Hsu
Chaniotakis
Pahre

# essential PHYSICS

$10^{19}$ $10^{20}$ $10^{21}$ $10^{22}$

Third Edition

PASCO
Education

Essential Physics Third Edition

Print Student Edition ISBN: 978-1-937492-13-7
Digital Student Edition ISBN: 978-1-937492-18-2
2019-06-01
EP-6323 textbook

2 3 4 5 6 7 8 9 - LSC - 22 21 20 19



**PASCO Scientific**
10101 Foothills Blvd.
Roseville, CA 95747

This book was written, illustrated, published, and printed in the United States of America.

We hope you find the Third Edition of *Essential Physics* a successful and engaging introduction to the exciting science of physics. As authors, we strove for clarity and completeness in a practical and useful physics course for all students. We drew inspiration and examples from renewable energy, technology, and engineering to connect the science of physics to both nature and the technology of our civilization.

As teachers ourselves, we place a lot of emphasis on *doing* physics because physics is *useful*. If physics is going to help you succeed later, and we know it can, someday you will need to apply what you learn to a new situation. This is why we chose to introduce many new concepts through investigations or other activities *before* we tell you the "answers" to what is going to happen. Figuring out the physics for yourself is experience that will help you solve the unfamiliar problems of the future.

Many people have strongly held misconceptions about physics, such as what we mean by action/reaction forces or energy conservation. No single course could cover *all* of physics, so we asked ourselves this: What are the most important concepts that you should understand and use? How could we best teach the *correct* explanations to replace misconceptions?

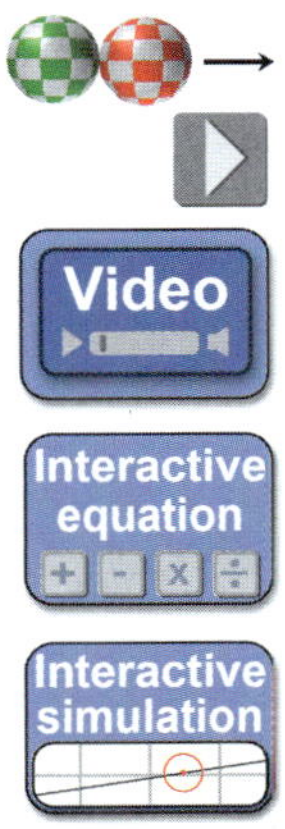


*Essential Physics* is state-of-the-art technology for learning using all the power of multimedia. Technology enhancements sprinkled throughout the electronic book will help you understand difficult or abstract ideas. Some graphics in the "book" are actually videos and many others are silent animations that explain a concept better than any static image could. Many of the lessons use interactive technology, such as simulations and interactive equations. Use these as "trial and error" ways to explore the meaning of an equation for yourself. Other lessons use simulations. These allow you to rapidly try things that would be difficult to do in real life. The *investigations* use both traditional physics equipment, such as force scales, and high-tech equipment, such as the *Smart Cart* Investigations provide direct experience with physics, such as measuring acceleration or the frequency of a sound wave.

To give you the quantitative skills you need for exams, and for real problems, every equation has practice problems and most e-Book pages have "test yourself" assessments at the bottom of the page. Use these to check your understanding. If you don't get it, then go back and see if you missed a key idea on the page.

The investigations, interactive elements, and lessons were developed simultaneously so they all work together seamlessly as a *learning system* for mastering physics. We wrote this for everyone who wants to know how this world of ours works. We hope you enjoy it!

Tom Hsu

2018

## About the authors

Dr. Tom Hsu is nationally known as an innovator in science education and has worked with more than 15,000 teachers over the past twenty years. The author of eight science programs in physics, physical science, and chemistry, Dr. Hsu also develops lab apparatus for teaching and learning. As an educator, Dr. Hsu taught high school physics as well as graduate and undergraduate courses at MIT, where he was nominated for the 1991 Goodwin Medal for excellence in teaching. He holds a Ph.D. in applied plasma physics from MIT and previously worked as an engineer for Eastman Kodak and Xerox. Dr. Hsu is the Chief Product Officer at PASCO Scientific, and was the founder of CPO Science in 1991 and cofounder of Ergopedia in 2003. Ergopedia joined PASCO in 2016.

Dr. Manos Chaniotakis, the cofounder of Ergopedia, Inc., is a former professor at MIT where for 18 years he developed and taught innovative, hands-on courses. A strong believer in *doing* science to learn, he enjoys teaching high school science, which he does at every opportunity. An expert in scientific instrumentation and control systems, Dr. Chaniotakis' designs for analytical chemistry sensors and meters are used by major companies worldwide. His vision for Ergopedia is to harness technology to create learning *systems* that unify curriculum, experiment, and teacher training. Dr. Chaniotakis holds a Ph.D. in plasma physics and fusion engineering from MIT.

Dr. Michael Pahre is an astrophysicist, formerly with the Smithsonian Astrophysical Observatory for 14 years, known for his research on elliptical galaxies, galaxy evolution, and infrared astronomy. Always interested in physics education, Dr. Pahre first taught high school physics as a Peace Corps volunteer in Ghana. He is the co-author of more than 150 research papers and the recipient of the 2001 Trumpler award for the outstanding doctoral dissertation in astronomy in North America. Dr. Pahre was a Hubble Fellow and a member of the team that developed the InfraRed Array Camera for the Spitzer Space Telescope. Dr. Pahre holds a Ph.D. in astronomy from Caltech.

**Editor in Chief**

**Freda Husic, M.A., M.S.**
*Director of Curriculum*
*PASCO Scientific*

**Contributing authors**

**J.J. Plank**
*Physics Specialist*
*PASCO Scientific*

**Jonathan Hanna**
*Physics Specialist*
*PASCO Scientific*

**Cathy Abbot, M.Ed.**
*Physics Teacher*
*Lexington High School*

**Joshua Roth, Ph.D.**
*Physics Teacher*
*Arlington High School*

**Jacalyn Crowe, Ph.D.**
*Physics Teacher*
*Lexington High School*

**William Konrad, M.Ed.**
*Physics Consultant*

**Brenda Weiss, Ph.D.**
*Physics Instructor and Post-Graduate Researcher*
*Univ. of California, Davis*

**Glenn Starkey**
*Physics Specialist*
*PASCO Scientific*

**Technical support and development**

**Michael Short, Ph.D.**
*Asst. Prof. Nuclear Science & Engineering*
*MIT*

**Sean Morton, M.Sc.**
*Lead Software Engineer*
*PASCO Scientific*

**Chris Murray**
*Lead Software Engineer*
*PASCO Scientific*

**Seva Khibkin**
*Software Engineering*
*Ergopedia*

**Marc Davidson**
*Engineering and Software*
*Ergopedia*

**Victor Youk**
*Dept. Electrical Engineering & Computer Science*
*MIT*

**Xola Ntumy**
*Dept. Electrical Engineering & Computer Science*
*MIT*

**Jemale D. Lockett**
*Dept. Electrical Engineering & Computer Science*
*MIT*

**Andreas Chaniotakis**
*Engineering*
*Univ. of Mass, Boston*

**Hunter Smith**
*Software Engineering*
*National Instruments*

**Graphic arts**

**James Travers**
*Director of Visual Arts*
*Ergopedia*

**Brennan Collins**
*Graphic Arts*
*PASCO Scientific*

**Kevin Young**
*Marketing Operations Manager*
*PASCO Scientific*

## Contributors


**Content specialists**

**Elizabeth Joy Toller**
Department of Physics
Massachusetts Institute of Technology

**Ravi Charan**
Department of Mathematics and Political Science
Massachusetts Institute of Technology

**Shannon Hope Harrison**
Department of Physics
Massachusetts Institute of Technology

**Luc Davidson**
Department of Electrical Engineering
Columbia University

**Robbie Gibson**
Harvard University

**Vincent Wenzel**
University of Illinois, Urbana-Champaign

**Kiara W. Cui**
Department of Chemical Engineering
Massachusetts Institute of Technology

**Melaní Suárez-Contreras**
*Spanish language consultant*
Mount Ida College

**Eli Workman**
Belmont High School

**Evan Greene**
Belmont High School

**Reviewers**

**Alan Gnospelius, M.Ed.**
Physics Teacher
Design and Technology Academy
San Antonio, Texas

**Joules Webb, M.Ed.**
Associate Director, Prefreshman Engineering Program
University of Texas at San Antonio

**William J. Montana**
Physics Instructor
All Saints Episcopal School
Tyler, Texas

**Beverly Trina Cannon, M.S.**
Physics Teacher
Kathlyn Gilliam Collegiate Academy
Dallas, Texas

**Melissa Pagonis, M.Ed.**
Physics/Earth & Space Science Teacher
Southwest Academy
San Antonio, Texas

**Karen Boone**
Physics Teacher
Hallettsville High School
Hallettsville, Texas

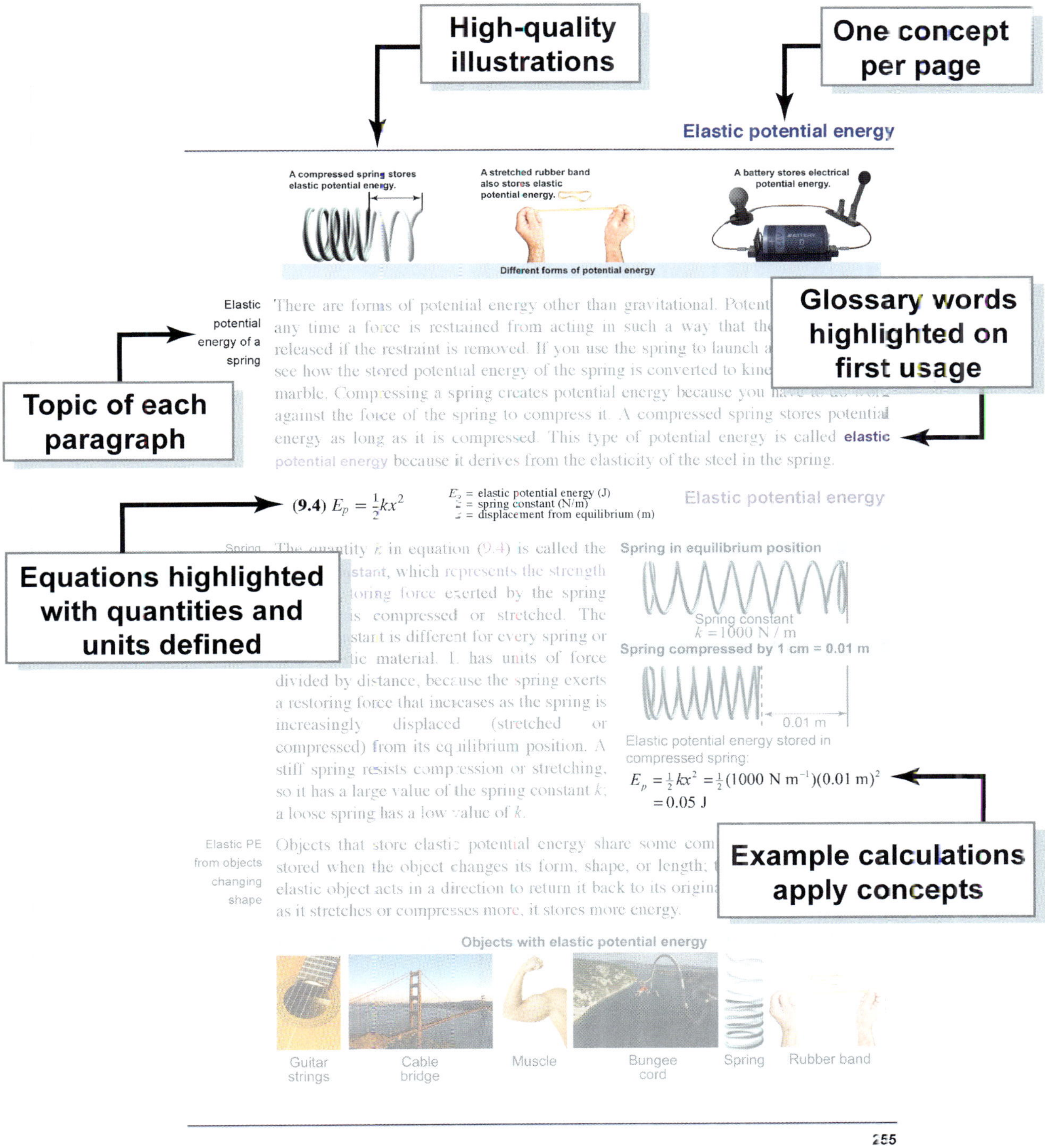

High-quality illustrations
One concept per page
Glossary words highlighted on first usage
Topic of each paragraph
Equations highlighted with quantities and units defined
Example calculations apply concepts
Elastic potential energy
A compressed spring stores elastic potential energy.
A stretched rubber band also stores elastic potential energy.
A battery stores electrical potential energy.
Different forms of potential energy
Elastic potential energy of a spring
There are forms of potential energy other than gravitational. Potent any time a force is restrained from acting in such a way that th released if the restraint is removed. If you use the spring to launch a see how the stored potential energy of the spring is converted to kine marble. Compressing a spring creates potential energy because you against the force of the spring to compress it. A compressed spring stores potential energy as long as it is compressed. This type of potential energy is called elastic potential energy because it derives from the elasticity of the steel in the spring.
(9.4) $E_p = \frac{1}{2}kx^2$
$E_p$ = elastic potential energy (J)
$k$ = spring constant (N/m)
$x$ = displacement from equilibrium (m)
Elastic potential energy
The quantity $k$ in equation (9.4) is called the stant, which represents the strength oring force exerted by the spring s compressed or stretched. The stant is different for every spring or ic material. $k$ has units of force divided by distance, because the spring exerts a restoring force that increases as the spring is increasingly displaced (stretched or compressed) from its equilibrium position. A stiff spring resists compression or stretching, so it has a large value of the spring constant $k$; a loose spring has a low value of $k$.
Spring in equilibrium position
Spring constant
$k$ = 1000 N / m
Spring compressed by 1 cm = 0.01 m
0.01 m
Elastic potential energy stored in compressed spring:
$E_p = \frac{1}{2}kx^2 = \frac{1}{2}(1000\ \text{N m}^{-1})(0.01\ \text{m})^2$
$= 0.05$ J
Elastic PE from objects changing shape
Objects that store elastic potential energy share some com stored when the object changes its form, shape, or length; elastic object acts in a direction to return it back to its origin as it stretches or compresses more, it stores more energy.
Objects with elastic potential energy
Guitar strings
Cable bridge
Muscle
Bungee cord
Spring
Rubber band
255

# Features of the e-Book

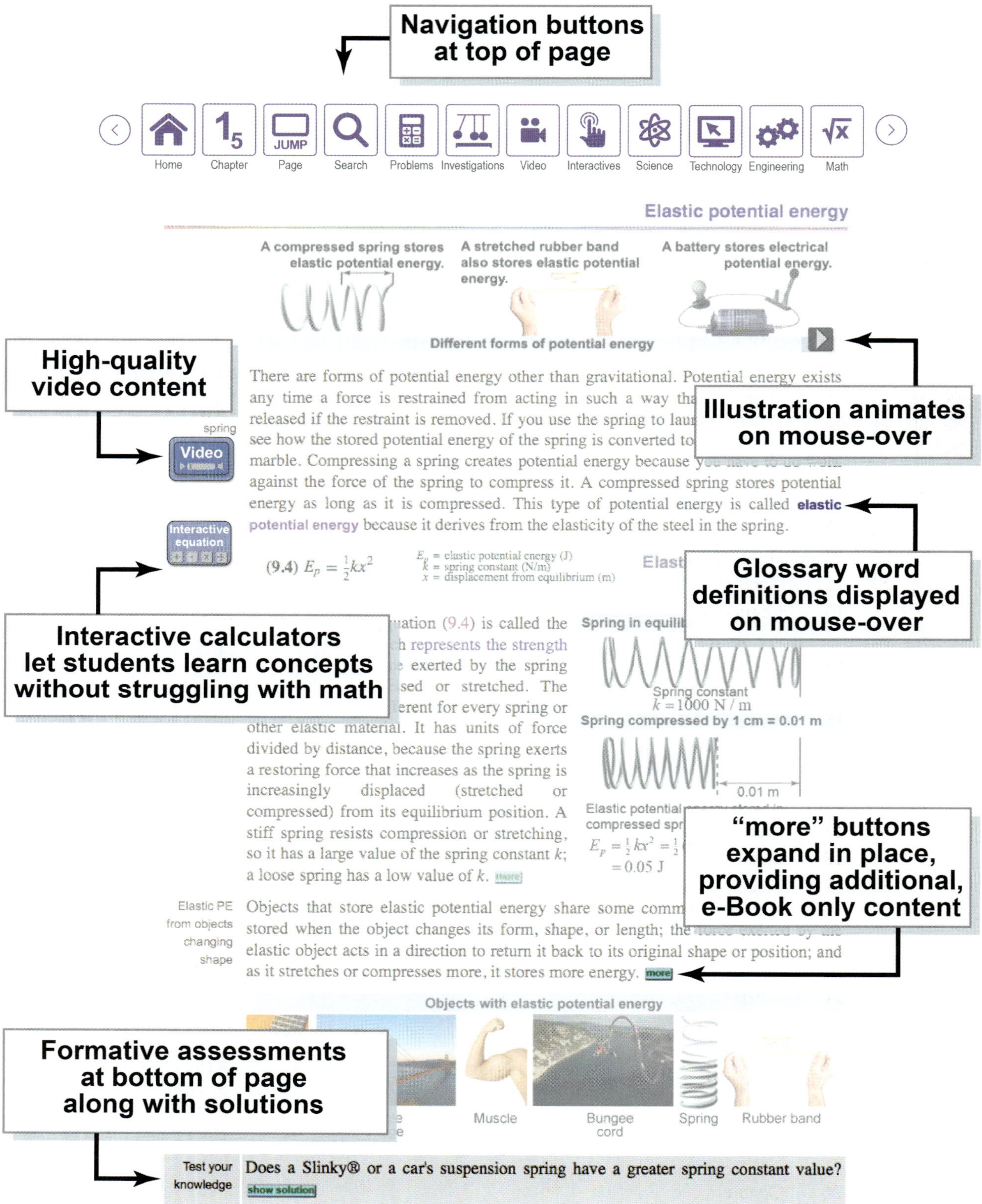

# Table of Contents

## Unit 4: Energy and Momentum (cont.)

## Unit 5: Waves and Sound

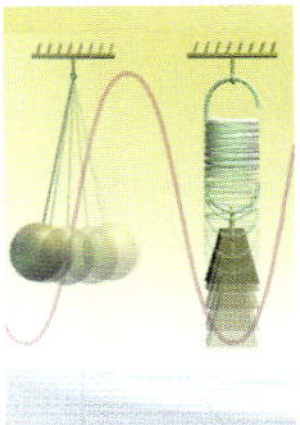

## Unit 6: Electricity and Magnetism

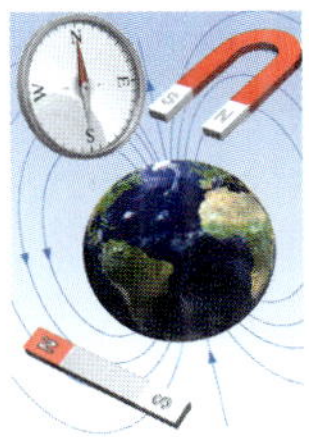



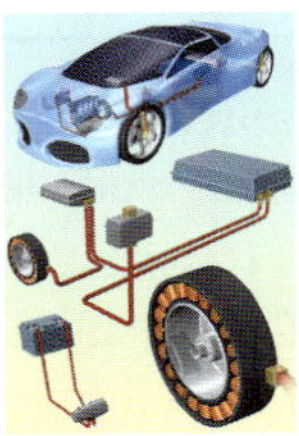

## Unit 7: Light and Optics

## Unit 8: Matter and Atoms

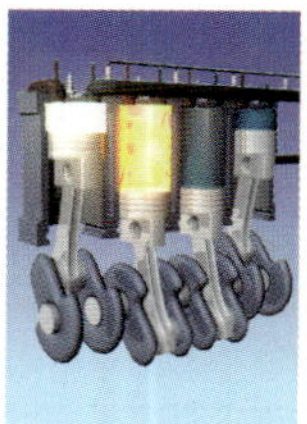



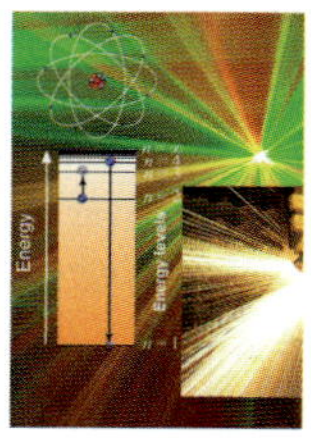



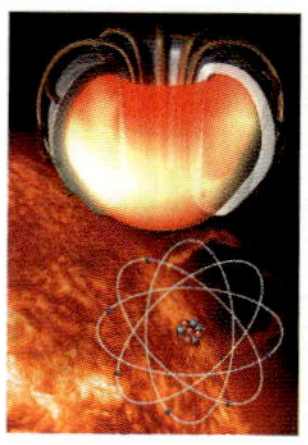

Hsu
Chaniotakis
Pahre

# essential PHYSICS

$10^{19}$ $10^{20}$ $10^{21}$ $10^{22}$

Third Edition

PASCO
Education

# Chapter 1

# Science of Physics

Physicists refer to 1905 as the *Annus mirabilis* (year of miracles) because of the unprecedented rethinking of the universe by the 25-year-old Albert Einstein. At that time, Einstein, a German by birth, worked as a patent clerk in the Swiss patent office. In that single year, revolutionary ideas poured forth from the young Einstein's mind. He not only finished his doctoral dissertation that year but also submitted four extraordinary papers for publication, each with a different and powerful new idea. Our understanding of matter, energy, space, and time changed forever. Any one of Einstein's 1905 papers would have won the Nobel Prize by itself. That four such contributions were made in a single year is a testament to the creative power of this genius's imagination and his willingness to leap with his mind into novel ideas and new ways of thinking.

Perhaps you think that creativity and imagination are the realm of painters and writers, but surely not *scientists*. If so, you would be mistaken! Truth turns out to be much stranger than fiction! Look up a painting called *The Persistence of Memory* created by the Spanish surrealist Salvador Dali in 1931. Dali's imaginary warped and drooping clocks were supposedly inspired by the fantastically *real* imagination of Albert Einstein. It was Einstein's creativity that prompted him to ask himself such questions as "What would light look like if we could see a light beam that was standing still? What would the world look like from a moving spaceship? Would the world look fundamentally different from how it appears when standing still?" It requires a creative mind to pose such questions and a powerful imagination to envision and explore the range of possible answers.

It was through a combination of logic and imagination that Einstein discovered that time and space are connected. In the third of his 1905 papers, Einstein proposed his theory of relativity. He described how time slows down and space contracts near the speed of light. In the theory of special relativity, Einstein gave us a realistic design for a *time machine* that could travel a million years into the future. Going forward in time and space takes a tremendous amount of energy—more than we can control with today's technology. Coming *back* would be another problem!

In 1915, Einstein extended his relativity theory to describe the connection among mass, space, and time—what we call *general relativity*. His creative idea was that the gravity caused by mass would *warp* the space around it, thereby causing the path of light to be deflected. This is strange indeed! Einstein's theory was successful in explaining the anomalous orbit of Mercury, but he became world famous when his prediction for the deflection of light by the Sun was confirmed in 1919. His theory has since been beautifully confirmed by many different experiments.

Chapter 1

# Chapter study guide

## Chapter summary

Physics is a study of the structure and interactions of matter and energy. Physics is concerned with fundamental rules for how matter interacts, so it typically deals with systems less complex than those addressed by either chemistry or biology. Inquiry lies at the heart of scientific methods, which test hypotheses and predictions against experimental data. The science of physics is closely linked to mathematics, engineering design principles, and the development of technology. Prominent historical and contemporary scientists have impacted both scientific thought and society in general.

## Learning objectives

By the end of this chapter you should be able to

- describe the major topic areas of physics and provide several real-world examples of each;
- describe the scientific method and provide examples;
- describe the impacts of a number of scientists on scientific thought and society;
- draw inferences from data to evaluate promotional claims;
- provide examples of the overlaps among physics, technology, engineering, and mathematics;
- draw connections between physics and careers; and
- describe the design parameters, and their trade-offs, for a household solar power installation.

## Investigations

1A: Forces and motion
1B: Musical sounds
1C: RGB color matching
Design project: Solar power

## Chapter index



## Important relationships

No equations are presented in this chapter.

## Vocabulary

**energy**
**force**
**atom**
**theory**
**experiment**
**average**
**quantitative**
**technology**
**photovoltaic cell**
**information**
**frequency**
**quantum physics**
**objectivity**
**scientific method**
**standard deviation**
**qualitative**
**additive primary colors (RGB)**
**matter**
**wave**
**hypothesis**
**reproducibility**
**error**
**standard error**
**engineering**
**subtractive primary colors (CMYK)**

## 1.1 - Science of physics

What is *physics?* Why is it important to know physics?

On a practical note, nearly all human technology is derived from physics, including buildings, bridges, electricity, lasers, noise-canceling headsets, LCD TVs, electric cars, cellphones, batteries, microwave ovens, and many more things you use every day. Physics includes the collected human knowledge, gathered over many centuries, of how things work. The mass of the electron, the properties of semiconductors, and Newton's laws of motion are examples of physics knowledge. *Physics will teach you the underlying principles behind how things work.*

Physics is also a *way of learning.* Science, of which physics is a part, is the most reliable way humans have ever devised for learning how our universe works. Newton's laws were discovered through creative thinking and careful observation. Physics blends creative thinking, the accurate language of mathematics, and rigorous observation to solve the deepest mysteries of nature.

### What is physics about?

**The universe**

The universe is *defined* to be everything that exists, including you. We believe "everything" is matter, energy, or information. **Matter** is "stuff" that takes up space and has mass. **Energy** is the quantity that causes matter to change and mediates how much change occurs. Energy is exchanged any time anything gets hotter, colder, faster, slower, or changes in any observable way. **Information** describes how matter and energy are arranged in time and space. For example, the words on your screen are pieces of information.

**What is physics?**

Broadly speaking, physics is concerned with the properties and interactions of matter and energy at the most fundamental level. Physics describes the basic forces, the nature of atoms and matter, and the processes by which matter and energy interact.

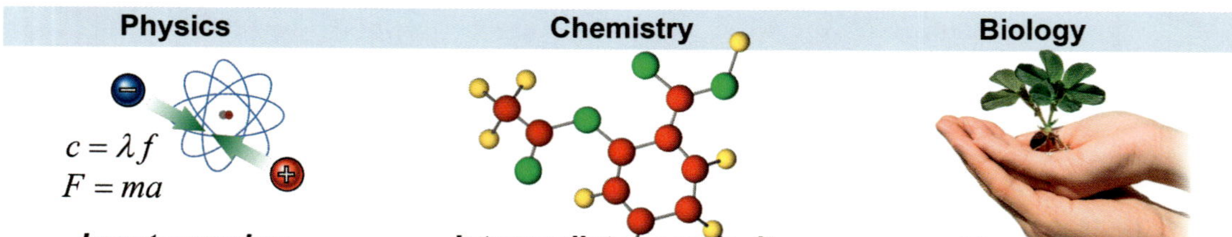


| ***Least complex*** | ***Intermediate complexity*** | ***Most complex*** |
|---|---|---|
| Fundamental rules for the interaction of matter, energy, and forces | Particles and forces combining in the diversity of compounds and reactions that is the material world. | Millions of compounds and innumerable reactions in complex living systems |

**How are chemistry and biology related to physics?**

One level higher in complexity is *chemistry.* Chemistry concerns how the basic particles and forces combine to create the trillions of different molecules that make up the diversity of matter we live in. Highest on the complexity scale is *biology.* Living organisms include millions of different molecules interacting with each other in extremely complicated systems. Biology is the most complex of the three basic disciplines of science. It may seem strange, but physics is the simplest.

# Investigation 1A Graphs of motion

**Essential questions** *What do graphs of motion look like?*

The Smart Cart is a device that displays its motion—position, velocity, and acceleration—on your computer in real-time while you move it! Look for the connection between the forces you apply to push the Smart Cart and how the Smart Cart's position and velocity change.

## Part 1: Testing the Smart Cart to see graphs of its motion

1. Open the experiment file **01A_GraphsOfMotion**, and then connect the Smart Cart to your software.
2. Start data collection and move the Smart Cart back and forth on the table.
3. Watch how the position and velocity graphs of the Smart Cart change as you move it.

*a. If you push the Smart Cart across the floor and let go, what happens to its velocity? What causes this?*
*b. If you roll the Smart Cart down a ramp, what happens to its velocity? What causes this?*
*c. Try to move the cart with constant velocity. What is the shape of the position graph for constant velocity?*
*d. How does the Smart Cart track its motion? Try spinning each wheel with a finger. Does each wheel track its motion? What is the smallest distance the Smart Cart can record?*

## Part 2: The position versus time graph

1. Go to Page 2 in the experiment file, which displays a graph of position versus time.
2. Use the Smart Cart to create the position versus time graphs shown.

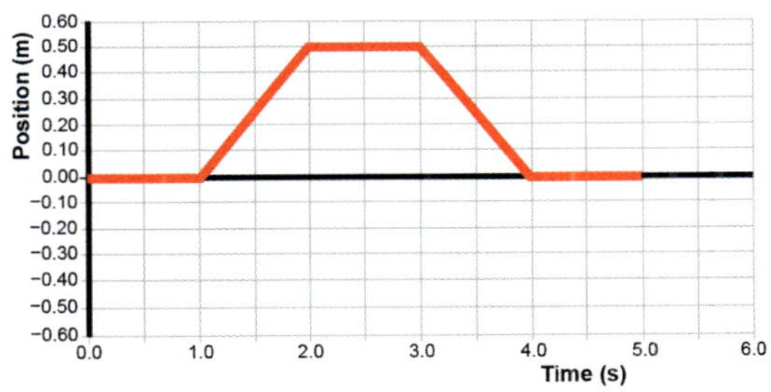


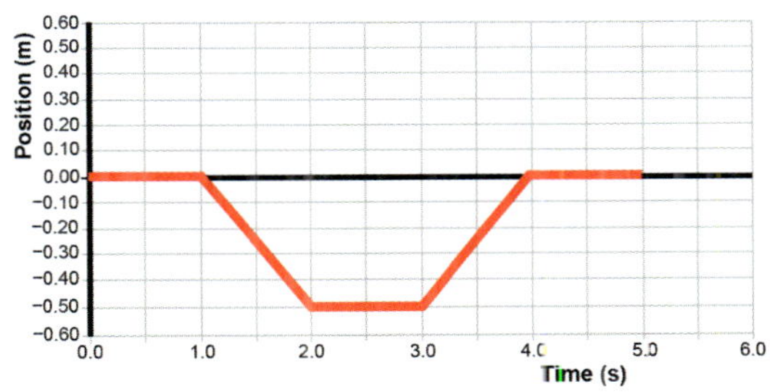


*a. How do you have to move the Smart Cart to create the left graph?*
*b. What is different about the Smart Cart's motion in the right graph?*

## Part 3: Testing the Smart Cart to see graphs of force

1. Attach the hook to the force sensor on the Smart Cart.
2. Go to Page 3 in the experiment file, which displays a graph of force versus time.
3. Loop a rubber band around the hook and pull the cart with the rubber band while collecting data.

*a. Is a pulling force positive or negative? If the cart moves in the direction of the pulling force, is the cart moving in a positive or negative direction?*
*b. Attach the rubber bumper to the cart and push the bumper. Is a pushing force positive or negative? If the cart moves in the direction of the pushing force, is the cart moving in a positive or negative direction?*
*c. Is it possible for the force on the cart and the motion of the cart to have opposite signs? For example, can there be a positive force on the cart while moving in the negative direction?*

**The physics of force and motion**

*Mechanics* is the branch of physics that deals with the "real" world of forces, objects, motion, and energy. Mechanics develops and applies mathematics to describe the physical world using concepts such as velocity, force, and acceleration. Acceleration is a challenging idea that eluded some of the brightest thinkers in history until the time of Galileo. Motion only changes through acceleration, leading us next to the concept of *force*.

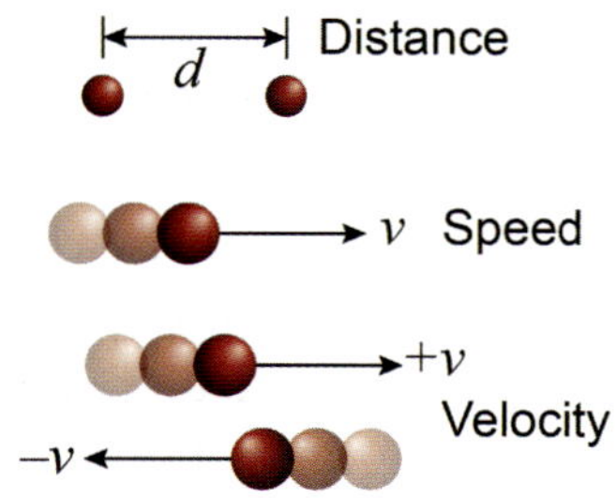


**Forces**

A **force** is an action that can make things move, change direction, or stop. We all have an intuitive notion of force. Physics makes the intuition *exact* by precisely describing how an object moves, or doesn't move, in response to the strengths and directions of forces acting on it. Newton's laws are the framework for understanding how forces cause accelerations and are caused by accelerations. For example, the force that accelerates a bicycle is *not* the force of the wheel pushing against the road. That force acts on the *road.* The reaction force acting back on the bicycle is what moves the bicycle because the reaction force acts on the bicycle.

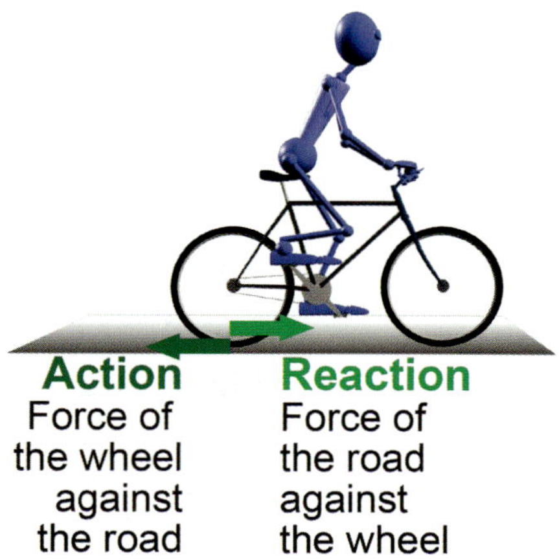


**Circular motion and extrasolar planets**

Physics concepts explored in forces and motion underlie cutting-edge advances, such as the discoveries of hundreds of planets orbiting around other stars, far from our Sun. Many of these *extrasolar planets* have masses similar to Jupiter, but they orbit their sun closer than Mercury is to our Sun. Some orbit as quickly as a few days!

**Planets around other stars**

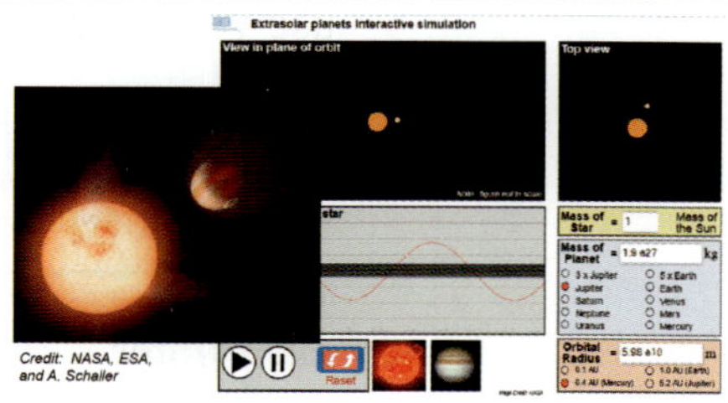


**Energy and you**

This is the age of energy, and mechanics helps us understand different kinds of energy. When you switch on an air conditioner, the equivalent energy output of ten working people travels through electrical wires from a power station 100 miles away. The engine of a moderate-sized car generates 150 horsepower—the equivalent energy output of 500 people! Modern energy use is clean, quiet, and simple. The human standard of living has risen hand-in-hand with the increase in available energy per person.

These tubes contain green algae being engineered to produce diesel fuel. Is this a future energy source?

### The physics of repeating motion

We observe things move in two fundamentally different ways. A car moving from one place to the next is well described by concepts of speed and position. The *vibration* of a guitar string, however, has additional complexity that needs additional physical concepts to describe. In the physics of waves, sound, and light, we will often use the term **frequency** to describe how quickly the vibration repeats.

### Examples of vibrations and waves

A guitar playing the note "A" oscillates back and forth at a frequency of 440 vibrations per second. In an acoustic guitar, the string vibrates the guitar top, which sets off a traveling oscillation in air, called a sound **wave**. Colors, cellphones, radio, and the ripples on a pond are other examples of *waves* and their applications.

### Waves and information

Sound waves carry both energy and information. Music is a form of information. When pressed, the keys of a piano create different frequencies of sound waves. A song is physically a complex "sonic image" created by your brain from patterns of frequency and loudness received by your ear. Guitars and pianos use vibrating strings of different length and tension to make different frequencies. Trumpets and pipe organs use vibrating columns of air.

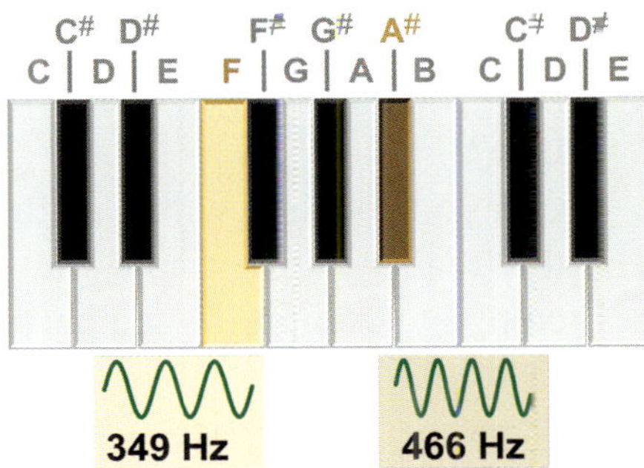

**Notes are frequencies of sound**

### What is light?

You see everything around you because it either emits or reflects light that then travels to your eyes. This is the physics of light and the technology of *optics*. Light travels through air, water, glass, or even empty space. Fundamentally, light is a wave in the electromagnetic field everywhere around us. Light ordinarily travels in straight lines, but it can be deflected by mirrors or lenses (such as in a telescope) to form images.

# Investigation 1B

## Musical sounds

**Essential questions**
*How do we understand sound?*
*How can sound contain so much information?*

Interactive simulation

For most people, sound creates one of the richest information flows into your brain. Sound is also a model for a whole class of physical phenomena involving *waves*. The creation, transmission, and recording of sound have driven a host of human inventions from the earliest musical instruments to mp3 players and electric guitars. This investigation explores some of the basic properties of sound waves.

### Part 1: The perception of frequency

1. Use the interactive keyboard to play and observe different **single-note** sounds.
2. The volume control on your computer sets the volume.

a. What property of sound corresponds to the perception of high or low pitch?
b. How long does a single oscillation of a 349 Hz sound wave take?
c. Can you come up with a mathematical relationship between the frequency of a sound and the time it takes to make one complete oscillation (the period)?

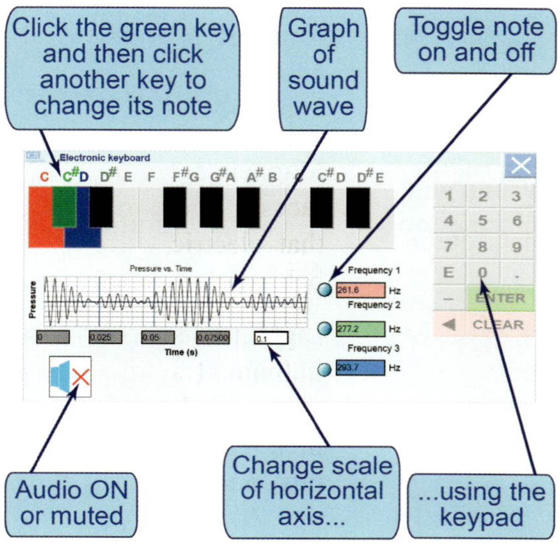


### Part 2: The richness of sound

1. Listen to two different frequencies of sound at the same time.
2. Play notes that are close together and that are far apart.
3. Play notes that have the same name but different frequencies.

a. Can you distinguish the two individual frequencies?
b. How does a two-frequency sound wave compare with a single frequency?
c. Are there combinations that sound *bad* and, if so, when do they occur?

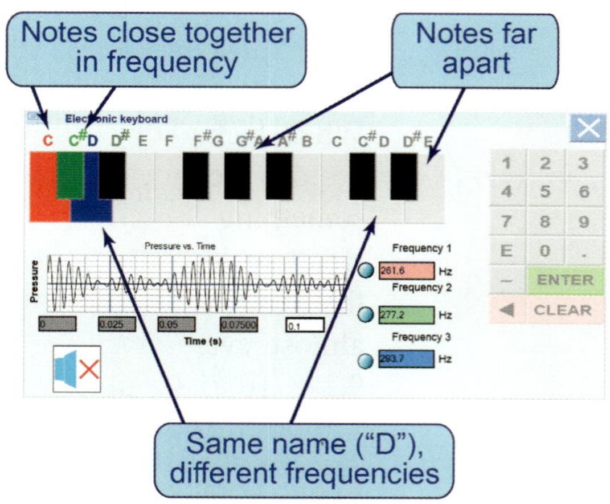

**The nature of electricity**

A single "D" battery can supply an astonishing 75,000 J of energy! If you could get all the energy out in one second that would be the equivalent of 100 horses, from a package that fits in the palm of your hand. Fortunately, 1.5 volt batteries can't deliver electrical power so quickly, or they would be *very* unsafe.

**Electric charge as a property of all matter**

Electricity exists because matter has a property called *electric charge*, which is normally hidden. Inside atoms are negatively charged particles called electrons and positively charged particles called protons. Under special circumstances, such as dragging your feet across a carpet on a dry winter day, you can sense electric charge in the form of a spark from static electricity. Lightning is another way to *see* the effects of electric charge—but hopefully not to experience it!

**Electricity and magnetism**

Magnets seem to have little to do with electricity but magnetism and electricity are two faces of the same fundamental phenomenon. In the late 19th century physicists discovered that electric current can cause magnetism. Electromagnetism is the operating principle behind electric motors, generators, computer disk drives, and countless other technologies.

**The idea of atoms**

Light is an electromagnetic wave and it interacts with individual atoms. What is an atom? Consider cutting a piece of aluminum foil into the smallest possible bits, then using a microscope to divide those bits into smaller bits, and the smaller bits into even smaller bits, and so on. At some point you come to a single **atom** of aluminum, which, at 0.00000001 m in diameter, is the very smallest particle of aluminum there can be. If you divide the atom further by separating some protons, neutrons, or electrons, you will no longer have aluminum at all but some other element.

**What is quantum physics?**

When technology was developed that allowed us to actually observe how electrons and protons behaved inside atoms, and how light interacted with single atoms, we found something completely new. **Quantum physics** was invented to explain the universe on very small scales. It is both a crowning intellectual achievement for its accuracy at describing the world of atoms and also a profound mystery because it causes us to completely rethink almost everything we know about how the world works. An electron with a mass of $9.1 \times 10^{-31}$ kg does not behave at all like a tiny marble that is simply a million trillion trillions times smaller than usual.

**Quantum physics and modern technology**

Many technologies in use—such as the MRI scanner in a hospital or a computer—explicitly depend on quantum physics. The laser in a laser cutting machine is based on stimulated emission, which is a quantum phenomenon. The quantum world is extraordinarily rich and surprisingly not well explored. A superconducting wire that carries electricity without friction is one potential quantum technology that may become reality in the decades to come.

# Section 1 review

The science of physics has two facets. Physics includes a process of thinking and problem solving that allows us to deduce reliable explanations for how things in the universe work. We can then apply what we know to solve practical problems. Physics is also the collected body of knowledge we have discovered about the universe and its underlying laws. Physics is roughly divided into four major branches. *Mechanics* deals with forces, objects, and motion and is the most directly observable branch of physics. Energy is the "fuel for change" that both enables changes, by generating forces, and also limits how much change is possible. Another branch of physics deals with *waves.* Waves are traveling oscillations, including sound and light, that carry both energy and information in the natural world and in technology. *Electricity and magnetism* is a third branch of physics. Both are phenomena that come from the arrangement and movement of the electric charges that make up matter. The fourth major branch of physics deals with the behavior of atoms themselves. This branch includes the *quantum theory*, which provides the fundamental explanation for how matter and energy behave on extremely small scales.

| Vocabulary words | energy, information, matter, force, frequency, wave, atom, quantum physics |
|---|---|

## Review problems and questions

1. Chemistry, biology, and physics are the sciences most commonly included in the course of studies for secondary students. Which of these sciences is considered the most fundamental? Which is considered the most complex?

2. Which of these seemingly separate topics in physics are actually deeply interconnected: electricity, harmonic motion, magnetism, and/or quantum physics?

3. Everything you do is done in complete accordance with the laws of physics. Every item of technology you use is based on the principles of physics. There are no exceptions. Think about physics in the context of your own activities and the technologies you use.
    a. Name two actions you did today and identify three aspects of physics that were directly involved in those actions.
    b. Name two items of technology you used today and identify three aspects of physics that were directly involved in how those technologies functioned.

4. Use each of the following words in two different sentences. The first sentence should use the word with the same meaning it has in physics. The second sentence should use the word differently from its physics meaning. For example, the sentence, "Electricity flows through wires and carries energy," uses the word *electricity* with the same meaning as electricity in physics. The sentence, "There was such electricity in the air from the crowd at the concert," uses the same word to mean *excitement,* which is a different meaning from physics.
    a. force
    b. energy
    c. magnetic
    d. wave

**What is *science*?**

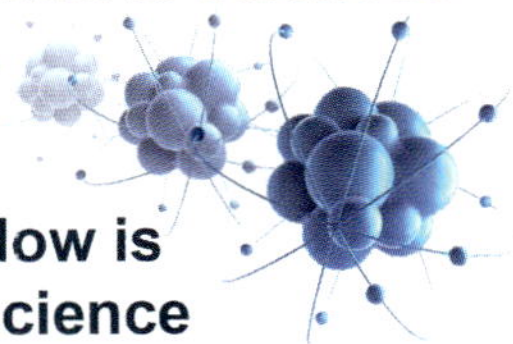

**How is science *different* from other kinds of human knowledge and thought?**

*Science* is the *systematic* study of the structure and behavior of the physical and natural world. *Systematic* refers to the system of logic by which scientific ideas are proposed, modified or rejected based on comparison with observational evidence.

Science is also a body of knowledge about the physical and natural world including elements such as *facts* and *theories*. *Science includes only that knowledge which can be tested and confirmed by comparison with objective, repeatable, and observational evidence.* If an idea cannot be supported or refuted by comparison with observational evidence, then that idea, true or not, cannot be part of *science*.

## Hypotheses and theories

**Limitations of science**

Science cannot answer ethical questions such as whether something is good or bad. As an example, the question of whether or not Earth's average temperature is increasing has a scientific answer based on measurable, objective data. The ethical question of whether it is *good* or *bad* if Earth's temperature increases, however, is *not answerable by science*. Questions of judgment, preference, ethics, or morality cannot be answered by science. These areas of thought *are important!* They are just not part of science.

**Hypotheses**

A **hypothesis** is a tentative explanation or conclusion about some aspect of the physical world. Hypotheses do not have to be right! The logic of science begins by *assuming* a hypothesis is correct and then seeking scientific evidence that either supports or refutes the hypothesis. Hypotheses are

1. *tentative* statements, which may or may not be correct,
2. *testable* statements, which may be evaluated by comparison with objective scientific evidence, and
3. necessary and fundamental steps in the construction of reliable scientific theories.

**Hypothesis:**

The Sun, Moon, and stars form a spherical shell that rotates around the Earth.
*Claudius Ptolemy (90–168 CE)*

**Scientific theories**

A scientific **theory** is a comprehensive and well-tested explanation of a natural or physical phenomenon. Unlike hypotheses, scientific theories are well established and *highly reliable* explanations that have been tested by multiple, independent researchers over a long period of time and under a wide range of conditions. Only hypotheses that have correctly explained a range of phenomena over a long time are incorporated into scientific theories.

**Theory:**

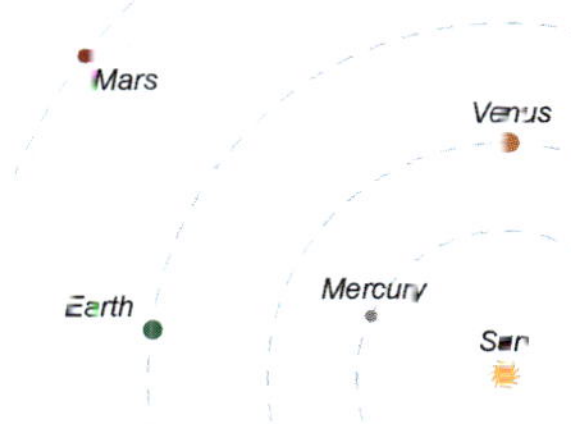


The Earth and planets move in elliptical orbits around the Sun.
*Nicolaus Copernicus (1473–1543 CE)*

**Other "theories"**

The word *theory* in science does not mean *hunch or guess.* Although scientific theories are always being tested, and occasionally proved wrong, *scientific theories* represent our best, most reliable knowledge about the natural or physical world.

## Scientific research and experiments

**What is research?**

Scientific *research* is the process by which multiple, independent people continuously test scientific theories by making new observations. One purpose of research is to find new phenomena that existing theories fail to correctly explain. Unexplained phenomena are often found in new areas of science, such as cosmology or quantum mechanics. Unexplained phenomena are also found when new technologies, such as particle accelerators, make different kinds of observations possible that could not be made before. Science grows as new observations lead to new hypotheses that augment or change existing scientific theories.

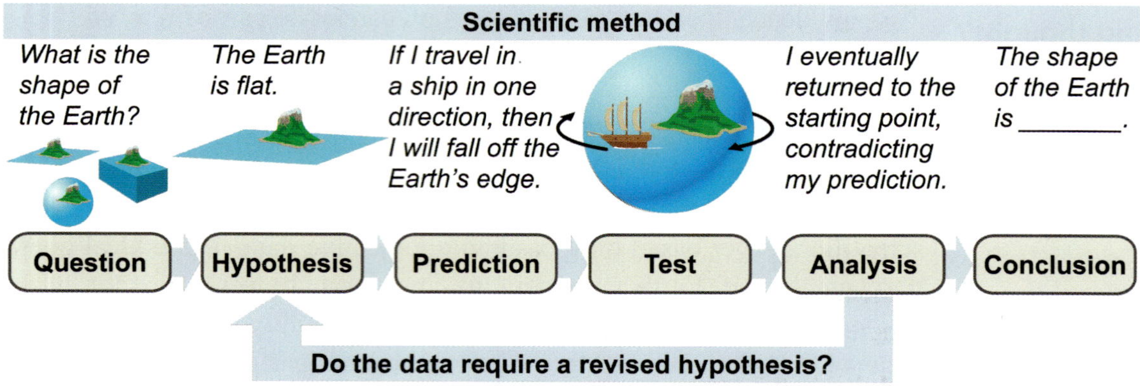


*New hypothesis: The Earth is a sphere.*

**What is scientific evidence?**

One outcome of research is to produce scientific evidence, such as drawings, measurements, data tables, graphs, or observations. The two most important characteristics of scientific evidence are that it be *objective* and *repeatable.* **Objectivity** means the evidence should describe only what actually happens as exactly as possible, without opinion, interpretation, exaggeration, embellishment, or bias. **Reproducibility** means that others who repeat the same experiment or make the same observation in the same way always observe the same results.

**Evidence from experiments**

An **experiment** is a controlled situation designed to collect scientific evidence on what happens under a controlled set of circumstances. A well-designed experiment changes one variable at a time so that any observed *effects* may be clearly associated with the variable that was changed. A poorly designed experiment is at best inconclusive, and at worst it supports erroneous conclusions. Experimental evidence should be critiqued by asking the following:

1. Is the experiment objective? Does it generate unbiased observations?
2. Do any observed effects result from changing the variable claimed, or could other variables have caused the effect?
3. Do other researchers repeating the experiment observe the same result?
4. Have the data been analyzed to understand the uncertainties in measurement?
5. Are the observed effects greater than the uncertainties in measurement for this experiment? If not, the experiment may be inconclusive.

**Control**

Well-designed experiments typically include a *control*, which is a standard set of conditions that are used for comparison to check or verify a result. For example, when testing the effects of a medicine on a sample of patients, the results might be compared to a control sample of patients who did not receive the medicine.

### If–then logic

Scientists use *logic* to evaluate whether observed evidence agrees or disagrees with testable predictions of a hypothesis. One type of logical argument is the *if–then* statement. This logic has the form "if ____ is true, then ____ must also be true." The diagram below shows how observations of the phases of Venus support the explanation that both Earth and Venus orbit the Sun.

**Scientific explanation:**
The observed phases of Venus occur because both Venus and Earth orbit the Sun.

**Evidence:** We observe Venus to have phases.

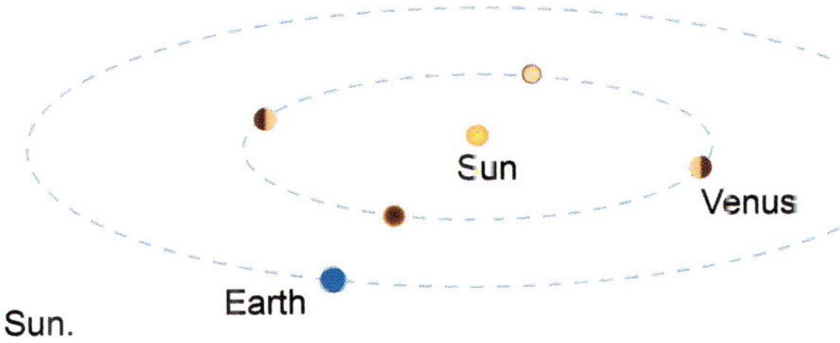


**Logic:** **If** Venus and Earth both orbit the Sun,
**then** Earth and Venus may at times be in the positions shown in the diagram relative to the Sun.

**If** Earth and Venus were in the positions on the diagram,
**then** the sunlit and shadowed sides of Venus would appear as phases like the Moon.

### Observational testing

The phases of Venus are an example of how *observational* testing can be used to evaluate a scientific explanation. The observations of Venus are both objective and repeatable and therefore satisfy the rules for acceptable scientific evidence.

### Logic and the round Earth

Multiple types of observational evidence may be needed. For example, the Earth appears generally *flat,* not round. Research and present evidence for the "flatness" of Earth and also evidence for "roundness." How does each piece of evidence support each claim? Which evidence shows that Earth cannot be flat? What logical argument would *convince* a 14th–century person that our planet is actually round?

**Earth does not appear "round." What evidence proves Earth cannot be "flat"?**

### Creating cause–and–effect hypotheses

Scientific explanations provide causes for the effects that we observe. For example, Newton stated that objects at rest start moving *because* of the action of unbalanced forces. The forces are the cause and the motion is the effect. Think of three "effects" for which you can identify the physical causes. Describe these to your class using the word "*because.*" For example, "I believe ____ happens *because* _____." You may not realize it but your statements are scientific hypotheses! They are tentative explanations of real effects. It may help to record your hypotheses.

### Accurate communication

*Clear communication* is crucial in science. Physics uses new vocabulary to ensure that what is said or written has the same meaning to everyone. For example, suppose someone offered to sell you a "strong" rope. What does "strong" mean? You might want to use the rope to tow a car, requiring a strength of 10,000 lb or more. The seller might mean strong enough that you cannot break it with your hands, a strength of 150 lb or so. Replacing the word "strong" with a scientific term—such as the rope having a "working load of 500 pounds"—avoids misunderstandings. The term "working load" has a clearly defined meaning.

**What does it mean to "analyze"?**

To *analyze* something means to intellectually take it apart and study both the pieces and how the pieces fit together. Scientific analysis takes a complex phenomenon and breaks it down into understandable parts that can each be explained. Consider the fact of *seasons.* Any explanation for why there are seasons must account for empirical evidence and observations such as:

1. The day is longer in the summer and shorter in the winter;
2. The average temperature is colder in winter than in summer; and
3. In the northern hemisphere the Sun appears at a lower angle in the sky in winter and higher in summer.

**Analyzing an explanation using logic and observation**

The explanation for seasons is that Earth's rotational axis is tilted about 23 degrees relative to the plane of its orbit, which you will learn more about on page 382. Logically, how does this explanation account for the evidence? Consider the angle of the Sun in the sky. If the explanation is true then logically (mathematically) it follows that:

1. An observer in Austin, Texas (30°N latitude) should see the Sun 53° away from vertical at noon in winter; and
2. The same observer should see the noon Sun 7° away from vertical in summer.

Observational evidence confirms that this is exactly the angle at which the Sun appears, therefore supporting the explanation.

**Tilt of the Earth's rotational axis causes the seasons**

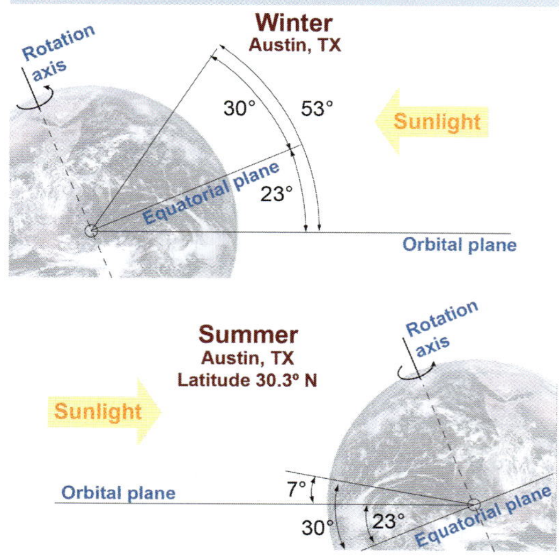


**Analyzing an explanation using experiments**

How could the solar angle explain the seasonal temperature change? Consider the following experiment. A solar cell of a certain size is held at different angles to a bright light. The light energy received by the solar cell is measured by recording the electric current produced at each angle. The experiment finds that the solar cell receives 40% less energy at 53° compared to when it is at 7°. Thus, experimental testing also supports the explanation of a tilted axis by accounting for another observed effect. In fact, the tilted axis theory accounts for every single observation and experimental prediction that can be tested. While technically this explanation is called a *theory* it is so well tested that it is accepted as *fact.*

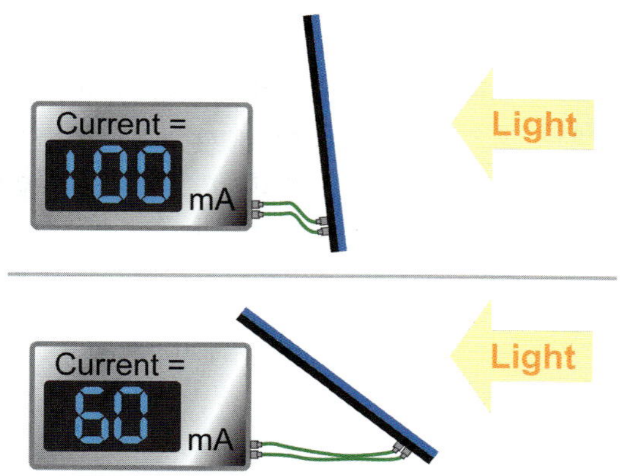


Measure the electric current in the solar cell for two different orientations relative to a bright light source.

**"Error" is not a mistake!**

In science, **error** is the unavoidable difference between a measurement and the true value of whatever is being measured. There is **always** error in every measurement, no matter how careful the experimenter or how elaborate the equipment. This is why it is important to take multiple measurements. It is unfortunate that the common use of the word error means mistake but *in the context of scientific data, the word error does not mean mistake!*

**Example of making a measurement**

Consider a measurement of the number of heads in ten flips of a coin. The table below shows the results of thirty measurements. Intutitively, you know the most likely outcome is 5 heads out of 10 flips. However, the outcome of a single measurement *could* be any number between 0 and 10. The most *likely* number is 5 but the data shows that out of 30 coin flips, 4 heads came up five times and 6 heads came up seven times.

Uncertainty in measurement

Experiment
Shake and drop 10 pennies and measure the number of heads.

Data
The experiment is repeated 30 times.

| Number of heads | | | | |
|---|---|---|---|---|
| 3 | 4 | 3 | 6 | 5 |
| 6 | 4 | 6 | 5 | 8 |
| 6 | 5 | 5 | 4 | 5 |
| 6 | 5 | 2 | 3 | 4 |
| 5 | 7 | 5 | 5 | 5 |
| 6 | 7 | 7 | 6 | 4 |

Average value = 5.06

Analysis

| Value | # |
|---|---|
| 1 | 0 |
| 2 | 1 |
| 3 | 3 |
| 4 | 5 |
| 5 | 10 |
| 6 | 7 |
| 7 | 3 |
| 8 | 1 |
| 9 | 0 |
| 10 | 0 |

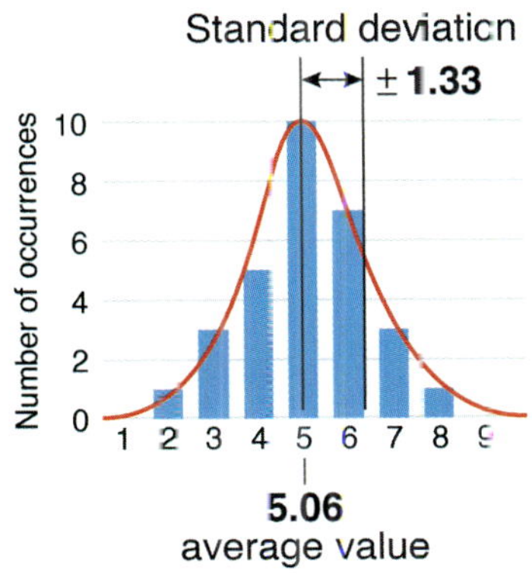


**The average**

All measurements have some variability (error) that is random, like flipping a coin. One way scientists account for random error is to take the **average** of several measurements. To calculate the average, you add up all the measurements and divide by the number of measurements you have. The average of the 30 coin-flip measurements is 5.06 - very close to 5. The average is a better estimate of the true value (5) because averaging smooths out random variability.

**Standard deviation**

Another reason for taking multiple measurements is *they provide an estimate of uncertainty* (or error). The solid red line on the graph shows the behavior of data that contains random variation. The difference between the average and the spread of the measurements is called the **standard deviation**. In the example the standard deviation is 1.33.

The standard error (error in the mean)

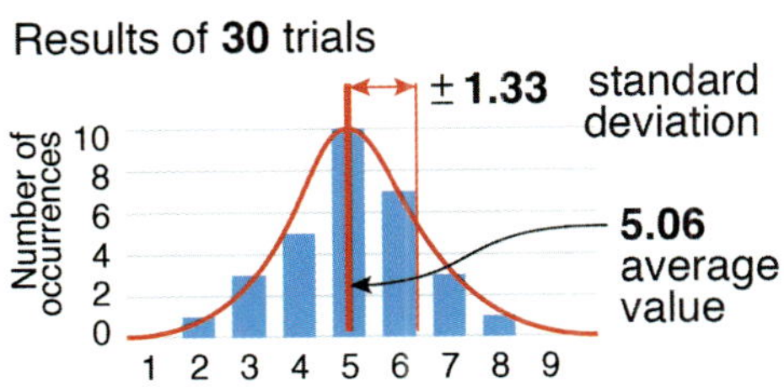


$$\text{Standard error} = \frac{\text{standard deviation}}{\sqrt{\text{number of measurements}}}$$

$$= \frac{1.33}{\sqrt{30}} = 0.24$$

**Result:** 5.1 ± 0.24

**The standard error**

The standard deviation is the uncertainty in a single measurement. The **standard error** in the average is smaller because many measurements are included. The standard error is the standard deviation divided by the square root of the number of measurements. A proper result states the average plus or minus the standard error.

## Comparing data: When are differences significant?

**Different numbers can be "the same"**

Scientific explanations make predictions that are tested by comparison with observational data. This brings up a **very** important question - *how can a scientist tell if the data is the same as the prediction when all the data has uncertainties?* The word "same" when applied to experimental measurements does not mean what it does in ordinary conversation. Two measurements are considered the same if their difference is less than the amount of uncertainty. *This is important to remember.*

Differences are not *significant* unless they are larger than the uncertainties in measurement.

**An experiment**

Any scientific use of data must consider the *significance* of differences between values. For example, a student decides to test a claim that heavier objects fall faster than lighter objects by dropping two stones of different weight. The time to fall is measured with a stopwatch (see the table at right). On average, the heavier stone is 0.02 s faster than the lighter stone. Does this evidence support the claim?

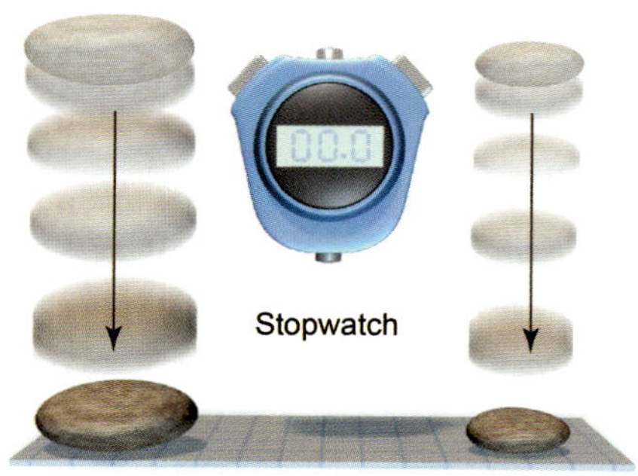


| Mass 1 Time (s) | | Mass 2 Time (s) |
|---|---|---|
| 0.68 | | 0.98 |
| 0.93 | | 0.81 |
| 0.79 | | 0.65 |
| 0.83 | | 0.77 |
| 0.67 | | 0.79 |
| **0.78** | **Average** | **0.80** |

**How to estimate the error**

The way scientists estimate the error is the following.

1. *Assume the average is the true value.*
2. Find the differences between each measurement and the average.
3. The standard deviation is the average difference between the measurements and the average.
4. The error is the standard deviation divided by the square root of the number of measurements.

Most calculators and spreadsheets have a function that calculates the standard deviation of a set of measurements. You can estimate the standard deviation as about 1/2 the difference between the average and the highest and lowest values.

**Significance**

Looking critically at this student's evidence, we see that the difference between the heavier and lighter rocks is less than the uncertainty. Therefore, the difference is *insignificant.* The two falling times are *identical* within the precision of the experiment. This experiment does *not* support the explanation because the experiment finds no *significant* difference between the heavy stone and the lighter stone.

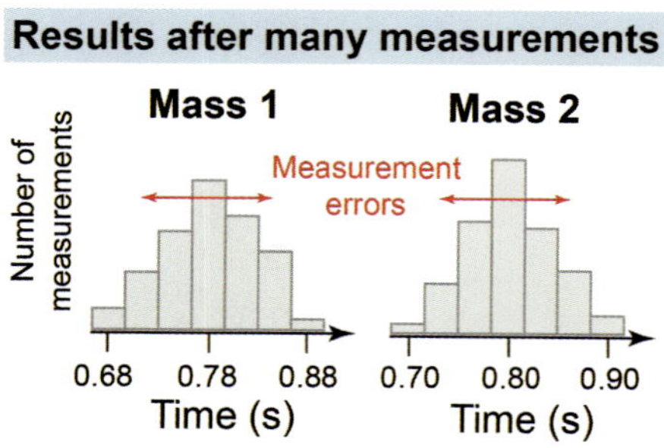


**Testing significance**

You cannot make a valid scientific conclusion without knowing the error in measurement. There are always differences between the calculated (exact) value of a prediction and any real measurement. The real question is whether the differences are significant compared to the error in measurement.

**Science and society**

Some scientific results are so profound that they can impact other scientists and society in ways that were never anticipated. Fundamental advances in science can challenge the assumptions of scientists and the public in general. New science encourages society to look at the world in a new light.

**Scientific developments in the physical understanding of the universe**

1927: Belgian mathematician Georges Lemaître proposes Big Bang theory.

1948: Ralph Alpher and Robert Herman estimate the temperature of the universe as a relic of the Big Bang.

1964: Arno Penzias and Robert Wilson use a microwave antenna to detect and measure the temperature.

Credit: NASA

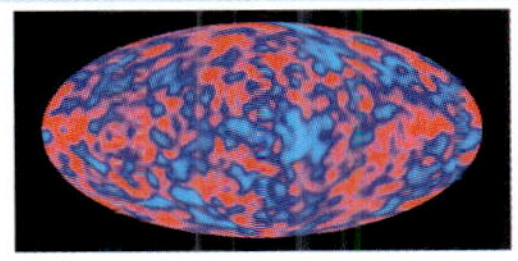

1992: George Smoot and John Mather measure small *variations* in the microwave background, which led to present-day clusters of galaxies.

**Impact of scientists on scientific thought**

In the scientific process, the results from one scientist's research can often lead to discoveries by others. A good example of this is the development of our understanding of the origins of the universe, which we now know as the Big Bang model. Mathematician Georges Lemaître first proposed the Big Bang, while Ralph Alpher and Robert Herman expanded upon the theory to predict a *temperature* for the present-day universe. Arno Penzias and Robert Wilson serendipitously discovered that temperature in 1964, and then in 1992 George Smoot and John Mather discovered tiny *variations* in the temperature. All these results contributed to our present-day understanding of how the universe formed and has evolved. Each scientist built on the work of previous scientists to advance scientific thought.

**Near-Earth objects**

**What do the dinosaurs and asteroids have in common?**

Sometimes advances by scientists in one field can lead to scientific research in another—and impact society, too. A geological team led by Luis and Walter Alvarez found high concentrations of iridium in rock layers corresponding to the time when many dinosaurs went extinct. Since only asteroids contain that much iridium, the team theorized that the dinosaur extinction was caused by the impact of a massive asteroid. Many teams of astronomers are now hunting for observational evidence of *near-Earth objects* that could conceivably collide with the Earth, because we wouldn't want to suffer the same fate as the dinosaurs. Some people ask, If we discover an asteroid on a collision course with Earth, then what should society do about it?

**Climate change**

Many scientists are currently engaged in research on changes in Earth's climate, such as global increases in temperature and rising sea levels. These scientific results have led to broad societal discussions about the impact of human activity on the Earth's climate. The research has motivated many people to explore alternative energy sources that generate few or no carbon emissions, such as solar and wind power.

# Section 2 review

The skills of scientific inquiry and problem solving are at least as important as physics facts and knowledge. The *scientific method* is the process of posing questions about how things work, proposing hypotheses to explain physical phenomena, and continually testing and refining hypotheses by comparison with observational *scientific evidence.* Scientific evidence must be *objective,* which means it must describe only what actually occurs. Popular or historical opinions and interpretations do not qualify as evidence. Scientific evidence must also be *repeatable,* which means others must be able to make the same observations and get the same results. *Experiments* are controlled situations specifically set up to collect scientific evidence.

| Vocabulary words | hypothesis, theory, objectivity, reproducibility, experiment, scientific method, error, average, standard deviation, standard error |
|---|---|

## Review problems and questions

1. Which of the following three questions *cannot* be answered by science?
   a. What is the age of planet Earth?
   b. What is the maximum speed of a cheetah?
   c. Are humans entitled to use all the natural resources of Earth?

2. Identify whether each of the following is scientific evidence. Give reasons for your choices.
   a. The speedometer on a car reads 57 miles per hour.
   b. A photograph shows the planet Venus at a specific position in the evening sky.
   c. A newspaper headline announces the discovery of a previously unknown fundamental particle.
   d. You notice that the temperature outside is 27°C.
   e. A teacher tells you that the atomic mass of carbon is 12.0 grams per mole.

3. Suppose you want to invent a better device for catching a mouse. List five different qualities or capabilities your invention must have. For each one, list one possible solution for how your device might achieve that quality or capability.

4. Is a scientific theory or hypothesis a more reliable explanation of how the physical world works?

5. Describe the control that could be used in each of the experiments described below.
   a. A drug company tests the effectiveness of a new pill for treating headaches.
   b. NASA tests various seeds to determine which types of grain germinate best in a weightless environment.
   c. Acousticians test the effectiveness of sound barriers of varying heights and materials in blocking highway noise.

6. Write one paragraph (of three to five sentences) that defines science. Your paragraph should include the concept of theories, hypotheses, and observational evidence.

Physics is not a subject that exists in isolation from everything else you learn. Historically, science has been closely intertwined with advances in technology, the engineering design process, and mathematics—and the connections will continue into the future. This book will help you draw those connections among the STEM (science, technology, engineering, and mathematics) disciplines.

## Language of physics

**Qualitative and quantitative**

Suppose someone offered to pay you "a lot" for your favorite shirt. The description "a lot" is a **qualitative** statement because it does not give a numerical value. But what is "a lot"? You need a **quantitative** answer to that question (such as $100) to make a decision. Physics involves both qualitative and quantitative explanations.

**Language of mathematics**

Mathematics is the language of quantitative relationships. It is a *language* with symbols and grammar, just like English, Spanish, Arabic, or any other language. While there are 6,000 spoken human languages, there is only *One* mathematical language—everyone on the planet speaks the same one. If you were born in the USA, English may seem easier than math, but that is because you have a lot more practice at it.

**Equations**

One of the goals of physics is to express natural laws in the *quantitative* language of equations. Equations are useful because they can be used to predict results—such as when, where, or how much—with numerical clarity. For example, the equation for speed is $v = d \div t$. If you travel 480 miles and it takes you 8 hours then this equation tells you that your average speed is $480 \div 8 = 60$ miles per hour. An equation defines the relationship between its variables for *all values those variables can have.*

**English**

The speed is the distance traveled divided by the time taken.

**Mathematics**

$$v = \frac{d}{t}$$

**Math icon in *Essential Physics***

You will use equations in math class as well as physics. The math icon in the e-Book will guide you to real-world physics applications of concepts you learn in math class.

The math icon

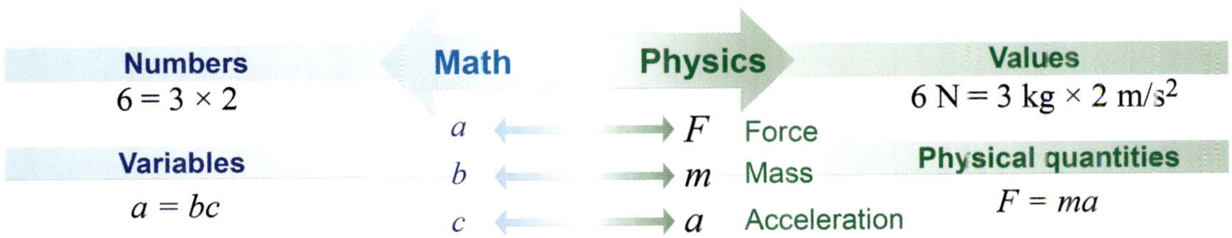


**An example of math and physics**

Physics teaches you how to use math in the real world. A math equation might say that six equals three times two, or $a = bc$. Physics uses a similar relationship that says that a force of six newtons is needed to accelerate a mass of three kilograms by two meters per second every second, or $F = ma$. The variable names are different but the techniques are the same. Throughout the book you will find examples of how techniques you learned in math class can be applied to physics and engineering.

# Engineering

**Engineering is the application of science**

**Engineering** is the application of science to design products or inventions that meet human needs. Engineering is very *creative.* A remarkable new product, such as a cellphone, has to be created in the mind of the engineer before it can be built of metal and plastic. Engineers invent **technology** that could make the future cleaner, safer, healthier, and more productive. Engineering also has a dark side because weapons are created by engineers, and some technologies have harmful side effects.

**We all use technology**

It is in our common human interest for everyone to understand the basic engineering design principles and trade-offs inherent in all technologies. This book features many *case studies* that show you how aspects of physics directly affect the design of technology.

**Engineering design**

An engineering problem typically involves designing a device that must perform a function. The design has constraints such as cost, energy consumption, reliability, or longevity. Engineering problems are fundamentally different from textbook problems because *there is no single right answer.* There may be many designs that achieve the goals. For example, more than a dozen different kinds of wind turbines are being tried as alternative energy sources. Different designs may best fit different constraints.

## Engineering Design Cycle

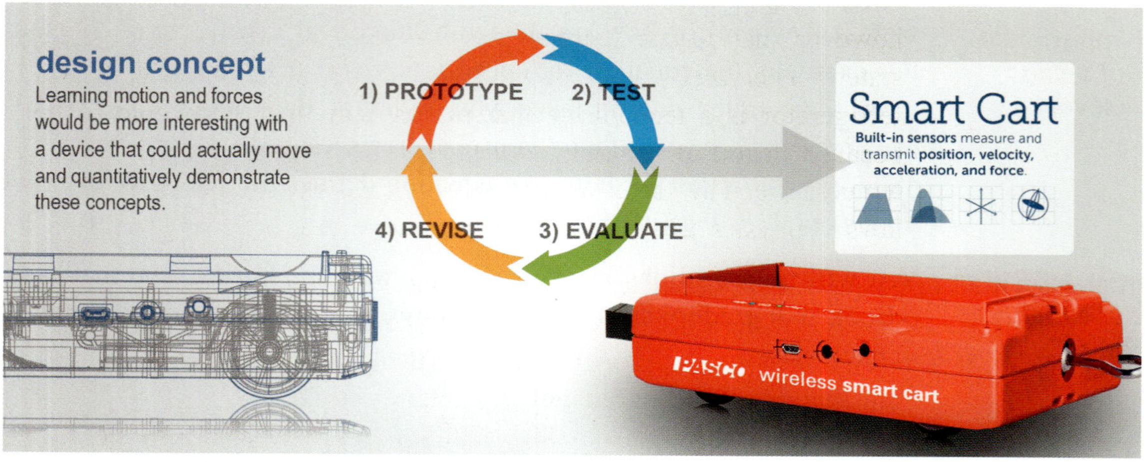


**The engineering method**

The *engineering method* typically starts with a design concept. Complicated goals are broken down into smaller goals that can each be solved with a *subsystem.* The design is made real with a *prototype,* a functional model that can be used to *test* and see how well things work. The results of tests are used to revise the design and make it better. The cycle of prototype–test–evaluate–revise is repeated many times before the final product is ready. Often the final product looks nothing like the initial concept! Much like the scientific method, the engineering method is a process of envisioning, testing, and improving a design until, after many iterations, it best solves the initial problem.

**Manufacturing**

Before a product such as a Smart Cart can be sold, it must be manufactured. Designing for manufacture is another interesting area of engineering. The manufacturing engineer creatively juggles aspects such as shape and color with processes such as molding or machining, and materials such as plastics or metals. A good design is durable, meets the need, and is attractive and practical to manufacture.

**What *technology* means**

Technology is everything in the built, or human-constructed, world. Every device, invention, material, or process created by human minds is technology, from a shovel to the space shuttle, and from hieroglyphics on a papyrus scroll to the Internet. In this course we will examine the fundamental physics behind many interesting technologies.

*Technology*
**Smartphone**

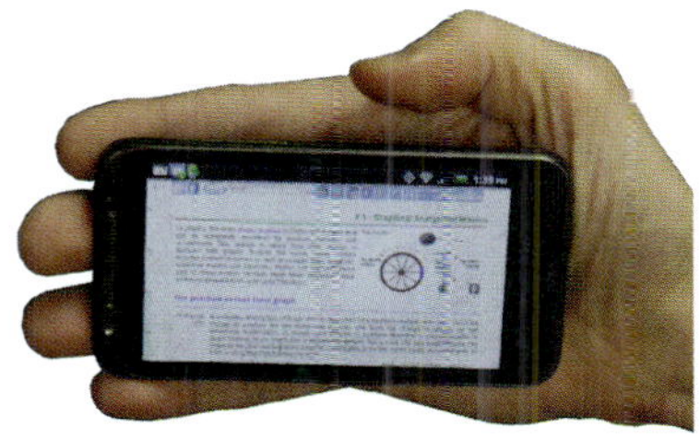

| *Subsystem* | *Physics* |
|---|---|
| Communication | Electromagnetic waves |
| Display | Optics, diodes |
| Touchscreen | Capacitance |
| Processor | Semiconductors |
| Memory | Semiconductors |
| Battery | Electrons, atoms |
| Case | Force, strength, mass |

**Physics and technology**

One book can hardly scratch the surface of the diversity of technologies created by human inventiveness. Instead, our goal is to show how fundamental concepts, such as electricity or waves, are employed in different technologies. We need creative new ideas to solve some of humanity's challenges—crises in medical care, food production, transportation, environmental degradation, and renewable energy.

**The importance of being clever**

All technologies obey all the laws of physics, with no exceptions. A clever approach, however, may lead to technologies that seem impossible. For example, the laws of physics state that thermal energy flows from hot to cold, or from higher to lower temperature. A refrigerator is a technology that makes heat flow from cold to hot. Warm food freezes because heat flows out of the food and into the warmer room, which is at a higher temperature. The secret of refrigeration (and air conditioning) is revealed on page 749. No laws of physics are violated but some very clever inventing has been done.

**Finding technology applications**

The technology and engineering icons in every chapter of the e-Book provide you with a quick way to see how physics concepts in the chapter are applied in various technologies. For example, digital imaging is a technology that we have all come to take for granted. Only 20 years ago a photograph was taken with chemical film, which had to be sent out and processed. Often you could not see your pictures for two weeks until the film came back! Today, you press the view button on a digital camera, or your cellphone, and the picture is taken instantly. You can take another if it did not come out right. How does the technology of digital imaging work? How are colors represented with numbers?

**Rainbows and light**

Where do all the colors of the rainbow come from? After a rainstorm, on the side of the sky opposite the Sun, you might see a rainbow if you're lucky. Light from the Sun scatters off of tiny water droplets in the air in such a way that the light is separated into its constituent colors, as you will learn on page 643. Rainbows are revealing something about the fundamental properties of light.

Where do the colors of the rainbow come from?

**White light**

The light from a lamp or the Sun may look white (or yellowish-white for the Sun), but white is actually a combination of colors! The perception of white comes from an equal blend of red, green, blue, and other colors, so that no single color dominates. Rather than waiting for a rainbow after the next rainstorm, you can demonstrate that white light contains all colors using a glass *prism*. The triangular glass shape of a prism *disperses* each color at a slightly different angle, producing the rainbow of colors.

Dispersion of white light by a glass prism

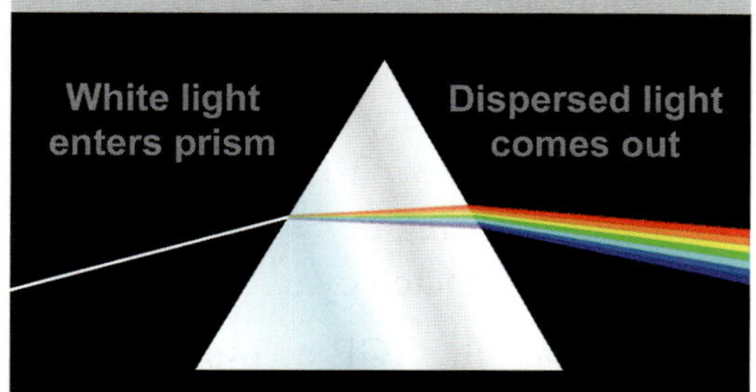


White light contains all the colors of the rainbow.

**ROYGBIV**

The order of the colors dispersed by a prism—or by a rainbow—is usually remembered by the acronym ROYGBIV. The letters stand for red, orange, yellow, green, blue, indigo, and violet. White light consists of all the colors, whereas blackness—as you can see on a dark, moonless night—is the absence of light.

**How do we sense color?**

If white light consists of a blend of colors, then how does the human eye or a digital camera sense that light? How about red light? Yellow light? Magenta? The sensation of color is actually a perception of the *energy* in light.

**Energy, color, and perception**

Within the range of *visible* light, lower energies appear red and higher energies appear blue. The human eye contains three color sensors that send nerve impulses to the brain. The lowest energy sensor responds most strongly to red light, the medium energy sensor to green light, and the highest energy sensor to blue light. There is some overlap in sensitivity, so *yellow* light stimulates the red and green sensors equally. Your brain "sees" yellow as an equal stimulus from the red and green sensors. A digital image tricks the brain into seeing yellow, and most other colors, using only varying amounts of red, green, and blue.

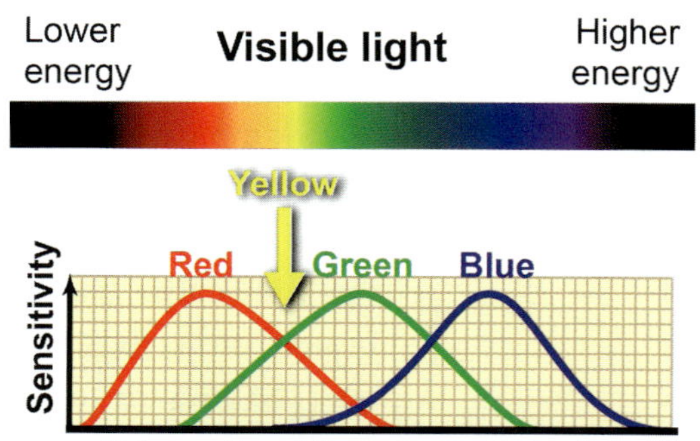


Schematic of the response for the three sensors in a human eye

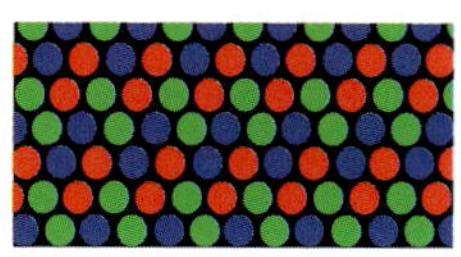

Pixels in a digital image fool the eye into seeing all colors using only red, green, and blue.

# Investigation 1C

## RGB color matching

**Essential questions** *How do computers and TV screens create color using only red, green, and blue?*

Interactive simulation

A computer or television screen produces a wide variety of colors using combinations of only red, green, and blue. RGB colors are used in HTML to create webpages, including this one. A color wheel is a simplified way of representing combinations of *two* of these primary colors. In this investigation you will construct combinations of all three primary colors in order to match various colors in a palette.

### Part 1: Creating simple colors

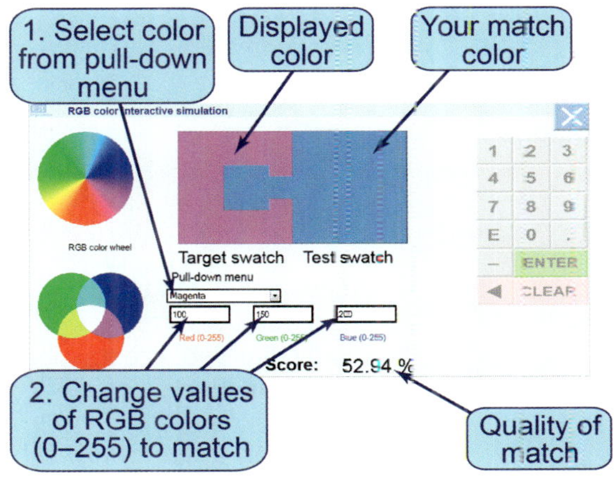


1. On your computer, launch the application in the electronic resources.
2. Match the first few colors using values of only zero or 255 for each primary color.
3. Select *magenta* from the pull-down menu. Enter values for red, green, and/or blue to match it.
4. Repeat to match cyan, yellow, white, and black.

a. What are the RGB colors for magenta? Cyan? Yellow?
b. What is the RGB color for white? Explain why by referring to a prism.
c. What is the RGB color for black?
d. Color wheels offer two different visual (yet scientific) explanations of how colors combine. Analyze each visual representation and describe how each explains how different colors of light combine to create other colors.

### Part 2: Creating complicated colors

1. Now use any value from zero to 255 for each color in the interactive simulation.
2. Start by matching light gray and dark gray. A good match has a "score" of greater than 95%.
3. Pick five different colors from further down the pull-down menu and find a good match for each one.

a. What is the RGB color for light gray? Dark gray?
b. Make a table of the names of the five colors you matched and their corresponding RGB values.
c. Select the color gold from the pull-down menu. Does the color appear to be a good match with what you think of as gold? Why or why not?
d. Critique how well the two different color wheels explain how colors combine based on your experiments with the interactive element.
e. Evaluate which of these two color wheel representations is more useful in helping you determine the red, green, and blue values to match the colors illustrated at right.

## Additive and subtractive primary colors

**Additive primary colors**

The three primary colors red, green, and blue are called the **additive primary colors (RGB)**. The additive color process occurs in the lamps inside a computer projector or an LED computer or television screen. Each has a collection of tiny lamps—some red, some green, and some blue—that illuminate in relative intensity to create any color in the rainbow. When two adjacent, tiny pixels on your LED screen show blue and red, your eye sees them combined together as magenta. If blue, red, and green all illuminate next to each other, then your eye sees white light.

**RGB additive primary colors**

On your computer screen, RGB primary colors add together to make other colors.

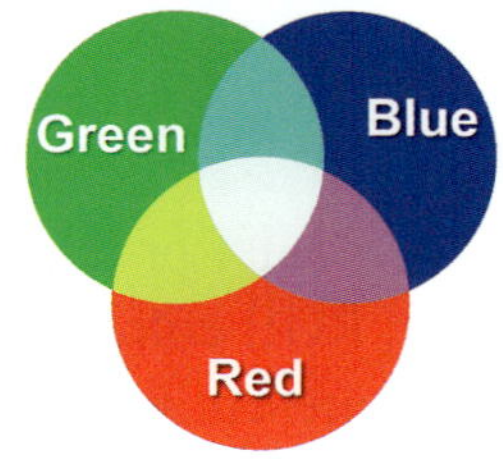


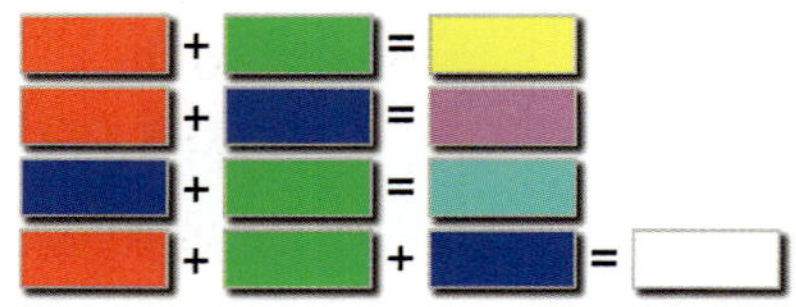

**Digital images**

This digital image is made up of many individual red, green, and blue pixels.

Digital images are made of tiny dots, or *pixels,* of color; a one megapixel image has one million dots. Each pixel in the image is captured by three tiny sensors in the camera, which, like the human eye, respond to red, green, or blue light. In turn, each sensor records a number from 0 to 255 corresponding to the intensity of the light it receives. A one megapixel digital image is a string of three million numbers that prescribe the intensity of red, green, and blue for each pixel. The maximum is assigned to 255 because computers use binary numbers and $2^8 = 256$.

**Digital information**

Digital (as opposed to analog) information can be stored reliably in computer memory. Digital files, such as images or music, can also be copied quickly, inexpensively, and without information loss; analog information degrades with each subsequent copy. On the other hand, many people have expressed concern about the security of digital personal information, the theft of copyrighted material, and the ease of accidental deletion.

**Reflected light**

How are the colors of the ink on a book page different from the colors on a TV screen? You see the light on the page *reflected* from the Sun or a light bulb. When printed, the red color on this page was created by a combination of pigments that reflect only magenta and yellow but absorb all the other colors within the white light shining on it.

**Subtractive primary colors (CMYK)**

Printed illustrations use cyan, magenta, and yellow pigments in the ink to create any possible color. Professional printers realized that they could not obtain dark blacks with only cyan, magenta, and yellow, so they added a fourth ink—black or "K." Cyan, magenta, yellow, and black (K) are known as **subtractive primary colors (CMYK)**. This is called a *subtractive* primary color process because the pigments *remove* the light upon reflection, rather than being sources that emit light.

**Subtractive primary colors (CMYK)**

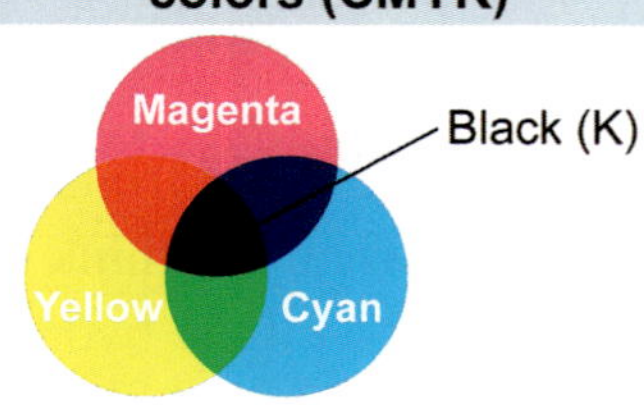


On this *printed* page, the CMYK pigments reflect cyan, magenta, or yellow colors while absorbing (*subtracting*) the other colors.

**Too good to be true?**

Have you ever heard claims about a product that will mop up any spill, reverse any dent in a car, or otherwise change your life? Whether or not you go into science or technology for a career, it is important for you to know how to evaluate product claims—to separate the reasonable claims from the far-fetched ones. If it sounds too good to be true, then it probably isn't true!

**Evaluating promotional claims**

An important step in evaluating the labeling of products or services is to *infer* whether their claims are reasonable according to the science you know. Reasonable claims must

1. be specific, observable, and testable, such as "reduces swelling up to 60%";
2. be supported by objective evidence from a reputable source, such as doctors, the FDA, the EPA, or independent lab tests;
3. be consistent with the laws of science, such as conservation of energy; and
4. not make unsupportable statements—such as "the ultimate."

Clark Stanley's Snake Oil Liniment

Have students in your class bring in toothpaste boxes. Each student verbally states the product's scientific claims from the packaging and then leads the class in discussing whether the claims satisfy the requirements above. Are the claims reasonable?

**Scientific evidence, not testimonials**

Late night television ads often feature "real people" saying why they like the product. Their stories are examples of *anecdotal* evidence and are generally not scientific because they have not followed standards of testing, analysis, and independent review. Anecdotal evidence may be *true*, but *empirical* evidence is far more reliable and harder to falsify.

**Creating energy for free**

One company recently claimed that their fuel cell product could create a substantial increase in gas mileage for cars. The U.S. Federal Trade Commission, however, examined the company's promotional claims and *inferred* that the product would violate conservation of energy—it would create energy for "free," out of nothing! The FTC pointed out that there was no scientific evidence supporting the promotional claims, and the company agreed to stop selling its product as an energy source.

**Claims about services**

Marketing or promotional claims about services should be judged by the same criteria as for products. A cleaning service claiming to be the "best, friendliest, and most reliable" should offer evidence to support the claim. For example, the company could state that an independent organization found that its cleaners had a 100% on-time arrival rate and offered the same cleaning services for the lowest price in town. These statements are empirical, testable, and objective.

***"Our cleaning service is the best, friendliest, and most reliable."***

**Fine print**

Sometimes an advertisement will "bury" relevant information about its product, hoping consumers won't bother to read the "fine print." Critique this product claim (at right): Is it reasonable? Why or why not?

***Our product will increase your car's gas mileage by 200%.****

**If you also attach a team of horses to tow your car.*

**Is this product claim reasonable?**

**Note taking**

Taking notes is an important part of learning because writing is a good way of remembering. But, *what should you write? Do you write every word?* Fortunately the answer is *no*. Written notes should *summarize* the key points, not capture every word. Ask yourself this: What is the most important concept? What words or techniques are new to me?

**Listening comprehension**

**Listening comprehension with *Essential Physics***

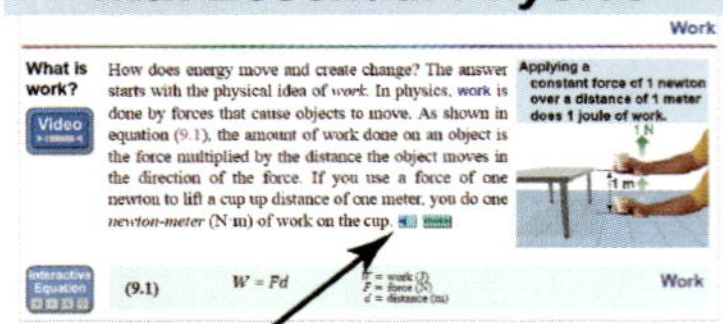

Press this button to hear the paragraph read out loud.

Sometimes you may find it useful to *hear* the text of *Essential Physics* read out loud, in order to learn new vocabulary or to better study the content of the book. The e-Book provides a button at the bottom of each paragraph that allows you to hear the entire text. Try it: Choose a paragraph in this chapter, listen to it, summarize it with one written sentence, and speak the sentence to a study partner and check your agreement.

**Outlining**

Many students find it useful to *outline* the important concepts in the book or in a verbal class presentation. *Essential Physics* has been written with "one idea per page," so the page titles can easily be used as main headings in an outline. Every paragraph on a page of *Essential Physics* also has its own short title that can form the subheadings in an outline. Vocabulary words, key illustrations, and equations can form additional subheadings for your outline.

**Outlining and *Essential Physics***

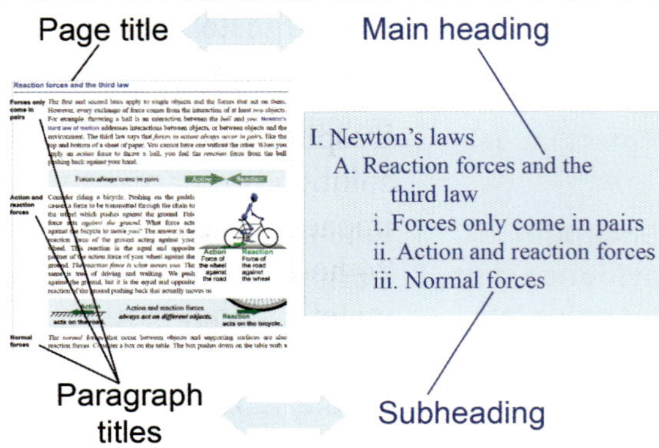


**Illustrations**

A picture is worth a thousand words, right? Science class is full of illustrations: diagrams, graphs, images, and even animations! These visual elements are not just for decoration; they have meaning. Choose any diagram in this book and ask yourself what message the diagram is trying to convey. Try writing down a one–sentence summary of the main point conveyed by each illustration you encounter—preferably using science words!

**Communicating to get more information**

What should you do if you need more information or don't understand a concept? Your classmates and teachers are some of your most important resources. If you have a question while studying after school, write it down to ask in class or to ask a friend. Try to use appropriate science vocabulary. Preparing a question to ask in class can help you—and other classmates—to grasp difficult concepts. It is even a good idea to write down a question or two every time you study!

### Energy use in the USA

On average, the USA uses about 4,000 terawatt–hours of electrical energy per year. About one-third of this energy powers our homes, including refrigerators, electric lights, tools, and appliances. Another one-third powers our businesses, schools, and stores. The remaining one-third powers our industries, such as car manufacturing, metal refining, textiles, machinery, and chemical manufacturing. Where does all the electrical energy come from?

**U.S. electrical energy consumption**

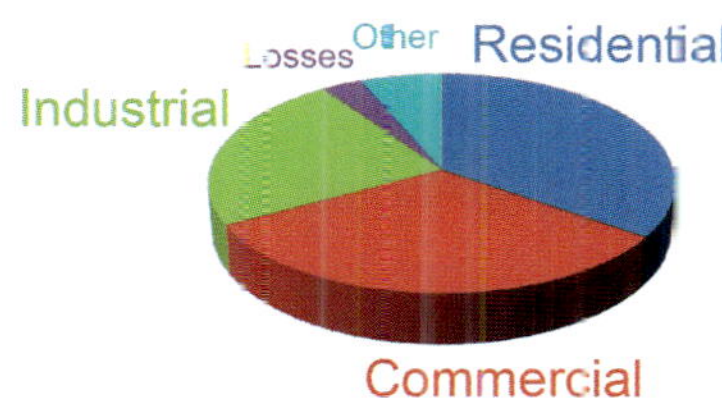


**U.S. electrical energy production**

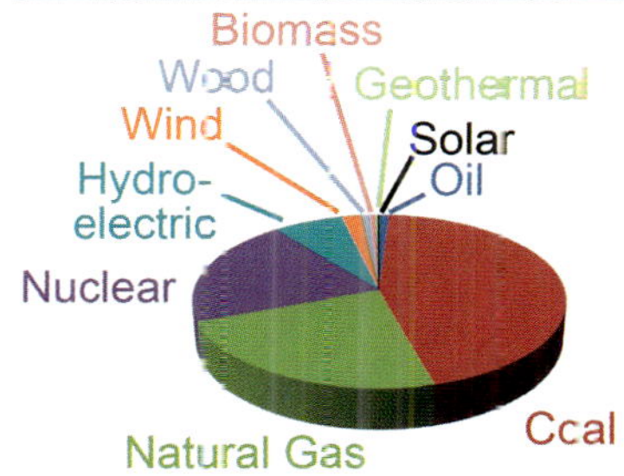


### Where does our energy come from?

Energy cannot be created but only transformed from one form into another. Today, about 70% of U.S. electrical energy is produced by generators that burn natural gas, coal, or oil. Most of the rest comes from nuclear reactors (20%) and hydroelectric dams (7%). A tiny fraction (<2%) comes from wind and an even tinier fraction (< 0.1%) comes from solar power. Energy consumption is plotted in the illustration at right. It is easy to infer from this pie chart that coal is the largest sector for energy production in the USA, because coal visually has the largest "wedge" in the pie chart.

### Potential of solar power

Despite its small contribution today, many believe expanding solar power offers the best solution to providing more electricity while also limiting the negative environmental impacts of energy production. The Sun emits an enormous quantity of energy. Over a 24 hour period, an average of 250 joules per second (or watts) of sunlight reach each square meter of the ground. The USA has a land area of 9.2 million square kilometers, or $9.2 \times 10^{12}$ $m^2$. If each square meter receives 250 J/s, the average energy in sunlight falling on the USA is 2,280 trillion J/s. *This is around 5,000 times the entire country's average electric power use!* So, at first glance, solar power appears to be an ideal, plentiful, and clean energy source.

### How a photovoltaic cell works

**How a photovoltaic cell works**

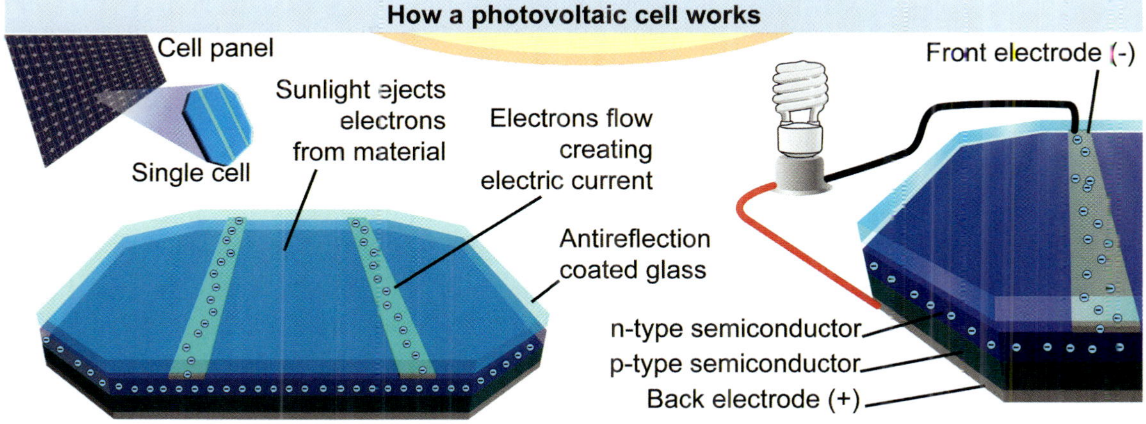


A **photovoltaic cell** (PV cell) transforms light directly into electrical energy. The basic physics of this process is fairly well understood. Light absorbed by a semiconductor *p-n junction* ejects electrons from certain atoms in the depletion region. The ejected electrons are collected and forced to flow through a circuit and carry electrical energy.

## Photovoltaic cells: the technology of solar power

**Typical solar array installation for a house**

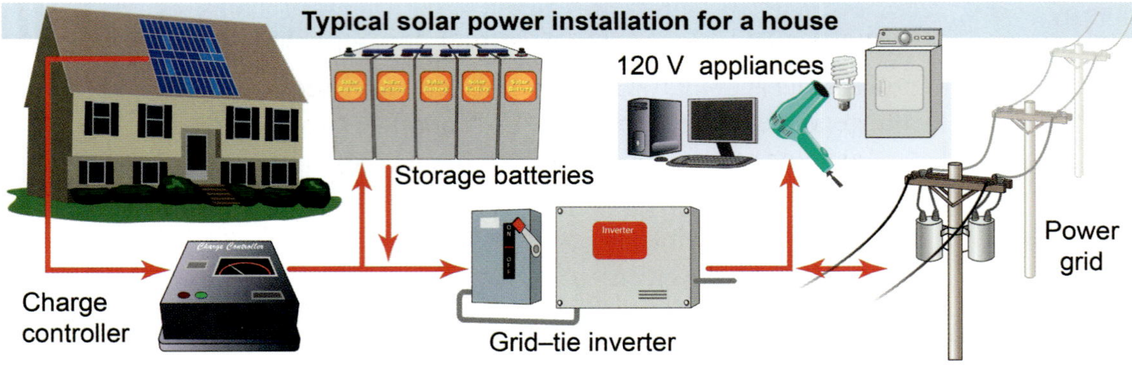


A typical solar power installation for a house requires a few electrical devices in addition to the solar panels. Since the Sun only shines by day, a battery is needed to store electrical energy for use at night. The solar panels may produce extra electrical power during sunny days, which can be sold back to the main electrical grid, but during cloudy days more power may need to be drawn from the grid. A grid–tie inverter controls the interactions between the outside power grid and the electricity produced by the solar panels.

| Design challenges for photovoltaic cells | |
|---|---|
| To catch the ejected electrons requires expensive nanofabrication techniques similar to how computer chips are made. | PV cells are expensive. |
| Ordinary PV cells convert only 12–18% of sunlight into electrical energy. The rest is converted to heat. | PV cells have low conversion efficiency. |
| Over time, UV light damages the microstructure of PV cells and can quickly reduce efficiency dramatically | PV cells are degraded by UV light. |
| PV cells are delicate and easily damaged by dirt, ice, wind, and rain. | PV cells are easily damaged by the environment. |
| PV cells do not work once the temperature gets too much over 30°C. | PV cells can easily overheat. |

**Thin silicon wafers are easily broken**

The most efficient *crystalline* PV cells are made from single-crystal silicon that has been sliced into extremely thin wafers. Like glass, silicon crystal is very brittle and easy to crack. The wafers used for PV cells are typically 180–220 μm thick, about the same as two sheets of paper! A single crack can break the electrical connection and ruin the PV cell. For this reason, the brittle silicon wafers are *encapsulated* in a tough, transparent resin. Like the front sheet, the encapsulant resin must be transparent to light and it must not constrain the thermal expansion or contraction of the wafer over a wide range of temperature.

**Thin-film photovoltaic materials**

Crystalline silicon is expensive to manufacture. A new form of PV technology called *thin-film* photovoltaics uses very thin layers of semiconductor materials, such as copper indium gallium selenide, deposited on a substrate. Thin-film PV cells are not as efficient but cost much less to manufacture.

**Design challenge**

Design a solar power array for the roof of a house that:

a. is sited at the latitude of your school, city, and/or town;
b. generates an average of 50 kWh per day, equivalent to 18,000 kWh per year (the average energy consumed by a household);
c. pays for itself within ten years, compared to the cost of grid power of \$0.12/kWh; and
d. delivers a peak energy of 70 kWh per day in July in order to run the house's air conditioning.

Designs are evaluated based on achieving these criteria at the lowest total cost.

**Identify need**

What is the need or problem that is addressed with a household solar power array? Which design criteria are the most important to meet? Write your explanations in your report.

**Simulating a solar power array**

In this interactive simulation, you will design a solar power system by modifying (and improving) an existing design that uses photovoltaic cells. Some of the parameters are fixed by typical commercial value: cost for PV arrays (\$300/m$^2$); installation cost for electrical hookups, such as the charge controller and grid-tie inverter (\$3,000); conversion efficiency from incident light into electrical energy (18%); and battery storage efficiency (100%). Because this is a simulation, the design process will not include the engineering process step of prototyping your design.

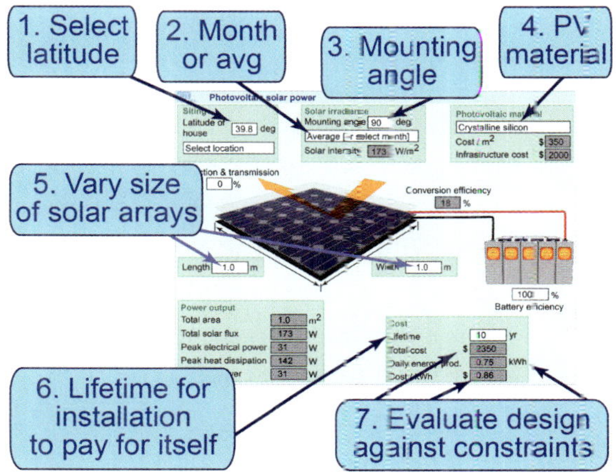


**Design**

Formulate testable hypotheses to address the first three design constraints. The parameters that should be varied are area (length and width) of the PV cells, mounting angle for the installation, month or year (start by using "Yearly Average"), and lifetime of the installation.

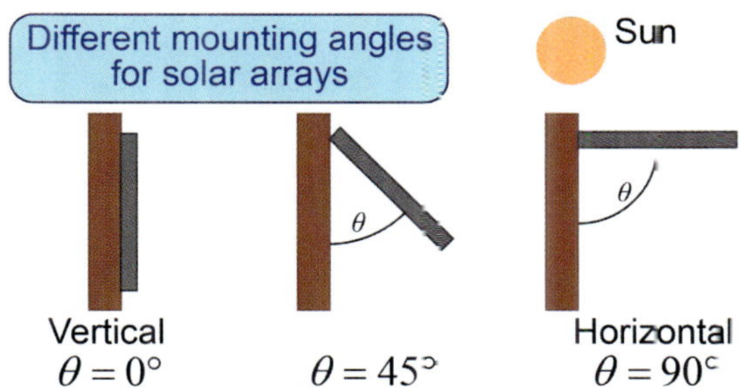


**Test**

In this simulation, you can vary the design parameters to determine how to optimize them. Change the design parameters to maximize or minimize the key design criteria: daily power production (in kWh), cost per kWh, and total cost.

**Evaluate**

Analyze your data to determine which solutions best meet the first three design criteria. Write down the parameters of your design and the resulting key design elements.

**Revise**

Now look at the power production for the month of July. Does your design meet the last of the four design criteria? If not, revise your design to improve it to meet this last criterion. How did your design change? Make a short presentation to your class of your final design along with data to support your design decisions. As a class, discuss the best design elements. Revise your design to incorporate any elements you might have missed.

Chapter 1

# Section 3 review

The study of physics develops a wide range of skills and enables us to live longer, richer lives. Physics students use both qualitative and quantitative statements to describe the natural world. The engineering design cycle uses physics (and other sciences) to develop technologies that meet specific human needs. Physics reveals the true nature of many commonplace phenomena, such as light and color. Physics provides a foundation for numerous careers, and scientific training can be used to evaluate promotional claims. Physics-based technologies range from the digital camera to photovoltaic panels that can reduce our dependence on fossil fuels.

| Vocabulary words | quantitative, qualitative, engineering, technology, additive primary colors (RGB), subtractive primary colors (CMYK), photovoltaic cell |
|---|---|

## Review problems and questions

1. Research the connection between physics and a career of your choosing. Interview a person who works in that career. Based on your research, describe at least three ways in which knowledge of the science of physics is needed for a person in that career to do his or her job. Present your findings in a five-paragraph essay by following these steps:
   - Be sure to take careful notes while fact-finding.
   - Organize and outline your notes before you begin writing.
   - Assign a topic to each paragraph.
   - Check your spelling, grammar, and punctuation after writing a first draft.
   - Finally, read your draft aloud (to a friend or family member or to the person you interviewed). Consider revising any sentences that are hard to read aloud or that confuse your audience.

2. Describe each of the following scientific observations as *qualitative* or *quantitative.*

   a. "The car was going pretty fast."
   b. "The water temperature was 80° when the crystals dissolved."
   c. "My backpack must weigh at least 20 pounds."
   d. "It took the butterfly a really long time to cross the field."

3. Your eyes contain three different color sensors.

   a. List the three colors that correlate with these sensors.
   b. Which of these colors of light corresponds to the highest energy, and which corresponds to the lowest energy?
   c. Do your eyes perceive the range of colors through an additive color process or a subtractive process?

4. This section of *Essential Physics* states that the engineering design process is different from the process of solving typical textbook problems. What is the evidence given to support this statement? Explain your conclusion in a complete sentence.

# Chapter review

## Vocabulary

Match each word to the sentence where it best fits.

### Section 1.1

| | |
|---|---|
| **atom** | **energy** |
| **force** | **frequency** |
| **information** | **matter** |
| **quantum physics** | **wave** |

1. In order for a car to stop, its brakes must apply a/an _______ to the wheels.
2. The stuff that makes up our bodies is called _______.
3. _______ is the field of study dealing with physical phenomena at the atomic and subatomic scale.
4. The basketball bounced up and down with a/an _______ of once every two seconds.
5. The arrangement of letters on this page and the black dots of ink that comprise them convey _______.
6. The smallest quantity of hydrogen is a/an _______.
7. A battery converts chemical _______ into electrical _______.
8. Moving one end of a bed sheet repeatedly up and down causes a/an _______ to travel across the length of the sheet.

### Section 1.2

| | |
|---|---|
| **experiment** | **hypothesis** |
| **objectivity** | **reproducibility** |
| **scientific method** | **theory** |

9. The steps of identifying a problem, suggesting a possible answer, predicting the observable effects whether or not the answer is correct, constructing an experiment to test the possible answer, and then evaluating the results are all part of the _______.
10. Various scientific advances of the early 20th century led to the quantum _______ of how matter behaves at the subatomic level.
11. How would you disprove the _______ that the Earth is flat?
12. If there are many unpredictable forces and energies affecting your laboratory equipment, then it is very difficult to perform a controlled _______.
13. When Donald wrote in his laboratory notebook his personal feelings about whether or not the gravitational force should exist, he had lost his _______ about the scientific topic.
14. When Max was able to describe his experiment so thoroughly that Maurice could get the same results, Max's experiments now had _______.

### Section 1.3

| | |
|---|---|
| **additive primary colors (RGB)** | **engineering** |
| **photovoltaic cell** | **qualitative** |
| **quantitative** | **subtractive primary colors (CMYK)** |
| **technology** | |

15. Examples of a/an _______ observation are 37 m, 9.37 s, and 100 mph.
16. Examples of _______ include a software program, an electronic device, or a fabrication process.
17. The _______ process may involve taking a set of physical principles to design and construct a device subject to a series of constraints.
18. A computer monitor combines the _______ to create any color in the rainbow.
19. Sunlight is converted into electricity using a/an _______.
20. Examples of a/an _______ observation are bluish, wet, and coarse.
21. The bluish green light reflected off the wall was created through a/an _______ process.

## Conceptual questions

### Section 1.1

22. An investigation uses a motion-sensing car to collect data. Which of the following is most likely to come *first* in the procedure?
    a. Move the car with your hand.
    b. Start the data collection system running.
    c. Compare the data with your actual motion.
23. Describe three ways in which you exerted forces earlier today to cause motion.

# Chapter review

## Section 1.1

24. Describe three ways in which you used energy earlier today.

25. Describe three kinds of waves (or periodic motion) that you experienced earlier today.

26. Describe three different ways in which you interacted with electricity or magnetism earlier today.

27. Describe three different ways in which you used optics earlier today to see light.

28. Describe three different devices that you used earlier today that were made possible, at least in part, by atomic physics.

29. Does physics apply to chemistry? Provide examples to defend your position.

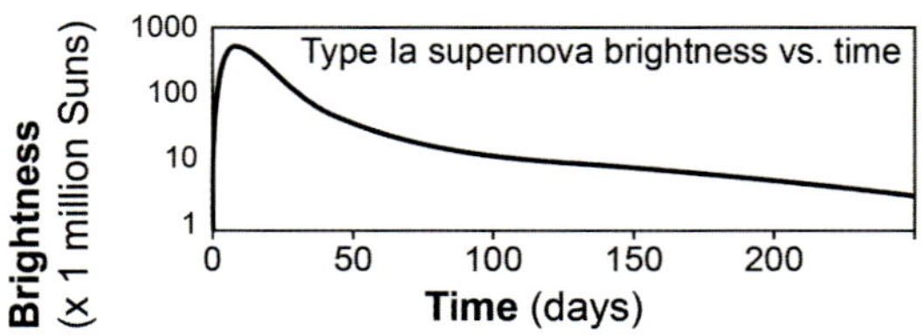


30. An astronomer observed a rare, bright, exploding star called a supernova and drew a conclusion about how its brightness changed over time. Are her observations repeatable? Why or why not?

31. Analysis of a physics experiment may involve equations with many variables and pages of mathematics. Analysis of a biology experiment might use simpler math. Does this make physics more complicated than biology? Explain your reasoning.

32. We observe that time moves forward, from the past into the future. We do not observe any examples of time moving backward. Now suppose that time could go backward, and that a backward–traveling time machine were possible. Think of a situation in which the rules of cause and effect are violated, therefore making backward time travel impossible in the universe as we know it.

## Section 1.2

33. What is the difference between a theory and a hypothesis? Which one might eventually lead to the other?

34. Are qualitative observations scientific?

35. Research and write a short report on the role of at least three contemporary scientists conducting research on climate change. Describe the impact that each has had on the development of scientific thought on the subject.

36. Define science using the vocabulary words theory, experiment, and hypothesis. Note the limitations of science.

37. Do the adjectives below describe hypotheses, theories, or both?
    a. tentative
    b. testable
    c. capable of being observed or supported by experimental evidence
    d. well established
    e. highly reliable
    f. based on natural physical phenomena
    g. subject to change

38. Professor Knowsalot proposes a clever hypothesis to explain something. He has not tested his hypothesis by comparing it with any real observations. Nonetheless, the professor is sure his hypothesis is correct, because he has never been wrong before.
    a. Is this hypothesis correct?
    b. When would this hypothesis be incorporated into a theory?
    c. Once his hypothesis became part of a theory, would Dr. Knowsalot be the only one who can test the theory?

39. Can scientific theories change, and if so, how?

40. Write down a statement you might give on a radio station to describe why you think light has energy. Your statement should use the word "because" and present at least one observable property of light that supports your statement.

| Causes | Effects |
|---|---|
| Icy roads<br>Strong winds<br>Bright light | Windshield glare<br>Dangerous skids<br>Forces on sail boats |

41. The chart above lists three things that occur and three potential consequences that may follow from them. Pair up the correct consequence with its cause by speaking (or writing) three sentences that have the form "______ is/are a cause of _____."

42. Describe the steps of the scientific method.

43. What happens when someone discovers a natural phenomenon or develops technology that contradicts an existing theory?

44. Your friend has a hypothesis that there is an angry monster with 20 teeth at the center of every black hole. Is this a good scientific hypothesis? Explain.

45. Relativity has been tested by many independent researchers and is well established as a good description of natural and physical phenomena. Is relativity a theory or hypothesis?

# Chapter review

## Section 1.2

46. A geologist suggested that an asteroid impact caused the extinction of the dinosaurs and predicted that iridium would be found at a particular sedimentary layer of rock. Searching for iridium is observational testing, not an experiment. Is this a valid method to obtain evidence to evaluate the geologist's prediction?

47. ❬ Describe the difference between a scientific law and a scientific theory. Give two examples of each and defend how you classify each of your examples.

48. ❬ Suzy said, "The color of the fire hydrant appears red to *my* eyes." Is this an objective statement?

49. ❬ Stephen Colbert once said that he wanted "to let the free market decide what the atomic weight of carbon is." Does his idea fit with the principles of scientific inquiry? Why or why not?

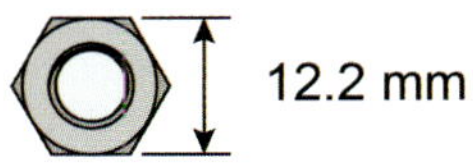


50. ❬ Ned measures the dimensions of a nut using a micrometer and writes down a single measurement in his notebook. What should Ned do to reduce the potential for errors in his measurement before moving on to the next step of his experiment?

51. ❬ Daniela wanted to improve the performance of her favorite race car. She designed and constructed a series of new cars, each with a different mass, engine, or aerodynamic shape. She then tested each car by driving it herself around a track to see which parameters improved the car's speed. What should she use as the *control* for her investigation?

52. ❬❬ Astronomers collect data by observing the sky with a telescope. Astronomers cannot do experiments that manipulate stars or galaxies. Do astronomers use the scientific method in their work? Can astronomers analyze scientific explanations in their work? Why?

53. ❬❬ Psychologists conduct an experiment to determine the effect of weather on mood. They compare the weather with the responses of a few people whom they asked how they feel that day. Is this an *objective* scientific experiment?

54. ❬❬ Research and write a short report on the role of at least three contemporary scientists conducting research on nuclear energy and power. Describe the impact of their work on society.

55. ❬❬ Geologists study the history of the Earth by digging down vertically through progressively older layers of sedimentary rock. How does the scientific method apply to this kind of research?

## Section 1.3

56. Give two examples of how mathematics is important in physics.

57. Describe the design method used in the engineering process.

58. Research a career that uses physics. Write a short report on the career, and include specific examples of how it uses physics.

59. A pizza delivery place claims that, to maintain freshness, they only start making your pizza after you order it. A pizza arrives six minutes after you order it. If it takes 8 minutes to cook a pizza, was the pizza made after your ordered it?

60. A barber advertises his haircuts using the following statements from customers:
"The best haircut I've ever gotten."
"The lowest price for a haircut anywhere in town."
"The nicest barber this side of the Rocky Mountains."
    a. Based on his advertisement, what kind of evidence do you infer he has to support his claims?
    b. Which of these claims about his services could conceivably be independently verified?

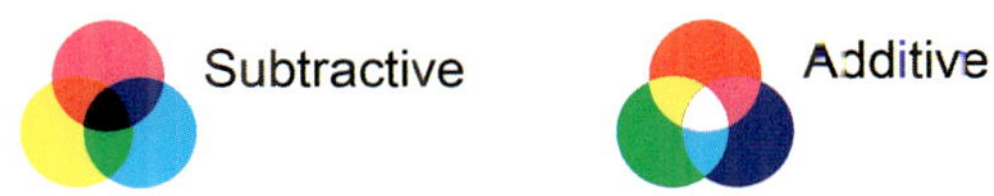


61. ❬ What is different about how additive and subtractive technologies create colors? Which applies to a computer monitor?

62. ❬ Explain how you can see ordinary, nonluminous objects in a classroom.

63. ❬ When the light from a light bulb bounces off the wall, are the colors best represented using an RGB or a CMYK color scheme?

64. ❬ Is black a color?

65. ❬ What combination of CMYK pigments can you use to create the color red?

66. ❬ Do advances in physics lead to developments in technology, or is it the other way around? Provide evidence to defend your position from things you have heard on TV or other media.

67. ❬ How are unforeseen problems addressed in the engineering design process?

68. ❬ List three kinds of technology you use everyday and list three physics concepts that go into each one.

### Section 1.3

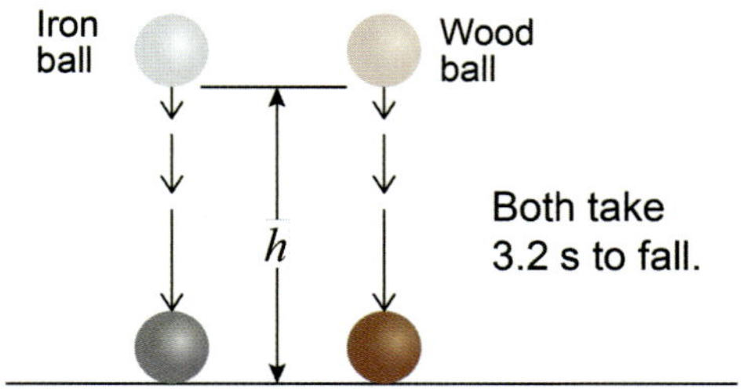


69. Sarah claims that heavier objects fall faster than lighter objects. To evaluate Sarah's claim, objects of the same size but different masses are dropped from the same height. Each takes 3.2 s to hit the ground.
    a. Is Sarah's experimental design effective for testing her claim? Explain.
    b. Is Sarah's claim supported by the evidence from the experiment?
    c. If Sarah's claim were true, what would happen when you tied two objects together and let them drop? Does this make sense?
    d. Propose a new claim that explains the rate at which objects fall and agrees with the evidence.

70. Devon claims that summer occurs when the Earth is closer to the Sun in its orbit, and winter occurs when the Earth is farther from the Sun in its orbit.
    a. When it is summer in the northern hemisphere, it is winter in the southern hemisphere. Using logic, show how this does or does not support Devon's claim.
    b. The Earth is closest to the Sun each year between January 2 and 5, at a distance of 147,098,290 km, and is farthest from the Sun between July 4 and 7, at a distance of 152,098,232 km. Does this evidence support Devon's claim?

71. Storing information digitally has advantages and disadvantages over older analog methods. Which of the following are true of digital information, such as images from a digital camera?
    a. Repeated recopying will result in lower quality.
    b. Digital material is easier to protect from copyright infringement.
    c. Relatively little storage space is required as compared to analog copies.
    d. Digital files can be transferred quickly and easily over long distances.
    e. Digital information is expensive to reproduce.

72. Two doctors advertise their services. One includes positive quotations from happy patients. The other says he has had the fewest complaints lodged against him at the state board of medicine over the past 10 years. Which promotional claim is more believable?

73. A company claims that its laundry detergent smells four times better than their leading competitor. After reading the fine print, you find that this is based off a survey of 50 people.
    a. Is their product claim based on empirical evidence?
    b. Provide a statement that the company could make about their laundry detergent that could be scientifically tested. Explain how it could be tested.

74. A medical doctor films an advertisement where she claims that a new pharmaceutical drug is effective at curing pneumonia. An enterprising reporter then uncovers the fact that the doctor has a paid position sitting on the board of directors of the company that produces and sells the new drug. How would you evaluate the claim made by the doctor?

### Quantitative problems

75. What kind of light is represented on a computer screen by an RGB value of
    a. (255, 255, 255)?
    b. (0, 0, 0)?
    c. (100, 100, 100)?
    d. (255, 0, 0)?
    e. (255, 0, 255)?

76. A company claims that their glue is so strong that 2 ml of their glue can attach a 1,000 lb object to a wall. You use this glue to hang up several pictures around your home, which all weigh less than 3 lb, yet all of them fall down after four days. Was this company's promotional claim correct?

77. A taxi service claims in its advertising that its service has the safest drivers in the city. This company has the second most car accidents of the four biggest taxi companies in the city. Of those same four companies, this one particular company has the most drivers who have been pulled over for speeding. Do you believe its claim?

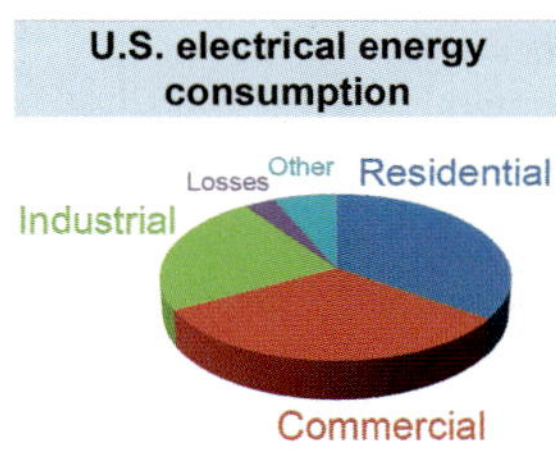


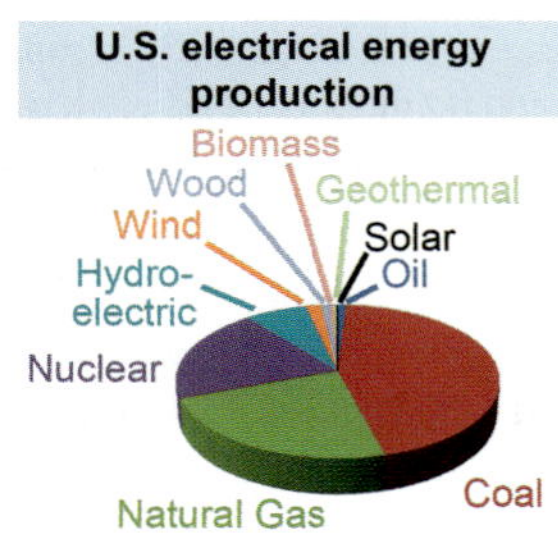


78. The graphs show electrical energy consumption and production.
    a. What group uses the most electrical energy?
    b. What sector produces the most electrical energy?
    c. How did you *infer* those two answers from a pie chart?
    d. If all commercial users agreed to cut their energy usage by 20%, which group would be the biggest user of electrical energy in the United States?

# Chapter review

## Standardized test practice

79. Which represents a valid approach to critiquing a scientific explanation?

A. using experimental testing
B. using observational evidence
C. analyzing the logical reasoning of the explanation
D. all of the above

80. Which is the best choice to refute a statement that Lake Toobin is round?

A. Argue louder than the person who made the statement.
B. Dismiss the person who said it as being uninformed.
C. Tell the person that water cannot have a round shape.
D. Design an experiment that might show that the lake is instead flat.

81. Which branch of science is primarily concerned with the properties and interactions of matter and energy on the most fundamental level?

A. psychology.
B. chemistry.
C. physics.
D. biology.

82. Which is the best description of how physics differs as a science from biology?

A. Physics is a quantitative science, whereas biology is a qualitative science.
B. Physics is mathematically based, whereas biology is observationally based.
C. Physics is concerned with the fundamental properties of matter, whereas biology is concerned with the interactions among and within complex, living organisms.
D. Physics deals with atoms and molecules, whereas biology does not.

83. What kind of light is represented on a computer screen by an RGB value of (255, 255, 0)?

A. white
B. blue
C. yellow
D. magenta

84. Within the scientific method, the statement, "The Moon is made of Swiss cheese," would best be described as a(n)

A. question.
B. hypothesis.
C. prediction.
D. conclusion.

85. Varian conducted an experiment in class that led him to conclude that Earth's gravity acted upward during his experiment. What is the best method to analyze his explanation?

A. Show him the page in a physics textbook that says that gravity acts downward, and tell him he must be wrong.
B. Attempt to repeat his experiment and critique any steps that appear flawed.
C. Turn your back on him.
D. Drop a book on his foot to show him that it falls downward.

86. A teacher says the following in a lesson:
"The Earth has two tides per day, not just one. Tides are caused by the gravity of the Moon acting on the oceans. The Moon, however, only passes overhead once per day. That explains the tide on the side of the planet *facing* the Moon. What explains the fact that there is another tide on the side facing *away* from the Moon?"
Choose the most appropriate notes you might take.

A. Earth has tides.
B. Earth, tides, there are 2 per day
C. Earth, tides, caused by Moon's gravity, 2 per day, why 2?
D. The Earth has two tides per day, not just one. Tides are caused by the gravity of the Moon acting on the oceans. The Moon, however, only passes overhead once per day. That explains the tide on the side of the planet *facing* the Moon. What explains the fact that there is another tide on the side facing *away* from the Moon?

87. Kassandra asserts that everything is composed of combinations of four elements: pizza, cereal, soda, and chocolate. Which is the best argument to refute her assertion?

A. There is not enough pizza in the world to make up one-fourth of the Earth.
B. Pizza cannot be an element because it is made from bread, cheese, and sauce.
C. None of those four elements contains matter in a gaseous state.
D. The ocean is blue but none of those four elements is blue.

88. Which is most likely to satisfy the criteria for scientific evidence?

A. a quote from an online encyclopedia about the results of a trial
B. the verdict of the jury in a civil trial
C. the fingerprints removed from objects at a crime scene
D. instructions by a judge to a jury as to what information they may use when deliberating

# Chapter 2 Physical Quantities and Measurement

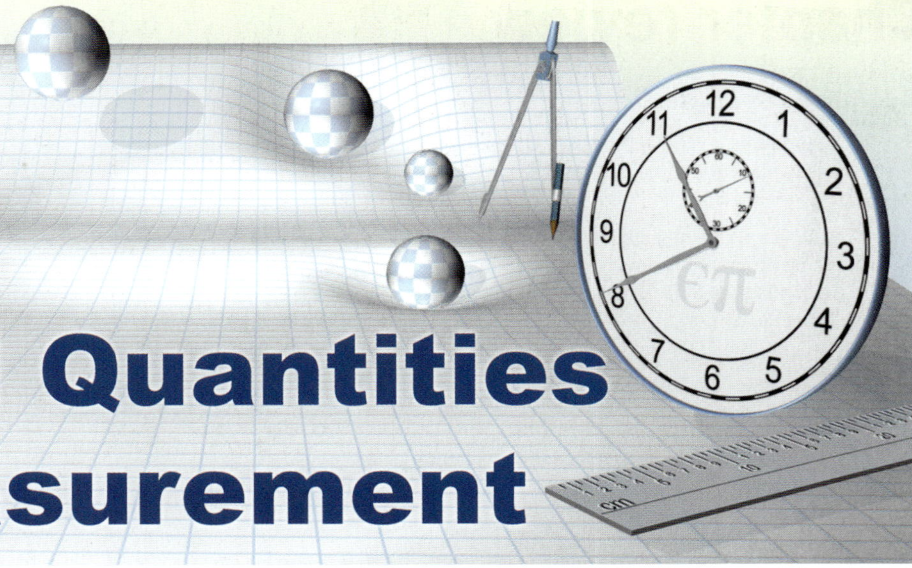

A young boy rockets off to Alpha Centauri, visits its famous zoo, and zooms back. He finds his twin sister has aged into an 80-year-old grandmother—even though he has hardly passed through adolescence. For now this is the stuff of science fiction. Or is it? After all, this *twin paradox* is a consequence of Albert Einstein's theories of relativity, and the paradox has been demonstrated by real experiments (though not with people!).

Einstein's theory postulates that physical constants such as the speed of light shouldn't depend on a particular observer's state of motion. This may seem odd, because many everyday experiences—such as *relative velocity*—*do* depend on your frame of reference.

Imagine that a baseball pitcher on board a moving train hurls a fastball toward the front end of his compartment at 80 miles per hour (mph). A bystander watching from the platform sees the ball fly past her at 170 mph. Why? *Because the train itself is passing her at 90 mph.* Einstein didn't disagree with this description of reality when the speeds involved were tiny fractions of the speed of light. But if the train were moving at nine-tenths the speed of light and the ball were thrown at another eight-tenths, the bystander would *not* see the ball exceed light speed (even though 0.8 plus 0.9 equals 1.7). Fast-moving (or *relativistic*) velocities add in a complicated way in Einstein's theory of relativity!

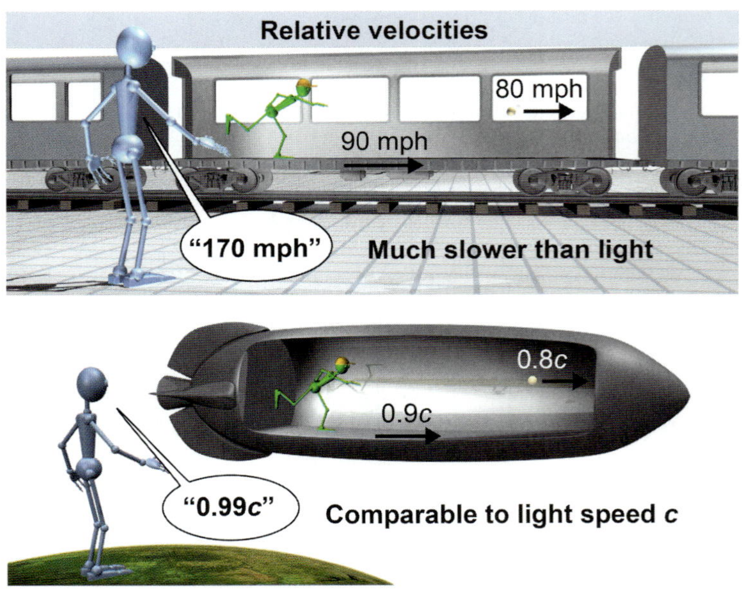


Even stranger: if the pitcher had turned on a flashlight, both he and the bystander would measure the speed of the light beam as the same 671 million mph! Speed is length divided by time. The pitcher and the bystander can only measure the same speed of light if moving "rulers" shrink or "clocks" tick more slowly. These strange effects aren't from the realm of science fiction; they impact our everyday technology, such as the clocks on global positioning system (GPS) satellites.

# Chapter study guide

## Chapter summary

Three fundamental quantities in physics are mass, distance, and time. All quantities are measured with both a value and its units. The units of the International System (SI) are used as the standard units for quantities such as the kilogram, meter, and second. Quantities can be converted from one set of units to another by using conversion factors. Scientific notation is useful for expressing quantities from microscopic to macroscopic scales. All measurements are expressed in a way that conveys their precision and uncertainty. When performing calculations, the precision of the result can never be better than that of the least precise quantity that contributes to the calculation. Physics often requires creating models—including equations and graphs—as representations to fit observational data. Algebra is a useful tool for constructing mathematical models of physical data.

## Learning objectives

By the end of this chapter you should be able to

- distinguish between fundamental and derived quantities;
- distinguish among mass, inertia, and weight;
- calculate surface area, volume, and density;
- use scientific notation and algebra to solve problems;
- convert units for fundamental and derived quantities;
- use the appropriate number of significant figures and decimal places when performing calculations;
- describe several causes and effects of uncertainties on measured data; and
- draw a graph of a set of data points, identify the dependent and independent variables, and measure the slope of the graph.

## Investigations

2A: Indirect measurement, estimation, and scale
2B: Graphical relationships
2C: Algebraic relationships

## Chapter index



## Important relationships

$$\rho = \frac{m}{V} \qquad S = 2ab + 2bc + 2ac \qquad S = 4\pi r^2 \qquad S = 2\pi rh + 2\pi r^2 \qquad S = \pi r^2 + \pi r\sqrt{r^2 + h^2}$$

$$V = abc \qquad V = \frac{4}{3}\pi r^3 \qquad V = \pi r^2 h \qquad V = \frac{1}{3}\pi r^2 h$$

## Vocabulary

| | | |
|---|---|---|
| **measurement** | **matter** | **mass** |
| **inertia** | **length** | **surface area** |
| **volume** | **density** | **scale** |
| **macroscopic** | **microscopic** | **temperature** |
| **scientific notation** | **exponent** | **accuracy** |
| **precision** | **decimal places** | **significant figures** |
| **conversion factor** | **dependent variable** | **independent variable** |
| **variable** | **model** | **linear regression** |
| **correlation coefficient** | **pie graph** | **bar graph** |
| **frequency table** | **error bars** | **box and whiskers plot** |

## 2.1 - Describing the physical universe

Three important quantities in physics are *mass*, *length*, and *time*. Mass describes the quantity of matter. Length describes a quantity of space, such as width, height, or distance. Time describes the flow of the universe from the past through the present and into the future. Like different languages, there are different *units* for each quantity.

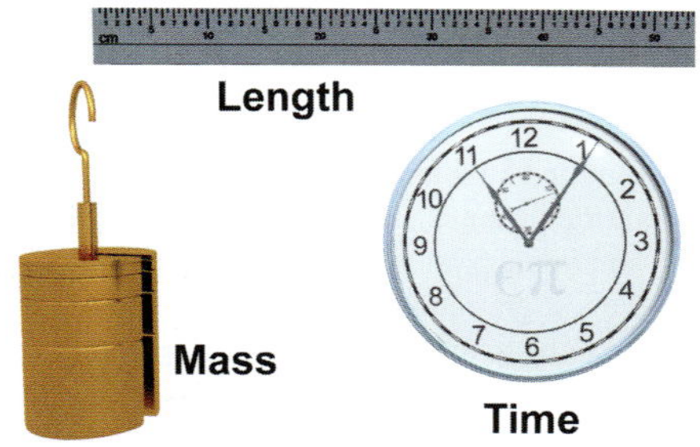


### Fundamental and derived quantities

**Fundamental quantities**

In ancient times, systems of measure were adopted locally. One kingdom might measure distances in farthings, another in stadia, and still another in multiples of the size of the king's hands. Over many centuries, civilizations developed common units of measure in order to facilitate communication, trade, and scientific development. The International System of Units (SI) contains seven fundamental quantities and units from which all other units can be derived. The three most important fundamental quantities for this chapter are length (or distance), mass, and time, and they are measured in units of the meter, the kilogram, and the second, respectively. These fundamental units are also referred to as *base units.* In later chapters of this book, you will learn about the other four fundamental quantities and their units.

**Fundamental units and physical quantities in the International System of Units**

| Physical quantity | Quantity symbol | Fundamental unit | Unit symbol |
|---|---|---|---|
| Mass | $m$ | kilogram | kg |
| Length | $x, d, l$ | meter | m |
| Time | $t$ | second | s |
| Electric current | $I$ | ampere | A |
| Temperature | $T$ | kelvin | K |
| Luminous intensity | $I_v$ | candela | cd |
| Amount of a substance | $n$ | mole | mol |

**Derived quantities**

How about speed? Speed is not a fundamental quantity—it is a *derived quantity*, representing distance divided by time. Fundamental quantities can usually be measured directly, such as with a meter rule, whereas derived quantities are often calculated from other measured quantities.

**Examples of derived quantities**

Area is a derived quantity because it is the product of length times length (because width is a length). Similarly, volume is a derived quantity, because it is length times length times length. How about miles per gallon? The word "per" should be a giveaway that one quantity is being divided by another. For example, mpg is length (miles) divided by volume (gallons or length cubed) and is a derived quantity.

**Measurements**

A **measurement** of a physical quantity involves a comparison with a standard of measurement. The standard of measure for the kilogram is one particular cylinder of platinum alloy held by the International Bureau for Weights and Measures (BIPM). If an object has a mass of 7.3 kg, then that means that its mass is 7.3 times larger than the BIPM's standard mass.

**Measured quantities have values and units**

A measurement of a *quantity*, such as mass, communicates how much there is. Measurements nearly always have two parts: a *value* and a *unit* (or *dimension*). Consider a quantity of mass of 2.5 kg. The number, 2.5, is the value and *kilograms* is the unit. All measurements need units because, without a unit, the value of the measurement cannot be understood. For example, describing a rock as having a mass of 2.5 makes no sense. You could be talking about a 2.5 g pebble or a 2.5 ton, car-sized boulder.

**mass = 2.5 kilograms**
Quantity — Value — Unit

Measurements of a quantity typically have both a value and a unit.

**Always seek clarification on units**

Whenever a value is communicated in science there should be some information to answer the question, *"What are the units of this quantity?"* For example, a sign for gasoline in Mexico may quote an equivalent price of USD$1.25. You must ask, however, whether that is *per gallon* or *per liter!* There will be some fine print on the sign that says "per liter." There are 3.78 liters in a gallon, so $1.25/liter = $4.72/gallon. *Always seek clarification on units if you are not given that information.*

**Dimensions**

The *dimensions* of a physical quantity are the fundamental quantities that comprise it. The dimension of speed is length divided by time, while the dimension of volume is length cubed (or length times length times length). The dimension of density is not mass divided by volume, because volume is not a fundamental quantity. Instead, the dimension of density is mass divided by length cubed.

**International standard units of measure**

The meter was originally defined as one ten-millionth of the distance from the North Pole to the Equator. The second was originally defined in terms of a day: one solar day = 86,400 s. As experiments became more accurate, both definitions became too imprecise. Today, one second is defined as the duration of 9,192,631,770 periods of the radiation corresponding to the transition between the two hyperfine levels of the ground state of the atom cesium-133. One meter is defined as the distance traveled by light in a vacuum during a time interval of 1/299,792,458 of a second. These modern definitions allow experimenters to compare measurements of length or time to within one part per billion.

**Old and new definition of the meter**

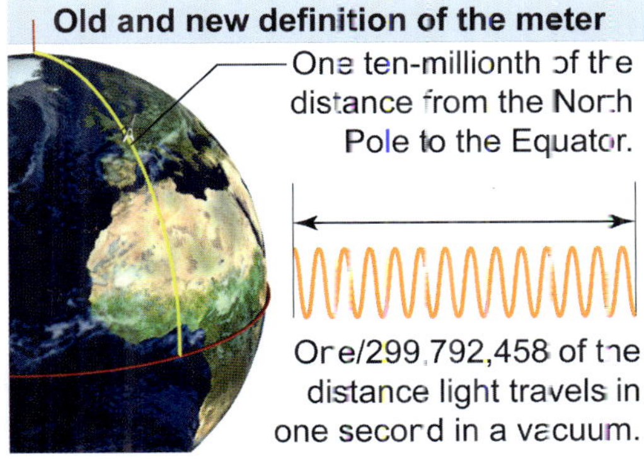

**Matter**

**Matter** has mass and takes up space. A solid rock is matter, but so is the air around you. Put your hand out the window of a moving car and you can feel the matter in the air pushing back against your hand as the motion of your hand pushes the air out of the way.

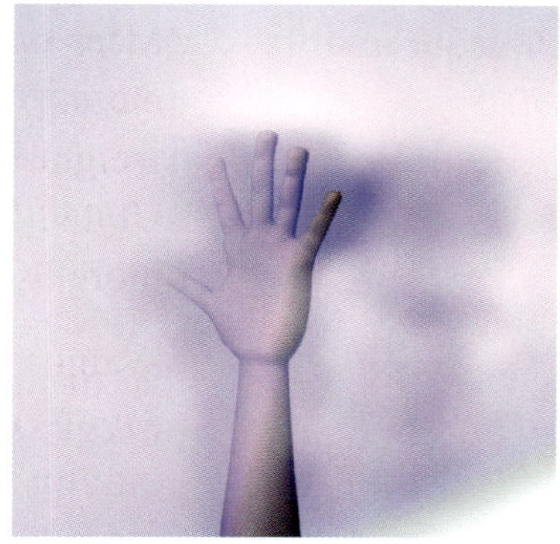

**Units of mass**

**Mass** measures the quantity of matter, such as in SI units of grams (g) and kilograms (kg). One kilogram is 1,000 grams, about the mass of the water in a 1 liter bottle.

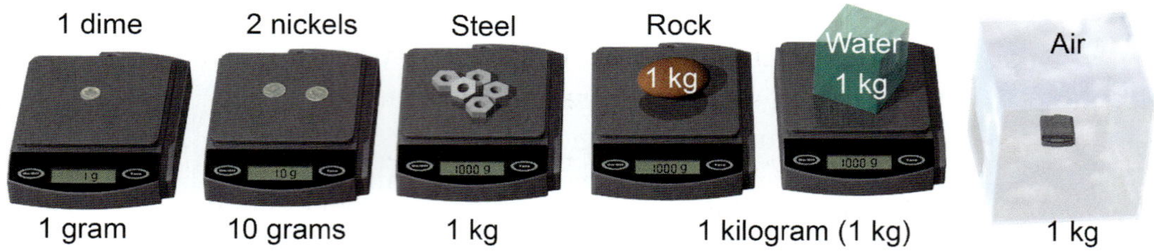


**Masses of some objects**

A fist-sized rock and a desk-sized volume of air both have a mass of 1 kg. The mass of a dime is around 1 g. The mass of a car might be 1,000 kg or more. The typical mass of an average adult human ranges from about 50 kg to 100 kg. An elephant might have a mass of 5,000 to 7,000 kg.

**Triple beam balance**

A *triple beam balance* is a commonly used instrument for measuring mass. Laboratory triple beam balances can usually measure masses between 0.1 g and 500 g. The large scale found in many doctor's offices is similar but larger, being used to measure weights from 10 to 350 lb (or masses from 4.5 to 159 kg).

**Balances**

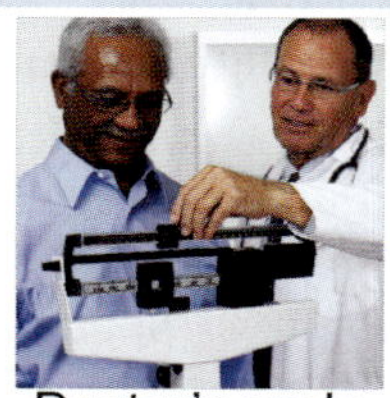
Doctor's scale

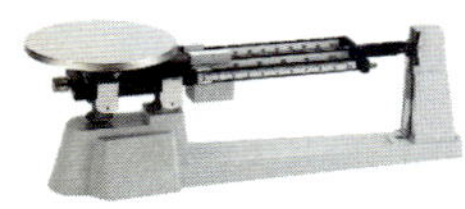
Triple beam balance

**How to read a triple beam balance**

The triple beam balance has three sliders, each with a different mass, often 10 g, 1 g, and 0.1 g. The sliders move on the three beams until they balance the mass on the measuring pan. The scale is balanced when the pointer is at the zero setting. It is a good idea to check *calibration* when first using a balance. The calibration is accurate if the pointer registers zero when the pan is empty and the sliders are all set to zero. To make a measurement, place the mass on the pan. Adjust the position of the *highest mass slider* to the highest value that does not push the pointer past zero. Then do the same for the middle mass slider. Finally, use the small mass slider to bring the scale into balance. The reading of a triple beam balance is the sum of the masses on all three slider scales.

**How to read a triple beam balance**

Adjust the position of the mass on each beam...

...until the scale is *just* balanced.

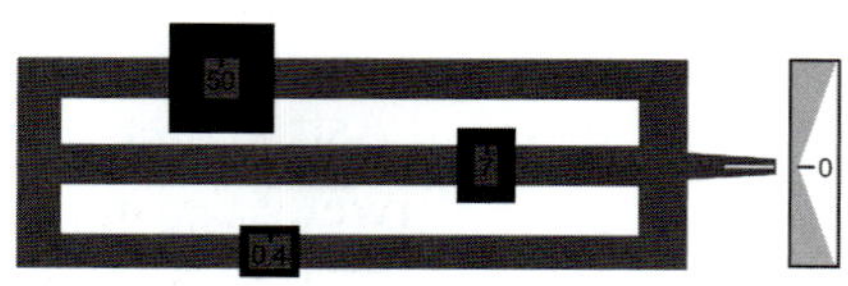


The total mass is the sum of the values on all three beams.

$$50 + 7 + 0.4 = 57.4 \text{ g}$$

**Weight and mass**

Mass produces the effects of *weight* and *inertia*. Weight is the force of gravity acting on mass. At the Earth's surface, 1 kg of mass weighs 2.2 lb. If your mass is 50 kg, then your weight is 50 × 2.2 = 110 lb at the Earth's surface. Mass is measured in kilograms. In the United States, weight is commonly measured in pounds. In physics, however, it will be more useful to express weight in *newtons,* the SI unit of force.

**Mass and weight are different**

Weight and mass are not the same. Mass (in kilograms) is the same on Earth, on the Moon, on Mars, or anywhere else. *Weight*, however, depends on the local strength of gravity. Your weight on the Moon is only 1/6 as much as it is on Earth even though your mass is the same. The Moon's gravity is weaker than Earth's gravity because the Moon has only 1/10 Earth's mass. Mass is an *intrinsic* property of matter because it does not depend on environmental conditions. Weight is an *extrinsic* property because it depends on the local strength of gravity, which is different in different places.

**Inertia and mass**

**Inertia** is the property of matter that resists changes in motion. Throwing a beach ball is easier than throwing a bowling ball even though both are the same size. The bowling ball, however, has more inertia because of its larger mass. Inertia plays an important role in physics.

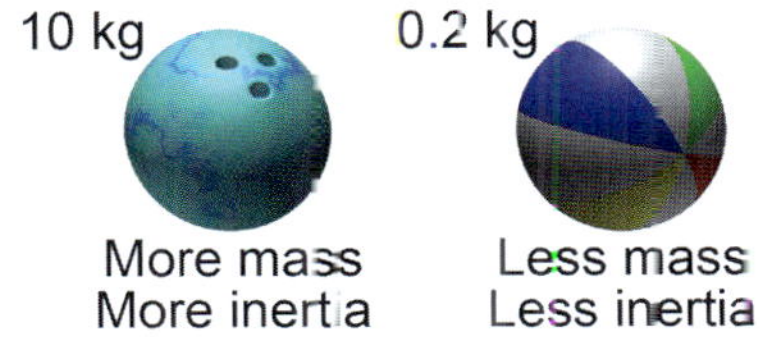


**Inertia and physics**

Think about skating across an ice rink. Once you get going your inertia keeps you going in the same direction, at the same speed, until something applies a force to stop you. Inertia explains why massive objects are hard to start moving and hard to stop once they get moving. For most situations, an object's inertia is equal to its mass. A 10 kg bowling ball has 50 times the inertia of a 0.2 kg beach ball.

Have both weight and mass

Same mass but *apparent* weightlessness

Same mass but virtually weightless

Same mass but less weight than on Earth

Credits: NASA, JPL, C. Kohlhase

**Weightlessness**

Astronauts orbiting the Earth can float above the floor of a spacecraft in *apparent* weightlessness. In reality, orbiting spacecraft are only a few hundred miles above Earth's surface and therefore still subject to 99% of Earth's surface gravity. The astronauts float because the floor is falling away beneath them! You experience the same state of *free fall* in the moments between jumping off a diving board and landing in the water. To truly become weightless, an astronaut would have to travel far from Earth. The first humans to experience this were the Apollo 8 astronauts who flew to the Moon in 1968.

**Space is measured in units of length**

Think about *space* as a quantity of its own, independent of any matter that might be present. For example, how *big* is the space inside your classroom. How do you describe precisely where objects are located within the room? The answers to both questions involve the quantity of *length*. **Length** is the measure of space between two points. Like mass, length is a fundamental quantity: length is the dimension of space itself. Objects, such as rocks or people, take up space and therefore *size* is described with length units such as meters, feet, and inches.

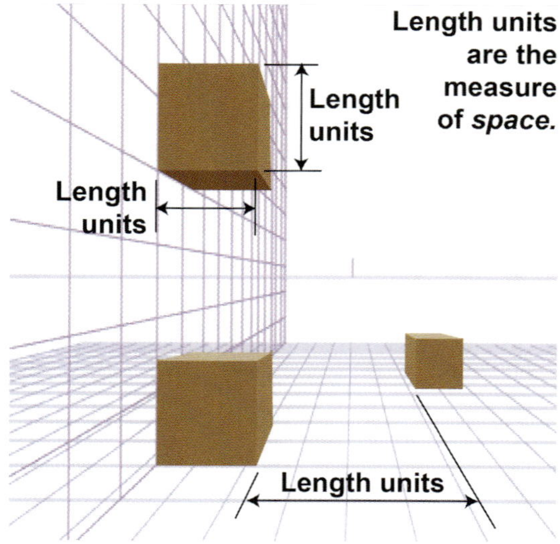


**Units for length**

There are thousands of different languages spoken among the seven billion people in the world. The concept of *length* is described in each one, often with a completely different word. As with languages, there are different units for describing the same length. For example, 1 meter, 39.37 inches, 3.28 feet, and 100 centimeters all describe the *same length*. There are two common systems of length units you need to know and understand. The English system uses inches (in), feet (ft), yards (yd), and miles (mi). The metric system uses millimeters (mm), centimeters (cm), meters (m), and kilometers (km). The light year is the distance light travels in one year—a very large distance!—and is suitable for measuring the distances between stars.

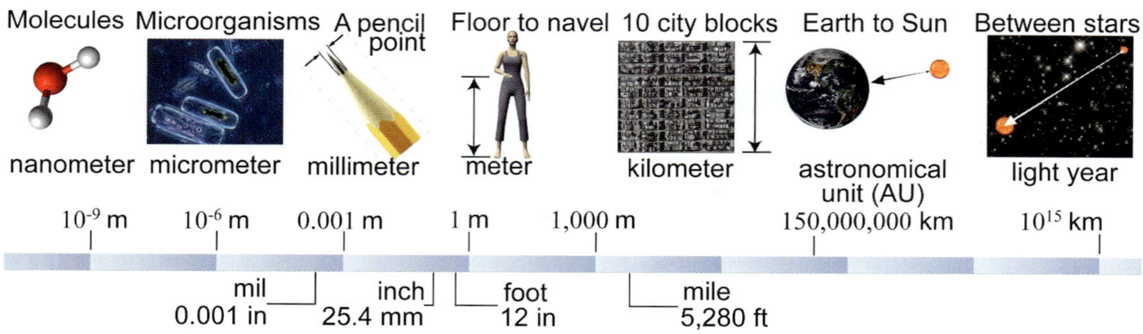


**Screw threads**

Screw threads provide an example of different length units. Nuts, bolts, and screws come in both English and metric threads that look similar but are not interchangeable. An M6 bolt has a diameter of 6 mm and each thread is 1 mm apart. A 1/4-20 bolt has a diameter of 1/4 inch and 20 threads per inch. A 1/4-20 screw will *not* thread into an M6 nut and vice versa. Because parts are made worldwide, and countries other than the USA use metric fasteners, American cars have a mixture of English *and* metric fasteners.

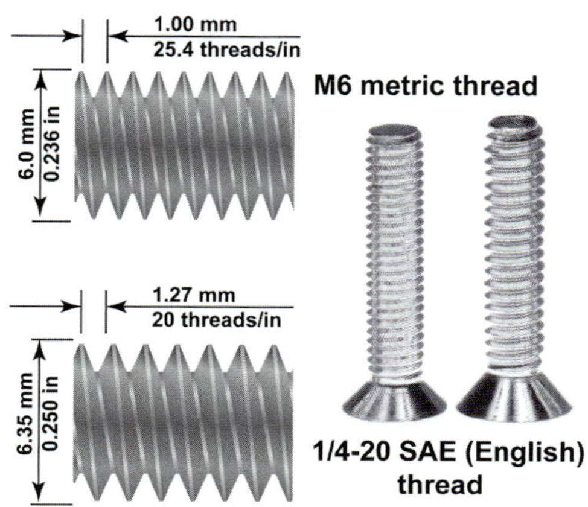

**Surface area**

The fundamental unit of length appears in other spatial quantities such as *surface area* and *volume*. **Surface area** describes how many square units it takes to completely cover a surface. For example, the top of a 3 m × 3 m board has a surface area of 9 $m^2$. A 3 m cube has a surface area of 54 $m^2$. A sphere with a diameter of 3 m has a surface area of only 28.3 $m^2$, considerably less than a cube of the same dimension. All units of surface area are units of length squared, such as square inches ($in^2$) and square kilometers ($km^2$).

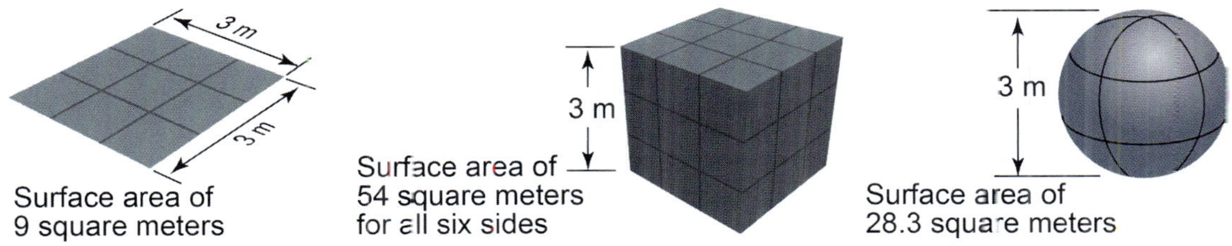


**Volume**

**Volume** measures an amount of space in units of length cubed. Examples include cubic meters ($m^3$), cubic centimeters ($cm^3$), and cubic inches ($in^3$). Volume is such an important concept that there are specific units for it, such as *liters* and *gallons*. Volume-specific units can always be related to the fundamental length units on which they are based. For example, one liter is defined as a volume of 1000 $cm^3$ and one gallon is 3,780 $cm^3$. For simple geometric shapes, the surface area and volume can be calculated by using formulas such as the examples in the diagram below.

Interactive equation

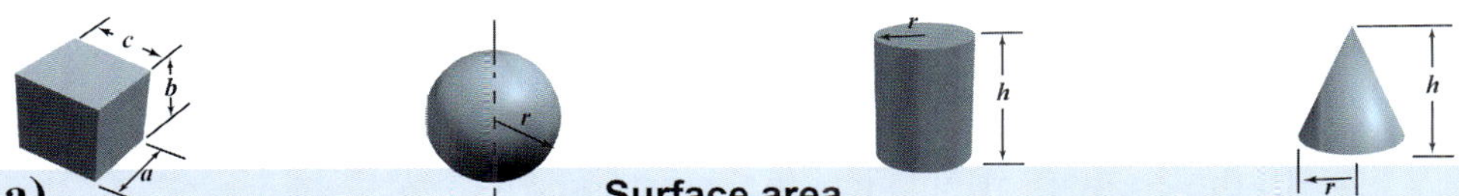


**(2.1a)** **Surface area**

$S = 2ab + 2bc + 2ac$ $\quad S = 4\pi r^2$ $\quad S = 2\pi rh + 2\pi r^2$ $\quad S = \pi r^2 + \pi r\sqrt{r^2 + h^2}$

**(2.1b)** **Volume**

$V = abc$ $\quad V = \frac{4}{3}\pi r^3$ $\quad V = \pi r^2 h$ $\quad V = \frac{1}{3}\pi r^2 h$

**Density**

Which has more mass, a block of steel or a block of wood the same size? Both blocks have the same volume but the steel block has more mass. The steel block has more mass because the *density* of steel is higher than the density of wood. The **density** of an object measures the concentration of matter in the object's volume. The units of density are units of mass divided by units of volume, such as kilograms per cubic meter ($kg/m^3$). Air has a low density, typically about 1 $kg/m^3$. Steel has a much higher density of around 7,800 $kg/m^3$.

**How to calculate density**

The equation for calculating density is given by equation (2.2).

Interactive equation

**(2.2)** $$\rho = \frac{m}{V}$$ **Density**

$\rho$ = density ($kg/m^3$)
$m$ = mass (kg)
$V$ = volume ($m^3$)

## Macroscopic and microscopic scales

**How do we explain what we observe?**

For thousands of years humans have wondered about the nature of electricity, light, and heat. Aristotle thought fire was an element, like water, air, and earth, and that its natural place was in the Sun, which explained why smoke rises. After a few thousand years of ideas, observations, and especially experiments, we now have an explanation that correctly fits all the observed facts. Electricity, light, and heat have their explanation in the *microscopic* structure of both matter and energy.

**Macroscopic scale**

Physicists use the concept of **scale** to describe the relative sizes of things. When you apply a force to push a cart up a ramp you can *see* the transfer of energy as the cart moves uphill. This is an example of work being done on the *macroscopic scale.* The **macroscopic** scale refers to things we can touch and sense directly. Rocks, shopping carts, dust specks, and planets are macroscopic objects. The macroscopic scale is the scale of ordinary life, from about 1/100th of a millimeter and larger.

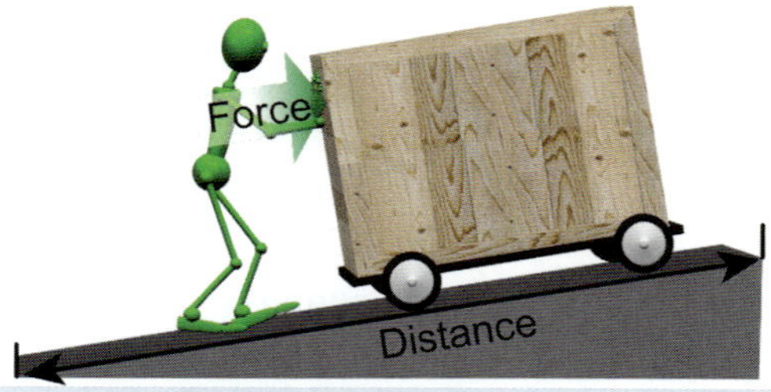


**Doing work on the *macroscopic* scale**

**Microscopic scale**

Hidden inside the macroscopic world is the *microscopic* world of atoms and particles. In physics, the **microscopic** scale refers to the size of an atom and smaller. Atoms are so small that even a dust speck contains trillions of atoms. A single atom has a diameter of around $10^{-10}$ m. In the context of physics, the word *microscopic* refers to things much, much smaller than are visible with an ordinary optical microscope.

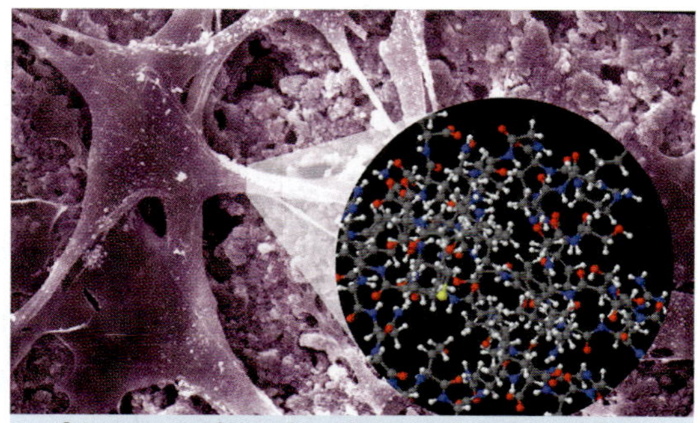

**Atoms and molecules make up matter on the *microscopic* scale**

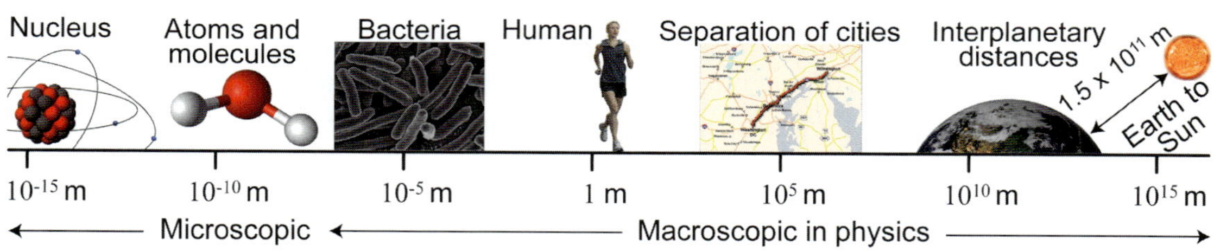


**The universe at two scales**

Electricity, heat, and light can only be fully understood by considering the behavior of matter and energy on the microscopic scale. For example, *electrons*, the tiny particles responsible for most observable aspects of electricity, typically move distances of $10^{-9}$ m between collisions with other electrons. *Photons*, the basic units of light, have energies of about $3\times10^{-19}$ joules. **Temperature** is a measure of the average kinetic energy of individual atoms (or molecules) moving around vigorously, even in solid matter.

**What is scientific notation?**

**Scientific notation** is a way of expressing a number as a *coefficient* multiplied by a *power of 10.* This is useful because many values in physics are either too large or too small to write as ordinary decimal numbers. The coefficient is a decimal number, greater than (or equal to) 1 but less than 10. Powers of 10 are $10^{-1} = 0.1$, $10^0 = 1$, $10^1 = 10$, $10^2 = 100$, and so on. For example, the number 1,500 is $1.5\times10^3$. The number 1.5 is the coefficient. The number $10^3$ is a power of 10. The small, raised numeral 3 is the **exponent**. For the number 1,500, this may seem like more trouble than it is worth. But consider a very large number, such as the speed of light, which is 3 hundred million or 300,000,000 m/s. In scientific notation this is $3\times10^8$ m/s, which is much easier to write without making a mistake.

| Scientific notation | Numbers larger than 1 | Numbers smaller than 1 |
|---|---|---|
| | Exponent | Exponent |
| | $1.5\times10^{3} = 1{,}500$ | $1.5\times10^{-3} = 0.0015$ |
| | Coefficient | Coefficient |

**Writing a number less than one**

For numbers less than one, scientific notation uses *negative* exponents. For example, the number 0.001 is $1 \div 1000 = 1 \div 10^3 = 10^{-3}$. A negative sign on the exponent of 10 does not mean that the number is negative! In scientific notation, negative exponents mean a value that is less than one. For example, the quantity 0.0025 m is $2.5\times10^{-3}$ m.

**Prefixes**

Scientific notation can also be represented by adding a *prefix* to the quantity. Two thousand meters is the same as $2\times10^3$ meters or 2 kilometers, where "kilo" means "one thousand" and is abbreviated "km." The tables below show some useful prefixes.

| Numbers larger than 1 | | SI prefix |
|---|---|---|
| 1 billion | $1\times10^9 = 1{,}000{,}000{,}000$ | giga |
| 100 million | $1\times10^8 = 100{,}000{,}000$ | |
| 10 million | $1\times10^7 = 10{,}000{,}000$ | |
| 1 million | $1\times10^6 = 1{,}000{,}000$ | mega |
| 100 thousand | $1\times10^5 = 100{,}000$ | |
| 10 thousand | $1\times10^4 = 10{,}000$ | |
| 1 thousand | $1\times10^3 = 1{,}000$ | kilo |
| 1 hundred | $1\times10^2 = 100$ | |
| ten | $1\times10^1 = 10$ | deca |
| one | $1\times10^0 = 1$ | |

| Numbers smaller than 1 | | SI prefix |
|---|---|---|
| 1 tenth | $1\times10^{-1} = 0.1$ | deci |
| 1 hundredth | $1\times10^{-2} = 0.01$ | centi |
| 1 thousandth | $1\times10^{-3} = 0.001$ | milli |
| 1 ten thousandth | $1\times10^{-4} = 0.0001$ | |
| 1 hundred thousandth | $1\times10^{-5} = 0.00001$ | |
| 1 millionth | $1\times10^{-6} = 0.000001$ | micro |
| 1 ten millionth | $1\times10^{-7} = 0.0000001$ | |
| 1 hundred millionth | $1\times10^{-8} = 0.00000001$ | |
| 1 billionth | $1\times10^{-9} = 0.000000001$ | nano |

**Scientific notation on a calculator or computer**

Computers and scientific calculators use scientific notation. Instead of the multiplication sign (×) a computer uses the letter "E" and a calculator uses either "E" or "EE." The letter "E" stands for *exponential,* which is another term for scientific notation. The diagram below shows how to enter scientific notation numbers on a calculator and computer. *Pay close attention to the order in which you use the +/– key when you need to enter a negative exponent on a calculator!*

Interactive equation

**Calculator**

| | | | Keys | Display |
|---|---|---|---|---|
| To enter | $6.5\times10^9$ | Press | 6 . 5 EE 9 | 6.5E9 |
| To enter | $2.6\times10^{-5}$ | Press | 2 . 6 EE 5 +/- | 2.6E-5 |
| | | or | 2 . 6 EE (-) 5 | |

**Computer**

$6.5\times10^9$ = `6.5E9`

$2.6\times10^{-5}$ = `2.6E-5`

## Time

**Historical time**

How things change over *time* is a basic question in all of science, including physics. Time is a fundamental quantity and there are two different ways to think about it. In your daily life, the word "time" usually means *historical time*, such as 2:25 pm. Historical time assigns a particular value to each successive moment of the past, present, and future. Clocks read historical time, as do calendars and bus schedules. Historical time has no known end, but scientists believe there is a beginning to historical time, roughly 13.6 billion years ago when the universe exploded forth in the Big Bang.

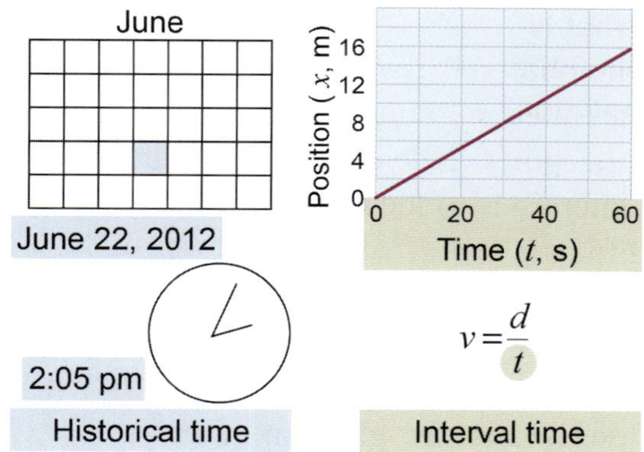


**Time intervals**

Historical time rarely appears in physics. Instead, physics deals mainly with time intervals, usually in *seconds*. A time interval is a quantity of time, such as 10 seconds or 3 hours. A stopwatch is a device that measures a time interval. Time intervals are independent of when they occur historically; ten seconds at 9:00 am on Tuesday is the same amount of time as ten seconds at 5:00 pm on Friday. This is important because the laws of physics cannot depend on what *time* it is!

**The second**

The *second* is the fundamental unit of time in both the English and the SI systems. A single second is about the time it takes to say "one thousand." There are 60 seconds in a minute, 3,600 seconds in an hour, and 86,400 seconds in one day. Prior to 1967, the second was defined as 1/86,400 of an average day. The standard definition of the second was changed in 1967 to be the time it takes a certain frequency of light emitted by a cesium-133 atom to perform 9,192,631,770 complete oscillations.

**Mixed units for time**

Physics calculations require that time be expressed in a single unit, usually seconds. Time, however, is often expressed in *mixed units*—using a combination of years, days, hours, minutes, and/or seconds. A race time of 2 hours, 20 minutes, and 9.2 seconds is an example of mixed units. *You cannot do calculations with time in mixed units!* Before calculating any quantity in physics all values for time must be in a single unit.

| Time equivalents |
|---|
| 1 millisecond (ms) = 0.001 s |
| 1 minute (m) = 60 seconds (s) |
| 1 hour (hr) = 60 m = 3,600 s |
| 1 day = 24 hr = 86,400 s |
| 1 year = 365.24 days = 31,579,200 s |

**Converting from mixed units**

Interactive equation

To convert time to hours, you convert each partial quantity—such as 20 minutes—to seconds. Then add up all the partial time intervals. Two hours is 7,200 seconds and 20 minutes is 1,200 seconds. Summing up the hours, minutes, and seconds in the example (at right) gives a race time of 8,409.2 seconds.

**Converting time in mixed units**

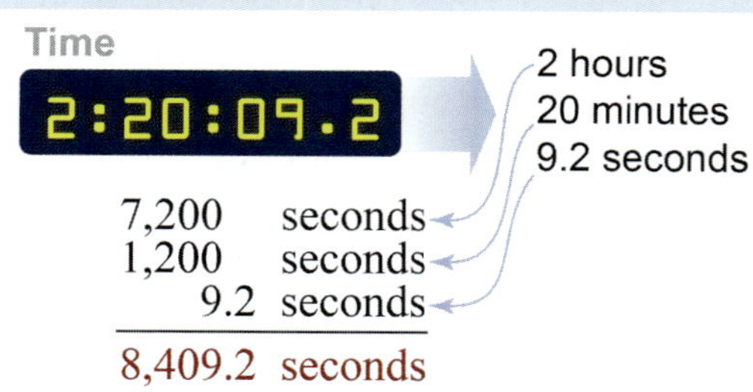

# Section 1 review

The observable world is described by mass, length, and time. *Mass* describes the quantity of matter and is measured in grams and kilograms. *Length* describes space, including concepts such as size or position. Units of length include centimeters, meters, and kilometers in the SI system. The concept of length extends in three dimensions to the concepts of *surface area* and *volume*. To fully understand the universe we need to look at different length scales. The *macroscopic scale* encompasses quantities and sizes that we can experience directly, ranging from about the size of a bacteria upward to the size of planets and larger. In physics, the *microscopic scale* refers to quantities and sizes comparable to individual atoms and smaller. Many macroscopic phenomena, such as temperature, can only be understood on the microscopic level. *Time* describes the progression of events that we observe to flow from the past, through the present, and into the future, and never in reverse. Units for measuring time include seconds, minutes, and hours.

| Vocabulary words | measurement, matter, mass, inertia, length, surface area, volume, density, scale, macroscopic, microscopic, temperature, scientific notation, exponent |
|---|---|

**Key equations**

$$\rho = \frac{m}{V}$$

| | Cube | Sphere | Cylinder | Cone |
|---|---|---|---|---|
| | $S = 2ab + 2bc + 2ac$ | $S = 4\pi r^2$ | $S = 2\pi rh + 2\pi r^2$ | $S = \pi r^2 + \pi r\sqrt{r^2 + h^2}$ |
| | $V = abc$ | $V = \frac{4}{3}\pi r^3$ | $V = \pi r^2 h$ | $V = \frac{1}{3}\pi r^2 h$ |

## Review problems and questions

1. Which of the following is closest to 1 kg in mass?
   (a) dime; (b) average size banana; (c) one liter of water; (d) average person

2. Which of the following is closest to one meter in length?
   (a) penny; (b) width of a door; (c) height of a room; (d) size of the Earth

3. What is the radius of a sphere that has a volume of one cubic meter?

4. A room measures 3 m wide, 4 m long, and 2.5 m high.
   a. What is the surface area of the room in square meters?
   b. If a certain paint covers 25 square meters per gallon, how many gallons will it take to paint the walls and ceiling, but not the floor?

5. Describe three objects that belong to the microscopic scale and three objects that belong to the macroscopic scale.

6. A basic concept in science is that things which happen, called *effects,* have causes and that the cause must precede the effect. Nothing happens *before* it is caused. Discuss how this might theoretically affect the possibility of time travel.

## 2.2 - Measurements

*Quantitative* means to express a measurement with a numerical value. Unfortunately, *it is impossible to measure the exact true value of any physical quantity.* All measurements contain some level of uncertainty. Suppose a balance measures the mass of a coin to be 1.00 g. This measurement does *not* tell you that the exact mass is 1.00 g because the specifications for the balance allow ±0.02 g of uncertainty! The measurement tells you that the exact mass of the coin lies between 0.98 and 1.02 g. Within the context of measurements, 1.0 and 1.000 are *different* because they imply different levels of uncertainty.

### Accuracy and precision

**Accuracy and precision**

In science, the words *accuracy* and *precision* have different meanings. **Precision** tells you the repeatability of a sequence of measurements of the same quantity. If you make three measurements of the mass of a coin on a balance precise to 0.1 g then each measurement will be within 0.1 g of the others. **Accuracy** tells you how close any measurement is to the true value of the quantity being measured. A meter ruler with small, millimeter divisions might be precise, but it would be inaccurate if the ruler were accidentally manufactured with a length of one yard instead of one meter.

**Difference between accuracy and precision**

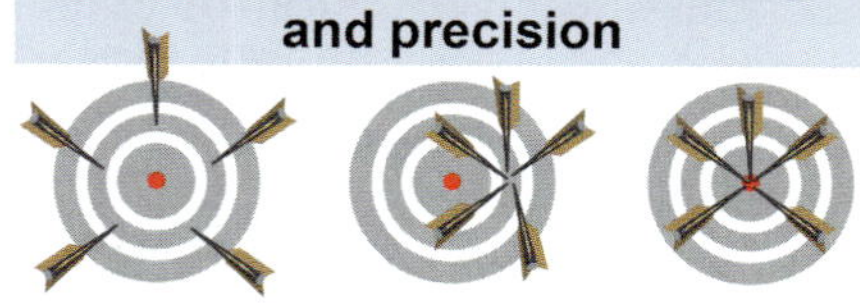


**Communicating uncertainty**

Every measurement of a physical quantity has an uncertainty associated with it. The uncertainty may be expressed explicitly with a plus-or-minus sign, such as ±0.001 m. A measurement of the length of a table precise to one millimeter could be written as 2.067 ±0.001 m.

$l = 2.067 \pm 0.001$ m (Measurement: 2.067; Uncertainty: ±0.001)

**Decimal places**

Precision determines the number of **decimal places** recorded for measurements. A meter stick is precise to 0.5 mm because the smallest gradations are every millimeter (0.001 m) and you can estimate to one half the smallest gradation. For example, the width of a table might be 885.5 mm. In meters this would be recorded with four decimal places, or 0.8855 m.

| One decimal place | Two decimal places | Three decimal places |
|---|---|---|
| 23.7 m | 0.07 J | 0.107 N |
| 1.1 s | 2.10 kg m | 0.007 m/s |
| 0.7 kg | 972.39 K | 2.100 mm |

**Decimal places and addition**

The results of a calculation can never be more precise than the precision of any of the individual measurements that went into it. When adding or subtracting, round off the final value to the least number of decimal places among the individual measurements.

**Decimal places and addition**

Measurement with fewest number of decimal places (1) → 23.6 m

```
   1.072  m
  23.6    m
+  0.19   m
---------
  24.862  m
```

Round final answer to one decimal place → **24.9 m**

**Claims about services**

One financial services company claims that it can provide the most *precise* measurement of a market index (such as the S&P 500® Index) at any given time, while another company claims it can provide the most *accurate* value. Assuming both claims are true, then which service is better? Most people will likely prefer the more accurate service, since knowing the value to five decimal places is unnecessary—but knowing how much money you have, within a factor of 2, is important!

### Significant figures

The number of **significant figures** (or digits) is a code that tells you the precision of a measurement. Count the number of significant figures starting from the left-most *nonzero* digit.

1. Include zeros to the right of the decimal.
2. Don't include zeros to the left of the decimal.

A force of 0.08 N has one significant figure, a time interval of 1.1 s has two significant figures, while a volume of 0.310 L has three significant figures.

| One significant figure | | Two significant figures | |
|---|---|---|---|
| 0.08 N | 70 m | 1.1 s | 0.026 mm |

| Three significant figures | |
|---|---|
| 23.7 m | 0.310 L |

### Estimating beyond the smallest gradation

Significant figures are not the same as decimal places. The length of the bolt in the diagram is measured with a meter stick to be 113.5 mm. This is the same as 11.35 cm or 0.1135 m. Each value has four significant figures even though they differ in digits to the right of the decimal point. The last digit (5) is the most uncertain because it was estimated by reading between the millimeter marks for 113 and 114. The bolt's true length could be anything between 113.25 and 113.75 mm.

**Estimating beyond smallest gradation**

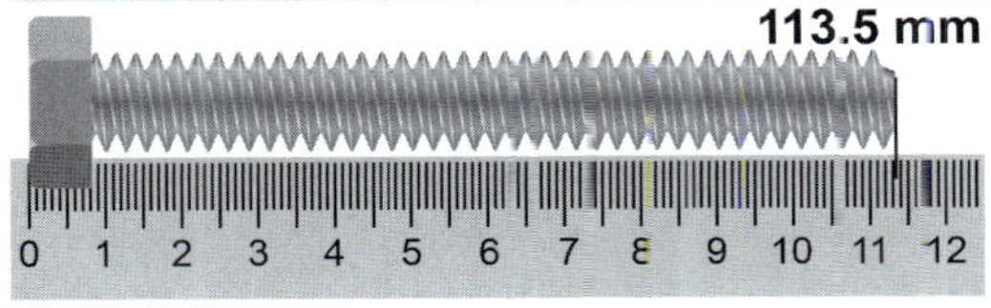


You may estimate length using one-half of the ruler's smallest subdivision.

### Significant figures and multiplication

A value that is the result of a calculation has the same number of significant figures as the least precise value (fewest significant figures) used in the calculation. In the example shown here, 4.37 m has three significant figures and is multiplied by 0.49 m, which has only two significant figures. The final answer is rounded off to two significant figures, or 2.1 m$^2$. *This rule is suspended when the result will be part of another calculation.* For intermediate results, one extra significant figure should be carried to minimize rounding errors in subsequent calculations. For this example, 2.14 would be used as the value to enter into the next calculation.

**Significant figures and multiplication**

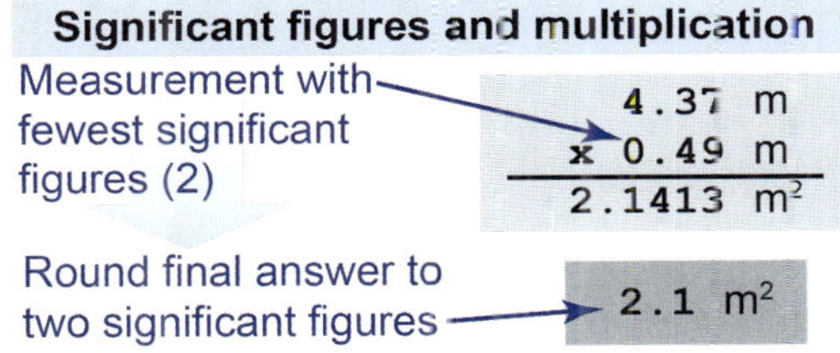


### Scientific notation

How many significant figures are in the value 216,500 m? There could be four, five, or six, depending on whether each of the last two zeros is a significant figure. Zeros to the left of the decimal may or may not be significant. Scientists solve this ambiguity by either quoting the uncertainty directly, such as ±0.1 m, or by using scientific notation. The same measurement written $2.1650 \times 10^5$ m tells you there are five significant figures. When a measured value is properly expressed using scientific notation, every digit that appears is a significant digit.

### Direct and indirect measurements

To measure the mass of a book, you might make a direct measurement by placing it on a scale and reading out the value in kilograms. But how do you measure the mass of a grain of rice? One approach would be to measure the mass of, say, 100 grains of rice and divide the result by 100. This is called an indirect measurement. Throughout all the sciences you may make both kinds of measurement, direct and indirect.

# Investigation 2A

## Indirect measurement, estimation, and scale

**Essential questions** *Is it possible for an answer to be "good enough"?*

When an astronomer says there are 200 billion stars in the Milky Way Galaxy, how precise do you think that number is? Could it actually be 201 billion? Did an astronomer actually count every single star? One website currently lists the world population as 6,840,507,003 people. Is that number correct *down to the single person?* One of the most common misunderstandings about science is that science produces *exact quantitative answers*. Except in rare cases there are no exact answers. Even the best formulas give answers that are only good enough, and very often *indirect* measurements or *estimation* provide the best answer that there can be.

### Part 1: Indirect measurement

**How thick is a single piece of paper?**

Sometimes an object to be measured is too small or is otherwise inaccessible. Use a metric ruler, triple beam balance, meter stick, and/or protractor to make the following indirect measurements and describe your methodology:

a. *What is the mass of a single grain of rice?*
b. *What is the thickness of a single piece of paper?*
c. *What is the height of a tree on the school grounds?*
d. *In which cases is a meter stick more appropriate? A metric ruler?*
e. *As a class, assemble your answers to parts a through c above. Use a calculator to determine the average values and the standard deviation for each. How well do your values agree?*

### Part 2: Atoms, size, and scale

The density of gold is 19,300 kg/m$^3$. The periodic table lists the mass of gold as 0.197 kg per $6.02 \times 10^{23}$ atoms.

a. *What is the mass of a single gold atom, in kilograms?*
b. *Estimate how many gold atoms are in a 1 m cube.*
c. *Assume gold atoms pack together as shown. What is the diameter (in meters) of a single gold atom?*

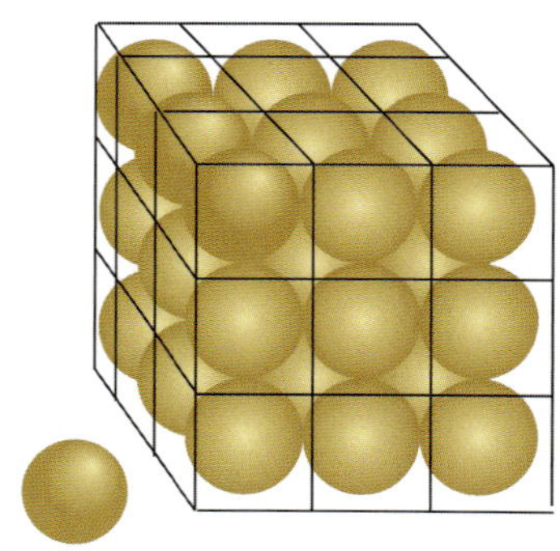


### Part 3: Estimating time

Physics deals with time scales ranging from the time it takes a single electron to move across an atom all the way up to the age of the universe. To compare times, for example, the times all must be represented in seconds. For each of the following questions, which is the *larger* quantity?

a. *one year or 100,000 s*
b. *the number of times a mosquito flaps its wings in 1 s or the number of seconds in 1 hr*
c. *one oscillation of the quartz crystal in your watch or computer or 1 ms*
d. *the time between your heartbeats or the time it takes a pencil to hit the ground falling from your desk*

**Measurement uncertainties**

It is impossible to measure *exactly* the value of any continuous variable. Exact values are only possible when *counting*, such as 311 people or 312 people. Counts are integers whereas continuous variables can have an infinite number of values. The word *error* is sometimes used to describe the fact that all real measurements are uncertain. In this context *error does not mean mistake!* The "error" in measurement is the difference between the measurement and the unknown "true value" of the quantity being measured.

**Why are measurements uncertain?**

The uncertainty in any macroscopic measurement is affected by many factors, including the limited precision of instruments. When using a meter stick, it is not possible to distinguish between 50.0 mm and 50.1 mm because the smallest gradation on the meter stick is 1 mm. Even an *accurate* measurement is uncertain to ±0.5 mm. Accuracy can also be affected by the environment—such as wind, humidity, or temperature—as well as the skill of the operator and the procedure for measuring. The uncertainty in microscopic measurements can result from fundamental quantum effects.

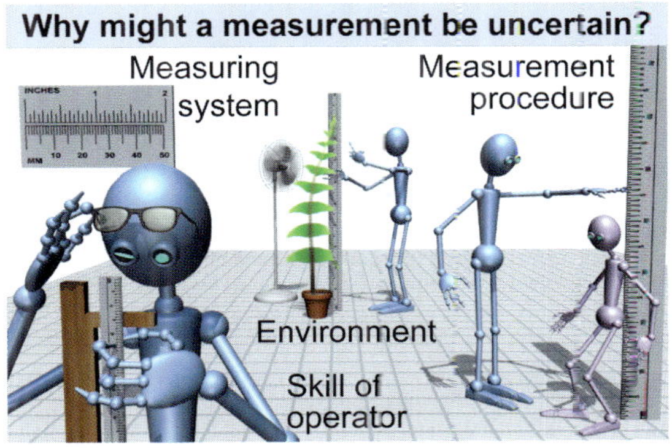


**What is the effect of measurement uncertainty?**

The effects of uncertainty depend on how a measurement is used. A GPS receiver determines position to within ±10 m. This precision is sufficient for navigational directions because roads and intersections are larger than 10 m, but it might be disastrous for steering a car because the width of a lane is smaller than 10 m. As another example, measuring a mass inaccurately can have serious consequences when calculating the load of an elevator or the payload for a rocket.

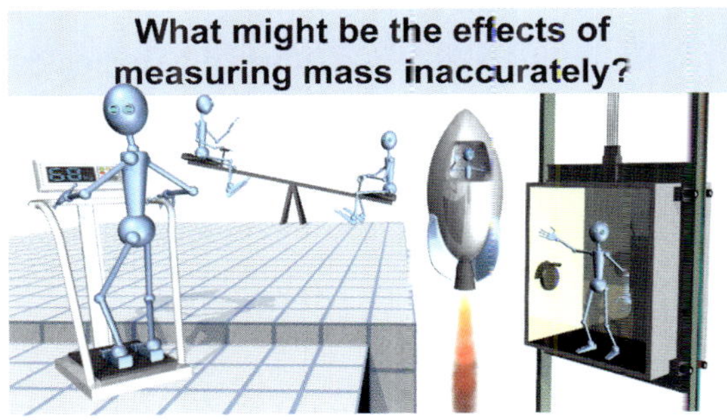


**Tolerances in engineering design**

In engineering design, defining the allowable uncertainties in dimensions or physical properties is often vital to ensuring that a product works. The *tolerance* gives the range of values that are either acceptable or known. In English units a dimension of 1.00 in is assumed to have a tolerance of ±0.010 in, whereas a dimension of 1.000 in is assumed to have a tolerance of ±0.005 in. Engineers select manufacturing processes with uncertainties that can meet or exceed the required tolerances. The "tighter" the tolerances, the more expensive it is to manufacture a part

## Safety in investigations

**Learning by doing**

*Doing* physics is a lot more interesting than just reading about it. Humans learned physics by doing it, and personal investigations in physics are an important part of the learning process. Some investigations involve interactive simulations on a tablet or computer. Other investigations use hands-on equipment, allowing you to gain direct familiarity with many physical concepts.

**Safety should be top priority**

Whenever you conduct an investigation or participate in a class demonstration, safety should be your top priority. You should be *proactive* to recognize potential safety hazards in every investigation. Consider the following:

- **Heat sources** can burn the skin and should not be brought near flammable materials such as clothing or drapes.
- **Light sources** can become quite hot. Light sources (including the Sun) can damage eyesight, so never look directly at a bright light source.
- **Moving objects**, such as cars rolling down ramps or masses swung around in circles, can hurt people and damage property. Always provide for a safe way for moving objects to stop, and conduct activities outside and away from other people when appropriate.
- **Sound sources** can damage hearing at high volume so always keep the volume low.
- **Electrical and electrostatic systems** carry the danger of electric shock if they are not handled and connected properly.
- **Field work** requires the same safety consideration as indoor lab work. Exercise the same safety precautions when conducting experiments or observations outdoors.

**Safety equipment**

You have only two eyes, and they have to last you for a lifetime, so protect them! Wear goggles or other approved eye protection when conducting physics investigations. Wear laboratory coats, disposable gloves, or other protective clothing—not just to protect your street clothing but to protect your skin should a liquid or chemical spill on you. Always know the location and operation of the fire extinguisher, eye washer, and safety shower in your classroom or laboratory space. All of these rules apply both indoors and outdoors.

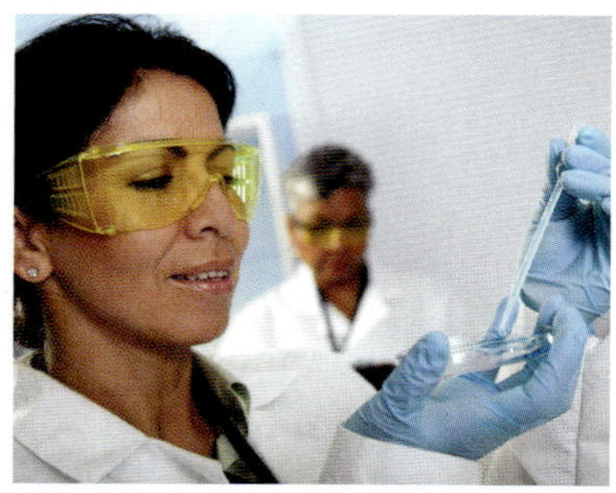

**Some other safety considerations**

Never conduct an experiment or investigation by yourself. Always work with at least one partner. Do not conduct experiments that you have not been authorized to perform. Read through the *complete* instructions for an investigation before beginning it. Ask your instructor for clarification if any part of a lab is not clear to you or if you believe any activity might be unsafe.

# Section 2 review

Measurements are important in all sciences, but providing only a value and quantity is not enough. Every measurement should also specify its *uncertainty*. Uncertainty can be expressed using a plus/minus sign (±), but it can also be expressed by using an appropriate number of significant figures or decimal places. These uncertainties propagate through to the answer when, for example, adding or multiplying. Uncertainties arise in measurements from factors such as the measuring device, the skill of the operator, the environment, and the procedure. The effects of errors can be important, especially when a product requires tight tolerances.

| Vocabulary words | accuracy, precision, decimal places, significant figures |
|---|---|

## Review problems and questions

1. How many significant figures do the following numbers have?
   203, 12.20, 0.05, −14.056, −0.0023, 100

2. You have five solid rods made of the same material. These rods all have the same cross-sectional shape (a 1 cm by 1 cm square) but are of different lengths. Below are some measurements you made of the length and mass of the rods.

| Length $x$ (cm) | Mass $m$ (g) |
|---|---|
| 1.8 | 2.5 |
| 2.2 | 3.4 |
| 3.7 | 4.9 |
| 4.2 | 5.9 |
| 5.3 | 7.1 |

   a. These measurements were taken using a ruler with gradations of 1 mm. Do these measurements match the precision of the instrument?
   b. As all of the rods are made from the same material, they should have the same density. Calculate the density for each of the five measurements. Are the densities the same to within the expected uncertainty? Based on the calculated densities, would you consider these measurements accurate?
   c. What could be some potential sources of uncertainty in the measurement of the length and mass of these rods?
   d. Your friend informs you that the ruler you used is missing the first two millimeters owing to wear and tear. This means that all of your measurements are 2 mm longer than they should be. Calculate the percentage error this causes in all five measurements of the length of the rods.
   e. How does the error caused by the shorter ruler change the error in the density calculation as the measured length becomes smaller?

3. A scientist has measured the weight of five objects of interest. Put these measurements in scientific notation: 450,000 g, 0.00089 g, 98.34 g, 2,340 g, and 0.0925 g.

Problem solving means using what you know to figure out something you don't know. This skill is useful in all careers, whether for valuing a company's stock or calculating the right size pipe to use in a new house. It is also a necessary skill to pass a physics test! One important problem solving skill is converting between two different sets of units, such as between feet and meters or between minutes and seconds. This book teaches you techniques that will help you apply your knowledge of physics to solve any problems you are likely to find in this book or on a physics exam.

## Converting units

**Comparing quantities**

Which is longer, 3,000 in or 100 m? The number 3,000 is larger than the number 100. Simply comparing values, however, gives you the wrong answer. To compare two lengths, both should be expressed in the same units. For example, by expressing the distance of 100 m in inches—100 m = 3,937 in—you can easily see that 100 m is longer than 3,000 in.

| Length conversion factors | | |
|---|---|---|
| $\frac{1\text{ in}}{2.54\text{ cm}}$ | $\frac{1\text{ mi}}{1{,}609\text{ m}}$ | $\frac{1\text{ light year}}{9.5\times10^{15}\text{ m}}$ |
| $\frac{100\text{ cm}}{1\text{ m}}$ | $\frac{1\text{ km}}{0.6215\text{ mi}}$ | $\frac{5{,}280\text{ ft}}{1\text{ mi}}$ |
| $\frac{1{,}000\text{ mm}}{1\text{ m}}$ | $\frac{1{,}000\text{ m}}{1\text{ km}}$ | $\frac{1\text{ micron}}{1\times10^{-6}\text{ m}}$ |
| $\frac{12\text{ in}}{1\text{ ft}}$ | $\frac{39.37\text{ in}}{1\text{ m}}$ | $\frac{1\text{ nm}}{1\times10^{-9}\text{ m}}$ |
| $\frac{1\text{ fathom}}{6\text{ ft}}$ | $\frac{1\text{ hand}}{101.6\text{ mm}}$ | $\frac{1\text{ mil}}{0.001\text{ in}}$ |

**Conversion factors**

Interactive equation

Often you will need to take a measurement in one set of units and express it in different units. This *conversion* is done using a relationship between units called a *conversion factor* (see table on the right). A **conversion factor** has the physical value of exactly one since the numerator and denominator represent the same distance. The *numerical* value of the conversion factor, however, is usually not 1.00 *because the lengths are expressed in different units.*

**Multiplying with conversion factors**

A conversion factor has a value of one whether it is used right-side-up or upside-down. When you use the conversion factor from feet to meters you divide by 3.28; when you use the same conversion factor from meters to feet you multiply by 3.28. *The correct calculation is the one in which all the units cancel except the ones you want.* Once you have the conversion factors arranged so that the units work out, then you know correctly which values to multiply and which to divide by. In this example, 30 m is 1,180.8 in.

**Only one of these calculations is correct. Can you see why?**

$$30\text{ m} = 30\,\cancel{\text{m}}\left(\frac{3.28\text{ ft}}{1\,\cancel{\text{m}}}\right)\left(\frac{1\text{ ft}}{12\text{ in}}\right) = 8.2\text{ ft}^2/\text{in}$$

$$30\text{ m} = 30\,\cancel{\text{m}}\left(\frac{3.28\,\cancel{\text{ft}}}{1\,\cancel{\text{m}}}\right)\left(\frac{12\text{ in}}{1\,\cancel{\text{ft}}}\right) = 1{,}180.8\text{ in}$$

$$30\text{ m} = 30\,\cancel{\text{m}}\left(\frac{1\text{ m}}{3.28\text{ ft}}\right)\left(\frac{12\text{ in}}{1\text{ ft}}\right) = 109.8\text{ m}^2\text{ in}/\text{ft}^2$$

**Tabulating data**

Units are important when organizing data into *tables*. A standard convention is to include the units in parentheses under the name heading of each column. For example, if a column includes length data in meters, then the heading of the column should be "Length" or "*l*" and directly below the word "length" should be the units in parentheses (m).

### Variables

A **variable** is a physical quantity, such as speed or time, that can have a value. In equations, variables are usually named with a single letter, such as $v$ for velocity. In this book, variables are written with *italics*. When there is more than one variable of the same type, such as two velocities, a subscript is added, such as $v_1$ or $v_2$. An important step in solving physics and engineering problems is to assign unique names to all the important variables!

### Dependent and independent variables

In an experiment, the experimenter—you!—will often physically change the value of one variable and then measure the resultant change in another variable. The variable you modify or manipulate is called the **independent variable**, while the variable that changes as a result is the **dependent variable**. A *controlled variable* is a quantity that is held constant in an experiment because variations in the quantity could conceivably affect the outcome.

### Purpose of graphing data

Graphs of data show patterns that are often hard to see in a data table. Throughout this course, we will use graphs of motion, waves, frequencies, and many other variables to demonstrate patterns. Computers are particularly useful at drawing graphs and other representations of data quickly. The tool we will use most is the $x$–$y$ graph.

### $x$–$y$ graphs

An $x$–$y$ graph is a visual representation of the relationship between two variables. Just as words are spelled a certain way, graphs are drawn by following certain rules. In the graph on the right, a student drops a pencil from counter height and measures its speed (the dependent variable) as a function of time (the independent variable). The independent variable (time) goes on the horizontal or $x$-axis, and the dependent variable (speed) goes on the vertical or $y$-axis. The graph shows a pattern of speed increasing with time. The mass and size of the pencil are controlled variables, because changing them *might* affect the measured speed.

**Speed vs. time**
(pencil dropped from a counter)

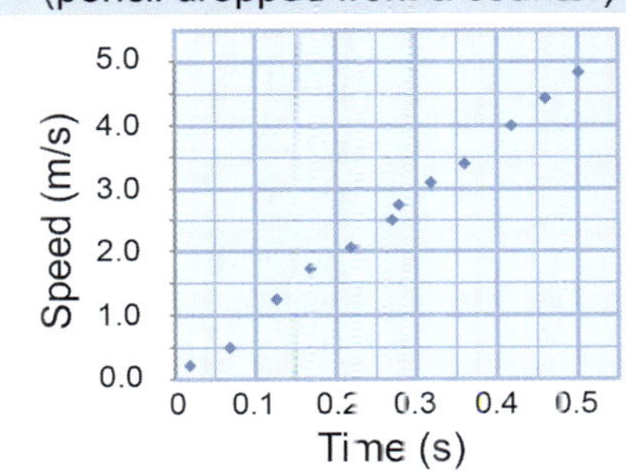


Note that the speed increases with time.

### Reading $x$–$y$ graphs

Graphs have a scale on both the $x$ and $y$ axes. The scale tells you the relationship between changes on the graph and changes in the variable itself. Be careful when interpreting graphs! A large visual change can sometimes mean only a small real change depending on the scale of the graph. The two graphs below show the same data. One looks like a big change whereas the other does not! The only difference between the two graphs is the *scale* of the $y$-axis.

**Which graph shows a greater change?**

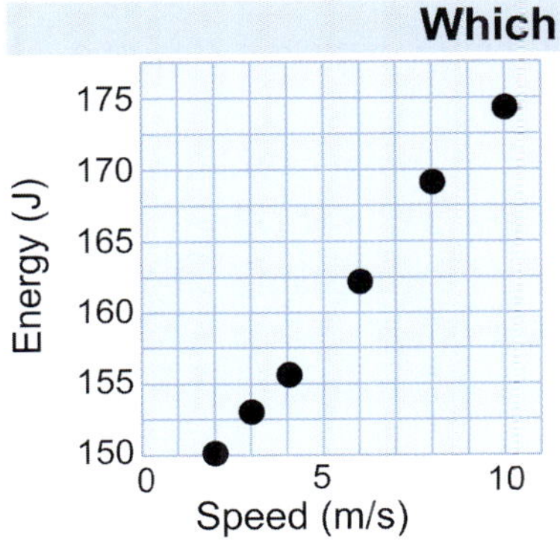


| Speed (m/s) | Energy (J) |
|---|---|
| 2 | 150 |
| 3 | 153 |
| 4 | 156 |
| 6 | 162 |
| 8 | 168 |
| 10 | 174 |

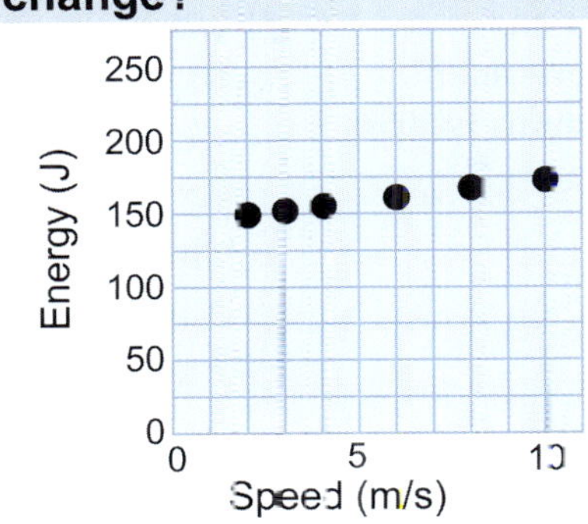

# Models and equations

### Models

In physics, a **model** is a relationship between variables that allows you to predict something unknown based on things that are known. For example, a model of motion allows you to predict the speed of a falling object from the height it has fallen. A graph is a simple form of model that relates two variables, such as speed and height. An equation, or formula, can be another form of model. Complex models can include many equations, computer spreadsheets, and the results of other models. Computer simulations of Earth's climate are a good example of this kind of model. Making models is important to science and engineering. A model allows us to evaluate how changes in one variable affect the others.

### Equations

An equation connects variables through the equals sign (=). An equation such as $v = d/t$ means that the speed $v$ is the same as the distance $d$ divided by the time $t$ for *all possible values of v, d, and t.* Not only are the values the same, but the *units* are also the same on the left and right sides of the equals sign. Distance divided by time therefore has the units of speed.

**Speed vs. height**
(marble dropped from different heights)

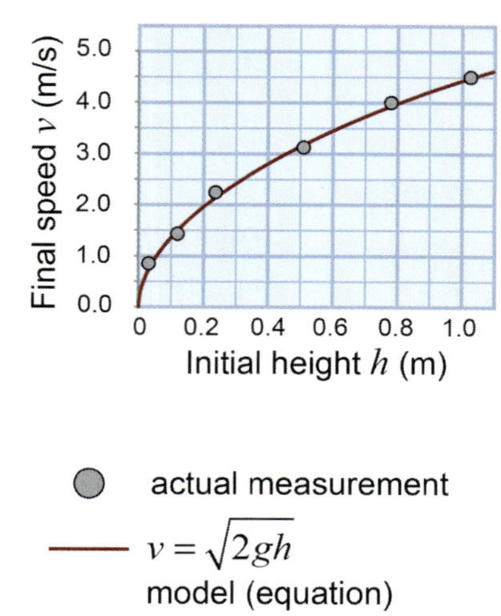


### Validating models with data

Some equations, such as $v = d/t$, define one variable in terms of others. Other equations are models that describe relationships among variables. These kinds of equation models are only useful if they also represent reality. The diagram on the right shows how an energy conservation model for the speed of a falling object compares with actual speed measurements made for a falling marble. The model used describes speed as $v = \sqrt{2gh}$.

### Graphical relationships

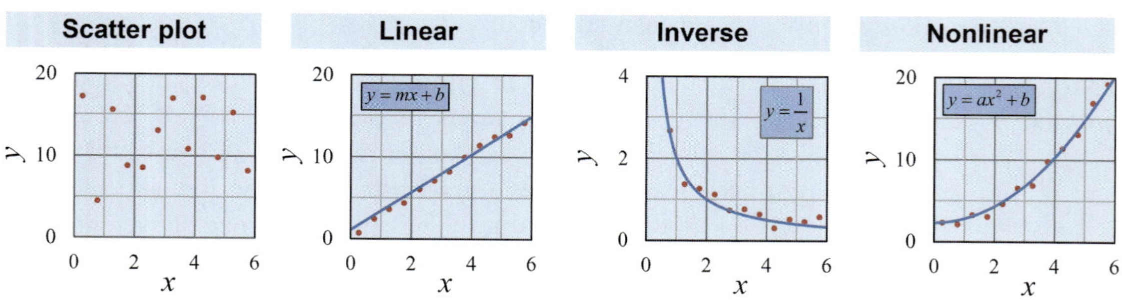


### Linear and nonlinear relationships

If the data on a graph indicate no relationship between the two plotted variables, this produces a scatter plot. If there is a relationship, however, then a graph will show a pattern represented by that relationship. Some kinds of mathematical relationships are linear (such as $y = mx + b$) or inverse ($y = 1/x$). There are also more general kinds of nonlinear relationships, such as a parabola ($y = ax^2$). In the graphs above, the data are plotted as points while the model—represented from these linear and nonlinear equations—are plotted as lines or curves.

**Finding the "best" straight line**

A theory or hypothesis often predicts a relationship between two variables. How do scientists evaluate whether data support or disprove a *relationship*? The answer is a form of "moving average" called a *regression*. The most common form of regression analysis is the **linear regression** in which the "best straight line" calculated for a set of x-y data used to explain the relationship between them. For example, the ideal gas law predicts that the pressure of a gas increases linearly as the temperature changes. The data below show the results of an experiment to measure the pressure and temperature over a range.

Linear regression

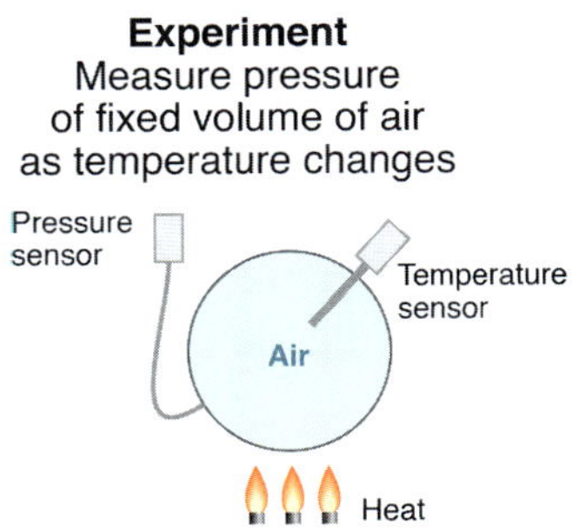


**Data**

| Temperature (°C) | Pressure (Pa) |
|---|---|
| 20 | 111 |
| 22 | 111 |
| 25 | 106 |
| 33 | 112 |
| 44 | 117 |
| 47 | 122 |
| 59 | 123 |
| 70 | 128 |

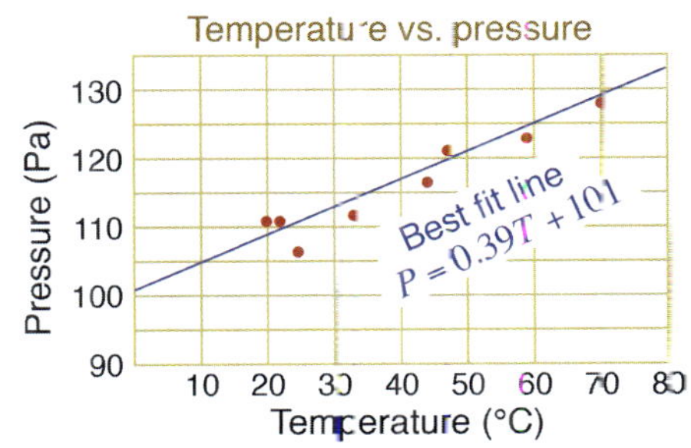


**Finding the best fit**

The blue line represents the best straight line that fits the data. In a regression analysis, "best" means the unique line that results in the smallest average difference between the line and each data point. It is easy to "eyeball fit" the best straight line with a ruler but, most data software has a function to calculate the best straight line. The diagram shows the calculation on a spreadsheet. The averages of x and y have a line on top and the Greek letter sigma, $\sum$, which means "the sum of" so $\sum x^2$ means "sum all the values of $x^2$."

Calculating the best fit straight line

| $x$ | $y$ | $x-\bar{x}$ | $(x-\bar{x})^2$ | $y-\bar{y}$ | $(x-\bar{x})(y-\bar{y})$ |
|---|---|---|---|---|---|
| 20 | 111 | -20 | 400 | -5.3 | 105.0 |
| 22 | 111 | -18 | 324 | -5.3 | 94.5 |
| 25 | 106 | -15 | 225 | -10.3 | 153.8 |
| 33 | 112 | -7 | 49 | -4.3 | 29.8 |
| 44 | 117 | 4 | 16 | 0.8 | 3.0 |
| 47 | 122 | 7 | 49 | 5.8 | 40.3 |
| 59 | 123 | 19 | 361 | 6.8 | 128.3 |
| 70 | 128 | 30 | 900 | 11.8 | 352.5 |

| $\bar{x}$ | $\bar{y}$ | $\sum(x-\bar{x})^2$ | $\sum(x-\bar{x})(y-\bar{y})$ |
|---|---|---|---|
| 40 | 116.3 | 2,324 | 907 |

Equation of a straight line $y = mx + b$

Calculate best-fit slope, m

$$m = \frac{\sum(x-\bar{x})(y-\bar{y})}{\sum(x-\bar{x})^2} = \frac{907}{2324} = 0.39$$

Calculate best-fit intercept, b

$$b = \bar{y} - m\bar{x} = 116.3 - (0.39)(40) = 100.7$$

**The coefficient of regression, R**

The "goodness" of the fit is described by the **correlation coefficient**, *R*, which varies from +1 to -1. A value of *R* = +1 indicates a perfect straight line with a positive slope (strong direct x-y relationship). A value of *R* = -1 describes a perfect straight line with a negative slope (strong inverse relationship). As *R* gets closer to zero, the linear relationship weakens. When *R* is equal to 0, the data do not have a linear relationship. The diagram below shows the patterns associated with different values for the correlation coefficient.

The correlation coefficient, R

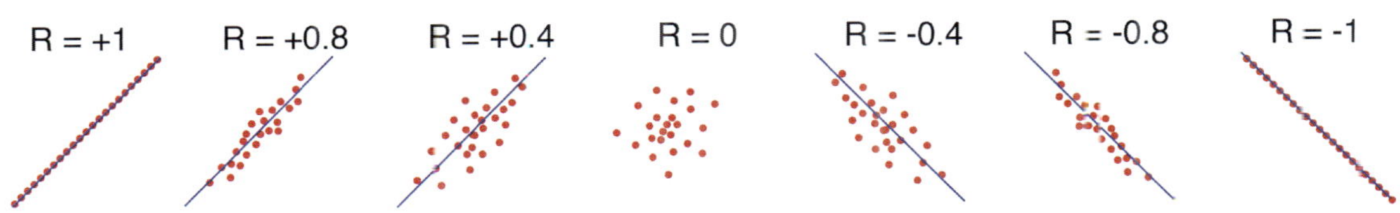

**Pie charts**

A **pie graph** is a good way to show the relative proportions of individual categories in relation to a whole. The size of each slice of the "pie" is proportional to its associated quantity divided by the whole. In the example below the "whole" is the production of electricity in the US. The pie graph makes it easy to see that chemical energy in the form of coal and gas are the source of almost two-thirds (64%) of the electric power generated in the US, with nuclear being the third largest contributor at 20%.

Pie graph
(fractions of a whole)

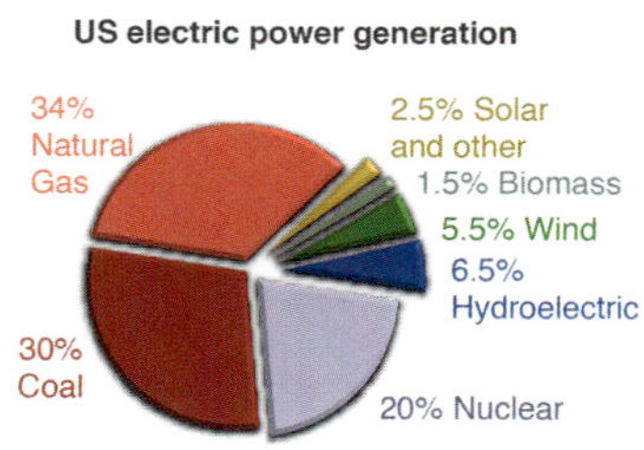


Bar graph
(data patterns over time)

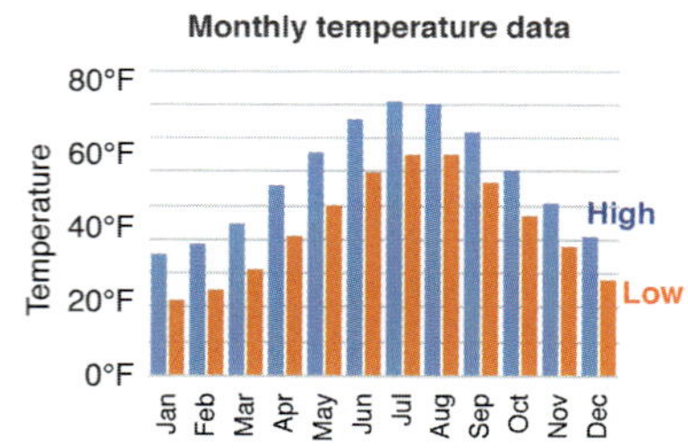


Bar graph
(frequency data)

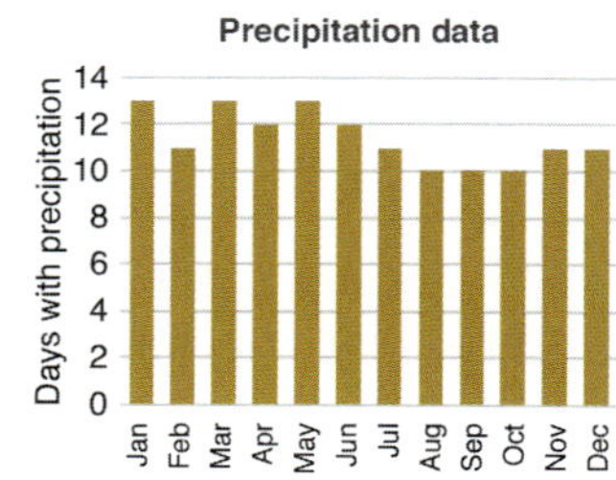


**Bar graphs**

A **bar graph** represents each data point by the height of the bar. Bar graphs are effective at showing a pattern in relatively small sets of data, such as the seasonal variation with temperature. Bar graphs lose their visual effectiveness when they have more than a dozen or so bars. Notice the example has two bars for each month showing the high and low temperatures. It is common to use bar graphs to compare the patterns in two or more related data sets, such as the average high and low temperatures.

**Frequency tables**

Many areas of research generate a lot of data that may be highly variable from point-to-point but has an overall pattern that is important to show. A good example is the number of days it rains or snows over a year. Plotting all 365 days separately does not describe the overall pattern that the average precipitation is approximately constant in the region described by the graph. A **frequency table** is a way to group data into bins than can be easier to represent on a graph. Combining many points into a bin helps smooth out fluctuations in the data to show the overall pattern.

Representing uncertainty on a graph

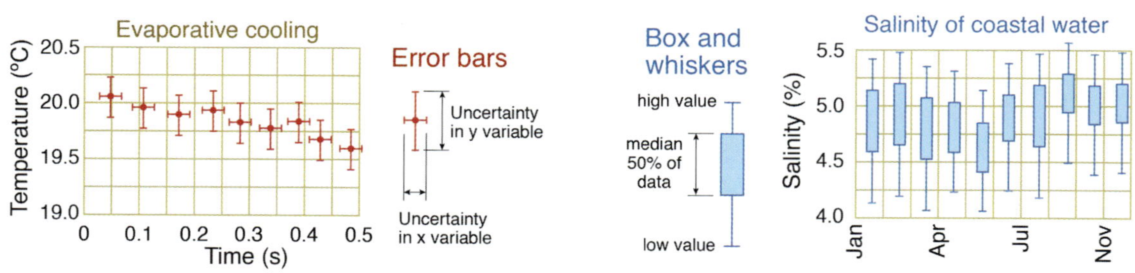


**Showing uncertainty on a graph**

To correctly interpret the results of data it is necessary to know the size of the uncertainties (error) in measurement. **Error bars** on each data point are one method used to indicate the error on a graph. The size of the error bars are proportional to the error in each data point and may vary from one point to the next. A **box and whiskers plot** is another way to represent uncertainty when the data sets are too large to plot every point. The box represents the media (middle) fifty percent of the data in each group and the whiskers represent the upper and lower twenty five percent.

# Investigation 2B Graphical relationships

**Essential questions**
*How do we recognize the "right" explanation?*
*How do graphs provide insight into the relationships among variables?*

Interactive simulation

The accuracy of a scientific theory is often assessed by comparing the predictions of theory with real observational data. For example, if a theory predicts how fast an object's speed changes, then actual measurements should agree with the theory to within the limits of experimental error. The data provided in the simulation show how some quantities for objects change under different conditions. The examples include an object moving at constant speed, an object dropped from rest, current through a resistor varying with resistance, and an object cooling over time.

## Part 1: Matching models to data

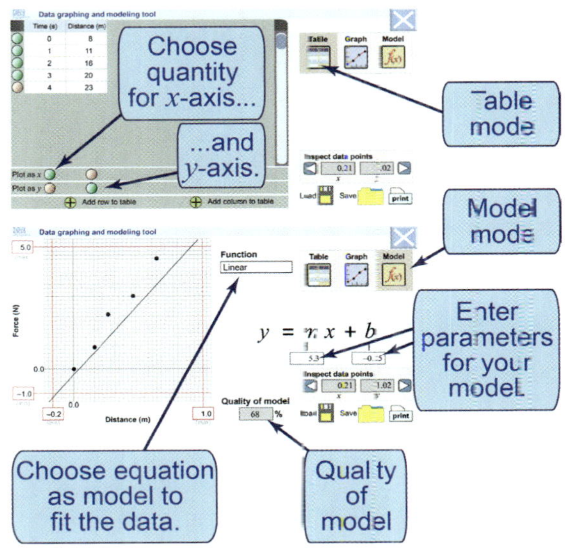


1. Plot time on the $x$-axis by selecting "$x$" for the column labeled "Time (s)." Plot distance on the $y$-axis for an object moving at constant speed by selecting "$y$" for the column labeled "Distance (m)."
2. Choose a model equation (line, parabola, exponential, or inverse) to try and fit the data.
3. Change parameters for that model to fit the data.

a. What model and parameters are a good fit to this distance versus time graph?
b. Are equal numbers of data points plotted above and below your model? Why is this important?
c. Repeat these steps for a graph of the object dropped from rest using data for height versus time. What model and parameters are a good fit to these data?
d. Repeat these steps for a graph of electric current versus resistance using data in the fourth and fifth columns. What model and parameters are a good fit to these data?
e. Repeat these steps for a graph of a cooling object using the temperature and time data. What model and parameters are a good fit to these data? Record your results on your assignment sheet.

## Part 2: Other ways to represent data: pie and bar charts

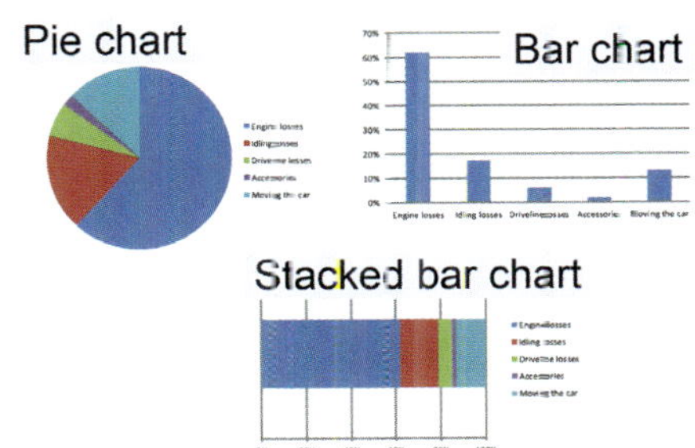


1. Use the data below to construct a pie chart, a bar chart, and a stacked bar chart. Be sure to label your charts!

a. In the pie chart, how can you tell visually which category uses the most power?
b. What is the advantage or disadvantage of representing these data with each of these kinds of charts?
c. When would you choose a pie or bar chart to plot data?
d. In what cases would you choose a line graph to plot data?

**Residential power usage by category**

| Space heating | Water heating | Lighting | Air conditioning | Refrigeration | Electronics | Washer and dryer |
|---|---|---|---|---|---|---|
| 32% | 13% | 12% | 11% | 8% | 5% | 5% |

## Problem–solving steps

**Four-step method for solving problems**

How do you start solving a physics problem? The technique on this page has four steps and will always get you started or even most of the way toward solving a problem:

- **Asked:** Determine what the problem is asking you for.
- **Given:** Determine what information you are given.
- **Relationships:** Write down the relationships you know that involve what is asked and what is given.
- **Solution:** Use the relationships and the given information to solve the problem.

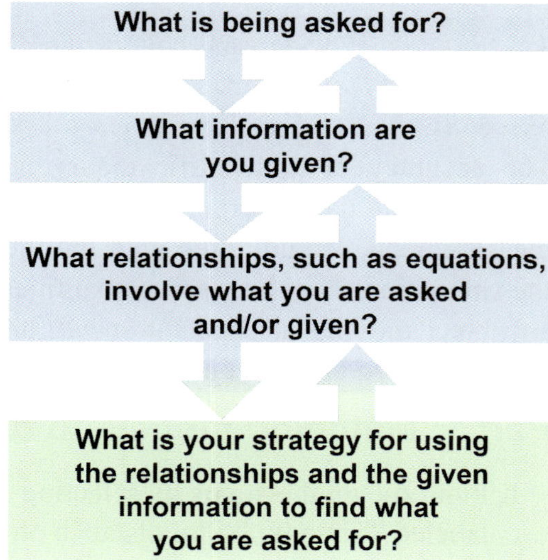


**What does the problem ask for?**

Determine what the problem is asking for. Be very specific. For example, when a problem asks how far an object has moved, the answer is a distance. It is easy to get sidetracked by looking for things you do not need, so *read the problem carefully*—twice, or as many times as you need—to make sure you understand what the problem is asking. *Assign variables to the unknowns.* For example, if the problem asks for a speed, the unknown is represented by the variable $v$. If there are two unknown speeds, then assign them variables $v_1$ and $v_2$.

**What information are you given?**

Identify all the information you are given in the problem. Information may be in the form of measurements, such as mass or length. Information may also be descriptive. For example, "at rest" means something is not moving and has a speed of zero. Some information may also be deduced from other information. For example, if an object is not moving, then the net force acting on it must be zero. If the net force were *not* zero, then the object would be accelerating and could not stay still. *Assign variables to the given quantities.* For example, if you are given a mass, then assign it the variable $m$. If there are two given masses, then assign them variables $m_1$ and $m_2$.

**What relationships do you know?**

Identify relationships, such as equations, that involve what you need to know and/or the information you are given. For example, if you want to know speed and are given distance and time, then you would write down the formula for speed that includes distance and time. Some problems require more than one relationship.

**How do you solve the problem?**

Apply the given information and relationships to solve the problem. Once you have collected the information and the relationships, you will often see how the problem can be solved. Sometimes the solution may require more than one step; you might use force to calculate acceleration, and then use acceleration to calculate speed.

**The most important step**

No matter how much you read about how to solve problems you will not become good at it until you start solving them. You can only learn how to solve problems by solving many of them. Don't worry if you make mistakes. Just keep on trying!

### Translating from English to math

Mathematics is the language of physics. Mathematics provides us with a concise and efficient way to describe physical phenomena. Consider *speed*, which is the distance you traveled divided by the time it takes. For example, if you go 120 miles in 2 hours, your speed is 120 miles ÷ 2 hours, or 60 miles per hour. If we write this in algebra, we let the variable $v$ stand for "the speed," $d$ stand for "the distance you go," and $t$ stand for "the time it takes." The sentence that defines speed translates to an equation that says the same thing.

**Translating from English to mathematics**

The speed is the distance traveled divided by the time taken.

$v \quad = \quad d \quad \div \quad t$

equation → $v = \dfrac{d}{t}$

### Rearranging relationships

The language of mathematics provides techniques for rearranging variables while preserving the relationship among them. As an example, try to rearrange the sentence defining speed in a way that says "the time taken is equal to...." The correct answer is that the time taken is equal to the distance traveled divided by the speed. If you have to travel 120 miles and your speed is 60 mph, it takes you 2 hours because 2 hr = 120 mi ÷ 60 mph. There is no way to know how to rearrange an English sentence that will provide this result. But there *is* a way to do it with algebra.

| Using numbers | | Using algebra | |
|---|---|---|---|
| Multiplying or dividing both sides by the same number preserves the equality. | $4 = \dfrac{12}{3}$ → | $v = \dfrac{d}{t}$ | Multiplying or dividing both sides by the same variable preserves the equality |
| | $3 \times 4 = \dfrac{12}{3} \times 3$ → | $t \times v = \dfrac{d}{\cancel{t}} \times \cancel{t}$ | |
| | $\dfrac{3 \times 4}{4} = \dfrac{12}{4}$ → | $\dfrac{t \times \cancel{v}}{\cancel{v}} = \dfrac{d}{v}$ → $t = \dfrac{d}{v}$ | |

### Solving equations

The key to using algebra is understanding the equals (=) sign. It means that whatever is on the left of the equals sign is the same as what is on the right. It also means that if you add something to the left, you have to add the same something to the right to preserve the equality. Learning what to add, subtract, multiply, or divide *on both sides of the equation* allows you to rearrange equations to turn the relationship you have into the one you want. The diagram illustrates how this works to solve the speed equation for time.

**Useful rules of algebra**

$$c(a+b) = ca + cb$$

$$\frac{a}{a} = 1$$

$$\frac{a+b}{c} = \frac{a}{c} + \frac{b}{c}$$

**Using algebra to solve an equation**

| | |
|---|---|
| $vt = \dfrac{d}{t}t$ | Multiply both sides by $t$. |
| $vt = \dfrac{d}{\cancel{t}}\cancel{t}$ | Cancel the $t$ from top and bottom. |
| $\dfrac{vt}{v} = \dfrac{d}{v}$ | Divide both sides by $v$. |

# Investigation 2C Algebraic relationships

**Essential questions** *How do you use algebra to solve physics problems?*

Interactive simulation

Mathematical equations are used to describe force and motion, work and energy, heat and temperature, and virtually every other topic in physics. Algebra is a basic technique in the physicist's toolkit, used to rearrange the variables in an equation to solve for one variable.

## Part 1: Using algebra to rearrange an expression

In this investigation, you will use algebra in a series of interactive simulations to solve for the desired variable. If you make a mistake, don't worry: You can choose a different step to replace your previous answer. Look for the traffic lights along the way: A red light indicates that you should try again!

1. Open the first interactive element that asks you to solve the equation $v = d / t$ for time $t$.
2. Choose the next step to solve the problem.
3. Inspect what happens algebraically to the expression because of your choice.

a. *What was your solution to the algebra problem?*
b. *Did you solve the problem in the minimum number of steps?*

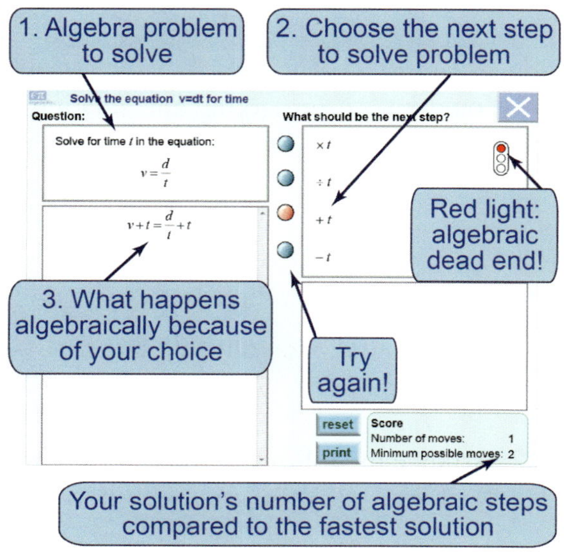


## Part 2: Solve two more problems

1. Using the second interactive simulation, solve the algebraic equation $x = x_0 + v_0t$ for the initial speed $v_0$.
2. Using the third interactive simulation, solve the equation $d = ½at^2$ for time $t$.

a. *What was your solution for the second algebra problem?*
b. *What was your solution for the third algebra problem?*

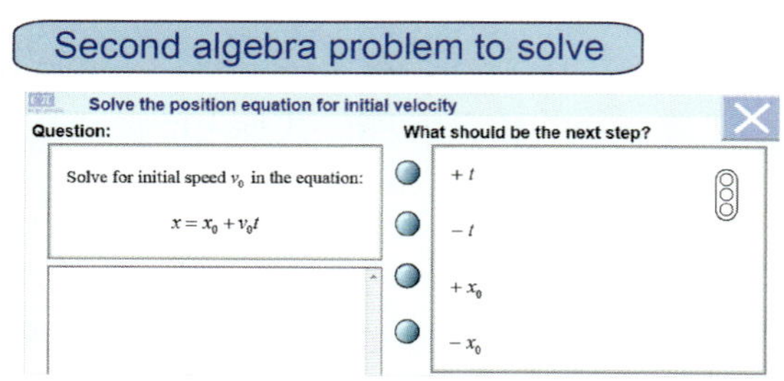


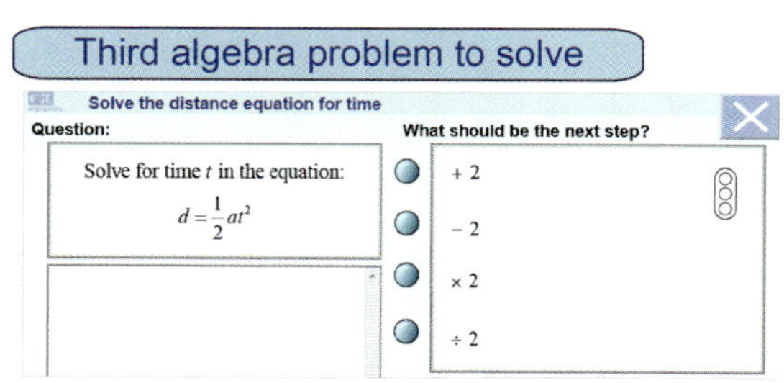

# Section 3 review

Quantities only have meaning if you know what *units* they are expressed in. For example, 22 does not tell you how much mass there is, but 22 kg does. Humans have invented many different units that are convenient for different applications. *Conversion factors*, such as 1 m = 3.28 ft, allow you to translate a quantity into a different set of units. *Scientific notation* is a convenient way of writing very large and very small numbers as a product of a base multiplied by a power of 10, such as $1.5 \times 10^3$. Many areas of physics involve *data*, which are measurements of one or more variables. A *graph* is a pictorial way of visualizing data to more easily recognize or demonstrate relationships among variables. Solving physics problems requires systematic thinking. Start by identifying what is *asked* and what information is *given*. The next step is to identify any known *relationships* that involve what is asked and/or given. Finally, the information and the relationships are used to *solve* the problem. The solution process often requires *algebra*. Algebra provides the mathematical and logical rules for combining and rearranging equations and variables while preserving both equality and underlying relationships.

| Vocabulary words | conversion factor, dependent variable, independent variable, variable, model, linear regression, correlation coefficient, pie graph, bar graph, frequency table, error bars, box and whiskers plot |
|---|---|

## Review problems and questions

1. Arrange the following lengths from largest to smallest.
   (a) 632 mm; (b) 35 in; (c) 1.1 m; (d) 2 ft

2. Arrange the following times from largest to smallest.
   (a) 1,500 s; (b) 500 min; (c) 2 hr; (d) 0.01 day

3. The graph shows the data for an experiment that used a rubber band to launch small model rockets of different masses straight up. The rubber band was stretched the same amount for each rocket. Answer the following questions based on the graph.

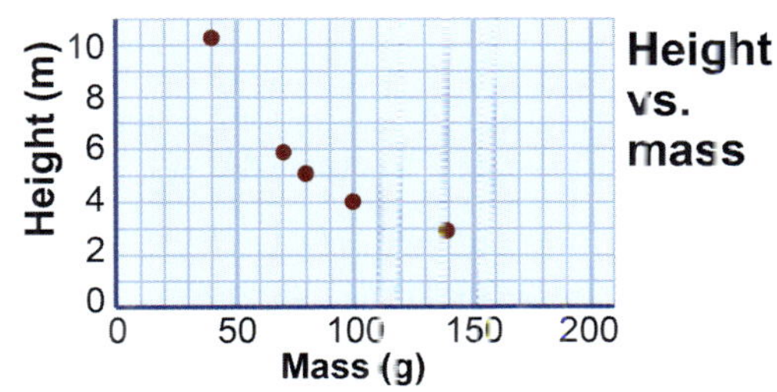


   a. What is the independent variable in this experiment and what is its range of values?
   b. What is the dependent variable in this experiment and what is its range of values?
   c. Does the relationship between height and mass appear directly or inversely proportional?
   d. How high would you expect a rocket of 120 g to fly?
   e. Suppose you observe a height of 7 m. Estimate the mass of the rocket.

4. Which *two* of the following statements are *not* true?

   a. $2x + y = 2(x + y)$
   b. $z(x - y) = (x - y)z$
   c. $z(x - y) = zx - yz$
   d. $xyz = (xy)(yz)$

Chapter 2

# Chapter review

## Vocabulary

Match each word to the sentence where it best fits.

### Section 2.1

| | |
|---|---|
| density | exponent |
| inertia | length |
| macroscopic | mass |
| matter | measurement |
| microscopic | scale |
| scientific notation | surface area |
| temperature | volume |

1. Mass, time, and _______ are three fundamental dimensions of the universe.
2. The relative size of things is in indicator of their _______.
3. Protein folding in the cell occurs at _______ scales.
4. A metal ball sinks in water because the metal's _______ is higher than that of water.
5. The celestial bodies of the Solar System are an example of the _______ scale of physics.
6. A car's gas tank holds 10 gallons of gasoline, which is a measure of its _______.
7. Room A is hotter than Room B because the former has a higher _______.
8. In the expression $8.7 \times 10^3$ kg, the number 3 is the _______ of 10.
9. A distance of 10 miles is an example of a _______ because it has a value and a unit.
10. A realtor said that the apartment for rent has 1000 square feet, which is a measurement of its _______.
11. Expressing a distance as $2.7 \times 10^{23}$ m uses _______.
12. The _______ of the boy is 150 kg.

### Section 2.2

| | |
|---|---|
| accuracy | decimal places |
| precision | significant figures |

13. The measurement 0.0789 m has three _______.
14. A measurement of 0.711 m has lower _______ than a measurement of 0.0063 m.
15. Samuil's throws lacked _______, because the water balloons he threw repeatedly missed Veronica.

### Section 2.3

| | |
|---|---|
| conversion factor | dependent variable |
| independent variable | model |
| variable | |

16. A relationship that connects two or more variables is an example of one kind of a/an _________.
17. The quantity (39.37 in/1 m) is a/an _______.
18. In an experiment to measure how much force is required to compress a spring by various lengths, the quantity force is a/an _______.
19. In an experiment, the height that a ball was raised above the floor was 10, 20, 30, and then 40 cm, while the mass of the ball was always 180 g. The quantity height is a/an _______.

## Conceptual questions

### Section 2.1

20. A cube and a sphere have the same volume. Which one stands *taller* when placed on the ground? Why?
21. Rank the three objects from smallest volume to largest volume. Assume all are solid with no interior voids.
    a. one kilogram of water
    b. one kilogram of wood
    c. one kilogram of aluminum
22. Give three examples each of fundamental and derived units in the SI.
23. Is volume a fundamental or derived quantity?
24. Describe three advantages of the SI over the English system of measurement.
25. Which has a higher density, a block of foam or a ball of clay?
26. List the following in order of increasing size.
    a. baseball  b. virus  c. gold atom
    d. ammonia molecule  e. tadpole  f. Sun-like star
27. List the following in order of increasing time duration.
    a. adult's resting heartbeat
    b. one flap of hummingbird's wings in flight
    c. one full rotation of the Earth
    d. one orbit of Mercury around the Sun
    e. the length of one class period

# Chapter review

## Section 2.1

28. What is the difference between weight and mass? Provide an example in your answer.
29. What is the difference between fundamental and derived units?
30. Which is longer, 1.23 mm or $2.34 \times 10^5$ µm?
31. Would you use an electron microscope or a meter stick to make measurements at macroscopic scales?
32. What metric unit of length is most useful for measuring the following lengths?
    a. height of a 5-year-old boy
    b. thickness of a piece of paper
    c. Diameter of a quarter
    d. Thickness of a dime
    e. Distance from New York, NY, to Sacramento, CA
33. Josephine measured the masses of a textbook and feather and concluded that the feather had more mass. How could you check her result?
34. « A block of foam and a ball of clay both have the same mass. Which one occupies a larger volume?

## Section 2.2

35. In case of a fire or other emergency in the classroom laboratory or outdoors during a field investigation, should you text message your friends, call your parents, or inform the instructor?

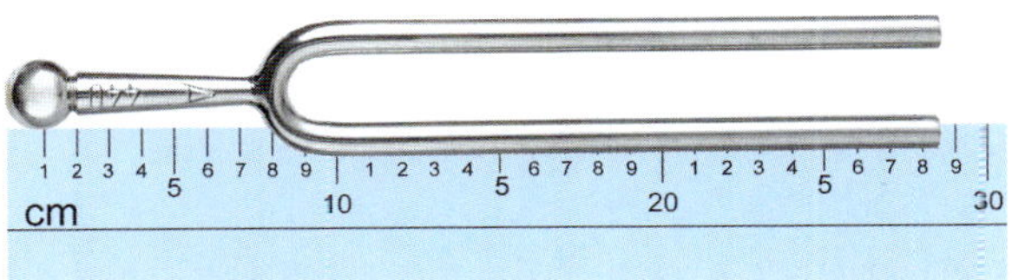


36. What is the length of this tuning fork as measured using this metric ruler?
37. Describe at least three different causes of uncertainties in measured data.

38. What is the difference between precision and accuracy? You may use the diagrams above in your explanation.
39. Write a short essay that is a persuasive argument either for or against a new initiative to increase the amount of material that is recycled at your school. Indicate which additional materials could and could not be recycled in this new program. Research details about the costs and benefits of recycling to include in your essay.

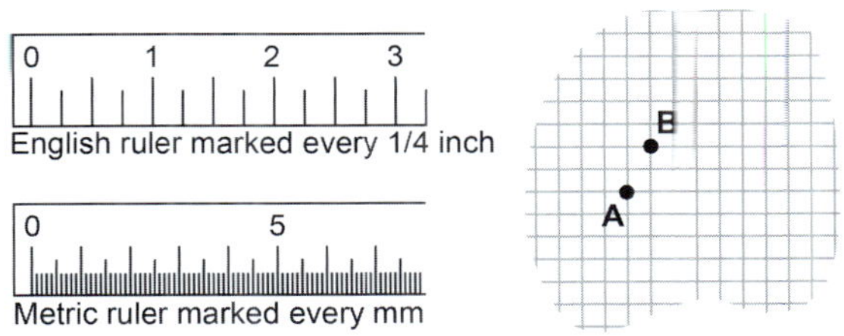


40. Which of the two tools pictured above would be best to measure the separation between points A and B on the graph above?
41. When you use water from the tap, should you be cautious in how much you use because fresh water is a resource that should be conserved?
42. How many decimal places are there in each of the following measurements? How many significant figures are there in each?
    a. 2.998 cm  b. 31.2 kg  c. 500 m
    d. 0.209 µm  e. 0.00030 s
43. Before the first day of conducting investigations in a classroom laboratory, which *three* of the following items should you know the location of and know how to operate?
    a. fire extinguisher  b. scientific calculator
    c. eye wash  d. safety shower
44. Which of the following are benefits of recycling aluminum cans? Select all that apply.
    a. Reusing it takes less energy than refining it from ore.
    b. Recycled aluminum has been sterilized.
    c. Recycling aluminum reduces refuse going into landfills.
    d. Recycled aluminum can be more expensive than aluminum that has never been used.
45. Which apparatus is the best choice to measure *mass*: a spring scale, a triple beam balance, or a digital hanging scale?

# Chapter review

## Section 2.2

46. What are some advantages of tabulating data over listing them in a single line?

47. Two lab partners were trying to decide what to do with a piece of paper that got some biowaste on it during a classroom investigation. Zareb thinks it is hazardous waste, so it should go into the biowaste bin. Doreen, however, thinks it should go into the paper recycling bin. Who is right?

48. When might a pie chart be a better choice than a line graph?

49. Chaia's family installed rain barrels to collect the rain water from the gutters of their house, in order to use the water at a later date to irrigate their vegetable garden. Is this a good idea?

50. Brendan's family was deciding how to dispose of a bottle of unused, liquid medicine. The container was plastic, and water was listed on the label as the largest ingredient by mass. How should the bottle be disposed of?

51. The smallest gradations on Mensah's meter rule are in millimeters. Which of the following are *not* appropriate measurements made with his measuring instrument? 0.767 m, 0.0095 m, 7.9 cm, 23.55 cm, 2.61 cm, 308 mm, 61.3 mm, 3.5 mm, $1.75\times10^1$ cm, and 6.0 mm

52. Construction workers are creating 100 miles of new highway. When they measure the mile markers, they accidentally space them one kilometer apart. (1 kilometer = 0.621 miles)
    a. What are the effects of the uncertainty resulting from their error?
    b. After placing the 100th mile marker sign, how far is their sign from where the 100th mile marker should have been placed?

## Section 2.3

53. What physical quantity, and what value for it, are implied when a physics problem says that a vehicle "starts from rest"?

54. After school for a week, Anthony measured the speed one particular species of ants moved and compared that to the ambient air temperature that day. In his investigation, what are the dependent, independent, and controlled variables?

55. Efi measured her height every 60 s for an hour and then plotted her height against time. Since her height did not change, is the value of the slope of her graph positive, negative, or zero?

56. In an investigation, Graciela measured the time a particular toy car took to reach the bottom of a ramp for different starting heights above the floor. What are the dependent, independent, and controlled variables in her investigation?

57. Can an equation be a model? Why or why not?

58. If a physics questions asks, "how long does it take," then what quantity are you asked to solve for?

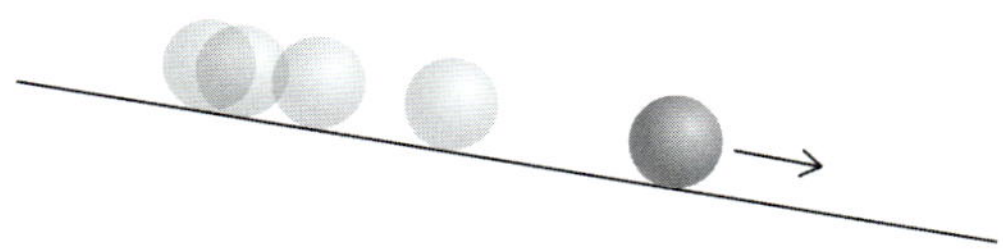

59. A ball, starting from rest, rolls down a hill. It gains speed all along the way. In the graph of speed versus time, is the slope positive, negative, or zero?

60. Ask five other members of your classroom for their height, age, and favorite number. Create a table to record their answers.

61. What conversion factor would you use to convert from inches to centimeters? From centimeters to inches?

62. What conversion factor would you use to convert ounces to pounds? Pounds to ounces?

63. Use words to express the equation $\rho = m/V$.

64. Shirley has a table of data for the number of Calories in a breakfast cereal that come from carbohydrate, fat, and protein. What kind of chart would be best to present her data?

65. In a typical plot of data taken in a laboratory investigation, which variable (dependent or independent) is plotted on the vertical axis? On the horizontal?

66. In the equation $a = v/r$, if $a$ is plotted against $v$ then what will be the shape of the graph? How about if $a$ is plotted against $r$?

67. Which is larger, one cubic yard or one cubic meter?

68. What conversion factor would you use to convert from inches to yards? From yards to inches?

69. Convert the following sentence into a mathematical equation: "Distance traveled is the product of speed and time."

70. Petra was paid for her work on the first day of the month, and then spent \$10 each day for the rest of the month. She then graphed the money she had left each day against the day of the month. Is the slope of her graph positive, negative, or zero?

# Chapter review

## Section 2.3

71. What conversion factors would you use to convert from seconds into days?

72. Give two examples of equations that are linear and two examples of equations that are nonlinear.

73. What conversion factors would you use to convert from miles per hour to meters per second?

## Quantitative problems

## Section 2.1

74. What is 270,000,000 in scientific notation?

75. How many seconds are there in 2 hours, 56 minutes, and 21 seconds (2:56:21)?

76. How many seconds are there in 01:39:20?

77. The fastest time in the Boston Marathon is 7,382 s. Express this time in hours, minutes, and seconds.

78. Jupiter's moon Io has an orbital period of 152,854 s, while Ganymede has an orbital period of 7.1546 days. Which period is longer?

79. An unknown substance has a mass of 670 kg and a volume of 782 $m^3$. Will it float in water? (Water has a density of 1,000 $kg/m^3$.)

80. When using a triple beam balance, one beam reads 30 g, another beam reads 100 g, and the last beam reads 8.5 g. What is the measured mass?

81. What is the sequence of buttons you need to press to enter the number $3.75\times10^{13}$ into a scientific calculator?

82. If 1 kg of feathers occupies 10 $m^3$ and 1 kg of bricks occupies 1 $m^3$, which has a greater density?

83. Express the answer to (4+8+15+16+23+42) in scientific notation.

84. What is 0.00000000000345 in scientific notation?

85. What is $8.945\times10^{12}$ in standard notation?

86. What is the surface area of the Earth? (The Earth's radius is 6,400 km.)

87. What is the sequence of buttons you need to press to enter the number $5.8\times10^{-8}$ into a scientific calculator?

$\rho = 19{,}300\ kg/m^3$ $\rho = 10{,}500\ kg/m^3$

88. Pure gold and silver have densities of 19,300 $kg/m^3$ and 10,500 $kg/m^3$, respectively. If a jeweler uses 0.25 $cm^3$ of metal to make a ring, then how much heavier (in grams) is the gold ring compared to the silver ring?

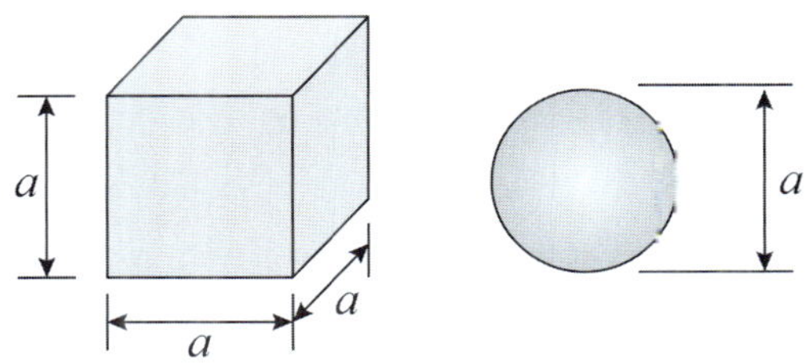


89. A cube has three equal sides each of length $a$, while a sphere has a diameter of $a$.

    a. Which has a larger volume, the cube or the sphere?
    b. Which has a larger surface area?

90. What is the average density of the Earth? (The Earth's radius is 6,400 km, and its mass is $6.0\times10^{24}$ kg )

91. Which is longer, one year, 8,897 hr, or $3.14\times10^7$ s?

## Section 2.2

92. How many people can squeeze into a car? Describe any assumptions you make in your estimation.

93. Estimate the number of ants in an anthill. Describe each step of your estimation and any assumptions you make.

94. How many atoms are there in your body? Describe any assumptions you make in your estimation.

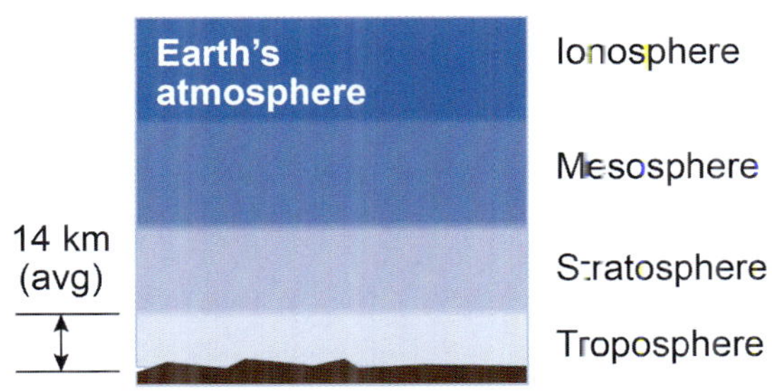


95. Estimate the mass of the Earth's troposphere, the lowest portion of the Earth's atmosphere. Describe any assumptions you make in your estimation. (The density of air is approximately 1.2 $kg/m^3$. The radius of the Earth is approximately 6,400 km. The Earth's troposphere is around 14 km high.)

# Chapter review

## Section 2.3

96. A particular service station in Canada sells gasoline for \$0.85 per liter. If your tank holds 14 gallons, then what is the cost to fill up?

97. Which is longer, 100 miles or 150,000 meters?

98. The summit of Mount Everest at 8,848 m is the highest point on the Earth's surface. If a small airplane can fly no higher than 30,000 ft, can it fly directly over the summit of Mt. Everest?

99. Convert the time 3:24:22 to seconds.

100. Convert 55 miles per hour into meters per second.

101. One cubic yard is equal to how many cubic feet?

102. On your 50th birthday, how many seconds have you lived?

103. What is the volume of a steel marble that has a mass of 50 g? (The density of steel is 7,800 kg/m$^3$.)

104. Which is longer, 10 million seconds or 100 days?

105. How long is a 5 mile road in centimeters?

106. One mil is 0.001 inch. How many millimeters is one mil?

107. One acre is 43,560 square feet. (There are 5,280 feet in a mile and 3.28 feet in a meter.)
    a. How many square miles is one acre?
    b. How many square meters is one acre?
    c. How many square kilometers is one acre?

108. Convert 1,500 revolutions per minute (rpm) into revolutions per second.

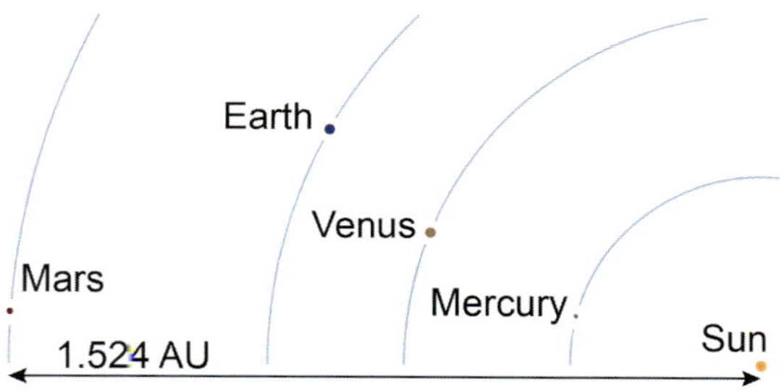


109. The average orbital radius of Mars is 1.524 astronomical units (AU) One astronomical unit is the radius of the Earth's orbit, or $1.5\times10^{11}$ m. What is the orbital radius of Mars in kilometers?

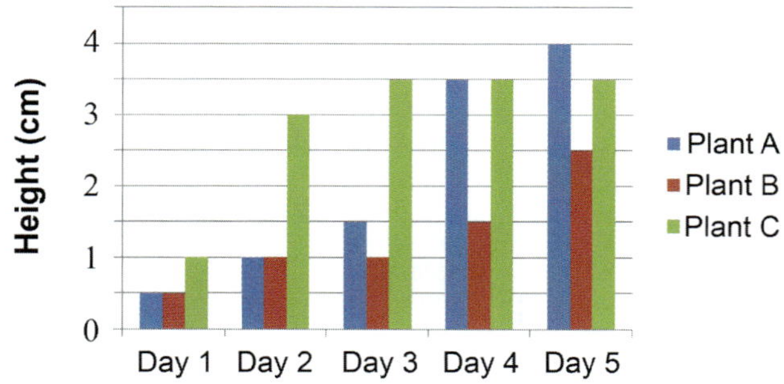


110. Measurements were made of the heights of three different plants over five days. The data collected are displayed in the chart above.
    a. Which plant grew the most over the five days the plants were measured?
    b. Which plant grew the most between day 1 and day 2?
    c. How could you *infer* those conclusions using the bar chart?
    d. Generate a new bar chart that plots the *change* in the height of each plant between days, i.e., between days 1 and 2, 2 and 3, 3 and 4, and 4 and 5.
    e. Which plant grew the most on any single day?
    f. A bar chart may not be the best way to represent these data. What kind of graph or chart would you use to plot the height of these plants? Explain.

111. Convert 293 cubic feet per hour into liters per second.

112. Refer to the table to answer the following questions.

| Position $x$ (m) | Speed $v$ (m/s) |
|---|---|
| 0.0 | 4.0 |
| 5.0 | 3.0 |
| 10.0 | 2.0 |
| 15.0 | 1.0 |
| 20.0 | 0.0 |

    a. Graph position and speed using speed as the dependent variable.
    b. Draw a trend line through the data points. What is the slope of your graph? Be sure to use the correct units in your answer.
    c. Use the graph to estimate the position when the speed is 2.5 m/s.
    d. Write an equation that is a model to represent the data.
    e. Use your graph to estimate the speed when the position is 6.8 m.
    f. Use the equation to estimate the speed when the position is 6.8 m.
    g. What is the range of values of the dependent variable?

# Chapter review

### Standardized test practice

113. Which of the following is *not* equivalent to 12.7 cm?

  A. $1.27 \times 10^3$ mm
  B. $1.27 \times 10^1$ cm
  C. $1.27 \times 10^{-1}$ m
  D. $1.27 \times 10^{-4}$ km

114. What is the volume of a sphere with a radius of five meters?

  A. $170\pi$ $m^3$
  B. 50 $m^3$
  C. $100\pi$ $m^3$
  D. $200\pi$ $m^3$

115. How many square meters are there in 560 $cm^2$?

  A. 5.6 $m^2$
  B. 0.56 $m^2$
  C. 0.056 $m^2$
  D. 0.0056 $m^2$

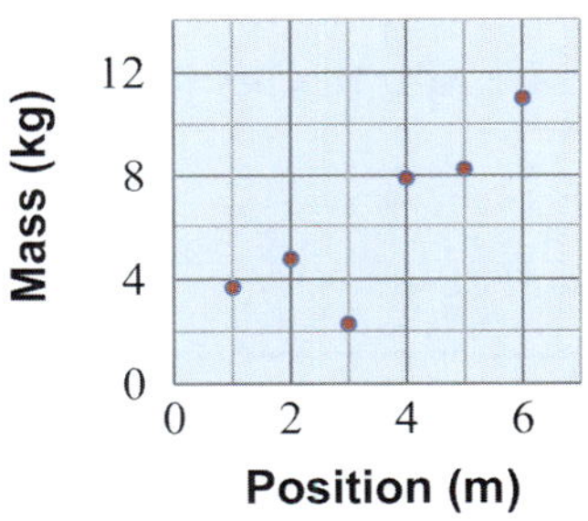


116. In the graph shown, which is the dependent variable?

  A. mass
  B. kg
  C. position
  D. m

117. In an investigation, a student collected data of the density of a liquid at various times. Which would be the best kind of graph to present the data?

  A. line graph
  B. pie chart
  C. bar chart
  D. stacked bar chart

118. One lunar cycle is about 30 days. Today is Morris's second birthday. Approximately how many complete lunar cycles has Morris been alive for?

  A. 60
  B. 24
  C. 15
  D. 182

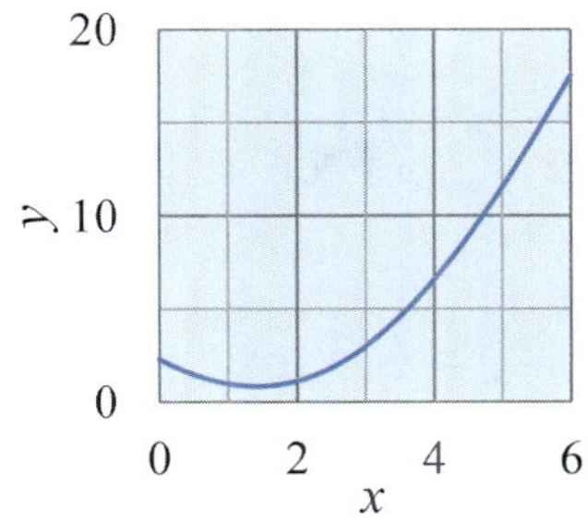


119. Which is the best description of the blue line in the graph above?

  A. scatter plot
  B. linear relation
  C. inverse relation
  D. nonlinear relation

120. How many seconds is 4 hours and 34 minutes?

  A. 16,440
  B. 9,650
  C. 13,470
  D. 12,740

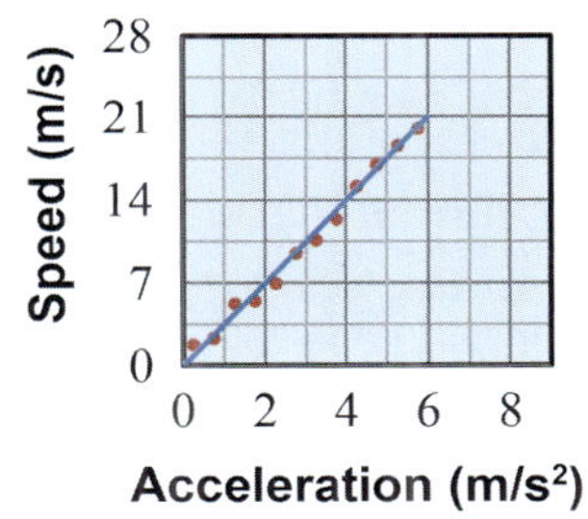


121. What is the slope of the blue line in the graph shown?

  A. 3.1 s
  B. 3.1 $s^{-1}$
  C. 3.5 s
  D. 3.5 $s^{-1}$

122. Which of the following equations is mathematically equivalent to $F = qE$?

  A. $q = E/F$
  B. $E = q/F$
  C. $q = F/E$
  D. $E = qF$

123. What is the volume of a cube with a side length of 3 km?

  A. 9 $km^3$
  B. 3 $km^3$
  C. 81 $km^3$
  D. 27 $km^3$

Chapter

# 3

# Position and Velocity

What happens when a policeman with a love for the outdoors gets a distress call from a hiker who has lost her way? If you're Mount Vernon, Washington, officer Tom Wenzl, you activate your cellphone's "geocaching" application and hit the trail. Hiker Brenda Johnston had called 911 after becoming disoriented in the early December evening darkness. In so doing, she had helpfully provided her precise coordinates to police dispatchers—not that she had any idea of her longitude and latitude, but her phone did, thanks to the Global Positioning System (GPS).

From Isaac Newton to Albert Einstein, physicists have dreamed of a day when they could pinpoint *where* an object is and *when*. GPS has turned that dream into a reality. Collect a sequence of such coordinates, and you can calculate an object's position, velocity, and acceleration at any moment. Going further, you can deduce the forces at work on that object.

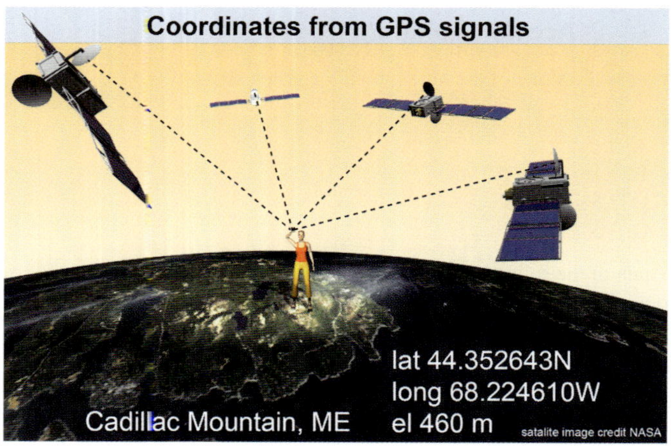


How does GPS work? At its heart lies a fleet of 31 satellites orbiting 20,000 km above ground—around four times as far from the center of the Earth as you are. When at least four of these satellites are above the horizon, a commercial GPS receiver can pinpoint your location on Earth with an accuracy of a few meters. Each GPS satellite constantly transmits *its* location and tags that information with a precise time. When your GPS unit receives that signal, it instantly "knows" how long the signal took to reach you. Divide that time interval by the speed of light, and you know the satellite's distance. Repeat the procedure with several other satellites, and your GPS-enabled device calculates your latitude, longitude, and elevation.

GPS receivers once were the sole province of secretive military personnel. Initially built by the U.S. Department of Defense, the system was invented to track and guide nuclear submarines and intercontinental ballistic missiles. Now GPS sensors are standard equipment in many cellphones. You can buy a satellite-navigation system for your car for the price of a small television.

Chapter 3

# Chapter study guide

## Chapter summary

Where are you? Where are you going? How fast? These are questions that are addressed *quantitatively* in physics using position, coordinates, displacement, speed, and velocity. An important distinction among some of these quantities is that some are scalars and some are vectors. In this chapter you will learn the equations of motion for displacement and velocity and use these to solve increasingly difficult problems—and how to tell from a graph whether an object is moving quickly or slowly.

## Learning objectives

By the end of this chapter you should be able to

- distinguish between position and displacement and between speed and velocity;
- solve multiple displacement problems in one and two dimensions;
- write down the equations of motion for constant velocity;
- solve one- and two-step problems using those equations;
- draw, analyze, and interpret position–time and velocity–time graphs; and
- describe motion in different frames of reference.

## Investigations

3A: Displacement
3B: Motion graphs

## Chapter index



## Important relationships

$$x_f = x_i + d \qquad v = \frac{d}{t} \qquad \vec{v} = \frac{\Delta x}{\Delta t} \text{ or } \vec{v} = \frac{\vec{d}}{\Delta t} \qquad v_{avg} = \frac{\Delta x}{\Delta t} = \frac{x_f - x_i}{t_f - t_i} \qquad x_f = x_i + vt$$

## Vocabulary

**position**
**vector**
**coordinates**
**slope**
**reference frame**
**origin**
**magnitude**
**speed**
**average velocity**
**inertial reference frame**
**displacement**
**scalar**
**velocity**
**instantaneous velocity**

## 3.1 - Position and displacement

This section looks at how physics describes and quantifies object locations. By the end of this section you should be able to apply the concepts of position, coordinates, vectors, and displacements to scenarios involving motion.

### Distance and position

**Are distance and position the same?**

Space has dimensions of length, and *distance* is the separation between any two points in space. Saying an object is 3 m away does not tell you where the object is *located*, however, because distance does not include any information about the starting position or the direction of the motion.

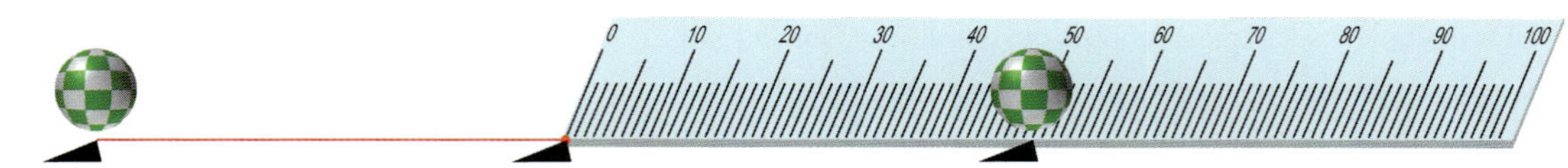


**Position ($x$) specifies location in a coordinate system relative to an origin**

**What is the meaning of *position?***

Interactive simulation

**Position** in physics means a unique location in space with reference to an *origin* and a *coordinate system.* Position has units of length but may be positive or negative depending on direction. For example, a meter stick defines a one-dimensional straight line with the origin at the zero mark. In this example *positive* positions are to the right of the origin and *negative* positions are to the left. A position of −0.50 m is a half meter to the *left* of the start of the meter stick. The *position* of −0.5 meters is different from a *distance* of 0.5 m because position implies an origin.

**What is the origin?**

The **origin** is a fixed reference point that you choose. In laboratory experiments you might choose the origin to be a particular point on the table or the top of a ramp. The key idea is that all position measurements are given relative to the *same origin point*; otherwise you would not know how to interpret them correctly. GPS devices use latitude and longitude to specify position on Earth's surface relative to zero longitude at the *prime meridian* in Greenwich, England.

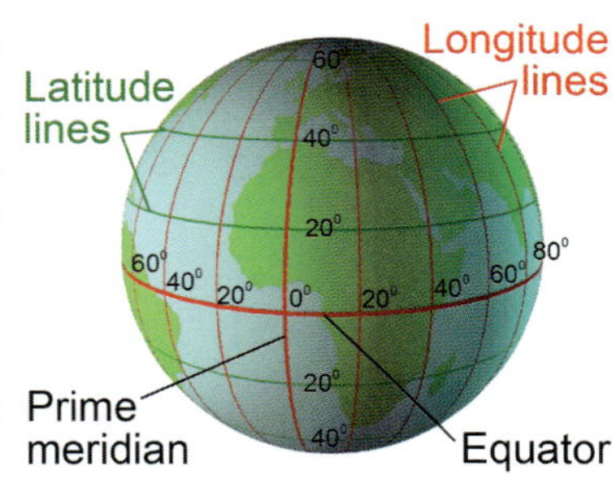


**How do we describe changes in position?**

A **displacement** is a change in position. Displacement can have positive and negative values, as can position. When negative is defined to be left of the origin, a negative displacement is to the left. A positive displacement is to the right. The position of an object is equal to its starting position plus any displacement.

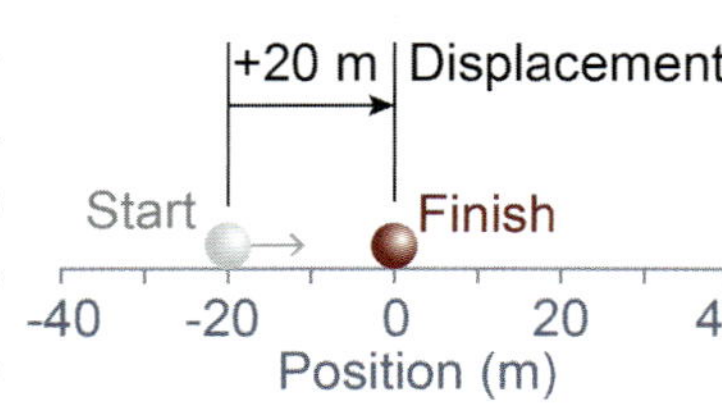


**Talking about displacement**

Describe an action that will take you from your seat to the classroom door *without* using the word "displacement." What other words or phrases mean the same thing? For example, you might say, "I should __________ about three meters to my left and two meters straight ahead." You don't need exact distances, just an estimate.

**What is a vector?**

A **vector** is a type of variable that includes information about direction as well as size. Unlike words such as "right" and "left," a vector includes direction information in a mathematical way so that vectors in different directions can be added and subtracted.

**What is a scalar?**

A **scalar** does not depend on direction and can be completely specified with a single value. Distance, mass, and temperature are all scalar variables. Each can be completely described by a single value that does not include directional information. A distance can be zero but cannot be negative! The same is true of mass.

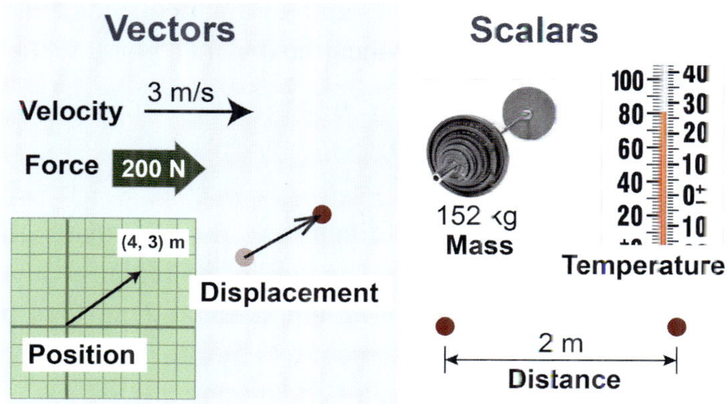


**Displacement is a vector**

*Displacement* is a vector even though *distance* is not. A *displacement* of 10 m tells you to move 10 m to the right in the coordinate system of the diagram below. A displacement of −10 m is a movement of 10 m to the left. With a vector, the positive and negative signs carry direction information.

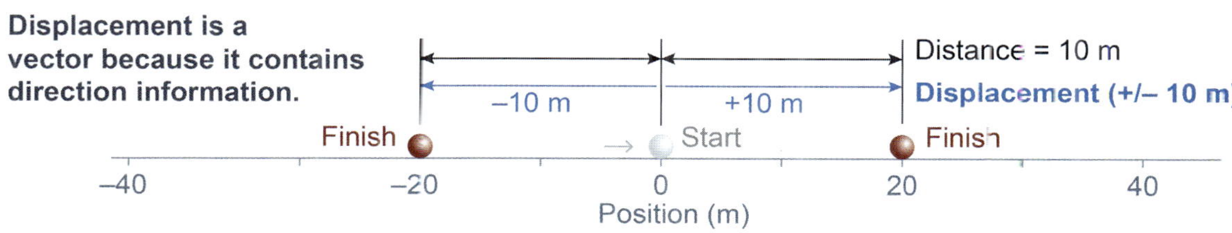


**The magnitude of a vector**

The **magnitude** of a vector is its "length" and is independent of its direction. For example, the displacements of +10 m and −10 m both have a magnitude of 10 m. Distance is the magnitude of both the position and displacement vectors. Other vectors such as force have magnitudes expressed in other units (such as newtons). The magnitude of any vector can be zero or positive but cannot be negative.

**What does two dimensional mean?**

Interactive simulation

Position and displacement may be used when mapping landmarks or plotting a course. A map is a two-dimensional surface that has length and width. The north–south axis provides one perpendicular reference line, usually aligned with the *y*-axis. The east–west axis provides the second perpendicular reference line, usually aligned with the *x*-axis. Where the two axes meet is usually the origin or fixed reference point, such as zero degrees latitude and longitude on a map of the world. Using these two axes, every point on the surface can be uniquely identified with two numbers, *x* and *y*. A map is *two* dimensional because it takes two numbers to uniquely specify any point.

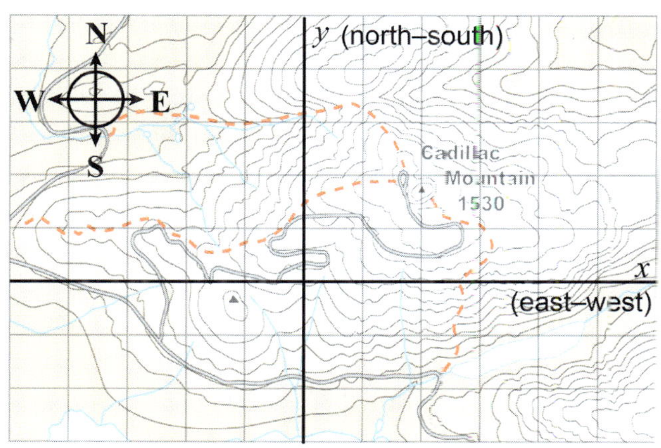

A flat map is a two-dimensional surface.

## Adding and subtracting displacements

**Relating position and displacement**

How the final position $x_f$ depends on the initial position $x_i$ and the displacement $d$ can be described by equation (3.1). This equation is a mathematical statement saying "where you end up is where you start plus how much you move." The *distance* you move is the absolute value of that displacement. If you changed position from +30 m (initial) to −30 m (final), then the distance you traveled is 60 m but your displacement is −60 m.

**(3.1)** $$x_f = x_i + d$$

$x_f$ = final position (m)
$x_i$ = initial position (m)
$d$ = displacement (m)

**Position**

**How are displacements added?**

In one dimension, displacements add just like positive and negative numbers. The final position is equal to the initial position plus all the successive displacements. For example, suppose an ant starts at 2 m, then moves a displacement of +7.1 m followed by another displacement of −5.5 m. What is its final position? The problem can be solved two ways: graphically and numerically.

1. To solve numerically, you add the displacements to the initial position, being careful to account for signs. For this example, the final position is
$$2 + 7.1 - 5.5 = 3.6 \text{ m}$$
2. To solve graphically, you draw the first displacement to scale starting from the initial position. Each successive displacement is drawn from the end of the previous displacement. The final position is at the end of the last displacement vector.

**Graphical addition**

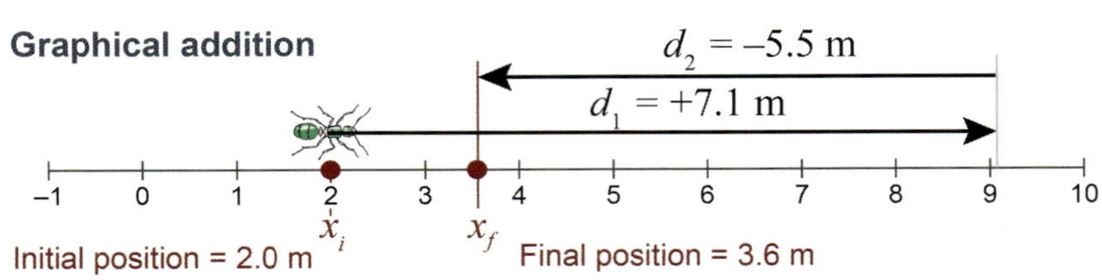


**Numerical addition**

| | |
|---|---|
| $x_i$ | 2.0 m |
| $+ d_1$ | + 7.1 m |
| $+ d_2$ | − 5.5 m |
| $x_f$ | 3.6 m |

**Which direction is positive and which is negative?**

In many problems either the north–south or the east–west axes are used to define directions. It is common to assign positive values to displacements to the east or north and negative values to displacements to the west or south. This is completely arbitrary! You can choose to define any direction as positive, making the opposite direction negative.

**Example problem**

An aircraft takes off from Dallas, flies 400 km north, lands, and then takes off again and flies 130 km south. What is the plane's final position relative to Dallas?

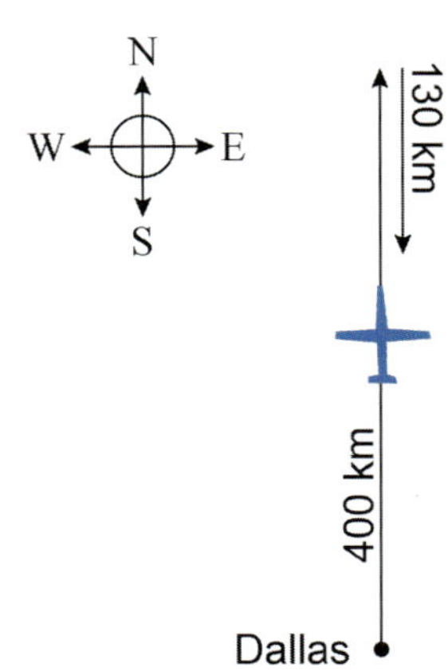


| | |
|---|---|
| *Asked:* | final position $x_f$ |
| *Given:* | $x_i$ = Dallas, $d_1 = +400$ km, $d_2 = -130$ km |
| *Relationships:* | $x_f = x_i + d$ |
| *Solution:* | $x_f = x_i + d$ |
| | $\rightarrow x_f = 0 + 400 \text{ km} - 130 \text{ km} = 270 \text{ km}$ |
| *Answer:* | The plane is 270 km north of Dallas. |

# Investigation 3A

## Displacement

**Essential questions** *How do we quantitatively describe movements in space?*

Interactive simulation

Within the next decade self-driving cars could make the leap from science fiction to dealer showrooms to people's driveways. How does a self-driving vehicle represent movements? In this investigation we look at one- and two-dimensional displacement vectors and how they are added to create motion paths.

### Part 1: Displacements in one dimension

The interactive simulation allows you to add multiple displacements in *one* dimension. You may also shift the origin. Use the simulation to solve the following problems.

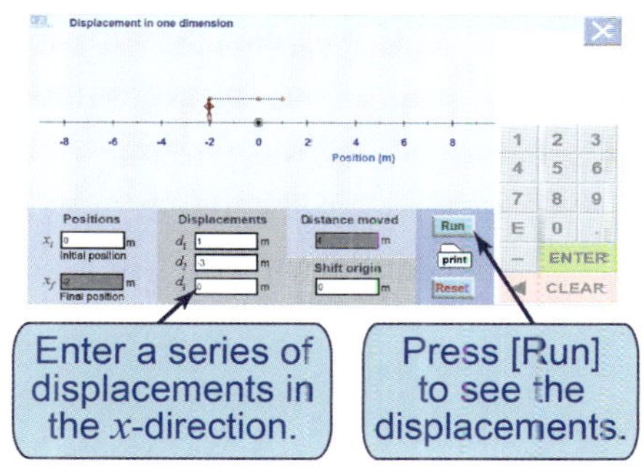


a. *A man moves 6 m east, stops, turns around, and moves 2 m west. What is his final position? How far did he travel?*

b. *A woman starts at position −10 m and has three successive displacements of −5 m each. What is her final position?*

c. *What displacement moves an object from an initial position of +25 m to a final position of −10 m?*

d. *A robot starts at the origin. Use three different displacements to move the robot to a final position of +8 m, where the second displacement must be negative.*

### Part 2: Displacements in two dimensions

The interactive simulation allows you to add multiple displacements in two dimensions. You may also reset the origin of the display. Use the simulation to solve the following problems.

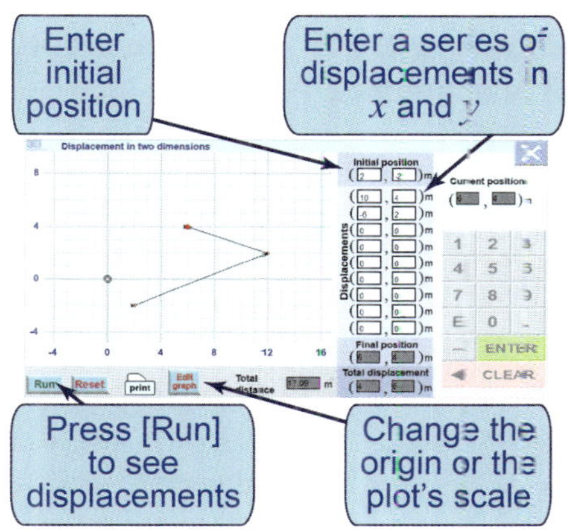


a. *What two perpendicular displacements will move an object from the origin to a position of (4, 3) m?*

b. *What is the final position of a robot that starts from (0, 0) m and makes displacements of (4, 4) m, (3, 3) m, and (−4, −2) m?*

c. *Create a path of four displacements of equal distance that move from a position of (1, 1) m to a position of (2, 2) m. Note that there are many possible solutions.*

d. *A student starts at the origin and ends up at a position 500 m north of the origin. She knows she walked 250 m straight west, but then she was blindfolded and led to the final position. Assume she walked a straight path for the hidden leg of the trip. What is the displacement vector of her movement?*

e. *What sequence of two displacements moves an object from (5, 5) m to (−5, −5) m while traveling a distance of exactly 20 m? How does this distance compare to the single displacement that connects the same starting and ending point?*

f. *Create a series of eight successive displacements that would program a robot to move in an octagonal path that is as close as you can get to approximating a circle. The robot should return to its starting point after the eighth displacement. What total distance does the robot move? Calculate the radius of the circle that has this distance as its circumference.*

## Two-dimensional coordinates

**Two-dimensional coordinates**

The same concepts of displacement, distance, and position apply in two- and three-dimensional space. In two dimensions each position requires two values, which we call the $x$- and $y$-coordinates. **Coordinates** are values that specify position relative to an origin. For example, the coordinates $(x, y) = (3, 1)$ m identify a point 3 m to the right and 1 m above the origin. Coordinates in physics have units of length, such as meters. When written as (3, 1) m, the first value is the $x$-coordinate and the second is the $y$-coordinate.

**What are displacements in two dimensions?**

Displacements in two dimensions are written in the same format as position coordinates, but they are interpreted as movements. For example, the displacement $d_1 = (3, 1)$ m describes movement along a straight line that takes you 3 m to the right and 1 m up *from your starting position.*

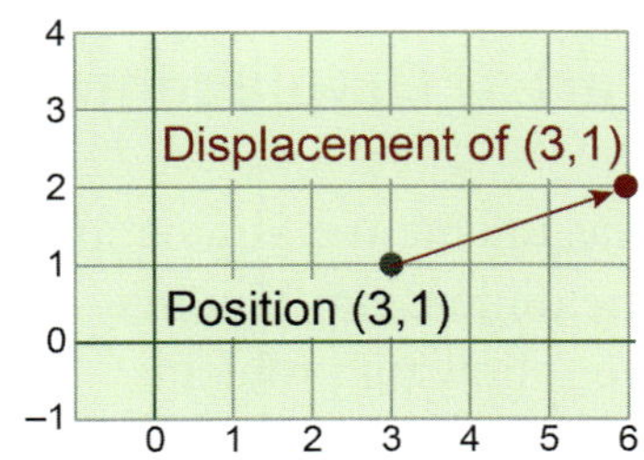


Equation (3.1) applies to two and three dimensional problems when using variables that are two- or three-dimensional vectors. For example, in two dimensions, equation (3.1) is actually two separate equations, one for the $x$-coordinates and an identical one for the $y$-coordinates.

**One-dimensional coordinates**

$$x_f = x_i + d$$

**Two-dimensional coordinates**

$$x_f = x_i + d_x \quad x\text{-coordinates}$$

$$y_f = y_i + d_y \quad y\text{-coordinates}$$

**How are 2-D displacements added?**

Two-dimensional (2-D) displacements can be added both graphically and numerically. In the graphical method, each successive displacement is drawn from the current position to the new position. This is called the "head-to-tail method," because the "tail" of each vector is drawn starting from the "head" of the previous vector. In the numerical method, $x$- and $y$-coordinates are added separately.

1. The final $x$ position equals the initial $x$ position plus $x$ values of each displacement.
2. The final $y$ position equals the initial $y$ position plus $y$ values of each displacement.

Add the following displacements starting from (0,0):
(5, 0) m
(0, 3) m
(−4, −1) m

**Graphical addition**
"Head-to-tail method"

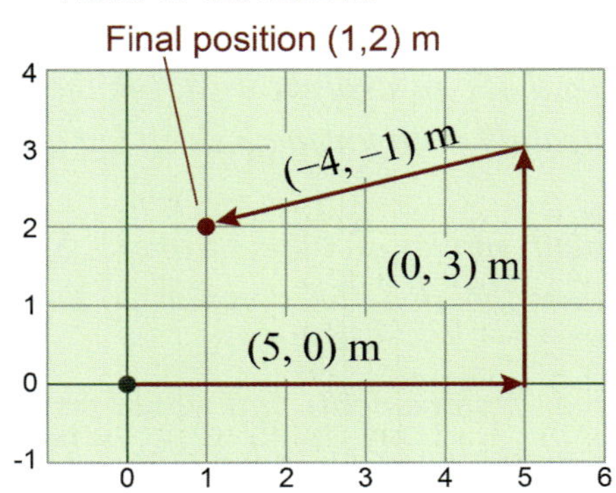


**Numerical addition**
Add $x$- and $y$-components separately

| | | | |
|---|---|---|---|
| $x_i$ | 0 m | $y_i$ | 0 m |
| $+x_1$ | 5 m | $+y_1$ | 0 m |
| $+x_2$ | +0 m | $+y_2$ | +3 m |
| $+x_3$ | −4 m | $+y_3$ | −1 m |
| $=x_f$ | 1 m | $=y_f$ | 2 m |

Final position (1, 2) m

# Section 1 review

Position describes a location in space relative to an origin. Coordinates describe a position in three dimensions, using a set of three numbers (such as $x$, $y$, and $z$) and a set of distance units (such as meters). An object's position is a vector: a quantity with both a value (magnitude) and direction. Displacement is also a vector: It describes a change in position. Vectors can be added graphically (with the "head-to-tail" rule) or numerically—by separately adding their $x$-, $y$-, and $z$-components.

| Vocabulary words | position, origin, displacement, vector, magnitude, scalar, coordinates |
|---|---|

## Review problems and questions

1. Classify each of the following as a *distance* or a *displacement*:
   a. Jodie sat 20 ft from the teacher's desk.
   b. Maurice drove his car a quarter-mile north of the train station.
   c. Yasmin hiked 6 km through the San Gabriel Wilderness.
   d. Ali kicked the ball 40 yards away.

2. In one dimension, positions and displacements can be positive, negative, or zero.
   a. What is the difference between a negative position and a negative displacement?
   b. Is it possible to undergo a negative displacement and end up at a positive position?

3. A dog initially located at position (3, 3) m in the $x$–$y$ plane undergoes the following displacements: $d_1 = (2, 1)$ m, $d_2 = (-4, -2)$ m, and $d_3 = (3, -2)$ m.
   a. What is the total displacement of the dog?
   b. What is the final position of the dog with respect to the origin?

4. A plane starts at airport A and undergoes the following displacements before landing at airport B: $d_1 = (+14 \text{ km}, +32 \text{ km})$ and $d_2 = (+18 \text{ km}, -11 \text{ km})$. Next, the plane returns directly to airport A.
   a. What displacement does the plane undergo during the return trip?
   b. What is the total displacement of the plane during the entire round trip?

5. *Step i:* A ball rolls from the origin on a one-dimensional coordinate axis and stops at ($x = 30$ m).
   *Step ii:* It next rolls from ($x = 30$ m) to ($x = -30$ m).
   *Step iii:* Finally, it rolls from ($x = -30$ m) to ($x = 0$ m).
   a. What is the ball's displacement during Step i?
   b. What is the ball's displacement during Step ii?
   c. What is the ball's displacement during Step iii?
   d. What is the total of the three displacements?
   e. Generalize about the displacement of any round trip (one that begins and ends at the same position).

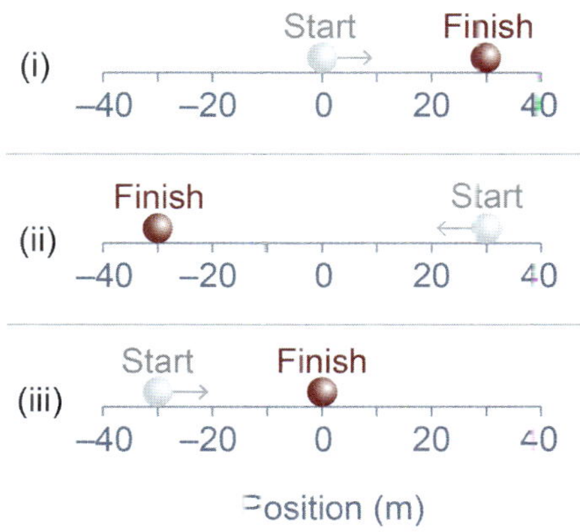

## 3.2 - Speed and velocity

Next time you are in a car, watch the speedometer and think about what it tells you. Is the reading on the speedometer different if you turn around and drive in the opposite direction at the same speed? What does the speedometer read when the car is backing up? A speedometer reads a scalar value—a speed. To describe forward and backward motion we need a *vector*—the velocity vector.

### Speed and velocity

**3-D and 1-D motion**

A complete three-dimensional (3-D) description of a car trip requires three coordinates: $x$, $y$, and $z$. The car's position *along a road*, however, can be described more simply by using a single value—such as mile post 167. When position can be described by a single value we say the description is *one dimensional.*

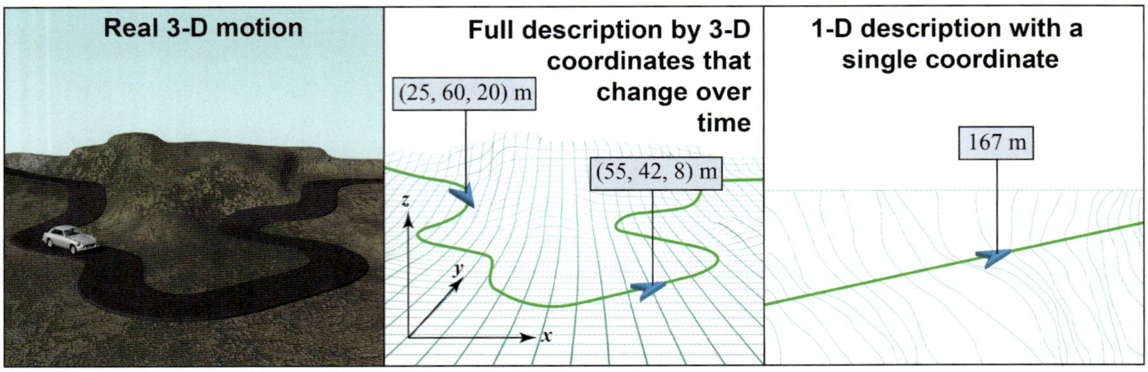


**What is speed?**

A one-dimensional (1-D) definition of **speed** is the ratio of distance traveled to time taken. If you go 120 miles in 2 hours your speed is the distance divided by the time, or 120 ÷ 2 = 60 miles per hour. In physics, we will typically use meters and seconds instead of miles and hours. Since 60 miles is 96,558 meters, and 1 hour = 3,600 seconds, 60 mph is the same speed as 26.8 meters per second (m/s).

Interactive equation

**Speed**

**(3.2)** $$v = \frac{d}{t}$$

$v$ = speed (m/s)
$d$ = distance traveled (m)
$t$ = time taken (s)

**What is velocity?**

Equation (3.2) is limited because a distance can be zero or positive but *not negative*, and therefore speeds are always positive. To account for moving *backward* we extend the concept of speed to include *velocity v.* Velocity is defined by equation (3.3):

Interactive equation

**Velocity**

**(3.3)** $$\vec{v} = \frac{\Delta x}{\Delta t} \quad \text{or} \quad \vec{v} = \frac{\vec{d}}{\Delta t}$$

$\vec{v}$ = velocity (m/s)
$\Delta x$ = displacement (m)
$\vec{d}$ = displacement (m)
$\Delta t$ = change in time (s)

**How are speed and velocity different?**

The symbol "Δ" translates to "the change in" and is pronounced "delta." Since $x$ is position, $\Delta x$ means "the change in position," which is identical to the displacement $d$. Equation (3.3) is a better definition because **velocity** is the change in position divided by the change in time. Because positions can be in front or behind, *velocity can be positive or negative depending on direction.* In short, velocity is a vector with both direction and magnitude—which can take on negative values—whereas speed is a scalar that represents only the *magnitude* of the velocity vector.

**What does the sign of velocity mean?**

Velocity can be positive or negative depending on how you define direction. If moving to the right is defined to be *positive*, then a negative velocity, such as −2 m/s, describes moving *to the left*. You should realize that this is a *choice* and not a rule of physics. You can choose *any* direction to be positive, including to the left instead of right, or down instead of up. You must be consistent with your choice, however, and not change it in the middle of solving a problem!

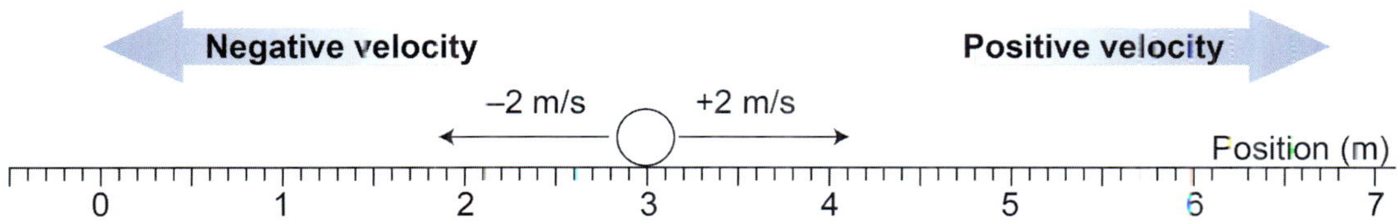


**When do I use velocity instead of speed?**

Mathematically, speed is the absolute value of velocity. Velocity has the same units as speed and has the same meaning in terms of describing *how fast* an object is moving. Velocity has the additional capability of indicating the direction. For one-dimensional motion, the direction of motion is described as positive or negative.

**What does “constant speed” mean?**

Many problems include the term *constant speed* or *constant velocity.* This means the value of the velocity $v$ does not change over time. For example, if a problem gives a “constant speed of 10 m/s,” then the speed stays 10 m/s for all times of interest in the problem. Of course, no real speed stays constant for very long; in many circumstances, however, it is a good approximation.

**Initial time**

In many physics problems, it is convenient to set the initial time $t_i$ to zero. In such cases, the time interval becomes $\Delta t = t$. The initial position (at time zero) may be written as $x = x_0$ or, alternatively, as $x_i$.

**Example with positive and negative velocities**

A cart travels to the right at a speed of 0.5 m/s for 15 s, then travels to the left at a speed of 0.3 m/s for 18 s. What is the final position of the cart if it started at $x = 0$?

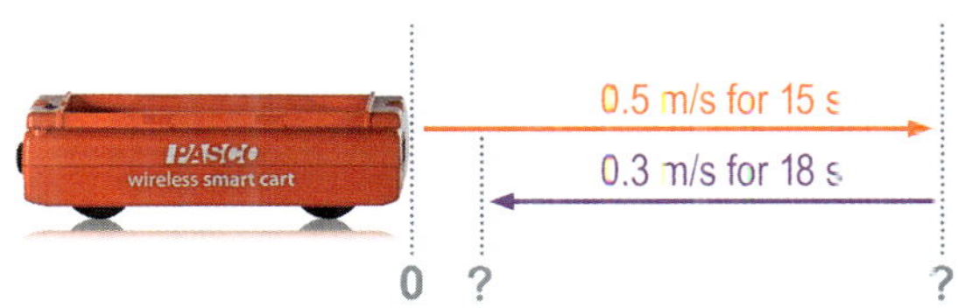


*Asked:* change in position $\Delta x$ (or final position since $x_i = 0$)

*Given:* $v = 0.5$ m/s for 15 s, then $v = -0.3$ m/s for 18 s. The second speed is expressed as a *negative velocity* because it is in the opposite direction.

*Relationships:* $\Delta x = v\,\Delta t$

*Solution:* This problem is solved by calculating two displacements and then adding them.

$$\Delta x_1 = v_1 \Delta t_1 = (0.5 \text{ m/s})(15 \text{ s}) = 7.5 \text{ m}$$
$$\Delta x_2 = v_2 \Delta t_2 = (-0.3 \text{ m/s})(18 \text{ s}) = -5.4 \text{ m}$$
$$\Delta x = \Delta x_1 + \Delta x_2 = 7.5 \text{ m} - 5.4 \text{ m} = 2.1 \text{ m}$$

Since the cart started at position $x_0 = 0$, its final position is:

$$x = 0 + \Delta x = 2.1 \text{ m}$$

*Answer:* The final position of the cart is +2.1 m.

## The language of physics: motion

**Everyday terms**

Physics uses specific meanings for words that may be used differently in conversation. You may be familiar with the following physics terms through things you learned before or from your own, everyday experiences. Below are different ways of describing each term in everyday usage.

| | |
|---|---|
| *speed* | How fast or slow, what the speedometer in your car reads, distance divided by time, miles per hour, kilometers per hour |
| *distance* | The space between two points; how far away something is; a measurement of length, height, meters, centimeters, feet, or inches |
| *position* | Where you are, the place occupied by a piece on a game board, a location on a map |
| *origin* | The place you start, the beginning, the reference point, (0, 0) |

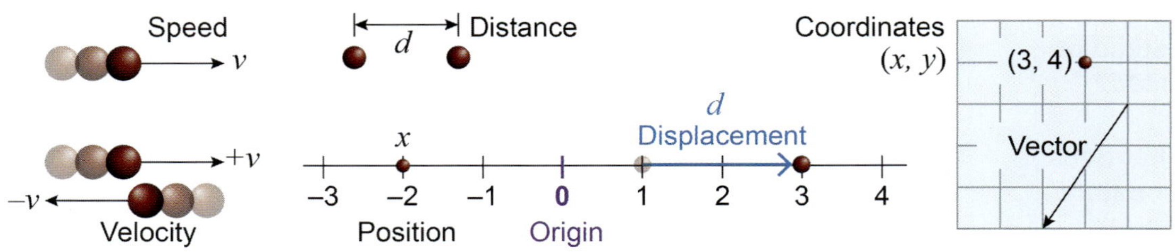


**Physics-specific terms**

Some physics terms are related to words used in conversation, but have deeper and more defined meanings in physics. These terms make talking about physics more accurate because the terms have very specific meanings. Below are ways of describing these terms along with examples for each.

| | |
|---|---|
| *velocity* | Speed with direction, 4 m/s *east*, 55 mph *north*, positive and negative speed |
| *displacement* | A movement between two points, a change in position that can be positive or negative, 30 m *south* |
| *coordinates* | The numbers that identify a specific place, $(x, y)$ or $(x, y, z)$, 30 km *north* and 8 km *east,* (8, 30) km |
| *vector* | A quantity that includes information about direction, 100 km *north* |

**Graphs are stories**

Graphs tell stories of how things change over time. Here is an example that might have produced the graph on the right.

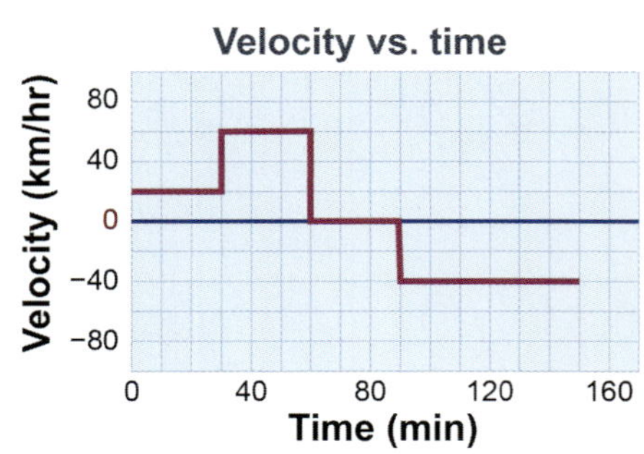


*Carmina drove through traffic for half an hour. Then traffic cleared and she drove at 60 km/hr for 30 min to reach a park in time to meet a friend. After staying for half an hour, Carmina drove home at 40 km/hr since traffic was light and she was in no hurry.*

Can you identify five ways in which the story matches the graph?

**What do I do with the velocity equation?**

The velocity equation appears in many physics problems in different ways. For example, you may be asked to

1. calculate a velocity given displacement or initial and final positions;
2. calculate a time given velocity and position or displacement; or
3. calculate initial position, final position, or displacement.

**How do I use the equation?**

Consider an object that moves with constant velocity $v$ over a time interval $\Delta t$, during which the object changes its position by $\Delta x$. The equation for velocity can be rearranged three ways. Each is useful for solving a different variation of problems involving velocity, displacement, and time. The equations are given two ways, in the notation of $\Delta x$ and $\Delta t$, and in the less precise form of $d$, $v$, and $t$. In either case, this is the same relationship expressed, or *solved*, in different ways.

Interactive simulation

**velocity** = $\dfrac{\text{change in position}}{\text{change in time}}$ → $v = \dfrac{\Delta x}{\Delta t}$ or $v = \dfrac{d}{t}$

**displacement** = velocity × change in time → $\Delta x = v\Delta t$ or $d = vt$

**change in time** = $\dfrac{\text{change in position}}{\text{velocity}}$ → $\Delta t = \dfrac{\Delta x}{v}$ or $t = \dfrac{d}{v}$

**An example problem**

A bicyclist rides due east for 3 min and 40 s at a velocity of 30 km/hr. How many meters does the cyclist travel?

*Asked:* distance $\Delta x$ in meters

*Given:* $v$ = 30 km/hr; $t$ = 3 min, 40 s

*Relationships:* $v = \Delta x / \Delta t \rightarrow \Delta x = v\Delta t$

*Solution:* Units must be made consistent.

Convert the velocity to m/s and the time to seconds.

$$\frac{30\text{ km}}{\text{hr}}\left(\frac{1{,}000\text{ m}}{\text{km}}\right)\left(\frac{1\text{ hr}}{3{,}600\text{ s}}\right) = 8.33\text{ m/s}$$

$$3\text{ min}\left(\frac{60\text{ s}}{1\text{ min}}\right) = 180\text{ s}$$

$$180\text{ s} + 40\text{ s} = 220\text{ s}$$

$$\Delta x = v\Delta t = (8.33\text{ m/s})(220\text{ s}) = 1{,}833\text{ m (or about 1,800 m)}$$

**Units**

Many real-world situations express quantities in different units, such as kilometers per hour (km/hr) or minutes. Before applying any equation the units must be consistent or the numerical answer may not be what you expect. This problem asks for the distance in meters. Therefore, it was best to convert all quantities to meters, seconds, or meters per second.

## Position versus time graph

**A driving trip**

A graph is a useful way to show motion in which velocity changes over time. Consider a trip between Houston and College Station, Texas, which are about 160 km (98 mi) apart. You drive at 100 km/hr (62 mph) for 1 hr, take a 24 min rest, then drive for another 36 min at the same speed. You travel 160 km over the course of 2 hr, so your *average* speed is 160 km ÷ 2 hr = 80 km/hr. Your car's speedometer, however, showed that you were driving at 100 km/hr. Did you drive at a speed of 100 km/hr or 80 km/hr?

**What does the position vs. time graph show?**

The *position versus time* (or $x$ vs. $t$) graph is a *graphical model* to represent motion. This model's graph shows position $x$ on the vertical axis and time $t$ on the horizontal axis. The origin is Houston and in the first hour your position changed from zero to 100 km. For the next 24 min the graph is flat—you were stopped. The final 36 min shows the last 60 km of movement.

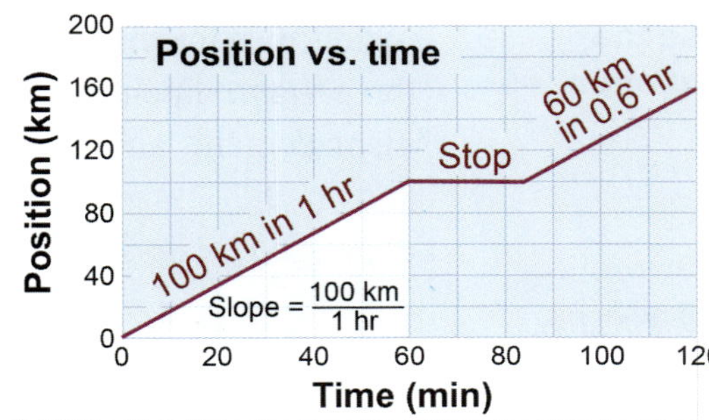


**How is velocity related to slope?**

The **slope** of a line is the ratio of rise over run, or change along the vertical axis divided by change along the horizontal axis. *The velocity is the slope of the line on a position versus time graph.* On the $x$ versus $t$ graph, staying at rest creates a flat line with a slope of zero. In the example, over the first hour the change in position is 100 km and the change in time is 1 hr, giving a velocity (slope) of 100 km/hr.

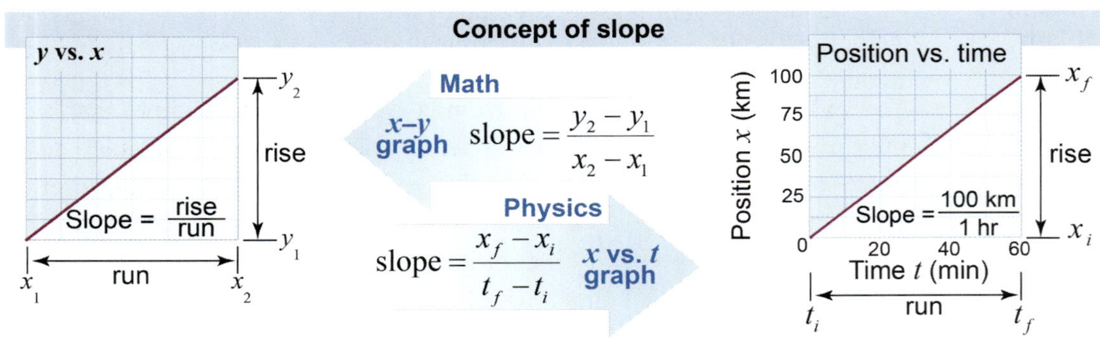


**How does the graph show moving backward?**

Suppose you turn around in College Station and head right back. The position versus time graph (below) shows the return trip as the downward sloping line starting at $t$ = 120 min. The negative slope here shows travel in the negative direction, *toward* the origin. A positive slope for the first part of the trip meant travel in the positive direction, away from the origin. In both cases the velocity is the slope of the position versus time graph—but during the return trip the velocity is *negative*.

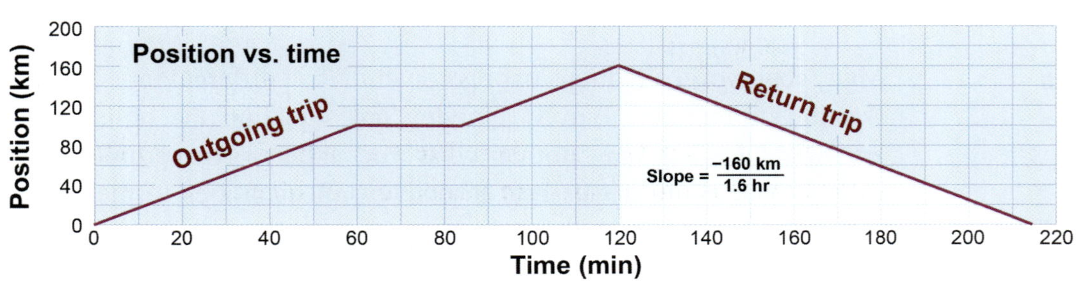

**Velocity vs. time graph for the car trip**

The velocity versus time (or $v$ vs. $t$) graph shows velocity on the vertical axis and time on the horizontal axis. This *graphical model* provides a quick visual history of how the velocity changes. The $v$ vs. $t$ graph can fool you, however, unless you remember that the $y$-axis is *velocity,* not position! In particular, a negative portion of a $v$ vs. $t$ graph means traveling in the backward direction.

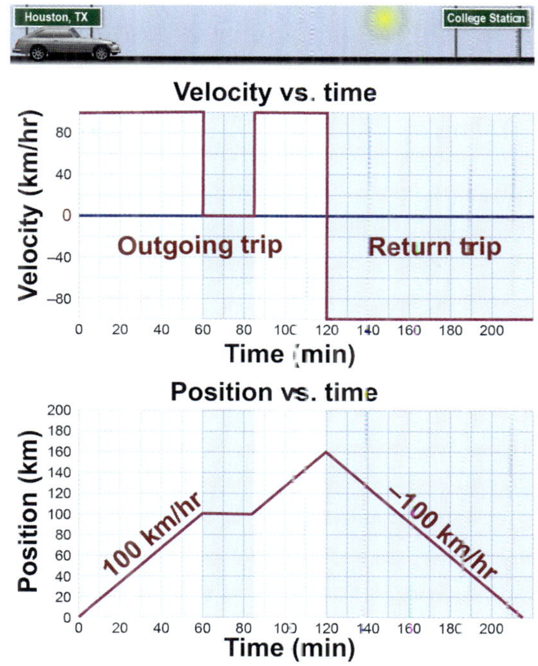


**Interpreting the $v$ vs. $t$ graph**

The diagram on the right shows the $x$ vs. $t$ and $v$ vs. $t$ graphs for a round trip between Houston to College Station. The motion is the same in both the graphs, but the way the motion *appears* on the two graphs is quite different:

a. Constant speed is a sloped line on the $x$ vs. $t$ graph but a flat line on a $v$ vs. $t$ graph.
b. The slope of the $x$ vs. $t$ graph is equal to the value of the $v$ vs. $t$ graph at the same time interval. In this example, during the first hour the slope on the $x$ vs. $t$ graph is 100 km/hr, and this is the value of the $v$ vs. $t$ graph during this hour.
c. On $v$ vs. $t$ graphs, zero velocity lies on the $x$-axis where the $y$ value is 0 km/hr.

**Distance from $v$ vs. $t$ graphs**

To find the distance traveled from a velocity versus time graph, we use a graphical interpretation of the equation $d = vt$. Consider the orange-shaded rectangle on the $v$ vs. $t$ graph below. This rectangle fills the area between the line representing velocity and the $x$-axis where $v = 0$. The area of a rectangle is length times height. On the $v$ vs. $t$ graph, length is equal to *time* and height is equal to *velocity*. Therefore, area on this graph equals velocity multiplied by time, which is the distance traveled. *This is an important result!* Area on a $v$ vs. $t$ graph represents *distance.*

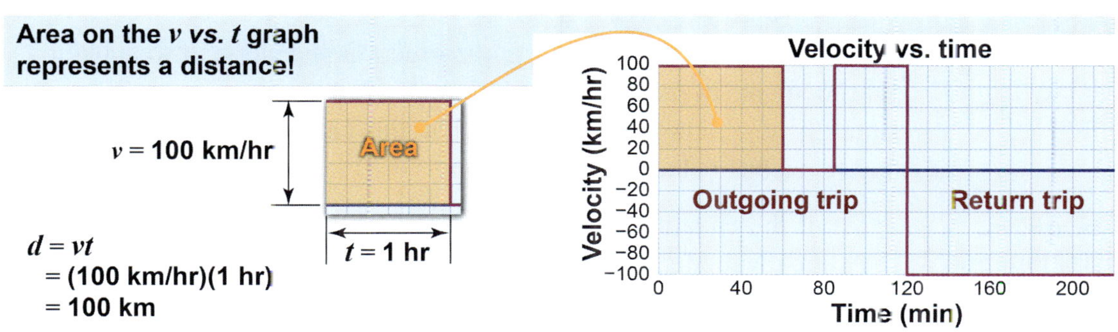


**Connecting the two graphs**

We have now learned that there are two important ways that position versus time and velocity versus time graphs are related to each other:

1. The slope of a position versus time graph is velocity,
2. The area under a velocity versus time graph is the total distance traveled.

## Average and instantaneous velocity

**What is the velocity for stop-and-go driving?**

Earlier in this chapter, velocity was defined as the rate of change of position with time, $v = \Delta x/\Delta t$. In many simple cases this is expressed as $(x_f - x_i)$ divided by $(t_f - t_i)$. But what happens when the velocity *changes* between the initial and final times, such as repeatedly speeding up and slowing down? What is the velocity of a car during stop-and-go driving?

**Average velocity**

The answer involves the difference between *average* and *instantaneous* velocity. The **average velocity** is the total displacement (or distance) traveled divided by the total time taken. In the figure shown, the total distance traveled is 55 m. The total time taken is 11 s. The average velocity is the ratio of the two: 55 m/11 s = 5.0 m/s. This calculation uses equation (3.4).

**Average velocity**

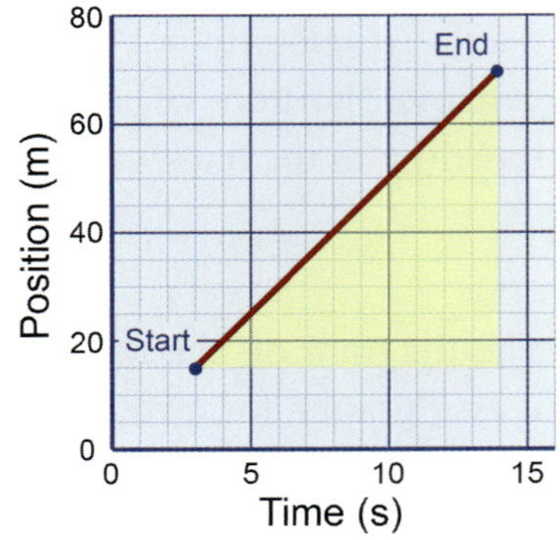


$$v_{avg} = \frac{x_f - x_i}{t_f - t_i} = \frac{70\text{ m} - 15\text{ m}}{14\text{ s} - 3\text{ s}} = 5.0\text{ m/s}$$

Interactive equation

**Average velocity**

$$\textbf{(3.4)} \qquad v_{avg} = \frac{\Delta x}{\Delta t} = \frac{x_f - x_i}{t_f - t_i}$$

$v_{avg}$ = average velocity (m/s)
$x_f$ = final position (m)
$x_i$ = initial position (m)
$t_f$ = final time (s)
$t_i$ = initial time (s)

**Instantaneous velocity**

Now imagine that the car is being driven through traffic—speeding up and slowing down. The position–time graph might be similar to the figure on the right. The **instantaneous velocity** is the velocity of the car *at any particular moment.* The instantaneous velocity may be faster than average or slower than average. The speedometer on a car's dashboard reads *instantaneous* speed, not the average speed over your entire trip.

**Instantaneous velocity**

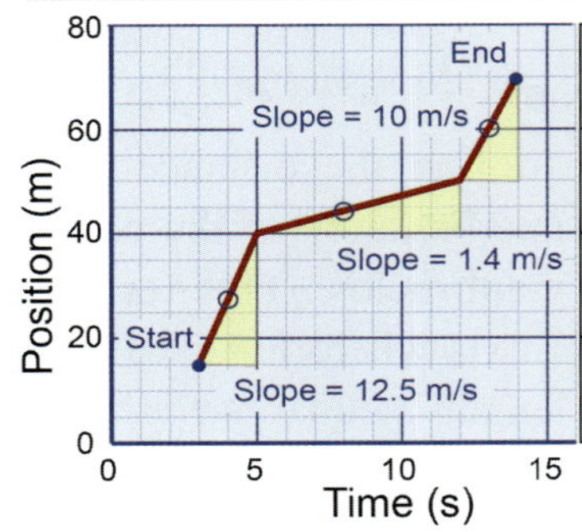


| Time | Instantaneous velocity |
|---|---|
| 13 s | 10 m/s |
| 8 s | 1.4 m/s |
| 4 s | 12.5 m/s |

**Instantaneous velocity from slope**

The instantaneous velocity at any moment is the slope of the position versus time graph at that moment. In the example, at time $t = 4$ s the instantaneous velocity is $v = 12.5$ m/s. At time $t = 5$ s, the graph shows that the car slows down. At time $t = 8$ s, the car's instantaneous velocity is $v = 1.4$ m/s.

**Average vs. instantaneous velocity**

Most of the equations in this chapter are used in the context of *average* velocity, because they are evaluated between two positions at two times. In a more complicated case, however, such as the trip from Houston to College Station, the average velocity differed from the instantaneous velocity because the car traveled, stopped, moved again, and then turned around. Is there any single correct velocity to describe that motion? The answer depends on the problem you are asked. The *average velocity* was 80 km/hr on the outbound trip whereas the *instantaneous velocity* was 100 km/hr while driving and 0 km/hr while stopped.

**Modeling motion**

If you know an object is moving with constant velocity $v$, can you predict the position of the object at any later time $t$? How? How do the equations we have developed so far lead to a model of constant-speed motion?

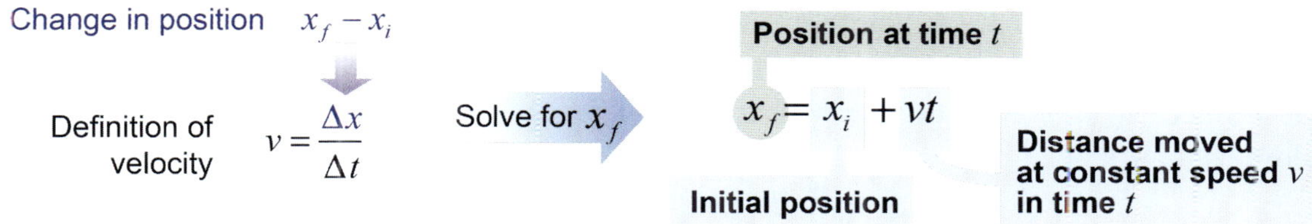


**The position model**

The answer is to solve the velocity equation for the displacement $\Delta x$, then break $\Delta x$ into the initial position $x_i$ and final position $x_f$. The result is equation (3.5), which is really just a restatement of equation (3.3).

Interactive equation

| | | | |
|---|---|---|---|
| **(3.5)** | $x_f = x_i + vt$ | $x_f$ = position at time $t$ (m)<br>$x_i$ = initial position at time $t = 0$ (m)<br>$v$ = velocity (m/s)<br>$t$ = time (s) | **Position**<br>constant velocity |

The position, time, and velocity in equation (3.5) are *instantaneous* values. This equation allows you to calculate the instantaneous position at any time when given the initial position and (constant) velocity.

**An example problem**

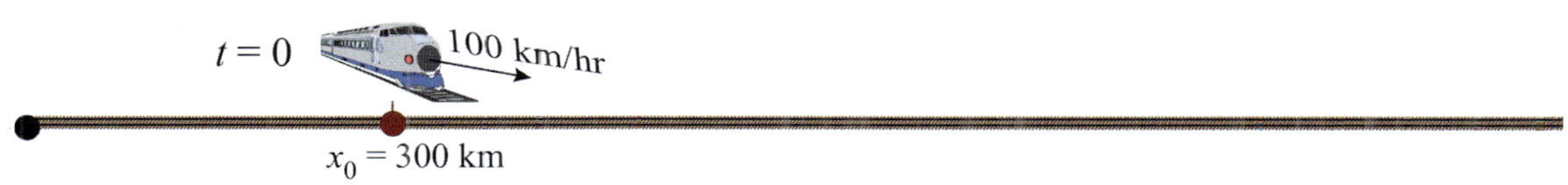


A train leaves a station located 300 km from the start of a track. How far is the train from the start of the track if it travels for 8 hr at 100 km/hr?

*Asked:* final position $x_f$ of the train

*Given:* $v = 100$ km/hr; $x_i = 300$ km; $t = 8$ hr

*Relationships:* $x_f = x_i + vt$

*Solution:* The units are km/hr and hours, which are consistent:

$$x = 300 \text{ km} + (100 \text{ km/hr})(8 \text{ hr}) = 1{,}100 \text{ km}$$

*Answer:* 1,100 km

**What if the velocity changes?**

You might think the limitation to constant velocity makes equation (3.5) useless in the real world. Nonetheless, complex motion can be accurately modeled by breaking up the time into intervals. The velocity is recalculated for each interval and is only considered constant during that interval. The final position for one interval becomes the initial position for the next interval. Many computer models use this approach, as do robots and self-navigating vehicles.

Multistep motion model

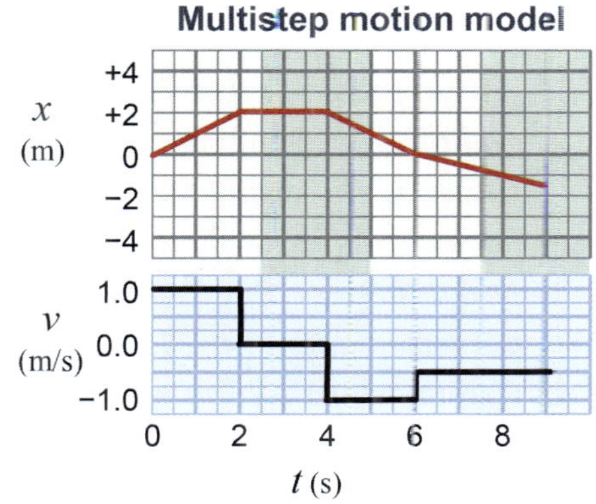

# Investigation 3B Motion graphs

**Essential questions** *How do we predict an object's position at a later time?*

Interactive simulation

Graphs and equations are valuable methods for describing the motion of an object. Position versus time and velocity versus time graphs can describe where an object is located, how fast it is going, and which direction it is headed. In this activity you will adjust the motion of a Smart Cart to match the velocity-time graphs below.

## Part 1: Matching the motion of a Smart Cart

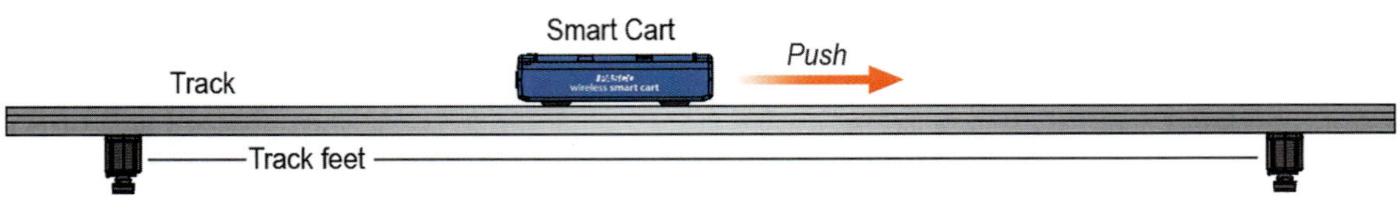


1. Set up your equipment like the picture.
2. Open the experiment file **03B_MotionGraphs**, and then power-on the Smart Cart and connect it wirelessly to the software.
3. Do the following for each velocity-time graph on pages 1 through 4 of the experiment file.
    a. Sketch a prediction for the corresponding position-time graph. Label the prediction.
    b. Find the page in the experiment file with the corresponding velocity-time match graph. Hide any data so the position-time graph is blank and only the velocity-time match graph is shown.
    c. Place the cart on the track and record data as you push, pull, roll, or use your hand to move the cart so that its velocity-time data matches the velocity-time match graph.
    d. Sketch the actual position-time graph in the same graph as your prediction.

*a. How does the position graph for a high positive velocity differ from a lower positive velocity?*
*b. How does the velocity graph for a high positive velocity differ from a lower positive velocity?*
*c. How does the position graph for a negative velocity differ from positive velocity?*
*d. How does the velocity graph for a negative velocity differ from a positive velocity?*
*e. Describe a situation for which the position versus time graph and the velocity versus time graph are both flat (zero slope) horizontal lines.*

*f. Go to page 5 in the experiment file and hide any data so the velocity-time graph is blank and only the position-time match graph is shown. Record data to match the position-time graph, and then describe the motion of the cart during each section shown in the graph to the right. Use terms such as forward, backward, at rest, fast, and slow.*
*g. Draw the resulting velocity-time graph. Label each section corresponding to the letters in the position-time graph above.*
*h. Use the slope tool in your software to find the slope of the position-time graph in each section A, B, C, and D. Record the slopes. How does the slope of the position time graph compare to the velocity recorded during the same period?*

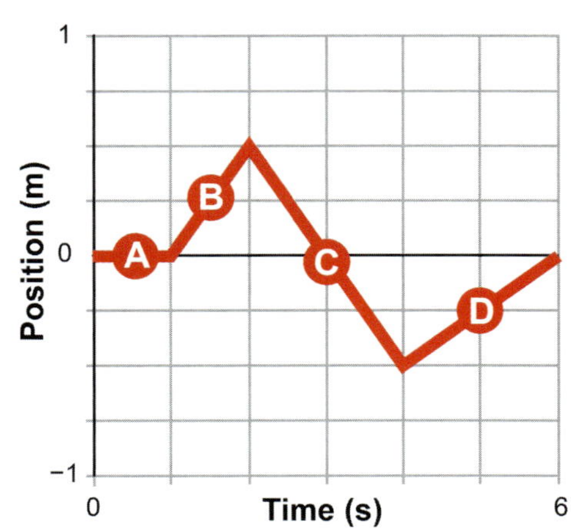

**Key questions**

Consider the questions on the right. These are not trivial questions. More than 2,000 years of thinking by many very smart people was necessary to come up with answers that are satisfactory.

*What does "at rest" really mean?*

*What is "zero" velocity?*

**What is speed on a moving train?**

To illustrate the point, imagine a person on a train throwing a dart at a speed of 10 m/s. The train car the person is in is moving forward at 10 m/s. What is the speed of the dart? An observer on the train measures the speed to be 10 m/s. An observer on the ground, however, measures the speed to be 20 m/s. Who is correct?

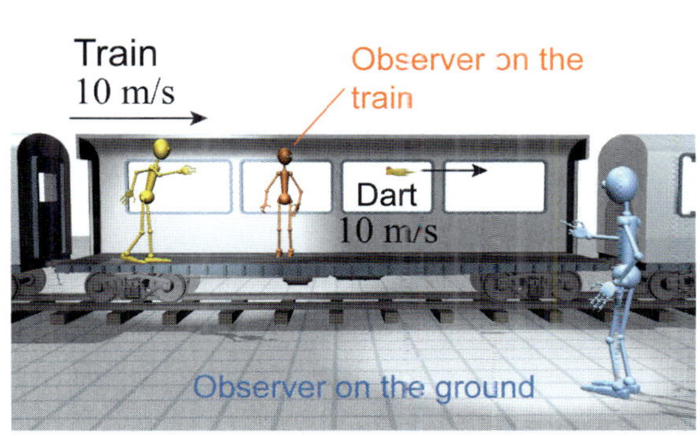


**Velocities depend on reference frames**

Both are correct because *there is no absolute "zero" velocity!* Whether an object is at rest or in motion depends entirely on your **reference frame**. Think of a reference frame as an imaginary "box" that you consider to be "at rest" for the purpose of making measurements. Velocities are measured with respect to your reference frame. The observer on the train might choose the inside of the train car to be the reference frame. In that reference frame the speed of the dart is 10 m/s. The observer outside might choose the ground to be the reference frame. In this reference frame the speed of the dart is 20 m/s.

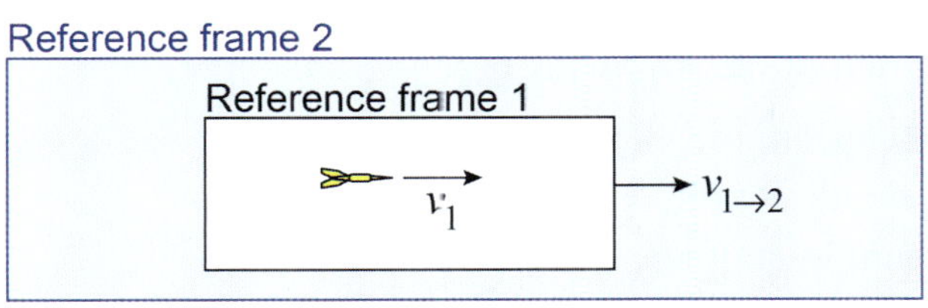


If $v_2$ is the speed of the same dart observed in reference frame 2, then

$$v_2 = v_1 + v_{1\to2}$$

where $v_{1\to2}$ is the velocity of reference frame 1 relative to reference frame 2.

**What is the velocity in another reference frame?**

The velocity of a moving object in one reference frame is equal to the velocity in a second reference frame *plus* the velocity of the second reference frame relative to the first. If this sounds confusing, consider the ground as the first reference frame. The moving train is the second. The velocity of the dart with respect to the ground (20 m/s) is the velocity of the dart with respect to the train (10 m/s) plus the velocity of the train with respect to the ground (another 10 m/s).

**Inertial frames**

An **inertial reference frame** is one in which the velocity is constant in both speed and direction. It was one of Einstein's brilliant insights that *no physical experiment can determine the velocity of an inertial reference frame without somehow looking "outside!"* This means that, if the train car were on a smooth track moving at constant speed (with no windows), then nothing a physicist could do *inside* the car could determine its motion relative to any external reference frame.

# Section 2 review

Velocity is the rate of change in position over time. In the case of one-dimensional motion, velocity can be positive or negative depending on the direction of motion. Speed is the absolute value of velocity. The graphs of position vs. time and velocity versus time are useful tools for analyzing motion. The slope of the *x* versus *t* graph is the velocity.

| Vocabulary words | speed, velocity, slope, average velocity, instantaneous velocity, reference frame, inertial reference frame |
|---|---|

| Key equations | $v = \frac{d}{t}$ | $\vec{v} = \frac{\Delta x}{\Delta t}$ or $\vec{v} = \frac{\vec{d}}{\Delta t}$ |
|---|---|---|
| | $v_{avg} = \frac{\Delta x}{\Delta t} = \frac{x_f - x_i}{t_f - t_i}$ | $x_f = x_i + vt$ |

## Review problems and questions

1. Describe in one sentence how the word *speed* relates to something you know. Your answer must use the word in its correct meaning for physics.

2. Use the word *position* in a sentence describing an experience you had. You must use the word so that its meaning is similar to what *position* means in physics.

3. The term *vector* is best described by which of the following?

   a. an arrow that represents an object's speed
   b. a quantity that has a magnitude and a direction
   c. the change in position from one point to another

4. The plot on the right shows the graph of position versus time of a vehicle. Answer the following questions:

   a. What is the total distance traveled by the vehicle?
   b. How far does the vehicle travel in the first 30 min?
   c. What is the average speed of the vehicle?
   d. What is the maximum speed of the vehicle?
   e. How long does it take for the vehicle to travel 20 km?

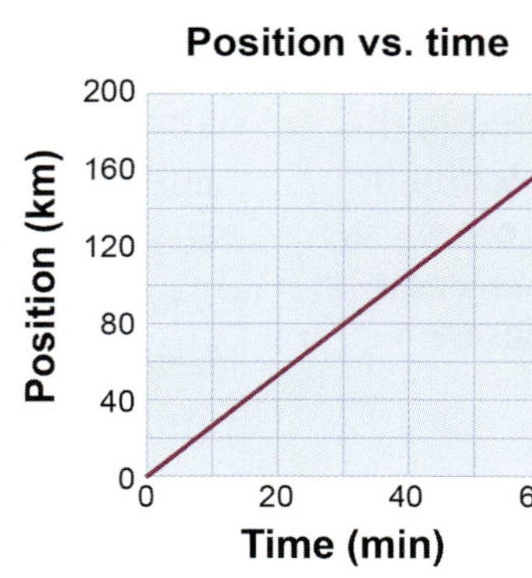


5. You start driving from Austin toward Houston with a constant speed of 120 km/hr (75 mph). After 1.5 hr you stop for 0.5 hr. Then you continue driving with a new constant speed $v_2$ and reach Houston 45 min later.

   a. If the total distance from Austin to Houston is 265 km, what is the speed $v_2$?
   b. What was your average speed during the trip?

Physics problems give you some information and ask you to apply physics to figure out something else. You need to analyze given information and think logically about how to get an answer. This section shows you how to analyze a problem that at first seems simple but is more complicated than it looks.

## How to start

**How do I start a problem?**

One of the most difficult steps in beginning a physics problem is *how to start.* That means translating diagrams and words into the language of math. *Mathematics is indeed a language!* To translate from English into math you need to know specific things.

a. What is the problem asking for?
b. What information are you given or do you know?
c. What relationships involve what is asked and/or given?

Two bicyclists approach each other on the same road. One has a speed of 5 m/s and the other has a speed of 8 m/s. They are 500 m apart. If they keep the same speeds, how long will it be before they meet each other?

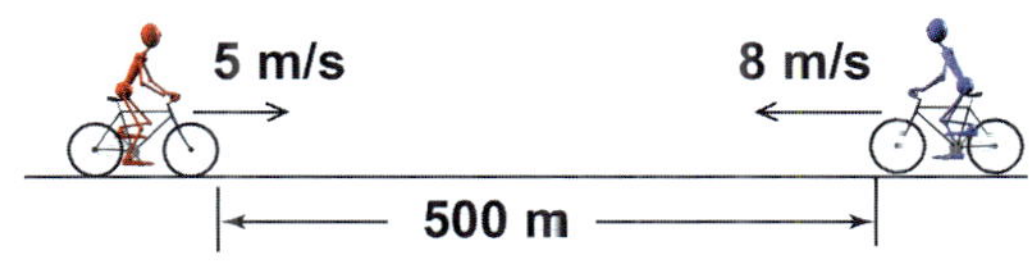


**What is the first thing to look for?**

The last sentence of the example problem tells you what the problem is asking for. *"How long will it be before..."* is a reference to *time.* This problem asks you to find a time. Choose the variable $t$ to represent time. The letter $t$ is a good choice. Mathematically, you are seeking a solution to the equation $t = ?$

**How do I organize my information?**

You are given two speeds and a distance. Because there are two speeds, you cannot use just $v$ for both. When there is more than one of the same type of variable a good strategy is to use *subscripts.* Make one bicycle "number 1" and the other "number 2." The speed of bicycle number one is $v_1$ (pronounced "vee one"). The speed of bicycle two is $v_2$.

**Naming variables**

$v_1$ = speed of bicycle 1

$v_2$ = speed of bicycle 2

$d$ = distance between bicycles

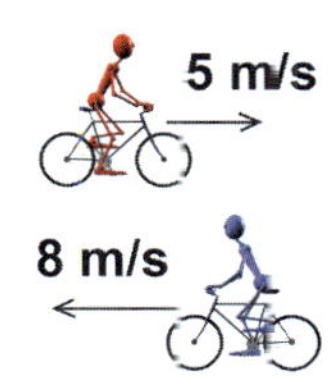


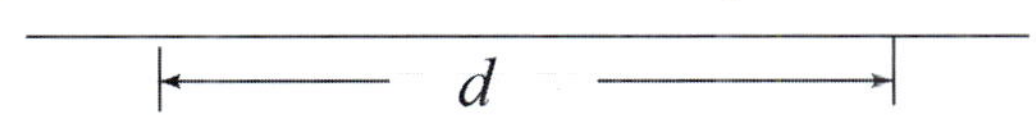


**How do I connect what I know and what I am given?**

Write down all the relationships you know that involve the types of information you are asked and given. For this problem you are asked for a time. You are given speeds and a distance. The relationship you need is one that relates speed, distance, and time. This relationship is $v = d/t$.

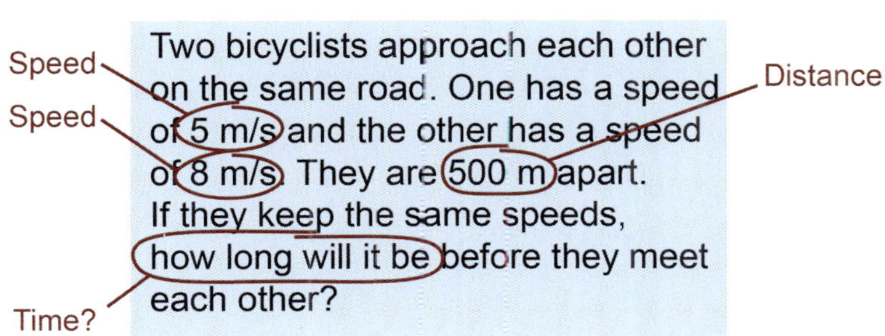


**Relationship**

Speed $v = \frac{d}{t}$ Distance / Time

## Using more than one equation

**Why "solve" an equation?**

How do you get time from speed and distance? The answer is to "solve" the equation for time. "Solving for $t$" means rearranging the equation into the form $t$ = ? in which "?" are the other variables and there are no $t$'s on the right of the equals sign. Using the rules of algebra, we can restate the same relationship as $t = d \div v$ or as $d = vt$.

### Solving the speed equation for distance or time

| Equation (2.1) | | | |
|---|---|---|---|
| $v = \frac{d}{t}$ | **Multiply both sides by $t$** $t \times v = \frac{d}{\cancel{t}} \times \cancel{t}$ | $d = vt$ | The distance traveled is the speed multiplied by the time. |
| | **Multiply both sides by $\frac{t}{v}$** $\frac{t}{\cancel{v}} \times \cancel{v} = \frac{d}{\cancel{t}} \times \frac{\cancel{t}}{v}$ | $t = \frac{d}{v}$ | The time is the distance traveled divided by the speed. |

**How do I use the equation?**

The general equation ($d = vt$) must be applied to the specific speed, time, and distance variables for your problem. Let's look at the equation for bicycle 1. We write the speed equation as $d_1 = v_1 t_1$. We do a similar thing for bicycle 2 to get a second equation $d_2 = v_2 t_2$.

**General relationship**

$d = vt$

**Applied to this problem**

5 m/s

$d_1 = v_1 t_1$

The distance traveled by bicycle 1 is the speed of bicycle 1 multiplied by the time for bicycle 1.

8 m/s

$d_2 = v_2 t_2$

The distance traveled by bicycle 2 is the speed of bicycle 2 multiplied by the time for bicycle 2.

**Is there enough information?**

We are given that $v_1$ = 5 m/s. We do not, however, know the distance $d_1$, so we cannot calculate a time. We have the same problem with the equation for bicycle 2. This brings up an important rule.

**One equation allows you to determine only *one* unknown value.**

**How many equations do I need?**

Looking at this problem, we see that there are *four* unknown values: $t_1$, $t_2$, $d_1$, and $d_2$. It is a fundamental rule that you need as many equations as you have unknown values. Four unknowns means that we need four equations to solve the problem. We have two equations; we need two more.

| | Unknown variables | | Known variables |
|---|---|---|---|
|  5 m/s Bicycle 1 | $t_1$ | $d_1$ | $v_1$ = 5 m/s |
|  8 m/s Bicycle 2 | $t_2$ | $d_2$ | $v_2$ = –8 m/s |

**What is contextual information?**

In most problems, it is assumed that you know information *implied* by the context of the problem, but not stated outright. Unless otherwise stated, you should assume the following:

1. There is zero friction, unless you are given friction information;
2. Velocities are constant, unless forces or accelerations are known; and
3. Initial position, time, and velocity are zero, unless you know otherwise.

Try to relate all the information you are given to the variables you chose. If you are stumped, think about what additional assumptions you might make that would make a problem solvable.

**How do I use the distance?**

The two bicycles are 500 m apart at the start, which is $t = 0$. To use this information, we need to relate it to the distance variables we defined. By the time they meet, the total distance traveled by both bicycles together has to be 500 m. In problem-specific variables this is written as $d_1 + d_2 = 500$ m. This is a third equation. Now we need a fourth equation because there are four unknowns.

... of 5 m/s and the other has a speed of 8 m/s. They are 500 m apart. If they keep the same...

$$d_1 + d_2 = 500 \text{ m}$$

The distance traveled by both bicycles individually must add up to 500 m total.

**Deducing implied information**

The last equation comes from reading the problem and recognizing that the bicyclists meet after traveling the same amount of *time*. Mathematically, that means $t_1 = t_2 = t$. Since the times are equal, they *do not need subscripts for us to tell them apart.* We can write the three equations in terms of a single time $t$.

| $d_1 + d_2 = 500 \text{ m}$ | $d_1 = v_1 t$ | $d_2 = v_2 t$ |
|---|---|---|
| The total distance traveled is 500 m. | The distance traveled by bicycle 1 | The distance traveled by bicycle 2 |

**How do I use multiple equations?**

We now know enough to solve the problem. Sentences from the problem and their mathematical equivalents are shown above. One unknown distance equation is $d_1 = v_1 t$. The other unknown distance equation is $d_2 = v_2 t$. This is true no matter what value $t$ has. The distance relationship tells you that $d_1 + d_2 = 500$ m. You can now *replace the distances with their equivalent velocities and times.* This gives you a single equation with a single unknown value. That unknown value is what you want because then you can calculate an answer.

### Solution

$$d_1 = v_1 t$$

$$d_1 + d_2 = 500 \text{ m} \;\rightarrow\; v_1 t + v_2 t = 500 \text{ m} \;\rightarrow\; (5 \text{ m/s} + 8 \text{ m/s})\, t = 500 \text{ m}$$

$$d_2 = v_2 t$$

$$(13 \text{ m/s})\, t = 500 \text{ m}$$

$$t = 38.5 \text{ s}$$

## Using consistent units

**Equations require consistent units**

Before you can use the information you are given in a physics problem, you must first check the units for consistency. This is very important because not paying attention to units will cause you to fail many physics tests!

*To use an equation every quantity must be in consistent units.*

For example, suppose a car is traveling 30 miles per hour. How long does it take the car to cross an intersection that is 18 meters across? Note that speed has units of miles per hour and distance has units of meters. These units are not consistent with each other. If you use the speed formula and calculate the time with speed in miles per hour and distance in meters *you will get the wrong answer!* To get the correct answer you have two choices. You can convert the speed of the car into meters per second to match the distance across the intersection in meters, or you could convert the distance across the intersection into miles to match the speed in miles per hour. *If you use this second choice your time will come out in hours!*

How long does it take a car moving at 30 mph to cross an intersection that is 18 m wide?

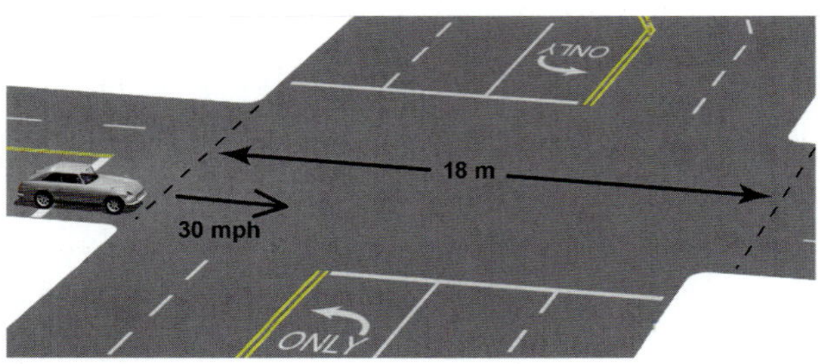


$$t = \frac{d}{v} = \frac{18\text{ m}}{30\text{ mph}} = 0.6\text{ s}$$

**Wrong—because units do not match**

**30 mph = 13.4 m/s**

$$t = \frac{d}{v} = \frac{18\text{ m}}{13.4\text{ m/s}} = 1.34\text{ s}$$

**Right—units are consistent!**

**Converting units of speed**

Speed has units of distance divided by time. When converting speeds, you need to use conversion factors for both time and distance. To convert the distance, you might use the conversion factor 1 mile = 1,609 meters. For time, you might use 1 hour = 3,600 seconds. The three calculations on the right show correct and incorrect ways to do the conversion. Which is correct?

**Which is the correct unit conversion?**

$$\left(\frac{30\text{ mi}}{\text{hr}}\right)\left(\frac{3{,}600\text{ s}}{1\text{ hr}}\right)\left(\frac{1{,}609\text{ m}}{1\text{ mi}}\right) = 173{,}772{,}000\text{ m s/hr}^2$$

$$\left(\frac{30\text{ mi}}{\text{hr}}\right)\left(\frac{1\text{ hr}}{3{,}600\text{ s}}\right)\left(\frac{1{,}609\text{ m}}{1\text{ mi}}\right) = 13.4\text{ m/s}$$

$$\left(\frac{30\text{ mi}}{\text{hr}}\right)\left(\frac{1\text{ hr}}{3{,}600\text{ s}}\right)\left(\frac{1\text{ mi}}{1{,}609\text{ m}}\right) = 0.00000518\text{ mi}^2/(\text{m s})$$

**Getting the units to cancel**

The correct conversion is the one in which all the units cancel out—except for meters in the numerator and seconds in the denominator. Once the conversion factors are arranged so that the units work out, all you have to do is multiply and divide the values as they appear in the numerator or denominator. In this example, the speed of the car is 13.4 m/s. Once the speed has been converted to units of meters per second it is consistent with the distance across the intersection in meters. To solve the problem you apply the formula $t = d/v$ to calculate that it takes the car 1.34 s to cross the intersection.

### What is relativity?

Nearly everyone has heard of *relativity*, but not many people know what it means. There are two different forms of relativity, both proposed by Albert Einstein. The theory of *special relativity* connects velocity, time, and space. The theory of *general relativity* presents a totally different way to understand gravity. Both theories may seem like science fiction, but they are well tested and part of the technology you use every day. GPS satellites incorporate general relativity, and virtually all optical fiber data networks must take account of special relativity.

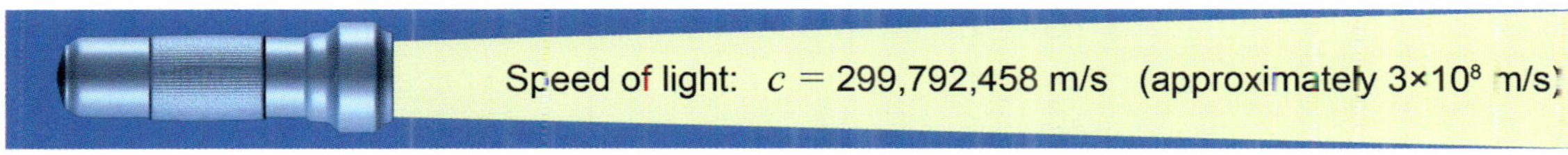


### Speed of light

The speed of light, $c$ = 299,792,458 m/s, is the fastest speed in our universe. Around 1900, Einstein asked himself what light would look like if he could see it when it wasn't moving. Instead of making the light stop, Einstein thought about traveling beside light at the same speed. You may have driven next to someone going the same speed on the highway. When you look out the window, the other car seems motionless relative to you. Einstein thought about whether it might be possible to travel at the same speed as light and about how light would appear if you could.

### Reference frames

Recall the dart thrown on a train from earlier in this chapter. If the person throws the dart at 10 m/s relative to the train, and the train is approaching you at 10 m/s, the dart gets 20 m closer to you every second. As far as you are concerned, you see the dart traveling 20 m/s toward you. The speed of the dart in *your* reference frame is the speed in its reference frame (the train) plus the speed of that reference frame with respect to you.

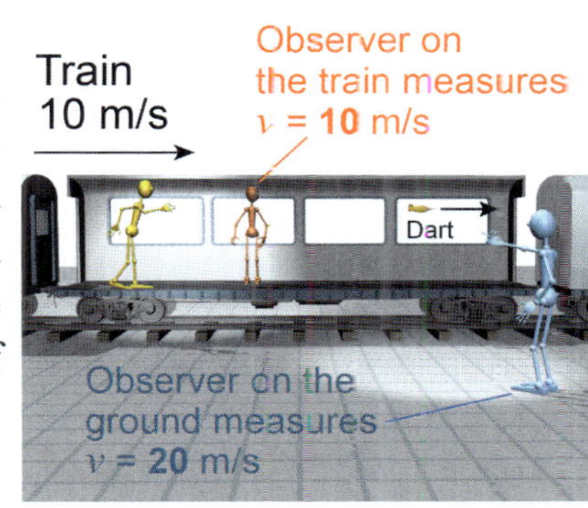


### Paradox of the speed of light

Einstein considered what the speed of light would be if the person on the train shined a flashlight at you as in the case of the thrown dart. Physicists at that time believed that the ground observer should measure a speed equal to the speed of light in the train's moving reference frame plus the speed of the train. But that is not what happens. The speed of light is 299,792,458 m/s *for both observers no matter the speed of the train or its direction, forward or backward!*

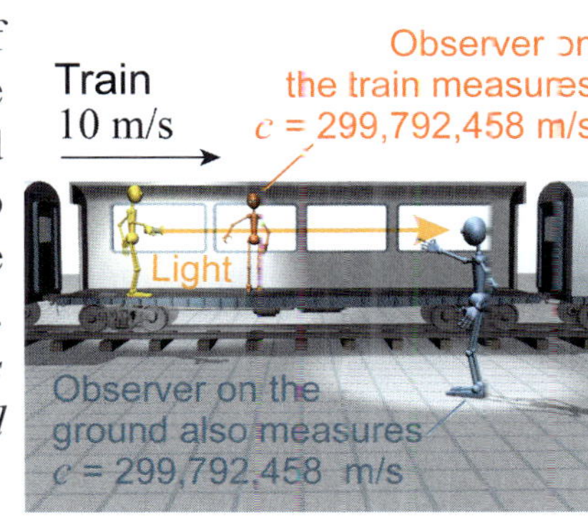


### Michelson–Morley experiment

In 1887, Albert A. Michelson and Edward W. Morley wanted to determine whether the speed of light changed as a result of the Earth's motion. Michelson and Morley made sensitive measurements of the speed of light both parallel and perpendicular to the orbital motion of the Earth. In effect, they used the Earth as a "train" moving at 29,800 m/s—Earth's orbital speed. They found the speed of light to be exactly the same for both directions! This result is not at all what they expected. How could it be explained?

## Time in moving frames of reference

**Why is the speed of light the same for all observers?**

Einstein thought deeply about what it meant for the speed of light to be constant to all observers. In the process, he came to surprising but inescapable conclusions that totally upended our notions of space and time. Prior to special relativity, time was a constant, flowing at the same rate for all observers. Einstein's thought experiments about light led him to realize that time and even space were not constants after all.

**Think about a light clock**

Imagine a clock that counts trips of a beam of light going back and forth between two mirrors. This type of clock is similar to how an atomic clock works today. Einstein placed his imaginary clock on a moving spaceship. A person standing next to the clock sees the light go up and down, back and forth. In the moving reference frame of the spaceship, the time it takes the light to make one trip is twice the distance between the mirrors divided by the speed of light.

**How does the clock appear in another reference frame?**

Next, consider a stationary observer outside the spaceship, watching from the ground. This stationary observer sees the light follow a zigzag path because the mirrors move with the spaceship. The zigzag path is longer than the simple up-and-down path seen by the observer inside the spaceship—*yet the speed of light must be the same to all observers, regardless of their motion.* Suppose it takes light one second to go between the mirrors. The speed of light must be the same for both people, yet the person on the ground sees the light travel a longer distance!

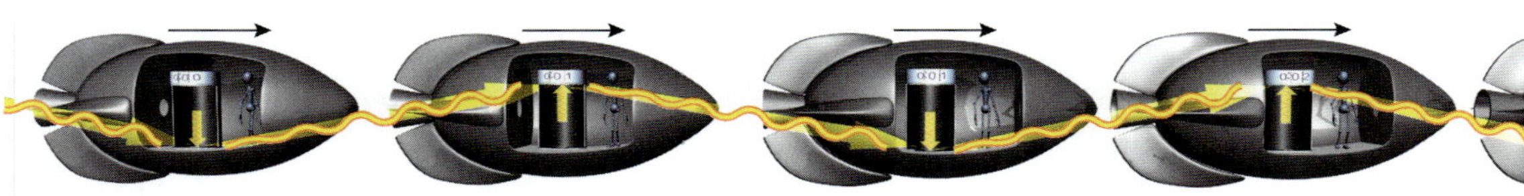

The observer on the ground sees the light take a longer path.
*The speed of light, however, must be the same for both observers!*

**Time runs at different rates for different observers**

The way to resolve this paradox is to realize that *one second on the ground is not the same as one second on the spaceship.* If one second of "ship time" is longer than one second of "ground time," then the problem is resolved. The ground observer divides a longer distance by a longer "second" and measures 299,792,458 m/s. The shipboard observer finds the same value based on the ship's time. The consequence of the speed of light being the same for all observers, Einstein's concluded, is that *time flows at different rates for observers in motion relative to each other.*

**Time runs slower in a moving reference frame**

A major consequence of the speed of light being constant is that time slows down for objects in motion (including people). If you move fast enough, the change in the rate at which time passes can be enormous. For a spaceship traveling at 99.9% of the speed of light, 22 years would pass on Earth while a single year passes on the spaceship. The closer the spaceship's speed is to the speed of light, the slower time flows for those aboard relative to those left behind on Earth.

**Consequences of special relativity**

Special relativity does not affect ordinary experience because the relative velocities need to be at least 100 million m/s before the effects become obvious. Relativistic effects, however, are observed every day in physics labs and in many technologies.

**Time dilation**

Time passes more slowly in a moving reference frame when compared to a stationary reference frame. In practical terms, clocks run slower on moving spaceships compared with clocks on the ground. By moving very fast, it is possible to travel into the future. If you were traveling in a spaceship at a speed of 0.9999*c* (or 99.99% of the speed of light), one year would pass for you while 100 years have passed on the ground. This effect is known as *time dilation.*

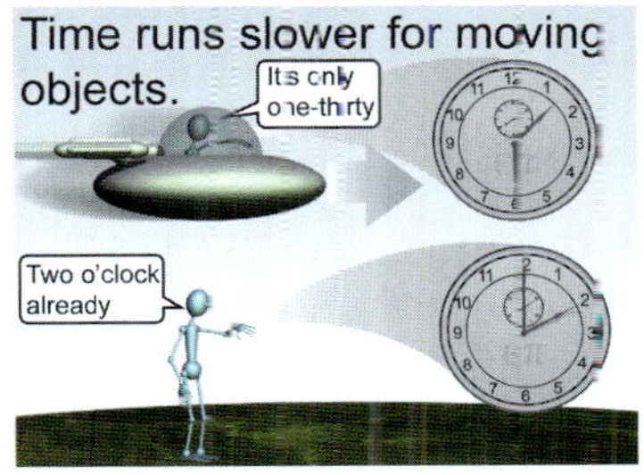

Time runs slower for moving objects.

**Space contraction**

Not just time, but also space itself, varies depending on the reference frame of the observer. The length of an object measured in a stationary reference frame is not the same as the length measured in a moving reference frame. An object does not get smaller or larger: *Space itself gets smaller for an observer moving near the speed of light.* Space contracts along (or parallel to) the direction of motion.

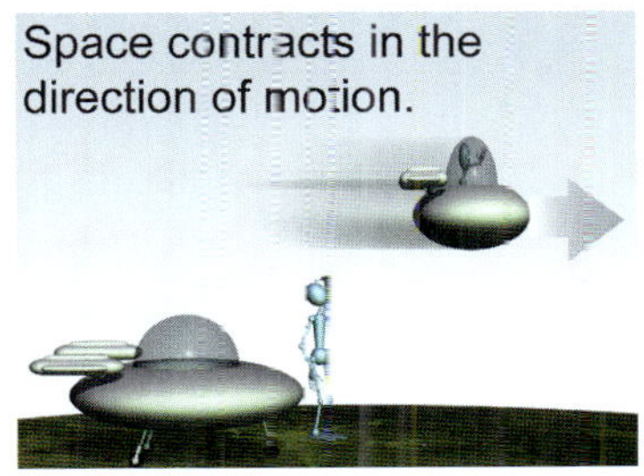
Space contracts in the direction of motion.

**Simultaneity becomes relative**

The concept of "simultaneous events" changes when one reference frame is moving relative to the other. Two events that are simultaneous to one observer may not be simultaneous to another who is moving relative to the first observer. The idea of "now" and "then" may no longer refer to the same point in time for observers in different reference frames.

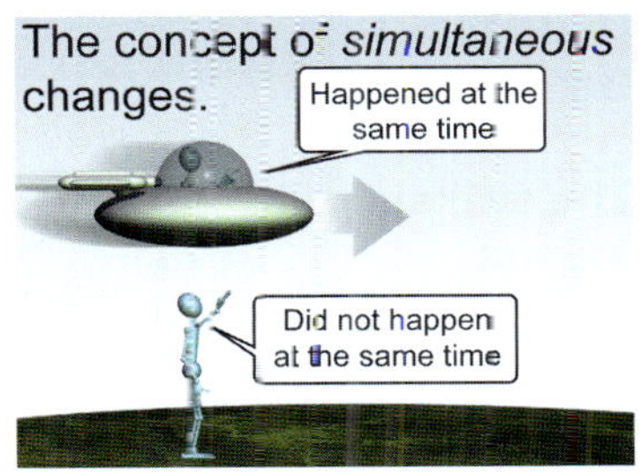

The concept of *simultaneous* changes.

**Increase in mass**

As the speed of an object approaches the speed of light, its mass increases as measured in any outside reference frame. The closer the speed of an object gets to the speed of light, the more of its kinetic energy becomes mass instead of motion. *Matter can never move faster than the speed of light because adding energy creates more mass instead of increasing an object's speed.*

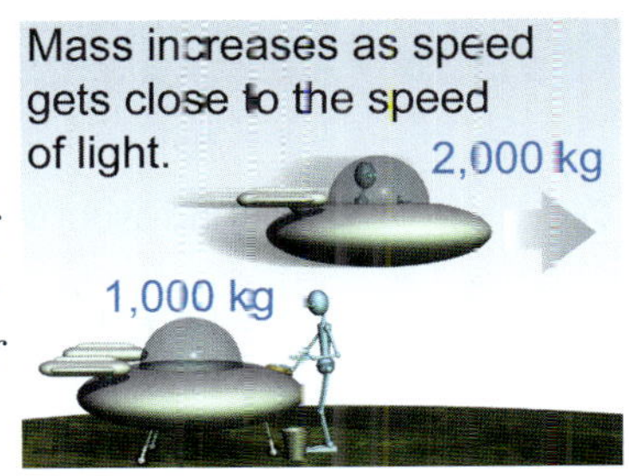

Mass increases as speed gets close to the speed of light.

**Frequency shift**

Because light is a wave, it has a frequency, and the frequency is affected by relative motion. Suppose a spaceship moving toward Earth emits red light with a frequency of $462\times10^{12}$ (462 trillion) waves per second. The same number of waves appear on Earth, but in a shorter time. If the ship were moving at 70% of the speed of light, the waves arrive at Earth at a rate of $630\times10^{12}$ (630 trillion) waves per second. *This is blue light!* The frequency of light emitted by a moving object becomes more blue ("blueshifted") when the object is moving toward you and more red ("redshifted") if it is moving away from you.

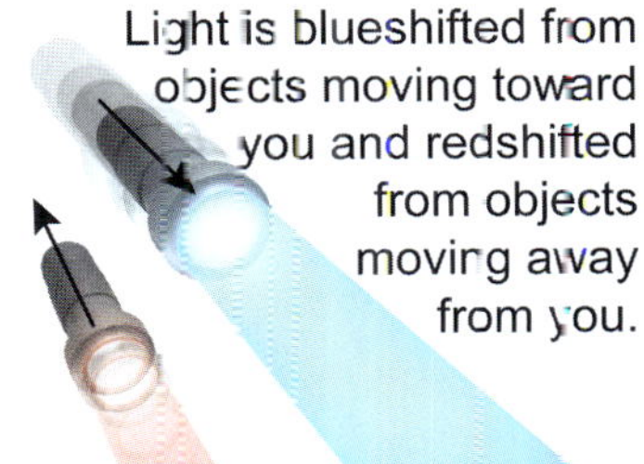
Light is blueshifted from objects moving toward you and redshifted from objects moving away from you.

## Simultaneity

**What does "at the same time" mean?**

The usual definition of *simultaneous* is "at the same time." Because time is not constant for all observers, whether two events are simultaneous depends on the relative motion of the observers. Einstein proposed that the meaning of "simultaneous" is that two equidistant events are simultaneous if *light* from both events reaches the observer at the same time. The whole concept of "at the same time" depends on the relative motion of different observers.

**A thought experiment**

Einstein imagined two lightning strikes hitting the front and back of a moving train. You are the observer on the ground watching the train from a distance. Because you are the same distance from both strike points, you see the two bolts of lightning hit the train at the same time. To you, the two lightning strikes are simultaneous because it takes the same amount of time for light from either event to reach you.

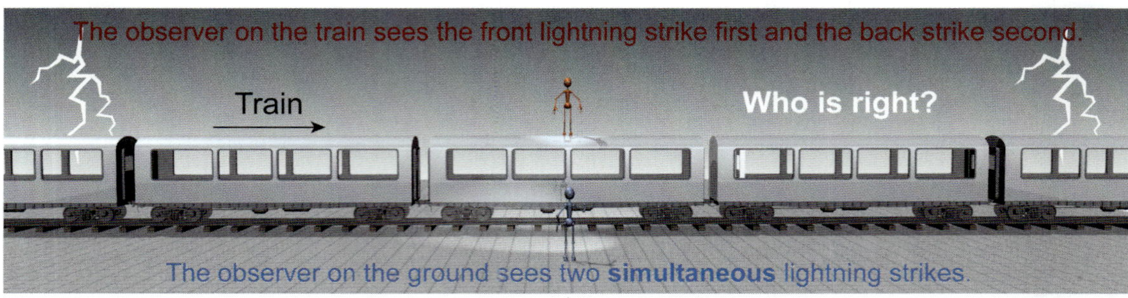


**The strikes are not simultaneous to the observer on the train**

If the train were at rest, then the train rider would also observe two simultaneous lightning strikes. The train, however, is *moving*. The observer in the center of the train sees the front of the train hit first, and then the rear, so the two lightning strikes are *not* simultaneous. The speed of light is unaffected by the motion of the train. But the observer is moving *toward* the point where lightning struck the front of the train. Light from the front strike is seen first because that light has a shorter distance to travel to reach the observer. The rear strike is seen second because the observer is moving *away* from the point where that lightning struck. Light from the rear strike has a *longer* distance to go to and reaches the observer a nanosecond after light from the front strike.

**Time, space, and causality**

Special relativity tells us that no information can travel faster than the speed of light. This has deep implications for *cosmology,* the study of the universe. The graph on the right shows space on the vertical axis and time on the horizontal axis. An *event* (A) occurs at a point in space and time. The *world lines* represent light spreading out at the speed of light. The event *never exists* for point (B) because light from (A) can never reach (B). Furthermore, point (D) experiences the event 1,000 years after point (C) because of the light travel time. To beings on a planet 165 million light years away, *dinosaurs still live on Earth* because that is the "now" they could see with their telescopes!

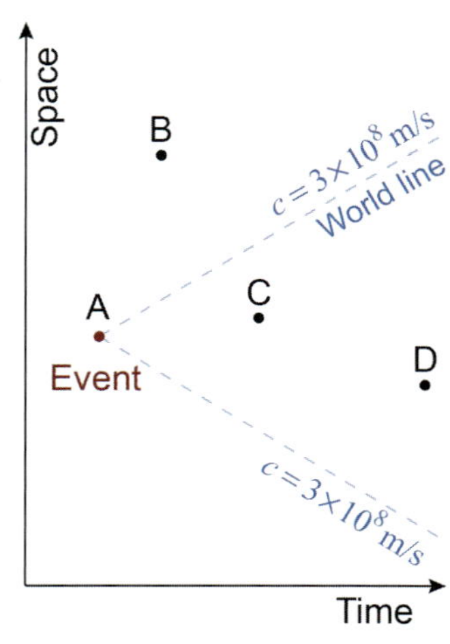

# Section 3 review

Solving physics problems usually involves translating the language of the problem into mathematical statements: identify what is asked for in the question; determine what quantities and other information are given to you; identify key relationships among the quantities; and solve the problem. In the process, it is important to assign variables to quantities, make clear and correct assumptions, and use consistent units.

## Review problems and questions

1. A student is biking to school. She travels 0.7 km north, then realizes something has fallen out of her bag. She travels 0.3 km south to retrieve her item. She then travels 0.4 mi north to arrive at school. The journey takes her 15 min.
   a. What is her total displacement?
   b. What is her average velocity?

2. Two students travel from school to their jobs at the bowling alley, which is 2.5 km away. One student rides a scooter at 9.0 m/s, and the other student rides a bike at 12 mi/hr.
   a. Who arrives at the bowling alley first?
   b. How long does the trip take for each of them?

3. Two cars are initially separated by 1.0 km and traveling toward each other, one at 30 mi/hr and the other at 30 km/hr. How far in kilometers from their respective starting points do the two cars meet?

4. A student was maneuvering the Smart Cart and generated the position versus time chart illustrated at right. Describe in words the motion of the cart for each segment of the chart.

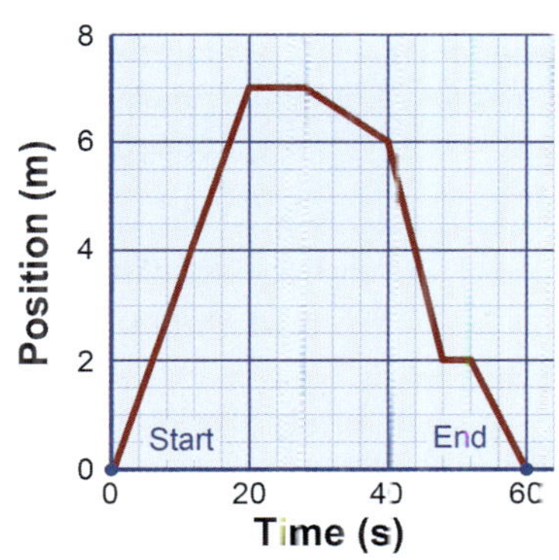


5. An airplane flew at a speed of 500 km/hr for 1.5 hr. It then hit a tail wind that boosted its speed to 650 km/hr for the next 750 km.
   a. How far did it travel during the first stage of its trip?
   b. How long did it take to complete the second stage (only) of its trip?
   c. How far did it travel during the entire trip?
   d. How long did it take for the entire trip?
   e. What was its average speed for the trip?

Chapter 3

# Chapter review

## Vocabulary

Match each word to the sentence where it best fits.

### Section 3.1

| | |
|---|---|
| **coordinates** | **displacement** |
| **magnitude** | **origin** |
| **position** | **scalar** |
| **vector** | |

1. When Ilana moved her pencil from the location (2, 4) cm on her paper to the location (6, 3) cm, the _______ she made was (+4, −1) cm.
2. _______ are values that together determine a unique position in space.
3. Marcel identified the ground beneath his feet as the _______ by labeling it with "0 meters."
4. Mass, speed, and temperature are examples of _______ quantities unlike the vector quantities such as velocity and force
5. An object's _______ is a vector that tells where it is relative to the origin.
6. A quantity that has both magnitude and direction is a/an _______.

### Section 3.2

| | |
|---|---|
| **average velocity** | **inertial reference frame** |
| **instantaneous velocity** | **reference frame** |
| **slope** | **speed** |
| **velocity** | |

7. Two people driving in cars at constant speed but in opposite directions on the highway are each traveling in a/an _______.
8. _______ can be calculated by dividing the change in position by the change in time.
9. _______ measures the steepness of a graph.
10. _______ has the same units as speed but is a vector, so it includes information about direction.
11. After Sinead finished a road trip, her GPS told her that her maximum speed was 58 mph. That means that at some point in her trip her _______ was 58 mph.
12. An office worker on the tenth floor of a building labels the origin of his _______ as zero meters at the bottom of his feet, despite being located 100 ft above ground.
13. During stop-and-go driving in a traffic jam, Darrius moved 5 km in one-half of an hour. His _______ was 10 km/hr.

## Conceptual questions

### Section 3.1

14. Which of the following sentences best matches the meaning of the equation $d = x_f - x_i$?
    a. The final position is the initial position minus the displacement.
    b. The initial position is the final position plus the displacement.
    c. The displacement is the initial position minus the final position.
    d. The displacement is the final position minus the initial position.
15. Latitude and longitude lines form a convenient coordinate system for specifying locations on Earth's surface. Where is the origin of this coordinate system? (Hint: The east–west origin and north–south origin both need to be identified.)
16. Are the following vector or scalar quantities?
    a. temperature
    b. computer memory space
    c. location from a GPS
    d. speedometer reading
17. What is the relationship between the scalar *distance* and the vector *displacement*?
18. Describe a situation in which someone moves a distance of 100 m yet has a displacement of zero.
19. ( What is the difference between position and displacement?
20. ( Give three phrases that are examples of displacements without using the word "displacement." These should be given as you would speak them to a friend. For example, "I ...."
21. ( Which of the following are vectors?
    a. distance
    b. position
    c. displacement
22. ( Can you add or subtract a distance from a displacement?
23. (( What factors would influence a person's choice of an origin and coordinate system for a given problem?

# Chapter review

## Section 3.2

24. A position versus time graph represents the motion of two objects. The line for object one has a steep slope and the line for object two has a shallow slope. Which object is moving faster?

25. What physical quantity is represented by area on a velocity versus time graph?

26. Your friend took a walk through a maze. She wrote down exactly how far she traveled along each straight path in the maze and at what times she turned. She always walked with a steady velocity along any straight path and only changed velocities when she turned.

    a. Explain how you would use her data to find her instantaneous velocity at any single point in the maze. Include the equation you would use.

    b. Explain how you would use her data to find her average velocity. Include the equation you would use.

27. Kristie walks once around a large circle at a constant speed. What is her average velocity for the whole walk? Explain using equations. Is Kristie's instantaneous velocity constant? Explain this second part in words.

28. Scott is chasing Damian along a track. Consider two frames of reference: Damian's moving frame of reference and the ground's fixed frame of reference. Is Scott's speed faster or slower in Damian's reference frame compared to the ground's frame?

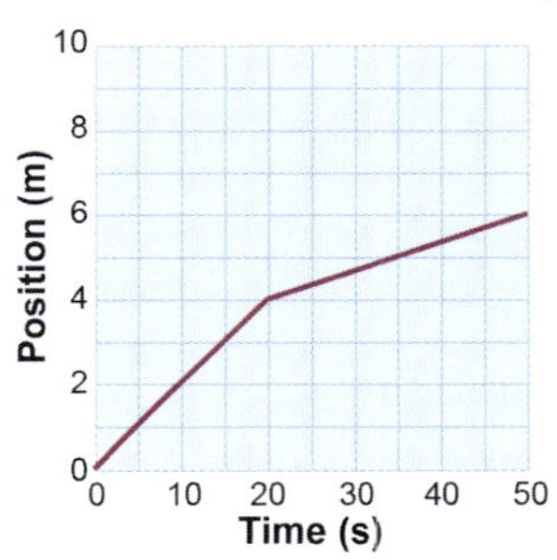


29. Without looking at the numbers, does the graph above show a greater speed at 10 s or at 40 s?
How can you tell?

30. The area on the velocity versus time graph represents distance. What does the area on a position versus time graph represent?

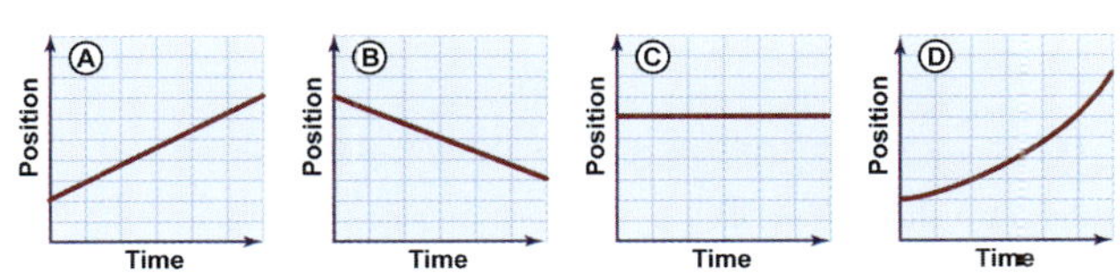


31. Which graph above shows the least amount of movement?

32. Describe three different ways you can travel 100 m in 10 s. Each different way must specify speed(s) and time(s).

33. Which of the following sentences best matches the meaning of the equation $d = vt$?

    a. The speed is the distance divided by the time.

    b. The time is the distance divided by the speed.

    c. The distance is the speed multiplied by the time.

    d. The distance is the speed divided by the time.

34. What does the slope in a position versus time graph represent?

35. What can you say about the shape of a position versus time graph representing accelerated motion?

36. How do you find distance traveled on a velocity versus time graph?

37. Why is the slope of a distance versus time graph equal to speed?

38. A position–time graph of Liz's trip to Boston shows a flat line starting at $x = 3$ hr and ending at $x = 5$ hr. What does this mean?

39. Janet is pulled over by a police officer for going faster than the speed limit. Janet, a high school physics student, argues that she didn't violate the law because her average velocity over the course of the last five miles was under the speed limit. The officer still gives Janet a ticket. What was the mistake in Janet's argument?

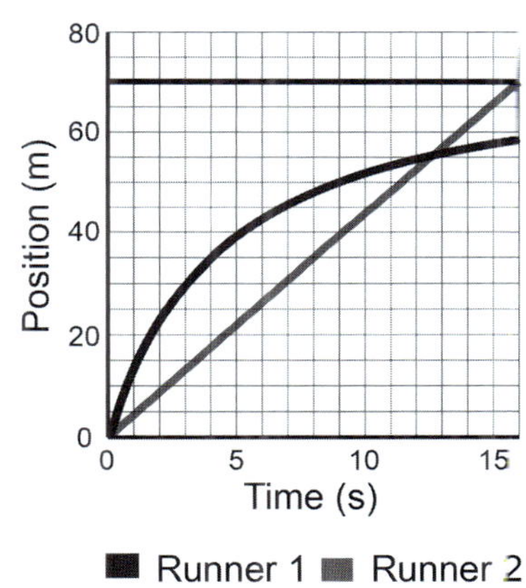


40. This position versus time graph shows a famous, mythic race. Can you tell which one? Explain and include a description of the race in your answer.

41. Give an example of a situation where an object moves in two or three dimensions but its motion can be characterized in one dimension?

## Section 3.2

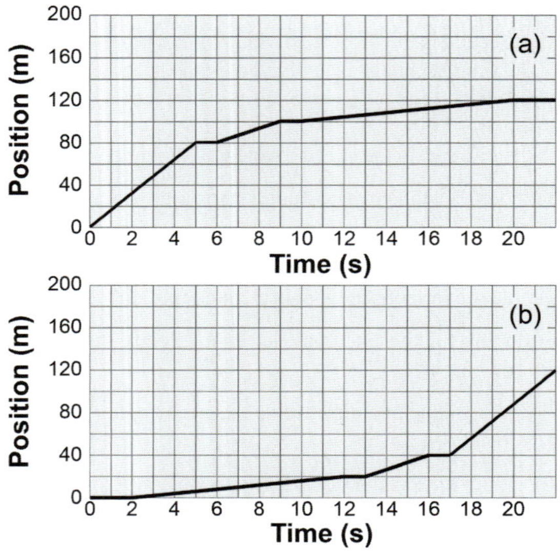


42. ⟪⟪ Which figure shows a driver getting stuck in the mud?

## Section 3.3

43. The equation $d = vt$ is applied to calculate the distance when time is given in seconds. If the speed is in units of centimeters per second, then what units will the calculated distance have?

44. Write down three forms of the equation that relates speed, distance, and time. Each form should solve for a different variable.

45. Which of the following equations best matches the meaning of the following sentence?
"The sum of two numbers is divided by 23."
   a. $(a + b)/23$
   b. $23(a + b)$
   c. $a + b - 23$
   d. $23/(a + b)$

46. ⟪ If $a^2 = b^2 + c^2$ then what is $c$ in terms of $a$ and $b$?

47. ⟪ As a solution to a physics problem, a student derived an equation for the distance traveled by a ball as $d = 7t^2$. Is this a reasonable answer? Why or why not?

48. ⟪⟪ A problem is given with three variables, $a$, $b$, and $c$. The relationships between the variables are
$$a = b + c$$
and
$$b = 2c$$
Is it possible to calculate a value for $a$ with the information given? Why or why not?

49. ⟪⟪ $A = BC^2$ and $A = DE$.
   a. Given $B$, $C$, and $E$, solve for $D$.
   b. Given $B$, $D$, and $E$, solve for $C$.

### Quantitative problems

## Section 3.1

50. You need to do a series of calculations involving the motion of an object. Another student hands you predefined parameters of the problem; you see that she has defined $x_i = -53.4$ m and $x_f = 71.6$ m. How might you redefine the problem so as to make it easier to solve?

51. ⟪ Trixie is driving laps at a racetrack. The track is circular and has a radius of 25 m. After completing 2 full laps, what is Trixie's displacement relative to her starting position?

52. ⟪ You walk 10 miles north, 20 miles south, and 5 miles north. Taking north to be the positive $y$ direction, define a position vector that describes your position on the $y$ axis.

53. ⟪⟪ Two robots compete in a game of tug-of-war. The orange robot first loses ground, until $x = -12$ m, then gains it until $x = 42$ m when the blue robot loses its grip. In order to win, the total displacement of a robot must be 50 m or more. Did the orange robot win the round?

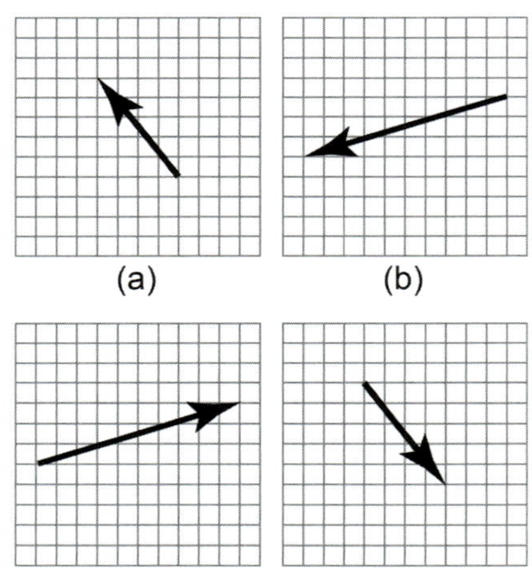

(a) (b) (c) (d)

54. ⟪⟪ Vector $\vec{X} = (3, 4)$ and vector $\vec{Y} = (-7, 1)$. Which diagram shows $\vec{Y} - \vec{X}$?

55. ⟪⟪ Starting at $\vec{x_i} = -5$ m, you walk right 10 m and then left for 3 m. Your final position is $\vec{x_f}$. Taking right to be the positive $x$ direction, describe a displacement vector $\vec{d_t}$ that characterizes the total change in position.

# Chapter review

## Section 3.2

56. A dog starts at the origin and runs forward at 6 m/s for 1.5 s and then turns around to fetch the ball by running backward at 7 m/s for 3 s. If the dog runs back to the origin at 4 m/s, then how much time has elapsed between the start and when he returns with the ball?

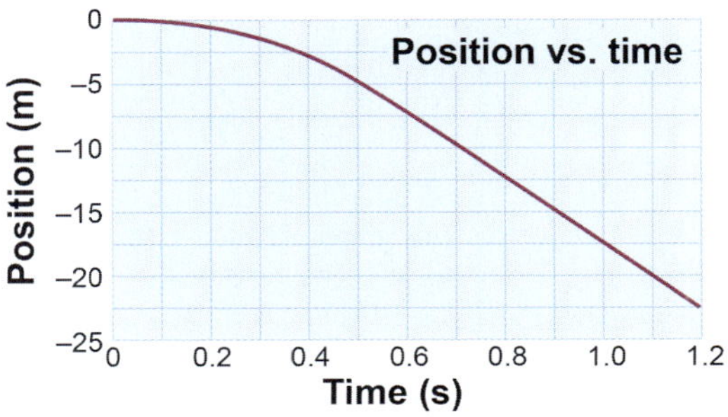


57. The position vs. time graph above represents the motion of a foam ball after being dropped. What is the instantaneous velocity at 0.8 s?

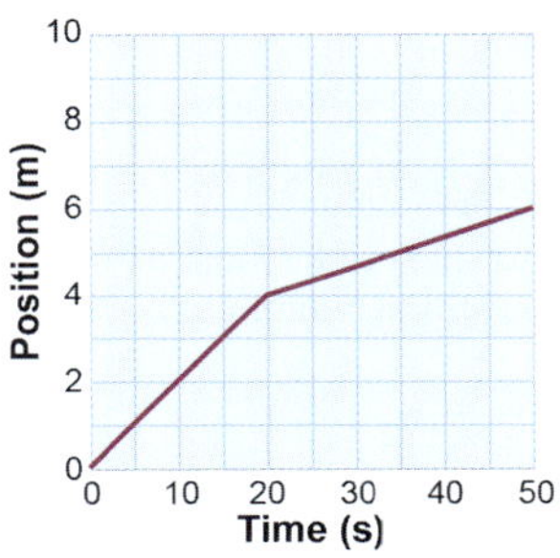


58. In the position versus time graph above, how far does the object travel in the first 35 s?

59. The cheetah, the fastest land animal, can sprint at 30 m/s. Could the cheetah get a speeding ticket on a highway where the speed limit is 55 mph?

60. A bicycle starts 50 m from an intersection. Ten seconds later the bicycle is 85 m from the same intersection. What is the slope of the bicycle's position versus time graph 5 s after it starts? Assume constant velocity motion.

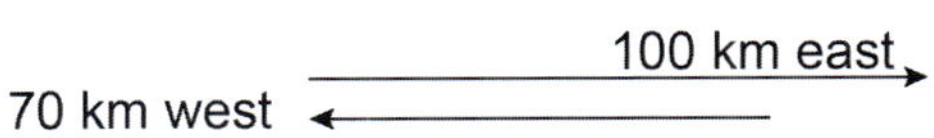


61. Mike drove 70 km due west and then 100 km due east.

   a. What is his total distance traveled?
   b. What is his total displacement?
   c. If Mike took 3 hr to complete the trip, what was his average speed?
   d. What was his average velocity over the 3 hr trip?

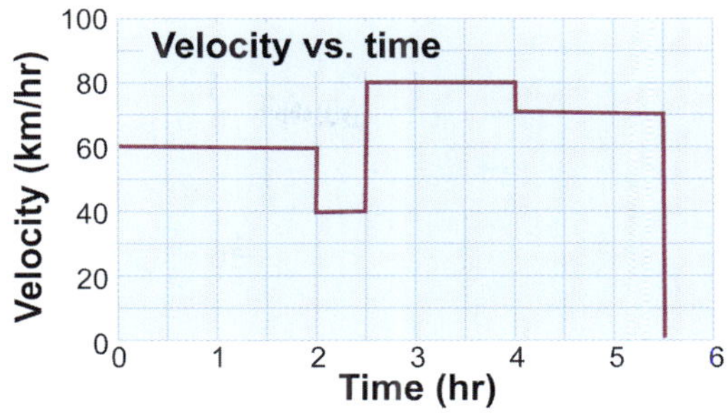


62. How far does the car travel between 1 hr and 3 hr according to the velocity versus time graph above?

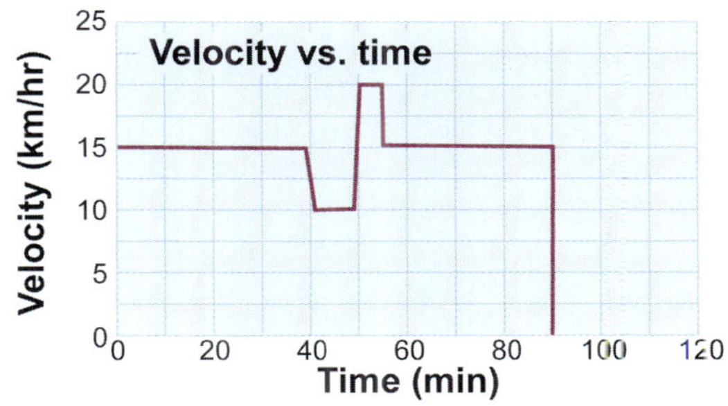


63. The velocity versus time graph above shows a bicycle trip. Use the graph to calculate the following.

   a. How many kilometers does the cyclist travel?
   b. What is the cyclist's final position?
   c. Which segment of the trip is most likely to have been downhill?

64. An object that starts at a position of 5 m and travels for 3 s at a velocity of −9 m/s ends up at what position?

65. How long does it take to ride 5 miles on a bicycle going 20 mph?

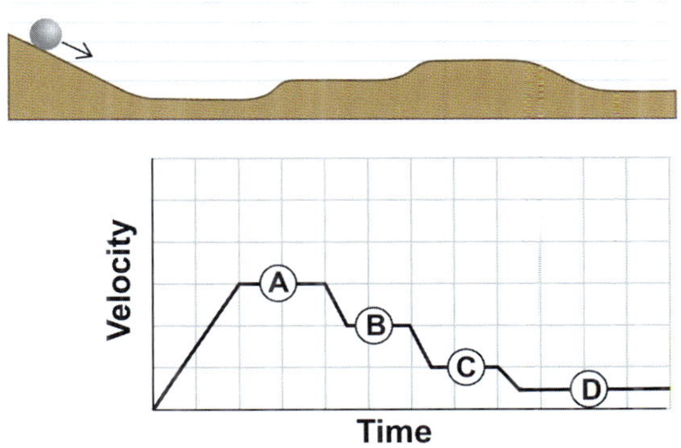


66. A ball rolls along the track shown. Use what you know about how speeds change on hills to determine which of the four places (A, B, C, or D) on the graph of velocity versus time cannot be correct.

### Section 3.2

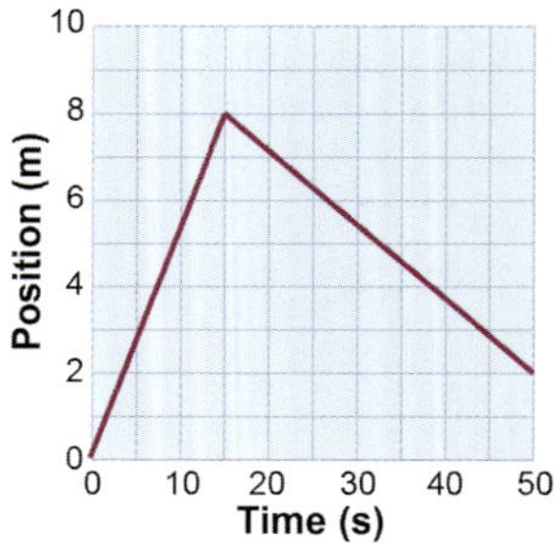


67. For the following questions, refer to the position versus time graph above.

    a. What is the average velocity from 30 to 40 s?
    b. What is the distance the object travels between 0 and 50 s? What is the ball's change in position?
    c. What is the object's velocity at $t = 5$ s?

68. Calculate the conversion factor between miles per hour and meters per second.

69. Marie is driving from the USA into Canada. She notices that all of the Canadian speed limit signs are in kilometers per hour; her odometer only displays miles per hour. How many kilometers per hour is she driving when the odometer reads 19 miles per hour?

70. Miguel solves a physics problem involving a car moving on the highway; the final velocity of the car in his answer is 3.2 m/s. Wanting to have an intuitive feel for his answer, he decides to convert to miles per hour. Calculate the velocity of the car in miles per hour.

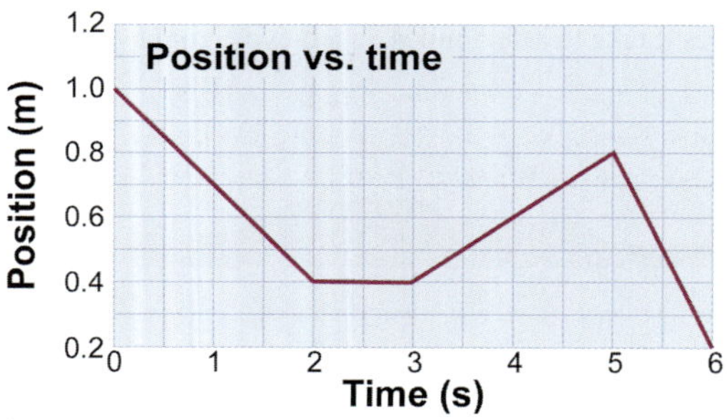


71. The graph shows the position of a robot car moving along a straight track for 6 s. Use the graph to answer the following questions.

    a. What is the car's speed 1 s after it starts?
    b. What is the maximum speed reached by the car?
    c. When does the car reach its maximum speed?
    d. What is the total *distance* the car travels between $t = 0$ and $t = 6$ s?
    e. What is the car's maximum *positive* velocity?

72. At the point when the race clock says 5:15.00 (315 s) one car is 500 m behind the starting line. By the time the clock reaches 5:56 (356 s) the car is 800 m in front of the starting line. What is the car's average velocity during this time? Assume the car moves in a straight line at constant speed in the positive direction.

73. Mitch wants to sketch a position versus time graph for a recent trip he took to New York City. During the trip, Mitch drove for one and one half hours at 60 miles/hr, stopped for one hour, and then drove at 50 miles/hr for the remaining two hours. Mitch stayed in town for three hours. Being in a rush on his way back, he drove at 95 miles/hr without stopping. Draw a position versus time graph of Mitch's trip.

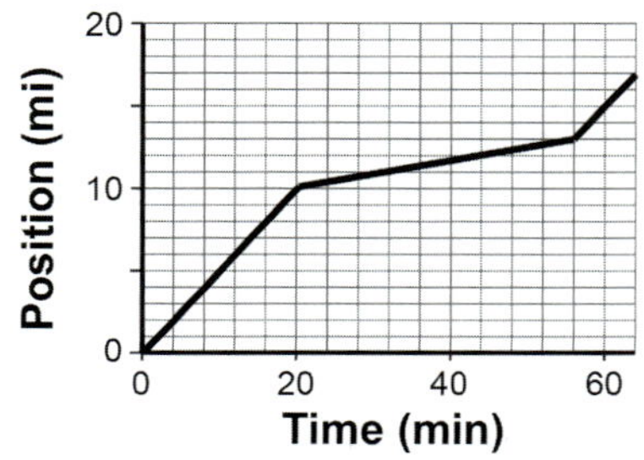


74. The figure shows a position versus time graph of Sylvia's trip to work.

    a. What is Sylvia's instantaneous velocity 12 min into the trip?
    b. What is it 36 min into the trip?
    c. What is it 60 min into the trip?

75. Which equation is the best match to the following sentence?
    "The initial position is the final position minus the product of velocity multiplied by time."

    a. $x_f = x_i + vt$
    b. $x_i = x_f - vt$
    c. $vt = x_f - x_i$
    d. $v = (x_f - x_i) \div t$

| | A | B | C |
|---|---|---|---|
| $x_i$ | 300 km | 0 km | 300 km |
| $x_f$ | 1,100 km | 1,200 km | 1,100 km |
| $t_i$ | 0 hr | 0 hr | 9 hr |
| $t_f$ | 4 hr | 4 hr | 1 hr |

76. A train travels along a 1,200 km track. At 9:00 am the train's position is 300 km from the eastern end of the track. At 1:00 pm the train's position is 1,100 km from the eastern end of the track. Which of the columns in the table above (A, B, or C) represents the best choices for the values of the variables needed to solve this problem?

# Chapter review

## Section 3.2

77. What is the average velocity of a man who runs forward at 8 m/s for 9 s and back again at 5 m/s for 3 s after an 8 s break?

78. Two cyclists that are 500 m apart start biking toward each other. They bike at speeds of 6 and 4 m/s.

    a. How long does it take for them to reach each other?
    b. How far does the slower biker travel?

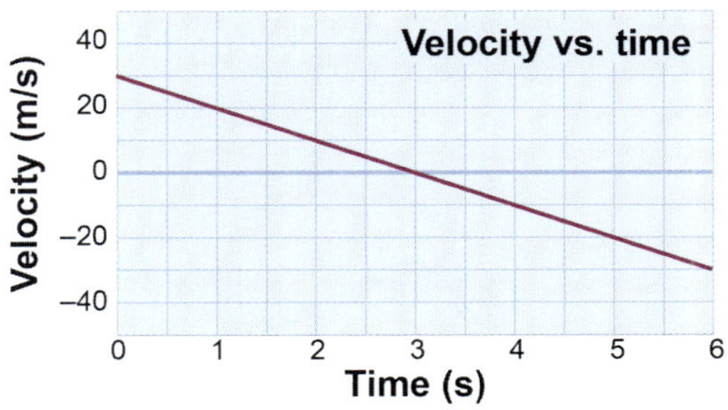


79. The velocity versus time graph above represents the motion of a ball that is thrown upward. Use the graph to calculate the maximum height the ball reaches.

80. A runner passes the 2,000 m mark of a race 380 s after starting. At what time (after starting) will the runner reach the 5,000 m mark if her speed remains constant?

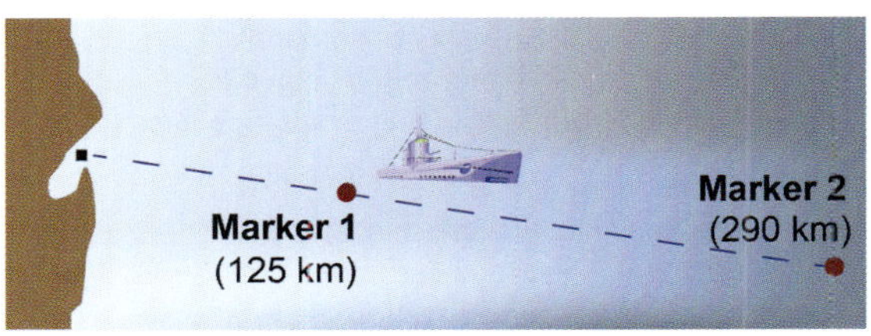


81. A submarine must reach a marker 290 km from shore 9 hr after leaving port. At what time will the submarine pass a marker that is 125 km from shore? Assume constant-velocity motion.

82. A person 200 m from a store walks directly away from the store at a constant velocity of 2.5 m/s for 65 s. How far is the person from the store now?

83. An asteroid is 180,000 km away from the Moon on a collision course. How much time will it take before the asteroid is 10 km away if it is moving at a constant velocity of 36,000 m/s? You may neglect the effects of the Moon's gravity.

84. A tennis ball is moving 30 m/s toward a player who swings a racquet at a velocity of 15 m/s toward the ball. What is the velocity of the ball relative to the racquet?

85. A sprinter running at 9 m/s is chasing a long distance runner who runs at 7 m/s. The long distance runner starts 200 m ahead. The sprinter, however, can only run for 2 min without stopping. Does the sprinter catch the distance runner?

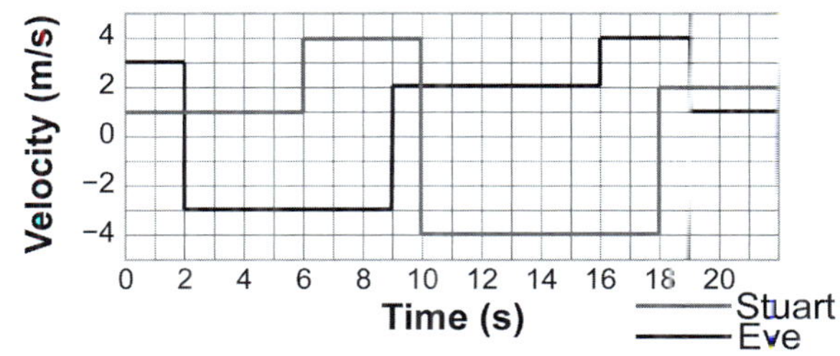


86. The figure shows the velocity versus time graph for a game of catch played between Stuart and Eve.

    a. Who has run farther after 6 s?
    b. Who has a greater total displacement from their starting position after 10 s?
    c. After 22 s, who has a greater displacement?
    d. Who has covered more distance after 22 s?

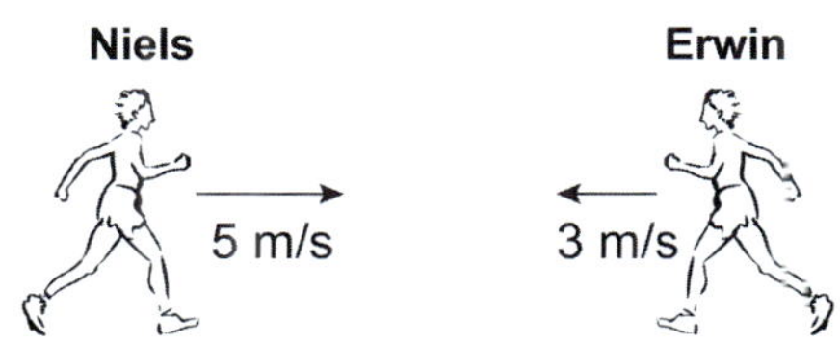


87. Niels is running 5 m/s toward Erwin, who runs 3 m/s toward Niels. From Erwin's reference frame, how fast is Niels moving toward him? From Niels's reference frame, how fast does Erwin move toward him?

88. Two people pass the 2,000 meter mark on a walking trail at the same time, going in the same direction. Jordan is jogging at 2.2 m/s and Annette is walking at 1.2 m/s. How much time passes before the two people are 100 m apart? Assume they maintain constant speed.

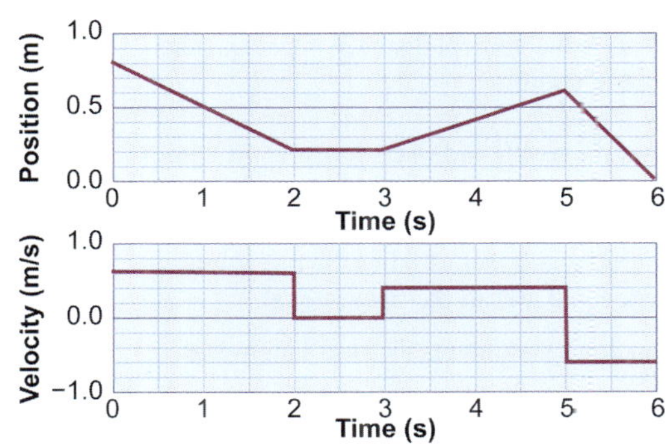


89. During which time interval does the velocity versus time graph *not* match the corresponding interval on the position versus time graph? Why?

## Section 3.3

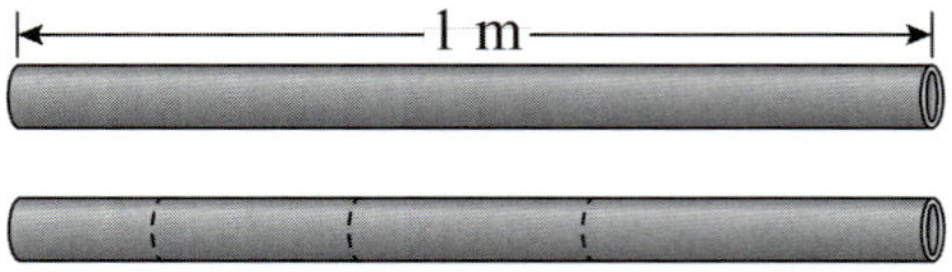

Four segments of different lengths

90. « A builder of musical chimes wishes to divide a 1 m length of pipe into four segments that have the following relationships between their lengths. Not wishing to have any waste, the builder knows it is possible to make three cuts that divide the pipe into exactly the four segments needed.

    a. The longest pipe is twice the length of the shortest pipe.
    b. The second longest pipe is 4/5 the length of the longest pipe
    c. The third longest pipe is 2/3 the length of the longest pipe.

    Calculate the lengths of each of the pipes.

Geometry of a stair

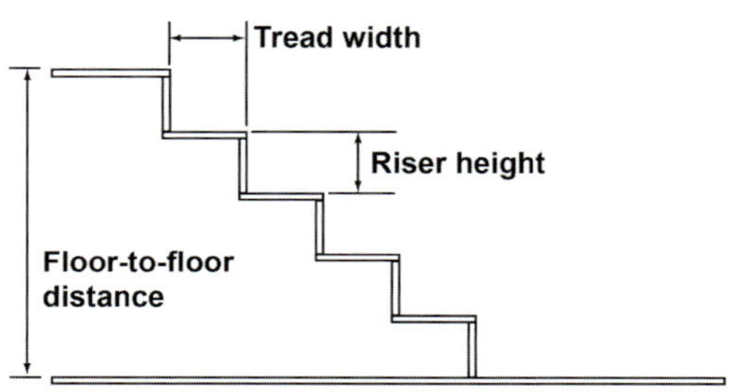


91. « The building code is a set of rules for how to design and build houses that are safe, strong, and comfortable. The rule for designing stairs is that twice the rise plus the run equals 25 in. The rise is the vertical distance between steps and the run is the tread width.

    A second rule is that the treads and risers must be the same for every step. The carpenter measures the height between floors and designs a stair that meets the rule by calculating a riser height and tread width that satisfies the rule while making all the risers equal and all the treads equal.

    a. Derive an equation that relates the riser height and tread width.
    b. Derive an equation that relates the distance between floors, the number of stairs, and the riser height.
    c. Suppose the distance between floors is 118 in. Use your equation to come up with two different stair designs that meet the rule but have riser heights no more than 9 in and tread widths no less than 9 in.

92. Given that
$$x_1 + x_2 = 4 \text{ m}$$
and that
$$x_1 x_2 = 4 \text{ m}^2$$
what are the possible values for $x_1$ and $x_2$?

93. ‹ You are given two algebraic equations: $y = 5x$ and $6y = x + 2$. Solve for $x$ and $y$.

94. « A skier leaves the lodge traveling at a constant speed of 12 m/s. Thirty minutes later, another skier leaves the same lodge traveling along the same trail. At what speed must the second skier travel to catch up with the first skier in 90 min?

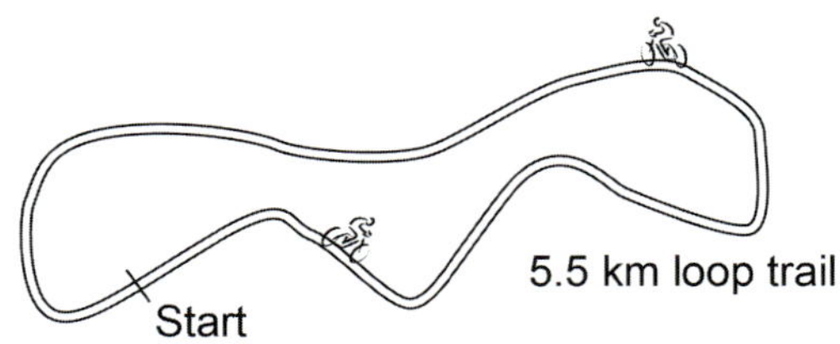


95. « Two cyclists are riding a trail that makes a 5.5 km loop. Both start at the same place at the same time. One cyclist rides at an average speed of 25 km/hr. The other cyclist rides ahead, maintaining an average speed of 29 km/hr. For how much time do the two cyclists ride before they meet up again?

96. ‹ Two boats are sailing directly toward each other at different speeds. One moves at 20 m/s and the other at 32 m/s. The boats start 2,500 m apart. How much time does it take before the boats collide?

97. « A car leaves a city traveling at a constant speed of 85 km/hr on a straight road. A second car leaves exactly one hour later traveling in the same direction. At what speed must the second car travel to catch up with the first car 250 km away from the city?

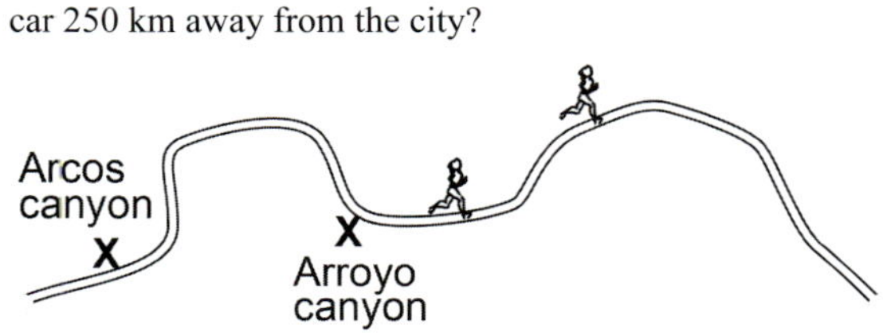


98. « Two hikers leave from different places on the same trail. Sonyia hikes at a constant speed of 4 km/hr starting from Arcos canyon at 1:00 pm. Marguarite starts from Arroyo canyon at 3:00 pm and hikes at 5 km/hr. If the two hikers meet up in 8 hr, then how far apart along the trail are Arcos and Arroyo canyons?

99. ‹ A swimmer travels at 0.8 m/s heading south across a small lake. At the same time, a second swimmer heads north from the opposite shore at a speed of 0.4 m/s. How wide is the lake if the two swimmers meet 12 min later?

# Chapter review

## Standardized test practice

100. A ball is +5 m from the origin and moves −7 m. What is the ball's position?

A. −2 m
B. 12 m
C. 7 m
D. 2 m

101. If Graham walks 5 m to the right over 6 s and then 3 m back to the left over the next 4 s, then what is his average velocity over the whole process?

A. 0.2 m/s
B. 0.33 m/s
C. 0.1 m/s
D. 0.8 m/s

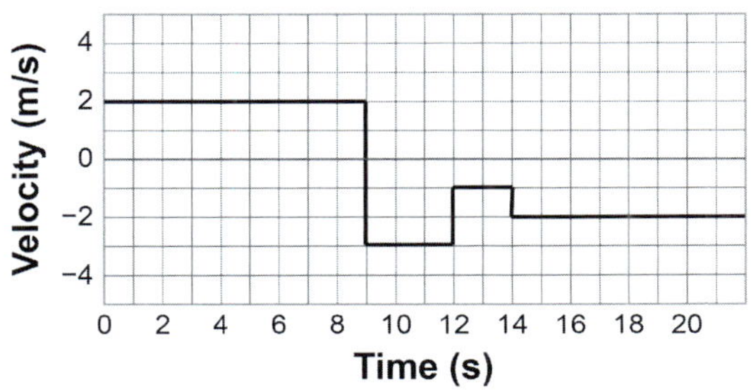


102. The figure shows a velocity versus time graph for a portion of Frank's warm-up run. What is Frank's total displacement?

A. 45 m
B. −9 m
C. −45 m
D. 9 m

103. What does the slope of a position versus time graph represent?

A. Acceleration
B. Time
C. Displacement
D. Velocity

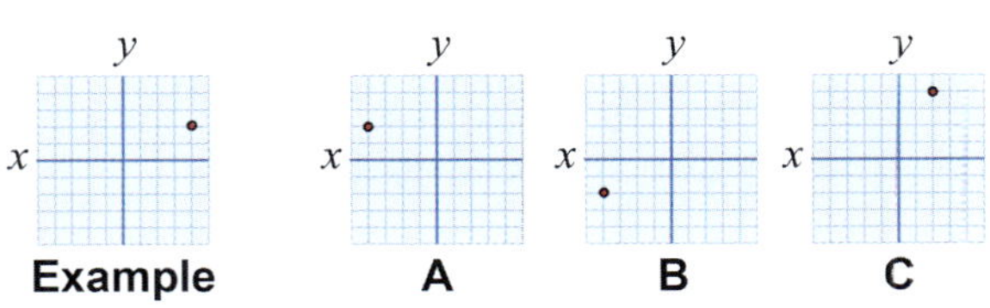


104. Which of the three (A, B, or C) points has its $x$ and $y$ coordinates swapped from the example?

A. Graph A
B. Graph B
C. Graph C

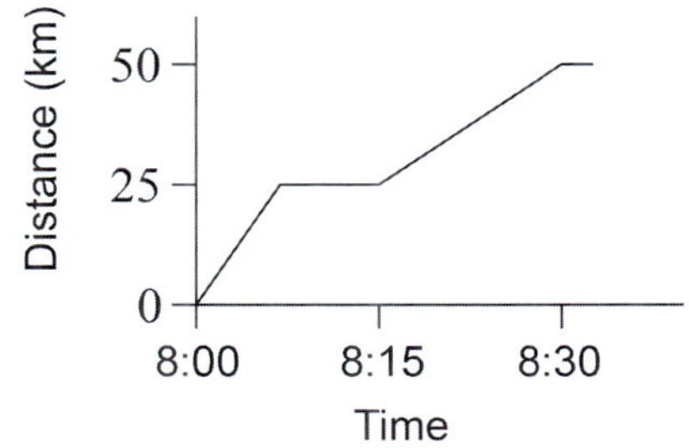


105. Calico is driving to work. Which of the following is a possible explanation of the position versus time graph of her drive?

A. Halfway to work, she stopped for breakfast. After eating, she realized she was late and sped the rest of the drive.
B. After going halfway, she had to return home to get something she forgot.
C. Halfway to work, Calico saw a police officer on the highway and slowed down. She completed her drive without incident.
D. Halfway to work, Calico was pulled over for speeding. After being issued a ticket, she drove the speed limit the rest of the way.

106. An ant starts at −5 m and moves 10 m left, 25 m right, 30 m left, and 5 m right. If the positive $x$-axis is to the right, what is the final position and total displacement of the ant?

A. position: −15 m; displacement: 70 m
B. position: 5 m; displacement: 10 m
C. position: 5 m; displacement: −70 m
D. position: −15 m; displacement: −10 m

107. Anand gets into an elevator and travels 50 m straight up in a time of 5 s. He then travels in the opposite direction, straight down, for 70 m in a time of 6 s. What is his average speed and velocity, respectively, over the 11 s period?

A. 10.9 m/s, 1.82 m/s
B. 10.9 m/s, −1.82 m/s
C. 21.8 m/s, 10.9 m/s
D. −10.9 m/s, 5.2 m/s

108. Which of the following is not a property of displacement?

A. Displacement can be negative.
B. Displacement describes a change in position.
C. Displacement is equal to the total distance traveled by an object in motion.
D. Displacement is equal to the final position of an object minus the initial position of the object.

# Chapter 4 Acceleration

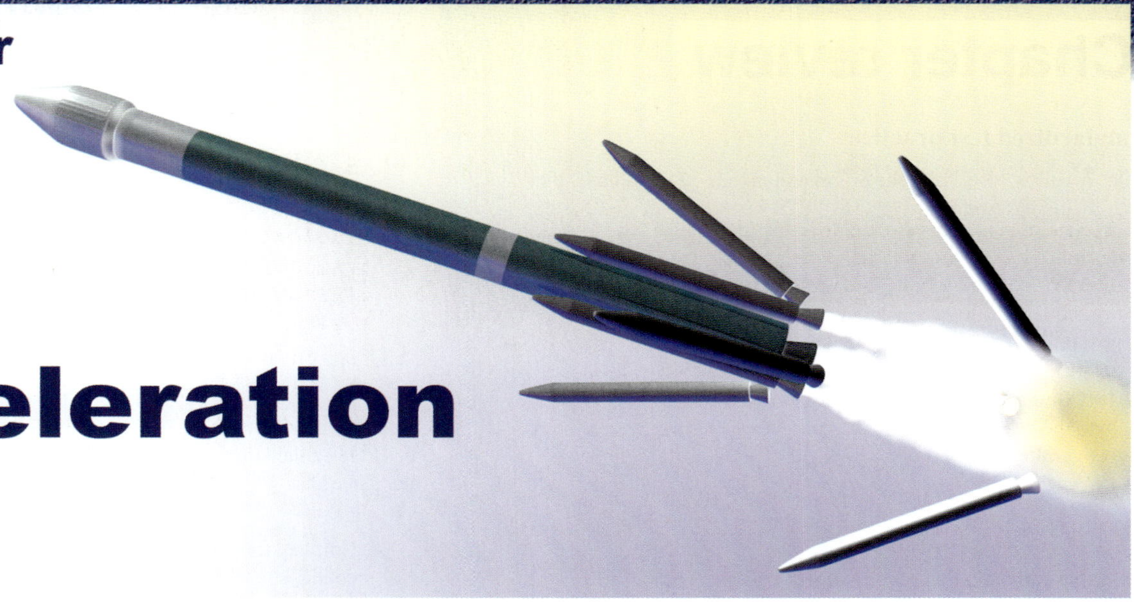

The stuff of legend, *g*-forces are endured by fighter pilots, astronauts, and roller-coaster riders. In a sense, however, they should be called *g-accelerations.* Acceleration is simply the rate at which your velocity changes. What does that feel like, and what can it do to you?

When described by car salespeople, acceleration is described in terms such as "zero to sixty [mph] in seven seconds." But "*g*'s" are a better measure. After all, when you drop a heavy book, the quantity *g* tells us how quickly it speeds up. And "*g*'s" tell us how heavy we feel—whether due to gravity or from being pressed against the seat in a fighter jet dodging enemy fire. At 2*g* we feel twice as heavy as normal; at 10*g* we feel *ten* times that force—enough to make many people lose consciousness. (In fact, *g*-forces first came to widespread attention because numerous WW I pilots passed out during dogfights.)

The term "*g*-force" usually makes one think of space travel, because an astronaut must achieve orbital speed within minutes of takeoff. In the 1950s and 1960s, U.S. astronaut candidates had to ride the Johnsville Centrifuge: a small metal capsule at the end of a 50 ft (15 m) rotor. Candidates briefly endured accelerations as great as 30*g*—similar to the sudden velocity change that you might experience in a car crash without an air bag. Requirements were stringent: Astronauts in the U.S. Mercury program would have to endure roughly 8*g* during takeoff and re-entry. Many capable pilots who had a history of heart trouble failed to qualify for space flight.

By contrast, Space Shuttle astronauts barely exceeded 2*g* during their missions. Your head accelerates more rapidly when you sneeze (about 3*g*), and your body undergoes the occasional 3*g* or 4*g* acceleration when you ride a typical roller-coaster. Sneezes and roller-coaster turns, however, are very short events. We still don't know how people will respond after exposure to even a few "*g*'s" over the long time periods that interplanetary space travel would require.

# Chapter study guide

## Chapter summary

This chapter takes the variables of position, time, velocity, and acceleration and weaves them together in a mathematical model that describes linear motion. This model is versatile and accurate enough to be used for everything from navigating submarines and spacecraft to moving toy robots and driverless vehicles. The model starts with the graphs of position versus time and velocity versus time. The concept of *slope* is applied to both graphs to derive two equations that describe position and speed in accelerated motion. These equations are then applied to free fall and other common situations.

## Learning objectives

By the end of this chapter you should be able to

- create and interpret $x$ vs. $t$ graphs for uniformly accelerated motion;
- create and interpret $v$ vs. $t$ graphs for uniformly accelerated motion;
- solve problems in one-dimensional motion involving position, velocity, time, and constant acceleration;
- solve a one-dimensional motion problem with two equations and two unknowns; and
- calculate speed and position for bodies in free fall.

## Investigations

4A: Acceleration
4B: A model for accelerated motion

## Chapter index



## Important relationships

$$a = \frac{\Delta v}{\Delta t}$$

$$v = v_0 + at$$

$$x = x_0 + v_0 t + \frac{1}{2}at^2$$

$$x = x_0 + \tfrac{1}{2}(v_0 + v)t$$

$$v^2 = v_0^2 + 2a(x - x_0)$$

## Vocabulary

**acceleration** **quadratic** **free fall**
**terminal velocity**

## 4.1 - Acceleration

How do we describe speeding up or slowing down? What is the difference between slowing down gradually and hitting a brick wall? Both these questions have answers that involve *acceleration.* Acceleration describes how velocity changes. Any change in velocity, including speeding up, slowing down, or turning, creates acceleration.

### What is acceleration?

**How do we describe changes in velocity?**

Almost nothing moves at constant speed for very long in everyday life. Even a car on cruise control speeds up and slows down by small amounts to compensate for hills. How do we describe *changes* in velocity, such as going from rest to moving, or from moving to rest? The answer is the concept of **acceleration**. Acceleration is defined in equation (4.1) as the rate of change of velocity.

**Acceleration** definition

$$a = \frac{\Delta v}{\Delta t} \qquad (4.1)$$

$a$ = acceleration (m/s$^2$)
$\Delta v$ = change in velocity (m/s)
$\Delta t$ = change in time (s)

Acceleration is a crucial concept in the physics of motion because acceleration, not velocity, is the result of applied forces.

**What does acceleration mean?**

Equation (4.1) describes how acceleration is the change in velocity ($\Delta v = v_f - v_i$) divided by the change in time ($\Delta t = t_f - t_i$). For example, if a car starts at rest and is moving at 30 mph, then 10 s later the car's *acceleration* is 3 mph/s, or three miles per hour *per second.* The speedometer increases by 3 mph each second for 10 s. Acceleration is the *rate at which velocity changes.*

**What are the units of acceleration?**

The units of acceleration are units of velocity divided by units of time. For a typical car a convenient unit would be miles per hour per second. A powerful sports car can accelerate from zero to 60 mph in 4 s. The change in velocity is 60 mph. The change in time is 4 s. The acceleration is 15 mph per second.

**What is m/s$^2$?**

Acceleration in this course will usually be expressed in SI units of m/s *per second*, or m/s$^2$. (This is sometimes written as m/s/s or m s$^{-2}$.) One meter per second squared means that the velocity changes by one meter per second each second.

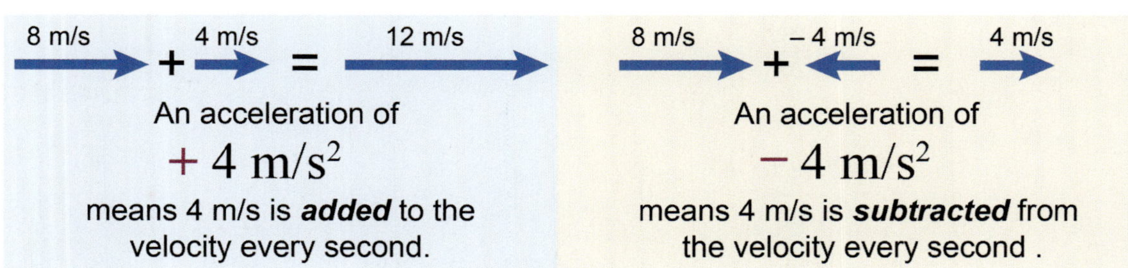


**Positive and negative acceleration**

Acceleration can be positive or negative. For example, an acceleration of +4 m/s$^2$ adds 4 m/s of velocity each second. A car starting from rest would move at 4 m/s after one second, 8 m/s after two seconds, 12 m/s after three seconds, and so on. A *negative* acceleration of –4 m/s$^2$ *subtracts* 4 m/s every second. A car moving at +40 m/s would be moving at 36 m/s after one second, 32 m/s after two seconds, 28 m/s after three seconds, and so on. A negative acceleration is sometimes called a *deceleration.*

**Acceleration on the position vs. time graph**

A moving object with an increasing velocity covers more distance during each new second than it covered in the previous second. That means the *slope* of the $x$ vs. $t$ graph must become steeper over time as the velocity increases. The changing slope is what we recognize as a curve.

*In general, acceleration results in curves on the x vs. t graph.*

**Why is the *x* vs. *t* graph a curve?**

Accelerated motion looks different on the $v$ vs. $t$ and $x$ vs. $t$ graphs. Constant acceleration means that the velocity changes by the same amount every second—which produces a straight line of constant slope on a $v$ vs. $t$ graph but results in a curved line on the $x$ vs. $t$ graph.

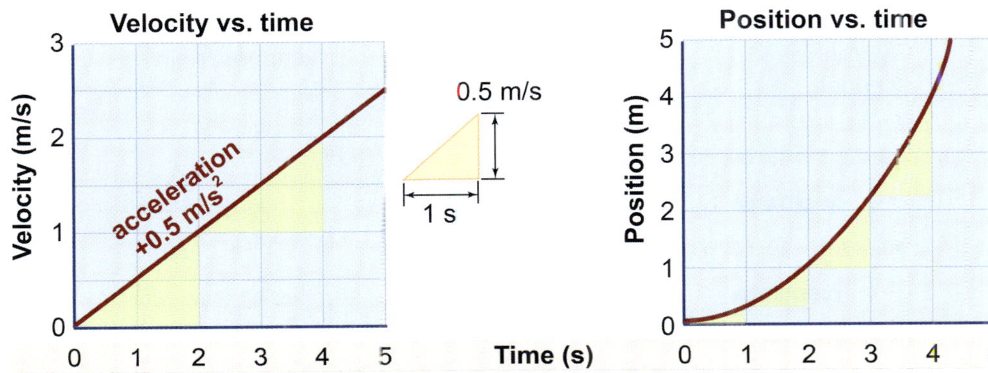


During accelerated motion, the slope of $x$ vs. $t$ changes over time, indicating changing speed.

**Reading acceleration from *v* vs. *t***

The acceleration equals the slope of the velocity versus time graph. In the example above, the velocity increases by 0.5 m/s in 1 s. This is equal to an acceleration of 0.5 meters per second *per second* or 0.5 m/s$^2$. A positive slope on a $v$ vs. $t$ graph means positive acceleration. A negative slope on a $v$ vs. $t$ graph means negative acceleration. *Do not confuse the sign of the slope of the v vs. t graph with the sign of the velocity itself!* You can have negative acceleration with positive velocity and vice versa.

**Constant acceleration versus constant velocity**

Constant velocity, or constant speed, both mean that the acceleration is *zero*. The position changes by the same amount in equal time intervals and the result is a flat $v$ vs. $t$ graph and a constant-slope $x$ vs. $t$ graph. Constant *acceleration* means that the velocity *changes* by an equal amount in equal time intervals. This means a constant-slope $v$ vs. $t$ graph and a curved $x$ vs. $t$ graph.

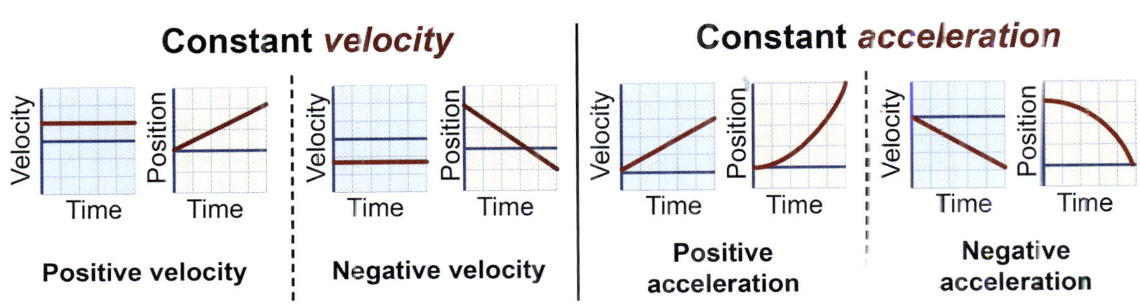


**Graphical model**

Graphical models for motion with constant acceleration are illustrated above and compared to graphical models for motion with constant velocity. You can quickly scan the shapes of these graphs and identify which kind of motion is being modeled in each case.

## Understanding acceleration

**What is acceleration?**

The concept of acceleration is not easy to understand. It took Newton's genius and the invention of a whole new form of mathematics (calculus) to fully define acceleration. Acceleration is important enough to understanding forces and motion that it is worth learning about it more deeply by discussing it with teachers and peers.

**Teacher-moderated peer discussion**

The diagrams below illustrate the concept of acceleration. Display these diagrams on a projector and discuss statements A–D that accompany each diagram. Use the questions below to focus the discussion in groups and with the teacher.

**A** An acceleration of ±5 mph/s means the speed changes by ±5 mph every second.

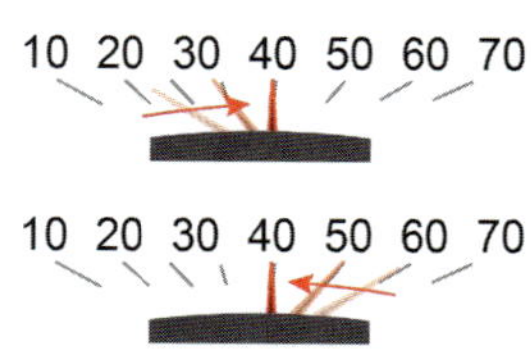


**B** Acceleration is the **rate** at which velocity changes.

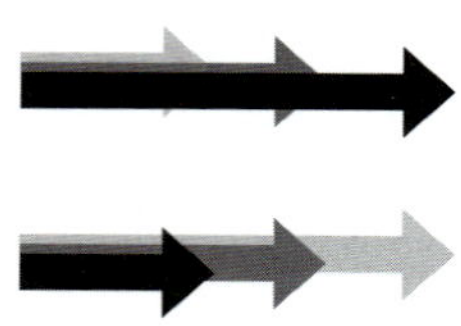

**C** Acceleration is the change in velocity divided by the change in time.

$$a = \frac{\Delta v}{\Delta t}$$

**D** Acceleration is the slope of the velocity versus time graph.

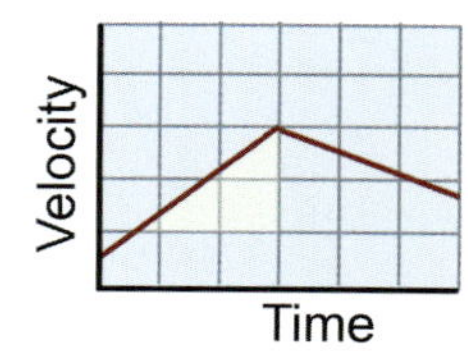


**Supporting questions for discussion**

1. How does statement A relate to the diagram below it?
   What does "+/− 5 mph/s" mean?
2. What does the word "rate" mean in the context of statement B? How do the arrows below represent the idea of a rate of change of velocity?
3. How does the equation represent statement C? Answer by translating each symbol and operation. For example the horizontal line means "divided by."
   Give numerical examples of a change in velocity divided by a change in time.
4. How does the diagram in D represent the text of statement D? What does the shaded triangle represent? What does it mean that the lines on the velocity versus time graph in diagram D go up, then down? How is that reflected in the concept of acceleration?

**Stories help to explain concepts**

Describe a personal experience that relates to diagram A using the word "acceleration." Your story should include both the upper and lower parts of the diagram. The following story provides an example.

*"If I push down on the gas pedal on a level road the car accelerates by about five miles per hour every second. From a stop it takes 12 s to get to 60 mph. If I hit the brakes gently, I can decrease the speed back to zero over 12 s, for an acceleration of −5 miles per hour per second."*

**Can $v = 0$ while $a \neq 0$?**

Acceleration is the rate at which velocity changes. Even when an object has zero velocity, it can still have a nonzero acceleration. Consider dropping a ball from your hand. The ball is not moving initially, and the instant your fingers release the ball it still has zero velocity. One tenth of a second later the ball is moving downward at −0.98 m/s. That means that the ball was accelerating from the moment it was released, even though its velocity was zero at that instant.

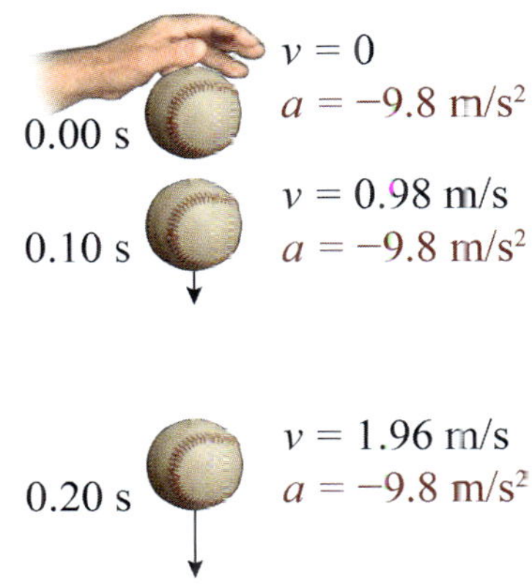


**Beware of signs!**

*Be very careful interpreting the signs of velocity and acceleration.* Many problems define positive as the expected direction of motion, such as *down* a ramp. In this case the acceleration of a car released from rest is also positive. Now think about what happens if the car is initially moving *up* the ramp. Is the acceleration still positive? How do we explain that the car moves upward, turns around, and rolls back down again?

**An uphill example**

The car in the diagram below has an initial upward velocity of −1 m/s. The constant downward acceleration adds +0.5 m/s to the velocity every second. The car's velocity starts negative then becomes 0.5 m/s *more positive* each second until $v = 0$. *At the car's highest point its velocity is zero.* After the turn-around, acceleration and velocity point in the same direction. Acceleration remains a constant +0.5 m/s$^2$ even though the sign of the velocity changes!

**Things to notice ...**

- The cart starts with an initial velocity of −1.0 m/s up the ramp.
- The acceleration is a constant +0.5 m/s$^2$ down the ramp.
- The position vs. time graph is curved when there is acceleration.

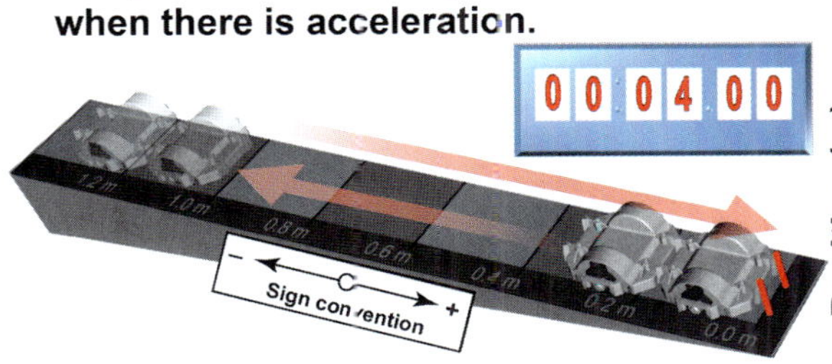


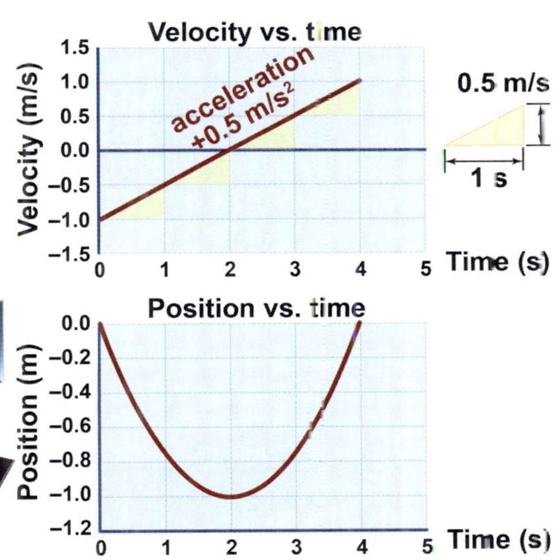


**When will motion reverse?**

Situations in which an object reverses direction always involve an acceleration that has the opposite sign from that of the velocity. For example, an acceleration of −1 m/s$^2$ adds −1 m/s to the velocity every second, so an initial positive velocity of +3 m/s becomes +2 m/s after one second, and then +1 m/s, 0 m/s, −1 m/s, etc. *Objects reverse direction when their velocity becomes zero and then change sign.* This fact is useful in solving many physics problems and in interpreting graphs.

## Determining acceleration

**How is acceleration determined?**

The acceleration of a moving object depends on two factors:

1. the amount the object's speed changes ($\Delta v$) and
2. the time over which the change occurs ($\Delta t$).

Unless it is otherwise stated, you should assume that the acceleration in a physics problem is *constant.* Thus the speed changes by the same amount every second.

**Calculating acceleration**

What is the acceleration of a cart that rolls down a hill if it starts at rest and reaches a speed of 1.2 m/s after 0.6 s?

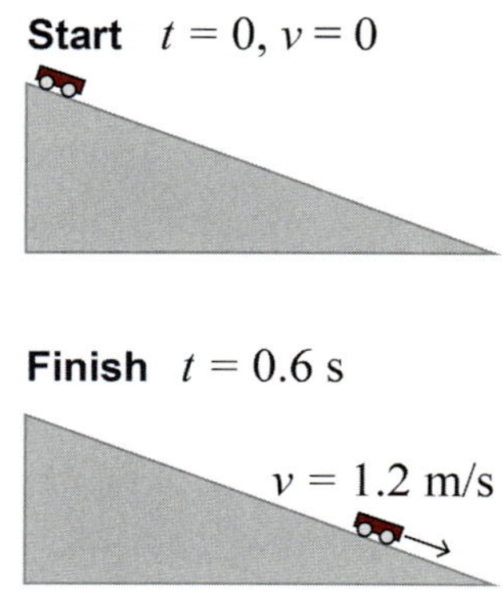


| | |
|---|---|
| *Asked:* | acceleration $a$ |
| *Given:* | change in speed of $\Delta v = 1.2$ m/s, interval of time $\Delta t = 0.6$ s |
| *Relationships:* | $a = \Delta v/\Delta t$ |
| *Solution:* | $a = 1.2 \text{ m/s} \div 0.6 \text{ s} = 2 \text{ m/s}^2$ |

**Calculating speed**

Another type of problem asks for the speed of an object given the acceleration and time. The change in speed $\Delta v$ for an accelerated object is

$$\Delta v = a\Delta t$$

What is the speed of an object that starts from rest and accelerates at a constant 2 m/s$^2$ for 10 s?

| | |
|---|---|
| *Asked:* | final speed $v$ |
| *Given:* | acceleration $a = 2$ m/s$^2$, time interval of $\Delta t = 10$ s, and initial speed $v_0 = 0$ (from rest) |
| *Relationships:* | $a = \Delta v/\Delta t \rightarrow \Delta v = a\Delta t$ |
| *Solution:* | We use the change in speed $\Delta v = v - v_0$ to rewrite the equation as $v - v_0 = a\Delta t \rightarrow v = v_0 + a\Delta t$ Then we calculate the final speed $v = 0 + (2 \text{ m/s}^2 \times 10 \text{ s}) = 20 \text{ m/s}$ |

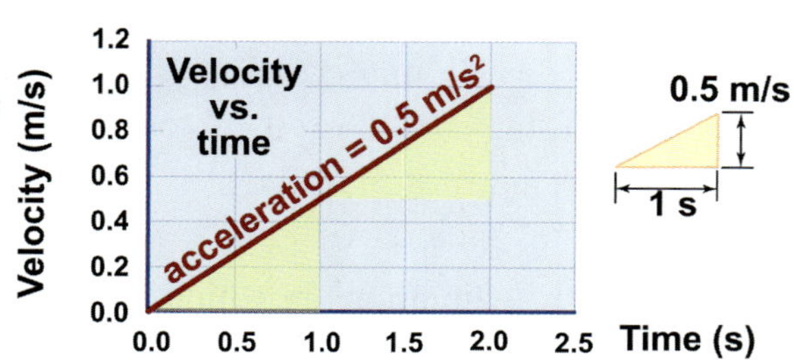


**Acceleration is slope on a $v$ vs. $t$ graph**

Acceleration causes a nonzero slope on the velocity versus time graph, because acceleration represents change in velocity over time. Mathematically, the slope of a $v$ vs. $t$ graph is the change in velocity divided by the change in time, which is the definition of acceleration. In the example above, velocity increases at a steady rate of 0.5 m/s each second, producing a straight-line graph with a slope of 0.5 m/s$^2$.

# Investigation 4A  Acceleration

**Essential questions** *What is acceleration and how does it relate to speed and velocity?*

A car rolling down a ramp accelerates. A car given an initial velocity *up* a ramp accelerates at the same rate! The Smart Cart allows us to quantitatively investigate acceleration and its relationship to speed and velocity.

## Part 1: Acceleration down a ramp

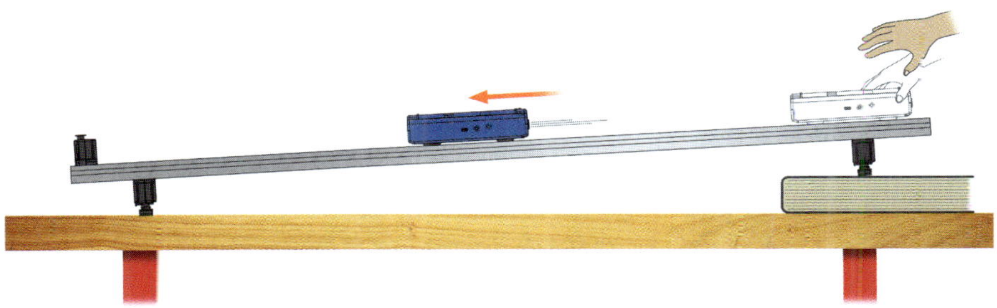

1. Open the experiment file **04A_Acceleration**, and then power-on the Smart Cart and connect it wirelessly to your software.
2. Set up the track at a low angle. Attach the End Stop at the bottom to catch the Smart Cart.
3. Hold the Smart Cart at the top of the ramp, Velcro® facing up the ramp. Begin collecting data and release the Smart Cart. Stop data collection just after the cart reaches the bottom of the track.

a. *Determine the acceleration in m/s² from the slope of the velocity graph. Does this value match with the acceleration recorded in the acceleration versus time graph?*

b. *What is the total elapsed time from the moment you release the Smart Cart until it reaches the bottom of the ramp?*

c. *Calculate the expected final velocity using acceleration from part (a) and time from part (b). Show your work, including units. What is the final velocity recorded on the velocity versus time graph?*

d. *Change the ramp angle and run the experiment again. Explain in one sentence the effect of increasing the angle on the acceleration of the Smart Cart.*

e. *How can you infer changes in acceleration from the Smart Cart velocity vs. time graph?*

## Part 2: Comparing acceleration up and down a ramp

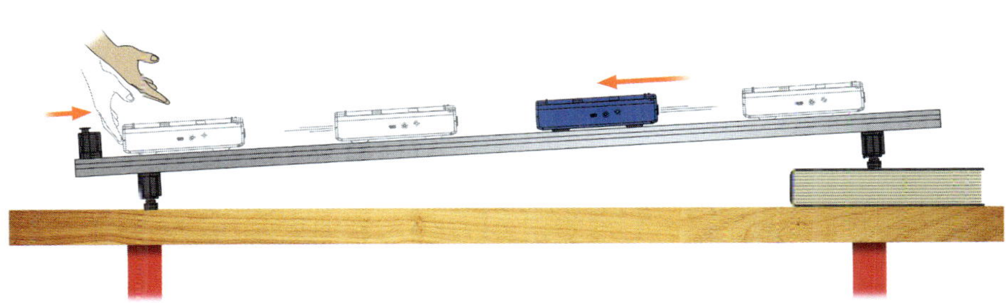

1. Reset the ramp angle to its original value. Set the Smart Cart at the bottom of the track, Velcro still facing up the ramp.
2. Start recording data. Give the Smart Cart a push so it rolls up the track and back down again. Stop recording data just before the cart reaches the bottom of the track.
3. Observe the velocity and acceleration graphs during a complete round trip.

a. *Describe the velocity during the motion. Does the velocity change sign?*

b. *At what point in the motion is velocity zero? How is this shown graphically?*

c. *Describe the acceleration during the motion. Does the acceleration change sign?*

d. *Determine the acceleration in m/s² from the slope of the velocity graph. Show data and units.*

e. *Compare the acceleration from Part 2 to the acceleration you calculated in Part 1. Do the accelerations have the same magnitude? The same sign? Explain.*

## Velocity in accelerated motion

**What is the velocity in accelerated motion?**

The equation defining acceleration can be rewritten as a *model* that describes how the velocity changes over time. The result is equation (4.2), which gives the instantaneous velocity $v$ at time $t$ under the assumption that the acceleration $a$ is constant.

Interactive equation

**Velocity**
for constant acceleration

**(4.2)** $$v = v_0 + at$$

$v$ = velocity (m/s)
$v_0$ = initial velocity (m/s)
$a$ = acceleration (m/s$^2$)
$t$ = time (s)

In this equation we set $\Delta t = t$ by assuming the motion starts at $t = 0$. The initial velocity $v_0$ is the velocity when $t = 0$.

**Determine the velocity**

A cart traveling at 1 m/s reaches a hill and accelerates down the hill at 0.5 m/s$^2$. What is the velocity of the cart 3 s after it starts accelerating?

| | |
|---|---|
| *Asked:* | *instantaneous* velocity $v$ |
| *Given:* | initial velocity of $v_0 = 1$ m/s, acceleration $a = 0.5$ m/s$^2$, and time $t = 3.0$ s |
| *Relationships:* | $v = v_0 + at$ |
| *Solution:* | $v = 1\text{ m/s} + (0.5\text{ m/s}^2)(3.0\text{ s})$ |
| | $= 2.5$ m/s |

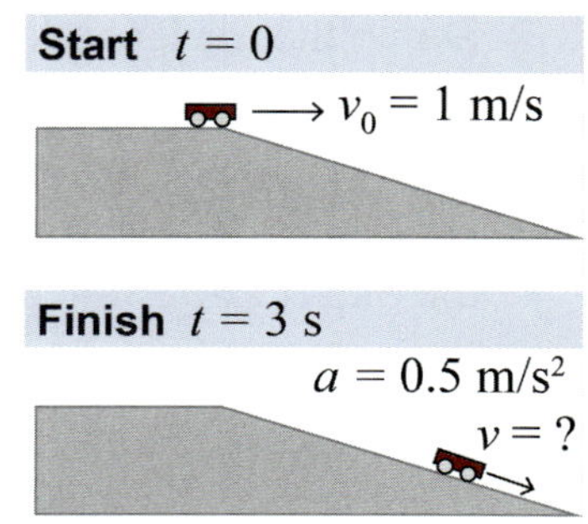


**Think about the sign of acceleration!**

In the above example the velocity and acceleration are in the same direction—both are positive. The speed increases from 1 to 2.5 m/s. Acceleration can also *decrease* an object's speed when the sign of the velocity is different from the sign of the acceleration. For example, an acceleration of −1 m/s$^2$ adds −1 m/s to the velocity each second. This would decrease a positive velocity. It would also make a negative velocity more negative—that is, faster in the negative direction.

**Acceleration opposite to velocity**

A cart traveling at 2 m/s along a level surface reaches an upward sloping hill and accelerates at −0.5 m/s$^2$. What is the velocity of the cart 3 s after it starts climbing the hill?

| | |
|---|---|
| *Asked:* | *instantaneous* velocity $v$ |
| *Given:* | initial velocity of $v_0 = 2$ m/s, acceleration of $a = -0.5$ m/s$^2$, and time of $t = 3$ s |
| *Relationships:* | $v = v_0 + at$ |
| *Solution:* | $v = 2\text{ m/s} + (-0.5\text{ m/s}^2)(3.0\text{ s})$ |
| | $= 0.5$ m/s |

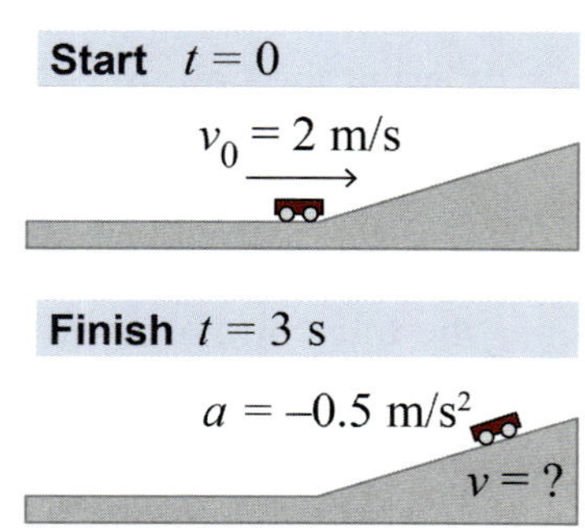


**Understanding equation (4.2)**

The model of motion given by equation (4.2) applies to the *instantaneous* velocity at time $t$. Translated to an English sentence the equation tells us that the velocity $v$ at time $t$ is the initial velocity $v_0$ plus the change in velocity due to acceleration $a$ applied every second for $t$ seconds. The assumption of constant acceleration means that the change in velocity each second is the same.

**How does position change with acceleration?**

The last step in building a model for motion is to develop a single equation that relates position, velocity, time, and acceleration. Consider a moving object with initial velocity $v_0$ that undergoes constant acceleration. At time $t$, the velocity has increased from $v_0$ to $v$. The distance the object travels between time $t = 0$ and time $t$ is the area shaded on the graph.

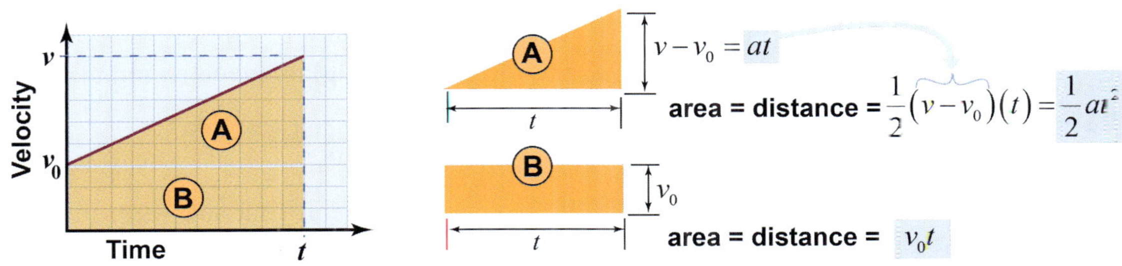


**The total distance traveled is the sum of the two areas.** $d = v_0 t + \frac{1}{2}at^2$

**How do we calculate the distance?**

The $v$ vs. $t$ graph breaks down into two shapes. The area of a triangle is ½ base × height. For triangle A this is $\frac{1}{2}(v - v_0)t$. We also know that the change in velocity is acceleration × time, so $v - v_0 = at$. Therefore, the area of triangle A is $\frac{1}{2}at^2$. The area of rectangle B is $v_0 t$. Area on a $v$ vs. $t$ graph equals distance, and adding the triangle to the rectangle gives us this result

$$d = v_0 t + \frac{1}{2}at^2$$

**Deriving the position from the distance**

One last step remains. The distance traveled is $d = x - x_0$. Substituting this expression yields equation (4.3), which relates position $x$ at any time $t$ to initial position $x_0$, initial velocity $v_0$, and acceleration $a$.

Interactive equation

**Position**
accelerated motion

$$x = x_0 + v_0 t + \frac{1}{2}at^2 \qquad (4.3)$$

$x$ = position (m)
$x_0$ = initial position (m)
$v_0$ = initial speed (m/s)
$a$ = acceleration (m/s$^2$)
$t$ = time (s)

**What does the equation mean?**

Equation (4.3) has three terms on the right-hand side, and each term has its own meaning. The first term is the initial position. The second term is the change in position the object would have had if it continued at constant initial speed $v_0$. The third term is the additional change in position resulting from changes in speed that come from acceleration. Note that, if the acceleration is zero, we get back $x = x_0 + vt$, the equation for constant velocity from the last chapter!

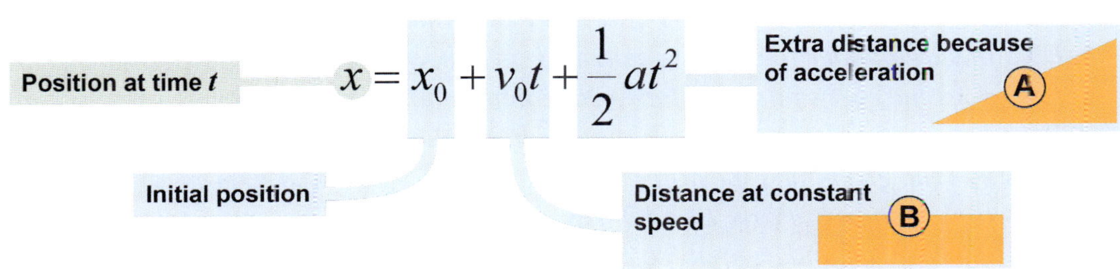

# Investigation 4B A model for accelerated motion

**Essential questions** *How can a velocity versus time graph be used to determine displacement?*

Interactive simulation

An object's position changes as it accelerates. Just like a moving object's position versus time graph can communicate information about its velocity and acceleration, graphs of velocity can communicate information about how an accelerating object's position is changing, given its initial velocity and position.

## Part 1: Determining displacement from motion graphs

1. Open the **04B_ModelsAcceleratedMotion** experiment file in your software, and then connect your Smart Cart using Bluetooth.
2. Set up your equipment like the picture with the plunger on the cart facing the end stop.

3. Push-in the cart's plunger two clicks, and then set the cart at the bottom of the track with the plunger resting against the end stop.
4. Start recording data, and then launch the cart up the track. Stop recording data just before the cart starts rolling back down the track.

*a. Sketch a copy of your graph and indicate which curve is velocity and which is position.*
*b. How can you tell from your graph that the cart was accelerating?*
*c. Use the area tool in your software to find the area under your velocity time graph.*
*d. Find the value of the cart's position when the last velocity data point was recorded.*
*e. If $x_i = 0$ m at the bottom of the track, what was the cart's total displacement $\Delta x$? How does the cart's total displacement compare to the area under its velocity-time graph?*

## Part 2: Extending to other graphs

1. Hide your Part 1 data so the graph is blank.
2. Repeat the same Part 1 procedure, except this time, after you launch the cart, let the cart bounce off the end stop several times. Record data for about 8 seconds as the cart bounces.

*a. Sketch a copy of your new graph and indicate which curve is velocity and which is position.*
*b. Find the area under the new velocity time graph, and the total displacement of the cart.*
*c. How does the cart's total displacement compare to the area under its velocity time graph?*
*d. Based on your results in Parts 1 and 2, what is the relationship between the area under an object's velocity time graph and the object's displacement?*
*e. The cart's motion in Parts 1 and 2 was mostly uniform (constant acceleration). Does the relationship between the area under an object's velocity time graph and the object's displacement extend to more erratic/random motion (variable acceleration)? Test the relationship by recording a third run of erratic/random velocity data, rolling the cart back-and-forth with your hand. Explain your results.*

### The equations of motion

We now have two equations that together make up a mathematical model of motion that includes position, velocity, acceleration, and time. Equation (4.2) describes velocity and equation (4.3) describes position.

| A model for accelerated motion | |
|---|---|
| (4.2) $v = v_0 + at$ | (4.3) $x = x_0 + v_0 t + \frac{1}{2}at^2$ |

### Using the equations to find distance

Some problems require both equations. For example, consider a car that starts from rest with a constant acceleration of 5 m/s$^2$. How far does the car go before it reaches a speed of 30 m/s (67 mph)? This problem asks for a distance but it gives you speed and acceleration. There are two unknown values: time and position. Therefore, to solve the problem you need both equations.

1. Equation (4.2) allows you to solve for the time in terms of the known final velocity and acceleration.
2. Equation (4.3) gives you the position.

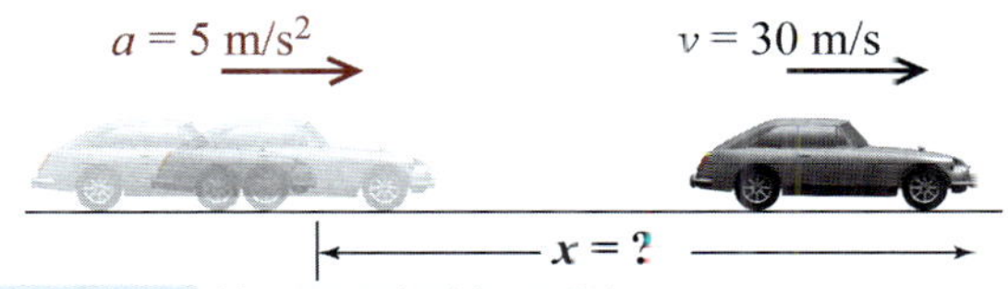


| | |
|---|---|
| Asked | You are asked for a distance. |
| Given | You are given final speed and acceleration. Assume initial speed and position are zero. |
| Relationships | (4.2) $v = v_0 + at$ (4.3) $x = x_0 + v_0 t + \frac{1}{2}at^2$ |
| Solution | From equation (4.2) $v - v_0 = at \rightarrow v = at \rightarrow t = \frac{v}{a}$ From equation (4.3) $x = \frac{1}{2}at^2 \rightarrow x = \frac{1}{2}a\left(\frac{v}{a}\right)^2 \rightarrow x = \frac{v^2}{2a}$ Substitute the values to get $x = \frac{v^2}{2a} = \frac{(30 \text{ m/s})^2}{2(5 \text{ m/s}^2)} = 90 \text{ m}$ |

### Using the equations to find acceleration

A car starts at rest at the top of a ramp. What is the acceleration if the car goes 1.4 m down the ramp in 1.6 s?

1. You are given position, time, and an initial velocity of zero. You only need equation (4.3) since there is only one unknown variable.
2. Solve the equation for acceleration to find that $a$ = 1.09 m/s$^2$.

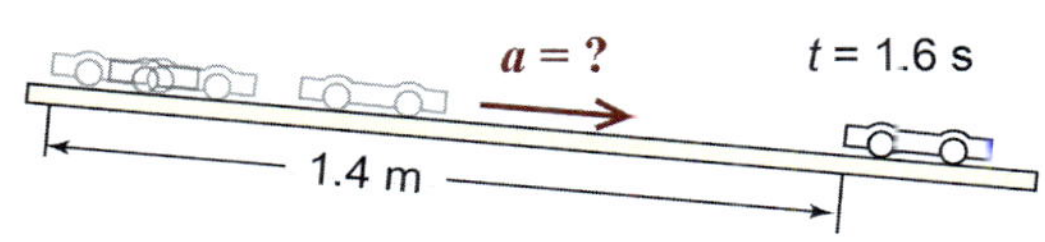


| | |
|---|---|
| Asked | You are asked for the acceleration. |
| Given | You are given the final position and time. Initial speed and initial position are zero. |
| Relationships | (4.3) $x = x_0 + v_0 t + \frac{1}{2}at^2$ |
| Solution | From equation (4.3) $x = \frac{1}{2}at^2 \rightarrow a = \frac{2x}{t^2} = \frac{2(1.4 \text{ m})}{(1.6 \text{ s})^2} = 1.09 \text{ m/s}^2$ |

### Using the equations to find time

A car starts from rest at the top of a ramp that creates an acceleration of 2.1 m/s$^2$. How much time does it take the car to travel 1 m?

1. You only need equation (4.3) since there is only one unknown variable.
2. Solve the equation for acceleration and substitute values to find that $t$ = 0.98 s.

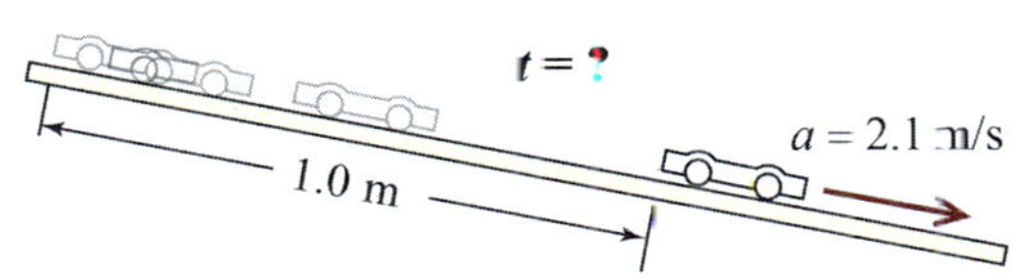


| | |
|---|---|
| Asked | You are asked for the time. |
| Given | You are given the distance traveled and the acceleration. Initial speed and position are zero. |
| Relationships | (4.3) $x = x_0 + v_0 t + \frac{1}{2}at^2$ |
| Solution | From equation (4.3) $x = \frac{1}{2}at^2 \rightarrow t = \sqrt{\frac{2x}{a}} = \sqrt{\frac{2(1.0 \text{ m})}{2.1 \text{ m/s}^2}} = 0.98 \text{ s}$ |

# Four equations of motion

**Why do we need two more equations?**

For constant acceleration, you have now learned two equations of motion: equations (4.2) and (4.3). These two equations can be used to solve a wide variety of motion problems in physics. In some problems, however, it might be more convenient to use two additional equations. These two additional equations aren't unique or different from the other ones, they are just more convenient to use in certain circumstances. As you will see below, we get the two new equations by manipulating the two equations you already know!

**Deriving the third equation**

Start with equation (4.3):

$$x = x_0 + v_0 t + \frac{1}{2}at^2$$

Assuming that we are starting at an initial time $t = 0$, acceleration is the change in velocity divided by the change in time or $a = (v - v_0)/t$. Use this equation to substitute for acceleration in equation (4.3) to get

$$\begin{aligned} x &= x_0 + v_0 t + \tfrac{1}{2}\left(\tfrac{v - v_0}{t}\right)t^2 \\ &= x_0 + v_0 t + \tfrac{1}{2}vt - \tfrac{1}{2}v_0 t \\ &= x_0 + \tfrac{1}{2}(v_0 + v)t \end{aligned}$$

This third equation of motion is useful when you don't know the acceleration.

| | | | |
|---|---|---|---|
| **(4.4)** | $x = x_0 + \frac{1}{2}(v_0 + v)t$ | $x$ = position (m)<br>$x_0$ = initial position (m/s)<br>$v_0$ = initial velocity (m/s)<br>$v$ = velocity (m/s)<br>$t$ = time (s) | **Position** when acceleration is not known |

**Deriving the fourth equation**

For the fourth equation of motion, start with equation (4.4) above, but rewrite it as

$$x - x_0 = \tfrac{1}{2}(v_0 + v)t \quad \Rightarrow \quad 2(x - x_0) = (v_0 + v)t$$

This time we want to remove the dependence on time. Rewrite the definition of acceleration to solve for time

$$a = \tfrac{v - v_0}{t} \quad \Rightarrow \quad t = \tfrac{v - v_0}{a}$$

and substitute this equation for acceleration to get

$$2(x - x_0) = (v_0 + v)\left(\tfrac{v - v_0}{a}\right) \quad \Rightarrow \quad 2(x - x_0) = \tfrac{v^2 - v_0^2}{a} \quad \Rightarrow \quad 2a(x - x_0) = v^2 - v_0^2$$

Rearranging these terms leads to equation (4.5), which is useful for solving problems when you don't know the time.

| | | | |
|---|---|---|---|
| **(4.5)** | $v^2 = v_0^2 + 2a(x - x_0)$ | $v$ = velocity (m/s)<br>$v_0$ = initial velocity (m/s)<br>$a$ = acceleration (m/s$^2$)<br>$x$ = position (m)<br>$x_0$ = initial position (m) | **Velocity** when time is not known |

Equations (4.2), (4.3), (4.4), and (4.5) are the four equations of motion.

| Equation | Initial conditions | Quantities you need to know or solve for | Quantities you *don't* need to know |
|---|---|---|---|
| (4.2) $v = v_0 + at$ | $v_0$ | $v$ $a$ $t$ | $x_0, x$ initial, final position |
| (4.3) $x = x_0 + v_0 t + \frac{1}{2}at^2$ | $x_0$ $v_0$ | $x$ $a$ $t$ | $v$ final velocity |
| (4.4) $x = x_0 + \frac{1}{2}(v + v_0)t$ | $x_0$ $v_0$ | $x$ $v$ $t$ | $a$ acceleration |
| (4.5) $v^2 = v_0^2 + 2a(x - x_0)$ | $x_0$ $v_0$ | $x$ $v$ $a$ | $t$ time |

**Understanding the four equations**

Equations (4.2), (4.3), (4.4), and (4.5) are useful for solving many physics problems. Each equation is a little bit different in terms of which quantities it uses—and which quantity it omits. Examine the table above to see how the first equation does not include a variable for position (either $x$ or $x_0$), the second equation does not include final velocity $v$, the third equation does not include acceleration $a$, and the fourth equation does not use time $t$.

**Which equation should I use?**

*Most motion problems have one quantity that you don't need to know.*

1. Write down the given quantities in the problem.
2. Write down the quantity for which you are asked to solve.
3. Determine which quantity you don't need to know.
4. Choose the equation based on #3.

**Determine the stopping time**

A car is moving at 20 m/s when the driver brakes and comes to a stop after traveling 30 m. If the car's deceleration were constant, how long did it take for the car to stop?

*Asked:* time $t$

*Given:* initial velocity of $v_0 = 20$ m/s, final velocity $v = 0$ m/s (comes to a stop), and displacement $x - x_0 = 30$ m

*Relationships:*

$$x - x_0 = \frac{1}{2}(v_0 + v)t$$

*Solution:* Solve the equation for time:

$$t = 2\left(\frac{x - x_0}{v_0 + v}\right) = 2\left(\frac{30\text{ m}}{20\text{ m/s} + 0\text{ m/s}}\right) = 3\text{ s}$$

**Highway on-ramp problem**

A car needs to accelerate from a standstill along a 150 m long on-ramp to merge into traffic moving at 25 m/s. What is the car's acceleration?

*Asked:* acceleration $a$

*Given:* initial velocity of $v_0 = 0$ m/s (standstill), final velocity $v = 25$ m/s, and displacement $x - x_0 = 150$ m

*Relationships:*

$$v^2 = v_0^2 + 2a(x - x_0)$$

*Solution:* Solve the equation for acceleration:

$$a = \frac{v^2 - v_0^2}{2(x - x_0)} = \frac{(25\text{ m/s})^2 - (0\text{ m/s})^2}{2(150\text{ m})} = 2.1\text{ m/s}^2$$

## Quadratic equations

**What it means to be quadratic**

Notice that the position equation has terms involving time and also time squared. Equations that involve a variable squared are called **quadratic** equations. An equation is said to be quadratic in $x$ if $x^2$ appears as the highest power of $x$ in the equation. Quadratic equations are special for two reasons:

1. There are special techniques for solving them.
2. There are *two solutions* for the squared variable.

**Quadratic equations**

| Math | | Physics |
|---|---|---|
| $ax^2 + bx + c = 0$ | | $x = x_0 + v_0 t + \frac{1}{2}at^2$ |
| **Solutions** | | **Solutions** |
| $x = \frac{1}{2a}\left(-b + \sqrt{b^2 - 4ac}\right)$ | $a \leftrightarrow \frac{1}{2}a$ | $t = \frac{1}{a}\left(-v_0 + \sqrt{v_0^2 + 2a(x - x_0)}\right)$ |
| $x = \frac{1}{2a}\left(-b - \sqrt{b^2 - 4ac}\right)$ | $b \leftrightarrow v_0$ | $t = \frac{1}{a}\left(-v_0 - \sqrt{v_0^2 + 2a(x - x_0)}\right)$ |
| | $c \leftrightarrow x_0 - x$ | |

**Understanding the math**

The variables $c$, $b$, and $a$ in the math formula are the coefficients of $x^0$ (= 1), $x^1 = x$, and $x^2$.The two solutions are *true for any values of the variables.* In physics, we will sometimes need the quadratic solutions above; but in some cases we can also use the easier method of *factoring* a quadratic equation to solve for the two solutions.

**An arrow shot directly upward comes back down and hits the ground in 5.0 s. The acceleration of the arrow was 9.8 m/s² downward. Calculate the initial velocity of the arrow.**

| | |
|---|---|
| **Asked** | Height and initial velocity |
| **Given** | Time $t = 5.0$ s |
| **Relationships** | $v = v_0 + at$ $\quad x = x_0 + v_0 t + \frac{1}{2}at^2$ |
| **Solution (setting up)** | First assign up = positive because there is both up and down motion, then use $a = -9.8\text{ m/s}^2$. Next, we can choose $x = x_0 = 0$ since the arrow starts at ground level at time $t = 0$ and finishes at ground level at later time $t$. |

**Solution**

$$x = x_0 + v_0 t + \tfrac{1}{2}at^2 \quad \rightarrow \quad 0 = v_0 t + \tfrac{1}{2}at^2$$

If we factor the $t$ from both terms we get

$$0 = t\left(v_0 + \frac{at}{2}\right)$$

This equation has two solutions:

$$t = 0 \qquad \text{and} \qquad v_0 + \frac{at}{2} = 0$$

From the second solution we find

$$v_0 = -\frac{at}{2} = -\frac{(-9.8\text{ m/s}^2)(5.0\text{ s})}{2} = 24.5\text{ m/s}$$

**Understanding the strategy**

To find the solution we start as usual by defining the initial position as $x_0 = 0$. We are also given that the arrow comes back down to $x_0 = 0$ again 5.0 s after it is fired upward. The position equation becomes a quadratic in time $t$, from which a $t$ can be factored from each term to give this intermediate equation:

$$0 = v_0 t - \tfrac{1}{2}gt^2 \qquad \rightarrow \qquad 0 = t\left(v_0 - \tfrac{gt}{2}\right)$$

There are two ways in which the result can be zero:

$$\text{if} \quad t = 0 \qquad \text{or if} \quad v_0 - \tfrac{gt}{2} = 0$$

The solution on the right gives us what we want, which is the value of $v_0$ when $t = 5.0$ s. The solution on the left is also realistic since it says the arrow was on the ground ($x = 0$) at the start when $t = 0$.

# Section 1 review

Acceleration is the rate at which velocity changes. Its SI unit is meter per second squared, or $m/s^2$. Acceleration is a vector. In one-dimensional motion, acceleration can be positive or negative (as well as zero). The slope of the velocity–time ($v$ vs. $t$) graph is the acceleration. The area under a velocity–time graph equals an accelerating object's displacement.

| Vocabulary words | acceleration, quadratic |
|---|---|

| Key equations | $a = \frac{\Delta v}{\Delta t}$ | $v = v_0 + at$ | $x = x_0 + v_0 t + \frac{1}{2}at^2$ |
|---|---|---|---|
| | $x = x_0 + \frac{1}{2}(v_0 + v)t$ | $v^2 = v_0^2 + 2a(x - x_0)$ | |

## Review problems and questions

1. Two friends each drive a car down Highway 101. During the same 10 s interval, Jodi's car moves at a constant 60 mph and Julie's car transitions from a speed of 20 mph to a speed of 30 mph.
   a. Whose car has the greater acceleration?
   b. Do you have to convert the given quantities into SI units in order to figure out which car has the greater acceleration?

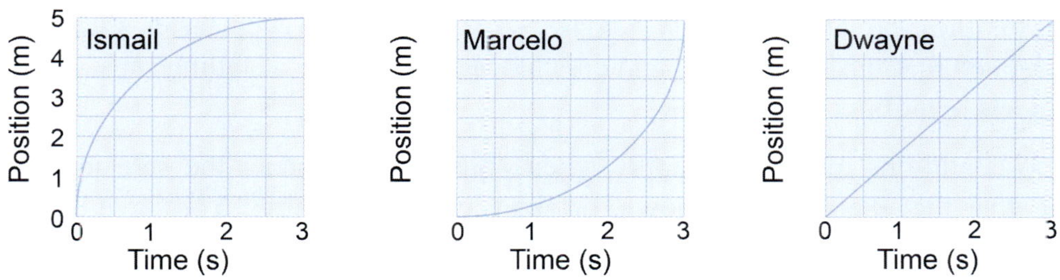


2. The position–time graphs of three sprinters are shown here. Which sprinter best matches each of the following statements?
   a. "I underwent positive acceleration during the three-second interval shown."
   b. "I underwent negative acceleration during the three-second interval shown."
   c. "I underwent zero acceleration during the three-second interval shown."

3. Nikita and Rachel are driving their new cars westward on the highway. At the very same instant, they cross the Weston town line (call it $t = 0$). Nikita's velocity is 20 m/s westward at that moment, while Rachel is going faster at 30 m/s (in the same direction). Each driver is accelerating westward: Nikita at $3\ m/s^2$, and Rachel at $1\ m/s^2$. They maintain these accelerations for 10 s (until $t = 10$ s).

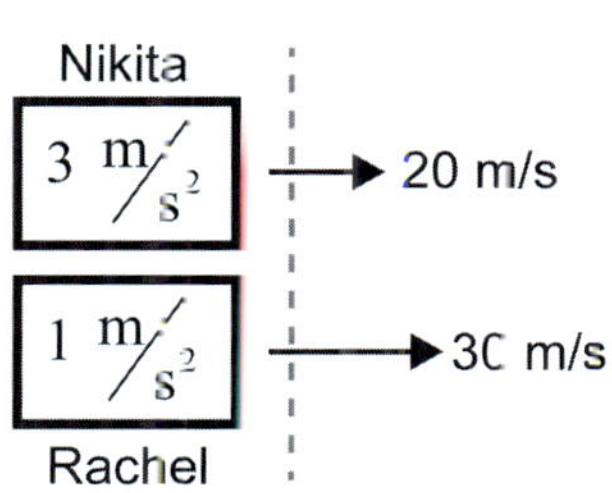


   a. At what time $t$ are Nikita and Rachel equally fast?
   b. At that time (when they are equally fast), who has gone farther from the town line?
   c. After 10 s has passed (that is, at $t = 10$ s ), who is farther from the town line?

## 4.2 - Gravity and free fall

**Free fall** is an idealized situation in which air friction is ignored and objects accelerate at a constant downward 9.8 $m/s^2$. Free fall does not mean objects necessarily move *down*. If you throw a ball upward, the ball is in free fall the moment it leaves your hand because, once you are no longer touching it, gravity is the only force acting on the ball (neglecting friction).

### Free fall

**The acceleration due to gravity**

If friction is negligible, a falling object near Earth's surface accelerates downward at a constant 9.8 $m/s^2$, or "one *g*." For example, after 3 s an object in free fall has a speed of 29.4 m/s, or nearly 66 mph! In real life, air friction typically becomes important at speeds greater than about 10 m/s. Smooth, compact objects such as marbles, however, really do accelerate at 9.8 $m/s^2$ even when dropped several stories.

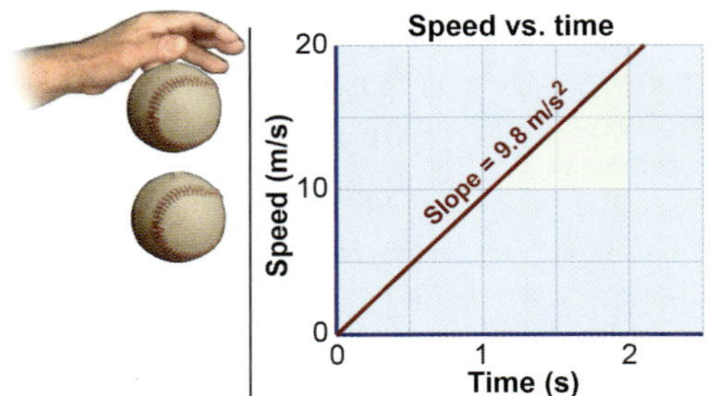

**When friction is negligible, freefalling objects accelerate downward at 9.8 $m/s^2$.**

**Galileo's famous critique of heavier objects falling faster**

The fact that a feather accelerates less than a falling stone was once explained by the hypothesis that heavier objects fall faster than light objects. Galileo critiqued this explanation through logic and was the first to explicitly realize that the true cause was air resistance. Galileo supposed that heavier objects *did* fall faster. Logically then, attaching a lighter object to a heavier one should hold the heavier object back somewhat, since the lighter object's natural rate of falling is less. But at the same time, adding the lighter object makes the heavier object *even heavier*, so it should fall even faster still! Galileo correctly reasoned that this was absurd and concluded that the rate of falling must be the same for all objects regardless of weight.

**Upward motion in free fall**

The conditions of free fall apply when objects move under the sole influence of gravity whether that motion is up, down, or sideways at any angle! You have probably heard the saying "what goes up must come down." We can now *calculate* this. An object launched upward accelerates downward at 9.8 $m/s^2$. If the ball is launched straight up, it comes straight back down. If the ball is launched at an angle, its velocity curves toward the ground by −9.8 m/s in the vertical direction every second.

**Examples of free fall**

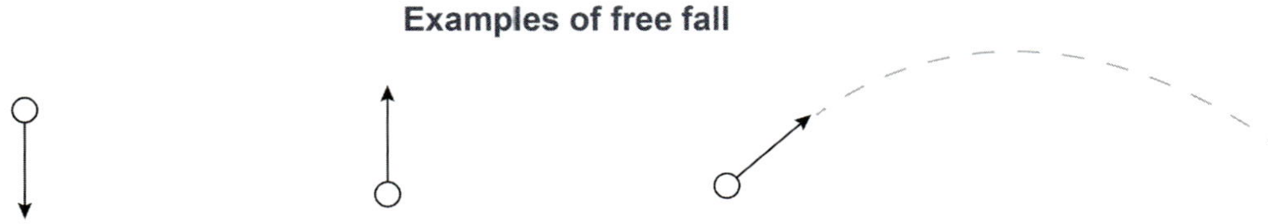

A ball falling down with negligible air friction

A ball thrown up with negligible air friction

A ball launched at any angle, even horizontally, as long as there is negligible air friction

**The sign of $g$**

In all free fall problems you will need to be careful regarding the *sign of g*. The acceleration of gravity is a constant $g = 9.8\ \text{m/s}^2$ *downward*. The value of $g$, however, can be positive or negative depending on how you define your coordinates.

- If *up* is defined to be positive, then $g = -9.8\ \text{m/s}^2$.
- If *down* is defined to be positive, then $g = +9.8\ \text{m/s}^2$.

**An example free fall problem**

Consider a ball that is dropped straight down from rest. Ignoring air friction, gravity causes a constant downward acceleration of $-9.8\ \text{m/s}^2$. Therefore, 9.8 m/s is subtracted from the ball's velocity each second. After 5 s of falling the ball's velocity is −49 m/s. After 10 s the velocity is −98 m/s, which is over 200 miles per hour! Notice that, in this example, negative acceleration causes the *speed* to increase, because both acceleration and velocity are in the same direction. The speed increased even as the velocity became more negative!

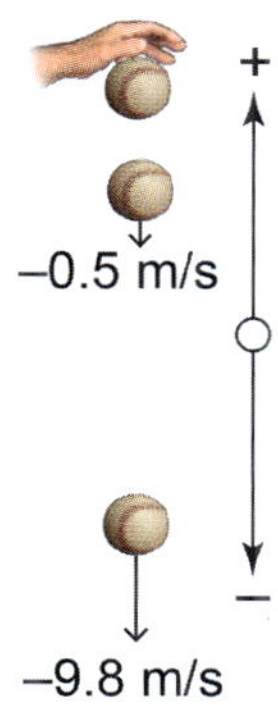


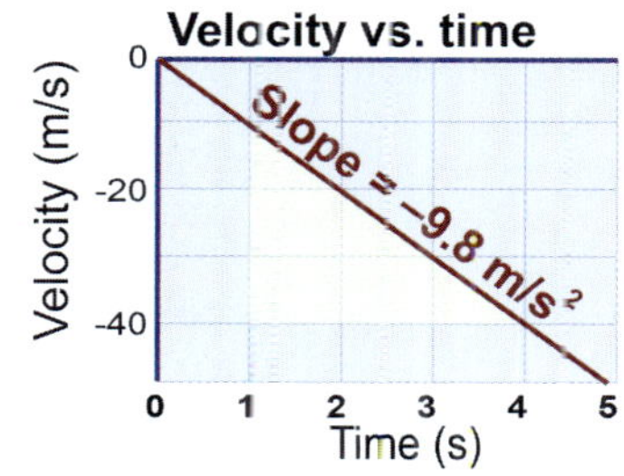


Ignoring friction, objects accelerate at 9.8 m/s² downward.

Model: $v = v_0 + at$

at 5 s

$$v = 0 + (-9.8\ \text{m/s}^2)(5\ \text{s}) = -49\ \text{m/s}$$

at 10 s

$$v = 0 + (-9.8\ \text{m/s}^2)(10\ \text{s}) = -98\ \text{m/s}$$

**Solving free fall problems**

In all free fall problems the first step is to choose coordinates and assign the value of $g$ to be positive or negative. Then rewrite the equations of motion, eliminate terms that are zero, and substitute $g$ for $a$. Finally, work out and execute a strategy for getting an answer.

**Problems that ask for height**

**How high should you drop a ball if you want it to hit the ground in exactly 1 s?**

| | |
|---|---|
| **Asked** | You are asked for a height. |
| **Given** | You are given the time of 1 s. |
| **Relationships** | $x = x_0 + v_0 t + \frac{1}{2}at^2 \quad v = v_0 + at$ |
| **Solution** | First assign down = positive because there isn't any upward motion, so $g = 9.8\ \text{m/s}^2$. |

Eliminate terms, substitute $a = g$, and assign $h$ = height:

$$x_0 = 0\ \text{m},\quad v_0 = 0\ \text{m/s} \quad x = h \Rightarrow h = \tfrac{1}{2}gt^2 = \tfrac{1}{2}(9.8\ \text{m/s}^2)(1\ \text{s})^2 = 4.9\ \text{m}$$

**How far does an object have to fall to reach a speed of 10 m/s (neglecting friction)?**

| | |
|---|---|
| **Asked** | You are asked for a height. |
| **Given** | You are given the final speed of 10 m/s. |
| **Relationships** | $x = x_0 + v_0 t + \frac{1}{2}at^2 \quad v = v_0 + at$ |
| **Solution** | First assign down = positive because there isn't any upward motion, so $g = 9.8\ \text{m/s}^2$. |

Eliminate terms involving $x_0 = v_0 = 0$, let $a = g$, then assign $x = h$:

$$v = gt \Rightarrow t = \frac{v}{g}$$

$$h = \frac{1}{2}gt^2 = \frac{1}{2}g\left(\frac{v}{g}\right)^2 = \frac{v^2}{2g} = \frac{(10\ \text{m/s})^2}{2(9.8\ \text{m/s}^2)} = 5.1\ \text{m}$$

**Initial assumptions**

In both example problems, we assume zero initial position and zero initial velocity. Unless a problem explicitly gives you an initial velocity or height, you should do the same. After eliminating terms involving $x_0$ and $v_0$ we see that the first problem has only one unknown, $h$, and therefore can be solved using only the position equation. The second problem requires both equations since there are two unknowns: $v$ and $t$.

## Free fall problems for objects thrown upward

**How do I model an object thrown upward?**

A ball thrown upward eventually comes back down again. The model for accelerated motion in equations (4.2) and (4.3) allows us to determine position (height) or velocity at all points in the motion:

(4.2) $v = v_0 + at$

(4.3) $x = x_0 + v_0 t + ½at^2$

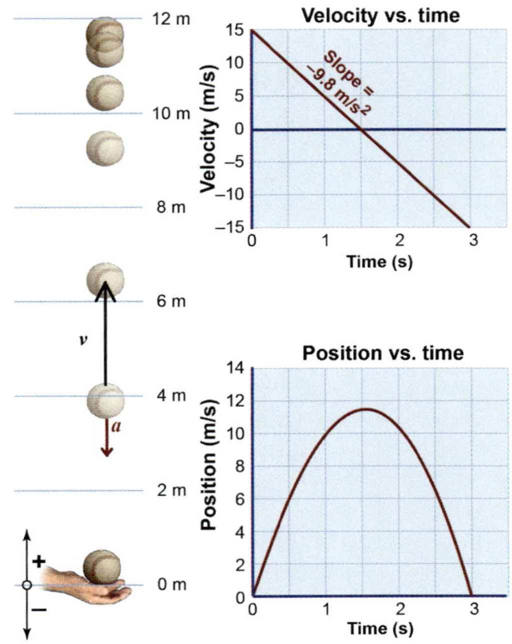


As an example, consider throwing a baseball straight *up* with a velocity of +15 m/s. The acceleration of gravity is –9.8 m/s$^2$, and so the ball's velocity changes by –9.8 m/s every second. After 1 s, the ball has a speed of only 5.2 m/s and by 2 s its velocity has reversed direction and the ball is heading downward at –4.6 m/s.

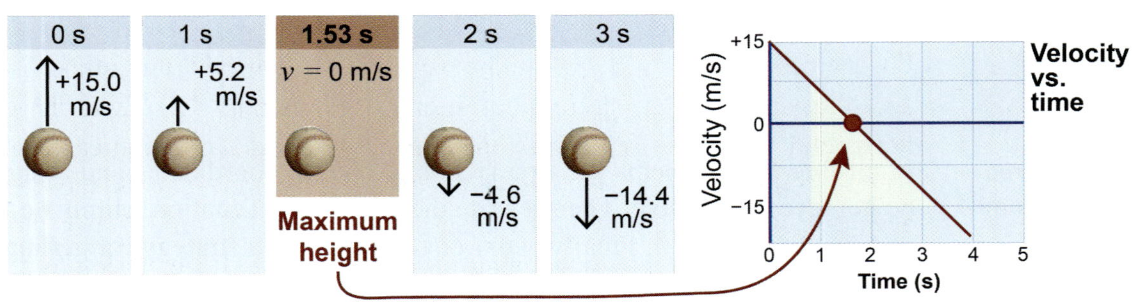


**What is the highest point?**

A common problem asks for the highest point reached when an object is thrown vertically upward. This problem is solved by recognizing that *the vertical velocity is zero at the highest point.* If the vertical velocity were not zero, then the object would continue to move upward. The velocity equation tells you the time it takes to reach a velocity of zero, while the position equation tells you the height the object reaches at that time.

| | |
|---|---|
| *Asked:* | height, which is the final position $x$ |
| *Given:* | initial velocity of 15 m/s, acceleration of –9.8 m/s$^2$ |
| *Relationships:* | $v = v_0 + gt$<br>$x = x_0 + v_0 t + ½gt^2$ |
| *Solution:* | $v = v_0 + at \rightarrow t = v_0/g$ = 15 m/s ÷ 9.8 m/s$^2$ = 1.53 s<br>$x = x_0 + v_0 t + ½gt^2 \rightarrow x$ = (15 m/s)(1.53 s) – ½(9.8 m/s$^2$)(1.53 s)$^2$<br>= 11.5 m |

**Particle model for motion**

Look more carefully at the illustration at the top of the page. The position of the baseball is plotted repeatedly every 0.25 s. This *particle model* of motion shows you that the speed was slowest at the top because successive ball images are close to each other. Objects are often replaced by small dots when creating a particle model.

### Acceleration is not always constant

Acceleration is seldom constant over long periods of time. In the next chapter we see that acceleration results from *forces*. In free fall, the force from *air resistance* gets large above about 10 m/s for falling objects such as balls. Air resistance dominates the motion of light objects with large surfaces—such as a sheet of paper. A flat sheet of paper falls slowly whereas a crumpled sheet falls quickly because of the relative strength of air resistance forces compared to gravity.

### The motion of a skydiver

The graph below shows the air speed of a parachute jumper who jumps out of a plane and falls freely for 28 s. At 28 s she opens her parachute. The vertical axis is speed (in meters per second) and the horizontal axis is time. Can you tell from the graph which part of the motion is accelerated and which is not?

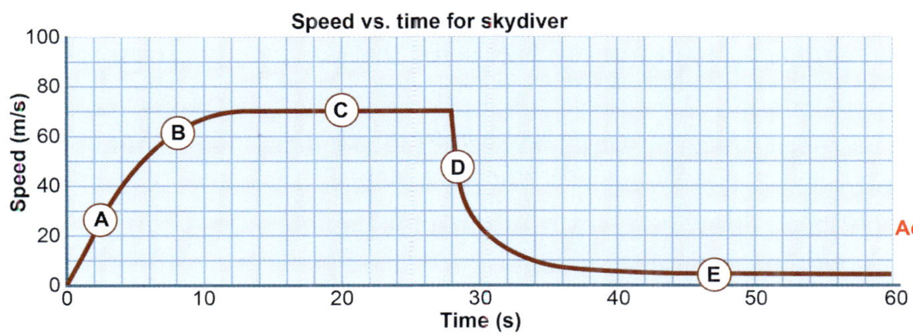


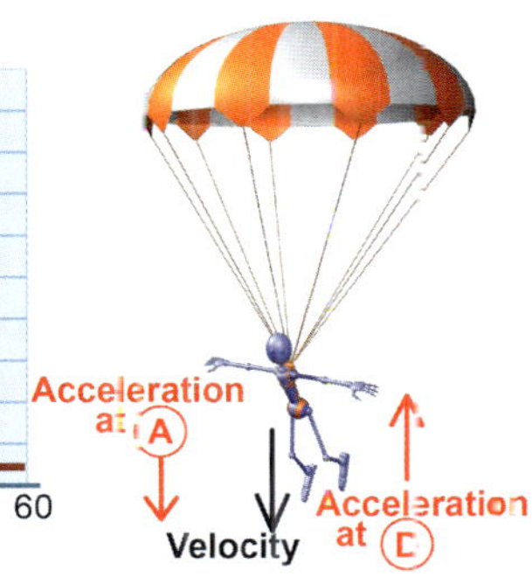


### Terminal velocity

For the first few seconds, before the parachute has opened, the jumper is in free fall with an acceleration of $g$ (A). Notice that after a few seconds the graph curves, becoming more horizontal (B). That is because the force of air resistance increases quickly with speed. Right around 70 m/s (C) the force of air resistance equals the force of gravity and acceleration is reduced to zero. For this skydiver, 70 m/s is her **terminal velocity**, the constant speed she reaches when air resistance reduces acceleration to zero.

### The effect of the parachute

An open parachute greatly increases air resistance and the jumper immediately slows down (D). She is still *accelerated*, but now the acceleration acts in a direction opposite to her velocity. Notice that the downward curve of the $v$ vs. $t$ graph is very steep at D just after the parachute opens. This is because the upward force of air resistance on an open parachute at 70 m/s is much greater than the downward force of gravity on the jumper. The area and shape of the parachute is designed to produce a safe terminal velocity of about 5 m/s (E) for a normal skydiver's weight.

### Terminal velocity and engineering

When an object falls in a medium—whether it is the air in the atmosphere or water in the ocean—it will reach a terminal velocity. How fast it reaches terminal velocity and the value of that terminal velocity depends on properties of the medium (air or water) and the *shape* of the falling object. When designing a spacecraft to re-enter the Earth's atmosphere, or an oil drilling device that will be lowered to the ocean floor, the terminal velocity of the medium is an important consideration.

Chapter 4

# Section 2 review

The term *free fall* describes any state of motion in which an object has a constant downward acceleration due only to gravity. Near Earth's surface, the magnitude of that acceleration is 9.8 $m/s^2$ (also known as one "g"). An object can be in free fall even if it is rising. When a rising projectile reaches its maximum height, its vertical velocity is momentarily zero (i.e., it is neither rising nor falling). A falling object reaches *terminal velocity* when the upward force of air resistance on the object equals the object's weight.

| Vocabulary words | free fall, terminal velocity |
|---|---|

## Review problems and questions

1. Which of the following objects are in free fall? (Assume that you can ignore air resistance.)
   a. a textbook lying on a desk
   b. a textbook that has been tossed upward and has left the student's hands but not yet touched any other object or surface
   c. a textbook that has been thrown downward from a tall cabinet
   d. a textbook held tightly by a running student

2. Josh tells Amanda that acceleration *must* be a negative number whenever an object is in free fall. "After all," he claims, "gravity pulls objects downward, and down is negative." Amanda replies that the acceleration in a free fall problem can be negative *or* positive. "You just have to make sure all related quantities, like displacement and velocity, are handled the same way," she adds. Who is right, and why? (Answer in complete sentences, and cite a specific passage in the text to support your claim.)

3. Zorgata drops her ball a distance of 1 m on the surface of planet Zogg. Izqif wants his ball to take *twice* as much time to fall as Zorgata's. What height should he drop it from? (Note that gravity is stronger on the surface of Zogg than it is on Earth.)
   a. 2 m
   b. 3 m
   c. 4 m
   d. You cannot say without knowing the value for *g* on planet Zogg.

4. During the Apollo 15 mission, astronaut David Scott performed a scientific experiment on the surface of the Moon. Scott dropped two objects at the same time from the same height. One was a 1.3 kg hammer; the other was a 0.03 kg feather. He dropped the two objects from a height of 1.6 m. The two objects fell, and *both* hit the Moon's surface 1.4 s after being released.
   a. What is the acceleration due to gravity on the Moon's surface, in units of meters per second squared?
   b. Were Scott's hammer and feather both in free fall?

# Chapter review

### Vocabulary

Match each word to the sentence where it best fits.

#### Section 4.1

| acceleration | quadratic |
|---|---|

1. The rate of change of velocity is _______
2. A/An __________ equation contains at least one term in which an unknown variable is squared.

#### Section 4.2

| free fall | terminal velocity |
|---|---|

3. The constant speed a falling object reaches when air friction and weight are equal is called the ______.
4. A thrown ball is approximately in ________ when it leaves your hand.

### Conceptual questions

#### Section 4.1

5. What does the graph of distance versus time look like when an object is accelerating?

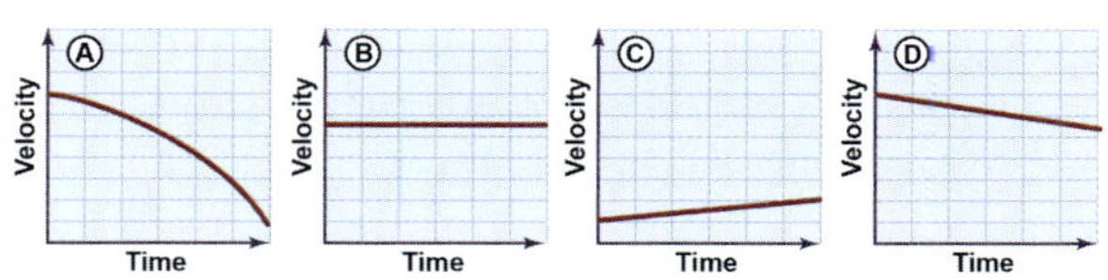


6. A motion-sensing lab car is used to explore motion in various different situations resulting in the four velocity versus time graphs shown. (There may be more than one correct answer for each question.)
   a. Which of these graphs shows an acceleration of zero?
   b. Which of the graphs shows constant acceleration?
   c. Which graph shows the smallest amount of distance traveled?
   d. Which of these graphs would not create a curved distance versus time graph?
   e. Which graph shows the greatest average acceleration?
   f. Which of the graphs most likely represents motion with an acceleration in the same direction as the velocity?

7. The position equation, $x = x_0 + v_0t + ½at^2$, has three terms. Write down the physical significance for each of the three.

8. In the equation $v - v_0 = at$, what does the term $v - v_0$ represent?

9. A student asks a friend the following question:
   "How is acceleration different from velocity?"
   Which of the following answers would best help the student understand acceleration? Why do you think so? Ask the question yourself to some friends and think about their answers.
   a. Acceleration has units of $m/s^2$ while velocity has units of m/s.
   b. How do you describe a change in speed from 10 to 20 mph that occurs over 2 s?

10. Write a sentence defining acceleration. Next, write a sentence that is equivalent to the equation that defines acceleration. Compare the two sentences. Is your definition of acceleration the same as the equation? Why or why not?

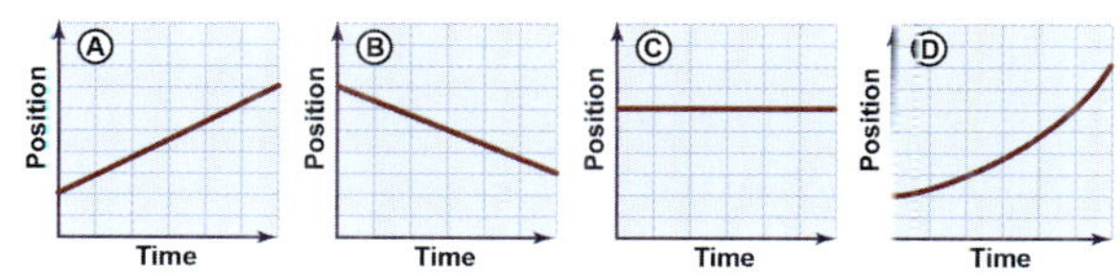


11. The four position versus time graphs above were generated by the motion of a small car with real-time technology (photogates) that sensed the motion of its wheels.
    a. Which of these is showing accelerated motion?
    b. Which graph would create a sloped (up or down) velocity versus time graph?
    c. Which of the graphs has a negative velocity?
    d. Which of the graphs shows an object at rest?
    e. Which of the graphs show an object whose equation of motion could be written as

    $$x = 15\text{ m} - (3\text{ m/s})t?$$

    f. Write down an equation for position versus time that could fit graph D.
    g. In which graph does time vary as the square root of position?
    h. In which graph does position not vary as a function of time?

12. If a bird flies at the constant velocity of 125 m/s, then what is its acceleration for the first 5 s?

13. Describe what constant positive acceleration looks like on an acceleration versus time graph, a velocity versus time graph, and a distance versus time graph.

14. Which of the following might be described by a positive initial velocity and a negative constant acceleration?
    a. The brakes are applied to a car traveling at 30 mph to reduce its speed to 20 mph.
    b. A car starts at rest and then begins to move backward with increasing speed.

# Chapter review

## Section 4.1

15. Use physics vocabulary of motion to describe the process of driving on an on-ramp onto a highway, merging with traffic, driving with the traffic, exiting at the next off-ramp, and then stopping at a traffic light.

16. ❬ Two students conduct an investigation to measure the acceleration of a ball rolling down a 1 m ramp inclined at 45°. One student releases the ball at the same time the second student starts the stopwatch. The students then exchange the (still running) stopwatch and the first student stops the clock when the ball reaches the bottom of the ramp. Evaluate and critique the design of this experiment.

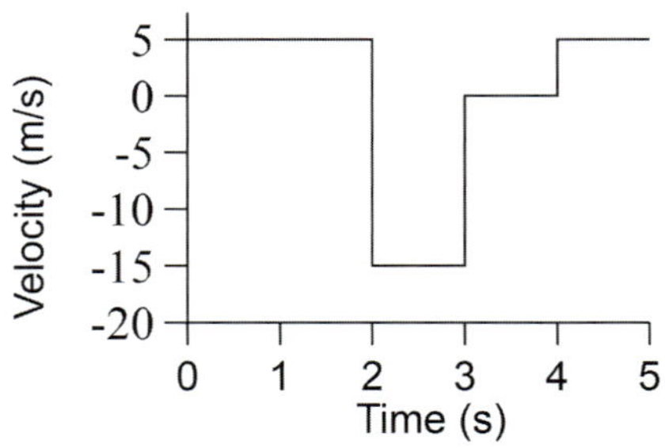


17. ❬❬ Draw the distance versus time graph that corresponds to the velocity versus time graph illustrated above. *(Be sure to look at the scale of the* y*-axis.)*

18. ❬❬ Two students conduct an investigation with a motion-sensing lab cart. Fai starts data collection then Moseph pushes the cart across the floor at roughly constant velocity. Data collection begins when the cart is still at rest and ends while the cart is still moving with constant speed. The graph shows zero acceleration, then positive acceleration, then zero acceleration again. Explain how the data correspond to the motion.

19. ❬❬❬ A physics problem has three unknown variables, $v$, $v_0$, and $x$. Both equations of accelerated motion apply. Is it possible to find a unique solution that assigns a value to each of the three unknown variables? Why or why not?

20. ❬❬❬ Assume that an object has negative acceleration.
    a. Is it possible that its speed can be increasing? How?
    b. What must the signs of velocity and acceleration be to ensure that speed is increasing?

## Section 4.2

21. Ask two friends to answer the following question:
    "Can there be acceleration but zero velocity? Give an example."
    Write down their answers. How do their answers compare with your own answer to the same question?

22. Two balls are dropped from rest and allowed to fall. If one ball is allowed to fall for 1 s and the other for 3 s, compare the distances they have fallen.

23. ❬ Say you drop two cannonballs out of a high window. The first cannonball is twice as heavy as the second. Which will hit the ground first if there is no air resistance?

24. ❬❬ Two balls are thrown at the same time with the same speed. One is thrown directly downward while the other is thrown straight up. Compare their speeds when they hit the ground.

25. ❬❬❬ Consider how far a falling object moves in each second during its fall. Assume the object is initially at rest. During the first second the object falls a certain distance. Between 1 and 2 s it falls a greater distance since its speed has increased even though the time interval is the same. How much farther does the object fall between 1 and 2 s compared to the distance it falls between 0 and 1 s?

### Quantitative problems

## Section 4.1

26. A car is traveling at 15 m/s. It goes down a hill where it accelerates at 3 m/s$^2$ for 4 s. How fast is it going at the bottom of the hill?

27. Silvia is driving at the speed limit of 25 m/s when she sees an ambulance approaching in her rear-view mirror. To move out of its way, she accelerates to 32 m/s in 3 s. What was her acceleration during those 3 s?

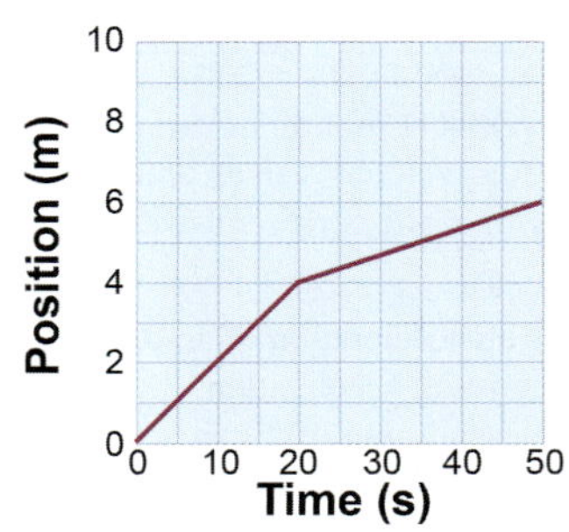


28. Answer the following questions based on the position versus time graph above.
    a. What is the average velocity over the whole 50 s interval shown?
    b. What is the maximum speed shown on the graph?
    c. What is the total distance traveled between $t = 0$ and $t = 50$ s?
    d. What is the final position at $t = 50$ s?
    e. How does the acceleration at $t = 10$ s compare with the acceleration at $t = 30$ s?

# Chapter review

## Section 4.1

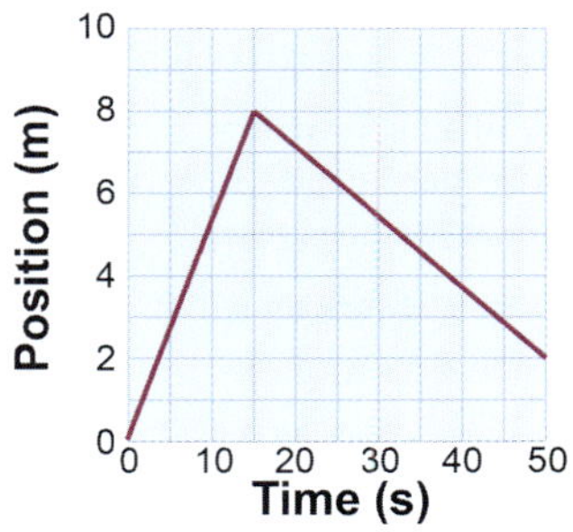


29. Answer the following questions based on the position versus time graph above.

    a. What is the average velocity over the whole 50 s interval shown?
    b. What is the maximum speed shown on the graph?
    c. What is the total distance traveled between $t = 0$ and $t = 50$ s?
    d. What is the final position at $t = 50$ s?
    e. How does the acceleration at $t = 10$ s compare with the acceleration at $t = 30$ s?

30. A car traveling at 15 m/s in a straight line accelerates for 5 s with an acceleration of 2 m/s$^2$. What is its final velocity?

31. How long should Pete's approach to his long jump attempt be if he wants to be at a speed of 9 m/s and can accelerate at 4 m/s$^2$?

32. Rachel is walking at 3 m/s when she walks into a glass door and stops in 0.05 s. What is her acceleration?

33. A zebra is at rest 60 m away from a charging lion running at 17 m/s. If the zebra accelerates at 2 m/s$^2$, will the lion catch it?

34. A submarine begins ascending to the surface of the ocean. It experiences an overall acceleration of 1.7 m/s$^2$ for the first 5 s. What is its velocity after 3.2 s?

35. What is the average acceleration of a cheetah that starts at rest and 3 s later is moving at a speed of 27 m/s? Is the acceleration greater or less than the acceleration of a $100,000 sports car that can go from 0 to 60 mph in 4 s?

36. Jane is riding her bicycle at 10 m/s when she begins to decelerate at 1.5 m/s$^2$.

    a. How long will it take Jane to stop?
    b. How far will she have traveled in this time?

37. Vinny is on a motorcycle at rest, 200 m away from a ramp that jumps over a gully. Calculate the minimum constant acceleration Vinny must have to get to the ramp in 8 s before his pursuers catch up with him.

38. At her most recent track race, Stella had an acceleration of 2.1 m/s$^2$ during the first 3 s of the race. What was her velocity after 2 s?

39. A speeding car traveling at 35 m/s passes a stationary police car. At the moment the car passes, the police car starts accelerating in the same direction with an acceleration of 3 m/s$^2$.

    a. How long does it take the police car to catch up with the speeder?
    b. What is the speed of the police car at the moment it catches up with the speeder?
    c. How far have both cars traveled when the police car catches up?
    d. Convert both car's speeds to miles per hour. Is it likely the police car would actually catch the speeder? Explain your reasoning.

40. A motorcycle has a maximum acceleration of 3 m/s$^2$ in either direction and a maximum speed of 25 m/s. How long does it take for the motorcycle to reach someone who is 1 km away?

41. A car travels 100 m while decelerating to 8 m/s in 5 s. What was its initial speed? What is the magnitude of the acceleration?

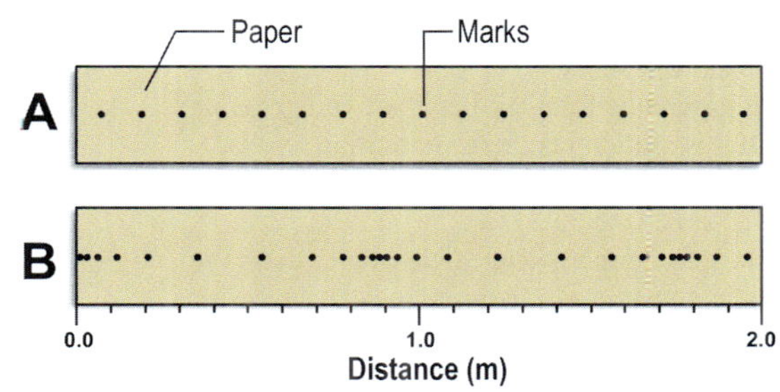


42. Imagine a cart moving with constant velocity over a piece of paper. If the cart makes a mark directly beneath it on the paper at a regular time interval, in this case every 0.1 s, the marks would look like diagram A above. This is known as a "ticker tape" diagram of motion.

    a. In the ticker tape diagram above, where does the cart have *positive* acceleration?
    b. Where does it have *negative* acceleration?
    c. Explain what features in diagram B you interpret as positive and negative acceleration.
    d. If the dots were spaced widely apart in the diagram, then what could you *infer* about the speed of the cart?

43. Linda is out jogging at a constant speed of 3 m/s. She jogs past Maggie, who is stationary but immediately takes off on her bicycle to try and catch up to Linda. Maggie accelerates on her bicycle at 2 m/s$^2$. How long does it take Maggie to catch Linda?

# Chapter review

## Section 4.1

44. A race car crosses the starting line traveling at 40 m/s. It then instantly starts to accelerate at 9 m/s$^2$. How far does it travel in 5 s?

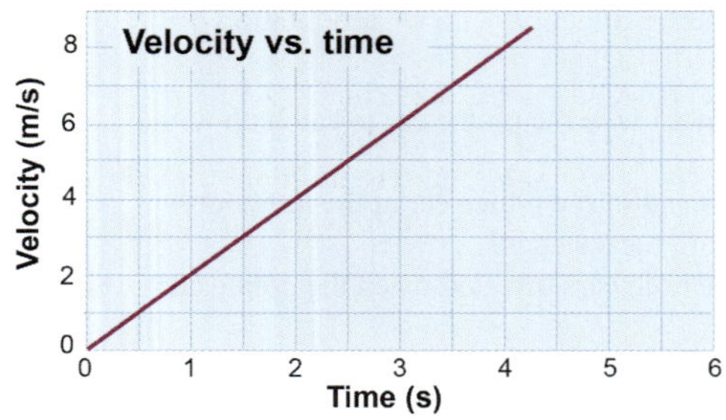


45. A student uses a robot cart as real-time technology to generate the velocity versus time graph above, which shows the cart increasing its speed steadily for 4 s. The area under the graph equals the distance the cart traveled.
    a. Generate a table of the total distance traveled for each of these four time intervals: 0 to 1 s, 0 to 2 s, 0 to 3 s, and 0 to 4 s. What pattern do you find?
    b. Generate a bar chart that shows the total distance traveled *during* each subsequent second (0 to 1 s, 1 to 2 s, 2 to 3 s, and 3 to 4 s). What pattern do you find in these distances?
    c. Make an inference from the bar chart you generated in (b): If this motion continues until time $t = 6$ s, then how far will the robot travel *during* the sixth second (from 5.0 to 6.0 s)?
    d. Generate a one-dimensional "ticker tape" diagram (such as in question 41) showing the cart's position from 0 to 4 s every 0.5 s. Plot a dot for the robot's position at each time, starting at the origin.
    e. Interpret the "ticker tape" diagram you generated in (d). How does the pattern of dots show you that the velocity of the robot is changing?

## Section 4.2

46. A ball is thrown straight up at 15 m/s.
    a. How long does it take to reach its highest point?
    b. How long does it take to hit the ground?
    c. Write an expression for height $x$ as a function of time $t$.
    d. Graph height $x$ versus $t$. Label the point of highest height and the time it reaches the ground.

47. From what height should you drop a ball so that it hits the ground at 40 m/s? How long will it take to hit the ground?

48. What is the average acceleration of a rocket that travels 400 m in 5 s if it starts from rest?

49. How long a runway does a plane need to accelerate from rest to a take-off speed of 50 m/s at a comfortable acceleration of $0.3g$ or 3.27 m/s$^2$? How long does it take for the plane to reach take-off speed?

50. A basketball player can jump 0.65 m off the ground. How fast must the player be going when she leaves the ground?

51. Bill throws a ball up in the air at 17 m/s. How long will it take for it to fall back to him?

52. Eve is standing at the top of a cliff. If she drops a rock off of the cliff and it takes 2.5 s to hit the ground, how tall is the cliff?

53. Four seconds after being launched, what is the height of a ball that starts from a height of 12 m with an initial upward velocity of 24 m/s?

54. A person drops a penny from a window that is 50 m high. Neglecting air friction, calculate the speed of the penny when it hits the ground. Do you think the penny would actually fall this fast or would air friction slow it down?

55. A juggler throws a ball upward at 9 m/s. One second later the juggler throws a second ball upward at the same speed. At what height do the balls reach the same position but are moving in opposite directions? (Hint: to solve this problem you must use the fact that $(t-1)^2 = t^2 - 2t + 1$.)

56. An arrow is launched straight up. What is the initial velocity the arrow must have to reach a height of 500 m? (Hint: At the highest point the arrow reaches, its velocity is $v = 0$.)

57. A train conductor sees an obstruction on the track 300 m ahead of the train. If the train is traveling at 24 m/s and it takes the conductor 0.5 s to apply the brakes, what must the train's acceleration be to stop before hitting the obstruction? Is this acceleration a reasonable value for a real train?

58. How far does a marble fall during its ninth second of free fall from rest?

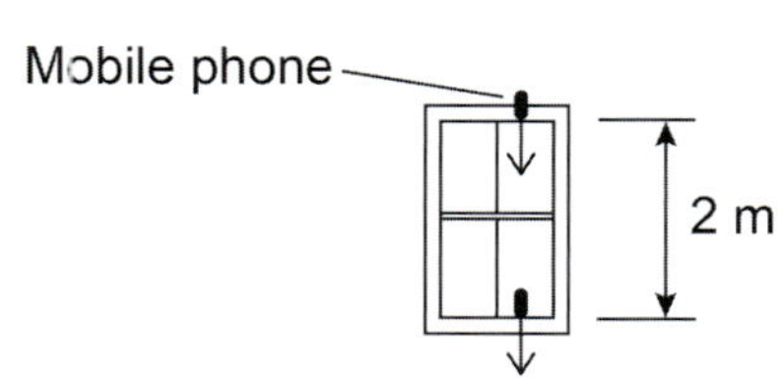


59. A person in a floor above you drops his mobile phone out the window. It takes 0.2 s for the phone to fall 2 m past your window. Estimate how far above you the person must have dropped the phone.

# Chapter review



### Standardized test practice

60. If the slope of a position versus time graph is decreasing, what does this indicate about velocity over time?

   A. Velocity is increasing.
   B. Velocity is decreasing.
   C. Velocity remains constant.
   D. Velocity is zero.

61. Ruth throws a baseball straight up at 20 m/s. What is the ball's velocity after 4 s?

   A. 10.2 m/s
   B. −19.2 m/s
   C. 16 m/s
   D. −76.8 m/s

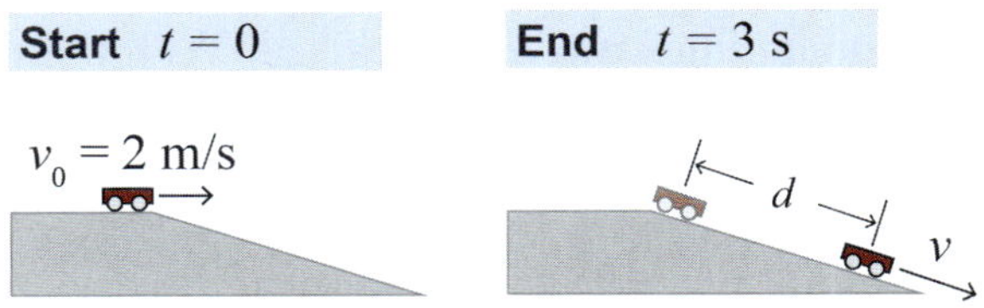


62. A cart is rolling along at a steady 2 m/s when it meets a slope. The cart accelerates at 1 m/s$^2$ as it rolls down the hill. How far will the cart move in 3 s?

   A. 6 m
   B. 5 m
   C. 15 m
   D. 10.5 m

63. If an object starts accelerating from rest at a rate of 34 m/s$^2$, how far does it travel during the first second?

   A. 20 m
   B. 17 m
   C. 34 m
   D. 26 m

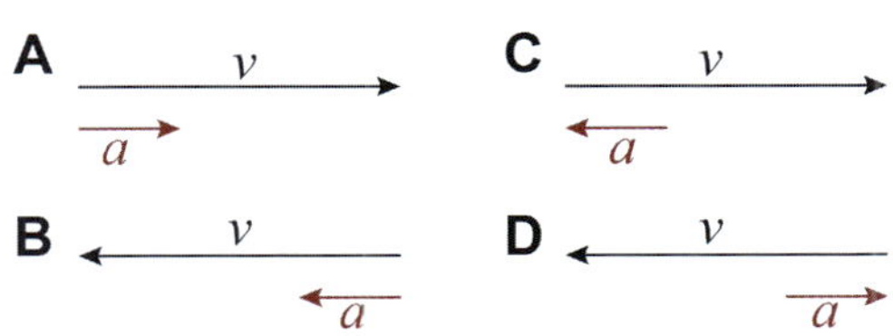


64. Which of the four diagrams best represents a negative velocity with a positive acceleration? Assume the positive direction is to the right.

   A. A
   B. B
   C. C
   D. D

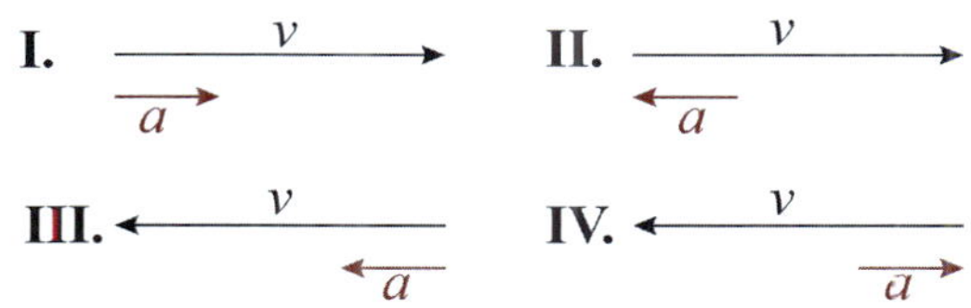


65. Four different objects are moving along a horizontal axis, each with a different combination of velocity and acceleration as shown here. How many of these objects are slowing down?

   A. one
   B. two
   C. three
   D. four

66. Jed drops a 1 kg box off of the Eiffel Tower. After 3 s, how fast is the box moving?

   A. 44.1 m/s
   B. 29.4 m/s
   C. 9.8 m/s
   D. 3.3 m/s

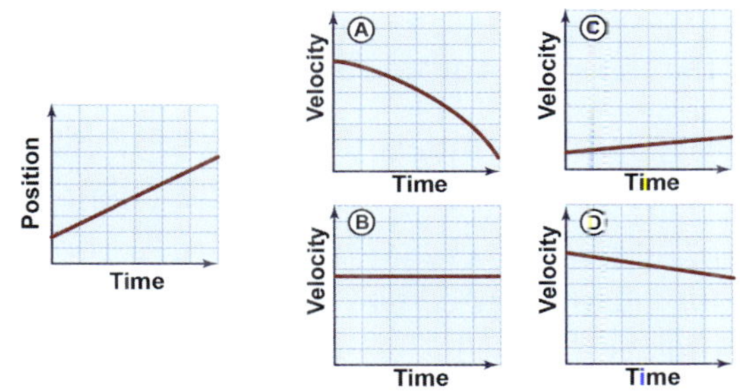


67. Only one of the four velocity–time graphs (above right) can possibly correspond to the single position–time graph (above left). Which is it?

   A. Graph A
   B. Graph B
   C. Graph C
   D. Graph D

68. You are driving a metric car that has a speedometer that displays in meters per second. When you look down at your speedometer, you see that it reads 15 m/s. Four seconds later, you look down and it reads 23 m/s. What is your average acceleration over those 4 s?

   A. 5.75 m/s$^2$
   B. 2 m/s$^2$
   C. 3.75 m/s$^2$
   D. 2.5 m/s$^2$

# Chapter 5 Forces and Newton's Laws

Many people consider sailing ships to be the finest examples of human ingenuity ever devised. With nothing but wind and sails for power, the elegant clipper ships of the mid-19th century carried passengers and precious cargoes across the Atlantic, making the 3,100 mile run in less than 14 days without a drop of fuel of any kind. Between 1840 and 1860 more than 350 fast clipper ships were built by American shipyards. Among many sailing records set by these ships, on March 1, 1854, the *McKay Lightning* traveled 436 nautical miles in 24 hours, the longest 24-hour run ever posted by a sailing ship up to that time.

For enthusiasts, sailing remains the ultimate technology for harnessing the forces of nature to cross the vast distances between continents. The current 24-hour distance record was set in 2009 by the French boat *Banque Populaire V*. The *Banque Populaire*, a 40 m, carbon-fiber trimaran, sailed an astounding 908 nautical miles in 24 hours at an average speed of 37.8 knots! A high-tech racing boat, the *Banque Populaire* set the 24-hour distance record while on the way to setting another record for the fastest trans-Atlantic crossing. With no source of power except the wind, the record-setting, west-to-east crossing was made in 3 days, 15 hours, and 25 minutes.

A variety of different and variable forces act on a sailboat as it travels through the water. The forces on a sail in a good wind can reach 100 N/m$^2$. For a large sailboat with 50 m$^2$ of sail, that comes to 5,000 N (or 1,200 lb) of force that can change from moment to moment. Most sailboats have a keel that extends down from the hull like a vertical fin. As the keel cuts through the water, its surface creates forces that keep the boat moving forward rather than sideways from the force of the wind. The keel of a large boat has a heavy weight at the bottom, helping to keep the boat upright against the tipping force from the sail. Tilting sideways (or "heeling") can increase drag from the water, thereby causing the boat to sail less efficiently.

Sailing requires expertise in manipulating the forces acting from wind and water. The wind might be blowing in one direction and the ocean current moving in a second direction, yet the sailboat crew probably wants to travel in a third direction. The crew can use the boom, for example, to vary the angle that the sail makes with the wind; they can also vary the number, size, and shape of the sails. You might think that sailing directly downwind would be fastest, but an expert sailor knows the highest speed is achieved by sailing at an angle of, say, 30°–60° from downwind in what is called a "broad reach." By using the "tack-and-turn" maneuver, a sailboat can even travel upwind.

Chapter
5

# Chapter study guide

## Chapter summary

Forces include gravity, friction, and forces from devices such as ropes or springs. Many forces may act on an object simultaneously. The *net force* is the vector sum of all forces acting on an object. The net force determines acceleration. Zero net force means zero acceleration. A nonzero net force causes acceleration equal to the net force divided by the object's mass. *Normal forces* act between objects and supporting surfaces. A *free-body diagram* is a sketch showing the location and direction of all forces acting on an object. Forces always exist in action–reaction pairs. Each force in a pair is equal in magnitude and opposite in direction to its partner and the two forces in a pair always act on different objects. The force from a spring is proportional to its deformation according to Hooke's law. *Friction* is a catch-all term for forces that act to oppose motion. *Kinetic friction* acts between moving surfaces, *rolling friction* acts between rolling surfaces, *viscous friction* applies to fluids, and *static friction* acts between surfaces with no relative motion.

## Learning objectives

By the end of the chapter you should be able to

- calculate the net force from any combination of one-, two-, or three-dimensional forces;
- draw a free-body diagram of a two-dimensional system of forces;
- identify and represent action–reaction, weight, friction, and normal forces on free-body diagrams;
- solve problems including forces from static, kinetic, and rolling friction; and
- solve problems involving forces from extension and compression springs.

## Investigations

5A: Newton's second law
5B: Hooke's law
5C: Static and kinetic friction

## Chapter index



## Important relationships

$$F_w = mg \qquad g = 9.8 \text{ N/kg} \qquad F_1 + F_2 + F_3 + \cdots = 0 \qquad a = \frac{F}{m} \qquad F = -kx$$

$$F_f = \mu_s F_N \qquad F_f = \mu_k F_N \qquad F_f = \mu_r F_N \qquad F_f = \frac{1}{2} c_d \rho A v^2$$

## Vocabulary

**force**
**weight**
**center of mass**
**friction**
**Newton's third law of motion**
**spring**
**brittle**
**coefficient of static friction**
**sliding friction**
**viscosity**
**newton (N)**
**free-body diagram**
**net force**
**Newton's first law of motion**
**reaction force**
**Hooke's law**
**material strength**
**coefficient of kinetic friction**
**rolling friction**
**drag coefficient**
**pound (lb)**
**normal force**
**equilibrium**
**Newton's second law of motion**
**spring constant**
**elasticity**
**static friction**
**lubrication**
**coefficient of rolling friction**

## 5.1 - Forces

Physics is based on the idea of *cause and effect*, which means that any change that we observe must have a cause. For example, the downward acceleration of a falling ball is *caused* by the force of Earth's gravity. The previous chapter described acceleration; in this chapter we will see that acceleration is caused by forces. You may not realize it, but you already know about one kind of force: weight, or the force due to Earth's gravity. In this section you will learn about forces and how to represent them visually on a *free-body diagram*.

### Forces

**What is a force?**

We all have an intuitive sense of the meaning of force. **Force** is an action represented by words such as *push* or *pull*. You probably exerted both pushing and pulling forces today without thinking about it. In physics, the concept of force is defined with strength (or magnitude), direction, and units.

**A conceptual definition of force**

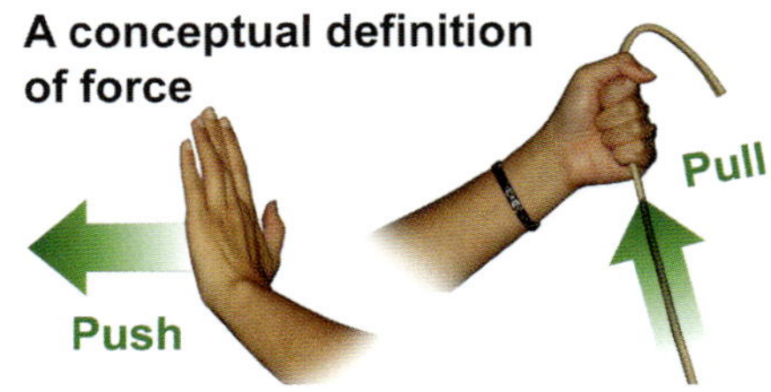


**Definition of 1 newton (1 N)**

One newton is the force required to change the velocity of a one kilogram object by one meter per second in one second.

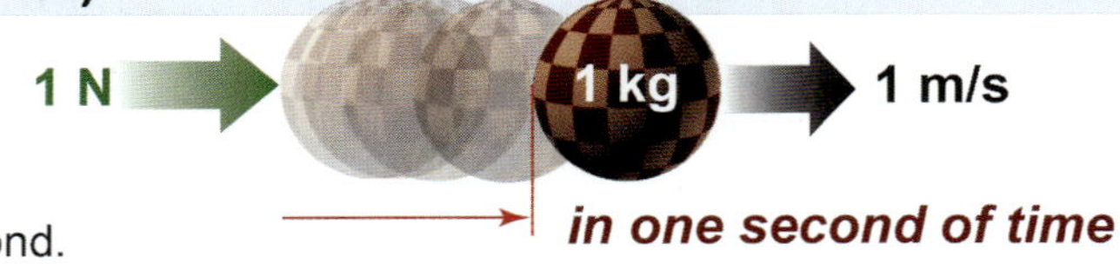


**How is force measured?**

The SI unit of force is the *newton*. One **newton (N)** is about the weight of a cellphone—a rather small force. You can easily exert a force of 100 N or more with one arm. Technically, one newton is the force required to change an object's velocity by one meter per second in one second. This is where the newton gets its fundamental units: One newton is the force required to accelerate one kilogram at one meter per second per second ($1 \text{ N} = 1 \text{ kg m/s}^2$).

**Connecting force and acceleration**

Forces are connected to acceleration—to changes in motion. An object at rest will stay at rest *unless* there is a force applied that causes the object to accelerate. Less obvious is that an object already moving will keep moving until a force is applied to change its velocity. Picture an ice skater on ideal frictionless ice. The skater could coast *forever* if there were neither friction nor other forces applied to slow her down or turn her. In other words, if no force is acting on her then her velocity will remain constant; i.e., her acceleration is zero.

**What are pounds?**

In your everyday life, force is often measured in *pounds*. One **pound (lb)** is equal to 4.448 N. If you weigh 100 lb you also weigh 444.8 N. In the grocery store, pounds are divided into ounces. One pound equals 16 ounces, so one newton is a just a bit less than 4 ounces. Many practical problems in life, such as buying a rope or building a house, will require you to understand forces in pounds. For most physics problems, however, the relationship between force and motion makes it much more convenient to use newtons.

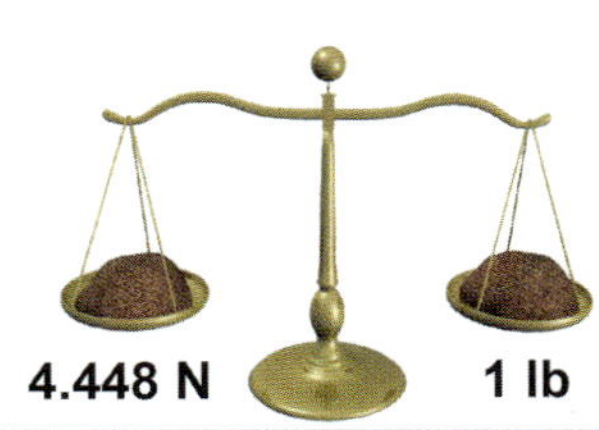


**4.448 newtons = 1 pound**

**What is weight?**

**Weight** is a force that results from the action of gravity on mass. At Earth's surface, gravity attracts all masses downward with a force of 9.8 N/kg. This gravitational force is called *weight*, which is calculated by multiplying its mass by the strength of gravity in newtons per kilogram. For example, a 10 kg object has a weight of 9.8 N/kg × 10 kg = 98 N.

Interactive equation

**Weight**

$$(5.1)\quad F_w = mg$$

$F_w$ = weight force (N)
$m$ = mass (kg)
$g$ = acceleration of gravity = 9.8 N/kg at Earth's surface

**The value of *g***

The value of 9.8 N/kg is used so often that it has its own symbol, $g$. When you see "$g$" in an equation such as $F = mg$, substitute the strength of gravity at Earth's surface $g = 9.8$ N/kg. The value of $g$ may also be given as $g = 9.8\ \text{m/s}^2$. This is equivalent because 1 N of force is defined as the force that causes a 1 kg object to accelerate at $1\ \text{m/s}^2$.

**Are weight and mass the same thing?**

Weight is not mass. Mass is measured in kilograms and grams. Weight is measured in newtons or pounds. The source of confusion is that we use weight to measure mass! When you place an object on an electronic balance, the sensor in the balance measures the force of gravity acting on the object. Since this force is proportional to mass, a calculation done inside the balance divides the force by 9.8 N/kg to read mass in kilograms, or by 0.0098 N/g to read mass in grams. A balance calibrated to work on Earth will not be accurate on the Moon or anyplace where gravity is not exactly the same as it is on Earth. In the extreme case, far from any planet or star, an object will still have mass, but it may be *weightless* because of the absence of gravity.

**How do I calculate weight?**

Equation (5.1) relates weight in newtons to mass in kilograms. If force is in newtons and $g$ is in newtons per kilogram, then mass must be in kilograms. Many physics problems require you to convert units, such as pounds to newtons. Do these conversions *before* substituting any values into the equation to calculate your results.

| Force conversion factors | | |
|---|---|---|
| $\frac{1\text{ lb}}{4.448\text{ N}}$ | $\frac{1\text{ lb}}{16\text{ oz}}$ | $\frac{1\text{ ton (US)}}{2{,}000\text{ lb}}$ |
| **Values for *g*** | | |
| $g$ = 9.8 N/kg | $g$ = 2.2 lb/kg | |

**Calculate weight**

If a person has a weight of 125 lb, then what is that person's mass in kilograms?

*Asked:* mass

*Given:* weight $F_w$ = 125 lb

*Relationships:* equation for weight: $F_w = mg$
conversion factor: (444.8 N)/(1 lb) = 1
acceleration due to gravity:
$g$ = 9.8 N/kg

*Solution:* Convert weight from pounds to newtons:

$$125\text{ lb} = (125\text{ lb})\left(\frac{4.448\text{ N}}{1\text{ lb}}\right) = 556\text{ N}$$

Calculate mass using weight equation:

$$F_w = mg \;\Rightarrow\; m = \frac{F_w}{g} = \frac{556\text{ N}}{9.8\text{ N/kg}} = 56.7\text{ kg}$$

**The meaning of a free-body diagram**

A **free-body diagram** is a geometric sketch of an object isolated from everything else except for the forces that act on it. A key idea in a free-body diagram is that every interaction the object has with the environment is *replaced by the force that interaction makes on the object.* That means every supporting surface, rope, spring, weight, and even friction appear on the diagram only as forces, not as sketches of the actual objects. In the example, the rope is replaced by the tension force $T$ exerted by the rope on the ball.

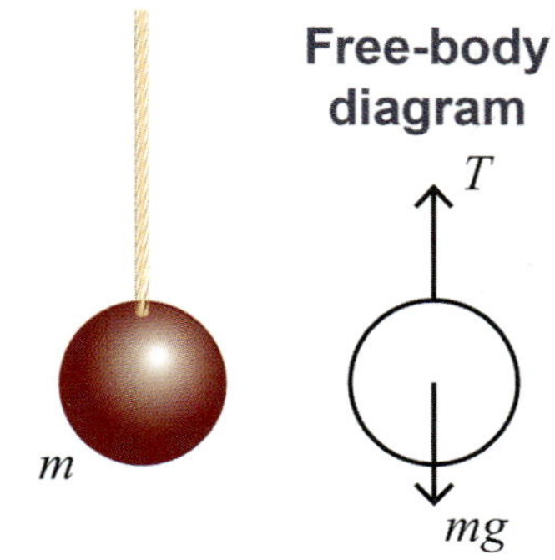


**Normal forces**

The "all" in "all the forces" means that, if you remove the object completely from its environment, the object behaves the same. If a ball is resting on a table, then the free-body diagram replaces the table with a force that allows the ball to remain at rest. The force from the table is called a **normal force**. Normal forces, also called *support forces,* are created at every point where two objects touch each other, such as between the ball and the table. In the example, the normal force is $F_N$.

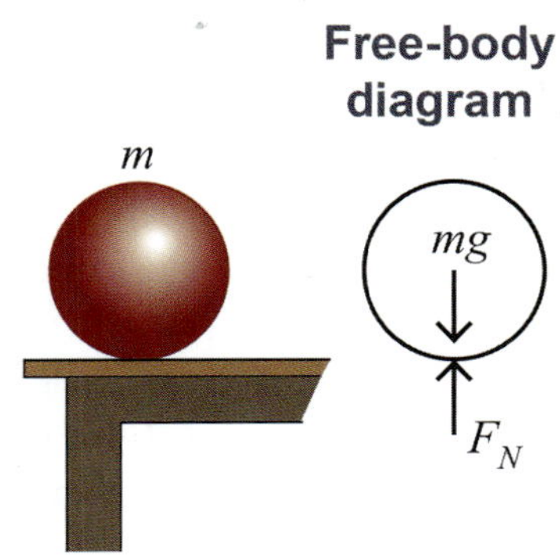


**How do I draw normal forces?**

Normal forces point *perpendicular* to the surface of contact between objects and always provide a *push*. If the surfaces are curved, then the normal force is along the line of contact between the surfaces. The diagram below shows some examples.

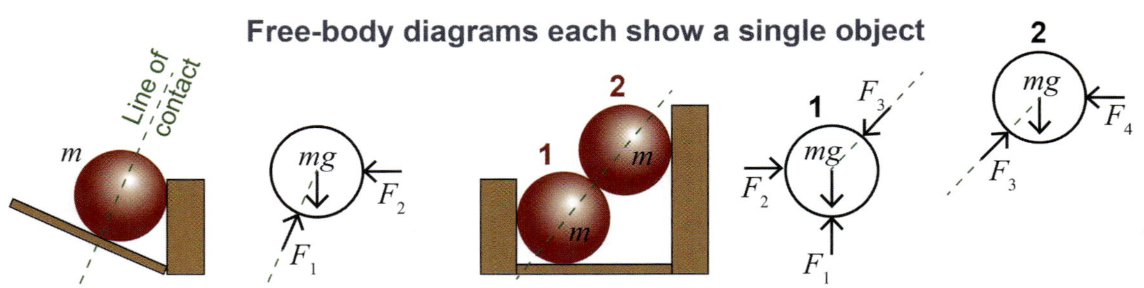


**How do I draw a free-body diagram?**

A free-body diagram needs to be neither *exact* nor artistic! A good free-body diagram is just an outline of the object that shows the approximate locations and assumed directions and *names* of the forces acting on each *single* object. The above-right diagram contains two balls, and each one gets its own free-body diagram. Equally important is that each unique force must have a unique name. In the example we use subscripts to name the four different forces $F_1$, $F_2$, $F_3$, and $F_4$.

**Which forces to include**

A tricky aspect of drawing free-body diagrams is to identify *all* the forces that act and to include only those forces that act *on the object*. The phrase "on the object" is important because there may be forces that come *from* the object but act *on* something else.

**Center of mass**

For rigid objects it is usually accurate to consider the weight force to act on an object's *center of mass*. The **center of mass** is the average location of all the mass contained in the object. For a solid, simple-shaped object like a cube, the center of mass is the geometric center. In contrast, if the mass is unevenly distributed within an object or the object is irregularly shaped, then the center of mass might be at another point that is not the geometric center. In certain shapes, such as a *ring*, the center of mass lies completely outside the object.

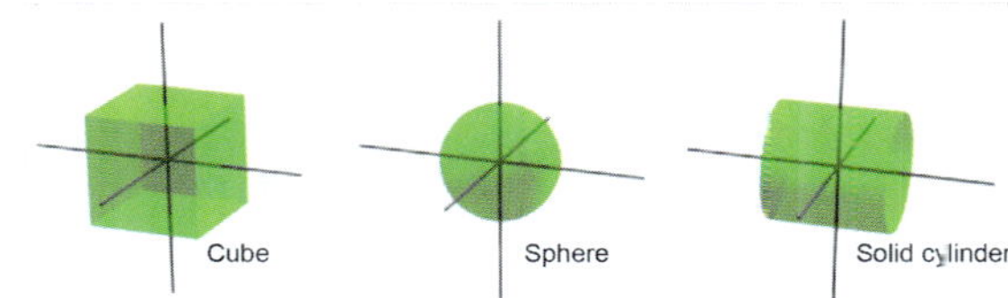


**For symmetric solid objects the center of mass is at the geometric center.**

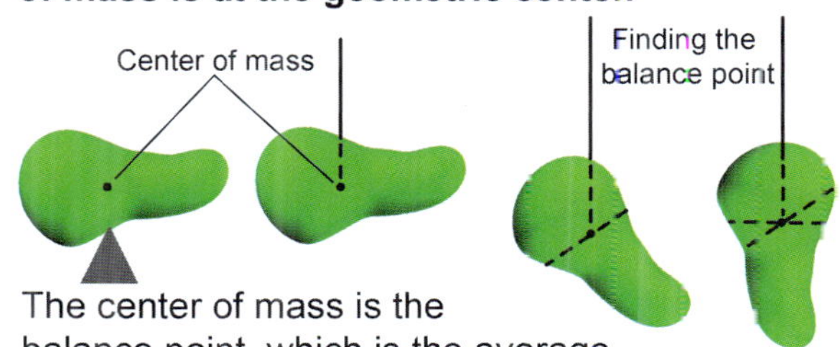


The center of mass is the balance point, which is the average position of all the mass in an object.

**Finding the center of mass**

You can find the center of mass of an irregular object by hanging it from a string. The center of mass will always lie directly below the line of the string. If you choose three different points, the intersection of the three lines (one for each string choice) shown in the diagram (above) locates the center of mass.

**Weight on free-body diagrams**

When constructing a free-body diagram, the weight force is drawn downward from the center of mass. The diagram on the right shows the force vector for weight drawn at the center of mass of the different 10 kg objects illustrated above.

When constructing free-body diagrams, draw the weight vector from the center of mass.

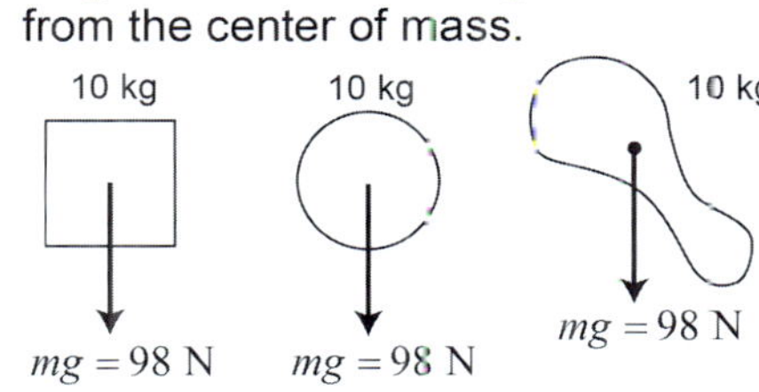


**An example problem**

A barbell with a mass of 50 kg is resting on the floor. Draw a free-body diagram of the barbell and calculate its weight in newtons.

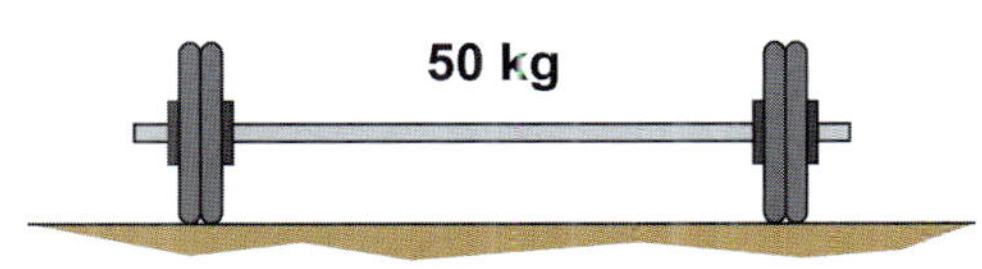


*Asked:* Calculate weight and draw a free-body diagram.

*Given:* mass = 50 kg

*Relationships:* $F_w = mg$

*Solution:* The weight force is drawn at the center of mass, which is in the middle of the barbell.

$$F_w = mg$$
$$= (50 \text{ kg})(9.8 \text{ N/kg}) = 490 \text{ N}$$

*Answer:* 490 N

**Free-body diagram**

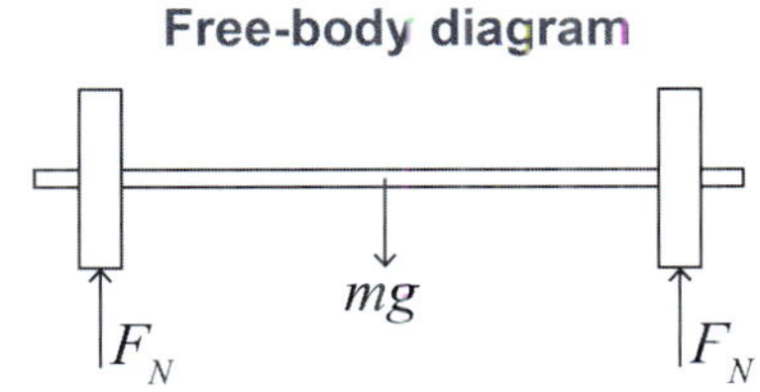

## Drawing free-body diagrams

**Are there different ways to draw free-body diagrams?**

Free-body diagrams are not art, but instead they are a language for describing forces and objects. As with other languages, there are different ways to represent the same information, or "say" the same thing. In this course we have chosen to draw free-body diagrams using the "outline" method. This is the way force diagrams appear in engineering applications, in practical building codes, and also in more advanced physics problems, such as those involving rotation. The other common method is the "point-mass" method, which is used in more theoretical books.

**An example**

Consider a block resting on the floor partially suspended by two springs.

**Problem**

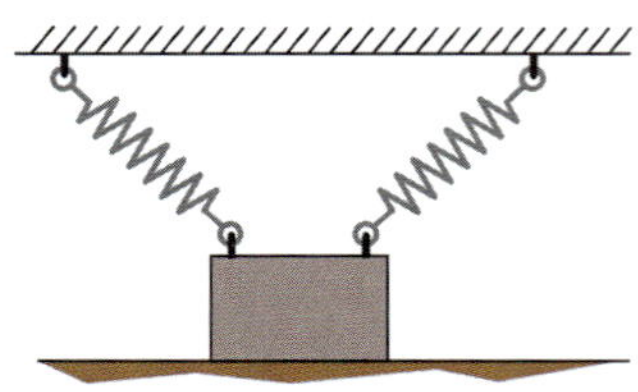

A block of mass $m$ sits on a floor partially suspended by two springs. Draw the free-body diagram of the block.

**Free-body diagrams**

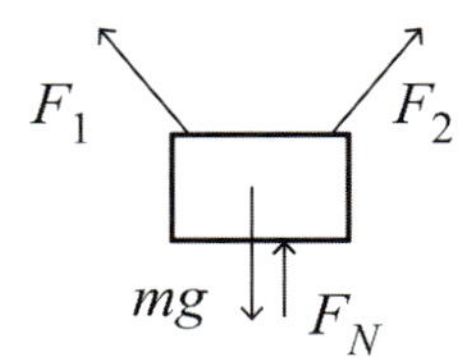


**Outline method**

- Draw an outline of the object.
- Draw forces pointing to or from the object.
- Draw forces about the location where they act.

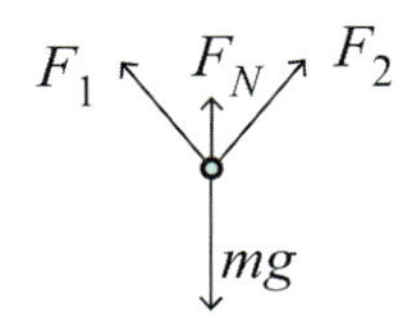


**Point-mass method**

- Draw object represented as a point.
- Draw all forces pointing away from the object.

Both free-body diagrams are correct as long as they correctly identify only the forces acting *on* the object and correctly label every force.

**Example problem with point-mass diagrams**

The diagram (below right) shows two masses connected by a rope that passes over two frictionless pulleys. Which of the three point-mass-style free-body diagrams best represents the forces acting on mass $m_1$? Explain why the other two are incorrect.

| | |
|---|---|
| *Asked:* | Choose the best free-body diagram for the forces acting on mass $m_1$. |
| *Given:* | three free-body diagrams |
| *Solution:* | There are three forces: weight $m_1g$, a normal force from the floor $F_N$, and the rope tension $T$, which equals the weight $m_2g$ of mass $m_2$. Only diagram (A) shows all three forces in the correct directions.<br>Diagram (B) omits the normal force from the floor.<br>Diagram (C) shows the tension $T$ acting *down* on mass $m_1$, which is incorrect because this force acts through the rope and should be pulling *up* on $m_1$. |

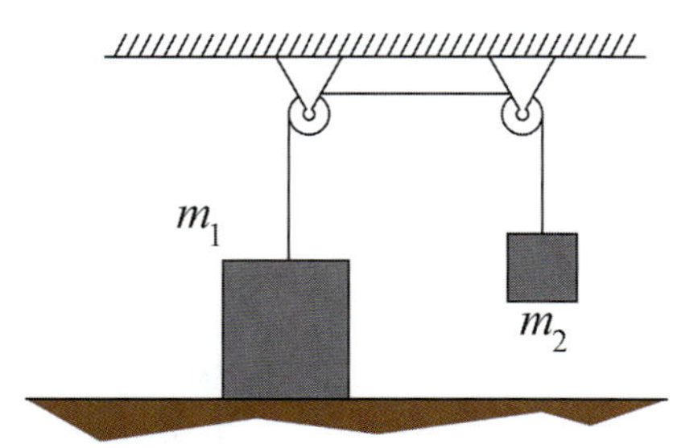


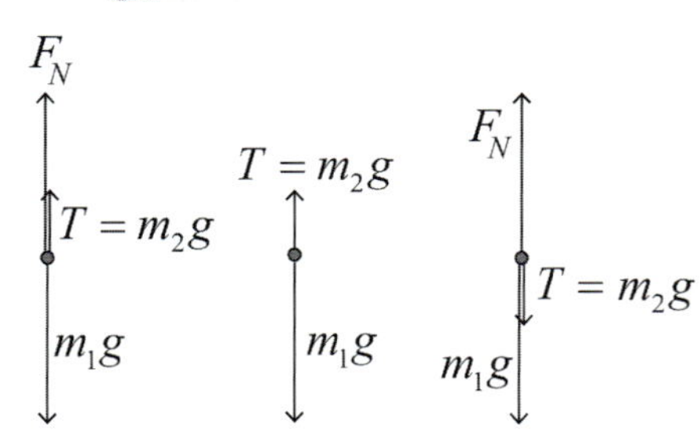


**(A)** **(B)** **(C)**

**What is a "net force"?**

Single forces do not always change an object's motion because *many* forces usually act on the same object at the same time. The **net force** is the total, or sum, of all forces acting on an object. Objects respond to the net force instead of individual forces. Forces may cancel each other out when multiple forces act in different or opposite directions. Stationary objects such as buildings and bridges stand still because of zero net force, not because there are no forces.

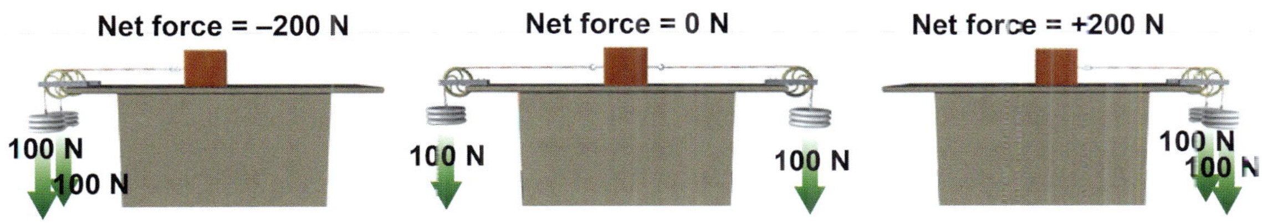


**Free-body diagrams**

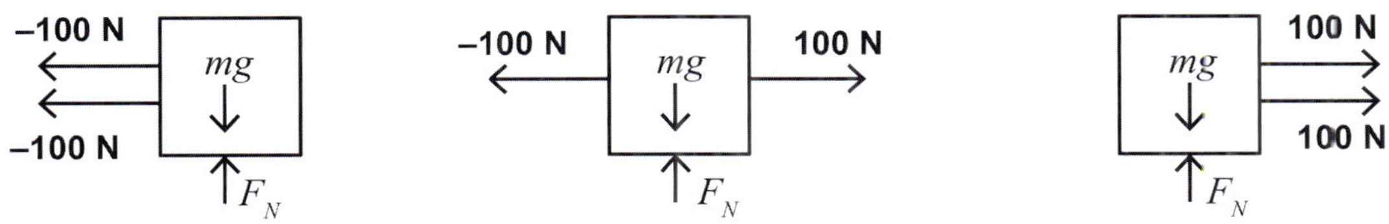


**Can the same forces make different net forces?**

Two 100 N forces act horizontally on the box in the diagram above. The net force is different in each of the three situations. The free-body diagrams show that the net force in the horizontal direction is −200 N to the left in one case, zero in another, and +200 N to the right in the third case. The net force in the vertical direction is zero.

**Force is a *vector***

Positive and negative signs indicate direction.

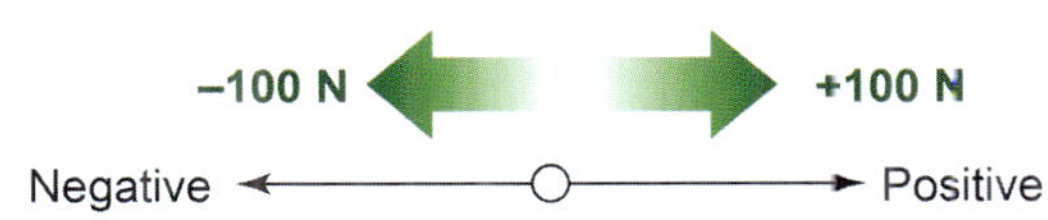


**Force is a vector**

The *direction* of a force matters when determining the net force. Force, like position, is a *vector*. In one-dimensional problems positive and negative signs are used to indicate direction. *Forces should be assigned positive and negative values that correspond to direction.*

**Which is positive and which is negative?**

When solving problems involving forces and free-body diagrams, *you must choose which direction is positive.* The opposite direction then becomes negative. Diagrams are often chosen such that right and up is positive and left or down is negative. The choice, however, is arbitrary and it may be easier for some problems to choose differently. Once you choose the positive direction, you must be consistent throughout the entire solution process.

**When can forces not be added?**

In most physics problems you will need to consider the net force separately in each of the perpendicular directions. For example, a 200 N *upward* force cannot cancel with a −200 N force to the *left* because these forces do not act along the same line. The net force may be different in different directions. In the example above, the vertical net force is zero but the net force in the horizontal direction may not be zero.

## Equilibrium and statics

### What is equilibrium?

In a *static* problem nothing moves, either in linear motion or rotation. Therefore, the *net force must be zero.* This condition is called **equilibrium**. Any object that remains at rest must be in equilibrium because, if there *were* any net force, then the object would accelerate and start moving. Equation (5.2) is the condition for equilibrium of forces.

**Equilibrium**

$F_1, F_2, F_3, \ldots$ = force (N)

$$\textbf{(5.2)} \qquad F_1 + F_2 + F_3 + \cdots = 0$$

### Equilibrium and vectors

*This is a vector equation!* That means that equation (5.2) holds separately for each direction relevant to the situation.

1. The total force in the $x$-direction must be zero.
2. The total force in the $y$-direction must be zero.
3. The total force in the $z$-direction must be zero.

### Drawing the free-body diagram

Equilibrium problems often involve finding one or more unknown forces using the fact that the net force is zero. A free-body diagram is *essential* for keeping track of the directions and positions of forces. In the diagram on the right two iron balls are hung from rods. The free-body diagrams include the weight of each ball and the forces from the rods.

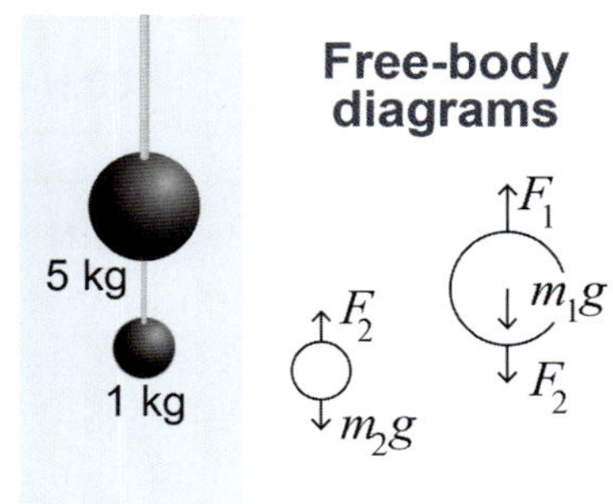


### How do I solve equilibrium problems?

1. Draw a free-body diagram including forces from ropes, attached supports, weight, and normal forces.
2. Assign positive and negative directions.
3. Use the equilibrium equation by setting the net force to zero independently for each direction. Each direction equation can be solved for one unknown force.
4. If there are still unknown forces then you may need more information.

### An example equilibrium problem

In many problems there may be a *symmetry* that causes some forces to be equal. In the example below, the forces from each rope must be equal because the acrobat is in the center, in a symmetrical position. That means that these forces can be assigned the *same* variable name $F$ instead of separate names such as $F_1$ and $F_2$. The name is the same because the forces are known to be equal and therefore have the same value.

### Rope tension for an acrobat

A 65 kg acrobat is suspended motionless between two rings as shown at right. What is the tension force in the rope holding up one of the rings if the acrobat is centered between them?

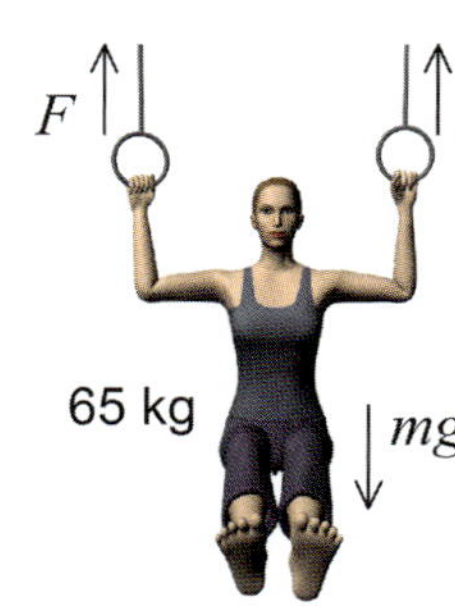


| | |
|---|---|
| *Asked:* | tension force $F$ |
| *Given:* | mass = 65 kg "at rest" |
| *Relationships:* | $F_w = mg$; net force = 0. |
| *Solution:* | $2F - mg = 0$ |
| | $\rightarrow F = \frac{1}{2}mg$ |
| | $= (0.5)(65\text{ kg})(9.8\text{ N/kg}) = 320\text{ N}$ |

# Section 1 review

A force is an action upon an object that can change the object's velocity. The SI unit of force is the newton (N). One newton is the force required to accelerate a 1 kg object by 1 m/s$^2$. Force is a vector: It has direction as well as magnitude (or strength). Weight is one example of a force. An object is in equilibrium when the net force (the vector sum of all external forces) on that object is equal to zero.

| Vocabulary words | force, newton (N), pound (lb), weight, free-body diagram, normal force, center of mass, net force, equilibrium |
|---|---|

| Key equations | $F_w = mg$ | $F_1 + F_2 + F_3 + \cdots = 0$ |
|---|---|---|

## Review problems and questions

1. Ariel's puppy weighs 15 lb. Heather's puppy weighs 50 N. Whose puppy weighs more?

2. Imagine that a spacecraft has been designed, built, and launched to explore the atmosphere of the planet Mars. To enter a circular orbit around Mars, the craft has to be slowed down. The spacecraft's reverse thruster (its "brakes") is supposed to deliver a force of 1,000 N whenever it is fired. The thruster, however, actually delivers 1,000 *lb* of force. Is it too weak or too strong?

3. The acceleration due to gravity at the surface of Mars is 38% of that on Earth. If a Mars rover has a mass of 950 kg on Earth, how much does it weigh on Mars?

4. A 100 kg astronaut weighs 500 N on a different planet.
   a. What is the astronaut's mass on that planet?
   b. What is the acceleration due to gravity on that planet?

5. A 120 lb woman sits on a 20 lb stool with her feet above the floor, as shown here. (Assume that the stool rests on a kitchen floor.)
   a. Is the woman in equilibrium? How about the stool?
   b. What is the net force on the woman? On the stool?
   c. Describe how to draw a free-body diagram for the woman.
   d. How strong is the normal force that the stool exerts upon the woman?

6. A student places a 15 kg box on a level floor.
   a. What is the net force on the box? What is the normal force on the box?
   b. The student now presses directly *down* on the box with a force of 30.0 N. What is the net force on the box? What is the normal force on the box?
   c. The student now pulls directly *up* on the box with a force of 80.0 N. What is the net force on the box? What is the normal force on the box?

## 5.2 - Newton's laws

Newton's three laws of motion tell us how objects move when they are acted on by net forces. The first law describes constant motion when there is *no* net force, the second law describes the acceleration of an object when it experiences a nonzero net force, and the third law describes how every action force has an equal but opposite reaction force.

### Newton's first law and friction

**Newton's first law**

**Newton's first law of motion** says that an object at rest or in motion remains in its identical state of rest or motion unless acted upon by a net force. Since inertia is the property of mass that resists change in motion, the first law is sometimes referred to as the *law of inertia.*

**What does the first law mean?**

It may be intuitive that an object at rest stays at rest without the action of a net force. But the first law additionally says that *an object in motion remains in the identical state of motion* unless acted upon by a net force. In real life, objects in motion slow down and eventually stop unless pushed or pulled constantly. They do *not* keep moving in their identical state of motion. How is this fact reconciled with the first law?

**Newton's first law**

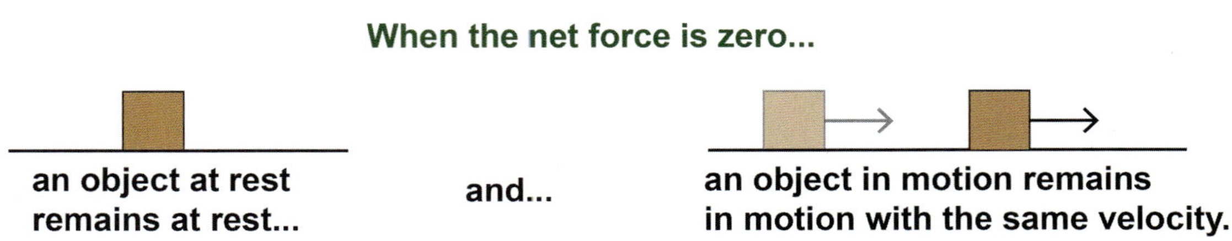


**What is friction?**

The first law is not violated because *friction* creates *another force* whenever motion occurs in the real world. **Friction** is a "catch-all" word used to describe any resistive force that may be caused by motion and always acts to resist motion. When a box slides across the floor, the sliding motion between the box and the floor results in a friction force that acts on the box to slow it down. The reason you must continually push a box to keep it moving at constant speed is to counteract friction. A box moving with constant velocity has *zero* net force acting on it because your applied pushing force cancels the force of friction.

**Effect of friction and the first law**

**How do I use the first law?**

The first law tells you how to approach problems that involve forces and motion. If things are at rest, and stay at rest, then the first law requires that *the net force must be zero.* This means that you can use the condition of equilibrium to find unknown forces. The same is true for situations of constant velocity. *Constant velocity is a condition of force equilibrium.* If the net force is *not* zero, then we need Newton's second law.

**What is the second law?**

**Newton's second law of motion** is probably the most widely used equation in all of physics. It says that the acceleration of an object is equal to the force you apply divided by the mass of the object. If you apply more force to an object, it accelerates at a higher rate. For a given force, an object with more mass will accelerate less. If the net force is *zero* then the acceleration is also zero, which means no change in speed or direction. The first law is therefore a special case of the second law, when $a = 0$.

Interactive equation

**Newton's second law**

$$(5.3) \quad a = \frac{F}{m}$$

$F$ = net force (N)
$m$ = mass (kg)
$a$ = acceleration ($m/s^2$)

**How do I use the second law?**

**The acceleration of an object equals the net force acting on the object divided by the object's mass.**

| Given | Find | Equations | Example |
|---|---|---|---|
| Net force and mass | Acceleration | $a = \frac{F}{m}$ | A net force of 100 N acts on a 5 kg mass, which starts at rest. What distance does the object move if the force acts in a straight line for 10 s? |
| | Velocity and/or position | $v = v_0 + at$ <br> $x = x_0 + v_0 t + \frac{1}{2}at^2$ | |
| Mass and acceleration | Net force | $F = ma$ | What force is required to cause a 100 kg sled to accelerate at 2 $m/s^2$? |
| Force and acceleration | Mass | $m = \frac{F}{a}$ | A 100 $m/s^2$ acceleration is observed for a rocket when its engine produces a force of 25 N. What is the mass of the rocket? |

**When to use each formula**

The three forms of the second law shown in the table above are each convenient for finding a different quantity. Note that the equations of motion are also included in the first row. In many problems you want to know the position and/or speed of an object given a force acting over some interval of time. To solve these problems the second law is used to find the acceleration, which is then used to determine the speed and position.

**Units and the second law**

As with all equations, *units* must be consistent. One newton equals one kilogram times one meter per second per second ($m/s^2$). When using the second law you must be sure to convert all information into consistent units. Speeds must be converted to meters per second. Masses in grams should be converted to kilograms and forces in pounds should be converted to newtons.

**Things to remember!**

1. Always remember that the $F$ appearing in the second law is the *net* force.
2. The second law is a *vector* relationship and the direction of acceleration $a$ is the same as the direction of the net force $F$.
3. If the motion is constrained, such as along a ramp, then the net force will always be along the allowed direction of motion.

**The acceleration is always in the direction of the net force.**

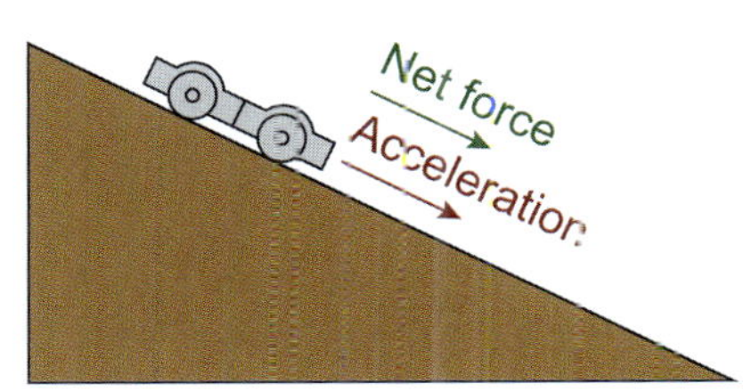

## Determining motion from forces

**Dynamics problems**

A *dynamics* problem involves accelerated motion and force. In dynamics problems, the second law is used to calculate the acceleration and resultant motion of an object from a known mass and known forces. For example, the acceleration of a rocket can be calculated from the thrust force of the engines and the mass of the rocket.

1. The acceleration is always in the direction of the net force.
2. Consistent units must be used for force, acceleration, velocity, distance, and time.

| Problem | Solution strategy |
|---|---|
| Find acceleration when given mass and force. | $a = \frac{F}{m}$ |
| Find velocity when given mass, force, and time. | $a = \frac{F}{m}$ ➡ $v = v_0 + at$ |
| Find position when given force, mass, and time. | $a = \frac{F}{m}$ ➡ $v = v_0 + at$, $x = x_0 + v_0 t + \frac{1}{2}at^2$ |

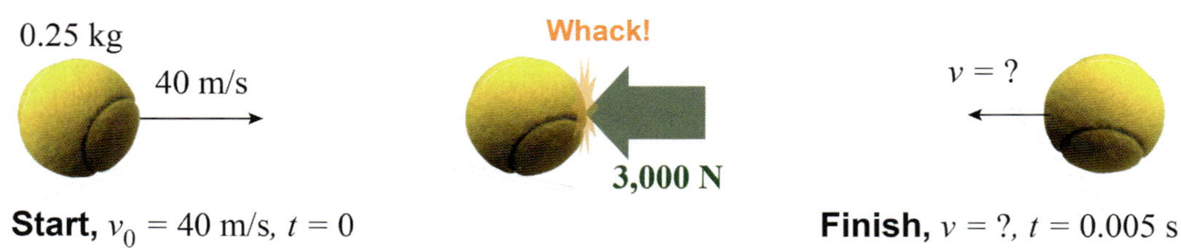


**A basic force and velocity problem**

A 250 g ball traveling to the right is hit by a racquet, which applies a force of 3,000 N to the left for 0.005 s. What is the velocity of the ball afterward?

*Asked:* final velocity $v$ of the ball

*Given:* $m = 0.25$ kg, $v_0 = +40$ m/s, $t = 0.005$ s, $F = -3{,}000$ N

*Relationships:* $a = F/m$, $v = v_0 + at$

*Solution:* Find the acceleration:

$$a = F/m = (-3{,}000 \text{ N}) \div (0.25 \text{ kg}) = -12{,}000 \text{ m/s}^2$$

Find the final velocity:

$$v = v_0 + at = (40 \text{ m/s}) + (-12{,}000 \text{ m/s}^2)(0.005 \text{ s}) = -20 \text{ m/s}$$

**Why do problems ask for the *minimum* force?**

In many problems you will see terms such as "minimum force" used. The word "minimum" is there because it is always possible to have a larger force acting partially in the wrong direction. The minimum force required to accelerate a 1 kg object at 1 m/s$^2$ is 1 N. *No net force smaller than 1 N can cause this acceleration.* You could, however, apply a 100 N force and still have a *net force* of 1 N if there were other forces acting, such as friction or normal forces.

**These situations all have a net force of 1 N and cause an acceleration of 1 m/s².**

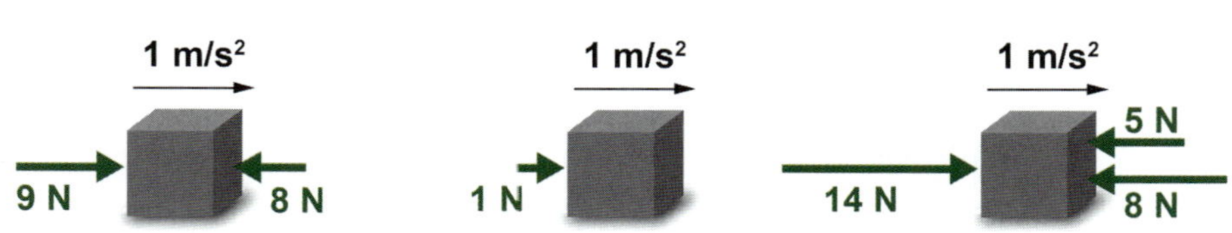

**How is the second law used to find a force?**

A good example of determining force from motion is the specification of aircraft engines. An aircraft must reach a minimum speed to take off and fly. The mass of the plane is known. The runway length is also known, as is the maximum comfortable acceleration for passengers. Newton's second law is applied to calculate how much thrust force must be supplied by the engines.

**Solution strategy for typical problems**

| Problem | Solution strategy |
|---|---|
| Find force when given mass and acceleration. | $F = ma$ |
| Find force when given mass, a change in velocity, and a change in time. | $a = \dfrac{\Delta v}{\Delta t}$ → $F = ma$ |
| Find force when given mass, a change in position, and a change in time. | $x = x_0 + v_0 t + \frac{1}{2}at^2$, $v = v_0 + at$ → $a$ → $F = ma$ |

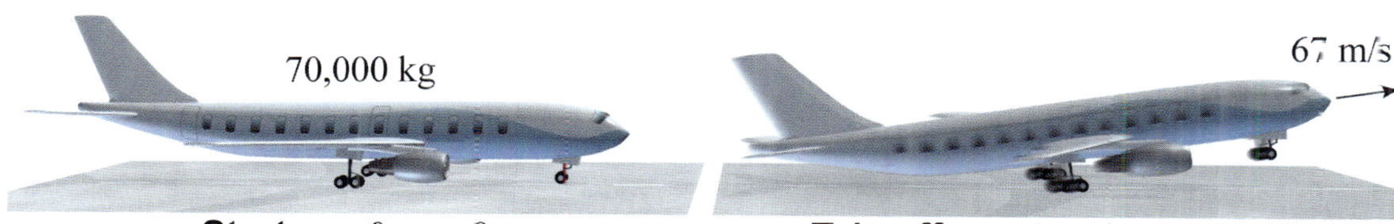


**Start,** $v = 0$, $t = 0$ **Takeoff,** $v = 67$ m/s, $t = 11$ s

**A problem giving velocity and time**

A 70,000 kg aircraft reaches a takeoff velocity of 67 m/s (150 mph) in 11 s. Calculate the minimum force required from the engines.

| | |
|---|---|
| *Asked:* | force $F$ |
| *Given:* | $m = 70{,}000$ kg, $v_f = 67$ m/s, and $t = 11$ s |
| *Relationships:* | $F = ma$, $a = \Delta v/\Delta t$ |
| *Solution:* | Find acceleration first: $a = \Delta v/\Delta t = (67 \text{ m/s}) \div (11 \text{ s}) = 6.1 \text{ m/s}^2$<br>Next, find force: $F = ma = (70{,}000 \text{ kg})(6.1 \text{ m/s}^2) = 427{,}000$ N |
| *Answer:* | 427,000 N |

**A problem giving position and time**

A 65 kg sprinter reaches a maximum speed of 11 m/s at a distance of 10 m from the start of a race. What average force must the sprinter's muscles create?

**Start**
$v_0 = 0$
$x_0 = 0$

**At 10 m**
$v = 11$ m/s
$x = 10$ m

| | |
|---|---|
| *Asked:* | force $F$ |
| *Given:* | $m = 65$ kg, $v_f = 11$ m/s, and $x = 10$ m |
| *Relationships:* | $F = ma$, $x = x_0 + v_0 t + ½at^2$, $v = v_0 + at$ |
| *Solution:* | Use both motion equations to find acceleration first:<br>$v = v_0 + at \rightarrow t = v/a$<br>$x = x_0 + v_0 t + ½at^2 \rightarrow x = ½at^2$<br>$x = ½a(v/a)^2 \rightarrow x = v^2/2a$<br>$a = v^2/2x = (11 \text{ m/s})^2 \div (2)(10 \text{ m}) = 6.05 \text{ m/s}^2$<br>Next, find force:<br>$F = ma = (65 \text{ kg})(6.05 \text{ m/s}^2) = 393$ N |
| *Answer:* | 393 N |

# Investigation 5A

## Newton's second law

**Essential questions** *How is an object's acceleration related to the net force acting on the object?*

When the forces acting on an object are unbalanced, the object accelerates. Newton's second law describes how an object's acceleration is related to the amount of net force acting on it. In this investigation you will explore this relationship.

### Force and acceleration

1. Open the **05A_NewtonsSecondLaw** experiment file, and then connect your Smart Cart using Bluetooth.
2. Set up the equipment like the picture. Be sure the track is level.

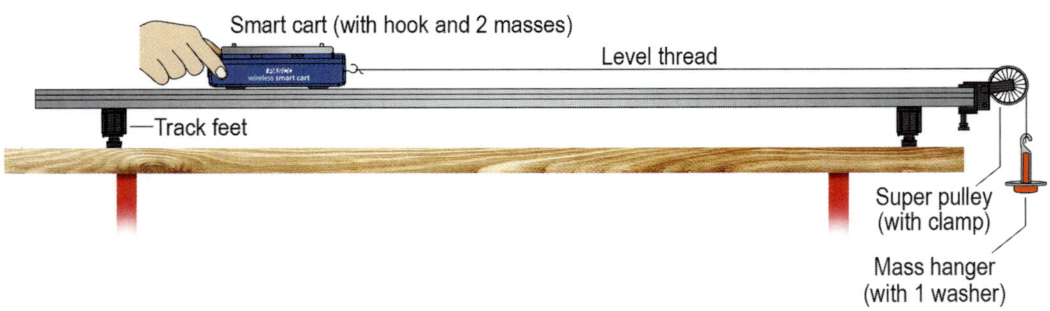


3. In your software, zero the Smart Cart force sensor while nothing is touching the hook.
4. Pull the cart to the end of the track, or until the mass hanger hangs just below the pulley, and then record data as you release the cart to roll freely down the track. Catch the cart before it hits the pulley.
5. Record five trials of data using the same steps, adding one more washer to the mass hanger before each trial: Trial 1 = 1 washer, Trial 2 = 2 washers, Trial 3 = 3 washers, and so on.
6. For each trial, find the cart's acceleration (slope of velocity graph) and average net force on the cart (net force = force measured by the sensor) while it was rolling freely down the track (only while it was rolling freely). Record your values in a table.

**Table 1: Average net force and acceleration data of a cart**

| Trial | Net Force (N) | Acceleration ($m/s^2$) |
|---|---|---|
| 1 | | |
| 2 | | |
| 3 | | |
| 4 | | |
| 5 | | |

7. Go to the second page of the experiment file and enter the data from the table in the experiment file. This data will appear in the graph on the same page.

*a. Sketch a copy of your net force versus acceleration graph. Be sure to label the axes.*

*b. How was the cart's acceleration different when the applied net force was greater? Support your answer with data.*

*c. What is the shape of your net force versus acceleration graph?*

*d. What is the slope of your net force versus acceleration curve? How does the slope value compare to the total mass of the cart plus any mass on it?*

*e. There are two common mathematical expressions for Newton's second law. One of these expressions is $F = ma$. How does your data support this mathematical relationship?*

**Forces only come in pairs**

The first and second laws apply to single objects and the forces that act on them. **Newton's third law of motion** addresses interactions between objects, or between objects and the environment. Every exchange of force comes from the interaction of at least *two* objects. For example, throwing a ball is an interaction between the *ball* and *you.* The third law says that *forces in nature always occur in pairs,* like the top and bottom of a sheet of paper. You cannot have one without the other. When you apply an *action* force to throw a ball, you feel the *reaction* force from the ball pushing back against your hand.

**Forces *always* come in pairs.**

**Action and reaction forces**

Consider riding a bicycle. Pushing on the pedals causes a force to be transmitted through the chain to the wheel, which pushes against the ground. This force acts *against the ground.* What force acts against the bicycle to move *you?* The answer is the **reaction force** of the ground acting against your wheel. This reaction is the equal and opposite partner of the action force of your wheel against the ground. *The reaction force is what moves you.* The same is true of driving and walking. We push against the ground, but it is the equal and opposite reaction of the ground pushing back that actually moves us.

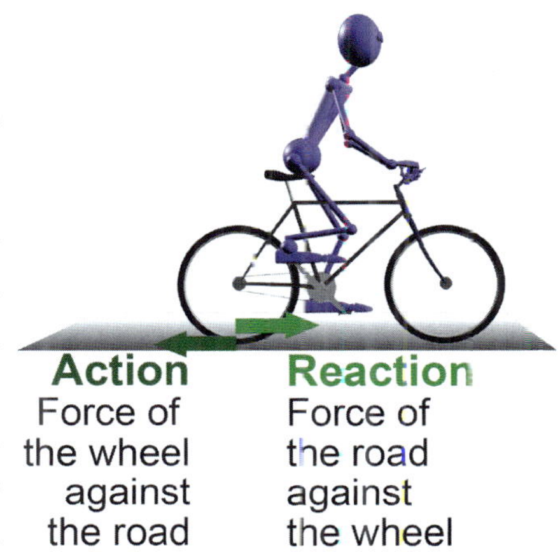


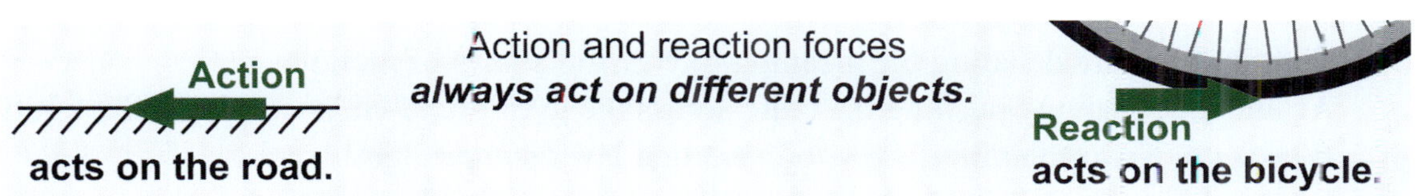


**Normal forces**

The *normal* forces that occur between objects and supporting surfaces are also reaction forces. Consider a box on the table. The box pushes down on the table with a force equal to its weight. That force acts on the table. The reaction force is the table acting back up on the box. The forces that hold objects up are reaction forces.

**Action and reaction labels**

Don't get confused about which force is the action and which is the reaction. The words *action* and *reaction* are just labels and it does not matter which is which.

1. All forces come in pairs. If one is the action, then the other is the reaction.
2. Action–reaction forces are *always* exactly equal in strength and opposite in direction.
3. The action and reaction forces always act on different objects, so they never cancel each other out. They also act *simultaneously* with each other.

# Section 2 review

Newton's three laws describe how motion is affected by forces. The first law says that, in the absence of a net force, an object at rest or in motion will continue to be in the identical state of rest or motion (that is, it will continue to have the same velocity). The second law states the relationship among net force, mass, and acceleration experienced by an object. The third law states that forces occur in pairs: For every "action" force (on a particular object), there is an equal but opposite "reaction" force operating at the *same* time, but on *another* object.

| | |
|---|---|
| **Vocabulary words** | **friction, Newton's first law of motion, Newton's second law of motion, Newton's third law of motion, reaction force** |

| | |
|---|---|
| **Key equations** | $a = \frac{F}{m}$ |

## Review problems and questions

1. Which of Newton's three laws of motion best applies to each statement?
   a. A 120 lb student sat on a stool, and the stool applied a 120 lb normal force to the student.
   b. The spacecraft cruised toward the North Star at a constant 500 m/s through the vacuum of interstellar space.
   c. Now that the moving van was full, its driver had to apply more force to reach highway speed before the on-ramp ended.

2. Two students are debating the meaning of Newton's laws of motion.
   Arjun: "See that guy pushing the crate across the gym floor? He's working hard, but the crate never speeds up. No acceleration! Newton's second law *can't* be right!"
   Buell: "I wonder. Newton's second law has to do with *net* force, which is the total force you are left with after you add up *all* the forces pushing on something. I think there is another force at work."
   Which student is correct, and why?

3. Rank the following objects by the magnitude of the net force that each experiences, from weakest to strongest:
   a. 150 g hockey puck that leaves a player's stick at 30 m/s and enters the goal one second later at 29 m/s
   b. 20 g marble that has been dropped 1 m above the floor (and is still falling)
   c. golf ball that is tapped with a 0.5 N force while resting on a table

A **spring** is a device specifically created to provide controlled amounts of force in many applications. The force exerted by a spring depends on how much it is deformed from its free length. The free length is the length of the spring when it is not exerting any force. The farther the spring is extended or compressed from its free length, the stronger the force the spring exerts on whatever is causing it to extend or compress.

## Springs and spring forces

**Hooke's law**

Springs come in many types. Three common types are shown below. In each, the restoring force is approximately proportional to the deformation $x$ of the spring relative to its free length. If you deform an ideal spring twice as far, it will produce twice the force. This relationship is known as **Hooke's law** and is mathematically expressed as equation (5.4). The force $F$ in equation (5.4) is the force that the *spring* exerts. It is equal in magnitude, but opposite in sign, to the force exerted by whatever deforms the spring, such as your hand.

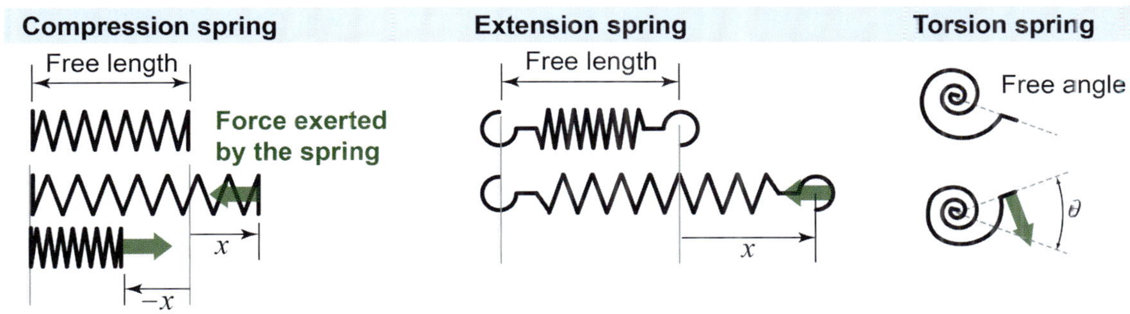


Interactive equation

**Hooke's law for springs**

$$(5.4)\quad F = -kx$$

$F$ = restoring force (N)
$k$ = spring constant (N/m)
$x$ = displacement from equilibrium (m)

**What is $k$?**

The quantity $k$ is called the **spring constant**, which is the magnitude of the force divided by the amount of deformation. The spring constant has units of force divided by distance. A spring constant of 1 N/m means the spring creates a force of one newton when stretched one meter. The spring constant is a property of the spring itself. A stiff spring deforms very little, even under a large force, and may have a value of $k$ = 100,000 N/m or higher. A loose spring takes very little force to deform and may have a value of $k$ = 100 N/m or less.

**A car suspension uses stiff (high-$k$) springs that move independently to allow the wheels to move up and down over bumps.**

**Why is there a negative sign in Hooke's law?**

The negative sign in equation (5.4) tells you that the force exerted by the spring acts in the opposite direction from the deformation. If you stretch the extension spring in the diagram above, the deformation $x$ is positive. The force is *negative*, indicating that it is to the left, opposite the deformation. If the deformation is negative, to the left, then the force is positive, to the right. *Be careful when using Hooke's law to use the proper sign for the deformation x.*

# Investigation 5B

## Hooke's law

**Essential questions** *How are force and displacement related when stretching a spring?*

The physics underlying a spring can be expressed by a straightforward relationship between how much the spring is extended (or compressed) and the restoring force the spring exerts. This relationship is known as Hooke's law. In this investigation you will measure force from the spring, and spring deformation (spring deflection) using the force and position sensors on the Smart Cart.

### Part 1: Extension and spring force

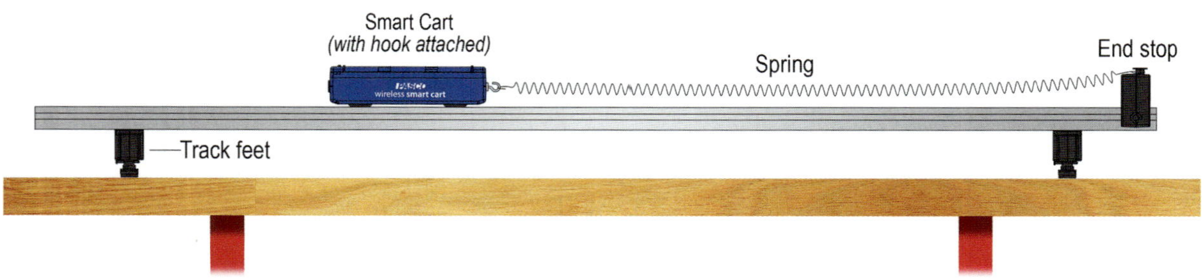


1. Set up the equipment like the picture using the loosest of the three springs.
2. Open the **05B_HookesLaw** experiment file in your software, and then connect your smart cart using Bluetooth.
3. In your software, zero the Smart Cart force sensor while nothing is touching the hook.
4. Begin recording data, and then roll the cart backward (extending the spring) about 10 cm and hold it in place. Click the check mark to record your first data point.
5. Keep pulling the cart backward and record four more data points at four different deformation lengths (any lengths will work as long as they are in order from shortest to longest).
6. Stop recording data, and then copy your data into a table. Sketch a copy of your graph.

### Part 2: Stiff and loose springs

1. Repeat the experiment using the medium and stiff springs.
2. Tabulate your data, and then sketch the graphs for each spring.

a. *When you stretch the stiffest spring by hand, how does it feel or respond that is different from the loosest spring? In supporting your answer, use data from your investigation.*
b. *How does the extension of the stiffest spring compare to the loose one for the same applied force?*
c. *What are the slope values of the three graphs? (Include units in your answer.)*
d. *What physical quantity do the slopes of your graphs represent? Why?*
e. *What is the spring constant of each of the springs? (Include units in your answer.)*
f. *Use your data to determine the force each spring would exert at these other deformations.*

| Deformation $x$ (m) | Loose Spring Force $F$ (N) | Medium Spring Force $F$ (N) | Stiff Spring Force $F$ (N) |
|---|---|---|---|
| 0.47 | | | |
| 1.31 | | | |
| 2.22 | | | |

**How do I calculate force with Hooke's law?**

Hooke's law ($F = -kx$) tells us the force exerted *by* an ideal spring when deformed a distance $x$ from its free length. The free length is the length of the spring when the deformation $x$ equals zero and the force $F$ also equals zero. Be careful that you use the *deformation* in Hooke's law. The deformation is the *change* in length, and not the actual length of the spring itself.

**Example problem**

A spring has a free length of 40 cm and is stretched to a length of 60 cm. What force does the spring exert if $k = 800$ N/m?

| | |
|---|---|
| *Asked:* | force $F$ |
| *Given:* | free length = 40 cm, extended length = 60 cm, $k = 800$ N/m |
| *Relationships:* | $F = -kx$ |
| *Solution:* | The deformation of the spring is the change in length, which is 60 − 40 = 20 cm. Then calculate force: $F = -kx = (-800 \text{ N/m})(0.2 \text{ m}) = -160 \text{ N}$ Note that we had to convert to meters before using Hooke's law because the spring constant is given in newtons per meter. |

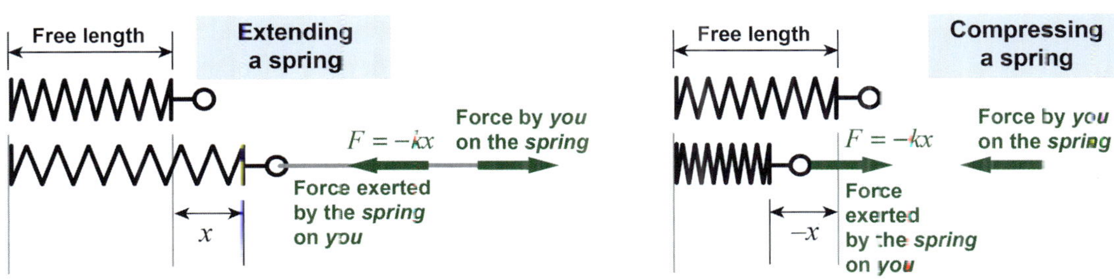


**The negative sign**

The negative sign in Hooke's law tells you that the force the spring exerts is opposite to the direction of the deformation. Consider a spring that is fixed at one end. If positive is to the right, then a positive displacement $x > 0$ means that the force exerted by the spring is in the negative direction. Conversely, a negative displacement means that the force exerted by the spring is in the positive direction.

A force of 50 N is applied to compress a spring that has $k = 1{,}000$ N/m. What is the deflection of the spring?

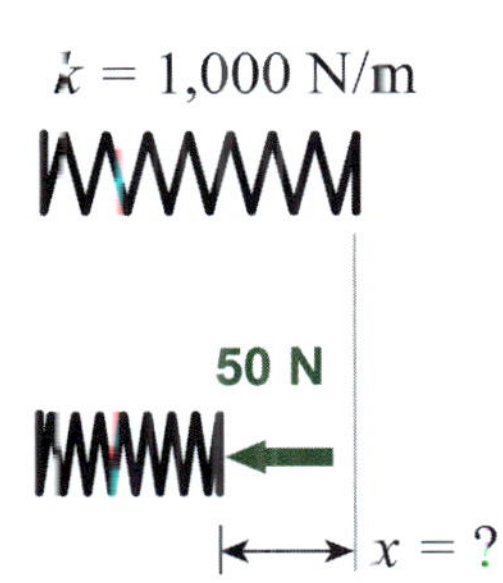


| | |
|---|---|
| *Asked:* | deflection $x$ |
| *Given:* | $k = 1{,}000$ N/m; $F = 50$ N |
| *Relationships:* | $F = -kx$ |
| *Solution:* | "Deflection" is another word for deformation or displacement. Solve Hooke's law for displacement: $x = -F/k = -(50 \text{ N}) / (1000 \text{ N/m}) = -0.050 \text{ m}$ |

**Solving spring problems**

Think about action–reaction forces when dealing with springs. Hooke's law gives the force exerted *by the spring*. An equal and opposite reaction force must be exerted by you to stretch the spring in the first place. That force acts on the spring!

## Generalization of Hooke's law

**Elasticity and brittleness**

A generalized version of Hooke's law describes the deformation of three-dimensional solids. The **elasticity** of a material describes the material's ability to bend or change shape without breaking. Rubber is very elastic and a rubber band can easily stretch 200% or more before breaking. On the opposite extreme, glass is not very elastic at all. Glass breaks before stretching even a few percent. Materials that break before stretching are called **brittle** and include glass, ceramic, and stone. Plastics and metals can fall in between the extremes of high elasticity and brittleness.

| Hooke's law for three-dimensional solids |
| --- |
| The deformation $x$ is proportional to the applied force $F$. |

$$x = \frac{F}{k}$$

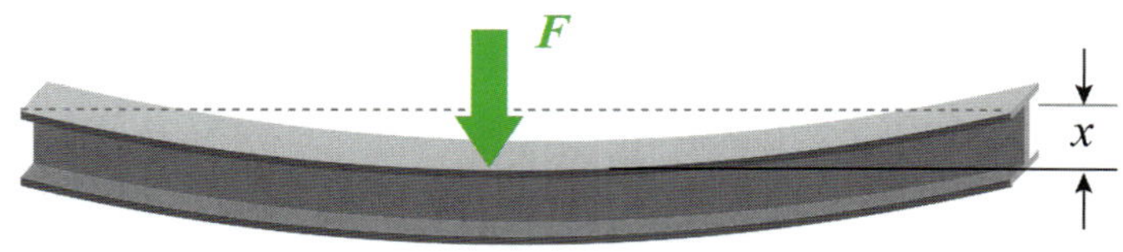


*Deflection is greatly exaggerated to show the relationship.*

**Elasticity and Hooke's law**

When you place a brick on a table, the table exerts precisely the right normal force up on the brick to maintain equilibrium (zero net force). The table does not solve the problem of how much force to exert using free-body diagrams! Instead, the table behaves like a very stiff *spring*. All materials have some degree of elasticity and a very sensitive instrument would show that the table surface deflects a tiny amount under the weight of the brick. Hooke's law applies and the force exerted on the brick increases as the deflection of the table increases. *The table continues to deflect until equilibrium is reached, at which point the upward force from the table exactly balances the downward force from the weight of the brick.*

| Spring | Three-dimensional solid |
| --- | --- |
| A brick compresses a spring until forces balance. | A table top deflects slightly under the force. The deflection increases until the forces in the vertical direction balance. |

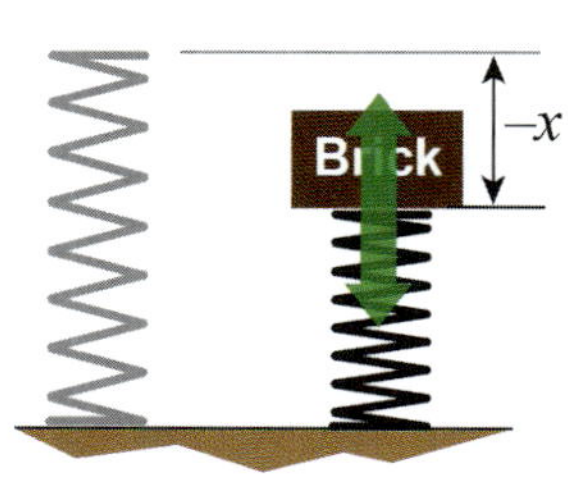


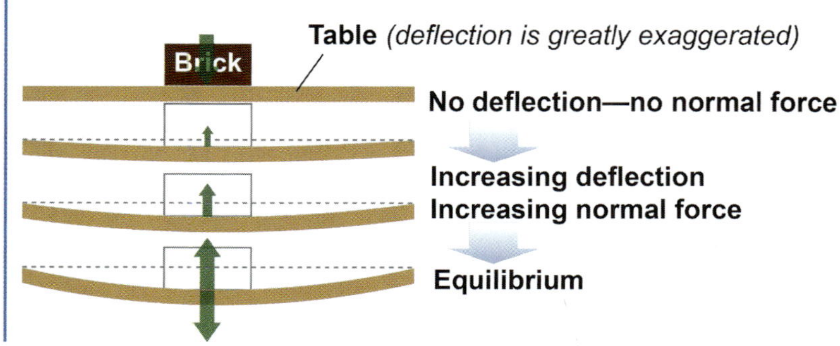


**Strength of materials**

When you talk about the *strength* of an object, such as a beam, what you are really describing is the size of its "spring constant" $k$. The **material strength** for an object or material is its ability to sustain force without permanent deformation or breaking. A large $k$ means a large force will cause very little deflection. For example, a *structural engineer* designing a building based on the International Building Code might specify that a floor deflect no more than 1/360 of its longest span when loaded at 700 N/m$^2$. The engineer chooses the size of the beams using Hooke's law!

# Section 3 review

When you stretch a spring, it pulls back on you; when you compress a spring, it pushes back. Moreover, the force exerted by the spring is generally proportional to the distance you push or pull it. This relationship is known as Hooke's law. The strength of a spring is quantified by its spring constant $k$. A three-dimensional solid, such as a tabletop, sags under the weight of a load until a spring-like force opposes the weight. In general, the sag distance is proportional to the force applied by the load. Brittle objects crack or disintegrate after bending a very small amount, whereas elastic objects can be bent or stretched by large amounts without breaking.

| | |
|---|---|
| **Vocabulary words** | **spring constant, spring, Hooke's law, elasticity, brittle, material strength** |
| **Key equations** | $F = -kx$ |

## Review problems and questions

1. Is the force $F$ in Hooke's law ($F = -kx$) the force applied *to* the spring by an outside agent, such as your hand?

2. What is the physical significance of the negative (minus) sign in Hooke's law?

3. Which of the following is a physically possible unit for the spring constant $k$? (More than one answer may be correct, and correct answers do not have to be in SI or metric units.)
    a. meters per second squared ($m/s^2$)
    b. pounds per inch (lb/in)
    c. newtons per millimeter (N/mm)
    d. meters per newton (m/N)

4. Chloe suspends a spring from a secure laboratory stand. Then, one by one, she attaches different metal objects to the spring. Each time, she measures and plots the metal object's weight $mg$ (in newtons) and the spring's displacement $x$ (in centimeters). She then plots her data as shown here. What is the spring constant $k$ for her spring (in SI units)?

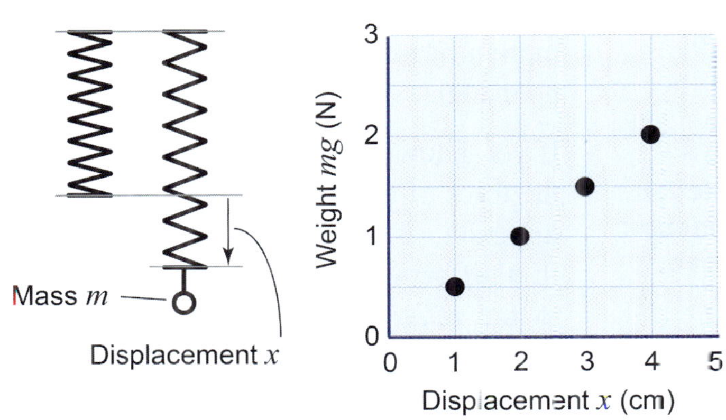

## 5.4 - Friction

Friction is not a specific force in the same way that gravity is a force between two masses. Instead, friction is a "catch-all" term that collectively refers to forces that may be caused by motion and that act to reduce motion. Friction transforms the energy of motion into thermal energy or the wearing away of moving surfaces. *Kinetic friction* describes the friction between sliding surfaces and *rolling friction* describes friction in wheels. *Static friction* describes the tendency of objects to stick to each other and not move until a minimum force is applied to "break things loose."

### Some causes of friction

**Friction** Everything that moves in the macroscopic world feels friction. We model the effect of friction as a force that is opposite in direction from actual motion, or the motion that *might* occur if there were no friction. Sliding, moving through air or water, or rolling are some forms of motion that generate friction that we can model this way.

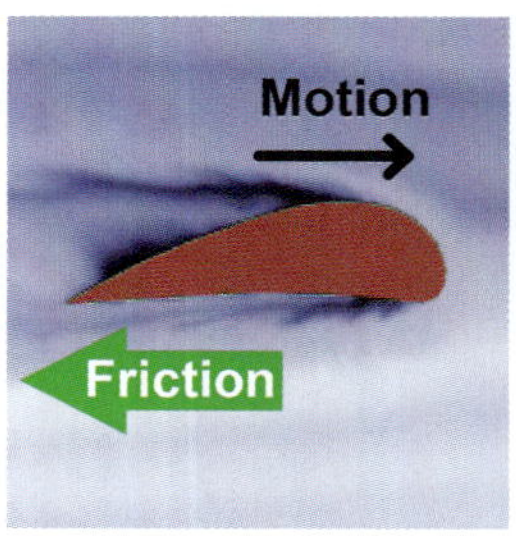


**Air friction**

comes from air being pushed aside or flowing around surfaces such as the body of a car or the wing of an aircraft.

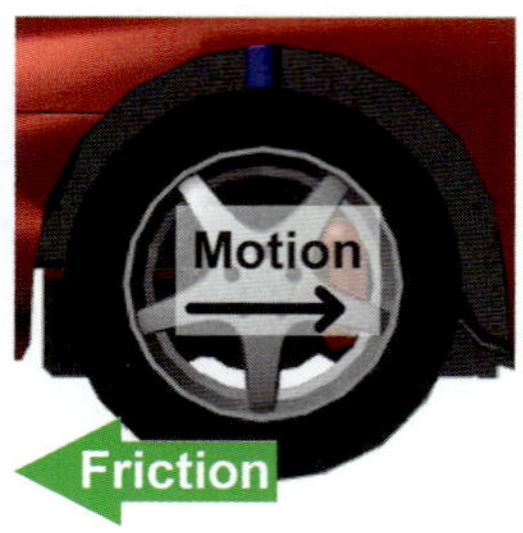


**Rolling friction**

comes from rolling contact between two surfaces, such as a wheel and the road.

**Sliding friction**

comes from sliding contact between two surfaces, such as the bottom of skis and a snow-covered hill.

**Viscous friction**

comes from liquids being displaced or forced to flow around or through objects such as pipes or boats.

**Where friction comes from** On a microscopic level friction comes from atomic-scale attraction between adjacent particles of matter. Slippery liquids such as oil can reduce friction but even the slipperiest oil cannot eliminate friction altogether. The only way to eliminate friction is to eliminate the relative motion of matter—such as by traveling through deep space where there is no air, and therefore no matter.

**The direction of friction forces** If you know the direction of motion, then the force of friction will be in the opposite direction. (More specifically, the force of friction between two surfaces is opposite to the *relative* motion between the two surfaces.) If a box slides to the left, the force of sliding friction points to the right. If an object is falling down, the force of air friction points up. If the direction of motion is unknown, the direction of friction forces can be tricky to determine.

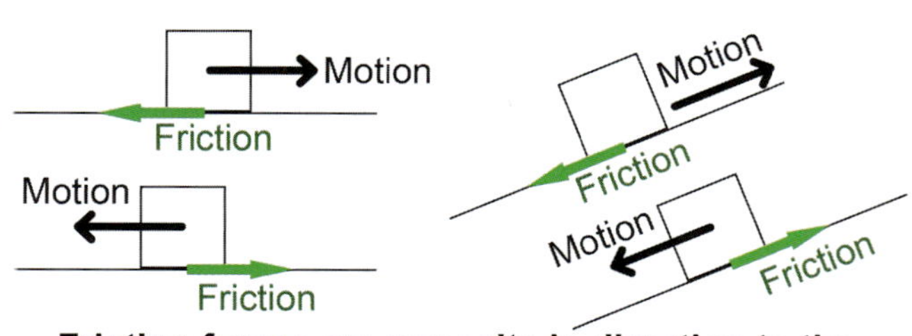


**Friction forces are opposite in direction to the (relative) motion that produces them.**

**Your sense of friction**

A board loaded with bricks takes more force to slide than an unloaded board. To account for the difference using a simple model, we assume that the friction force $F$ is proportional to the normal force acting between the sliding surfaces. As an equation, the friction force equals the normal force $F_N$ times a constant called the *coefficient of friction* $\mu$, which depends on the materials of the two surfaces.

A board

**Which takes more force to keep moving?**

A board with 3 bricks on it

**Definition of static friction**

**Static friction** occurs when two surfaces *could* move relative to each other but are not yet moving. The maximum force of static friction is modeled with equation (5.5).

**Static friction**

$$(5.5) \quad F_f = \mu_s F_N$$

$F_f$ = maximum force of static friction (N)
$\mu_s$ = coefficient of static friction
$F_N$ = normal force (N)

The **coefficient of static friction** $\mu_s$ is a number generally between 0 and 1. The subscript $s$ in $\mu_s$ identifies the friction as *static*.

**Direction of static friction**

Like other forms of friction, the direction of static friction depends on what other forces are present. The direction of static friction is

1. parallel to the contact surface and
2. opposite to the net of all other forces acting.

**An example**

A 10 N wood board is on a table. How much force does it take to make the board slide if $\mu_s = 0.2$?

*Asked:* friction force $F_f$
*Given:* $\mu_s = 0.2$, $F_N = 10$ N
*Relationships:* $F_f = \mu_s F_N$
*Solution:* The board will *just* slide when $F = F_f$, so
$F_f = (0.2)(10 \text{ N}) = 2$ N

Free-body diagram

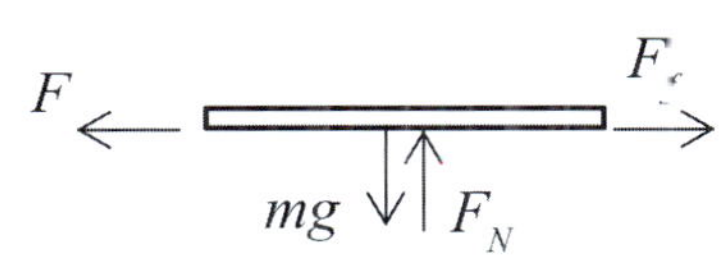


**Maximum force of static friction**

The coefficient of static friction $\mu_s$ predicts the *maximum* force of static friction. The word maximum is very important because the actual force of static friction is equal and opposite to the net of all other applied forces up to the maximum value. In the example above, the coefficient of static friction for wood sliding on wood is 0.5 and the weight of the board is 10 N. The *maximum* force of static friction would be 5 N. The actual force of static friction could be anything between 0 and 5 N. If the net applied force is 3 N, static friction creates a resisting force of 3 N.

**Typical values of the coefficient**

The table shows typical values for $\mu_s$. Coefficients vary greatly with conditions. A rubber tire on wet concrete has less than 1/3 of the friction compared to dry concrete. Real friction forces are highly variable. Numbers in tables are only *rough estimates!*

| Materials in contact | | $\mu_s$ |
|---|---|---|
| Rubber | Dry concrete | 1.0 |
| Rubber | Wet concrete | 0.3 |
| Wood | Wood | 0.5 |
| Wood | Concrete | 0.6 |
| Steel | Dry steel | 0.8 |
| Steel | Oiled steel | 0.16 |
| Steel | Teflon® | 0.04 |

**Sliding friction**

Once an object starts moving the strength of the friction force usually drops, because it takes less force to keep surfaces sliding than it does to break them free. The **coefficient of kinetic friction** $\mu_k$ is the ratio of the force of **sliding friction** (or kinetic friction) to the normal force between the sliding surfaces. The subscript $k$ in $\mu_k$ refers to kinetic. The table below gives some representative values for both static and kinetic friction.

**Kinetic friction**

$$(5.6)\quad F_f = \mu_k F_N$$

$F_f$ = force of kinetic friction for dry sliding (N)
$\mu_s$ = coefficient of static friction
$F_N$ = normal force (N)

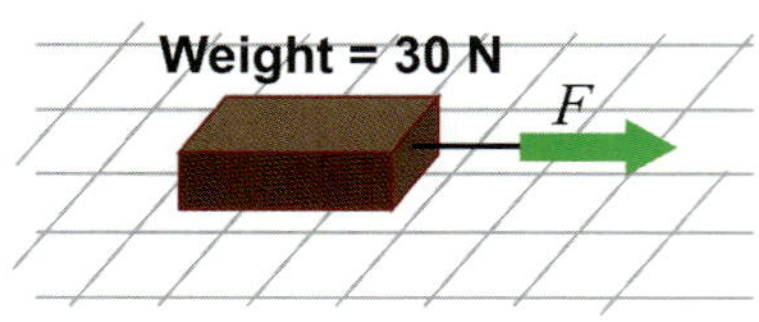


**Free-body diagram**

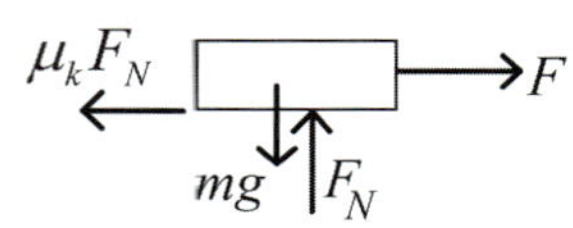


***y*-direction**

$$F_N = mg$$

***x*-direction**

$$F = \mu_k F_N = \mu_k mg$$

**An example problem**

What force is needed to drag a 30 N brick across a wood table at constant speed? Brick has a surface that is similar to that of concrete, so from the table we find $\mu_k = 0.5$.

*Asked:* friction force $F_f$

*Given:* $\mu_k = 0.5$, $F_N = 30$ N

*Relationships:* $F_f = \mu_k F_N$

*Solution:* The board will slide at constant speed when $F = F_f$, so

$$F_f = (0.5)(30\text{ N}) = 15\text{ N}$$

A force of 15 N is required.

| Materials in contact | | $\mu_s$ | $\mu_k$ |
|---|---|---|---|
| Rubber | Dry concrete | 1.0 | — |
| Rubber | Wet concrete | 0.3 | — |
| Wood | Wood | 0.5 | 0.3 |
| Wood | Concrete | 0.6 | 0.5 |
| Wood | Aluminum | 0.3 | 0.2 |
| Metal | Wood | 0.2–0.6 | — |
| Felt | Aluminum | 0.26 | 0.22 |
| Steel | Dry steel | 0.80 | 0.50 |
| Steel | Oiled steel | 0.16 | 0.09 |
| Steel | Teflon® | 0.04 | 0.04 |

**Lubrication**

The coefficient of kinetic friction for dry steel sliding against steel is 0.5, which is much greater than the coefficient for steel on oiled steel, which is 0.09. **Lubrication** is the technology of using substances such as *oil* to reduce friction. Oil-lubricated surfaces have much lower friction because a thin layer of oil prevents the surfaces from touching each other. Notice that the coefficient for steel sliding on Teflon® is also very low. Teflon is a special type of polymer invented by DuPont that is very slippery even when dry. Because Teflon is also chemically inert, Teflon-coated wearing surfaces are used in many artificial joints to replace injured knees and hips.

**Useful friction**

Friction is both useful and wasteful. Without the friction between the soles of your shoes and the floor, you could not walk. Engineers design rubber compounds and tread patterns to provide static friction between car tires and the road even in wet or icy conditions. The brakes in your car or bicycle rely on friction to provide stopping force. Friction can also be wasteful. Only 13% of the energy from a gallon of gasoline goes into the forward kinetic energy of an average car. The remaining 87% is transformed into nearly useless heat through friction and the inefficiency of the engine.

# Investigation 5C

## Static and kinetic friction

**Essential questions** *What determines the force of friction?*

Friction is everywhere and can be either helpful or wasteful depending on the situation. In this investigation you will test models of friction against actual measurements to get a sense of how accurate these friction models are.

### Coefficient of static and kinetic friction using a friction block

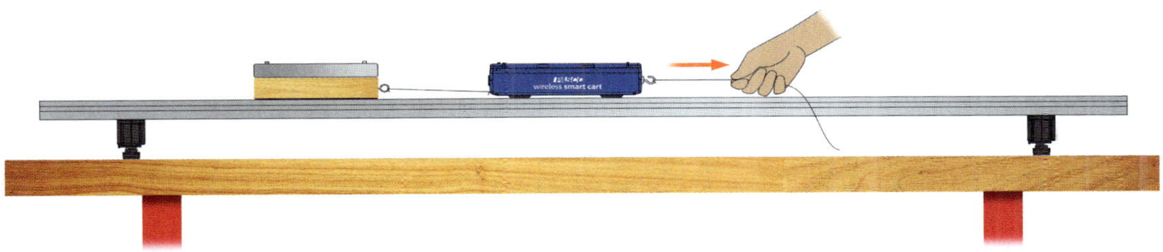


1. Open the experiment file **05C_Friction**, and then connect the Smart Cart to the software using Bluetooth.
2. Set up the equipment like the picture. Zero the Smart Cart force sensor while nothing is touching the hook.
3. Start data collection, and then very slowly pull on the string, increasing the force you exert until the block starts to slide. Once the block is sliding, keep pulling at a constant speed until you get to the edge of the table, and then stop data collection.
4. Record the mass of the block in a table.
5. Repeat this activity two more times, each time adding a 250-g mass on top of the block.
6. For each trial, use the graph tools to find the maximum force exerted just as sliding was about to start (static friction force), and the average force while the block was sliding (kinetic friction force). Record the values for each trial in the table.

a. *Draw two free-body diagrams of the friction block: one representing the moment just before it began to slide (static friction), and one representing the time when it was sliding at a constant speed (kinetic friction). Label all the forces acting on the block in each diagram, including friction.*

b. *The coefficients of static and kinetic friction are defined as the ratios of each frictional force to the normal force of the block. Write two equations: one for the coefficient of static friction, and one for the coefficient of kinetic friction–both in terms of the frictional force and the mass of the block.*

c. *Determine the coefficients of static friction and kinetic friction for each of the trials you performed. Enter these values in the table. Calculate the average for each.*

d. *The model for static friction treats $\mu_s$ as constant, even as the mass of the friction block was increased. Do your data support this model? Explain your answer.*

e. *Compare your average values to the tabulated values for the coefficients of static and kinetic friction in section 5.4 of your text. Calculate the percent error of your average values using the accepted values in the table.*

### Friction and wheels

**Rolling friction** occurs when two surfaces are in rolling contact with each other. Wheels are found throughout human technology because rolling motion usually results in much lower friction than sliding. Rolling friction is more complex than static or kinetic friction because the radius of the wheel matters as well as the nature of the contacting surfaces. This is because rolling friction comes from deformation and larger wheels deform proportionally less than smaller wheels. The force of rolling friction is approximately given by equation (5.7).

**Rolling friction**

$$F_f = \mu_r F_N \quad (5.7)$$

$F_f$ = force of rolling friction (N)
$\mu_r$ = coefficient of rolling resistance
$F_N$ = normal force (N)

### Calculating the coefficient

The **coefficient of rolling friction** $\mu_r$ depends on the size of the wheel. The parameter $b$ is used to calculate a value for $\mu_r$ if you know the radius of the wheel in meters. The coefficient of rolling friction is $\mu_r = b/R$, where $b$ and $R$ are both measured in meters. This causes the units of $b$ and $R$ to cancel to make $\mu_r$ dimensionless. The table below gives some values for wheels of 5 and 50 cm radius. You can see that the coefficient of rolling friction drops by a factor of 10 when the wheel radius increases from 5 to 50 cm.

**Rolling friction**

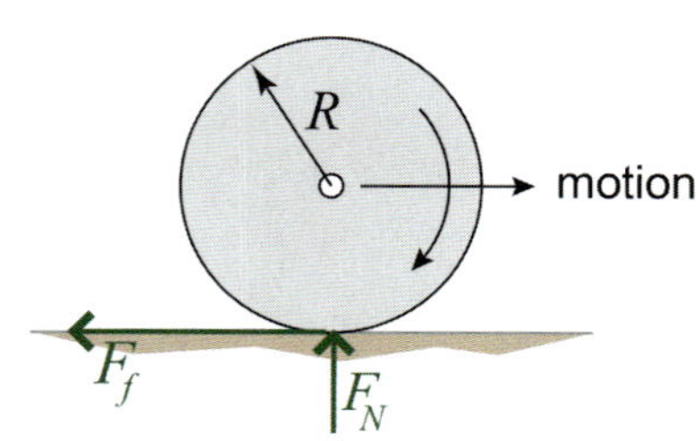


$$\mu_r = \frac{b}{R}$$

The coefficient of rolling friction is $b$ (from the table) divided by $R$ in meters.

| Coefficients of rolling friction | $b$ (m) | $\mu_r$ (5 cm wheel) | $\mu_r$ (50 cm wheel) |
|---|---|---|---|
| Steel on steel | 0.0005 | 0.010 | 0.001 |
| Wood on steel | 0.0012 | 0.024 | 0.002 |
| Wood on wood | 0.0015 | 0.030 | 0.003 |
| Plastic on steel | 0.002 | 0.040 | 0.004 |
| Hard rubber on steel | 0.008 | 0.16 | 0.016 |
| Hard rubber on concrete | 0.01–0.02 | 0.2–0.4 | 0.02 - 0.04 |
| Rubber on concrete | $1.5 \times 10^{-5}$–0.035 | 0.3–0.7 | 0.03 - 0.07 |

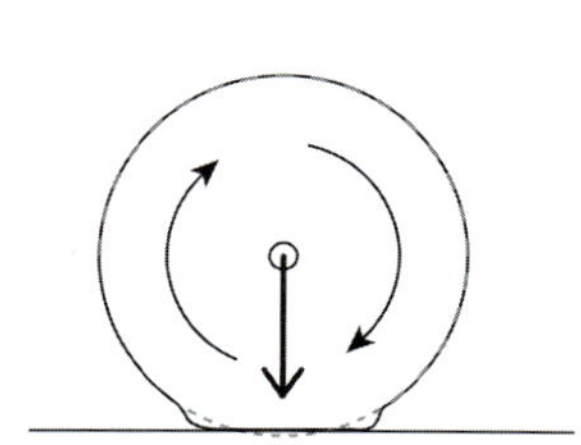

One cause of rolling resistance is the small deformation of the wheel as it contacts the ground.

### An example problem

A bicycle has two wheels 74 cm in diameter. What is the force of rolling friction on a level road if the weight of the bicycle and rider is 700 N?

*Asked:* friction force $F_f$

*Given:* $R = 0.74$ m, $F_N = 700$ N

*Relationships:* $F_f = \mu_r F_N$, $\mu_r = b/R$

*Solution:* Assuming rubber on concrete, $\mu_r = b/R = (0.035) \div (0.74) = 0.047$, so

$$F_f = \mu_r F_N = (0.047)(700 \text{ N}) = 33 \text{ N}$$

### Limitations of the friction models

The models for static, kinetic, and rolling friction are useful for making estimates. These models are *not* rigorously obeyed natural laws such as $F = ma$. Friction models are at best *rough approximations* of the actual friction in real-world situations. Friction is very dependent on the precise conditions of the surfaces, their temperature, and the presence of oils and dirt.

### Air resistance

The largest inefficiency in transportation vehicles, such as cars, trucks, and planes, involves overcoming *air resistance*. In boats, friction forces are even larger because water is both denser and more *viscous* than air. Friction from fluids, such as water or air, is caused by two factors:

1. the *shear* resistance of the fluid from viscosity and
2. the effort required to push the fluid out of the way.

### What is viscosity?

The **viscosity** of a fluid describes its resistance to flow. "Thick" fluids such as honey have a high viscosity and flow very slowly. "Thin" fluids such as water have a low viscosity and flow much faster under the same conditions. Viscosity depends strongly on temperature and is a major factor in lubricating oils for engines. The SI unit for viscosity is pascal-second (Pa s), but a commonly-used unit is the *poise*, which is equal to 0.1 Pa s. The specification "10W30" in a motor oil means the viscosity of the oil at 0°C ("W" means "winter") is no more than 3.1 poise while the viscosity at 100°C is no less than 0.1 poise.

### What causes air resistance?

For vehicles moving through water or air, the force of air resistance is dominated by inertial effects—essentially the force needed to push the fluid out of the way. The amount and rate at which air or water must be pushed out of the way depend on the shape and size of the object and also on the velocity of the object. When an object moves faster through a fluid, two factors contribute to increasing the air resistance:

1. more fluid must be displaced per second and
2. fluid must be accelerated out of the way more rapidly.

### Equation for fluid friction

For a given shape, fluid friction increases as the square of the speed. Doubling the speed of a car from 30 to 60 mph increases the air resistance by a factor of 4.

**Fluid resistance**

$$F_f = \frac{1}{2} c_d \rho A v^2 \tag{5.8}$$

$F_f$ = friction force (N)
$c_d$ = drag coefficient
$\rho$ = fluid density (kg/m$^3$)
$A$ = cross-sectional area (m$^2$)
$v$ = speed (m/s)

### What is the drag coefficient?

Equation (5.8) gives the friction force on an object moving at speed $v$ through a fluid of density $\rho$. The **drag coefficient** $c_d$ is a geometrical shape factor that describes the relative ease with which an object moves through air or water. An aerodynamic shape has a low drag coefficient. For example, $c_d$ is 0.04 for an airfoil. A blunt cube, in contrast, has a high drag coefficient $c_d = 1.05$. This means that, at the same speed, the force of air resistance on a cube is 26 times greater than on an airfoil with the same cross-sectional area!

**Drag coefficient for various shapes**

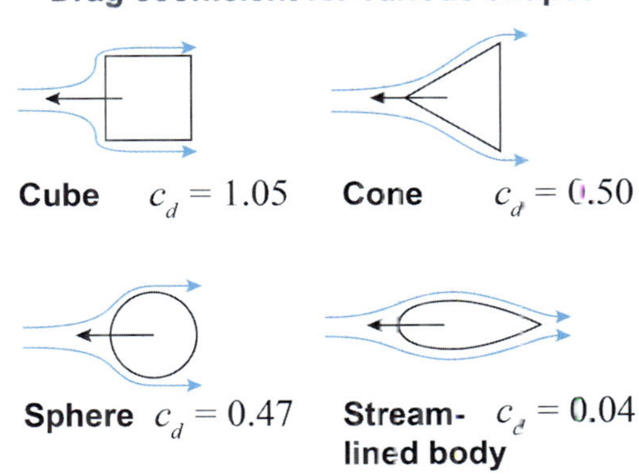


### Drag on a sprinter

When sprinter Usain Bolt set the world record of 9.58 s for the 100 m sprint at the World Championships in 2009, his position was tracked every 0.1 s using a laser velocity guard device. Physicists published calculations using these data to show that an astonishing 92% of Bolt's effort went into overcoming drag. They estimated his body's drag coefficient at $c_d = 1.2$, within the typical human range of 1.0–1.3.

# Section 4 review

The term *friction* covers a wide range of forces, all of which either oppose motion *or* oppose the forces that would cause motion. *Static friction* prevents an applied force from making an object slide. *Kinetic friction* creates a force that acts opposite to the direction of sliding motion. We model the force of friction as a coefficient times the normal force acting between moving surfaces. The concept of friction also extends to air resistance and viscosity, both of which resist an object's motion through a fluid.

| Vocabulary words | static friction, coefficient of static friction, coefficient of kinetic friction, lubrication, sliding friction, rolling friction, coefficient of rolling friction, viscosity, drag coefficient |
|---|---|

| Key equations | $F_f = \mu_s F_N$ | $F_f = \mu_k F_N$ | $F_f = \mu_r F_N$ | $F_f = \frac{1}{2} c_d \rho A v^2$ |
|---|---|---|---|---|

## Review problems and questions

1. A mover is pushing on a 40 kg crate that rests upon a hardwood floor. Her efforts proceed as follows:
   - I. She applies a horizontal 120 N force. The crate does not move.
   - II. She next applies a 160 N force. The crate remains motionless.
   - III. She increases her push to 200 N. The crate just begins to slide.
   - IV. She returns to a 160 N push. The crate slides at a constant speed.

   Questions:
   - a. What is the maximum static friction force (in newtons)?
   - b. What is the kinetic friction force (in newtons)?
   - c. What is the coefficient of static friction?
   - d. What is the coefficient of kinetic friction?

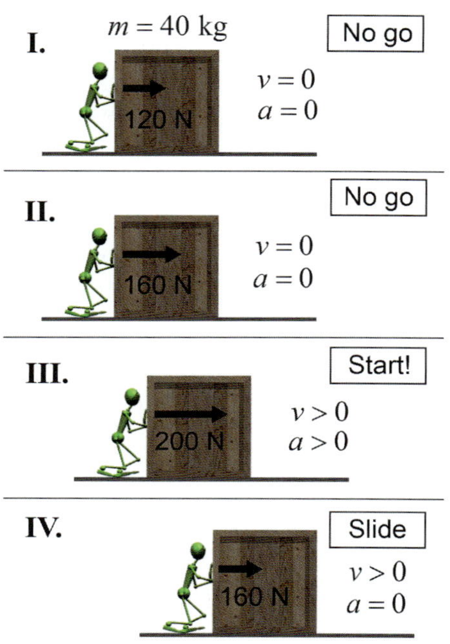


2. A skydiver jumps from an airplane and falls toward Earth. His instructor tells him to open his parachute. When he does so, the force of air resistance pushes upon his parachute, and the cords pull up on him. His fall slows down and gradually he reaches terminal velocity. The diagrams show three moments during this process; the lengths of the force vectors indicate how strong the forces are.
   - a. Why does the skydiver slow down when the parachute is opened?
   - b. Which of the three diagrams corresponds to terminal velocity, and why?

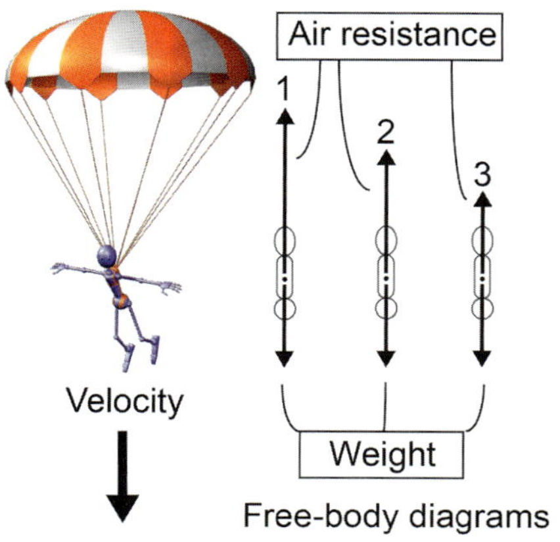

# Chapter review

## Vocabulary

Match each word to the sentence where it best fits.

### Section 5.1

| | |
|---|---|
| center of mass | equilibrium |
| force | free-body diagram |
| net force | newton (N) |
| normal force | pound (lb) |
| weight | |

1. The _______ is the vector sum of forces acting on an object.
2. A system is in __________ when there is no change over time in its motion.
3. The ______ is a force perpendicular to a surface
4. The supermarket often sells meat and vegetables by the ______, a unit of force often associated with weight.
5. The force of gravity acting on an object is the _______.
6. A/An __________ represents the location, magnitude and direction of all forces acting on a single object isolated from its environment.
7. Weight is a force that acts through the _________ of an object.
8. An action that may cause the motion of a body to change is called a/an _______.

### Section 5.2

| | |
|---|---|
| Newton's first law of motion | Newton's second law of motion |
| Newton's third law of motion | friction |
| reaction force | |

9. "To every action there is always an equal and opposite reaction" is a statement of _______.
10. A force that acts to resist motion is called _______.
11. A/An _________ occurs when one object exerts a force on another object.
12. Another name for _______ is the law of inertia.
13. According to _______, the acceleration of an object is directly proportional to the net force exerted on it.

### Section 5.3

| | |
|---|---|
| Hooke's law | brittle |
| elasticity | material strength |
| spring | spring constant |

14. A material that can bend easily has a lot of _______.
15. The _______ measures how stiff a spring is and how much force it takes to deform it.
16. The ability of an object to sustain force without breaking is described by its _______.
17. A/An _______ is a device that can store energy when it is deformed.
18. A material that, when bent even just a little, will break is very _______.
19. The restoring force of a spring is proportional to the distance it is compressed or extended is _______.

### Section 5.4

| | |
|---|---|
| coefficient of kinetic friction | coefficient of kinetic friction |
| coefficient of rolling friction | coefficient of static friction |
| drag coefficient | lubrication |
| rolling friction | sliding friction |
| static friction | viscosity |

20. _______ is the friction force that occurs when two surfaces are in rolling contact.
21. In order to reduce the friction among moving parts in the engine of a car, _______ is used.
22. _________ is a frictional force between two objects that are not moving relative to one another.
23. The force of ______ opposes the motion of objects in motion.
24. The ratio of the rolling friction force to the normal force is the ________.
25. An airfoil has a smaller _______ than a cube, which is one reason why an airfoil is used for vehicles that are designed to operate at high speeds.

### Section 5.4

| | |
|---|---|
| coefficient of kinetic friction | coefficient of kinetic friction |
| coefficient of rolling friction | coefficient of static friction |
| drag coefficient | lubrication |
| rolling friction | sliding friction |
| static friction | viscosity |

26. It is easier to swim through water than gelatin, because gelatin has a higher _______.
27. The _______ is a ratio of the friction between two surfaces in contact that are not moving relative to each other to the normal force between them.
28. The _______ is a ratio of the amount of friction between two surfaces in contact that are sliding relative to each other to the normal force between them.

## Conceptual questions

### Section 5.1

29. Where is the center of mass of a doughnut? Is it inside or outside the doughnut? How do you know?

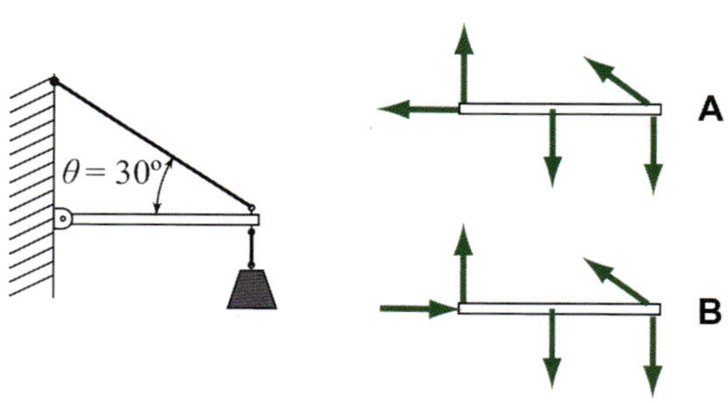


30. Which of the two sketches is the correct free-body diagram of the horizontal arm holding the weight? Explain why the other is incorrect.
31. What quantities sum to zero when a system is in equilibrium?
32. You and your friend want to balance a teeter-totter, but your friend is much heavier than you. How can you balance it?
33. What is the difference between mass and weight? What units of the metric system are used to describe them? What United States customary units are used?
34. When you walk, you push on the ground. This exerts a force on the ground. What exerts a force on *you* to propel you forward?

### Section 5.2

35. Two balls of equal mass collide. One of them experiences a 12 $m/s^2$ acceleration at one moment. What acceleration does the other ball experience at the same moment?
36. A 50,000 kg plane is flying at a constant velocity of 130 m/s. What is the net force that is acting on it?
37. Can an object be moving when the net force acting on it is zero? If so, explain how this can be true.
38. What is the reaction partner to the weight of an object sitting on a table? Is the reaction force the normal force of the table acting upward to hold the object up? If not, then what is the correct reaction force?
39. Explain why action and reaction forces cannot cancel each other out.
40. Jamayla and Iris are ice skating and standing next to each other. Jamayla pushes Iris and each then moves away from the other. Express what they just did using vocabulary from Newton's laws.
41. If every action has an equal and opposite reaction, then when you jump and gravity pulls you back down, it also pulls the Earth up to you. Why can't you see this motion?
42. If you add mass to a car on a ramp the acceleration of the car does not increase proportional to the increase in mass. Why not?
43. The tendency of a moving body to keep moving and a body at rest to stay at rest is described by the first of which set of principles?
44. Why does it hurt more to kick a bowling ball than a beach ball?
45. Which of Newton's laws best explains why it hurts to punch someone?
46. If, according to Newton's third law, every reaction has an equal and opposite reaction, then how can motion ever occur? Wouldn't every force be canceled out by its reaction force?

# Chapter review

## Section 5.2

47. Ask two friends to answer the following question:
"Can there ever be a single isolated force?"
Ask them to explain why or why not. Write down their answers. How do their answers compare with your own answer to the same question?

## Section 5.3

48. In the lesson on springs, your teacher described the force exerted by a spring. Based on your teacher's statements, why is there a negative sign in Hooke's law?

49. Why is a spring used in a force scale?

50. A spring with a spring constant of $k$ is cut in into three equal-length pieces. What is the spring constant of each of the pieces?

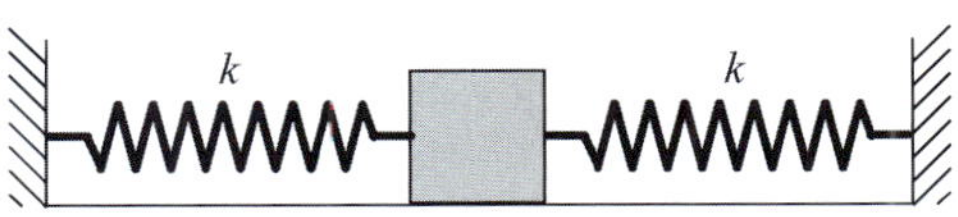


51. A mass is connected to two springs as shown above. Each spring has a spring constant of $k$. What is the effective spring constant that the mass experiences?

52. Does a spring scale measure weight or mass? Why?

53. How would you calculate the spring constant of an unknown spring?

54. Ask two friends to define the meaning of the "spring constant." Write down their answers. How do their answers compare with your own definition?

## Section 5.4

55. Does adding a lubricant such as oil between sliding surfaces raise or lower the coefficient of friction?

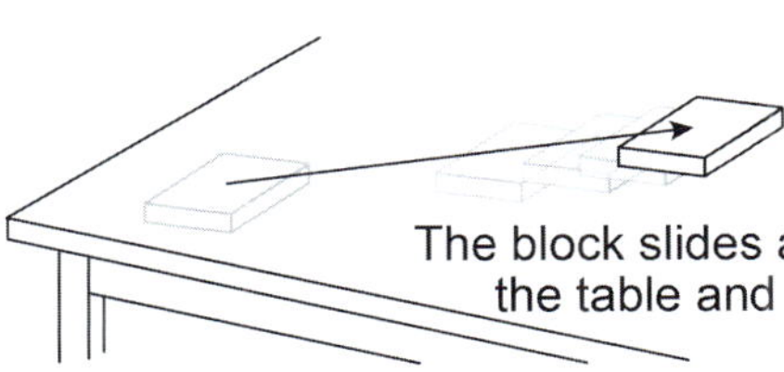

The block slides across the table and stops.

56. Sylvie applied a quick push to a motionless friction block, which then slid part way across a table and came to a stop. What kinds of friction were acting on the friction block in her activity?

57. Why does rubbing your hands together warm them up?

58. Why is it difficult to walk on surfaces such as ice that have a low coefficient of static friction?

59. Consider the model $F_k = \mu_k F_N$. Would a block with a large surface area or a small surface area experience a greater force of kinetic friction, assuming that the blocks have identical mass?

60. A friction block consists of a piece of wood with slotted masses on top of it. If you double the total mass of the block plus slotted masses, how does the coefficient of static friction between the friction block and concrete change?

61. How do you determine the sign of the force of static friction?

62. How do you determine the sign of the force of kinetic friction?

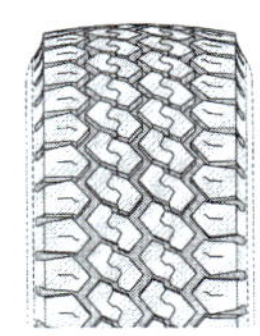

**All street tires have grooves called treads. Racing tires do not have treads.**

63. The purpose of tires is to *increase* static friction between the wheels and the road so that there is no slipping or skidding. With deformable materials such as rubber, the amount of friction, or *traction*, increases as more rubber surface actually contacts the road. Explain why street-legal tires have treads that *decrease* the amount of rubber touching the road whereas racing car tires are smooth with no treads.

64. Why is it far more efficient to pull a sled with the rope slanting upward instead of downward? Draw a diagram to solve this problem by showing the forces from both an up-slanted and a down-slanted rope.

65. Why do we say that friction models such as $F = \mu F_N$ are only approximations? The models can be used to calculate a friction force to five decimal places with any ordinary calculator.

66. A team of physicists studied the world-record-setting sprint by Usain Bolt in 2009. The physicists showed that Bolt exerted a roughly constant horizontal force against the frictional drag forces acting against him. The physicists said the "[drag] force causes a reduction of his acceleration, so his speed tends to a constant value (terminal speed)." Explain why Bolt reaches a *terminal speed* by describing the forces acting on him.

# Chapter review

## Quantitative problems

### Section 5.1

67. Which contains more matter: 100 lb of stone or 100 N of stone? Express both quantities in kilograms to justify your answer.

68. What is the weight of a 20 kg sack of cement?

69. There are 16 ounces in a pound. Calculate the equivalent of a one ounce force in newtons.

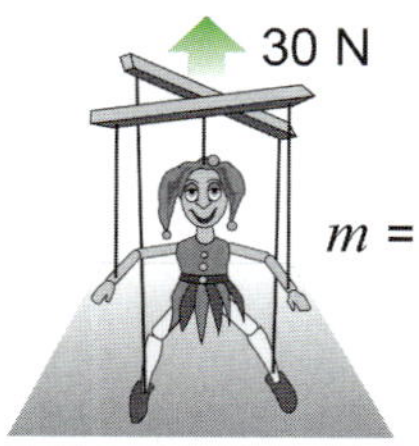


70. What is the normal force between a 5 kg puppet and the floor if the puppet is attached to a string that applies 30 N of force upward?

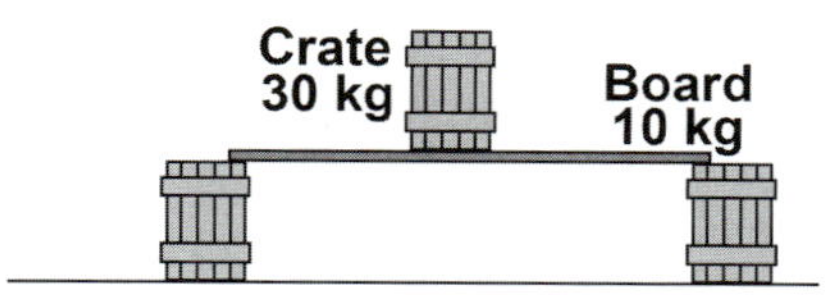


71. Draw a free-body diagram of the board shown in the diagram above with all the forces acting on it. You may assume the crate is at the center.

72. You have 10 kg of coconuts.

    a. How many kilograms of bananas have the same mass?
    b. How many pounds is this?

73. Forces are often classified by the number of "$g$"s. For example, a $1g$ force is equal to an object's weight, a $2g$ force is twice an object's weight, and so on. If a person has a mass of 50 kg, then what force would be equivalent to $7g$'s? Give your answer in pounds and newtons. This is approximately the force a pilot can withstand before losing consciousness.

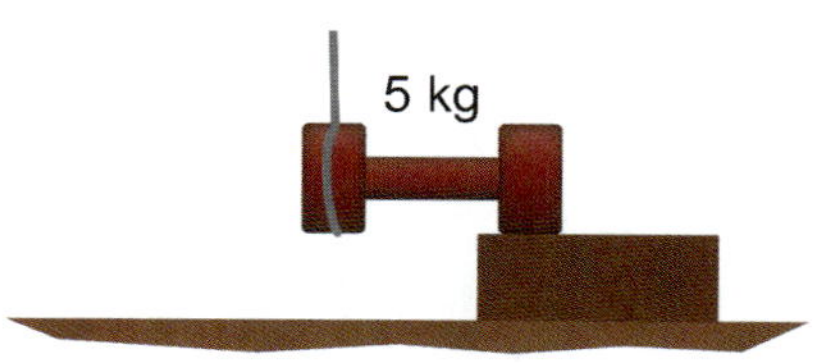


74. Draw a free-body diagram of the dumbbell shown in the picture above. Be sure to include normal forces. Assume the dumbbell is not moving.

75. What is the weight of a 50 kg woman at the top of Jupiter's atmosphere, where $g$ = 24.8 N/kg? Give your answer in both newtons and pounds.

### Section 5.2

76. What is the mass of a man who accelerates 4 m/s$^2$ under the action of a 300 N net force?

77. What is the normal force when a 40 kg person jumps and is accelerating at 3 m/s$^2$ upward while in contact with the ground?

78. What is the force required to accelerate a 180,000 kg blue whale at 3 m/s$^2$?

79. There is a tug-of-war game going on. There are 5 people on each team with an average mass of 60 kg. What is the acceleration if there is a net force of 400 N to the right?

80. How far will a 600 kg boat travel in 12 s if there is a constant 900 N force on it and it starts from rest?

81. A 10 N net force is applied to an object, which then accelerates at 4 m/s$^2$. What is the mass of the object?

82. A cart is moving at 9 m/s on a frictionless surface. Kayla pushes the front end of the cart with a force of 60 N, while Kristofer pushes from the back with a force of 60 N. How fast is the cart moving after 5 s?

83. What net force is required to give a 4 kg rock an acceleration of 2 m/s$^2$?

84. Bathroom scales read the normal force that is exerted against the floor. What would a scale read when a 100 kg man is in an elevator accelerating upward at 1.2 m/s$^2$? What would it read when the man is accelerated downward at 1.8 m/s$^2$?

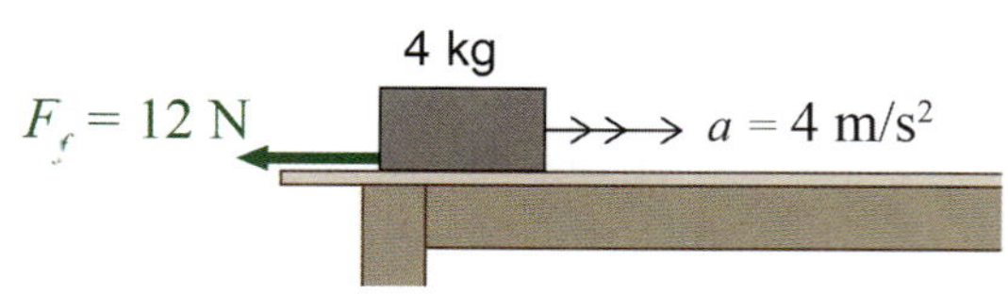


85. Draw a free-body diagram for a 4 kg box sitting on a table. The box is accelerating at a rate of 4 m/s$^2$ to the right, and there is a friction force of 12 N.

86. One important part of testing a car is seeing how fast it can brake. A 1,100 kg car is traveling at 15 m/s when the brakes are suddenly applied. How far a distance will it travel if the brakes can exert a force of 6,000 N?

# Chapter review

## Section 5.3

87. What is the magnitude of a force that is required to keep a spring with a spring constant of 500 N/m compressed 0.3 m from its starting point?

88. How far must a 100 N/m spring stretch to hold up a 4 kg box?

89. What is the spring constant of a spring that exerts a 56 N force after it is stretched 0.16 m?

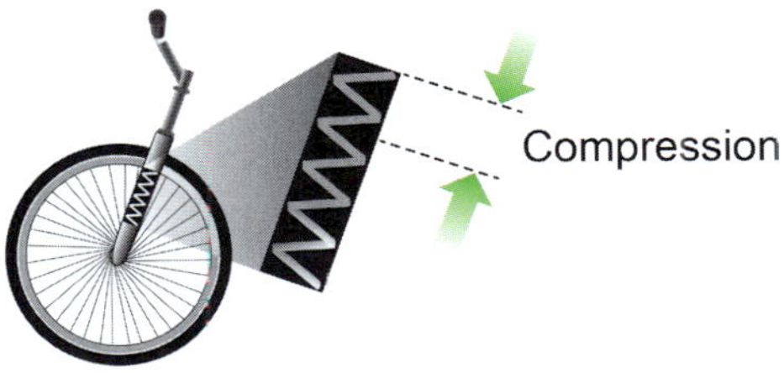


90. How much does a front suspension system that has a spring constant of 50,000 N/m compress when it supports forces of 1,500 N from a bike and rider?

91. What is the force exerted by a spring where $k = 400$ N/m when it is stretched 0.6 m in the positive direction?

92. How far does a spring stretch to support a 50 N weight if it has a spring constant of 3,000 N/m?

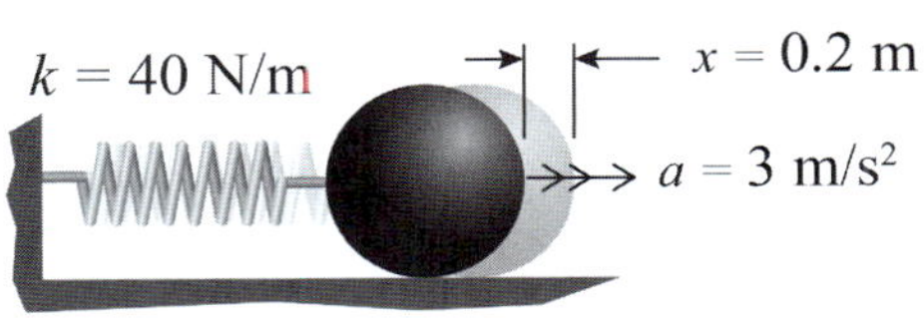


93. An object of unknown mass is connected to a spring with $k = 40$ N/m. When the spring is compressed 0.2 m, the object accelerates at 3 m/s$^2$. What is its mass?

94. Resistance bands used for exercise are essentially large rubber bands that act similarly to a spring. If Madison can supply a maximum force of 500 N how far will she be able to stretch a resistance band that has $k = 400$ N/m?

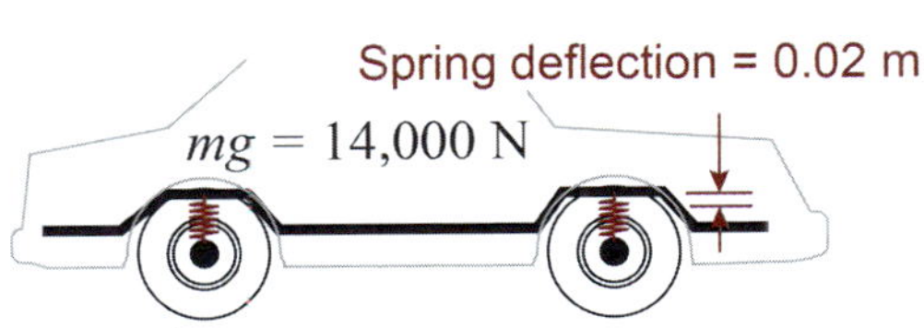


95. A 14,000 N car is supported by four springs. If each spring is deflected 0.02 m, what is the spring constant?

96. A 400 N force causes a spring to compress 0.2 m. What is the spring constant of the spring?

97. When a bow is at its most stretched it can accelerate a 0.5 kg arrow at 80 m/s$^2$. If the spring constant is $k = 100$ N/m, how far has the bow been stretched?

98. When testing a spring, you find that it takes 20 N of force to compress the spring by 10 cm. What is the spring constant for this spring?

99. How far does a 5,000 N/m spring stretch when supporting a 300 N weight?

100. You hang a 10 N mass from a scale and find that this makes the spring stretch 0.04 m. What is this spring's spring constant?

## Section 5.4

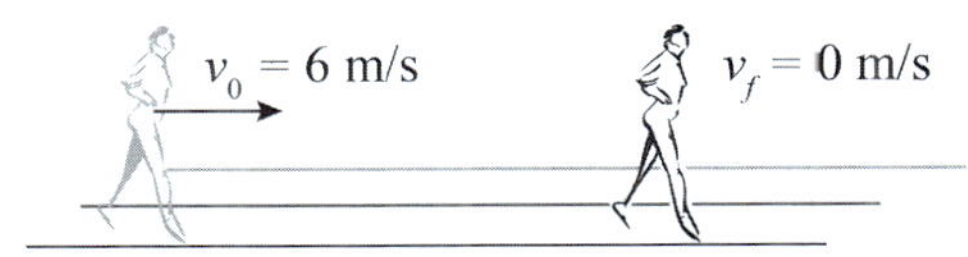


101. The coefficient of kinetic friction between newly waxed floors and socks is 0.12. How far does Laura slide if she has an initial speed of 6 m/s?

102. What is the magnitude of the acceleration caused by the force of kinetic friction acting on a box sliding across a level table? The coefficient of friction is $\mu_k = 0.3$.

103. How far does a box travel until it is stopped by friction if it has an initial velocity of 8 m/s and there is a coefficient of friction of 0.102?

104. How many 20 N bricks should you add to increase the friction force on a cart by 80 N if the coefficient of friction is 0.8?

105. A 12 kg computer is sitting on a table and the coefficient of static friction between the computer and the table is 0.7. If the computer is pushed with a horizontal 30 N force, what is the actual force of static friction between the computer and the table?

106. A box of oranges lies in the back of a truck. The coefficient of static friction is 0.5. If the truck accelerates from 20 m/s to 10 m/s in 3 s, will the box slide?

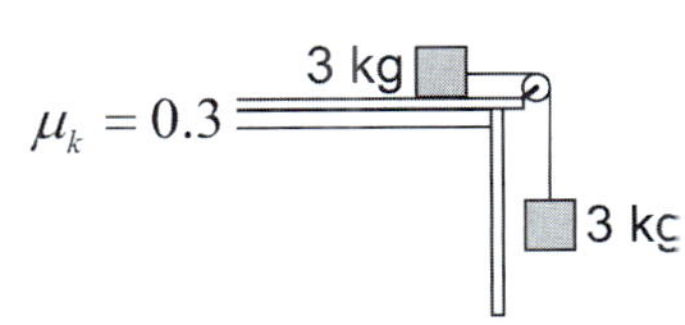


107. Two identical 3 kg masses are connected by a string that passes over a frictionless pulley as shown in the drawing above. If the coefficient of friction is 0.3, what is the acceleration of the hanging block?

### Section 5.4

108. A 100 N wood friction block is on top of concrete. What is the net force on the block when sliding on the concrete if there is a 75 N force pushing it?

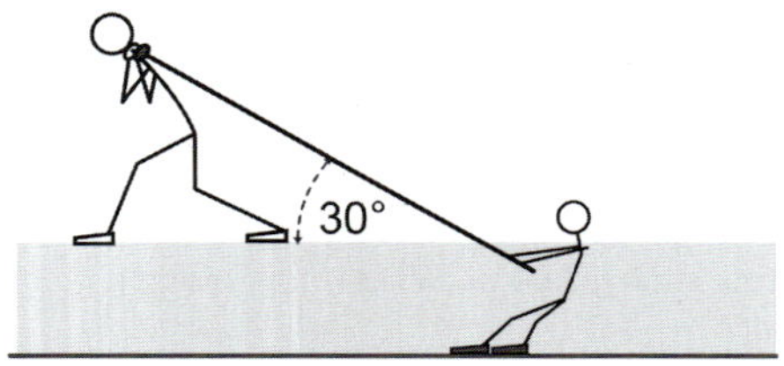


109. Katie weighs 600 N, wears wooden shoes, and is being pulled along the wood floor of the orchestra pit by her mother using an 800 N force at a 30° angle as shown. Find the force of friction between Katie and the floor and the net force acting on Katie. Assume that the coefficient of kinetic friction is 0.3 for wood-on-wood.

110. A 10 kg block and a 5 kg block are resting on each other on a flat surface. The heavier block is pushed with a 400 N force.
    a. Draw the free-body diagram of the 10 kg block.
    b. What is the net force acting on the 10 kg block? Give the $x$ and $y$ components.
    c. What is the net force acing on the 5 kg block in each of the $x$ and $y$ directions?

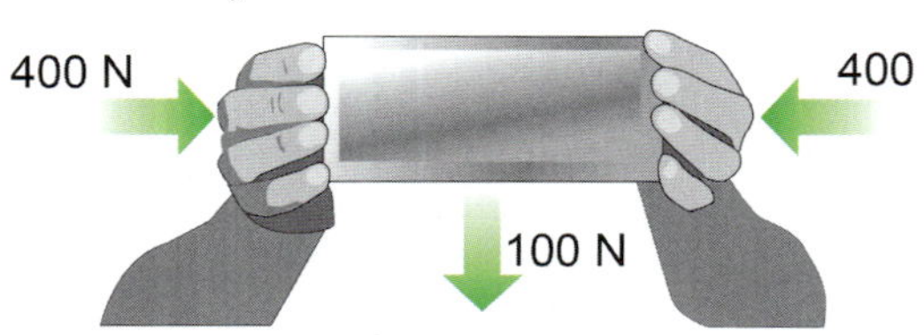


111. A 100 N tin box is prevented from falling by holding it on both sides. If the pushing forces are 400 N each, what is the coefficient of friction?

112. How much work does the kinetic friction force do on Jesse, who weighs 720 N, as he slides for 4 m on a surface with a coefficient of friction of 0.4?

113. By what factor does stopping distance change if the coefficient of friction is doubled?

114. If the friction force is 450 N when the coefficient of friction is 0.3, what is the normal force?

115. What is the coefficient of kinetic friction if a car going at 20 m/s stops in 50 m?

116. Which of the following sentences is closest to the meaning of the equation $\mu_k = F_f / F_N$?
    a. The force of friction is the coefficient of friction divided by the normal force.
    b. The coefficient of friction is the ratio of the friction force to the normal force.
    c. The normal force creates a friction force when multiplied by a coefficient of friction.

117. A 3 kg rubber block is resting on wet concrete. The coefficient of static friction is 0.3. What is the minimum force that must be applied to the block to make the block begin to accelerate?

118. A crate of physics textbooks is sliding across the room at 18 m/s. What is the coefficient of friction that causes it to come to a stop in 12 s?

119. A 4 kg box sits in the back of your 1,400 kg car with a coefficient of static friction between the box and your car of 0.4. The car is traveling at 15 m/s. What is the minimum stopping distance that can be obtained without the box sliding?

120. The police find skid marks that are 80 m long left by a 1,300 kg car. The coefficient of kinetic friction between the tires and the road is 0.6. How fast was the car going before the brakes were applied? If the speed limit is 60 mph, was the driver speeding?

121. A cart is propelled by an engine so that when it is on a surface with a coefficient of kinetic friction of 0.3 it travels at constant velocity. What is the acceleration of the cart on a surface with a coefficient of 0.2 assuming the force is the same?

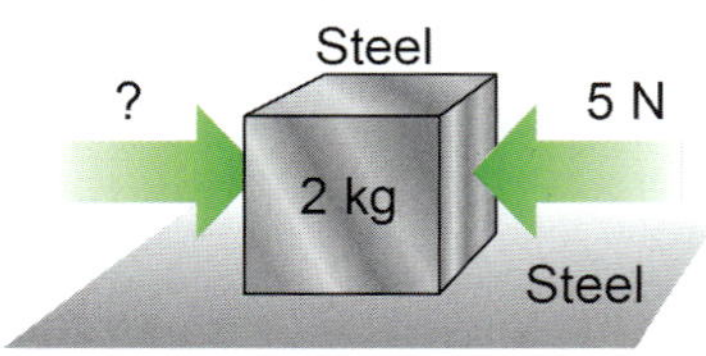


122. How much force is required to start moving a 2 kg block of steel on dry steel if there is a 5 N force opposing you? How much is required to keep it from slowing down assuming the opposition force remains? (For steel on dry steel, $\mu_s = 0.8$ and $\mu_k = 0.5$.)

123. The coefficient of friction between a 0.17 kg puck and and the ice is $\mu_k = 0.15$. If the puck leaves a hockey stick traveling at 20 m/s, what is its speed when it reaches the goalie 15 m away? How much time does the goalie have to react? Assume the puck travels on the ice the whole time.

# Chapter review

## Standardized test practice

124. What is the acceleration due to gravity on Jupiter if someone who weighs 600 N on Earth weighs 1400 N on Jupiter?

A. 26 $m/s^2$
B. 19 $m/s^2$
C. 21 $m/s^2$
D. 23 $m/s^2$

125. What is the mass of something that weighs 300 N on Earth?

A. 31 kg
B. 15 kg
C. 2940 kg
D. 600 kg

126. How much force does it take to accelerate a 34 kg box at a rate of 4 $m/s^2$?

A. 136 N
B. 68 N
C. 8.5 N
D. 340 N

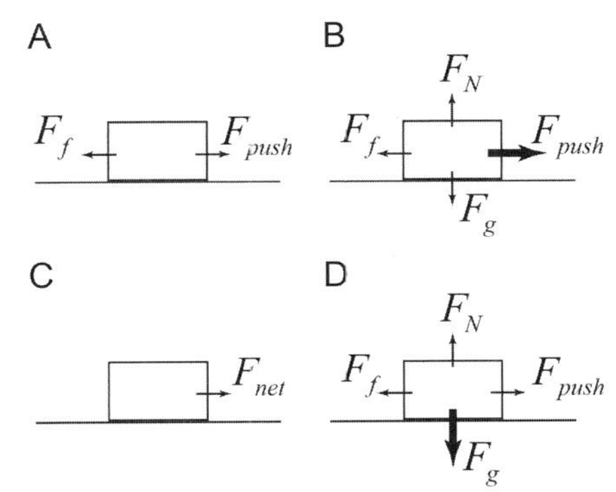


127. A woman pushes a box on a table. There is friction, but the box still accelerates. Which diagram above shows the correct free-body diagram for this situation?

A. A
B. B
C. C
D. D

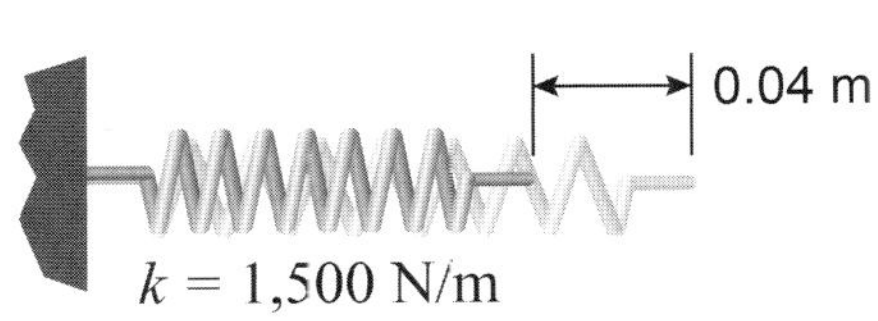


128. What force must be applied to the spring in the diagram above to create the displacement shown?

A. 6 N
B. 60 N
C. 600 N
D. 37,500 N

129. If you can pull on a spring with 360 N of force and the spring has a spring constant of 400 N/m, how far will it stretch?

A. 0.9 m
B. 0.8 m
C. 0.7 m
D. 0.6 m

130. If you pull a 8 kg box across the floor with a net force of 50 N, how fast will it accelerate?

A. 5.5 $m/s^2$
B. 9.8 $m/s^2$
C. 3.2 $m/s^2$
D. 6.25 $m/s^2$

131. A 5 kg box is lying at rest on a frictionless surface. If you push it with a constant net force of 10 N, how far will it travel over the first 4 s?

A. 16 m
B. 8 m
C. 4 m
D. 2 m

132. If someone weighs 711 N on Earth, what is that person's mass?

A. 71.1 kg
B. 70.0 kg
C. 72.6 kg
D. 73.5 kg

133. How much would a 100 kg man weigh on the surface of the Sun, where the acceleration due to gravity is 274 $m/s^2$?

A. 10,000 N
B. 36,985 N
C. 27,400 N
D. 98 N

134. A 2 kg box is moving at 5 m/s on a frictionless plane. When will it stop?

A. after 1 second
B. after 10 second
C. after 20 seconds
D. never

135. Jim passes a 1.5 kg basketball to his teammate with a force of 60 N. Find the acceleration of the basketball.

A. 40 $m/s^2$
B. 40 m/s
C. 90 $m/s^2$
D. 0.025 $m/s^2$

Chapter

# 6

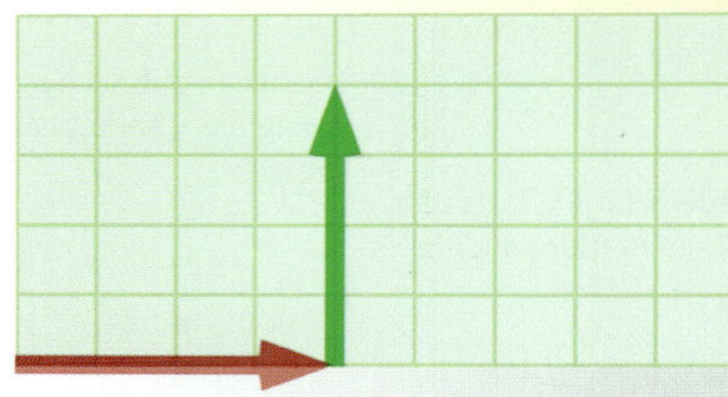

# Motion in Two and Three Dimensions

Imagine what life would be like if we could sleep in our car on the way to school or work!

If you have driven a long trip, you might have wondered if the almost-mindless task of driving couldn't be done better by a smart car than by a sleep-deprived human driver. An automated car could also allow elderly passengers to get around town without worrying about their aging eyesight and slower reaction times. Robotic cars could even eliminate the societal costs of drunk drivers. The challenge of automated cars, however, is that they must navigate safely in a complex, potentially dangerous, and rapidly changing environment.

On November 3, 2007, eleven design teams and their mechanical progeny assembled at the mothballed George Air Force Base in Victorville, California, for the third competition of the DARPA Grand Challenge. The prize was $2 million for the vehicle that could complete a 96 km (60 mi) urban road course in under six hours. Each robot vehicle had to recognize and obey all traffic regulations. Each vehicle had to safely execute maneuvers such as lane changes and merges with other vehicular traffic. The urban challenge was much more difficult than the two previous robot vehicle races in that vehicles had to recognize other vehicles and make "intelligent" decisions in real time based on what they recognized.

Six of the designs successfully completed the course, although two did so outside of the six-hour time limit. The winning entry averaged 14 mph during the race. As a sign of the continuing challenges of the new technology, some of the unsuccessful entries were eliminated because of collisions or because they froze in a traffic circle—a confusing traffic pattern even for human drivers! The New York Times summarized the event as demonstrating "that the state of the art in robotics has reached the point where the most sophisticated autonomous vehicles can now drive comfortably and safely on a city course while surrounded by traffic and other obstacles."

Several states have now passed laws permitting driverless cars. The Nevada statute "defines an *autonomous vehicle* to mean a motor vehicle that uses artificial intelligence, sensors and global positioning system coordinates to drive itself without the active intervention of a human operator." The complexity of a fully autonomous vehicle containing those three subsystems—artificial intelligence, sensors, and GPS—means that the field is ripe for future innovation and technological advances. When do you think you will first encounter an autonomous car on America's roads?

# Chapter study guide

Chapter 6

## Chapter summary

If we were points confined to a straight line along a single coordinate axis, then distance and speed might suffice to describe all possibilities and there would be no need for vectors. Fortunately, the universe is three dimensional and much more interesting. Vectors are a fundamental part of the language of physics because they allow us to describe three-dimensional behavior. This chapter describes how to use vectors, add and subtract vectors, and solve problems with vectors. Position and displacement are vectors that describe location and changes in location. Velocity and acceleration vectors describe motion. The force vector describes the three-dimensional character of forces. Vectors are useful in solving many real-world problems, such as projectile motion of a soccer ball kicked through the air, motion of a car rolling down a ramp, and control of a robot maneuvering through a maze.

## Learning objectives

By the end of this chapter you should be able to

- find the magnitude and components of a force, displacement, velocity, or acceleration vector;
- represent and perform calculations with force, displacement, velocity, or acceleration vectors in Cartesian and polar forms;
- convert between Cartesian and polar vectors;
- find the resultant of two or more vectors both graphically and by components;
- apply the technique of breaking down a two- or three-dimensional problem into separate one-dimensional problems; and
- solve two-dimensional motion problems, including projectile motion and motion down a ramp.

## Investigations

6A: Vector navigation
6B: Projectile motion
6C: Acceleration on an inclined plane
6D: Graphing motion on an inclined plane

## Chapter index



## Important relationships

$$\vec{F} = (F_x, F_y, F_z)$$

$$F_x = F\cos\theta \qquad F_y = F\sin\theta$$

$$F = \sqrt{F_x^2 + F_y^2}$$

$$\theta = \tan^{-1}\left(\frac{F_y}{F_x}\right)$$

$$\vec{a} = \frac{\Delta\vec{v}}{\Delta t}$$

$$x = v_{x0}t \qquad v_x = v_{x0}$$

$$y = v_{y0}t - \frac{1}{2}gt^2 \qquad v_y = v_{y0} - gt$$

$$a_{ramp} = \left(\frac{h}{L}\right)g$$

$$a_x = g(\sin\theta - \mu_r\cos\theta)$$

## Vocabulary

| | | | |
|---|---|---|---|
| **vector** | **vector diagram** | **magnitude** | **scalar** |
| **resultant vector** | **component force** | **component** | **resolution of forces** |
| **sine** | **cosine** | **tangent** | **radian (rad)** |
| **displacement** | **polar coordinates** | **Cartesian coordinates** | **compass** |
| **velocity** | **speed** | **acceleration** | **trajectory** |
| **projectile** | **range** | **inclined plane** | **ramp coordinates** |

## 6.1 - Force vectors

Suppose someone told you to push a box with a force of 200 N (about 45 lb). What do you do? A moment's thought and you realize that you need to know what *direction* to push or pull. Do you lift the box (upwards force) or push it to the left, right, forward, or backward? Force is a vector because a complete description of a force includes both strength and direction. This section describes the force vector and also the mathematics for how vectors are used to solve two- and three-dimensional problems that involve direction as well as strength.

### Properties of vectors and scalars

**Vectors**

A **vector** is a type of variable that includes information that indicates direction as well as strength or size. Directional words such as "left" or "up" are fine for conversation, but until they are expressed as vectors they are not that useful in solving complex problems, such as navigating a robot vehicle. A vector includes direction information in very specific mathematical terms so that vectors in different directions can be added, subtracted, or multiplied. Directional information for a vector is specified very precisely compared to our typical descriptions of direction in everyday conversation.

**Vector diagrams**

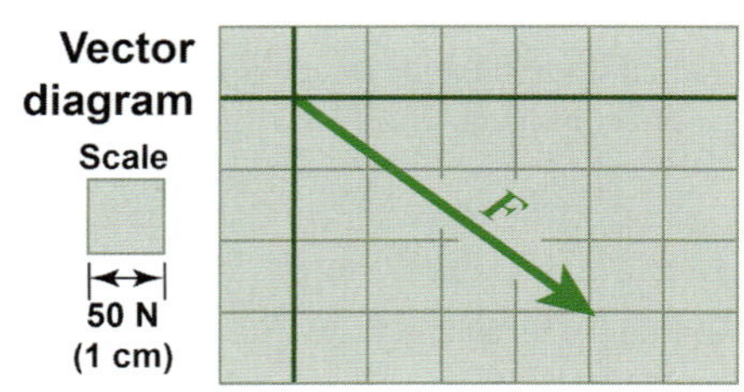


A **vector diagram** is a type of graph in which the perpendicular axes represent the *x*- and *y*-directions the vector can have. When representing force vectors, the *scale* of the diagram relates length on the graph to the strength of the force. For example, suppose we choose a scale of 1 cm = 50 N. A 250 N force to the right is represented as an arrow starting at the origin and extending to the right with a length of 5 cm. A 250 N upward force has a length of 5 cm pointing up along the *y*-axis. The arrowhead points in the direction of the force.

**Magnitude of a vector**

The **magnitude** of a force vector is its "length" on the diagram—or, equivalently, the strength of the force in newtons. The force vector in the example below has a magnitude of 200 N and points in the $+x$ direction, whereas the displacement vector shown has a length of 5 units and therefore a magnitude of 5 m. The magnitude of a vector can either be zero or have a positive value, such as 200 N or 5 m. Magnitude cannot be negative.

**Vectors and scalars**

As you learned on page 73, vectors are different types of variables from scalars. A **scalar**, such as mass or temperature, does not depend on direction and can be completely specified with a single value. A vector always includes direction and usually requires more than one value to describe it. To differentiate between the two types of variables, vectors are often written with a small arrow over them, such as $\vec{F}$.

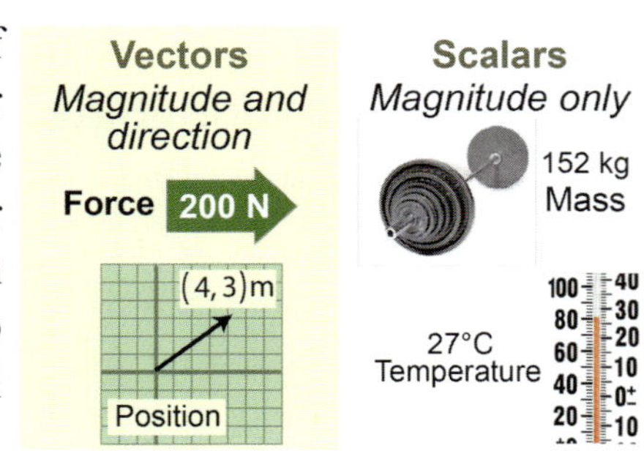


**Vectors in and out of the page**

When a vector quantity, such as force $\vec{F}$, is directed into or out of the page, the symbols at right are used to denote its direction. These symbols use the "arrow" metaphor: A vector coming out of the page looks like an arrowhead, while a vector going into the page looks like the crossed feathers at the end of an arrow.

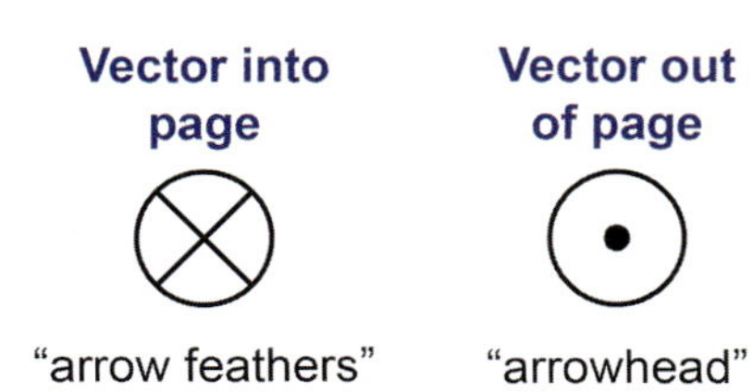

**Multiple forces**

Most real situations involve two or more forces acting at the same time. When multiple forces act on an object, the resulting motion depends on the net force, which is the sum of the forces. As an example, consider a block at rest on a flat table. The block is pulled by a 3 N force and a 4 N force that are perpendicular to each other.

**How do you add vectors in different directions?**

In what direction will the box move? Intuitively, you know the box moves at an angle in between the two forces. How do you add forces that are in different directions? Simple addition gives 4 N + 3 N = 7 N, but that is the wrong answer because the forces do not act in the same direction. Forces can only be added in this way if they have the same *line of action*, meaning that they are either in the same direction or in exactly opposite directions. The forces in this example are at right angles to each other, so they do not have the same line of action.

**How do you add forces in different directions?**

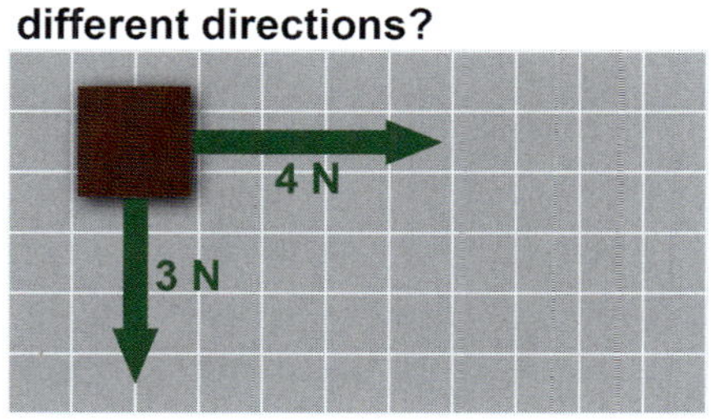


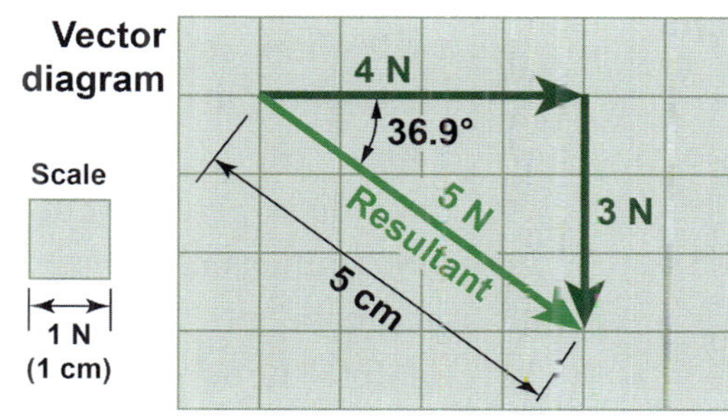


**Resultant vector**

This is a two-dimensional problem and must be solved by thinking in two dimensions. A **resultant vector** is a single vector that represents the net sum of two or more individual vectors. Here, two force vectors add to create a resultant force vector.

**Add vectors "head to tail"**

To find the resultant of the two forces in the example, the 4 N force is drawn to scale on a vector diagram, starting from the origin. To add the second force, the 3 N vector is drawn from the end of the 4 N vector rather than from the origin. The 3 N force vector ends at coordinates of (4,–3) N on the scale diagram. The resultant force is the bold green vector drawn from the tail of the first vector, at the origin, to the head of the second vector at (4,–3) N on the diagram. Measuring the length of the arrow yields a value of 5 cm = 5 N, so the resultant of these two forces is a single force of 5 N at an angle of −36.9°. The block accelerates in the direction of the 5 N resultant force.

**Adding vectors "head to tail"**

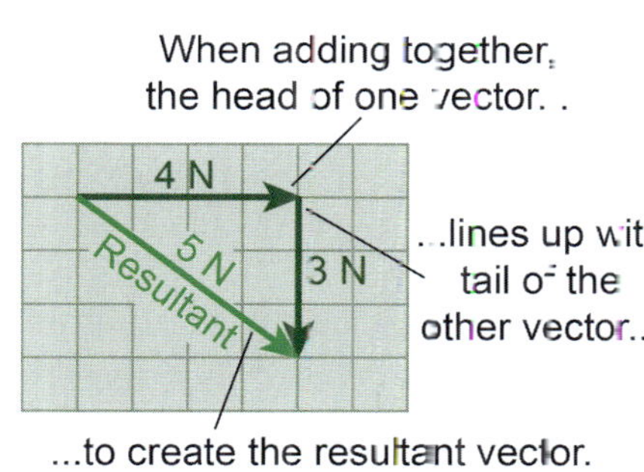


**Subtracting forces**

Subtracting a force is the same as adding the force in the opposite direction. To find the negative of a force, draw the same length as the original force along the same line of action, but pointing away from the origin in the opposite direction. For example, the opposite force from (4,–3) N is (–4,3) N. Notice that each of the $x$ and $y$ values changed sign. Both forces have a magnitude of 5 N, but they are in opposite directions.

**Components of a vector**

Our last example points to a very powerful approach to working with vectors. A resultant 5 N force at −36.9° has the same effect as the combination of a 4 N force in the $x$-direction and a −3 N force in the $y$-direction. *It is much easier to work with forces along the x- and y-axes than to work with forces at arbitrary angles.* Forces along the $x$-, $y$-, or $z$-axis that together represent another force are called *components*. In this example, the **component** along the $x$-axis is a 4 N force, which is also called the $x$-component. The −3 N force is the $y$-component because it is along the $y$-axis. Any vector, at any angle, can be *resolved* into its components. The easiest and most accurate way to add, subtract, and even multiply vectors is by performing the operations separately on the components of the vectors in each direction.

**The general form of a vector**

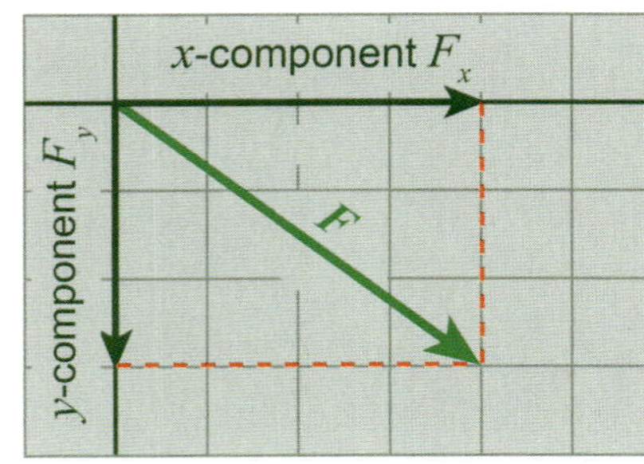


Consider an arbitrary force vector $\vec{F}$ with magnitude $F$. The **component force** of the vector in the $x$-direction is $F_x$ and is found by drawing a perpendicular line from the head of the vector to the $x$-axis. The $y$-component $F_y$ is similarly found by drawing a perpendicular line to the $y$-axis. (For a three-dimensional vector there would also be a $z$-component $F_z$ found in a similar way.) We write the total force vector by arranging the $x$-, $y$-, and $z$-components in the order ($x$, $y$, $z$) inside parentheses like equation (6.1).

**Force vector** by components

$$\vec{F} = (F_x, F_y, F_z) \tag{6.1}$$

$F_x$ = $x$-component of force (N)
$F_y$ = $y$-component of force (N)
$F_z$ = $z$-component of force (N)

**Components and direction**

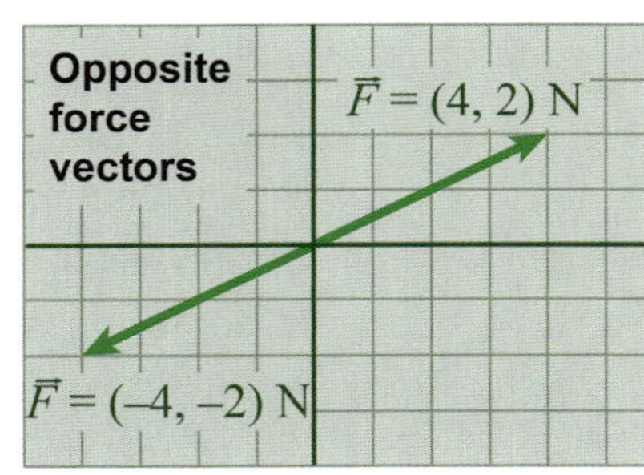


The components of a force vector are also force vectors themselves, and they can have positive and negative values. For example, the force vector $\vec{F_1} = (+4,+2)$ N points in the opposite direction from the vector $\vec{F_2} = (-4,-2)$ N. When solving problems, draw a diagram to determine the sign of vector components. By using positive and negative values, a two-component $(x, y)$ vector can describe all possible directions on a plane. The three-component ($x$, $y$, $z$) vector can uniquely describe all possible directions in space.

In this book, we are referring to components of a vector (e.g., $\vec{F_x}$) as *vector components*, because they can take negative values and those negative values imply direction. This is reinforced graphically in order to make resolving vectors into their components very clear, such as in the first illustration on this page.

**Resolution of forces**

The process of breaking a force vector into its components is called **resolution of forces**. When a vector has been written in the form $\vec{F} = (F_x, F_y)$, we say the vector has been *resolved* into its components. For two-dimensional problems such as this, only the $x$- and $y$-components are used.

**Vector triangle**

The properties of right triangles are the key to resolving a vector into components. In the vector diagram for a 10 N force, the $x$-component is the projection of the vector along the $x$-axis. The $y$-component is the projection along the $y$-axis. These projections are the sides of the right triangle formed by $F$, $F_x$, and $F_y$. The hypotenuse of the triangle is the original force $F$.

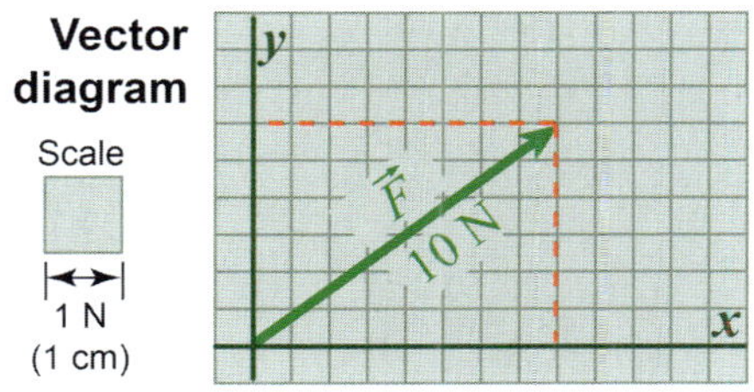


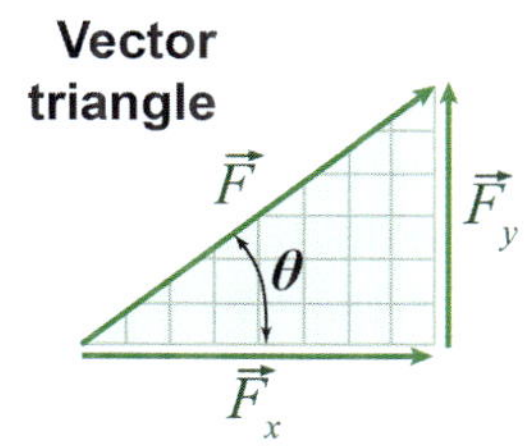


$\vec{F}$ is the hypotenuse

$\vec{F}_x$ is the adjacent side ($x$-component)

$\vec{F}_y$ is the opposite side ($y$-component)

**Using the trigonometric functions to find the components of force**

The ratios of the sides of a right triangle depend only on the angle and are called *sine*, *cosine*, and *tangent*. In the example above, the force $F$ makes an angle $\theta$ with the $x$-axis. The cosine of the angle is the ratio of the adjacent side divided by the hypotenuse: $\cos\theta = F_x/F$. The sine of the angle is the ratio of the opposite side divided by the hypotenuse, or $\sin\theta = F_y/F$. The tangent of the angle is the ratio of the opposite side divided by the adjacent side, or $\tan\theta = F_y/F_x$.

**Trigonometry**

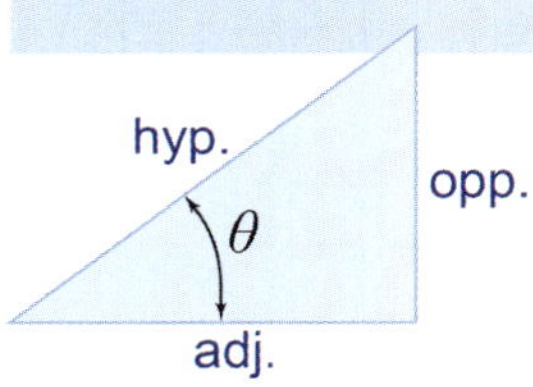


| Math (Right triangle) | Physics (Vector triangle) |
|---|---|
| $\sin\theta = \dfrac{\text{opp.}}{\text{hyp.}}$ | $\sin\theta = \dfrac{F_y}{F}$ |
| $\cos\theta = \dfrac{\text{adj.}}{\text{hyp.}}$ | $\cos\theta = \dfrac{F_x}{F}$ |
| $\tan\theta = \dfrac{\text{opp.}}{\text{adj.}}$ | $\tan\theta = \dfrac{F_y}{F_x}$ |

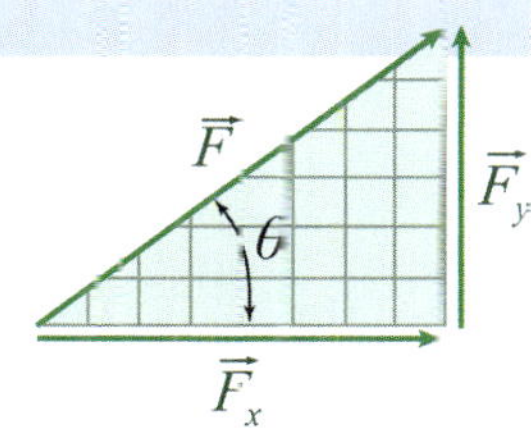


Interactive equation

**Force vector**
$x$- and $y$- components

$$F_x = F\cos\theta \qquad (6.2)$$
$$F_y = F\sin\theta$$

$F$ = magnitude of force (N)
$F_x$ = $x$-component of force (N)
$F_y$ = $y$-component of force (N)
$\theta$ = position angle (degrees)

**Equation for components**

In the equations above one assumes that the angle $\theta$ is measured from the $x$-axis. If instead the angles are measured from the $y$-axis, then the sine and cosine functions in these equations will swap places. The sine and cosine relationships can be solved for the components $F_x$ and $F_y$ in terms of the magnitude and angle of the original force. Because the sides of a right triangle can never be longer than the hypotenuse, the values of sine and cosine are always between 0 and 1 for angles that lie between 0° and 90°. Sine and cosine can be negative for other angles but can never have a value greater than +1 or less than −1.

**A force of 50 N acts on a cart at an angle of 30° to the horizontal. Find the horizontal and vertical components of the force.**

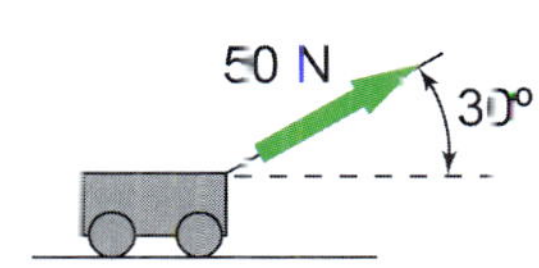


| | |
|---|---|
| **Asked** | $x$ and $y$ components of force |
| **Given** | magnitude and angle of force |
| **Relationships** | $F_x = F\cos\theta$ $\quad F_y = F\sin\theta$ |
| **Solution** | |

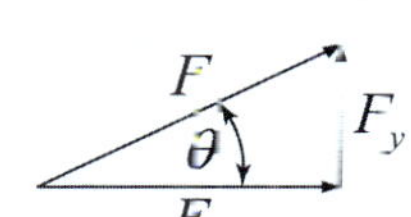


$F_x = F\cos\theta = (50\text{ N})\cos(30°) = \boxed{43\text{ N}}$
$F_y = F\sin\theta = (50\text{ N})\sin(30°) = \boxed{25\text{ N}}$

## Adding and subtracting component vectors

**Adding vectors by components**

To add two vectors, you add their $x$-, $y$-, and $z$-components separately. The $x$-component of the resultant is the sum of the $x$-components of each separate vector; the same is true for the $y$- and $z$-components. To subtract two vectors, you subtract the components. This is both faster and more accurate than drawing vector diagrams. Working in components is the easiest way to do calculations with vectors.

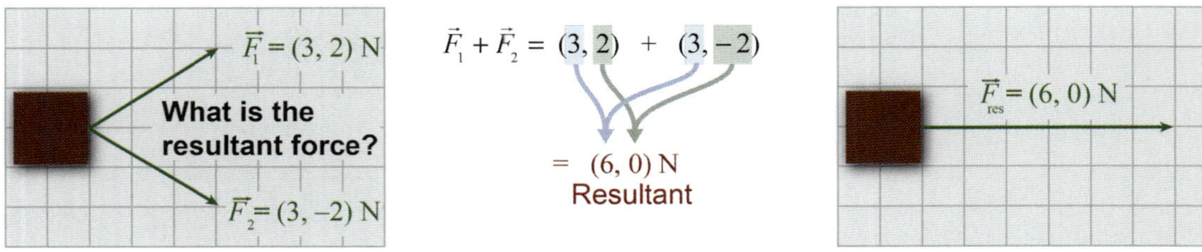


**An example of adding components**

Consider a tanker pushed by two tugboats as shown. Tugboat A pushes with a force of 40,000 N and tugboat B pushes with a force of 10,000 N. What is the net force on the tanker? Will it go straight north or will it move at an angle? The answer to both questions requires adding the two forces from the tugboats. Before this can be done, we draw the vector triangles and find the components of each force. The $x$- and $y$-axes are aligned north–south and east–west.

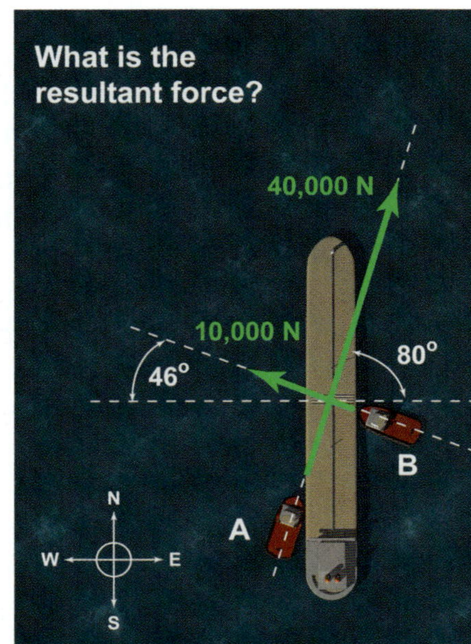


Force components for tugboat A:

$$F_x = +(40,000\ \text{N})\cos 80° = +6,946\ \text{N}$$
$$F_y = +(40,000\ \text{N})\sin 80° = +39,392\ \text{N}$$

Force components for tugboat B:

$$F_x = -(10,000\ \text{N})\cos 46° = -6,946\ \text{N}$$
$$F_y = +(10,000\ \text{N})\sin 46° = +7,193\ \text{N}$$

**Solving the problem**

The force from tugboat A is resolved into +6,946 N in the $x$-direction and +39,392 N in the $y$-direction. Tugboat B is much less powerful, and its 10,000 N force resolves into an $x$-component of −6,946 N and a $y$-component of +7,193 N. The forces in the $y$-direction add to a $y$-component of +46,585 N. The $x$-components add up to zero! The net force on the tanker is $F_{net} = (0, 46,585)$ N. Since there are no forces acting in the $x$-direction, the tanker will go straight north. Clearly, the captain of tugboat B knew enough physics to adjust his angle to compensate for the action of the stronger tugboat A.

**Problem-solving strategy**

How do you solve problems involving vectors, including force, position, speed, and others?

Follow this problem-solving strategy:

1. Resolve each vector into its $x$-, $y$-, and $z$–components.
2. Add or subtract components separately for $x$, $y$, and $z$ as needed by the problem.
3. Put the components back together to arrive at the full vector solution.

Remember that subtracting a vector is the same as adding the negative of the vector.

**Pythagorean theorem**

When you know the $x$- and $y$-components of a vector, you know two sides of the vector triangle. The magnitude is the hypotenuse, which you can calculate using the Pythagorean theorem. This theorem states that the square of the hypotenuse is the sum of the squares of the other two sides of the triangle.

**Finding the magnitude using the Pythagorean theorem**

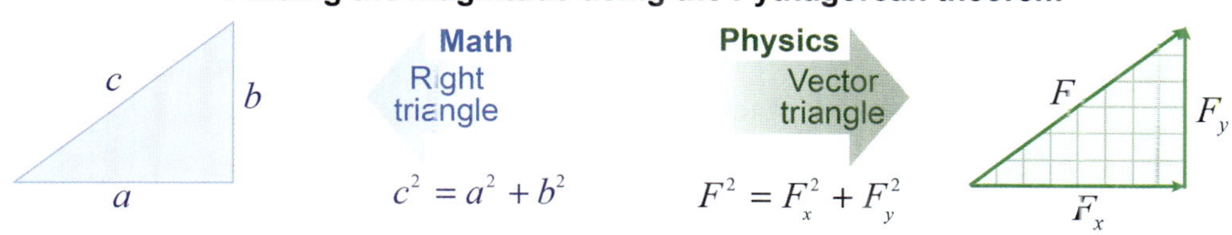


Using the vector triangle in the diagram, the Pythagorean theorem says $F^2 = F_x^2 + F_y^2$. If you take the square root of both sides of this equation, then you have solved for the magnitude $F$ in terms of the components, as given in equation (6.3).

Interactive equation

$$\text{(6.3)} \qquad F = \sqrt{F_x^2 + F_y^2}$$

$F$ = magnitude of the force (N)
$F_x$ = $x$-component of force (N)
$F_y$ = $y$-component of force (N)

**Magnitude** of a vector

**Finding the angle**

To find the angle of a vector from its components you can use the *tangent* function in reverse. The *inverse tangent* is a function that gives you back the angle if you know the ratio of the sides of a right triangle. Mathematically, the inverse tangent of a number $n$ is written as $\tan^{-1} n$. When the number is the ratio of the $x$- and $y$-components of a vector, the inverse tangent gives you the angle the vector makes with the $x$-axis, as given in equation (6.4).

Interactive equation

$$\text{(6.4)} \qquad \theta = \tan^{-1}\left(\frac{F_y}{F_x}\right)$$

$\theta$ = angle (degrees)
$F_x$ = $x$-component of force (N)
$F_y$ = $y$-component of force (N)

**Angle** from components

**Radians and degrees**

For example, suppose you have a vector $\vec{F}$ = (+6,+8) N and wish to know the angle this force makes with the $x$-axis. Applying the inverse tangent results in an angle of +36.9°. When using a calculator, carefully check whether it is set to degree or *radian* mode. One **radian (rad)** is about 57.3°, so a 30° angle is the same as 0.52 rad.

**Finding the angle using the inverse tangent**

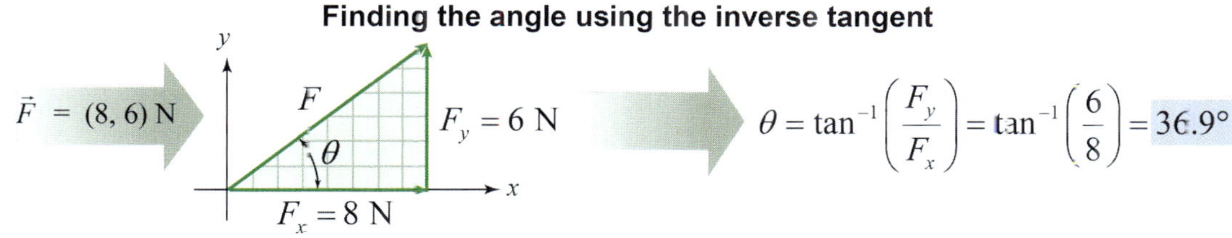


**Three-dimensional vectors**

For any vector in $(x, y, z)$ form, you can find the magnitude using the three-dimensional version of the Pythagorean theorem. The square of the magnitude is the sum of the squares of all three coordinate values. That tells us that the magnitude of force $(F_x, F_z, F_z)$ is the square root of the sum of the squares: $F_x^2 + F_y^2 + F_z^2$.

## Net force and free-body diagrams

**Why draw a free-body diagram?**

An accurate free-body diagram is essential to solving all but the simplest force problems. Keep in mind that *free-body diagrams are not supposed to be art!* Draw simple outlines of objects—adding too many details does not help you solve the problem and may even make a diagram hard to interpret. The important thing is to get the general shape of the object, identify and sketch all the forces in the right places, and assign variable names so you don't confuse yourself.

**Drawing an accurate free-body diagram**

**Three boxes are stacked on top of each other as shown and pulled with a force of 80 N at a 30° angle. Draw the free-body diagram of the 15 N box.**

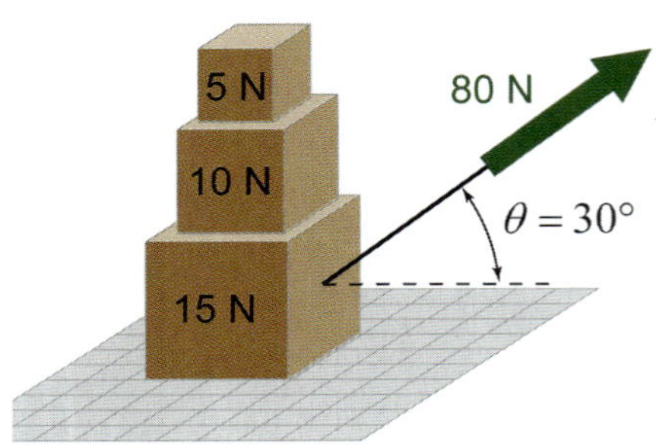


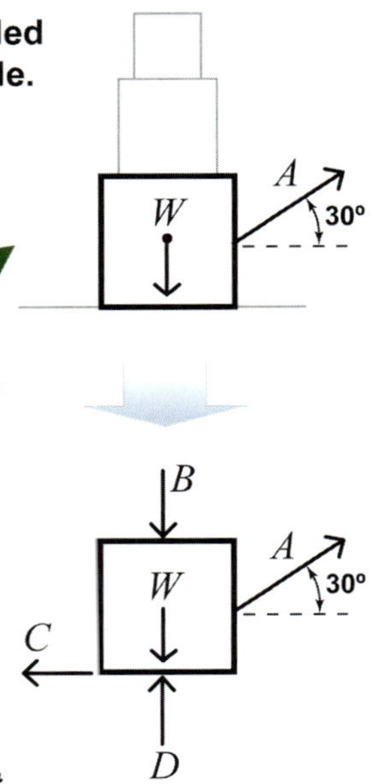


**Include only forces that act *on* the object.**

***Do not include forces the object may exert on other objects!***

1. Draw the outline of the object isolated from all other objects.
2. Draw arrows for forces applied from ropes, springs, pushes, or other external sources. Arrows should be in approximately the right places and in the direction you believe the force acts.
3. Draw an arrow for the weight acting down from the center of mass.
4. Add a normal force acting perpendicular to every surface where the object contacts another object or touches a surface.
5. If the object can slide or move, draw arrows for any friction forces you might expect.
6. Assign a name and/or value to every force.

**Drawing component forces**

One of the main reasons for drawing a free-body diagram is to uniquely assign names and directions to each force, especially unknown forces. You may wish to draw the component forces when a problem includes forces at angles other than 90°. This will help you keep track of the subscripts when there are several forces.

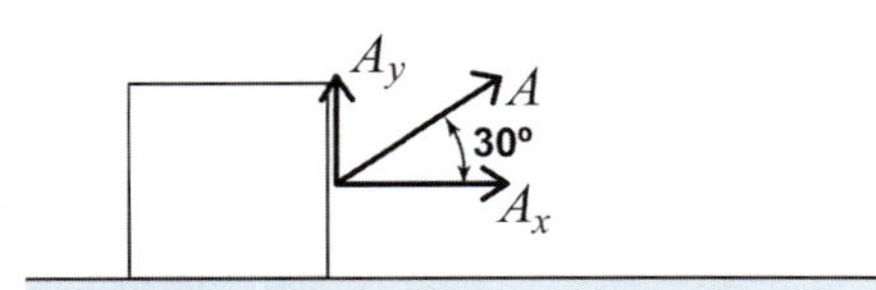


**Drawing component forces**

**Positive and negative directions**

In free-body diagrams you assign positive and negative directions to forces. When you draw an arrow for an unknown force on a free-body diagram you are explicitly choosing the direction of the force. Don't worry if you don't choose correctly! After you solve the problem, if the force has the opposite sign from what you chose, it means the force is in the opposite direction from the one in your diagram.

**Find the *x* and *y* net force**

In almost all force problems the next step after drawing the force diagram is to find the net force in each component direction. A good way to accomplish this is to make a table that lists the equation for the forces in $x$ and $y$ separately. Put the table right next to the free-body diagram so you have a ready reference for identifying each force with the name and direction you assigned. *Forces that point in a negative direction on the diagram should have negative signs in the equation!*

# Section 1 review

Vectors are quantities that specify direction as well as amount. Adding vectors is necessary in order to determine the resultant, or net, force on an object (the force that appears in Newton's second law). Vectors can be added graphically using the "head-to-tail" method. Vectors can be expressed in terms of components; they also can be expressed in terms of magnitude and angle. To add two vectors mathematically, first add their $x$-components to obtain the $x$-component of the vector sum; then repeat this procedure with $y$ (and, in the case of three dimensions, $z$ as well). A free-body diagram can help you identify all of the forces acting on an object of interest.

| | |
|---|---|
| **Vocabulary words** | **vector, vector diagram, magnitude, scalar, resultant vector, component force, component, resolution of forces, sine, cosine, tangent, radian (rad)** |
| **Key equations** | $\vec{F} = (F_x, F_y, F_z)$    $F_x = F\cos\theta$, $F_y = F\sin\theta$    $F = \sqrt{F_x^2 + F_y^2}$    $\theta = \tan^{-1}\left(\frac{F_y}{F_x}\right)$ |

## Review problems and questions

1. Using each force vector's $x$- and $y$-components, calculate the sum $\vec{F_1} + \vec{F_2}$ indicated here. Use component arithmetic and the graphical "head-to-tail" method. Express your sum graphically and with numbers.

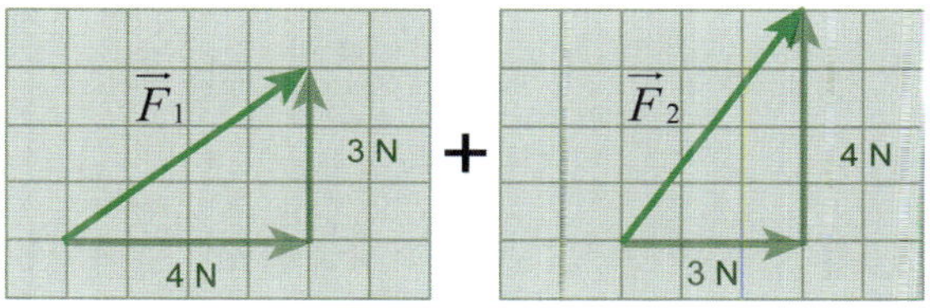


2. Calculate the *magnitude* of the vector $\vec{F_1} + \vec{F_2}$ from the previous problem.

3. Using each force vector's $x$- and $y$-components, calculate the sum $\vec{F_1} + \vec{F_2}$ indicated here. Use component arithmetic and the graphical "head to tail" method. Express your sum graphically and with numbers.

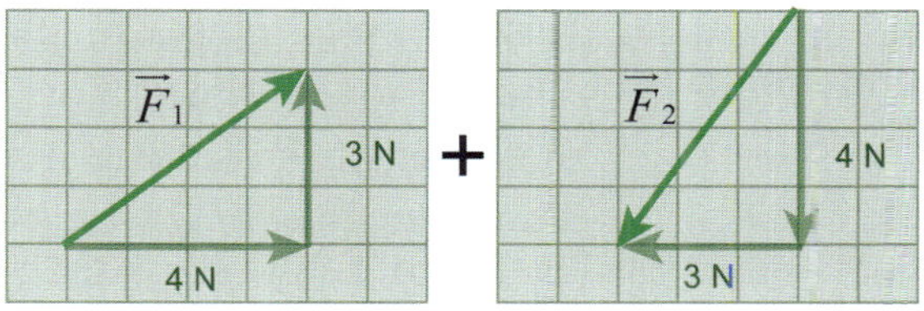


4. Calculate the *magnitude* of the vector $\vec{F_1} + \vec{F_2}$ from the previous problem.

5. Calculate the magnitude of $\vec{F_1}$ from the first problem in this review. Next, calculate the magnitude of $\vec{F_2}$. Finally, calculate the magnitude of the sum $\vec{F_1} + \vec{F_2}$. Does the magnitude of the vector sum equal the sum of the two individual vectors' magnitudes? Explain.

In Chapters 3 and 4, we used position, velocity, and acceleration to describe motion in one dimension. In this section we broaden our scope to include the full, three-dimensional vector character of these quantities. Position, velocity, and acceleration are vectors; the tools we learned with one-dimensional motion can now be extended to three dimensions. The key to working in three dimensions is to use vector components to separate complex, three-dimensional problems into three separate but solvable one-dimensional problems—one along each coordinate axis.

### Displacement vector

**Adding displacements**

Suppose you walk 10 m east, turn, go 5 m north, then turn and go 5 m west. Where do you end up? To answer the question, you must treat each leg of the walk as a separate *displacement* vector. The **displacement** vector describes a change in position, such as a move from one place to another. The position at the end of the walk is the result of adding the three displacements together. This resultant vector is also the final position vector. The illustration below shows the calculation done graphically and by components, resulting in $\vec{d}$ = (+5,+5) m. Note that the westward displacement vector has a negative $x$-component because the positive $x$-axis points east. Displacements are often given in compass coordinates in which the $x$-axis is east–west and the $y$-axis is north–south.

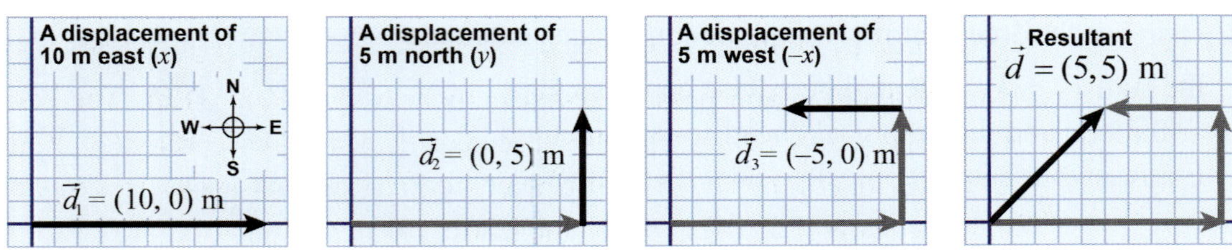


Scale: 1 m

$$\vec{d} = \vec{d}_1 + \vec{d}_2 + \vec{d}_3 = (10, 0)\text{ m} + (0, 5)\text{ m} + (-5, 0)\text{ m} = (10-5,\ 5)\text{ m} = (5, 5)\text{ m}$$

**Displacement by components**

Most real displacements are not purely north, south, east, or west. Instead, they are at angles that can change over time. For purposes of calculation, however, it is useful to represent a displacement by components *as if it were two separate displacements in x and y*. For example, moving 7 m at 30° north of east puts you in the same place as moving 3.5 m north and then 6.06 m east. The displacement is written in component form as $\vec{d}$ = (+6.06,+3.50) m, even though the *actual* motion follows the diagonal path.

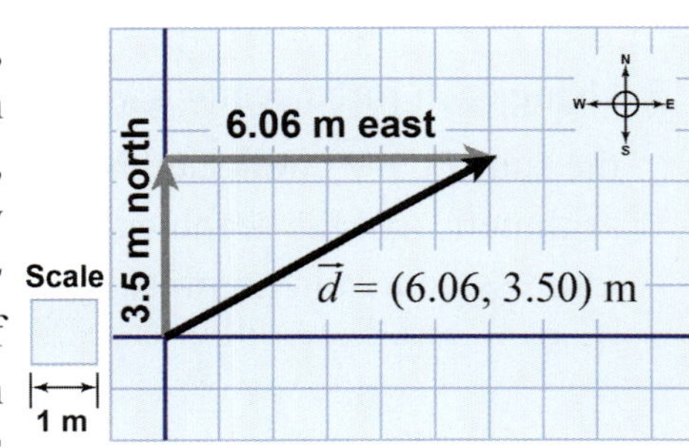


**Vector model for displacement**

The illustrations above depict displacement using a *vector model*, where each displacement is represented by a vector. In the vector model, more than one vector can be added *graphically* using the head-to-tail method or *algebraically* by adding the components.

**Calculating components**

Calculating the components of a displacement vector is the same as for the force vector. If the magnitude of the displacement is $d$, then the components are $d_x = d\cos\theta$ and $d_y = d\sin\theta$.

**$x$- and $y$-components of displacement**

$$d_x = d\cos\theta$$
$$d_y = d\sin\theta$$

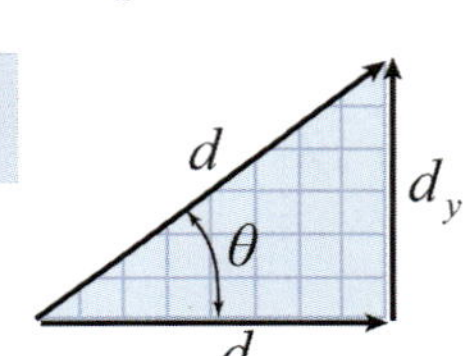

**Polar vectors**

Describing a vector as 5 m at an angle of 36.9° is one way of expressing both magnitude and direction. This is called a *polar vector* because it uses the resolution of components in which every point is specified with a radius (magnitude) and angle. To follow convention, the 5 m displacement (illustrated at right) is given an angle of −36.9°, because angles increase from 0° on the $x$-axis to 90° on the $y$-axis. Clockwise angles are negative and counterclockwise angles are positive.

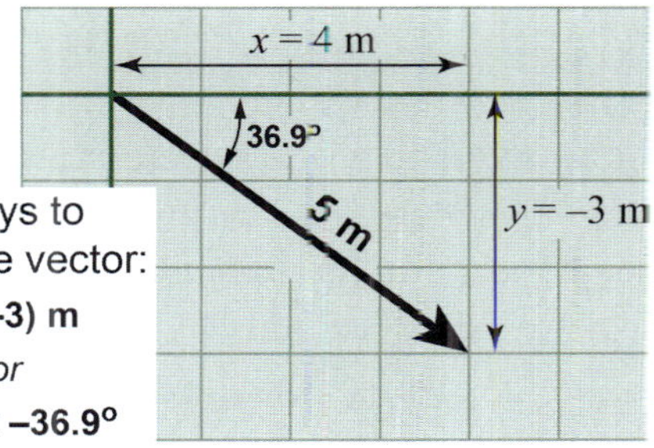


Two ways to write the vector:
**(4, −3) m**
*or*
**5 m at −36.9°**

**Vector diagram**

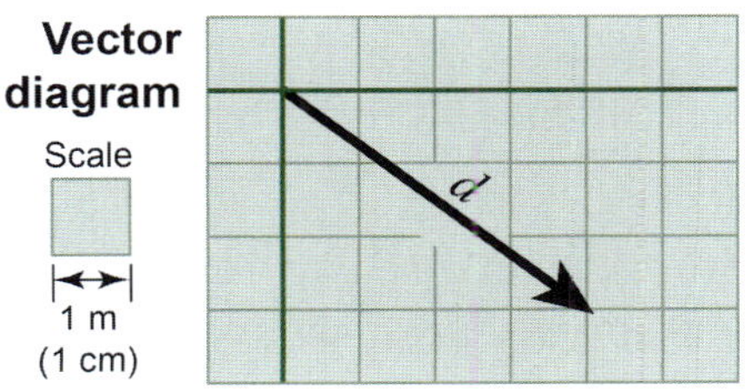


| **Cartesian ($x$–$y$ components)** | **Polar ($r$, $\theta$ values)** |
|---|---|
| $\vec{d} = (4,-3)$ m | $\vec{d} = 5$ m at $-36.9°$ |

**Different ways to represent vectors**

Vectors can be written in more than one way using different coordinate systems. The most common representation is using **Cartesian coordinates**, which uses the vector components along the $x$-, $y$-, and possibly $z$-axes, such as $\vec{d} = (+4,-3)$ m. The coordinate system that uses polar vectors is called **polar coordinates**. The vector can also be represented by using a vector diagram. The 5 m displacement of our example is shown above in a vector diagram, as well as using polar and Cartesian coordinates.

**Compass bearings**

Most navigation is done using compass coordinates. A **compass** divides direction into 360°—a full circle. *Compass coordinates are rotated from standard geometry coordinates because they use north instead of east as a reference.* (In mathematics class you will usually measure angles starting from the horizontal direction, rather than the vertical direction!)

In compass coordinates

- angles increase in the clockwise direction (starting from north) and
- 0° is north, which is usually the $y$-axis in a two-dimensional system.

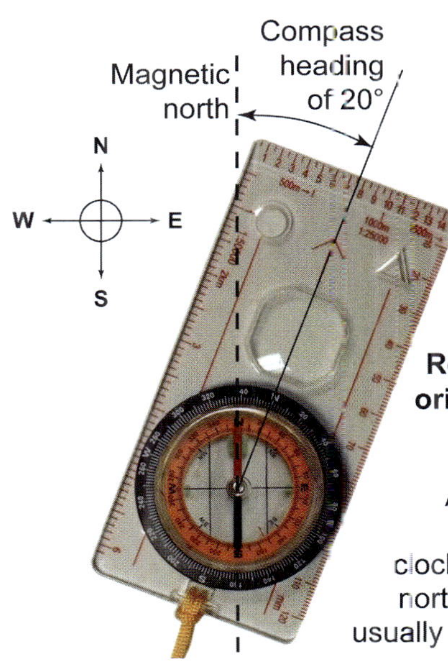


**Reading an orienteering compass**

Angles are measured clockwise from north, which is usually the $y$-axis.

**Example for a compass heading**

A compass heading of 20° is 20° clockwise from north. This is the same as a heading of +70° from the $x$-axis. Compass headings range from 0–360°; a compass heading of 210° is 210° clockwise from north, for example. You should *always* draw a vector diagram that puts the vector in the right quadrant to figure out components from compass angles.

# Investigation 6A Vector navigation

**Essential questions** *How does a robot use vectors to navigate?*

Interactive simulation

Vectors are the fundamental way in which robotic technology operates. Each motion of a robot is described by displacement, velocity, and acceleration vectors. In this investigation you will solve different navigation challenges using two-dimensional vectors to solve a series of mazes.

## Part 1: Navigating a maze

To solve the challenge, find a series of vectors that move from the green circle (start) to the red circle (finish) without crossing any of the walls of each maze. You may make up to 20 displacements.

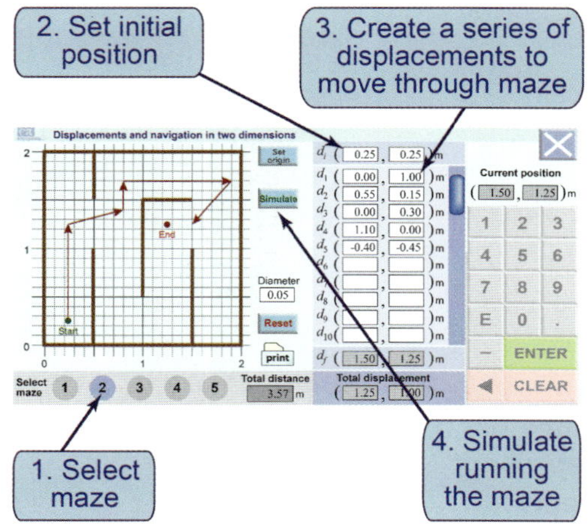


1. Choose the first maze and try to solve it on the screen with as few vectors as possible.
2. Choose the second maze. Work out a solution.

a. *What is the minimum number of vectors that can solve each of the four mazes?*

b. *What is the minimum total distance you can achieve that solves each of the four mazes?*

c. *Maze 4 has a circular path. Do you think more than 20 vectors would allow you to find a solution with a shorter distance? Why or why not? Estimate with your group how much shorter you believe you could get. Could you find a course that is 10% shorter with an unlimited number of vectors? How about 20% shorter?*

d. *What changes would you have to make in order for an actual robotic car to navigate a real maze like the ones in the simulation? How will the displacements be different and why?*

e. *Discuss how the accuracy and precision of the robotic car's motion affects the programming needed to successfully solve the maze without hitting a wall.*

## Part 2: Force, inertia, and a race course

1. In this interactive simulation, two players race through a course to the finish line.
2. During their turns, players may apply a force of 1 N in any direction (or choose zero force instead).
3. Using the setup button, you may choose a different race course or switch to one-player mode.

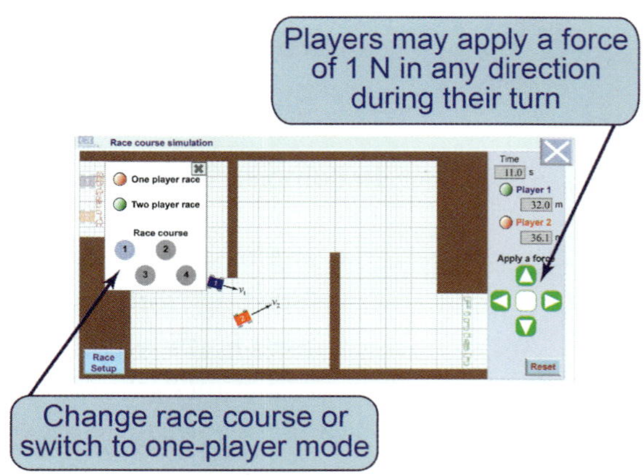


a. *How does an upward applied force change the velocity of the cart?*

b. *Are there negative forces applied in this activity? Explain.*

c. *Explain the role of inertia in this activity.*

**The velocity vector**

How quickly a displacement occurs depends on the velocity vector $\vec{v}$. The velocity vector is the directional analog of speed. The velocity vector can be represented in polar or Cartesian coordinates, just as for displacement. As an example, suppose a car is moving at 5 m/s on a road that makes an angle of 36.9° east of north. One second after passing through the origin, the car's position has changed by 5 m along the 36.9° line.

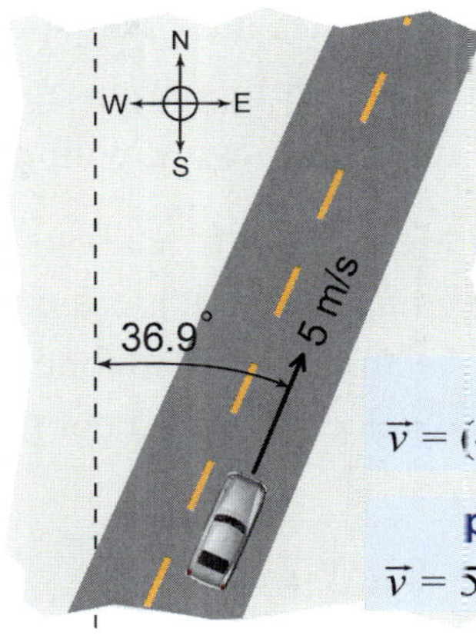


**Equivalent ways to describe a velocity vector:**

***x–y* components**

$\vec{v} = (4, 3)$ m/s

**polar coordinates**

$\vec{v} = 5$ m/s at 36.9°

**Velocity, time, and displacement**

The displacement of the car has components given by $\vec{d} = (+3,+4)$ m. In Chapter 3 we defined **speed** as distance divided by time. The vector equivalent is that ***velocity** is displacement divided by time.* Time is a scalar; when you divide a vector by a scalar, each component is divided separately—and the result is also a vector. Therefore

$$\vec{v} = \frac{\vec{d}}{t} = \left(\frac{d_x}{t}, \frac{d_y}{t}\right) = \left(\frac{3\text{ m}}{1\text{ s}}, \frac{4\text{ m}}{1\text{ s}}\right) = (3,4)\text{ m/s}$$

For every 5 m the car moves along its actual path, it changes its position by +3 m in the $x$-direction and +4 m in the $y$-direction. The car actually moves along the diagonal, but we can mathematically treat the motion as components along $x$ and $y$.

**Velocity in two and three dimensions**

All the relationships between among, speed, and time also exist in the same form among displacement, velocity, and time. In three dimensions the single vector equation $\vec{v} = \vec{d}/t$ is actually three separate equations, one for each direction $x$, $y$, or $z$. For two-dimensional problems, there are two equations, one for $x$ and one for $y$.

**How scalar speed relates to the vector velocity**

| Scalar | Vector | | *x*-axis | *y*-axis | *z*-axis |
|---|---|---|---|---|---|
| speed = distance/time<br>$v = \frac{d}{t}$ | velocity = displacement/time<br>$\vec{v} = \frac{\vec{d}}{t}$ | Three dimensions<br>$\vec{v} = (v_x, v_y, v_z)$ | $v_x = \frac{d_x}{t}$ | $v_y = \frac{d_y}{t}$ | $v_z = \frac{d_z}{t}$ |
| | | Two dimensions<br>$\vec{v} = (v_x, v_y)$ | $v_x = \frac{d_x}{t}$ | $v_y = \frac{d_y}{t}$ | |

**Using vector equations**

All of the techniques for solving problems we used for one dimension apply to two and three dimensions as well. The main difference is that there are more variables and more equations. In one dimension, there was just one position variable $x$, one velocity variable $v$, one acceleration variable $a$, and one set of equations. In two dimensions there are two position components ($x$ and $y$), two velocity components ($v_x$ and $v_y$), two acceleration components ($a_x$ and $a_y$), and two separate sets of equations for the $x$- and $y$-directions. In three dimensions there are three each. That's a lot of components and equations! Despite the challenge in keeping track of all these components, the problems are conceptually similar to what you have already done.

## Resolving component velocities

**Velocity in polar coordinates**

Velocities are often specified in polar form, giving a speed (magnitude) and an angle, such as $\vec{v}$ = 5 m/s at an angle of 36.9° east of north. Polar vectors are easy to understand but difficult to work with mathematically.

**Interpreting velocity vectors**

Most of your analysis and problem solving will be done with components in $(x,y)$ form, which are much easier to work with. Therefore you will often need to resolve the *velocity vector* into components. This is done with a vector triangle just as with force and displacement, except that the axes represent perpendicular velocities.

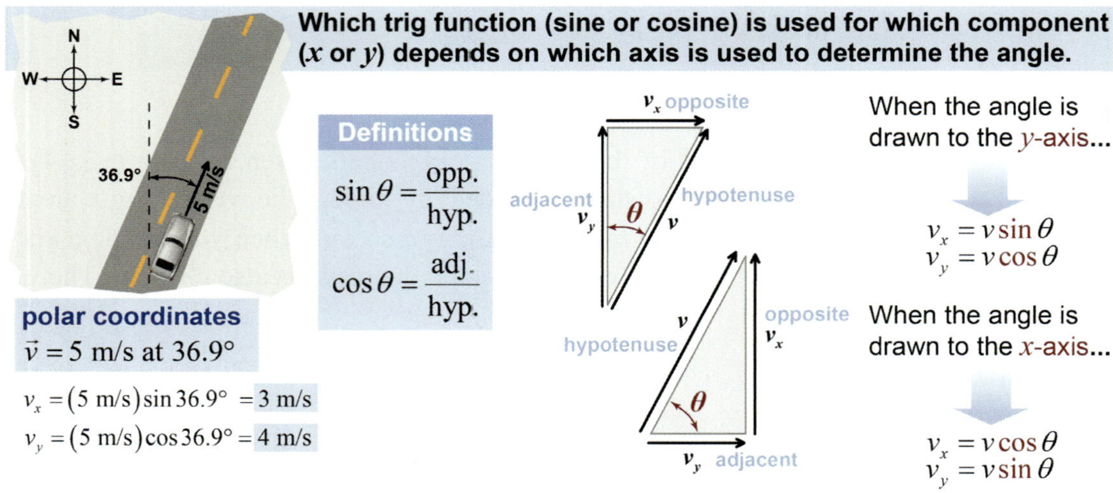


**Choosing sine or cosine**

Notice that, in this example, the angle is given relative to the $y$-axis. If the vector triangle is drawn using the 36.9° angle, the $x$-component is the opposite side and therefore $v_x = v \sin\theta$. Because the angle can be drawn to either the $x$-axis or $y$-axis, it is a good idea to always sketch the vector triangle to determine whether the $x$-component uses a sine or cosine ratio relative to the speed. In this example, the $x$-component uses the sine and therefore the $y$-component of velocity is given by $v_y = v \cos\theta$.

**Angles with the *x*-axis**

If you prefer, you can choose to make all angles relative to the $x$-axis. If the angle with the $y$-axis is 36.9°, the complementary angle with the $x$-axis is 53.1° because the axes always make a 90° right angle. The components in this case have sine and cosine reversed because now the $x$-component is the adjacent side of the angle: $v_x = v \cos\theta$ and $v_y = v \sin\theta$.

**Celestial sphere**

The celestial sphere is a coordinate system used by astronomers to describe the night sky. The celestial sphere is usually plotted with north up but *east to the left*! Why? Lie on your back on the ground with your head pointing north. Your left hand—not your right hand—will be pointing east, because the celestial sphere is reversed with respect to standard map coordinates. As a result, position angles in astronomy are measured from north through east, or counterclockwise. Compare the celestial coordinates with the figure at the top of this page, where the position angle is measured clockwise!

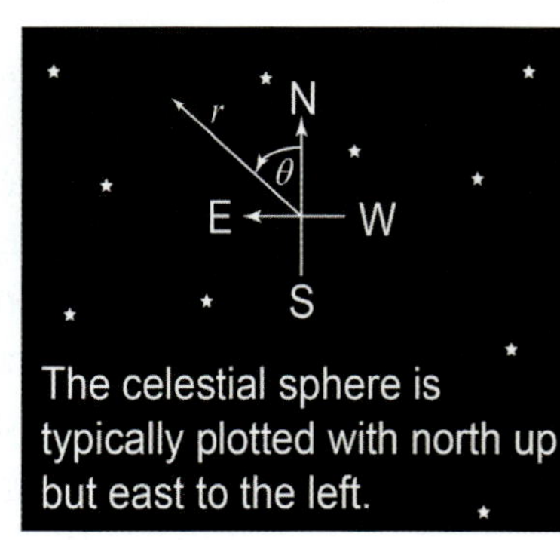

The celestial sphere is typically plotted with north up but east to the left.

### Adding relative velocities

There are many situations in which the velocity of an object is a combination of two or more velocities. An example is a boat moving with a certain velocity relative to the water of a river. The water is moving with a different velocity. The velocity of the boat relative to the land is the sum of the boat's velocity relative to the water plus the water's velocity relative to the land.

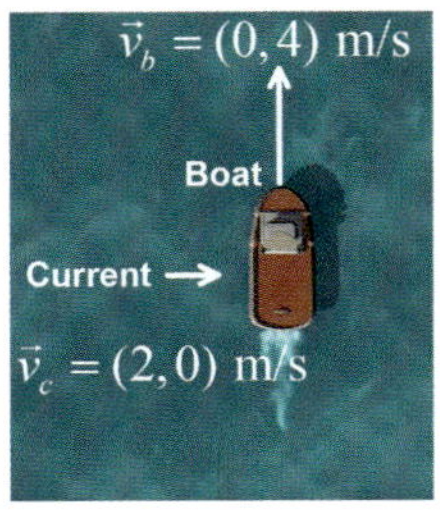


A boat steers north at 4 m/s relative to the water in a current flowing 2 m/s east. What is the velocity of the boat relative to the land?

**Solution**

Velocity vectors may be added by components just like other vectors.

$$\begin{aligned}\vec{v} &= \vec{v}_b + \vec{v}_c \\ &= (0,4)\text{ m/s} + (2,0)\text{ m/s} \\ &= (2,4)\text{ m/s}\end{aligned}$$

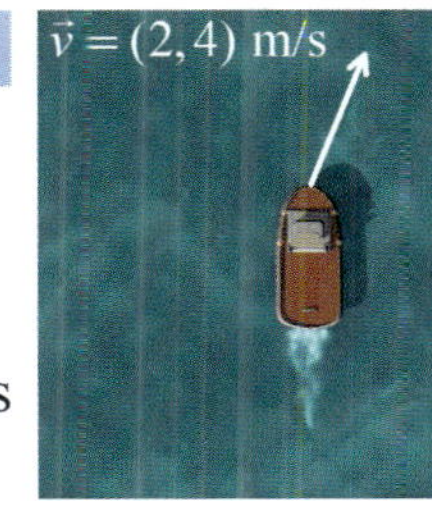


### Add velocities by components

Velocity vectors are added by components, just like force and displacement vectors. Add the $x$-components separately and the $y$-components separately to calculate a resultant velocity. The illustrations above and below show the *vector model* of motion for velocity.

### Velocity of an airplane in a crosswind

Consider a commuter airplane flying due north with a velocity of 400 km/hr. A strong and steady high-altitude wind blows from the west at 80 km/hr. The resultant velocity of the plane is $\vec{v}$ = (+80,+400) km/hr. Relative to the land, the plane is flying at an angle of 11.3° instead of true north. The effect of adding velocities is very real and a concern for anyone navigating a plane or boat. What is the actual speed of the plane with respect to the ground when the effect of the wind velocity is included? We calculate the magnitude of the net velocity (the speed) by drawing the vector triangle. The magnitude is given by the Pythagorean theorem:

$$v^2 = v_x{}^2 + v_y{}^2.$$

**Solution**

Add the two velocity vectors to find a resultant velocity:

$$\begin{aligned}\vec{v} &= \vec{v}_p + \vec{v}_w \\ &= (0,400)\text{ km/hr} + (80,0)\text{ km/hr} \\ &= (80,400)\text{ km/hr}\end{aligned}$$

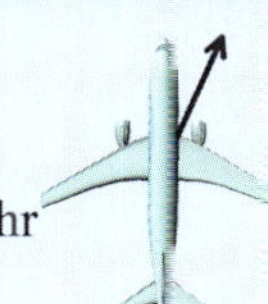

Resultant velocity
$\vec{v} = (80,400)$ km/hr

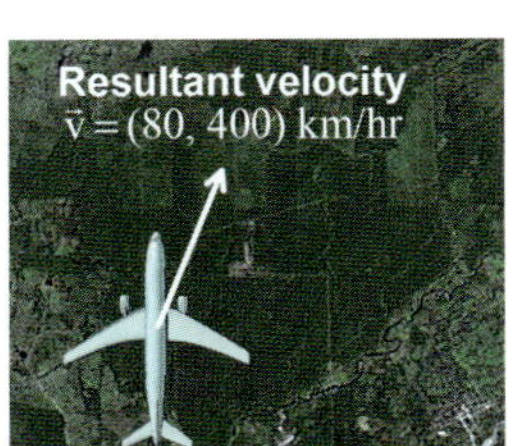


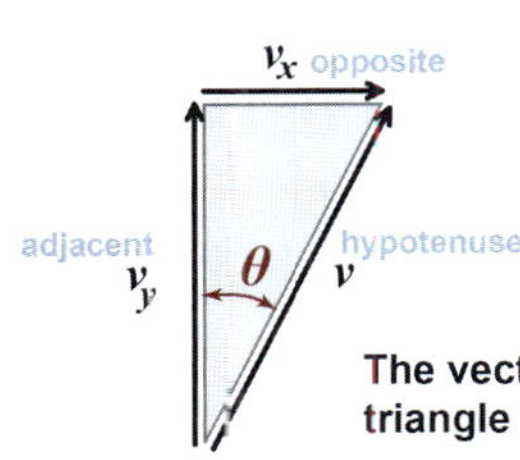


The vector triangle

**Magnitude (speed)**

Use the Pythagorean theorem

$$\begin{aligned}v &= \sqrt{v_x^2 + v_y^2} \\ &= \sqrt{80^2 + 400^2}\text{ km/hr} \\ &= 408\text{ km/hr}\end{aligned}$$

**Angle**

Use the inverse tangent

$$\begin{aligned}\theta &= \tan^{-1}\left(\frac{v_x}{v_y}\right) \\ &= \tan^{-1}\left(\frac{80\text{ km/hr}}{400\text{ km/hr}}\right) \\ &= 11.3°\end{aligned}$$

### Displacement from added velocities

Relative to the land, the plane is moving at a speed of 408 km/hr at a position angle of 11.3°. Because the component velocities add as squares, the actual speed of the plane is not that much greater than its northward component alone. If the pilot continued to steer straight north, however, then over a 2 hr flight the easterly displacement $d_x = v_x t$ would be 160 km. She will miss the airport! What should she do to compensate?

## Acceleration vector

**The acceleration vector**

Force, displacement, and velocity are vectors so it should be no surprise that acceleration is also a vector. Objects may accelerate and change their velocity in any direction, $x$, $y$, or $z$, including any combination of $a_x$-, $a_y$-, or $a_z$-component accelerations. **Acceleration** is the change in velocity over the change in time.

**Acceleration** vector definition

$$\text{(6.5)} \qquad \vec{a} = \frac{\Delta \vec{v}}{\Delta t}$$

$\vec{a}$ = acceleration (m/s$^2$)
$\Delta v$ = change in velocity (m/s)
$\Delta t$ = change in time (s)

**Components of acceleration**

This equation is really shorthand for three component equations, one for each direction.

$$\vec{a} = \frac{\Delta \vec{v}}{\Delta t} \quad \text{really means} \quad a_x = \frac{\Delta v_x}{t}, \quad a_y = \frac{\Delta v_y}{t}, \quad a_z = \frac{\Delta v_z}{t}$$

**Understanding acceleration vectors**

The acceleration vector can be difficult to understand because it represents the rate of change in the velocity vector—which itself represents the rate of change in the position vector. Consider the following two cases of constant acceleration where the acceleration vector points in a different direction from the velocity.

**Accelerating at right angles to initial velocity**

What does the motion look like if an object has an acceleration in one direction but not in another? In the example on the lower left, the initial velocity is equal to $\vec{v} = (0, 10)$ m/s north, while the acceleration vector points due east at $\vec{a} = (1, 0)$ m/s$^2$. After one second, the velocity vector equals $\vec{v} = (1, 10)$ m/s; after two seconds, $\vec{v} = (2, 10)$ m/s. Notice that the $y$-component of the velocity stays the same, because there is no acceleration in the $y$-direction! This means that the speed is slowly increasing and the direction is turning more to the east every second.

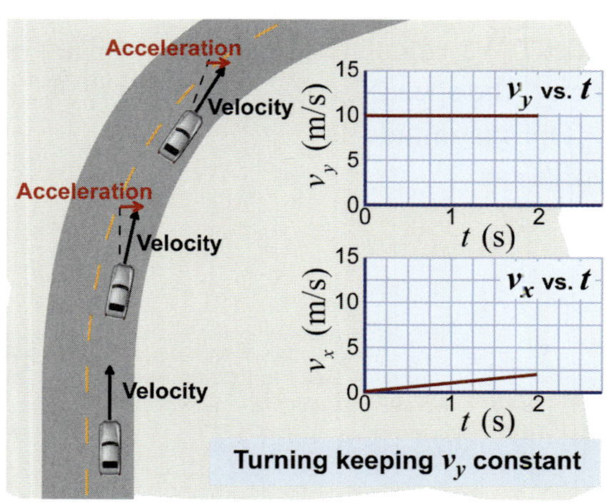


Turning keeping $v_y$ constant

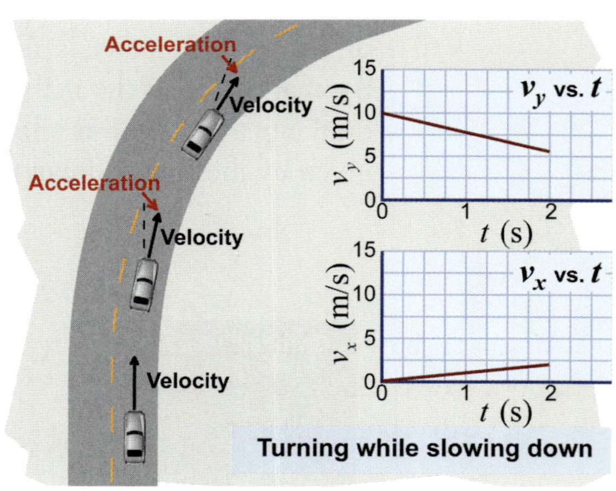


Turning while slowing down

**Slowing down while turning**

In a more realistic scenario, the car starts with the same velocity $\vec{v} = (0, 10)$ m/s, but it slows down while turning (above right). The acceleration $\vec{a} = (1, -2)$ m/s$^2$ acts to slow the car down in the $y$-direction. After one second the velocity is $\vec{v} = (1, 8)$ m/s, while after two seconds $\vec{v} = (2, 6)$ m/s. Notice that the $y$-component of the velocity decreases by 2 m/s each second. The speed decreases over time and the direction changes.

**Comparing models**

The illustration above depicts the motion using both *graphical* and *vector* models. Do you find that one is one more informative than the other?

# Section 2 review

When studying motion in two dimensions, we need to specify two numbers to describe a displacement, velocity, or acceleration vector. The two numbers can be $x$ and $y$ (Cartesian) coordinates or a magnitude and an angle (polar coordinates). The resultant (sum) of two or more vectors can be calculated by separately adding the vectors' Cartesian coordinates. When a boat is rowed across a flowing river, or an airplane flies in windy air, the vehicle's velocity must be added to the current or wind vector to predict how the motion will appear from the ground.

| Vocabulary words | displacement, polar coordinates, Cartesian coordinates, compass, velocity, speed, acceleration |
|---|---|
| Key equations | $\vec{a} = \frac{\Delta \vec{v}}{\Delta t}$ |

## Review problems and questions

1. A programmable toy car has to reach the target shown here. Which of the following vector sums will displace the car from its starting point to the target?
   a. (1, 5) m + (2, −3) m + (3, 1) m
   b. (3, 1) m + (2, −2) m + (1, 3) m
   c. (1, 4) m + (2, −2) m + (3, 3) m

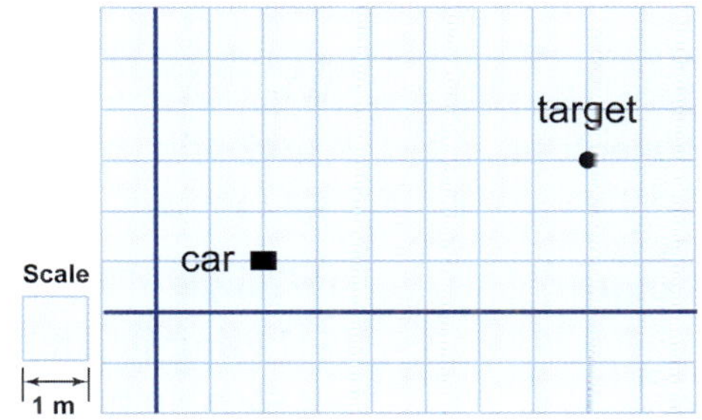


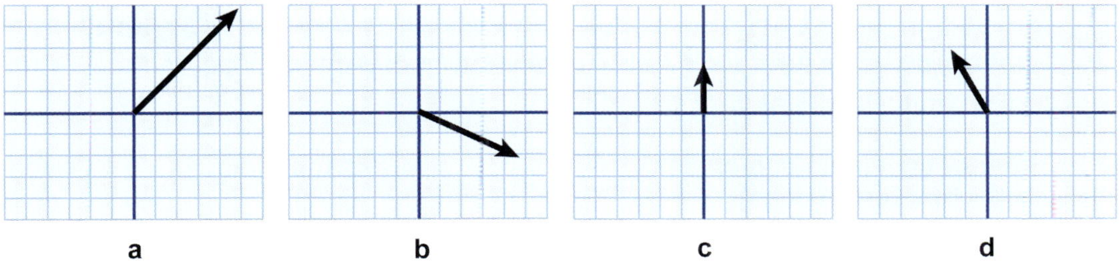


2. Match each of the labeled vectors (**a** through **d**) with one of the following expressions, given in polar coordinates. (Each square on the grid represents one meter of displacement.)
   (i) (2 m, 90°) (ii) (5 m, −30°) (iii) (7 m, 45°) (iv) (3 m, 120°)

3. A rowboat rests on the southern bank of a river 20 m wide flowing west to east. The rowboat is moored 60 m west of a treacherous waterfall. Jamie rows the boat at a speed of 2 m/s northward *with respect to the flowing water*. The current flows at a rate of 4 m/s eastward, which is the water's velocity *with respect to the ground*. Jamie enters the boat and begins to row.
   a. Sketch a diagram of the scenario (as seen from above), with the velocity vectors oriented correctly and drawn to scale.
   b. What is the boat's velocity (in meters per second) with respect to the ground? (Provide components as well as the speed and approximate compass heading.)
   c. Can Jamie reach the northern riverbank before the current carries him into the waterfall?

## 6.3 - Projectile motion and inclined planes

Throw a ball across a football field. Slide down a hill on a sled. How can you describe the motion in each case using physics? Throwing a ball is an example of *projectile motion*, while sledding on a hill is an example of *motion down an inclined plane*. Both are good applications of the vector equations of motion, because they are intrinsically two- or three-dimensional physics problems.

### Projectile motion

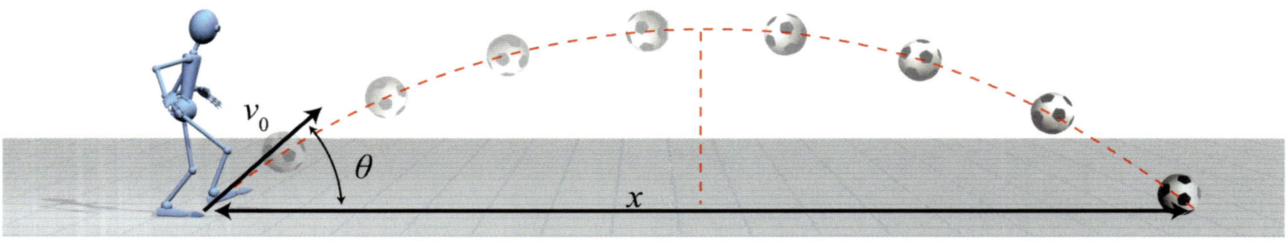


**Trajectory**

Imagine kicking a soccer ball toward a distant goal. The ball starts moving upward at an angle. Then it moves along a curved path called a **trajectory**, gradually redirecting its velocity vector so it points downward again, back toward the ground. The trajectory of the soccer ball is a mathematical curve called a *parabola* that is the result of the fact that gravity accelerates moving objects *down* but not sideways.

**Projectiles**

A **projectile** is any object that is moving and affected only by gravity. If we ignore friction for a moment, the soccer ball is an example of a projectile.

**Range**

The **range** of a projectile is the horizontal distance it covers before landing on the ground at its starting height. When using standard coordinate axes $x$ and $y$, the range is the distance measured in the $x$-direction.

**Independence of motion in two dimensions**

Underlying the physics of projectile motion is a powerful concept: *that the motion in each direction (x and y) can be analyzed independently of the other!* The equations of accelerated motion in vector form look just like they did in one-dimensional motion—except that there are separate equations for $x$ and $y$!

| Vector equations of motion (two dimensions) | | $x$-component equations | $y$-component equations |
|---|---|---|---|
| Position | | Position | Position |
| $\vec{x}=\vec{x}_0+\vec{v}_0t+\frac{1}{2}\vec{a}t^2$ | $\vec{x}=(x,y)$ | $x=x_0+v_{x0}t+\frac{1}{2}a_xt^2$ | $y=y_0+v_{y0}t+\frac{1}{2}a_yt^2$ |
| Velocity | $\vec{v}=(v_x,v_y)$ | Velocity | Velocity |
| $\vec{v}=\vec{v}_0+\vec{a}t$ | $\vec{a}=(a_x,a_y)$ | $v_x=v_{x0}+a_xt$ | $v_y=v_{y0}+a_yt$ |

**What the subscripts mean**

There are a lot of subscripts, so it is worth explaining them. In one dimension, a subscript "0" as in $x_0$ means "initial position"; therefore, in vector components $x_0$ means the $x$-component of the initial position and $y_0$ means the $y$-component of the initial position. The variables $x$, $v_x$, and $a_x$ refer to the time-dependent $x$-components of position, velocity, and acceleration. The variables $y$, $v_y$, and $a_y$ refer to the time-dependent $y$-components of position, velocity, and acceleration. Time $t$ is a scalar, so there is only one value for time.

**Gravity pulls downward**

How can we simplify these equations of motion for projectiles? We use the fact that *gravity accelerates down*; therefore, $a_y = -g$ and $a_x = 0$! The equations of motion simplify a great deal because all the terms that include $a_x$ become zero. Also, to make the math simpler, we choose the initial position to be zero, so $x_0 = 0$ m and $y_0 = 0$ m.

**Equations of motion in the *x*-direction**

Let's first examine the motion in the $x$-direction. Since there is no acceleration in $x$, the component of velocity in the $x$-direction *is constant*. The projectile moves with constant velocity along $x$! The $x$ position changes linearly, as shown in equation (6.6).

**(6.6)**

$$x = v_{x0}t$$

$$v_x = v_{x0}$$

$x$ = position in $x$-direction (m)
$v_{x0}$ = initial velocity in $x$-direction (m/s)
$v_x$ = velocity component in $x$-direction (m/s)
$t$ = time (s)

**Projectile motion**
$x$-component

**Equations of motion in the *y*-direction**

Now look at the equations of motion in the $y$-direction. The acceleration in $y$ has a constant value of $-g$. As we learned in Chapter 4, constant (nonzero) acceleration results in velocity that changes linearly, as shown in equation (6.7). Position in the $y$-direction is a function of $t$ and $t^2$, as also shown in equation (6.7).

**(6.7)**

$$y = v_{y0}t - \frac{1}{2}gt^2$$

$$v_y = v_{y0} - gt$$

$y$ = position in $y$-direction (m)
$v_{y0}$ = initial velocity in $y$-direction (m/s)
$v_y$ = velocity component in $y$-direction (m/s)
$t$ = time (s)
$g$ = acceleration due to gravity = 9.8 m/s$^2$

**Projectile motion**
$y$-component

Equations (6.6) and (6.7) are the equations of projectile motion when gravity is acting in the negative $y$-direction. The initial velocity given to the projectile—soccer ball, cannonball, etc.—has components $v_{x0}$ in the $x$-direction and $v_{y0}$ in the $y$-direction.

**Projectile motion of a hit baseball**

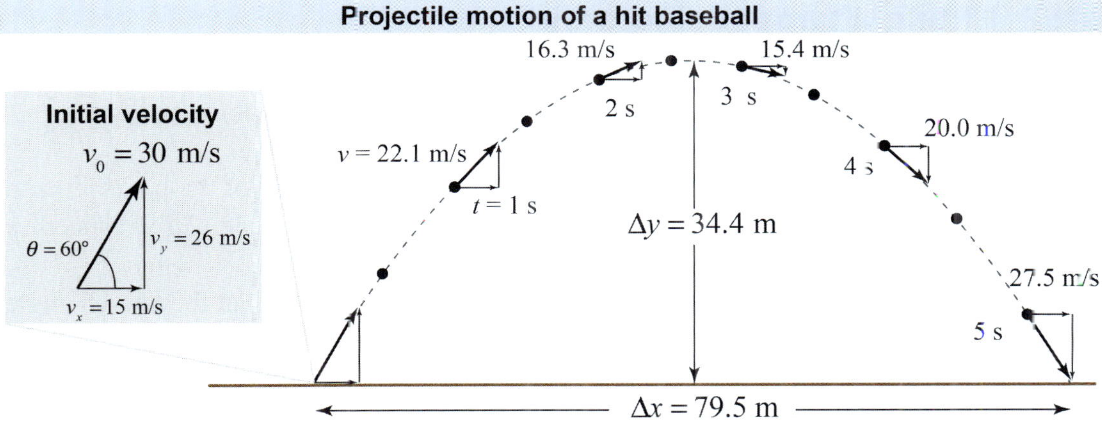


**Trajectory of a baseball**

Consider a baseball player hitting a ball with an upward velocity of 30 m/s at an angle of 60°above horizontal. The ball's initial velocity has components of 15 and 26 m/s in the $x$- and $y$-directions, respectively. Applying equations (6.6) and (6.7) predicts the trajectory illustrated above—with the magnitude of the velocity calculated from $v^2 = v_x^2 + v_y^2$.

**Particle model**

In the illustration of the fly ball above, the position of the ball is depicted at fixed time intervals (every 0.5 s) to create a *particle model* of its motion. The ball positions are closest to each other at the top of the trajectory, which means that the speed is slowest there. The ball positions are furthest apart near the ground where the speed is fastest.

# Investigation 6B Projectile motion

**Essential questions** *What is the angle that maximizes the range of the projectile? How are the projectile motion equations used to calculate a projectile's range?*

Projectile motion occurs whenever a moving object is under no force except gravity. The equations of motion for each coordinate axis are different—and independent of each other—so *you can analyze them separately*. Combined, the equations are used to describe the range of a projectile.

## Part 1: Maximizing range

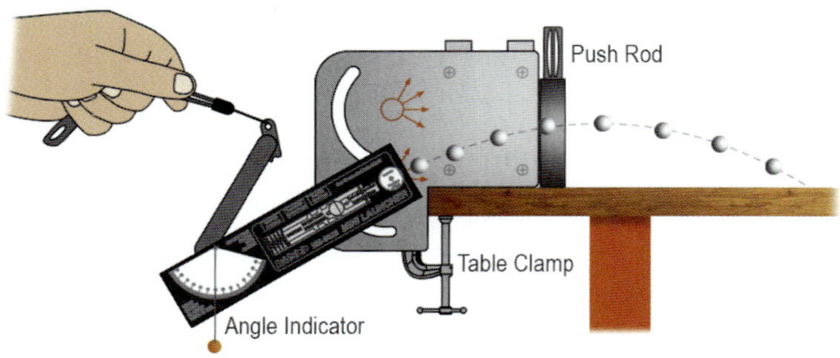


1. Set up your equipment like the picture so the projectile will land on the lab table.
2. Press-in the projectile TWO clicks using the push rod, and then pull up on the string to launch it.
3. Launch the projectile at different angles such as $\theta = 20°$, $25°$, $30°$, $35°$, and so on. Include $30°$ and $60°$ in your data.
4. Measure the range of the projectile for each angle. Tabulate your data.
5. Use the same initial speed $v_0$ (two clicks) for every launch.

a. *What angle $\theta$ launched the projectile furthest? Provide a conceptual explanation for why you think this angle gives the maximum range.*

b. *Provide a conceptual explanation for why the range was approximately the same when $\theta = 30°$ and $\theta = 60°$.*

## Part 2: Hitting a target using projectile motion

1. Adjust the launcher to launch the projectile horizontally ($\theta = 0°$) onto the floor.
2. Launch the projectile using THREE clicks instead of two.
3. Measure the range of the projectile along the floor and the height of the launcher, and then use those values and the projectile motion equations to calculate the projectile's initial speed $v_0$ when leaving the launcher. Show your work.
4. Adjust the launcher so the projectile will land on the table again. Set the launch angle to 60°.
5. Use the range equation (6.10) and your values from the previous steps to calculate the new range of the projectile. Show your work.
6. Place an empty cup at your calculated range and attempt to launch the projectile into the cup using THREE clicks.

a. *Did the projectile land in the cup first try? If not, what factors may have caused the projectile to miss the cup?*

b. *Neglecting air resistance, in which direction was the net force acting on the projectile while it was in flight? In which direction was the projectile accelerating when it was in flight?*

c. *Describe how the projectile's x-direction velocity and y-direction velocity changed during its flight.*

**Trajectory of a cannonball**

Imagine you are firing a cannon, where the initial velocity has components $v_{x0}$ = 15 m/s and $v_{y0}$ = 20 m/s. What trajectory will the cannonball follow? To answer this question, we will use the equations of motion and graphs of the equations. For projectile motion of this cannonball, how do the equations of motion in the $x$-direction differ from the equations of motion in the $y$-direction?

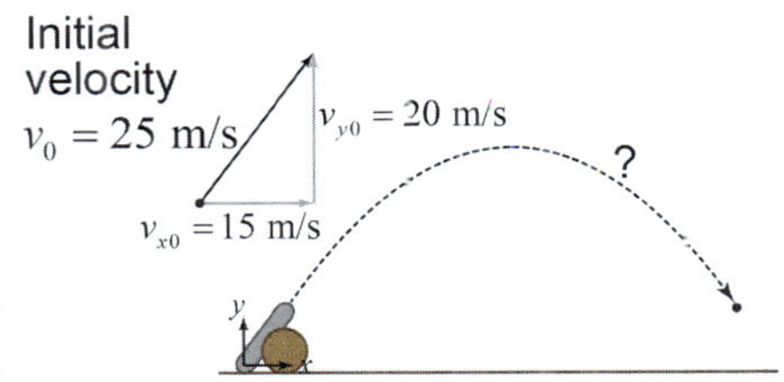


What trajectory will the cannonball follow?

**Comparing motion in the two directions**

The basic difference is that gravity acts only in the $y$-direction, not in the $x$-direction. This is an important observation, because it means that motion is *accelerated* in the $y$-direction, but it is *unaccelerated* in the $x$-direction.

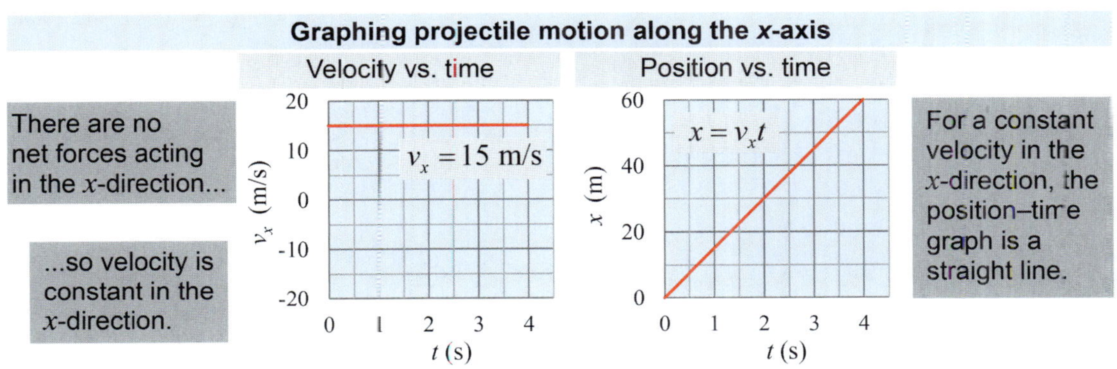


**Projectile motion in the $x$-direction**

In the $x$-direction, there is no acceleration, so the velocity in the $x$-direction is constant at $v_x$ = 15 m/s, which is graphed above left. If $v_x$ is constant, then the position in the $x$-direction is given by $x = v_{x0}t$ = (15 m/s)$t$, which is a straight line graph of constant slope (above right).

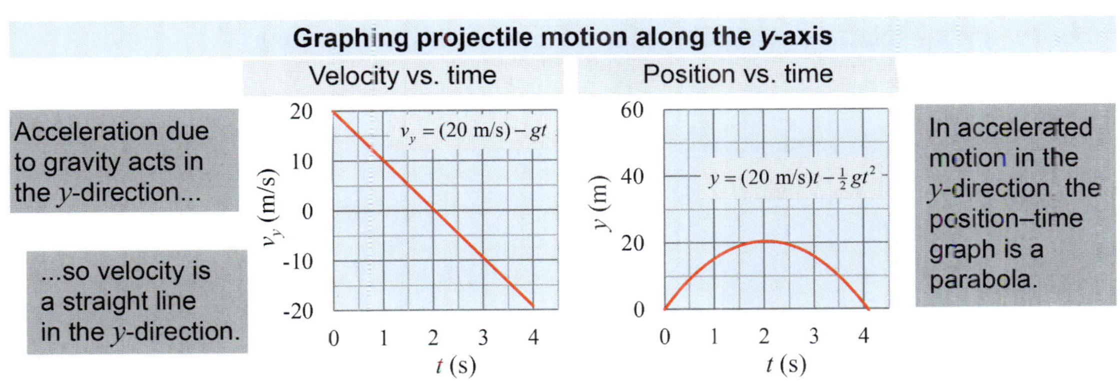


**Projectile motion in the $y$-direction**

In the $y$-direction, the motion has a constant acceleration due to gravity $g$. The velocity is $v_y = v_{y0} - gt$, which is a straight line of constant slope (above left). The position in the $y$-direction is $y = v_{y0}t - ½gt^2$, which is a parabola (above right).

**Parabolic motion**

These equations of motion and their graphs highlight a fundamental property of projectile motion: The position versus time graph in $y$ is a parabola, whereas the same position versus time graph in $x$ is a line.

## Range of a projectile

**An example of a projectile**

Suppose a ball is kicked upward with an initial velocity of 10 m/s at an angle of 30° from the horizontal. How far does it travel before it hits the ground? At what angle of elevation should you kick the ball for it to reach the maximum distance? Both questions are answered by applying the equations of motion in two dimensions. The equations of motion are the same as before (equations 6.6 and 6.7), except that we are now given the magnitude of the initial velocity $v_0$ instead of the individual velocity components. Using the projection angle, we get the initial individual velocity components of $v_{x0} = v_0 \cos\theta$ and $v_{y0} = v_0 \sin\theta$, leading to the modified equations given below.

**Vector triangle for the initial velocity**

$$\sin\theta = \frac{\text{opp.}}{\text{hyp.}} \qquad \cos\theta = \frac{\text{adj.}}{\text{hyp.}}$$

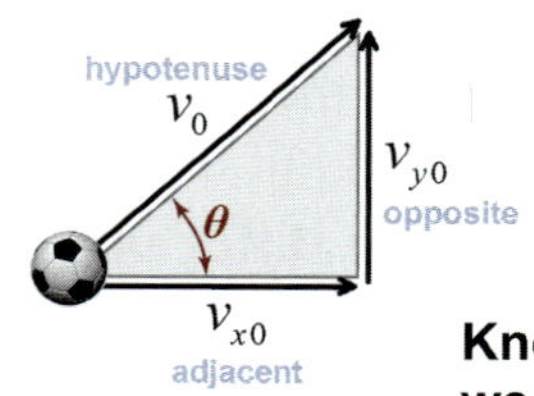


$$v_{x0} = v_0 \cos\theta$$

$$v_{y0} = v_0 \sin\theta$$

**Known quantities, since we are given $v_0$ and $\theta$**

**Projectile motion** *x*-coordinate equations

$$\textbf{(6.8)} \quad x = v_0 t \cos\theta \qquad v_x = v_0 \cos\theta$$

$x$ = position in $x$-direction (m)
$v_0$ = initial speed (m/s)
$\theta$ = projection angle (degrees)
$t$ = time (s)
$v_x$ = velocity component in $x$-direction (m/s)

**Projectile motion** *y*-coordinate equations

$$\textbf{(6.9)} \quad y = v_0 t \sin\theta - \tfrac{1}{2} g t^2 \qquad v_y = v_0 \sin\theta - gt$$

$y$ = position in $y$-direction (m)
$v_0$ = initial speed (m/s)
$\theta$ = projection angle (degrees)
$t$ = time (s)
$v_y$ = velocity component in $y$-direction (m/s)

**Solving the equations**

The *range* of a projectile is the horizontal distance it covers before hitting the ground. That means $x = v_0 t \cos\theta$ is the equation we want to solve. The initial speed $v_0$ is known, and the angle $\theta$ is also known, while the time $t$ is not known. So we use the $y$ equations to solve for time $t$. The $y$-component equation for position is a *quadratic*, which means there are two solutions. Factoring the equation gives the following.

**From the *y*-component equation for position**

Quadratic equation for time $t$

$$y = v_0 t \sin\theta - \tfrac{1}{2} g t^2$$

$$0 = v_0 t \sin\theta - \tfrac{1}{2} g t^2 = t\left(v_0 \sin\theta - \tfrac{1}{2} g t\right)$$

Two solutions for time

$t = 0$ → **Ball kicked off the ground**

$t = \dfrac{2 v_0 \sin\theta}{g}$ → **Ball hits the ground again**

The first solution is for $t = 0$; when the ball is first kicked at the start. The second solution is the one we want because it describes the *second time* the ball touches the ground. Solving the second solution for time yields $t = 2v_0 \sin\theta_0 / g$. Substituting this into $x = v_0 t \cos\theta$ yields the range equation:

**Range equation**

$$\textbf{(6.10)} \quad x = \frac{2 v_0^2 \sin\theta \cos\theta}{g}$$

$x$ = range (m)
$v_0$ = initial speed (m/s)
$\theta$ = launch angle relative to horizontal
$g$ = acceleration of gravity (9.8 m/s$^2$)

**Calculating the range**

Now that we have the range equation (6.10), we can calculate how far a soccer ball goes when kicked upward at an angle of 30° with an initial velocity of 10 m/s. If we use the values $v_0$ = 10 m/s and $\theta$ = 30° to plug into equation (6.10), then we find that the range of the ball is 8.8 m, which is approximately 30 ft.

**Calculating the range**

$$x = \frac{2v_0^2 \sin\theta\cos\theta}{g} = \frac{2(10\text{ m/s})\sin(30°)\cos(30°)}{9.8\text{ m/s}^2} = 8.8\text{ m}$$

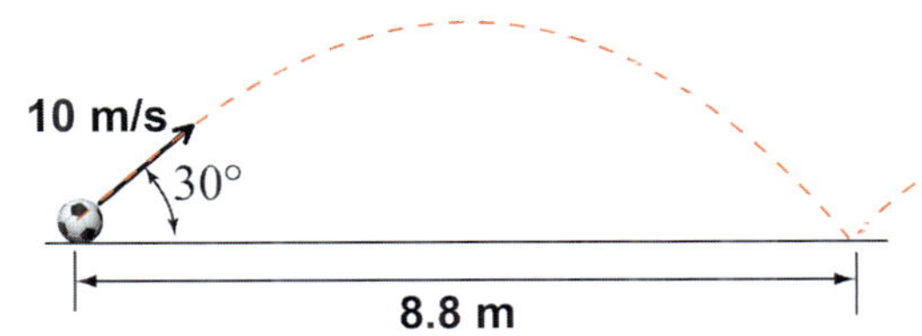


**Investigating the equation**

Looking at the equation we see that the speed appears as $v^2$, which tells us that the range increases with the square of the speed. If you launch at twice the speed, the ball will go four times as far. Kicking the ball four times faster gets you a factor of 16 increase in range! At speeds much over 10 m/s, *air resistance* becomes impossible to ignore, so the range equation is not accurate at higher speeds!

**Maximizing range**

In the range equation (6.10), the range depends on the initial projection angle $\theta$ through the product of sin $\theta$ and cos $\theta$. When you multiply sine and cosine together, the combined function is zero when either sine or cosine is zero. That means that the range is zero at 0° (when the ball is kicked along the ground) and at 90° (when kicked straight upward). In the absence of air resistance, the range is maximum at 45°—halfway between 0° and 90°.

**An example problem**

Not all projectile problems start with an angle. Suppose you launch a marble horizontally off a cliff at a speed of 10 m/s. If the cliff is 10 m high, how far will the marble travel before it hits the ground?

The first step in the solution is to write down the equations of motion in the $x$- and $y$- directions, omitting any terms that are zero. In this problem, $x_0$ = 0 and $v_{y0}$ = 0. The marble starts at $y_0 = h$ with initial velocity $v_{x0}$ in the $x$-direction. The $y$ equation for position has only one unknown (time), so solve it for the time. The time is substituted back into the position equation in the $x$-direction to calculate the answer. The marble travels 14.2 m before it hits the ground.

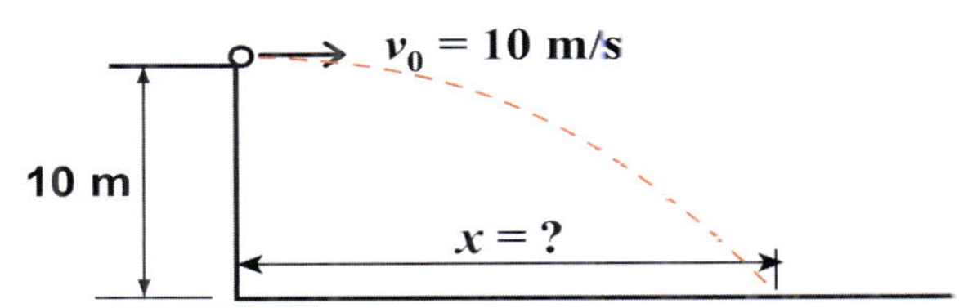


| | |
|---|---|
| **Asked** | Distance in the horizontal direction |
| **Given** | Initial velocity in the $x$-direction; initial position in the $y$-direction; (assume) zero initial velocity in the $y$-direction. |
| **Relationships** | $v_x = v_0$   $v_y = -gt$ <br> $x = v_0 t$   $h = h - \frac{1}{2}gt^2$ |
| **Solution** | Use the equation in $y$ to solve for time until the marble hits the ground ($y = 0$). |

$$0 = h - \tfrac{1}{2}gt^2 \Rightarrow t = \sqrt{\frac{2h}{g}}$$

Calculate the distance from the $x$ equation:

$$x = v_0 t = v_0\sqrt{\frac{2h}{g}} = (10\text{ m/s})\sqrt{\frac{2(10\text{ m})}{9.8\text{ m/s}^2}} = 14.2\text{ m}$$

# Investigation 6C

## Acceleration on an inclined plane

**Essential questions**

*What is the acceleration of an object down an inclined plane?*
*What do the motion graphs of an object down an inclined plane look like?*

Galileo Galilei used *inclined planes* for his quantitative experiments into the nature of position, time, and acceleration. Galileo took advantage of the property that objects roll faster down steeper ramps. In this investigation, you will use the slope of a velocity versus time graph to determine how acceleration varies with the *inclination* of a ramp.

## Part 1: Acceleration down a ramp

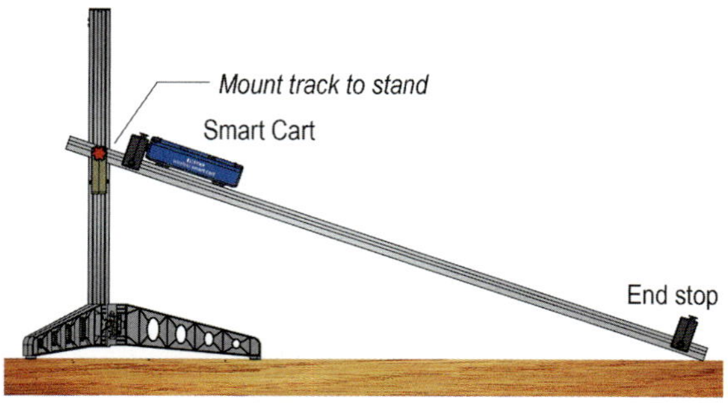


1. Set up the track to act as an inclined plane with height around 30 cm.
2. Open the experiment file **06C_AccelerationOnAnInclinedPlane**, and then power-on the Smart Cart and connect it wirelessly to your software.
3. Begin data collection and release the Smart Cart down the track. Stop data collection just after the cart reaches the bottom of the track.
4. Use your software to fit a curve to each graph as the cart moves down the track. Sketch the shape of the graphs.

a. *What is the motion equation you are fitting to the position vs. time graph? The velocity vs. time graph?*
b. *Explain the connection between the slope of the velocity vs. time graph and the acceleration.*

## Part 2: How does acceleration vary with the ramp's inclination?

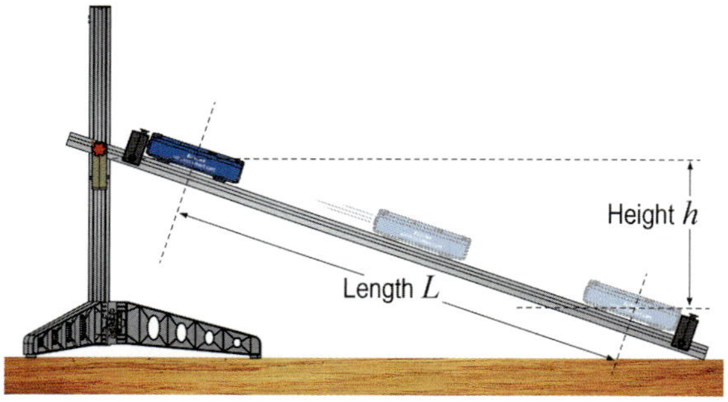


1. Measure the height $h$ and length $L$ of the ramp.
2. Record data and release the Smart Cart.
3. Use the resulting graphs to measure the acceleration of the Smart Cart down the inclined plane.
4. Repeat for four more values of $h$, lowering the height around 2-3 cm each time. Enter your values in a table.
5. Calculate the inclination of the ramp ($h/L$) for each run and enter the values in the table.

a. *How does the acceleration vary as the inclination of the ramp increases?*
b. *Go to Page 2 of the experiment file and graph acceleration versus inclination. Fit a line to the graph. What is the value of the slope of a line through your data (with units)?*
c. *Is this slope similar to any physical constant you have learned about in Physics? Explain.*
d. *Prepare a written report for your investigation (parts 1 and 2). In addition to addressing the above questions, describe the procedure, results, and conclusion or interpretation of your investigation.*

**Inclined planes and ramps**

You may have ridden a bicycle on a hill and noticed the substantial acceleration going down and the extra force required to go up. A hill is a natural example of an **inclined plane**, which is a flat but sloped surface also called a *ramp*. Because of the angle, a fraction of the object's weight acts downward, along the ramp. The fraction increases as the angle increases and explains why you accelerate faster down a steep hill than a shallow one. The relationship between acceleration and ramp steepness is given in equation (6.11).

**Interactive equation**

**Acceleration down a ramp**

$$\textbf{(6.11)} \quad a_{ramp} = \left(\frac{h}{L}\right)g$$

$a_{ramp}$ = acceleration down the ramp (m/s$^2$)
$h$ = vertical height of the ramp (m)
$L$ = distance along the ramp (m)
$g$ = acceleration due to gravity (m/s$^2$)

**Acceleration varies with the steepness of a ramp**

If you ever drove over a high mountain pass, you probably saw signs on the descent warning truck drivers about high speeds. It may seem obvious, but trucks (and cars) can rapidly accelerate to high speeds on steep downhills if their drivers are not careful. The underlying phenomenon is found in the equation above: The acceleration due to gravity increases as the slope of the inclined plane (or ramp) increases. The slope of the hill is given by the ratio ($h/L$) in equation (6.11).

**Acceleration down a ramp**

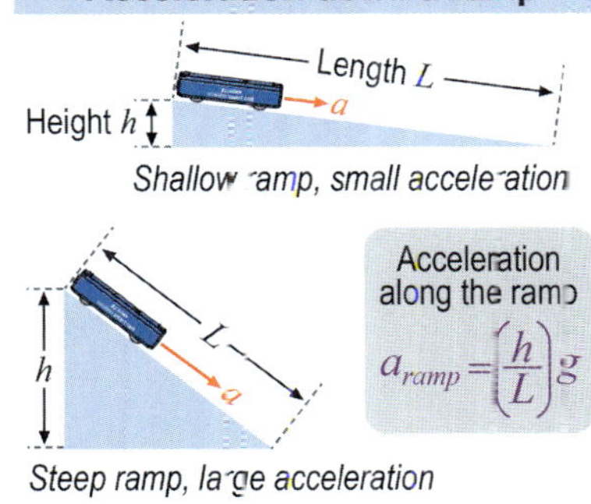


**Galileo's inclined plane experiments**

Galileo probably did not drop masses off the tower at Pisa to demonstrate the properties of gravitational acceleration. He did, however, construct inclined planes to do so! Why use ramps instead of dropped masses? Galileo did not have reliable time pieces available that could measure short time intervals. By using ramps, he *slowed down the motion* so that it could be measured accurately using his time pieces.

**Revisiting Galileo's inclined planes**

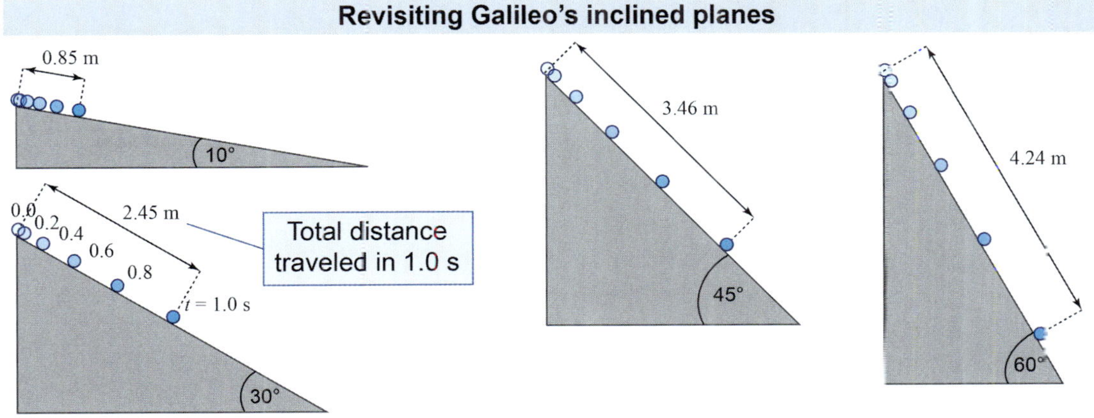


**Revisiting Galileo's experiments**

If we look back on Galileo's experiments using a particle model illustrated above, we see that the total distance traveled per second increases as the ramp becomes steeper. Since the points are plotted every 0.2 s for all the ramps, the ball is moving fastest where the points are furthest apart. The ball is moving slower on the shallow ramp than on the steep ramp. Since we know that acceleration is the change in velocity per unit time, the acceleration must be larger along the steeper ramp.

## Forces along a ramp

**Ramp coordinates**

Consider a car (or cart) moving on a ramp that makes an angle $\theta$ relative to the horizontal direction. There are forces both parallel and perpendicular to the ramp surface. It is convenient to define **ramp coordinates** in which the $x$-axis is along the direction of motion and parallel to the ramp. The $y$-axis is then the direction of the normal force.

**Ramp coordinates**

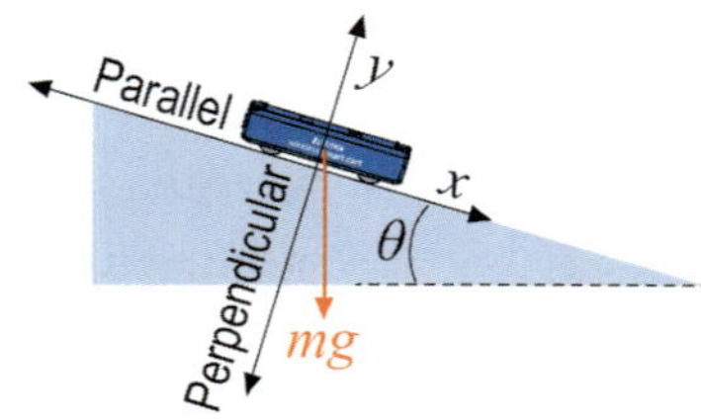


**Acceleration along the ramp**

The car's motion depends on the net force in the $x$-direction (along the ramp). Because the ramp is at an angle, the weight force is not vertical in ramp coordinates but is tilted at the ramp angle. A component of the car's weight equal to $mg\sin\theta$ lies parallel to the ramp surface. This component causes the car to accelerate down the ramp at $a_x = g\sin\theta$.

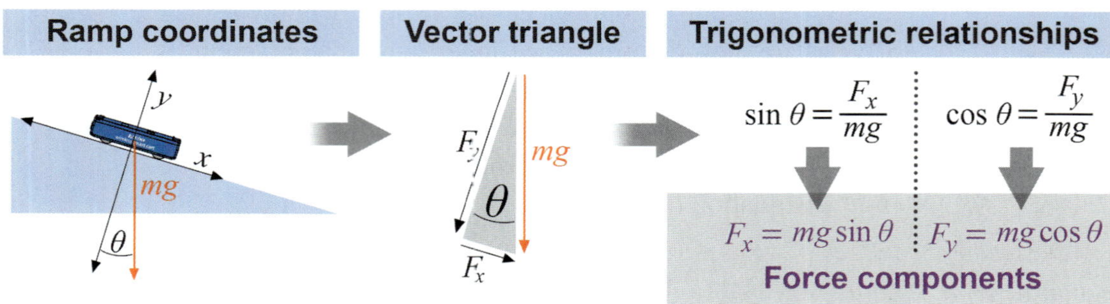


Interactive equation

| | | | |
|---|---|---|---|
| **(6.12)** | $a_x = g\sin\theta$ | $a_x$ = acceleration along ramp (m/s$^2$)<br>$g$ = acceleration of gravity = 9.8 m/s$^2$<br>$\theta$ = ramp inclination angle (degrees) | **Acceleration** in ramp coordinates |

**Acceleration versus angle**

The table on the right shows how the acceleration of the car varies with the angle of the ramp. The acceleration varies as the sine of the ramp angle. The sine of 0° is 0, and the acceleration of a horizontal ramp is also zero. The sine of 90° is 1, and a vertical ramp has an acceleration of 9.8 m/s$^2$, the same as free fall.

**Acceleration vs. angle (no friction)**

| Angle | sin $\theta$ | Acceleration (m/s$^2$) |
|---|---|---|
| 0° | 0.000 | 0 |
| 5° | 0.087 | 0.85 |
| 10° | 0.174 | 1.70 |
| 20° | 0.342 | 3.35 |
| 30° | 0.500 | 4.90 |
| 45° | 0.707 | 6.93 |
| 60° | 0.866 | 8.49 |
| 90° | 1.000 | 9.80 |

**Ramp as a simple machine**

Notice that for shallow angles the acceleration is small. This means that the force along the ramp is also small. This works to great advantage, because you can push a load up a ramp using a force smaller than the weight of the load. The ratio $F_x/mg$ is the sine of the inclination angle. At an angle of 10° the sine has a value of 0.174, so it only takes 0.174 N of force applied parallel to the ramp to raise each newton of weight. Used this way, the ramp is a *simple machine*, which you will learn more about on page 347.

# Investigation 6D

## Graphing motion on an inclined plane

**Essential questions** *What is the acceleration of an object down an inclined plane?*
*What do the motion graphs of an object down an inclined plane look like?*

Interactive simulation

Graphs of position versus time and velocity versus time show the nature of the motion of an object, such as a block sliding down a frictionless ramp. Unaccelerated motion (i.e., with constant velocity) will have one set of shapes for those graphs, whereas accelerated motion will show different graphical relationships. Physicists often use properties of a graph to measure fundamental physical quantities. In this investigation you will answer this question: How can you use motion down a ramp graphically to measure the acceleration due to gravity $g$?

### Part 1: Position–time and velocity–time graphs

1. Set the initial height $h_0$ to 200 m.
2. Set the inclination angle $\theta$ to 30°.
3. Check the boxes to graph both position and velocity versus time.
4. Run the simulation.

*a. Describe the shapes of the position–time and velocity–time graphs.*

*b. Using the equations of motion, explain why each graph has that shape.*

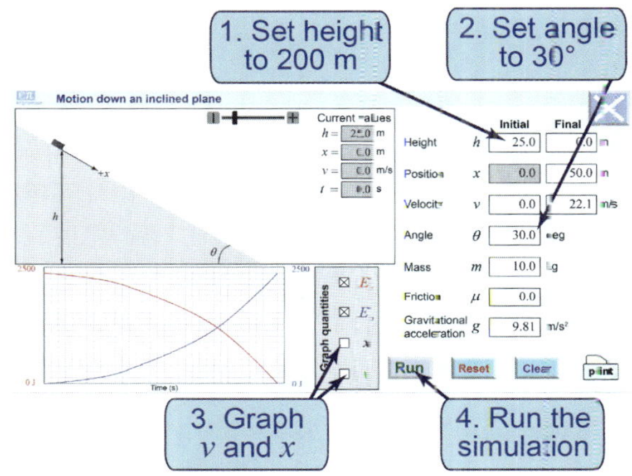


**What are the shapes of the position–time and velocity–time graphs?**

### Part 2: Measuring the acceleration due to gravity

1. Devise a procedure to measure $g$ using the simulation, a table, a graph, and $a = (h/x)g$.
2. Your procedure should use at least five different values of $\theta$.
3. Your procedure should include graphing your data, drawing a trend line through the data points, measuring the slope of the line, and using that slope as part of calculating $g$.

*a. For your written report, write down your procedure, noting the steps required for collecting data, graphing data, making calculations, and arriving at the answer.*

*b. What is the value of the slope of a line through your data? (Remember to include units!)*

*c. What is the physical meaning of the graph's slope? Use velocity, time, and acceleration in your answer.*

*d. In your written report, note your conclusion for the value of the acceleration due to gravity and explain any discrepancies from the commonly accepted value.*

## Friction on an inclined plane

**Finding the friction force**

The inclined plane is an excellent laboratory for studying friction during motion, because the acceleration can easily be controlled by changing the angle. Consider a ramp that makes an angle $\theta$ with the horizontal. From the vector triangle, the normal force is the $y$-component of the car's weight, or $mg\cos\theta$.

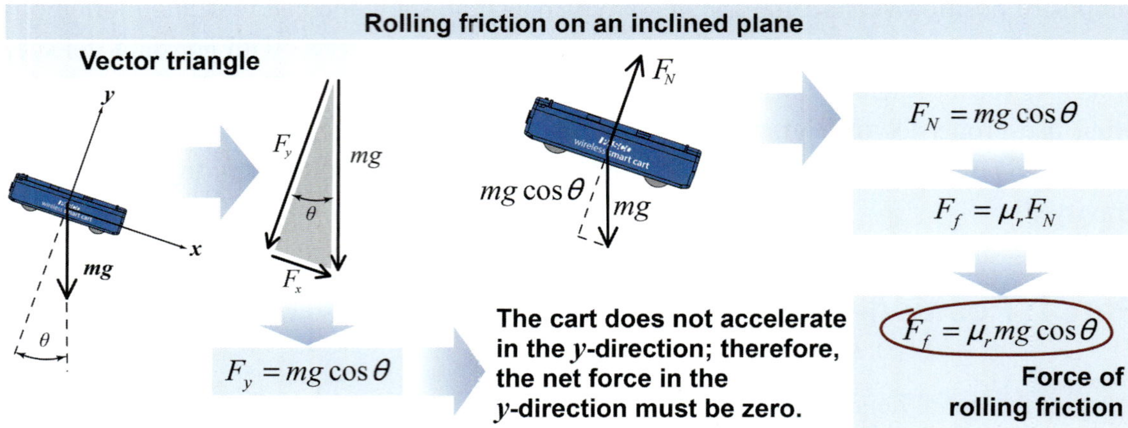


**Net force and acceleration**

The acceleration of the car is due to the net force acting parallel to the ramp. This includes the $x$-component of the car's weight and rolling friction. You can see both forces in the equation for the net parallel force (below). The term with $\sin\theta$ is the $x$-component of the car's weight. The term $\mu_r\cos\theta$ is the effect of rolling friction. Applying Newton's second law gives us an equation for the acceleration of the car. Friction lowers the acceleration of the car by $g \times \mu_r\cos\theta$.

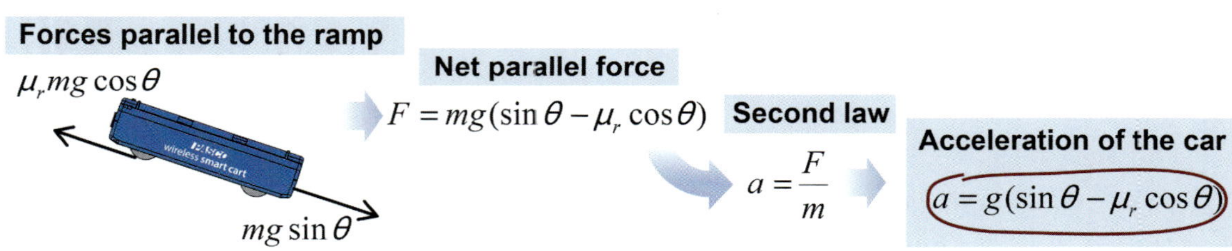


**Acceleration with rolling friction**

$$\textbf{(6.13)} \quad a_x = g(\sin\theta - \mu_r\cos\theta)$$

$a_x$ = acceleration along ramp (m/s$^2$)
$g$ = acceleration of gravity = 9.8 m/s$^2$
$\theta$ = ramp angle (degrees)
$\mu_r$ = coefficient of rolling friction

**Mass and acceleration**

Notice that the mass of the car does not appear in equation (6.13)! As with free fall, adding mass creates extra force from weight—but also increases inertia, which cancels out the effects of added weight. Similarly, rolling friction depends on the normal force and hence weight, so adding mass increases friction but does not change the acceleration.

**An example**

The table on the right shows the acceleration as the coefficient of rolling friction increases from 0.0 to 0.2. Notice that, when $\mu_r$ = 0.1, the car does not accelerate even at a ramp inclination angle of 5°. When $\mu_r$ = 0.2 the car does not start moving when on a ramp with an angle of 10°.

**Acceleration for different coefficients of rolling friction**

| | Acceleration (m/s$^2$) | | | |
|---|---|---|---|---|
| Angle | $\mu_r = 0$ | $\mu_r = 0.05$ | $\mu_r = 0.1$ | $\mu_r = 0.2$ |
| 0° | 0 | 0 | 0 | 0 |
| 5° | 0.85 | 0.37 | 0 | 0 |
| 10° | 1.70 | 1.22 | 0.74 | 0 |
| 20° | 3.35 | 2.89 | 2.43 | 1.51 |
| 30° | 4.90 | 4.48 | 4.05 | 3.20 |
| 45° | 6.93 | 6.58 | 6.24 | 5.54 |
| 60° | 8.49 | 8.24 | 8.00 | 7.51 |

**Combining physics and engineering**

The design of the Smart Cart provides a good example of physics and engineering together. A tool to help study the fundamentals of physics and mechanics, the Smart Cart underwent several iterations and engineering cycles before being ready for classroom use.

While there were many aspects to the Smart Cart design, the primary focus was to build a cart that could roll with virtually zero friction, measure forces applied to it, and simultaneously measure its motion, both linear and rotational; All while working without the use of cords.

**Measuring motion in one-dimension**

The Smart Cart measures its one-dimensional motion using an optical encoder system. This system detects rotation in the cart's wheels by measuring light reflected from an encoder wheel attached to the axle. Because the encoder system is accurate, very lightweight, and completely frictionless, it was a perfect tool for the Smart Cart to measure motion in one-dimension.

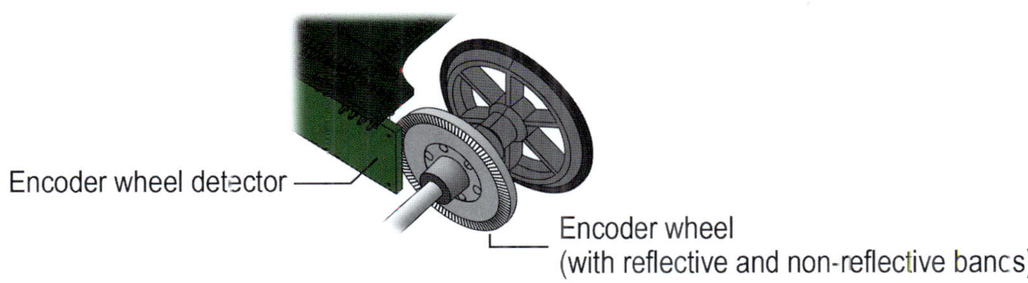


**How the encoder works**

On the Smart Cart's encoder wheel are bands of reflective and non-reflective surfaces. Next to the encoder wheel is a light emitter and detector system. As the cart's axle spins, light from the emitter reflects only from the reflective bands on the encoder wheel and hits an array of detectors. This light appears to the detectors as tiny flashes as the encoder wheel spins. These flashes are converted to digital signals, or "states," and sent to the cart's on-board computer, each state change representing the smallest measurable rotation in the encoder wheel. Depending on the order and number of the state changes, the cart's computer can determine if it is moving forward or backward, and the distance the cart has traveled.

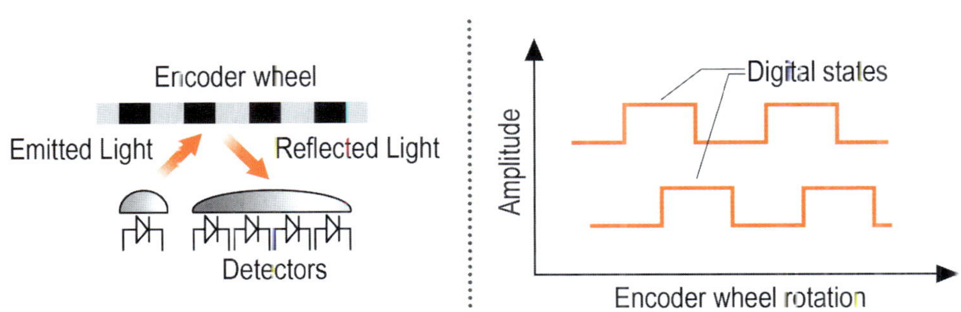

### What is the smallest movement the Smart Cart can sense?

The smallest position measurement the Smart Cart can make is dependent on the distance traveled by the cart between state changes. The detectors register a total of 816 state changes every time the encoder wheel makes one full rotation. Since the encoder wheel is attached to cart's axle, the cart wheels and the encoder wheel both make one full rotation at the same time. With the circumference of the cart's wheel being 95.6 mm, the smallest measurable movement is (95.6 mm)/(816 pulses) = 0.117 mm.

816 states/rotation

$$\frac{1 \text{ rotation}}{816 \text{ states}} \times \frac{95.6 \text{ mm}}{1 \text{ rotation}} = 0.117 \text{ mm}$$

### Converting from position to velocity and acceleration

The only direct measurement made by the Smart Cart's encoder system is position. To make the cart a more useful tool in mechanics experiments, it was necessary to have the cart also measure its one-dimensional velocity and acceleration. To do this, software engineers developed a simple algorithm that uses the encoder's position measurements to calculate velocity values: take the difference between two neighboring position data points $\Delta x$, and divide by the time elapsed between them $\Delta t$.

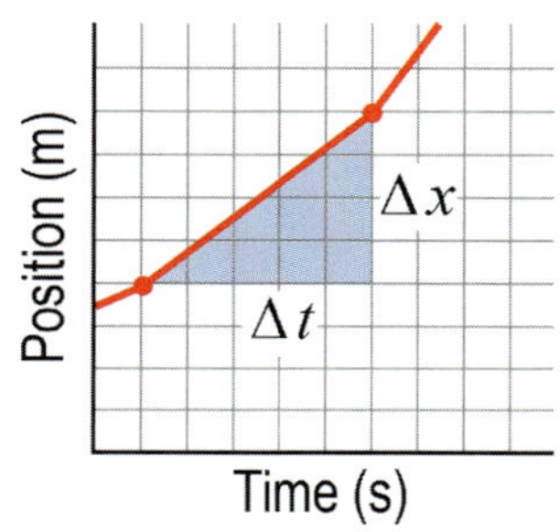


$$v = \frac{\Delta x}{\Delta t}$$

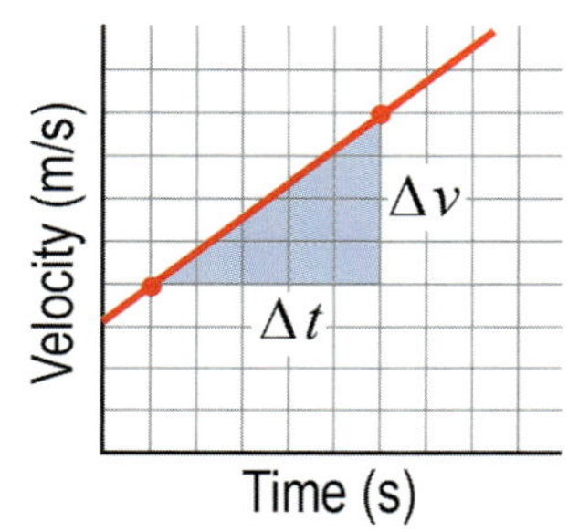


$$a = \frac{\Delta v}{\Delta t}$$

Similarly, acceleration measurements are calculated using the difference between two neighboring velocity data points $\Delta v$, and the time elapsed between them $\Delta t$.

### Measuring force

A major part of mechanics is the study of forces and their effects on motion. It was critical that the Smart Cart have the ability to measure forces applied to it while simultaneously measuring its motion. Built-in to the front of the Smart Cart is a force sensor capable of measuring one-dimensional push or pull forces up to ±100 N. This force sensor is built from an aluminum beam with a strain gauge mounted to the back of it. As force is applied to the front of the beam, the area of the beam under the strain gauge will flex slightly, changing the shape and resistance of the elements inside the gauge. By monitoring the differential voltage across the strain gauge, a calibrated value for force can be determined.

Smart Cart force beam (strain gauge on back)

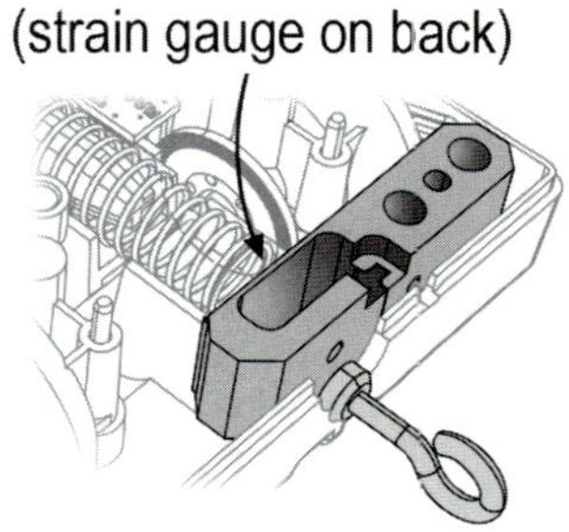

### Three-axis accelerometer and gyroscope

The Smart Cart also has a three-dimensional accelerometer and three-dimensional gyroscope built-in to its electronics, capable of measuring the cart's acceleration in the $x$, $y$, and $z$-directions as well as its angular velocity along the same axes. The addition of these measurements make the cart a very useful tool for studying dynamics in two and three-dimensions as well as rotational mechanics.

# Section 3 review

Projectiles move horizontally at uniform velocity and vertically, accelerated by gravity. The motions can be analyzed separately, and this is the key to solving problems, such as a projectile's range or time of flight. Motion along a ramp is accelerated by gravity if the ramp is inclined: the steeper the angle, the greater the acceleration. Rolling friction reduces this acceleration, but is less significant as the ramp steepens.

| Vocabulary words | trajectory, projectile, range, inclined plane, ramp coordinates |
|---|---|

| Key equations | | | |
|---|---|---|---|
| | $x = v_{x0}t$ | $y = v_{y0}t - \frac{1}{2}gt^2$ | |
| | $v_x = v_{x0}$ | $v_y = v_{y0} - gt$ | |
| | $x = v_0 t\cos\theta$ | $y = v_0 t\sin\theta - \frac{1}{2}gt^2$ | $x = \dfrac{2v_0^2\sin\theta\cos\theta}{g}$ |
| | $v_x = v_0\cos\theta$ | $v_y = v_0\sin\theta - gt$ | |
| | $a_{ramp} = \left(\dfrac{h}{L}\right)g$ | $a_x = g\sin\theta$ | $a_x = g(\sin\theta - \mu_r\cos\theta)$ |

## Review problems and questions

1. A soccer ball is kicked and leaves the ground with an initial speed $v_0$ at an angle $\theta$. Assume that the initial speed is fixed but the angle can vary. Match each quantity in the left-hand column with one quantity in the right-hand column:

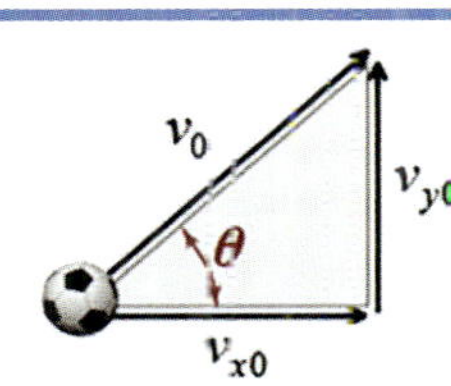


| time ball is in air | largest when $\theta = 0°$ |
|---|---|
| horizontal speed | largest when $0 < \theta < 90°$ |
| horizontal distance | largest when $\theta = 90°$ |

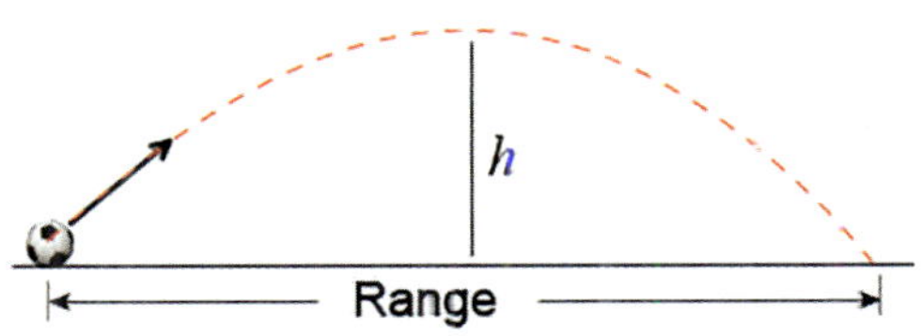


2. A soccer ball is kicked at an upward angle $\theta$ between 0° and 90°. Its initial vertical ($y$) velocity $v_{y0}$ is 19.6 m/s. It follows a parabolic trajectory and lands at the end of its range.
   a. How long (how much time) is the ball in the air?
   b. How high does the ball get?
   c. If the ball's range is 60 m, is the angle $\theta$ less than, equal to, or greater than 45°?

3. A small toy car is placed upon a ramp that is 2 m long. The car is released (with zero initial velocity) from the top of the ramp. Assume that friction is negligible.

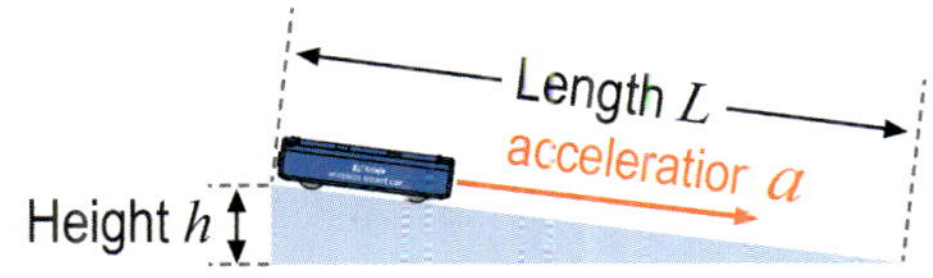


   a. How high is the top if the car takes 2.0 s to reach the bottom?
   b. Now suppose that you want the car to go from top to bottom in just one second (and you are still releasing it, not pushing it, so its initial velocity remains zero). In other words, you want to divide the travel time by two. Will you achieve this by doubling the height, or will you have to raise the ramp by a different amount?

Chapter 6

# Chapter review

## Vocabulary

Match each word to the sentence where it best fits.

### Section 6.1

| | |
|---|---|
| component | component force |
| cosine | magnitude |
| resolution of forces | resultant vector |
| sine | tangent |

1. ________ is the ratio of the adjacent side over the hypotenuse.
2. The process of breaking a force down into its components is called ___________.
3. The _______ of a force vector is its strength in newtons.
4. The _______ differs from the sine and cosine because it does not depend on the hypotenuse of a right triangle.
5. The ________ is the single vector that is the sum of two or more vectors.
6. The part of a force that lies on the $x$-axis is a ______
7. _______ is most useful to solve for the vertical component of a force.
8. For the vector (6,2) N, two newtons is the _______ in the $y$-direction.

### Section 6.2

| | |
|---|---|
| Cartesian coordinates | acceleration |
| compass | displacement |
| polar coordinates | speed |
| velocity | |

9. A ship's navigator would use a/an ________ to determine a direction when plotting a course.
10. In the expression $v_0 \cos \theta$, the variable $v_0$ is the ________.
11. The average _______ vector is the change in velocity divided by the change in time.
12. A force that is described as 50 N at 30° is an example of a vector in ________.
13. The average __________ vector is calculated by dividing the displacement by the time.
14. A change from one position vector to another is called a/an _________.
15. A force that is 12 N in the $x$-direction and 6 N in the $y$-direction is an example of using _______ to describe the vector.

### Section 6.3

| | |
|---|---|
| inclined plane | projectile |
| ramp coordinates | range |
| trajectory | |

16. A/An __________ is a moving body traveling only under the influence of gravity.
17. An example of ________ would be the horizontal distance a soccer ball moves between being kicked and touching the ground again.
18. Another term for ramp is _______.
19. A kicked soccer ball follows a parabolic path called its _______.
20. _______ is the rotated reference frame where the $x$-direction is along the surface of an inclined plane.

## Conceptual questions

### Section 6.1

21. Is it possible for a single 100 N force to have zero effect in the $x$-direction? If so, then describe how this might be possible.
22. Is it possible for three forces to have a resultant of zero even if all three forces have different magnitudes? If so, then explain or sketch how this might be possible.
23. What is the $\sin^{-1}(\sin 30°)$?

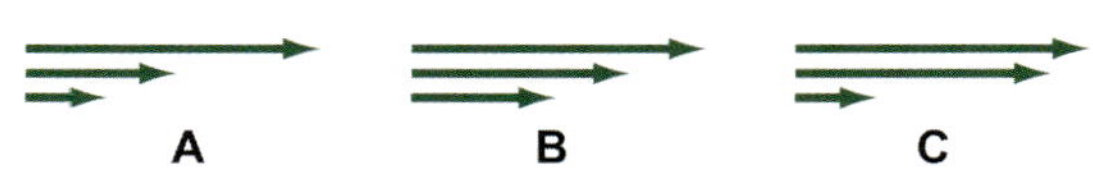


24. Which of the three vector diagrams would best represent force of 2 N, 4 N, and 8 N?

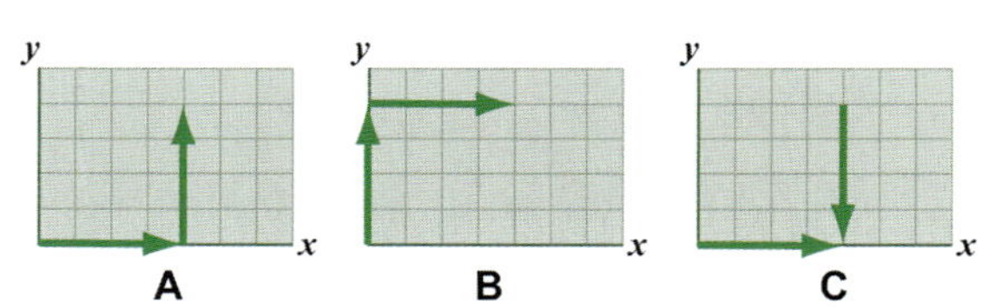


25. Which of the three diagrams above does *not* correctly show the addition of the two vectors on a vector diagram.
26. Describe how to transform a force that is in $x$–$y$ components into a force of the same magnitude and opposite direction.

# Chapter review

## Section 6.1

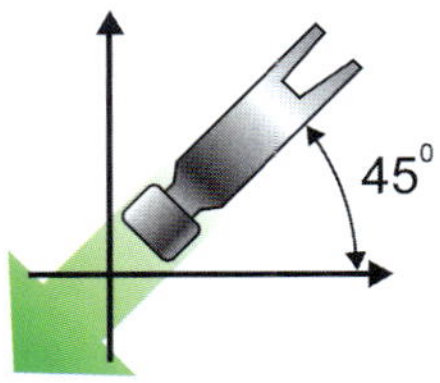


27. ( Is the force of the hammer striking a board greater in the $x$- or $y$-direction when a hammer strike is directed at a 45° angle?

28. ( Describe how to transform a force that is given in polar coordinates into a force of the same magnitude and opposite direction.

## Section 6.2

29. Your teacher described the difference between displacement and distance. Based on your teacher's statements, do you think there can be four displacements of equal distance that add up to no change in position? If so, then explain how this might be done.

30. Explain what is different between two successive displacements of (0, 5) m and (5, 0) m and a single displacement of (5, 5) m?

31. Describe the difference between how a compass defines angles and how angles are defined on the standard $x$–$y$ graph you might find in a math book.

32. ( Explain how the vector equation $\vec{d} = \vec{v}t$ can represent one, two, or three equations.

33. ( What does it mean when we claim that the $x$- and $y$-components of a velocity vector are independent of each other?

34. (( Given that Newton's second law $\vec{F} = m\vec{a}$ is a vector equation, describe the relationship between the directions of force and acceleration that the second law implies.

35. (( Is it possible for the acceleration vector and the velocity vector to be perpendicular to each other? If so, describe a situation where this might occur and the kind of motion that might result.

## Section 6.3

36. Name a real-time technology that can generate a *particle model* chart that represents an object's motion?

37. Why do we choose the coordinate axes $x$ and $y$ to be horizontal and vertical instead of aligning them at the same angle as the initial velocity?

38. In the absence of friction, explain why the horizontal component of a projectile's velocity is a constant.

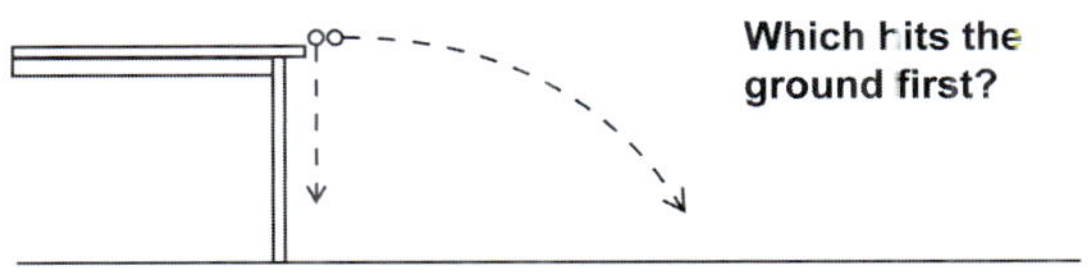


39. One marble sits motionless at the edge of a table. A second marble is fired horizontally along the tabletop at the first one using a marble launcher. When the second marble strikes the first, one marble falls straight down and the other follows a curved path as shown in the diagram above. Compare when the two marbles hit the floor. Explain your reasoning.

40. It is possible to adjust the angle of a ramp until a car just rolls down with constant speed and zero acceleration. Explain why there is no acceleration.

41. How is carbon paper useful in collecting data in a projectile motion investigation?

42. If Galileo's descendants were to use a real-time technology to generate a ticker tape diagram (such as in Chapter 4 review question 41) for constant velocity motion along a horizontal surface, then what would the chart look like?

43. Name one or more advantages of using real-time technology—such as a ticker tape—instead of Galileo's methods to generate a chart of motion down an inclined plane?

44. ( Why is the trajectory of a soccer ball kicked in the air curved and not straight? Can you think of an example of projectile motion that does have a straight trajectory?

45. ( A real-time graphing technology for a moving car, such as the Smart Cart, might measure the position of the wheels over time. How then does such as real-time system generate a chart of velocity versus time?

46. ( Why is it useful to tilt the coordinate system to align with the angle of a ramp?

47. (( When you add mass to a real car on a ramp, there is a measurable increase in acceleration. It is not large, maybe a few percent, but the effect is real and repeatable. Why?

48. ((( Consider a projectile launched upward at an angle between 0° and 90° from the horizontal. As the launch angle increases, what happens to the total time of flight for the projectile?

## Section 6.3

49. Consider the range equation.
    a. If the velocity doubles, what effect does this have on the range?
    b. Explain how the equation describes the range at angles of 0° and 90°.

50. Why is white paper (e.g., from a roll of white craft paper) placed underneath carbon paper when conducting a trajectory experiment with a falling marble, as in Investigation 6B on page 188?

51. You and your lab partner are given a marble and a trajectory apparatus ramp that acts as a marble launcher. You will compete with other lab groups to see who can launch their marble the furthest. With what projection angle should you set up the end of your marble launcher ramp?

52. What are the advantages of using real-time technology to generate graphs of motion (position–time or velocity–time graphs) down an inclined plane over how Galileo collected his data?

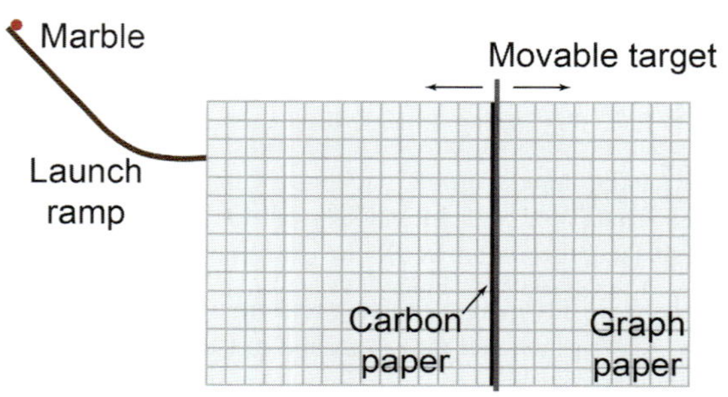


53. The figure above depicts a marble on a launch ramp that strikes a vertical, movable target. The carbon paper leaves a mark on the target each time a marble hits it. Describe how you would use this trajectory apparatus to graph the shape of the marble's trajectory after it leaves the launch ramp.

### Quantitative problems

## Section 6.1

54. What is the net force that is the result of combining a 30 N force at a 45° angle above the horizontal and a 45 N force 37° above the horizontal?

55. At what angle is a force directed if its horizontal component is 10 N and the vertical component is 15 N?

56. What is the range of magnitudes that can be attained from the combination of forces with magnitudes of 7 and 5 N?

57. What is the resultant of the following calculation adding two vectors:

$$\vec{F} = \vec{F}_1 + \vec{F}_2$$

where $\vec{F}_1 = (a, b)$ and $\vec{F}_2 = (c, d)$?

58. Calculate the magnitude of the force with components of $F_x = 12$ N, $F_y = 16$ N, and $F_z = 15$ N.

59. What is the horizontal component of a force with a magnitude of 100 N that is directed 60° above the horizontal?

60. Calculate the magnitude of a force given by $\vec{F} = (5, 12)$ N.

61. What is the vertical component of a 64 N force directed 53° above the horizontal?

62. Find the components of the resultant of the following forces: 15 N up, 8 N down, 4 N left, 20 N right.

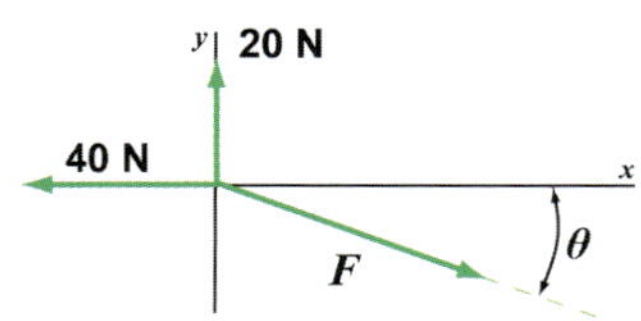


63. Determine the force $\vec{F}$ required to make the net force zero in the diagram above.

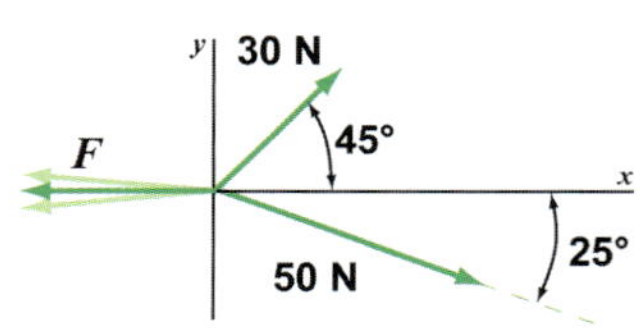


64. Determine the force $\vec{F}$ required to make the net force zero in the diagram above.

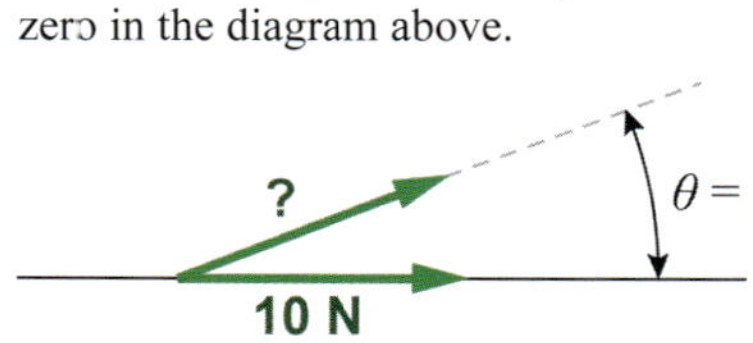


65. What magnitude force at a 17° angle above the horizontal will produce a 10 N horizontal component force?

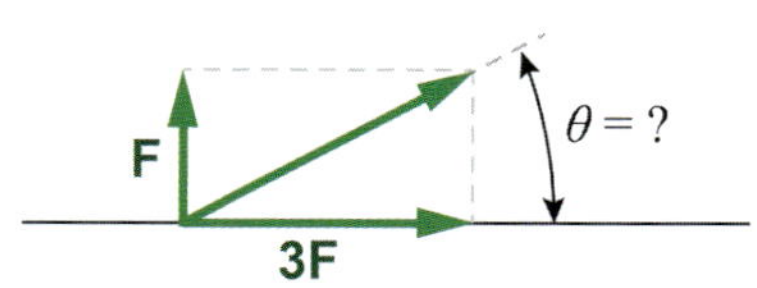


66. At what angle should a force be applied so that its $x$-component is three times that of its $y$-component?

# Chapter review

## Section 6.2

67. You go 3 m east and 4 m north in 5 s. Your friend goes 5 m at a 36.9° angle east of north in 5 s. Who has the greater average speed? Who has the greater average velocity?

68. How far have you walked from your starting position if you walk 12 m north, then 18 m east, then 9 m west?

69. How far east have you traveled if you ran 80 m directly northeast?

70. What is the magnitude of your acceleration if your horizontal velocity changes from 10 to 13 m/s and your vertical velocity changes from 2 to 6 m/s in 4 s?

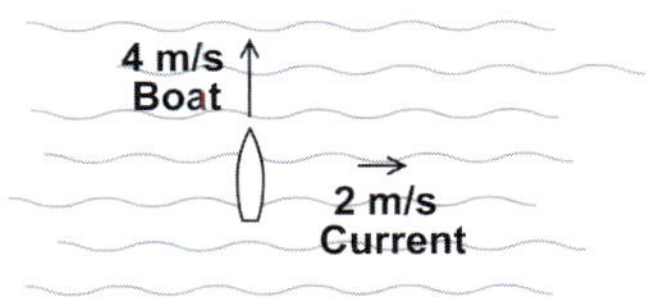


71. A boat is traveling at 4 m/s north relative to the water on a river that is flowing east at 2 m/s.
    a. What is the boat's velocity relative to the land?
    b. How far downstream does the boat drift in 10 s?
    c. How long does it take the boat to move 100 m across the river?

72. How long does it take a car traveling 18 m/s at a 40° angle west of north to go 900 m west?

73. What is the velocity of a bike with components of 8 m/s west and 15 m/s north?

74. At what angle is your displacement after going 9 miles east and 15 miles north?

75. What are the components of a 20 m/s velocity that is directed 10° above the horizontal?

76. Your friend's house is 4 miles away to the east and 7 miles away to the south. If you run there in a direct line, what angle does your velocity vector make relative to a line pointing straight east?

77. John, at third base, throws a ball to Sam, who is at second base 90 ft away. Sam then throws it to Mike, who is at first base 90 ft away. What is the ball's displacement?

78. If a marathon (26.2 miles) follows a straight course and the winner goes 12 miles north during the race, how far west did he travel?

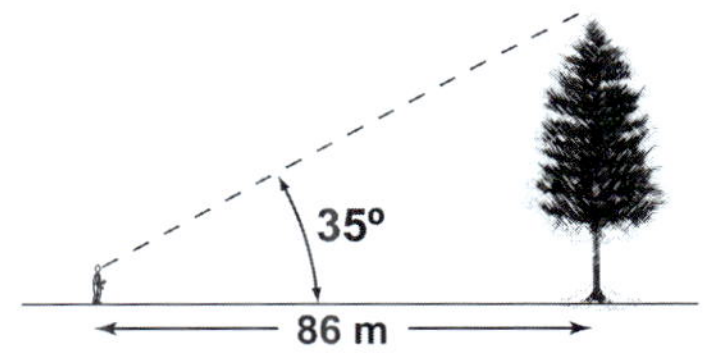


79. You are in the woods and use a plumb bob to determine that your line of sight to the top of a tall tree is 35°. How tall is the tree if it is 86 m away from you?

80. If a compass reads a heading of 74°, what angle does this direction make with the east ($x$-axis)?

81. A spacecraft making a course correction has a velocity of $\vec{v} = (100, 300, 100)$ m/s. The spacecraft turns on its engines to create an acceleration of $\vec{a} = (2, 5, 5)$ m/s$^2$ for 5 s. What is the velocity vector of the spacecraft after the engine burn?

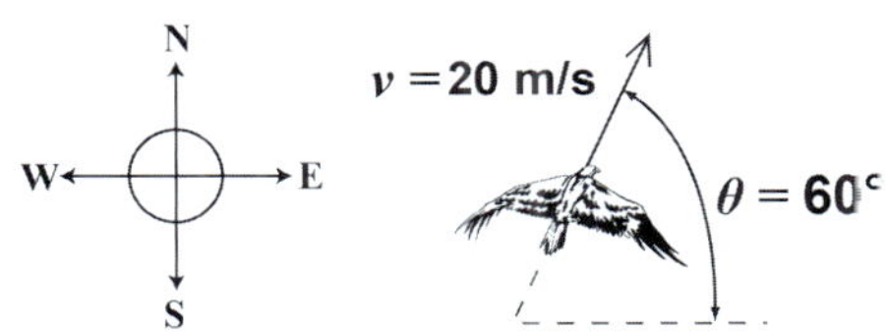


82. A bird is flying at a velocity of 20 m/s in a direction 60° north of east.
    a. Calculate the bird's velocity vector in component form.
    b. How long does the bird have to fly to get 100 km north?
    c. How far east has the bird traveled during this time?

83. How much does your speed change if you start moving at (2, 4) m/s and have an acceleration of 4 m/s$^2$ in the north direction for 8 s?

84. What are the components of your velocity if you started at 10 m/s southwest and then accelerated west at 3 m/s$^2$ for 2 s?

85. A swimmer can swim 3 m/s in still water and heads directly to the opposite bank of a 30 m wide river. How far downstream will he be pushed by a perpendicular 2 m/s current?

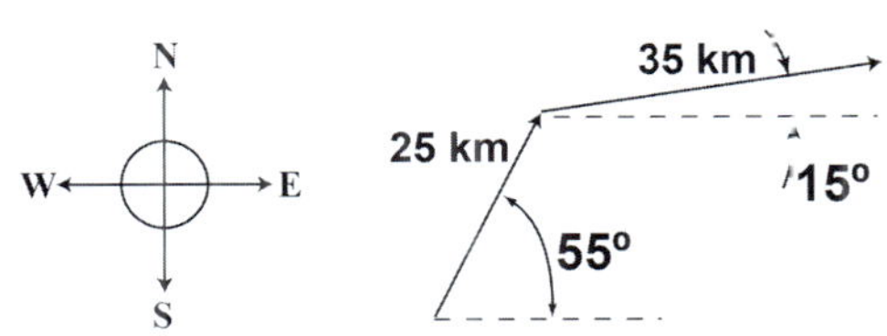


86. Calculate the $x$- and $y$-components of the two displacements shown and the resultant in component form.

# Chapter review

## Section 6.2

87. A robotic cart is located at a position of (4, 8) m relative to the origin. It has been programmed to move in displacements of (15, 16) m and (23, 42) m. Where will it end up relative to the origin?

88. ❮❮ What is the velocity of an object that travels 80 m north and 20 m west in 3 s?

89. ❮❮ A plane can fly 100 m/s in still air with no wind. To what angle should the plane be steered if the pilot wants to head straight north in a crosswind that is blowing 10 m/s to the east?

90. ❮❮ A rocket plane flying on a velocity vector of (0, 500, 0) m/s needs to turn to the east. How long must the plane maintain a sideways acceleration of (5, 0, 0) $m/s^2$ before its velocity vector has moved 10° to the east?

## Section 6.3

91. A baseball leaves the bat with a velocity of 55 m/s. The ballpark fence is 120 m away. Does the ball reach the fence if it leaves the bat traveling upward at an angle of 30° to the horizontal?

92. The Empire State Building is 381 m tall.
    a. How long would it take for a marble to fall from the top of the Empire State Building to the ground, neglecting air friction.
    b. What initial velocity would you have to give the marble to make it hit the ground in 5 s?
    c. What initial velocity would you have to give the marble to make it hit the ground in 10 s?

93. A volcano throws a stone 9,000 m into the air. What is the minimum initial velocity of the stone? In reality, air friction is significant and the actual velocity may be higher.

94. What ramp angle would produce an acceleration of 2 $m/s^2$ for a cart rolling down? Assume no friction and that the angle is measured with respect to the horizontal.

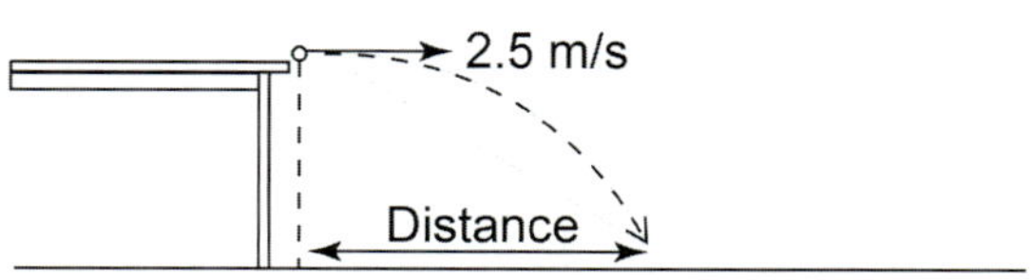


95. ❮ A marble is shot horizontally from a tabletop with an initial speed of 2.5 m/s using a marble launcher. If the table is 0.8 m high, how far will the marble travel along the floor before it lands?

96. ❮❮ A cart rolls down a 2.5 m frictionless ramp in 1.25 s after starting from rest at the top. What is the angle of the ramp?

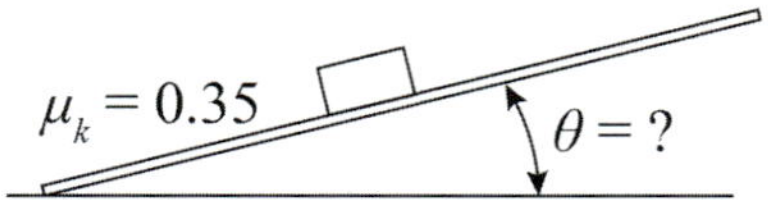


97. ❮ If the coefficient of static friction is 0.35 for the ramp shown, at what angle should you set a ramp so the block just begins to slide down on its own?

98. ❮❮ On your computer, use the interactive simulation of projectile motion on page 188 to determine the projection angle needed to hit a target 60 m away when the initial speed of the projectile is 25 m/s.

99. ❮❮ Determine the tension in a rope used to tow a 90 kg skier up a 15° incline while accelerating the skier at 0.5 $m/s^2$. The coefficient of kinetic friction is 0.2.

100. ❮❮ A projectile can have the same range at different launch angles. Use the range equation to determine a different launch angle at which the calculated range is the same as it is for an angle of 53°. (*Hint:* $2 \sin\theta \cos\theta = \sin 2\theta$.)

101. ❮❮ A slippery plastic block is resting on a ramp that makes an angle of 20° with the horizontal. The coefficient of static friction between the block and the ramp is 0.2. The coefficient of kinetic friction is 0.15.
    a. Will the block start sliding on its own?
    b. If you found that the block slid in part a, calculate its acceleration. If not, then calculate the minimum angle the ramp must have to start the block sliding.

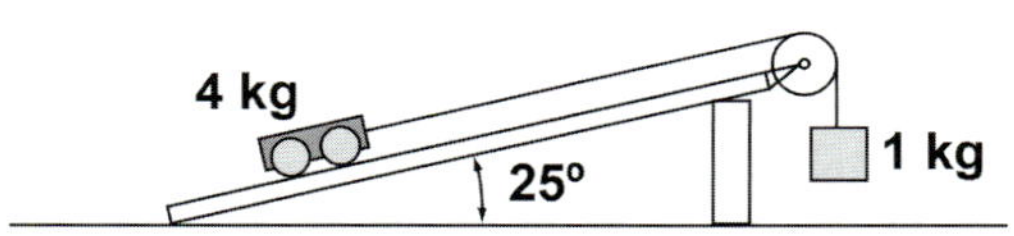


102. ❮❮ A 4 kg car with frictionless wheels is on a ramp that makes a 25° angle with the horizontal. The car is connected to a 1 kg mass by a string that passes over a frictionless pulley.
    a. Draw a free-body diagram of the car.
    b. In which direction does the car move: up or down the ramp?
    c. Calculate the acceleration of the car and mass using Newton's second law.

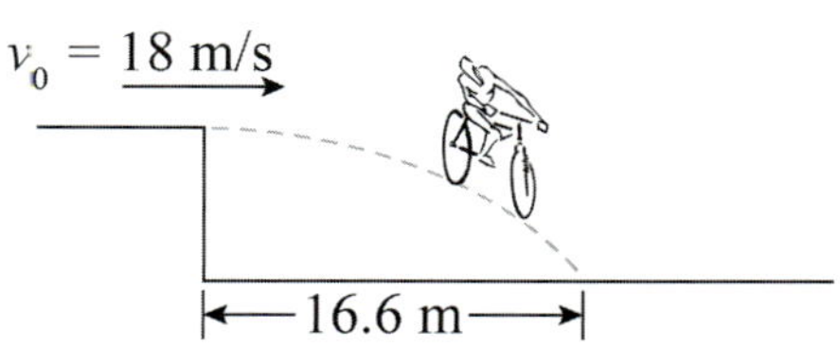


103. ❮❮ A cyclist jumps off a low cliff at an initial velocity of 18 m/s. The cyclist travels 16.6 m before touching down. How high is the cliff?

# Chapter review

### Standardized test practice

104. A projectile has an initial velocity of 17 m/s. The projectile has an $x$-component velocity of 15 m/s and a $y$-component velocity of 8.0 m/s. What is the range of the projectile?

   A. 1.6 m
   B. 13 m
   C. 24 m
   D. 28 m

105. An object is placed on a frictionless inclined plane with a height of 3.0 m. It has an acceleration of 3.7 m/s$^2$. What is the length of the inclined plane?

   A. 1.1 m
   B. 7.9 m
   C. 11 m
   D. 36 m

106. In writing a report for Investigation 6C, which of the following statements is best for a student to use in order to maintain an objective tone and formal style?

   A. The Smart Cart sped up more and more when my buddy made the ramp steeper.
   B. Keeping the ramp low was a lousy idea.
   C. Height mattered a lot.
   D. The acceleration of the Smart Cart increased as the height it was released from increased.

107. Which shape best describes the trajectory of a projectile?

   A. straight line
   B. semicircle
   C. triangle
   D. parabola

108. Which is the best choice of materials or equipment for determining where an object strikes the ground?

   A. mud
   B. clear plastic transparency
   C. carbon paper on white paper
   D. photogate timer

109. A marble is launched at a speed of 9 m/s at an angle of 25° above the horizontal. If the experiment is done on a level floor, then how far away will the marble land if friction is neglected?

   A. 1.58 m
   B. 3.16 m
   C. 6.33 m
   D. 6.98 m

110. A projectile travels a horizontal distance of 25.0 m in 3.00 s. What was the magnitude of the total initial velocity?

   A. 6.37 m/s
   B. 8.33 m/s
   C. 14.7 m/s
   D. 16.9 m/s

111. A projectile with a total velocity of 10 m/s is in the air for 0.5 s. What was its initial $x$-component velocity?

   A. 2.4 m/s
   B. 3.2 m/s
   C. 5.0 m/s
   D. 9.7 m/s

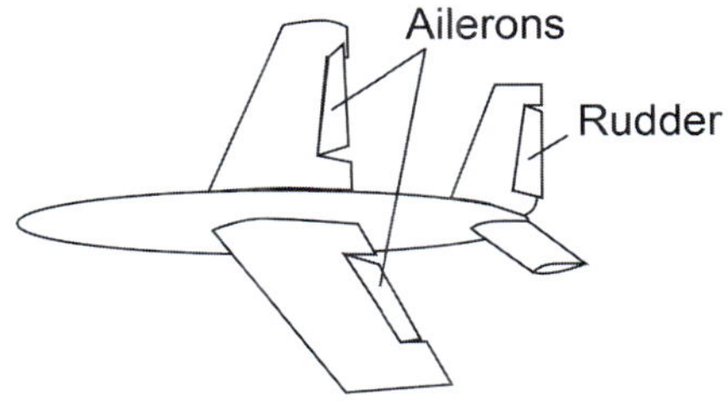


112. A flight instructor makes the following statement: "The aircraft's velocity vector can change direction through the use of the controls that move the ailerons on the wings, the rudder on the tail, or both."
   Which of the following best summarizes what the instructor is telling you?

   A. how to speed the plane up
   B. how to turn the plane
   C. how to slow the plane down
   D. how to increase the plane's altitude

113. "Two velocity vectors at right angles to each other cannot be added as if they were along the same line."

   The term "at right angles" used above means that

   A. the angle is between 0° and 90°.
   B. the angle between the vectors is 90°.
   C. the vectors are given as speeds and angles.
   D. the angles are the correct kind of angles for velocity vectors.

114. Which of the following is the best example of a graphing technology?

   A. photogate timer that displays the time an object crossed it
   B. digital odometer on a car
   C. seismograph for sensing earthquakes
   D. robotic pen for signing legal documents remotely

# Chapter 7 Circular Motion

On October 15, 1997 a massive Titan IVB/Centaur rocket launched the *Cassini–Huygens* space probe toward Saturn. Once in interplanetary space, *Cassini–Huygens* reached speeds of 13,000 m/s or 29,000 mph. No Earth-bound vehicle comes close to this extraordinary velocity. After traveling for nearly seven years—and three and a half *billion kilometers*—the spacecraft successfully arrived on schedule at Saturn on June 30, 2004. The Earth orbits the Sun at a speed of 29,780 m/s, while Saturn orbits at 9,670 m/s. By comparison, a bullet travels at 240 m/s (540 mph). *Cassini–Huygen*'s navigators had to launch while moving 124 times as fast as a bullet, and hit a target moving 40 times as fast as a bullet, after traveling 3.5 billion kilometers. The *Cassini–Huygens* mission is an extraordinary feat of navigation and it required precise control of the spacecraft's motion through time and space.

*Cassini*'s mission planners calculated that a direct flight from Earth to Saturn would burn too much fuel and also pass up opportunities to explore other planets. The flight engineers devised a clever but complex trajectory that included two flybys of Venus, a flyby of Earth, and a fourth flyby of Jupiter. Each flyby served two purposes. First, each was an opportunity for the spacecraft to make new scientific observations. Second, each close approach allowed a gravitational "slingshot" maneuver providing extra velocity. Only after the four "gravity assists" was *Cassini–Huygens* able to reach Saturn.

***Cassini–Huygens* path to Saturn**

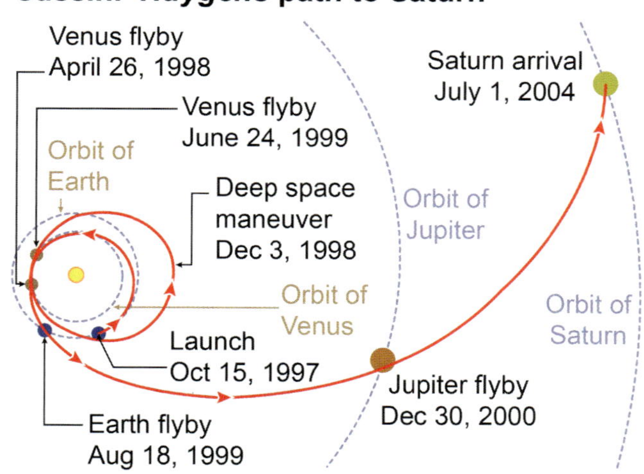


How does physics help us to understand how to send spacecraft to other planets? This chapter is about how force and motion—displacement, velocity, and acceleration—work in the two-dimensional case of circular motion and orbits. The techniques of this chapter are the very same techniques that *Cassini–Huygen*'s engineers and scientists used to calculate and control the spacecraft over its incredible journey.

# Chapter study guide



## Chapter summary

In the Solar System, the Milky Way Galaxy, and throughout the universe, most objects don't travel in straight lines; they move in circular or elliptical orbits around a more massive object. Circular motion describes not only how satellites orbit the Earth but how wheels roll and what forces you feel when riding on a rotating amusement park ride. When an object moves in a circle, its velocity constantly changes direction. In order to change an object's velocity, something else has to apply a force. That *centripetal force* always points toward the center of the circle, which is why the force due to the Sun's gravity keeps the planets moving in near circles around it.

## Learning objectives

By the end of this chapter you should be able to

- draw vectors of velocity, acceleration, and force for circular motion;
- calculate angular velocity and centripetal acceleration and force;
- draw free-body diagrams for circular motion problems;
- describe the historical development of the concept of the gravitational force;
- calculate gravitational force and describe how it depends on mass and distance; and
- describe properties of planetary orbits and summarize Kepler's laws.

## Investigations

7A: Circular motion
7B: Orbits
7C: Extrasolar planets

## Chapter index



## Important relationships

$$\omega = \frac{\Delta\theta}{\Delta t} \qquad v = \omega r \qquad a_c = \frac{v_t^2}{r}$$

$$F_c = \frac{mv_t^2}{r} \qquad F = G\frac{m_1 m_2}{r^2} \qquad R = \frac{Gm}{v^2} \quad v = \sqrt{\frac{Gm}{R}}$$

## Vocabulary

**angular velocity**
**centripetal acceleration**
**orbit**
**black hole**
**radian (rad)**
**law of universal gravitation**
**orbital period**
**centripetal force**
**satellite**
**escape velocity**

## 7.1 - Circular motion

The Earth orbiting the Sun, a turning wheel, and a spinning top are all examples of *circular motion*. The concepts of position, velocity, and acceleration used to describe *linear motion* take slightly different forms when used to describe circular motion.

### Angular velocity

**How does physics describe rotating speeds?**

The rate at which a rotating object spins is called its **angular velocity**, which is represented by the Greek letter $\omega$ ("omega"). To understand angular velocity, consider an object with a rotational position given by its *angle* $\theta$. Angular velocity describes the amount that angle changes per unit time. If the angle is measured in degrees then angular velocity is expressed in degrees per second. If the angle is measured in full turns (360°), then angular velocity might be in rotations per second or cycles per second. The angular speed of motors is often given in revolutions per minute (rpm).

**Interactive equation**

**Angular velocity**

**(7.1)** $$\omega = \frac{\Delta\theta}{\Delta t}$$

$\omega$ = angular velocity (rad/s)
$\Delta\theta$ = change in angle (rad)
$\Delta t$ = change in time (s)

***What is a radian?***

For the purpose of angular speed, a **radian (rad)** is a more natural unit of angle than a degree. One radian equals approximately 57.3°. Radians are *dimensionless* because radians are a ratio of lengths. One radian is the angle formed by wrapping one radius of a circle around the circumference. There are $2\pi$ (about 6.28) radians in a full circle, which makes $2\pi$ rad = 360°.

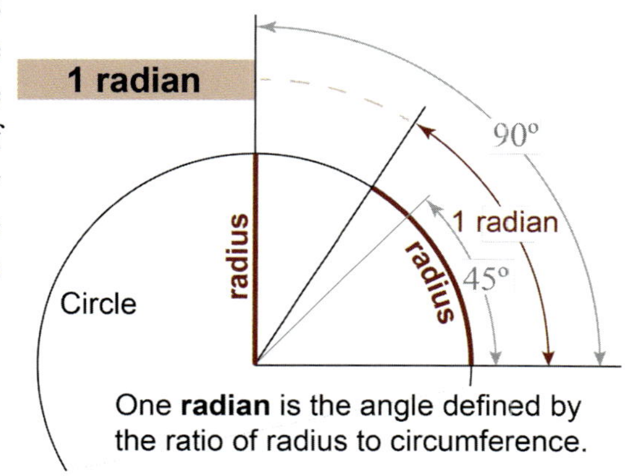


One **radian** is the angle defined by the ratio of radius to circumference.

**Positive and negative**

The sign of the angular velocity depends on direction. If counterclockwise rotation is defined to be positive then clockwise rotation is negative.

**What are the units of $\omega$?**

Radians are *pure numbers* without units in the sense that meters or seconds are units. Expressed in radians per second, angular velocity has units of 1/s or $s^{-1}$. If you encounter an angular velocity expressed in units of 1/s, then interpret the value as "radians per second."

**Angular velocity of the Earth**

The Earth rotates once every 24 hours. What is its angular velocity in radians per second?

| | |
|---|---|
| *Asked:* | angular velocity $\omega$ |
| *Given:* | Earth makes one full rotation ($2\pi$ rad) every 24 hours. |
| *Relationships:* | $\omega = \Delta\theta/\Delta t$; one day = $(24\text{ hr})\left(\frac{60\text{ min}}{1\text{ hr}}\right)\left(\frac{60\text{ s}}{1\text{ min}}\right) = 86{,}400$ s |
| *Solution:* | $\omega$ = (1 rotation)/(1 day) = ($2\pi$ rad)/(86,400 s) = $7.27 \times 10^{-5}$ rad/s |
| *Answer:* | $7.27\times10^{-5}$ rad/s. Note that the units "radian" are not necessary, because the radian is dimensionless. |

**Rolling motion**

A rolling wheel has both linear and angular velocity. If the wheel is not slipping, the two velocities are related by geometry, because the circumference of a circle is $2\pi$ times the radius $r$. A wheel rotates through an angle of $2\pi$ radians (a full rotation) as it moves forward by a distance of $2\pi r$ (one circumference).

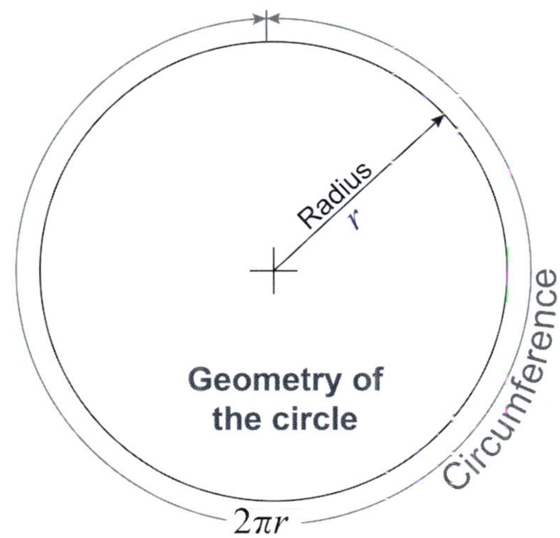


**What is the linear velocity?**

The linear velocity is the circumference divided by the time it takes to make one turn, i.e., $v = 2\pi r/t$. For one rotation, the quantity $2\pi/t$ is the angular velocity $\omega$. The linear velocity is therefore equal to $\omega$ multiplied by the radius $r$. Linear and angular velocity are related to each other by the equation

Interactive equation

**(7.2)** $$v = \omega r$$

$v$ = linear velocity (m/s)
$\omega$ = angular velocity (rad/s)
$r$ = radius (m)

**Linear velocity** from angular velocity

**Why did old bicycles have such large wheels?**

The radius $r$ that appears in equation (7.2) means that larger wheels have a higher linear velocity than smaller wheels for a given angular velocity. Early bicycles had very large wheels because the larger wheels would create a higher linear velocity. Unfortunately, they were very difficult to ride and quite unstable! Modern bicycles use gears and chains so that the pedals can turn at a different angular velocity from the wheels.

**Linear speed of a rolling wheel**

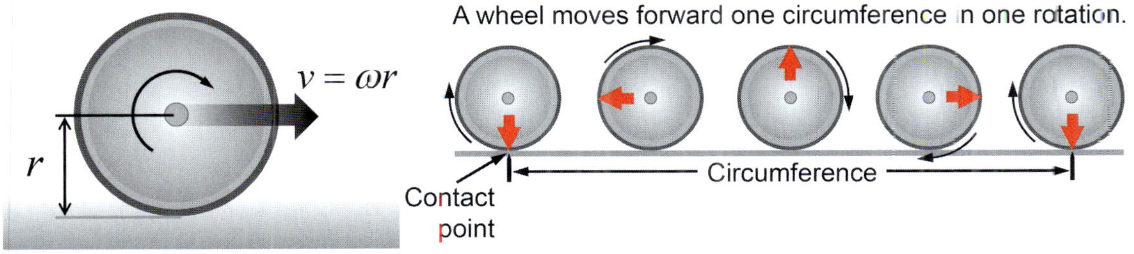


**Rotation of a rolling wheel**

A car has wheels with a radius of 30 cm. What is the angular velocity of the wheels in both radians per second and revolutions per minute when the car is moving at 30 m/s (67 mph)?

| | |
|---|---|
| *Asked:* | angular velocity $\omega$ calculated in two units: rad/s and rpm |
| *Given:* | radius $r$ = 0.3 m; velocity $v$ = 30 m/s |
| *Relationships:* | linear velocity $v = \omega r$; one revolution is $2\pi$ = 6.28 radians |
| *Solution:* | In rad/s, $\omega = v/r$ = (30 m/s)/(0.3 m) = 100 rad/s.<br>In rpm, $\omega$ = (100 rad/s)×(60 s/min)×(1 revolution/6.28 rad) = 955 rpm. |
| *Answer:* | $\omega$ = 100 rad/s = 955 rpm |

# Investigation 7A

## Circular motion

**Essential questions** *How are radius, velocity, acceleration, and force related in circular motion?*

Interactive simulation

Many objects—from spinning wheels to the orbits of satellites around the Earth—undergo circular motion. For an object to remain in circular motion, it must experience a centripetal force (and acceleration). In this investigation, you will explore the properties of circular motion at scales similar to that of a mass swung around horizontally at the end of a string.

### Part 1: Directions of the velocity, force, and acceleration vectors

1. Set $m = 5.0$ kg, $r = 5.0$ m, and $v = 5.0$ m/s.
2. Play the simulation, and then pause it at various positions around the circle.
3. Sketch the velocity, force, and acceleration vectors for at least five positions distributed around the circle.

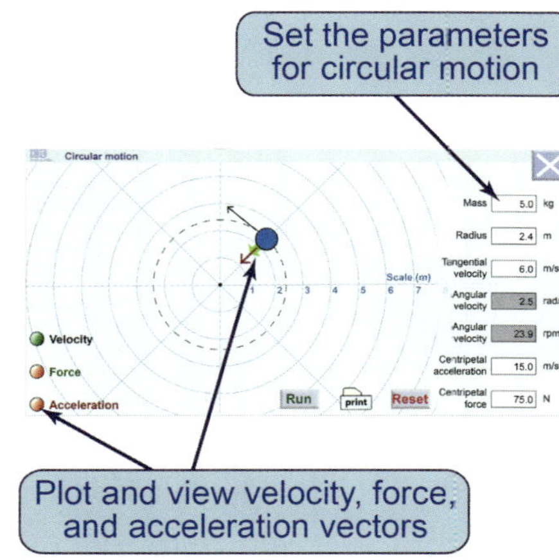


a. Which vector quantity or quantities are radial and which are tangential? Are the radial quantities pointed inward (toward the center) or outward?
b. Do the lengths of the velocity, acceleration, or force vectors change around the circle?
c. Notice that the angular velocity is exactly 1 rad/s. Why?

### Part 2: Approximating a mass swung overhead

1. Set $r = 1.0$ m and $m = 0.3$ kg.
2. Calculate the tangential velocity needed to spin the object around once per second, and enter that into the simulation.

a. How much force is needed to maintain this object in circular motion?
b. Compare that force with the force required to hold the object motionless against the force of gravity.
c. Lengthen the string to $r = 2.0$ m. Does it now require more or less force than before to maintain the object in circular motion with the same angular velocity?

### Part 3: Variation of velocity with radius for circular motion

1. Hold the force constant at 10 N and the mass constant at 2 kg, but vary the length of the string from $r = 1$ m to 5 m.
2. Record the velocity and radius for each case.

a. Graph $v$ (on the vertical axis) against $r$ and describe the shape of your graph.
b. Graph $v^2$ against $r$, describe the shape of your graph, and measure its slope (including units).

**Tangential velocity in circular motion**

Imagine whirling a small weight around your head on a string. If the string broke, the weight would fly off and no longer move in a circle. This is a consequence of Newton's first law: A moving body will continue in a straight line if no net force acts on it. What direction will it travel? *An object undergoing circular motion has a velocity that is tangential to the circle.* If the string were to break, then the object would continue moving in the direction of its velocity —upward in the illustration.

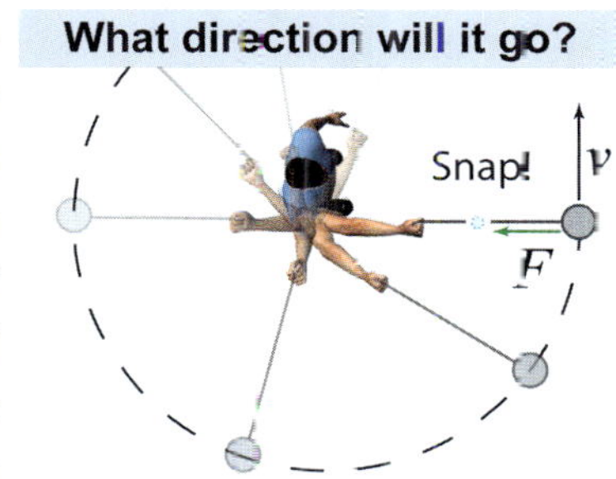


**Centripetal acceleration**

What keeps the object in circular motion over your head? The string exerts an inward force that causes an inward acceleration that continually bends the object's direction of motion toward the center of the circle. In the illustration, the velocity of the object changes direction as it moves from point A to point B. The *change in velocity* is represented by the vector $\Delta v$, which is perpendicular to the direction of velocity. Since acceleration is the rate at which an object's velocity changes, the **centripetal acceleration** acts in the direction of the vector $\Delta v$, which is toward the center of the circle. Centripetal acceleration always points toward the center of the circle and is perpendicular to the object's velocity.

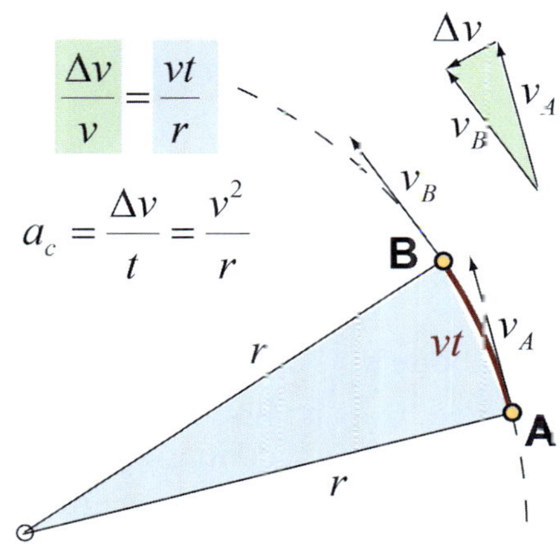


**An equation for centripetal acceleration**

How are acceleration and velocity related quantitatively in circular motion? Look again at the illustration above. Between points A and B the object moves a distance $d = vt$, while the change in its velocity is $\Delta v$. The blue- and green-shaded triangles are similar; therefore, the ratio $\Delta v/v$ is the same as the ratio $vt/r$. Acceleration is defined as $\Delta v/t$; therefore, the centripetal acceleration is given by equation (7.3).

Interactive equation

**Centripetal acceleration**

$$(7.3) \quad a_c = \frac{v_t^2}{r}$$

$a_c$ = centripetal acceleration (m/s$^2$)
$v_t$ = tangential speed (m/s)
$r$ = radius of circle (m)

**Centripetal force**

Whether a force changes a moving object's speed or its direction depends on the force's *direction*. A force in the direction of motion causes the object to change speed. A force *perpendicular* to the motion causes the object to change its path from a line to a circle —without changing speed. The force (or combination of forces) that points towards the center of a circle and is perpendicular to an object's motion is called a **centripetal force**. Some source of centripetal force is required to keep an object in circular motion, whether it is a planet in orbit or a race car going around a curved track. We get the equation for centripetal force by combining equation (7.3) and Newton's second law.

Interactive equation

**Centripetal force**

$$(7.4) \quad F_c = \frac{mv_t^2}{r}$$

$F_c$ = centripetal force (N)
$m$ = mass (kg)
$v$ = tangential speed (m/s)
$r$ = radius of circle (m)

## Centrifugal "force"

**Effect that is felt as centrifugal "force"**

Think about the last time you were in a car that drove around a corner quickly. You may have noticed a "force" pushing you toward the side of the car away from the corner. The faster the car rounds the corner, the stronger the perception of being "flung" outward. This effect is commonly known as *centrifugal force.*

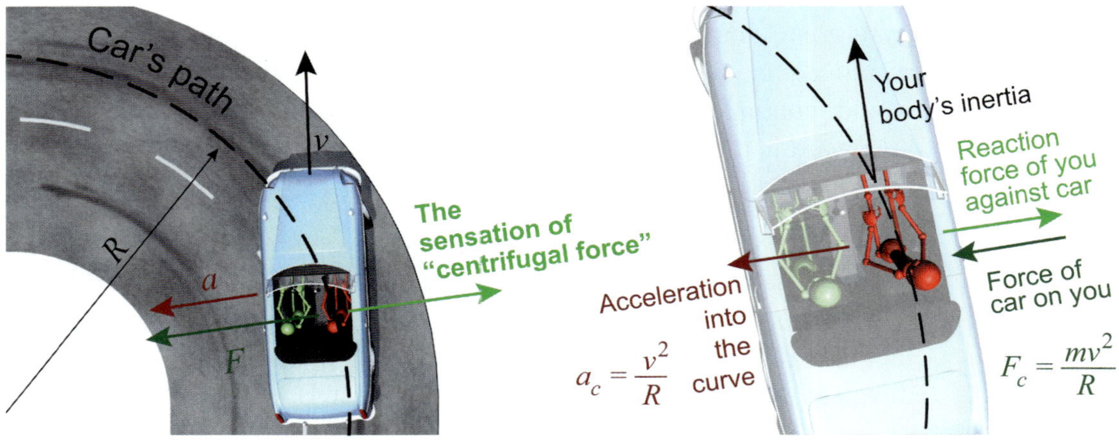


**Your body's inertia**

There is really not a force pushing you outward against the side of the car. What you feel is the structure of the car forcing your body to redirect its motion into a curve, rather than continuing in a straight line. The interior walls of the car exert a sideways force on you that points toward the center of the curve. The *reaction force* is your body pushing back against the side of the car. Although often called "centrifugal force," this sensation of being pushed radially outward is actually a consequence of Newton's first law (inertia) in a situation where your motion is being forced into a circle.

**What does *"high-g"* mean?**

While centrifugal "force" is not a real force, centrifugal effects are found throughout nature and technology. For example, high-performance-aircraft pilots often push the limits of human endurance when they execute "*high-g*" maneuvers. A typical high-*g* maneuver is a tight radius turn at high speed. Consider a plane traveling at 223 m/s (500 mph) that makes a turn with a 500 m radius. A radius of 500 m means that the plane's turn is 1 km across, which seems gradual. In reality, the centripetal acceleration is 100 m/s$^2$! This is 10.2 times the acceleration of gravity ($g$ = 9.8 m/s$^2$) and this turn would be a "10.2g" maneuver if attempted.

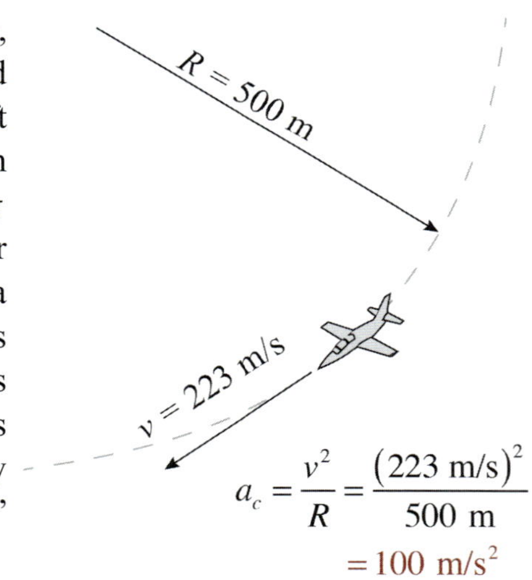


**Why are high-*g*'s dangerous?**

At 1*g*, your body feels its normal weight. In a 10.2*g* turn, however, the force acting on the body of a 70 kg pilot is nearly 7,000 N or 1,600 lb. At this acceleration, the heart cannot pump blood to the brain and an actual human pilot would lose consciousness.

**What kinds of problems use centripetal force?**

Problems involving centripetal acceleration often ask for the speed or radius at which the centripetal acceleration meets a specified value. For example, the safe radius for a curve on a roadway depends on the speed of the cars on that road. High-speed roads require large-radius curves. The *redline* for a car engine is related to the centripetal acceleration applied to the piston by the rotating crankshaft. If the engine spins too fast the force required to accelerate the piston breaks the connecting rod in a spectacular (and expensive) engine malfunction.

**Example problem using centripetal acceleration**

The Federal Highway Association guidelines suggest a maximum safe sideways acceleration in a turn of 0.1*g*, or 1 m/s². What is the minimum radius at which a civil engineer should design a curve on a road intended for cars traveling at up to 30 m/s (67 mph)?

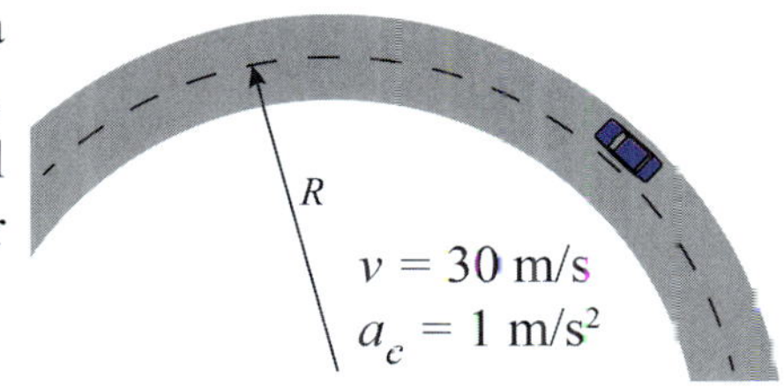


*Asked:* radius $R$

*Given:* $v = 30$ m/s, $a_c = 1$ m/s$^2$

*Relationships:* $a_c = v^2/R$

*Solution:* Solve the equation for the radius:

$$a_c = v^2/R \rightarrow R = v^2/a_c$$

$$R = (30 \text{ m/s})^2 \div (1 \text{ m/s}^2) = 900 \text{ m}$$

The minimum radius is 900 m (about 3,000 ft).

**Example problem using centripetal force**

A high-performance car engine has a piston and connecting rod that have a combined reciprocating mass of 500 g. The connecting rod can safely sustain a force of 20,000 N. The radius of the crankshaft is 5.0 cm (0.050 m). What is the maximum safe operating speed of the engine in revolutions per minute?

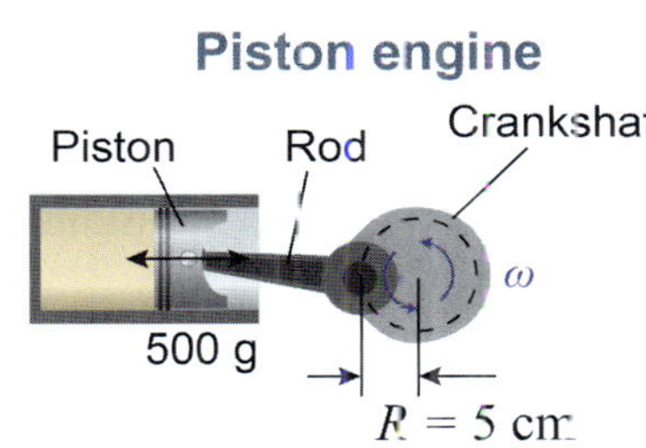


*Asked:* rotational speed $\omega$ in rpm

*Given:* $m = 0.500$ kg, $F = 20{,}000$ N, $R = 0.050$ m

*Relationships:* $F_c = mv^2/R$, $v = \omega R$, 1 rad/s = 9.55 rpm

*Solution:* First, put the force equation in terms of angular speed $\omega$:

$$F_c = \frac{mv^2}{R} \quad \rightarrow \quad F_c = \frac{m(\omega R)^2}{R} = m\omega^2 R$$

Next, calculate $\omega$:

$$\omega = \sqrt{\frac{F_c}{mR}} = \sqrt{\frac{20{,}000 \text{ N}}{(0.500 \text{ kg})(0.050 \text{ m})}} = 894 \text{ rad/s}$$

$$894 \text{ rad/s}\left(\frac{9.55 \text{ rpm}}{1 \text{ rad/s}}\right) = 8{,}500 \text{ rpm}$$

*Answer:* 8,500 rpm

# Section 1 review

Circular motion can involve rotating, rolling, or orbiting objects. Any object moving in a circle is undergoing centripetal acceleration, which changes the direction of the velocity vector, but not its magnitude (speed). A centripetal force—directed toward the center of the circle—is required to maintain an object in circular motion.

| Vocabulary words | angular velocity, radian (rad), centripetal force, centripetal acceleration |
|---|---|

| Key equations | | |
|---|---|---|
| | $\omega = \frac{\Delta\theta}{\Delta t}$ | $v = \omega r$ |
| | $a_c = \frac{v_t^2}{r}$ | $F_c = \frac{mv_t^2}{r}$ |

## Review problems and questions

1. Radians and degrees are related by $\pi$ rad = 180°. Perform the following conversions and draw each one of these angles on a circle.
   a. Convert 45° to radians
   b. Convert 0.5236 radians to degrees
   c. Convert 270° to radians
   d. Convert 7.85 radians to degrees
   e. Convert 585° to radians

2. A wheel spins at a rate of 30 revolutions per minute (rpm). What is the angle that a point on the wheel makes in one second? Give your answer in degrees and radians.

3. A race car is moving with a speed of 200 km/hr on a circular section of a race track that has a radius of 300 m. The race car and the driver have a combined mass of 800 kg.
   a. What is the magnitude of the centripetal acceleration felt by the driver?
   b. What is the centripetal force acting on the car?

4. A bicycle moves with a speed of 30 km/hr. If the wheels of the bicycle have a radius of 35 cm, what is the angular speed of the wheels?
   Give your answer in rad/s and degrees/s.

5. A 62 kg student rides a Ferris wheel that has a diameter of 50 m and makes one complete rotation every 35 s.
   a. What is the angular velocity of the wheel (in radians per second)?
   b. What is the linear velocity of the student (in meters per second)?
   c. What is the centripetal force acting on the student?

You learned in elementary school that the Moon orbits the Earth and the planets orbit the Sun because of *gravity*. This section explains *why* by showing you how the law of universal gravitation bends the motion of the planets and the Moon into circular or nearly circular orbits.

## The apple and the Moon

**Sir Isaac and the apple**

The story goes that Sir Isaac Newton deduced the law of gravitation upon observing an apple fall from a tree. Newton saw that the apple was accelerated as it fell. Thus, he knew there must be a force causing the acceleration. Following Newton's thinking, let us call this force "*gravity.*" Experience shows that an apple falling from a tree that is twice as high will be similarly accelerated. It is natural to deduce, as Newton did, that the force of gravity reaches at least as high as the tallest trees.

**What was Newton's idea?**

Newton's truly brilliant insight was to question whether the same force of gravity we experience on Earth might reach much further than the tallest trees. Might it reach all the way to the Moon? If so, then the orbit of the Moon about the Earth could be a consequence of the *same* force that caused the apple to fall. While this may now seem obvious, in Newton's time it was not. The motions of "heavenly bodies" such as the Moon were divine and not subject to the understanding of humans, nor to laws deduced by mere human minds. Historians of science mark Newton's explanation of orbits through the law of gravitation as *the* turning point in Renaissance thought that began the scientific revolution and expanded our understanding of the universe.

**How is a falling apple like the Moon?**

How can "falling like an apple" cause the Moon to orbit? Suppose we fire a cannon up into the sky. The cannonball eventually falls to Earth due to the downward acceleration of gravity. As we increase the velocity of our imaginary cannon, the cannonball travels further and further before hitting the ground.

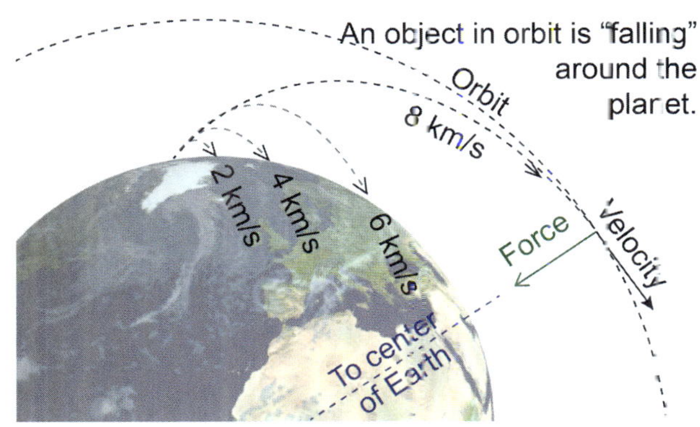


**Newton's intuitive leap**

Newton reasoned that, given sufficient velocity, the curvature of the "falling" cannonball would match the curvature of the planet itself. In essence the "ground" would fall away at the same rate as the cannonball. Newton realized that this path is precisely what we call an *orbit.* This intuitive leap led to a concept that gravity is a *universal force* that any object with mass (such as the Earth) exerts on any other object with mass (such as you or the Moon).

**What was Newton's idea about gravity?**

Newton correctly deduced that an attractive gravitational force exists between all objects that have mass. The gravitational force is a universal force that exists between any two objects that contain mass. The strength of the gravitational force depends on the mass of the objects and the square of the distance between them. The equation Newton derived for gravity is called the **law of universal gravitation**.

Interactive equation

**Law of universal gravitation**

$$(7.5)\quad F = G\frac{m_1 m_2}{r^2}$$

$F$ = force (N)
$G$ = gravitational constant = $6.67\times10^{-11}$ N m²/kg²
$m_1$ = mass of object #1 (kg)
$m_2$ = mass of object #2 (kg)
$r$ = distance between centers (m)

**What does the law mean?**

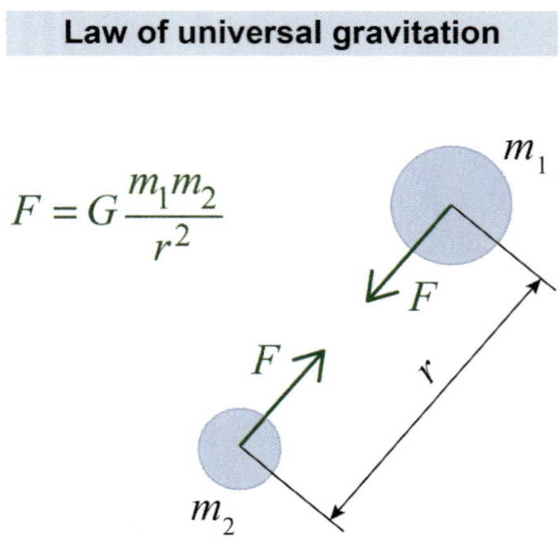


Equation (7.5) says that the attractive force between two masses ($m_1$ and $m_2$) is proportional to the mass of both objects multiplied together. If one object doubles in mass, then the force between the objects doubles. If both objects double in mass, then the force of gravity between them is multiplied by four. The force is inversely proportional to the square of the radial distance $r$ between the masses. This means that increasing the distance by a factor of 2 decreases the force by $(1/2)^2 = 1/4$.

**What is *G*?**

The gravitational constant $G$ in equation (7.5) has the value $6.67\times10^{-11}$ N m²/kg². The value of $G$ is the same everywhere in the universe.

**What is the reaction partner of weight?**

Gravitational forces always come in pairs, just like all other forces. If the *action* force is the force acting on $m_1$ from $m_2$, then the *reaction* force is the force acting on $m_2$ from $m_1$. Similarly, consider the weight of an apple in your hand. If the apple's weight is the action force, the reaction partner is the force of the apple pulling back on the Earth itself—*not the normal (support) force from your hand holding up the apple!*

**What direction is the force of gravity?**

The force of gravity between two objects is always directed along the line connecting their centers of mass. As objects move, the force of gravity changes its direction to stay pointed along the line of those centers.

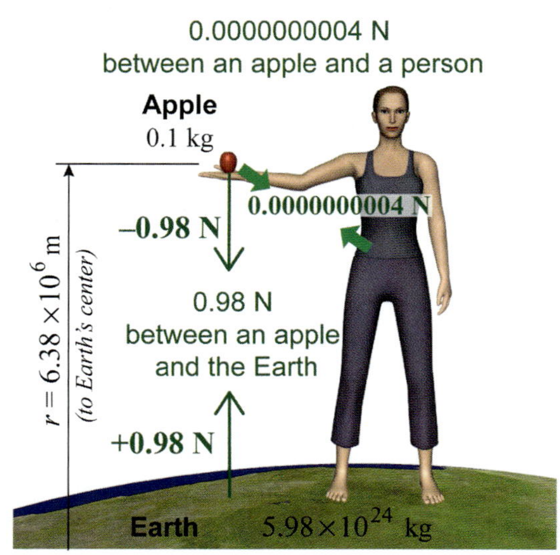


**Gravity**

The attractive gravitational force between ordinary objects is *tiny* because gravity is a relatively weak force. It takes a planet-sized mass to create gravity strong enough to feel. The attraction between you and a 0.1 kg apple is 0.0000000004 N whereas the attraction between Earth and the apple is 0.98 N. Earth's force is larger because it has a mass of $5.98\times10^{24}$ kg, compared to a person with a mass of 60 kg.

**What is a satellite?**

A **satellite** is a natural or technological object gravitationally bound to a larger object, such as the Moon is bound to the Earth. The planets of the Solar System are technically satellites of the Sun. Gravity provides the centripetal force that bends the path of a satellite into an **orbit**. An orbit is a closed path, such as a circle, which the satellite repeats. Artificial satellites in orbit around Earth include the *Hubble Space Telescope*, the *International Space Station*, and hundreds of global positioning and communications satellites.

**Orbits and planets**

Mathematically, orbits are ellipses; a *circle* is a type of ellipse in which all points are equally far from one center. Most of the planets, including Earth, orbit the Sun in nearly perfect circles. A satellite in a circular orbit moves with constant speed. Comets by comparison orbit the Sun in highly elliptical orbits. Comets move fastest at *perihelion,* when closest to the Sun and slowest at *aphelion,* when farthest from the Sun. Comets may come closer to the Sun than Mercury at perihelion and be farther than Neptune at aphelion.

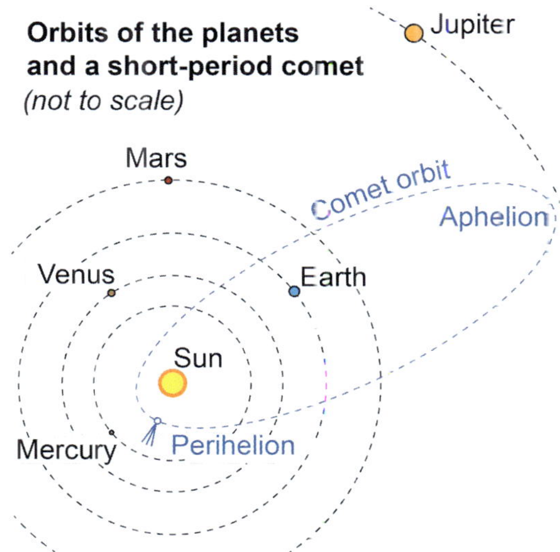


**What orbits what?**

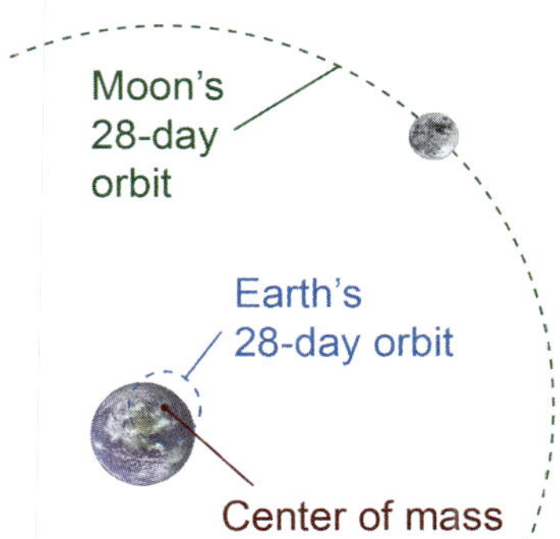


The gravitational force that attracts the Moon to the Earth creates an equal and opposite reaction force that attracts the Earth to the Moon. The Earth and Moon *both* orbit the center of mass of the Earth–Moon system. In fact, part of the definition of a *planet* is that the center of mass of the system including the planet and all its moons be within the main planet. The Earth–Moon center of mass is about 2/3 of the way to Earth's surface on the Moon-ward side.

**Does gravity bend all paths into orbits?**

Not all objects moving in a gravitational field go into orbit! The path taken by an object depends on the ratio of its tangential velocity to its radial velocity. If the tangential velocity is too high, the acceleration of gravity is not enough to bend the object's path into a closed orbit. The object moves along a *hyperbolic* path that does not return. If the tangential component of velocity is too low, the object's trajectory makes it crash into the planet.

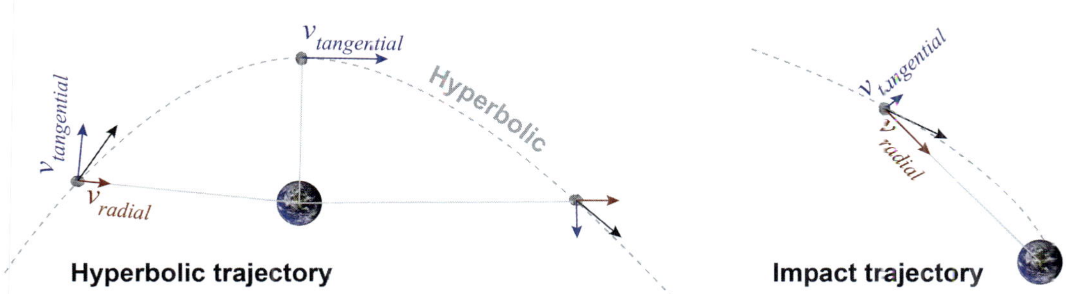


**Hyperbolic trajectory** **Impact trajectory**

## The orbit equation

**When does an object orbit?**

To determine whether the path of a moving object will be an orbit, consider a satellite of mass $m_1$, moving with linear velocity $v_1$, near a planet with a much larger mass $m$. When is the gravitational force sufficient to move the satellite in a circular orbit with radius $R$? To solve the problem, recall the equation for the centripetal force required to cause an object to move in a circular path of radius $R$.

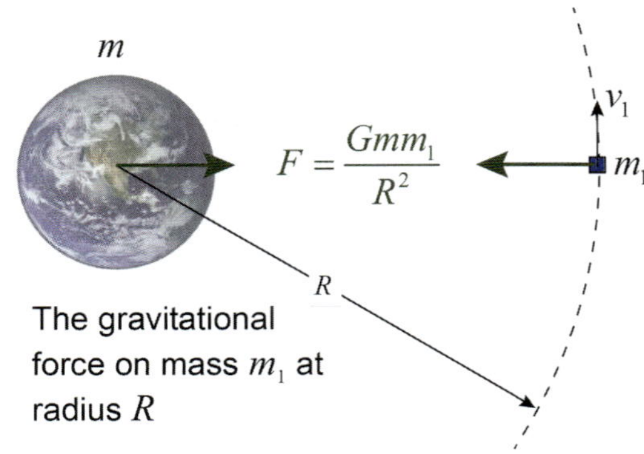


The gravitational force on mass $m_1$ at radius $R$

**Orbit conditions**

To derive the orbit equation (7.6) we set the gravitational force at radius $R$ equal to the centripetal force needed to keep a mass $m_1$ moving in a circle.

**Deriving the orbit equation**

Gravitational force at radius $R$ $\quad \frac{Gmm_1}{R^2} = \frac{m_1 v_1^2}{R} \quad$ Centripetal force to make $m_1$ move in a circle of radius $R$ with velocity $v_1$.

$$(7.6) \qquad R = \frac{Gm}{v^2} \qquad v = \sqrt{\frac{Gm}{R}}$$

$R$ = orbital radius (m)
$G$ = gravitational constant = $6.67 \times 10^{-11}$ N m$^2$/kg$^2$
$m$ = mass of central object
$v$ = satellite orbital speed (m/s)

**Orbit equation** for circular orbits

**What does the orbit equation mean?**

The radius of a circular orbit depends on the mass of the planet divided by the square of the satellite's velocity. The higher the velocity, the *smaller* the orbital radius. This is because it takes a stronger force to hold a faster moving object in a circular path, and gravitational force decreases with distance. A satellite with double the velocity will orbit at 1/4 the radius. As equation (7.6) shows, the velocity a satellite needs to maintain a particular orbit does *not* depend on the satellite's mass, but only on the mass of the planet.

**An example problem**

Calculate the orbital velocity of a 4,000 kg satellite that orbits at a radius of 42,000 km.

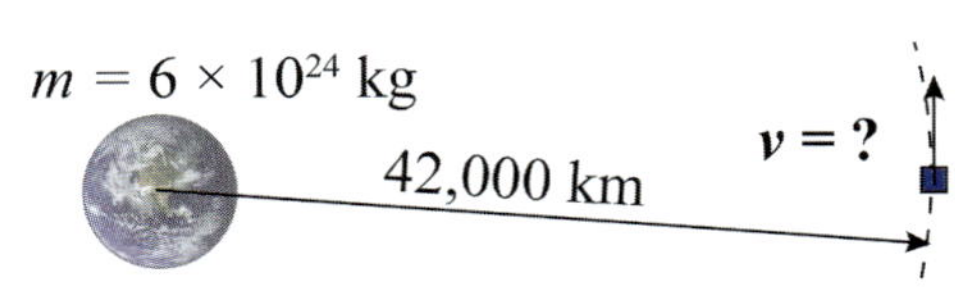


*Asked:* orbital velocity $v$

*Given:* $R = 42{,}000{,}000$ m, $m = 6 \times 10^{24}$ kg, $m_1 = 4{,}000$ kg

*Relationships:* $v = \sqrt{Gm/R}$

*Solution:*

$$v = \sqrt{\frac{Gm}{R}} = \sqrt{\frac{(6.67 \times 10^{-11}\ \text{N m}^2/\text{kg}^2)(6 \times 10^{24}\ \text{kg})}{42{,}000{,}000\ \text{m}}}$$

$v = 3{,}090$ m/s (6,900 mph)

**Orbital velocities**

Notice that the orbital velocity is quite high. The velocities are even greater at lower altitude orbits. The orbital velocity of the *International Space Station* is 7,700 m/s (27,700 mph) at an average altitude of 414 km above Earth's surface.

# Investigation 7B Orbits

**Essential questions**

*How does velocity relate to a planet's orbital radius?*
*How do you send a satellite from Earth to another planet's orbit?*

Interactive simulation

The planets in the Solar System orbit the Sun in ellipses—elongated circles—although many of the planets have nearly circular orbits. As a planet is placed further from the Sun, it experiences a smaller acceleration due to the Sun's gravity; the planet's velocity will therefore change with distance from the Sun to maintain a stable orbit. In this investigation you will determine the velocities of different planetary bodies in the Solar System and correlate these with their distances from the Sun.

## Part 1: Orbital velocities of the planets

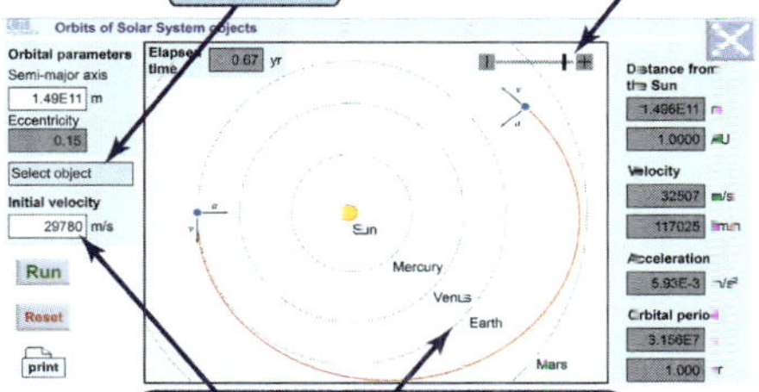


1. Using your computer, click on the interactive simulation in the electronic resources to conduct the investigation. Select *Earth*.
2. Vary the orbital velocity *v* to find the best match for Earth's true orbit. Record Earth's minimum and maximum velocities and average distance from the Sun (semi-major axis).
3. Repeat for the other seven planets.

a. *Write a hypothesis about which planets move faster or slower, outer planets or inner planets.*
b. *From your results in the simulation, does Jupiter move faster or slower than the Earth? Does this evidence support or contradict your hypothesis?*
c. *Using the simulation results, how does orbital velocity vary with distance from the Sun? Do these data support your hypothesis?*
d. *Graph average orbital velocity v against distance from the Sun r. Is it a straight line?*
e. *Devise a new graph of v against a different quantity derived from r that will be a straight line. Measure its slope (including its units).*
f. *Using digital resources such as the Internet, research the average orbital speed for the eight planets. Compare the values with those you obtained from the simulation.*

## Part 2: Transfer orbits

To send a satellite from one planet to another, it is necessary to change the satellite's velocity by using a trajectory called a *transfer orbit*.

**Transfer orbit to Mars**

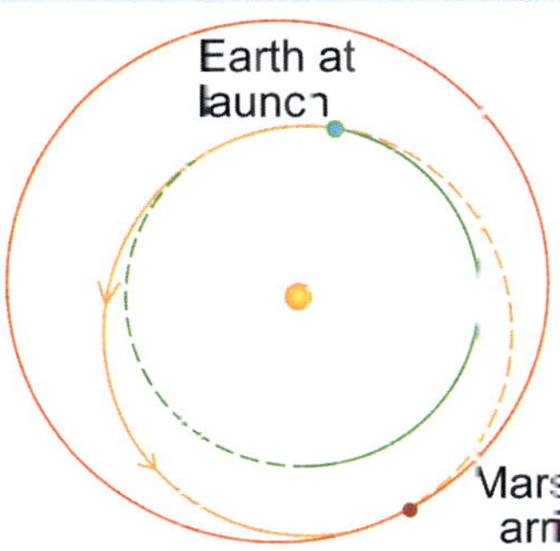


1. Select Earth. Determine the initial velocity necessary to propel a satellite initially from Earth's orbit to just reach Mars's orbit.

a. *Compare the average velocity for Earth's orbit (from Part 1) with the velocity necessary to reach Mars's orbit. How much does the velocity have to change to reach a Martian orbit?*
b. *What provides this change in velocity of the satellite?*
c. *How much "elapsed time" does it take for this satellite to reach Mars? Compare this length of time to typical missions of the Space Shuttle and International Space Station. (Use digital resources to find out the typical mission lengths.)*

## Satellite orbits

**The orbital period**

The time it takes to complete a full orbit is called the **orbital period**. Earth's orbital period is 31,558,149.5 s or 365.256 days. The orbital period of the *International Space Station* in its low-Earth orbit is 92 min and 50 s, which means the space station orbits the globe in just over an hour and a half. If you know the orbital velocity $v$ and the orbit radius $R$, then the orbital period for a circular orbit is given by the equation

$$\text{orbital period} = 2\pi\sqrt{\frac{R^3}{Gm}}$$

Orbit velocity

$$v = \sqrt{\frac{Gm}{R}}$$

Distance for one orbit

$$C = 2\pi R$$

Time to travel distance $C$ at speed $v$.

$$t = \frac{C}{v} = \frac{2\pi R}{\sqrt{\frac{Gm}{R}}} = 2\pi\sqrt{\frac{R^3}{Gm}}$$

Period of a circular orbit

**Geostationary orbit**

A satellite in *geostationary orbit* completes one orbit in exactly one day, so its position remains directly over the same place on Earth's surface. Weather and communications satellites use geostationary orbits because a dish antenna on the ground can receive a signal 24 hours a day by pointing in a fixed direction. If the satellite were not in a geostationary orbit, a dish antenna would have to move around to track the satellite. As of December 31, 2012, there were 404 satellites in geostationary orbit from more than 30 different countries, including 60 from the USA.

**Where is a geostationary orbit?**

A geostationary satellite must have an orbital period of 24 hours (86,400 s) as well as a velocity that satisfies the orbit equation. The combination of these two conditions determines the radius of the geostationary orbit, which is 42,300 km from the center of the Earth. That is an *altitude* of 35,920 km after subtracting 6,380 km for the radius of Earth itself.

**Polar orbits**

Geostationary satellites orbit above the equator. Above 70° latitude, however, a geostationary satellite appears less than 10° above the horizon. At this low angle reception is difficult. There is no possible orbit in which a satellite hovers directly over a pole, but there is a way for a satellite to spend *most* of its time over a polar region: using a polar, highly elliptical orbit (or HEO).

**How do polar orbits enhance coverage?**

The orbital velocity at aphelion in an HEO is slower than the velocity at perihelion by the ratio of $R_p/R_a$. By orienting the orbit above the north or south pole, the satellite spends most of its time above the pole where it is moving slowest. A communications satellite in a north polar HEO offers line-of-sight coverage of the northern hemisphere for most of its orbital period. A pair of satellites in north polar HEO and another pair in south polar HEO provide continuous coverage for high-latitude regions of the globe.

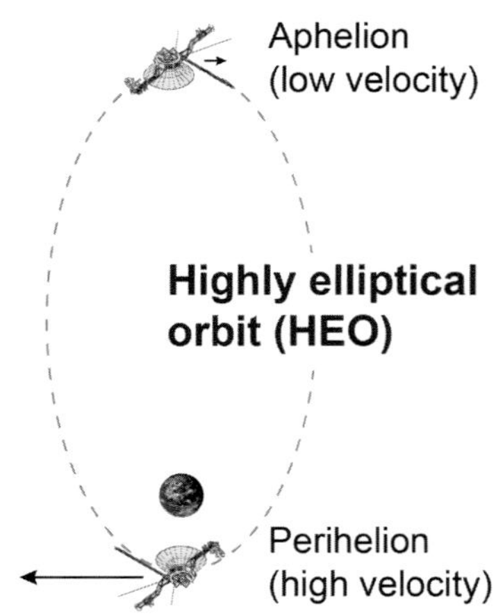

# Investigation 7C Extrasolar planets

**Essential questions** *How do planets and stars influence each other's motion?*

Interactive simulation

Are there planets around other stars? Science fiction writers have been pondering this question for many years. In the past two decades, astronomers have detected many *extrasolar planets*—planets around stars other than the Sun—using a *Doppler technique*. Astronomers detect tiny wobbles in the velocities of nearby stars in order to infer the presence of orbiting planets. Many planets discovered in this way are as massive as Jupiter but are very *hot*, because they orbit at very small radii from their star—closer than Venus or Mercury. Why have astronomers found so many *hot Jupiters* with this technique, but no Earth-like planets?

## Part 1: What planets in our Solar System could be detected?

The *Doppler velocity* for an astronomical object is how fast it is moving toward or away from you *along the line of sight*. The large mass of a star will cause its planets to move around in large orbits, but the planets themselves will also cause their parent stars to wobble a little. The mutual forces form an action–reaction pair! Astronomers using the Doppler technique try to detect the component of the star's velocity along the line of sight.

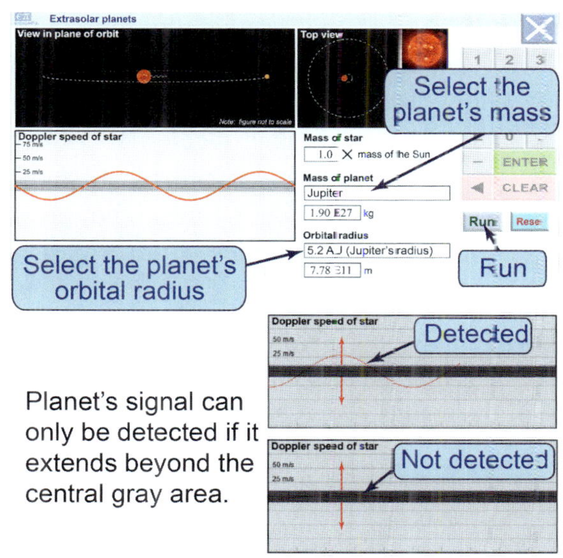


**What planets in our Solar System can be detected at their orbital radii?**

1. Select a planet with the mass and orbital radius of Jupiter.
2. Play the orbit simulation. Can this planet be detected at this orbital radius?
3. Repeat for each of the eight planets in the Solar System at that planet's orbital radius. (You may have to research the planet's orbital radii!) Tabulate your results.

a. *If you were an astronomer on another star and studying our Solar System, which planets could you detect?*

b. *Inspect the orbital period in the upper-left graphic. If you were an astronomer, how long would you have to collect scientific data to detect Jupiter?*

c. *Why can certain planets be detected with this technique while others cannot?*

## Part 2: Hot Jupiters and super-Earths

1. Try detecting a Jupiter-mass and "super-Earth"-mass planet (five times the mass of the Earth) at various orbital radii.

a. *At what orbital radii can a "super-Earth" be detected, if any?*

b. *Do you expect this Doppler technique to reveal Earth-like planets? Why or why not?*

c. *Why is a typical extrasolar planet discovered using this technique referred to as a "hot Jupiter"?*

## Kepler's laws and the birth of modern science

**Ptolemy's universe**

The Sun rises in the east, sets in the west, and appears to circle around the Earth. The planets, the Moon, and the stars have similar apparent motion. Early theories of astronomy put Earth at the center of the universe with celestial objects circling around it. Ptolemy (AD 90–AD 168) is credited with the model illustrated in the *Flammarion* on the right in which an explorer reaching the edge of Earth peers out from under the first, and lowest, celestial sphere containing the sky, clouds, and air.

Medieval concept of astronomy

**The birth of modern astronomy**

Nicolaus Copernicus (1473–1543) correctly deduced that the planets revolve around the Sun, which is called the *heliocentric model*. Nonetheless, Copernicus was not widely believed at any time during his life. Danish nobleman Tycho Brahe (1546–1601) made the first accurate and systematic measurements of the positions of stars and planets over many years, aided by his assistant Johannes Kepler (1571–1630). A brilliant mathematical thinker, Kepler laboriously worked with Brahe's data long after Brahe's death, and in 1609 he published the first of what are now known as Kepler's laws of planetary motion. In Kepler's laws, the gravitational force between the Sun and a planet determines the properties of the orbit of the planet.

**Kepler's laws**

The orbits of the planets around the Sun are ellipses with the Sun at one focus.

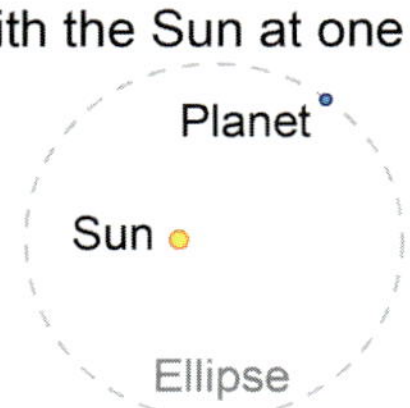


A line from the Sun to any planet sweeps out equal areas in equal times.

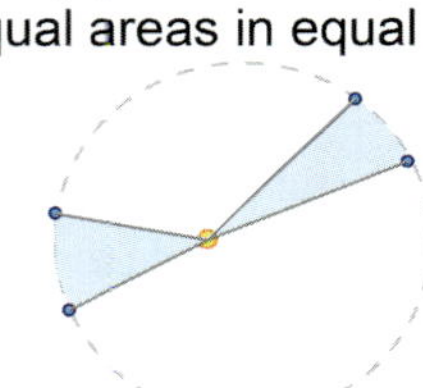

The square of the orbital period is proportional to the cube of the orbit's radius.

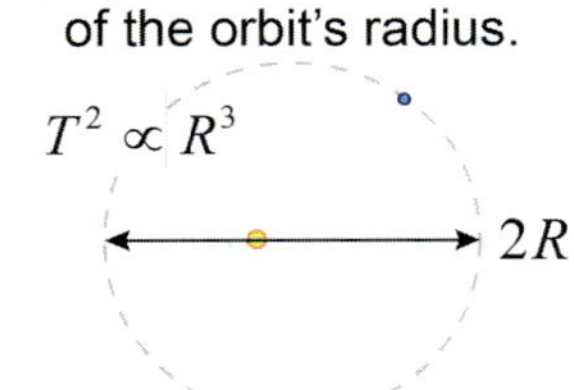


**Galileo and the telescope**

In 1610 Italian philosopher Galileo Galilei turned his newly made telescope upon Jupiter and observed four "stars" near the planet. Over a period of a few days these "stars" changed their positions, as Galileo's notes in his journal show clearly. Galileo correctly deduced that these were in fact not stars orbiting Earth, but four moons orbiting Jupiter. These *Galilean moons* provided the first clear observational evidence that not all celestial objects circle around the Earth.

**Galileo's observations of Jupiter's moons**

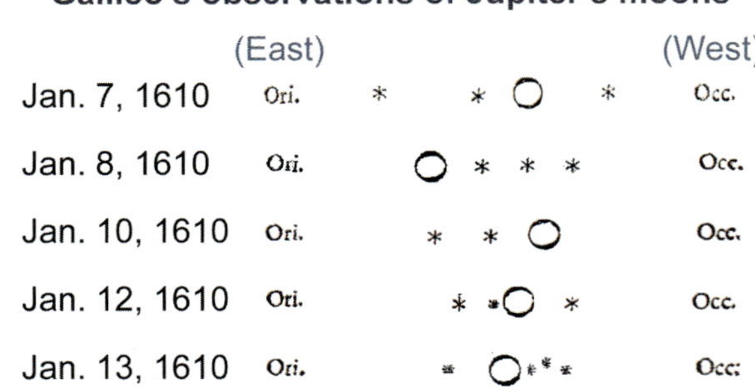


**The emergence of scientific thinking**

Though it took 200 years to become accepted by society, the Sun-centered solar system forever changed how humans think. Prior to the Renaissance, the correctness of ideas was based on tradition and reasoning—even incorrect reasoning. Kepler, Galileo, Newton, and others ushered in a new way of thinking where truth is based on what is actually seen, heard, and demonstrated. This is the essence of the modern view of science: that truth is what actually happens, not what people *think* should happen.

**Why is the same mass *m* in both equations?**

The mass $m$ that creates weight ($F = Gmm_1/R^2$) is the same mass that represents resistance to being accelerated in Newton's second law ($a = F/m$). Why should this be so? *Inertia has nothing to do with gravity!* Why should an object's resistance to acceleration be the exact same property that determines an object's interaction with gravity? Why should it be the same "$m$" in both formulas? The results of every experiment tell us that the "$m$" is the same, but the experiments do not explain *why.*

**The general theory**

Einstein thought the coincidence was a clue that gravity, space, and time were connected. His theory of general relativity was published in 1915. It has been called the greatest feat of human thought ever accomplished. General relativity describes gravity in a fundamentally different way from Newton's law of gravitation. According to Einstein, the presence of mass changes the shape of spacetime itself. In general relativity, an object in orbit is moving *in a straight line through curved space!* The force we call gravity is an effect created by the curvature of space and time.

**A thought experiment**

Imagine a boy and girl jumping into a bottomless canyon, with no air friction. As they fall, they throw a ball back and forth. The girl sees the ball go straight to the boy and vice versa. An observer at rest watching them fall, however, sees the ball follow a curved zigzag path back and forth.

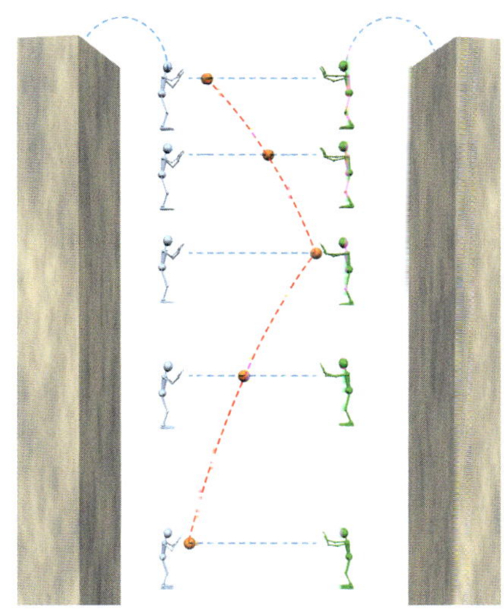

**The view from a falling reference frame**

Now, imagine the boy and girl throwing the ball but enclosed in a windowless box falling with them. From inside the box, they see the ball go straight back and forth. To them, *the ball moves as if there were no gravity.* Remember, gravity would cause the ball to follow a curved path.

**An accelerated reference frame**

Next, imagine the boy and girl throwing the ball back and forth in the same box but in deep space where there is no gravity. This time the box is *accelerating* upward. When the boy throws the ball it does not go straight but drops in a parabola toward the floor. This happens because the floor is accelerating upward and pushing the girl with it. The girl calculates the path of the ball and finds it to be exactly the same as if gravity was pulling it down.

Motion in an accelerated reference frame is indistinguishable from motion under "gravity"

**The equivalence principle**

The *equivalence principle* says that no experiment the boy or girl can do will distinguish whether they are feeling the force of gravity or they are in a reference frame that is accelerating. Any result from *any* experiment is exactly the same whether the experiment is done where there is a gravitational field of 9.8 N/kg or in a reference frame that is accelerating at 9.8 $m/s^2$.

## Curved spacetime

**Reconciling equivalence and special relativity**

The theory of special relativity was discussed at the end of Chapter 3. It tells us that the speed of light is the same for all observers whether or not they are moving. To make the equivalence principal true for experiments that measure the speed of light, two things must be true:

1. Space itself must be curved.
2. The path of light must be deflected by gravity, even though light has no mass.

**Flat space**

What is *curved space?* To understand it without mathematics, consider rolling a ball along a flat sheet of graph paper on a level table. The ball rolls along a straight line. The flat graph paper is like "flat space." Flat space is what you would consider "normal." In flat space, parallel lines never meet and the three angles of a triangle add up to 180°.

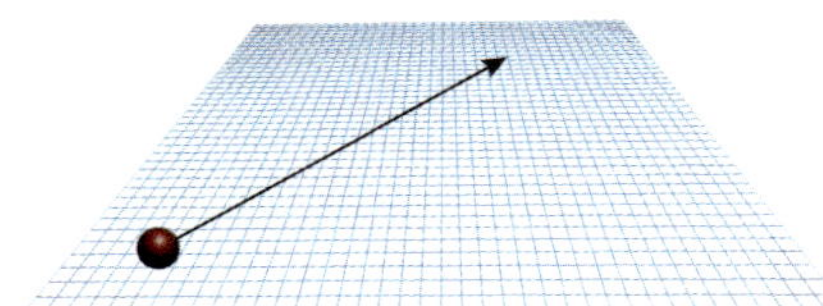

The path of a particle in flat spacetime with no gravity.

**Curved space**

A large mass, such as a star, causes space to curve in the region nearby. Imagine a graph paper printed on a sheet of rubber. Imagine a heavy mass representing the effect of a star. The "star" deforms the flat space of the graph paper. If you roll a ball along this "curved" graph paper, it bends as it passes by the "well" created by the mass.

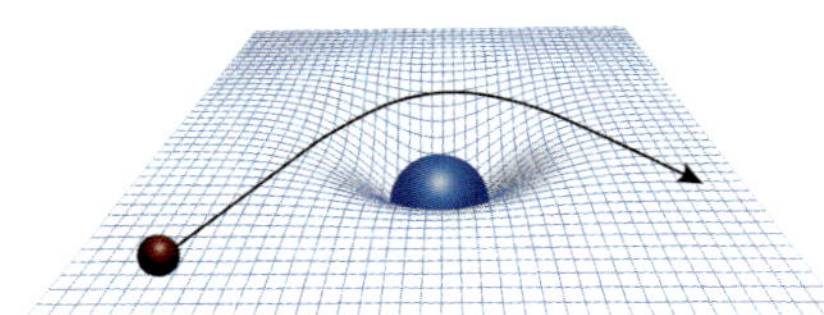

The path of a particle in **curved** spacetime

**Properties of curved space**

Graph paper is two dimensional, but the effect of mass on three-dimensional space is similar. In curved space, parallel lines may meet and the angles of a triangle may not add up to 180°!

**Looking at 3-D space in two dimensions**

From directly overhead, the deformed graph paper looks square. If you could only look straight down, the path of the ball appears to be deflected by a *force* pulling it toward the center. You could create an equation describing the motion of the ball as accelerated by the *force of gravity.* As Newton did, you would be right. The effect of curved space is identical to the force of gravity.

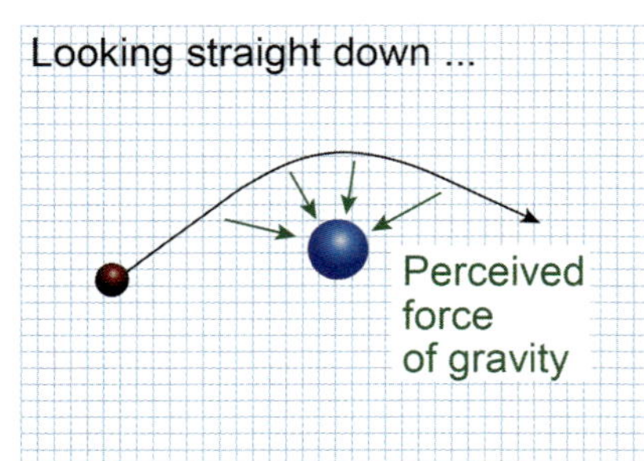


**When is a straight line a circle?**

Close to a large mass, the curvature of space causes straight lines to become *orbits.* A planet in an orbit is actually moving in a straight line through curved space. This is a strange way to think, but the experimental evidence tells us it is the *right* way to think. Einstein was not a household name when general relativity was published, and few physicists could follow the math to understand the theory. Convinced that he was right, however, Einstein predicted that light from distant stars would be bent by passing near the Sun. What made Einstein famous was Sir Arthur Eddington's 1919 confirmation of the bending of starlight by the Sun.

**Who thought of a black hole?**

The equations of general relativity are so complex that even Einstein thought they might have no exact solutions. Later in 1915, however, German physicist Karl Schwarzschild found an exact analytical solution, which described a point mass with infinite density. Although Schwarzschild's solution was mathematically interesting, no one took it very seriously for nearly 40 years. The term "black hole" was coined by American physicist John Wheeler in 1967 while jokingly trying to describe what Schwarzschild's theoretical *singularity* might look like. Wheeler had a gift for colorful phrases and the name stuck, although virtually no one believed it was anything real. A point mass with infinite density that stopped time and sucked in mass and light like a cosmic vacuum cleaner was an *absurd* idea. Then, in 1972, astronomers actually *found one!*

**Escape velocity**

To understand a black hole, consider a rocket leaving Earth. If the rocket does not go fast enough, the Earth's gravity eventually pulls it back to the ground. The minimum speed a rocket must go to escape the planet's gravity is called its **escape velocity**. The escape velocity from Earth's surface is approximately 11,200 m/s. Light emitted from Earth escapes easily to space because Earth's escape velocity is far less than the speed of light (299,792,458 m/s).

**What is a black hole?**

A **black hole** is an object that is both small enough and massive enough that its escape velocity equals the speed of light. The *event horizon* is the radius at which the speed of light equals the orbital velocity. An object outside the event horizon may escape if it has sufficient velocity. Inside the event horizon nothing can escape. To be a black hole, a very large mass must be compressed into a very tiny space. For example, a black hole with Earth's mass would have an event horizon the diameter of a dime.

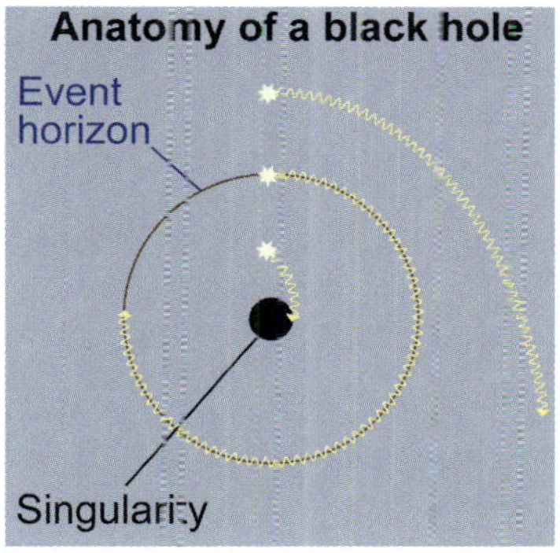


**Real black holes**

In a sense, the region inside the event horizon is outside our universe because no matter, energy, or information can get out. At the center of the event horizon may be a *singularity*, Schwarzschild's point of infinite density. No one knows what is inside, but astronomers have found *many* black holes in the universe. While you cannot see a black hole itself, any mass that falls into one loses so much potential energy that the area around the black hole gives off incredibly bright radiation. Astronomers believe supermassive black holes are at the core of most galaxies. Our own Milky Way contains a monster black hole with a mass of approximately four million times that of the Sun. The Galactic Center—the core of our Galaxy—is an inhospitable place for life as whole stars are consumed in a whirling maelstrom of radiation and fury. In the past two decades, astronomers have even been able to track *individual* stars as they orbit around the center of the Galaxy!

**A dream deferred**

Human exploration of Mars has been a dream of writers and scientists for many years. The dream drew inspiration from colorful reports that astronomers had seen irrigation canals or seasonal vegetation on Mars. Today, of course, we know that the planet is colder, dryer, and windier than the harshest desert on Earth: No farmers, canals, or crops are anywhere to be seen. But is Mars entirely lifeless? Could it support a human outpost? To find out, many believe, we must set foot on its dusty soil. And that can't happen until we know what it takes to bring people to Mars—along with many tons of supplies and shelter.

Credit: NASA

**Mechanical Martians**

With the help of unmanned spacecraft, in the past half-century we've learned quite a bit about reaching Mars and working there. The first such spacecraft simply cruised past the "red planet," snapping pictures along the way. The next generation entered into orbit around Mars, ultimately mapping its entire surface. A third generation reached the planet's surface, telling us about its environment and showing us its rocky plains. Now a fourth generation of mechanical Martians is roving about the red planet, exploring craters and searching for signs of past or present life.

**Getting there**

Interactive simulation

As you learned in the investigation on orbits, the Earth (and everything on it!) circles the Sun at a speed of roughly 30,000 m/s (67,000 mph). But you need to travel at an even higher speed to overcome the Sun's gravity and reach the more distant orbit of the red planet. The *Spirit* rover got the needed boost from the launch rocket, from booster rockets that fired shortly after launch, and from much smaller engines on the spacecraft itself.

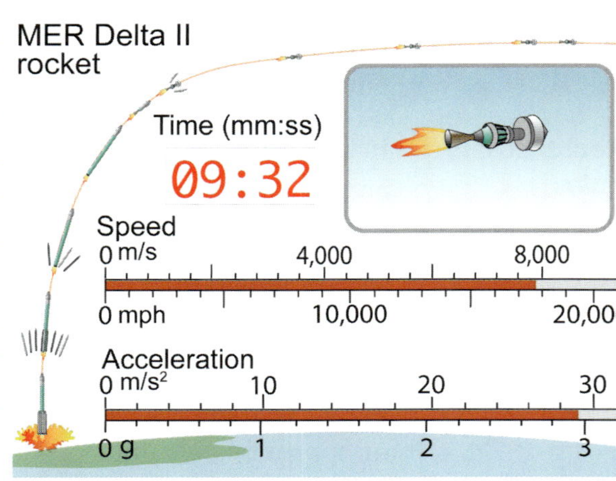


**Time to reach Mars**

How long does it take to reach Mars? The answer is that it depends on the exact alignment between Mars and the Earth at the time of launch. Typical travel times are between 130 and 300 days, where the longer travel times are for heavier payloads. That's a long trip, and it's only one way!

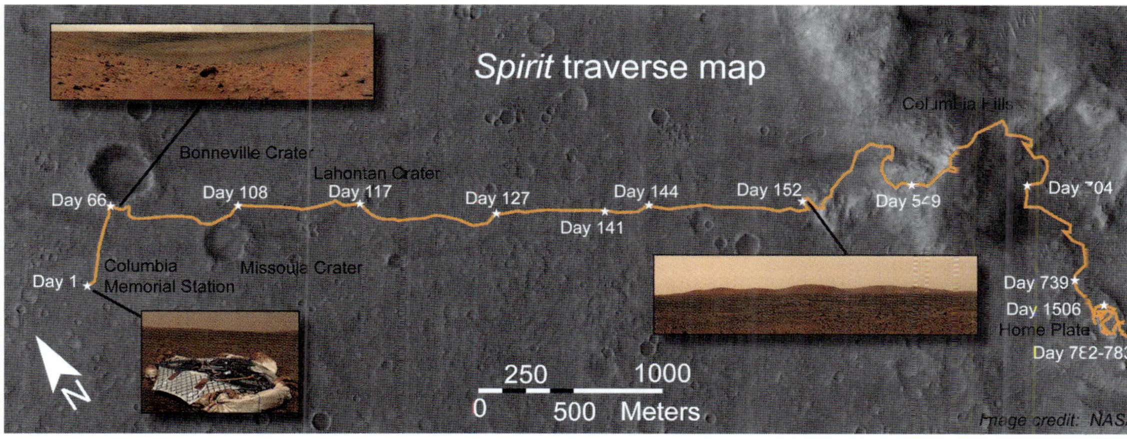


**_Spirit_ and _Opportunity_**

Launched in mid-2003, *Spirit* and *Opportunity* were twin 180 kg rovers sent to explore opposite sides of Mars. *Spirit* operated for six years before going silent; its traverse map is shown above. *Opportunity* has lasted ten years and was still exploring at the time of publication. This is an impressive lifespan, considering that NASA had only required each rover to last for 90 days! Powered by sunlight, the rovers have both found compelling evidence that Mars once held large bodies of water—evidence such as salt deposits and layered sediments. A third, much larger rover, *Curiosity*, reached Mars in 2012.

**The next step?**

There are many challenges involved in getting unmanned spacecraft to Mars. (In fact, many of the 40 missions attempted so far have ended in failure, with some crashing shortly after launch and others falling silent just as they reached their destination.) But these challenges pale beside the requirements for a successful human expedition. For starters, you would need to supply astronauts with enough air, food, and water to make the nine month journey to Mars—roughly fifty tons of supplies!—and additional supplies that will be consumed while on the red planet. Air and water can be partially recycled, but their weight is still many times more than each crew member.

Credit: NASA/JPL and C. Kohlhase

**Other challenges for Mars travel**

Once the astronauts land on Mars, you would have to protect them from constant bombardment by high-speed subatomic particles known as cosmic rays. This potentially fatal radiation is 100 times more intense on Mars than on Earth. Finally, you've got to bring these interplanetary pioneers home. The surface gravity on Mars is only 40% as strong as it is on Earth. It's possible that settlers could make some of the needed rocket fuel from materials found on Mars. If not, then some of the payload brought to Mars would have to include fuel for the return flight.

# Section 2 review

The law of universal gravitation specifies the mutual force of attraction between two objects. The law states that the force of gravity is proportional to the product of the two objects' masses and inversely proportional to the square of the distance between them. Gravity is the force that keeps satellites and planets in their orbits, which can be circular or elongated (elliptical). If a planet orbits a star, its orbital speed and period depend on the *star's* mass and the distance between the two bodies: The *planet's* mass is irrelevant! A black hole is an object whose escape velocity exceeds light speed; it can be studied by the effects it has on other objects, such as the radiation emitted by any mass that falls into it.

For more information to research the historical development of the concept of the gravitational force, see

*Gravity's Arc: The Story of Gravity from Aristotle to Einstein and Beyond* by D. Darling,
*On The Shoulders Of Giants* by Stephen J. Hawking (ed.), and
*Gravity's Engines: How Bubble-Blowing Black Holes Rule Galaxies, Stars, and Life in the Cosmos* by C. Scharf.

| | |
|---|---|
| **Vocabulary words** | **law of universal gravitation, satellite, orbit, orbital period, escape velocity, black hole** |
| **Key equations** | $F = G\dfrac{m_1 m_2}{r^2}$ $\quad$ $R = \dfrac{Gm}{v^2}$ $\quad$ $v = \sqrt{\dfrac{Gm}{R}}$ |

## Review problems and questions

1. Using the formula for the universal law of gravitation, calculate the strength of the gravitational force between two spheres. Each sphere has a mass of 100 kg, and the distance between the centers of the spheres is 2 m. Compare the force to the weight of either sphere. How easy do you think it would be to measure this gravitational force?

2. A 15.0 kg dog rests on Earth's surface. Since Earth is nearly spherical, assume that you can pretend that Earth's mass is all at the planet's center. Earth's radius is 6,370 km, and its mass is $5.98 \times 10^{24}$ kg.
   a. What is the strength of the gravitational attraction between the dog and the Earth?
   b. Calculate the ratio of this force to the dog's mass (in kilograms).
   c. What is the significance of the ratio you just computed?

3. Planet Zorg has two moons named Kwazu and Daewok. Kwazu and Daewok have the same mass, but Kwazu is three times as far from Zorg as Daewok is.

   a. Which moon feels a stronger gravitational pull from Planet Zorg?
   b. How many times stronger is the force on that moon when compared to the force on the other (that is, what is the ratio)?

# Chapter review

### Vocabulary

Match each word to the sentence where it best fits.

#### Section 7.1

| angular velocity | centripetal acceleration |
|---|---|
| centripetal force | radian (rad) |

1. _______ is expressed in radians per second.
2. When measuring angles, the _______ is an alternative unit to a degree.
3. The force that causes an object to move in a circle is the _______.
4. Although acceleration usually causes a change in speed, _______ causes a change in direction.

#### Section 7.2

| black hole | escape velocity |
|---|---|
| law of universal gravitation | orbit |
| orbital period | satellite |

5. A consequence of Einstein's theory of general relativity is that there can be a gravitational singularity called a/an _______.
6. The Moon is the Earth's natural _______.
7. The _______ relates the two objects' masses and the distance between them to how much they are attracted to each other.
8. Many artificial satellites _______ the Earth and relay information across the globe.
9. If you fire a rifle straight up in the air, the bullet will return to Earth because the bullet is not fired fast enough to exceed the Earth's _______.
10. It takes 687 Earth days for Mars to move once around the Sun. The _______ of Mars is therefore 687 days.

### Conceptual questions

#### Section 7.1

11. Ask two friends to tell you what a *radian* is. Compare their answers with your own understanding of radians. Did you learn anything new from your friends' answers?
12. Is a radian larger or smaller than a degree?
13. Which of the following equations best represents the same meaning as the following sentence: "The angular velocity is the linear velocity divided by the radius of the motion."
    a. $v = \omega r$
    b. $v = \omega / r$
    c. $\omega = v / r$
    d. $\omega = r / v$
14. When a satellite orbits a planet in circles, its velocity is _______ to its orbit and its acceleration is _______ to its velocity.
    a. tangential, parallel
    b. parallel, perpendicular
    c. perpendicular, parallel
    d. tangential, perpendicular
15. What are the units of angular velocity? Do these units have dimensions such as length, mass, or time?
16. How does centripetal acceleration change when the radius decreases and the angular speed remains the same?
17. What is the SI unit of angular frequency?
18. Imagine that you build a toy car. For it to work well, you want all four wheels to have the same linear velocity. (Otherwise, some will be trying to get ahead of the others!) Does this mean that all four wheels must have the same radius?
19. Which has a higher linear velocity when rolling, a big wheel or a small wheel with the same angular velocity?
20. A roller coaster car in an amusement park goes around a tight semi-circular turn. Describe the source and direction of the force that causes the car to move in a circular path.
21. Why are radians useful?

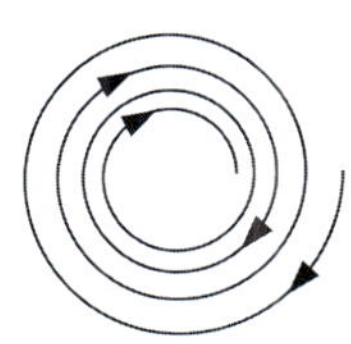

Consider moving on a spiral path at constant speed.

22. Say you are moving inward in a spiral at constant speed. Does the centripetal force needed for this increase or decrease as you move closer to the center?
23. Consider a purely centripetal force, such as gravity. If an object is moving in a circular path, does the centripetal force cause the object's *speed* to change? Explain why or why not.

# Chapter review

## Section 7.1

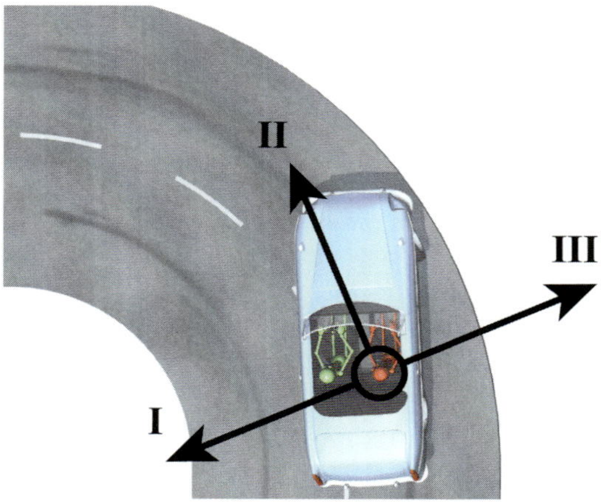


24. At a single moment in time, a passenger (circled) sits in the right-hand seat of a sports car. The car is turning sharply to the left, and the passenger feels as if she were being pushed against the door to her right.
    a. Which of the three labeled vectors (I, II, or III) shows the direction of the passenger's *velocity*?
    b. Which of the three labeled vectors (I, II, or III) shows the direction of the passenger's *acceleration*?
    c. Which of the three labeled vectors (I, II, or III) shows the direction of the *centripetal force* that acts upon the passenger?

## Section 7.2

25. Research the history of the discovery of the law of universal gravitation on the Internet or at a library. Prepare a two-minute oral report using three vocabulary words from this chapter. Make sure to include when and where the law was published and by whom.

26. Write the equation for the law of universal gravitation

$$F = G\frac{m_1 m_2}{r^2}$$

in words. Describe how the magnitude of the force changes as objects move closer and farther from one another.

27. Using the law of universal gravitation, explain why we clearly feel the gravity of Earth but don't notice the gravitational pull of the Sun, even though the Sun is much more massive than the Earth.

28. When Newton first proposed the idea of the law of universal gravitation, a man named Robert Hooke claimed that Newton stole the idea from him. Research Robert Hooke and his main contributions to science. Write a short report on what you learned. Follow a standard format for citations in your paper.

29. Astronauts conduct microgravity experiments aboard the *International Space Station* orbiting at an altitude of 416 km above Earth's surface.
    a. At this altitude what is the relative strength of Earth's gravity compared to its strength at the surface?
    b. Explain how astronauts can be virtually "weightless" even when they feel gravity nearly equal to that on the surface.

30. ‹ The law of universal gravitation determines your weight on Earth's surface. To an excellent approximation, we can assume Earth's mass is all concentrated at the planet's center. Your weight is then determined by your mass, Earth's mass, and Earth's radius.
    a. Suppose the Earth shrank to three-quarters of its present-day radius but retained all of its mass. Would your weight increase, decrease, or stay the same?
    b. Now suppose the Earth somehow got rid of one-third of its mass but maintained the same radius. Would your weight increase, decrease, or stay the same?
    c. Suppose the Earth's mass doubled *and* its radius doubled, too. Would your weight increase, decrease, or stay the same?

31. ‹‹ A geostationary satellite is a spacecraft that remains above a single spot on Earth, such as a city that receives its radio signals. Geostationary satellites orbit Earth at an altitude of roughly 36,000 km (that is, 42,300 km from the center of our planet).
    a. Suppose that Earth's day length was 12 hours (rather than 24 hours) at our time in history. Would a geostationary satellite's altitude be higher than, lower than, or equal to 36,000 km?
    b. Suppose that Earth had formed, nearly five billion years ago, with twice as much mass. Would a geostationary satellite's altitude be higher than, lower than, or equal to 36,000 km?

32. ‹‹ Consider Jupiter moving in its elliptical orbit around the Sun according to Kepler's laws. If Jupiter were suddenly to become twice as massive, then the gravitational force the Sun exerts on Jupiter would double, according to the law of universal gravitation. How would this increase in force change Jupiter's orbit?

# Chapter review



## Section 7.2

33. The International Astronomical Union in 2006 adopted new definitions for planets and dwarf planets: "(1) A *planet* is a celestial body that (a) is in orbit around the Sun, (b) has sufficient mass for its self-gravity to overcome rigid-body forces so that it assumes a hydrostatic equilibrium (nearly round) shape, and (c) has cleared the neighborhood around its orbit.
(2) A *dwarf planet* is a celestial body that (a) is in orbit around the Sun, (b) has sufficient mass for its self-gravity to overcome rigid-body forces so that it assumes a hydrostatic equilibrium (nearly round) shape, (c) has not cleared the neighborhood around its orbit, and (d) is not a satellite.
(3) All other objects, except satellites, orbiting the Sun shall be referred to collectively as *Small Solar System Bodies*."
    a. Is the Moon a dwarf planet? Why or why not?
    b. If the dust and rocks near a Solar System object have been swept up into its rings, is the object more likely to be a planet or a dwarf planet?
    c. Write one or more sentences communicating what you think is meant by "hydrostatic equilibrium" as it is written in the IAU's text.
    d. Provide an example of an object that is *not* in hydrostatic equilibrium and communicate why you think so using one or more sentences.
    e. The Earth bulges out slightly along its equator. Is the Earth a planet?
    f. Research and write a one-page analysis of the IAU's decision to reclassify Pluto as a dwarf planet. Cite specific data corroborating or challenging this decision.

## Quantitative problems

## Section 7.1

34. What is the largest diameter of circular pipe around which you can just wrap a 10 m length of string?

35. If a wheel has a radius of 2 m, how far does it roll in two complete rotations?

36. Robbie is swinging his favorite yo-yo in circles on a meter-long string.
    a. His classmate, Lucy, measures the instantaneous velocity of the yo-yo as 12 m/s. What is the centripetal acceleration of the yo-yo?
    b. The mass of the yo-yo is 60 g. What is the centripetal force on the yo-yo?

37. David is swinging his 0.5 m long sling and exerting a centripetal force of 1,250 N. Soon he will hurl the 250 g rock in the sling at Goliath. What is the velocity of the rock as it leaves the sling?

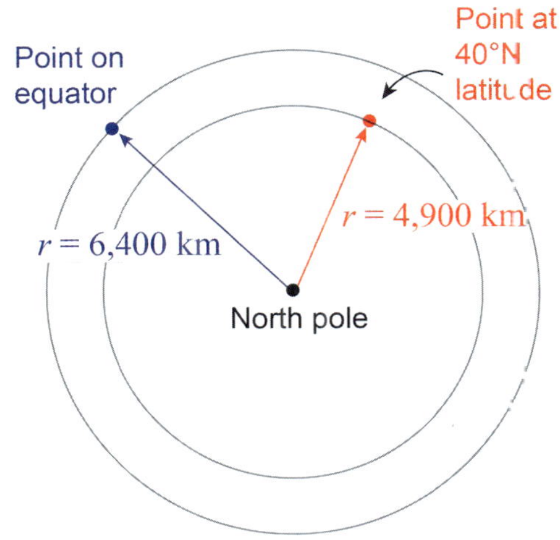


38. Seen from a point above the north pole, the Earth spins (rotates) in a counterclockwise direction, and every point on its surface (other than the poles) traces a circular path around the planet's axis. The radii of these paths are indicated here for two locations.
    a. With what approximate speed (in meters per second) does the equatorial spot go around the Earth's spin axis?
    b. With what approximate speed (in meters per second) does the spot at 40°N latitude go around the Earth's spin axis?
    c. Speculate on the reason why the USA launches most of its space probes from Florida and the Europeans launch most of theirs from the northern coast of South America.

39. On a calm day, a windmill blade 4 m long turns 0.75 rad in 3 s. What is the linear velocity of the tip of the blade?

40. A dragster is speeding down the track at 150 m/s. Its rear wheels are 2 m in diameter, and its front wheels are 40 cm in diameter. What are the angular velocities of the front and rear wheels, respectively?

41. A 1,400 kg car traveling at 35 m/s enters a curve with a radius of 100 m.
    a. How much centripetal force does it take to keep the motion of a car following the curve?
    b. Calculate the ratio of the centripetal force from Part a to the car's weight.
    c. Will the car make it around the curve? Explain.

42. John is swinging a 100 g mass on a string around his head. On one end of the string is a spring scale that says he is exerting 5 N of centripetal force. Attached to the mass is a speedometer that tells him the mass is moving at 4 m/s. How long is the string?

# Chapter review

## Section 7.1

43. A boy on a swing set has a speed of 4.5 m/s and a centripetal acceleration of 8.1 m/s$^2$ at the bottom of his swing. How long are the ropes of the swing?

44. An amusement park ride features a vertical cylinder 8 m in radius with a horizontal floor. Riders stand on the floor with their backs against the inner surface of the cylinder. The cylinder spins, completing one revolution every 4 s. Once spinning, the axis of the cylinder rotates so it tilts up to 45 degrees. Even as the cylinder tilts sideways, the riders don't fall off.

    a. Why don't they fall off?
    b. What direction is the force exerted by the cylinder on the riders?
    c. What is the angular velocity of the riders?
    d. What is the linear velocity of the riders?
    e. What centripetal acceleration do the riders experience?
    f. Compare this to the acceleration of gravity ($g$).

## Section 7.2

45. What is the angular velocity of Mercury's orbital motion? (Mercury makes one orbit of the Sun every 88 Earth days.)

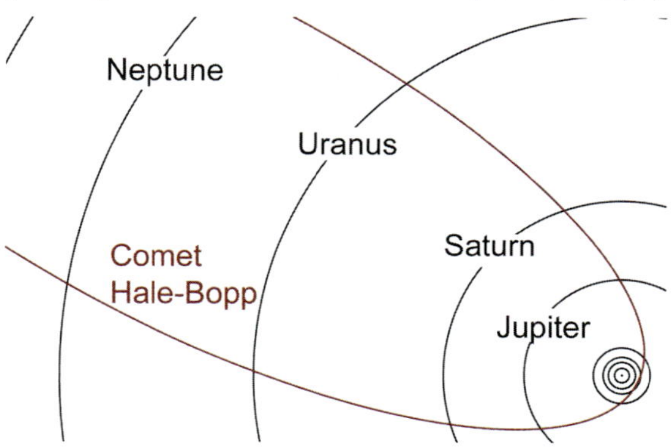


46. Comets move in highly elliptical orbits. For example, the aphelion, or farthest point of Comet Hale-Bopp's orbit is 371 AU from the Sun. The perihelion, or closest approach, is 0.91 AU, within the Earth's orbital radius!

    a. Use the orbit equation to calculate the velocity of Comet Hale-Bopp at aphelion.
    b. Use the orbit equation to calculate the velocity of Comet Hale-Bopp at perihelion.

47. The Moon is Earth's only natural satellite. It is much smaller and less massive than Earth, with a radius of $1.74\times10^6$ m and a mass of $7.35\times10^{22}$ kg.

    a. What is the acceleration due to gravity on the surface of the Moon?
    b. What is your weight on the Moon? (1 lb = 4.45 N, and 1 kg weighs 2.21 lb on Earth.)

48. The mass of the Sun is $2\times10^{30}$ kg. Earth's average orbital radius is $1.52\times10^{11}$ m.

    a. Use the orbit equation to calculate Earth's average orbital velocity.
    b. Suppose Earth increased its velocity by 50%. Calculate the new radius of the planet's orbit.
    c. Compare your answer to Part b with Venus's orbital radius of $1.08\times10^{11}$ m. Given that Venus has a surface temperature hot enough to melt lead, speculate on the possibility for life if Earth had this orbital velocity.

49. Suppose that you lived on a planet with Earth's mass ($5.97\times10^{24}$ kg) but only half its radius (that is, a radius of 3,189 km instead of 6,378 km). Suppose, too, that your own mass was 75 kg.

    a. What would your weight be on the surface of this hypothetical planet? (State your answer in both newtons and in pounds.)
    b. What would the acceleration due to gravity be on the surface of this made-up planet? (State your answer in newtons per kilogram and compare it to the value on Earth's surface.)

50. Compared to our $1.99\times10^{30}$ kg Sun, a 70 kg person on Earth ($1.52\times10^{11}$ m away from the Sun) is very small and light. What is the attractive force due to gravity between the Sun and a person who is on Earth?

51. The Sun has a mass of $2.0\times10^{30}$ kg. Jupiter has a mass of $1.9\times10^{27}$ kg. It orbits $7.5\times10^8$ km away from the Sun. Assume its orbit is circular.

    a. What is the gravitational force between Jupiter and the Sun?
    b. How fast does Jupiter orbit around the Sun?
    c. What is Jupiter's angular velocity?
    d. How may radians does Jupiter travel in one Earth year? How many degrees?

52. The orbit of the dwarf planet Haumea is at an orbital *semi-major axis* of 43 astronomical units (AU), i.e., 43 times larger than the Earth's orbital radius. Using this information and Kepler's laws, what do you predict is the orbital period of Haumea?

53. On your computer, use the interactive simulation of orbits on page 219 to determine the *change in velocity* needed to launch a satellite in a transfer orbit from Earth to Venus.

54. The planets of our Solar System orbit the Sun. The Sun has a mass of $1.99\times10^{30}$ kg. Mars, a planet $2.37\times10^{11}$ m from the Sun, has a mass of $6.42\times10^{23}$ kg. What is the linear velocity of Mars as it orbits the Sun? Assume Mars has a circular orbit.

# Chapter review

## Standardized test practice

55. How many rad/s are equivalent to an angular velocity of 490°/min?

A. 0.23 rad/s
B. 0.05 rad/s
C. 0.14 rad/s
D. 0.19 rad/s

56. Tomás faces due north. He then turns *counterclockwise* until he is facing due east. How many radians did he turn?

A. $\frac{1}{2}\pi$ rad
B. 6.28 rad
C. $\frac{3}{2}\pi$ rad
D. 0.75 rad

57. What is the linear velocity of a rolling wheel with a diameter of 4 m, rotating at $2\pi$ rad/s?

A. 6.3 m/s
B. 12.6 m/s
C. 25.1 m/s
D. 4.0 m/s

58. A rolling ball is rotating at 5 rad/s. If the ball has a radius of 0.5 m, how far does it travel in 3 s?

A. 30 m
B. 0.83 m
C. 7.5 m
D. 45 m

59. A wheel on a watermill turns halfway every 5 s. What is the wheel's angular velocity?

A. $5\pi$ rad/s
B. $\pi/5$ rad/s
C. 2.5 rad/s
D. 25 rad/s

60. Assume 1,200 N is the maximum friction force between the tires and the ground for an 1,100 kg car. Without skidding, how fast can the car go around a curve with a radius of 200 m?

A. 25 m/s
B. 20 m/s
C. 15 m/s
D. 10 m/s

61. Say you and your friend each have a mass of 80 kg. You are floating motionless in space, 1.5 m apart from each other. How much gravitational force is there between you?

A. $2.9\times10^{-6}$ N
B. $1.9\times10^{-7}$ N
C. $2.4\times10^{-9}$ N
D. $3.6\times10^{-9}$ N

62. What is the gravitational force between the proton in a hydrogen atom (mass of $1.7\times10^{-27}$ kg) and the electron in a hydrogen atom (mass of $9.1\times10^{-31}$ kg)? Assume they are $2.5\times10^{-11}$ m apart.

A. $1.6\times10^{-46}$ N
B. $3.5\times10^{-46}$ N
C. $6.4\times10^{-46}$ N
D. $8.9\times10^{-46}$ N

63. How many degrees are equal to 8 radians.

A. 386°
B. 424°
C. 342°
D. 458°

64. What is your centripetal acceleration when you run at 11 m/s around the curve of an Olympic track, which has a radius of 36.5 m?

A. 0.30 m/s$^2$
B. 1.2 m/s$^2$
C. 3.3 m/s$^2$
D. 2.4 m/s$^2$

65. At what speed would a 1 kg satellite orbit if it were 1,400 km above the Earth's surface? (Earth has a mass of $6.0\times10^{24}$ kg and a radius of 6,400 km.)

A. 7,200 m/s
B. 230,000 m/s
C. 130,000 m/s
D. 68,000 m/s

66. Which of the following best explains why Mars takes longer than Earth to orbit the Sun.

A. Mars has farther to go (its orbit is larger).
B. Mars orbits more slowly (its speed is lower).
C. Both A and B
D. Neither A nor B

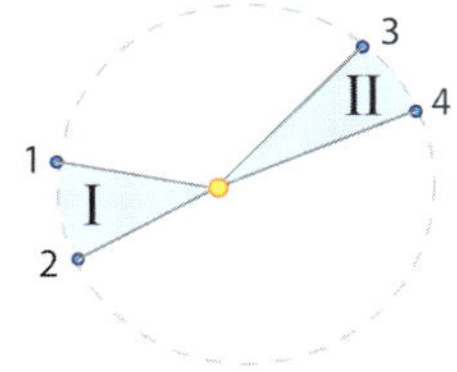


A line from the Sun to any planet sweeps out equal areas in equal time intervals.

67. One planet's elliptical orbit is shown here, along with a statement of Kepler's second law. The time taken to go from 1 to 2 equals the time to go from 3 to 4. Which of the following best completes the following sentence?
The planet is closer to the Sun in _______ and moves faster in _______.

A. Stage I, Stage I
B. Stage I, Stage II
C. Stage II, Stage II
D. Stage II, Stage I

Chapter

# 8 Static Equilibrium and Torque

Throughout history performers and athletes have performed astounding feats of balance and acrobatics. Among the more audacious feats is *tightrope walking* (or *funambulism*), in which a performer balances on a tightrope or cable and walks across a high place. Although the practice of *jultagi*—a traditional Korean art of tightrope walking—may date back many centuries, it was a French acrobat who first caught the public imagination in the west.

On June 30, 1859, at 5:00 pm French stunt master Charles "the Great" Blondin amazed onlookers by being the first human to walk a tightrope above the mighty Niagara river just below the falls on the border between Canada and the USA. The Great Blondin's rope was 1,100 ft (340 m) long and 160 ft above the churning water. When he reached the middle of the span, Blondin was so self-confident that he lowered a rope to the cruise ship *Maid of the Mist* carrying spectators below him and pulled up a bottle of water to drink! Soon refreshed, Blondin resumed walking toward the Canadian shore. Then he stopped, steadied his balancing pole, and *did a backward somersault, landing back on the rope!* A famous crowd pleaser, the Great Blondin later repeated his feat while riding a bicycle, pushing a wheelbarrow, having his hands and feet handcuffed, and carrying his manager on his back.

A *tightrope* walker uses a long pole to balance. While heavy to carry, the pole can be easily shifted side-to-side to supply small torques that rotate the walker slightly left or right, allowing the walker's center of mass to stay directly above the rope. The long poles increase the walker's total *moment of inertia* or rotational inertia, which makes it more difficult for the walker to tip over—and hence easier for him or her to stay upright.

By comparison, *slack rope* walkers do not use poles but use body motion to keep their center of mass above the rope and centered between the rope's support points. The tension in a slack rope is provided by the walker, whereas a tightrope is strung tightly whether or not a person is walking on the rope. A spin-off activity called *trickling* is even being developed as a competitive sport!

Chapter 8

# Chapter study guide

## Chapter summary

Architects and engineers often design structures for which the desired net acceleration is zero. After all, you don't want a building that will move this way and that! For a structure to remain at rest in a state of static and rotational equilibrium, the net force and net torque on the structure must be zero. Applying the equilibrium conditions is an important step in the structural design process. In this chapter you will learn about the basic concepts of static and rotational equilibrium and apply them to address engineering design problems.

## Learning objectives

By the end of this chapter you should be able to

- define static and rotational equilibrium;
- calculate torque and define line of action;
- apply the equilibrium condition to solve force and torque problems;
- summarize the design process for structures;
- differentiate among tension, compression, and shear;
- describe different kinds of reactions provided by structural elements such as pins, rollers, and fixed supports;
- apply the equilibrium conditions to solve structural design problems; and
- define stress and tensile strength.

## Investigations

8A: Static equilibrium
8B: Structural design

## Chapter index



## Important relationships

$$\sum \vec{F} = 0$$

$$\sum \tau = 0$$

$$\tau = r \times F$$

## Vocabulary

**torque**
**fulcrum**
**stress**
**line of action**
**lever**
**tensile strength**
**lever arm**
**safety factor**

Newton's laws form the foundation for understanding the interaction of forces with objects—such as buildings, bridges, and people. All three laws apply to all situations, all the time. For buildings, bridges, and all other structures that are not supposed to move, the *net* force is zero—otherwise they would be accelerated. This chapter considers the special case of *static equilibrium* in which net force is zero. Although there are always forces acting on bridges and buildings, the *net* force is always zero in a static situation.

**Newton's first law**
An object at rest remains at rest and an object in motion remains in motion with the same speed and direction unless acted upon by a net force.

**Newton's second law**
$\vec{a} = \frac{\vec{F}}{m}$ The acceleration of an object is equal to the net force acting on the object divided by the mass of the object.

**Newton's third law**
Forces always exist in pairs because every force (action) acting on one object creates an equal and opposite force (reaction) acting on another object.

## Equilibrium of forces

**Static problems and equilibrium**

In a static problem all the forces acting on an object are in equilibrium and the net force is *zero*. While equilibrium technically includes the condition of *constant velocity*, we usually apply equilibrium analysis to things at rest—or things we wish to stay at rest, such as buildings and structures.

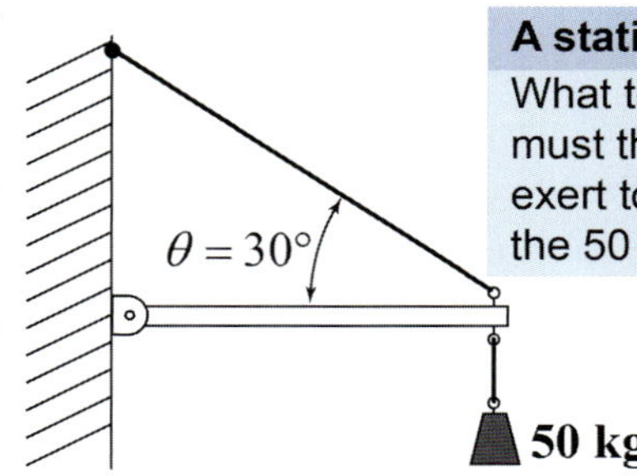


**A static problem**
What tension force must the rope exert to balance the 50 kg mass?

**How do I start an equilibrium problem?**

The starting point for any static problem is the assumption of zero net force *in every direction.* In mathematical language the capital Greek letter *sigma* (Σ) translates as "*the sum of.*" The equilibrium of forces is usually written as $\Sigma F = 0$ (as in equation 8.1), which translates to "the sum of the forces is zero." Since force is a vector, this equation applies separately to all three directions.

**(8.1)** $$\sum \vec{F} = 0$$

$F$ = force vector (N)

**Force equilibrium**

**How do I use the equilibrium equation?**

This compact equation is really three separate equations, one for each direction:

$\Sigma F_x = 0 \rightarrow$ the sum of forces in the $x$-direction is zero.
$\Sigma F_y = 0 \rightarrow$ the sum of forces in the $y$-direction is zero.
$\Sigma F_z = 0 \rightarrow$ the sum of forces in the $z$-direction is zero.

Only a single equation is necessary for 1-D problems, but two equations $(x, y)$ are required for 2-D problems.

**Free-body diagrams**

Every force has a direction, magnitude, and point of action, so a *free-body diagram* is essential to show an object isolated from everything else except for the forces that act on it. A good free-body diagram organizes the information you need in a picture that helps you keep track of the direction, name, and strength of each force.

**An example problem**

To illustrate how to solve equilibrium problems, consider a 50 kg mass hanging from a rope between two walls. What is the tension in each rope in the illustration below, when both ropes make angles of 45° with the vertical? For such equilibrium problems, follow these steps:

1. Draw the free-body diagram of the mass, replacing the rope with the forces exerted at the proper angles.
2. Resolve all forces into $x$- and $y$-directions. Use subscripts "$x$" and "$y$" to keep track of force components.
3. Apply the equilibrium equation to the $x$- and $y$-directions to solve the problem.

Once you have a proper free-body diagram, the key step is to resolve all forces into their $x$- and $y$-components before setting the net force in each direction to zero.

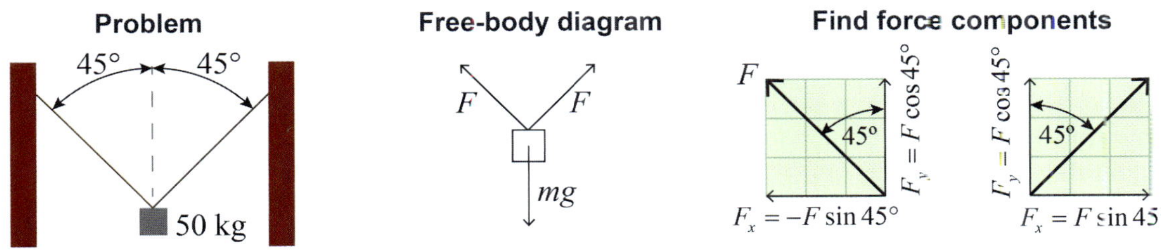


**Force equilibrium**

$$\sum \vec{F} = 0$$

**Solution**

$$\sum F_x = 0 \quad \rightarrow \quad -F\sin 45° + F\sin 45° = 0$$

$$\sum F_y = 0 \quad \rightarrow \quad 2F\cos 45° - mg = 0$$

$$F = \frac{mg}{2\cos 45°} = \frac{(50\text{ kg})(9.8\text{ N/kg})}{2(0.7071)} = \mathbf{346\text{ N}}$$

**Symmetry**

In this example, the forces on the left and right ropes are assumed to be equal, and that is why both get the same name, $F$. This is a good assumption because the problem is *symmetric:* The left and right forces are reflections of each other. The direction of each force is different but the *magnitudes* are the same, because both ropes make the same angle with the vertical. Symmetry is often useful in solving equilibrium problems.

**Solution**

As shown above, the force $F$ is found by applying the equilibrium condition in the $y$-direction. The vertical (upward) components of the rope tension, $2F\cos 45°$, balance the vertical (downward) weight of the mass, $mg$.

**Can all force problems be solved?**

The problem at right appears at first to be easy to solve using force equilibrium. The pin holding the beam to the wall, however, may exert reaction forces ($R_x$, $R_y$) in both the $x$- and $y$-directions, so there are *three* unknown forces—and there is no symmetry we can exploit. A problem with three unknowns needs at least three equations. This problem must be solved using the additional concept of *rotational equilibrium.*

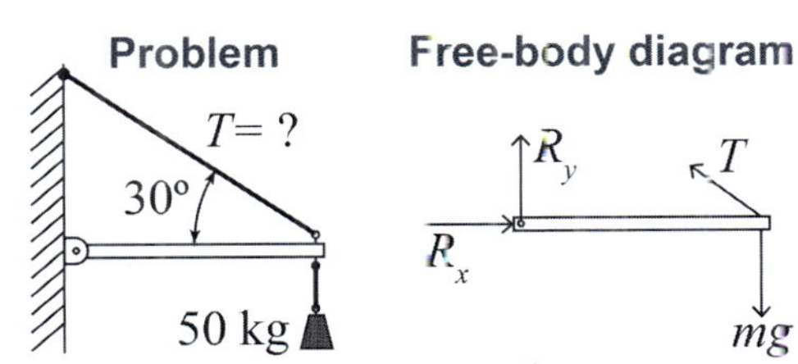


**Force equilibrium equations**

$$\sum F_x = 0 \quad \rightarrow \quad -T\cos 30° + R_x = 0$$

$$\sum F_y = 0 \quad \rightarrow \quad T\sin 30° + R_y - mg = 0$$

**3** unknown forces, **2** equations

**What is torque?**

Consider the two forces acting on the box in the diagram on the right. The net force is *zero* because the forces are equal and in opposite directions. The box rotates, however, because there is a *torque* created by these forces. A **torque** is a twisting action that may be created by forces depending on where they act.

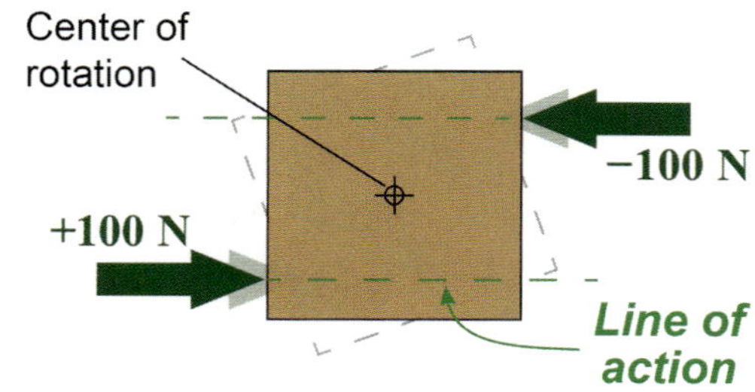


Net force = +100 N − 100 N = 0 N

**Although these two forces add to produce *zero net force*...**

**...their lines of action differ, which creates a torque and causes the box to rotate.**

**When does a force make a torque?**

A force creates torque about a particular center point when its *line of action* does not pass through the center point. The **line of action** is an imaginary line in the direction of a force passing though the force's point of application on an object. Both forces in the illustration at right create torque.

**How do I calculate torque?**

Torque is usually represented by the lower case Greek letter $\tau$ ("tau"). The torque $\tau$ is the length of the lever arm $r$ times the magnitude of the force $F$.

Interactive equation

| (8.2) | $\tau = r \times F$ | $\tau$ = torque (N m)<br>$r$ = distance from center to line of action (m)<br>$F$ = force (N) | **Torque** |
|---|---|---|---|

**Lever arm**

The **lever arm** is the perpendicular distance from the center of rotation to the line of action of the force. The torque created by a particular force depends on the length of the lever arm. A longer lever arm means larger torque even through the force may not change. When the line of action passes *through* the center of rotation, however, the lever arm is *zero*. In this case the torque is also zero *no matter how large a force is applied.*

**What are the units of torque?**

The units of torque are force times distance, or newton-meters. A torque of 20 N m is created by a force of 100 N acting with a lever arm of 0.2 m. Because torque is a product of two variables, it is possible to create the same torque with different forces. For example, a 20 N force applied with a lever arm of 1 m also produces a torque of 20 N m.

**The torque from a force**

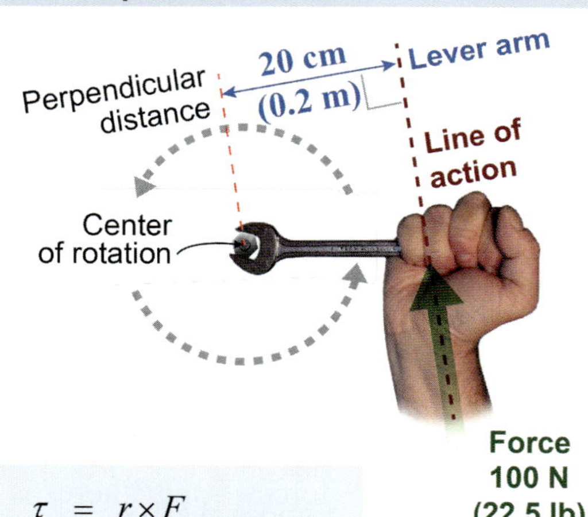


$$\begin{aligned}\tau &= r \times F\\ &= (0.2\ \text{m})(100\ \text{N})\\ &= \mathbf{20\ N\ m}\end{aligned}$$

**Are there other units for torque?**

In English units, torque is usually expressed in *pound-feet.* A torque of one pound-foot represents the twisting action of a one pound force with a lever arm of one foot.

**Can a force make different torques?**

*The same force can produce a different torque around a different center of rotation.* The torque from any force depends on the line of action of the force relative to the location of the center of rotation. The drawing to the right shows the same +10 N force creating torques of +20, +10, and −20 N m when applied at three different places with different centers of rotation.

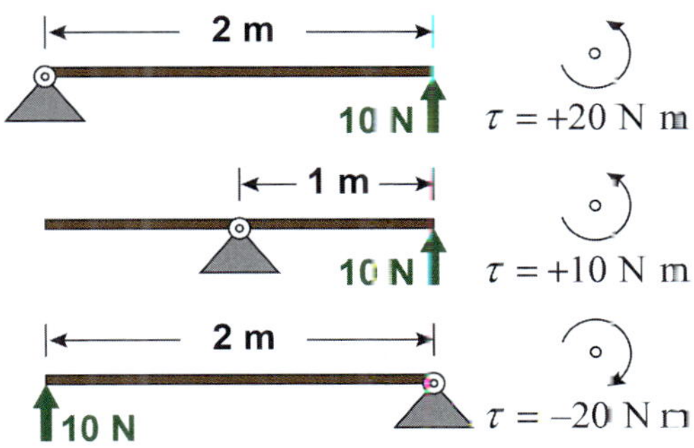


**Sign convention for torque**

Torques can be negative or positive depending on whether they tend to cause clockwise or counterclockwise rotation about any given center. By convention, a *positive* torque tends to increase the angle with the $x$-axis, which means *counterclockwise* rotation. A *negative* torque tends to decrease the angle with the $x$-axis, which means *clockwise* rotation.

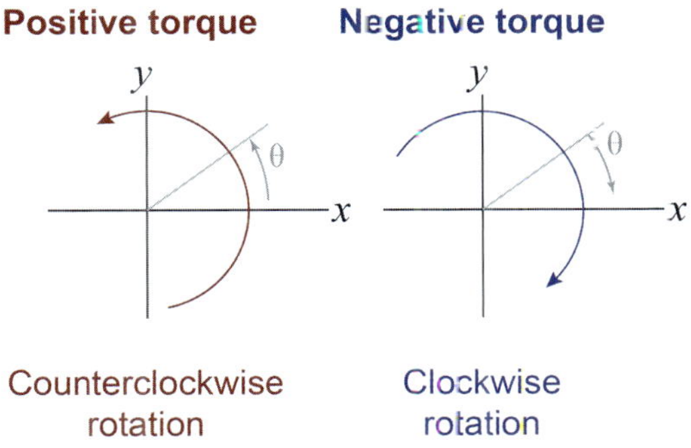


**How do torques add?**

Like forces, torques add as vectors. The *net torque* is the sum of all torques acting around a particular center. Because torque depends on the choice of center, it is absolutely crucial that all torques be determined around the *same* center point in order to add them. In the example at right of a balancing board, the net torque from the three forces is zero.

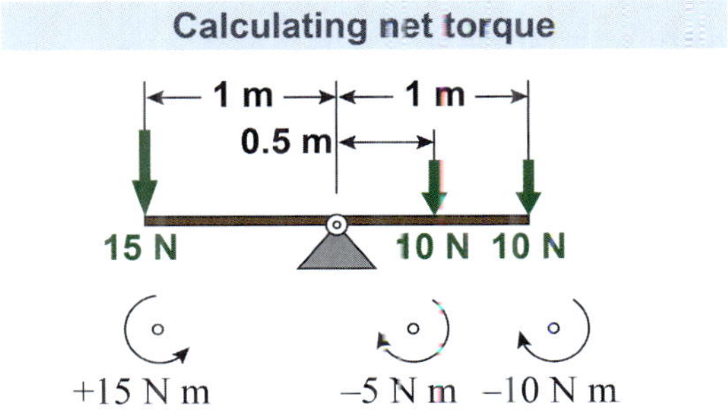


$$\tau_{net} = (+15 - 5 - 10) \text{ N m} = 0 \text{ N m}$$

**How do I choose the "center"?**

Torque can be determined around *any* point in space; *it does not have to be the actual center of rotation.* You are free to choose the "center" to be anywhere. This is very useful for the following reason:

An object in equilibrium has zero net torque around *any* center!

If the net torque were not zero around some point in space, the object would rotate around that point. It follows that any object that is not moving must have zero net torque around all possible choices of "center." For static equilibrium problems it is most convenient to choose the "center" to be the place where one or more unknown forces act. This is useful because the torque is *zero* for any force whose line of action passes through the center.

## Rotational equilibrium

**What is rotational equilibrium?**

An object is in *rotational equilibrium* when the net torque on the object is zero about any point (center). In the notation we used for forces, this condition is stated by equation (8.3), which translates to "the sum of the torques is zero."

**(8.3)** $$\sum \tau = 0$$

$\tau$ = torque (N m)

**Rotational equilibrium**

**Solving a lever problem**

In the problem below, a 5 kg mass is 2 m to the left of the center of a see-saw. Where should the 8 kg mass be placed so the see-saw balances? A see-saw is an example of a **lever**, which is a board free to rotate around a point called the **fulcrum**. This is a typical problem of rotational equilibrium.

Interactive equation

**Where should the 8 kg mass be placed so that the see-saw is in equilibrium?**

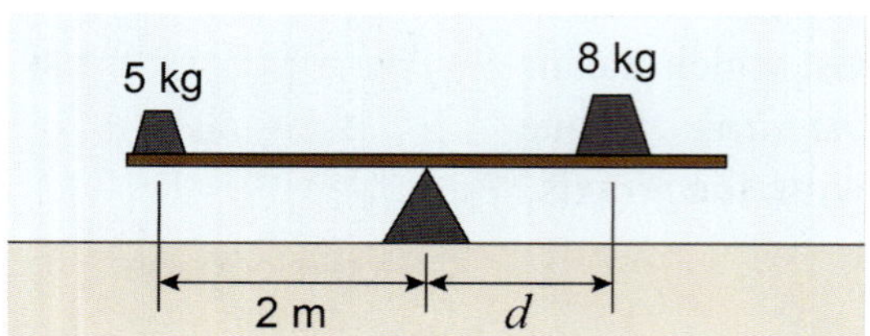


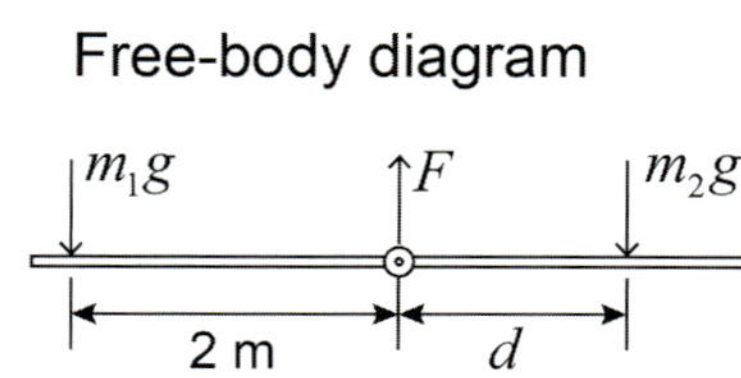


**Problem-solving steps**

1. Draw the free-body diagram and choose a center of rotation.
2. Draw each force at approximately the right distance. Label the known and unknown distances.
3. Calculate the torque created by each force, being careful to assign positive and negative directions with respect to the center.
4. Solve the problem by setting the net torque to zero.

**Calculate the torque from each force**

$\tau_1 = (+2\text{ m})m_1 g$ — Counterclockwise torque (positive)

$\tau_2 = -dm_2 g$ — Clockwise torques (negative)

**Rotational equilibrium** $\sum \tau = 0 \rightarrow \sum \tau = (+2\text{ m})m_1 g - dm_2 g = 0$

$$d = \frac{(+2\text{ m})m_1 \cancel{g}}{m_2 \cancel{g}} = \frac{(+2\text{ m})(5\text{ kg})}{(8\text{ kg})}$$

**Solution** $d = 1.25$ m

**What are the actual torques?**

The 5 kg mass placed 2 m from the fulcrum creates a positive torque of +98 N m. But 8 kg at 1.25 m creates a negative torque of −98 N m. The heavier mass sits closer to the fulcrum than the lighter mass to create equal and opposite torques that cancel out in equilibrium. This is consistent with what you know about see-saws.

# Investigation 8A Static equilibrium

**Essential questions** *How do we predict when something will move or not?*
*How do we determine forces we cannot measure?*

Interactive simulation

Objects remain at rest *only* when the net force and the net torque are zero. The converse is also true: if an object remains at rest, then *you know the net force and the net torque must be zero.* The first statement predicts whether an object will remain at rest or begin to move, while the second statement is used to determine unknown forces.

## Part 1: A simply supported beam

1. This interactive model allows you to place up to four masses on a beam that is supported at its ends. You can adjust the placement of the masses by entering distances for each one.
2. [Reset] clears all the masses and distances.
3. The [Force] or [Torque] button toggles between displaying the force or torque diagrams below the bar.

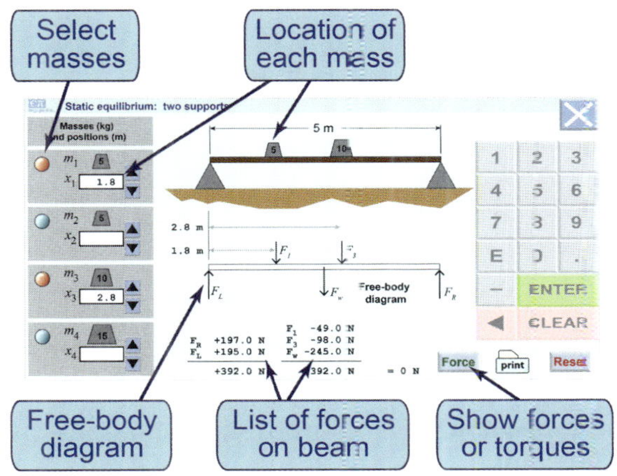


a. *What is the relationship between the upward and downward forces on the free-body diagram?*

b. *What is the relationship between the clockwise and counterclockwise torques when measuring torques?*

c. *Use the masses to create an equilibrium in which a force scale under the left support will measure a force of close to 300 N.*

## Part 2: Find the unknown mass using static equilibrium

1. Set up two Tripod Stands with a cross rail and two Universal Hooks
2. Make a string a little longer than 1 meter with three loops and a weight as shown. The string on the right of the weight should be horizontal and the string on the left side should make an angle of 30° - 50°.
3. Record the tension in the string for several different angles (adjust the hook positions)

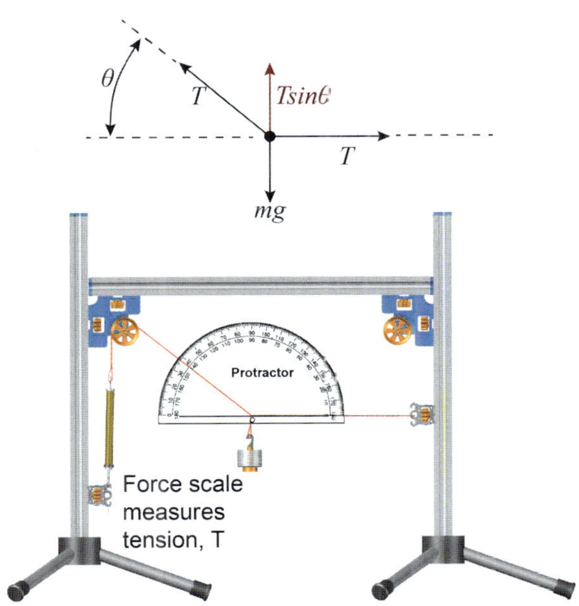


a. *Draw a free-body diagram for the knot above the hanging mass.*

b. *Assume one side of the string is horizontal therefore exerts only an x-component of force on point A. What is the equilibrium condition for forces in the vertical direction?*

c. *Derive an equation for the tension in the string in terms of the angle, θ, and the weight, mg, of the hanging mass.*

d. *Use the equation you derived in part (c) to predict the tension. How does this agree with your spring scale measurements?*

## The general case for static equilibrium

**Two conditions of static equilibrium**

The general condition for static equilibrium has two parts, both of which must be satisfied:

1. The sum of forces must be zero in all directions ($\Sigma F = 0$).
2. The sum of torques must be zero around any choice of center ($\Sigma \tau = 0$).

The force equilibrium equation (8.1) can result in up to three equations, one for each coordinate direction. The torque equilibrium equation (8.3) adds at least one additional equation for each different choice of the center of rotation.

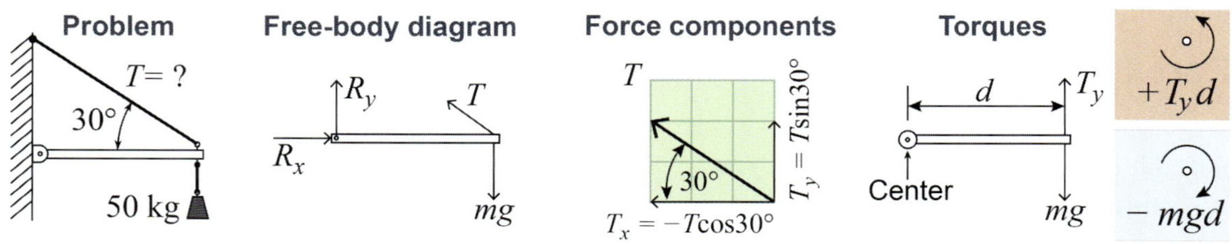


**Force equilibrium equations**

$$\Sigma F_x = 0 \quad \rightarrow \quad -T\cos 30° + R_x = 0$$

$$\Sigma F_y = 0 \quad \rightarrow \quad T\sin 30° + R_y - mg = 0$$

**The torque equation is the key!**

We can now solve the problem of the mass hanging from the beam. The two force equilibrium equations in $x$ and $y$ do not provide a solution. If we choose the center, however, to be the point where the beam attaches to the wall, then the two reaction forces ($R_x$, $R_y$) contribute zero torque since their line of action passes through the center. The same is true of the $x$-component of the tension $T_x$.

**Rotational equilibrium equation**

$$\Sigma \tau = 0 \quad \rightarrow \quad T_y d - mgd = 0 \quad \rightarrow \quad T = \frac{mg}{\sin 30°} = \frac{(50\text{ kg})(9.8\text{ N/m})}{(0.5)} = 980\text{ N}$$

**Finding the solution**

The two vertical forces create opposite torques. To be in rotational equilibrium, these torques must sum to zero. The resulting equation has only a single unknown value and can be solved directly for the tension $T$. Note that the problem is easiest to solve when the center is chosen to eliminate the torque from the unknown reaction forces $R_x$ and $R_y$.

**What other forces create torques?**

Any type of force may exert a torque. Torques can be created by friction and also by normal forces and reaction forces. Normal forces act perpendicular to surfaces of contact. Friction forces act parallel to surfaces of contact. The example of a board resting on a chair and the floor shows both kinds of forces.

**Free-body diagram**

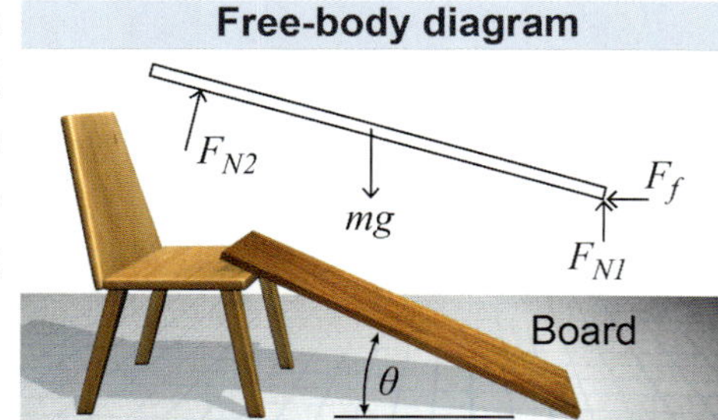

# Section 1 review

Static equilibrium means that both the net force is zero in all directions and the net torque is zero around any center. Torque is the "twisting" equivalent of force. Torque is force times distance, which has SI units of newton-meters and has units of pound-feet in the English system. The torque created by a force is the magnitude of the force multiplied by the lever arm, which is the perpendicular distance from the center to the line of action of the force. The general procedure for solving static problems is to create a free-body diagram, resolve all forces into components, and then apply the equilibrium conditions for both forces and torques.

| | | | |
|---|---|---|---|
| **Vocabulary terms** | **torque, line of action, lever arm, fulcrum, lever** | | |
| **Key equations** | $\sum \vec{F} = 0$ | $\sum \tau = 0$ | $\tau = r \times F$ |

## Review problems and questions

1. Which of the following gives the correct force equilibrium equations for the problem on the right?
   a. $T_x - F = 0,\ T_y - mg = 0$
   b. $T_y - F = 0,\ T_x - mg = 0$
   c. $T - F - mg = 0$
   d. $T + F + mg = 0$

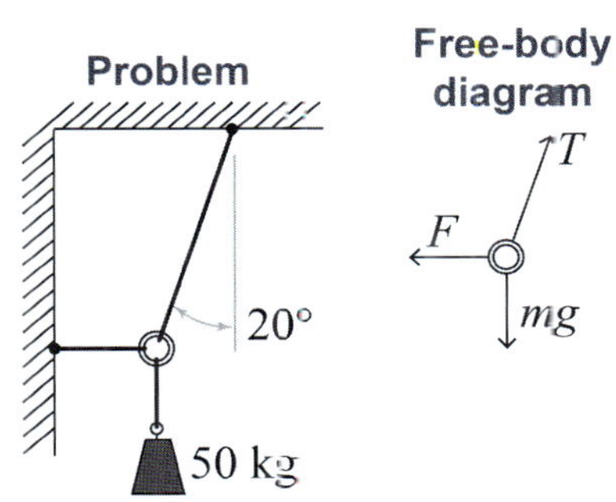


2. What is the torque created by the force drawn in the diagram?
   a. 25 N m
   b. −25 N m
   c. 100 N m
   d. −100 N m

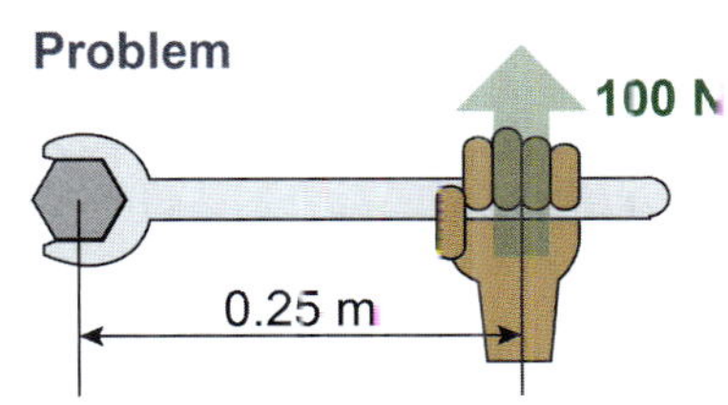


3. Which of the following equations is a correct statement of rotational equilibrium for the problem on the right?
   a. $m_1g + m_2g + T = 0$
   b. $m_1gd_1 + m_2gd_2 = 0$
   c. $m_1gd_1 - m_2gd_2 = 0$
   d. $m_1gd_1 + m_2gd_2 + T = 0$

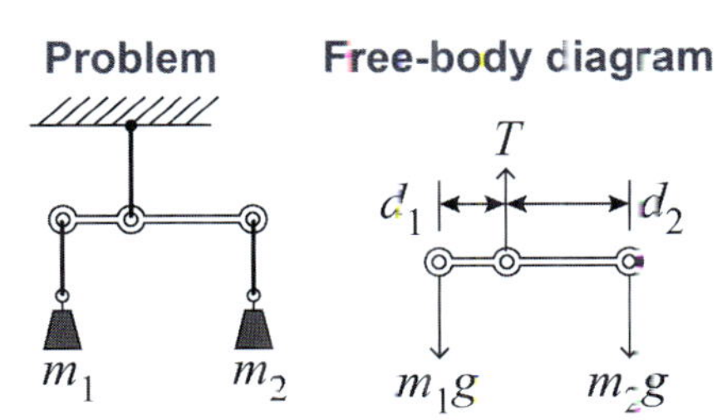

## 8.2 - Structures and design

Humans began building structures long before we could write down equations to solve physics problems. The ideas in this chapter were developed by people trying to understand why some buildings and bridges fell down while others did not. There is a crucial point here: We design something to succeed by *designing it not to fail.* You *must* think about how something might fail in order to design it to be strong enough to succeed. When designing buildings, bridges, and other structures, engineers must ensure that their designs are strong enough not only for the expected loads of normal operation but also for the highest possible loads.

### The design process

**How are structures designed?**

The process of designing and evaluating a structure follows a well-defined procedure.

1. Envision a *design concept* that satisfies all the constraints, such how much weight the structure needs to support.
2. Create a *schematic design* that provides details for the concept, such as length, height, angle, width, and materials used.
3. Calculate the forces—using physics! —acting on each part of the structure under the design loads.
4. Evaluate which design choice(s) can handle the calculated forces and meet other project requirements such as cost, weight, and environmental factors.
5. Modify or refine the design based on these results, then go back to steps 3 and 4 to recalculate forces and reevaluate design choices, respectively.
6. Keep repeating steps 3–5 until all parts of the structure are satisfactory.

Consider an example where a physics tutor wants to suspend a sign off the side of a building. After calculating the force exerted by the sign, she must evaluate different ropes, cables, and chains to determine which can support the load.

**Schematic design:**
Use a link and cable at 30° to the wall.

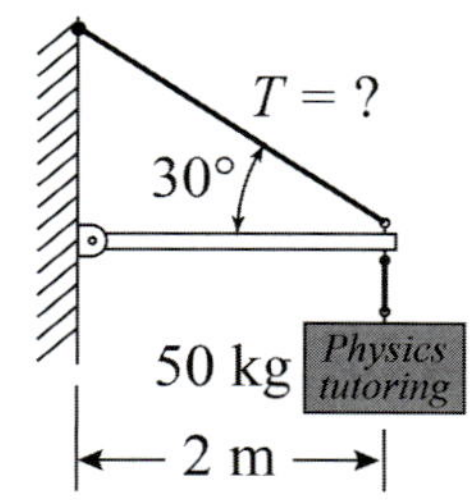


**Calculate forces:**

$$T = \frac{mg}{\sin 30°}$$

$$= \frac{(50 \text{ kg})(9.8 \text{ N/s})}{0.5}$$

$$= 980 \text{ N}$$

Assume a safety factor of 5. This means designing for a tension of 4,900 N (1,100 lb).

**Evaluate design choices:**

| Material | Working load (N) | |
|---|---|---|
| 3/16 in rope | 2,800 | |
| 1/4 in rope | 3,500 | |
| 3/8 in rope | 6,800 | |
| 1/16 in steel cable | 2,100 | |
| 1/8 in steel cable | 8,900 | |
| 5 mm chain | 3,700 | |
| 6 mm chain | 6,500 | |
| 8 mm chain | 9,800 | |

**Safety factors**

Engineers always apply a **safety factor** to all calculated forces when evaluating specifications of real materials. The safety factor helps account for the approximations of all force calculations, wear and tear, and variations in conditions, such as wind loads. When injury might result from failure, safety factors are often between 10 and 20. Note that four choices meet the required strength for the cable in the example: 3/8 in rope, 1/8 in steel cable, and the 6 mm and 8 mm chains.

**What makes up a structure?**

Some elements of structures are made with rigid materials, such as concrete, wood, glass, or metal. Rigid materials can withstand compression and shear forces. Other elements of structures are flexible, such as ropes, cables, or chains. Flexible elements can support only tension forces, but they offer the advantages of being light and allowing parts of a structure to move.

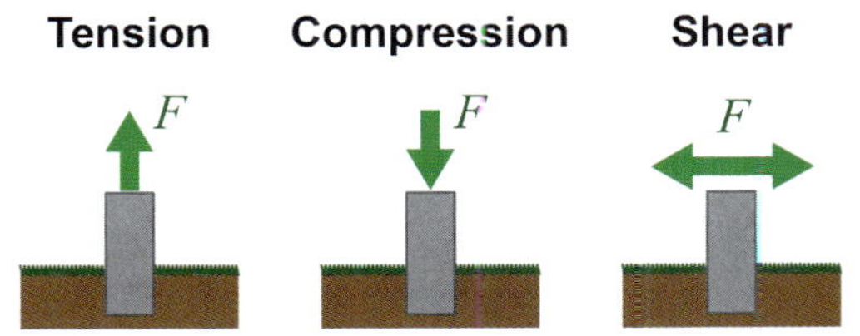


Rigid bodies can transmit all three forces.

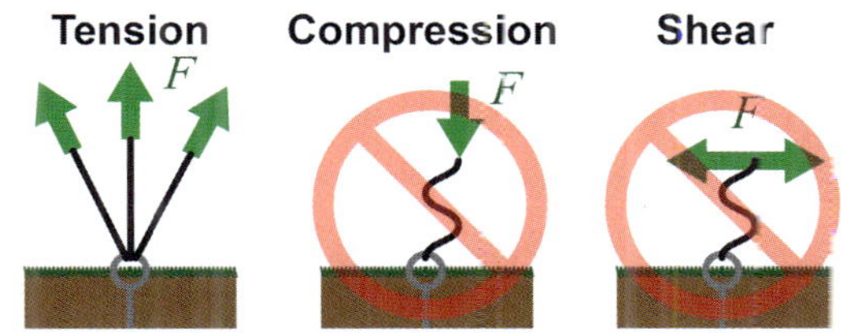


Rope-like elements transmit only tension.

**How do tension forces behave?**

The force from a cable, rope, or chain lies along the direction of its length and is a *tension* force. There are no reaction forces at the attachment point other than the tension force. The tension force transmitted by a "perfectly frictionless" rope is the same everywhere along the rope. For example, if you were to cut a rope and hold the cut ends you would feel the same tension force on either hand. *Ropes cannot transmit compression or sideways forces!* Wherever a rope (or cable or chain) is connected, you may assume the tension force on the connection points away from the connection, along the rope. Ropes only have tension if a force is applied at *both* ends, such as during a game of "tug-of-war."

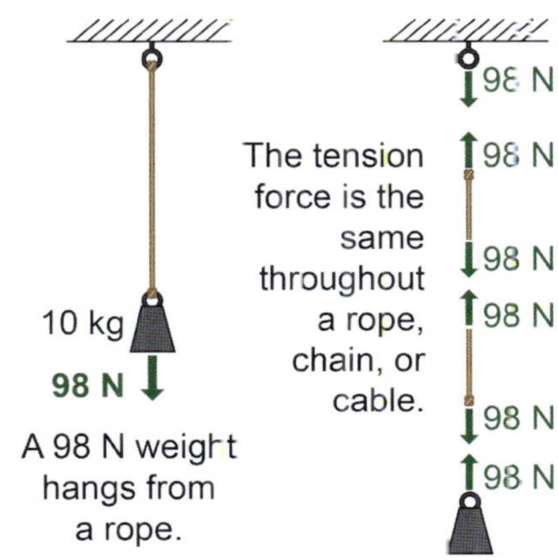


**Posts and beams**

Another common structural element is a *beam.* A beam can transmit both compression and tension forces. A vertical beam is often called a *post.* Posts and beams can be connected in different ways and the forces they transmit depend on how they are connected. "Post-and-beam" timber framing is a style of house construction that has been used for many centuries.

A simple structure

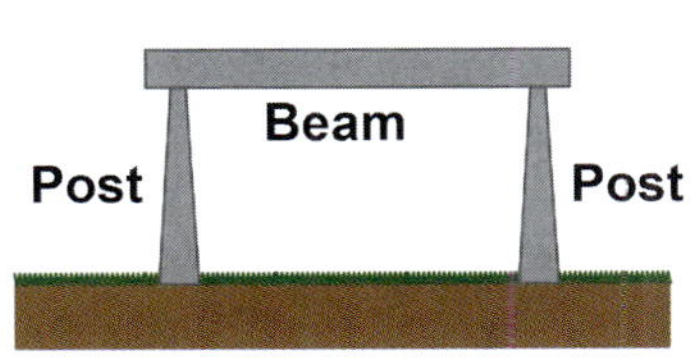


A truss

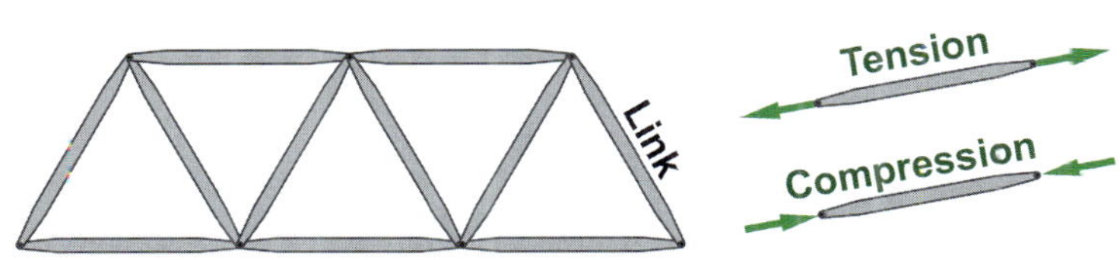


**Links and trusses**

A *link* is a section of a beam or post that has pinned connections at both ends. Very strong, lightweight structures called *trusses* are made from links. A link can transmit either a tension or a compression force along its length. Where two links either join or meet a wall, each exerts a force on the other in the direction of the connecting link.

**Reaction forces**

Forces from supports or other parts of the structure are often called *reactions*. Identifying reaction forces is one of the most important steps to solving an equilibrium problem. Start by isolating one or more elements of the structure, alone, in a free-body diagram. Every interaction of each isolated element with other elements or supports is replaced by forces and/or torques. Reactions occur wherever the isolated element touches any other element or any support.

**Frictionless and roller contacts**

A roller or frictionless surface exerts only a normal force at the point of contact. There is only one unknown force and it has a known line of action.

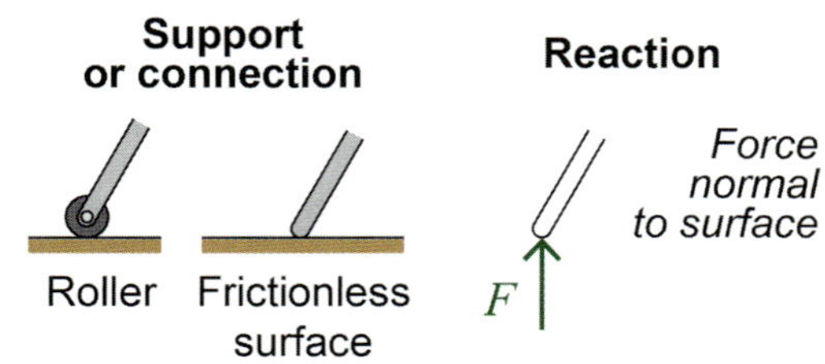


**Ropes and links**

A rope or link exerts a force with a known line of action and therefore also has only one unknown force.

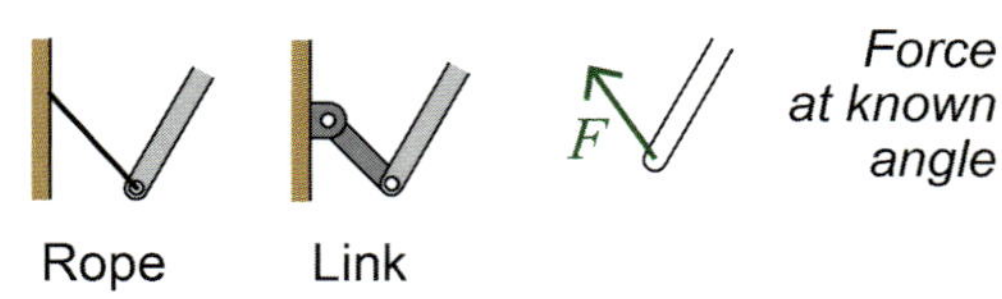


**Rough surfaces**

A rough (frictional) surface or sliding connection includes the force of friction parallel to the surface and a normal force. There is only one unknown force if the coefficient of static friction $\mu_s$ is given.

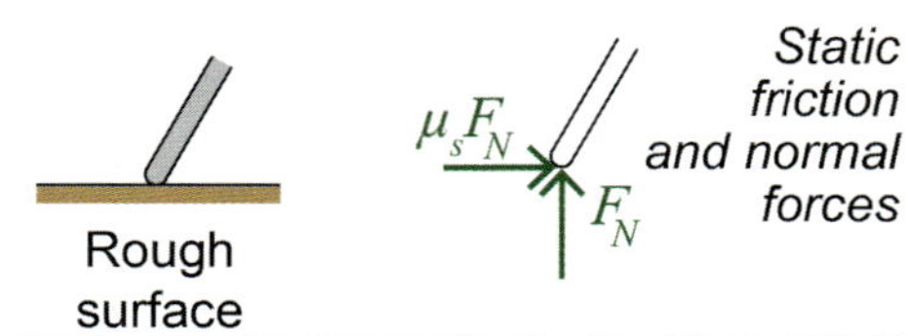


**Pin supports**

A frictionless *pin* connection exerts reaction forces in two directions ($x$ and $y$), so there are two unknown forces. The pin indicates that the connection is free to rotate; as a result, there is no *reaction torque* generated by the support. *Many problems are approximately solved by assuming a pinned connection.*

**Fixed supports**

The most complex connection is a fixed support that can transmit both forces and torques. As with forces, Newton's first law also applies to torque. Any torque created by the structure on the support creates an equal and opposite reaction torque that acts on the structure. There are two unknown forces and one unknown torque.

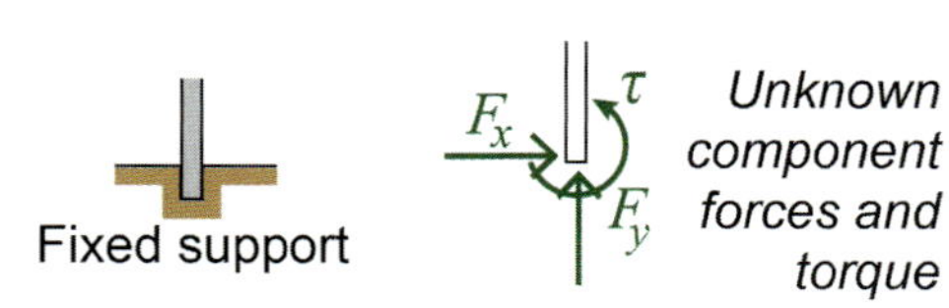


**Using the diagrams**

These diagrams show what kinds of support reactions are created for all of the problems you will encounter in this course—and some for future study too! The fixed support is the most difficult because there are three unknowns: two forces and a torque. Static problems with fixed supports are addressed in more advanced courses.

**Solving structure problems**

Most structure problems involve at least two equations. At least one equation is force equilibrium, and the other equation is often rotational equilibrium.

As an example, consider a 60 kg person who is standing off-center on a 40 kg scaffold suspended by chains. What is the tension force in each chain?

**Problem**

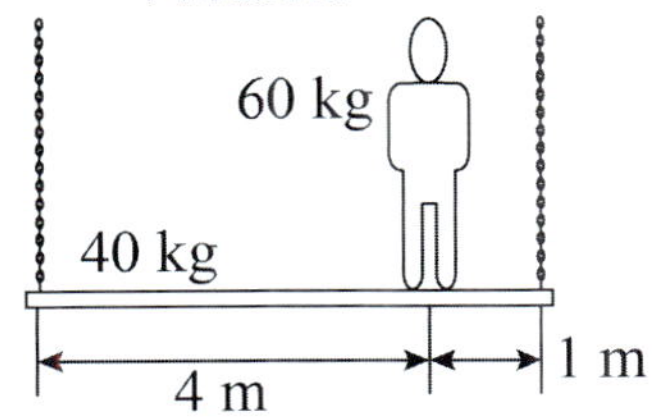


**Free-body diagram**

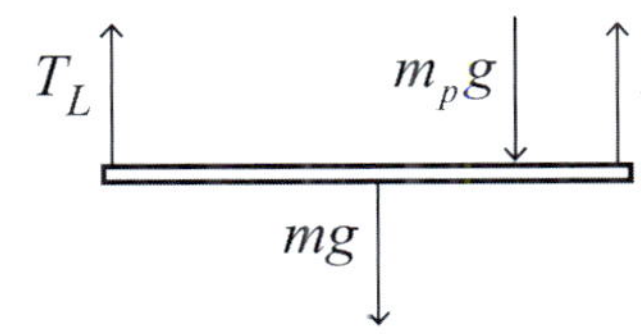


**Free-body diagram**

The free-body diagram includes two reaction forces from chains, which are treated the same as ropes. The tension forces are not equal because the person is standing off-center. The weight of the scaffold $mg$ is assumed to act at the center, which is 2.5 m from either end. Let the weight of the person be $m_p g$.

**The force equation**

The equation for force equilibrium tells us that the sum of forces in the vertical direction must be zero. (This is true in the horizontal direction, too, but no forces act in either horizontal direction in this problem, so we need not analyze the $x$-axis.)

**Sum of forces = 0**

$$\Sigma F_y = 0 \longrightarrow T_L - mg - m_p g + T_R = 0$$

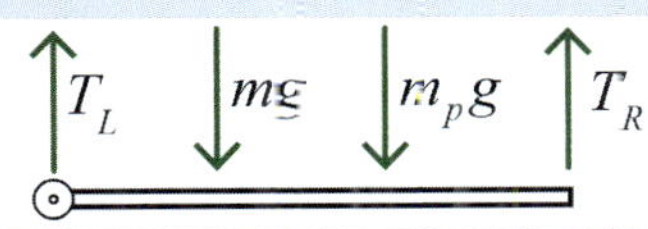


**Sum of torques = 0**

$$\Sigma \tau = 0 \longrightarrow (5\text{ m})T_R - (2.5\text{ m})mg - (4\text{ m})m_p g = 0$$

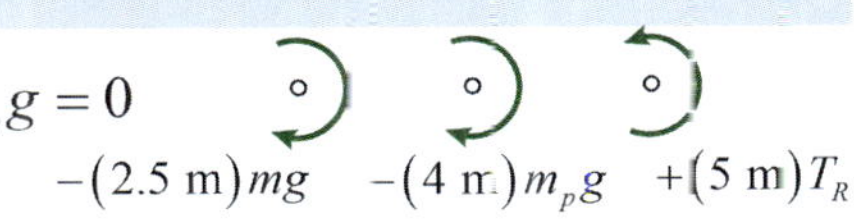


**The torque equation**

The best choice for a center about which to calculate torques is one of the chain attachments. It does not matter which one, so we chose the left support. The sum of torques around this center must be zero when in rotational equilibrium.

**Solution**

The torque equation has only one unknown and can therefore be solved for $T_R$. *Once $T_R$ is known, the force equation is used to solve for $T_L$.*

**From the torque equation**

$$T_R = \frac{1}{5}(2.5mg + 4m_p g) = \mathbf{670\ N}$$

**From the force equation**

$$T_L = mg + m_p g - T_R = \mathbf{310\ N}$$

**Problem-solving strategy**

Solving static equilibrium problems starts with determining forces from reactions, weights, and loads. The next step is to apply the equations for equilibrium of forces and equilibrium of torques. You need one equation for each unknown variable. Often, as in this example, one of the equations can be solved directly for one of the variables. The other equations can then be used to solve for the remaining variables.

# Investigation 8B

## Structural design

**Essential questions** *How does an engineer design something to be "strong enough"?*

Interactive simulation

We design buildings and structures to stand up by making sure they don't fall down. That may seem obvious, but it really isn't! You *cannot* design something to stand up under every possible circumstance. The best you can do is to define the possible ways something might fall down and make it strong enough not to fall down under each of your possible scenarios. In practical terms, you design for success by preventing failure!

### Design a suspension bridge that will not fail

The interactive model simulates a suspension bridge with steel support cables and concrete towers. There are many ways that a bridge can fail, but this simulation looks only at two: tensile failure of the cables and compression failure of the towers. The simulation defaults to a bridge with a length similar to that of the Bronx-Whitestone Bridge in New York City but with design parameters that fail.

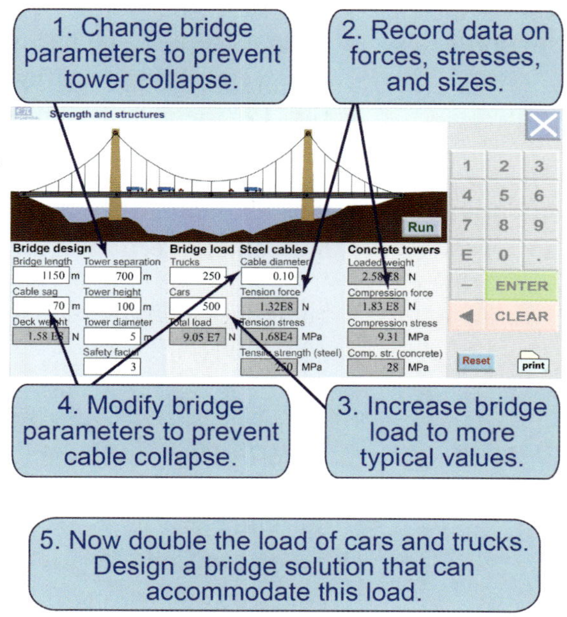


What parameter(s) are a good predictor of failure of the cables? Of the towers?

1. What parameter(s) must you change to prevent the bridge towers from collapsing?
2. Record the data on loads, stresses, and sizes.
3. Now increase the load on the bridge to a busy day during rush hour, such as 250 trucks and 500 cars.
4. How must you change the bridge design to prevent the bridge cables from collapsing? Record the data on loads, stresses, and sizes.
5. Double the load of cars and trucks. Come up with a solution so that the bridge does not fail.

a. *What parameter is a good predictor of failure of the towers? Explain how you know and propose a reason or model for why the towers behave this way.*

b. *What parameter is a good predictor of failure of the cables? Explain how you know and propose a reason for why the cables behave this way.*

c. *What parameter is a good predictor of failure of the towers? Explain how you know and propose a reason or model for why the towers behave this way.*

d. *Devise and carry out a virtual experiment to determine whether the height of the towers has an effect on failure. What is your conclusion? Is there an effect? Propose an explanation using the principles of static equilibrium and describe the component forces that are present to justify your conclusion.*

e. *List at least three additional ways that a suspension bridge might fail.*

f. *Propose at least one design attribute for each of the three failure modes you identified in the last step that would prevent that failure mode from occurring. Explain in one or two sentences why your proposed design feature addresses the failure mode you are trying to prevent.*

**When does a structural element fail?**

The most obvious failure occurs when a part breaks, but failure also occurs when a part deforms past the point where the material "springs back" after the load is removed. Consider a load hanging from a steel cable (as in the graph at right). As the load is increased, the cable stretches like a very stiff spring. Up to a point the steel is *elastic*, meaning that it springs back when the load is removed. When the load reaches its *yield* limit, the steel deforms permanently. If the load continues to increase, eventually the steel will break.

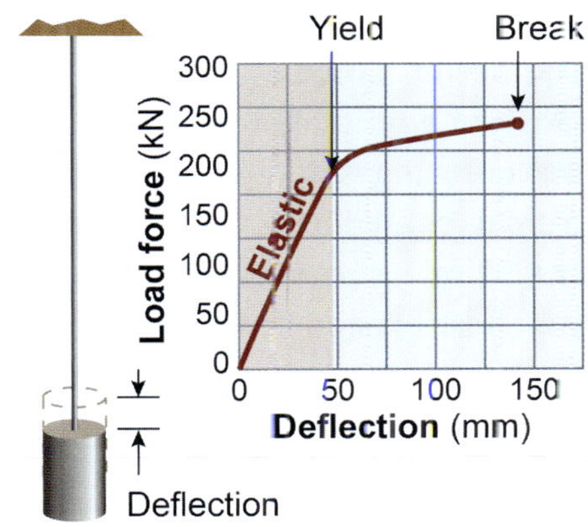


**Stress and tensile strength**

Materials fail when the *stress* reaches a certain level. **Stress** is force divided by cross-sectional area, which has units of pressure: N/m$^2$ (pascals or Pa) in SI and lb/in$^2$ (psi) in English units. The **tensile strength** is the value of stress at which a material typically fails in tension. The tensile strength of plain steel is 250 million N/m$^2$ (MPa) or 25,000 N/cm$^2$. A cable with 1 cm$^2$ area can withstand 25,000 N force before breaking, whereas a thicker 4 cm$^2$ cable can hold up to 100,000 N of force.

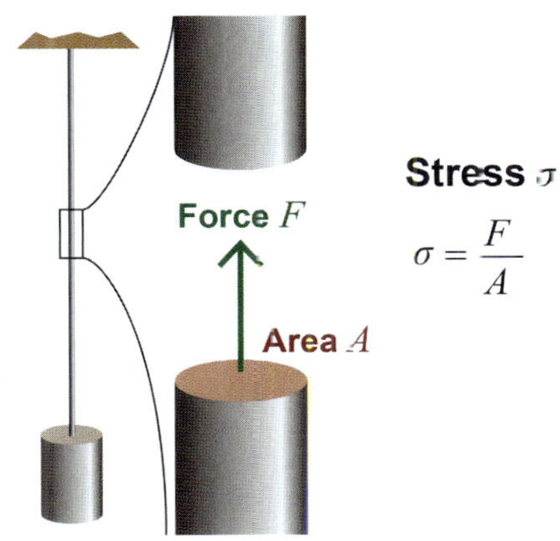


**Bending failure**

Material may also fail in *bending*. Consider what happens as a beam bends. Material at the center is not deformed at all. Material at the bottom of the beam is *compressed*. Material at the top of the beam is extended and therefore in *tension*. The beam fails when the tension stress exceeds the material's tensile strength at the most highly stressed area, which is the part of the beam farthest away from the center.

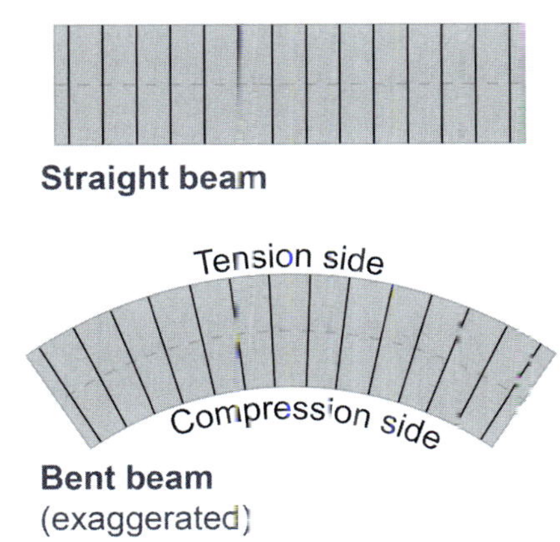


**Why are I-beams shaped that way?**

I-beams are often used as structural beams in buildings and bridges. The I-beam makes the most efficient use of material by providing the maximum strength for the minimum weight. In an I-beam, most of the material is placed where the stress is highest—at the top and bottom surfaces. For example, a 100 kg square steel bar 5 m long fails at a load of 440 kg. The same 100 kg of steel formed into an I-beam can handle a load of 2,340 kg!

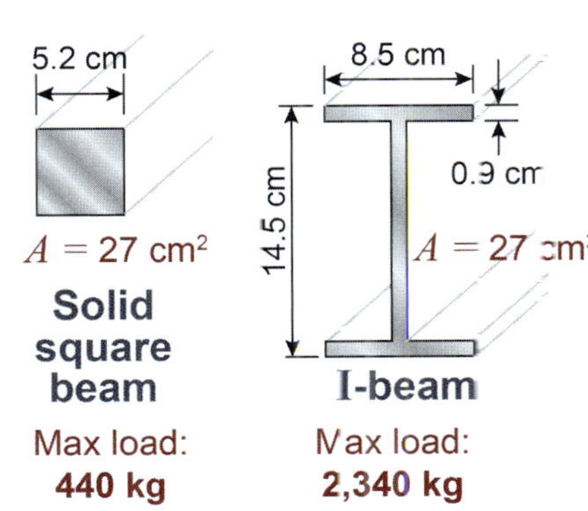

# Section 2 review

The design process consists of a cycle of creation, analysis, and revision, in which a human need is to be satisfied with real-world materials. The design process makes use of physics while also accounting for cost and ensuring safety in unusual circumstances. A bridge or building will include a safety factor so that its critical components can withstand 10–20 times the forces expected in regular use; this provides some protection against rare but dangerous events. Extremely flexible structural elements such as ropes and chains only bear tension (pulling forces). Stiffer materials (such as beams) can withstand stress and shear (caused by pushing) as well as tension.

| | | |
|---|---|---|
| **Vocabulary words** | **safety factor, stress, tensile strength** | |
| **Key equations** | $\sum \vec{F} = 0$ | $\sum \tau = 0$ |

## Review problems and questions

1. A 35 kg lighting fixture hangs from a chain in a restaurant dining room. The builder chose a 4-mm-gauge chain with a breaking strength of 5,000 N, for this application. What safety factor does this imply?

2. A steel cable is used to lift a 24,000 kg container from a cargo ship. The tensile strength of steel is $2.50 \times 10^8$ N/m$^2$.
   a. What is the minimum cross-sectional area of the cable that can support this load? (Do not include a safety factor.)
   b. What is the diameter of this cable?

3. A 60 kg acrobat stands upon a 3-m-long, 20 kg beam that is hanging from two chains. He stands a distance $d_p$ from the right-hand chain. When computing torques, consider the right-hand end of the beam to be the center of rotation.

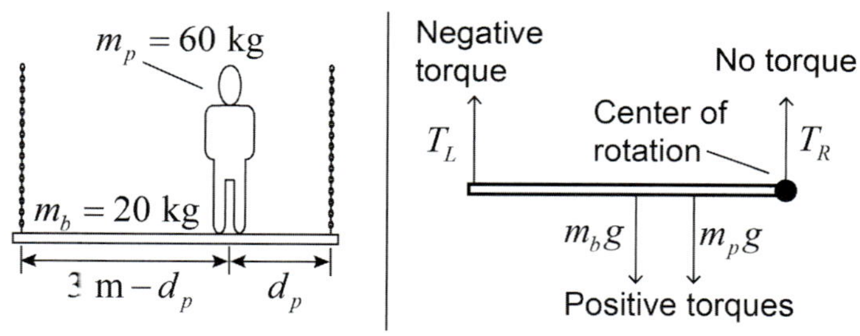


   a. Calculate the tension in each of the two chains when $d_p = 1.5$ m (that is, when the acrobat is standing on the center of the beam).
   b. The acrobat moves a half-meter toward the right-hand chain (that is, to $d_p = 1.0$ m). Calculate the tension in each of the two chains.
   c. How did $T_L$ and $T_R$ each change?
   d. *Predict* the values of $T_L$ and $T_R$ if the acrobat were to slide another 0.5 m to the right.
   e. Test your predictions by performing the necessary calculations.
   f. Suppose that a safety factor of 20 were required, *no matter where the acrobat stood upon the beam.* What minimum breaking strength (in newtons) would be required for each chain?

# Chapter review

## Vocabulary

Match each word to the sentence where it best fits.

### Section 8.1

| fulcrum | lever |
|---|---|
| line of action | torque |

1. A _______ is a simple machine that uses torque to multiply linear forces.
2. The point about which a lever rotates is the _______.
3. _______ is measured in units of newton meters.
4. The _______ goes through an object and points where a force goes.

### Section 8.2

| safety factor | stress |
|---|---|
| tensile strength | |

5. If the applied force per unit area exceeds a material's _______, then it may break.
6. An engineer was tasked to design a footbridge to support a load of 5000 N, so he designed it to withstand a force of 50,000 N instead to account for a _______ of 10.
7. The amount of applied force per unit area is the _______ on the material.

## Conceptual questions

### Section 8.1

8. Ask two friends if they can find the difference in the following two statements.

   Statement 1: "A torque results from a force applied in a way that could cause an object to rotate."
   Statement 2: "A torque is the result of an object's rotation under the action of a force."

   Which one is more correct? Write down their answers. How do they compare with your own answer?

9. Which sentence comes closest to the meaning of the following equation?

$$\sum \vec{F} = 0$$

   a. The product of all forces equals zero.
   b. The sum of all forces equals zero independently in every direction.

10. Julietta claims that an object in rotational equilibrium cannot spin. Javier disagrees and says that an object in rotational equilibrium *can* spin. Who is right, and why?

11. Describe a situation in which a force is applied that results in zero torque.

12. Michael wishes to loosen a bolt and is holding a wrench that is attached to the bolt. Which of the following changes would enable him to exert a greater torque? (More than one choice may be correct.)

    a. using a wrench with a longer handle
    b. applying a lubricant to the bolt
    c. pushing or pulling harder on the wrench

13. You are designing a see-saw (teeter-totter) for a playground. Which kind of equilibrium should the see-saw *always* maintain: force equilibrium, rotational equilibrium, both, or neither?

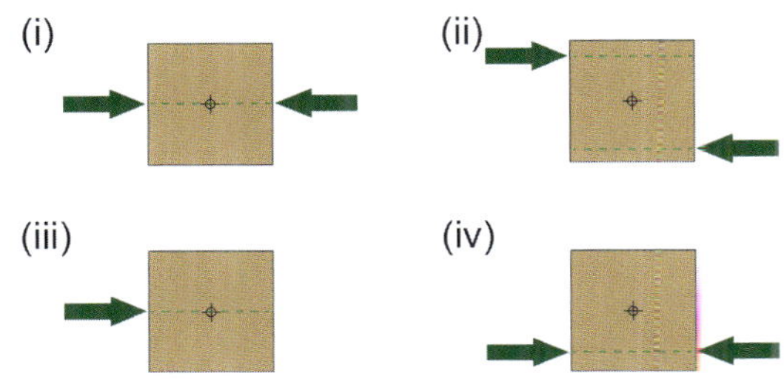


14. Each of the forces shown here (as arrows) acts upon the block with a strength of 100 N. The center of rotation and line(s) of action are shown in each scenario.

    a. Which of the four scenarios (i, ii, iii, or iv) shows the block in *force* equilibrium? (More than one may be correct.)
    b. Which of the four scenarios (i, ii, iii, or iv) shows the block in *rotational* equilibrium? (More than one may be correct.)

### Section 8.2

15. Which of the two following statements do you agree with? Write down your answer and explain your reasoning.

    Statement 1: "Engineering is a kind of science, and that means there is always a single right answer to a problem."
    Statement 2: "Engineering is different from science. In science, a problem has just one right answer, but in engineering there might be several good solutions."

    Then ask two friends which statement each of them agrees with and why. Record their answers and compare them with your own.

16. The number known as a *safety factor* is never assigned a value less than one. Why?

### Section 8.2

17. Thick rope may be strong enough to keep a sailboat from breaking free of a dock, even in rough weather. But rope is not used for the handle of a hammer or wrench. Using the language of statics and engineering, explain why.

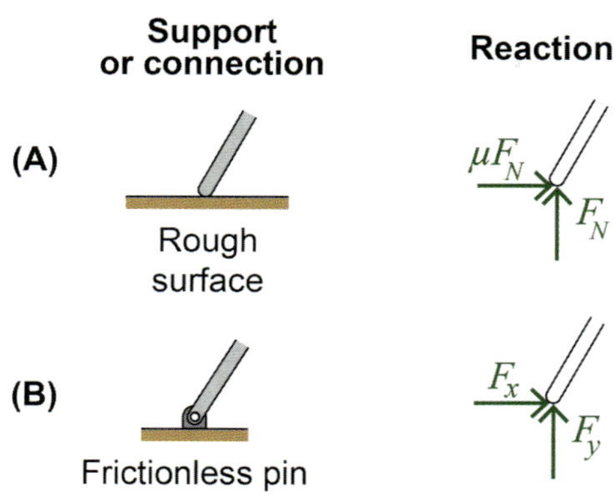


18. Two approaches (A and B) for supporting a tilted beam are shown here. Each offers an upward support force and a force pushing to the right. Suggest one advantage and one disadvantage for each of these two approaches.

## Quantitative problems

### Section 8.1

19. A door of width 90 cm is held open by a door stop 55 cm from the hinge. The door stop will exert up to 125 N of force to resist the door from closing. Rather than removing the doorstop, Geoffrey pushes on the end of the door to close it. How much force must he exert on the door to move it?

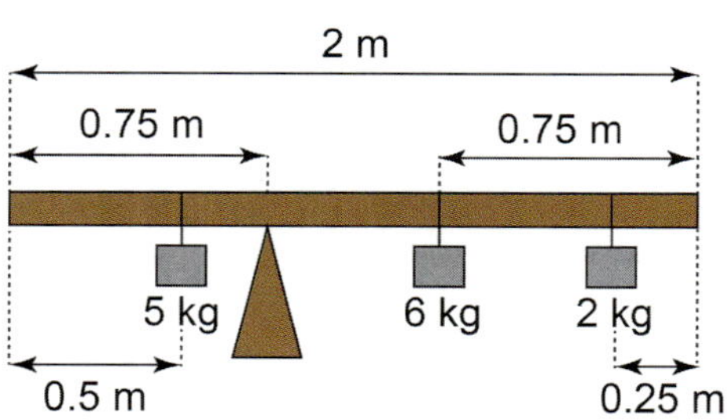


20. You have a balance beam of length 2 m with negligible mass. The fulcrum is 0.75 m from the left end. There is a 5 kg weight hanging 0.5 m from the left end, a 2 kg weight hanging 0.25 m from the right end, and a 6 kg weight hanging 0.75 m from the right end. You have a spare 5 kg weight. Where would you hang it to balance the beam?

21. You have a very long balance beam. You have placed a 50 kg box 2.1 m to the left of the fulcrum, and a 30 kg box 3.5 m to the right of the fulcrum. Does the balance beam move? Explain using one of Newton's laws.

22. If a 5 kg weight is 3 m from the fulcrum of a lever, how far away from the fulcrum on the other side must you place a 2 kg weight to balance the lever?

### Section 8.2

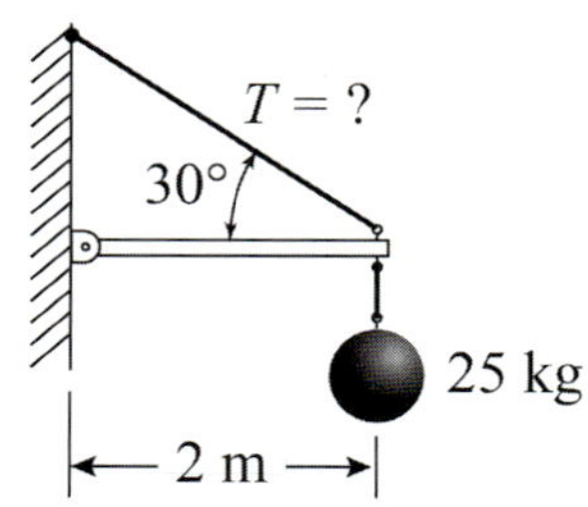


23. Calculate the tension in the cable that makes a 30° angle as shown in the diagram above.

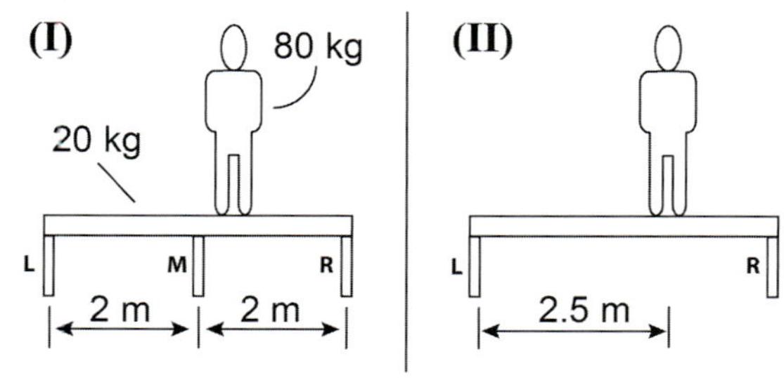


24. In Case I, an 80 kg acrobat stands on a 20 kg, 4-m-long balance beam, which is supported by three equally spaced posts (marked L, M, and R). She stands 2½ m from the left end of the beam. At a later time, in Case II, she returns to the same position, but the middle post is missing.
    a. In which case does the L post provide a greater support force, Case I or Case II, or are they equal in both?
    b. In which case does the R post provide a greater support force, Case I or Case II, or are they equal in both?
    c. Which case has the largest *sum* of all of the support forces, Case I or Case II, or are the sum equal?
    d. In Case I, which post exerts the largest support force, L, M, or R?
    e. In Case II, which post exerts the largest support force, L or R?
    f. Now suppose that only two posts hold the beam up at its ends (as in Case II), but the acrobat can stand *anywhere* along the beam's length. If a safety factor of 10 is called for, how much compression (in newtons) must either post be able to bear?

# Chapter review

### Standardized test practice

25. You see balance beam with a length of 1 m. The fulcrum is 0.25 m from one end and there is a 10 kg bag of flour 0.75 m from the same end. How much torque does the bag create?

    A. 24.5 N m
    B. 49 N m
    C. 73.5 N m
    D. 196 N m

26. A 5 kg box sits on one end of a balance beam of length 1 m. The fulcrum is in the middle. How far from the fulcrum would you have to place a second 10 kg box to balance the system?

    A. 0.25 m
    B. 0.5 m
    C. 0.75 m
    D. 1 m

27. A 35 kg lighting fixture hangs from a church ceiling by a metal cable rated for a tension of 1,000 lb. What is the cable's safety factor? (1 lb is 4.45 N.)

    A. 4
    B. 7.9
    C. 13
    D. 28

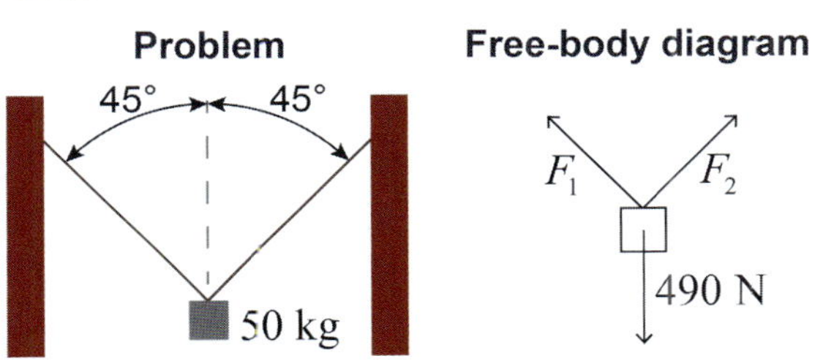


28. A 50 kg mass is suspended from two identical cables, each at an angle of 45° from the vertical. Which statement correctly describes the *magnitudes* of the forces $F_1$ and $F_2$ exerted by the respective cables upon the hanging mass?

    A. $F_1$ and $F_2$ are equally strong, and each has a magnitude that is less than 245 N.
    B. $F_1$ and $F_2$ are equally strong, and each has a magnitude that equals 245 N.
    C. $F_1$ and $F_2$ are equally strong, and each has a magnitude that is greater than 245 N.
    D. You cannot choose between A, B, and C without more information.

29. Which of the following expressions most nearly equals one pound-foot of torque in SI units? (Use the following conversion factors: 1 ft = 0.30 m and 1 lb = 4.45 N.)

    A. 0.07 N m
    B. 1.3 N m
    C. 4.75 N m
    D. 14.8 N m

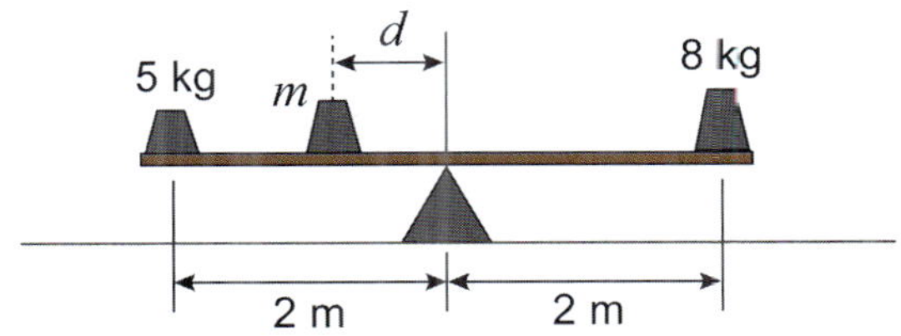


30. A lightweight beam, 4 m long, is supported by a fulcrum at the midpoint. A 5 kg mass rests on the beam's left end, and a 8 kg mass rests on the right end. A third object, of mass $m$, is to be placed a distance $d$ to the left of the fulcrum in order to level the beam. Which mass-and-distance pair will do this?

    A. $m$ = 12 kg and $d$ = 0.5 m
    B. $m$ = 10 kg and $d$ = 1.0 m
    C. $m$ = 12 kg and $d$ = 1.0 m
    D. $m$ = 3 kg and $d$ = 1.0 m

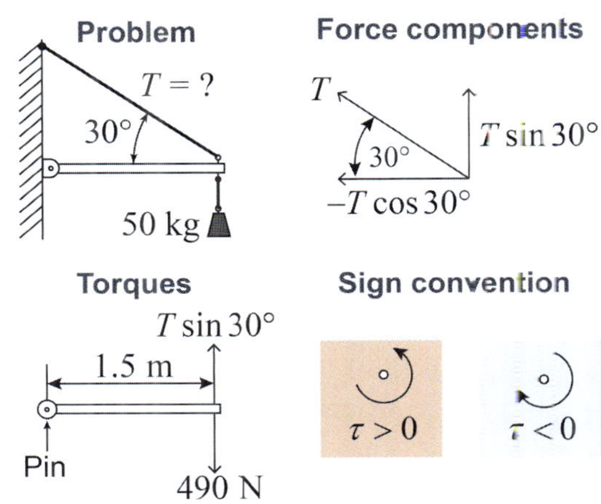


31. A lightweight beam, 1.5 m long, is attached to a wall with a frictionless pin at one end and a cable at the other. The cable length allows the beam to hang horizontally, as shown. A 50 kg mass is suspended from the end of the beam that is attached to the cable. Which of the following equations best describes rotational equilibrium about the frictionless pin? (Assume that the beam's mass is negligible.)

    A. $T \sin 30° + 490 \text{ N} = 0$
    B. $T \sin 30° - 490 \text{ N} = 0$
    C. $T \cos 30° + 490 \text{ N} = 0$
    D. $T \cos 30° - 490 \text{ N} = 0$

32. Which of the following statements correctly describes the forces and torques on objects in statics problems?

    A. The net force upon an object of interest is zero, but the net torque may be nonzero.
    B. The net torque is zero, but the net force may be nonzero.
    C. The net force and the net torque upon an object of interest must both be zero.
    D. Neither the net force nor the net torque can be zero.

# Chapter 9 Work and Energy

Energy is a word you hear every day, whether in the news or on people's lips. Nations fight over oil for its energy. We expend enormous resources trying to draw natural gas from miles beneath the Earth's surface for its energy. We accept some risk in order to benefit from the energy released in nuclear reactions. This chapter is about the fundamental nature of energy: what it is and how it is measured and used. Though it cannot be tasted, touched, or weighed in its pure form, energy is the fundamental essence of the universe and the natural currency of change. Anything that changes only does so through the exchange and transformation of energy —whether it be a change in speed, color, temperature, or any other physical and observable quantity.

The average American today uses a thousand times as much energy as their parent's grandparents did. The energy required to run your air conditioner might be the equivalent energy output of ten people working full-tilt. This energy came to the appliance through electrical wires; the electricity coming into your house may have originated in a power station a hundred miles away. The engine of a moderate-sized new car generates 150 horsepower—roughly the equivalent energy output of 150 horses working simultaneously! Modern energy use is clean, quiet, and simple. Compare turning the ignition key and pushing the gas pedal to having to connect, feed, and clean up after 150 horses every time you need to travel.

Ours is the age of energy. Computers, electric lights, and cars are all energy-intensive technologies. The rising human standard of living goes hand in hand with a steady increase in available energy per person. Where does it end? The total energy in the universe is the same today as it has always been; energy cannot be created or destroyed, only transformed from one form into another.

Why then are we worried about energy running out? Modern technology relies heavily on chemical energy in the form of gas, oil, and coal for transportation, heat, and the electricity that is the lifeblood of the modern world. In recent years we have recognized that energy conversion from fossil fuels has undesirable side effects—the release of enormous quantities of carbon dioxide that impact the Earth's climate. Today's emerging energy technologies may transform tomorrow's cars to use diesel fuel from bioengineered algae, electricity from solar cells, or hydrogen fuel from water. Energy isn't scarce, but *specific* forms of energy are limited and some have troubling ecological impacts.

Chapter 9

# Chapter study guide

## Chapter summary

Everything in the universe changes through the movement and transformation of *energy*. Work is a form of energy. Kinetic energy is the energy of motion and potential energy is the stored energy of position. Power is the *rate* at which work is done or energy is transformed. Light, heat, and electricity are *macroscopic* forms of energy that we can experience directly. Renewable energies are those forms of energy that nature replenishes at least as fast as we can use them, such as solar, hydroelectric, and wind power.

## Chapter index



## Learning objectives

By the end of this chapter you should be able to

- describe different types of energy;
- calculate work given force and distance for one-dimensional movements;
- calculate potential and kinetic energy in joules;
- solve one-step problems involving power, work, energy, and time;
- describe electrical energy in terms of amperes and volts; and
- understand the meaning of *intensity* in the context of light.

## Investigations

9A: Work and the force versus distance graph

## Important relationships

$$W = Fd$$

$$E_k = \frac{1}{2}mv^2$$

$$E_p = mgh$$

$$E_p = \frac{1}{2}kx^2$$

$$P = \frac{\Delta E}{\Delta t} = \frac{W}{\Delta t}$$

$$P = IV$$

## Vocabulary

**mechanical energy**
**kinetic energy**
**reference frame**
**power**
**volt (V)**
**intensity**
**renewable energy**
**work**
**potential energy**
**elastic potential energy**
**watt (W)**
**electric current**
**light**
**joule (J)**
**gravitational potential energy**
**spring constant**
**ampere (A)**
**radiant energy**
**horsepower (hp)**

## 9.1 - Energy

Energy is a fundamental essence of our universe but also a slippery concept to get a grip on. Energy measures the capacity for change. Energy can create changes in speed, height, or temperature—or even turn matter from liquid into gas or back. Anything that changes over time by moving, turning on, or even living changes through the flow and transformation of energy. Without energy *nothing can change*—not speed, height, temperature, or even color. Understanding the flow of energy through matter opens a powerful window into understanding how things work.

### Forms of energy

**Mechanical energy**

**Mechanical energy** is energy that comes from position or motion. Gravitational potential energy, elastic potential energy, rotational energy, and ordinary kinetic energy are examples of **mechanical energy**.

**Radiant energy**

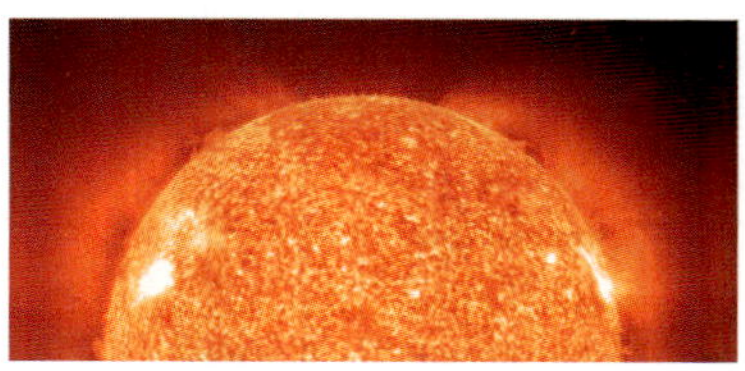

**Radiant energy** includes visible light, microwaves, radio waves, x-rays, and other forms of electromagnetic waves. Nearly all the energy on Earth ultimately comes from radiant energy that originates in the Sun.

**Nuclear energy**

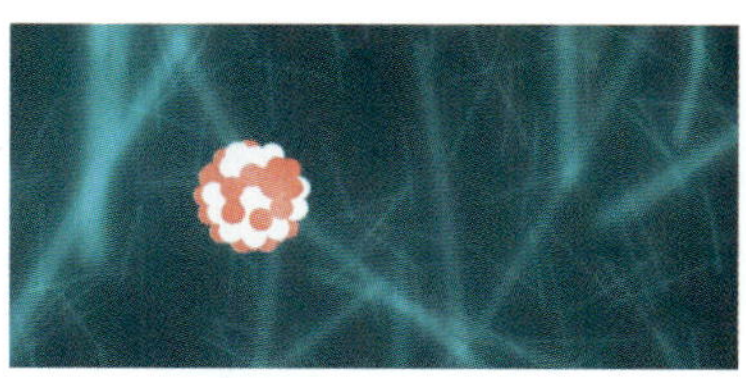

**Nuclear energy**, sometimes called atomic energy, is energy contained in matter itself. Nuclear energy can be released when atoms are changed from one element into another, such as in a nuclear reactor or in the core of the Sun.

**Electrical energy**

**Electrical energy** moves in the form of electric currents that flow in response to electrical voltages, such as in a battery or wall socket. Electrical energy is used whenever we turn on an appliance such as a lamp or coffee maker.

**Chemical energy**

**Chemical energy** is stored in the bonds between the constituent atoms in molecules that make up most matter. Chemical energy can be released by rearranging the atoms into different molecules, such as by burning natural gas to produce water and carbon dioxide.

**Thermal energy**

**Thermal energy** is another word for heat. Thermal energy is energy that is attributed to an object's temperature. Hot objects have more thermal energy than cold objects.

**Internal energy of gases**

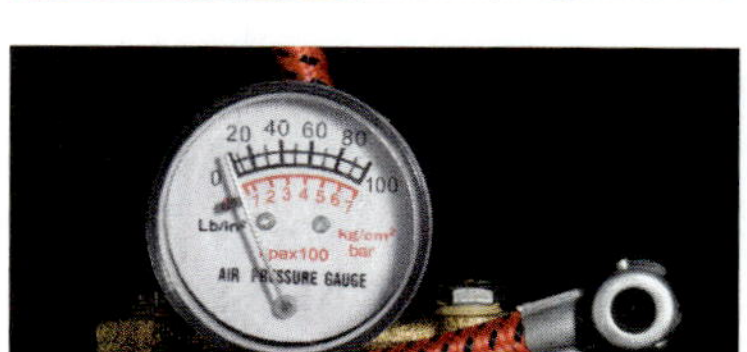

**Pressure** is a consequence of thermodynamic energy—a form of mechanical energy at the microscopic level—that is important in gases and liquids. It takes work to inflate a tire; some of the work is stored as energy in the form of high-pressure air inside the tire.

**What is work?**

Video

How does energy move and create change? The answer starts with the physical idea of *work*. In physics, **work** is done by forces that cause objects to move. As shown in equation (9.1), the amount of work done on an object is the force multiplied by the distance the object moves in the direction of the force. If you use a force of one newton to lift a cup up a distance of one meter, you do one *newton-meter* (N m) of work on the cup.

**Applying a constant force of 1 N over a distance of 1 m does 1 J of work.**

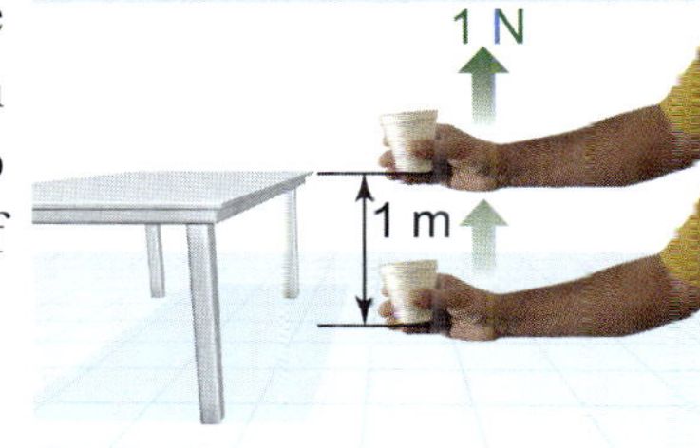


Interactive equation

**Work**

| | | |
|---|---|---|
| **(9.1)** | $W = Fd$ | $W$ = work (J)<br>$F$ = force (N)<br>$d$ = distance (m) |

**Does "work" mean other things?**

The meaning of the word "work" is different in physics from how we use the word in everyday life. You might complain that you *work* too hard on your homework or that you have to go to *work* early in the morning. In physics, work is the product of force exerted and distance moved, and it can have a precise value such as 10 N m. Work is done whenever energy is transferred from one object to another.

**One *joule* of energy is equivalent to one *newton-meter* of work**

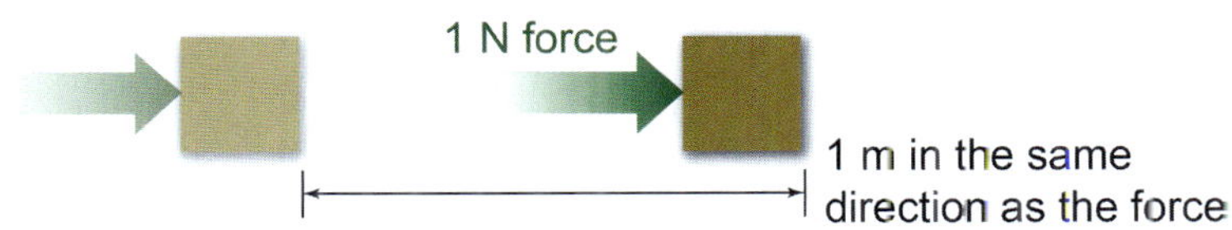


**What are the units of work?**

In the SI system, one **joule (J)** is defined as one newton-meter (N m). The unit of work—the joule—is also the basic unit of energy. Throughout physics and throughout all of science—including chemistry, biology, and earth sciences—you will find the joule as a standard unit of energy. It is not, however, the *only* unit of energy. Calories (food), British thermal units (heat), and kilowatt-hours (electricity) are other units of energy that are commonly used.

**Two ways to do 10 J of work**

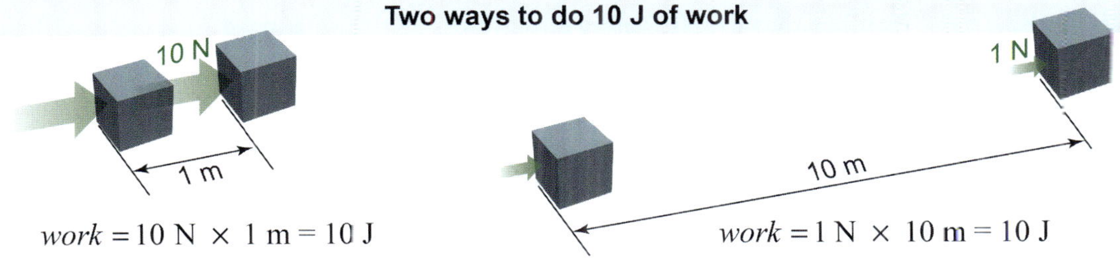


**Is work the same as energy?**

Work provides an operational definition for energy: *Energy is the ability to do work*. If you have 10 J of energy, you have the ability to produce a force of 10 N for a distance of 1 m, or 1 N for 10 m. A system that has 1 J of energy can exert any combination of forces and distances whose product is equal to or less than 1 N m. We say "less than" because the amount of energy available limits the amount of work that can be done. No system can produce any work that exceeds its available energy.

# Kinetic energy

**Thinking about kinetic energy**

**Kinetic energy** ($E_k$ or KE) is the energy of motion. Any object that has mass and is moving has kinetic energy *because* it is moving. When you catch a ball, your hand applies a force over some distance (your hand recoils a bit) to stop the ball. That force multiplied by the distance represents the transfer of the ball's kinetic energy to your hand by doing work on your hand. Now think about catching balls of different masses and speeds. You can probably guess that a massive or a fast-moving ball has more kinetic energy—it is harder to stop!—than a lighter, slower ball.

**Kinetic energy**

$$E_k = \frac{1}{2}mv^2 \qquad (9.2)$$

$E_k$ = kinetic energy (J)
$m$ = mass (kg)
$v$ = speed (m/s)

**How is kinetic energy related to mass?**

Kinetic energy is calculated using equation (9.2) above. The kinetic energy of a moving object is proportional to mass. If you double the mass, you double the kinetic energy. For example, a 2 kg ball moving at a speed of 1 m/s has 1 J of kinetic energy according to the equation. A 4 kg ball moving at the same speed has 2 J of kinetic energy, or twice as much. This is a *linear* relationship.

**KE is proportional to mass.**

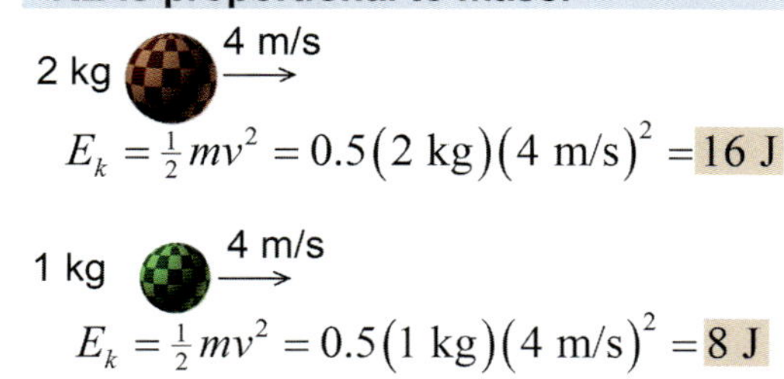


**Kinetic energy increases as speed squared**

According to equation (9.2) kinetic energy depends on the *square* of the speed of a moving object. Consider a 2 kg ball traveling at 1 m/s with 1 J of kinetic energy. The same ball moving at 3 m/s has 9 J of kinetic energy. If you multiply the speed by 3, then the kinetic energy is multiplied by a factor of $3^2 = 9$. This is an example of a *nonlinear* relationship.

**KE is proportional to speed *squared*.**

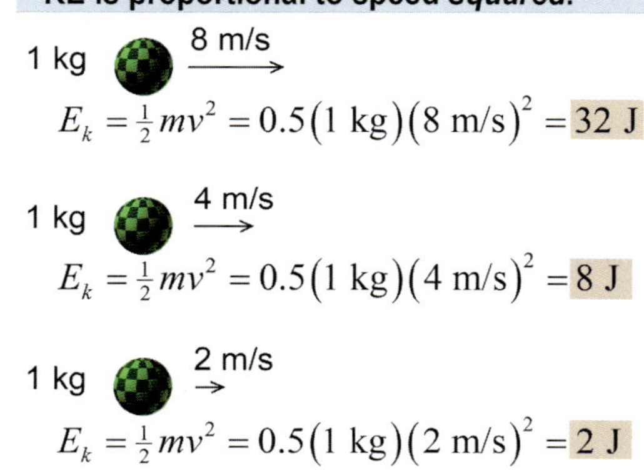


**Kinetic energy and braking distance**

The fact that kinetic energy increases with the square of the speed has implications for the stopping distance of a car. As a car brakes, work is done to transform the car's kinetic energy into thermal energy. Work is force multiplied by distance. Assuming that the braking and road conditions result in a relatively constant stopping force, then the distance it takes the car to stop is proportional to the initial kinetic energy. At 30 mph a car can stop in about 15 m. When the speed is doubled to 60 mph (and the kinetic energy increases by a factor of 4) it takes four times as much distance to stop.

**Because of kinetic energy, braking distance increases as the square of a car's speed.**

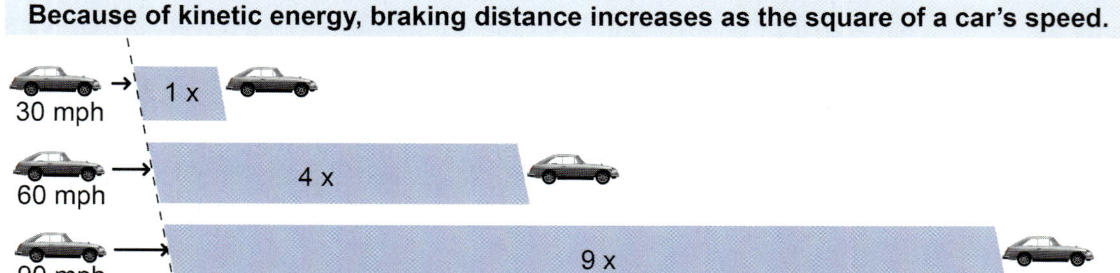

**Potential energy is energy of position**

**Potential energy** is the energy of position. The word "potential" describes the fact that potential energy is not active. It is stored energy that has the ability to become active. A good example is to compare the energy of a brick on the table to that of a brick on the floor. By virtue of its being higher than the floor, the brick has more potential energy when it is on the table than it does when it is on the floor. The brick on the table, however, is not using its energy by moving, heating up, or doing anything else.

A brick has more potential energy on the table than on the floor.

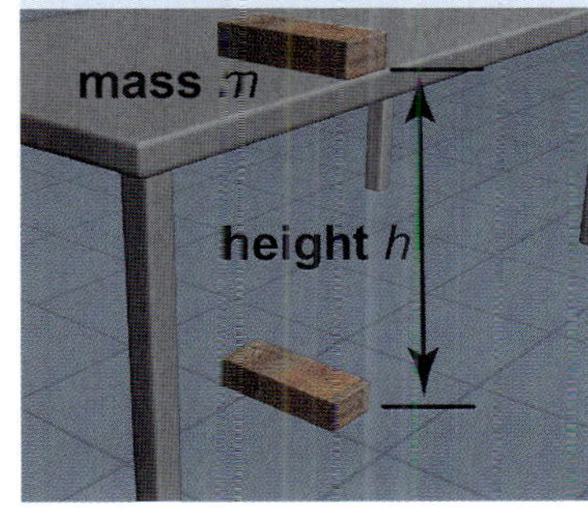


**Equation for potential energy**

How much potential energy does the brick have? If the brick has more mass, the force of gravity gives it more weight and more potential energy. If it has more height, it can fall farther and also has more potential energy. By these arguments we deduce that potential energy depends, at least, on weight and height. Checking the units supports that conjecture—the product of weight times height has units of force times distance, the same as work. The equation for **gravitational potential energy** ($E_p$ or PE) is exactly what we surmised. The potential energy of any object of mass $m$ raised by a height $h$ is given by equation (9.3).

Interactive equation

**Gravitational potential energy**

**(9.3)** $$E_p = mgh$$

$E_p$ = potential energy (J)
$m$ = mass (kg)
$g$ = acceleration of gravity
$h$ = change in height (m)

**Gravitational potential energy**

Video

Potential energy exists because there is a force that we must work *against* to raise the brick from the floor to the table. In this case the force we must work against is gravity. Gravity still acts on the brick once it is on the table and therefore the work we do to raise the brick is stored as potential energy, which *can be recovered by letting the brick fall back down again.* An important concept is that potential energy represents stored work that may be recovered when the energy becomes active—such as by changing an object's height.

**Potential energy and work done for weightlifting**

The current weightlifting world record of 263 kg was set by Hossein Rezazadeh of Iran in 2004. (a) How much gravitational potential energy did the barbell have if he held it 2.0 m above the floor? (b) How much work did he perform to lift the barbell from the ground to the overhead position?

| | |
|---|---|
| *Asked:* | (a) potential energy $E_p$ of the barbell overhead and (b) work $W$ done by the weightlifter to lift the barbell overhead from the floor |
| *Given:* | barbell mass $m$ = 263 kg, barbell final height $h$ = 2.0 m |
| *Relationships:* | $E_p = mgh$, $g = 9.8\ \text{m/s}^2$ |
| *Solution:* | (a) $E_p = mgh = (263\ \text{kg})(9.8\ \text{m/s}^2)(2.0\ \text{m}) = 5{,}155$ J. (b) As given in Part (a), the change in potential energy of the barbell from the floor ($E_p = 0$) to its overhead position is 5,155 J. The minimum work done by the weightlifter is the same as the change in PE. |
| *Answer:* | (a) Gravitational PE is 5,155 J; (b) minimum work done is 5,155 J. |

**Where is zero height?**

Hold a ball above the ground. What is the ball's potential energy before you drop it? It depends! The ball is not at zero height—if it was it couldn't have fallen because it would have already been on the ground. For that matter, even the ground is not truly zero height because you could dig a hole and let the ball fall farther into the hole. Depending on your choice of *coordinate system*, the ball's height—and hence its potential energy—varies. From the perspective of potential energy, where is "zero height"? Is there even such a thing?

**Potential energy depends on height.**

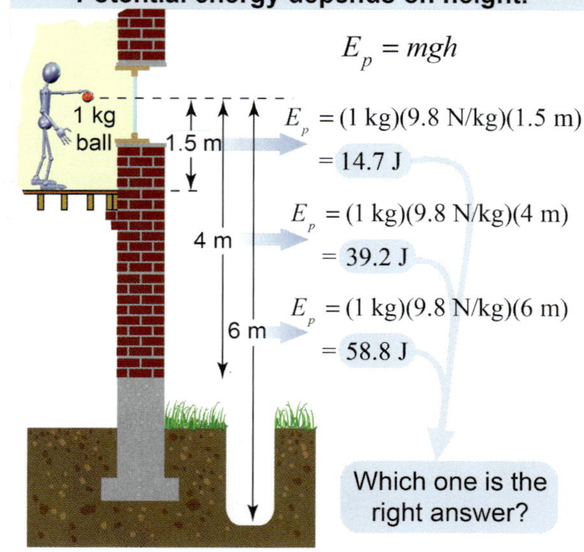


**Potential energy is relative**

The answer to that question is that the energy of a system depends on the **reference frame** in which things are measured. For example, if the reference frame is the room, a 1 kg ball that is 1 m from the floor has a potential energy of 9.8 J. If the reference frame is the ground two stories down, the potential energy is 68.6 J. The actual position of the ball is the same but the choice of reference frames is different.

**Why reference frames can be chosen**

The choice of a reference frame (such as where "zero" is) is arbitrary because it is *differences* in energy that matter, not absolute values. A ball that falls 1 m loses 9.8 J of potential energy. It does not matter if the ball falls from 1 m to zero or from 7 m to 6 m. The amount of potential energy lost or gained from a *change* in height is independent of where zero is defined. We are therefore free to choose any reference frame that makes the calculation easiest! The only rule is that the reference frame should not change in the middle of a problem.

**Is kinetic energy independent of reference frame?**

You might think kinetic energy is independent of reference frames. A ball at rest in your hand has zero velocity, doesn't it? In your reference frame, which is fixed to the Earth, the ball's kinetic energy is zero. Consider, however, that Earth is moving through space at 29,780 m/s. *Relative to the Solar System*, the ball on the table has a kinetic energy of $443 \times 10^5$ J! Like potential energy, kinetic energy also depends on the reference frame in which speed is measured.

**Kinetic energy and reference frames**

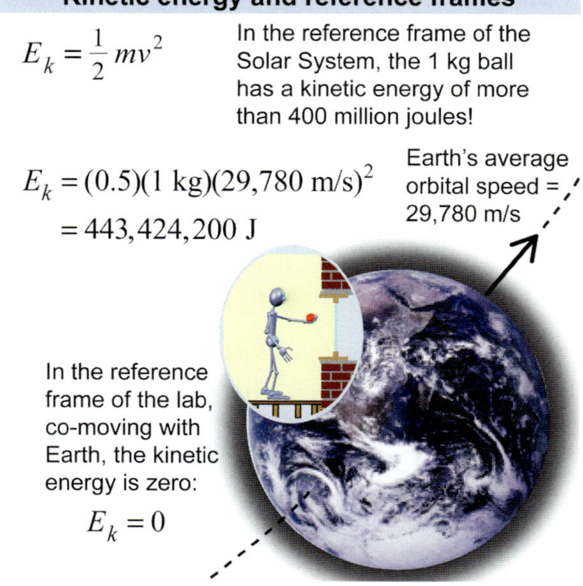

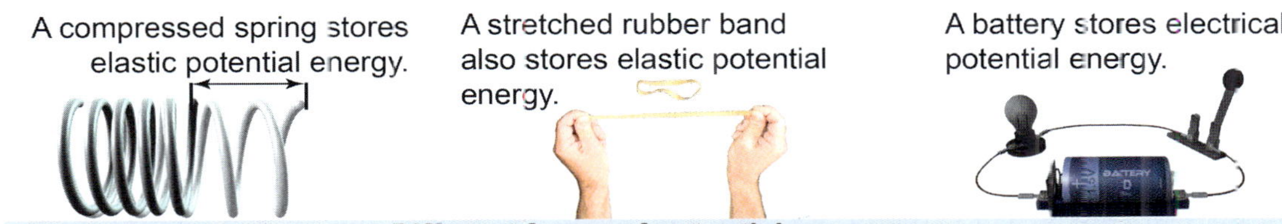


**Different forms of potential energy**

**Elastic PE of a spring**

Video

There are forms of potential energy other than gravitational. Potential energy exists any time a force is restrained from acting in such a way that the energy can be released if the restraint is removed. If you use the spring to launch a marble you can see how the stored potential energy of the spring is converted to kinetic energy of the marble. Compressing a spring creates potential energy because you have to do work against the force of the spring to compress it. A compressed spring stores potential energy as long as it is compressed. This type of potential energy is called **elastic potential energy** because it derives from the elasticity of the steel in the spring. It can be calculated by using equation (9.4).

Interactive equation

**Elastic potential energy**

$$\textbf{(9.4)}\quad E_p = \frac{1}{2}kx^2$$

$E_p$ = elastic potential energy (J)
$k$ = spring constant (N/m)
$x$ = displacement from equilibrium (m)

**Spring constant**

The quantity $k$ in equation (9.4) is called the **spring constant**, which represents the strength of the restoring force exerted by the spring when it is compressed or stretched. The spring constant is a property of the spring itself and is different for every spring or other elastic material. It has units of force divided by distance, because the spring exerts a restoring force that increases as the spring is increasingly displaced (stretched or compressed) from its equilibrium position. A stiff spring resists compression or stretching, so it has a large value of the spring constant $k$; a loose spring has a low value of $k$.

**Spring in equilibrium position**

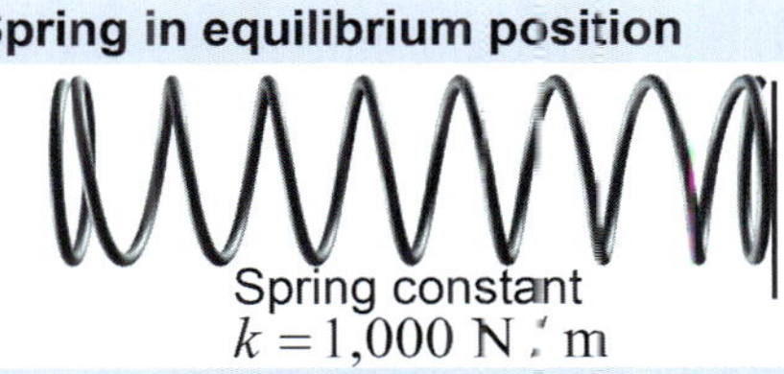


**Spring compressed by 1 cm = 0.01 m**

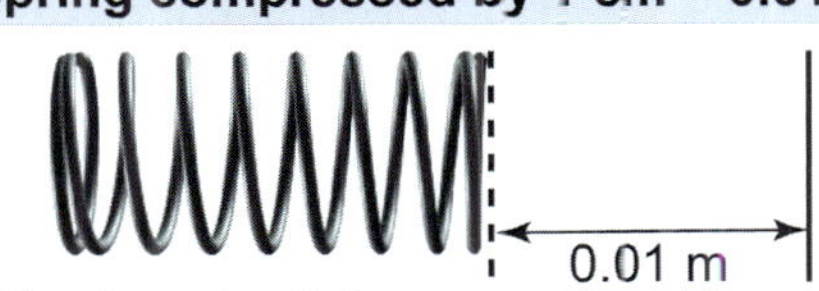


Elastic potential energy stored in compressed spring:

$$E_p = \tfrac{1}{2}kx^2 = \tfrac{1}{2}(1{,}000\ \text{N/m})(0.01\ \text{m})^2 = 0.05\ \text{J}$$

**Elastic PE from objects changing shape**

Objects that store elastic potential energy share some common features: Energy is stored when the object changes its form, shape, or length; the force exerted by the elastic object acts in a direction to return it back to its original shape or position; and as it stretches or compresses more, it stores more energy.

**Objects with elastic potential energy**

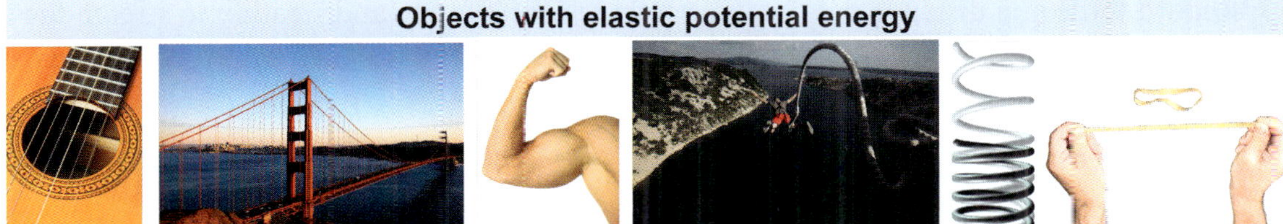

Guitar strings — Cable bridge — Muscle — Bungee cord — Spring — Rubber band

# Investigation 9A

## Work and the force versus distance graph

**Essential questions** *How are force, distance, and work related graphically?*

To lift a mass, you must do work on it against the gravitational force. Similarly, when you stretch a spring, you do work on it to overcome its restoring force. How does the work done on these objects relate to the forces applied to them and the distances they move or stretch? In this investigation you'll find out by using a Smart Cart to apply those forces so that you can measure both force and distance.

## Part 1: Lift a mass

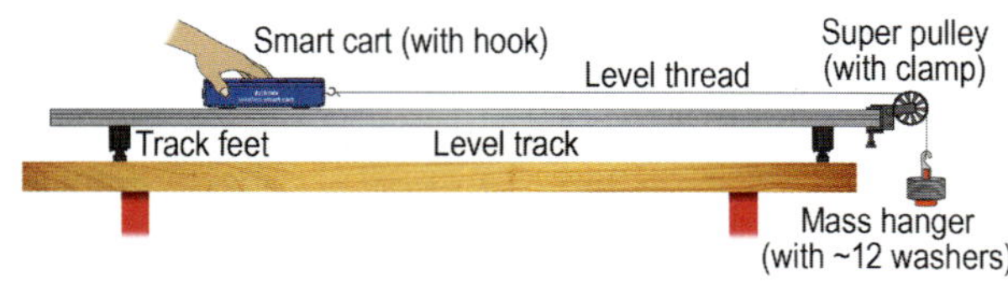


1. Open the **09A_ForceDistanceGraph** experiment file in your software, and then connect your Smart Cart using Bluetooth.
2. Set up the equipment as in the picture. Be sure the track is level.
3. In your software, zero the Smart Cart force sensor while nothing is touching the hook.
4. With the thread reattached to the cart, position the cart so that the mass hanger is suspended just above the floor.
5. Record force and distance data as you move the cart slowly and steadily to lift the mass a distance of 30 cm, as measured by the markings on the track. Then stop recording data.

*a. Sketch a copy of your force versus distance graph. Be sure to label the axes.*
*b. Use the statistics tool in your software to determine the average (mean) force applied to lift the mass.*
*c. Determine the actual distance you lifted the mass during data recording.*
*d. Use these two values to calculate the work done on the mass to lift it, including units.*
*e. Use the statistics tool in your software to find the area under your force versus distance graph, including units. How does this area compare to the work you calculated above? What is the geometric shape of the shaded area?*
*f. Did the work done on the mass to lift it change its energy? If so, explain which form of energy changed and whether it increased or decreased.*

## Part 2: Stretch a spring

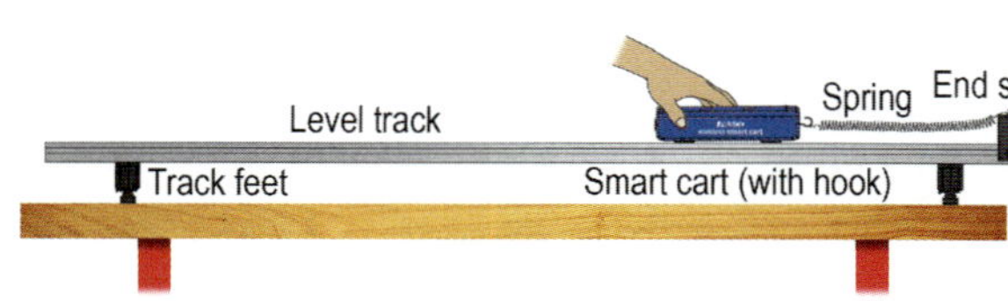


1. Modify your equipment as shown in the picture. Use the stiffest of the three springs.
2. Position and hold the cart so that it is just starting to separate the spring's coils, and then use your software to zero the Smart Cart force sensor.
3. Record force and distance data as you move the cart slowly and steadily to stretch the end of the spring a distance of 30 cm, as measured by the markings on the track. Then stop recording data.

*a. Sketch a copy of your force versus distance graph. Be sure to label the axes.*
*b. Was the force constant? Would it be appropriate to use the formula $W = Fd$ to calculate the work done on the spring? Explain.*
*c. Use the statistics tool in your software to find the area under your force versus distance graph, which is the work done on the spring to stretch it.*
*d. Did the work done on the spring to stretch it change its energy? If so, explain which form of energy changed and whether it increased or decreased.*

# Section 1 review

Energy can take many different forms: mechanical, radiant, nuclear, electrical, chemical, and thermal. Mechanical energy includes kinetic energy and potential energy (both gravitational and elastic). Work and energy are closely related, because energy is the ability to do work, and doing work on an object changes its energy. Potential energy is usually calculated in terms of a reference frame, in which a position of zero energy must be defined. Energy and work are both measured in joules (J).

| Vocabulary words | mechanical energy, work, joule (J), kinetic energy, potential energy, gravitational potential energy, reference frame, elastic potential energy, spring constant |
|---|---|

| Key equations | | |
|---|---|---|
| | $W = Fd$ | $E_k = \frac{1}{2}mv^2$ |
| | $E_p = mgh$ | $E_p = \frac{1}{2}kx^2$ |

## Review problems and questions

1. What forms of energy are found in
   a. the Sun?
   b. a car's engine?

2. A 30 kg boy sitting on a 10 kg go-kart wants to travel down a hill of height 10 m, starting from rest.
   a. What is the initial potential energy of the boy and go-kart together?
   b. What is their initial kinetic energy?

3. A 1,000 kg car traveling 15.0 m/s brakes and comes to a stop after traveling 20.0 m.
   a. What is the car's initial kinetic energy?
   b. What is the car's final kinetic energy?
   c. How much work does it take to stop the car?
   d. How much constant force is applied in bringing the car to a stop?

4. A baseball outfielder catches a fly ball traveling 25.0 m/s with his gloved hand. When he catches the 0.140 kg baseball, the outfielder's hand recoils by 10.0 cm. How much force did the outfielder's hand exert in catching the ball?

5. A horizontal spring with $k = 100$ N/m is compressed by 10 cm by a 100 g mass.
   a. How much elastic potential energy does the compressed spring store?
   b. How much elastic potential energy would the compressed spring store if it were compressed the same distance by a 300 g mass?

## 9.2 - Flow of energy

All processes in both nature and technology occur through a continuous flow of energy. Scientists learn a great deal by following the flow and transformation of energy. A system can transform its available energy into speed, height, heat, or even mass. The system must have energy to begin with, and any energy transformed into one form depletes other forms by the same amount. This section examines how energy is transformed through technologies such as the photovoltaic cell, which converts radiant energy into electrical energy.

### Energy flow diagrams

**Mechanical energy**

Consider the sewing machine, one of the most revolutionary devices ever invented. Clothing is a basic human need and prior to the sewing machine all clothes were stitched together by hand. Hand sewing requires considerable skill and takes a long time. A single shirt might take several hours to sew by hand. The same shirt can be sewn on a machine in six minutes with perfect, uniform stitches.

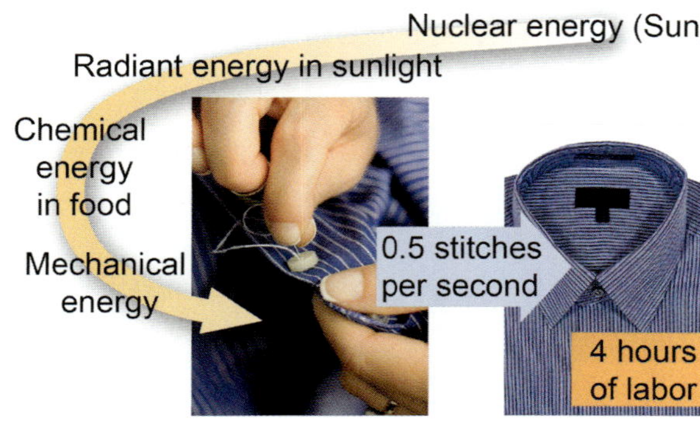


**Energy flow in muscle work**

Sewing requires mechanical work—forcing a needle through cloth. To estimate the energy involved, assume a weight of 20 N for your forearm and hand. Your hand moves about 40 cm to make a stitch so each stitch represents about 8 J of muscle work. If you average a stitch every 2 s that is an energy flow of 4 J/s. That muscular energy comes from food which derives ultimately from sunlight, so the flow of energy through your muscles traces its way back to nuclear energy in the sun.

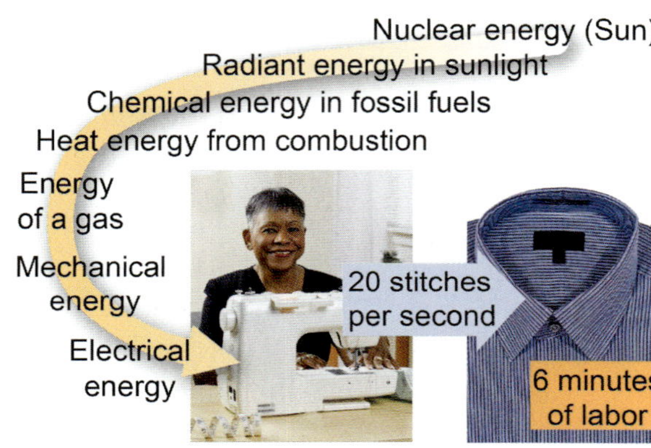


**Energy flow in technology**

Transforming energy from one form to another is depicted in an *energy flow diagram* or described as an *energy chain*. The energy used by the sewing machine comes from an electric motor that converts electrical energy to mechanical energy. Two-thirds of U.S. electricity is generated from burning coal or natural gas. The fossil fuel itself came from decaying plants. Burning fuel transforms chemical energy into thermal energy, and the heat produced is used to boil water into steam. The steam turns a turbine that converts the thermodynamic energy of the gas into mechanical energy; the turbine drives a generator to produce electrical energy. The electrical energy we use is ultimately derived from nuclear energy in the Sun, transported to the Earth as radiant energy, accumulated by plants over millions of years, then stored in fossil fuels.

**Thinking about doing work**

In everyday conversation the words energy and power are often used interchangeably, but the true meanings are different. *Energy* is the ability to do physical work and has units of joules. However, the same amount of work can be done fast or slow. For example, lifting a 1 kg ball by 1 m takes a minimum of 9.8 J of work. You could do it slow, taking a whole minute, or fast, in one second. The work done is the same but there must be a difference between fast and slow.

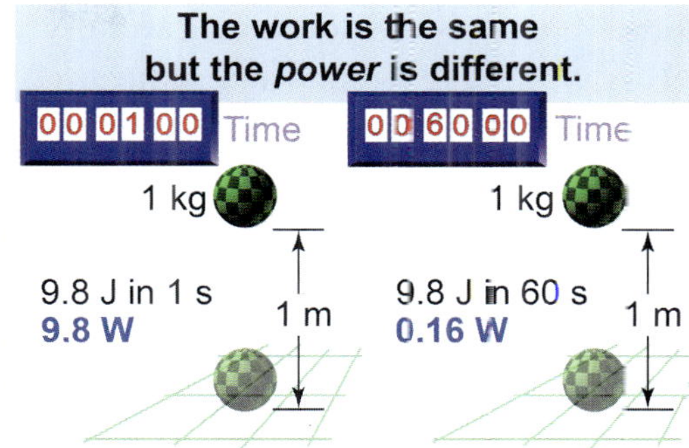


**The rate of doing work**

The difference is the *rate* at which the work is done. The rate at which work is done, or the rate at which energy is transferred, is called **power**, and is calculated using equation (9.5).

Interactive equation

**Power**

$$(9.5) \qquad P = \frac{\Delta E}{\Delta t} = \frac{W}{\Delta t}$$

$P$ = power (W)
$\Delta E$ = change in energy (J)
$W$ = work done (J)
$\Delta t$ = time duration (s)

Power is an indication of the level of "effort" required to perform a given amount of work. If you want to move a boulder a few meters, then one person can do it in a few minutes using only a modest amount of power. If you want to move the boulder the same distance but *more quickly* (in a few seconds!), then you will need much more power—that can only be delivered by a team of horses or a front end loader machine.

**Power is measured in watts**

Whereas work and energy are measured in joules, power is measured in joules *per second*. A power of one joule per second (J/s) is one **watt (W)**. The watt is named in honor of James Watt, a Scottish engineer who developed the first practical steam engine and thereby provided the power for the industrial revolution. Doing 9.8 J of work in 60 s requires 0.16 W—approximately the power output of a small mouse. Doing the same 9.8 J of work in 1 s, as shown in the figure above, requires 9.8 W of power, which is 60 times greater.

**Energy used by a light bulb**

How much energy does an incandescent light bulb rated at 100 W use in one hour?

*Asked:* energy $\Delta E$ used by the light bulb

*Given:* power of the light bulb $P = 100$ W; time that bulb is on $\Delta t = 1$ hr

*Relationships:* power $P = \Delta E/\Delta t$

*Solution:* Time is usually expressed in seconds, but we are given it in hours. Convert time to seconds:

$$\Delta t = 1\ \text{hr}\left(\frac{60\ \text{min}}{1\ \text{hr}}\right)\left(\frac{60\ \text{s}}{1\ \text{min}}\right) = 3{,}600\ \text{s}$$

Solve for $E$ by multiplying the power equation by the elapsed time $\Delta t$:

$$P \times \Delta t = \frac{\Delta E}{\Delta t} \times \Delta t \quad \Rightarrow \quad \Delta E = P\Delta t$$

$$\Delta E = P\Delta t = (100\ \text{W})(3{,}600\ \text{s}) = 360{,}000\ \text{J}$$

*Answer:* 360,000 J

**Volts**

If you look carefully at a battery, it will tell you how many volts—an indicator of electrical potential energy—the battery produces. A battery that produces 1 **volt (V)** provides half the electrical potential energy (per ampere of current) of a battery that produces 2 V. You normally can't feel the electrical potential energy from ordinary batteries because the energy is too low to push electric current through your skin. You have probably seen *lightning*, however, which is a spectacular example of high voltage—many millions of volts!—that can send enough electrical energy through your body to kill you.

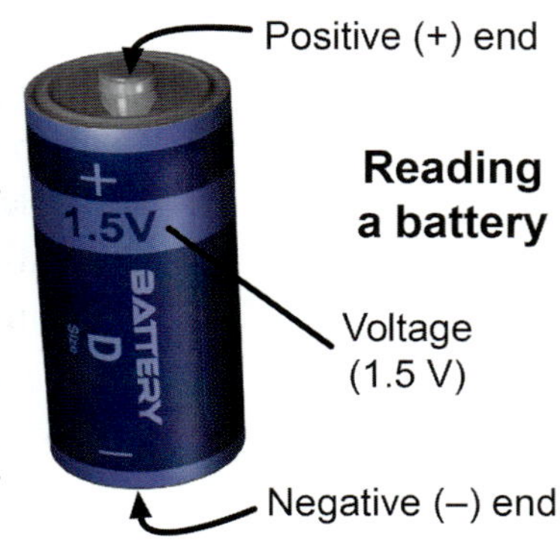


**Electric current and the ampere**

To make potential energy active, something has to move. A falling mass converts its gravitational potential energy into kinetic energy. In electricity, an **electric current** flows, usually in wires, to carry electrical energy and make it useful. Most electric current is the movement of electrons. One **ampere (A)** of electric current represents the flow of a large number of electrons per second.

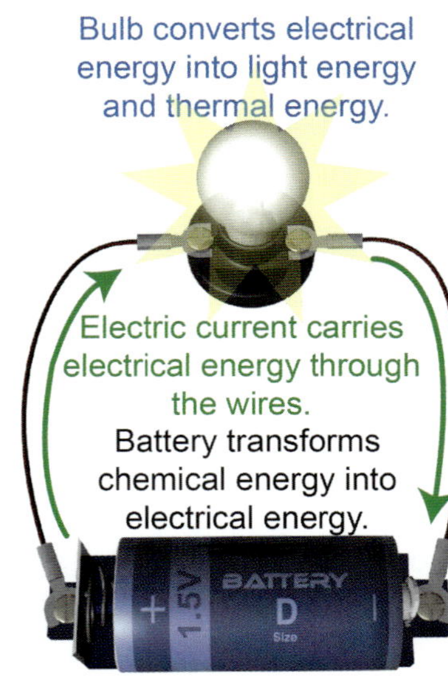


**Current is how electrical energy is transformed**

Electric current is invisible, although it is possible to see its effects in cases such as lightning. When you turn on an appliance or a light bulb, electric current transforms electrical potential energy into other forms of energy such as light and thermal energy. If it flows through a motor, electrical energy is transformed into mechanical energy.

**Electrical power**

Electrical power measures how much electrical energy flows through a circuit per unit time (or per second). Electrical power is greater if the voltage across the circuit is higher (such as with high-voltage power lines) or if there is more electric current through the circuit. Electrical power is the product of electric current and voltage.

Interactive equation

**Electrical power**

$$(9.6)\quad P = IV$$

$P$ = electrical power (W)
$I$ = electric current (A)
$V$ = voltage or potential difference (V)

**Volts, amperes, and power**

Volts and amperes (sometimes called "amps") are useful because together they determine how much electrical power flows. Voltage is an indication of how much electrical *potential* energy a circuit has, but it is more useful to think about *joules per second* (power) instead of energy. If you have 9 V, this means that 1 A of current carries 9 W of power. The wall outlet in your classroom is 120 V, so it delivers 120 W of power per amp of current—a lot of power from a little wire. This is why electricity is so useful!

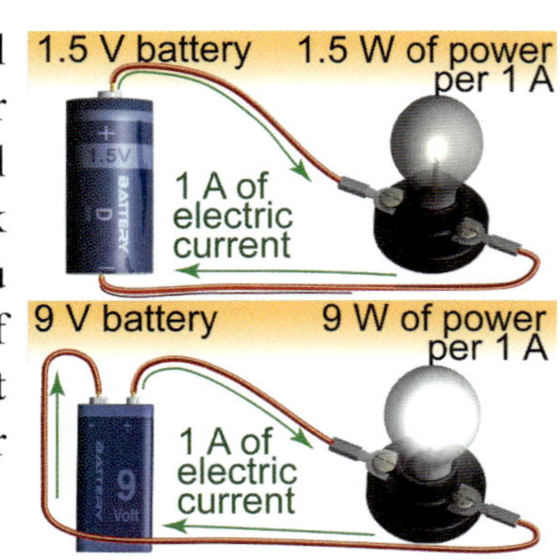

**Radiant energy**

**Light** is a form of *radiant* energy that is all around us. **Radiant energy** has no mass and can travel unimpeded through the vacuum of space. Visible light is a small part of the whole spectrum of radiant energy that includes microwaves, infrared radiation, and ultraviolet light—as well as more exotic forms such as x-rays and gamma rays. The human eye responds to a narrow band of energy that starts in the dull red, then proceeds through orange, yellow, green, blue, and violet. Violet is the most energetic light you can see. Light of higher energy falls into the ultraviolet (UV) range and it can give you a sunburn, but you cannot see it with your eyes.

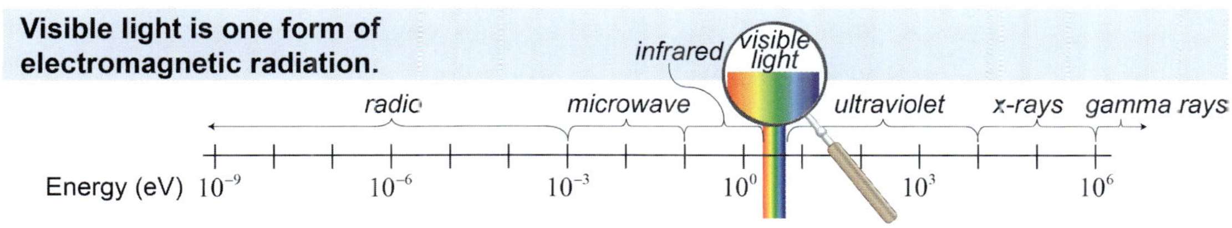


**Power and radiant energy**

Radiant energy always travels at the speed of light, which is 300,000,000 m/s. Nothing in the known universe can travel faster than light. Since light is energy that is always moving, it is best described in terms of its power. An old-fashioned, incandescent 100 W light bulb produces a power output of about 2 or 3 W of visible light; the remaining input power is transformed into thermal energy. The Sun produces an incredible amount of power—about $3.8 \times 10^{26}$ W.

**Intensity**

**Intensity** is power per unit area. An intensity of 1 watt per square meter (1 $W/m^2$) means 1 W of light power crosses or lands on each square meter of surface. When we say a light is *bright* or *dim*, what we really mean is that the *intensity* of the light is either low (dim) or high (bright). The intensity of bright sunlight can be as high as 600 $W/m^2$. Thus sunlight delivers the energy of six 100 W light bulbs onto every square meter of the Earth. This is the reason why solar energy has great potential as a renewable energy source.

**Intensity and the human eye**

Your eye is an incredibly sensitive detector of light energy. The light receptor cells in the back of your retina can detect an intensity of $10^{-9}$ $W/m^2$, one billionth of a watt per square meter. Since the area of the retina is very small, this corresponds to a power of less than a trillionth of a watt!

**Color and energy**

Apart from whether the light is bright or dim, light has an intrinsic energy that produces the perception of *color* in your eye. The sensation of color is created by the brain in response to the differing energy content in the light. The lowest energy light humans can see is dull red. The highest energy light you can see is violet. A human eye is most sensitive to green light, which is in the middle of the visible energy spectrum. What we perceive as *white* light is actually an even mixture of light of different energies.

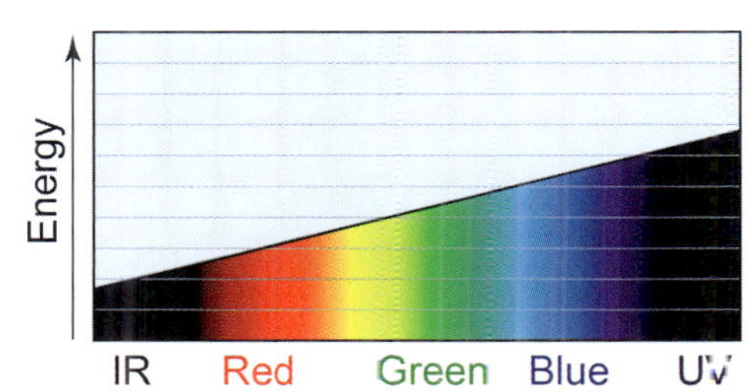


**We perceive light of different energy as different colors.**

### Power in everyday uses

When you think of how "strong" something is, you are often thinking about how much power it produces. Many appliances and cars provide power in units of **horsepower (hp)**, which is equal to 746 W. The electric motor in a typical washing machine is ½ hp or 373 W, about the same as the power output of a very fit athlete during competition. The motor in an electric saw is about 1.5 hp. A small car engine operates at around 100 hp. A blue whale can develop 500 hp, or 370,000 W. It takes 0.13 hp to power a standard 100 W bulb. How many horses would it take to power your house?

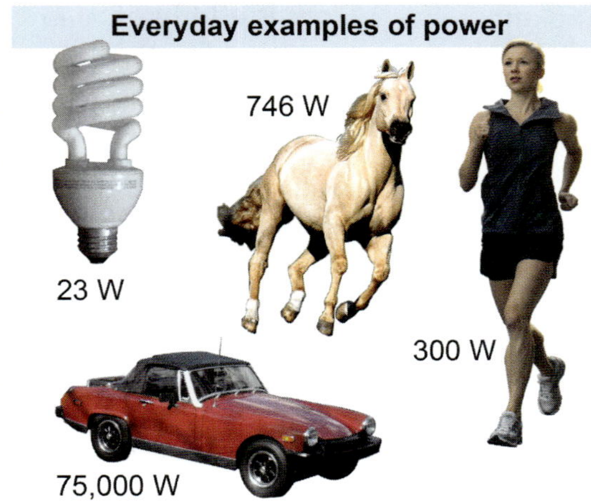


### Radiant power and light bulbs

An old-fashioned incandescent 100 W light bulb produces a power output of about 2 W of visible light. The other 98 W of electrical power used by the bulb become heat or infrared light (which we cannot see). Typical compact fluorescent light bulbs use less than one-fourth the electrical power, only 23 W, to produce the same power output of visible light. Fluorescent bulbs therefore lose around 21 W of input energy as heat, while incandescent bulbs lose around 98 W of input energy as heat. Power consumption and heat loss are the reasons why many people now use fluorescent bulbs.

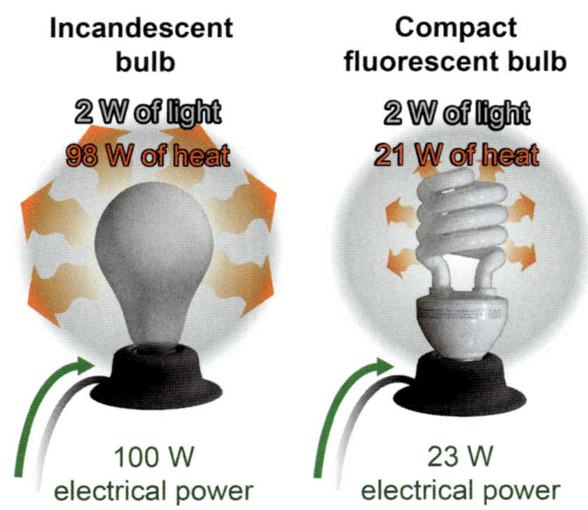


### The "100 W" light bulb

When is 100 W not really 100 W? The answer may be right over your head. What is sold in stores as a "100 W" rated compact fluorescent light bulb actually uses a power of around 23 W. The "100 W" classification of the bulb comes from the fact that the 23 W modern compact fluorescent bulb produces the same output radiant energy per second as an old-style 100 W incandescent light bulb—around 2 W of power in the form of light. So today, a 23 W bulb is still *called* a 100 W bulb even though the real input power is quite different! This strange historical anachronism could have been avoided if light bulbs had been rated by the light intensity (or light power) they *produced*, rather than by the electrical power they *consumed*.

**Energy and food**

What does physics have to do with food? The answer is far more important today than it was even 50 years ago. A healthy diet has a daily energy content of 2,000 Calories, or 8.4 million joules. This energy must come from food, and the food required to feed Earth's human population has increased dramatically. Consider that there were about 1 billion people on Earth in 1800. By 1927 the population had doubled to 2 billion. By 1960 there were 3 billion people. By 1974 there were 4 billion, and by 2012 the world population was 7 billion and growing. Yet the amount of land and sunlight available for farming is the same today as it was 200 years ago.

**Green revolution**

Since 1950 the world population has almost tripled. Yet, according to the U.S. Geological Survey, the amount of land and water devoted to growing food has actually *decreased* because of urbanization and soil degradation. How do we feed more people with less farmland? The answer is a widespread change from traditional methods of farming to more energy-intensive, high-yield agriculture. This largely took place between 1950 and 2000, a period that has been dubbed the "green revolution."

**Home electricity bill**

Your family's monthly home electricity bill shows how much electrical energy you used in the past month (measured in kilowatt-hours or kWh) and charges for the generation and transmission of that electricity. Electricity *generation* costs are for the power plant—hydroelectric, coal-fired, natural gas, etc.—that produced the power in the first place. Electricity *transmission* costs cover the high-voltage power lines, the power lines on your street, the transformers, and so on. In the example at right, the delivery costs (9.4 cents/kWh) are actually higher than the generation costs (7.0 cents/kWh)!

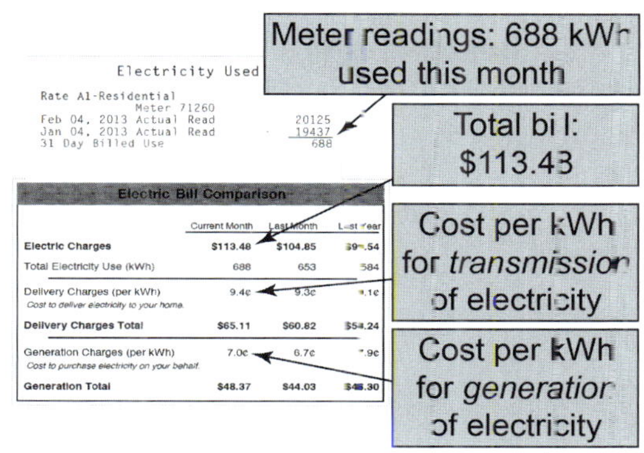


**Compare your bill**

Have class volunteers bring in a copy of their most recent electricity bills. Each student can extract and communicate the information in his or her bill, such as energy used per month, cost per kilowatt-hour, and change in consumption compared to the previous month.

**Energy guide labels**

Major home appliances are sold with an "EnergyGuide" label that indicates their expected energy consumption per year. This standardized labeling compares the energy usage of a particular appliance with similar models produced by the same or other companies. In the example at right, this refrigerator will cost approximately $67 to operate per year and consumes somewhat more energy than the average of other models.

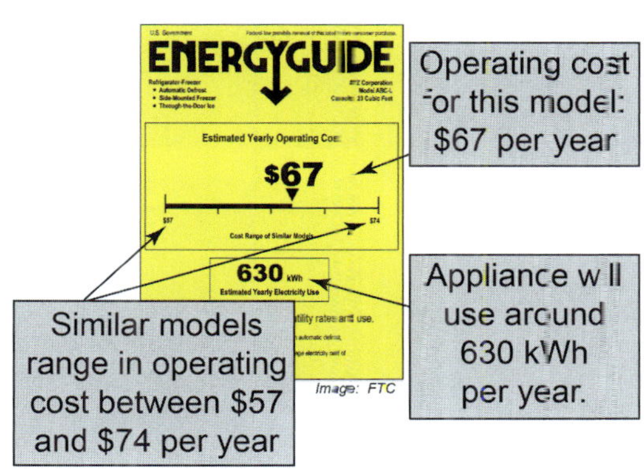


**Compare clothes dryers**

Research clothes dryers either on the Internet or in a local store to compare the energy consumption of each appliance. In class, each student should verbally summarize the content of a label, such as its annual energy consumption, yearly operating cost, and how it compares to similar models. Which models are the most energy efficient?

**What is renewable energy?**

Energy is a topic of concern to us all. We increasing desire a secure, affordable, and unlimited energy supply, but the environmental side effects of fossil fuels and nuclear power have become major concerns. *Renewable energy* may be the key to meeting our energy needs in an ecologically sustainable way. What do we mean by *renewable energy?* Is any "natural" form of energy renewable? The answer is no. Nature created the fossil fuels that we burn to supply most of our energy needs today, but it took millions of years to build up these reserves of coal, oil, and natural gas that civilization will deplete in only one or two centuries. **Renewable energy** includes forms of energy that we do not destroy, deplete, or disrupt faster than nature can restore or replenish them.

**Energy from sunlight**

**Solar energy** derives power from the radiant energy in sunlight. Sunlight can be used to heat water for household use; it can be used to create steam, which then powers a turbine; and it can directly generate electric current when absorbed by a photovoltaic cell.

**Gravitational potential energy**

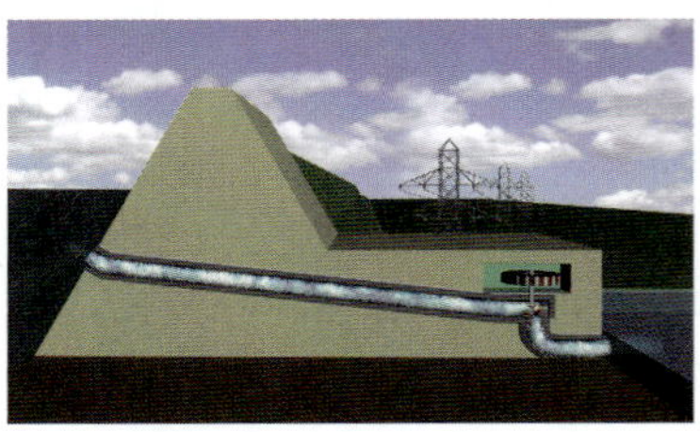

**Hydroelectric energy** refers to the mechanical energy of water flowing from higher places to lower ones. For centuries, this energy has crushed grain and spun lathes in mills built alongside rivers and waterfalls; today it is used mainly to generate electricity.

**Mechanical energy from moving air**

**Wind energy** has long been used to pump water or grind grain. Wind energy is today increasingly being tapped to generate electricity. Ultimately, wind derives its power from sunlight and, to a lesser extent, from Earth's rotation.

**Thermal energy**

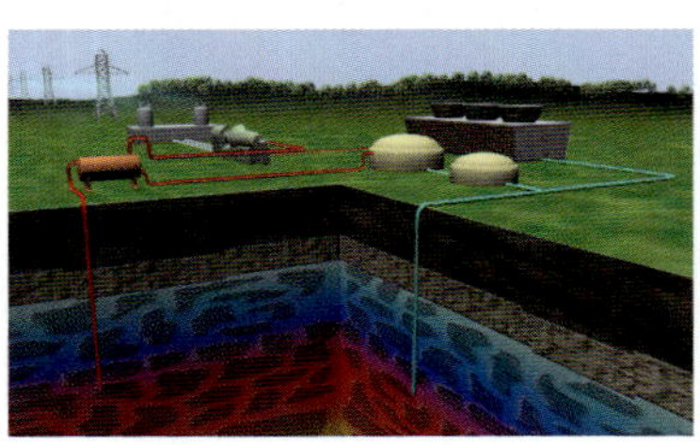

**Geothermal energy** is thermal energy and pressure contained within underground bodies of water heated by *magma*—molten rock that rises from deep within the Earth. Hot springs and geysers are two natural geothermal phenomena.

**Mechanical energy from ocean tides**

**Tidal energy** ultimately comes from the gravitational interaction among the Earth, the Moon, and the Sun. This interaction causes ocean water to flow back and forth twice daily. Tides involve tremendous amounts of mechanical energy—energy that is hardly tapped today.

**Nuclear (fusion) energy**

**Nuclear fusion** derives energy through nuclear reactions that combine lighter elements into heavier elements. Nuclear fusion is the energy source for the Sun and stars. A practical fusion reactor, however, has yet to be developed. (The opposite process, nuclear *fission*, powers today's nuclear plants.)

**Renewable energy and careers**

Putting the world on a path toward renewable energy will require both brilliant new ideas and also careful re-engineering of existing technologies. Physics underlies all aspects of renewable energy, from the invention of new types of solar cells to schemes for beaming microwave energy to Earth from orbiting solar power satellites. There are many exciting careers in these fields awaiting you. The examples below should provide you a start for researching the connection between physics and your future career.

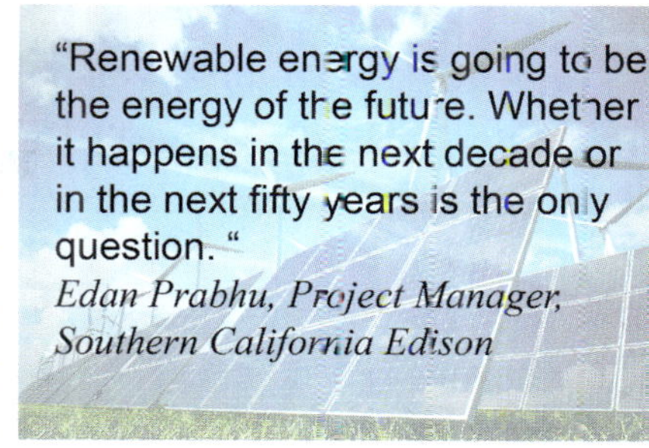

"Renewable energy is going to be the energy of the future. Whether it happens in the next decade or in the next fifty years is the only question."

*Edan Prabhu, Project Manager, Southern California Edison*

**Environmental engineer**

The ebb and flow of the tides release daily more than five times the total energy consumption of the human race. Environmental engineers are working on ways to convert some of this energy to useful electricity. Tidal power devices have to withstand harsh weather, the ceaseless beating of waves, and the corrosive effects of salt water. Demonstration power plants are working in several areas of the world, but more ideas and better technologies are still needed.

**Electrical engineer**

The rate at which wind and solar sources produce electricity goes up and down constantly and depends upon weather and season. Demand for electricity also goes up and down. How do we build and manage the complex network needed to match ever-changing power capacity to an ever-changing demand? Design of a world-spanning smart power grid is an enormously complex task awaiting tomorrow's electrical engineers.

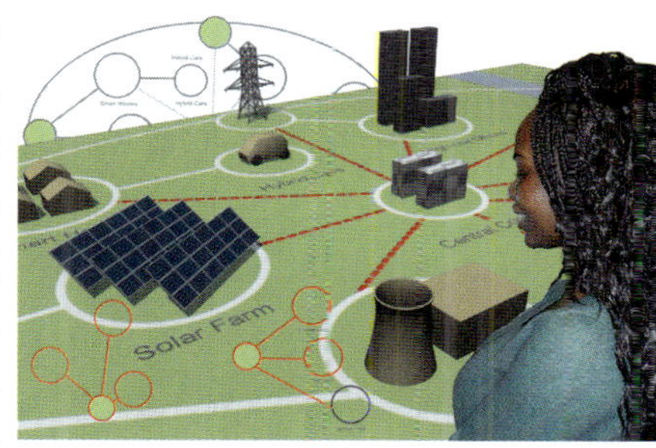


**Materials scientist**

Photovoltaic cells use the photoelectric effect to transform sunlight into electric current. The process, however, is only 14% efficient and very expensive. Are there better ways to convert sunlight into electricity? Engineers and materials scientists are working toward building a synthetic "leaf" that uses a nanotechnology form of *photosynthesis*.

**Architect**

Architects can design structures that heat, cool, and light themselves with a minimum of fuel or electricity. How? By channeling sunlight, breezes, and geothermal energy into interior spaces that are insulated against sudden temperature changes. Thermodynamics, optics, and fluid dynamics all support this increasingly valued specialty.

# Section 2 review

The process of making everyday items often requires the transformation of energy through many different forms, which can be followed as a flow of energy. Power is a basic quantity that measures the change in energy per unit time and has units of watts (W). Power is measured for many technologies all around us, such as light bulbs, batteries, automobile engines, washing machines, or stereo speakers. Voltage (measured in volts), electric current (measured in amperes), and power (measured in watts) are the most useful quantities for understanding the energy consumption of electric devices. Radiant energy spans the full electromagnetic spectrum, but our eyes can only see one small part of it called visible light.

| | |
|---|---|
| **Vocabulary words** | **power, watt (W), ampere (A), volt (V), electric current, radiant energy, intensity, light, horsepower (hp), renewable energy** |

| | | |
|---|---|---|
| **Important equations** | $P = \frac{\Delta E}{\Delta t} = \frac{W}{\Delta t}$ | $P = IV$ |

## Review problems and questions

1. Describe the energy flow that goes into producing a gold ring. Which step in the process do you think requires the most energy?

2. A 40 kg wheeled cart needs to be moved to the top of a platform that is 1.0 m high.
   a. How much power is required to pick up the cart and place it on the platform in 3.0 s?
   b. How much power is required to roll it 20 ft up a ramp, a process taking 20 s?
   c. How much power is required to roll it 40 ft up a shallower ramp, a process that also takes 20 s?

3. An immersion water heater is rated at 1,000 W. If it is plugged into a 120 volt outlet, how much current does it draw?

4. The cost of electricity in Mount Pleasant is 5 cents per kilowatt-hour.
   a. How much does it cost to operate a 100-W-rated incandescent light bulb for 24 hours?
   b. How much does it cost to operate a 100-W-rated fluorescent light bulb for 24 hours?
   c. How much does it cost to operate each for an entire year?
   d. Which one generates more light?

5. From the standpoint of energy and power, what are the advantages for a car with high horsepower? What are the disadvantages?

6. Why does ultraviolet light cause sunburn but infrared light does not?

# Chapter review



## Vocabulary

Match each word to the sentence where it best fits.

### Section 9.1

| | |
|---|---|
| elastic potential energy | gravitational potential energy |
| joule (J) | kinetic energy |
| mechanical energy | potential energy |
| reference frame | spring constant |
| work | |

1. _____ is done on an object when a force acts on the object as it moves through a distance.
2. A/An _____ is the amount of energy that is sufficient to produce a force of 1 N that acts for 1 m.
3. Global positioning satellites provide a/an ________ from which you can determine your position on the Earth and the current time.
4. A compressed spring has ________ equal to the work that the spring could do if it expanded again.
5. The _______ goes up by a factor of 4 if the speed increases by a factor of 2.
6. The ________ is a quantity that indicates the strength of the restoring force exerted by an elastic material when it is stretched.
7. Water at the top of a dam has ________ as a result of its height above the bottom of the dam.
8. Energy due to motion and energy due to position (such as height) are both categorized as kinds of _______.

### Section 9.2

| | |
|---|---|
| ampere (A) | electric current |
| horsepower (hp) | intensity |
| light | power |
| radiant energy | renewable energy |
| volt (V) | watt (W) |

9. ________ is the kind of energy that is carried by electromagnetic waves.
10. Seven-hundred and forty-six watts is equal to one _________.
11. _________ is a form of energy that can be replenished as rapidly as it is used.
12. An indicator of electric potential is the ________.
13. A/An ________ is the SI unit for electric current.
14. The flow of charged particles is called ________.
15. The _______ of light describes how many watts of power fall cross one square meter of area.
16. An energy flow of one joule per second is called one ________.
17. The rate at which work is done is called ______.

## Conceptual questions

### Section 9.1

18. How much work does it take to hold a dumbbell motionless over your head for 10 s?
19. What physical quantity is measured by a spring scale?
20. String and nylon thread will stretch when pulled with a moderate force, but only a small amount. If you apply the same force to a spring and it stretches much further than the string and thread, how do the spring constants of the string and thread compare to the spring?
21. How do you account for the mass of the hook when using a hooked lab mass?
22. Two apartment mates are arguing over the potential energy of a 10 kg TV that hangs on the wall. The first person claims that the television is 2 m from the floor so its potential energy is 10 kg × 9.8 N/kg × 2 m = 196 J. The other claims that since they are on the second floor, the TV is 12 m above the ground so the potential energy is 10 kg × ×9.8 N/kg × 12 m =1,176 J. Who is right or are they both right? Explain.
23. ☾ For a classroom investigation, a teacher hangs a heavy bowling ball by a rope from the ceiling so that the ball hangs at a height just a little lower than the teacher's nose. The teacher faces the ball, steps backward a few steps and draws the ball toward her until it just touches the tip of her nose. She then asks her class if she will remain uninjured when she releases the bowling ball—knowing it will swing away then reverse and swing back toward her nose again. How should you answer her?
24. ☾ What happens to the energy "lost" to friction?
25. ☾ Compare the change in potential energy between a ball that falls from 10 m to 5 m in height and another ball of the same mass that falls from 7 m to 2 m in height.

## Section 9.1

26. For examples (a)–(f) below, match the event with the correct form of energy.

| | |
|---|---|
| I. kinetic | II. gravitational potential |
| III. elastic potential | IV. thermal |
| V. electrical | VI. chemical |

a. ___ Ice melts when placed in a cup of warm water.
b. ___ Campers use a tank of propane gas on their trip.
c. ___ A car travels down a level road at 25 m/s.
d. ___ A bungee cord causes the jumper to bounce upward.
e. ___ The weightlifter raises the barbell above his head.
f. ___ A spark jumps from the girl's finger to the doorknob after she scuffs her feet on the wool rug.

27. Which of the following statements correctly describes the relationship between work and energy?
    a. Energy is the stored ability to do work.
    b. Work must be done in order for energy to be transferred between objects.
    c. Work must be done in order for energy to change from one form to another.
    d. All of the above.

28. Are these statements about the spring constant true or false?
    a. The spring constant is a measure of the stiffness of the spring.
    b. The spring constant tells you how many newtons of force it takes to stretch the spring one meter.
    c. If a spring stretches easily, it will have a high spring constant.
    d. The spring constant of a spring varies with the amount of stretch or compression of the spring.

29. The Smart Cart is a piece of dynamics demonstration equipment that displays in real time its position, velocity, acceleration, and the force applied to it.
    a. Explain how you would use it to measure kinetic energy.
    b. Explain how you would use it to measure work done on it.

30. The force of hot expanding gas from exploding gunpowder propels a cannonball out of the barrel of a cannon. How might the cannon be modified so that the cannonball leaves at a higher speed without increasing the amount of gunpowder used?

31. Ask two friends if they can find the difference in the following two statements.

    Statement 1: "A force that causes no change in an object's position also does no work."
    Statement 2: "A force that does no work may not cause any change in an object's position."

    Are both correct? In not, which is more correct? Write down their answers. How do they compare with your own answer?

## Section 9.2

32. Is it possible to have very little power but use a lot of energy? How?

33. Compare water flowing in a hose to electricity. Pressure in the pipe causes water to flow through the hose. What electrical quantity acts like pressure and causes electric current to flow?

34. Which is the correct statement about a 12 V battery?
    a. 12 A of electric current can flow from the battery.
    b. Each ampere of electric current that flows carries 12 W of power.
    c. The battery contains 12 times as much stored energy as a 1 V battery.

35. Which of the following energy flow diagrams might describe the path of the energy that powers the electric light in your classroom?
    a. nuclear energy > thermal energy > mechanical energy > electrical energy > light energy
    b. chemical energy > electrical energy > nuclear energy > light energy
    c. mechanical energy > chemical energy > electrical energy > thermal energy > light energy

36. Ask two friends if they can find the difference in the following two statements.

    Statement 1: "A runner with more power can expend energy more quickly to accelerate."
    Statement 2: "A runner with more energy uses more power when accelerating."

    Which one is more correct? Write down their answers. How do they compare with your own answer?

# Chapter review

### Section 9.2

37. Research and describe three energy flow diagrams for converting solar energy into electrical energy.

38. List five ways in which food you buy in a grocery store has an "energy content" other than the actual calories your body gets from eating the food.

39. How does the Sun's energy reach us?

40. Mercury thermometers, once common in science classrooms, have been largely replaced due to safety concerns. The "fire diamond" and the global harmonization system (GHS) are two methods used to label a mercury hazard. Locate and reproduce images of both labels and evaluate their relative strengths and weaknesses.

41. Identify the only career in the following list that does not require any knowledge of physics: optometrist, petroleum engineer, astronomer, industrial engineer, grocer, robotics designer, architect, and physical therapist.

42. Research a physics-related career and develop a class presentation. The presentation should use digital media (text, graphics, audio, etc.) to illustrate your findings and provide interest.

## Quantitative problems

### Section 9.1

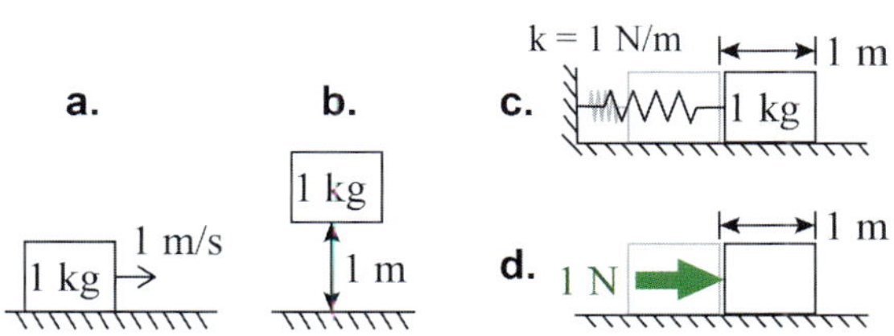


43 Rank the following scenarios in order of most energy to least energy.

   a. 1 kg block moving at 1 m/s on the ground
   b. 1 kg block at rest 1 m above the ground
   c. 1 kg block at rest on the ground but attached to a spring with $k = 1$ N/m that is stretched 1 m
   d. the work done to push the block a distance of 1 m with a force of 1 N

44. What is the elastic potential energy of a rubber band with a spring constant of 25.0 N/m if it is stretched by 10.0 cm from its original length?

45. If a box's mass is cut in half, but the box is raised to a height four times higher, how does its potential energy change?

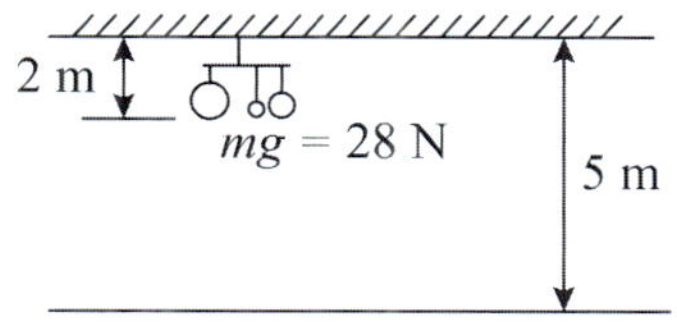


46. A mobile that weighs 28 N is hanging 2 m below a ceiling that is 5 m high. What is its potential energy with respect to the floor, the ceiling, and a point at the same height as the mobile?

47. What is the kinetic energy of a 6 kg bird moving at a speed of 15 m/s?

48. What is the largest force that an energy source containing 100 J can exert continuously over a distance of 25 m?

   a. 0.25 N  b. 4 N  c. 25 N  d. 2,500 N

49. At what height would a 15 kg bird have a potential energy of 2,400 J?

50. If you double the speed and mass of an object how does its kinetic energy change?

51. A 1 kg ball and a 9 kg ball have the same $E_k$. Compare their speeds.

52. A small airplane has a kinetic energy of 5,000,000 J when it travels at 100 m/s. What is its mass?

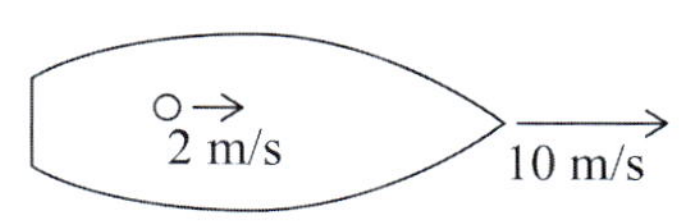


53. A ball with a mass of 150 g rolls due north along the deck of an ocean liner at a speed of 2.0 m/s. The ocean liner is also moving north, with a speed of 10.0 m/s relative to a nearby island.

   a. What is the kinetic energy of the ball as measured from the reference frame of the ocean liner?
   b. What is the kinetic energy of the ball as measured from the reference frame of the island?

54. How far is a spring extended if it has 1.0 J of potential energy and its spring constant is 1,000 N/m?

55. How much work do you do when you push a 400 kg car for 6 m with a force of 300 N? What is the car's final speed if all your work became kinetic energy?

56. Relative to sea level, how much total mechanical energy does a plane with mass of 1,000 kg have when it moves at 235 m/s at an altitude of 2 km above sea level?

# Chapter review

## Section 9.2

57. A light bulb is connected to a 120 V household circuit. How much current does the light bulb draw if it uses
    a. 100 W of power?
    b. 60 W of power?
    c. 150 W of power?

58. In the Empire State Building Run-up, participants compete to see who can run up the 320 m Empire State Building the fastest. The current record holder, Paul Crake, who has a mass of 64 kg, covered this distance in 9 min and 33 s. What is the minimum average output power that his body produced while doing this?

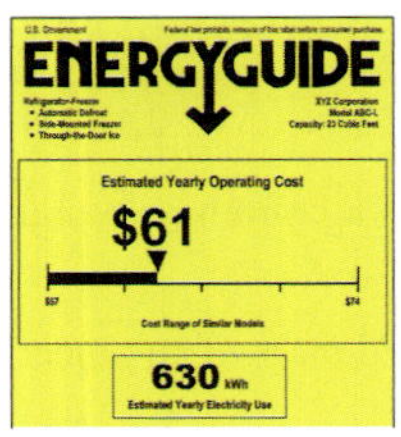


59. Does the refrigerator with the EnergyGuide label shown here use more or less energy than comparable appliances?

60. A man rides a bicycle that is connected to an electrical generator. If he rides as hard as he can, his body can produce a mechanical power of 500 W, but the generator is only 40% efficient at converting mechanical energy into electrical energy.
    a. How many 100 W incandescent light bulbs can he power?
    b. How many 100-W-*rated* compact fluorescent light bulbs can he power?

61. A small electric motor produces a force of 5 N that moves a remote-control car 5 m every second. How much power does the motor produce? Give your answer in watts and horsepower.

62. A fully charged cellphone battery contains 20,000 J of stored energy. If the cellphone uses 2 W of power, how long will the battery last? Give your answer in seconds, minutes, and hours.

63. In 2009, Usain Bolt set the world record of 9.58 s in the 100 m sprint. A published physical model of his race showed that he exerted approximately a constant horizontal force of 816 N and that 75,200 J of his work was used to overcome drag from the air.
    a. What was his average speed?
    b. How much total work did he do in the race?
    c. What was his average power output?
    d. What was his efficiency?

64. The *Boston Globe* reported on an electricity rate change:
    "[The electric utility] NStar filed for electric rate cuts yesterday that would save the average residential customer about $6 a month starting in January.... Counting both the cost of energy and delivery-oriented charges, the typical homeowner using 500 kilowatt-hours monthly would pay... less in Cambridge, NStar said.... Cambridge Electric's power rate is slated to go from 6.351 to 5.626 cents [per kilowatt-hour], and the delivery cost from 7.282 to 6.782 cents [per kilowatt-hour]."
    a. How much energy is consumed monthly by the typical consumer referred to in the story?
    b. Does electric generation or delivery cost more per kilowatt-hour in Cambridge?
    c. What will be the new cost per kWh for electricity generation plus delivery?
    d. For typical consumers, what will be the change in their total electric bill owing to the rate change?
    e. Do you think that consumers will like or dislike this rate change? Support your argument by communicating evidence found in, or inferred from, the article.

65. What is the maximum height a 5 hp engine could push a 10 kg box in 5 s? Assume you could attach the engine to an ideal mechanical device with perfect (100% efficient) transfer of energy.

66. Suppose you have a solar energy conversion system with a sunlight collecting area of 10 $m^2$. On a cloudless day sunlight has an intensity of about 600 $W/m^2$.
    a. How much energy is collected in 1 hr?
    b. A single electric light bulb uses 25 W of power. How long could the collected energy keep the bulb lit?
    c. A very efficient cabin uses about 400 W on average over 24 hr. If the solar conversion to electricity is 15% efficient, can this collector supply the average power draw for the cabin?

67. Imagine that a 1,000-acre (400-hectare) farm field can grow enough corn to produce 1.4 million liters ($1.4\times10^6$ L) of ethanol, an alternative automobile fuel, over the course of a year. Each liter of ethanol can produce roughly 20 million joules ($2\times10^7$ J) when burned. Now imagine that 20 *trillion* joules ($2\times10^{13}$ J) of energy had to be used to irrigate, fertilize, transport, and process all of this corn. Assuming that corn can be grown on this field indefinitely, would you consider this a renewable form of energy? Why or why not?

# Chapter review

## Standardized test practice

68. Anthony is standing on the top of a building 10 m high holding a 7 kg bowling ball. Mildred dug a 2-m-deep hole next to the base of the building. What is the gravitational potential energy of the bowling ball relative to the bottom of the hole?

    A. 137.2 J
    B. 548.8 J
    C. 686.0 J
    D. 823.2 J

69. How much energy is stored in a spring with a spring constant of 500 N/m if it is compressed a distance of 0.4 m?

    A. 80 J
    B. 100 J
    C. 20 J
    D. 40 J

70. A dock worker pushes a 50 kg crate up a 1-m-high, 3-m-long ramp. Ignoring friction, how much work did he do?

    A. 150 J
    B. 490 J
    C. 1,470 J
    D. 1,960 J

71. If you push a car with a constant force for 13 m and it gains 91 J of energy, with what force did you push it?

    A. 7 N
    B. 3 N
    C. 10 N
    D. 16 N

72. If an object is moving at a speed of 5 m/s and has kinetic energy of 600 J, what must its mass be?

    A. 81 kg
    B. 60 kg
    C. 96 kg
    D. 48 kg

73. How much energy does it take to accelerate a 90 kg object from rest to 13 m/s?

    A. 3,681 J
    B. 7,605 J
    C. 6,943 J
    D. 9,810 J

74. The kilowatt hour (kWh) is commonly used to measure electricity usage. How many joules is a kilowatt hour?

    A. $1\times10^3$ J
    B. 1 J
    C. $6\times10^3$ J
    D. $3.6\times10^6$ J

75. Three 5 V batteries are wired in series to power a light bulb. If 5 A of current is flowing through the circuit, how much power is the light bulb drawing?

    A. 20 W
    B. 25 W
    C. 75 W
    D. 100 W

76. A 300 W motor in a bowling alley lifts a 6 kg bowling ball directly upward at a constant speed. At what speed does the ball rise?

    A. 1.00 m/s
    B. 5.10 m/s
    C. 7.07 m/s
    D. 10.0 m/s
    E. 17.3 m/s

77. Which of the following is the most likely energy flow diagram for the electric lighting in one's house?

    A. nuclear energy > thermal energy > light energy
    B. mechanical energy > light energy > pressure energy
    C. light energy > electrical energy > nuclear energy
    D. chemical energy > electrical energy > light energy

78. A 4 kg electric car has a 9 V battery powering the motor. If 1.5 A runs through the circuit to run the motor at a constant power and it is accelerating from rest, how fast will the car be moving after 10 s?

    A. 0.34 m/s
    B. 8.22 m/s
    C. 16.43 m/s
    D. 67.5 m/s

79. An object 4 m off the ground has 300 J of gravitational potential energy. How much gravitational potential energy does it have relative to the bottom of a hole 3 m deep?

    A. 300 J
    B. 600 J
    C. 900 J
    D. 1,200 J

# Chapter 10 Conservation of Energy

"In this house, we obey the laws of thermodynamics!" So proclaimed Homer Simpson when he found out that his daughter Lisa had built a perpetual-motion machine in season six of *The Simpsons*. Even resourceful Lisa Simpson cannot create a perpetual-motion machine in a world ruled by the laws of physics. But that hasn't stopped generations of inventors from trying.

It's no wonder that perpetual-motion machines obsess so many minds. It costs money to fill up your car's gas tank every week, and watches need to have their batteries replaced. Wouldn't it be great if such machines could be kept in motion without having to refill or recharge them? Think of the savings, convenience, and environmental benefits.

Some of the ideas behind perpetual-motion machines *are* legitimate. Many designs are meant to change energy from one form (potential energy, say) into another (such as kinetic energy). A number of wheeled contraptions supposedly can keep rolling because masses slide or pivot along their spokes, exerting repeated torques while transforming energy into potential or kinetic form.

It isn't hard, however, to see the flaw in such reasoning. Take a simple, U-shaped track and release a toy car or marble from one end. In the absence of friction, the toy or marble would roll endlessly from end to end. But that never happens in the real world. In fact, you usually can hear energy leaving the system in the form of sound. The law of conservation of energy states that the sound or heat generated by friction must come at the expense of some of the system's mechanical energy, with the car or marble eventually coming to a rest at the bottom of the track.

Newton's first law states that an object in motion remains in motion in the absence of a nonzero net force. Objects within the nearly empty voids between galaxies may be able to remain in motion for billions of years—a time that seems eternal when compared to a human lifespan—but an object with that motion does no work. In order to do work, such a system must be acted upon by an unbalanced force, which means that its subsequent motion would not continue unchanged forever.

# Chapter study guide

## Chapter summary

Energy powers modern society. Most energy that we use, however, has been transformed from one form of energy to another. Is energy lost in this process of transformation? No, although energy can be converted into other forms, such as thermal energy (or heat) and radiated away. Energy conservation is a fundamental principle in physics: Energy cannot be created or destroyed, only transformed from one form to another. On a practical level, energy conservation allows us to solve physics problems—such as free fall or motion down a ramp—more easily than by simply applying the equations of motion. On a deeper level, energy conservation is behind many important industrial and technological applications in society, such as power generation and the automobile engine. In this chapter, you will learn the principles of energy conservation and how to apply it to better understand the physics behind many phenomena and applications.

## Learning objectives

By the end of this chapter you should be able to

- define energy conservation and describe examples of its application;
- apply the principles of open and closed systems to energy conservation;
- solve energy conservation problems, including conservation of kinetic and potential energy;
- define the work–energy theorem and apply it to solve problems;
- explain the role of friction in energy conservation; and
- calculate the efficiency of a system or energy transformation.

## Investigations

10A: Inclined plane and the conservation of energy
10B: Work and energy
10C: Springs and the conservation of energy
10D: Work done by friction

## Chapter index



## Important relationships

$$E_{initial} = E_{final}$$

$$W_{net} = \Delta E_k$$

$$\Delta E = 0$$

$$\Delta E = Q + W$$

$$\eta = \frac{E_{out}}{E_{in}}$$

## Vocabulary

**system**
**law of conservation of energy**
**friction**
**closed system**
**state**
**efficiency**
**open system**
**work–energy theorem**

One of the foundational laws of physics is that the total amount of energy in a closed system does not change. As events happen, such as a ball rolling downhill, energy changes from one form to another. *Nonetheless, the total energy remains constant.* Any process that requires a certain amount of energy will *not* occur if there is insufficient energy available. This section addresses the law of conservation of energy and how this law governs all processes in nature.

## Systems and interactions

**What is a system?**

If we want to understand the speed of a falling ball, we might include the ball's initial height, mass, and speed, as well as the force of gravity. The color of the sky or exact materials inside the ball are information we exclude because these details do not matter to what we are trying to understand. *A group of interacting objects and influences is called a* ***system****.* We choose a system to include only the objects and influences—such as forces—that are important to what we are investigating. Choosing a system frees us from having to consider *everything,* which would be an impossible task.

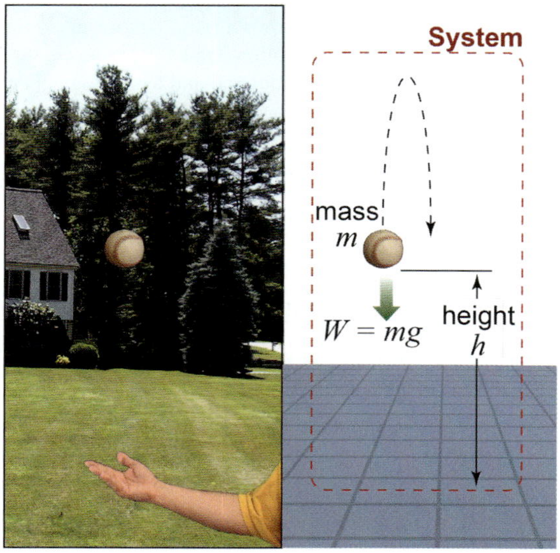


**Open and closed systems**

A system can be thought of as everything within an imaginary shell, or system boundary, that completely encloses the behavior you want to understand. If we allow matter or energy to cross the boundary then we say the system is *open.* A cup of hot coffee on a table is an open system. In a few hours the thermal energy leaks out into the room and in a few days the water evaporates away, leaving only the coffee residue in the bottom of the cup. Both matter and energy have left the system. If no matter or energy or force can cross the boundary then we say the system is *closed.* Closed systems are useful to think about, because in a **closed system** the total energy remains constant. The total energy at one time will be the same as the total energy at some later time. In an **open system**, however, the energy can change because the surroundings can add or remove energy from it.

**Open system**

Energy and/or matter can cross the system boundaries.

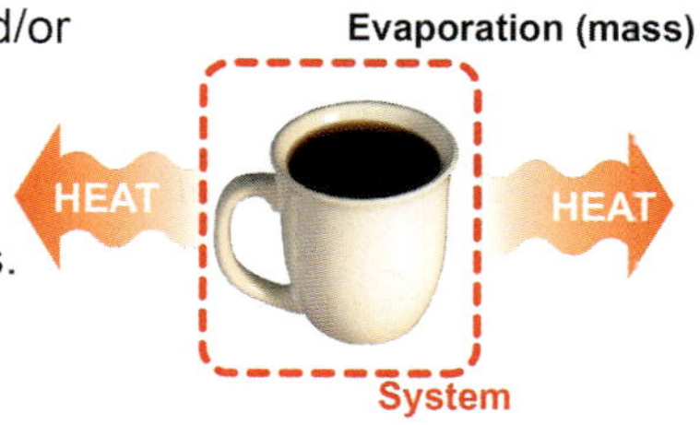


**Closed system**

No matter or energy is allowed to cross the boundries of the system.

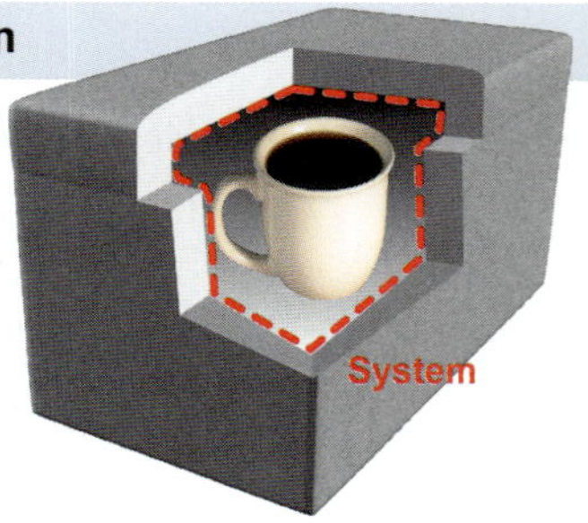


**Approximating a closed system**

While there is no such thing as a truly closed system, it is often a very good approximation to *assume* a closed system as a starting point for an investigation. The analysis is much easier and we get the "big picture" of what happens. We can start with a simplified model and then add details based on comparing the model's predictions with actual data.

**What is the big idea of energy conservation?**

The total energy in a closed system remains constant. Any change in the system that increases one form of energy, such as kinetic energy, can only do so if another form of energy decreases by the same amount. This is a powerful way of looking at changes. Changes involve the exchange of energy. Any change that increases a system's energy in one form can only occur in parallel with one or more changes that decrease the system's energy in other forms.

Video

**Law of conservation of energy**

*Energy cannot be created or destroyed—only transformed from one form into another. The total energy in a closed system remains constant.*

**What kinds of energy are involved?**

The **law of conservation of energy** applies to the *total* energy of a system, and not to any one form of energy. Some or all of the energy can be transformed from one form to another, as long as the *total* energy remains constant. "Constant" means that the total energy in a closed system at any time is the same as the system started with.

**How is energy conservation used?**

To see how this is useful for understanding how things change, consider a car on a roller coaster. Let the system be the roller coaster and car, including the effect of gravity. If we assume the system is closed, then the energy at the start is all potential because the car is at rest on top of the hill.

**Conservation of energy on a rollercoaster**

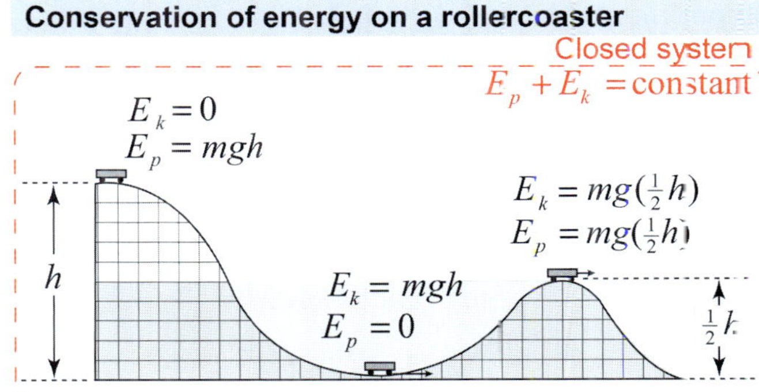


**What kind of problems can we solve?**

At some later time the car is rolling fast at the very bottom of the hill. How fast can the car go? Energy conservation tells us that *if* potential energy and kinetic energy are the only forms of energy in the system, *then* the car's kinetic energy gain can only come from potential energy lost. The kinetic energy at $h = 0$ must equal the lost potential energy. Therefore, we can set $\frac{1}{2}mv^2$ equal to $mgh$ and solve for the speed at the bottom of the hill. Since the mass cancels from both sides, conservation of energy tells us that the speed of the car is completely determined by the change in height $h$.

**Conservation of energy**

$$E_{initial} = E_{final} \quad (10.1)$$

$E_{initial}$ = initial total energy (J)
$E_{final}$ = final total energy (J)

**Using energy conservation**

Equation (10.1) is a good way to apply the law of conservation of energy to solving physics problems. In a closed system the total energy at the start is the same as the total energy at any later time or in any other configuration of the system.

**"Conserving" energy in daily life**

In physics, the term "conservation" means something *different* from what it does in conversation. When we talk about "conserving" energy in everyday life, we mean to *use less energy*. When you leave a room, you turn off the lights to "conserve" energy. When a truck is loading, it will turn off its engine to "conserve" fuel. When people speak of "running out of energy" what they are really referring to is that certain forms of energy, such as fossil fuels, are limited resources. "Using" these resources does not destroy the energy; instead, the energy is turned into heat, chemical change, and other forms.

**States of a system**

Conservation of energy is usually applied between two different *states* of a system before and after a change. In the context of physics a **state** is a particular configuration of all the elements in the system that has a certain energy. The change rearranges one or more elements in the system, usually transforming or transferring some energy. If the system is closed, the total energy before the change is the same as the total energy after the change. The distribution of energy in the system before and after the change determines what changes are possible.

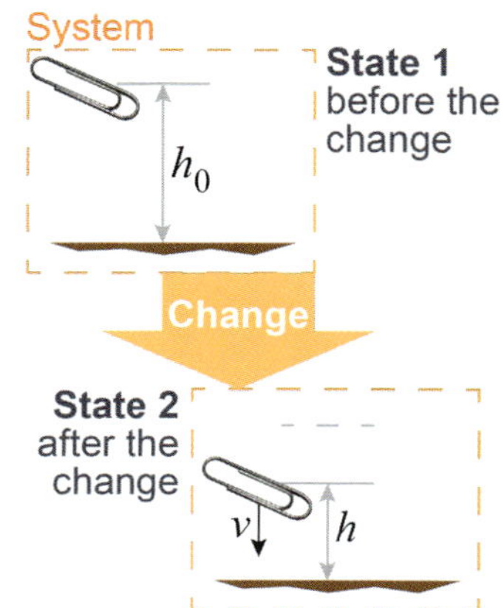


**Galileo's thought experiment**

Popular legend holds that Italian scientist Galileo Galilei dropped two balls, one ten times more massive than the other, from the top of the Leaning Tower of Pisa to demonstrate that, in the absence of air resistance, the acceleration of gravity is independent of the mass of the object. Historians dispute that Galileo actually performed this experiment, since only Galileo's secretary claimed in writing that he had done so. Even if Galileo never did the experiment, we can use conservation of energy and a few paper clips to test Galileo's hypothesis.

**A real experiment**

Which falls faster, 1 paper clip or a group of 10 paper clips together?

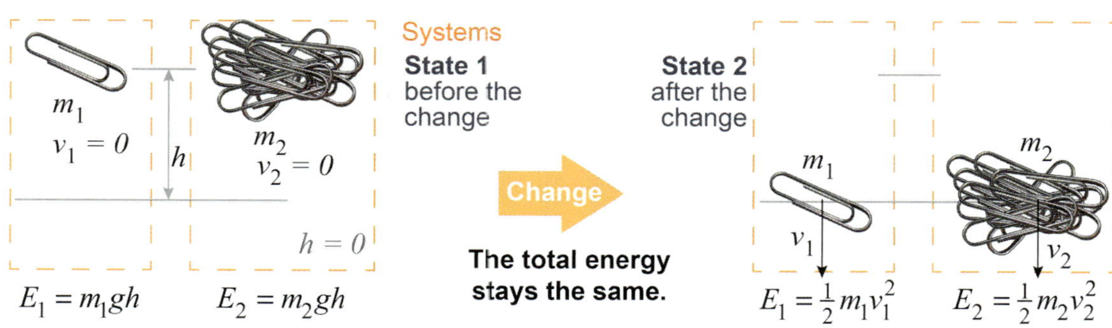


**The results: Mass does not matter**

Before the change, neither mass is moving, so there is only potential energy. After the change, the height is zero, so there is only kinetic energy. For each system we set the energy before the change equal to the energy after the change and solve the resulting equations for the speed $v$. The result is *the same for both!* In the absence of friction, the speed depends only on the change in height $h$. Galileo was correct in his conclusion.

| Energy before the change | = | Energy after the change | |
|---|---|---|---|
| $m_1gh$ | = | $\frac{1}{2}m_1v_1^2$ | ➡ $v_1 = \sqrt{2gh}$ |
| $m_2gh$ | = | $\frac{1}{2}m_2v_2^2$ | ➡ $v_2 = \sqrt{2gh}$ |

**What do the energies tell you?**

Things tend to speed up, or gain kinetic energy, as they move downward, such as when they roll down a hill or fall. When friction can be neglected, conservation of energy is used to relate speed and height. If the mass of the moving object is known, then:

1. potential energy ($mgh$) tells you *height,* and
2. kinetic energy ($\frac{1}{2}mv^2$) tells you speed.

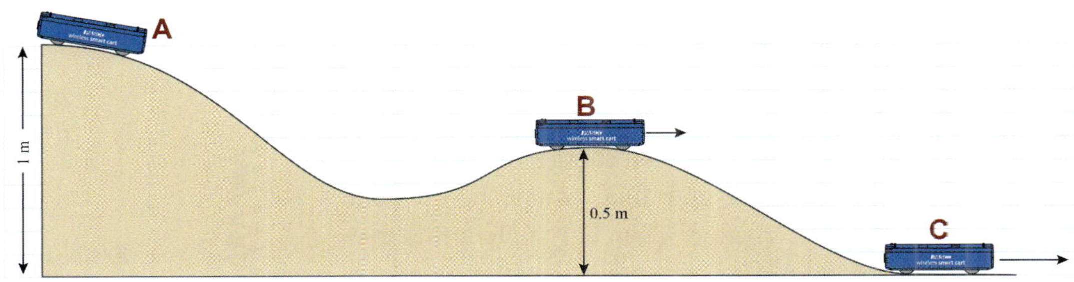


**How to apply energy conservation**

Consider a car that is released from rest at the top of a hilly track. If there is no friction, then we can assume a closed system. At any point, the total mechanical energy of the car rolling down the track is the same as the total mechanical energy it has at the start—i.e., potential energy $mgh$ plus kinetic energy $\frac{1}{2}mv^2$. Apply energy conservation between two different points by naming variables, writing down the relevant forms of energy at both points (step 1), and then eliminating terms that are zero (step 2). If the problem is to find the speed of the car at point B, then the resulting equation is solved for the speed at B (step 3).

**Problem-solving strategy**

| Step | Point A | | Point B | Solution |
|---|---|---|---|---|
| 1 Write down relevant forms of energy. | $mgh + \frac{1}{2}mv^2$ | = | $mgh + \frac{1}{2}mv^2$ | |
| 2 Eliminate any terms that are zero. | $mgh_A + \frac{1}{2}mv^2$ (0) | = | $mgh_B + \frac{1}{2}mv^2$ | |
| 3 Solve for the variable you want. | $mgh_A$ | = | $mgh_B + \frac{1}{2}mv^2$ | $v = \sqrt{2g(h_A - h_B)}$ |

**Height change determines speed**

The reduction in potential energy from A to B is $mg(h_A - h_B)$; therefore, the speed of the car at point B depends only on the difference in height between A and B and not on the path the car takes. The fact that there is a valley between A and B makes no difference. The gain in kinetic energy equals the loss in potential energy:

$$v_B = \sqrt{2g(h_A - h_B)} = \sqrt{2(9.8\ \text{m/s}^2)(1.0\ \text{m} - 0.50\ \text{m})} = 3.1\ \text{m/s}$$

**Choosing where $h = 0$**

It is usually convenient to define the zero of potential energy ($h = 0$) to be the lowest point in the problem. At point C the height is defined to be zero, so at this point all of the car's initial potential energy is converted to kinetic energy. The result is that the car is fastest at C because it has lost more potential energy and gained more kinetic energy:

$$v_C = \sqrt{2gh_A} = \sqrt{2(9.8\ \text{m/s}^2)(1.0\ \text{m})} = 4.4\ \text{m/s}$$

## Problems involving springs

**How do I include a spring?**

A spring allows for a third form of energy: elastic potential energy. In problems with springs, the total energy has three terms: one each for gravitational potential energy, elastic potential energy, and kinetic energy.

**Dropping a ball onto a spring**

Consider a 2.0 kg ball that drops onto a vertical spring with a spring constant $k$ of 1,000 N/m. From what height did the ball drop if the spring compresses by 25 cm?

**Problem-solving strategy**

To solve the problem we follow the same three steps. First, name the variables and write down the total energy of the system in the initial and final states, before and after the change. Then eliminate terms that are zero. The remaining equation contains the solution.

**Total energy**

$$E_{total} = mgh + \tfrac{1}{2}mv^2 + \tfrac{1}{2}kx^2$$

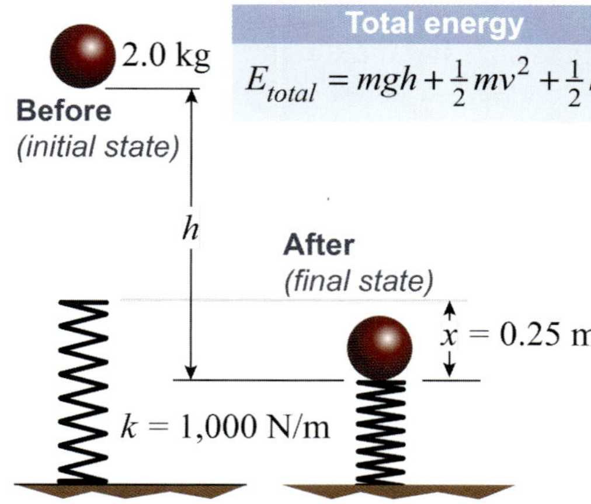


| | Initial state | | Final state |
|---|---|---|---|
| 1 Write down relevant forms of energy. | $mgh + \frac{1}{2}mv^2 + \frac{1}{2}kx^2$ | = | $mgh + \frac{1}{2}mv^2 + \frac{1}{2}kx^2$ |
| 2 Eliminate any terms that are zero. | $mgh + \frac{1}{2}m\cancel{v}^2_0 + \frac{1}{2}k\cancel{x}^2_0$ | = | $mg\cancel{h}_0 + \frac{1}{2}m\cancel{v}^2_0 + \frac{1}{2}kx^2$ |
| 3 Solve for the variable you want. | $mgh$ | = | $\frac{1}{2}kx^2$ |

**Solution**

$$h = \frac{kx^2}{2mg}$$

$$h = \frac{kx^2}{2mg} = \frac{(1{,}000 \text{ N/m})(0.25 \text{ m})^2}{2(2.0 \text{ kg})(9.8 \text{ m/s}^2)} = 1.6 \text{ m}$$

**Springs and speed**

A horizontal spring is used to launch a 2.0 kg ball. The spring is compressed by 0.25 m and has a spring constant $k$ of 1,000 N/m. What is the maximum speed of the ball?

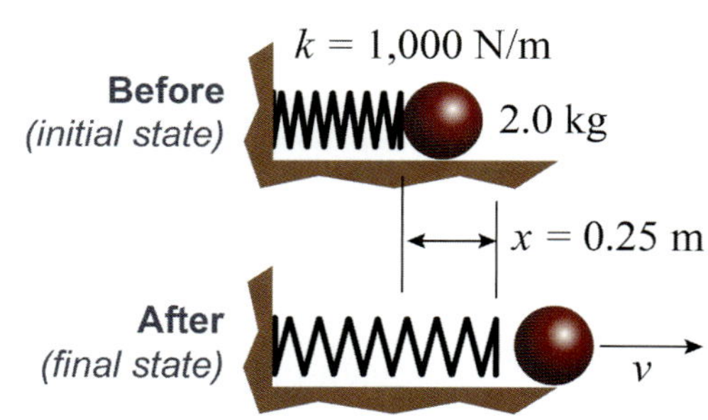


| | |
|---|---|
| *Asked:* | speed $v$ |
| *Given:* | $k = 1{,}000$ N/m; $x = 0.25$ m; $m = 2.0$ kg |
| *Relationships:* | total mechanical energy: $E_{total} = mgh + ½mv^2 + ½kx^2$ |
| *Solution:* | |

| | Initial state | | Final state |
|---|---|---|---|
| Write down relevant forms of energy. → | $mgh + \frac{1}{2}mv^2 + \frac{1}{2}kx^2$ | = | $mgh + \frac{1}{2}mv^2 + \frac{1}{2}kx^2$ |
| Eliminate zero terms. → | $mg\cancel{h}_0 + \frac{1}{2}m\cancel{v}^2_0 + \frac{1}{2}kx^2$ | = | $mg\cancel{h}_0 + \frac{1}{2}mv^2 + \frac{1}{2}k\cancel{x}^2_0$ |
| Solve for the variable you want. → | $\frac{1}{2}kx^2$ | = | $\frac{1}{2}mv^2$ |

**Solution**

$$v = x\sqrt{\frac{k}{m}}$$

$$v = (0.25 \text{ m})\sqrt{(1{,}000 \text{ N/m})(2.0 \text{ kg})} = 5.6 \text{ m/s}$$

*Answer:* $v = 5.6$ m/s

# Investigation 10A

## Inclined plane and the conservation of energy

**Essential questions** *What law governs the energy transformations of motion on an inclined plane?*

Interactive simulation

This investigation uses a cart on a frictionless track to explore how gravitational potential energy and kinetic energy change as motion changes. For example, if you want to design a roller coaster to reach 30 mph, how high must it be at the start?

### Part 1: Changes in energy for motion down an inclined plane

1. Set up your equipment as in the picture. The plunger on the cart should be facing the end stop, and the ramp should have an incline of 15°.

2. Open file **10A_ApplyingConservationOfEnergy**, and connect your Smart Cart using Bluetooth.
3. Explore the motion of the cart on the ramp. Start recording data with the cart at the bottom of the ramp, then roll it to the top of the ramp and release it. Stop recording when the cart reaches the bottom. Answer the questions for the time the cart was rolling down the ramp.

a. *Sketch position and velocity. What are the shapes of the graphs? Why do you get these shapes?*
b. *Sketch kinetic energy and potential energy. What are the shapes of the graphs? Why?*
c. *What is the sum of the kinetic and potential energies at the top of the ramp when the cart is released? At the bottom? How are their changes related to each other?*
d. *What is the speed at the bottom of the ramp? Try adding mass to the cart and take another set of data. How does the speed change in each case? Why?*

### Part 2: Designing a roller coaster that reaches 30 mph

1. Convert 30 mph into units of m/s. Make this the final speed of the block at the bottom of the ramp.
2. Set the mass of the roller coaster to $m = 2{,}000$ kg.
3. Vary the simulation parameters—such as the vertical height $h_0$ or inclination angle $\theta$—to produce a speed of 30 mph at the bottom of the ramp.
4. The fastest roller coaster in the United States reaches a maximum speed of 128 mph. Use the interactive simulation to estimate the height of this roller coaster.

a. *How does changing the steepness of the roller coaster (while keeping the same initial height) affect the speed at the bottom of the ramp? Why?*
b. *How does changing the mass of the roller coaster change its speed at the bottom? Why?*
c. *Research the* actual *height of the roller coaster in step #4. How well does your estimate agree with it?*

## Friction and open systems

**What happens with friction?**

In our initial treatment of energy conservation we assumed 100% conversion of mechanical energy among various forms. In reality, every transformation of energy diverts some energy into heat, wear, or other forms gathered into the catch-all basket of *frictional losses*. When a system has changed, any energy diverted into frictional losses reduces the energy available for mechanical forms, such as kinetic, gravitational potential, and elastic potential energy.

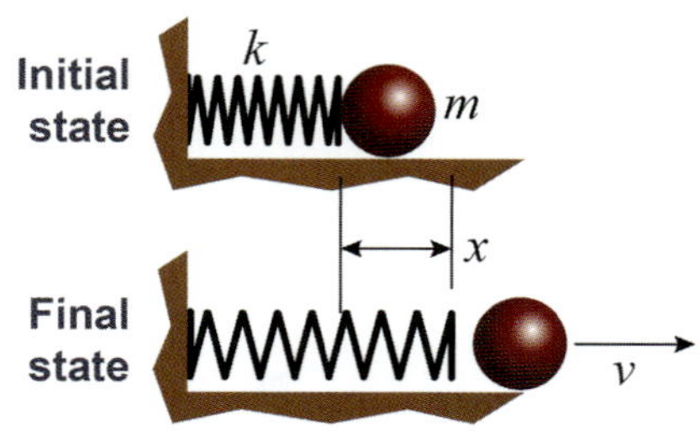


**Initial state** | **Realistic final state**

$$mgh + \tfrac{1}{2}mv^2 + \tfrac{1}{2}kx^2 = mgh + \tfrac{1}{2}mv^2 + \tfrac{1}{2}kx^2 + \textbf{losses}$$

**In real processes, some energy is always diverted into unusable forms called "frictional losses."**

**Is energy actually lost?**

The law of conservation of energy states that energy can be neither created nor destroyed, only transformed. If we counted *every* form of energy, then the law of energy conservation would be perfectly obeyed. Unfortunately, the forms of energy that become "frictional losses" are not reflected in the observable properties of the system, such as speed, height, or length. Strictly speaking, if only observable forms of energy, such as kinetic energy, are counted then the system is *open* and frictional energy crosses the boundary, out of the system. That is why we loosely call frictional energy "losses."

**How good is the approximation?**

The *closed system* is always an approximation in the macroscopic world. Nonetheless, the approximation is extraordinarily useful for analyzing the physics of an incredible variety of natural and technological systems! If the frictional losses are small—often less than 10%—then the approximation can be very good. The predictions of a model based on closed-system energy conservation often will accurately reflect real experiments. Even when friction is present and important, the ideal, frictionless case provides a "best-case" estimate of changes in a system.

**How useful is an ideal calculation?**

This is an important point: *Energy conservation sets an upper limit on what changes are possible.* If we solve the spring problem on page 284, then we find that frictionless ideal energy conservation predicts a speed of 5.6 m/s.

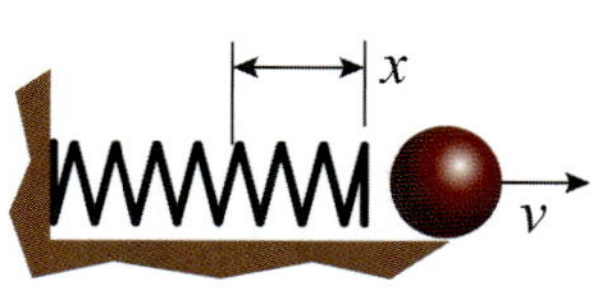


Frictionless *(ideal)* energy conservation

$$\tfrac{1}{2}kx^2 = \tfrac{1}{2}mv^2 \longrightarrow v = x\sqrt{\frac{k}{m}} = (0.25\text{ m})\sqrt{\frac{1{,}000\text{ N/m}}{2.0\text{ kg}}} = 5.6\text{ m/s}$$

**What is a "maximum" value?**

Any frictional losses would reduce the kinetic energy of the ball in the final state of the system. Therefore, 5.6 m/s is the *maximum possible* speed the ball could have *if* friction could be reduced to insignificance. When a problem on a physics exam asks for the "maximum" speed or height, the problem is actually asking you to approximate the system as ideal and frictionless.

# Section 1 review

The law of conservation of energy states that energy cannot be created or destroyed—only transformed. More specifically, the total energy remains constant in a closed system. An object's mechanical energy is defined as the sum of its potential and kinetic energies, and it may remain constant in the absence of "frictional losses."

| Vocabulary words | system, closed system, open system, law of conservation of energy, state |
|---|---|

| Key equations | $E_{initial} = E_{final}$ | $E_p + E_k = \text{constant}$ |
|---|---|---|

## Review problems and questions

1. You shoot a 2.0 kg basketball toward the hoop with an initial total energy of 500 J. (Neglect air friction on the ball.)
   a. When the ball reaches the top of its arc, what is its total energy?
   b. When the ball is just about to hit the rim, what is its total energy?
   c. What principle are you demonstrating?

2. Panel A shows a ball shortly after being tossed upward. Panel B shows the same ball at an instant on its way back down. Assume that air resistance can be ignored. Which statement is true?
   i. The potential and kinetic energy are both greater in Panel A than in Panel B.
   ii. The potential and kinetic energy are both greater in Panel B than in Panel A.
   iii. The potential energy is greater in Panel A, but the kinetic energy is greater in Panel B.
   iv. The potential energy is greater in Panel B, but the kinetic energy is greater in Panel A.

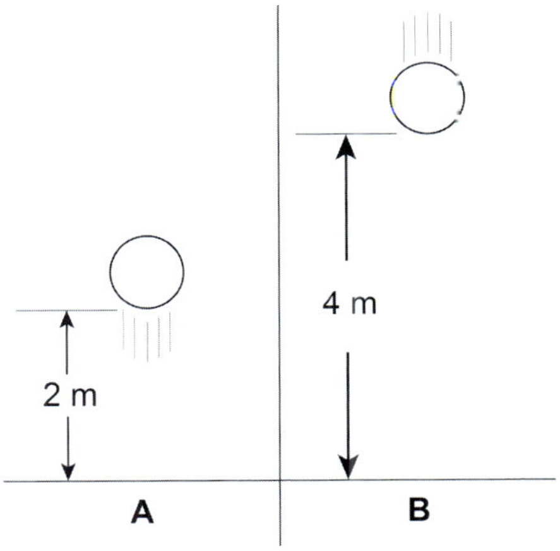


3. A roller-coaster cart, initially stationary at position *a*, is given a gentle push to the right. As it glides along the track, it passes through positions *b*, *c*, and *d*. Assume that friction and air resistance can be ignored.
   i. At which of the four positions is gravitational potential energy greatest?
   ii. At which position is the cart moving fastest?
   iii. At which position(s) is/are potential and kinetic energy equal?

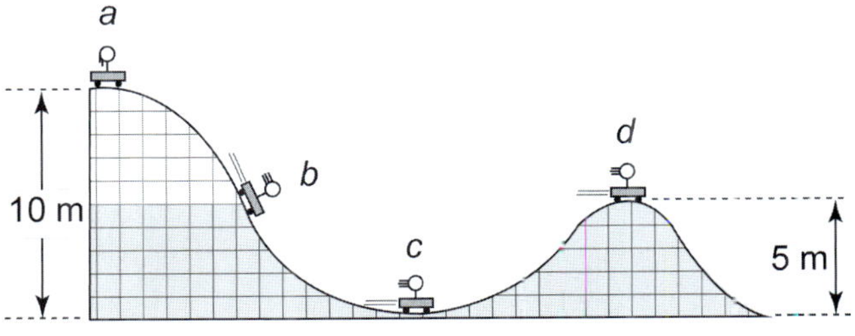


4. A ball is dropped from two heights, one four times as high as the other. What is the ratio of the speeds in the two cases just before the ball hits the ground? (Assume that air resistance can be ignored.)

## 10.2 - Work and energy transformations

Up to now, we have considered the energy *within* a system before and after a change. But energy can also enter or leave an *open* system. Mechanical energy enters or leaves a system through *work,* the action of forces. Work done *on* a system increases the system's total energy, whereas work done *by* a system decreases its energy. This section connects work done with the conservation of energy.

### Work and energy

**How do forces change the energy of a system?**

As you learned on page 257, work is a form of energy. The work done by a force is the force multiplied by the distance moved in the direction of the force. Consider a system containing an uncompressed spring at its free length. A force acts from *outside the system* to compress the spring a distance $x$. The final energy of the system is the initial energy it started with plus the work done *on* the system by the external force.

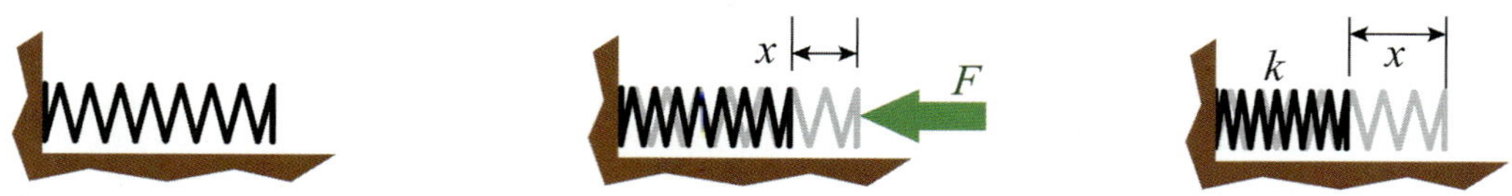


| Initial energy | | Work done | | Final energy |
|---|---|---|---|---|
| $mg\cancel{h}_0 + \frac{1}{2}m\cancel{v}_0^2 + \frac{1}{2}k\cancel{x}_0^2$ | + | $W$ | = | $mg\cancel{h}_0 + \frac{1}{2}m\cancel{v}_0^2 + \frac{1}{2}kx^2$ |
| | | $W$ | = | $\frac{1}{2}kx^2$ |

**The final energy of a system is the initial energy plus work done on the system.**

**What if the system does work?**

Again, consider the system to consist only of the spring. If the spring is now used to launch a ball, then the system does work on something outside the system: the ball. When a system exerts a force that does work outside the system, then the final energy is the initial energy *minus* the work done *by* the system.

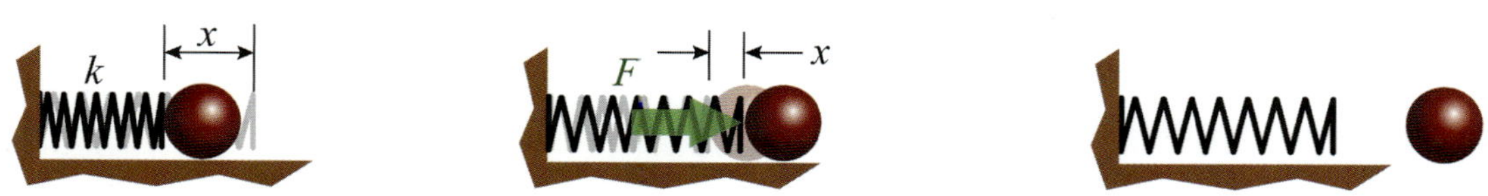


| Initial energy | | Work done | | Final energy |
|---|---|---|---|---|
| $mg\cancel{h}_0 + \frac{1}{2}m\cancel{v}_0^2 + \frac{1}{2}kx^2$ | − | $W$ | = | $mg\cancel{h}_0 + \frac{1}{2}m\cancel{v}_0^2 + \frac{1}{2}k\cancel{x}_0^2$ |
| $\frac{1}{2}kx^2$ | − | $W$ | = | 0 |

**The final energy of a system is the initial energy minus work done by the system.**

**Work and conservation of energy**

If all forces act *inside* the system, then the total energy of the system remains constant because all the energy lost by one part of the system is gained by another part. If forces act *outside* the system, then the energy of the system either increases when work is done on the system or decreases when the system does work on the outside environment. Notice how the initial and final energies are not equal for the spring system illustrated above because there are forces acting outside the system.

**Forces that change speed**

The major energy transformation in many situations is between forces doing work on a system and the kinetic energy of the system. The **work–energy theorem** states that the net work done on an object is always equal to its change in kinetic energy, as given in equation (10.2).

**Work–energy theorem**

$$(10.2) \quad W_{net} = \Delta E_k$$

$W_{net}$ = net work done (J)
$\Delta E_k$ = change in kinetic energy (J)

**What does the work–energy theorem mean?**

Equation (10.2) says that the *net* work done on a system equals the system's change in kinetic energy. In this context, the word "net" means two things:

1. The work is done by the net force, not by individual forces that may cancel or partially cancel each other; and
2. the net work includes only the work done by the component of the net force that acts in the direction of the object's motion.

The first requirement should be familiar since objects respond to the net force. When a net force does net work on an object, the object accelerates according to Newton's second law and therefore gains kinetic energy. The second requirement is new and important. Other elements within a system may constrain the motion of an object so it does not move in the direction of the net force. An equivalent statement is that work is only done by the component of the net force in the direction of motion.

**Positive and negative work**

The net work in equation (10.2) can be positive or negative. Consider a car on the road. A reaction force acts on the car created by the car's engine applying a force against the road. The reaction force does *positive* work on the car, and the car's kinetic energy increases. Work is positive when it is done *on* a system.

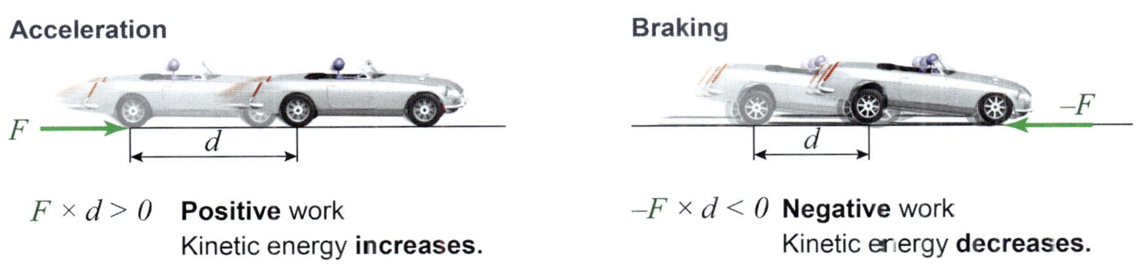


**When is work negative?**

The car can also have its kinetic energy decreased by the driver applying the brakes. Brakes resist the rotation of the car's wheels. This resistance creates a reaction force from the road that acts back against the car, opposite to the direction of motion. The quantity $Fd$ is now *negative* and kinetic energy therefore decreases. In the context of the work–energy theorem, the net work can be positive or negative. When the net work is positive, the kinetic energy of the system increases. When the net work is negative, the kinetic energy of the system decreases.

**Multistep processes**

Sometimes forces do work that becomes stored potential energy. This happens when you stretch a rubber band. If that stretched rubber band is later used to launch a paper airplane, then the work done by the rubber band on the airplane is what increases the plane's kinetic energy.

# Investigation 10B Work and energy

**Essential questions** *How is the work done on a system related to its change in energy?*
*How is the efficiency of a system calculated?*

Interactive simulation

In this investigation you will explore the relationship between work and kinetic energy by measuring the work done on a cart to accelerate it, and then comparing that work to the cart's change in kinetic energy.

## Part 1: Work and kinetic energy when force is constant

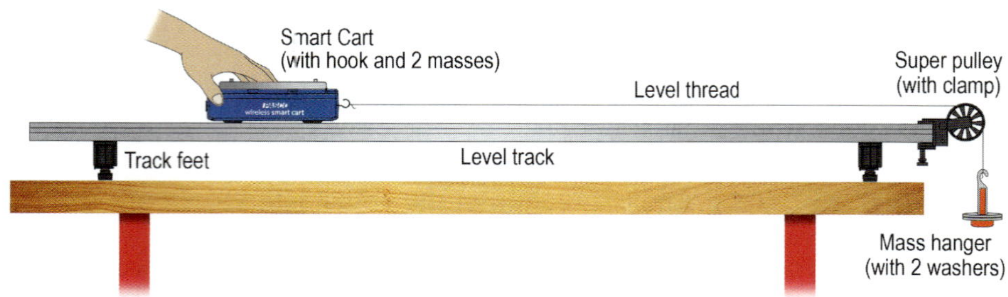


1. Set up the equipment like the picture. Open the experiment file **10B_WorkEnergyTheorem**, and connect to the Smart Cart with Bluetooth.
2. Unhook the thread from the cart. Measure the mass of cart and two 250−g masses.
3. Zero the cart's force sensor while nothing is touching the hook.
4. Reattach the thread, and then roll the cart all the way to the end of the track opposite the pulley and hold it in place. Start recording data and release the cart.
5. Catch the cart before it hits the pulley. Data collection automatically stops after 1.5 s.
6. Use your data to determine the net work $W_{net}$ done on the cart ($W_{net}$ = area under your force versus position data), and the final speed of the cart $v_f$. Record both values in a table.

## Part 2: Work and kinetic energy when force is not constant

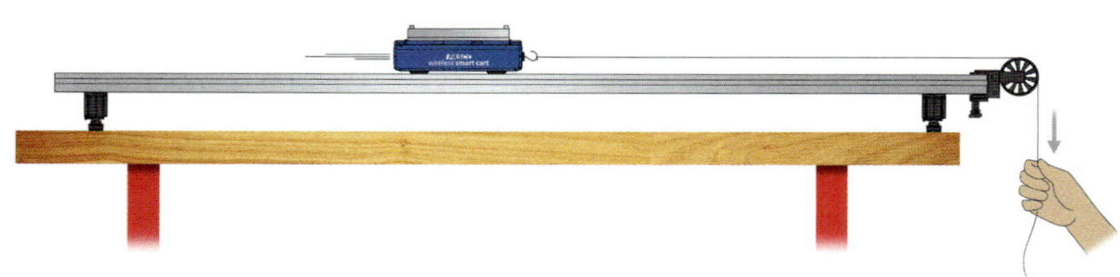

1. Unhook the hanging mass and have a group member hold the loose end of the thread in their hand at the other end of the track (no tension in the thread). Zero the force sensor.
2. Start recording data, and then gently pull on the string so the cart rolls down the track. Catch the cart before it rolls into the pulley.
3. Use your data to determine, and record, the work done on the cart, and the cart's final speed.

a. *Use the work-energy theorem to derive an equation for the final speed of the cart $v_f$ in terms of mass $m$ and net work $W_{net}$. Assume the initial speed of the cart is zero. Show your work.*

b. *Use the equation you derived and your measured value for $W_{net}$. Calculate and record the theoretical final speed of the cart for Part 1 and Part 2.*

c. *How does the theoretical final speed of the cart compare the experimental (measured) value in both parts? If the values are different, what do you think caused the difference?*

d. *How efficient was the work done to change the cart's kinetic energy in each part? Explain the significance of the values you obtain. (Efficiency = $\Delta E_k / W_{net} \times 100\%$)*

e. *If you do 0.621 N·m of work on a cart (same mass as the cart in this experiment) to accelerate it from rest, what would the cart's final speed be if the efficiency of the work was 90.3%?*

**Applying the work–energy theorem**

"If you can read this, you're too close." Have you ever seen a bumper sticker that says this? It is used to discourage *tailgating*—the act of driving too close to the moving vehicle in front of you. How close is too close? The work–energy theorem holds the key to this practical physics question.

**Solving an example problem**

Suppose a car is traveling at 30 m/s (67 mph). The maximum force of traction between the road and a car is typically about $0.8mg$, or 80% of the car's weight. How much distance does the car need to stop? In a real situation there would be other factors, such as driver reaction time, but in this physics problem assume zero reaction time. The work done by the reaction force from braking decreases kinetic energy to zero.

**Stopping distance**

**Work–energy theorem:**

| Work done by friction | Change in $E_k$ |
|---|---|
| $W = 0.8\,mgd$ | $\Delta E_k = \frac{1}{2}mv_i^2$ |

| | |
|---|---|
| *Asked:* | distance $d$ |
| *Given:* | $v = 30$ m/s; $F = 0.8mg$ |
| *Relationships:* | $W = Fd$; $E_k = ½mv^2$ |
| *Solution:* | Set the work done equal to the change in kinetic energy: |
| | $Fd = ½mv^2 \rightarrow 0.8mgd = ½mv^2$ |
| | $d = v^2/(2\times0.8g) = (30\text{ m/s})^2/(2\times0.8\times9.8\text{ m/s}^2) = 57$ m (190 ft) |

**How does speed affect stopping distance?**

We can draw a very important point from the solution to the problem above: Stopping distance is proportional to the *square* of velocity, or $d \propto v^2$. *Twice the speed means four times the stopping distance!* This means that a car initially traveling at 60 mph needs four times the stopping distance compared to a car traveling 30 mph. And that's not all. If you see a hazard ahead it takes time—*your reaction time* $t_r$— to apply the brakes. Since distance equals the product of speed and time, this will add $v_i \times t_r$ to the amount we calculated earlier. At 30 m/s, each one-half second of reaction time adds another 15 m of stopping distance, or almost 50 ft.

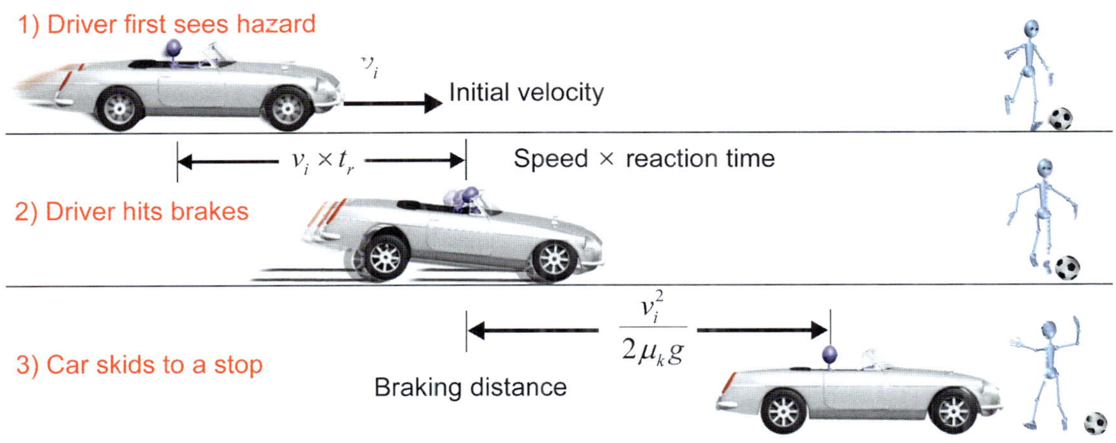


**Stopping distance and friction**

When accounting for friction between tires and the road for a skidding vehicle, the force of sliding friction is $\mu_k mg$, not $mg$ (or $0.8mg$ as in the problem above). The equation for stopping distance is then $d = v^2/(2\mu_k g)$. If the coefficient of kinetic friction $\mu_k$ is small, as is the case for a wet or icy surface, then the stopping distance becomes very large!

## Relating force and work to energy transformations

**Force as an agent of change**

A very fundamental way to think of *force* is as the active agent through which energy moves. Energy is constantly being transformed from one form to another in every second of our existence. When work is done by a force, that work *transfers* energy from one object to another or *transforms* the energy from one form to another. Work is the means through which energy is transformed into and out of mechanical forms, such as potential and kinetic energy.

**Expressing changes in energy**

Consider a pendulum swinging back and forth. The pendulum has both kinetic energy and potential energy. At different points in its motion the energy is all of one form or the other, but the balance continually shifts. When the swinging mass is at the highest point in each swing, the energy is all potential. When the mass is lowest (and moving the fastest) the energy is all kinetic. The energy moves back and forth through the action of the net force, which is a component of the gravitational force.

**Energy of a pendulum**

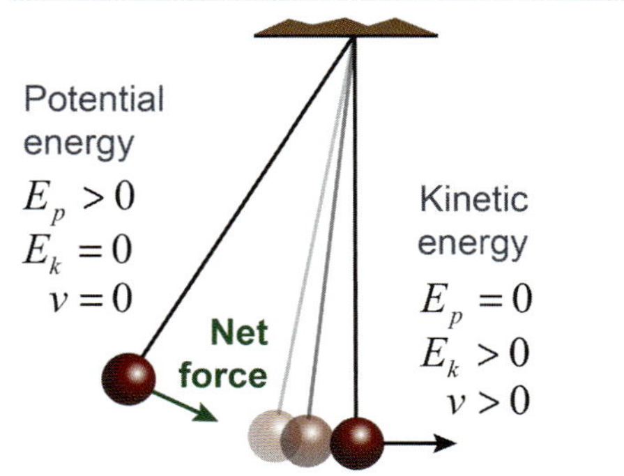


**Energy changes in closed systems**

In a closed system, if one kind of energy increases in value, then one or more other kinds of energy must decrease by the same amount to keep the total energy constant. Another way to express the law of energy conservation is by using energy *changes*: The total change in energy for a closed system must equal zero.

**(10.3)** $\Delta E = 0$

$\Delta E$ = change in total energy of system (J)

**Energy conservation** alternative formula

**What about other forms of energy, such as heat?**

For an *open* system, the energy equation is written as equation (10.4) below, for which $\Delta W$ is the work done *on* the system. Equation (10.4) says that the change in total energy of a system equals the heat added $Q$ plus the work done on the system. Many systems convert some energy into heat through friction. On the next page we will see that many energy technologies, such as a car engine, convert heat into work. This general version of the law of conservation of energy extends the work–energy theorem and is known as the *first law of thermodynamics*. We will learn more about it on page 734 in Chapter 25.

**(10.4)** $\Delta E = Q + W$

$\Delta E$ = change in energy (J)
$Q$ = heat added (J)
$W$ = work done on the system (J)

**First law of thermodynamics**

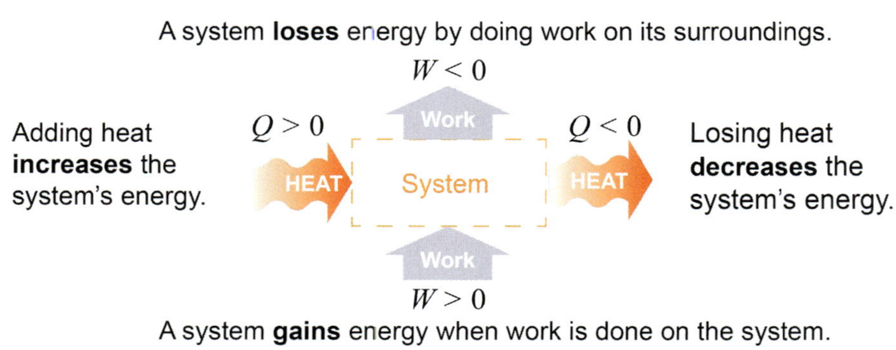

**Hydroelectric power plants**

Energy technology, such as a power plant, does not *create* energy. A power plant converts energy in one form, such as the potential energy of elevated water, into another form, such as electrical energy. A *hydroelectric power plant* converts the gravitational potential energy of water in the reservoir into electrical energy that can be sent to your home. The hydroelectric plant's reservoir of water is stored behind a dam so that there can be a larger vertical height (or *head*) between the reservoir and the turbines in the power plant.

**Design of a hydroelectric power plant**

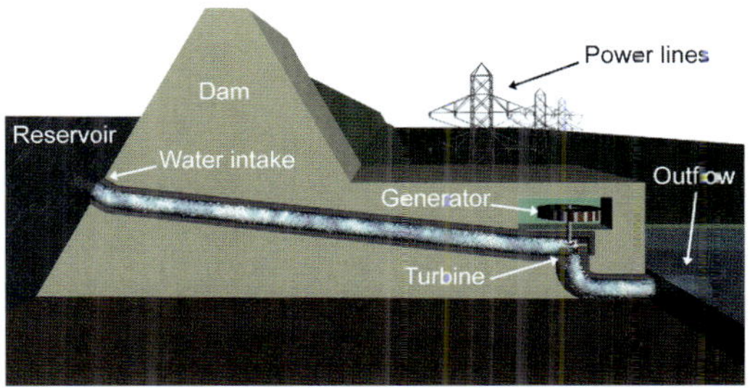


**How does hydropower work?**

The basic idea of the hydroelectric dam is that the water is allowed to drop a substantial distance, converting potential energy into kinetic energy. The kinetic energy of the water is then used to turn a propeller-like machine called a *turbine.* The turbine's kinetic energy does work on an electric generator connected to the turbine by a metal shaft. The generator turns the work done into electrical energy. Hydroelectric power is clean, efficient, and stable. There are not enough places, however, that are geographically suited to large-scale dams. Approximately 6.4% of the electricity generated in the USA is from hydropower, which makes up 68% of all renewable energy generation.

**Piston engine**

The first successful internal combustion engine was created by the German inventor Nikolaus Otto in 1867. Otto's engine is often called a "four-stroke" engine because a full operating cycle takes four strokes of the piston, two up and two down. Only one of the strokes produces power.

**How does an engine work?**

1 Intake

2 Compression

3 Expansion

4 Exhaust

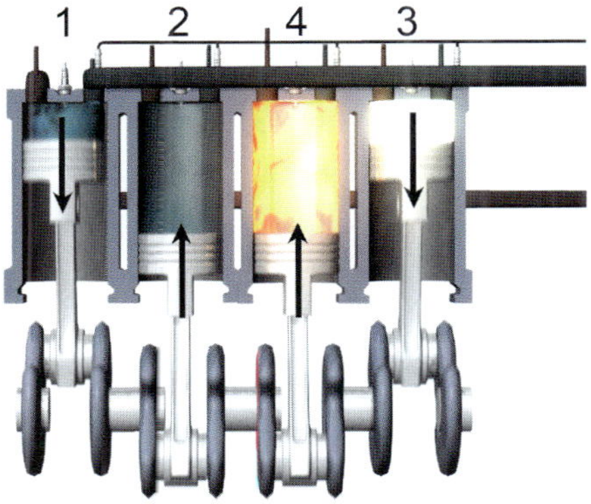


The cycle begins with a downward *intake* stroke—in which the piston pulls air and fuel into the cylinder. Next comes an upward *compression* stroke, where the piston compresses the air–gas mixture. When the gas is most compressed, the spark plug fires in the *power* stroke; the resulting internal explosion causes expansion of the gas that forces the piston downward. The last part of the cycle is the upward *exhaust* stroke, in which the piston expels the spent fuel and air mixture out of the engine (and then eventually out the car's tailpipe).

**Where does the energy come from?**

The internal combustion engine is a marvelously compact and efficient technology. A device half the size of a person can produce the equivalent power of 200 large horses. Essentially, the engine converts chemical energy in gasoline into heat. The heat does work as the gas trapped inside the cylinder expands. The force produced by the moving piston in effect converts the heat energy into mechanical energy by turning a crankshaft that is ultimately connected to the wheels of a vehicle.

# Investigation 10C Springs and the conservation of energy

**Essential questions** *Can energy conservation be used to predict the behavior of a system?*

A system consisting of a cart, attached to a spring, rolling freely on an inclined track includes three forms of energy. By carefully modeling the flow of energy among potential, kinetic, and elastic forms, we can accurately predict how far down the track the cart will roll before it is stopped by the spring.

## Building a mathematical model and testing it

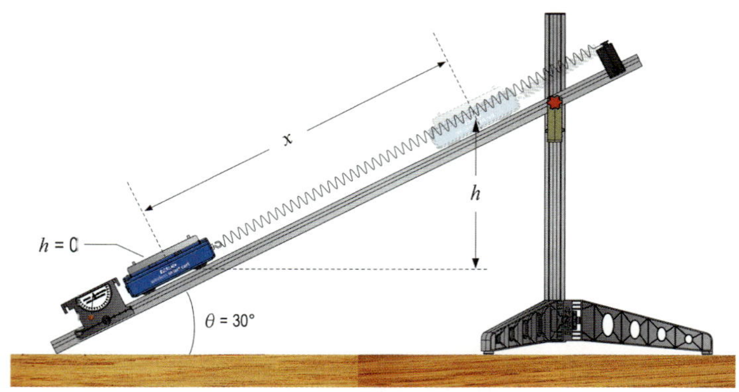


1. Set up your equipment as in the picture with two 250-g cart masses on the cart.
2. Given the setup, derive an expression which predicts the maximum distance the cart rolls down the track after being released from the top of the track (top is where the spring is unstretched).
3. Write your expression in terms of spring extension $x$, mass $m$, spring constant $k$, and gravitational acceleration $g$. *NOTE: Because the track is inclined 30° a helpful substitution can be made for height: $h = x/2$.*
4. Measure the mass of the cart plus masses, and measure the spring constant of the spring. Use those values and your expression to predict the maximum distance the cart will roll down the track.
5. Open the file **10C_EnergyTransformations**, and then connect your Smart Cart using Bluetooth.
6. Hold the cart at the top of the track (where the spring is unstretched), start data collection, and then release the cart. Catch the cart at the top of its rebound.

*a. How does your experimental value for maximum distance $d_{exp}$ compare to your prediction? What may have caused differences between them?*

*b. Use your experimental value for $d_{exp}$, your motion data, and energy equations to complete the table.*

**Table: Motion and energy data for a cart and spring system on a 30°-inclined track**

| Position | Spring extension $x$ (m) | Cart speed $v$ (m/s) | Cart height $h = (d_{exp} - x)/2$ (m) | Cart potential energy $E_p$ (J) | Cart kinetic energy $E_k$ (J) | Spring potential energy $E_s$ (J) | Total energy $E_T$ (J) |
|---|---|---|---|---|---|---|---|
| When the cart is released | | | | | | | |
| At maximum speed down | | | | | | | |
| At maximum distance | | | | | | | |
| At the top of the rebound | | | | | | | |

*c. How did the total energy of the cart and spring system change as the cart rolled down the track?*

*d. Create a bar chart to display the kinetic, gravitational potential, elastic potential, and total energies at each of the four positions shown in the table. What pattern do you see?*

*e. Calculate the efficiency of the system for two scenarios:*

- *From the point of release to the furthest point reached on the track*
- *From the point of release to the rebound position when the cart was caught.*

**Examples of friction**

Friction often converts kinetic energy into forms that are not easily converted back into motion. Kinetic friction on a skidding car or a sliding baseball player acts to oppose the motion, transforming kinetic energy into unrecoverable heat. A skydiver acted on by air resistance and a swimmer subject to drag are also examples of systems that do work on their surroundings through the action of a frictional force, losing mechanical energy in the process.

**Examples of friction**

**How to include friction**

In problems involving friction, work done against friction is energy that cannot appear in other forms. When you set up the conservation of energy equation, the work done against friction appears in the final state, after the change has occurred.

**Example problem**

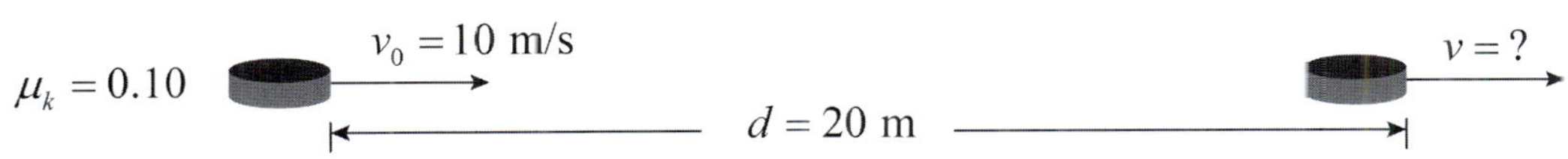


Write down relevant forms of energy, then eliminate zero terms.

| Initial state | | Final state | |
|---|---|---|---|
| $mgh_0 + \frac{1}{2}mv_0^2$ | = | $mgh_0 + \frac{1}{2}mv^2 + F_f d$ | Work done against friction |

For example, a 170 g hockey puck with an initial speed of 10 m/s slides with a coefficient of friction of $\mu_k = 0.10$. What is its speed after moving 20 m?

*Asked:* final speed $v$

*Given:* coefficient of kinetic friction $\mu_k = 0.10$, mass of puck $m = 0.17$ kg, initial velocity $v_0 = 10$ m/s, distance traveled $d = 20$ m

*Relationships:* $F_f = \mu_k mg$, $W = Fd$, $E_k = ½mv^2$

*Solution:* $\frac{1}{2}mv_0^2 = \frac{1}{2}mv^2 + \mu_k mgd \Rightarrow v^2 = v_0^2 - 2\mu_k gd \Rightarrow v = \sqrt{v_0^2 - 2\mu_k gd}$

$$v = \sqrt{(10\text{ m/s})^2 - 2(0.10)(9.8\text{ m/s}^2)(20\text{ m})} = 7.8\text{ m/s}$$

**Work with no change in energy**

When work is done against friction the net change in kinetic energy can be *zero*! A swimmer expends 2,500 J of energy swimming 25 m against a friction force of 100 N.

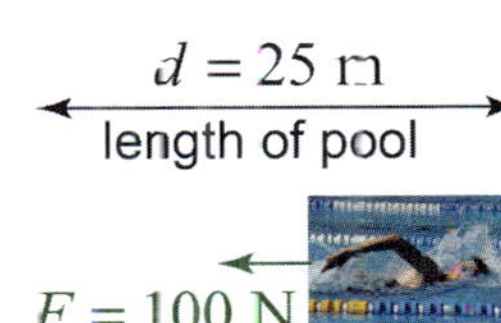


a. How much work is spent overcoming friction?

b. What is the change in the swimmer's kinetic energy?

*Asked:* (a) the work $W$ done against friction and (b) the change in kinetic energy $\Delta E$

*Given:* $\Delta E = 2{,}500$ J, $F_f = 100$ N, $d = 25$ m

*Relationships:* $W = Fd$, $\Delta E = Q_{\text{friction}} + W$

*Solution:* (a) The work done against friction is $W = Fd = (100\text{ N})(25\text{ m}) = 2{,}500$ J.
(b) The change in energy is $\Delta E = Q + W = 2{,}500\text{ J} - 2{,}500\text{ J} = 0$.
There is no change in kinetic energy since all the work done by the swimmer (2,500 J) is spent overcoming friction.

### Comparing real and theoretical speed for a falling object

How fast will the beach ball fall?

How about the bowling ball?

Imagine dropping a beach ball from the top of a building. At first, its measured speed would agree very well with the theoretical formula $v = \sqrt{2gh}$ we derived from energy conservation, but at some point it cannot fall any faster. It's not that energy conservation is wrong; it's that we neglected to account for air friction slowing down the beach ball's descent. The *real* speed of the beach ball is slower than the *theoretical* speed because of friction.

### Efficiency

The **efficiency** of a process describes how well the process transforms input energy into output energy. When you drop a ball, the input energy is the ball's potential energy, while the output energy is its kinetic energy.

### Friction and a falling wood ball

At speeds up to 10 m/s, the process of falling is close to 100% efficient—the potential energy lost by reducing height is converted to kinetic energy. At speeds above 10 m/s, however, the efficiency gradually diminishes as some potential energy is spent overcoming air friction. This is shown in the graphical model of a falling wood ball (right) You can see the actual speed of the ball is increasingly different from the model with no friction. At the terminal velocity of 33 m/s, the efficiency becomes *zero!* Lost potential energy is spent overcoming friction.

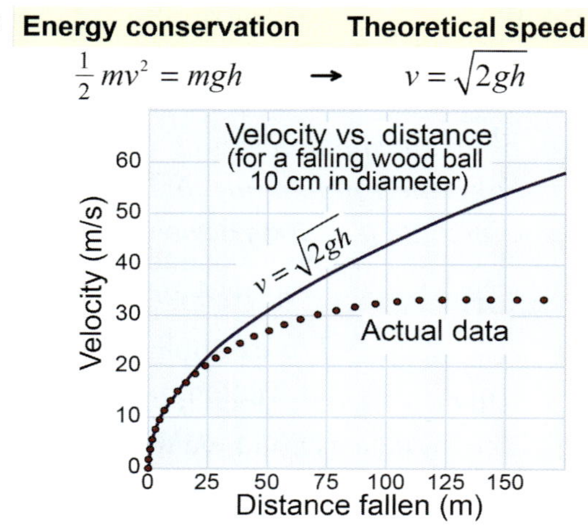


Interactive equation

**Efficiency**

$$\eta = \frac{E_{out}}{E_{in}} \qquad (10.5)$$

$\eta$ = efficiency
$E_{out}$ = output energy of a system (J)
$E_{in}$ = input energy of a system (J)

### Definition of efficiency

Efficiency is the ratio of the output energy of a system to its input energy. Since work and energy are directly related, the definition of efficiency can also be expressed as the ratio of the work performed by a system to the work input to the system. If a man does 2,000 J of work pushing a box up a ramp, and the box gains 1,200 J of potential energy, then this process is only 60% efficient because the man spent 40% of his work overcoming friction.

### Efficiency in technology

Efficiency is a very important parameter in technology. The best solar cells commercially available convert only 15% of the power in sunlight into electrical power. Typical rechargeable batteries are 20% efficient, which means that you get only 20 J of output energy for every 100 J of input energy you spend charging the battery. A solar power system that can provide electricity overnight requires both solar cells *and* batteries. Together, the system efficiency is 15% × 20% or only 3%! That means as much as 97% of the original solar energy ends up as waste heat. Many scientists and engineers are working to improve the efficiency of both solar cells and batteries.

# Investigation 10D Work done by friction

**Essential questions** *How is friction included in models of motion?*

The work done against friction reduces the amount of energy available for transformation into other forms. This is a significant way in which real systems are "*open*." This investigation explores friction in a typical braking situation in which work done against friction reduces kinetic energy.

## Part 1: Rolling friction

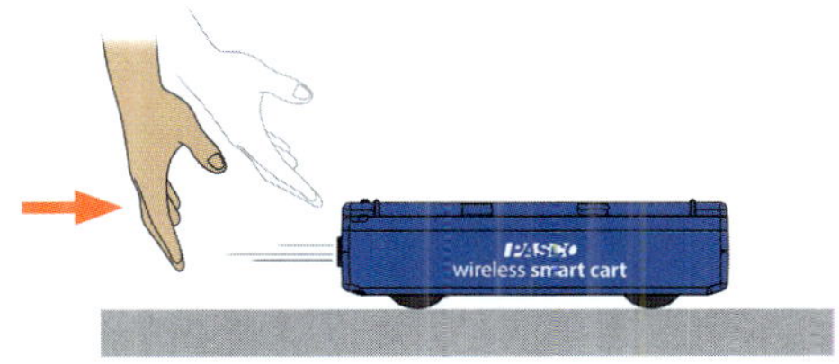


1. Open the **10D_FrictionAndLosses** experiment file in your software, and then connect your Smart Cart using Bluetooth.
2. Set the cart on a level surface. Start data collection, and then give the cart a gentle push.
3. Observe the cart's position, velocity, and acceleration graphs as the cart is pushed and as it rolls freely to a stop.

a. *Write an equation expressing conservation of energy for the cart as it rolls to a stop.*
b. *Solve the equation for the coefficient of rolling friction* $\mu_r$.
c. *Calculate* $\mu_r$ *using your equation, and the data from your position and velocity graphs.*

## Part 2: Kinetic friction

1. Attach a friction block to the plunger-end of the Smart Cart with a short string.
2. Record data as you give the cart a push, pulling the friction block behind it. There should be no slack in the string when you push.
3. Observe the graphs for the motion of the cart and friction block as they freely come to rest.

a. *Draw separate free-body diagrams representing the cart and friction block as they slow down.*
b. *Write a conservation of energy equation that includes the kinetic energy of the cart and friction block and the work done by friction at the start and end of the sliding motion.*
c. *Solve the equation for the coefficient of sliding friction* $\mu_k$.
d. *Calculate* $\mu_k$ *using the measured masses and data from your graphs.*
e. *Create an equation model to predict the stopping distance.*
f. *Add mass to the friction block, and then predict the stopping distance.*
g. *Test your prediction with the cart using the same initial velocity as before.*
h. *Explain any differences between your prediction and measurement.*

## Part 3: Combining the two

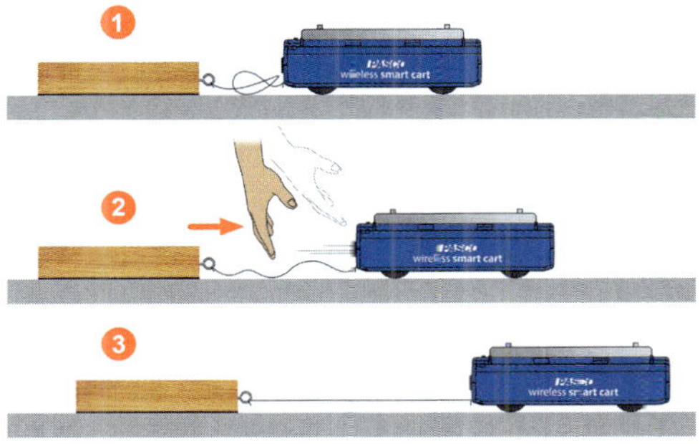


1. Move the mass to the cart and reattach the friction block with 1/2 m of string. Arrange for the cart and block to roll freely before starting to move the block.
2. Observe the graphs as the cart is pushed, rolls freely, catches the friction block, and drags it until both come to rest.

a. *Sketch a copy of your velocity graph and identify where on the graph (i) positive work is done on the cart and (ii) the cart does positive work on its environment (the friction block and surface).*

### Where does your trash go?

Technology makes our lives more productive and easier, and it also allows us to create enormous amounts of waste. Count how many trash bins go out on an average street every week. It takes energy, natural resources, and manufacturing labor to make a plastic cup, bottle, or cardboard box—all of which get thrown away. Where does it go after trash pickup? As much as 34% of solid waste gets either recycled or composted, about 12% gets burned, but 54%—more than half—is dumped into landfills. That is a lot of garbage that ends up in landfills!

**Where does municipal solid waste go?**

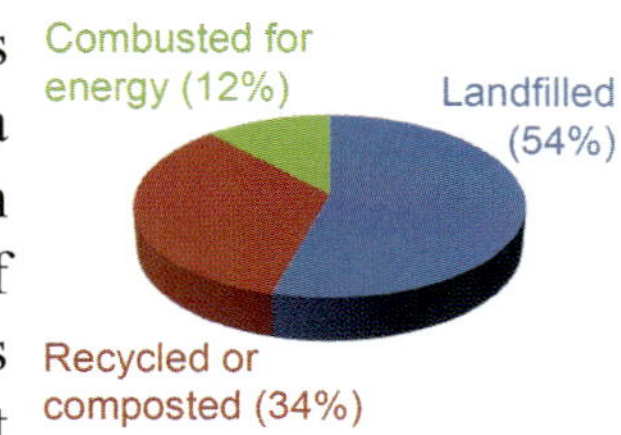


### Reduce, reuse, and recycle

The mantra of "reduce, reuse, and recycle" is a reminder to all of us of how to do our part to reduce the waste that would otherwise end up in a landfill. *Reduce* waste by not using as much packaging. *Reuse* items such as plastic take-out containers instead of throwing them away. As much as possible, *recycle* items that cannot be reused. Plastic containers have a triangular recycling symbol with a number inside; those with appropriate numbers can be placed in special recycling bins rather than in the trash. Electronic devices should be taken to centers that reprocess the components. Used oil should be recycled at a service station. Household chemicals and fluorescent light bulbs should be disposed through your town's special hazardous waste pickup days or at town recycling centers.

### Recycling aluminum cans

Nearly 500 billion beverage cans are produced worldwide every year, most of which are made out of aluminum. Manufacturing aluminum from raw materials such as bauxite ore requires a great deal of energy, since the melting point of aluminum is 660°C (1,220°F). It takes approximately 800 kJ (800,000 J) of energy—enough to power a 100 W light bulb for more than two hours—to produce an aluminum can that has a mass of only 15 g. A *recycled* aluminum can, however, requires 95% less energy to manufacture. Recycling of aluminum is not just about conserving a natural resource; it's also about reducing our use of energy.

### Recycling paper

Look at the sheet of paper in your notebook or on the printed page of this textbook. It took around 100 kJ of energy to produce that single sheet of paper directly from wood. Recycling the paper, however, saves one-third of that energy. Reducing your use of paper, reusing paper when possible (such as using old sheets as scratch paper), and recycling all will reduce the energy required to create the paper we use every day. It will also mean that fewer trees need to be felled to provide the paper we need.

### In the laboratory

In your classroom, do your part in the proper disposal and recycling of materials.

1. **Reduce**: Take only those materials you need to do your investigation,
2. **Reuse**: Use materials more than once where possible,
3. **Recycle**: Dispose of materials in the appropriate locations so that as much as possible can be recycled.

### What are hazardous materials?

A hazardous material is *any substance that poses an unreasonable risk to life, the environment, or property when not properly contained.* The Occupational Safety and Health Administration defines a hazardous *chemical* as any substance that would be a risk to employees if they were exposed to it in the workplace. Common to both of these definitions is that a hazardous material can pose a risk when not handled or stored properly; when proper procedures are followed, however, that risk is minimized. The most important part of handling hazardous materials is to know the protocol for handling them *before* you try to do so. Learn before you act!

### How are hazardous materials classified?

**"Fire diamond": identifying hazardous materials for emergency response**

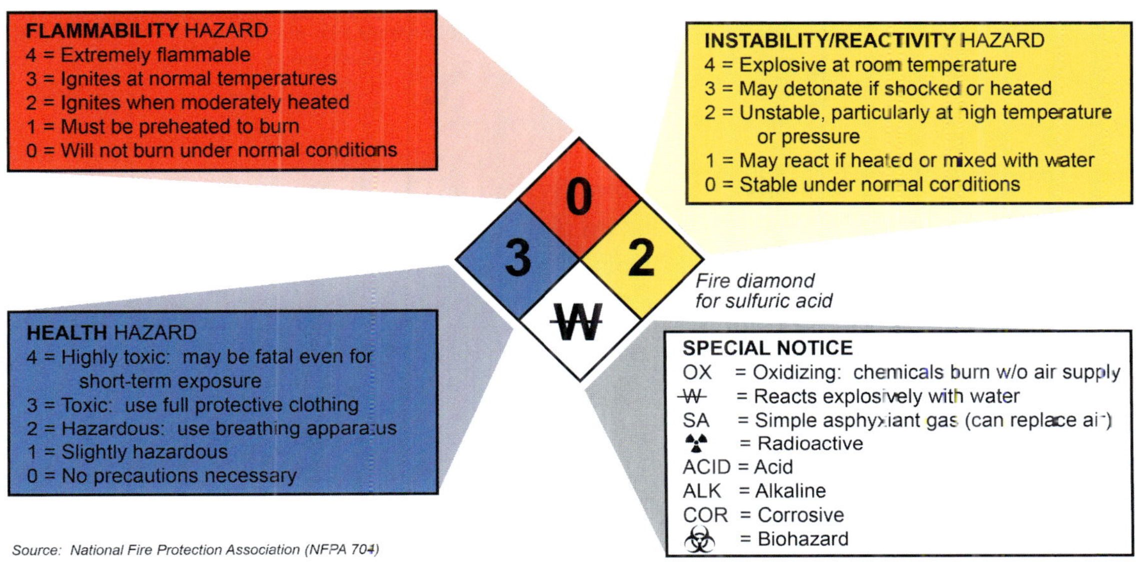


*Source: National Fire Protection Association (NFPA 704)*

Materials can pose many different types of hazards to life, property, and the environment. You may have seen before one kind of classification: the fire hazards typically associated with transportation of hazardous chemicals. These "fire diamonds" can be recognized easily on the sides of trucks and containers as a set of four diamonds in red, blue, yellow, and white. Each diamond contains a number or abbreviation inside it indicating the degree to which the chemical is a hazard in that way. Zero (0) means that the substance does not pose a hazard under normal conditions; the maximum four (4) indicates that the substance poses an extreme hazard. These standardized symbols have been adopted based on the National Fire Protection Association's standard #704 (known as NFPA 704).

### Hazardous waste disposal

Although hazardous chemicals are not part of this curriculum, you should still know how to dispose of any chemicals or other waste you may generate in your laboratory. Never dispose of hazardous chemicals in the regular trash or pour them down the sink; special containers are provided instead. Look for posters, product labeling, or safety data sheets for disposal instructions or ask your teacher.

### Household hazardous waste

Household hazardous waste is composed of post-consumer products that cannot be discarded in the normal waste stream—items that cannot be thrown away in the trash can. Oil-based paint, motor oil, pesticides, fluorescent bulbs, rechargeable batteries, and some electronics (such as mobile phones, computers, and televisions) are all examples of products that should never be dumped into a garbage can. Many cities, towns, and counties have collection locations and times when they accept household hazardous waste.

**What is the GHS?**

### New global harmonization system (GHS) labeling for hazardous materials

**Health hazard**

Carcinogen
Mutagenicity
Reproductive toxicity
Respiratory sensitizer
Target organ toxicity
Aspiration toxicity

**Flame**

Flammables
Pyrophorics
Self-heating
Emits flammable gas
Self-reactives
Organic peroxides

**Exclamation mark**

Irritant (skin and eye)
Skin sensitizer
Acute toxicity
Narcotic effects
Respiratory tract irritant
Hazardous to ozone layer (non-mandatory)

**Gas cylinder**

Gases under pressure

**Corrosion**

Skin corrosion/burns
Eye damage
Corrosive to metals

**Exploding bomb**

Explosives
Self-reactives
Organic peroxides

**Flame over circle**

Oxidizers

**Environment** (non-mandatory)

Aquatic toxicity

**Skull and crossbones**

Acute toxicity (fatal or toxic)

*Source: Occupational Safety & Health Administration*

The Globally Harmonized System of Classification and Labeling of Chemicals (GHS) is a new, worldwide standard for identifying and labeling hazards on the data sheets and packaging for chemicals. There are nine different categories of hazards in the GHS, as illustrated above. The GHS standard allows people who handle these materials to have both illustrations and text that allow rapid identification of the hazards associated with the substance. These regulations provide safety guidelines that you should always follow when handling the chemical.

**How is this information labeled on a product?**

Products should be labeled with the product name, pictograms of the hazards, appropriate signal words, hazard statements, precautionary statements, and supplier information. Precautionary statements come in four categories: prevention, response, storage, and disposal. Both the packaging box and the individual products should be labeled. If you pour out some of the chemical and put it into another container (such as a spray bottle for a cleaning agent), then you should place the label onto the new container, too.

### Product labeling with GHS information

**What are safety data sheets?**

A safety data sheet (SDS)—previously called a material safety data sheet (MSDS)—is an important document to provide workers with clear instructions for the safe handling, storage, and disposal of a chemical. A SDS provides workers with information on first aid, toxicity, reactivity, flammability, protective equipment, and spill-handling procedures for the chemical. Schools and research facilities should keep these forms in a labeled binder that is easily accessible in case of an accident. When you work with chemicals, make sure you know where these forms are kept and be familiar with their contents for any chemicals you work with.

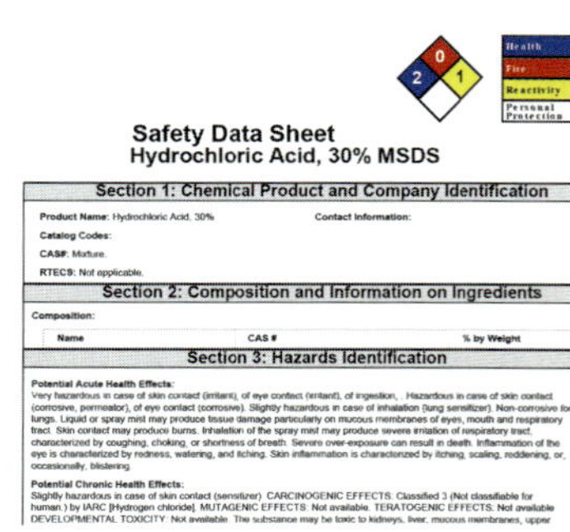

# Section 2 review

The work–energy theorem states that the *net* work done on an object equals that object's change in *kinetic* energy. Work also can change an object's gravitational potential energy. Energy changes can be represented by the symbol $\Delta E$, and they can be positive or negative. When acting on moving objects, friction transforms some kinetic energy into thermal energy. Air friction (also called air resistance) can prevent a falling object from converting all of its potential energy into kinetic energy. *Efficiency* divides the energy one gets out of a system by the energy that one puts in. Efficiency can never be greater than 100%, and often is a great deal less.

| | | |
|---|---|---|
| **Vocabulary words** | **work–energy theorem, friction, efficiency** | |
| **Key equations** | $W_{net} = \Delta E_k$ | $\Delta E = 0$ |
| | $\Delta E = Q + W$ | $\eta = \frac{E_{out}}{E_{in}}$ |

## Review problems and questions

1. A 10 kg go-kart driven by a 30 kg boy starts from rest and rolls down a hill of height 10 m.
   a. What is the predicted speed of the boy and go-kart at the bottom of the hill?
   b. The boy's actual speed at the bottom of the hill is measured to be 10 m/s. Does this agree with your prediction? Why or why not?

2. A man applies 500 N of horizontal force to push a 100 kg wooden crate 20 m across a floor.
   a. What is the work that the man performs upon the crate over the 20 m distance?
   b. When sliding, the crate experiences a friction force of 200 N in the direction opposite its motion. What is the work that the friction does upon the crate over the 20 m distance?
   c. What is the net work done on the crate?
   d. What is the change in the crate's kinetic energy?
   e. Calculate the efficiency with which the man's work is turned into the crate's kinetic energy.

3. A moving hockey stick hits a stationary 160 g hockey puck ($m = 0.16$ kg) and applies a horizontal force $F$ to the puck. The two objects remain in contact for the first 0.25 m (25 cm) of the puck's motion. The puck then slides away at a speed $v$ of 20 m/s. What force $F$ is delivered to the puck?

4. A weightlifter grabs a 50 kg barbell resting on the floor, raises it 2 m, then holds it in place.
   a. What is the barbell's kinetic energy before being lifted?
   b. What is the barbell's kinetic energy when the lift is complete?
   c. What is the barbell's change in kinetic energy?
   d. What is the net work done on the barbell?
   e. How much work does the weightlifter perform on the barbell?
   f. Do your answers to questions d and e match?
   g. Is the weightlifter's pull the only force acting upon the barbell while it is being lifted?

# Chapter review

### Vocabulary

Match each word to the sentence where it best fits.

#### Section 10.1

| | |
|---|---|
| **closed system** | **law of conservation of energy** |
| **open system** | **state** |
| **system** | |

1. The ________ says that energy stays constant over time.
2. We choose the objects and influences in a/an ______ to focus on what we are trying to investigate and exclude things that are irrelevant.
3. A particular configuration of all the elements in a system is its _______.
4. A/An ________ is one in which energy can flow into or out of it over time.
5. A/An ________ is one in which there is no net flow of energy into or out of it.

#### Section 10.2

| | |
|---|---|
| **efficiency** | **friction** |
| **work–energy theorem** | |

6. The _______ of a system describes how effectively the system transforms input energy into output energy.
7. The amount of work you do on a basketball is converted to its energy of motion through the _______.
8. When a car rolls down a ramp, not all of the potential energy becomes kinetic energy because ______ transforms some energy into heat and wear.

### Conceptual questions

#### Section 10.1

9. According to physics, energy can never be created or destroyed and the energy content of the universe is constant. So why are smart people worried about "conserving energy lest we run out"?
10. Ask two friends the meaning of the word "conservation" in each of the following sentences.

    Sentence 1: "Potential and kinetic energy obey a conservation law that applies to all physical processes."
    Sentence 2: "Many towns practice ecological conservation by keeping certain areas of land wild and undeveloped."

11. Your friend brags that his ball can bounce to a higher height than it was initially dropped from. Is he telling the truth?
12. A roller coaster car begins at rest at the top of the first hill. Draw a sketch of a roller coaster with a second hill where the roller coaster cannot reach the top. Draw a sketch of a different roller coaster with a second hill that the roller coaster can get over.
13. A bowling ball and a tennis ball are dropped separately in such a way that both have the same kinetic energy when they hit the ground. Were they dropped from the same height or a different height? If the latter, which one was dropped from a higher point?
14. Give a five-slide oral presentation describing the difference between the law of conservation of energy and the everyday use of the term "energy conservation."
15. You are at an amusement park to conduct a field investigation.
    a. How can you estimate the maximum height of a roller coaster if you are only given a radar gun?
    b. Alternatively, your teacher wants you to use the roller coaster to *demonstrate* energy conservation, and she allows you to use both a radar gun and a meter stick. Explain how you would demonstrate energy conservation.
16. Two balls are dropped, one from twice the height of the other. What is the ratio of the speeds each ball reaches when they hit the ground?
17. From a roof, you throw three equal-mass balls each with the same initial speed: one straight down at the ground, the second straight up, and the third horizontally out from the building. When each of the three balls hit the ground, which one has the most kinetic energy?

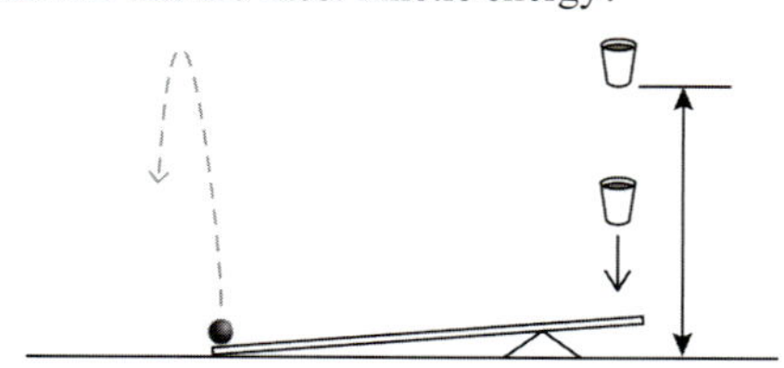

18. A bucket of sand is dropped several meters onto a lever that launches a ball upward.
    a. What should be included in the system you choose to apply energy conservation to find the height the ball reaches?
    b. Is the system open or closed?
    c. What might make the approximation of a closed system inaccurate?
    d. How would an open system affect the height the ball reaches compared to an ideal closed system?

# Chapter review

## Section 10.1

19. Explain how measuring the height of an icy hill and the speed of a sledder at the top and bottom of the hill can be used to demonstrate energy conservation.

20. Prove mathematically, using energy arguments, that objects of different mass should fall at the same rate in the absence of friction.

## Section 10.2

21. A thrown ball follows a curved path called a *trajectory* as it first goes up and then comes down again. At what point in the trajectory does the ball have maximum kinetic energy? At what point does it have a maximum potential energy?

22. Linda claims that she can defy the law of conservation of energy by riding an elevator and gaining potential energy. Is Linda right? Explain.

23. In a closed system where only kinetic and electrical energy are changing, if the kinetic energy is increasing, what is happening to the electrical energy?

24. State the law of conservation of energy using delta notation.

25. Why is it harder to rub two pieces of sandpaper together than to rub two pieces of notebook paper together?

26. When friction removes kinetic energy from an object, what does it typically transform it into?

27. When a falling object has reached terminal velocity, what is its efficiency in converting potential energy to kinetic energy?

28. Suppose you have a half can of unused, oil-based house paint. Can you dispose of the paint by dumping it down a storm drain at the road side? If not, then how should you properly dispose of the paint?

29. When does a pendulum have the highest gravitational potential energy? When does it have the highest kinetic energy?

30. Name two real-life examples of energy transformations, and explain what types of energy are being transformed.

31. Describe how you used the work–energy theorem to determine the efficiency of the work done by the elastic band in Investigation 10B on page 290.

32. In a closed system where kinetic, potential, and thermal energy are changing, the kinetic energy decreases by 54 J, and the thermal energy increases by 13 J. What is the change in the potential energy?

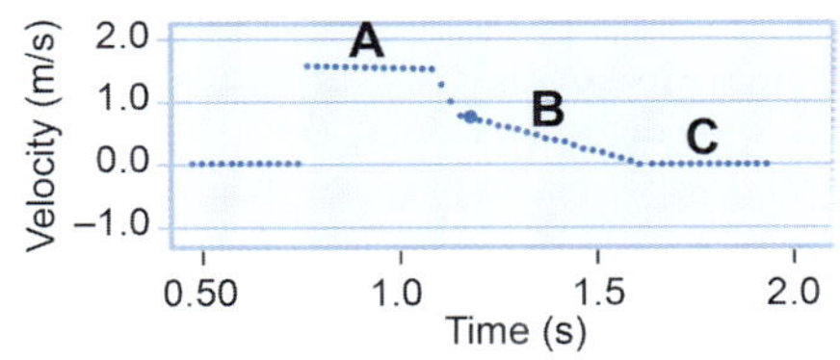


33. This graph was generated from a real-time graphing technology for the motion of a rolling car.
    a. Which portion of the graph (A, B, and/or C) corresponds to static friction? Kinetic friction? Rolling friction?
    b. In which portion of the graph (A, B, and/or C) does velocity change with a sign opposite to the changes in time?
    c. In which portion of the graph (A, B, and/or C) is the slope negative?
    d. Write down a symbolic relationship for velocity as a function of time for portion B of the graph.
    e. Using your equation in the previous question and/or the chart, describe how time varies as a function of velocity for portion B.

34. Explain why energy is still conserved when friction causes an object to "lose" energy.

35. Two friends, Noor and Meera, are pushing boxes. Noor has an efficiency of 50% in pushing her box. Meera pushes her box on a path that is twice as long as Noor's but has half the friction. Assume Meera pushes with the same amount of force as Noor. Is Meera more or less efficient than Noor?

## Quantitative problems

### Section 10.1

36. A 1.0 kg brick falls off a ledge of height 44 m and lands on the ground 3.0 s later.
    a. Find the final velocity of the brick using the equations of motion.
    b. Find the final velocity of the brick using conservation of energy.

37. A frictionless roller coaster with a mass of 200 kg starts 15 m above the ground with a speed of 10 m/s. When it is 5 m above the ground what is its speed?

# Chapter review

## Section 10.1

38. ❮❮ A 20 kg chair has 250 J of potential energy relative to the ground. If the chair is dropped from its position, what is its speed when it strikes the ground?

39. ❮❮ A 60 kg diver drops from a ledge that is 50 ft above the water. What is the speed of the diver upon hitting the water? Give your answer in meters per second and miles per hour.

40. ❮❮ A large bird with a mass of 1.0 kg is flying at a height of 10.0 m at a speed of 10 m/s. What is the mechanical energy of the bird?

41. ❮❮❮ A 60 kg diver jumps off a diving board upward with an initial velocity of 5 m/s. The diving board is 10 m higher than the water. What is the diver's speed when just entering the water after the dive?

## Section 10.2

42. ❮ A 6 kg box is traveling at a speed of 10 m/s. Suddenly, it comes across a rough patch in the ground, which exerts a friction force of 40 N on the box. How far will the box go before it stops?

43. ❮ A boy walks up a 10 meter tall snow-covered hill and sleds down. When he reaches the bottom, the boy is moving at 7 m/s. What is the efficiency of the boy's sled ride?

44. ❮ John weighs 80 kg, and eats a 1,000 J candy bar. If his body perfectly transforms the energy in the chocolate into kinetic energy, what is the fastest he can run?

45. ❮❮ How much distance does it take to stop a car going 30 m/s (67 mph) if the brakes can apply a force equal to one half the car's weight?

46. ❮❮ Consider a rubber ball whose bounce has an efficiency of 80%. That means 80% of the kinetic energy before the bounce remains kinetic energy after the bounce. (The rest of the energy is converted to thermal energy.)
    a. If the ball drops from a height of 5 m, how high does it bounce back on the first bounce?
    b. How high does the ball bounce on the second, third, and fourth bounces?

47. ❮❮ You can only apply a 200 N force but you need to make a 20 kg wagon of mulch travel at a speed of 10 m/s. How far do you have to push the wagon before its speed reaches 10 m/s if it starts from rest? You may assume no friction.

48. ❮❮ A pitcher for a baseball team can throw a fastball that travels at 45 m/s (100.6 mph). The pitcher throws a ball directly upward with the same energy that he uses to throw a fastball over the plate. Assume that the ball does not experience friction.
    a. How high does the baseball go?
    b. What is the speed of the baseball when the pitcher catches it?

49. ❮❮ A 10 kg bowling ball sits at the top of a 10 m hill and then slides down its icy hillside.
    a. What is the speed of the bowling ball when it reaches the bottom of the hill?
    b. What is the change in kinetic energy of the system as the bowling ball travels from the top of the hill to the bottom of the hill?
    c. What is the bowling ball's mechanical energy at the top of the hill?
    d. What is the bowling ball's mechanical energy at the bottom of the hill?

50. ❮❮ Juan, who is 60 kg, falls from a height of 20 m on a strange planet that has a large amount of air resistance and a gravity exactly equal to Earth's. When he reaches the ground, he is traveling 2 m/s.
    a. How efficient was his jump at converting potential into kinetic energy?
    b. How much energy do Juan's surroundings gain as he jumps?

51. ❮❮❮ Nina, whose mass is 60 kg, is traveling on a bike at 15 m/s at the top of a 3.0 m hill. Nina continues down this hill without pedaling. Assume friction is negligible.
    a. What energy transformation is occurring as she travels down the hill?
    b. How fast is she traveling when she reaches the bottom of the hill?

52. ❮❮❮ A 1,500 kg car begins at rest. A force of 2,000 N is applied to the car for a 20 m stretch in order to accelerate it. At the end of the 20 m, the car is going 5 m/s, and the road gains 21,250 J of thermal energy.
    a. Is energy conserved?
    b. Why does the road gain thermal energy?

# Chapter review

## Standardized test practice

53. George does 18 J of work to lift a 1 kg box at a constant speed. If he drops it, how fast will the box be going when it hits the ground?

A. 18 m/s
B. 1.8 m/s
C. 6 m/s
D. 36 m/s

54. A 0.70 kg ball is placed on a vertical 200 N/m spring that is compressed 40 cm. When the spring is released, how high above its starting point will the ball go?

A. 2.3 m
B. 5.8 m
C. 80 m
D. 0.40 m

55. A 15 kg ball is thrown straight upward with a speed of 20 m/s. What is the maximum height the ball reaches?

A. 14.0 m
B. 19.8 m
C. 20.4 m
D. 40.8 m

56. A 2.0 kg ball, traveling at 1.0 m/s, rolls off of a cliff of height 3.0 m. If the ball lands directly on top of an unstretched spring with a spring constant of 40 N/m, how far does it compress the spring?

A. 0.22 m
B. 0.68 m
C. 1.7 m
D. 1.4 m

57. Dennis drops a quarter from a skyscraper with floors spaced 3 m apart. If the quarter hits the ground at 25 m/s, how many floors above ground is Dennis? (Ignore air drag.)

A. 20 floors
B. 10 floors
C. 2 floors
D. 32 floors

58. A man accelerates a crate across a rough, level floor with an efficiency of 50%. Which is true?

A. Energy is not conserved, because the efficiency is less than 100%.
B. Part of the man's work goes into the kinetic energy of the crate, while the other part of his work is done against friction.
C. The energy gained by the crate equals the work done by the man.
D. The work done by the man equals the distance the crate moves multiplied by the friction.

59. Kate pushes a box at a constant speed with a constant force of 30 N up a frictionless incline 3 m long. If the box weighs 90 N, how high off the ground does she push it?

A. 0.5 m
B. 1.0 m
C. 1.5 m
D. 2.0 m

60. For a typical car, approximately 65% of the energy in the fuel is radiated away as heat, 13% is output as work in moving the car, 10% is spent overcoming friction, 7% is spent idling rather than moving, and 5% runs accessories (e.g., the heater). What is the efficiency of the typical car?

A. 13%
B. 20%
C. 23%
D. 30%

61. A 500 kg roller coaster starts from rest at a height of 40 m. What is its speed when it is 13 m off the ground?

A. 28 m/s
B. 16 m/s
C. 23 m/s
D. 9 m/s

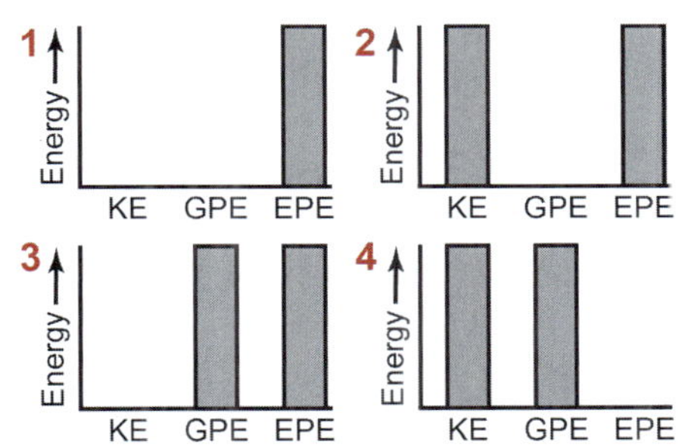


62. Which bar graph might represent the kinetic energy (KE), gravitational potential energy (GPE), and elastic potential energy (EPE) of a pole vaulter just before she leaves the ground at the beginning of the vault?

A. 1
B. 2
C. 3
D. 4

63. A 70 kg woman who wants to climb up the 900 m face of the El Capitan vertical rock formation in Yosemite ponders her fate were she to fall without using a safety rope. How fast would she be traveling at the bottom when she hits the ground were she to fall from the top of the cliff's face?

A. 13.6 m/s
B. 93.9 m/s
C. 132.8 m/s
D. 17,640 m/s

# Chapter 11 Momentum and Collisions

In most of today's world, modern life is unimaginable without the automobile. This century-old invention has liberated millions from having to live in or near their workplaces. Along with telecommunications and electricity, it has accelerated economic growth, lifting millions out of poverty. The automobile has also helped the physically challenged to live fuller lives.

The undeniable advantages of automobiles have come at a terrible price—and arguably an avoidable one. From developing nations such as India to first-world mainstays such as the USA, traffic fatalities far outnumber deaths from war and terrorism. A key goal of modern automobile design is to make cars safer in a collision.

That wasn't always the case, some say. When Ralph Nader published *Unsafe at Any Speed: The Designed-In Dangers of the American Automobile* in 1965, cars did relatively little to protect their occupants if a crash occurred. In physics terms, they did little to reduce the rate at which an occupant came to a stop. Air bags were all but nonexistent, body frames rigid, and car interiors riddled with hazards. Fortunately, automobile design no longer ignores the laws of momentum.

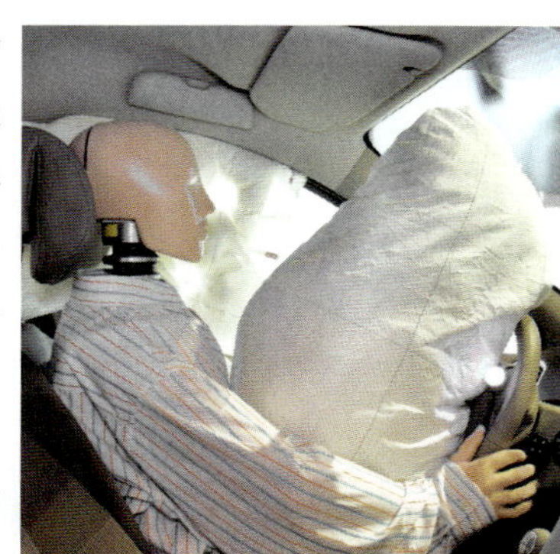

A combination of inertia and motion, momentum can only be transferred by enduring a force. Imagine riding in a car that hits a brick wall at 30 mph (13 m/s). Impart your body's momentum to the dashboard in 1/20th of a second, and you almost certainly will leave the scene in an ambulance. Encounter an air bag, and your collision may last a leisurely quarter-second —often enough to let you walk away unharmed.

Progress doesn't end with air bags, which now protect riders from side impacts as well as from dashboards and steering wheels. Materials scientists, physicians, and engineers are working to develop car bumpers that protect a car's occupants as well as any pedestrians who might be struck. And automated, driverless cars, which could avoid collisions altogether, appear to be right around the corner. Understanding the physics of momentum transfer promises to reduce the number of lives lost every year to car accidents.

# Chapter study guide

## Chapter summary

Why do cars have crumple zones and airbags? How does rocket propulsion work? How can you predict where two pool balls will go after they collide? These are problems that involve momentum, impulse, and the conservation of momentum. Momentum is a physical description of *mass in motion*; the higher the mass or velocity of an object, the more momentum it has. The conservation of momentum is a fundamental law of physics—akin to energy conservation—that is essential for understanding collisions, whether between two cars or between two subatomic particles.

## Learning objectives

By the end of this chapter you should be able to

- describe momentum and impulse and solve problems using them;
- apply impulse to real-world problems, including situations involving cushioning and impacts;
- define momentum conservation and describe applications of it;
- use momentum conservation to solve noncollision problems;
- distinguish among elastic, inelastic, and perfectly inelastic collisions;
- apply momentum conservation to solve collision problems; and
- determine whether a collision is elastic using energy arguments.

## Investigations

Design a crash barrier
11A: Conservation of momentum
11B: Inelastic collisions
11C: Elastic collisions

## Chapter index



## Important relationships

$$\vec{p} = m\vec{v}$$

$$J = \Delta p = F\Delta t$$

$$m_1 v_{i1} + m_2 v_{i2} = m_1 v_{f1} + m_2 v_{f2}$$

$$\frac{1}{2}m_1 v_{i1}^2 + \frac{1}{2}m_2 v_{i2}^2 = \frac{1}{2}m_1 v_{f1}^2 + \frac{1}{2}m_2 v_{f2}^2$$

## Vocabulary

**momentum**
**collision**
**impulse**
**inelastic collision**
**law of conservation of momentum**
**elastic collision**

## 11.1 - Momentum and impulse

Momentum is a property of an object in motion, whether it is a person, car, truck, or spaceship. A car traveling down the road has momentum; it requires force between the tires and road to change its velocity. A big truck traveling at the same velocity has much more momentum, because it is more difficult to change its velocity. Momentum is a quantity that describes the tendency for an object in motion to remain in motion.

**Calculating momentum**

The **momentum** of a moving object is its mass multiplied by its velocity. The higher an object's mass or velocity, the more momentum it has. A truck has more momentum than a car moving at the same speed because the truck has more mass. A car has more momentum while speeding on the highway than when moving slowly in heavy traffic.

Interactive equation

**Momentum**

**(11.1)** $\vec{p} = m\vec{v}$

$\vec{p}$ = momentum (kg m/s)
$m$ = mass (kg)
$\vec{v}$ = velocity (m/s)

**Momentum is a vector quantity**

Momentum is the product of mass and velocity; since velocity is a vector quantity, momentum is also a vector quantity. Momentum in one dimension can take on either positive or negative signs to indicate its direction. For example, a 10 kg ball moving at 2 m/s to the right has a momentum of +20 kg m/s, and the same ball moving to the left has a momentum of –20 kg m/s. Momentum has units of mass multiplied by speed—or kilogram meters per second (kg m/s).

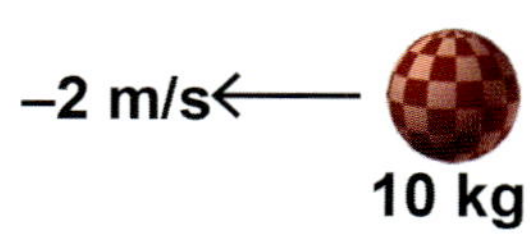


momentum $p = -20$ kg m/s

**Momentum is mass multiplied by velocity and has a *direction*.**

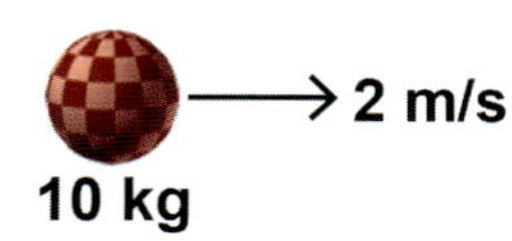


momentum $p = +20$ kg m/s

**Why use "p" for momentum?**

The symbol for momentum is a lowercase $p$ or vector $\vec{p}$. Historically, momentum was identified with the *persistence* of an object to continue in its original speed and direction of motion. While we call it "momentum" today, Newton called it "impetus" in 1760. Impetus is derived from the Latin word *petere,* which means to go toward.

**Can different objects have the same momentum?**

Objects with different masses can have the same momentum. For example, a truck moving slowly at 1 m/s can have the same momentum as a small car moving quickly at 20 m/s. Both have the same "persistence of motion" described by Newton.

**Two different ways to have the same momentum**

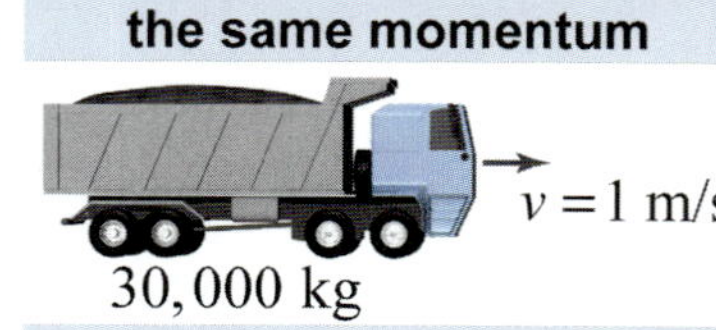


$p = mv = 30{,}000$ kg m/s

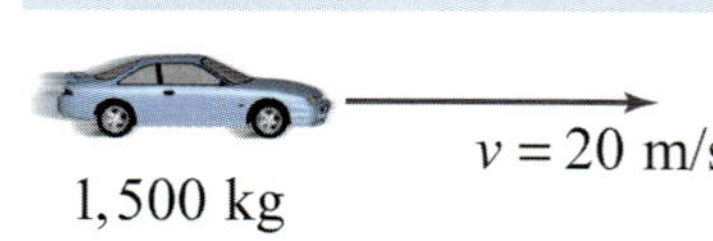


Since the velocity of an object depends on the reference frame, its momentum will also depend on the reference frame. A speeding car will have a large momentum when observed by a pedestrian but zero momentum in the driver's reference frame.

**Momentum and Newton's first law**

Momentum is an important quantity in physics. In the quantum world, momentum is more fundamental than either mass or velocity taken separately. Newton's first law might be more accurately named the law of *momentum*, because it is really momentum that remains unchanged when the net force is zero.

**Momentum and inertia in common usage**

In everyday language, a speeding train might be described as having a lot of inertia—or a lot of momentum. The terms inertia and momentum are often used interchangeably to describe a moving object that is difficult to stop.

**How is momentum different from inertia?**

In physics, inertia and momentum have distinct meanings, but they sometimes overlap. Inertia is a property of an object that is dependent on the object's mass alone—it is an object's resistance to change in velocity. A bicycle has the same inertia whether it moves at 10 m/s or is stationary. It does not, however, have the same momentum in both cases. Since momentum depends on both mass and velocity, momentum will be greater when the bicycle is moving at 10 m/s than when it is stationary.

**Momentum comes from motion**

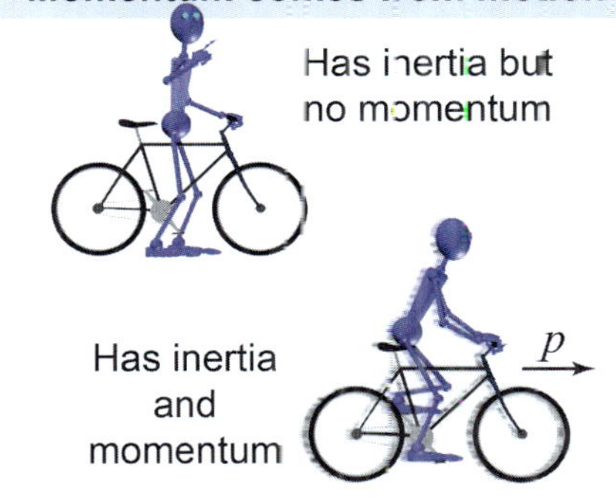


**Momentum is a vector quantity, with magnitude and direction.**

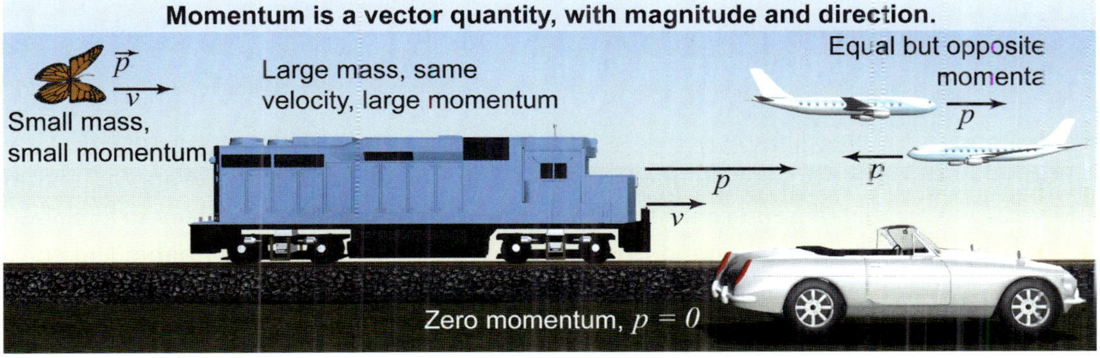


**Momentum is a vector**

Momentum is the product of mass and velocity, and since velocity is a vector quantity, momentum is a vector quantity too. As a vector, momentum has both magnitude and direction. A fluttering butterfly will have a small momentum because its mass is small. A train, traveling at the same velocity, will have much more momentum than the butterfly because its mass is greater. In the illustration above, the train's momentum vector is therefore drawn much longer than the butterfly's. The two airplanes in the figure have momentum vectors pointing in opposite directions because their velocities are in opposite directions. A stationary car has a momentum vector of magnitude zero.

**Adding momentum vectors**

**Adding momentum vectors**

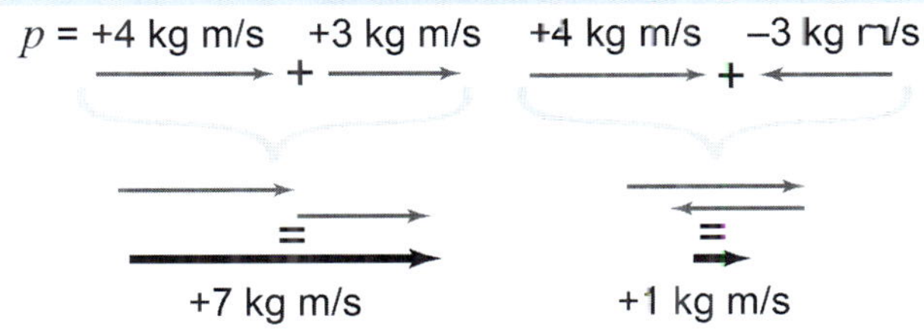


Momentum vectors add in the same way as any other vectors we've encountered. As we learned on page 74, vectors can be added in one dimension and in more than one dimension. The figure on the right presents an example of graphical addition of vectors in one dimension. Adding $\vec{p} = +10$ kg m/s to $\vec{p} = +3$ kg m/s gives the resultant vector of $\vec{p} = +13$ kg m/s. Similarly, adding $\vec{p} = +10$ kg m/s to $\vec{p} = -3$ kg m/s gives the resultant vector of $\vec{p} = +7$ kg m/s.

## Impulse and Newton's second law

**How Newton expressed his second law**

We often think of Newton's second law as $F = ma$. But Newton originally expressed his second law in a way that relates to momentum. Newton defined force as the rate of change of momentum—the change in momentum divided by the time interval. Just as velocity is the rate at which displacement changes, force can be expressed as the rate at which momentum changes.

$$F = \frac{\Delta p}{\Delta t}$$

$F$ = force (N)
$\Delta p$ = change in momentum (kg m/s)
$\Delta t$ = change in time (s)

**Force**
rate of change of momentum

**Impulse is the product of force and time interval**

Newton's definition of force in terms of momentum leads to a new quantity: *impulse*. Multiplying both sides of $F = \Delta p/\Delta t$ by $\Delta t$ gives $F\Delta t = \Delta p$. The product of force and the duration of time the force is applied is called **impulse** and is usually denoted with the letter $J$. An impulse applied to an object causes a change in its momentum, $\Delta p$, equal to $F\Delta t$.

Interactive equation

$$\text{(11.2)} \quad J = \Delta p = F\Delta t$$

$J$ = impulse (kg m/s)
$\Delta p$ = momentum (kg m/s)
$F$ = force (N)
$\Delta t$ = time (s)

**Impulse**
force applied over time interval

**Impulse and average force**

The equation above assumes that the applied force $F$ is constant. However, most impulses are from variable forces. For our purposes an acceptable approximation for equation 11.2 is that $F$ represents the *average force* acting on an object to change its momentum. This approximation modifies the equation to address both variable and constant forces.

**Impulse and applied force**

The same impulse can be applied with different forces. Braking gently applies a small average force, whereas slamming on the brakes applies a large average force. Both stop the car, so both apply the same impulse. They are different because braking gently stops the car over a *long* time $\Delta t$, while slamming on the brakes does it over a short time. In both cases the car changes its momentum from a large value to zero!

**Impulse is the product of force and the time interval it is applied.**

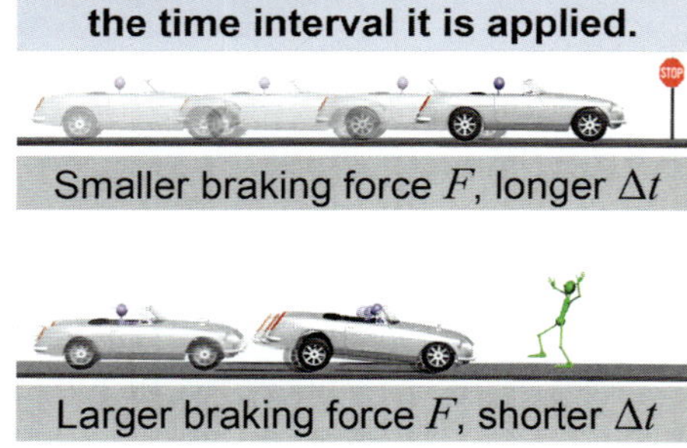


**Cushioning**

Falling on a hard surface causes a rapid change in momentum in a short time interval $\Delta t$. As a result, the force is large. *Cushioning* increases the time interval over which a force is applied, without changing the total impulse. Falling on a cushion spreads the impulse over a longer time interval as the cushion collapses. This dramatically decreases the impact force. One hurts, and the other doesn't! The same thing happens when you bend your legs while landing. Bending your knees extends the time the impulse is applied, reducing the force you experience.

**Cushioning increases impact time.**

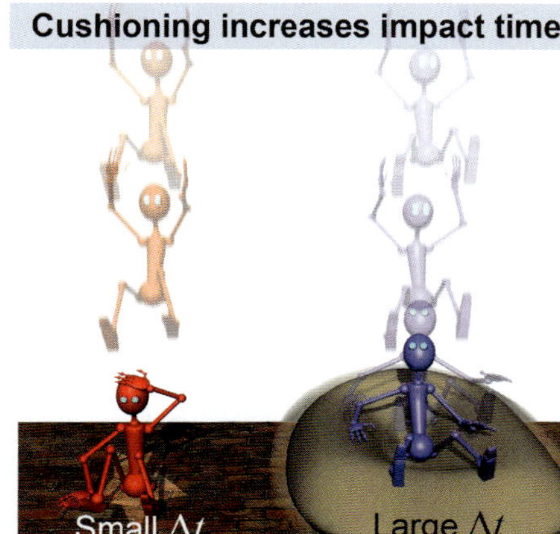

**Changes in momentum**

The momenta of things around us often change. As a car approaches an intersection and slows down, its momentum decreases. A train departs from the station and its momentum increases as its accelerates. A soccer player's velocity may change direction as she weaves between defenders, and because of this her momentum frequently changes direction.

**Impulse is change in momentum**

As we learned on the previous page, impulse $J$ is equivalent to a change in momentum: $J = \Delta p = p_f - p_i$. Imagine a car moving down the street at a velocity of 10 m/s. The driver brakes to a stop; the car's velocity and hence its momentum have changed. The change in momentum, or impulse $J$, is the final momentum minus the initial momentum. In the illustration at right, the change in momentum is −20,000 kg m/s. In most problems you will encounter, an object will change its velocity but not its mass. In such cases, a change in velocity means a change in momentum—and that means an impulse has been applied to the object.

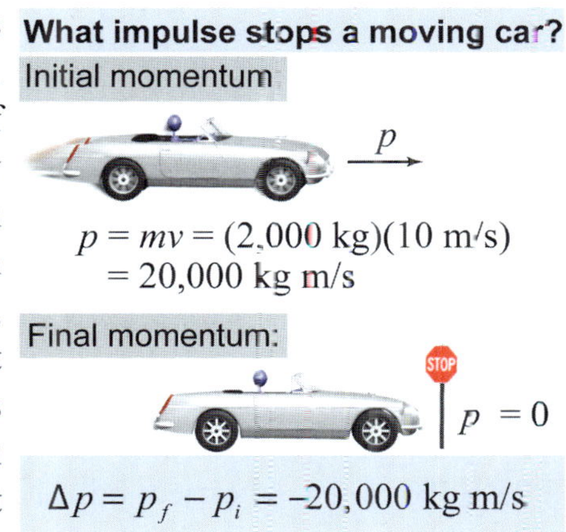


Interactive equation

$$J = \Delta p = p_f - p_i$$

$J$ = impulse (kg m/s)
$\Delta p$ = change in momentum (kg m/s)
$p_f$ = final momentum (kg m/s)
$p_i$ = initial momentum (kg m/s)

**Impulse**
change in momentum

**Units of impulse**

Impulse has the units of newton seconds (N s). But since impulse is equal to the change in momentum, it also has the same units as momentum, or kilogram meters per second (kg m/s), because 1 N s = 1 kg m/s.

**Applying impulse**

An impulse can be applied in many different ways. A driver can step on the gas pedal to apply a positive impulse, whereas the brake pedal delivers a negative impulse. Even when slowing down, there are many different ways to deliver the same impulse.

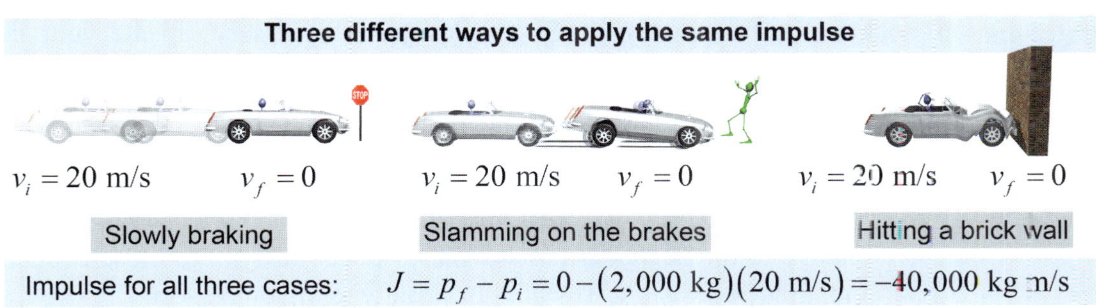


**Three different ways to apply the same impulse**

The illustration above shows three examples of how to apply the same impulse to slow down a car. The driver can brake slowly, slam on the brakes, or get into an accident (such as by hitting a brick wall). No matter how his momentum is reduced, the total impulse applied to his car is the same: His momentum changes from 40,000 kg m/s to zero. Even though the impulse applied is the same, the three examples look very different. Why? In each case a different force is applied for a different length of time, but in each case the product of these two quantities equals the change in momentum of the car.

## Design a crash barrier

**Design challenge**

Construct a crash barrier designed to minimize the impact force experienced by a cart in a collision.

*Test equipment:* Smart Cart with bumper, dynamics track with feet and end stops.

*Barrier materials:* straws, popsicle sticks, toothpicks, newspaper, craft paper, bathroom tissue, aluminum foil, duct tape, cotton balls, thread, and cardboard.

*Other constraints:* maximum impact force is less than 10.0 N, maximum mass is 50 g, maximum length is 12 cm, and the front of the crash barrier is flat.

**Change in momentum and force**

When a car impacts a wall, the wall exerts a strong force on the car to abruptly change the car's momentum over a very small time interval. Common features on highways are crash barriers built to reduce the force of incidental impact from cars. Crash barriers reduce the force during a collision by lengthening the time interval over which the impact occurs.

**Test different materials and configurations**

Set up your test equipment. Use the **DC_CrashBarrier** experiment file to record data. Use your test station to test different barrier materials and configurations you might use in your design. Vary one or two parameters: padding material, shape, and so on. Write a testable hypothesis on the effect of each parameter.

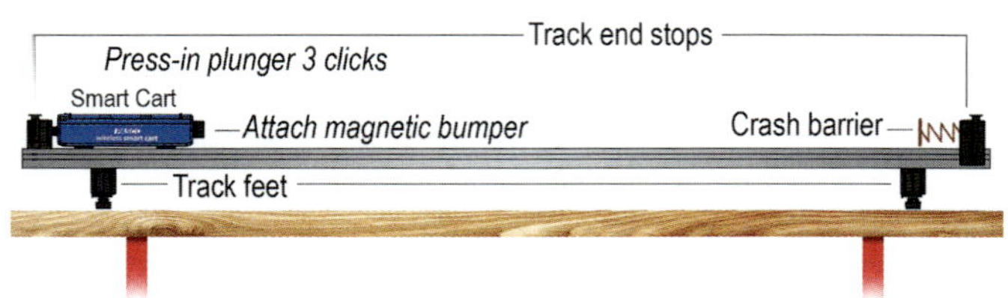


**Design**

Design the crash barrier based on test results. Describe your design with words and sketches.

**Prototype**

Construct a prototype of your design. If necessary, update your design to include any changes made during prototyping as a result of unforeseen problems.

**Test**

Test your prototype. Before running the test, identify what you will be evaluating at the conclusion of each test, such as what parts of the barrier did not perform as expected.

**Evaluate**

Analyze the tests to determine which design elements produce the best performance relative to the design criteria. What evidence led you to those conclusions? Document the strengths and weaknesses of your prototype in your report.

**Revise**

Based on your evaluation, redesign your barrier using the design elements that best meet the evaluation criteria. Then prototype and test it in preparation for the final evaluation.

**Final performance evaluation**

Your teacher will evaluate each submitted barrier by measuring its mass and length (less than 50 g and shorter than 12 cm), and launching a cart across a level track (using the cart plunger pressed-in three clicks) into the flat front of the crash barrier. The peak force must not exceed 10.0 N. Afterward, each student group will present to the class a short summary of their design and performance. As a class, discuss what design elements performed best.

### Football and cars

How many "$g$'s" does a football player's brain undergo when ramming an opponent? Does the front end of a car always need to take such a beating? These questions may seem unrelated, but they are unified by the physics of impulse and cushioning.

### Concussion and contact sports

Play a sport such as football, ice hockey, or lacrosse, and sooner or later your head will almost certainly experience an abrupt impulse. Recent studies suggest that the human brain has a 75% chance of being concussed when it experiences a $100g$ acceleration (which corresponds to a force of roughly 1,400 N!). In the hope of reducing concussion risk, new football helmets are being designed with flexible face masks and foam layers of varying densities. If the helmet's cushioning can *increase* the amount of time the head is decelerated, then the forces imparted on the player's brain are *reduced.* Padding likewise reduces the forces felt by other parts of a contestant's body.

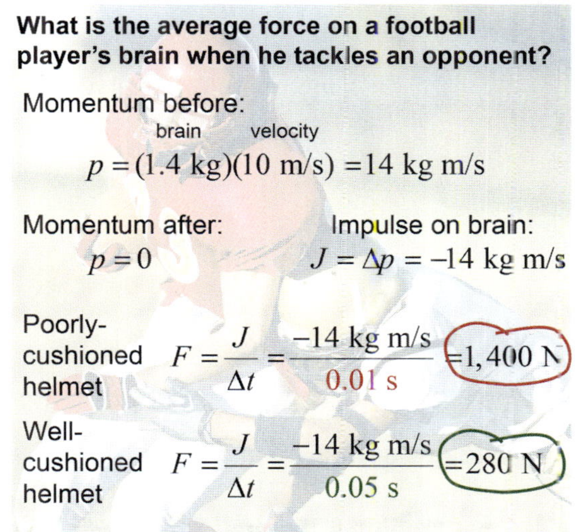


### Extending car crashes

The same reasoning lies behind many features of today's automobiles. While it might seem odd to make a car crash last *longer*, increasing its duration is exactly what improves the odds of survival for the vehicle's occupants. Consider a collision that abruptly stops your 35 mph (15.6 m/s) car. If your body mass were 60 kg, your initial momentum would have been roughly $p_i$ = 936 kg m/s. You might have transferred that momentum to the steering wheel in a dangerous 100 milliseconds (0.1 s) if it were not for several life-saving technologies such as the crumple zone. The corresponding force of 9,360 N is *approximately the weight of your car*, and it might be sufficient to kill you. The front end of the car, however, is designed to collapse in the event of a crash—rather than remain rigid—thereby extending the time of the collision. The force imparted on the passenger in the case of a crash is reduced. An easily crushed engine compartment is not a flaw but a well-designed, possibly life-saving feature.

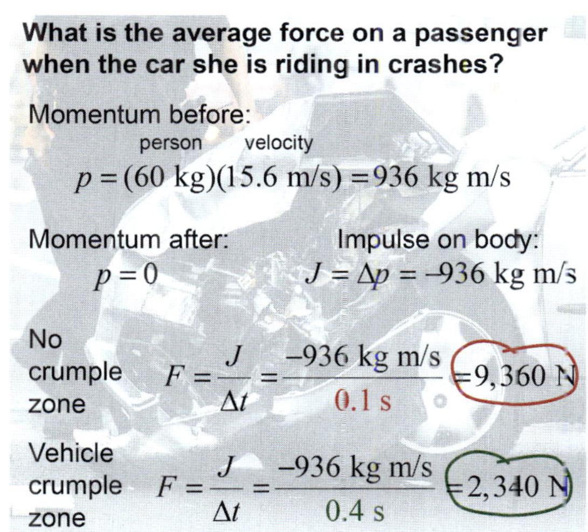


### Seat belts and airbags

Anything that increases the time over which a collision occurs will reduce the deceleration of the occupants and decrease the forces imparted on them. Seat belts and airbags are both vehicle technologies that are designed to do just that.

# Section 1 review

Momentum is a vector quantity: the product of mass and velocity. Impulse (also a vector) is the change in an object's momentum. Impulse can be expressed as the product of an applied force and the force's duration. Cushioning increases the time interval for momentum transfer, and this reduces the forces felt in car crashes and contact sports.

| | | |
|---|---|---|
| **Vocabulary words** | **momentum, impulse** | |
| **Key equations** | $\vec{p} = m\vec{v}$ | $J = \Delta p = F\Delta t$ |

## Review problems and questions

1. State one similarity and one difference between inertia and momentum.

2. A hockey puck that has a mass of 170 g travels with a speed of 30 m/s.
   a. What is the momentum of the puck?
   b. What impulse must be imparted on the puck by a player who wishes to change the puck's direction by 180°, while keeping the puck moving at the same speed?

3. Julieta kicks a stationary 0.2 kg soccer ball. The ball leaves her foot at a speed of 20 m/s.
   a. What impulse was delivered to the ball?
   b. The ball and Julieta's foot experience equal-but-opposite forces, each 40 N strong. How long (in seconds) were the ball and foot in contact?

4. Marco strikes a punching bag with 80 N of force for 0.05 s. Marcela strikes a body bag with 60 N of force for 0.1 s. Who delivers a greater amount of impulse?

5. A ball with a mass of 200 g is thrown straight down at the floor. It strikes the floor at a speed of 10.0 m/s and bounces straight up again with a speed of 6.0 m/s. What is the change in the ball's momentum?

6. A 1,200 kg car is heading due north at 14.0 m/s. It collides with an identical car heading due south at 14.0 m/s. What is the combined momentum of the cars before the collision?

7. Which of the following has greatest inertia? Which has greatest momentum?
   a. 6,000 kg elephant charging at 11 m/s
   b. 200 g bullet fired at 300 m/s
   c. 18,000 kg fire engine parked on the street

8. Sean, whose mass is 60 kg, is riding on a 5.0 kg sled initially traveling at 8.0 m/s. He brakes the sled with a constant force, bringing it to a stop in 4.0 s. What force does he apply?

9. A 1000 kg car is traveling north on a highway at 20 m/s. What is the car's momentum when observed by another car traveling in the *opposite* direction at 20 m/s?

Just as the total amount of energy within a closed system remains constant, so too does the total amount of momentum. Unlike energy, however, momentum is a vector. This means that conservation of momentum has several interesting and novel implications for subjects ranging from rocketry to billiards, car crashes, and contact sports.

## Conservation of momentum

**Momentum and interactions**

Imagine two roller skaters who are stationary. They each have zero momentum. If they push their arms against each other, they will move apart. One skater acquires a momentum to the left, while the other acquires an equal but opposite momentum toward the right. Although each skater's *individual* momentum changed, *the sum of their momenta did not change!*

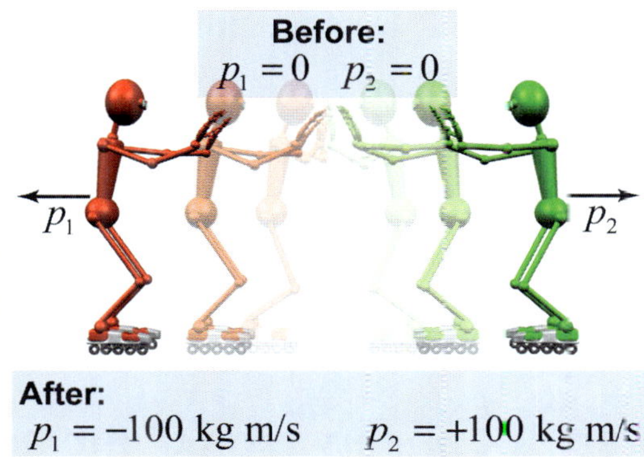


**Action–reaction forces and momentum**

Why is this so? Look at these free-body diagrams for the two skaters. Each skater is in equilibrium along a vertical axis—the weight is canceled out by the support force from the floor. When pushing off, however, each skater feels a net force along the *horizontal* axis. Newton's third law tells us that these two pushes are action–reaction forces: equal in strength but opposite in direction.

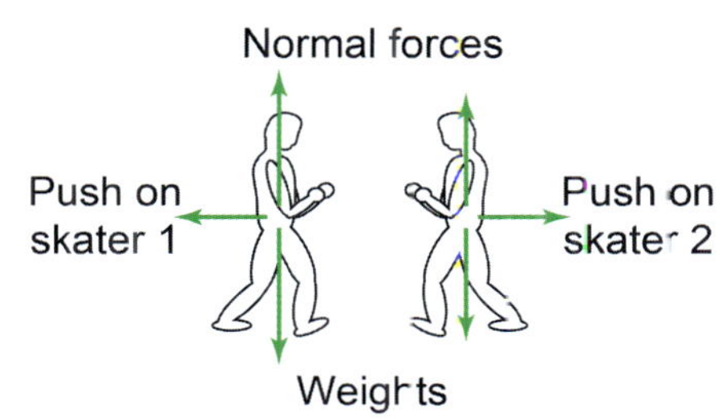


**Consider the impulse**

If the two skaters are equally massive, pushing off each other causes equal and opposite accelerations. Now suppose the skaters have different masses. They still feel equal and opposite forces, and these forces act over the same time period, the time interval $\Delta t$ of their pushing.

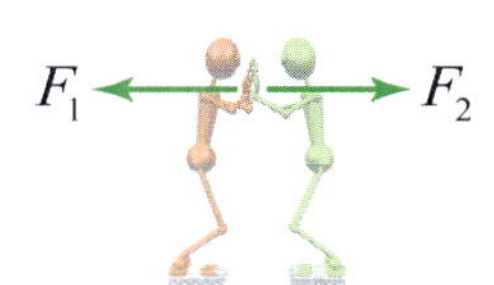


Robots apply equal and opposite forces to each other...

$$\vec{F}_1 + \vec{F}_2 = 0$$

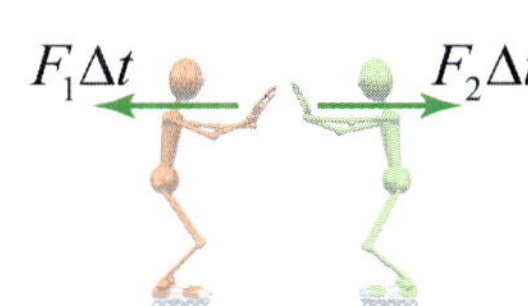


...in contact for the same time...

$$\vec{F}_1 \Delta t + \vec{F}_2 \Delta t = 0$$

...which gives each robot the same change in momentum:

$$\Delta \vec{p}_1 + \Delta \vec{p}_2 = 0$$

**Momentum is conserved!**

**Conservation of momentum**

Since force multiplied by time equals impulse, the two skaters experience equal and opposite impulses $\Delta p$. In general, if there are no external forces acting on a system, impulses come in pairs that cancel each other out, leaving no *net* change in the momentum of the system. For a closed system, this implies the **law of conservation of momentum**:

**Momentum conservation**

***The total momentum of a closed system remains constant, as long as no net outside force acts upon that system.***

Objects *within* the closed system—such as the two skaters—are free to exchange momentum between them, but the *total* momentum cannot change.

# Investigation 11A

## Conservation of momentum

**Essential questions** *How does momentum change for objects in an isolated system?*

The law of conservation of momentum is a powerful generalization of Newton's third law. For an isolated system, the total momentum of all the objects inside is constant. In this investigation, you will explore the conservation of momentum for two carts that are subject to no outside net force. The only catch is that the carts have a compressed spring inserted between them! Is momentum conserved for this system?

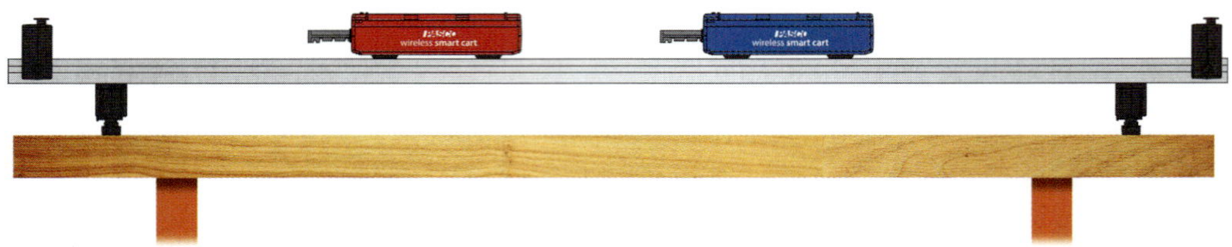

## Part 1: Conservation of momentum for spring-loaded carts

1. Open the experiment file **11A_ConservationOfMomentum**, and then power-on the Smart Carts and connect them wirelessly to your software.
2. Set a red cart and a blue cart on a level track, as shown.
3. Press the plunger on the blue cart until it loads at its third position. Set both carts in the middle of the track with the plunger on the blue car touching the red cart.
4. Begin collecting data and then tap on the plunger release trigger to launch the carts.
5. Use your software to find the velocity of each cart after the launch.
6. Run the experiment for different combinations of masses for the two carts. Use your data table to record the mass and velocity for each combination.

| $m_{red}$ (g) | $v_{red}$ (m/s) | $p_{red}$ (kg m/s) | | $m_{blue}$ (g) | $v_{blue}$ (m/s) | $p_{blue}$ (kg m/s) |
|---|---|---|---|---|---|---|
| 250 | | | | 250 | | |
| 500 | | | | 250 | | |
| 750 | | | | 250 | | |
| 250 | | | | 500 | | |
| 500 | | | | 500 | | |
| 750 | | | | 500 | | |
| 250 | | | | 750 | | |
| 500 | | | | 750 | | |
| 750 | | | | 750 | | |

a. *Describe the velocities when the masses of the two carts are equal.*
b. *Describe the velocities when the red cart has more mass than the blue cart.*
c. *Describe the velocities when the blue cart has more mass than the red cart.*
d. *Evaluate the data in your table. What quantity can you construct or calculate that is equal and opposite for the two carts after they are released? How is this the most logical conclusion to draw from your data?*
e. *If the two carts together are considered a closed system, what is the net force on the system? What is the change in the system's momentum after being released? Use appropriate equations to explain how these two questions are related to each other.*
f. *What similarities are there in the velocities and momenta when the masses of both carts are equal?*
g. *What is the difference in velocity when one mass is twice that of the other? When one mass is three times the other?*
h. *What is the difference in momentum when the masses are equal? When one cart has twice the mass of the other? When one cart has three times the mass of the other?*

### Putting momentum conservation to work

The law of conservation of momentum is a powerful tool for understanding how objects move after they interact. A rocket exhaust, a car crash, and the collision of billiard balls all involve objects exchanging momentum through action–reaction forces.

$$m_1v_1 + m_2v_2 = m_1v_1 + m_2v_2$$

### Momentum conservation

Imagine a system with two objects that can only move to the left or right. They may or may not collide; they may or may not stick together if they do so. It doesn't matter: The total momentum of *any* closed system remains unchanged. This rule applies as long as any outside forces on the system add up to zero.

### A strategy for momentum problems

To solve momentum problems, set up a table showing the momentum of each object, both before and after the objects interact. (Subscripts are often used to identify the two objects or to indicate initial and final values.) If the motion is along a single axis, decide which direction corresponds to positive velocity. Then be sure to assign plus or minus signs accordingly.

**How to solve momentum problems**

1. Calculate all the known momenta for objects *before* the interaction.
2. Calculate all the known momenta for objects *after* the interaction.
3. Equate the *total momentum* before the interaction to the system's total momentum after the interaction.
4. Solve for the unknown quantity.

Typically you can calculate most of the momenta right away and state a number in kilogram-meters per second. At least one momentum value, however, will contain an unknown variable: the quantity you are trying to solve for. You may need to rearrange the equation.

### Recoil of an astronaut who throws a wrench

A motionless 100 kg astronaut is holding a 2 kg wrench while on a spacewalk. To get moving, the astronaut throws the wrench forward at a speed of 5 m/s. How fast does the astronaut move backward?

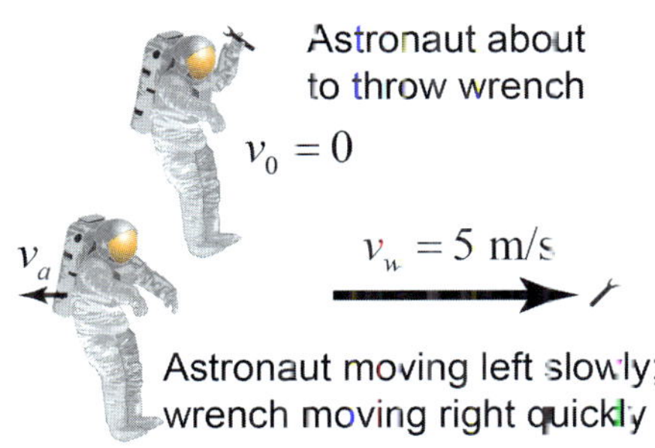


*Asked:* astronaut's final velocity $v_a$

*Given:* masses $m_w = 2$ kg (wrench) and $m_a = 100$ kg (astronaut); initial velocity $v_0 = 0$ (both); final velocity $v_w = 5$ m/s (wrench)

*Relationships:* momentum $p = mv$; momentum conservation

*Solution:* Momentum before equals momentum after:

$$m_a \cancel{v_0} + m_w \cancel{v_0} = m_a v_a + m_w v_w \Rightarrow 0 = m_a v_a + m_w v_w$$

Solve for the astronaut's final velocity:

$$v_a = -\frac{m_w v_w}{m_a} = -\frac{(2\text{ kg})(5\text{ m/s})}{(100\text{ kg})} = -0.1\text{ m/s}$$

*Answer:* The astronaut moves 0.1 m/s backward (negative direction).

**The puzzle of propulsion**

Many people think that rockets work by the propellant pushing against the air or the launch pad. If this were true, then how would rockets still work in outer space? There is nothing in the vacuum of outer space to push against! Space travel is made possible by the physics of momentum conservation.

**Action–reaction forces in rocketry**

Rockets are propelled "forward" when exhaust gases stream out "backward." Since those gases have mass, they have momentum—momentum that they lacked until they were blasted out of the exhaust nozzle. So the rocket has a forward (or positive) momentum while the exhaust has a backward (or negative) momentum. Since the rocket and exhaust form a closed system, their total momentum must be conserved.

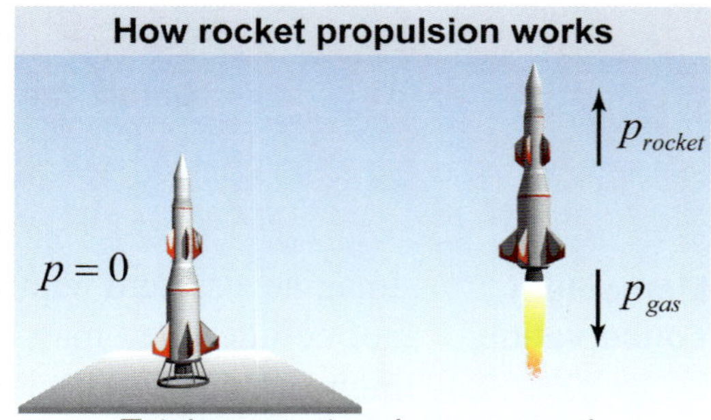


Total momentum is conserved:
Total momentum always adds to zero!

**Impulse in rocket propulsion**

If the change in momentum is zero, then the total impulses must be zero because impulse is the change in momentum. The exhaust gases were given a *negative* (or backward) impulse, while the rocket itself received a *positive* impulse because of the law of conservation of momentum. And *that* means you can push a rocket forward as long as you have something (such as exhaust gases) to toss in the opposite direction—no launch pad, atmosphere, or other obstacle is required.

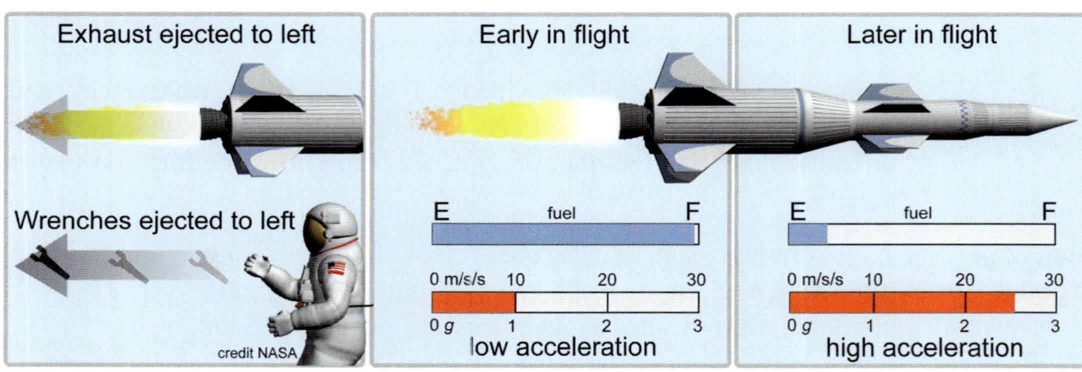


**Increasing acceleration**

A typical, modern rocket experiences an *increasing*, not constant, acceleration. Why? In a rocket, the force of *thrust* (which is the product of the exhaust velocity and the mass-loss rate, measured in newtons) is more or less constant until the fuel is nearly used up. The mass of the rocket *decreases* over time as the fuel is used up and expelled out the nozzle. Now combine these two ideas: A *constant* force divided by a *decreasing* mass implies an *increasing* acceleration from Newton's second law, $a = F/m$. That is exactly what we see in the first few minutes of a rocket launch. Not only does the rocket gain speed, it does so at an increasing rate.

**Parameters for modern rockets**

A modern cargo-carrying rocket might have a mass of 100,000 kg when empty and 500,000 kg on the launch pad, full of the fuel it will use for the first stage of its journey. Exhaust typically exits such rockets at many thousands of meters per second, with hundreds or thousands of kilograms flowing out each second.

# Section 2 review

Newton's third law states that any two interacting objects push or pull each other with equal (and opposite) force. One key consequence of Newton's third law is that the total momentum of any closed system is conserved (does not change). If two objects push each other away, their individual momenta change by equal amounts, though in opposite directions, resulting in a zero net change in the momentum of the system. Conservation of momentum forms the basis for understanding rocketry, jet propulsion, and the recoil motions of guns and cannons.

| Vocabulary words | law of conservation of momentum |
|---|---|

## Review problems and questions

1. The law of conservation of momentum applies to *closed systems.* What does the term "closed system" mean in this context?
   a. No force of any kind can act upon any object within the system.
   b. No net force can act upon any object within the system.
   c. No net force may operate upon the system from outside the system.

2. Jaime and Ayesha stand motionless while wearing roller skates. They gently press on each others' hands, palms up. Each friend then rolls away at a speed of 0.8 m/s. Which statement accurately compares their masses?
   a. Jaime and Ayesha have equal masses.
   b. Jaime is more massive than Ayesha.
   c. Jaime is less massive than Ayesha.
   d. There is not enough information to compare their masses.

3. During a spacewalk, Dmitri finds himself floating 3 m from his space station's airlock. He only has one minute of air left. He has one detachable 10 kg toolkit, which he can toss to propel himself toward safety.
   a. How fast will he need to move to reach the airlock before he runs out of air?
   b. Dmitri and his space suit together have a mass of 100 kg. How fast must he throw the toolkit to propel himself toward the airlock at the required speed (found in the previous step)?

4. During the spacewalk described above, Dmitri tosses his detachable 10 kg toolkit. The statements below refer to Dmitri and the toolkit during the throw. Decide whether each statement is true or false.
   a. The magnitude of the force is the same for each.
   b. The duration of the event is the same for each.
   c. The magnitude of the acceleration is the same for each.
   d. The magnitude of the impulse is the same for each.

Collisions happen in many ways: A cue ball strikes other pool balls, a hiker brushes past branches, a baseball player strikes a fastball, one car rear-ends another car, the space shuttle docks with the space station, or a baker pounds bread dough. People do not usually refer to all of these events as collisions, but in all these cases one object comes in contact with another object. In physics, a **collision** is any such interaction that causes one or more objects to change its velocity. As you will learn in this section, all collisions can be categorized as *elastic*, *inelastic*, or *perfectly inelastic*. In the illustration at right, you can describe *qualitatively* the outcome of each type of collision based on how high the ball bounces. How can we describe collisions *quantitatively* using physics?

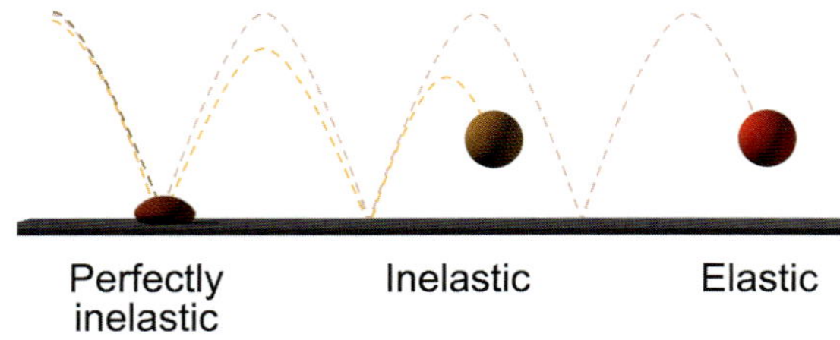


## Conservation of momentum in collisions

**Momentum is conserved in all collisions**

As we learned on page 315, momentum is always conserved for a closed system. The law of conservation of momentum applies to all different kinds of collisions, whether the two objects bounce off each other or stick together.

**Collisions involving a stationary target**

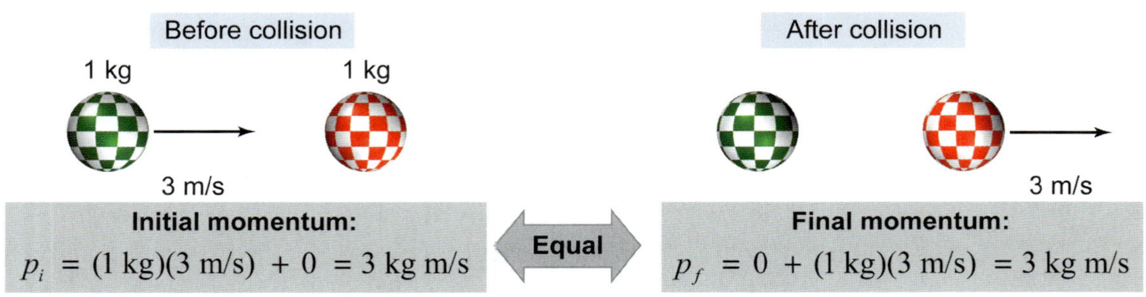


In the illustration above, two balls with masses of 1 kg collide. The red ball is stationary before the collision, and the green ball is stationary after the collision. How can we check that momentum is conserved? Add together the momenta of the objects before the collision and compare this to the momenta of the objects after the collision. Because the red ball is stationary, the total momentum before the collision is that of the green ball: $p = (1\ \text{kg})\times(3\ \text{m/s}) = 3\ \text{kg m/s}$. After the collision the green ball is stationary, so the total momentum is now simply that of the red ball: $p = (1\ \text{kg})\times(3\ \text{m/s}) = 3\ \text{kg m/s}$. We can see that momentum is conserved, because it did not change as a result of the collision.

**Using momentum conservation to analyze collisions**

Whenever analyzing a collision, calculate the momentum for all the objects before the collision and equate that to the momentum of all the objects after the collision. If one object has an unknown momentum, then you can determine it by comparing the *total* momentum before the collision to the *total* momentum after the collision.

**How to solve collision problems**

1. Calculate all the known momenta for objects *before* the collision.
2. Calculate all the known momenta for objects *after* the collision.
3. Equate the *total momentum* before the collision to the total momentum after the collision.
4. Solve for the unknown momentum.

**Collision between two objects**

Imagine a collision between two objects with masses $m_1$ and $m_2$. The two objects have initial velocities $v_{i1}$ and $v_{i2}$ and final velocities $v_{f1}$ and $v_{f2}$. Momentum conservation for these two colliding objects can be written as

Interactive equation

**Conservation of momentum (two objects)**

$$(11.3)\quad m_1 v_{i1} + m_2 v_{i2} = m_1 v_{f1} + m_2 v_{f2}$$

$m_1$ = mass of object 1 (kg)
$v_{i1}$ = initial velocity of object 1 (m/s)
$v_{f1}$ = final velocity of object 1 (m/s)
$m_2$ = mass of object 2 (kg)
$v_{i2}$ = initial velocity of object 2 (m/s)
$v_{f2}$ = final velocity of object 2 (m/s)

**Kinetic energy is lost in inelastic collisions**

There are two basic types of collisions in physics: elastic and inelastic. In an **inelastic collision**, some of the initial kinetic energy of the objects is transformed into heat and/or works to deform the shape of the objects. Auto collisions are nearly always inelastic, because of the damage caused to the cars. In the special case of a *perfectly inelastic collision*, the two objects stick together after impact.

**Perfectly inelastic collisions**

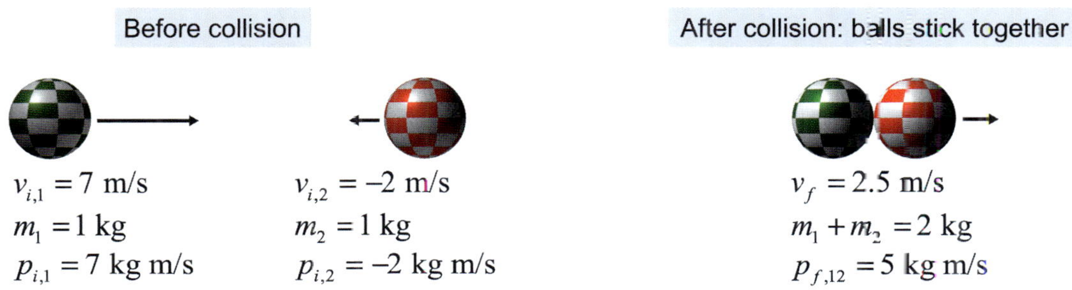


**Solving perfectly inelastic collisions**

A perfectly inelastic collision is depicted in the illustration above. These collision problems are solved in the same way as any other collision problem, using the conservation of momentum. Moreover, in the perfectly inelastic collision case the *final* velocities of the two objects are set to be equal—because the objects stick together!

**Final velocity in a perfectly inelastic collision**

A 100 kg hockey player, moving at 2 m/s, collides head-on with a 75 kg hockey player moving at 1 m/s. After impact, they become entangled and slide together. What is their velocity after impact?

*Asked:* final velocity $v_f$ after collision

*Given:* $m_1 = 100$ kg, $m_2 = 75$ kg,
$v_{i1} = +2$ m/s, $v_{i2} = -1$ m/s,
$v_{f1} = v_{f2} = v_f$, since they move together after the (inelastic) collision

*Relationships:* Conservation of momentum: $m_1 v_{i1} + m_2 v_{i2} = m_1 v_{f1} + m_2 v_{f2}$

*Solution:* Equate the final velocities: $m_1 v_{i1} + m_2 v_{i2} = (m_1 + m_2) v_f$.
Divide both sides by $(m_1 + m_2)$:

$$\frac{m_1 v_{i1} + m_2 v_{i2}}{m_1 + m_2} = \frac{(m_1 + m_2) v_f}{m_1 + m_2}$$

Cancel terms to solve for $v_f$ and substitute the values:

$$v_f = \frac{m_1 v_{i1} + m_2 v_{i2}}{m_1 + m_2} = \frac{(100 \text{ kg})(2 \text{ m/s}) + (75 \text{ kg})(-1 \text{ m/s})}{100 \text{ kg} + 75 \text{ kg}} = +0.71 \text{ m/s}$$

*Answer:* $v_f = +0.71$ m/s, in the same direction as the 100 kg player.

# Investigation 11B Inelastic collisions

**Essential questions** *How can we predict the outcome of an inelastic collision?*

In this investigation you will explore how the total kinetic energy and momentum of a closed system involving two carts is affected by a perfectly inelastic collision between the carts.

## Perfectly inelastic collisions

1. Set a red cart and a blue cart on a level track with the Velcro® ends facing each other, as shown. Adjust the track feet to make sure the track is as level as possible.

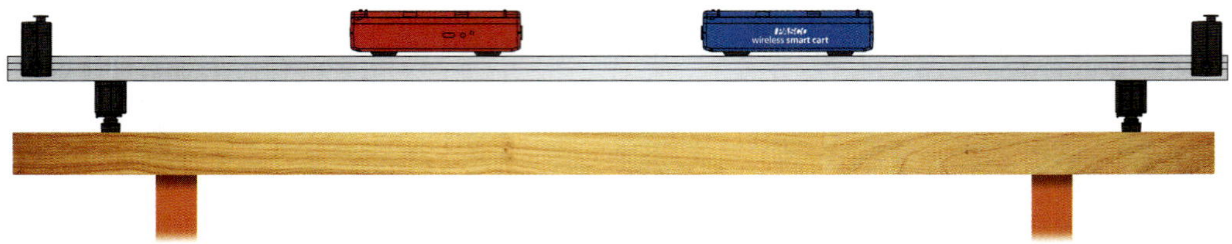

2. Open the experiment file **11B_InelasticCollisions**, and then power-on the Smart Carts and connect them to your computer using Bluetooth.
3. Begin collecting data with both carts stationary, and then push the red cart into the blue cart. Stop recording data after the carts have collided.
4. Use your software to find the velocity of each cart before the collision, and the final velocity of the two carts after the collision.
5. Run the experiment for different combinations of masses for the two carts (0.25 kg = mass of empty cart). Use your data table to record the velocities for each combination.
6. Calculate the total system momentum and kinetic energy before and after each collision. Record the values in your table and examine the table for patterns.

**Table: Velocity after the perfectly inelastic collision**

| Before collision | | | | | | After collision | | | |
|---|---|---|---|---|---|---|---|---|---|
| $m_1$ (kg) | $v_1$ (m/s) | $m_2$ (kg) | $v_2$ (m/s) | $p_{tot}$ (kg m/s) | $E_{tot}$ (J) | $m_{tot}$ (kg) | $v_{tot}$ (m/s) | $p_{tot}$ (kg m/s) | $E_{tot}$ (J) |
| 0.25 | | 0.25 | 0 | | | 0.50 | | | |
| 0.25 | | 0.75 | 0 | | | 1.00 | | | |
| 0.75 | | 0.25 | 0 | | | 1.00 | | | |

*a. Describe the velocities before and after the collision when masses are equal.*
*b. Describe the velocities (before and after) when the blue cart has more mass.*
*c. Describe the velocities (before and after) when the red cart has more mass.*
*d. As a result of the collision, does the total momentum of the system increase, decrease, or remain the same? Does the total kinetic energy of the system increase, decrease, or remain the same?*

**Energy is conserved in elastic collisions**

In an **elastic collision**, kinetic energy is conserved as well as momentum. An example of a perfectly elastic collision occurs when an ideal (frictionless) rubber ball bounces off a floor and reaches the same height from which it was initially dropped. A nearly-elastic collision occurs in billiards when a fast-moving cue ball strikes another ball, causing the cue ball to stop in place and the target ball to move off in the same direction. Real collisions are rarely perfectly elastic however, the amount of kinetic energy lost may be so small that it is often a good approximation to assume perfect elasticity.

**Conservation of energy for elastic collisions**

$$\frac{1}{2}m_1 v_{i1}^2 + \frac{1}{2}m_2 v_{i2}^2 = \frac{1}{2}m_1 v_{f1}^2 + \frac{1}{2}m_2 v_{f2}^2 \quad \text{(11.4)}$$

**Solving elastic collision problems**

Elastic collision problems typically involve two equations: conservation of momentum and conservation of kinetic energy. The momentum equation involves the masses and velocities before and after the collision. The energy equation involves the masses and the velocities *squared* before and after the collision. The squared velocities make the algebra of solving momentum problems a little more challenging. In problems involving two and three dimensions, momentum must be conserved separately in each direction. Kinetic energy is a scalar however, and there is typically only one kinetic energy equation.

**Is this collision elastic?**

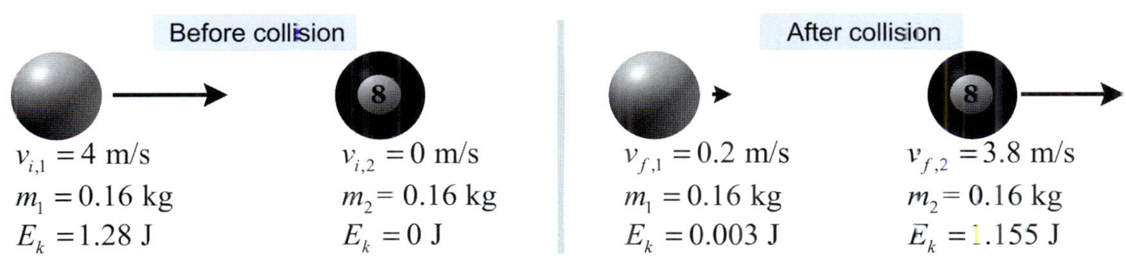


A 0.16 kg cue ball traveling at 4 m/s strikes a stationary 0.16 kg eight ball. After the collision, the cue ball travels at 0.2 m/s while the eight ball travels at 3.8 m/s. Is this an elastic collision? Why or why not?

*Asked:* whether it is an elastic collision; i.e., is kinetic energy conserved?

*Given:* masses of the balls, $m_1 = m_2 = 0.16$ kg; initial speed $v_{i1} = 4$ m/s of the cue ball; initial speed $v_{i2} = 0$ m/s of the eight-ball; final speed $v_{f1} = 0.2$ m/s of the cue ball; final speed $v_{f1} = 3.8$ m/s of the eight ball

*Relationships:* conservation of energy for two objects in an elastic collision:

$$\frac{1}{2}m_1 v_{i1}^2 + \frac{1}{2}m_2 v_{i2}^2 = \frac{1}{2}m_1 v_{f1}^2 + \frac{1}{2}m_2 v_{f2}^2$$

*Solution:* Compare the kinetic energies of the balls before and after the collision:

$$\tfrac{1}{2}(0.16\text{ kg})(4\text{ m/s})^2 + \tfrac{1}{2}(0.16\text{ kg})(0)^2 \stackrel{?}{=} \tfrac{1}{2}(0.16\text{ kg})(0.2\text{ m/s})^2 + \tfrac{1}{2}(0.16\text{ kg})(3.8\text{ m/s})^2$$

$$1.28\text{ J} + 0\text{ J} \neq 0.003\text{ J} + 1.155\text{ J}$$

*Answer:* The collision is not elastic because some kinetic energy is lost. (Since only a small amount of the kinetic energy is lost, the collision is "nearly elastic.")

# Investigation 11C Elastic collisions

**Essential questions** *How can we predict the outcome of an elastic collision?*

Interactive simulation

In an inelastic collision kinetic energy is lost, but momentum is conserved. Is the same true for perfectly elastic collisions? In this investigation, explore how the total kinetic energy and momentum of a closed system is affected by a perfectly elastic collision.

## Part 1: Simulation of elastic collisions between balls

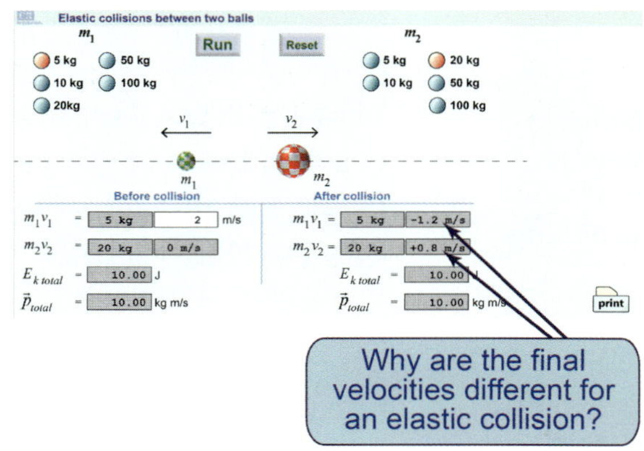


1. Run the simulation using $m_{\text{green}} = 5$ kg, 10 kg, and 50 kg with $m_{\text{red}} = 5$ kg, and $m_{\text{green}} = 5$ kg with $m_{\text{red}} = 10$ kg and 50 kg.
2. For each combination, record in a table the masses and initial and final values for velocities, $p_{\text{total}}$ and $E_{\text{k total}}$.
3. Examine the table for patterns in the data.

a. *Describe the velocities before and after the collision when: the masses are equal; the green ball has less mass; the green ball has more mass.*

b. *If this had been an inelastic collision investigation, how would your energy values differ?*

c. *As a result of the collision, does the total momentum of the system increase, decrease, or remain the same? Does the total kinetic energy increase, decrease, or remain the same?*

## Part 2: Elastic collisions between carts

A collision between two carts is nearly elastic, but not completely. How well can a physical situation approximate a perfectly elastic collision? Explore how the total kinetic energy and momentum of this closed system is affected by an elastic collision between the carts.

1. Attach a magnetic bumper to the front of each smart cart, then set a red cart and a blue cart on a level track with the magnetic bumpers facing each other, as shown. Adjust the track feet to make sure the track is as level as possible.

2. Open the experiment file **11C_ElasticCollisions**, and then power-on the Smart Carts and connect your Smart Carts to your computer using Bluetooth.
3. Begin collecting data, gently pushing the red cart toward the blue cart and allowing them to collide. Stop recording data once carts have collided.
4. Run the experiment for different combinations of masses for the two carts. Use a data table to record the velocity for each combination.
5. Examine the table for patterns in the data as in Part 1.

**Car-crash forensics**

A passenger car and a pickup truck have collided. Their bumpers lock and the pair skid together across the pavement. Police, insurance agents, and lawyers will analyze the scene in hopes of determining who was at fault. They can determine the post-collision speed from the distance that the two vehicles slid. They also use the direction the vehicles moved after the collision to determine the relative momentum each vehicle had before the accident. The law of conservation of momentum tells the reconstruction team the relationship between the vehicles' initial speeds and the distance the vehicles traveled after colliding.

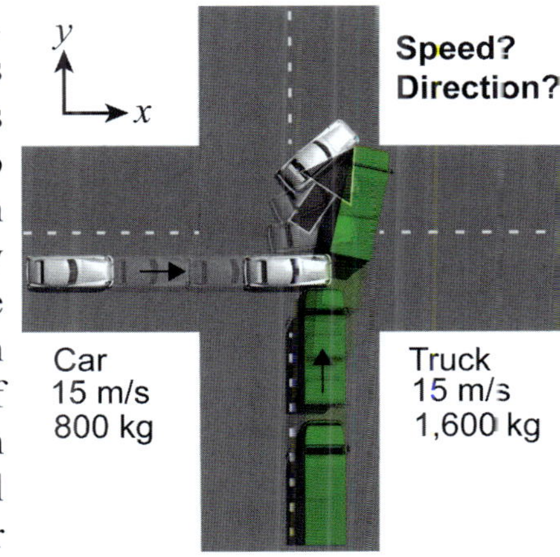


**Who was driving faster before the collision?**

To determine the relative speeds of the vehicles before the collision, the team uses momentum conservation. Since momentum is a vector quantity, *the total momentum along each axis must be considered separately.*

**Add momentum vectors, not velocities**

| Analyzing a perfectly inelastic collision in two dimensions | |
|---|---|
| 1. Calculate the initial momentum for each object. | 5. Use the Pythagorean theorem or graphical vector addition to get the total momentum of the system. |
| 2. Resolve each object's momentum into $x$- and $y$-components. | 6. Use trigonometry or graphical vector addition to get the direction of motion for the final system. |
| 3. Add all the $x$-components to get the final momentum in the $x$-direction. | 7. Divide momentum by total mass to get the final velocity of the system. |
| 4. Repeat Step 3 for the $y$-direction. | |

Let's follow the procedure above for the car–truck collision. Multiply each vehicle's mass and velocity to calculate its momentum. The car's momentum is entirely along the $x$-axis, while the truck's momentum is along the $y$-axis. We use the Pythagorean theorem to calculate the total momentum. Since the total momentum is conserved, we equate the total momentum before the collision to the final momentum of the vehicles afterward. We divide the final momentum by the total mass to calculate their mutual velocity after colliding.

**Step-by-step solution of the car–truck crash**

Step 1:

$$p_{car} = (15 \text{ m/s})(800 \text{ kg}) = 12{,}000 \text{ kg m/s}$$
$$p_{truck} = (15 \text{ m/s})(1{,}600 \text{ kg}) = 24{,}000 \text{ kg m/s}$$

Step 2:

$$p_x = 12{,}000 \text{ kg m/s (car)} + 0 \text{ (truck)}$$
$$p_y = 0 \text{ (car)} + 24{,}000 \text{ kg m/s (truck)}$$

Steps 3 and 4:

$$p_x = 12{,}000 \text{ kg m/s}$$
$$p_y = 24{,}000 \text{ kg m/s}$$

Step 5:

$$p_{tot} = \sqrt{(12{,}000)^2 + (24{,}000)^2} \text{ kg m/s} = 26{,}800 \text{ kg m/s}$$

Step 6:

$$\theta = \tan^{-1}\left(p_y / p_x\right) = 63^\circ$$

(or measure with protractor)

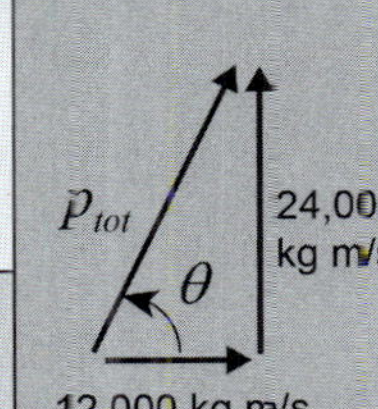


Step 7:

$$v_{fin} = \frac{26{,}800 \text{ kg m/s}}{(800 \text{ kg} + 1{,}600 \text{ kg})} = 11 \text{ m/s}$$

# Section 3 review

There are two basic categories of collisions, elastic and inelastic. Momentum is conserved in both types of collisions. Kinetic energy, however, is conserved only in elastic collisions. Kinetic energy isn't really lost during inelastic collisions; it is converted into other forms of energy, such as thermal energy or sound. In a *perfectly inelastic* collision, the colliding objects stick together and move as a single body.

| Vocabulary words | collision, inelastic collision, elastic collision |
|---|---|

| Key equations | $m_1 v_{i1} + m_2 v_{i2} = m_1 v_{f1} + m_2 v_{f2}$ | $\frac{1}{2} m_1 v_{i1}^2 + \frac{1}{2} m_2 v_{i2}^2 = \frac{1}{2} m_1 v_{f1}^2 + \frac{1}{2} m_2 v_{f2}^2$ |
|---|---|---|

## Review problems and questions

1. Rudolf is traveling at 3.0 m/s in his toy bumper car when he has a rear-end collision with Marcel, whose bumper car is initially at rest. The mass of each boy and his bumper car is 100 kg. As a result of the collision, Marcel glides off at 3.0 m/s in the direction that Rudolf had been going.
   a. What is Rudolf's velocity after the collision?
   b. Is the collision elastic or inelastic?

2. A collision takes place on a hockey rink. Sonja ($m$ = 80 kg) slides into Yoon-Hee ($m$ = 60 kg), who is initially at rest. Sonja skates into Yoon-Hee at 3 m/s, and as a result Yoon-Hee is ejected with the same velocity, in the direction that Sonja had been going.
   a. What is Sonja's velocity after the collision?
   b. Is the collision elastic or inelastic?

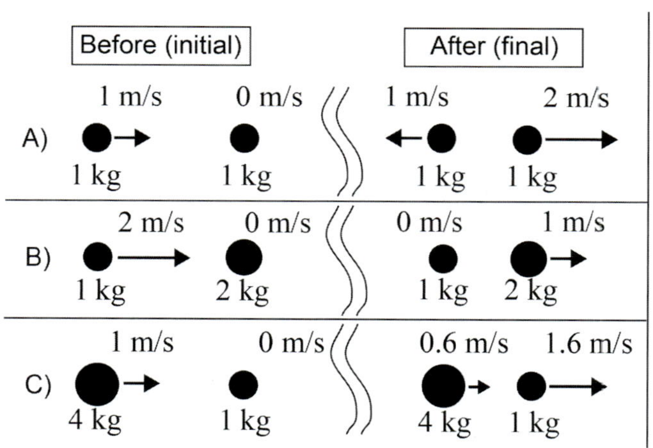


Match A, B, and C with

I. Collision is impossible (system kinetic energy higher after collision than before).

II. Collision is impossible (momentum is not conserved).

III. Collision is elastic.

IV. Collision is inelastic.

3. The graphic above shows three collision scenarios (A, B, and C). Each collision is between two objects that only can move horizontally. Match each scenario with one of the numbered statements (I, II, III, or IV).

# Chapter review

### Vocabulary

Match each word to the sentence where it best fits.

#### Section 11.1

| impulse | momentum |
|---|---|

1. A/an _______ can be thought of as either the product of force and duration time or as change in momentum.
2. An object's _____ is a product of its mass and velocity.

#### Section 11.2

| law of conservation of energy | law of conservation of momentum |
|---|---|

3. The _______ explains how it is possible for a rocket to speed up in empty space even if no external force is exerted on it.

#### Section 11.3

| collision | elastic collision |
|---|---|
| inelastic collision | |

4. A/An _______ is an interaction between objects that causes one or more of them to change velocity.
5. Momentum is conserved in all collisions, but kinetic energy is only conserved in a/an _______.
6. A collision where the objects stick together after impact is called a/an _______.

### Conceptual questions

#### Section 11.1

7. Describe two different scenarios where the same object has identical momentum.
8. A rubber ball and a clay ball of equal mass are dropped from the same height. The rubber ball bounces back up, and the clay ball sticks to the floor.
   a. Which ball has the greatest momentum just before hitting the floor, or are they equal?
   b. Which ball experiences a greater change in momentum as a result of the collision, or are they equal?
   c. Which ball experiences a greater change in kinetic energy as a result of the collision, or are they equal?
9. For a given running speed, does it take more force to slow down if you are wearing a backpack or not?
10. Philip is chasing his dog. He has four times more mass than the dog, but he can only run half as fast as the dog. What is Philip's momentum, compared to his dog's?
11. Which sentence below best expresses the meaning of the following equation?

$$F\Delta t = \Delta p$$

   a. Force multiplied by time equals momentum.
   b. The net force on an object multiplied by the change in momentum equals the impulse.
   c. The force on an object divided by the momentum is equal to the time duration.
   d. The change in momentum equals the net force exerted on the object multiplied by the time interval.
12. Describe two situations in which it is possible to double the mass of an object without changing the momentum of the object.
13. A stunt man jumps from a burning building and comes to rest at ground level. Consider his impulse during landing if he lands on a soft mat, on a concrete surface, or in a safety net. Which of the following is true?
   a. His impulse is greatest if he lands on a soft mat.
   b. His impulse is greatest if he lands on a concrete surface.
   c. His impulse is greatest if he lands in a safety net.
   d. The impulse is the same in all three cases.
   e. There is zero impulse in all three cases because he comes to rest.
14. Stuart pushes a table with a force of 100 N. Is the impulse different if he pushes for 1 s or 10 s?
15. ❬ If you want to increase an object's momentum by the greatest amount possible, should you double the mass or the velocity while keeping the other the same?
16. ❬❬ Are you better off in an accident where the hood is deformed or where the car is not damaged at all but bounces backward?
17. ❬❬ When designing an egg drop container as on page 312, what are the comparative advantages of using a roll of white craft paper, as opposed to sheets of newspaper or bathroom tissue?
18. ❬❬❬ Is impulse a vector quantity?
19. ❬❬❬ Newton's first law states that an object tends to stay in motion unless acted on by an external force. But braking—an internal force—causes a car to stop. Explain this apparent paradox.

# Chapter review

## Section 11.1

20. ❮❮ How is it possible that two objects with the same mass could have the same kinetic energy but different momenta?

## Section 11.2

21. When two identical pucks collide on a frictionless surface, the impulses they receive are always equal in magnitude and opposite in direction.
    a. Why must this be true? Explain this using Newton's third law of action and reaction.
    b. If the pucks are of unequal mass, and the surface is not frictionless, then are the impulses on the pucks from the collision still equal?
22. Explain how you would use two frictionless carts with a spring between them to demonstrate momentum conservation.
23. Give a five-slide oral presentation describing how the concept of momentum conservation applies to a jet engine. Your presentation should use the following terms: "exhaust velocity," "mass," and "forward momentum."
24. ❮ Two ballistic carts have a compressed spring between them, such as in Investigation 11A on page 316. If one cart is much more massive than the other, which one will have a greater speed after the carts are released?
25. ❮❮ What physical principle would suggest it is a bad idea to jump forward off of a skateboard without putting one foot on the ground while you jump?
26. ❮❮ A stationary package on a frictionless surface explodes into two chunks. One chunk has mass $M$, and the other chunk has mass $2M$.
    a. What is momentum of the system before the explosion?
    b. What is momentum of the system after the explosion?
    c. The smaller chunk moves due north with a speed of 10 m/s. Describe the motion of the larger piece.
27. ❮❮❮ Two students are arguing about the following situation:
    *A girl is traveling at constant speed in a frictionless cart. She has a supply of baseballs with her. What happens to the velocity of the cart if she starts dropping the baseballs over the side?*
    Jamal argues that the velocity of the cart does not change. Samuel says that momentum is conserved, so the cart must speed up as its mass decreases. Who is right?
28. ❮ How can you demonstrate the law of conservation of momentum using a billiards table and any additional equipment you need?

## Section 11.3

29. Jorge is conducting an investigation into perfectly inelastic collisions using equipment where two carts collide with each other. He can set up the carts either to bounce off each other or to stick together upon impact. Which setting should he use?
30. ❮ Two football players collide head on. The momentum of each player changes after the collision. Is momentum conserved?
31. ❮ How does solving an elastic collision problem differ from solving an inelastic collision problem?
32. ❮❮ An action hero jumps out of a stationary helicopter onto a car moving on an icy lake. What happens to the velocity of the car when the hero lands on it?
33. ❮❮ The bumpers on cars, as well as some cars' engine compartments, are designed to collapse in a collision. Why is this a useful design?
34. ❮❮ Two unknown objects collide. If one of the objects is stationary before the collision, can both objects be stationary afterward? Can only one be stationary afterward?
35. ❮❮ If two objects collide and one is stationary before impact, can only one of the two objects be stationary after impact? Justify your answer.
36. ❮❮ When two pucks undergo a perfectly inelastic collision, which statement below is always true about the total kinetic energy $E_k$ of the pucks?
    a. $E_k$ after the collision will be less than $E_k$ before the collision.
    b. $E_k$ after the collision will be greater than $E_k$ before the collision.
    c. $E_k$ after the collision will be zero.
    d. $E_k$ after the collision will equal $E_k$ before the collision.
37. ❮❮❮ In an elastic collision, a tennis ball hits a rigid wall and bounces back with a momentum that is equal in magnitude and opposite in direction. The wall is rigidly fixed to its surroundings. Is momentum conserved? Explain.
38. ❮❮❮ Two identical balls collide head on, while traveling at the same speed, but in opposite directions. The two balls come to a complete stop as a result of the collision. Is the collision elastic or inelastic? Why or why not?

# Chapter review

## Quantitative problems

### Section 11.1

39. What is the change in momentum for a 5,000 kg ship in outer space that experiences no net force over a 1 hr period?

40. A constant net force of 20 N is applied to a 32 kg car, causing it to speed up from 4.0 to 9.0 m/s. How long is the force applied?

41. A golfer tees up a 45 gram golf ball and hits it with her driver. The club head is in contact with the ball for 9.0 milliseconds (ms), and the velocity of the ball immediately after impact is 79 m/s.
    a. What is the impulse on the ball?
    b. What is the impulse on the club head due to the collision?
    c. What was the average force on the ball during the collision?

42. ❬ Calculate the momentum of a 75 kg person running at 10 m/s.

43. ❬ Calculate the momentum of a jet moving at 700 mph with a mass of 20,000 kg.

44. ❬ Calculate the magnitude of the change in momentum in each of the following examples.
    a. A 1,200 kg car accelerates from rest to a speed of 20 m/s.
    b. A 75 g pellet moving at 18.0 m/s hits a tree and lodges in it.
    c. A 250 g rubber ball hits the floor at 10.0 m/s and rebounds at 6.0 m/s.
    d. A 60 kg student jogging at 3.0 m/s speeds up to 6.5 m/s.

45. ❬ A green ball with a mass of 10.0 kg is heading due east at 50 m/s. A red ball with a mass of 25 kg is heading due west at 15 m/s. Calculate the total momentum of this system.

46. ❬❬ Which moves faster after being accelerated from rest, (a) a 500 kg boat that has been imparted with a 10,000 N s impulse or (b) a 780 kg boat imparted with a 14,000 N s impulse?

47. ❬❬ Consider two bowling balls, each with a mass of 4.0 kg. The red ball is moving east at 2.0 m/s. The blue ball is moving west at 1.0 m/s. Calculate the total momentum of the system.

48. ❬❬ A 1.5 kg tennis ball is slammed by a player who changes its direction by 180°, keeping its speed of 30 m/s the same. What impulse is required to redirect the ball?

49. ❬❬ A wind-up toy car has a mass of 66 g. The car is wound, and then released, going from being at rest to a constant velocity of 0.8 m/s. If the internal force of the winding on the tires is 0.1 N, what is the duration of the impulse imparted by the winding on the car.

50. ❬❬ A baseball player hits a 2 kg ball that is approaching with a velocity of 45 m/s. After making contact with the bat, the ball has a velocity of 40 m/s in the opposite direction. Seconds later, the ball is caught by an outfielder. Who imparts a greater impulse on the ball, the batter or the outfielder?

51. ❬❬ A 100 kg stunt woman falls from a three-story building that is 9.9 m high. If she falls into a net, which slows her down over the course of 1 s, what force did she experience while landing?

52. ❬❬ A train moves at 40 m/s heading due west. A young child with a mass of 20 kg, disregarding safety, is running down the hallway of the train opposite to the direction of travel, at 4 m/s. What momentum does the child have relative to an observer on the ground? The train has a mass of 100,000 kg.

53. ❬❬ An impulse $J$ is applied to a ball that is initially at rest, giving it a resulting momentum $p_1$ and kinetic energy $E_{k1}$.
    a. If the impulse is doubled to $2J$, what is the resulting momentum $p_2$?
    b. If the impulse is doubled to $2J$, what is the resulting kinetic energy $E_{k2}$?

### Section 11.2

54. ❬❬ A 3,000 kg truck is traveling 28 m/s through a rainstorm. At the end of its journey, it has collected 500 kg of water and is now moving 24 m/s. Was momentum conserved?

55. ❬❬❬ A 15,000 kg rocket traveling at +230 m/s turns on its engines. Over a 6.0 s period it burns 1,000 kg of fuel. An observer on the ground measures the velocity of the expelled gases to be −1,200 m/s.
    a. What is the resulting speed of the rocket after the burn?
    b. What is the acceleration of the rocket during the burn? Express your answer in meters per second squared and in "$g$"s.
    c. What impulse is applied to the rocket?

56. ❬ A car and an attached trailer have a combined mass of 3,000 kg. The car and trailer are moving on the highway at a velocity of 25 m/s. After going over a small bump in the road, the trailer detaches from the car. If the car's speed after separation is 27 m/s and its final momentum is 35,000 kg m/s, what are the masses of the car and trailer, respectively?

## Section 11.2

57. A ball with a mass of 0.5 kg is traveling at 25 m/s and collides with a ball at rest that has a mass of 2 kg. After the collision, both balls are traveling 5 m/s in the same direction. Was momentum conserved?

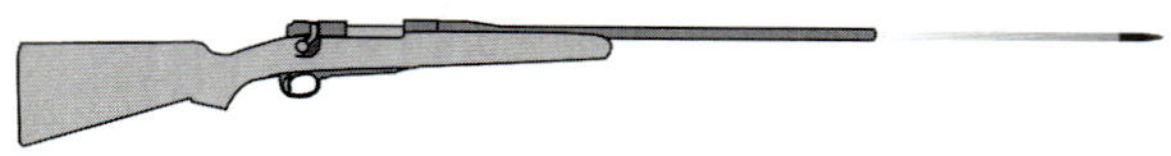

58. Calculate the recoil (backward momentum) and velocity of a 4.0 kg rifle that fires a 20 g bullet that travels 1.0 m in $2.15 \times 10^{-3}$ s.

59. Two carts connected with a compressed spring are placed somewhere along a frictionless track of 4.0 m in length. One cart has a mass of 150 g, and the heavier cart has a mass of 275 g. The spring is released and the two carts move off toward the two ends of the track, reaching the ends at the same instant.
    a. Which cart is initially closer to its end of the track, the light cart or the heavy cart?
    b. If the heavy cart gains a velocity of 2.0 m/s, what is the velocity of the light cart?
    c. How far from its end of the track is the heavy cart placed initially?
    d. How much momentum is produced by the release of the spring?
    e. How much kinetic energy is produced by the release of the spring?
    f. Energy is always conserved. Where does the kinetic energy of the carts come from?
    g. If the compression of the spring is increased, which of these answers a–f will change?

## Section 11.3

60. A 10,000 kg railroad car traveling north at 10 m/s collides with a 5,000 kg rail car also moving north but at an unknown speed. After the collision, the two cars lock together and move north at 8 m/s. How fast was the second car moving before the impact?

61. A 2.0 kg puck is moving east at 5.5 m/s. It catches up to and collides with a second identical puck moving due east at 3.0 m/s. The collision is perfectly inelastic.
    a. What is the resulting velocity of the pucks?
    b. What is the initial kinetic energy $E_{ki}$ of the system?
    c. What is the *change* in kinetic energy, $\Delta E_k$, of the system as a result of the collision?
    d. If the mass $m$ is doubled, but the initial velocities are unchanged, does the resulting velocity increase, decrease, or remain unchanged?

62. A 2,000 kg car moving at 10 m/s collides head-on with a 2,500 kg car moving in the opposite direction at 15 m/s. The two cars are locked together after impact.
    a. Is this an elastic or an inelastic collision? Why?
    b. What is the speed of the cars after impact?
    c. Calculate the kinetic energies of the cars, both before and after impact.
    d. What fraction of the kinetic energy was lost during the impact? Where did the energy go?

63. A stationary 165 kg football player is tackled by a 178 kg player running at 8 m/s.
    a. How fast are they moving after the collision?
    b. What is the impulse imparted on the stationary player?
    c. What is the impulse imparted on the moving player?

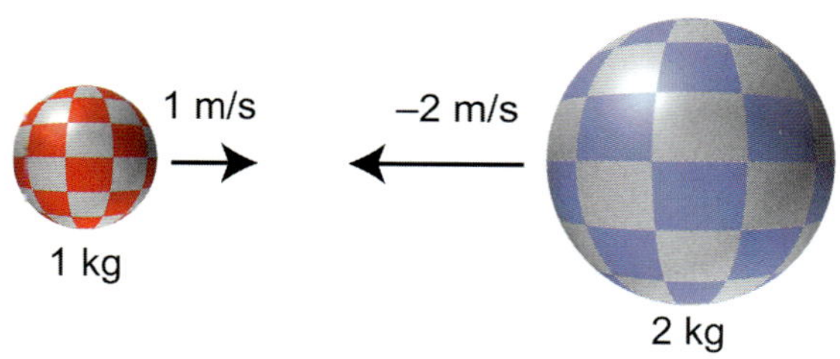


64. In an elastic collision, a 1.0 kg ball moving at 1.0 m/s collides with a 2.0 kg ball moving at −2.0 m/s. The 2.0 kg ball transfers all of its momentum to the 1.0 kg ball.
    a. What velocity does the 1.0 kg ball have after the collision?
    b. What is the initial kinetic energy of the system?
    c. What is the final kinetic energy of the system?

65. In an elastic collision involving two balls where one is stationary before the collision, can you make any generalizations about the speed and direction of the two balls after the collision? Consider the situations where (a) the moving ball is more massive than the stationary one; (b) the stationary ball is more massive than the moving ball; and (c) the two balls have equal masses.

66. Two clouds collide and form another, more massive cloud. One cloud is stationary, while the other is traveling at 1 m/s. After the collision, the new, combined cloud travels with a velocity of 0.25 m/s. What is the ratio of the masses of the two original clouds?

# Chapter review

## Standardized test practice

67. Jimi (80 kg) and Tony (100 kg) stand motionless on ice skates, facing each other at arm's length on a smooth, frozen pond. Which of the following actions will give Tony the greatest speed?

    A. Tony pushes Jimi with 100 N of force for 0.1 s.
    B. Jimi pushes Tony with a force of 60 N for 0.2 s.
    C. Tony pushes Jimi with 200 N of force for 0.08 s.
    D. Jimi pushes Tony with a force of 250 N for 0.04 s.

68. A 1,500 kg car sliding on frictionless ice at 15 m/s hits a stationary, 2,500 kg minivan. The two vehicles are locked together after impact on the ice. What is their speed after impact?

    A. 0 m/s
    B. 5.6 m/s
    C. 7.5 m/s
    D. 15 m/s

69. A truck with a mass of 2,000 kg and moving at 15 m/s collides with a stationary passenger car of mass 1,000 kg. The two cars stick together after impact. Assuming that they are free to move afterward, what will the momentum of the truck+car object be?

    A. 2,000 kg m/s
    B. 3,000 kg m/s
    C. 15,000 kg m/s
    D. 30,000 kg m/s

70. An 800 kg car with 200 kg of passengers and cargo is moving at 10 m/s. Suddenly, the driver applies the brakes and the car skids to a stop. What happens to the momentum that the car (and its contents) had before braking?

    A. It is transferred to the pavement (and the Earth).
    B. It is destroyed.
    C. It is transferred to the passengers and cargo.
    D. It is turned into sound.

71. A spacewalking astronaut is stranded a few meters from the space-station door. His jet pack is broken, but he can take it off and toss it in order to propel himself toward the door. In what direction should he toss it?

    A. toward the door
    B. away from the door
    C. toward the Earth
    D. The answer cannot be determined from the information provided.

72. Which of the following equipment is the best choice for investigating elastic or nearly elastic collisions?

    A. a marble launcher that causes two marbles to collide with each other
    B. two sliding carts on a frictionless track that stick together after impact
    C. ball of soft clay that flattens when it hits the ground
    D. crash test cars with front ends that crumple upon impact to absorb all the initial kinetic energy

73. Two spring-loaded ballistic carts, one of 200 g and the other 800 g, are released and the lighter cart is observed to move at +1 m/s afterward. What is the velocity of the other cart?

    A. −0.25 m/s
    B. +0.25 m/s
    C. −4 m/s
    D. +4 m/s

74. In an inelastic collision, what quantities are conserved?

    A. momentum only
    B. kinetic energy only
    C. momentum and kinetic energy
    D. neither momentum nor kinetic energy

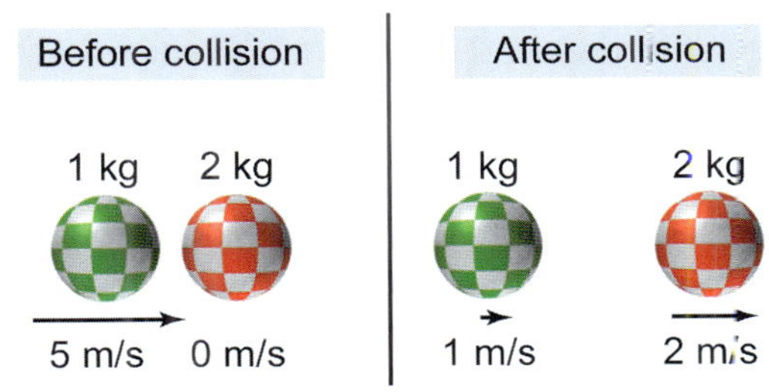


75. One sphere of mass 1 kg is moving at 5 m/s to the right until it smacks into a stationary, 2 kg sphere. After the collision, both spheres travel to the right: the 1 kg sphere at 1 m/s and the 2 kg sphere at 2 m/s. What kind of collision took place?

    A. an elastic collision
    B. an inelastic (but not perfectly inelastic) collision
    C. a perfectly inelastic collision
    D. The answer cannot be determined from the information provided.

76. Two friends stand on roller skates, facing each other within arm's reach. Jimi pushes gently on his friend Tony's upraised palms. What happens?

    A. Jimi remains motionless and Tony rolls backward.
    B. Jimi rolls backward and Tony remains motionless.
    C. Jimi rolls forward and Tony rolls backward.
    D. Jimi and Tony each roll backward.

# Chapter 12 Machines

If you wanted to get around town 200 years ago you had to walk, ride a horse, or use a horse-drawn vehicle. Animals are not a very convenient transportation option; they require stabling, food, and veterinary care. The invention of the bicycle in the 19th century revolutionized personal transportation, allowing people to travel far more easily with an inexpensive, reliable vehicle.

In early bicycles the pedals were connected directly to the wheel. One turn of the pedals advanced the rider forward one circumference of the wheel. To move at a reasonable speed, this required a very large wheel. Although they provided good speed, large-wheel bicycles were difficult to get on or off and their high center of mass made them quite unsafe. Upon hitting a rut in the road an unlucky rider often rotated face-first into the ground!

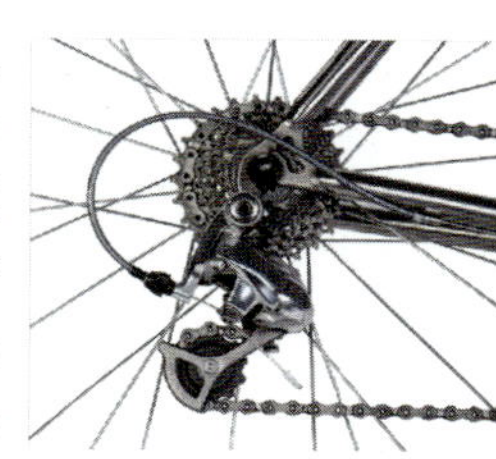

By the late 19th century, bicycles began to use a drive train with a chain and sprockets. Pedals turn the front sprocket, which turns the smaller rear sprocket using a chain. By giving the front sprocket more teeth than the rear sprocket, each turn of the pedals could produce four or five turns of the rear wheel. Initially called the *safety bicycle* the new design quickly caught on. In 1949, Italian racer and inventor Tullio Campagnolo introduced the modern *derailleur*, which allows the rider to change the gear ratio between the front and rear sprockets while moving.

The development of the bicycle impacted society beyond allowing workers who could now live in a suburb to commute to work. This "freedom machine" empowered women, giving them mobility and the freedom that came with it. Suffragette Susan B. Anthony once remarked, "I think [the bicycle] has done more to emancipate women than anything else in the world."

Today, bicycles come in a wide variety of models, from road bikes suitable for racing and mountain bikes for off-road riding to sturdy hybrids suitable for commuting in cities on potholed roads. Cycling enthusiasts compete in a host of different categories, including endurance events and single-speed bike races. Many of the early bicycle designs in the 19th century cost a worker the equivalent of several months' salary. Bicycles today can be bought at virtually any quality and price, with children's bikes costing less than a tank of gas.

# Chapter study guide

## Chapter summary

A machine helps us to accomplish a task by changing the direction and/or magnitude of the force we apply. All the constituent parts of machines fall into six categories of simple machines: levers, pulleys, wheels and axles, ramps, wedges, and screws. A compound machine, such as a bicycle or the human leg, is composed of two or more simple machines that work together. Real-world machines have an efficiency of less than 100% because some of the input work is lost to friction.

## Learning objectives

By the end of this chapter you should be able to

- define mechanical advantage, ideal mechanical advantage, and efficiency of a machine;
- describe the six types of simple machines;
- calculate the mechanical advantages of each type of simple machine; and
- define a compound machine, calculate its total mechanical advantage, and provide several examples.

## Investigations

12A: Levers
12B: Pulleys
12C: Ramps and inclined planes
Design a wind turbine power plant

## Chapter index



## Important relationships

$$MA = \frac{F_o}{F_i} \qquad MA_{lever} = \frac{L_i}{L_o} \qquad MA_{wa} = \frac{r_w}{r_a} \qquad MA_{ramp} = \frac{L_{ramp}}{h_{ramp}}$$

$$MA_{wedge} = \frac{L}{h} \qquad MA_{screw} = \frac{2\pi L}{p} \qquad MA_{ideal} = \frac{d_i}{d_o} \qquad \eta = \frac{W_o}{W_i}$$

$$GR = \frac{\text{output turns}}{\text{input turns}} = \frac{\text{input teeth}}{\text{output teeth}}$$

## Vocabulary

**machine**
**input force**
**fulcrum**
**tension**
**efficiency**
**gear**
**wedge**
**biomechanics**
**simple machine**
**output force**
**input arm**
**pulley**
**ideal mechanical advantage**
**gear ratio**
**screw**
**mechanical advantage**
**lever**
**output arm**
**block and tackle**
**wheel and axle**
**ramp**
**compound machine**

## 12.1 - Simple machines and the lever

What do a bicycle, a pulley, a pair of scissors, a door hinge, and the human body have in common? You might not think of calling them all by the same name, but all are examples of *machines*. A **machine** is an assembly of moving parts through which forces act. A machine might change the direction of an applied force, change the magnitude of a force, or change both direction and magnitude. Pedals, gears, and levers are mechanical machines. Bones and muscles form biological machines. The same laws of physics apply to all machines, whether mechanical or biological.

**What do these have in common?**

### Six types of simple machines

**Simple machines**

A **simple machine** is a mechanical system through which an applied *input force* directly creates an *output force* through a single motion. A simple machine has no source of energy except the instantaneous work done by the input force. The wheel of a bicycle is an example of a simple machine. The pedal and crank system on a bicycle is another example of a different simple machine.

**The six types of simple machines**

There are six types of simple machines: the lever, the pulley, the wheel and axle, the inclined plane, the wedge, and the screw. The physics of how they operate divides the six types into two groups: the levers and the inclined planes. You might not immediately see the relationship between a car engine and a simple lever. Nonetheless, the operation of any machine—no matter how complex—can be analyzed in terms of combinations of the six different simple machines. A car engine includes screws, ramps, levers, gears, and wheels.

**Levers**

As a child you may have played on a lever called a "see-saw." Levers are used to lift heavy objects, tighten lug nuts on a car's wheel, or turn the gears on a bicycle. Levers work through torque and rotational motion. The pulley and the wheel and axle are two other simple machines related to the lever.

**Inclined planes**

It takes more force to lift a heavy load straight up compared to moving the same load up a ramp, even when the total height change is the same. A ramp is an example of an inclined plane, another simple machine. Inclined planes work by dividing forces acting along the direction of motion from forces acting perpendicular to the direction of motion. The wedge and screw are two other types of simple machines that are based on inclined planes.

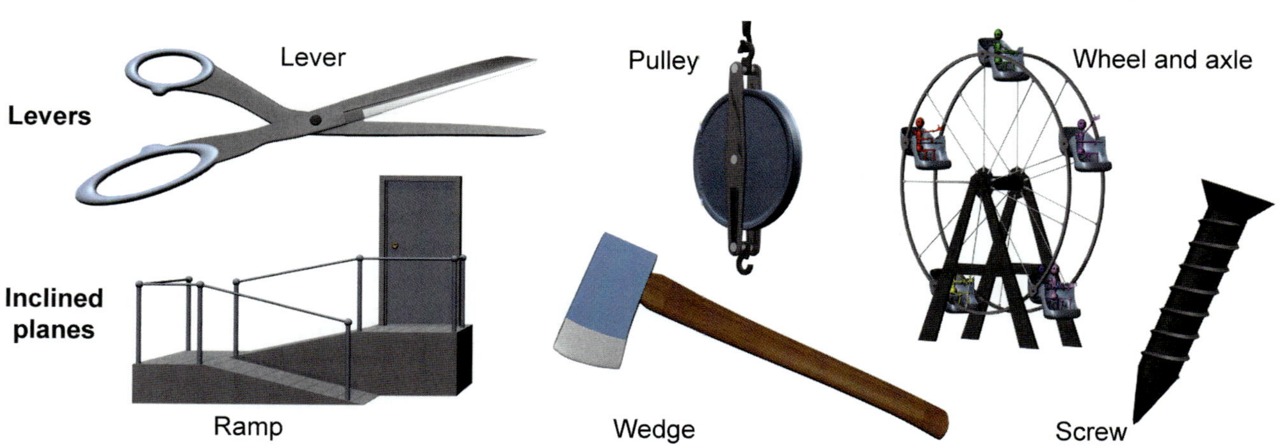

**Machines and force**

Consider using a board and a log to lift a heavy boulder. The board and log together create a simple machine (a lever) that converts your **input force** into a more powerful **output force** capable of lifting the boulder. As you apply an input force downward on one end of the board, this lever applies a larger output force *upward* on the boulder at the other end. In terms of forces, the lever both magnifies the input force and changes its direction.

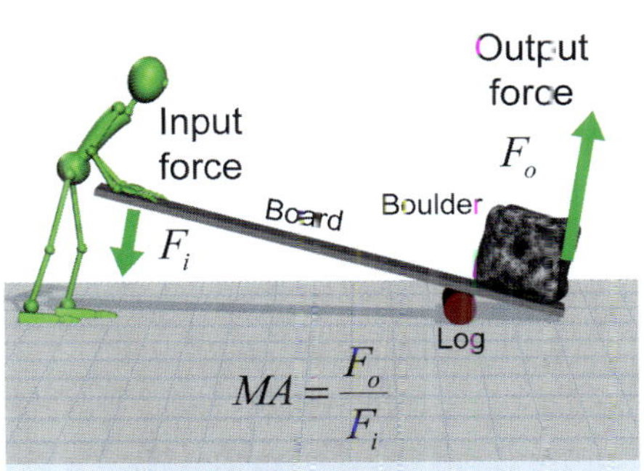

Mechanical advantage

**Mechanical advantage**

The **mechanical advantage** tells you the "magnification factor" relating the strength of the output force to the strength of the input force. Mathematically, the mechanical advantage ($MA$) is the output force divided by the input force. A mechanical advantage of 5 means that the output force is 5 times larger than the input force.

Interactive equation

**Mechanical advantage**

$$MA = \frac{F_o}{F_i} \quad (12.1)$$

$MA$ = mechanical advantage
$F_o$ = output force (N)
$F_i$ = input force (N)

**What do the values of *MA* mean?**

When $MA = 1$, the magnitude of the output force is equal to the input force—the machine has no mechanical advantage. A mechanical advantage greater than one ($MA > 1$) means the output force is larger than the input force. Mechanical advantage can also be less than one! The overall mechanical advantage of a bicycle is less than one because a bicycle trades lower output *force* for greater output *distance*.

**Mechanical advantage of Prince Ludwig's contraption**

Inventor Prince Ludwig designed a hand-cranked machine to lift water from a stream to the kitchen window. Igor, Ludwig's assistant, counted 27 gears, 7 levers, 12 wheels, and 3 ramps in the machine. Igor found that the machine required 200 N of applied force to lift 100 kg of water. What is the mechanical advantage of Prince Ludwig's machine?

| | |
|---|---|
| *Asked:* | mechanical advantage $MA$ |
| *Given:* | input force $F_i = 200$ N; mass of water lifted, $m = 100$ kg |
| *Relationships:* | $F_w = mg$, $MA = F_o/F_i$ |
| *Solution:* | The output force that is lifted is the weight of the 100 kg of water: $$F_w = mg = (100\text{ kg})(9.8\text{ m/s}^2) = 980\text{ N}$$ The mechanical advantage of Prince Ludwig's contraption is therefore $$MA = \frac{F_o}{F_i} = \frac{980\text{ N}}{200\text{ N}} = 4.9$$ |
| *Answer:* | The mechanical advantage is 4.9. Note that mechanical advantage compares the input and output forces, regardless of how the machine is constructed inside! |

# Investigation 12A Levers

**Essential questions** *How can a lever be used to multiply force?*
*How does the work done vary with levers?*

The lever is a simple machine with many applications. The most common use is to lift heavy objects, but not all lever designs increase the force applied! In this investigation you will use a see-saw to determine the ways in which a lever can be used to increase force.

## Part 1: Building a first class lever

1. Build the first class lever as shown.
2. Hang one mass at 20 cm and the other at 40 cm from the fulcrum on opposite sides.
3. Adjust the masses until the lever "balances."
4. Use a spring scale to measure and record the weight of each mass and also record the distance *d* between each mass and the fulcrum.
5. Set up levers in which the ratio of input force to output force (mechanical advantage) is 3, 4, 1/2, and 1/4

a. *Propose a model of the form:*
*input force = (?) × output force.*

b. *Use your data to test your model. How do the measured and predicted ratios compare?*

c. *Suggest an explanation for any differences.*

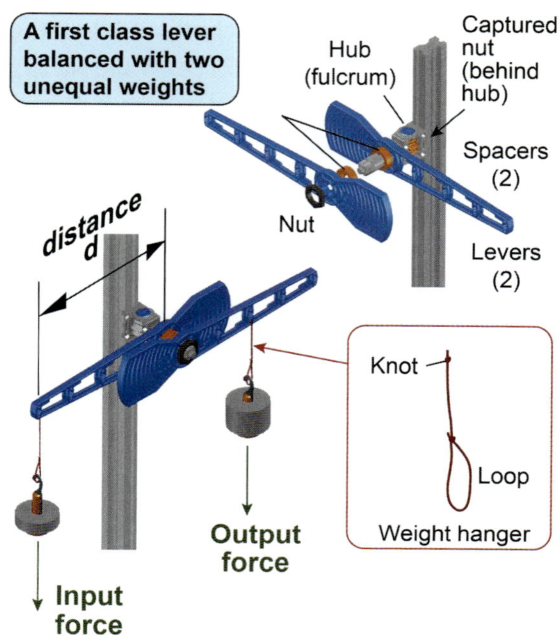


## Part 2: Building a second and third class lever

There are three fundamental ways to configure a lever. The second and third class levers have the input and output force on the same side of the fulcrum.

1. Design and construct a second class lever with a mechanical advantage of three. Use the hanging weight as the output force (load) as shown in the diagram. Measure and record the actual mechanical advantage using a spring scale.
2. Design and construct a third class lever with a mechanical advantage of 1/3. Use the hanging weight as the output force (load) as shown in the diagram. Measure and record the actual mechanical advantage using a spring scale.

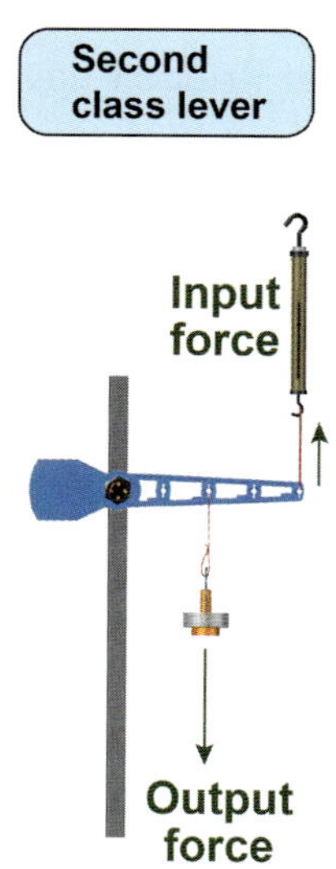

a. *How are the forces applied different from the forces applied in the first part? (Hint: Force is a vector!)*

b. *For this part and the previous one, draw diagrams to describe the three different configurations for the relative positions of the fulcrum, applied force ("effort"), and resistance force ("load").*

**How does a lever work?**

A **lever** is a rigid beam that can rotate around a point called the **fulcrum**. In the example on the right, the input force on one side is in equilibrium with the load force on the other side. The **input arm** is the distance between the fulcrum and the input force, while the **output arm** is the distance between the fulcrum and the load force.

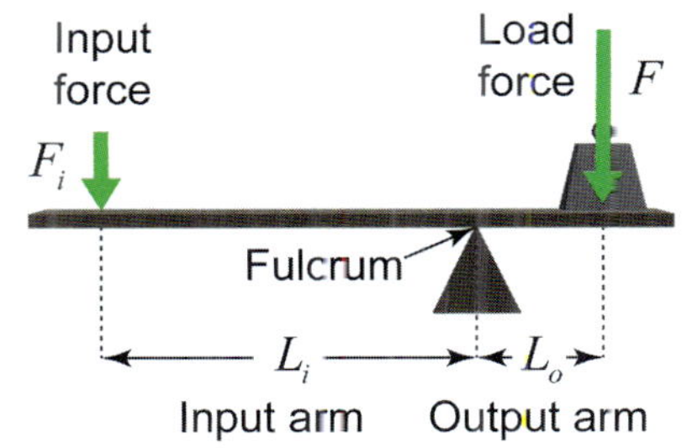


**Mechanical advantage of a lever**

The key to understanding how a lever works is that the input torque on one side must balance with the torque from the load force on the other side. As we learned in Chapter 8, torque is the product of force applied and distance from the fulcrum. The input force produces a torque $\tau_i = +L_iF_i$ while the torque from the load is $\tau_o = -L_oF_o$. To be in rotational equilibrium, the net torque must be zero:

**Torques for a lever**

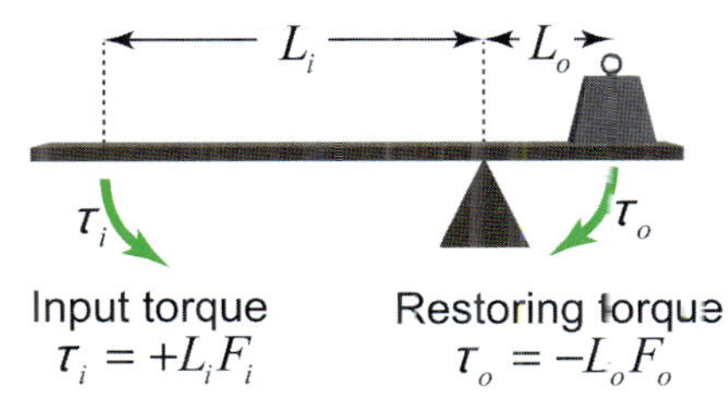


$$\tau_{net} = \tau_i + \tau_o = +L_iF_i - L_oF_o = 0$$

or $L_iF_i = L_oF_o$. Rearranging terms leads to the mechanical advantage of a lever in equation (12.2).

Interactive equation

$$\text{(12.2)} \quad MA_{lever} = \frac{L_i}{L_o}$$

$MA_{lever}$ = mechanical advantage
$L_i$ = input arm (m)
$L_o$ = output arm (m)

**Mechanical advantage** of a lever

**Lever's *MA* depends on location of fulcrum**

The great fact about levers is that the mechanical advantage depends only on the relative lengths of the input and output arms—in other words, the lever's mechanical advantage depends on where you place the fulcrum! If the fulcrum is placed close to the output force, then $L_o > L_i$ and $MA > 1$. To increase the mechanical advantage of a lever, it is only necessary to move the fulcrum of the lever even closer to the output force (the "load").

**Setting the fulcrum's location for a lever**

Where should Michellin place the fulcrum when using a meter stick to construct a lever with $MA = 4$?

| | |
|---|---|
| *Asked:* | location of the lever's fulcrum, i.e., the length of the output arm, $L_o$ |
| *Given:* | total length of lever, $L = 1$ m; mechanical advantage $MA_{lever} = 4$ |
| *Relationships:* | $MA_{lever} = L_i/L_o$; and length of meter rule, $L = L_i + L_o$ |
| *Solution:* | Solve for the length of the input arm: $L_i = MA_{lever} \times L_o = 4L_o$ Use this to substitute for $L_i$: $L = L_i + L_o = 4L_o + L_o = 5L_o \Rightarrow L_o = \frac{1}{5}L = \frac{1}{5}(1\text{ m}) = 0.2\text{ m}$ |
| *Answer:* | The fulcrum should be placed 0.2 m from the output end of the meter stick. |

## Uses of levers

**Forces and distances**

A lever can allow you to multiply your force to move a heavy object, but at what price? The lever is a machine for doing work that enables a tradeoff between forces applied and distances moved. If you make a lever with a long input arm, then you will also have to move the input arm by a large distance to move the load. Doubling the length of the input arm means that you can use half the force, but you also have to move the input arm twice as far.

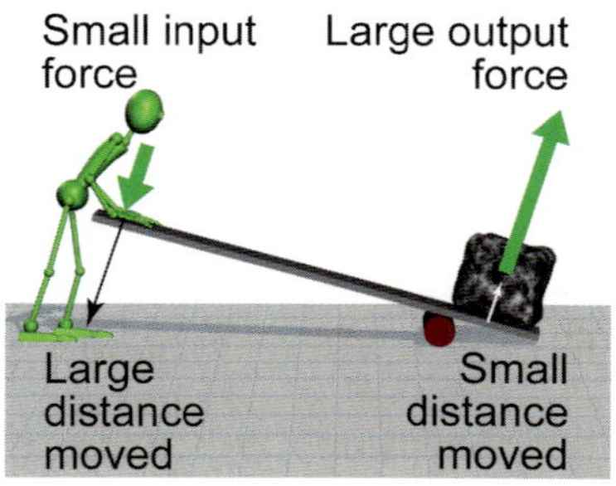


**Removing nails with a claw hammer**

Nails driven into wood are difficult to remove because there is a strong friction force holding them in place. A claw hammer has a "V"-shaped wedge that is useful for removing nails. The claw hammer acts as a lever with a large input arm but short output arm. The input force from your arm is multiplied substantially to deliver a large force to remove the nail. As a tradeoff, however, you must swing the handle through a large motion in order to move the nail a short distance.

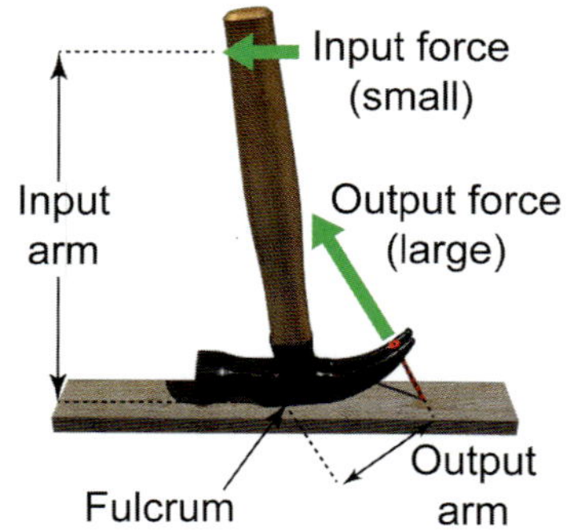


**Nutcracker**

Most nuts have hard shells that are difficult to crack. A nutcracker is a lever that multiplies the force of your squeezing hand to break the shell. Your hand must squeeze through a larger distance than the crushing motion on the shell. Now look at the diagram more closely: Can you see a difference between the nutcracker and hammer in the arrangement of the forces and the fulcrum?

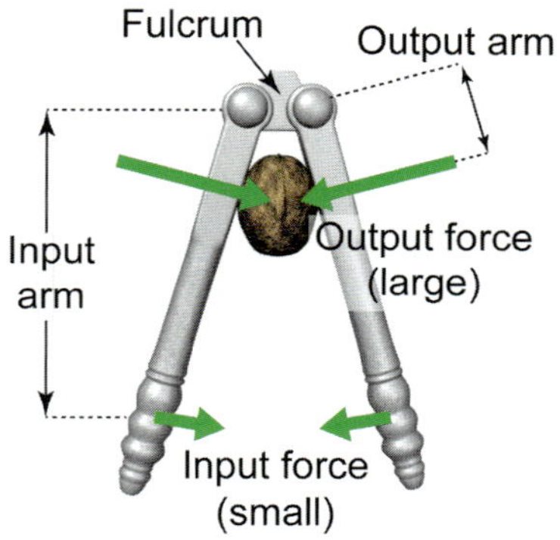


**Three different classes of levers**

As you may have discovered in Investigation 12A, there are three different classes of levers based on the location of the input and output forces relative to the fulcrum. The fulcrum lies between the input and output force for a first-class lever. The input force is further from the fulcrum than the output force for a second-class lever, whereas the output force is further for a third-class lever. All levers fall into one of these three classes.

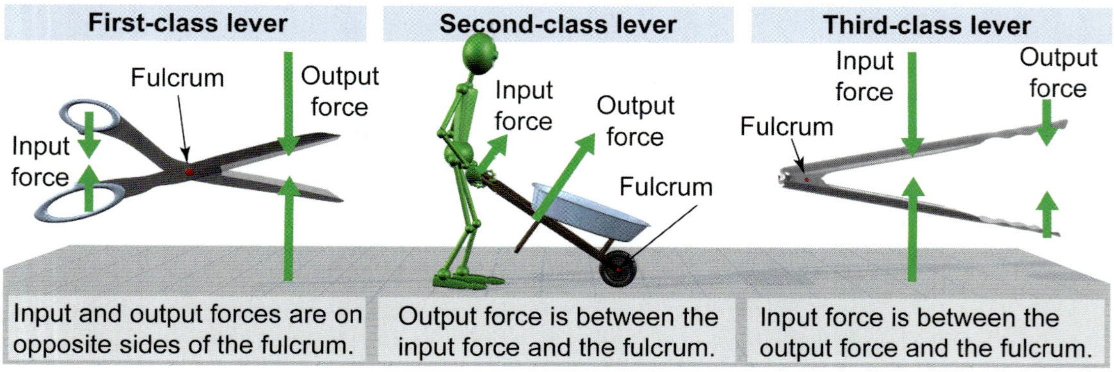

# Section 1 review

There are many kinds of machines that we use everyday to do work or help us accomplish a task. All machines are fundamentally built up of combinations of six categories of *simple machines*: lever, pulley, wheel and axle, ramp, wedge, and screw. The mechanical advantage of a machine indicates how it changes the input force into an output force. The lever is a simple machine that multiplies force by the ratio of the input arm to the output arm.

| | |
|---|---|
| **Vocabulary words** | **machine, simple machine, mechanical advantage, input force, output force, lever, fulcrum, input arm, output arm** |
| **Key equations** | $MA = \frac{F_o}{F_i}$ $\quad$ $MA_{lever} = \frac{L_i}{L_o}$ |

## Review problems and questions

1. Is a car a machine? Is it a *simple* machine? Why or why not?

2. How are first-, second-, and third-class levers different?

3. The Mega-Rontastic Machine, resembling a fancy broom, is advertised on late-night television as a device to help you do household chores with less effort. Maximillian bought one and tested its mechanical advantage to see whether it actually worked as advertised. What measured values of mechanical advantage would justify the product claims about the Mega-Rontastic Machine?

4. Prince Ludwig II creates his own version of a machine that will lift a 25 kg bucket of water. Igor tries out the device and determines that, to lift the water by 2 m, he must pull the input rope by 10 m while applying 50 N of force. What is the mechanical advantage of Prince Ludwig II's machine? (Assume that no energy is lost to friction.)

5. Astrid has two slotted masses and a meter ruler. She wants to hang the 100 g from one end of the meter ruler (at the 0 cm mark), hang the 300 g mass from the other end (at the 100 cm mark), and place the meter ruler on a knife edge to balance the lever. At what mark on the ruler should she place the knife edge? The mass of the ruler itself is negligible.

6. Zing is a safecracker who wants to lift a 150 kg safe full of loot, but he can only exert 490 N of force. He has a 60-cm-long steel rod as well as a rock he can use as a fulcrum.
   a. What minimum mechanical advantage does he need?
   b. How far from the safe should the rock be placed to provide this minimum mechanical advantage?
   c. What torque does Zing apply to lift the safe?
   d. What is the net torque about the fulcrum?

Is the wheel on your school bus a lever? Is the rope and pulley used to lift a piano through a second story window a lever? The answer to both questions is yes! The lever is not just one type of simple machine: It is also a category that includes the pulley and the wheel and axle. In this section, you will learn why levers, pulleys, and wheels are related to each other.

## Ropes and pulleys

**Tension in a rope**

To understand pulleys, we first need to understand the strings or ropes that pass around the pulleys. When you pull a rope with a force, you create **tension** in the rope. The rope carries that force all along its length so that the tension is the same everywhere along the rope. But the rope only has tension if *both ends* of the rope have a force applied—there is no tension if you drag the rope along the ground! Tension is a kind of force, so it has units of newtons (N). In rope and pulley diagrams, the rope's tension is usually denoted by $T$.

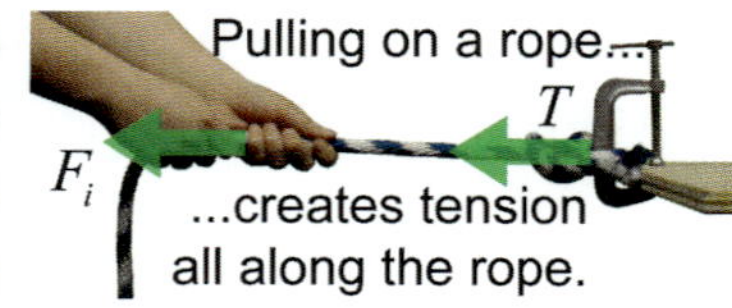


**What does a pulley do?**

A **pulley** is a wheel with a groove around the circumference to hold a rope or some similar flexible cord. A pulley is a machine that changes the direction of the force exerted by the rope. By using a pulley, you can lift a heavy object by pulling in a more convenient direction, such as down instead of up.

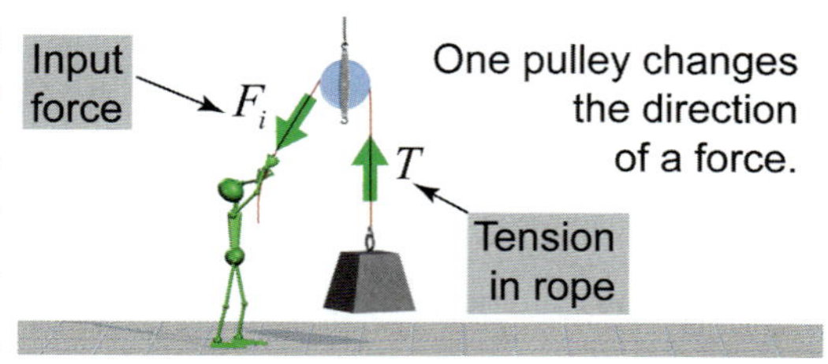


**Changing magnitude of a force**

Although changing the *direction* of a force might be helpful, we often want a machine that changes the *magnitude* of a force, too, in order to lift something heavy. How can we use pulleys to lift heavy objects? As we will learn soon, *combinations* of pulleys can be used to change the magnitude of a force, not just its direction.

**Showing the direction of tension**

Earlier in the book, we learned how to draw free-body diagrams to identify all the forces acting on an object. The tension in a rope or string is also a force, so tension must also be identified and labeled on free-body diagrams. The illustration at right shows the direction of the tension force on two blocks. Notice that the tension vector always acts along the direction of the rope, and it always pulls.

**Tension in free-body diagrams**

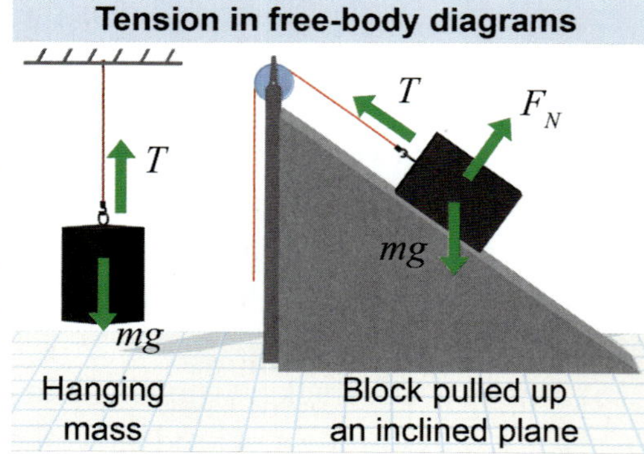


**Other kinds of tension**

A rope is not the only kind of material that can exhibit tension! In a more general sense, tension is an outward force on an object that acts to stretch it (and so is the opposite of compression). You can create tension in both a rubber band and a steel beam by pulling on them—although you have to pull harder on one than the other. The tension in those cases is a force that acts to resist the stretching. How can a bug called a water strider seemingly "walk" on water? The weight of the bug acts to expand the surface area of the water. The *surface tension* of the water resists the bug's weight, applying an upward force to keep it on top of the water!

# Investigation 12B Pulleys

**Essential questions** *How do you use pulleys to lift a heavy object?*
*Does a system of pulleys change the work required to lift an object?*

Pulleys can be used to change the direction of the force needed to lift an object and also to *reduce* the force needed to lift it! The *block and tackle* is a simple machine that can be used to lift heavy objects. In this investigation, you will combine two or more pulleys to form a block-and-tackle system. In each case, you will measure how far you have to pull on a string to lift the mass by 20 cm and record the force required to do so.

## Part 1: The mechanical advantage of a block and tackle

1. Set up the block and tackle machine as shown. The output force is the weight of the bottom pulley block and weights.
2. Configure the block and tackle with 1, 2, 3, 4, 5, and 6 support strings. For each configuration, measure and record the input force and the output force.

a. *What is the rule that tells you the mechanical advantage of a pulley system?*

b. *What was the largest mechanical advantage you could construct? Why is this the limit of this system of pulleys?*

c. *Was the actual measured mechanical advantage greater or less than you think it should have been? Propose an explanation for any differences you observed.*

d. *Is it possible to build an infinite mechanical advantage using an infinite number of pulleys? Why or why not? What limits they realistic advantage of a machine made with blocks and pulleys?*

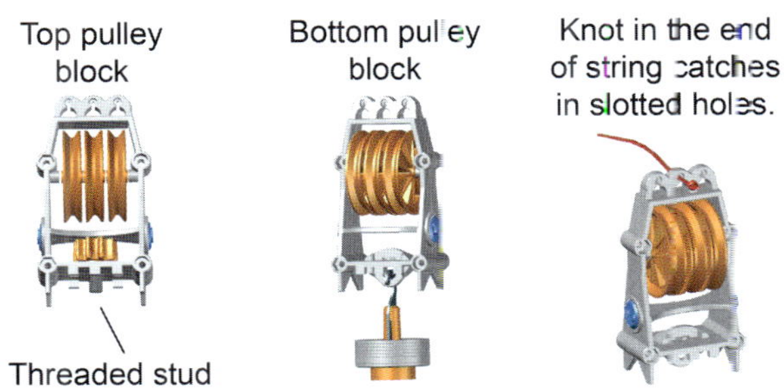


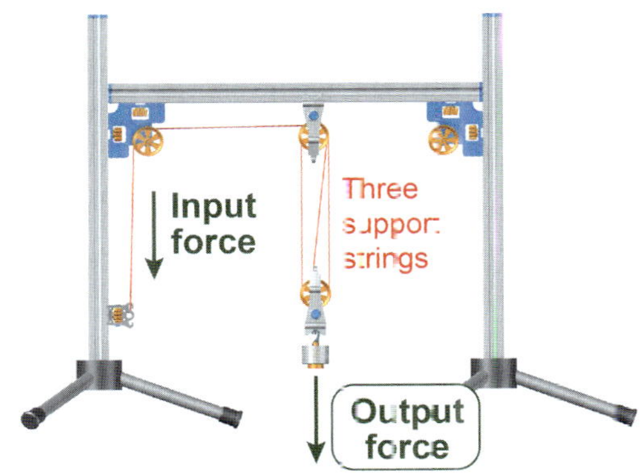


## Part 2: Mechanical work

1. Repeat the block and tackle setups for each mechanical advantage you found.
2. Measure and record the distance the load (output) rises when you pull the input string 30 cm.
3. Calculate and record the input and output work for each machine.

a. *How do the input and output work compare?*

b. *Do the input and output work agree with the principle that the output can never exceed the input?*

c. *Calculate the efficiency of your machine for different mechanical advantages.*

d. *Can you explain why the efficiency decreases as the mechanical advantage increases?*

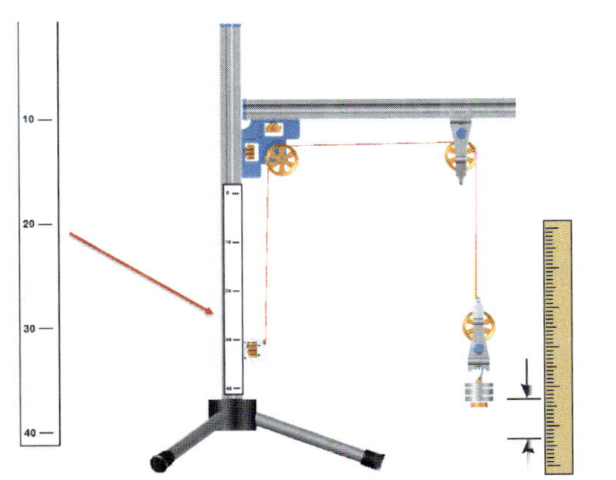


**The input distance**
Make a scale for the side of the stand. Make marks for 10, 20, 30, and 40 cm.

**The output distance**
Measure the height the load rises with a ruler.

## Combining pulleys

### What is the mechanical advantage of a pulley?

A single pulley has a mechanical advantage of one because the input and output forces are equal. Combinations of pulleys are used together to achieve mechanical advantages different from one. Modern elevator drives often use pulleys and cables to create mechanical advantage as high as 15.

### How to double your force with pulleys

To create a mechanical advantage pulleys must be arranged so that the load is supported by *more than one strand of the same rope.* Since the tension in a rope is the same everywhere, two strands of the same rope supporting a load means that the tension force is applied to the load *twice.* The diagram on the right shows two different ways to use pulleys so that the output force is twice the input force. In both cases, notice how the load is supported by two strands of the same rope—each applying tension $T$.

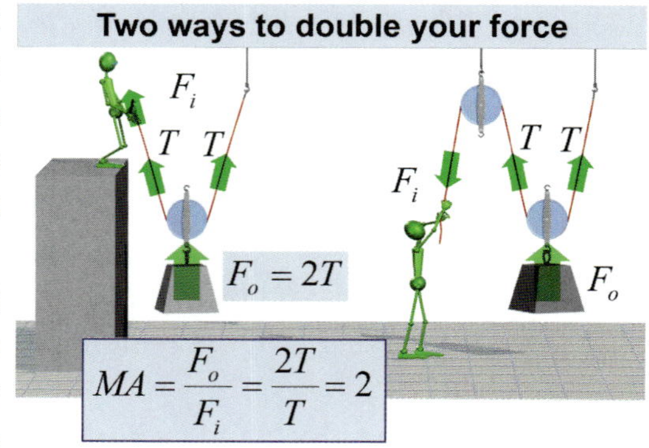


### Block and tackle

The **block and tackle** is a compound machine using two or more pulleys to support a load from a single length of rope. With a block and tackle, it is possible to arrange three pulleys so that a single input force applied to the rope is applied three times to the load, creating a mechanical advantage of three. When analyzing the input and output forces in a block and tackle, the crucial step is to identify how many times the tension in the rope is applied to the load. *The mechanical advantage of a block and tackle is equal to the number of rope tension forces directly supporting the load.*

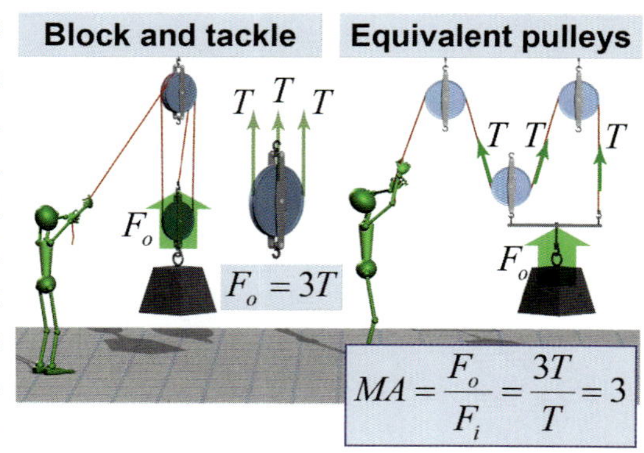


### Force and distance

Energy (work) must be conserved and the block and tackle enables a tradeoff between force and distance that is similar to the tradeoff for levers. The block and tackle above has a mechanical advantage of three, which means that the output force is three times larger than the input force. But it also means that you have to pull the rope by three meters for every one meter you lift the load vertically upward. You do the same amount of work, whether you pull a load using one or three pulleys. With three pulleys, however, you must pull the rope three times as far—while exerting only one-third of the input force—when compared to using only one pulley.

### Mechanical advantage of a block-and-tackle system

What is the mechanical advantage of the block and tackle system in the illustration at right?

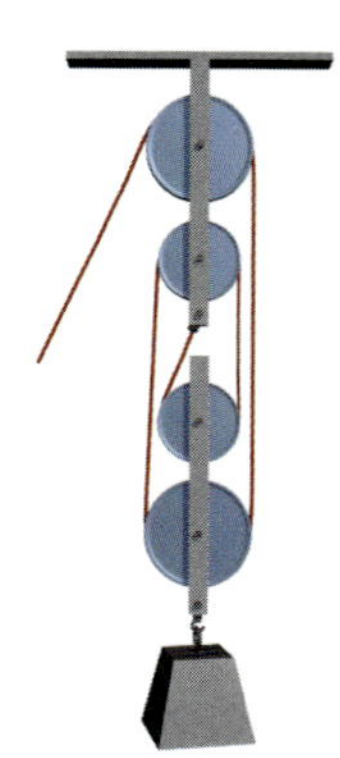

| | |
|---|---|
| *Asked:* | mechanical advantage $MA$ |
| *Given:* | four pulley system (as shown on right) |
| *Solution:* | There are four strands of the same rope supporting the mass. (Can you identify each of them?) The mechanical advantage is therefore four. |
| *Answer:* | $MA = 4$ |

**Efficiency**

Because of friction, real machines cannot convert 100% of applied input work into output work. The **efficiency** of a system measures the fraction of input work converted to output work. Since input and output distances are often fixed by geometry, efficiencies less than 100% tend to reduce output forces. For example, suppose an input force of 1 N is applied to a machine with *MA* = 10. The *ideal* output force should be 10 N, but an actual output force might only be 9 N.

Friction causes you to work harder to generate the same output work.

Interactive equation

**Efficiency of a machine**

$$(12.3)\qquad \eta = \frac{W_o}{W_i}$$

$\eta$ = efficiency
$W_o$ = output work (J)
$W_i$ = input work (J)

**Ideal mechanical advantage**

In an ideal machine all of the input work is converted into output work, so $F_i d_i = F_o d_o$. Since the ratio of the output force to the input force is the mechanical advantage, we can rearrange the terms to determine the *MA* of an *ideal* machine:

$$MA = \frac{F_o}{F_i} = \frac{d_i}{d_o}$$

To determine the **ideal mechanical advantage** of a machine one assumes an efficiency of 100% (i.e., there are no energy losses). The ideal *MA* is the ratio of the distance moved at the machine's input to the distance moved by the output.

Interactive equation

**Ideal mechanical advantage**

$$(12.4)\qquad MA_{ideal} = \frac{d_i}{d_o}$$

$MA_{ideal}$ = ideal mechanical advantage
$d_i$ = input distance (m)
$d_o$ = output distance (m)

**Measuring forces or distances**

The ideal mechanical advantage of a system is determined from the input and output *distances* moved. For the (actual) mechanical advantage, however, you measure the input and output *forces*. To determine efficiency, you need both force *and* distance!

**Efficiency of a pulley system**

A block and tackle lifts a mass of 0.25 kg by 15 cm. The input string is pulled by 45 cm with a force of 0.95 N. What is the efficiency of the system?

*Asked:* efficiency $\eta$ of the pulley system

*Given:* mass $m$ = 0.25 kg, input distance $d_i$ = 0.45 m, input force $F_i$ = 0.95 N, output distance $d_o$ = 0.15 cm

*Relationships:* efficiency $\eta = W_o/W_i$; work $W = Fd$

*Solution:* The output force is the weight of the mass:

$$F_o = F_w = mg = (0.25\text{ kg})(9.8\text{ m/s}^2) = 2.45\text{ N}$$

The efficiency is the ratio of the output work to the input work:

$$\eta = \frac{W_o}{W_i} = \frac{F_o d_o}{F_i d_i} = \frac{(2.45\text{ N})(0.15\text{ m})}{(0.95\text{ N})(0.45\text{ m})} = 0.86 = 86\%$$

*Answer:* The efficiency is $\eta$ = 86%.

**Why are wheels useful?**

The invention of the wheel around 4,000 BC was a major advance in the development of civilization. What about the physics of wheels makes them so useful? Consider a wheel turning a drum to raise a bucket of water. The radius of the wheel is larger than the radius of the drum. The difference in radius creates a mechanical advantage. A small input force applied to turn the outer rim of the wheel creates a larger output force on the rope.

**Components of a wheel and axle**

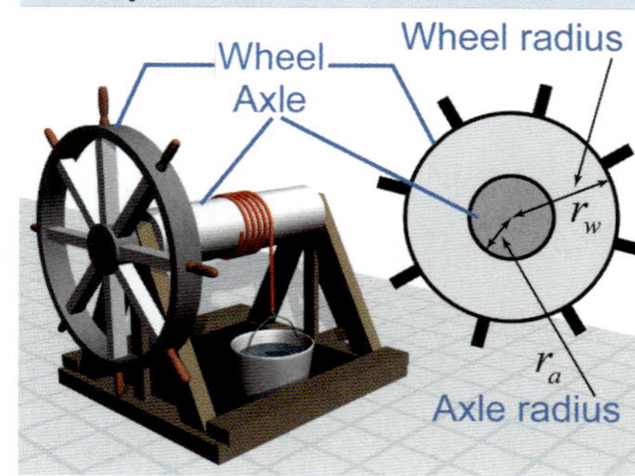


**Wheel and axle as a lever**

The physics of a **wheel and axle** is similar to that of a *first-class lever*, for which the center of the axle acts as the fulcrum. The input force is typically applied to the outer rim of the wheel, while the output force is exerted on the outer rim of the axle. The mechanical advantage is the ratio of the wheel radius (input arm) to the axle radius (output arm). In the example the axle radius is 1 cm and the wheel radius is 3 cm giving a mechanical advantage of 3. In actual wheels the ratio is much larger. The radius of a typical bicycle wheel is 35 cm while the axle radius is 0.4 cm, making a mechanical advantage of 87.5!

**Wheel and axle** **Equivalent lever**

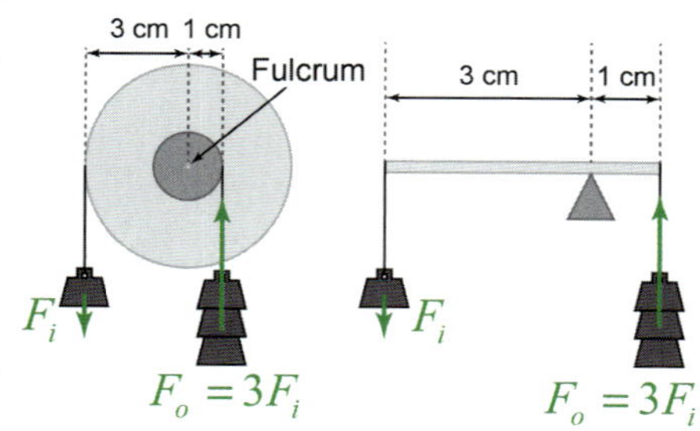


For both: $MA = 3$

Interactive equation

$$\textbf{(12.5)} \quad MA_{wa} = \frac{r_w}{r_a}$$

$MA_{wa}$ = mechanical advantage
$r_w$ = wheel radius (m)
$r_a$ = axle radius (m)

**Mechanical advantage** of a wheel and axle

**Wheels and friction**

Wheels greatly reduce the force needed to overcome friction when pulling heavy objects. Consider sliding a crate along the floor. You have to apply a force equal to or greater than the friction force between the bottom of the crate and the floor. With a wheel, the coefficient of rolling friction is much lower. There *are* high forces at the small contact area where the axle is supported in the cart. Although sliding friction occurs in this place, the contact area is small, protected, and can be lubricated to reduce friction.

Frictional force is reduced by using a wheel. It takes less force to move a heavy object.

**Applying force to the axle**

It is also possible to apply the input force closer to the axle, rather than at the rim of the wheel. This causes the output force to be *smaller* than the input force. This is the case with the bicycle. When you ride a bicycle, the input force applied by the chain to the axle turns the rear wheel. A small motion at the axle produces a large motion on the outer rim of the wheel. The bicycle's use of the wheel and axle *trades force for speed.* You can pedal slowly, yet you still travel fast!

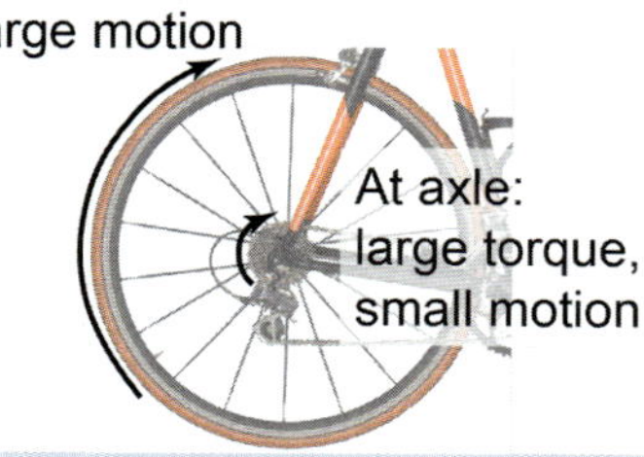


Applying torque to the axle trades off torque for speed.

**Gears change rotational speed**

A **gear** is a wheel with teeth that interlock with matching teeth on a mating gear to transmit rotating motion and torque. Gears can transmit much higher forces than smooth wheels. Gears are used to change rotational forces and speeds in many machines, including automobile transmissions. Related to the gear is the *sprocket*, which is found in the drive train of a bicycle. A sprocket is a toothed wheel that engages a chain instead of another gear.

**Automobile transmission**

Gears are used to change the rotating speed in the engine

**Gear ratios**

Interlocking gears with different numbers of teeth turn at different rotational speeds. The **gear ratio** is the ratio of turns of the output gear to turns of the input gear. The gear ratio in the example is 16 ÷ 8 = 2. Each turn of the input gear moves 16 teeth through the point of contact, forcing the output gear to turn twice. The gear ratio is the inverse of the ratio of teeth. Note that, like two touching wheels, interlocking gears always rotate in *opposite* directions.

Two gears | Equivalent wheels

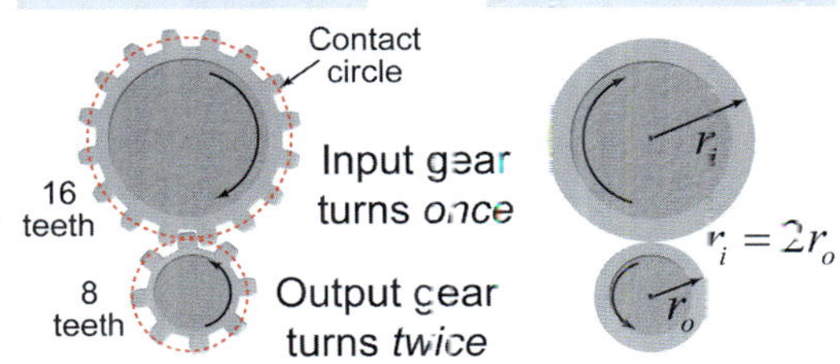


$$\text{Gear ratio} = \frac{\text{output turns}}{\text{input turns}} = \frac{r_i}{r_o} = \frac{\text{input teeth}}{\text{output teeth}} = 2$$

**Gear ratio**

Interactive equation

$$GR = \frac{\text{output turns}}{\text{input turns}} = \frac{\text{input teeth}}{\text{output teeth}}$$

$GR$ = gear ratio

**Mechanical advantage of two gears**

The input gear's teeth apply a force to the output gear's teeth at the point of contact. At this point of contact, the input force equals the output force (if 100% efficiency is assumed). If the *radii* of the input and output gears differ, then the input and output *torques* will differ, too, because torque is force times radius. The mechanical advantage of two gears is the ratio of the output to input torque, which is equal to $MA = r_o/r_i$ or $1/GR$. As you can see in the illustration at right, the $MA$ is also the ratio of the number of output teeth to the number of input teeth.

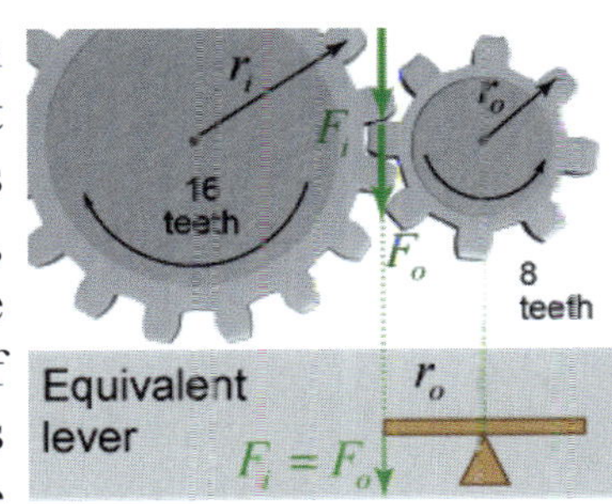


$$MA = \frac{\tau_o}{\tau_i} = \frac{r_o \cancel{F_o}}{r_i \cancel{F_i}} = \frac{\text{output teeth}}{\text{input teeth}}$$

***MA* of two gears**

Interactive equation

$$(12.6) \qquad MA_g = \frac{\tau_o}{\tau_i} = \frac{\text{output teeth}}{\text{input teeth}}$$

$MA_g$ = mechanical advantage
$\tau_o$ = output torque (N m)
$\tau_i$ = input torque (N m)

**Using gears to change torque and speed**

Small electric motors turn at high speeds but produce low torque. Many robotic applications require *low* speed and *high* torque. Multiple gear ratios are used to both reduce the speed and increase the torque. The four gears in the diagram make an overall mechanical advantage of 16.

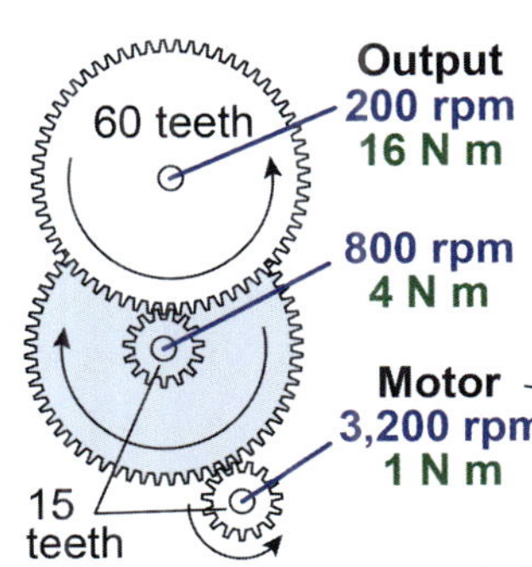

*FIRST Tech Challenge 2012*

# Section 2 review

The pulley and the wheel and axle are two other categories of simple machines related to the lever. Pulleys can be used to redirect a force, and combinations of pulleys used in block-and-tackle systems can significantly multiply force. The wheel and axle multiplies force by the ratio of the wheel to axle radius. When the axle drives the wheel, the output force is reduced as a tradeoff for increased speed. Losses such as friction cause real machines to have less than 100% efficiency and lower output forces than expected from their ideal mechanical advantage.

| Vocabulary words | tension, pulley, block and tackle, efficiency, ideal mechanical advantage, wheel and axle, gear, gear ratio |
|---|---|

| Key equations | |
|---|---|
| | $\eta = \frac{W_o}{W_i}$ $\quad$ $MA_{ideal} = \frac{d_i}{d_o}$ |
| | $MA_{wa} = \frac{r_w}{r_a}$ $\quad$ $MA_g = \frac{\tau_o}{\tau_i} = \frac{\text{output teeth}}{\text{input teeth}}$ |

## Review problems and questions

1. Describe the measurements and calculations that you need to make to determine the mechanical advantage of a
   a. lever,
   b. block and tackle,
   c. wheel and axle.

2. A force is applied to a large gear with 51 teeth. That large gear then turns a smaller gear with only 17 teeth.
   a. What is the gear ratio of the two gears?
   b. What is the mechanical advantage of the two gears?
   c. If the small gear instead were to turn the large gear, what is the gear ratio?
   d. Likewise, what is the mechanical advantage when the small gear turns the large gear?

3. In a car, when the steering wheel is turned it causes the steering column (the axle) to turn. Why wouldn't you want to design the car to simply have a *steering axle*? Use forces in your answer.

4. In a winch, a crank with a radius of 50 cm is turned to cause a rope to wrap around a drum with a radius of 10 cm.
   a. What is the ideal mechanical advantage of this machine?
   b. If the winch had friction that caused the efficiency to be 60%, what would be its mechanical advantage?

Did you walk up a simple machine today? This might sound like a strange question, but in all likelihood you did. An inclined plane—or ramp—is a type of simple machine. If you walked up a wheelchair access ramp, or even walked on a road up a hill, your path was a simple machine! Ramps are machines because less force is needed to move up a ramp than is needed to climb directly up the same vertical distance. Ramps are used to move cars onto auto carriers, move dump trucks out of mining pits, or move vehicles on roads over mountain passes.

## Ramps as simple machines

**Why are ramps useful?**

Consider two ways to lift a heavy dresser onto a truck. Lifting it vertically requires a force equal to the weight of the dresser. A better idea is to use a *ramp*. A **ramp** uses an angled surface to divide the weight force into two directions: parallel and perpendicular to the ramp. The perpendicular force is supported by the normal force from the ramp. The dresser can now be moved along the ramp by overcoming only friction and the (much smaller) parallel force.

An inclined plane uses an angled surface to divide forces into perpendicular and parallel components.

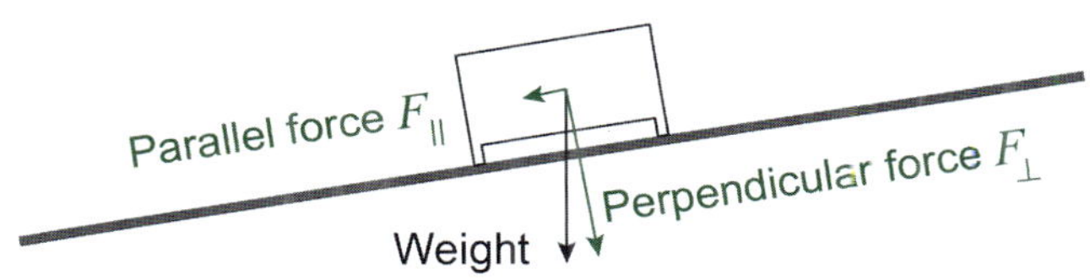


**Mechanical advantage of a ramp**

It takes a little experimentation to discover that the steeper the slope of the ramp the more force is required to move up it. How can this intuition be converted into an equation for mechanical advantage of the ramp? The input work done by using the ramp is force times distance, or $F_iL$, where $L$ is the length of the ramp. The output work represents what would happen if you had no machine to help you, or $F_oh$, where $h$ is the vertical height it is being lifted. The gravitational potential energy gained by the dresser will be the same *mgh* whether you use a ramp or lift it directly into the truck. Since changes in energy are equivalent to work done, both ways of doing work are equal, or $F_iL = F_oh$. Mechanical advantage is the ratio of output force to input force, so rearranging terms leads to equation (12.7).

**Mechanical advantage of a ramp**

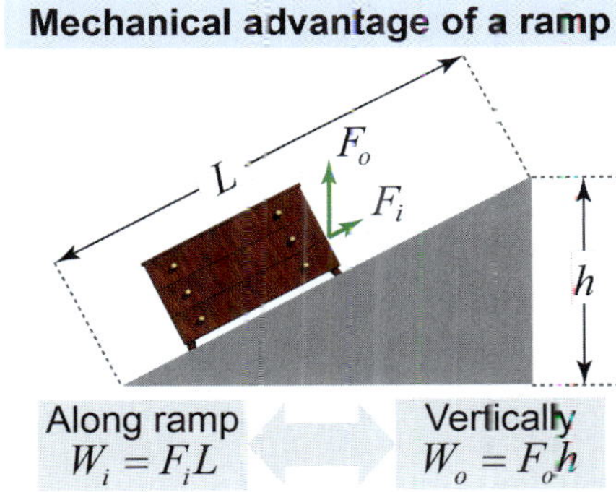


Equal work done because both increase $E_p$ by the same amount

$$MA_{ramp} = \frac{F_o}{F_i} = \frac{L}{h} = \frac{\text{ramp length}}{\text{ramp height}}$$

$$\textbf{(12.7)} \quad MA_{ramp} = \frac{L_{ramp}}{h_{ramp}}$$

$MA_{ramp}$ = mechanical advantage
$L_{ramp}$ = length of ramp (m)
$h_{ramp}$ = height of ramp (m)

**Mechanical advantage** of a ramp

# Investigation 12C Ramps and inclined planes

**Essential questions** *How does a ramp change the force required to move an object uphill?*

When the Egyptians built their massive pyramids, they faced the engineering challenge of how to lift the heavy stone blocks vertically into position. The Egyptians may have moved the blocks up ramps (or inclined planes) constructed along the side of the pyramid. How would using a ramp make their job easier? In this investigation you will measure the force required to move the Smart Cartup a ramp to a height of 30 cm. How does the force vary? How about the work done?

## Force and work required to move the cart up a ramp

1. Open the **12C_InclinedPlanes** experiment file in your software, and then connect your Smart Cart using Bluetooth. Zero the cart's force sensor while nothing is touching the hook.
2. Set up your equipment as in the picture. The plunger on the cart should be facing the end stop, with the hook facing up the ramp. Start with an incline of at least 20°.

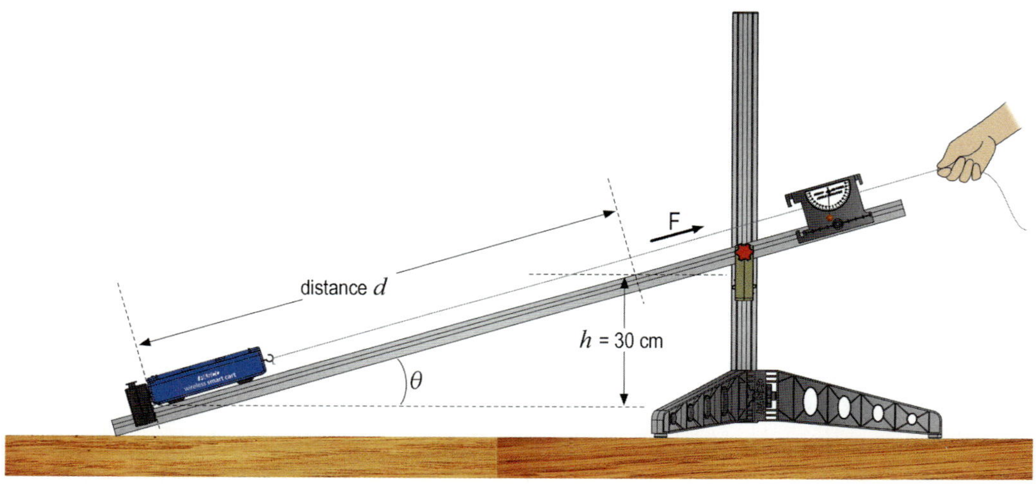


3. Attach the cart to the end of a string.
4. Use tape to mark a location on the ramp 30 cm vertically higher than the starting position of the cart.
5. Measure the average force $F$ required to move the cart at a slow constant speed to the final position. Measure the distance the cart traveled along the ramp.
6. Repeat for at least two other inclinations of the ramp. Tabulate your results.

*a. Is more force needed to move an object up a steep ramp or a shallow ramp?*

*b. Using the force applied and the distance moved, calculate the work done on the cart to move it in each case. Explain your results.*

*c. How can you calculate the mechanical advantage of the ramp using the values in your investigation?*

*d. Calculate the mechanical advantage for each inclination of the ramp and include the results in your table.*

*e. Measure the mass of the Smart Cart, and then calculate a theoretical value for the amount of work needed to lift the cart 30 cm. How does this calculated value compare to the work values in your table? Explain why they are different and/or similar.*

**What is a wedge?**

Do you like to cook? If so, then you regularly use a knife, which is also a *wedge*—a class of simple machines. A **wedge** has a tapered cross section: thick on one end and thin on the other. The thin end is often inserted into a crevice to split the object when an input force is applied to the thick end. The input force causes two output forces to be generated normal to the two angled faces of the wedge.

**A wedge is a simple machine.**

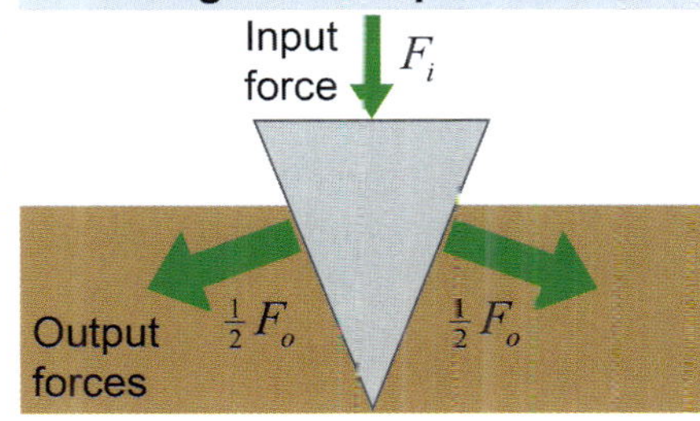


**A wedge is composed of inclined planes**

Divide a wedge along its plane of symmetry and what do you get? You get two inclined planes! The physics behind the forces acting on a wedge is similar to the physics of a ramp because both are types of inclined planes. Similar to two back-to-back ramps, the mechanical advantage of a wedge is the ratio of the length of the sloped side to the width of the thick end of the wedge, as calculated from equation (12.8).

**Two inclined planes constitute a wedge.**

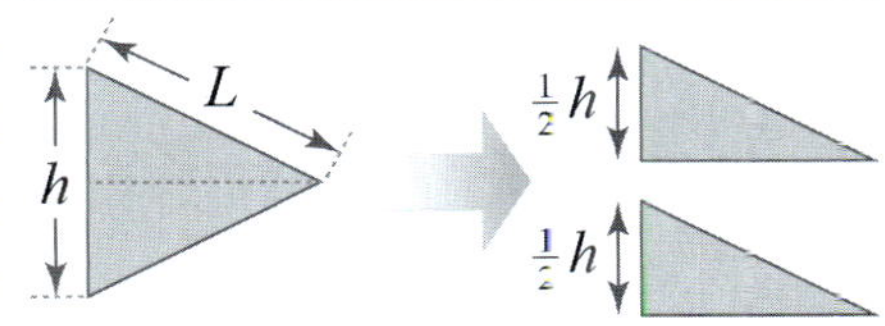


Mechanical advantage of a wedge $MA_{wedge} = \frac{L}{h}$

**Mechanical advantage of a wedge**

$$(12.8)\quad MA_{wedge} = \frac{L}{h}$$

$MA_{wedge}$ = mechanical advantage
$L$ = length of wedge (m)
$h$ = maximum thickness of wedge (m)

**Uses of wedges**

Wedges have many uses and are not just for splitting wood. Wedges can be used to hold objects in place, such as in a doorstop. The tines at the end of a fork are each a wedge. The sharp edge of a knife is a thin wedge forcing apart the material being cut. A fairly small downward force on a sharp blade can easily cut food while it takes a much larger force if the knife edge is blunt.

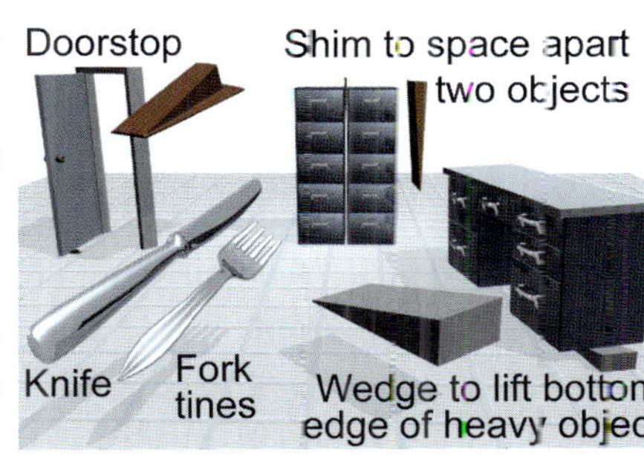


**Building a backyard pyramid**

Tony needs to move a 500 kg block to a height 5 m above the ground. Tony can push the block on rollers up a ramp with a force of 1000 N. How long must his ramp be?

*Asked:* length $L$ of the ramp

*Given:* input force Tony can apply, $F_i = 1000$ N; height of the ramp, $h = 5$ m; mass of the block, $m = 500$ kg; acceleration of gravity, $g = 9.8$ N/kg

*Relationships:* mechanical advantage: $MA = F_o/F_i$; for a ramp: $MA_{ramp} = L/h$

*Solution:* The output force Tony is working against is gravity, $F_o = mg$.
We have two expressions for mechanical advantage (above), so equate them and solve for the length $L$ of the ramp:

$$MA = \frac{F_o}{F_i} = \frac{L}{h} \Rightarrow L = h\frac{F_o}{F_i} = (5\text{ m})\frac{(500\text{ kg})(9.8\text{ N/kg})}{1{,}000\text{ N}} = 24.5\text{ m}$$

*Answer:* The ramp needs to be 24.5 m long.

**Spiraling ramps**

How do you drive a car up to the fifth level of a city parking garage? Chances are good that there is not enough space around the parking garage to fit a straight, quarter-mile long ramp! Instead, architects will design some parking garage ramps to spiral around. You get off the ramp when you get to the fifth floor. Parking ramps such as the one at right can be wound around the shape of a cylinder to conserve space in a densely populated city environment.

**The screw as a spiraling ramp**

You have used many of these spiral ramps before, although you might not have realized it! The threads of a **screw** resemble a spiraling ramp. The threads wrap around a central cylinder, just as the parking garage ramp in the illustration wraps around a central cylinder. You can make the shape of a spiral ramp yourself using a triangular piece of paper wrapped around a pen. Ramps, wedges, and screws are therefore all related simple machines: They all use the principles of an *inclined plane*.

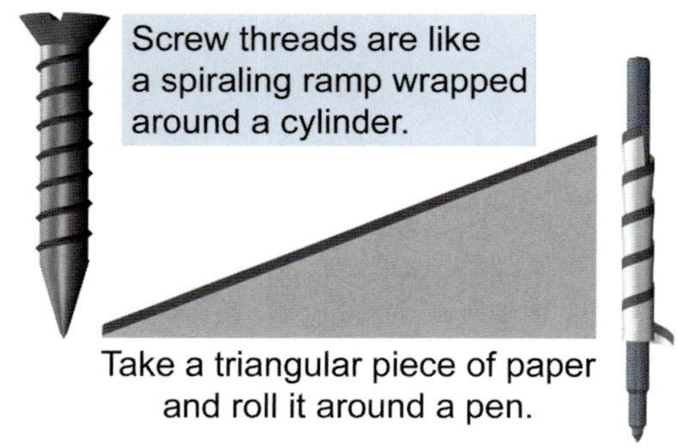


The result resembles a screw thread!

**Mechanical advantage of a screw**

The key property of a screw is its pitch $p$, which is the vertical distance between each successive thread. When you turn the screw around once, the screw will travel a distance $p$ into the material. The output work done is then $F_o \times p$. What is the screw's mechanical advantage? The input force is usually applied to the screw using another device, such as a screwdriver. When your hand has turned the outside of the screwdriver handle once, it has traveled a circumference of $2\pi L$, where $L$ is the radius of the handle. The input work done is therefore $F_i \times (2\pi L)$. Since the input work equals the output work, the mechanical advantage of the screw can be derived:

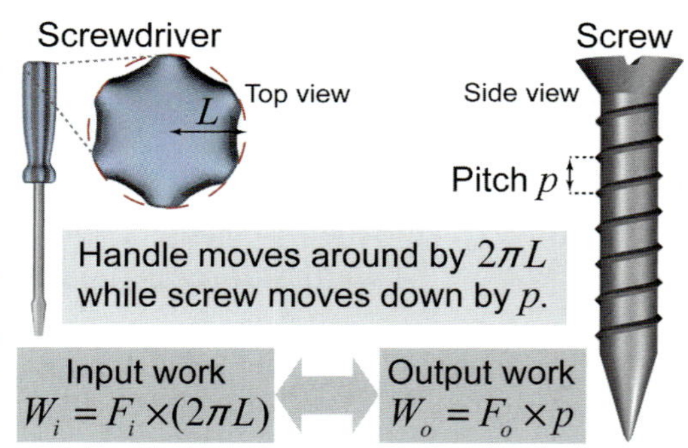


$$MA_{screw} = \frac{F_o}{F_i} = \frac{2\pi L}{p}$$

$$F_i \times (2\pi L) = F_o \times p \;\Rightarrow\; \frac{\cancel{F_i} \times (2\pi L)}{\cancel{F_i} \times p} = \frac{F_o \times \cancel{p}}{F_i \times \cancel{p}} \;\Rightarrow\; MA_{screw} = \frac{F_o}{F_i} = \frac{2\pi L}{p}$$

**Mechanical advantage** of a screw

$$\textbf{(12.9)}\quad MA_{screw} = \frac{2\pi L}{p}$$

$MA_{screw}$ = mechanical advantage
$L$ = radius of screwdriver handle (m)
$p$ = pitch of screw (m)

# Section 3 review

Three types of simple machines—ramps, wedges, and screws—are related to each other because each contains an inclined plane. Less force is required to move an object up a ramp than is needed to lift it vertically, but the object must be moved a further distance. A wedge is composed of two ramps back to back, and it is often used to split a material, such as an axe through wood. A screw consists of a spiraling ramp that wraps around a cylinder.

| | |
|---|---|
| **Vocabulary words** | **ramp, wedge, screw** |

| | | | |
|---|---|---|---|
| **Key equations** | $MA_{ramp} = \frac{L_{ramp}}{h_{ramp}}$ | $MA_{wedge} = \frac{L}{h}$ | $MA_{screw} = \frac{2\pi L}{p}$ |

## Review problems and questions

1. Describe the measurements and calculations that you need to make to determine the mechanical advantage of the following:
    a. a ramp
    b. a wedge
    c. a screw

2. Is it easier to use a screwdriver with a wide handle or a narrow handle?

3. The blade of an axe has a length of 15 cm and a mechanical advantage of 12. What is the maximum thickness of the blade?

4. A screwdriver with a handle that is 4.0 cm in diameter is used to drive in a metric screw with a pitch of 2.5 mm.
    a. What is the mechanical advantage of this combination?
    b. If you want to increase the mechanical advantage to 90, what pitch should you use?

5. Billy Bob is trying to decide whether to push a 50 kg dresser up a 4.5 m ramp or lift it directly into the back of a truck. The floor of the truck is 1.5 m above the surface of the street.
    a. What is the ideal mechanical advantage of the ramp?
    b. How much force is required to lift the dresser vertically upward into the back of the truck?
    c. If the ramp were frictionless, how much force is needed to push the dresser up it at constant speed?
    d. Billy Bob realizes that the ramp has a fairly rough surface. When he pushes the dresser up the ramp, he has to apply 225 N of force. What is the mechanical advantage of the ramp?
    e. How much work is required to lift the dresser vertically into the back of the truck?
    f. How much work is required to move the dresser up the ramp?
    g. What is the efficiency of the ramp?

It is not surprising that a car is a machine. But what kind of machine is a car? Is it a lever? Does it contain a wheel and axle? A better question to ask is this: *What types of machines does it contain?* A car is a *compound machine*, which means that it is made up of two or more simple machines that work together in order to accomplish its tasks.

What type of machine is this? A lever? Wheel and axle? Screw? Ramp?

## Combining more than one machine

**Compound machines**

Earlier in this chapter you worked with a compound machine, although you might not have realized it! The block and tackle is a compound machine made up of two or more pulleys connected together. A **compound machine** is a combination of two or more simple machines that work together to accomplish a task.

**How does a crane work?**

The crane was an important machine developed during the classical period and used to construct the Parthenon and other ancient monuments. With a modest-sized device, one man could lift an object that would normally require the strength of several men—and he could do so with a more convenient posture! One early crane design developed by the Greeks was the *trispastos*, which was made up of two simple machines: the pulley and the wheel and axle. This design was later enhanced by Roman engineers and was used extensively across the Roman Empire.

**Early Greek crane design**

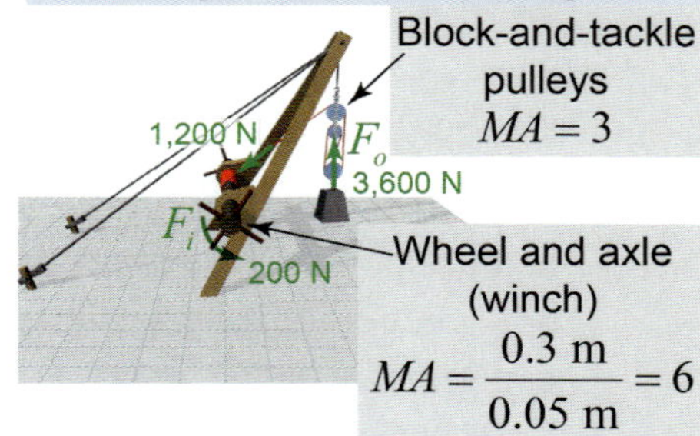


**Modern tower crane**

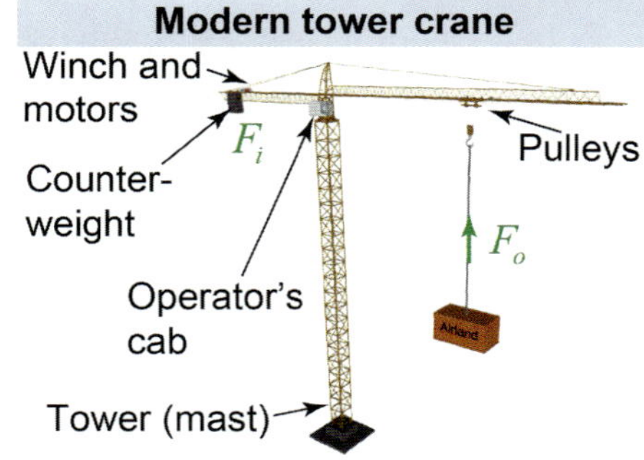


The block and tackle contained three pulleys (the origin of the prefix "tri" in trispastos), which had a combined mechanical advantage of 3. The user turned a winch, which is a form of the wheel and axle and might have had a mechanical advantage of around 6. By combining these two simple machines together into the crane, a mechanical advantage of 18 was achieved!

**Modern tower cranes**

Modern tower cranes use these same principles: An electric motor drives the winch and a series of pulleys multiply the force above the object being lifted. These cranes are used to lift heavy materials to the top of a building under construction. They are also used in ports to load containers onto ships. To avoid tipping over when lifting such heavy loads, a tower crane must employ a counterweight on the opposite side.

**Mechanical advantage for a compound machine**

A compound machine is made up of two or more simple machines that vary the forces applied. The individual mechanical advantages of each of those simple machines are multiplied together to determine the mechanical advantage of the compound machine. To determine the mechanical advantage of the Greek crane design in the illustration above, just multiply the individual mechanical advantages: $MA_{pulleys} \times MA_{winch} = 3 \times 6 = 18$!

**How does the body move?**

Your body is a machine composed of many individual simple machines. When you move your limbs, your bones act as the levers, the joints are the fulcrums, the contraction of muscles provides the input forces, and the motions of your arms and legs are the output forces. **Biomechanics** is the study of how the body moves.

**Biceps and triceps**

Video

Your forearm moves up and down as a result of the contraction of two muscles in your upper arm called the biceps and triceps. When the biceps contracts, it lifts your forearm upward—pivoting about a fulcrum located within your elbow joint. The biceps attaches to the forearm at a location closer to the fulcrum than the load on the arm (such as the weight of the dumbbell shown in the figure), so the biceps acts as a third-class lever. When the triceps contracts, it causes your arm to move downward in the opposite direction. The triceps attaches to the bones *on the opposite side of the elbow from the forearm*, so the triceps acts as a first-class lever. The biceps and triceps must apply large input forces close to the elbow to create small output forces at the end of the arm. *Your forearm has a mechanical advantage of less than one.*

**Levers and the human forearm**

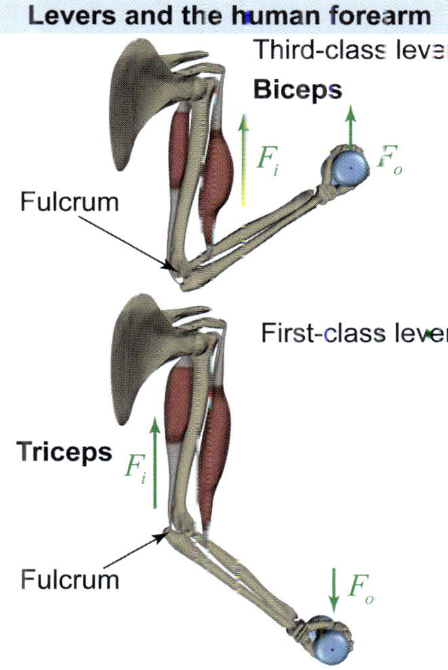


**Trading force for speed**

Why is it useful to have an arm with $MA < 1$? Your arm moves through a much larger range of motion than the small motions of your muscles. Your muscles trade off force for distance—or trade off force for speed, which is distance over time. That's why you can swing your arm around fast!

**Kicking a ball**

When you kick a ball, your leg acts as a combination of *three levers*: the upper leg pivoting about your hip joint, the lower leg pivoting about your knee, and your foot pivoting about your ankle. Each of these three lever motions is controlled by a muscle on the front of your leg. In all three cases, the input force is applied closer to the fulcrum than the output force—they are all acting as third-class levers.

When you swing your leg backward while preparing to kick, your leg uses three other muscles that act in the opposite direction. Where do you think those three muscles are located?

**Kicking a ball combines three levers**

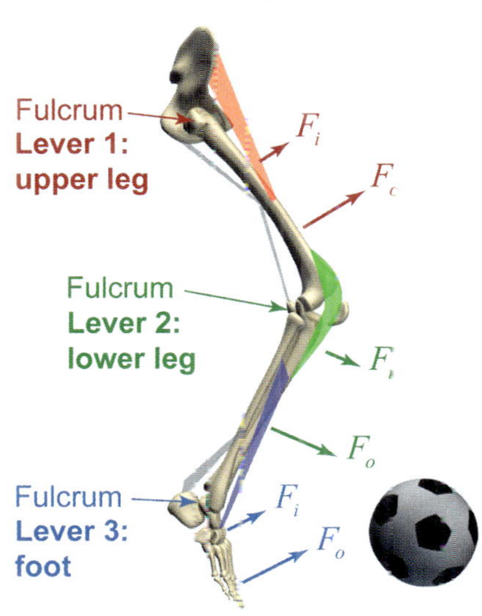

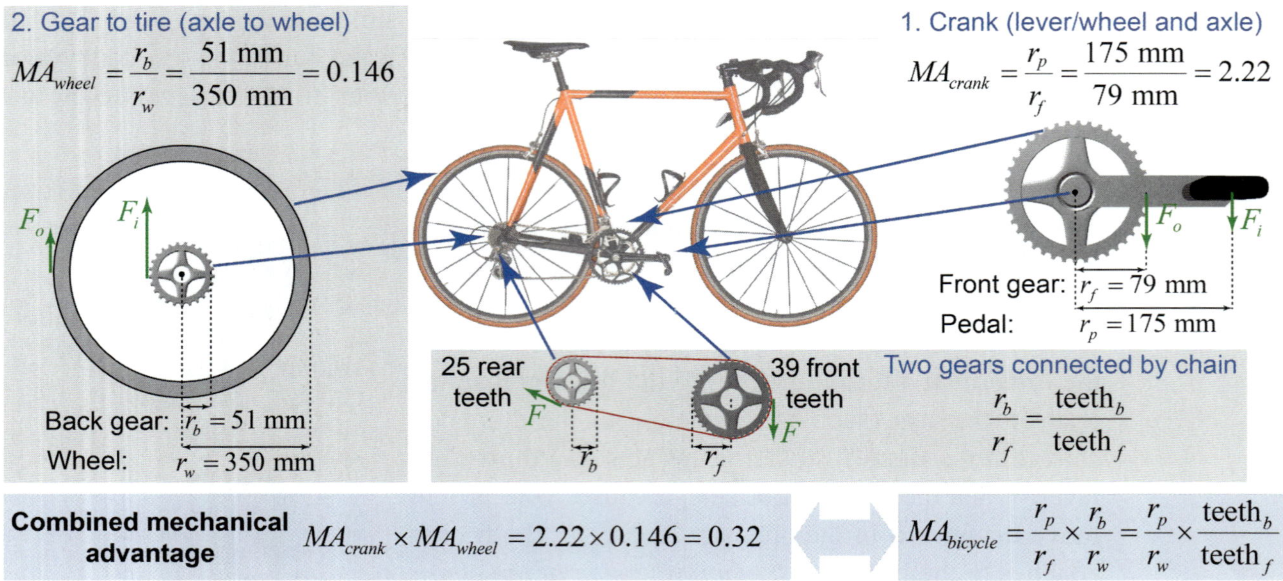


**How does a bicycle work?**

When you pedal a bicycle, you cause the rear wheel to turn. How is the motion of your feet converted into forward motion of the bicycle? The moving parts of the bicycle that connect the pedals to the wheel are called the *drive train.*

**A crank is a lever**

The crank is a lever integrated into a wheel-and-axle simple machine. Pushing down the pedal provides the input force and the output force turns the front gear. The mechanical advantage of the bicycle crank is the ratio of the radius of the pedal to the radius of the front gear, or $MA_{crank} = r_p / r_f$.

**Wheel and axle**

The back wheel is another wheel-and-axle simple machine, but this time the input force turns the back gear (the axle) and the output force is exerted by the tire on the outside of the wheel. The mechanical advantage of the back wheel is the ratio of the radius of the back gear to the radius of the wheel, or $MA_{wheel} = r_b / r_w$.

**Combined mechanical advantage**

The chain connects the front gear to the back gear such that the force is the same everywhere in the chain. Thus the combined mechanical advantage of a bicycle is the product of the mechanical advantages of the crank and the back wheel:

$$MA_{bicycle} = MA_{crank} \times MA_{wheel} = \frac{r_p}{r_f} \times \frac{r_b}{r_w}$$

Since the teeth are the same size on the front and back gears—the same chain runs over both gears!—the ratio of the radii of the back gear to the front gear, $r_b / r_f$, is the same as the ratio of their number of teeth. We can substitute the number of teeth for the ratio of the gear radii to get the mechanical advantage of a bicycle:

$$MA_{bicycle} = \frac{\text{pedal radius}}{\text{wheel radius}} \times \frac{\text{number of back gear teeth}}{\text{number of front gear teeth}}$$

**Mechanical advantage** of a bicycle

**Changing gears**

The mechanical advantage of a bicycle depends on the *ratio of the number of teeth in the back and front gears*. Now we can see why most modern road bicycles have more than one gear! When you change gears on a bicycle, you are changing its mechanical advantage.

*There is enough energy in high altitude winds to power civilization 100 times over; and sooner or later, we're going to learn to tap into the power of wind and use it to run civilization.*

— Ken Caldeira, Carnegie Institution for Science

**Energy dependence**

*Work, power,* and *energy* are three words that invoke modernity. Without energy, we'd have no way to do the work that we need to survive: building shelter, staying warm, gathering and cooking food, obtaining and purifying water, and so on. Doing those things by hand, or even with the help of animals, condemns people to short and precarious lives, dependent on nature's whims. By replacing human and animal muscles with machines, the industrial revolution made it possible for billions of people to live long lives of relative comfort.

**Danger ahead**

But our modern lifestyle may be living on borrowed time. After all, most of the power supplying our farms, cars, and factories comes from burning fossil fuels—and fossil fuels contain solar energy that took millions of years to store. Scientists disagree on how much coal, oil, and natural gas remains, but at some point—possibly during your lifetime—we will need to use more energy to obtain these resources than they give back.

**Advantages of wind power**

The vital yet controversial role that energy plays in our lives underlies the passion with which people pursue *renewable energy*—energy from natural processes that can be harnessed to power our modern lifestyle without being destroyed, disrupted, or depleted. Renewable energy sources include sunlight, wind, geothermal energy (hot springs and geysers), and moving water (waterfalls, rivers, ocean waves, and tides). Of these, wind holds much promise in the eyes of many conservationists: It poses relatively few hazards; it is abundant; and it can be harvested on many scales, from tiny backyard windmills to enormous offshore turbines.

**Limitations of wind power**

Windmills have been used for centuries to mill grain and pump water, as exemplified by the dike system in the Netherlands. Windmills have dotted the landscape for generations. But that, say some, is part of the problem. Modern wind turbines are as tall as a football field is long. Many find them rather noisy. They may pose a significant hazard, some say, to birds and bats. But worst of all, they're inefficient. Wind turbines typically operate at full power only 10% or 20% of the time; in other words, they have a *low duty cycle.* Even at an altitude of 100 m above the ground, wind loses lots of energy to friction with the ground. And, as sailors know well, low-altitude winds often stop blowing altogether.

**High-altitude resources**

At very low altitudes, wind loses energy from friction with the ground. Reach up a few hundred meters, however, and the wind blows more vigorously and more often. In the past few years, this untapped resource has spawned a new industry: high-altitude wind power. In this fledgling industry, high-tech kites, blimps, and gliders are taking the place of towering windmills.

**Energy from the jet stream**

The ultimate goal of wind power might be to tap into the *jet stream*—the river of air that circles the globe at temperate latitudes, powered not only by solar heating but by the Earth's rotation. Some 10 km above ground, this atmospheric current can shorten (or lengthen!) the travel times of cross-continental passenger jets by an hour or more. Above several cities, the jet stream could provide at least 10 kW to a turbine with a cross-sectional area of just one square meter—a windmill you probably could rest on your bed. A power output of 10 kW can provide for the domestic needs of five to ten small households. Tokyo, Seoul, and New York are among the Northern Hemisphere cities blessed with strong, fairly reliable winds at altitudes of 10 km.

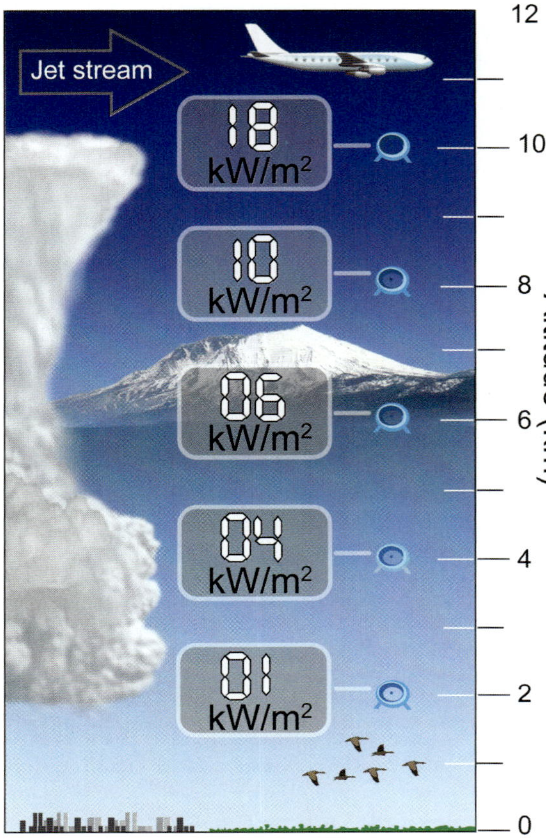


**Climbing the atmosphere**

The jet stream may be wind power's ultimate goal, but in the meantime wind power can make meaningful gains by climbing just a few hundred meters. Humans have taken their first practical steps toward power generation at these more modest altitudes. Altaeros Energies, a Massachusetts-based company, has recently launched helium-filled vessels that carry small but conventional wind turbines. In tests, this company's engineers deployed a 10 m prototype over a frigid winter landscape in Maine. The floating windmill ascended to an altitude of 100 m and generated electric current, which was carried to the ground by cables. Altaeros plans to make airborne turbines that can operate 300 m up, where winds can typically provide substantially more power than they do for fixed-height generators.

**Design challenge**

Design a wind turbine installation that

- is located in your area (state or region);
- produces 200 megawatts (MW) average power;
- has a minimum 30 m clearance between the bottom of the blades and the ground (or water);
- pays back its cost in less than 10 years by selling electricity to the grid at $0.07 per kilowatt-hour (kWh); and
- produces electricity at the minimum possible cost.

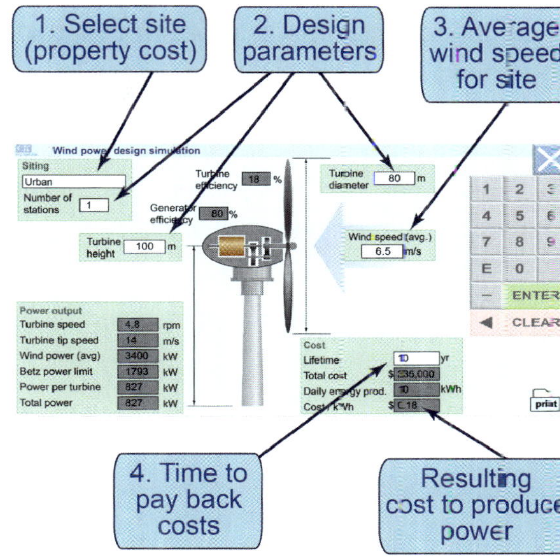


Which are the design *criteria*? Which are the design *constraints*? Write your answers in your report.

**Simulation**

In this interactive simulation, you will modify (and improve) an existing design for a wind turbine installation to produce electricity. You can vary the diameter of the turbine blades, the height of the tower holding the rotors, the number of turbines, and the wind speed for your location. Note that average wind speeds vary significantly not only across the United States but also within individual states.

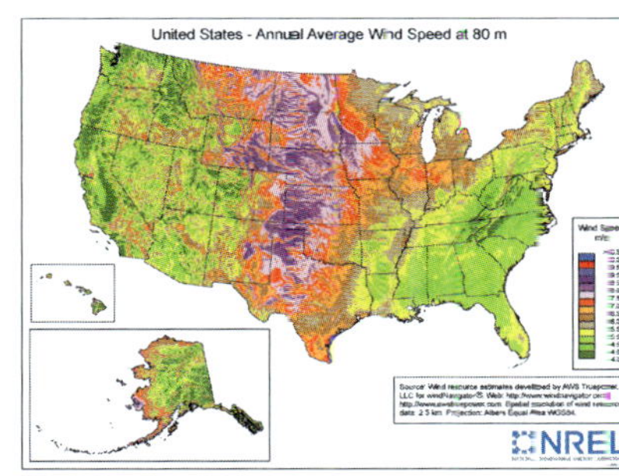


**Design and prototype**

In the simulation it is assumed that you must purchase the land (urban, suburban, or rural sites) or lease it from the government (offshore, such as for the Cape Wind project off the Massachusetts coast); the costs will change depending on your siting choice. Use resources such as the maps at right to determine the typical wind speed for your potential sites.

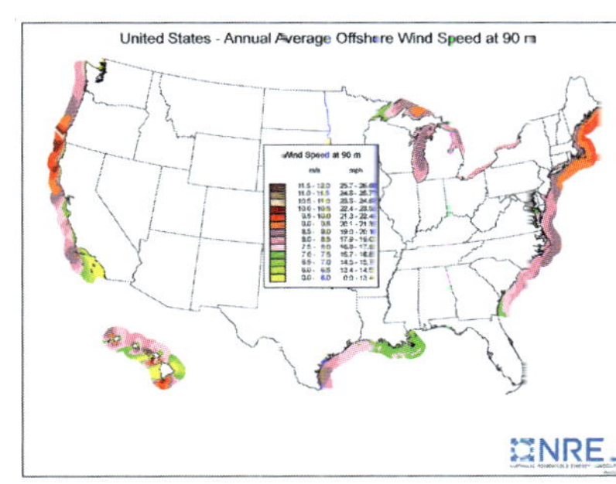


**Test**

Investigate how the power produced and cost per kilowatt-hour vary with the design parameters (diameter, height, number of turbines, siting costs, average wind speed, and life expectancy). Identify the controlled variables in your investigation.

**Evaluate**

Evaluate your design to determine which solutions best meet the design criteria. Your design will pay back its cost over its lifetime when the cost per kilowatt-hour equals the average cost in your area. Research the average cost of electricity in your area (or state), which is usually provided in dollars per kilowatt-hour ($/kWh). A monthly home electricity bill is one way to determine the local cost of electricity.

**Revise**

How might you improve your design? Revise your design and evaluate it.

# Section 4 review

A compound machine is a machine that is made up of two or more simple machines that work together. All compound machines, from a Roman crane to the modern automobile, can be broken down into different components that fall within one of the six kinds of simple machines. Even the human body can be modeled as a compound machine; each joint in the human arm and leg acts as either a first- or third-class lever, depending on which muscle is being used. The bicycle is a compound machine. Its drivetrain is composed of a front wheel-and-axle system called the crank, a chain connecting the front and back gears, and a rear wheel-and-axle system.

| Vocabulary words | compound machine, biomechanics |
| --- | --- |

## Review problems and questions

1. During the dissection of a human cadaver, a medical student makes the following measurements: The biceps are attached 1.25 cm below the elbow; the triceps are attached 0.8 cm above the elbow; the upper arm is 30 cm long; and the distance between the elbow and middle fingertip is 45 cm.
   a. What was the ideal mechanical advantage of the forearm when contracting the arm?
   b. What was the ideal mechanical advantage of the forearm when extending the arm?
   c. If, during a range of motion extending the arm, the upper arm had a mechanical advantage of 0.15, what was the combined mechanical advantage of the full arm?

2. A block and tackle with a mechanical advantage of 3.0 is used to pull a boat up a ramp. The ramp is 1.5 m high and 5.0 m long.
   a. What is the combined mechanical advantage of this arrangement?
   b. How much input force must be applied to obtain the needed output force of 3,500 N?

3. When early humans were hunting large game, why would small mechanical advantages of their legs be advantageous?

4. A bicycle's crank has two front gears, with 39 and 52 teeth. The rear cassette has five gears, with 15, 17, 19, 22, and 25 teeth.
   a. What are all the possible gear ratios for the bicycle?
   b. What are all the possible mechanical advantages?

5. Nadja has the multispeed bicycle described in the last problem and is stopped at a traffic light.
   a. When the light turns green, what mechanical advantage should she use at first, low or high? What gear ratio, low or high? Which gears should she use? Why?
   b. As she goes faster and faster, what gear ratio and mechanical advantage should she use, low or high? Which gears should she use? Why?

# Chapter review

## Vocabulary

Match each word to the sentence where it best fits.

### Section 12.1

| | |
|---|---|
| fulcrum | input arm |
| input force | lever |
| machine | mechanical advantage |
| output arm | output force |
| simple machine | |

1. When a wedge is pounded into a log using a sledgehammer, the sledgehammer provides the _______.
2. The ramp, screw, and lever are each a different kind of _______.
3. The ratio of the output force to the input force is called the _______.
4. When you use a crowbar to remove a nail from a piece of wood, the distance from the crowbar's pivot point to the nail is the _______.
5. A tricycle is a/an _______, because its drive train changes the magnitude and direction of a force.
6. The balance point for a see-saw is called its _______.
7. When a person turns the wheel of a wheel and axle system, the axle delivers the _______.
8. A rigid object that rotates about its fulcrum is called a/an _______.

### Section 12.2

| | |
|---|---|
| block and tackle | efficiency |
| gear | gear ratio |
| ideal mechanical advantage | pulley |
| tension | wheel and axle |

9. A rope and pulley system that uses more than one pulley is called a/an _______.
10. The gear and rear tire of a bicycle is an example of a _______.
11. A wheel with 24 teeth was turned, causing a wheel with 8 teeth to turn. The system has a/an _______ of three.
12. A toothed wheel is called a/an _______.
13. The ratio of the radius of a wheel to the radius of its axle is the _______ of that wheel-and-axle system.
14. During a tug-of-war competition, the _______ is the force exerted by the two teams along the rope.
15. Real-world machines generate less output force than ideal machines because the _______ is usually less than 100%.

### Section 12.3

| | |
|---|---|
| impulse | ramp |
| screw | wedge |

16. The switchbacks of a trail up the side of a mountain are an example of a/an _______.
17. A triangular doorstop placed partially under a door to keep it open is an example of a/an _______.
18. When you turn a/an _______, it converts the rotational motion into translational motion into the material.

### Section 12.4

| | |
|---|---|
| biomechanics | compound machine |
| microphone | |

19. A device that combines a lever, pulley, and screw is an example of a _______.
20. _______ is the field of science to study how various parts of the body move.

## Conceptual questions

### Section 12.1

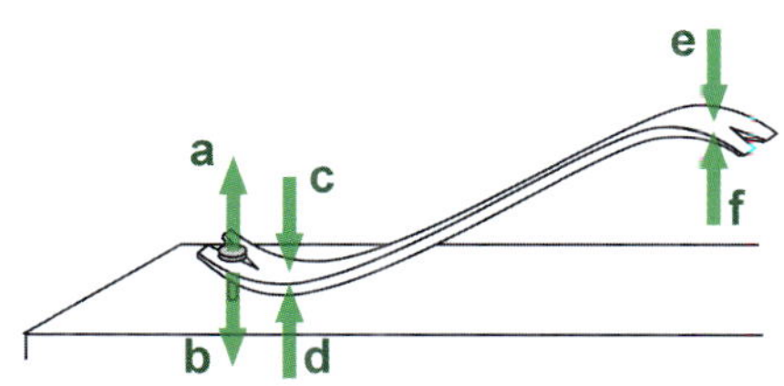


21. The diagram shows three pairs of action–reaction forces acting between a crowbar, a board, a nail, and a hand (not shown) pushing the right end of the crowbar. Consider the crowbar as a simple machine (lever).
    a. Which force is the input force on the lever and which is the output force?
    b. Where is the fulcrum located?
22. What is the difference between a simple and compound machine?
23. Define mechanical advantage in your own words.
24. Name four simple machines.

## Section 12.1

25. In conducting an investigation where you need to use a variety of different masses, compare the advantages of slotted masses and a small pail with sand poured into it.

26. A slotted lab mass typically has a wedge removed from it (the "slot"). How does the "missing" slot affect the mass printed on the slotted lab mass?

27. For the largest mechanical advantage, would you place a lever's fulcrum closer to the input or output force?

28. How can you determine to which class a lever belongs?

29. Using the equivalence of torques for a lever in equilibrium, find the mechanical advantage of a lever.

## Section 12.2

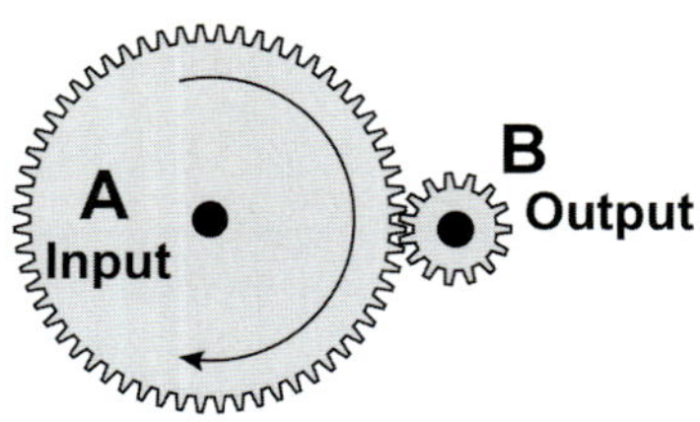


30. Gear A has more teeth than gear B.
    a. If gear A is the input gear, is the mechanical advantage greater than one or less than one?
    b. If gear B is the input gear, is the gear ratio greater than or less than one?

31. When constructing a block-and-tackle system for an investigation such as 12B on page 341, why is it better to use a spool of string or nylon thread rather than an elastic band to pass over the pulleys?

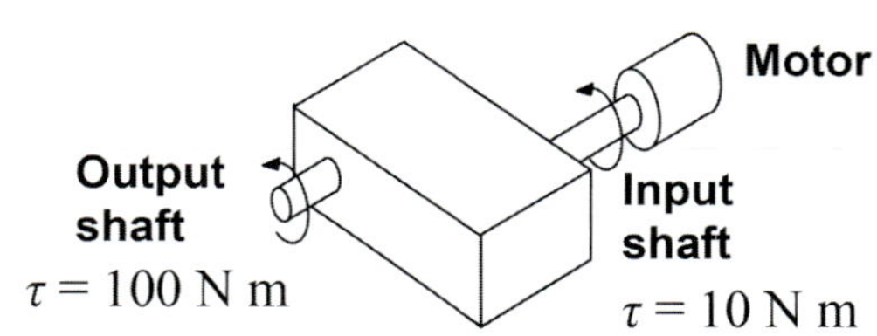


32. Irini has a motor that can provide 10 N m of torque. She plans to use gears to make a machine to produce 100 N m of output torque. Will the output shaft of her machine turn faster or slower than the input shaft connected to the motor? Why?

33. Rosa slides a block across the table at constant speed against the force of friction. What is the efficiency at which her input work is transformed into mechanical energy of the block?

34. Describe the relationship between the gear ratio and the mechanical advantage for a set of gears.

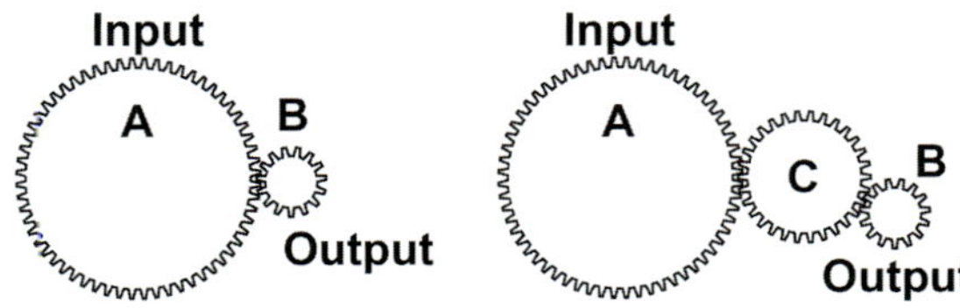


35. Gear A has 52 teeth and gear B has 15 teeth. Gear A is used to drive gear B. Gear C has 30 teeth and is inserted between gears A and B. (In this case, gear A remains the input gear and gear B remains the output gear.)
    a. Does the gear ratio change when gear C is added?
    b. What happens to the direction of rotation of gear B?

36. A system of gears has a mechanical advantage greater than one. You apply a force to the input gear and your friend applies a force to the output gear so that neither gear spins.
    a. Compare how much force each of you provides.
    b. Compare how much torque each of you provides.

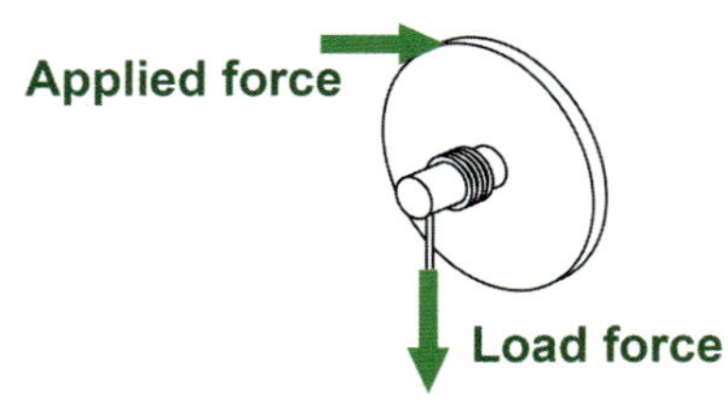


37. Suppose you use a wheel and axle to hold steady a load that is attached to the axle. The axle and wheel are held up by supports that are not shown.
    a. Does the torque from your hand on the wheel balance with the torque from the load?
    b. Does the strength of the force from your hand on the wheel equal the strength of the force exerted by the load?
    c. If either does not balance, then explain what ensures that the system remains in static equilibrium.

## Section 12.3

38. Bruce needs to lift an 80 kg dishwasher from street level to the bed of a moving van, 1 m higher. He plans to put the dishwasher on a wheeled dolly and then to push it up a ramp. He has two ramps to choose from, one 5 m long and the other 8 m long. Assume there is no friction.
    a. Which ramp (if either) requires Bruce to push with greater force?
    b. Which ramp (if either) requires Bruce to do a greater amount of work?

# Chapter review



## Section 12.3

39. Conceptually speaking, how can you change a ramp into a wedge? Into a screw? What do the three devices have in common in terms of mechanical advantage?

40. Tricia is building a birdhouse and having trouble driving a screw into the wood. Which of the following would make this task easier? (She has to do it by hand and can only deliver a certain amount of force to her screwdriver.)

    a. a screwdriver with a thicker handle
    b. a screw with more threads per inch
    c. both a and b
    d. neither a nor b

41. A wedge is useful for splitting wood because it exhibits mechanical advantage, which increases whatever force you can deliver. Suppose that your hammer strikes a wedge with a force of 100 N, and the wedge splits a wooden block in two. Suppose too that the wedge pushed the two halves of the block apart from each other with 1,000 N of force. Explain why this does not violate any conservation laws.

## Section 12.4

42. The mechanical advantage (MA) of a compound machine is the _______ of the MAs of the individual machines.

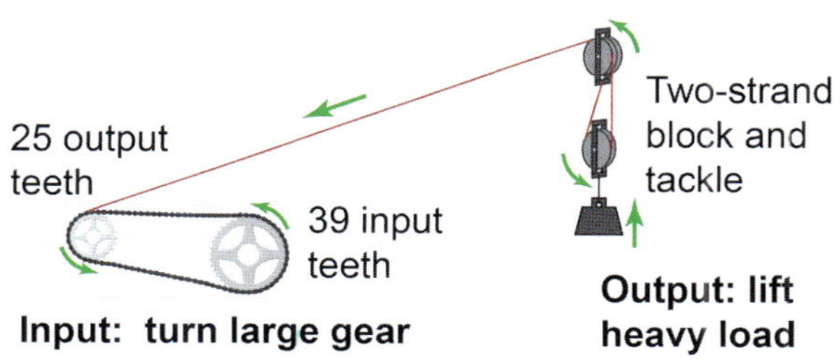


43. The compound machine shown here uses a bicyclist to turn a large gear, which turns a smaller gear (with fewer teeth). The smaller gear pulls a cord that uses a block and tackle to lift a heavy load. Which of the following changes will *increase* the mechanical advantage of this compound machine? (More than one answer may be correct. Note that the number of teeth is proportional to the gear radius, because the gears share a common chain.)

    a. increasing the radius (and therefore the number of teeth) of the large gear
    b. increasing the radius (and therefore number of teeth) of the small gear
    c. using a block and tackle with a mechanical advantage of three
    d. replacing the block and tackle with a simple pulley that hangs above the load

## Quantitative problems

## Section 12.1

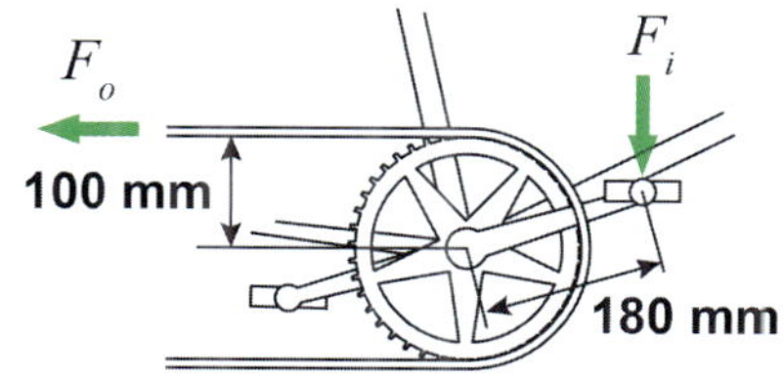


44. The diagram shows the arrangement of a bicycle crank and pedals. How much force is transmitted to the chain if 500 N of force is applied to the pedal? What is the mechanical advantage of the crank–chain machine?

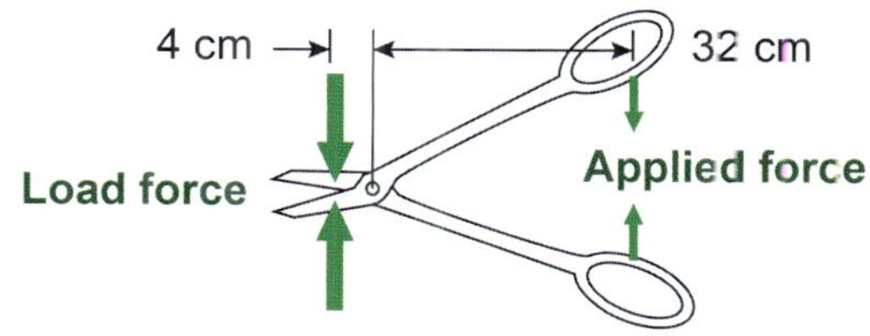


45. Shears are special scissors for cutting metal. Calculate the mechanical advantage of the shears shown in the diagram. If 30 lb of force is applied to the handles, how much force is applied to the metal between the blades?

46. Janelle has a machine with a mechanical advantage of 2.5. She inputs a force of 20 N. What is the output force?

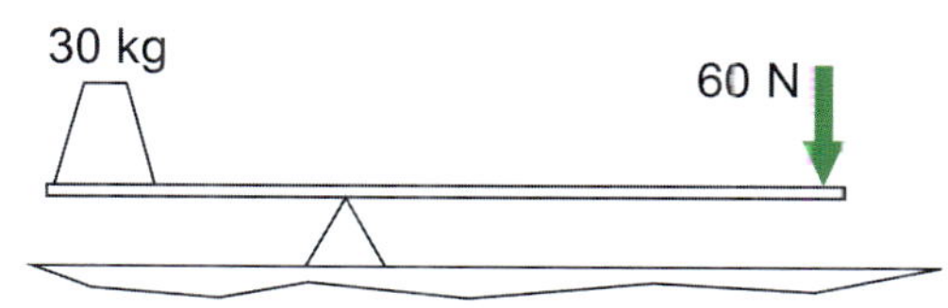


47. Sofia has placed a 30 kg mass on the end of a first-class lever. At the other end, she applies a 60 N force to balance the lever. What is the mechanical advantage of her lever?

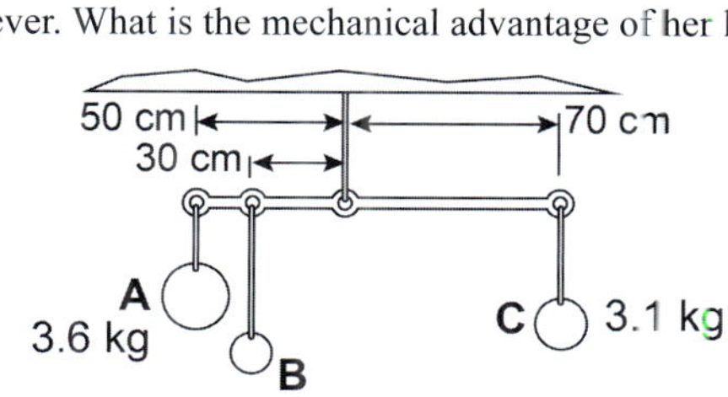


48. A lamp designer creates a ceiling fixture with three globes suspended from a rod. What should be the mass of globe B so that the fixture will balance and hang with the rod horizontally?

49. Stanley has placed a 50 kg weight at the end of a 4 m lever. How far should the fulcrum be from the weight so that he can lift the weight with only 49 N of force?

# Chapter review

## Section 12.2

50. Louise kicks a 10 N soccer ball up a hill. It rolls over the top of the hill, which is elevated 20 m above the ground she is standing on, and comes to a stop halfway down the other side of the hill where there is a wall. That wall's elevation is 10 m below the top of the hill.
    a. If she kicks with 400 J of energy, what is the efficiency?
    b. How can she kick the ball to improve her efficiency?
    c. If she does so, what is the maximum efficiency she can achieve?

51. A set of pulleys has a mechanical advantage of 1.5 and an ideal mechanical advantage of 2. If you wish to lift a 30 N weight by 10 m, how far do you have to pull and how hard?

52. Jared lifts a 100 N mass by 10 m. He does so using a wheel and axle, providing an input force of 10 N and turning the wheel for 200 m.
    a. What is the mechanical advantage?
    b. What is the ideal mechanical advantage?
    c. What is the efficiency?

53. An engine is attached to a set of gears with a gear ratio of two. Those gears drive an axle, which is connected to wheels. The ratio between the wheels' radius and the axle's radius is 10.
    a. The engine provides 100 N m of torque. How much torque do the wheels provide?
    b. The engine provides 50 N of force. How much force do the wheels provide?
    c. Use your answers to a and b to calculate the radius of the wheels.

54. Larissa and Kasey built a block-and-tackle assembly to lift a 15 kg box. They had to pull their end of the rope with an input force of 120 N. Furthermore, they moved their end a distance of 3.0 m to lift the box 150 cm. What was the *efficiency* of the block-and-tackle assembly?

55. Gears A, B, and C have 10, 20, and 30 teeth, respectively.
    a. If A is the input and B is the output, what is the gear ratio?
    b. If A is the input and C is the output, what is the gear ratio?
    c. If B is the input and A is the output, what is the mechanical advantage?
    d. If B is the input and C is the output, what is the mechanical advantage?

## Section 12.3

56. You place a cart with a mass of 460 g on a ramp that is 3.4 m long and 0.8 m high.
    a. What is the mechanical advantage of this ramp?
    b. Draw a free-body force diagram of the cart.
    c. Using your diagram, calculate the force it would take to push the cart up the ramp it is placed on.
    d. What direction is the cart most likely to travel once you let it go? Use your free-body diagram to help you answer.

57. A man is pushing his 75 kg dresser into a moving van up a ramp with a force of 175 N. The ramp is 5.7 m long and 1.1 m high.
    a. Draw a free-body diagram of the dresser.
    b. Assuming there is no friction, how much force does the man have to apply to keep the dresser from slipping down the ramp?
    c. What is the acceleration of the dresser when the man pushes with a force of 175 N? Does the dresser move up or down the ramp? Use your free-body diagram to help you answer.
    d. What is the mechanical advantage of the ramp?

## Section 12.4

58. On your computer, use the interactive simulation of a wind turbine on page 688 to calculate the difference in cost per kWh (ten-year average) between a 120 m high, 80 m diameter turbine in 8 m/s average wind speed that is located in a rural versus urban site. What about the sites makes the costs different?

59. The ancient Greek crane shown on page 357 has a block and tackle with a mechanical advantage of three and a winch (wheel-and-axle assembly) with a mechanical advantage of six. Which one of the following changes would *most* increase the MA of the entire crane?
    a. Increase the radius of the winch handle from 30 cm (0.3 m) to 45 cm (0.5 m).
    b. Decrease the radius of the winch axle from 5 cm (0.05 m) to 2.5 cm (0.025 m).
    c. Replace the block and tackle with one that supports the load with four strands instead of three.
    d. Replace the block and tackle with one that supports the load with five strands instead of three.

# Chapter review

## Standardized test practice

60. Scott has a machine with a mechanical advantage of five. How much force does he need to apply to create an output force of 120 N?

    A. 5 N
    B. 24 N
    C. 125 N
    D. 6,000 N

61. A lever with an output arm of length 40 cm has an object on the end of it. Larisa pushes on the 80 cm input arm with 100 N of force to balance the lever. What is the mass of the object?

    A. 2.0 kg
    B. 10.2 kg
    C. 20.4 kg
    D. 204 kg

62. Rafi uses a pulley system to lift a 15 N weight by 6 m. If he pulls his end of the rope a distance of 3 m, which of the following would be the force he uses? (Assume the pulley is 100% efficient.)

    A. 5 N
    B. 15 N
    C. 25 N
    D. 30 N

63. Katarina pushes a 41 kg crate up a ramp 7.5 m long. In going from the bottom of the ramp to the top, she raises the crate 1.5 m while doing 1,000 J of work. Which of the following is most nearly the efficiency of the ramp?

    A. 0.20
    B. 0.30
    C. 0.40
    D. 0.60

64. Which of the following simple machines provides the *least* mechanical advantage?

    A. A lever 1 m long whose fulcrum is 80 cm from the input force and 20 cm from the load.
    B. A block-and-tackle assembly that lifts a 1,000 N load when you provide a 200 N input force.
    C. A wedge with 80-mm-long sides and a base 10 mm wide.
    D. All three machines are equally efficient.

65. Which of the following pieces of equipment is the best choice for measuring *force*?

    A. spring scale
    B. triple beam balance
    C. meter stick
    D. friction block

66. Which of the following is the best choice to provide the widest variety of different values of mass?

    A. a set of seven hooked lab masses, each having a different mass
    B. a set of five slotted masses, each with a different mass
    C. a hanging plastic bottle with water
    D. a set of 25 identical marbles

67. Cristin and Adele built a block-and-tackle assembly to lift a 20 kg motor. They had to pull their end of the rope with an input force of 100 N. Furthermore, they had to move their end a distance of 2.5 m to lift the motor just 50 cm. What was the *ideal mechanical advantage* of the block-and-tackle assembly?

    A. 0.05
    B. 0.5
    C. 2.5
    D. 5

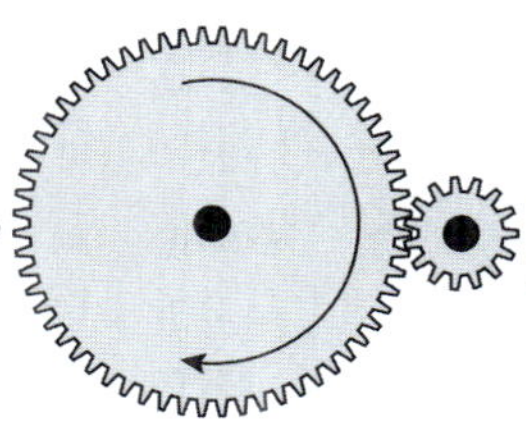


68. A 15-tooth gear makes 20 turns. How many turns will the 52-tooth gear make?

    A. 2.60
    B. 3.67
    C. 5.77
    D. 10.00

69. Suppose that a compound machine is made by connecting a wheel-and-axle assembly with a mechanical advantage of 4.0 to a block-and-tackle assembly with a mechanical advantage of 5.0. What will be the mechanical advantage of the machine as a whole?

    A. 0.8
    B. 1.25
    C. 9.0
    D. 20

70. Suppose that a compound machine is made by connecting a wheel-and-axle assembly with an efficiency of 0.80 to a block-and-tackle assembly with an efficiency of 0.50. What will be the efficiency of the machine as a whole?

    A. 0.3
    B. 0.4
    C. 1.3
    D. 1.6

# Chapter 13 Angular Momentum

How can you prevent a spacecraft from spinning out of control? How does a smartphone, tablet computer, or hand-held controller for a video game sense that it has been tilted or turned? The answer is a gyroscope, and the physics behind it is the conservation of angular momentum.

The heart of a gyroscope is a *rotor*, which is a disk that spins about an axle connected to the frame. Inexpensive, commercially available gyroscopes, such as the one at right, are mounted inside a fixed frame that consists of two metal rings. You start the gyroscope by spinning the rotor fast, often by pulling a string wound around the axle. Anything that rotates possesses *angular momentum*—just as anything moving in a straight line possesses linear momentum. The angular momentum from the gyroscope's spinning motion keeps it from falling over. You can even balance a spinning gyroscope on the tip of a pencil!

**How a gyroscope works**

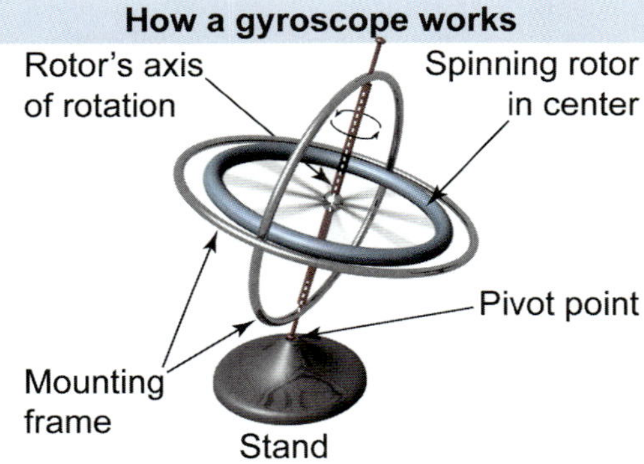


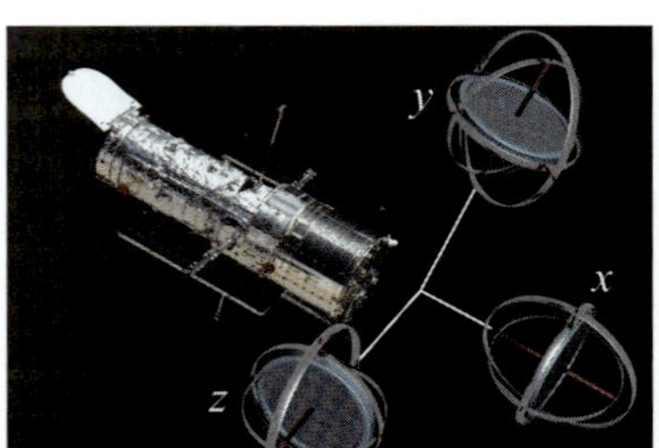

Three gyroscopes are needed to maintain a spacecraft's orientation.

When spacecraft leave the Earth's surface, they cannot use the Earth's magnetic field to tell direction. They use gyroscopes instead! When a spacecraft tilts in one direction, sensors in a gyroscope detect the spacecraft's tilt with respect to the desired orientation of the spinning rotor—and the spacecraft's attitude (its orientation relative to an inertial frame of reference) can be adjusted to compensate. Three dimensions (the *x*-, *y*-, and *z*-axes) are needed to define the spatial coordinate system or *inertial reference frame*, so a spacecraft needs at least three gyroscopes to maintain its orientation.

Many smartphones and hand-held video game consoles measure the rotation or tilt of the device using accelerometers and gyroscopes, but not the clunky ones pictured above. Instead, they use a *vibrating structure gyroscope* as part of a tiny chip called a microelectromechanical system (MEMS). These solid state gyroscopes are also used in two-wheeled, self-balancing personal transporters, such as the Segway® PT.

# Chapter study guide

## Chapter summary

When bodies move in a straight line they have momentum; when they rotate they have angular momentum. Rotation is a fundamental property of many objects around us, from the rolling wheels of a car to the rotation of the Earth about its axis. Everything that has mass has rotational inertia, which is the resistance of an object to changing its state of rotation. Rotational inertia depends not just on mass but also on how that mass is distributed relative to the axis of rotation. Rotating objects also possess rotational energy in addition to their linear kinetic energy.

## Learning objectives

By the end of this chapter you should be able to

- define angular momentum and calculate its value for a rotating object;
- describe rotational inertia and calculate the moment of inertia for objects with simple shapes;
- explain why objects can change their rotational velocities by applying the conservation of angular momentum;
- define center of mass and apply it to practical situations;
- define and calculate rotational energy; and
- explain why rolling objects of different shapes are accelerated differently.

## Investigations

13A: Rotational inertia
13B: Conservation of angular momentum
13C: Center of mass
13D: Rolling down an inclined plane

## Chapter index



## Important relationships

$$L = r \times mv$$

$$I = mr^2$$

$$L = I\omega$$

$$E_r = \frac{1}{2} I\omega^2$$

## Vocabulary

**axis**
**translation**
**angular momentum**
**center of mass**
**rotation**
**rotational inertia**
**linear momentum**
**rotational energy**
**revolution**
**moment of inertia**
**conservation of angular momentum**
**precession**

Spinning objects tend to keep spinning for the same reason moving objects tend to keep moving with the same speed and direction. A spinning bicycle wheel has *angular momentum* as a result of its rotational motion. Angular momentum resists changes in rotational speed or orientation, similar to the way linear momentum resists changes in linear speed or direction. If you spin a heavier wheel at the same speed, or spin the same wheel faster, it will have more angular momentum and be more difficult to stop or turn. Angular momentum plays a key role in the behavior of objects from the smallest microscopic scale inside the atom to the largest macroscopic scales in the dynamics of the Solar System and the Milky Way Galaxy.

**Objects with angular momentum**

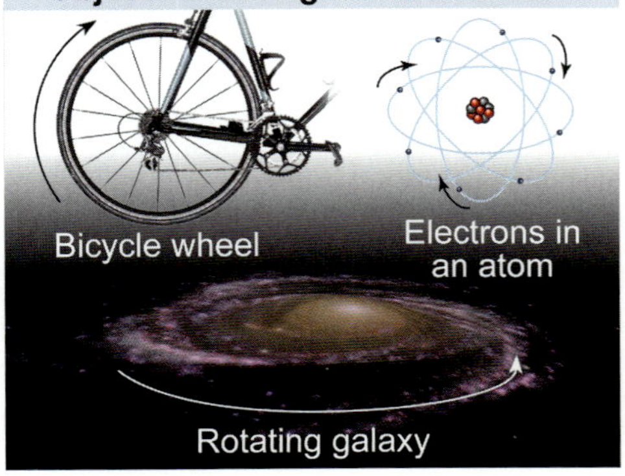


## Describing rotational motion

**Translational and rotational motion**

Objects may move from one place to another while they are simultaneously rotating. A good example is a rolling ball. The ball has *translational motion* that takes it across the floor. **Translation** is a type of motion that causes a change in position. While the ball is *translating* it is also *rotating*. When we analyze complex motion we may treat translational and rotational motions independently and associate momentum and energy with each.

**Translational and rotational motion**

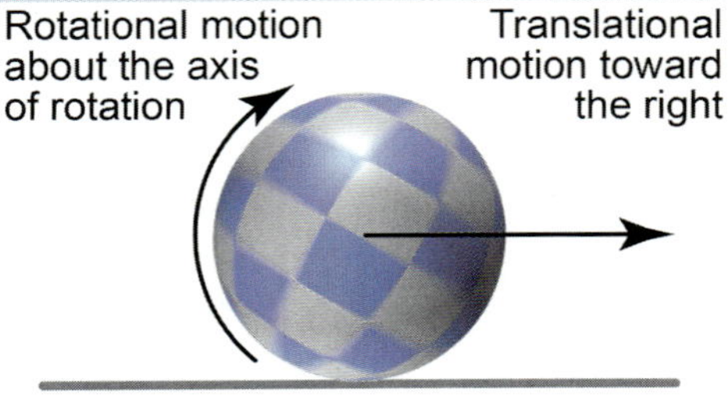


**Axis of rotation**

The **axis** of rotation is an imaginary line around which rotation occurs. A spinning top rotates about a vertical axis—a straight line—passing through the center of the top. The axis of rotation for a bicycle wheel is the axle of the wheel, which is horizontal when the bicycle is being ridden.

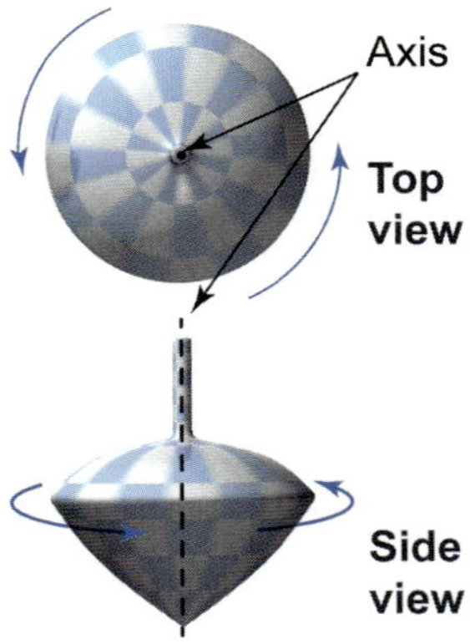


**Why does the axis matter?**

Many properties of rotational motion, such as *angular momentum,* depend on the location and orientation of the axis of rotation. The same object rotated around a different axis will have different rotational kinetic energy and angular momentum, even if the mass and rotational speed stay the same!

**Rotation and revolution**

A **rotation** describes a motion in which the axis *lies within the object*, such as the motion of the spinning top or bicycle wheel. A **revolution** refers to a rotational motion about an axis that is *outside the object.* As an example, Earth undergoes both rotation and revolution. Earth makes a full rotation about its north–south axis every 24 hours, creating day and night. Earth makes a full revolution (orbit) around the Sun once per year. The axis of Earth's orbital motion passes through the Sun and is perpendicular to the plane of the Solar System.

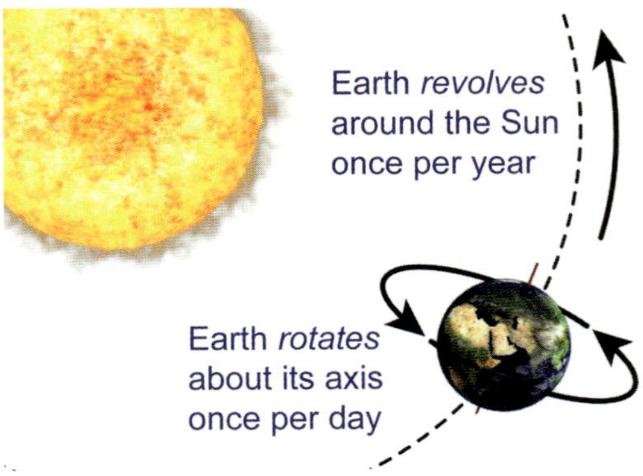

### Starting and stopping a Ferris wheel

An object's *rotational inertia* is its resistance to changing its state of rotation—increasing or decreasing its angular velocity. A wheel with low rotational inertia is easy to spin. An amusement-park Ferris wheel starts and stops very slowly because the heavy steel structure has a large rotational inertia. Racing cars and bicycles have very light aluminum-alloy wheels to reduce rotational inertia.

**Why does it take time for a Ferris wheel to get up to speed or come to a stop?**

### Rotational inertia

**Rotational inertia** depends not only on mass and rotational speed but also on *how mass is distributed around the axis of rotation.* Consider a solid cylinder and a hoop with the same mass $m$ rotating at the same angular velocity $\omega$. Now think about a small element of mass inside each object. Mass element (B) is moving faster than mass element (A) because point (B) is farther from the axis of rotation than point (A). A hoop has more rotational inertia than a cylinder of equal mass *at the same angular velocity* because the mass in the hoop is moving faster because of its greater distance from the axis of rotation.

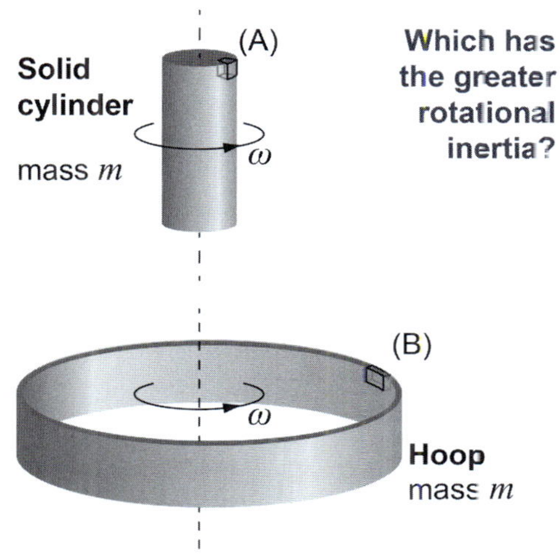


### Moment of inertia

The quantity of rotational inertia possessed by an object called its **moment of inertia** and is identified with a capital letter "$I$" in equation (13.1). The units of $I$ are kg m$^2$. Notice that the moment of inertia varies with the *square* of the radius. That means that mass that is twice as far from the center of rotation contributes *four times* as much rotational inertia.

**Moment of inertia** of a point mass

$$I = mr^2 \quad (13.1)$$

$I$ = moment of inertia (kg m$^2$)
$m$ = mass (kg)
$r$ = radius (m)

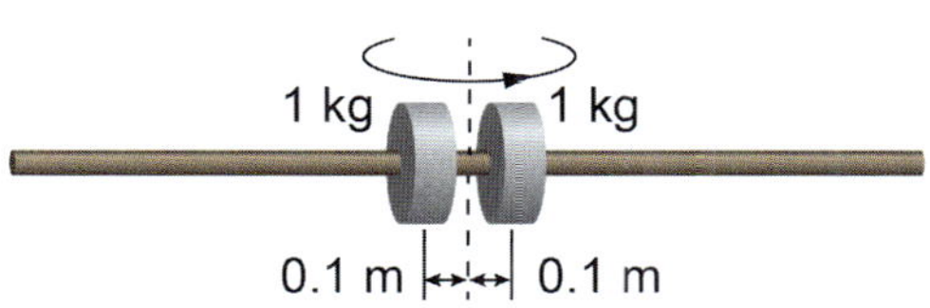


$$I = mr^2 = 2\times(1\text{ kg})(0.1\text{ m})^2 = 0.02\text{ kg m}^2$$

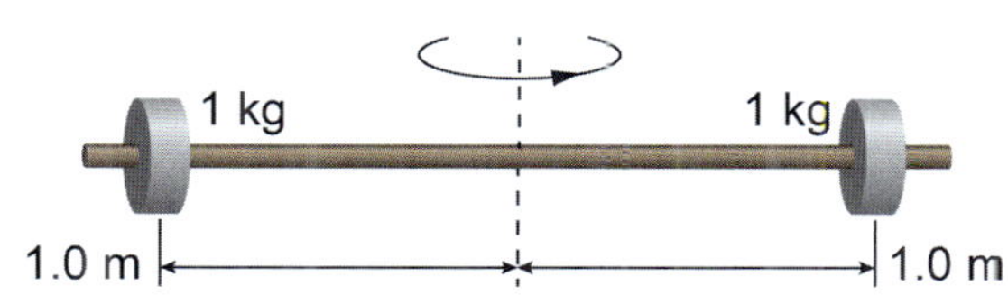


$$I = mr^2 = 2\times(1\text{ kg})(1.0\text{ m})^2 = 2\text{ kg m}^2$$

### How does the moment of inertia affect motion?

To get a sense of how mass distribution changes rotational inertia, consider a light wooden pole with two 1 kg iron masses. When the masses are 0.1 m from the center, the moment of inertia is 0.02 kg m$^2$. When the masses are 1.0 m from the center, the moment of inertia is a hundred times larger, even though the total mass is identical! Consequently, it takes a hundred times more torque to change the rotational speed when the masses are spread out.

# Investigation 13A Rotational inertia

**Essential questions** *What physical effects result from rotational inertia?*

Rotational inertia can affect the motion of objects that are not "rotating" in the sense of continually turning. A good example is "tipping over." When an object tips over it is momentarily rotating about the point in contact with the ground. The rotational inertia about that axis determines how quickly the object tips! In this investigation, you will see how some objects are far easier to balance than others because they have different amounts of rotational inertia. For example, is it easier to balance a meter stick or a pencil upright on the palm of your hand?

## Part 1: Which has more rotational inertia?

1. Hold a pencil upright in the palm of your hand and try to balance it.
2. Repeat for a broom, a meter rule, a lollipop (with the sphere both at the top and the bottom), and a long-handled screwdriver (also both ways).
3. Tabulate your results as to which ones are easy or difficult to balance.

*a. If gravity can rotate the object rapidly, such that you cannot keep the object balanced, then does the object have a little or a lot of rotational inertia?*

*b. Which objects are easier to balance: those with most of the mass closest to or farthest from your hand?*

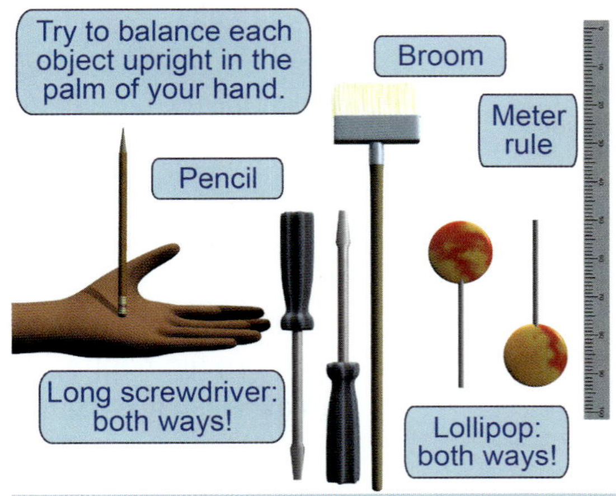


Which ones are the easiest to balance? Is it easier to balance when most of the mass is located close to your hand?

## Part 2: Rotational inertia of the meter rule with added mass

1. Attach a large blob of clay near one end of the meter rule (around the 10 cm mark).
2. By creating a model based on the physics of rotational inertia, predict when it will be easiest to balance the ruler: when the clay end is nearest to your hand or when the clay end is furthest from your hand.
3. Try both cases and write down which one is easier.

*a. Which way made it easier to balance the meter rule?*

*b. Does the meter rule have more rotational inertia when the clay is located close to your hand or far away? Why might this be the case?*

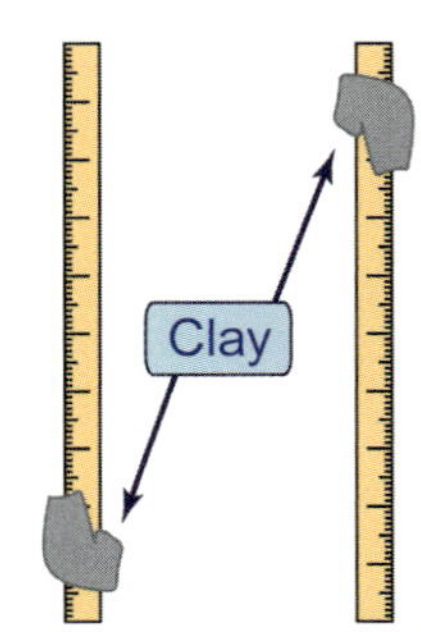


Which way is easier to balance?

### Linear vs. angular momentum

A rolling ball has both linear momentum and *angular momentum.* The linear momentum is the product of mass and translational velocity, as you learned on page 308. Angular momentum is a different kind of momentum, with different units that obeys a separate conservation law. Any moving mass has both angular momentum and linear momentum. The quantity of angular momentum depends on the choice of center of rotation as well as the mass and velocity.

**Rolling ball has linear momentum and angular momentum.**

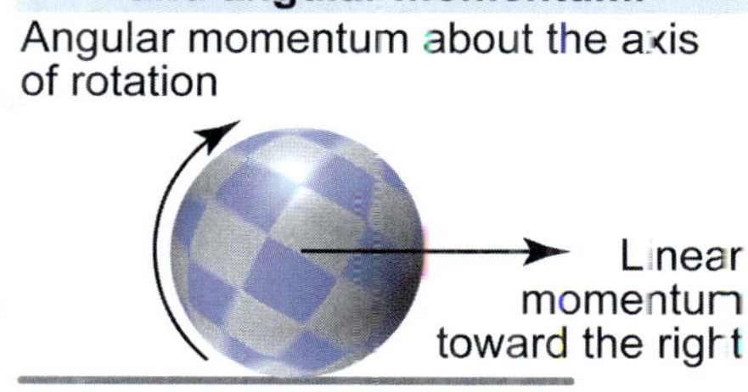


### Angular momentum

**Angular momentum** is associated with the movement of mass about a particular axis of rotation. Imagine spinning a mass attached to a string around over your head. At any given moment, the mass $m$ is moving with a velocity $v$, so it has a linear momentum $p = mv$. As it spins over your head, the same mass has an *angular momentum* (denoted with the letter $L$). Angular momentum is equal to the radius of rotation times the linear momentum, or $L = r \times mv$. Note that the same mass moving at the same velocity may have a *different* angular momentum about a different center.

**Momentum of a mass spun around on a string**

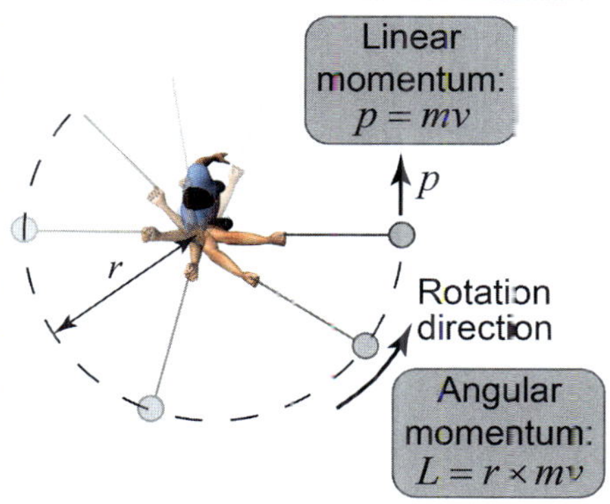


Interactive equation

$$L = r \times mv \tag{13.2}$$

$L$ = angular momentum (kg m$^2$/s)
$r$ = radius from axis (m)
$m$ = mass (kg)
$v$ = velocity (m/s)

**Angular momentum** for a point mass

### Units of angular momentum

Angular momentum and linear momentum have different units! The units of linear momentum are kg m/s. When we calculate angular momentum we multiply linear momentum by distance, so angular momentum has units of kg m$^2$/s.

### Rotational and orbital angular momentum

An object can have multiple quantities of angular momentum around multiple axes. For example, Earth has *rotational* angular momentum because of its rotation about its north–south axis. Earth also has *orbital* angular momentum because the planet revolves around its orbital axis passing through the Sun.

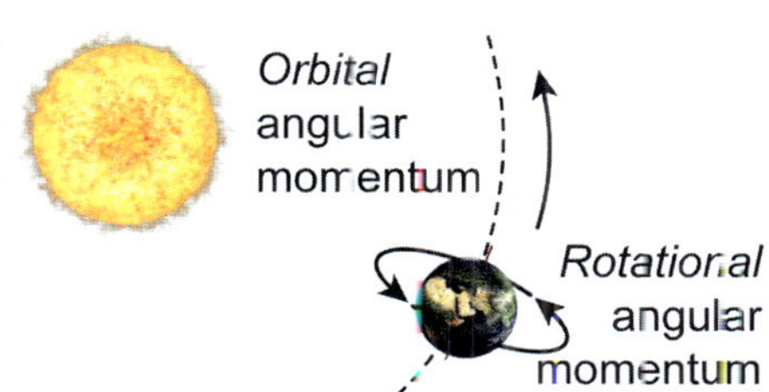


### Calculating angular momentum

Calculate the angular momentum of a 100 g mass revolving in a circle once per second at the end of a 0.75 m string.

| | |
|---|---|
| *Asked:* | angular momentum $L$ |
| *Given:* | mass $m$ = 100 g = 0.1 kg, radius $r$ = 0.75 m, $\omega$ = 1 rev/s |
| *Relationships:* | angular momentum $L = rmv$ |
| *Solution:* | The mass moves $2\pi r$ every second, so its velocity is<br>$v$ = distance/time = $(2\pi)(0.75\text{ m}) \div (1\text{ s}) = 4.7$ m/s<br>Its angular momentum is therefore<br>$L = rmv = (0.75\text{ m})(0.1\text{ kg})(4.7\text{ m/s}) = 0.35\text{ kg m}^2/\text{s}$ |
| *Answer:* | The angular momentum is 0.35 kg m$^2$/s. |

**Another equation for angular momentum**

Angular momentum obeys a separate conservation law from that of linear momentum. To best understand this conservation law we need to rewrite the equation for angular momentum. Recall that linear momentum is the product of mass and velocity, $p = mv$. We want a similar form of equation for the angular momentum in terms of the angular velocity $\omega$. We start with equation (13.2) then use the fact that the linear velocity and angular velocity are related by $v = \omega r$.

Equation (13.2)

$$L = r \times mv \quad \leftarrow \text{Linear and angular velocity: } v = \omega r$$

$$L = rm(\omega r)$$

$$L = mr^2 \omega \quad \leftarrow \text{Moment of inertia: } I = mr^2$$

Equation (13.3)

$$L = I\omega$$

**Where does moment of inertia come in?**

If we regroup the terms, we find that the moment of inertia $I$ in the angular momentum equation appears in the same place as the mass $m$ in the linear motion equation. The result is equation (13.3), which tells us that angular momentum $L$ is the product of moment of inertia $I$ and angular velocity $\omega$. Moment of inertia is the rotational analog of mass.

Interactive equation

**(13.3)** $L = I\omega$

$L$ = angular momentum (kg m$^2$/s)
$I$ = moment of inertia (kg m$^2$)
$\omega$ = angular velocity (rad/s)

**Angular momentum** and moment of inertia

**Conservation of angular momentum**

Recall that Newton's first law says the linear momentum of an object remains constant when the net force is zero. A similar rule holds for angular momentum.

**The angular momentum of a rotating body remains constant when there is *zero net torque* acting on the body.**

**Ice skaters and platform divers**

Figure skaters and divers make use of angular momentum to perform feats of rotational agility. A skilled skater can increase her rate of spin to as much as five revolutions per second by pulling her arms close to her body. Platform divers tuck in their arms and legs while tumbling then open up their bodies to slow their rotation before hitting the water. Both techniques employ conservation of angular momentum.

**How does the ice skater spin faster?**

An ice skater sets up a fast spin by spinning slowly, with her arms and one leg extended. In the absence of any net torque, the skater's total angular momentum is conserved. To spin faster, the skater reduces her moment of inertia $I$ by pulling in her arms. To conserve $L$, her angular velocity $\omega$ increases!

**An ice skater spins faster because of angular momentum conservation.**

**Ice skater tucks in her arms and legs, which causes her to spin faster.**

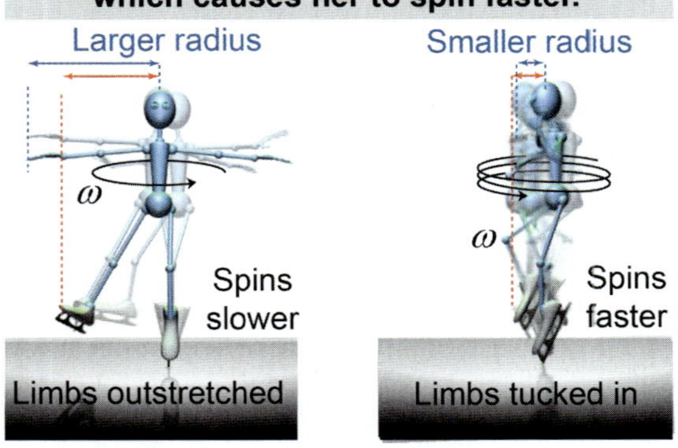


**How does it work?**

Skater's angular momentum $L = I\omega$ must remain constant for both cases.

| Skater's limbs are outstretched: | Skater's limbs are tucked in: |
|---|---|
| **Large $I$** | **Small $I$** |
| **Small $\omega$** | **Large $\omega$** |

# Investigation 13B Conservation of angular momentum

**Essential questions** *Why does conservation of angular momentum cause objects to spin faster or slower?*

Interactive simulation

When an ice skater executes a spin, she will usually start with her arms and one leg extended. As she spins, she tucks in her limbs, causing her to spin faster. Why does this happen? In this investigation you will explore two similar examples where rotating objects speed up as the radius of rotation decreases.

## Part 1: Spinning in an office chair

1. Sit in an office chair that swivels, holding a small barbell (1–3 kg) in each hand.
2. Spin yourself around with both arms outstretched.
3. Pull your arms into your body.

*a. How does your angular velocity change as you pull your arms in or extend them?*

*b. Why does this happen?*

## Part 2: Conical pendulum

*Safety warning: keep the pendulum bob away from people and objects that could be hit.*

1. Attach a 1 m string to a pendulum bob and thread through the hole in one end of a meter stick.
2. Attach the meter stick to a table with this hole extended away from the table.
3. Attach a small piece of tape to the string approximately 1/3 of the distance above the pendulum bob.
4. Mark a circle on the floor centered below the hole in the meter stick. Measure and record its radius.
5. Gently set the bob in circular motion as a conical pendulum with a radius matching the circle. Measure the time for ten revolutions using a stopwatch. Calculate the angular velocity $\omega_i$.
6. While the bob is still moving, quickly pull up on the string until the tape marker reaches the hole. Measure the new radius of the conical pendulum's motion and the time for ten revolutions. Calculate the angular velocity $\omega_f$.

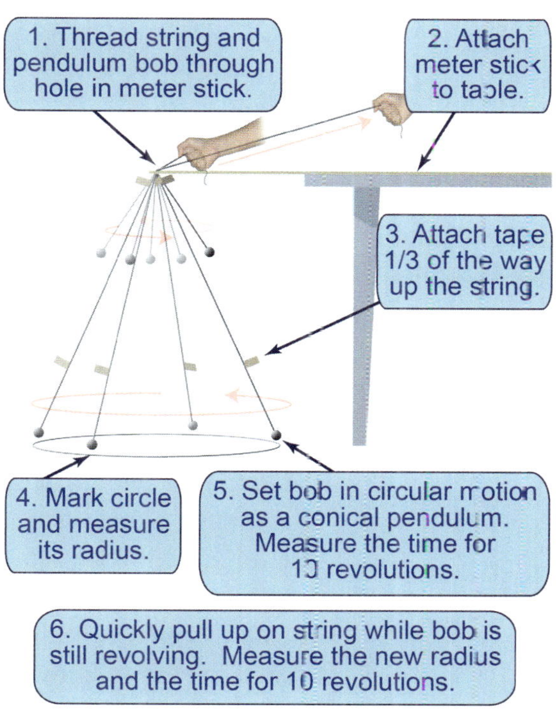


*a. How does the moment of inertia of the spinning bob change as you pull up on the string? Why?*

*b. How does the angular velocity change as you pull up on the string? Why?*

*c. Are your results (roughly) consistent with conservation of angular momentum? Explain.*

**Center of mass**

The **center of mass** is the average position of all the mass contained in an object (see page 137). That means that there is an equal quantity of mass on either side of any plane splitting an object through the center of mass regardless of which plane is chosen. The center of mass is sometimes called the "center of gravity" because it represents the average point at which the force of gravity acts on an object.

**Center of mass is the average position of the mass in an object.**

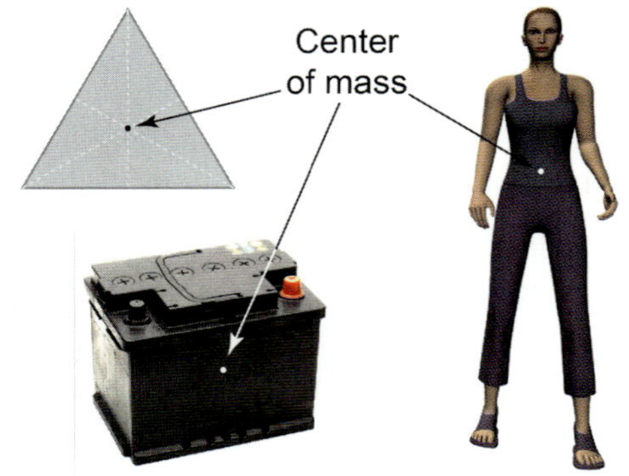


**Why is the center of mass important?**

When an object is rotating and translating at the same time we often separate the motions into two parts using the center of mass as a reference.

1. The linear motion is the translation of the center of mass. The linear momentum is the object's mass times the velocity of the center of mass.
2. The rotational motion is about the center of mass. The angular momentum is defined relative to an axis passing through the center of mass.

**Center of mass can be outside an object**

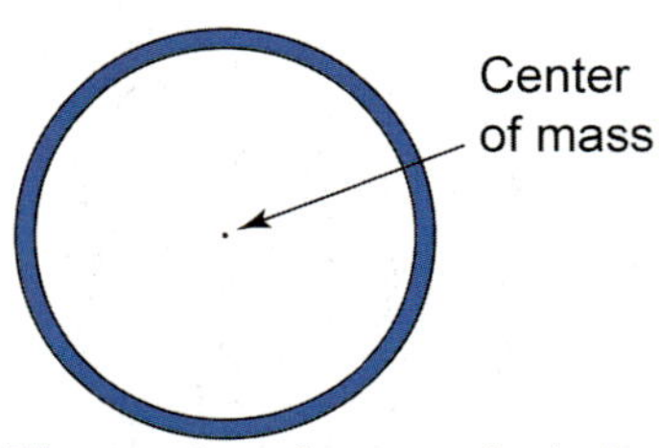


The center of mass of a hula hoop is located at its center.

For geometrical objects, such as a solid sphere or cube, the center of mass is located at the object's geometrical center. While some irregularly shaped objects may still have their center of mass located inside them, others can have the center of mass outside their body. Where is the center of mass of a hula hoop located? The answer is at its center, even though no mass is located there!

**Gravity acts on the center of mass**

The beautiful feature of the center of mass is that it allows us to simplify a complex situation. *The force of gravity can be assumed to act on the center of mass,* as if all the mass of the object were concentrated at that point.

**Stability and the center of mass**

The center of mass represents the *balance point*: If you placed the object on a pencil tip located directly below the center of mass, the forces should balance! Imagine a truck driving horizontally along a steep slope. Will it tip over? The force of gravity acts on the center of mass of the truck; gravity creates a torque about the truck's wheelbase. If the truck's center of mass is located vertically above or inside the wheelbase, the torque will act to keep the truck upright. But if the truck's center of mass is located beyond the wheelbase, the torque exerted by gravity will cause it to tip over. This is why the heaviest cargo should be loaded near the bottom of the truck!

**Which truck will tip over?**

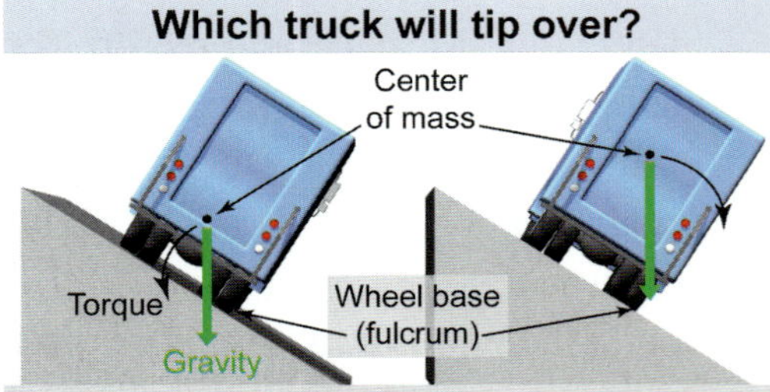


If the center of mass is outside the truck's wheel base, then the force of gravity will create a torque that tips over the truck!

# Investigation 13C Center of mass

**Essential questions** *How do we find the center of mass of an object?*

What is the physics behind the balance beam event in gymnastics? Gymnasts must keep their *center of gravity* directly above the balance beam—or else they fall off! Every object (including a gymnast) has a center of gravity, which is the average position for all the mass in an object. In this investigation, you will find the center of mass of an irregularly shaped piece of cardboard. Once you find it, can you balance the cardboard on the tip of a pencil?

## Part 1: Find the center of mass by hanging

1. Cut out an irregularly shaped piece of cardboard and use a pen or other sharp object to create three widely spaced holes near the edges of the cardboard.
2. Using string, hang the object using one hole, trace the line of the string, and use a ruler to extend the line across the object.
3. Repeat for the other two holes. Locate the center of mass at the intersection of the lines.

a. *Why does drawing a line below the hole pass through the center of mass?*

b. *Why is the intersection of the three lines the center of mass?*

c. *Try to balance the cardboard on the tip of a pencil. Where must you place the pencil?*

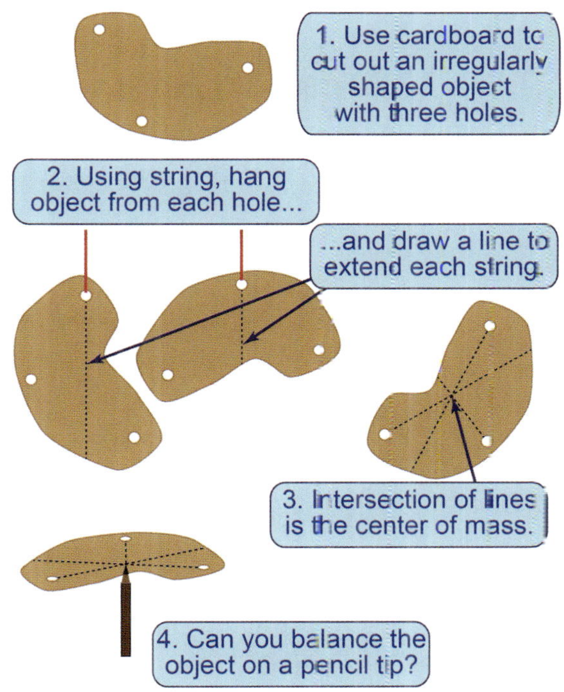


## Part 2: Find the center of mass using a plumb line

1. Insert a pencil through one hole to hang the object.
2. Using string, hang a plumb line from the pencil and trace the path of the string on the cardboard.
3. Repeat for the other two holes. Locate the center of mass at the intersection of the lines.

a. *Did you get the same center of mass as in Part 1?*

b. *What happens to a gymnast when her center of mass is not located directly above her feet? Use the equipment to demonstrate what happens.*

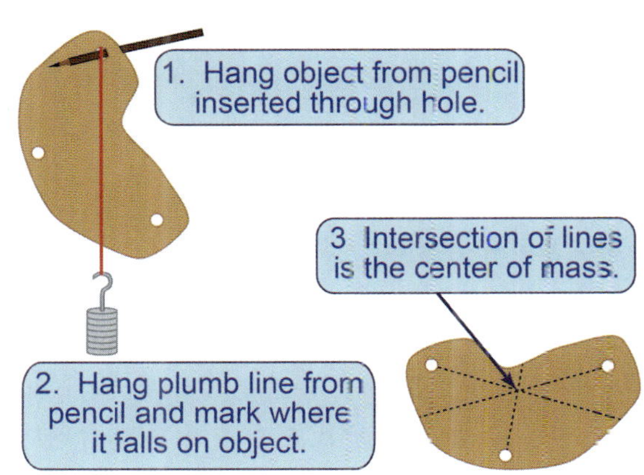

## Rotation and athletics

**Rotation about the center of mass**

Tumbling and spinning are key parts of many different sports. When athletes spin in the air or on ice, they rotate around their center of mass—which is typically located in the lower torso of a person when standing upright. Athletes may control their moment of inertia by tucking in their limbs; since angular momentum $L = I\omega$ is conserved, reducing the moment of inertia $I$ leads to an increase in the athlete's angular velocity $\omega$.

**Diving and tumbling**

An advanced diver may do several somersaults in the air. As he jumps off the diving board, he begins tumbling. By tucking his body in tightly, he will complete the somersaults more quickly—and he can do more of them before straightening out again and entering the water! Throughout the dive, his body rotates around his center of mass. The same technique is used in gymnastics, such as in performing multiple somersaults in midair in the vault event.

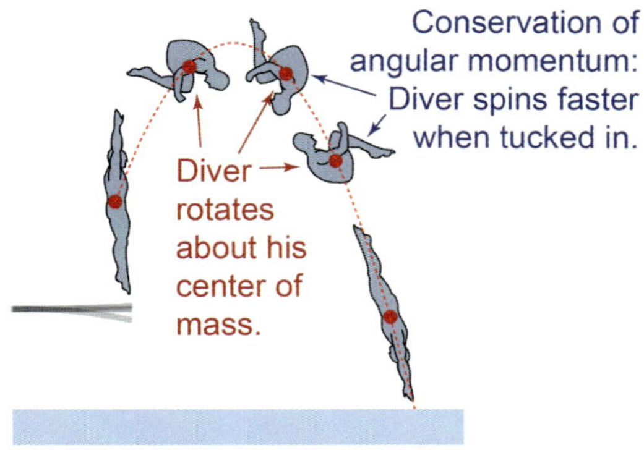


**High jumping**

In track and field, high jumpers rotate their bodies about their center of mass—and over the bar—using a technique called the "Fosbury flop," named after American Dick Fosbury, the 1968 Olympics gold medalist. When the jump is properly executed, a high jumper's center of mass actually passes *under* the bar, allowing the athlete to gain a little bit of extra height!

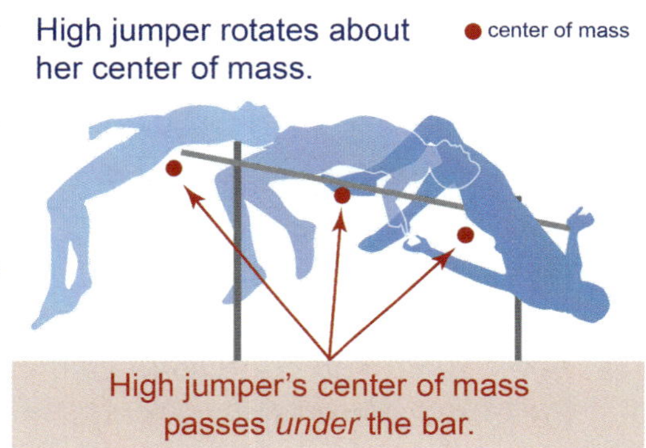


**Collisions and rotation**

Two figure skaters will "collide" with each other slightly offset—by an arm's length—so that they lock arms and spin together afterward. Why do they spin around instead of collapsing in a heap, as they would in a head-on collision? As they come together, each has an angular momentum relative to their *combined* center of mass. Both skaters have angular momenta spinning in the *same* direction around the center of mass, so their angular momenta add, not cancel. Since their combined angular momentum is conserved, after the "collision" they will continue to spin about their mutual center of mass.

**Angular momentum of two ice skaters**

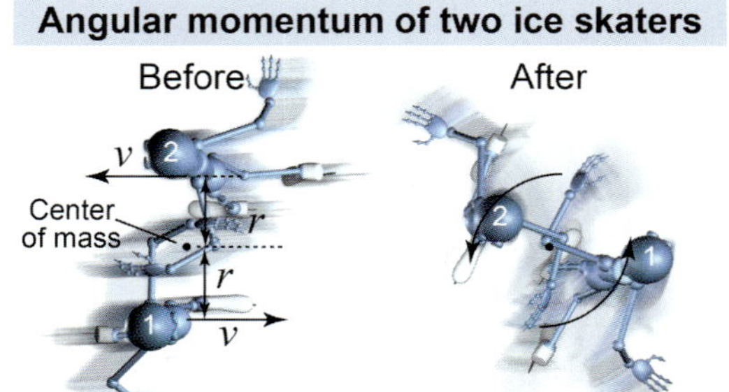


| Angular momentum is *positive*: $L_{tot} = L_1 + L_2 = mvr + mvr$ | Angular momentum is conserved: **Skaters spin together!** |
|---|---|

# Section 1 review

Motion can be translational, rotational, or both. Rotating objects move about an axis of rotation that can be located either inside or outside their body. Just as a mass has inertia that resists changes in its linear motion, the same mass has rotational inertia that resists changes in its rotational motion. Angular momentum is the rotational analog of linear momentum, is expressed by the letter $L$, and is conserved in the absence of any net external torque. Every object has a center of mass, which can be located either inside or outside its body. Many athletes take advantage of rotation about their center of mass, and conservation of angular momentum, in performing impressive feats.

| Vocabulary words | axis, rotation, revolution, translation, rotational inertia, moment of inertia, angular momentum, linear momentum, conservation of angular momentum, center of mass |
|---|---|

| Key equations | $I = mr^2$ | $L = r \times mv$ | $L = I\omega$ |
|---|---|---|---|

## Review problems and questions

1. When you throw a spinning plastic disk, such as a Frisbee®, why can it fly so far without tipping over sideways?

2. The *geographic center* of the contiguous 48 states in the USA is the two-dimensional mean (or average) location of the land mass. How could you use a piece of poster board to locate the geographic center of the USA?

3. A tall office worker doesn't like to lean over to reach things, so he decided to load all his files into the top drawer of his filing cabinet, leaving the other drawers empty. Is this a good or bad idea? Why?

4. Where is the center of mass of the letter "L" located?

5. Why does a tightrope walker's horizontal pole usually droop at the ends?

6. Judy is conducting an investigation by spinning washers (with a total mass of 100 g) attached to a string over her head. When the washers follow a path of 30 cm in radius, she spins them at one revolution per second.
   a. What is the angular velocity of the washers in radians per second?
   b. What is the linear velocity of the washers?
   c. What is the angular momentum of the washers?
   d. She then pulls on the string and shortens the radius at which the washers are rotating to 10 cm. What is the new angular momentum of the washers?
   e. What is the new angular velocity of the washers?

## 13.2 - Rotational dynamics

Newton's first law says that an object at rest remains at rest and an object in motion remains in motion with the same velocity when the net external force is zero. The *rotational analogy* to Newton's first law involves rotational inertia, angular velocity, and torque:

**When the net torque is zero, an object at rest remains at rest and a rotating object continues to rotate at the same angular velocity.**

**Which one is harder to twist?**

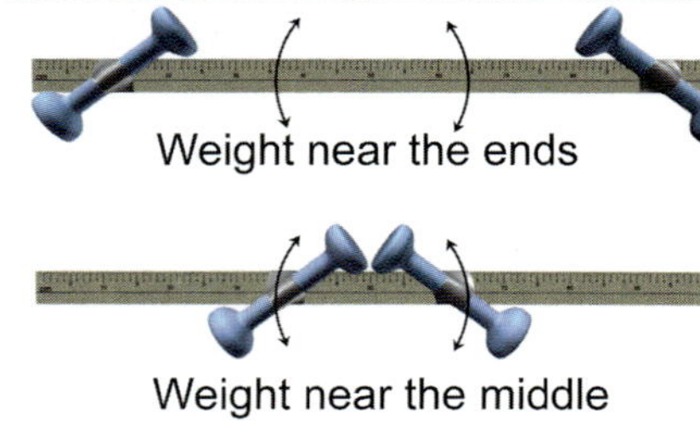


Attach two dumbbells to a meter rule. Is it harder to twist the weighted ruler if the dumbbells are located close to the center or at the ends? When the dumbbells are located near the ends, the weighted ruler has a larger *moment of inertia*—or more *rotational inertia*—which means that it is more difficult to twist. Moment of inertia is an important quantity in rotational dynamics. It helps us answers questions such as whether a sphere or a cylinder (with equal radii) will roll downhill faster.

### Moment of inertia

**Moment of inertia and rotational inertia**

In the previous section, we learned about the rotational inertia of a Ferris wheel. The *moment of inertia* of an object is a measurement of its resistance to changes in its rotational state. Moment of inertia $I$ is calculated as the product of mass times the square of the distance from the rotational axis, and the term is often used interchangeably with "rotational inertia." The moment of inertia depends strongly on how far the mass is located from the axis of rotation, because $I$ depends on $r^2$. If you move the mass twice the distance from the axis, the moment of inertia increases by a factor of 4! Since the equation for moment of inertia is $I = mr^2$, its units are kg m$^2$.

**What is the rotational inertia of an orbiting "point mass"?**

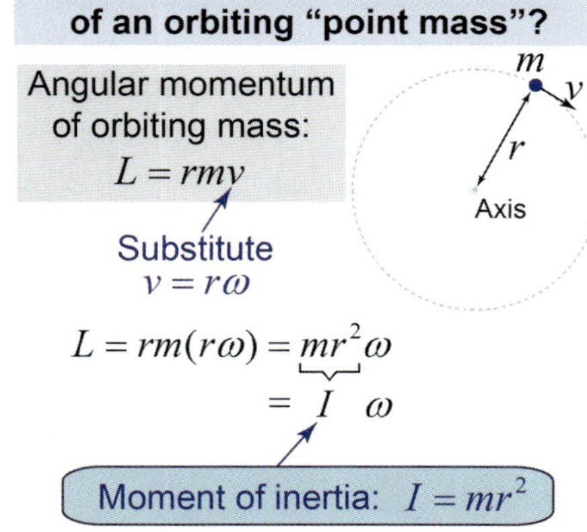


**Using the equation**

Think back to the example of the rotational inertia of a Ferris wheel on page 367. A large wheel has more rotational inertia, because moment of inertia $I$ is directly proportional to the square of the radius $r$ in equation (13.1). A massive wheel also has more rotational inertia, because $I$ is proportional to mass $m$ in equation (13.1).

**Moment of inertia of two masses combined**

What is the moment of inertia of two masses (in the figure at right) as they rotate about the center of a massless rod?

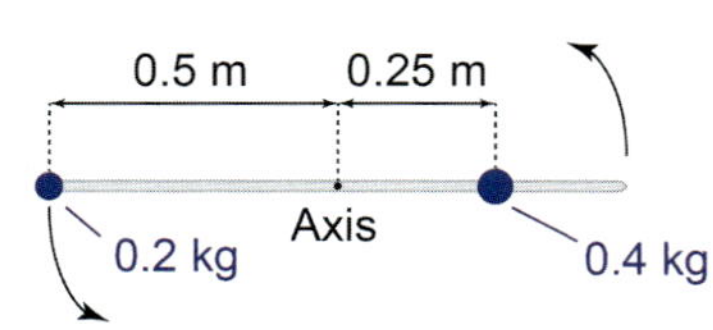


*Asked:* combined moment of inertia $I$

*Given:* masses $m_1 = 0.2$ kg and $m_2 = 0.4$ kg; radii $r_1 = 0.5$ m and $r_2 = 0.25$ m

*Relationships:* moment of inertia $I = mr^2$

*Solution:*

$$\begin{aligned} I &= I_1 + I_2 = m_1 r_1^2 + m_2 r_2^2 \\ &= (0.2\text{ kg})(0.5\text{ m})^2 + (0.4\text{ kg})(0.25\text{ m})^2 \\ &= 0.075\text{ kg m}^2 \end{aligned}$$

**Point masses and rotation**

In Newtonian mechanics you can often simplify a situation by replacing an object with a point mass located at its center of gravity. Now think about the rotation of a disk about its center axis: Can we replace the disk with a point mass at its center? No! The disk possesses a moment of inertia about the axis that *depends on how the mass of the disk is distributed in space*. In rigid-body rotation, we cannot replace objects with point masses because the shape of the object is important.

**Are the rotational properties of these two objects equivalent?**

Disk of mass $m$ rotating about its center

Point mass of mass $m$ rotating about its center

**Moment of inertia for distributions of mass**

Every object has a center of mass; likewise, every object has a moment of inertia when it is rotated about a particular axis. The illustration below shows the moment of inertia for several geometrical shapes. Look at the two cases of the rotation of a rigid rod: The moment of inertia is different depending on which axis you use! The moment of inertia is an important quantity to understand when designing a device that rotates!

**Moments of inertia for various geometrical shapes**

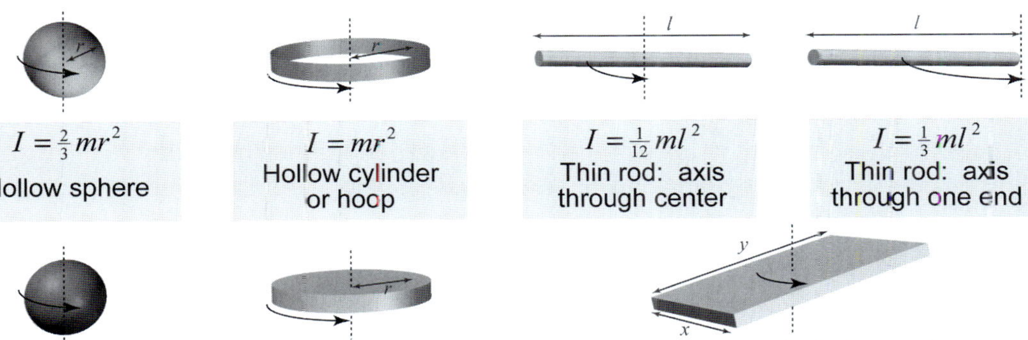


| $I = \frac{2}{3}mr^2$ Hollow sphere | $I = mr^2$ Hollow cylinder or hoop | $I = \frac{1}{12}ml^2$ Thin rod: axis through center | $I = \frac{1}{3}ml^2$ Thin rod: axis through one end |
|---|---|---|---|
| $I = \frac{2}{5}mr^2$ Solid sphere | $I = \frac{1}{2}mr^2$ Solid cylinder or disk | $I = \frac{1}{12}m(x^2 + y^2)$ Thin rectangular sheet | |

**Angular momentum and moment of inertia**

The moment of inertia is a convenient quantity for describing the rotational motion of rigid bodies. As we saw on the previous page, angular momentum can also be expressed as the product of moment of inertia and angular velocity, or $L = I\omega$.

**Comparing moments of inertia**

A hollow and a solid sphere, each with a radius of 10 cm and mass of 0.3 kg, are rotating at 2 rad/s about an axis passing through their centers. Which has more angular momentum?

*Asked:* angular momenta of a hollow sphere, $L_h$, and a solid sphere, $L_s$

*Given:* radius $r = 0.1$ m, mass $m = 0.3$ kg, angular velocity $\omega = 2$ rad/s

*Relationships:* angular momentum $L = I\omega$; moments of inertia of hollow sphere, $I_h = \frac{2}{3}mr^2$, and solid sphere, $I_s = \frac{2}{5}mr^2$

*Solution:*

$$L_h = I_h\omega = \frac{2}{3}mr^2\omega = \frac{2}{3}(0.3\text{ kg})(0.1\text{ m})^2(2\text{ rad/s}) = 0.004\text{ kg m}^2/\text{s}$$

$$L_s = I_s\omega = \frac{2}{5}mr^2\omega = \frac{2}{5}(0.3\text{ kg})(0.1\text{ m})^2(2\text{ rad/s}) = 0.0024\text{ kg m}^2/\text{s}$$

*Answer:* The hollow sphere has a larger angular momentum.

## Rolling motion and rotational energy

### Rolling

When a ball rolls across the ground *without slipping*, the faster it moves across the ground, the faster it rotates about its center. The linear velocity $v$ and angular velocity $\omega$ of a rolling ball are directly related to each other by the equation $v = \omega r$, as we learned on page 209. Every time the ball rotates one complete turn (or $2\pi$ radians), it has moved forward by a distance of $2\pi r$ (the ball's circumference).

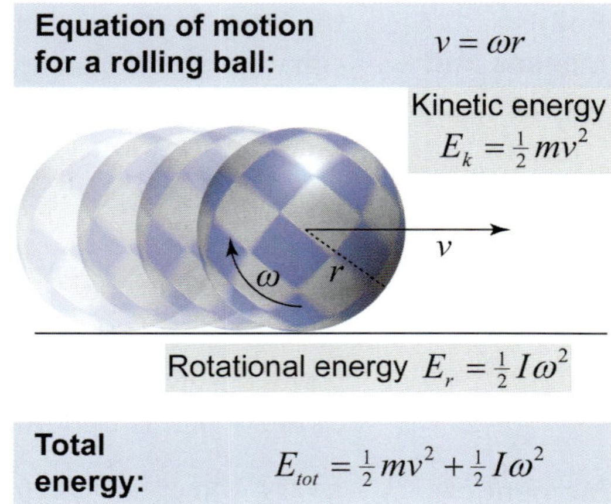


### What is the energy of a rolling ball?

When the ball is rolling, it has both kinetic energy and **rotational energy**, which is sometimes called the kinetic energy of rotation. Linear kinetic energy is given by $\frac{1}{2}mv^2$, while rotational (kinetic) energy is given by $\frac{1}{2}I\omega^2$. The equations look similar: You just substitute $I$ for $m$ and $\omega$ for $v$!

**Rotational energy**

**(13.4)** $$E_r = \frac{1}{2}I\omega^2$$

$E_r$ = rotational energy (J)
$I$ = moment of inertia (kg m$^2$)
$\omega$ = angular velocity (rad/s)

### Energy of rolling motion

When a ball is rolling across the ground, it has both kinetic energy and rotational energy. Its total energy is the sum of the two:

$$E_{tot} = E_k + E_r = \frac{1}{2}mv^2 + \frac{1}{2}I\omega^2$$

How does the energy of a rolling ball compare with the energy of a *skidding* ball—i.e., a ball that is *sliding* across a frozen lake—moving at the same velocity? When the ball is sliding, it has no rotational energy, so its total energy is less than the rolling ball that moves at the same velocity.

### Comparing translational and rotational motion

Linear motion and rotational motion may appear different, but the mathematical representations for both are very similar. Whereas linear motion uses distances and velocities, rotational motion uses angles and angular velocities. Mass is used in linear motion but moment of inertia is used in rotational motion.

**Comparison between linear and rotational motion**

| Linear motion | | | Rotational motion | | |
|---|---|---|---|---|---|
| **Quantity** | **Variable or equation** | **Units** | **Quantity** | **Variable or equation** | **Units** |
| Position | $x$ | m | Position angle | $\theta$ | rad |
| Velocity | $v = \Delta x/\Delta t$ | m/s | Angular velocity | $\omega = \Delta\theta/\Delta t$ | rad/s |
| Mass | $m$ | kg | Moment of inertia | $I$ | kg m$^2$ |
| Momentum | $p = mv$ | kg m/s | Angular momentum | $L = I\omega$ | kg m$^2$/s |
| Kinetic energy | $E_k = \frac{1}{2}mv^2$ | J | Rotational energy | $E_r = \frac{1}{2}I\omega^2$ | J |

# Investigation 13D Rolling down an inclined plane

**Essential questions**

*What is moment of inertia (or rotational inertia)?*
*What physical quantities determine the rotational inertia of an object?*

Galileo demonstrated the principle that objects are all accelerated at the same rate by gravity, regardless of their mass. But do all objects *roll* downhill with identical acceleration? In this investigation, you will roll a number of different objects down a ramp and determine which one—if any!—is accelerated the most. Each of the objects has a *different distribution of mass* within it; some are hollow whereas others are solid, for example. Does the distribution of mass within an object change its rotational properties?

You will need the following materials:

- a stiff board (wood, foam, etc.) and a support to make a ramp;
- a meter rule, to use as the starting gate;
- different shapes of rolling objects, such as a solid metal can, which has relatively solid contents (not soup), a hollow metal can, a solid ball, and a hollow ball; and
- additional examples of the same geometrical shapes, but having different masses and radii.

## Which object rolls downhill the fastest?

1. Prop up a stiff board with a support to form a ramp with an inclination angle of 15°–30°.
2. Line up the meter rule across the board to use it as a starting gate, and place two (or more) objects behind it.
3. Lift up the meter rule to start the objects rolling and note which one reaches the bottom first.
4. Repeat for all four geometrical shapes: a solid sphere, a hollow sphere (ball), a solid cylinder, and a hollow cylinder (ring).

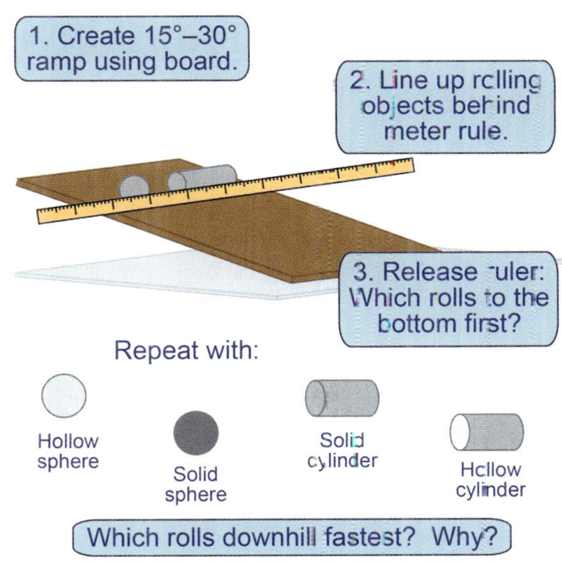


a. *Which of the four geometrical shapes reaches the bottom first? Last? List them in order from fastest to slowest.*
b. *What is the equation for the moment of inertia of each geometrical shape?*
c. *Explain why you think that the objects reached the bottom in that order.*
d. *For two objects of the same shape but different radii, will they reach the bottom at the same time? Design and carry out an experiment to test your prediction.*
e. *For two objects of the same shape but different masses, will they reach the bottom at the same time? Design and carry out an experiment to test your prediction.*

## Rolling downhill

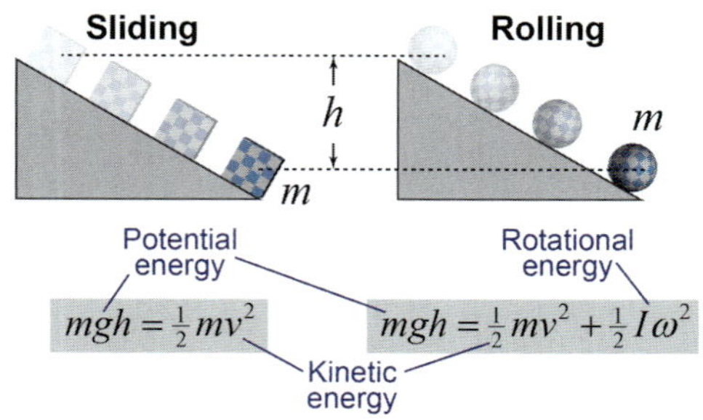


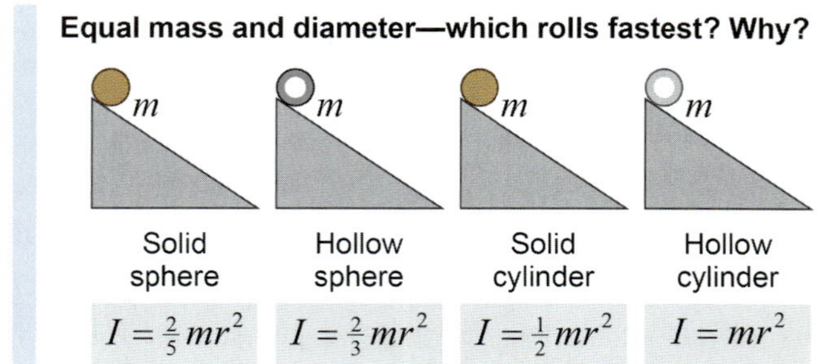


### Conservation of energy rolling downhill

How fast is a rolling ball moving when it reaches the bottom of a hill? Just as in the case of a block sliding down a frictionless hill, the energy of a rolling ball is conserved. We can use energy conservation to determine what kind of rolling objects will reach the bottom first!

### Which has a higher velocity?

A ball and block at the same height start with the same gravitational potential energy: $E_p = mgh$. At the bottom of the hill their final energies will also be equal. Whereas the sliding block has only kinetic energy, *the rolling ball has both linear kinetic energy* ($\frac{1}{2}mv^2$) *and rotational kinetic energy* ($\frac{1}{2}I\omega^2$). The rolling ball has a lower velocity because it has *less* linear kinetic energy than the sliding block. The difference is the rolling ball's rotational kinetic energy.

### Moment of inertia and rolling

In the picture above, four different objects with the same mass and diameter roll down identical ramps. Which one reaches the bottom first? To answer the question, consider your results from the investigation in which you rolled differently shaped objects down a hill. What did you find?

### Mathematics behind rolling down an inclined plane

If you start with the conservation of energy equation, substitute for the angular velocity using $\omega = v/r$, and rearrange the terms, you get

$$mgh = \frac{1}{2}mv^2 + \frac{1}{2}I\omega^2 = \frac{1}{2}mv^2 + \frac{1}{2}I\left(\frac{v}{r}\right)^2 = \frac{1}{2}v^2\left(m + \frac{I}{r^2}\right)$$

Simplifying the expression further gives

$$mgh = \frac{1}{2}v^2\left(m + \frac{mI}{mr^2}\right) = \frac{1}{2}mv^2\left(1 + \frac{I}{mr^2}\right)$$

From this equation you can solve for the velocity of the ball or cylinder when it reaches the bottom of the inclined plane:

$$v = \sqrt{\frac{2gh}{1 + \left(I/mr^2\right)}}$$

As the moment of inertia $I$ increases, the velocity of a rolling object will decrease. This makes sense because a higher moment of inertia means more energy is in rotational kinetic energy—leaving less for linear kinetic energy. The solid sphere reaches the bottom first because it has the smallest moment of inertia and therefore the highest translational kinetic energy!

**What causes the tides?**

The tides are a twice daily rise and fall of the ocean level. What causes the tides in the Earth's oceans? The complete answer requires understanding the *center of mass* of the Earth–Moon system.

**Center of mass of the Earth–Moon system**

Even though the Earth is much more massive than the Moon, the Moon does not orbit around the center of the Earth. Instead, the Earth and the Moon both *orbit around their combined center of mass*. The system's center of mass is located within the Earth itself, approximately 1,600 km below the surface of the Earth. The Earth travels in a nearly circular orbit around the system's center of mass, as does the Moon—although the Moon's orbit is much larger! (Note that the figure is not shown to scale; the Moon is located much further away.)

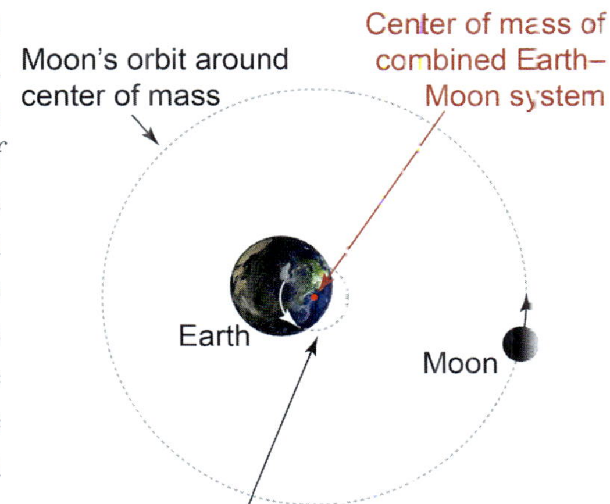


Earth orbits around the Earth–Moon system's center of mass

**Gravity of the Moon causes one high tide**

Consider the side of the Earth facing the Moon. The Moon's gravity exerts an attractive force on the water in the ocean. Unlike solid land, liquid water can flow in response to the Moon's feeble gravity. As a result, the ocean surface rises a few meters in the "Moonward" direction. This causes one of the high tides on the Earth. But there are *two* tides per day; the second tide is trickier to explain.

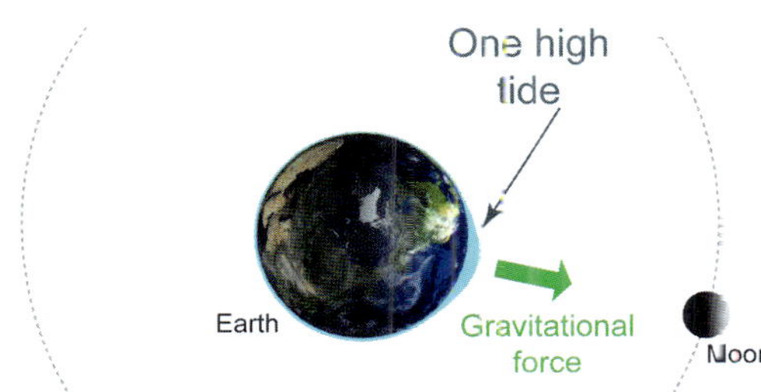


The Moon's gravitational pull creates a bulge in the Earth's oceans in the direction toward the Moon.

**Centrifugal "force" causes the other high tide**

The second tide is due mainly to Earth's 28 day orbit around the Earth–Moon system's center of mass. This 28 day rotation creates a "centrifugal" effect that tends to move ocean water radially outward and away from the Moon. The effect is similar to what happens if you swing a bucket of water around your head in a circle fast enough that the water stays in the bucket. This centrifugal effect is the cause of the second high tide—the tide that faces away from the Moon.

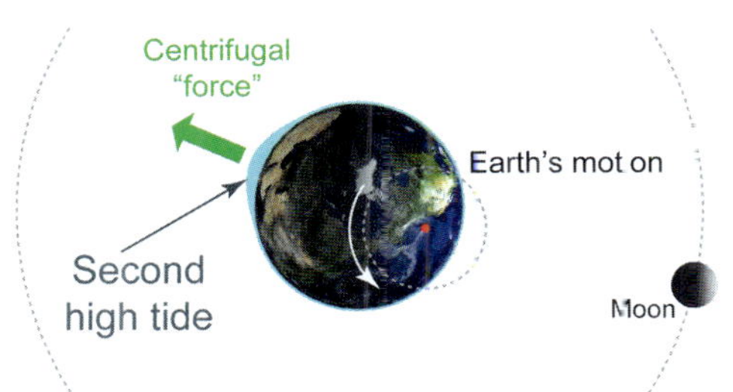


Earth's motion around the system's center of mass causes a centrifugal "force" acting on the Earth's oceans *opposite* the Moon.

**Earth's rotation and the tides**

A more accurate picture is that the ocean surface is distorted into a slightly oval shape while the Earth rotates under the oval's surface. Both tides lag slightly behind the passing of the Moon overhead due to friction and inertia of huge amounts of water sloshing back and forth between the continents. High tide occurs when a point on Earth's surface passes through a bulge of the "oval." Low tide occurs when that point passes in between the two bulges. Tides are highest near the equator and lowest near the poles.

## The seasons and precession

### What causes the seasons?

It is a common misconception that the seasons are caused by Earth being closer or farther from the Sun at different places along the planet's "elliptical" orbit. This is *not* the cause of the seasons. Mathematically, a *circle* is a type of ellipse. Earth's orbit is so nearly circular that the small variation in distance from the Sun does not explain the large temperature changes that occur with the seasons. Nor does ellipticity explain why summer in the northern hemisphere occurs at the same time as winter in the southern hemisphere. This could not be true if summer were caused by Earth being closer to the Sun.

### Tilt of the Earth's axis

The seasons are caused by the orientation of Earth's rotational axis relative to the plane of its orbit around the Sun. The Earth rotates about its axis once every 24 hours, and it orbits around the Sun once every year. Earth's rotational axis, however, is not perpendicular to its *orbital plane*. The Earth's *rotational* axis is tilted by 23.4° relative to the *orbital axis* about which it revolves around the Sun.

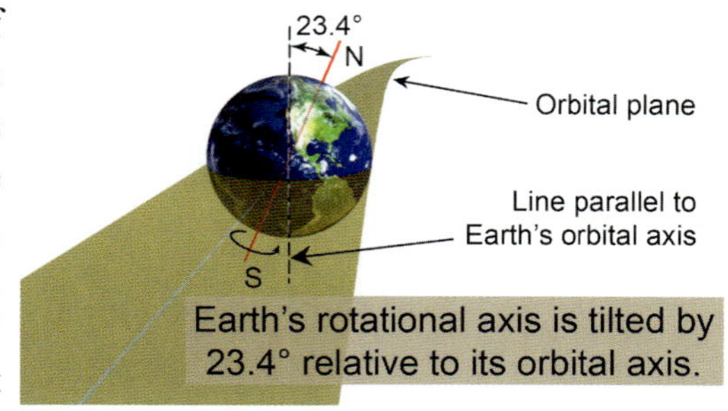

Earth's rotational axis is tilted by 23.4° relative to its orbital axis.

### Seasons are caused by the tilt of the Earth's axis

Because of the tilt of the Earth's axis, in January the North Pole tilts 23.4° away from the Sun. This reduces the intensity of sunlight creating winter in the northern hemisphere. Six months later, in June, the North Pole tilts toward the Sun. This increases the intensity of sunlight, making summer. The opposite is true in the southern hemisphere.

The tilt of Earth's axis also causes polar regions of our planet to have extremes of day and night near the summer and winter solstices. Near the North Pole at the summer solstice, the Sun never sets! This creates 24 hours of daylight. At the same time but near the South Pole, the Sun never rises, creating 24 hours of night.

**Seasons are caused by the tilt of the Earth's rotational axis.**

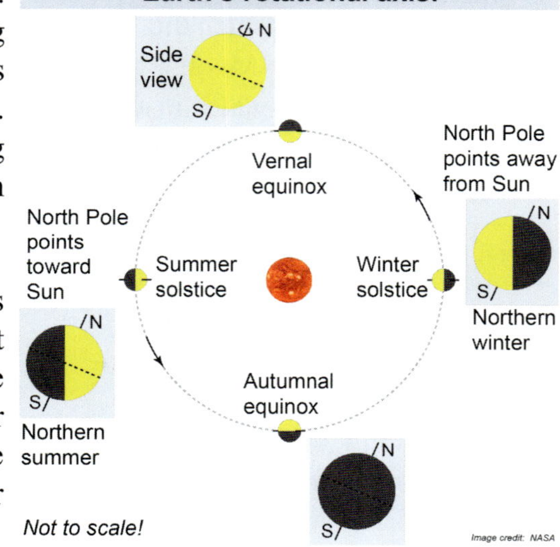


### Precession of the Earth's axis

A strange thing happens to the Earth's axis of rotation: It rotates slowly around the Earth's *orbital axis,* tracing a 46.8° wide circle on opposite sides of the sky. It takes a full 26,000 years for a complete cycle of this **precession** of the Earth's axis.

### Precession of a gyroscope

You don't have to wait 26,000 years to see precession at work. Just use a gyroscope, which you can buy for less than $10! Tilt the gyroscope axis with respect to the vertical—or even point it horizontally. Start the rotor spinning and place one end on its stand. The gyroscope's rotational axis will precess about the vertical. This precession occurs as a result of the interaction between the angular momentum of the gyroscope and the torque exerted by the gravitational force.

Set the rotor spinning while gyroscope points horizontally.

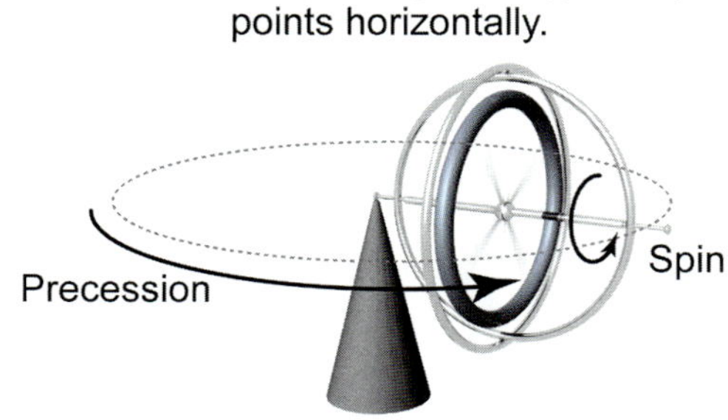

Gyroscope's axis will precess!

# Section 2 review

Rotational inertia is at the heart of the rotation of rigid bodies. The moment of inertia of an object represents its rotational inertia about a particular axis. Moment of inertia depends not just on mass but also on the *distribution* of that mass. Even if a solid sphere and a hollow sphere have the same mass, they have different moments of inertia because their mass is distributed differently. Rolling objects possess not just kinetic energy but also rotational energy (sometimes called the kinetic energy of rotation). When rolling downhill, objects will be accelerated differently, depending on their moment of inertia.

| | |
|---|---|
| **Vocabulary words** | **rotational energy, precession** |

| | |
|---|---|
| **Key equations** | $E_r = \frac{1}{2} I\omega^2$ |

## Review problems and questions

1. In bowling, usually the bowling ball will initially slide down the lane without spinning. Partway down the lane, however, the ball begins to roll. What happens to the velocity of the ball when it begins to roll?

2. What is the moment of inertia for each of the following? (Mass of the Earth: $6.0\times10^{24}$ kg. Mass of the Moon: $7.3\times10^{22}$ kg. Radius of the Earth: $6.4\times10^{6}$ m. Radius of the Moon: $1.7\times10^{6}$ m. Radius of the Earth's orbit around the Sun: $1.5\times10^{24}$ m. Radius of the Moon's orbit around the Earth: $3.9\times10^{8}$ m.)
   a. Moon's orbit around the Earth
   b. Earth's orbit around the Sun
   c. rotation of the Moon about its axis
   d. rotation of the Earth about its axis

3. A beginning bowler pushed a 6 kg bowling ball so that it rolled down the lane without slipping at 2 m/s. How much work did the bowler do on the ball to start it rolling?

4. Will a hoop or a solid disk be accelerated more when rolling down the same inclined plane? (Assume that the two objects have the same radius.)

5. You are considering two different designs for the wheel that forms part of a wheel-and-axle simple machine. Although both designs have the same mass, one puts most of the mass near the rim while the other distributes the mass evenly throughout the wheel. What are the advantages or disadvantages of each design if the input force will be applied to the axle?

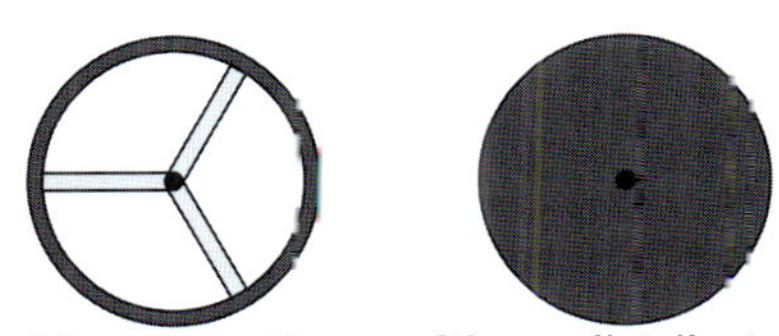

Chapter 13

# Chapter review

## Vocabulary

Match each word to the sentence where it best fits.

### Section 13.1

| | |
|---|---|
| **angular momentum** | **axis** |
| **center of mass** | **conservation of angular momentum** |
| **linear momentum** | **moment of inertia** |
| **revolution** | **rotation** |
| **rotational inertia** | **translation** |

1. The product of mass and velocity is also called _______.
2. The motion of a spinning wheel is called _______.
3. It is important to have your _______ located directly over your feet or you might fall over!
4. It is hard to stop a spinning carousel because it has a lot of _______.
5. The central line about which a gyroscope rotates is called its _______ of rotation.
6. When a spaceship moves from here to there the motion is called _______.
7. Any object that is spinning or rotating has _______.
8. _______ holds in the absence of net, external torques.
9. The rotational analog to mass is _______.

### Section 13.2

| | |
|---|---|
| **precession** | **rotational energy** |

10. A ball *rolling* downhill possesses _______ while a ball *sliding* downhill does not.
11. A horizontally spinning gyroscope slowly spins around a vertical axis in a phenomenon called _______.

## Conceptual questions

### Section 13.1

12. How does rotational inertia differ from linear inertia?
13. Compare and contrast *rotation* and *revolution* from the point of view of angular motion. Then give an astronomical example of each.
14. Two objects have the same mass. Must they have the same rotational inertia?
15. For each of the following, determine whether the motion is rotational or translational. If rotational motion is included, then identify whether it is rotation, revolution, or both.
    a. a car wheel when the car is being driven on the highway
    b. an electric motor lifting a projector screen
    c. a falling rock
    d. a gyroscope on a table
    e. a car wheel when the driver slams on the brakes and the car skids as a result
    f. an orbiting satellite
    g. the Earth in its usual motion
16. Based on the units of each of the following quantities, determine whether each is a measurement of linear or angular momentum.
    a. 10 kg m/s
    b. 13 kg $m^2$/s
    c. (15 kg/s)(10 m)
    d. (13 kg m)(11 m/s)
    e. (11 $m^2$/s)(5 kg/m)
17. Write a sentence explaining the relationship between rotational inertia and the moment of inertia. Then write a similar sentence for inertia and mass.
18. A spinning object is modified so that it doubles its moment of inertia but its angular velocity does not change. What happens to its angular momentum?
19. If a gyroscope were to fall over, would its angular momentum change?

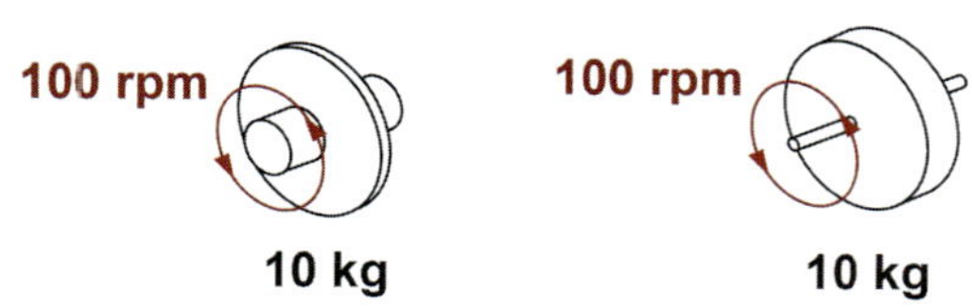


20. The two shapes shown have the same mass, diameter, and rotational speed. Which is true and why?
    a. Both have the same angular inertia because they have the same mass and speed.
    b. The thin wheel has more angular inertia because its mass is concentrated near the axis of rotation.
    c. The thick wheel has more angular inertia because its mass is distributed farther from the axis of rotation.
21. For each of the following objects undergoing rotational motion, describe where the axis of rotation is.
    a. a planet rotating
    b. a planet revolving
    c. a pencil stood on end and falling over
    d. a spinning top
    e. a satellite orbiting above the equator

# Chapter review



### Section 13.1

22. If you double an object's rotational speed, what happens to its angular momentum?

23. A small, dense object is spun around on a string at 10 m/s. If you triple the length of the string (the radius of the object's "orbit") *without* changing the speed, what happens to the object's angular momentum?

24. A ferris wheel takes a long time to get up to speed. If you double its radius without changing its mass or engine, will it take more or less time to get up to speed? Explain.

25. Which is easier to balance, a carrot with its thick end up or a carrot with its thick end down? Explain.

26. ❨ The radius $r$ in the equation

$$L = r \times mv$$

refers to

a. the minimum radius of the object.
b. the maximum radius of the object.
c. the distance between mass element $m$ and the axis of rotation.
d. the distance from the axis of rotation to the outermost part of the object.

27. An object orbits the Earth at a radius beyond the Earth's atmosphere.

a. What force keeps the object from flying off into space?
b. Does that force exert any torque?
c. Is the object's angular momentum conserved?

28. A figure skater jumps into the air and contracts her arms so that her spinning speeds up in midair.

a. Does her angular velocity change?
b. Does her moment of inertia change?
c. Does her angular momentum change?
d. Does she exert any torque?

29. ❨ When you pick up a heavy object, why do you usually lean your torso backward?

### Section 13.2

30. Why would toymakers build children's tops with relatively large moments of inertia, even though a large moment of inertia makes it harder to start something spinning?

31. Jeremiah has a thin metal hoop and a solid metal disk, each 40 cm in radius. Suppose that he conducts an experiment and finds that their moments of inertia are equal. Which object is heavier?

32. ❨ Suppose that Eleftheria has two metal disks, each with a radius of 40 cm and a mass of 400 g. An experiment reveals that Disk A has a greater moment of inertia than Disk B. What does this prove?

33. A spinning gyroscope wheel completes 10 full rotations per second. Suppose that the rate is increased to 20 rotations per second. By what factor does that multiply the wheel's rotational energy?

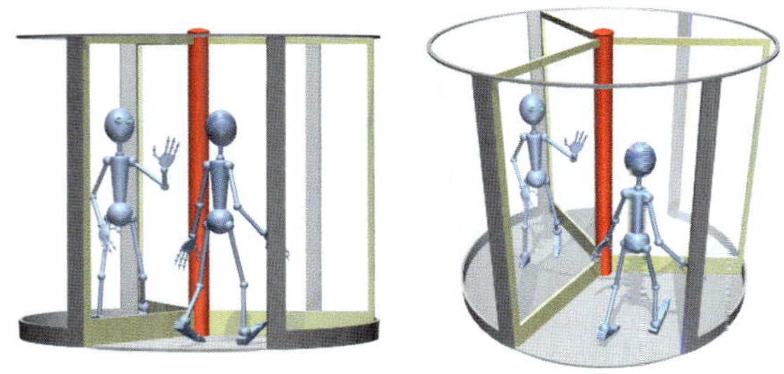

34. A revolving door generally consists of three glass panels 120° apart, all attached to a central axis and revolving within a partial cylinder. Visitors must step into the space between two panels and push on one in order to pass through the door. State one possible advantage and one possible disadvantage to building a revolving door with a large moment of inertia.

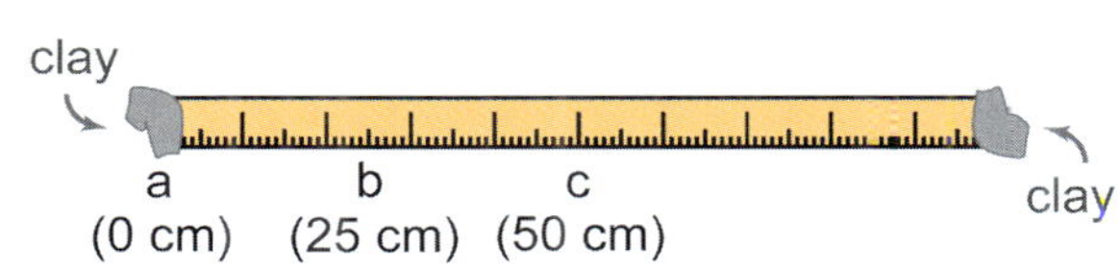


35. A student has attached two equally massive lumps of clay to opposite ends of a lightweight but stiff meter stick. She plans to twirl the stick end over end while holding it at each of the three labeled positions. Which position will be easiest to twirl? Hardest? (Assume that the mass of the meter stick itself can be ignored.)

36. ❨ Suppose a ball is placed at the top of a ramp and released from rest. It rolls to the bottom in 2 s. Next, the ramp is lubricated and the ball is released from the top once more. This time, it slides to the bottom without spinning. Will it take less than 2 s, more than 2 s, or 2 s to reach the bottom this time?

Chapter 13

# Chapter review

## Quantitative problems

### Section 13.1

37. An object has a moment of inertia about an axis of rotation of $I = 30$ kg m$^2$. What is its angular momentum $L$ if it rotates with an angular velocity of $\omega = 2$ s$^{-1}$?

38. Lizzie is spinning on a chair so that she has a moment of inertia of 10 kg m$^2$. Her angular momentum is 5 kg m$^2$/s.
    a. What is her angular velocity?
    b. If she withdraws her arms, making her moment of inertia 5 kg m$^2$, what is her new angular velocity?

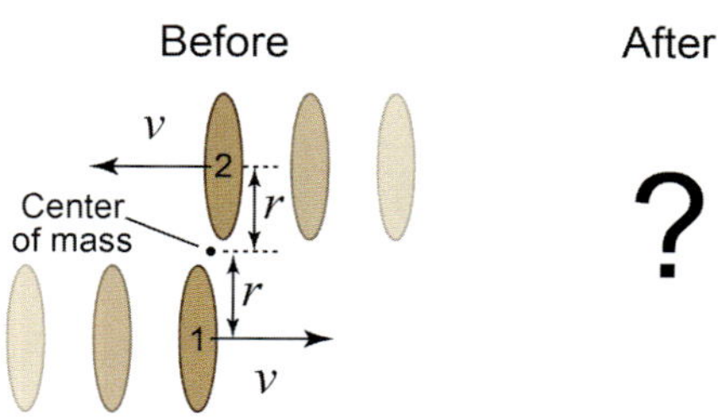


39. Two figure skaters of equal mass move toward each other, as shown here in this view from above. As they pass one another they join hands. Let $v$ be the *speed* with which each moves upon the ice, and $r$ the closest distance between each skater's head and their shared center of mass.
    a. What is the total *linear* momentum of the pair before the skaters join hands? After?
    b. What is the total *angular* momentum of the pair before the skaters join hands? After?
    c. Use your answers to the previous two questions to *qualitatively* describe how the skaters would appear to move, as seen by a spectator.

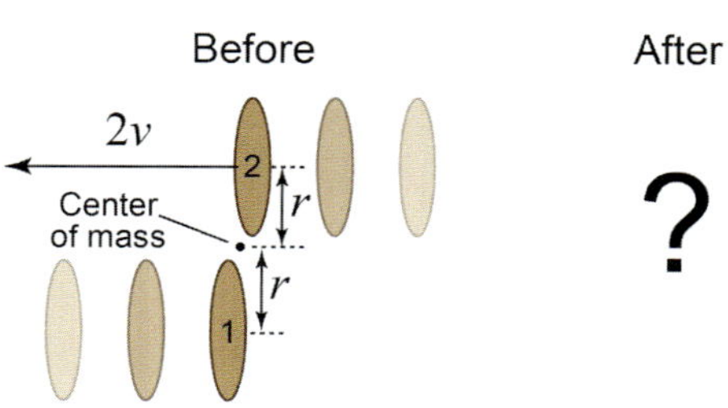


40. Two equally massive skaters face each other on the ice. Skater 1 is motionless and Skater 2 approaches with a *speed* of $2v$. The skaters join hands when Skater 2 reaches Skater 1.
    a. What is the skaters' total linear momentum before joining hands? After?
    b. What is the skaters' total angular momentum before joining hands? After?
    c. Use your answers to the previous two questions to *qualitatively* describe how the skaters would appear to move from a spectator's point of view.

41. A 1,000-kg communications satellite orbits the Earth at an altitude of 500 km, completing one revolution every 90 minutes. What is its angular momentum? (The Earth's average radius is approximately 6,400 km.)

### Section 13.2

42. A solid, 50 kg cylinder with a radius of 0.25 m is rotating with an angular velocity of 60 s$^{-1}$.
    a. How much rotational energy is stored in this rotating cylinder?
    b. If that energy could be entirely used to lift the cylinder, how high would it go?

43. An engineer wishes to modify an existing flywheel design so that the device can store more rotational energy. Which of the following changes will increase rotational energy by the largest factor? (The original design calls for a solid, 50 kg cylinder with a radius of 0.5 m, rotating with an angular velocity of 60 s$^{-1}$.)
    a. increasing the radius to 0.75 m
    b. increasing the mass to 100 kg
    c. replacing the solid cylinder with a hollow cylinder *and* reducing the mass to 25 kg
    d. doubling the angular velocity

44. For which of the following astronomical motions is the amount of rotational energy the greatest: the Earth's rotation about its axis or the Moon's revolution about the Earth?
The mass of the Earth and Moon are $6.0 \times 10^{24}$ kg and $7.3 \times 10^{22}$ kg, respectively. The radius of the Earth is $6.4 \times 10^6$ m. The Moon's orbital radius is $3.9 \times 10^8$ m. (Assume that the Moon completes one revolution every 28 Earth days and that the Earth is a solid sphere of uniform density.)

45. Suppose that a sphere or cylinder is released from rest at the top of a ramp. At the top, it has a gravitational potential energy of *mgh*. When it reaches the bottom, this energy will be converted entirely to kinetic energy and rotational energy (assuming friction and air resistance are negligible). For each of the following shapes, what fraction of the original potential energy becomes rotational energy?
    a. solid sphere
    b. hollow sphere
    c. solid cylinder
    d. hollow cylinder

# Chapter review

## Standardized test practice

46. Which pair of mass and length gives the lowest moment of inertia for a uniform metal bar spun around its end?

   A. 5 kg, 5 cm
   B. 10 kg, 5 cm
   C. 5 kg, 10 cm
   D. 10 kg, 10 cm

47. Which statement best describes the Earth's motions?

   A. The Earth rotates about its axis and the Sun.
   B. The Earth revolves about its axis and the Sun.
   C. The Earth revolves about its axis while rotating around the Sun.
   D. The Earth rotates about its axis while revolving around the Sun.

48. The formula for the angular momentum of a point mass is $L = mvr$. Which of the following quantities can be the SI unit for angular momentum?

   A. kg m/s
   B. kg $m^2$/s
   C. kg m/$s^2$
   D. kg $m^2$/$s^2$

49. A quarter-kilogram (0.25 kg) tetherball is attached to a pole by a 2 m length of rope. If the tetherball circles the pole once in 0.2 s (and the rope is horizontal), how fast does it travel?

   A. 10 m/s
   B. 20 m/s
   C. 31 m/s
   D. 63 m/s

50. A quarter-kilogram (0.25 kg) tetherball is attached to a pole by a 2 m length of rope. If the tetherball circles the pole once in 0.2 s (and the rope is horizontal), what is its angular momentum?

   A. 25 kg $m^2$/s
   B. 31 kg $m^2$/s
   C. 63 kg $m^2$/s
   D. 78 kg $m^2$/s

51. An ice skater sets herself spinning in place with her arms outstretched. She then pulls her arms in close to her body. Which of the following can be explained by the law of conservation of angular momentum?

   A. She appears smaller.
   B. She spins faster.
   C. She eventually stops spinning.
   D. None of the above.

52. An object's center of mass can be outside of the object itself. Which of the following objects does not contain its own center of mass?

   A. calculator
   B. eraser
   C. pencil
   D. paper clip

53. Wheel 1 has a larger moment of inertia than Wheel 2. Which statement best explains this fact?

   A. Wheel 1 is harder to start spinning.
   B. Wheel 1 is harder to stop spinning.
   C. Both a and b are correct.
   D. Neither a nor b is correct.

54. Two identical metal spheres are placed at the top of a wooden ramp and then released. Sphere 1 has been oiled, and it slides down without rolling. Sphere 2 has not been oiled, and it rolls down, since friction prevents it from sliding. Which sphere will reach the bottom of the ramp first? (Assume that there is no air resistance.)

   A. Sphere 1 will reach the bottom of the ramp first.
   B. Sphere 2 will reach the bottom of the ramp first.
   C. The two reach the bottom at the same time.
   D. You cannot tell without more information.

55. Two metal spheres are placed at the top of a wooden ramp. The spheres have identical masses and radii. Sphere 1 is hollow, with its mass all contained in a thin outer shell. Sphere 2, by contrast, is solid with its mass uniformly distributed throughout its volume. The two spheres are released at the same time and roll toward the bottom of the ramp. Which reaches the bottom first?

   A. Sphere 1 will reach the bottom of the ramp first.
   B. Sphere 2 will reach the bottom of the ramp first.
   C. The two will reach the bottom at the same time.
   D. You cannot tell without more information.

56. Which of the following is a reason for a student to use a computer instead of pen and paper to write her investigation report?

   A. It is easy to make changes on the computer version of her report and print out a new copy, but it takes much longer to recopy the pen and paper version.
   B. The computer copy is easier to read than most people's handwriting.
   C. With the computer, it is easier to share drafts with other members of her group when submitting a joint report.
   D. All of the above.

# Chapter 14 Harmonic Motion

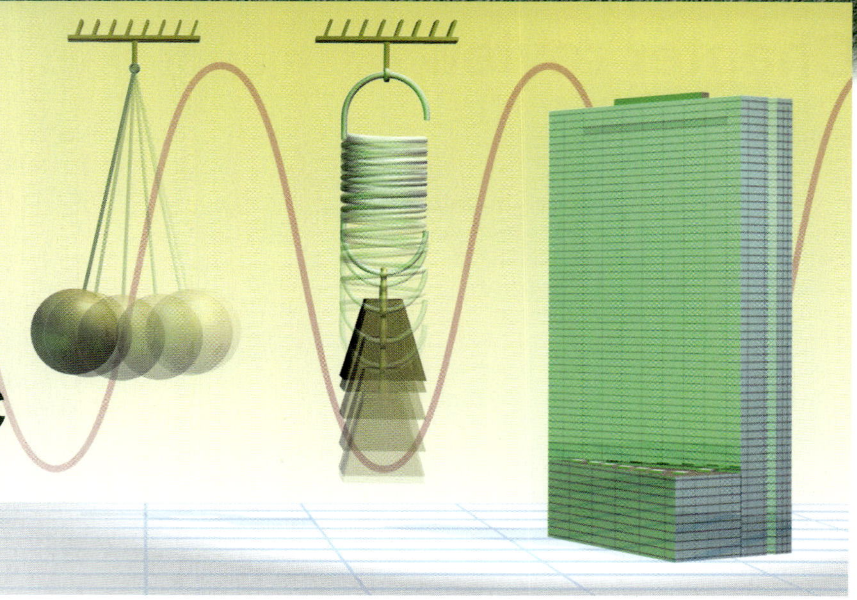

*"Would the Hancock ever, really, have fallen down? Nobody knows, but nobody was willing to take the risk."*

—Robert Campbell, in *The Boston Globe*, March 3, 1995

Why would a modern skyscraper in Boston's posh Back Bay neighborhood fall down? And what does that have to do with harmonic motion? The answer is as simple as pushing a child on a swing. After all, not many grandmothers can lift a ten-year-old boy overhead. Instead, put the child into a swing and let grandma push him a few dozen times. Soon he flies as high as grandma can reach.

As with many things, timing is key. A child on a swing is a form of pendulum with a natural frequency of motion. Deliver an impulse at the system's natural frequency, and soon you will have a lot of energy stored in the system's back-and-forth motion, or *oscillation*. That is what grandma does for her grandson, and that is what the wind did to the John Hancock Tower in Boston when it was completed in 1976.

Push a skyscraper to one side, and it tends to flex. Elastic forces then tend to straighten the building out. But when the wind repeatedly pushed the Hancock Tower at its natural frequency (0.14 Hz, or once every 7 s), the building continued to wobble, making occupants seasick and threatening to age its support structure prematurely. What was worse was that the building not only bent—it *twisted*, too. The two motions of bending and twisting resonated with each other, storing a large amount of vibrational energy in the process.

The solution came in the form of two 300-ton boxes attached to the Hancock's frame by springs and fluid-filled "shock absorbers." Occupying the tower's 58th floor, the boxes float on oil; because of their inertia, they remain nearly motionless as the building dances around them. If the Hancock lurches or twists, the springs and shocks push or pull it back into alignment. The system is tuned to the building's natural frequency and transforms vibrational energy into easily dissipated heat. Similar *tuned mass dampers* now operate in dozens of structures. One, a 730-ton pendulum in Taiwan's half-kilometer-tall Taipei 101, has become a popular tourist attraction.

# Chapter study guide

## Chapter summary

The pendulum is an example of an *oscillator,* which has motion that repeats in *cycles* with properties of frequency, period, and amplitude. Period and frequency are inversely related. Most oscillators have a *natural frequency* at which *resonance* occurs. The frequency and period of a simple pendulum depend on its length. An object oscillating vertically on a spring is another example of an oscillator, one whose frequency and period depend on the object's mass and the spring constant.

## Learning objectives

By the end of this chapter you should be able to:

- describe and measure period, frequency, and amplitude for a wave;
- determine period, amplitude, and frequency from a graph;
- calculate frequency and period from each other;
- describe the meaning of resonance and provide examples;
- explain the principles behind an oscillating pendulum and calculate its period; and
- explain the principles behind a mass oscillating vertically on a spring and calculate its period.

## Investigations

14A: Oscillators
14B: Damping and shock absorbers
14C: Resonance

## Chapter index



## Important relationships

$$f = \frac{1}{T}$$

$$f = \frac{1}{2\pi}\sqrt{\frac{k}{m}}$$

$$T = 2\pi\sqrt{\frac{L}{g}}$$

## Vocabulary

| | | |
|---|---|---|
| oscillator | cycle | period |
| frequency | hertz (Hz) | amplitude |
| damping | phase | periodic force |
| resonance | | |

## 14.1 - Concepts of harmonic motion

As we look at moving things, we see two classes of motion. *Linear motion* goes from point to point without repetition. The concepts of Chapter 2, such as velocity and acceleration, arise from thinking about linear motion. This chapter deals with *harmonic motion*, which is motion that repeats over and over in identical patterns called *cycles*. A pendulum swinging back and forth is an example of harmonic motion. While the familiar concepts of position, velocity, and acceleration still apply, there are also new concepts that are unique to harmonic motion: frequency, period, amplitude, and resonance.

### Cycles and oscillators

**Cycles**

The *cycle* is the building block of harmonic motion. A **cycle** is a unit of motion that repeats over and over. All harmonic motion is a repeated sequence of cycles. The cycles of three common examples of harmonic motion are shown in the figure at right.

Examples of oscillators

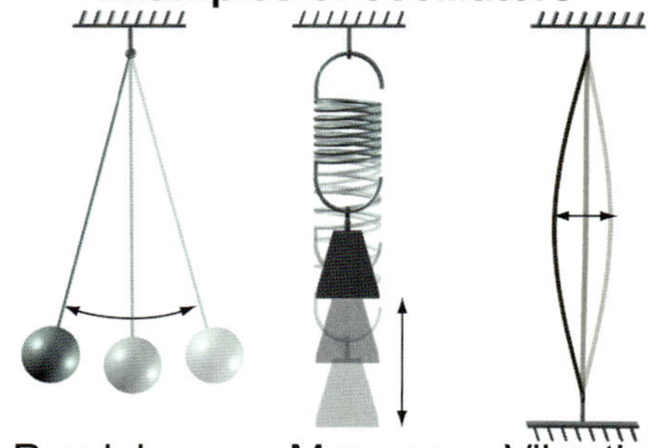


**Oscillators**

A system in harmonic motion is called an **oscillator**. Examples of oscillators you might find in your physics lab include a pendulum, a mass hanging from a spring, and a vibrating rubber band.

**Natural oscillators**

Examples of oscillators are found throughout nature. A tree swaying in the wind is an oscillator. The electrons inside atoms as they absorb light can be thought of as oscillators. The tides form a planet-sized oscillator driven by the gravity of the Moon acting on the oceans.

**Clocks are cycle counters**

If you have a pendulum with a cycle one second long, you can count time in seconds by counting cycles of the pendulum. Counting cycles of an oscillator is the basis for nearly all clocks. Modern quartz watches and atomic clocks also count cycles, and the latter do so to an accuracy of better than 1 s in every 1,400,000 years!

**Earth's cycles**

A few of Earth's cycles

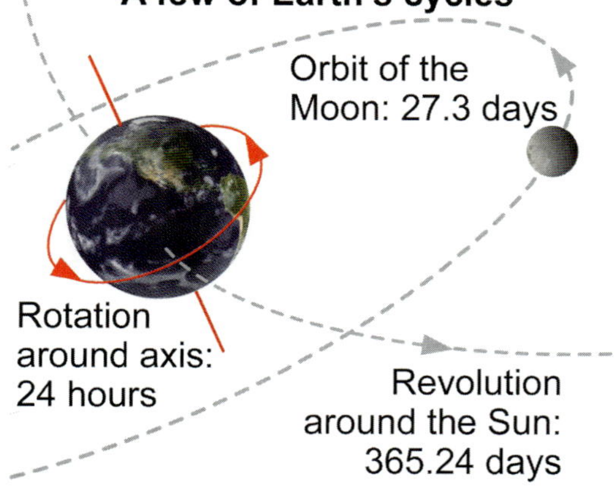


An *orbit* and a *rotation* are both cycles because they are repeating motions. Both are the basis for calendars. The Earth–Sun system has an orbital cycle of one year. The Earth–Moon system has a orbital cycle of approximately one month. Earth itself has several cycles. Earth rotates on its axis once a day, creating the 24-hour cycle of day and night. There is also a wobble of the Earth's axis, moving the orientation of the North and South Poles around by hundreds of miles every 22,000 years. .

**Weather cycles**

There are cycles in weather. The El Niño and La Niña are oscillations in ocean currents. These huge currents carry heat from the equator to the poles and their oscillations produce fierce storms approximately every decade. Much of our planet's ecology is created by the cycles that drive climate and weather.

**Why harmonic motion occurs**

Harmonic motion is a characteristic of what physicists call *stability.* Consider the two diagrams below. The ball in the valley will oscillate when disturbed. The ball on the right will not. Both start from an equilibrium point but the *stable* system has *restoring forces* that pull the system back toward equilibrium. Oscillation occurs whenever a system has restoring forces that give it stability and act to return it to the equilibrium position. Once you know what to look for, you can predict when harmonic motion is likely to occur.

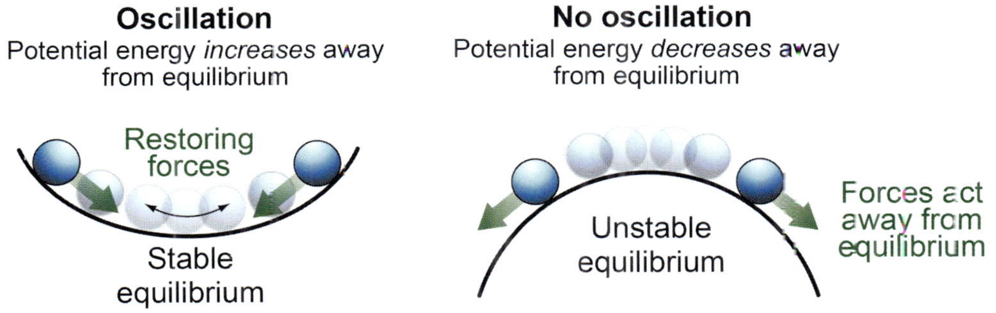


**Energy and stability**

Another way to look at a system is to consider how its potential energy changes. Any system for which the potential energy *increases* away from equilibrium will be stable and will oscillate. Any system for which the potential energy *decreases* away from equilibrium will be unstable. A system for which the potential energy does not change is *neutral* —neither stable nor unstable.

**Inertia**

To understand the effect of inertia, think about the ball in the valley. The ball is raised up a little to one side and then released. When it reaches the bottom of the valley the net force on the ball may be zero, but its motion causes it to keep going past equilibrium. Now the restoring force pulls back, eventually reversing the motion. When the ball gets back to the center, its motion—now going in the opposite direction—causes it to overshoot again. The tendency for the ball to overshoot its equilibrium position every time is caused by its inertia. All systems that oscillate on their own have some type of restoring force, as well as some property that acts like inertia, causing the motion to overshoot the equilibrium position.

**Stability and aircraft**

The concept of stability is critical to flight. Wind gusts often pitch a plane slightly up or down. Consider an aircraft with fins in the front versus one with fins in the back. If the fins are in the front, any slight pitch of the plane results in a force that causes the plane to rotate *more.* This is called *positive feedback* and it makes the plane unstable. When the fins are in the back, air flowing over the tail creates forces that push the plane back toward level flight. These restoring forces create *negative feedback*, reducing pitch and creating more stable flight. The farther back the tail is from a plane's center of mass, the more stabilizing is the effect.

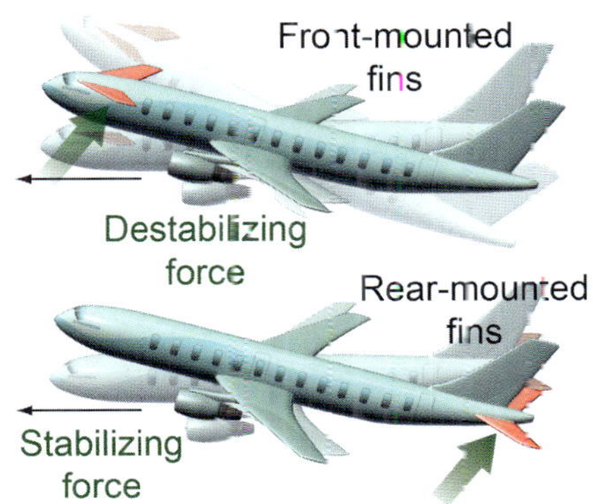

## Frequency and period

**Period**

The time for one cycle is called the **period**. The graph below shows position versus time for a mass oscillating up and down on a spring. Notice that one complete cycle occurs from 7 to 14 s; if we start counting the cycle at 7 s, the mass returns to the same position at 14 s. This graph is typical of *simple* harmonic motion. By *simple* we mean that there is a single period (or single frequency) that describes the motion.

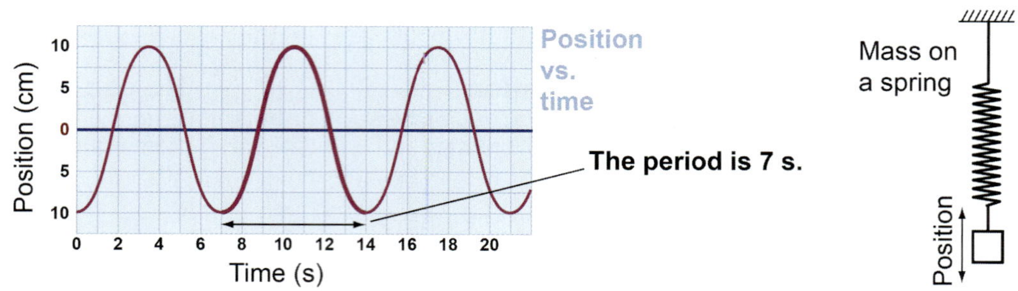


**Frequency**

The **frequency** of an oscillator tells you the number of cycles it completes each second. You experience a wide range of frequencies in your environment. A human heartbeat might have a frequency of ⅔ cycle per second when you first awake in the morning, 1–1.5 cycles per second for your normal activity, and 2–3 cycles per second when you are exercising. A plucked rubber band might have a frequency of 50 cycles per second. The sound of the musical note "A" has a frequency of 440 cycles per second.

**Unit of frequency is the hertz (Hz)**

The unit of frequency is called the **hertz (Hz)**, where one hertz corresponds to one cycle per second. A frequency of 440 cycles per second is usually written as 440 hertz, or abbreviated 440 Hz. The hertz is a unit that is the same in English and metric measurement systems.

**Period and frequency**

Frequency and period are the inverse of each other and are related by equation (14.1). A swing with a period of 2.5 s has a frequency of 1 ÷ 2.5 = 0.4 Hz. Large oscillators (such as a swing) tend to have low frequencies. Small or stiff oscillators (such as a guitar string) tend to have higher frequencies.

Interactive equation

**Frequency and period**

$$f = \frac{1}{T} \qquad (14.1)$$

$f$ = frequency (Hz)
$T$ = period (s)

**Converting between frequency and period**

A sprinter's heart beats once every 0.33 s. What is the frequency of her heartbeat?

*Asked:* frequency $f$ of the heart beat

*Given:* period of beating heart, $T = 0.33$ s

*Relationships:* $f = 1/T$

*Solution:* Use the relationship between period and frequency:

$$f = \frac{1}{0.33 \text{ s}} = 3.0 \text{ Hz}$$

*Answer:* The frequency is 3.0 Hz.

# Investigation 14A Oscillators

**Essential questions** *How do we understand motion that repeats in cycles?*

Interactive simulation

A pendulum and a mass on a spring are both good examples of *oscillators,* or systems that exhibit repetitive behavior. This investigation looks at how motion characteristics such as period and frequency are affected by physical variables such as mass, length, and spring constant.

## Part 1: Period of a pendulum

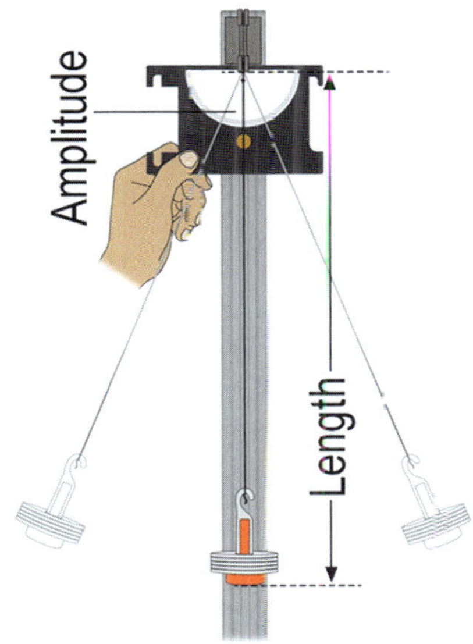


1. Assemble the pendulum as shown, with 5 washers on the mass hanger.
2. Set the pendulum swinging and observe the motion. Do not let the swinging mass hanger hit the stand.
3. Hold the angle indicator behind the string to measure the amplitude of the motion.
4. With a stopwatch, measure the time it takes to complete 10 full cycles.
5. Change the amplitude, mass, and string length and see how each variable affects the period of your pendulum. Tabulate your data.

a. *Describe how you determined one full cycle of the pendulum.*
b. *How does the period of the pendulum depend on length, mass, and amplitude? Support your answers using the data.*
c. *Propose a design for a pendulum that has a period of 2.0 s.*
d. *How did you choose the number of trials for each variable?*

## Part 2: Mass and spring oscillator

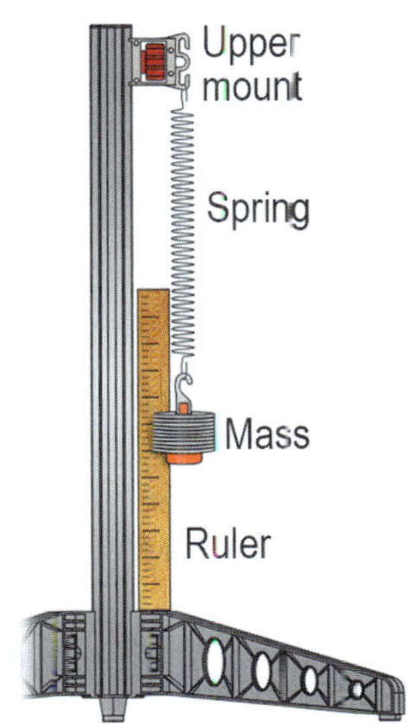


1. Set 6 washers on the mass hanger. Attach the mass and stiff spring. Place the meter rule against the stand. Note the position of the top washer in its equilibrium position.
2. Displace the mass 5 cm and release it. Record the time to complete 10 oscillations.
3. Repeat the experiment and record data for different masses and amplitudes.
4. Replace the stiff spring with the medium spring (different spring constant) and set 6 washers on the mass hanger.
5. With a stopwatch, measure and record the time to complete 10 oscillations.
6. With a spring scale, measure and record the force needed to extend each spring 10 cm. Calculate the spring constants. The spring constant is $k = F/x$ where $F$ is in newtons and $x$ is in meters.

a. *How did you determine one full cycle of the motion?*
b. *How does the period of the mass/spring oscillator depend on mass and amplitude? Your answer should be supported by the data.*
c. *Explain the answer to part (b) using Newton's second law, $F = ma$.*
d. *How does the period of the mass/spring oscillator depend on the spring constant? Your answer should be supported by the data.*
e. *In step #3 above, what were the independent, dependent, and controlled variables?*

## Amplitude and energy

**Equilibrium**

Most oscillators have a resting state, or *equilibrium* position. At equilibrium the net force is zero. A system in equilibrium remains in equilibrium until some outside force disturbs it. Any force that disturbs the system's equilibrium adds energy and the added energy is what causes the oscillations. The energy in the oscillations is therefore *additional* to any energy the system has in equilibrium.

**Amplitude**

The **amplitude** describes how far an oscillator moves away from equilibrium during each cycle. Amplitude is measured in units that match the oscillation. For a mass on a spring, the amplitude is the maximum distance the mass moves up or down away from its equilibrium position. With other oscillators the amplitude might be a voltage or a pressure. The key idea is that amplitude always describes the *maximum displacement from equilibrium.* This is usually *half* the distance from the highest to the lowest point. Notice that positive and negative values are used to represent motion on either side of equilibrium. The graph in the figure shows the mass on the spring moving between +10 cm and −10 cm, and thus with an amplitude of 10 cm.

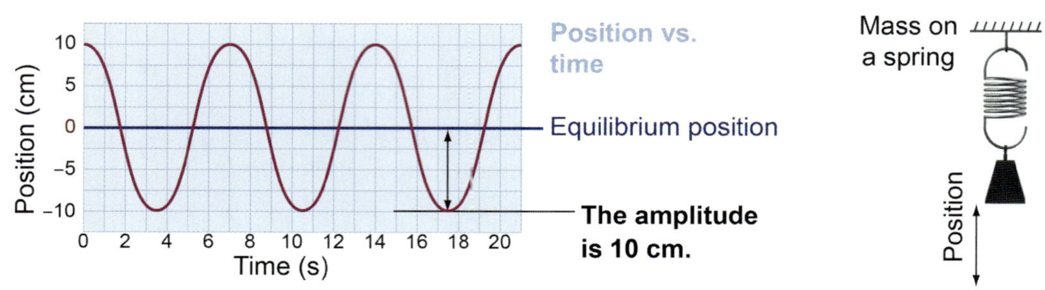


**Energy in an oscillator**

Harmonic motion involves *energy that oscillates among different forms.* Consider the mass on the spring. The system has kinetic energy because there is moving mass. The system also has elastic potential energy because the oscillation stretches a spring. The kinetic energy is largest when the mass moves through the equilibrium position. This is when its speed is greatest, before the force from the spring starts to slow it down again. The system has the most elastic potential energy when it reaches the highest or lowest point in its cycle, when its kinetic energy is zero. All the kinetic energy has been converted to elastic potential energy at these points. The continuous transformation of energy back and forth between two or more forms is a characteristic of most oscillators.

**Friction and damping**

Friction converts kinetic energy into heat and wear. As the energy of an oscillator decreases from friction, the amplitude decreases. The frictional decrease in amplitude is called **damping**. Over time, damping reduces the speed of an oscillating mass on a spring until it gradually comes to a stop. The graph of position versus time (on the right) shows the reduction of amplitude over many cycles owing to damping. In most situations, even though the amplitude decreases, *the frequency stays the same.* The peaks on the graph occur at the same time intervals.

**Damping reduces amplitude**

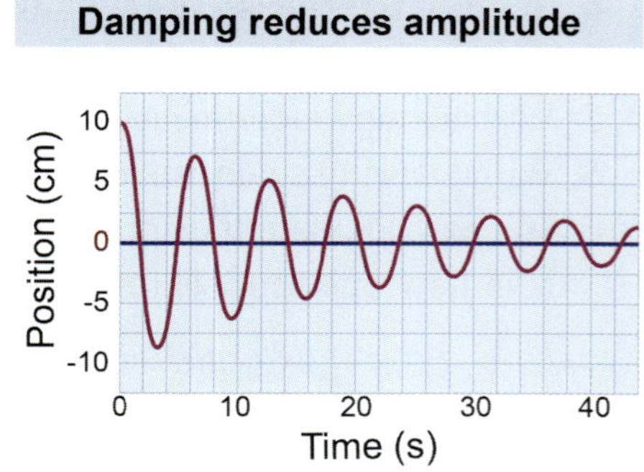

**A familiar example of a pendulum**

If you have ever used a swing you have first-hand experience of harmonic motion. A swing is physically a large pendulum, a mass suspended below a pivot point by a rope or chain, allowing the mass to oscillate back and forth. The equilibrium point is directly below the pivot, where the swing hangs at rest. The period is the time it takes to complete one full back-and-forth swing. The amplitude is the maximum distance the swing moves from its resting (center) position.

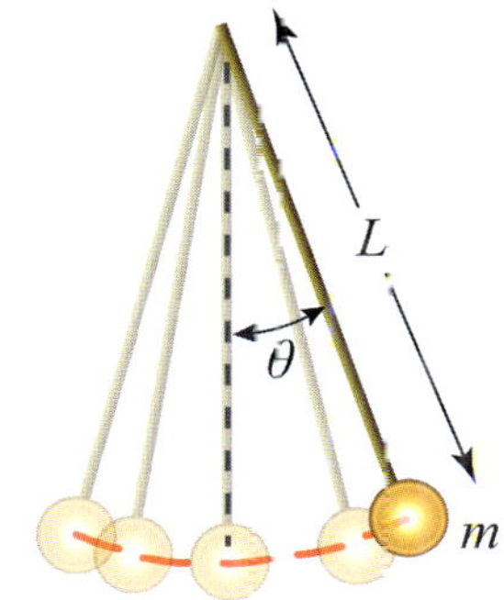


**Restoring force and energy of a pendulum**

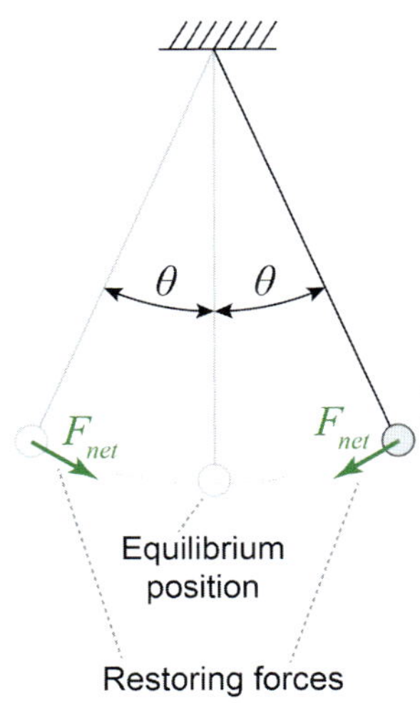


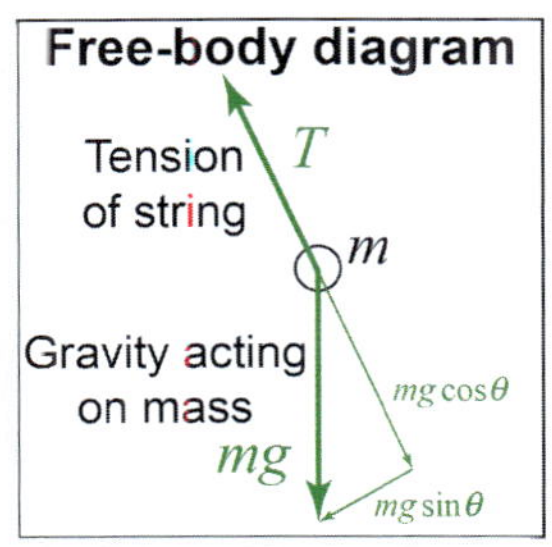


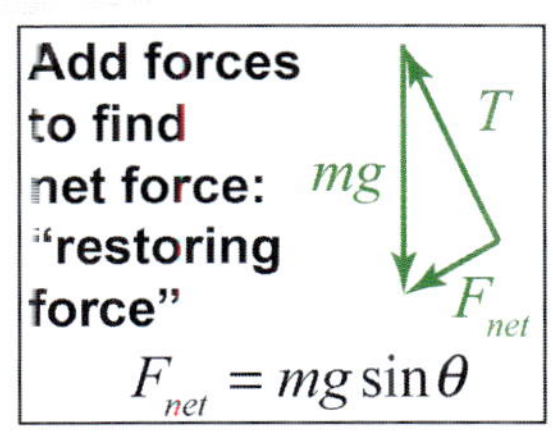


The weight $mg$ acts vertically down while the tension $T$ acts along the string. When the string makes an angle $\theta_{\max}$ to the right, the forces along the direction of the string cancel, leaving a net force $F_{net}$ that points down and to the left—back toward equilibrium. When the string makes the same angle to the *left*, the net force points down and to the right. The net force pushes the bob back toward equilibrium.

**The cycle and period**

As the bob moves back toward equilibrium, the restoring force goes to zero. At the center, the restoring force is zero but the pendulum keeps moving because of inertia. Past the center the force is reversed, and the bob slows, reverses direction, and moves back toward equilibrium to begin the next cycle. The pendulum crosses equilibrium twice in each cycle—once moving to the left and once to the right.

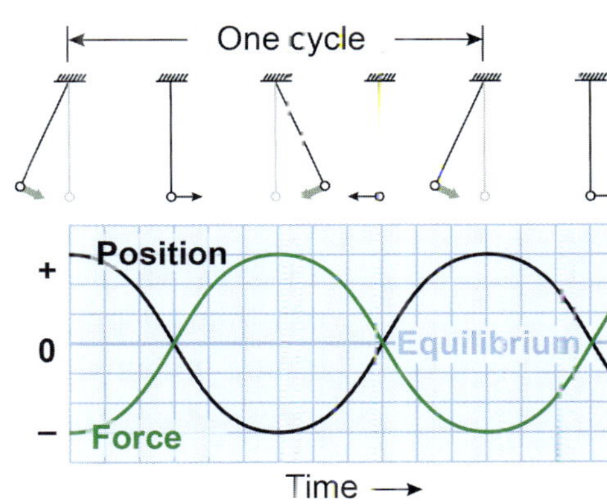


**Energy of the pendulum**

The total energy of the pendulum remains constant even though its energy changes form. At the highest point of the oscillation, it has only potential energy. As the bob moves through the equilibrium position, all of its energy has been converted into kinetic energy.

**Observations**

A pendulum is described by three variables: the mass of the bob, the length of the string, and the amplitude of the motion. Experimentally, we observe that the pendulum's period $T$

1. depends on the square root of the string length, $T \propto \sqrt{L}$ ,
2. does *not* depend on its amplitude, and
3. does *not* depend on its mass.

Why does the period depend on length but not on amplitude or mass? What is the physics behind it?

## Mass and spring oscillator

**What is a spring–mass oscillator?**

Consider a mass on a frictionless surface. The mass is connected to a spring that is fixed at one end. If the mass is displaced we have a situation similar to that of the pendulum. The spring supplies a restoring force that always pushes (or pulls) the mass back toward its equilibrium point. Understanding these systems is important because many real physical systems behave like spring–mass oscillators, including musical instruments, geological formations in an earthquake, and even atoms in a solid.

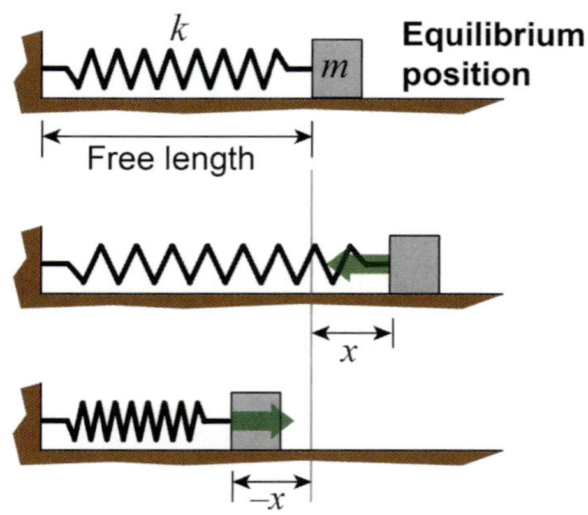


**How does the oscillator behave?**

When we observe a mass and spring oscillator in the lab we find the following:

1. The frequency increases when the spring gets stiffer (that is, has a higher $k$).
2. The frequency decreases when the mass increases.
3. Like the pendulum, the frequency is virtually independent of amplitude.

**The vertical mass and spring**

A way to create a nearly frictionless mass and spring oscillator is to hang the spring vertically. The spring stretches until at equilibrium the upward force from the spring is equal to the weight of the hanging mass. This "pre-stretched" state is the equilibrium position for the vertical system.

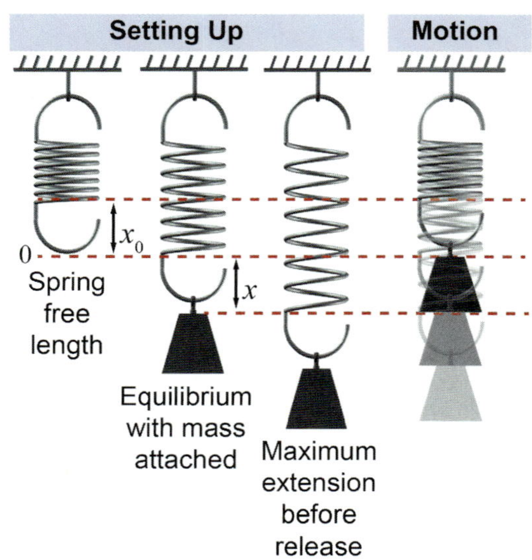


**What is the amplitude?**

If the mass is pulled down below this "pre-stretched" equilibrium position and released, the system oscillates up and down. The amplitude of the oscillation is measured relative to the equilibrium position. The oscillation exchanges elastic potential energy stored in the spring with kinetic energy of the moving mass.

**Does gravity matter?**

Although there are also changes in height, gravitational potential energy does not contribute to the oscillation! You can see this by looking at the free-body diagram of the hanging mass. The force of gravity (weight) stays constant and does not change direction, no matter what position the mass is in. Gravity *does not create the restoring force*, and therefore gravitational potential energy is a "spectator" and does not contribute to the oscillation.

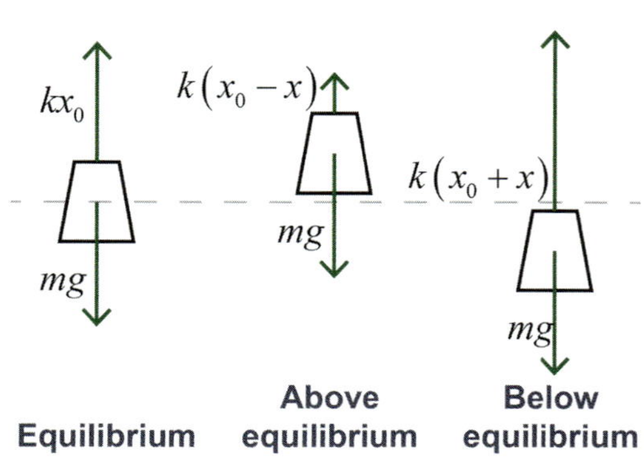

# Investigation 14B

## Damping and shock absorbers

**Essential questions** *How are oscillators used?*

Interactive simulation

When you extend or compress a spring, most of the work you do is converted into stored *elastic potential energy* in the spring. If the mass attached to the spring is allowed to move, this stored energy causes the spring and mass to oscillate. The spring in a car's shock absorber allows the wheel to go up and down over uneven roads independently from the whole car. The resultant oscillations, however, are undesirable for passengers! In this investigation, you will use an interactive simulation to see how frictional *damping* is used to reduce oscillations while still allowing wheel movement.

### Part 1: Springs in a car's suspension

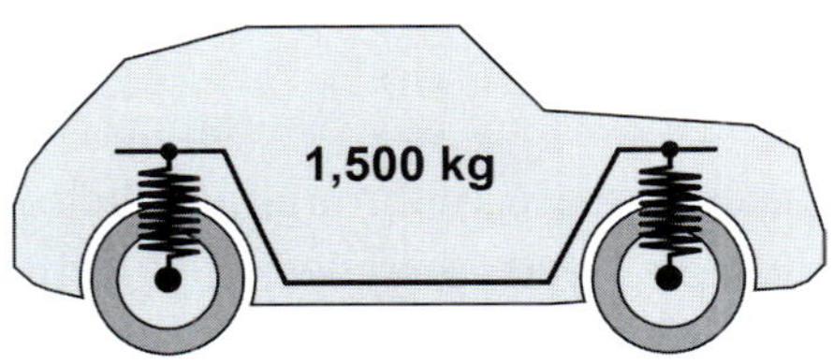


**Suspension springs in a typical car**

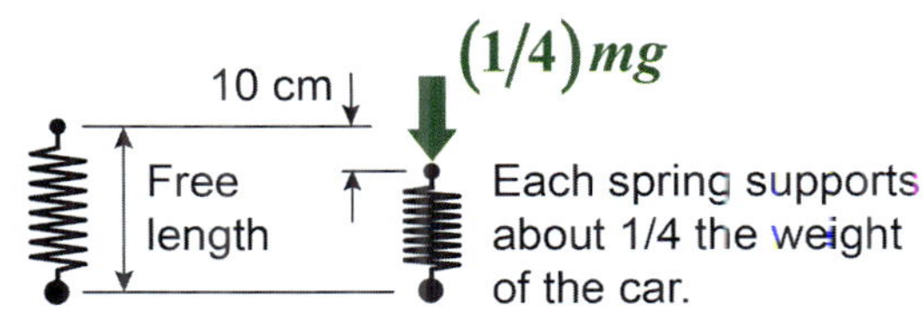


Consider a 1,500 kg car chassis that supports its mass evenly among four suspension springs, one for each wheel, that compress 10 cm when the car is at rest.

1. Draw a free-body force diagram of the chassis showing the gravitational and elastic forces that are in equilibrium.
2. Use the diagram to calculate the spring constant $k$ that is required to support one-fourth of the mass of the car when depressed by 10 cm.
3. Is this a stiff or a loose spring? Do large or small values of the spring constant $k$ correspond to a stiff or loose spring?
4. Use the interactive simulation to model that spring by setting the mass, spring constant, and displacement. How much energy is stored in the spring?

### Part 2: Shock absorbers and damped oscillations

This interactive simulation models the response of a damped car-suspension spring. You can set the initial displacement from equilibrium, $x_0$, spring constant $k$, and mass $m$, as well as the damping constant $b$.

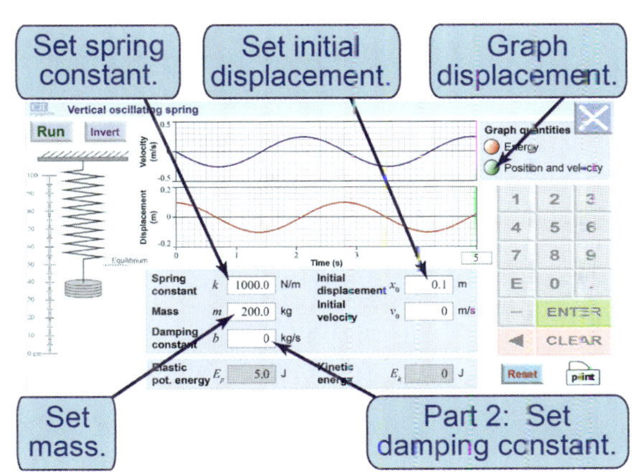


1. Let the damping constant $b = 0$ and give the spring a displacement of 10 cm, simulating a wheel hitting a pothole. Describe the resulting motion of the 1,500 kg car. Is this desirable? Why or why not?
2. Set $b = 100$ kg/s. How does this affect the period and amplitude of the motion? Graph the resulting motion for $b = 100$ and 25,000 kg/s.
3. Adjust the damping constant $b$ until you achieve two goals. First, the wheel should be able to move down as much as possible to accommodate potholes. Second, the subsequent oscillations should damp away quickly. What value of $b$ is optimal?
4. What happens to the damped energy lost by the spring?

## Phase

**What is the cycle of a wheel?**

A rotating wheel returns to the same place every 360°, giving rotation a cycle, just like harmonic motion. A useful difference is that each cycle of circular motion always has a "period" of $2\pi$ rad or 360°. For this reason, the $2\pi$ (360°) cycle of circular motion is used to represent the periodicity of all forms of harmonic motion.

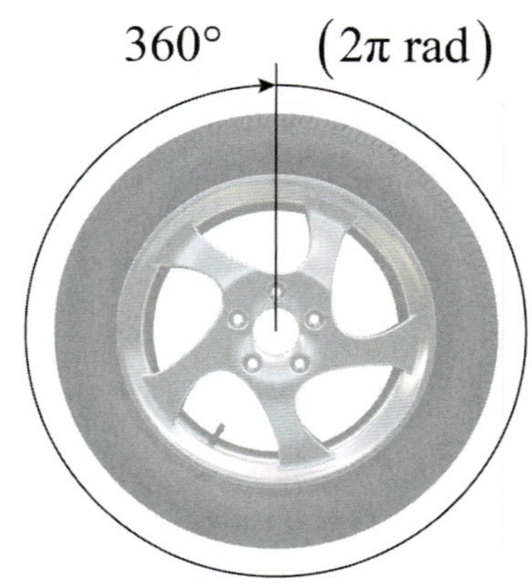


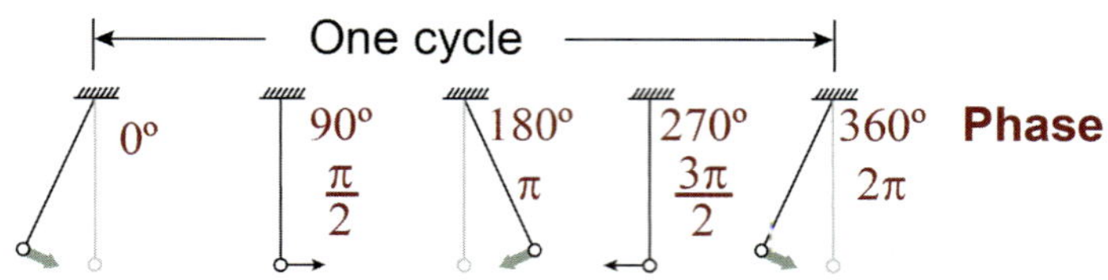


**The concept of phase**

The fraction of a full circle is a useful idea to describe where any oscillator is within its full cycle. For example, a pendulum that is one-quarter of the way through its cycle is described as having a *phase* of 90° or $\pi/2$. The word **phase** refers to where an oscillator is at a particular moment in relation to its full cycle. If we let one cycle be 360°, then one-quarter of a cycle is 90°. In radians, a full cycle is $2\pi$ rad, and therefore a quarter-cycle is $\pi/2$ rad.

**In-phase motion**

Two oscillators may have the same period but different phases. For example, if you start two identical pendulums together, their position versus time graphs would look like the diagram on the left. These pendulums are *in phase* because each is always at the same position at the same time.

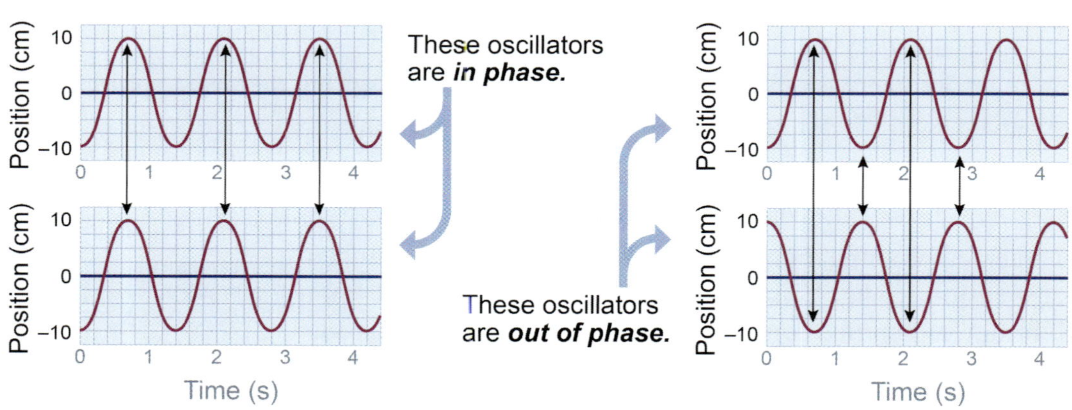


**Out-of-phase motion**

Consider the graphs if one pendulum is started when the other is halfway through its cycle. When the first pendulum is at its extreme left, the second is at its extreme right. The graphs of position versus time have the same cycle, but they are *out of phase* with each other. The second pendulum is always 180° ($\pi$) behind the first. The graph shows the lead of the first pendulum as a *phase difference.* These two pendulums are out of phase by 180°, or one-half cycle. Oscillators may differ in phase by any amount between 0° and 360° (0–$2\pi$ rad).

# Section 1 review

*Harmonic motion* repeats in identical patterns called *cycles*. An *oscillator,* such as a pendulum, is a system that features repeating cycles. The *period* of an oscillator is the time to complete one full cycle. The *frequency* is the number of cycles per second. The *amplitude* of an oscillation is the maximum displacement from equilibrium.

| Vocabulary words | oscillator, cycle, period, frequency, hertz (Hz), amplitude, damping, phase |
|---|---|
| Key equations | $f = \dfrac{1}{T}$ |

## Review problems and questions

1. The Earth exhibits several different types of harmonic motion. Describe at least three examples of harmonic motions associated with our planet.

2. A swing is observed to move back and forth 12 times in 30 s.

   a. What is the period of the motion?
   b. What is the frequency of the motion?

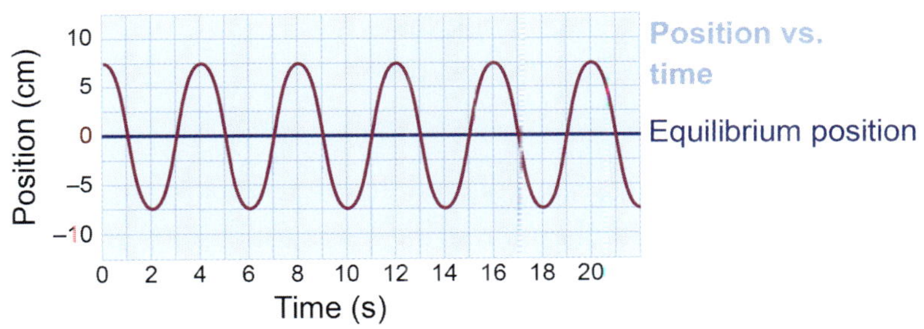


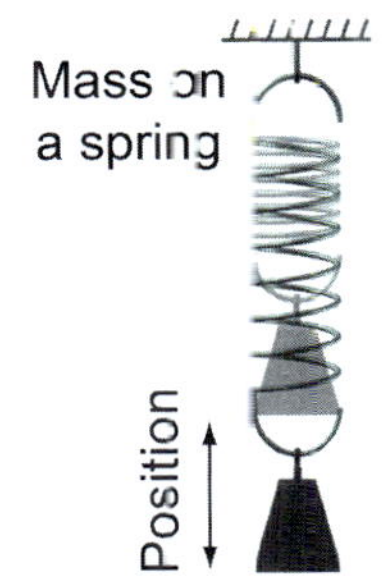


3. Determine the following information from the graph above.

   a. How many cycles occur in 10 s?
   b. What is the amplitude of the motion?
   c. What are the period and frequency of the motion?

4. Systems undergoing harmonic motion contain energy that oscillates between different forms. What forms of energy are changing in the systems described below?

   a. a simple pendulum
   b. a mass on a spring that is sliding on a frictionless, horizontal surface

5. A simple pendulum and a mass on a spring are both moving in simple harmonic motion. In which of these systems does the period of the harmonic motion depend on mass? On amplitude?

6. If an oscillator has a frequency of 60 Hz, how many complete oscillations take place in 2 s?

7. A quartz crystal in a watch has a period of 0.00350 s. What is its frequency in hertz?

If you experiment with a pendulum you observe a curious fact: It always oscillates with the same frequency. For example, a certain pendulum completes one cycle every 1.5 s (0.67 Hz). Every time you set the same pendulum swinging, it *always* swings with a frequency of 0.67 Hz and never changes its frequency. The frequency at which a system tends to oscillate is called its *natural frequency.* Everything that can oscillate has a natural frequency, and most systems have more than one. The natural frequency depends on the balance between the strength of restoring forces and the amount of inertia in the system.

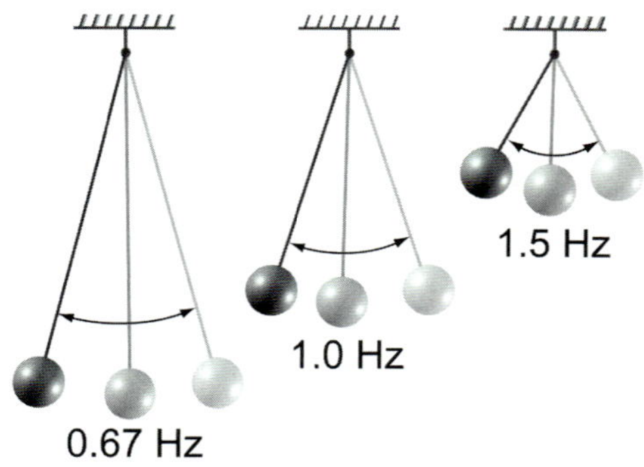


## Natural frequency

**Using the natural frequency**

The natural frequency is useful because many inventions are designed to work at a specific frequency. For example, a guitar string that plays the note middle C is *tuned* to have a natural frequency of 262 Hz. The process of tuning is really a process of adjusting the natural frequency of a vibrating string oscillator. Watches, computers, and many devices rely on the precise natural frequency of quartz crystal oscillators. A computer that advertises a speed of 2.6 GHz has an internal quartz clock that oscillates at a frequency of 2.6 billion cycles per second.

Tuning a guitar changes the natural frequency of the string by changing its tension.

**Effect of restoring force on natural frequency**

If the restoring forces in a system are strong, then the system accelerates quickly and tends to have a higher natural frequency. A steel guitar string takes much more force to deflect than a rubber band of the same length, because rubber is a much weaker material than steel. The guitar string has a much higher natural frequency because of the higher restoring force created by steel compared to rubber.

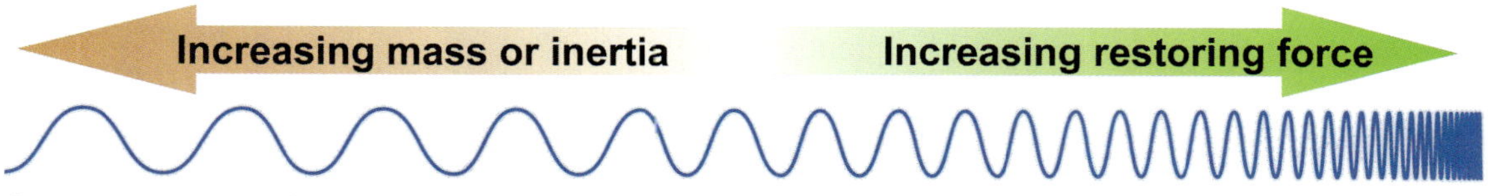


**Effect of inertia**

A system with more inertia is harder to accelerate and responds more slowly. Adding inertia therefore lowers the natural frequency. The bass strings on a guitar are wound with extra wire to make them heavier. The increased mass creates more inertia and causes the strings to oscillate more slowly, which means that the natural frequency has been lowered. Tying a steel nut onto a rubber band will produce the same effect. The added inertia of the steel nut greatly lowers the natural frequency of the system.

**How can we calculate the period of a pendulum?**

Consider a pendulum released from a distance $x$ from center so the string makes an angle $\theta$ with the vertical. The restoring force on the pendulum bob is a component of the gravitational force:

$$F_{restoring} = mg\sin\theta = mg\frac{x}{L}$$

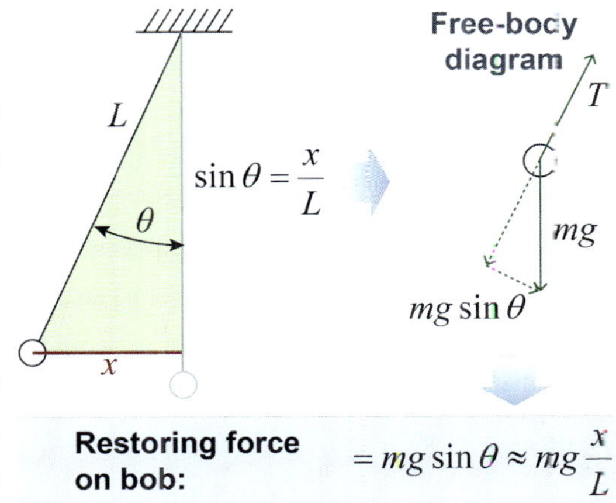


From Newton's second law we know that the acceleration $a$ is the force divided by the mass. This tells us the maximum acceleration of the pendulum bob:

$$a = \frac{F}{m} \quad \rightarrow \quad a_{\max} = g\frac{x}{L}$$

**Estimating the period**

The pendulum bob has zero *tangential* acceleration at the center, so we approximate the average acceleration $a$ as one half the maximum, or

$$a = \frac{gx}{2L}$$

From the equations for accelerated motion, the displacement $x$ is given by $x = \frac{1}{2}at^2$ for an object moving with constant acceleration $a$ starting from rest. Solving for the time, $t$, gives an estimate of the time it takes the pendulum to move the distance $x$:

$$x = \cancel{x_0} + \cancel{v_0 t} + \tfrac{1}{2}at^2 \;\Rightarrow\; t = \sqrt{\frac{2x}{a}} \;\Rightarrow\; t = \sqrt{\frac{2x}{gx/2L}}$$

**What does the estimate tell us?**

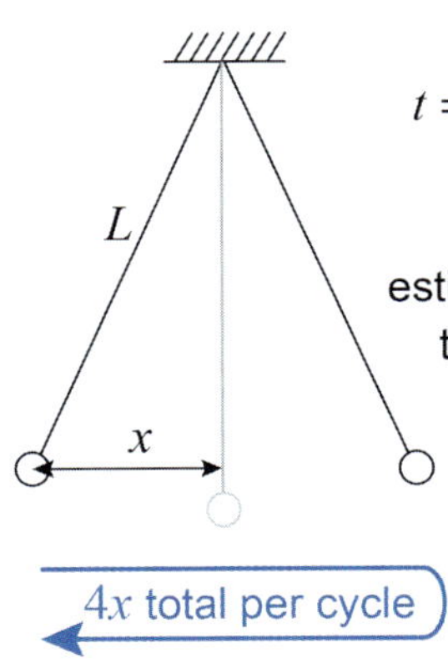


$$t = 2\sqrt{\frac{L}{g}}$$

This is the estimated time time it takes to move $x$, which is 1/4 cycle.

When we simplify the fraction under the square root, the amplitude $x$ cancels out! To get the period of the pendulum, we multiply by four because the pendulum moves approximately $4x$ in one cycle. We are left with a simple formula that tells us that the period of a pendulum depends only on gravity and *the square root of the length of the string!*

**How does this compare to an exact calculation?**

Estimate: $t_{period} \approx 8\sqrt{\frac{L}{g}}$ Actual solution: $t_{period} = 2\pi\sqrt{\frac{L}{g}}$

A better calculation of the period gives a factor of $2\pi$ instead of 8 but the dependence on the square root of $L/g$ is the same. More importantly, the analysis predicts that the period of a pendulum is independent of mass and amplitude, depending only on the square root of the string length, in good agreement with observations.

## Natural frequency of a mass and spring oscillator

**How do we analyze a spring–mass oscillator?**

Consider a mass $m$ moving on a horizontal frictionless surface. The mass is connected to a spring with spring constant $k$ that is fixed at one end. Although the force approach could be used to determine the natural frequency of this oscillator, the energy approach provides a more direct and insightful way to treat the mass and spring.

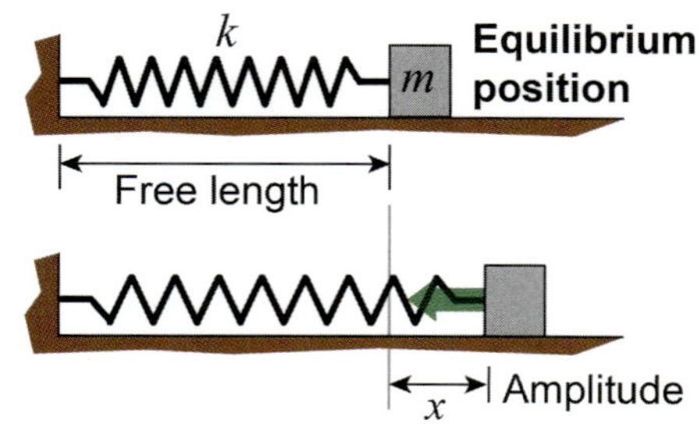


**Energy of the oscillator**

At maximum amplitude the spring is extended a distance $x$. At this point the velocity $v$ of the mass is zero, and the energy of the system is entirely elastic potential energy. A quarter-cycle later the mass reaches its maximum velocity $v_{max}$ when the spring is at its free length and the restoring force is zero. Here, the energy of the system is entirely kinetic. Conservation of energy says that the elastic potential energy at $x$ equals the kinetic energy at $x = 0$. As the mass moves back and forth the energy oscillates between elastic potential energy in the spring and kinetic energy of the moving mass.

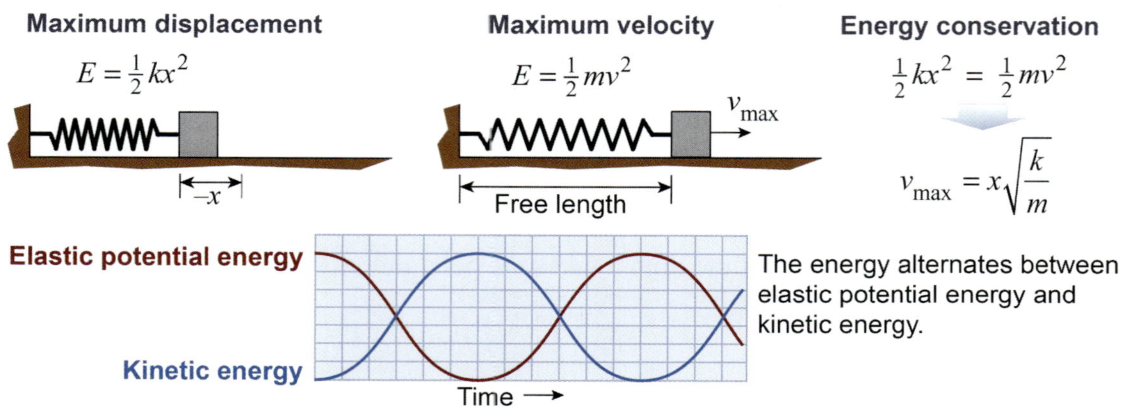


**Estimating the period and frequency**

The period is the time it takes the mass to complete a full cycle. The average speed of the moving mass may be estimated as ½$v_{max}$. Since the distance traveled is $4x$, we can divide this by the average speed to calculate the period. This estimated result is shown below along with the exact calculation. The estimated frequency $f_{est}$ is the inverse of the period.

$$t = \frac{d}{v} \quad \rightarrow \quad T = \frac{4x}{\frac{x}{2}\sqrt{\frac{k}{m}}} \quad \rightarrow \quad T \approx 8\sqrt{\frac{m}{k}} \qquad T = 2\pi\sqrt{\frac{m}{k}}$$

$$f = \frac{1}{T} \quad \rightarrow \quad f_{est} \approx \frac{1}{8}\sqrt{\frac{k}{m}} \qquad f = \frac{1}{2\pi}\sqrt{\frac{k}{m}}$$

**What does this mean?**

The frequency of a mass–spring oscillator depends on the square root of the ratio of the spring constant $k$ divided by mass $m$. A spring that is four times stiffer ($4 \times k$) will have twice the natural frequency. Increasing the mass by a factor of 4 reduces the frequency by a factor of 2. Note that the only difference between the exact calculation and the estimation is that the factor of 8 becomes $2\pi$.

**Periodic forces**

The connection between force and motion is more complex for harmonic motion than it is for linear motion. The difference is that *forces can be periodic*. A **periodic force** is a force that repeats in cycles—like the repeated push–wait–push–wait–push—that you use to get a swing going. Of course, Newton's laws still apply, but the *frequency* of a periodic force is a new variable that can make a huge difference. When the frequency of the force matches a system's natural frequency, even a small force can produce a surprisingly large oscillation. The effect is called *resonance*, and it is found throughout both nature and technology.

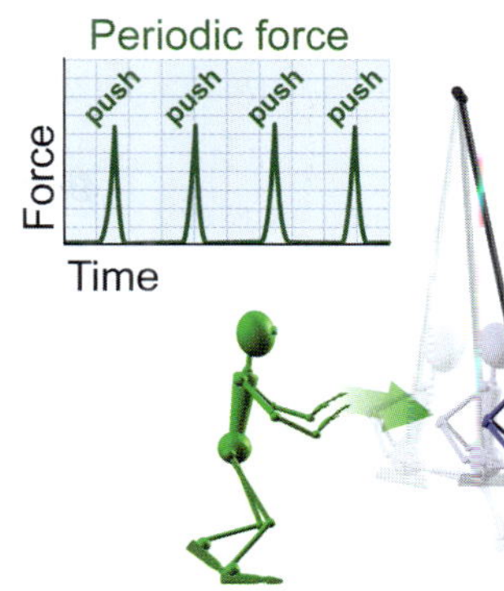


Over time, a small periodic force at the natural frequency can create a large-amplitude motion.

**Example of a periodic force**

Think about pushing someone on a swing. A swing is a pendulum and it has a natural frequency. To create a large amplitude you supply a small push every time the swing reaches the end of its cycle. In the language of physics, your repetitive pushes are a *periodic force* at the *natural frequency* of the swing. Over many pushes, the swing builds up a large amplitude of motion even though any single push does not do much by itself.

**Resonance**

When the frequencies of the force and the system are matched, each push comes at just the right moment and the amplitude increases dramatically. This behavior is called **resonance**. Resonance occurs when the frequency of a periodic force matches the natural frequency of a system. Although the physical laws are the same as for linear motion, resonance is a unique behavior of harmonic motion.

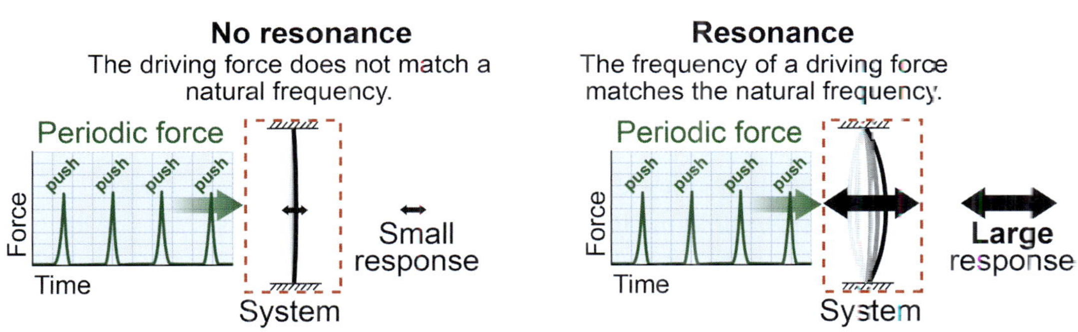


**Resonance, energy, and technology**

A resonant system accumulates energy with every cycle of the applied force. A system in resonance is a very efficient *energy accumulator*. This is useful in technology because *the energy is concentrated at the natural frequency!* If you want a musical instrument to have a vibration of 100 Hz, you create something physical that has a natural frequency of 100 Hz. By doing this, only vibrations at a frequency of 100 Hz will have a large amplitude. This is the principle behind musical instruments and modern communications technologies. When you *tune* the FM stereo in your car what you are doing is adjusting the resonance frequency of your car's receiver to match the frequency transmitted by a particular station.

**Forced resonance?**

Video

In 1940, the Tacoma Narrows Bridge in Washington State vibrated and twisted amid 40 mph winds, resulting in its dramatic collapse caught on video (at left). Was this a real-world example of *forced resonance*, where the oscillatory frequency of the wind matched the natural frequency of the bridge? Watch the video and decide for yourself

# Investigation 14C Resonance

**Essential questions** *What is resonance?*

Interactive simulation

When *periodic forces* are applied to a system that can oscillate, the resulting motion can vary tremendously. If the frequency of the force matches a natural frequency of the system, a *very* large amplitude response can occur. The extra-large response is what we call *resonance.*

## Part 1: Finding natural frequency

1. Mount the lever to the top of the stand.
2. Hang the stiff spring from the lever arm, and then hang the mass (12 washers) from the spring.
3. Attach a ruler to the stand centered on the hanging mass.
4. Pull the mass down slightly to start it oscillating.
5. Measure the time for 10 oscillations.

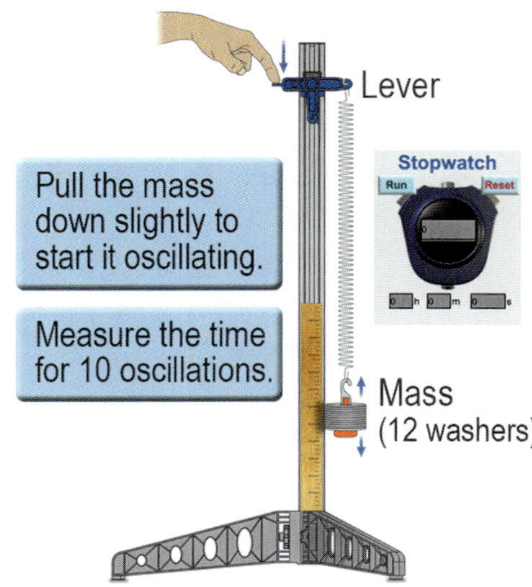


a. *What is meant by the oscillator's "natural frequency"?*

b. *What are the values of your oscillator's natural period and frequency?*

c. *Describe and then apply a method to estimate the error bars on your frequency measurement.*

d. *Describe the meaning of the terms phase and amplitude.*

## Part 2: Creating resonance

1. Set the timer to beep at a frequency of between 0.1 and 3 Hz. (Convert between frequency and period first!)
2. While using one hand to hold the top of the stand in place, use your other hand to press down on the lever on each beep.
3. Have your partner estimate the amplitude of the motion using the ruler.
4. Measure and tabulate the oscillation amplitude for at least ten frequencies ranging between 0.1 Hz and three times the natural frequency. One measurement should be at the frequency you measured in Part 1.
5. Graph the oscillation amplitude versus the frequency of the periodic force.

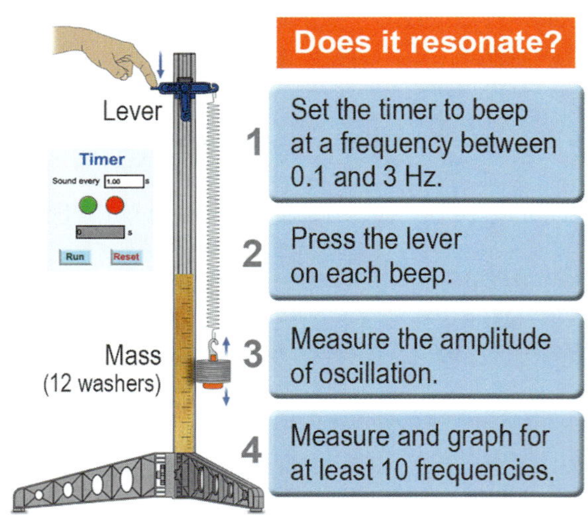


a. *In your own words, define resonance by referring to the motion you just observed and the graph of your data.*

b. *At what frequencies is the oscillator in resonance? At what frequencies is it out of resonance?*

c. *Describe the flow and storage of energy in the system at resonance compared to frequencies that are not at resonance.*

d. *Prepare a written report for your investigation (parts 1 and 2). In addition to addressing the above questions, describe the procedure, results, and conclusion (or interpretation) of your investigation.*

# Section 2 review

Most systems have a *natural frequency* at which they oscillate when disturbed; many have several such natural frequencies. Systems may exhibit an especially large amplitude of motion at their natural frequency. *Resonance* describes the large-amplitude response that occurs when an oscillatory system is driven with a force that matches its natural frequency. The natural frequency of a simple pendulum depends on the acceleration due to gravity ($g$) and the length of the pendulum but not upon the mass of the bob or the amplitude. The natural frequency of a mass-and-spring assembly depends upon the spring constant and the mass of the attached object but not upon $g$ or the amplitude.

| | |
|---|---|
| **Vocabulary words** | **periodic force, resonance** |
| **Key equations** | $T = 2\pi\sqrt{\frac{L}{g}}$     $f = \frac{1}{2\pi}\sqrt{\frac{k}{m}}$ |

## Review problems and questions

1. Describe the relationship among resonance, amplitude, force, and natural frequency.

2. Josué and Dalia are building a small pendulum to serve as a primitive stopwatch for a ball-and-ramp experiment. They hang a dense metal nut from a lab stand with string. They discover that their pendulum has a period $T$ of ½ second.
   a. How long is the pendulum?
   b. Suppose that the partners now want their pendulum to swing back and forth once every second, not every half-second. Josué argues that they should double the length of the string. Dalia argues that the string should only be 50% longer. Which partner (if either) is right?

3. Tracie and Bob are performing a series of experiments with small, dense metal weights hanging from springs (which are attached to sturdy lab stands). They find a mass-and-spring combination that bobs up and down once every 2 s, after being gently stretched and then released. Their teacher asks them to double the system's frequency. Which of the following changes might achieve this?
   a. Replace the spring with a stiffer (harder-to-stretch) spring.
   b. Replace the metal weight with a lighter (less massive) object.
   c. Either a or b will increase the frequency.
   d. Neither a nor b will increase the frequency.

4. A 1.0 kg mass hangs on a spring with a spring constant of $k = 158$ N/m. The mass is pulled down 10 cm and released, causing the system to move in simple harmonic motion with a natural frequency of 2.0 Hz and an amplitude of 10 cm. What is the new frequency if the mass, spring constant, and amplitude are changed as described below?
   a. The mass and spring constant are doubled, and the amplitude is unchanged.
   b. The mass is quadrupled, and the spring constant and amplitude are unchanged.
   c. The spring constant is quadrupled, and the mass and amplitude are unchanged.
   d. The amplitude is doubled, and the mass and spring constant are unchanged.

# Chapter review

## Vocabulary

Match each word to the sentence where it best fits.

### Section 14.1

| | |
|---|---|
| **amplitude** | **cycle** |
| **damping** | **frequency** |
| **hertz (Hz)** | **oscillator** |
| **period** | **phase** |

1. A video runs at 30 frames/s, therefore its _______ is 30 Hz.
2. The amount of time that an oscillator takes to complete one cycle is known as the _______.
3. An oscillation that repeats seven times per second can be written as 7 _______.
4. The maximum displacement of an oscillator is its _______.
5. After a car hits a pothole it may bounce up and down, but those oscillations decrease over time because of _______.
6. Each time an oscillator returns back to its beginning state it has completed one _______.
7. A pendulum is an example of a/an _______.

### Section 14.2

| | |
|---|---|
| **periodic force** | **resonance** |

8. When Sally pushed Jenny on the swing with a frequency that matched the swing's natural frequency, they were demonstrating _______.
9. When a girl pushes a boy on a playground swing in a regular pattern that helps him swing higher, the girl is applying a _______.

## Conceptual questions

### Section 14.1

10. Ask two friends the meaning of the word "cycle" in each of the following sentences.

    Sentence 1: "The cycle of a pendulum is one complete back-and-forth motion."
    Sentence 2: "Fashion tends to go through cycles where styles are in favor for a while, become unpopular, then come back to being in favor again."

    Have them compare the two meanings. How are they similar? How are they different?

11. ❨ What feature about a pendulum allows it to be used as a clock?
12. ❨ If you hang a dumbbell from a vertically suspended spring and start it oscillating up and down, eventually it will stop oscillating. Why?
13. ❨ When measuring the length of a pendulum, what part of the clamp holding it should you use as the top of the pendulum?
14. ❨ What are two ways to measure and describe a pendulum's maximum displacement from equilibrium?
15. ❨ An AM radio station broadcasts at 1,050 kHz. How many times per second does its signal oscillate?
16. ❨ A pendulum takes 2 s to complete one cycle of oscillation. What is the frequency of the pendulum's motion?
17. ❨ Describe the different forms of energy involved when a pendulum oscillates back and forth.
18. ❨❨ Imagine placing a spherical metal ball on a small round bowl. Why is it easy to place the ball motionless in the middle of the upright bowl but difficult to place it motionless in the middle of the upside-down bowl?
19. ❨❨ Two pendulums are swinging back and forth. The first pendulum makes two complete oscillations in the time it takes the second pendulum to make only one complete oscillation. Which pendulum has a longer period? Which one has a higher frequency?
20. ❨❨ Arrange the following in order of increasing period.
    a. motion of the Earth around the Sun
    b. rotation of the Earth about its axis
    c. motion of the Moon around the Earth to complete one lunar cycle
21. ❨❨ You are using a stopwatch to measure the period of a pendulum. After releasing the pendulum, you find that it reaches maximum velocity for the first time at $t_1 = 0.5$ s. What is the pendulum's period?
22. ❨❨ Arrange the following in order of increasing frequency.
    a. motion of the Earth around the Sun
    b. rotation of the Earth about its axis
    c. motion of the Moon around the Earth to complete one lunar cycle
23. ❨❨ What is the relationship between a pendulum's amplitude and the length of the arc the pendulum traces out during one complete swing?

# Chapter review

## Section 14.1

24. ‹‹ Describe the energy of a mass oscillating at the end of a spring.

25. ‹‹ Can a simple harmonic oscillator be in motion but have no internal forces acting on it?

26. ‹‹‹ Based on your understanding of the motion of a pendulum, what can you say about the acceleration of the pendulum bob during oscillation at (a) equilibrium and (b) maximum displacement?

27. ‹‹‹ How does gravity impact the oscillation of a mass suspended vertically from a spring?

## Section 14.2

28. ‹ You are pushing a child on a swing set. The child wants to go higher on the swing. How can you make this happen?

29. ‹ Describe the relationship between the displacement and acceleration of a swinging pendulum.

30. ‹‹ A pendulum has been oscillating for a minute and you begin to push it at a frequency of 1.7 Hz. The natural frequency of the pendulum is 1 Hz. Will your pushes increase or decrease the energy in the pendulum?

31. ‹‹‹ You are trying to increase the natural frequency of a mass on a spring. Should you add or remove mass to increase the natural frequency?

32. ‹‹‹ In 1665, Christian Huygens had two pendulum clocks standing on the floor of his room. He noticed that, no matter what state he started them in, they would always eventually reach a state in which the two pendulums swung *out of phase* or opposite to each other. What was going on?

33. ‹‹‹ Centuries ago soldiers crossing a bridge learned not march together in time. Instead, they all broke stride. Why break stride?

34. ‹‹‹ Describe how the phenomenon of resonance is related to playing a guitar.

35. ‹‹‹ Using primary sources, such as research papers or videos, draw diagrams to distinguish among three possible explanations for the collapse of the Tacoma Narrows Bridge: forced resonance, Strouhal vortices, and aeroelastic fluttering.

## Quantitative problems

## Section 14.1

36. An oscillator has a period of 13 s. Will it complete an integer number of full cycles of oscillations after 1.3 or 130 s? Why?

37. ‹ In cycles per hour, what is the frequency of (a) the second hand of a watch, (b) the minute hand, and (c) the hour hand?

38. ‹ The frequency of an annoying tapping sound suddenly gets twice as fast. What has happened to the period of time between tapping sounds?

39. ‹ What is the frequency of a pendulum that completes 20 cycles in 45 s? What is its period?

40. ‹ A mass suspended vertically from a spring is oscillating at 1.1 Hz. Find the period of oscillation.

41. ‹ A child on a pogo stick oscillates with an amplitude of 11 cm. What is the total distance the child travels during one period?

42. ‹ An oscillator triples in frequency. What happens to its period?

43. ‹ How many times does the Moon orbit around the Earth in the time that the Earth orbits once around the Sun?

44. ‹‹ A surfer estimates that the waves at a particular beach have an average height of 2.3 m, measured peak to trough. She also estimates that a buoy moves up and down at a rate of two cycles in 8 s. Using her estimates, find the total distance traveled in 12 s by a seagull riding the waves.

45. ‹‹ The angular frequency of an oscillator is $\omega = 6$ rad/s. Find the period.

46. ‹‹ The period of a simple harmonic oscillator is 2 s. Find the angular frequency.

47. ‹‹ A marathon runner makes 180 steps per minute. What is the frequency at which his feet hit the ground? At 20 miles he starts to walk at 90 steps per minute. What is his new pace frequency?

48. ‹‹ A 3 kg mass is hanging vertically from a spring suspended from the ceiling. The mass is pulled 10 cm from its equilibrium position and released. Find the speed of the mass as it passes through the equilibrium position. The spring constant is $k = 400$ N/m. (Assume no friction.)

49. ‹‹ A 1.2 m pendulum is pulled 13° away from equilibrium and let go. What is its frequency of oscillation?

# Chapter review

## Section 14.1

50. A mass suspended vertically from a spring oscillates with a frequency of 2.3 Hz. The amplitude of oscillation is 10.2 cm. Find the total distance the mass travels in 3.0 s.

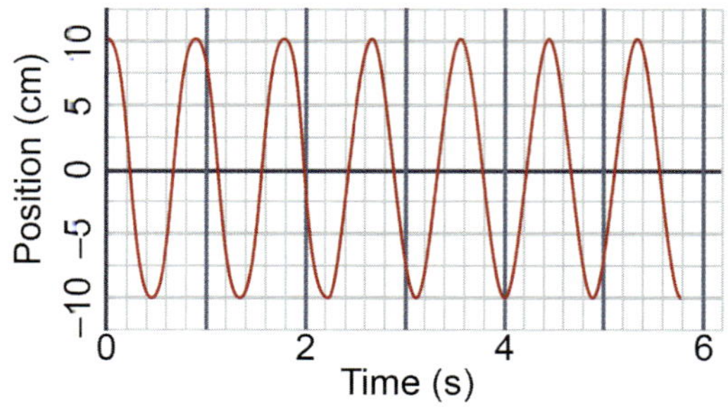


51. The graph above represents the motion of an oscillating mass that is suspended vertically from a spring.
    a. What is the amplitude of the oscillation?
    b. What is the period of the oscillation?
    c. What is the frequency of the oscillation?

52. While running, a car's engine might operate around 1,000 revolutions per minute.
    a. What is the engine's frequency in hertz?
    b. What is the period of one engine cycle (or revolution)?

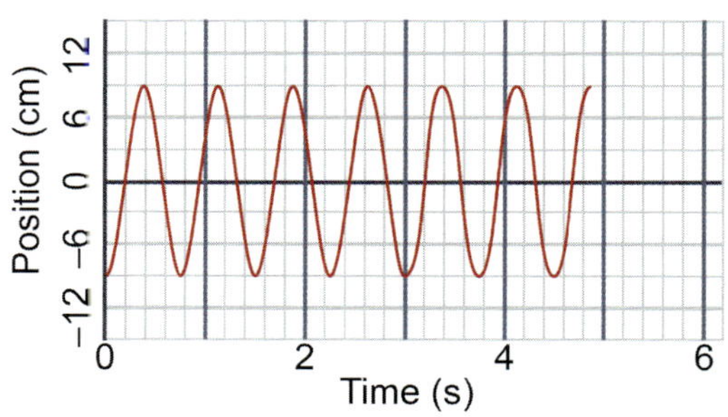


53. Using the position versus time graph of a harmonic oscillator, find the following:
    a. the period of oscillation
    b. the amplitude of oscillation
    c. the frequency of oscillation
    d. the frequency of oscillation if the period were halved

54. A pendulum has a kinetic energy of 1.6 J at the bottom of its swing. If the mass of the bob on the end of the pendulum is 3.1 kg, what height does the pendulum reach? (Assume that there is no friction.)

55. A 1.8 kg mass oscillates on a spring hanging from the ceiling. If the speed of the mass is 1.4 m/s as it moves through its equilibrium position, what is the potential energy of the system when totally compressed?

56. A mass of 1.33 kg oscillates at the end of a spring hanging vertically from the ceiling. The frequency of oscillation is 1.2 Hz and the total energy of the system is 0.001 J. Find the oscillator's amplitude. (Assume that there is no loss due to friction.)

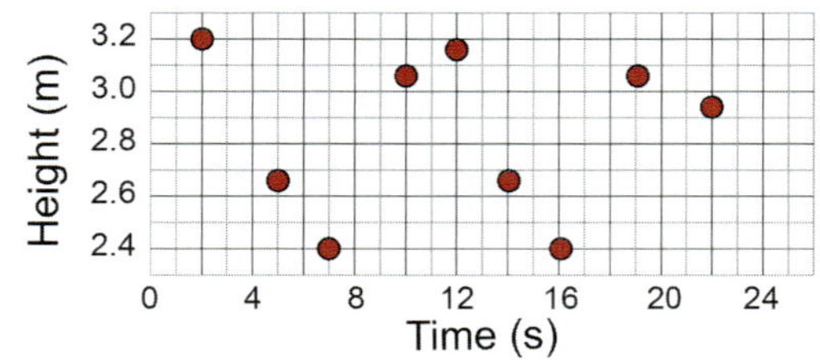


57. The graph shows data collected by observing the height of a buoy floating on the ocean over a time interval of 30 s.
    a. Is this an example of harmonic motion? If so, what are the period and amplitude?
    b. Had the observations continued, at what time would the buoy have reached its next maximum height?

## Section 14.2

58. A pendulum is oscillating at its natural frequency of 0.7 Hz. If you change the pendulum so that the period is now three times greater, how has the natural frequency changed?

59. A 0.75 kg mass on a spring oscillates at 3.1 Hz with a maximum displacement of 8.0 cm. Determine both the maximum velocity of the mass and its total energy.

60. A 1.3 kg mass on a horizontal, frictionless surface is attached to a spring and oscillates back and forth. Its maximum restoring force is 6.0 N. A similar system has a 2.0 kg mass and experiences a 4.0 N maximum restoring force. Which system has the greatest maximum acceleration?

61. The period of a rubber band oscillating at resonance is 2.5 s. If the period changes to 3.7 s and the rubber band oscillates at resonance, what happens to the natural frequency?

62. A mass of 1.6 kg hanging vertically from a spring with $k = 175$ N/m is pulled by a periodic force with a frequency of 1.4 Hz. Is the system at resonance?

63. A pendulum of length $L = 0.22$ m is oscillating. A cat swipes at the pendulum with a period of $T = 1.3$ s. Are the cat's applied forces resonant with the pendulum?

64. For a given oscillator, the position and acceleration at time $t = 3$ s are $x = 7$ cm and $a = -0.7$ m/s$^2$. Find the frequency of oscillation.

# Chapter review

## Standardized test practice

65. Which of the following is *not* an example of harmonic motion?

   A. a ball rolling down a hill
   B. pistons in a running car engine
   C. a swinging pendulum
   D. waves in a pond spreading outward from a stone that was dropped

66. What is the period of a wave that has a frequency of 10 Hz?

   A. $(v/10)$ s
   B. $(10/c)$ s
   C. 0.1 s
   D. $10v$ s

67. Which is *not* an example of stable equilibrium?

   A. a small mass suspended vertically from a spring
   B. a ball at the top of a hill
   C. a marble at the bottom of an empty bowl
   D. a pendulum at rest

68. Planet X orbits Black Hole Y every 2,000 yr. In revolutions per year, what is the frequency of Planet X's orbit?

   A. $10^{-6}$ rev/yr
   B. $5\times10^{-4}$ rev/yr
   C. 5 rev/yr
   D. $2\times10^{3}$ rev/yr

69. Two physical systems that can oscillate are (1) an object hanging from a string or wire (i.e., a pendulum) and (2) an object attached to a spring. Any such object has a specific *natural frequency*. Which of these systems changes its natural frequency when the object's *mass* changes?

   A. both
   B. neither
   C. only the pendulum
   D. only the spring system

70. Which of the following is an example of resonance?

   A. a pendulum swinging back and forth
   B. a small mass vertically suspended from a spring
   C. the flickering of a fluorescent light
   D. the musical note "B flat" played on a trumpet

71. Which is the best choice of equipment to measure the period of oscillation of a pendulum?

   A. your heartbeat
   B. meter stick
   C. triple beam balance
   D. stopwatch

72. The highest and lowest points of a mass oscillating on a spring are 20 cm apart. What is the amplitude of the movement?

   A. 10 cm
   B. 20 cm
   C. 30 cm
   D. 40 cm

73. Which is *not* an example of the natural frequency of a system?

   A. an electric motor spinning at 10 Hz
   B. a pendulum swinging at 2 Hz
   C. a mass on a spring oscillating at 4.7 Hz
   D. a boy on a swing going back and forth at 0.3 Hz

74. When constructing the support for a simple pendulum, which is the best choice for how to attach the pendulum's string to the stand?

   A. Attach the string to the stand using duct tape.
   B. Attach the string to the stand using wood glue.
   C. Clamp the string to a pin in the stand.
   D. Tie a loop of string around the pin in the stand.

75. Three pendulums are shown here, each labeled with its natural frequency. Which statement best describes the relationship between pendulum length and pendulum *period*?

   A. The greater the length, the greater the period.
   B. The greater the length, the lesser the period.
   C. Length is independent of period.
   D. One cannot tell from the information provided.

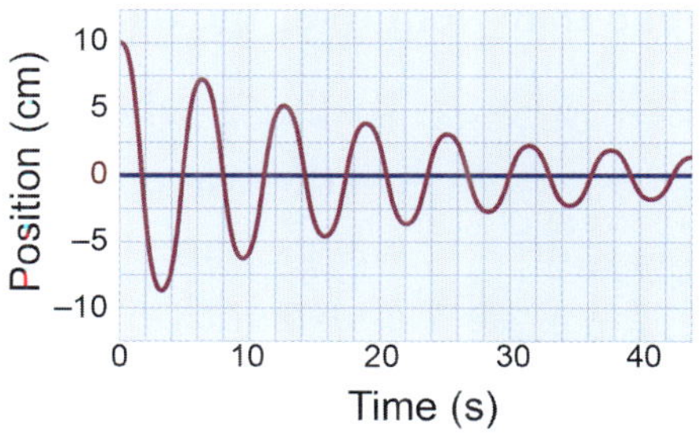


76. The position of an oscillating mass is graphed here as a function of time. Which quantity is changing as time passes?

   A. period
   B. frequency
   C. amplitude
   D. none of the above

# Chapter 15 Waves

In 1952, a magazine writer claimed that supernatural forces were causing ships and planes to disappear in a place the writer named the *Bermuda Triangle*—a triangle bracketed by Miami, Puerto Rico, and the island of Bermuda. Subsequent investigations showed that the losses, while tragic, were statistically unremarkable. Nevertheless, ships at sea *can* disappear suddenly with hardly a trace. The culprit, in many cases, is a *rogue wave:* a mountain of water that can break a ship apart in a single blow.

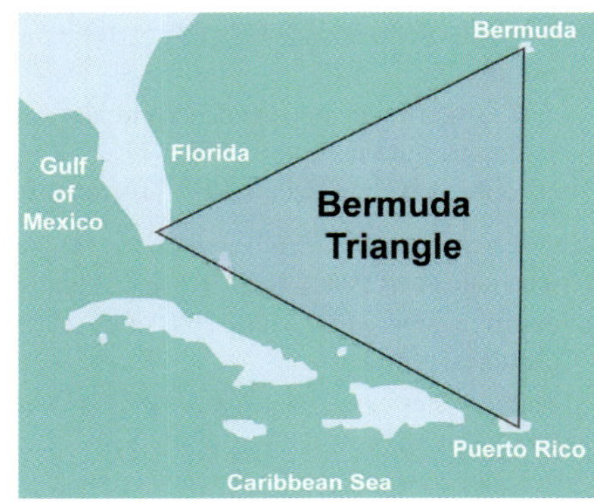


"Most people don't survive encounters with such waves," according to Sebastian Junger, author of *The Perfect Storm.* One sailor who did survive was English adventurer Beryl Smeeton. Recalling a rogue wave off of Cape Horn, she wrote, "The whole horizon was blotted out by a huge grey wall... a wall of water with a completely vertical face, down which ran white ripples, like a waterfall." The wave flipped her 46-ft (14-m) boat and stripped it of its masts, but the crew survived. Others have not been so lucky. In 1976, the oil tanker *Cretan Star* reported being "struck by a huge wave that went over the deck." The ship was never heard from again.

To those who have survived to tell the tale, the appearance of a rogue wave seemingly violates natural law. But in fact rogue waves are the result of a basic physics concept called *constructive interference.* When the crests of several unrelated waves arrive in one place simultaneously, their heights are added—even if they may have largely canceled one another out in other nearby locations. Research suggests that these waves can also be amplified by ocean currents that crowd their crests together. Such currents frequent Africa's Cape Horn and the Gulf Stream, an immense flow of warm water that crosses the Atlantic Ocean from the Caribbean Sea to Western Europe. Fortunately, rogue waves are rare!

# Chapter study guide

Chapter 15

## Chapter summary

In the previous chapter, you learned about many different kinds of oscillators: pendulums, masses on springs, and so on. What is the difference between an oscillator and a *wave*? A wave is an oscillation that *travels through space*. Waves are found all around us in nature, musical instruments, technology, and medical and industrial applications. Waves can reflect, refract, diffract, resonate, and interfere with each other. In this chapter, you will learn about the physical characteristics of waves and the behaviors associated with wave propagation.

## Learning objectives

By the end of this chapter you should be able to

- describe waves and wave pulses and provide examples;
- define amplitude, frequency, wavelength, speed, and phase for a wave and identify each graphically;
- solve problems involving the speed, frequency, and wavelength of a wave;
- distinguish between transverse and longitudinal waves and provide examples of each;
- describe wave propagation in various types of media;
- describe wave behaviors of reflection, refraction, diffraction, and resonance and provide examples of each;
- describe constructive and destructive interference of waves and provide examples of each;
- describe standing waves and resonance and provide examples of each; and
- describe medical applications of waves.

## Investigations

15A: Waves
15B: Wave interactions
15C: Interference
15D: Wavelength and standing waves

## Chapter index



## Important relationships

$$v = f\lambda$$

## Vocabulary

| | | |
|---|---|---|
| **wave** | **wavelength** | **transverse** |
| **polarization** | **longitudinal** | **crest** |
| **trough** | **wavefront** | **reflection** |
| **concave** | **convex** | **refraction** |
| **diffraction** | **absorption** | **superposition principle** |
| **constructive interference** | **destructive interference** | **node** |
| **antinode** | **mode** | **standing wave** |

## 15.1 - Waves

Think about dropping a stone into a pond on a very calm day. As the stone breaks the surface of the water, the surface oscillates up and down—in harmonic motion. But something else happens to the water surface: *Ripples form and spread out.* Everything the ripples touch also oscillates up and down with the same frequency. An oscillation that travels is a **wave**, and waves are the subject of this section. Both sound and light are waves. There are even gravity waves created when black holes crash into each other.

### The importance of waves

**Why are waves important?**

Waves are an essential way by which energy travels. Think about the example of a stone falling into a pond. The ripple causes the water and objects floating in it to move up and down some distance away. Where did their energy of motion come from? The answer is that *it came from the stone and was carried by the wave.* When the stone hit the water surface some of its kinetic energy was converted to waves. The waves spread out over the surface of the pond, dispersing the kinetic energy of the stone over a much broader area of space than was directly touched by the stone itself.

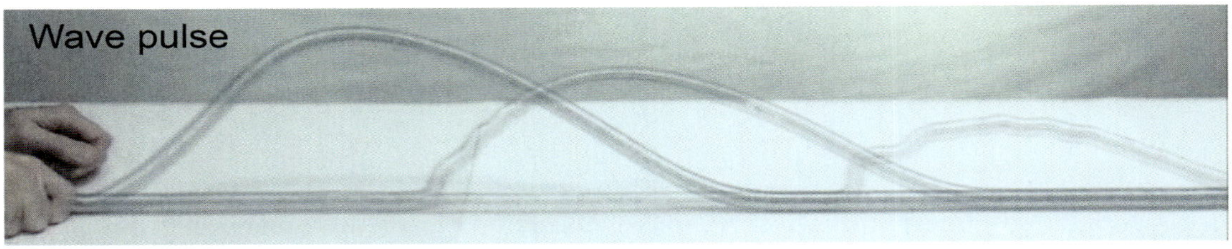


**Wave pulses**

A fundamental reason for why waves are important is that any disturbance that releases energy often produces waves. The waves spread the energy out and propagate the disturbance through space, affecting other regions, which may be quite far away. A *wave pulse* on a long spring is a good example. To make a wave pulse on a spring, disturb one end by rapidly jerking it up and down once. The disturbance quickly moves away from your hand and travels along the spring. Areas of the spring far away from your hand are affected as the wave pulse reaches them. As the pulse moves, the energy of the disturbance is spread along the spring and also dissipates through friction. The wave pulse gets smaller until the spring is at rest again.

**Earthquakes**

This aspect of waves is frighteningly displayed by earthquakes. In an earthquake a tremendous amount of elastic potential energy is released when stressed rock deep underground suddenly slips and realigns itself. That energy is largely released as seismic waves that oscillate the ground up, down, and sideways. Just as a wave pulse moves along a spring, seismic waves race away from the earthquake epicenter at the speed of sound. When the seismic waves interact with matter, energy can be released—which can topple buildings 100 miles away during a powerful earthquake! Architects must design buildings to withstand the energy from oscillations (or shaking) caused by earthquakes.

# Investigation 15A Waves

**Essential questions** *What is a wave and what are the properties of waves?*

Interactive simulation

A wave is a traveling form of energy that carries oscillations from one place in space to another. Sound and light are both waves and share characteristics of frequency and wavelength with familiar water waves. This investigation will use a simulation of water waves to characterize the properties of frequency, amplitude, and wavelength.

## Part 1: Match a wave's properties

1. Open the interactive simulation. You will create a mathematical model of a wave (red line) to match the blue waves representing water.
2. Adjust the amplitude and wavelength to match the blue wave.
3. Run and Pause the waves. Adjust the frequency until the red circle bobbing in your model matches the bobbing of the floating ball.

a. Describe how changing the amplitude changes the wave.
b. Describe the effect of changing the wavelength.
c. Describe the effect of changing the frequency.
d. What are the frequency, amplitude, and wavelength of the blue wave?
e. Draw a graph showing the amplitude and wavelength of this wave.
f. Calculate the speed of the wave. Show your work.

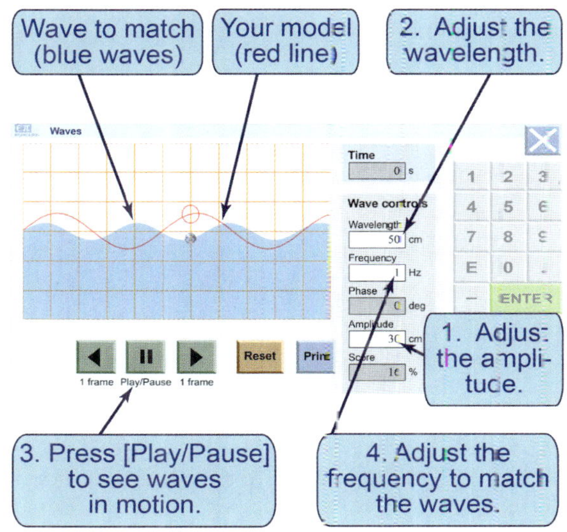


## Part 2: Transverse and longitudinal waves

1. Hold one end of a Slinky® spring (or other long spring) and have your partner hold the other end. Stretch the spring a little bit so that it is not slack.
2. Create *transverse* waves by moving your hand side to side.
3. Create *longitudinal* waves by moving your hand sharply toward your partner.
4. Repeat the above steps, but this time using a wave motion rope or other heavy string.

a. What are the differences between these two types of waves? Describe the characteristics of each in words.
b. Can you make both types of waves on both pieces of equipment? Why or why not?
c. Can you create waves of different velocities with the spring or rope? If so, how?

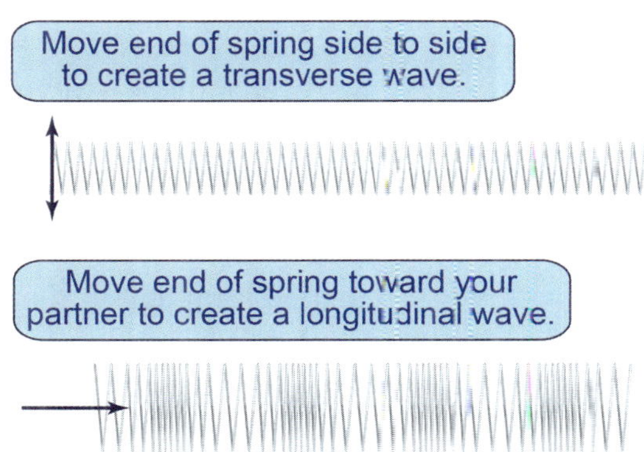

**Waves in time and space**

Oscillations have cycles and so do waves. Waves spread out and move, so their cycles extend over both *time* and *space.* Consider a water wave on the surface of a pond. If you watch one single location in space as *time* goes by, the water oscillates up and down at the frequency of the wave. A ball floating at this location oscillates up and down over time. Next, consider freezing the entire surface of the pond at a single instant of time. In the frozen surface there is also an up–down repeating pattern that repeats in *space.*

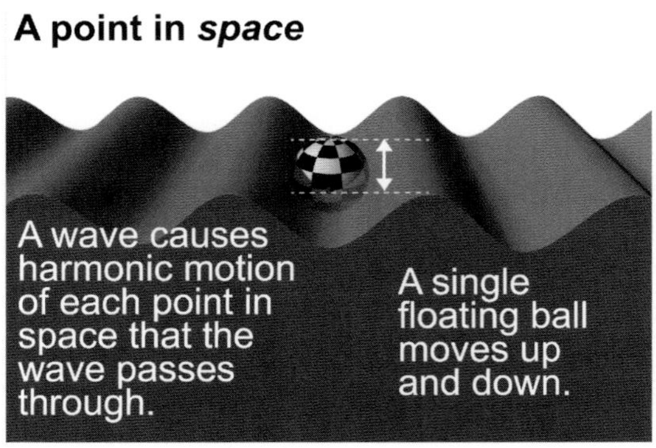


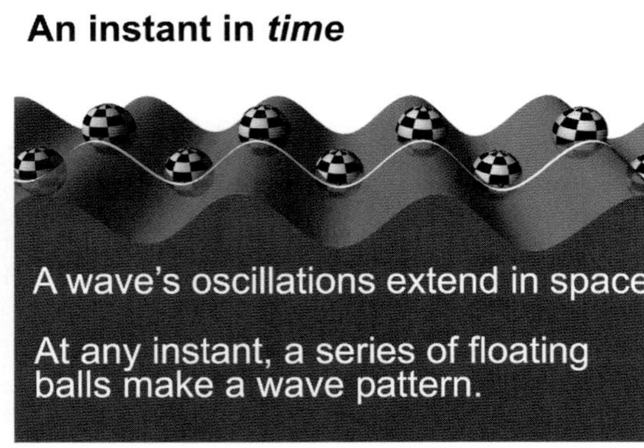


**Wavelength**

At any moment of time, the cycles of a single wave are described by its *wavelength.* The **wavelength** is the spatial length of one complete cycle of the wave at one instant of time. Wavelength is a new property of waves that does not exist for a stationary oscillator such as a pendulum.

**Measuring wavelength**

If you could "freeze" a wave in place, you could measure the wavelength as the distance between two successive crests (or peaks) of the wave. Alternatively, wavelength can be described as the distance between two successive *troughs* (or lowest points) of the wave.

**Frequency**

The frequency of a wave is just like the frequency of an oscillator. A wave with a frequency of 10 Hz repeats itself 10 times per second *at every place in space the wave reaches.* This allows waves to carry *information* over great distances. When you listen to a guitar, the sound wave carries information about the vibration of the string from the instrument to your ear. A light wave carries information about color and space to your eyes. Electrical waves in wires and light waves in optical fiber cables carry Internet and television information. A microwave carries cellphone conversations. In nature and in human technology, waves carry energy and information from one place to another. The information could be sound, color, pictures, commands, or virtually anything else.

**Amplitude**

The amplitude of a water wave is the maximum amount the wave causes the water to rise (or fall) compared to its average resting level. While the amplitude of a water wave is measured as a height in meters, the amplitudes of other waves may have different units. For example, a sound wave is a pressure oscillation and therefore its amplitude is measured in force per unit area, not distance. Most waves become smaller with increasing distance from the source, just as ripples decrease far from the splash of a pebble. Even though the *amplitude* of the wave decreases, the *frequency* remains the same.

**Mechanical waves**

Sound and water waves are *mechanical waves*, which means that they propagate via the oscillation of matter in a medium. The medium can be a solid, liquid, or gas. As you will learn in Chapter 22, light is an *electromagnetic wave* that can propagate in a vacuum —without any medium whatsoever.

**Speed of a wave**

The speed of a wave is not quite the same as the speed of a moving object. Think about dropping a stone into a pond in which a ball is floating a few meters away. When the ripples reach the ball, the ball oscillates up and down. The speed of the wave is the speed at which the ripples spread out from where the stone fell. Physicists use the word *propagate*, which means to "spread out and grow." The speed of a wave is the speed at which the *wave* propagates, or spreads itself out.

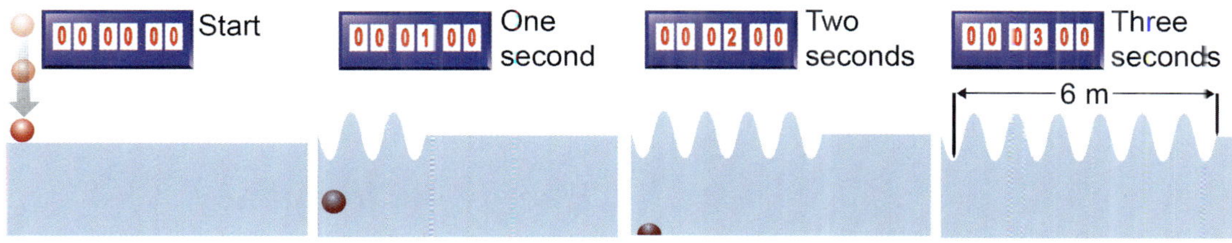


**Speed, frequency, and wavelength**

As a wave moves forward, it advances one wavelength with each complete cycle. Speed is distance divided by time, and therefore the speed of a wave is its wavelength $\lambda$ divided by the period $T$ of its cycle. Since the frequency is the inverse of the period ($f = 1 / T$) we usually write the speed of a wave in terms of frequency and wavelength. The result is true for sound waves, light waves, and even gravity waves. Frequency multiplied by wavelength is the speed of the wave.

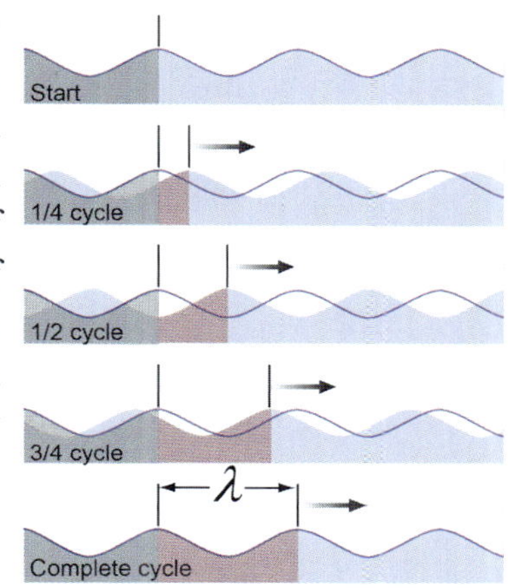


$$v = \frac{d}{t} = \frac{\lambda}{T} = \lambda f$$

Interactive equation

**Speed, frequency, and wavelength of a wave**

$$(15.1) \quad v = f\lambda$$

$v$ = speed of the wave (m/s)
$f$ = frequency (Hz)
$\lambda$ = wavelength (m)

**Waves have a wide range of speeds**

Waves have a wide range of speeds. Lab-sized water waves are fairly slow; a few miles per hour, or 0–5 m/s, is typical. Deep ocean waves, such as tsunamis, can be much faster, reaching 600 mph (268 m/s) or more—as fast as a jet airliner! Light waves are extremely fast—300,000 kilometers *per second,* which is $3\times10^8$ m/s or 671,000,000 mph. Sound waves travel at about 343 m/s in air—faster than water waves but much slower than light.

**Calculating the speed of a wave**

Two men use a long spring to create a wave. The wavelength is 2 m and the frequency is 2 Hz. How fast is the wave traveling along the spring?

*Asked:* speed $v$ of the wave ("how fast")

*Given:* wavelength $\lambda = 2$ m, frequency $f = 2$ Hz

*Relationships:* $v = f\lambda$

*Solution:* Insert the values into the equation:

$$v = f\lambda = 2\ \text{Hz} \times 2\ \text{m} = 4\ \text{m/s}$$

*Answer:* The wave travels at 4 m/s.

**Sound waves in solids**

Sound waves travel faster through water (1,500 m/s) than they do through air (343 m/s). Sound waves also generally travel faster through solids than through liquids. A seismic wave may travel at 2,000 to 8,000 m/s through the Earth's crust.

**Waves are a form of energy**

Waves are a form of moving energy. Waves may move *through* matter, but a wave is not matter itself. For example, a water wave is *not* the water, which exists independently of the wave. A water wave is a form of pure energy. When a water wave moves across a pond surface, the wave's energy causes the matter to respond—by moving up and down. Once the wave passes, the matter returns to equilibrium again. Waves transfer energy from one location to another.

**Energy and frequency**

The energy of a wave increases with frequency. The three waves in the diagram below have the same amplitude and different frequencies. The wave with the higher frequency transfers more energy. This is true for almost all waves, including water waves, sound, and light.

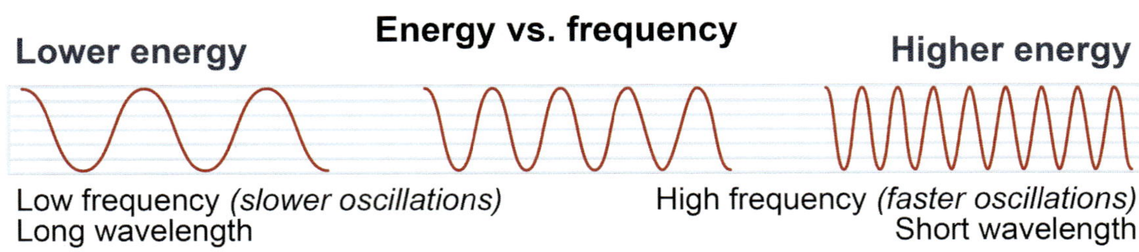


**Energy and amplitude**

The energy of a wave also increases with amplitude. If two waves have the same frequency, the wave with the larger amplitude transfers more energy. With water waves, larger amplitude means that the wave lifts the water a greater distance above its equilibrium level. With sound waves, larger amplitude means louder sound. With light waves, larger amplitude means brighter light.

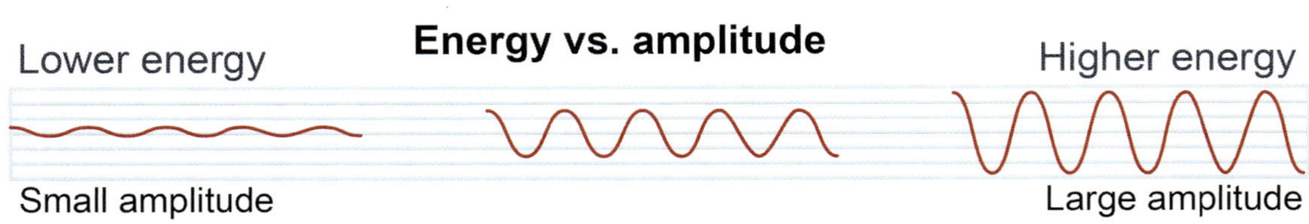


**Amplitude decreases over space**

As a wave spreads out, its amplitude decreases. This can happen for two reasons. One is damping, a process in which friction reduces the wave energy over time. But there is also a second reason that amplitude decreases. As a wave propagates, its energy may spread out over a larger area. That leaves less energy in any given portion of the wave. This is the main reason why ripples get smaller and smaller as they spread. It is also the reason why light gets dimmer as you get far from a bulb and sound gets fainter far from its source.

Amplitude tends to decrease over time and space.

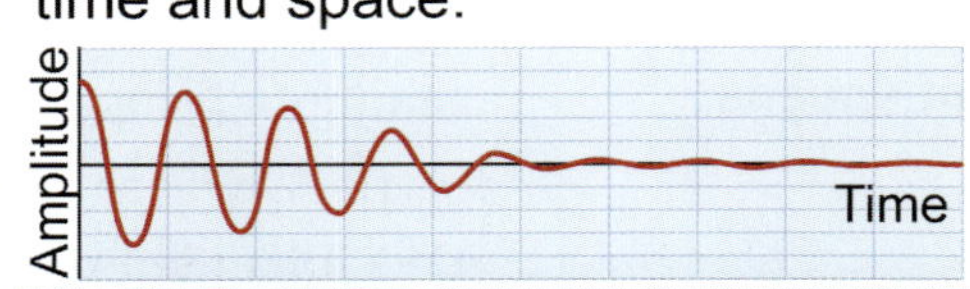

**Waves in three dimensions**

Space is three dimensional and waves can cause oscillations in all three dimensions as well as *scalar* oscillations that have no direction. How a wave oscillates relative to its direction of motion is one important way to classify waves. To explain this we define the forward dimension as the direction the wave moves. The other two dimensions—up–down and left–right—are both perpendicular to the direction the wave moves.

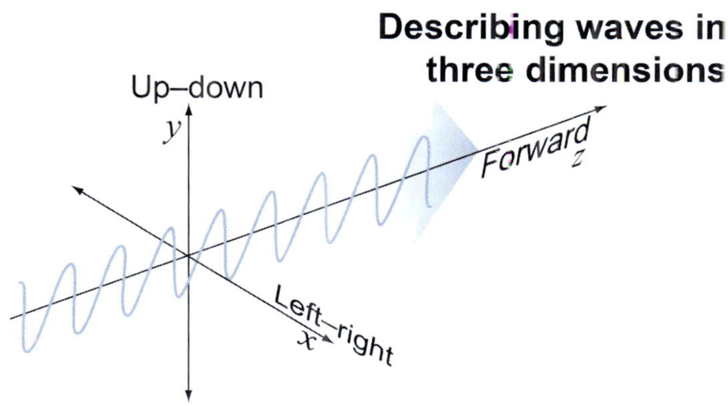


**Transverse waves**

A **transverse** wave causes oscillations perpendicular to the forward direction the wave moves. Waves in a stretched string are transverse waves because the wave moves along the string and the oscillations are up and down, perpendicular to the line of the string. Light is also a transverse wave, although the explanation for *why* will have to wait until Chapter 22 when we discuss electric and magnetic fields.

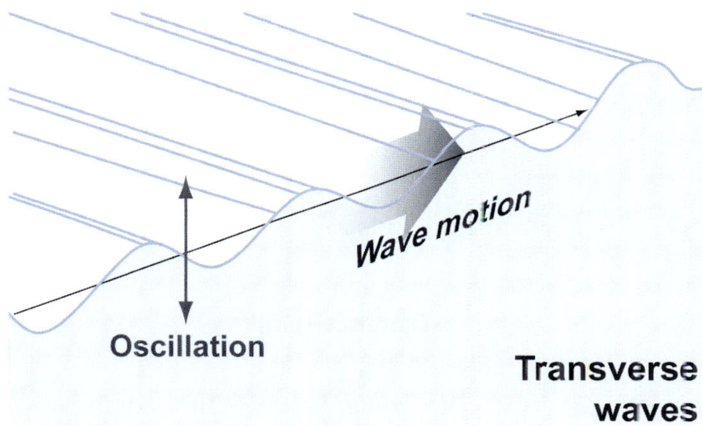


**Polarization**

A spring can be used to create transverse waves. Shaking a long spring up and down versus side to side demonstrates the property of *polarization* that is common to all transverse waves. **Polarization** describes the direction of the oscillation in a plane perpendicular to the direction the wave moves. A transverse wave has a *polarization* because there are two directions perpendicular to the motion of the wave. For example, a wave on a spring moving in the $z$-direction could be polarized in the horizontal $x$-direction, the vertical $y$-direction, or any other direction in between.

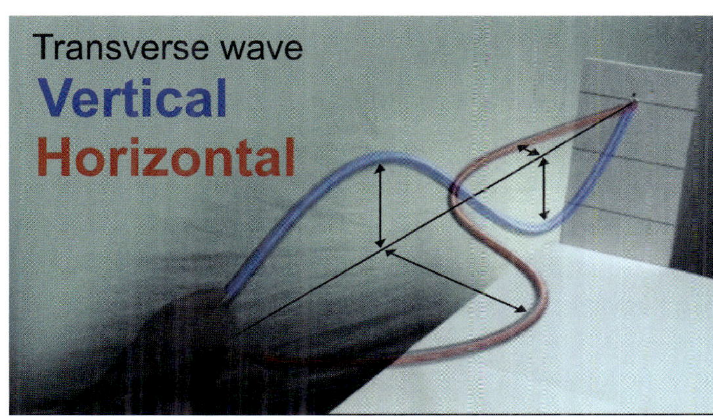

**Longitudinal waves**

If you rapidly move the end of a Slinky® spring forward and backward again, a *longitudinal* compression wave moves along the spring. A **longitudinal** wave is either a *scalar* oscillation or an oscillation of the medium that moves back and forth in the same direction as the wave travels. The compression wave on the Slinky® moves along the length of the slinky. The oscillations are the "bunching" and "stretching" that move along through the spring as the wave travels. Sound is another example of a longitudinal wave.

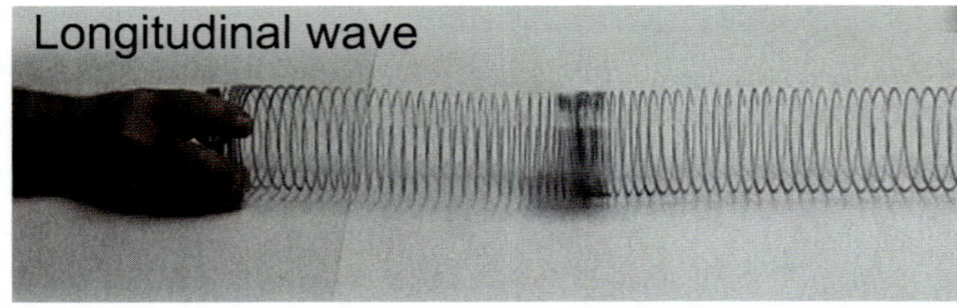


Longitudinal compression wave

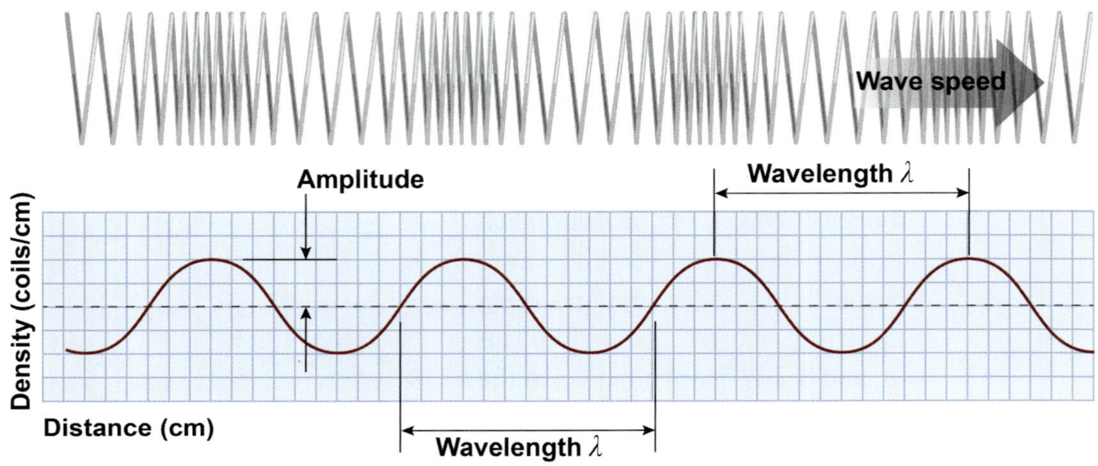


**Amplitude and wavelength**

The *amplitude* of the compression wave is the difference between the maximum "bunching" and the average, resting spacing of the spring. The amplitude of any longitudinal wave is the difference between the average and the maximum displacement away from average of the parameter that is oscillating. For the Slinky® spring the oscillation is in the density of coils per centimeter. For a sound wave the oscillation will be of air pressure. The *wavelength* of a longitudinal wave is the length of one complete cycle of the amplitude oscillation. The graph shows the wavelength as related to the motion of the spring. This is the same as the distance between one maximum compression and the next maximum compression.

**Perpendicular and parallel oscillations**

The key distinction between transverse waves (such as *light*, which you will learn about in Chapter 22) and longitudinal waves (such as sound, which you will learn about in Chapter 16) is the relationship between the direction of the oscillation and the direction the wave travels. Oscillations in transverse waves are *perpendicular* to the direction the wave travels—meaning that they are rotated 90° away from the wave's direction of motion. Oscillations in longitudinal waves are either *parallel* to the direction the wave travels—such as the compression wave on the Slinky®—or they are oscillations of a scalar variable, such as pressure.

# Section 1 review

Waves are oscillations that travel. Waves have amplitude and frequency just like oscillators, but waves have the additional property of *wavelength.* A wave moves one wavelength forward in each cycle, so the speed of a wave is its wavelength divided by its period. This is equivalent to frequency times wavelength. A wave carries energy that is proportional to both amplitude and frequency. The higher the amplitude, the higher the energy at a given frequency. At equal amplitudes low-frequency waves have less energy than high-frequency waves. Transverse waves cause oscillations that are perpendicular to the direction of the wave's motion, whereas longitudinal waves cause oscillations that are parallel to the wave's motion.

| | |
|---|---|
| **Vocabulary words** | **wave, wavelength, transverse, polarization, longitudinal** |
| **Key equations** | $v = f\lambda$ |

## Review problems and questions

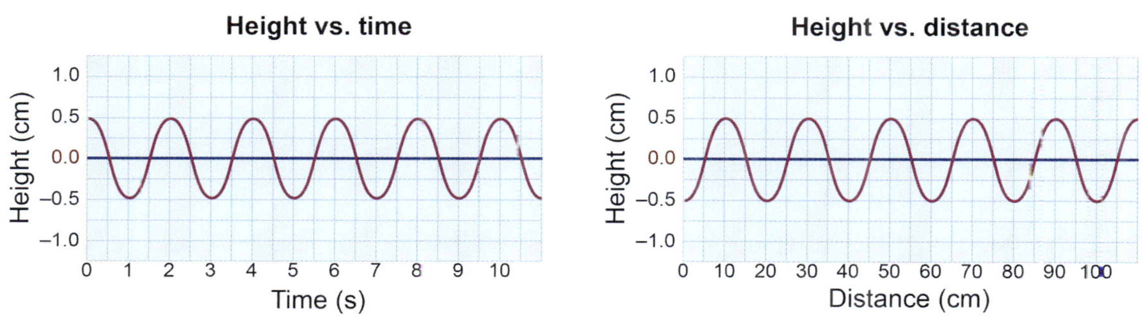


1. The figure above shows a graph of the oscillation of a single point on a transverse wave as a function of time and a second graph of the waveform at a given instant as a function of distance. Use the graphs to answer the following questions.
   a. What is the frequency of the wave?
   b. What is the wavelength of the wave?
   c. What is the amplitude of the wave?
   d. What is the speed at which the wave propagates?

2. A water wave has a frequency of 2 Hz and a wavelength of 1.5 m. What is the speed at which this wave travels?
   a. 0.75 m/s
   b. 1.5 m/s
   c. 2.0 m/s
   d. 3.0 m/s

3. a. If you are using a water tank in an investigation into waves, is a still image or a video a better choice for measuring wavelength?
   b. How about for measuring frequency?
   c. Amplitude?
   d. Velocity of the wave?

## 15.2 - Wave propagation

The word *propagate* means *to spread out and grow.* Waves propagate outward from their source, carrying both energy and information. The direction and speed at which a wave propagates depend on many factors, such as the shape of the wave and the medium through which the wave moves.

### How waves propagate

**How does a wave propagate?**

Sound and water waves propagate through matter because of connections. Any disturbance in one place causes a disturbance in the matter immediately adjacent, which disturbs the matter adjacent to *that,* and so on. Consider a stone falling into water. Some water is pushed aside where the stone crosses the surface (left). The higher water pushes the water next to it out of the way as it tries to flow back down to equilibrium (center). The water that has been pushed then pushes on the water in front of it, and so on. The wave spreads through the interaction of each bit of water with the water immediately next to it (right).

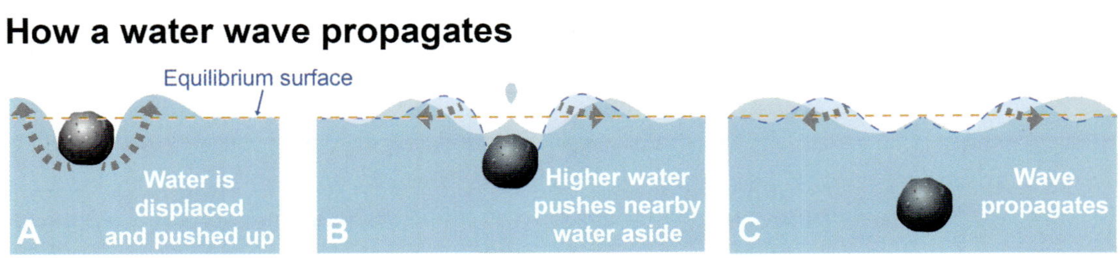


**Crests, troughs, and wavefronts**

Water waves propagate in two dimensions along a surface. Think of the expanding circular ripples spreading out on the water from the drop of a stone. Each circle represents a **crest** or **trough** of the expanding wave. A crest represents all the high points of one cycle. A trough represents all the low points. We often draw the crests on diagrams to show how a wave moves. The crest is sometimes called a **wavefront**. A wavefront describes the two- or three-dimensional shape of the crest of a wave.

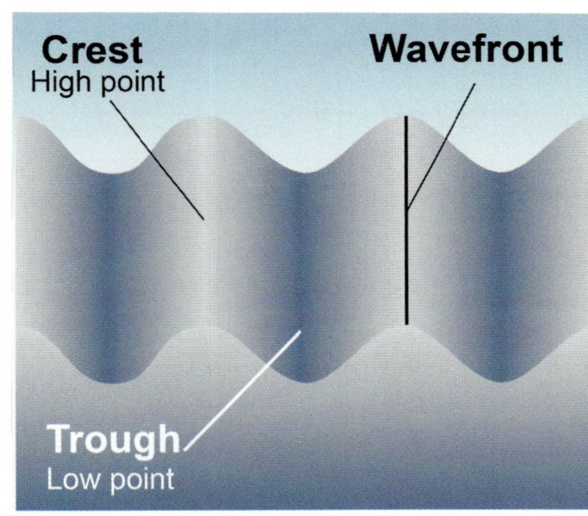


**Plane waves and circular waves**

*Waves always move perpendicular to their wavefronts*. This is easiest to see with circular waves and plane waves. The wavefronts of a plane wave are straight lines and the wavefronts move perpendicular to the lines. The wavefronts of a circular wave are circles and the wave moves radially outward from the center. At each point on the circle the wave expands in the direction perpendicular to the wavefront. This property of waves is very important to technology. If you want to control the direction of a wave, then you do it by changing the shape of its wavefront.

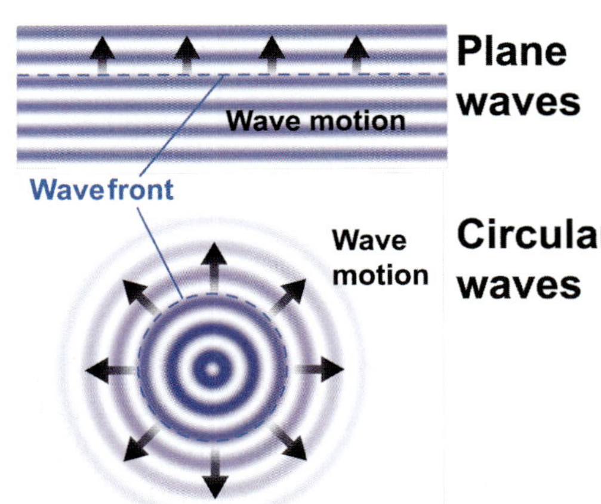

# Investigation 15B Wave interactions

**Essential questions** *What happens when a wave hits an obstacle?*

Interactive simulation

Waves interact with matter in four fundamental ways: reflection, refraction, diffraction, and absorption. This simulation allows you to see how each of these interactions affects waves of different frequency and wavelength.

## Part 1: Investigate reflection

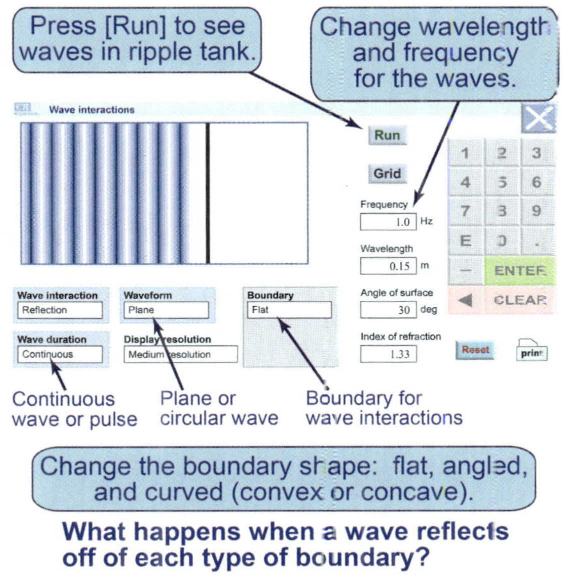


The simulation defaults to continuous plane waves reflecting from a flat surface.

1. Press [Run] to watch the waves propagate.
2. Change the wavelength and/or frequency (using the keypad) and press [Run] to see the new simulation.
3. Investigate three different boundaries: angled wall; curved, *concave* wall; and curved, *convex* wall.

a. Sketch a plane wave reflecting from a straight wall. How does the direction of the wave change?
b. Sketch a plane wave reflecting from an angled wall. How does the direction of the wave change?
c. Sketch a plane wave reflecting from a concave wall and a convex wall. Which one diverges *the wavefronts? Which one* converges *the wavefronts?*

## Part 2: Investigate refraction, diffraction, and interference

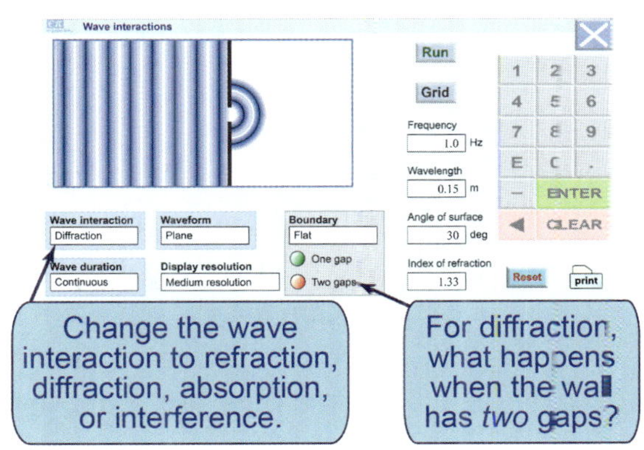


1. Investigate refraction of plane waves for two cases: a flat boundary and an angled boundary.
2. Investigate diffraction of plane waves for three cases: around a half wall and through a flat wall with either single or double gaps in its middle.
3. Investigate diffraction further by varying the wavelength for the single-gap wall.
4. Investigate absorption using a flat boundary.
5. Investigate interference using two circular waves.

a. Sketch the wavefronts refracted by the flat boundary. What happens to the direction and wavelength?
b. Sketch wavefronts refracted by an angled boundary. What happens to direction and wavelength?
c. Sketch the wavefronts for a plane wave passing by a half wall. Describe what you observe using the term *diffraction*?
d. Sketch the wavefronts passing through the single gap.
e. Sketch the wavefronts passing through the double gap. Describe what you observe using the term *interference*.
f. How does varying the wavelength change the way the waves diffract?
g. Sketch the wavefronts before and after they interact with the absorbing material. What happens to the wave? Propose a hypothesis to explain what happens to the energy of the wave.
h. Sketch the wavefronts for two circular waves. Where do the waves cancel each other out?

## Reflection

**Reflection**

**Reflection** causes a wave to change direction and may also change the shape of its wavefront. Reflection occurs for both transverse and longitudinal waves. A plane wave encountering a straight boundary reflects in a new direction while keeping the same waveform. The same is true of a circular wave. A circular wave reflecting from a straight boundary will still be a circular wave.

**Reflection from a fixed boundary**

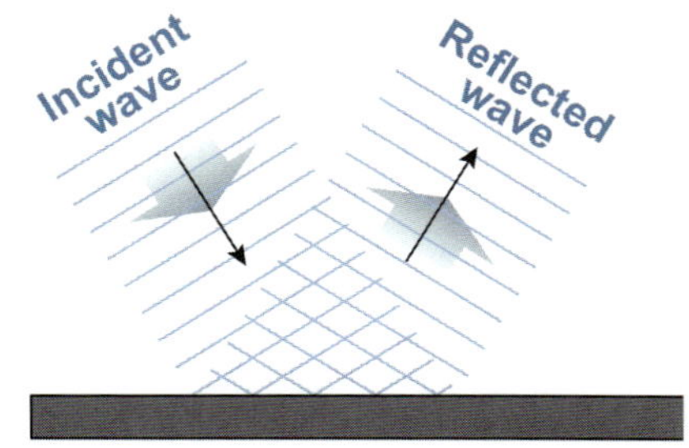


**Boundaries**

Reflection occurs at *boundaries* where conditions change, such as the edge of a pool or the wall of a room. The end of a long, snaky spring can also be considered a boundary for waves on the spring since it represents a change of conditions.

**Fixed boundary for a spring**

**Open boundary for a spring**

**Boundary types**

The kind of reflection that occurs at a boundary depends on whether the boundary is *fixed* or *open.* A fixed boundary does not move in response to the wave. A transverse wave pulse on a spring reflects onto the opposite side of the spring when encountering a fixed boundary. The free end of a spring is an open boundary. The end moves in response to a wave traveling along the spring. A wave pulse reflects back on the same side of the spring when encountering an open boundary.

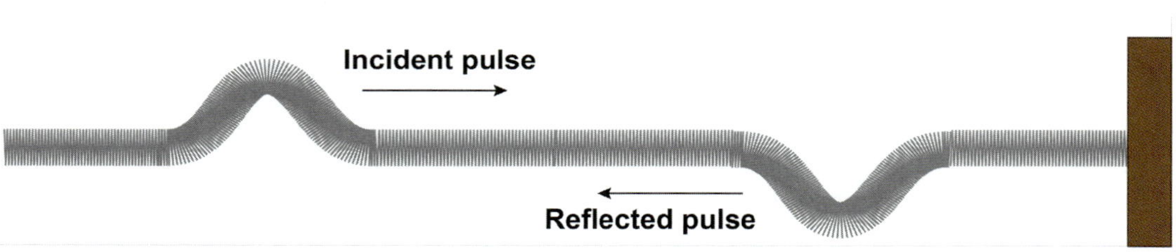


**Curved boundaries**

Curved boundaries can change the shape of a wave, altering its direction. A **concave** reflector can turn a plane wave into a circular wave that converges to a point. A circular wave reflecting off the same shape can turn into a plane wave. A **convex** reflector can turn a plane wave into a circular wave that diverges from a point. The convex shape can also turn a circular wave into another circular wave with a different curvature. Curved boundaries are extensively used in communications technology such as satellite dish receivers.

**Concave reflector**

**Convex reflector**

**Refraction**

Some boundaries divide regions where conditions change, such as the depth of water. A plane wave in water that crosses a depth boundary changes its direction. This process is called **refraction**. Refraction is the process by which a wave changes direction as its wavefront is altered by passing *through* a boundary. A plane wave passing through a straight boundary remains a plane wave but changes direction as shown in the diagram.

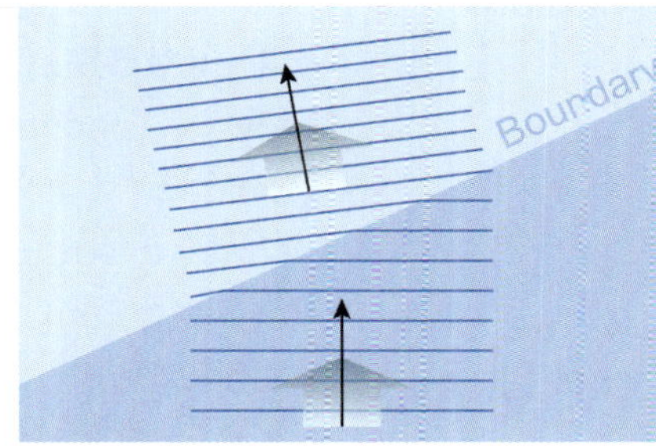


**Refraction**
occurs when a wave changes direction crossing a boundary.

**What causes refraction?**

Refraction occurs when the speed of the wave is different on the two sides of a boundary. Consider three points on the crest of a plane wave crossing a boundary between deep and shallow water. Water waves travel slower in shallow water compared to deep water. Point A reaches the boundary first and starts to move slower. Point B hits next and point C hits last. The wavefront over the shallow section has a different angle than the original wave. A wave moves in the direction perpendicular to the wavefront, and therefore the change in the wavefront causes a corresponding change in direction of the wave as it crosses the boundary.

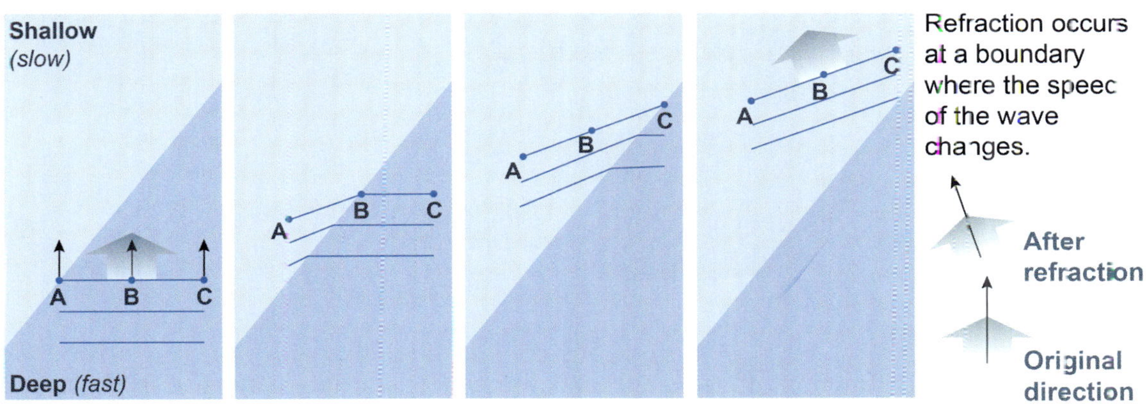


Refraction occurs at a boundary where the speed of the wave changes.

**Do the wavelength or the frequency change?**

Refraction usually changes the wavelength as well as the direction. Note in the diagram that the wavefronts get closer together on the slow side of the boundary. This occurs because the frequency of the wave does *not* change. The same number of crests per second must enter the boundary as leave. If the speed $v$ gets slower and the frequency $f$ stays the same, then, according to the equation $v = f\lambda$, the wavelength $\lambda$ must decrease.

**Is refraction useful?**

Refraction occurs with all types of waves, including transverse and longitudinal waves, and is important in many technologies. In optical systems, such as in cameras and telescopes, curved refracting surfaces bend light waves to create images. Ultrasound imaging detects the changes in tissue density within the human body by measuring the refraction of very high frequency sound waves.

**A paradox**

Sound and light are both waves, but they behave differently when they encounter obstacles. Consider a speaker and a lamp behind a corner. You cannot see the lamp but you can hear the sound! Why? Why does a sound wave bend around a corner but a light wave does not!

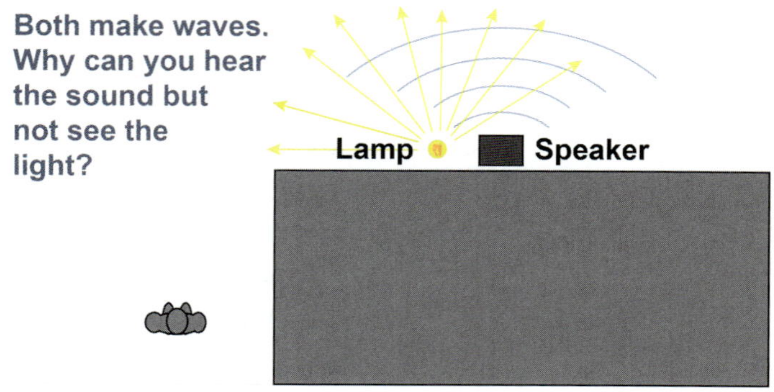


**The answer to the paradox**

The answer is *diffraction.* **Diffraction** is a property of waves that allows a wave to bend around an obstacle, such as a corner. Sound waves diffract around the corner because the wavelength of sound is typically a few centimeters. This wavelength is comparable to the size of the corner itself. Light waves also diffract *slightly* around the corner but the typical wavelength of a light wave is on the order of $10^{-5}$ cm. This is far smaller than the scale of the corner itself, and the amount of diffraction is imperceptible.

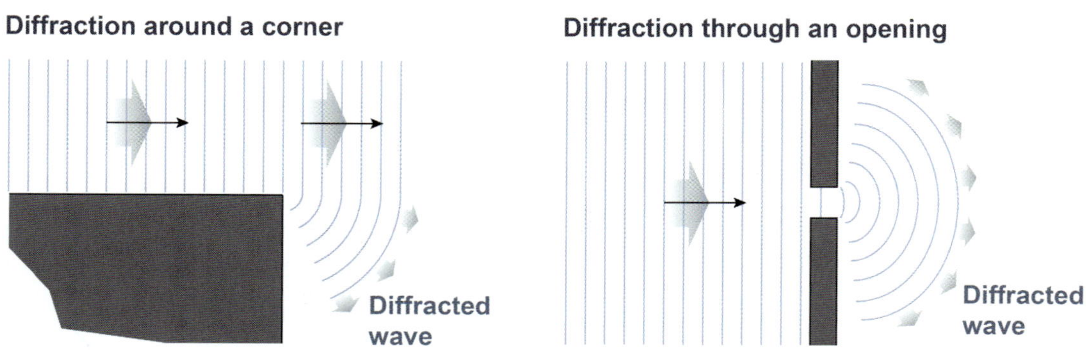


**What does diffraction do?**

Diffraction often changes the direction and shape of a wave. A plane wave passing a corner is diffracted so that its wavefronts become nearly circular at the edges. The same plane wave passing through a slot spreads out into a circular wave on the other side of the slot. Diffraction of sound explains why you can hear someone in another room even if the door is open only a crack. Diffraction causes a sound wave to spread out from the crack. This also occurs for light waves but you typically have to look at fairly small systems or very fine details to see the effects of diffraction of light.

**Is diffraction important?**

Diffraction has many implications for technology. Radio waves have wavelengths that can be tens of meters to kilometers long. This allows radio signals to bend around obstacles such as mountains. Cellphone transmissions, in contrast, use much shorter wavelengths, typically 6–12 cm. As a result, cellphone waves diffract (spread) less, which is why good reception typically requires a direct line of sight from the phone to the nearest tower.

**How do waves change as they move?**

When a wave travels through matter some of the wave's energy is dissipated through processes we loosely call "friction." This loss of energy is more properly called *absorption*. **Absorption** describes the transformation of wave energy into other forms of energy that occurs when a wave travels through matter. Absorption at some level affects virtually all waves, including sound, light, water, and even seismic waves. One of the few exceptions is an electromagnetic wave, such as visible light, traveling through a pure vacuum. Light waves can travel through the vacuum of space across the known universe without significant absorption.

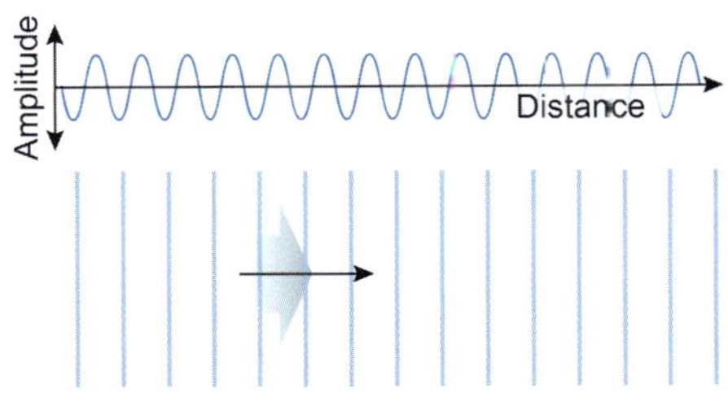

**Wave traveling with no absorption**

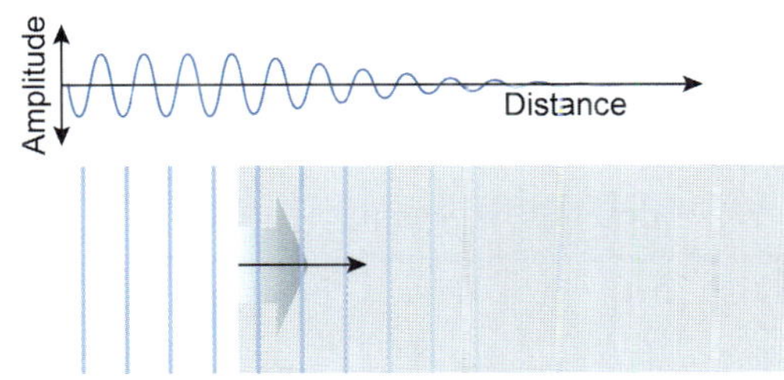

**Wave traveling *with* absorption**

**What changes in absorption?**

As a wave is absorbed, the amplitude of the wave decreases. Except in special cases *the frequency remains the same.* How quickly the amplitude decreases depends on both the type of wave and the medium the wave is moving through. For example, a sponge absorbs water waves within a fraction of a wavelength, but it transmits sound waves through several centimeters. Theaters often use thick heavy curtains to absorb sound waves so that the audience cannot hear backstage noise. The tinted glass in sunglasses absorbs some energy from light but still passes enough to see by.

**Can amplitude decrease for other reasons?**

Wave amplitude may decrease for reasons other than absorption. For example, when a circular ripple spreads out, the amplitude of the ripple decreases with increasing distance from the center. This decrease mostly results from the wave's energy being distributed over a larger area as the wave expands. Circular and spherical waves share this characteristic because both distribute wave energy over an increasing area as they spread.

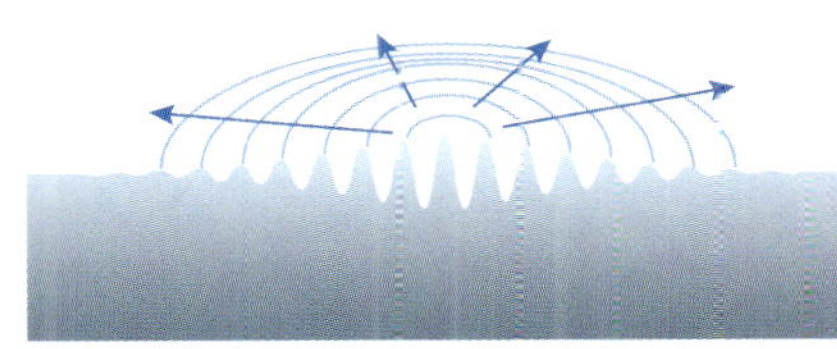
Ripples decrease in amplitude because wave energy is spread out over a larger and larger circle as the wave expands.

**What happens to absorbed energy?**

The energy of a wave is transferred to the material that absorbs the wave. With water waves this effect is dramatically illustrated by the power of hurricanes. When a large wave hits the shore, all of its energy is released quickly. This energy may be enough to move cars and trucks and demolish roads and houses. With other types of waves the absorbed energy is often converted to heat.

# Section 2 review

Wave propagation is the passage of wave energy from one location to another. As waves propagate, they can change amplitude, direction, or wavelength, depending on the kind of wave and the material through which it travels. Wavefronts are extended wave crests, such as ripples on a pond. Waves always travel at right angles to wavefronts. Waves can be reflected, refracted, diffracted, or absorbed at a boundary or when encountering an obstacle. Absorption can rob a wave of its energy, but a wave's amplitude can also decrease simply because the wave is spreading out over an ever-larger area (as when ripples spread from a pebble dropped into water).

| Vocabulary words | crest, trough, wavefront, reflection, concave, convex, refraction, diffraction, absorption |
|---|---|

## Review problems and questions

1. Define the following as one of the wave–boundary interactions, using *reflection, refraction, absorption,* and *diffraction* each once.
    a. Tarmac heats up on a sunny day.
    b. A magnifying glass enlarges an image.
    c. Waves curve around a boulder in the water.
    d. A shout echoes off of a building.

2. How is it that you can hear people inside another room even when the door is only opened very slightly?

3. A dozen classmates line up shoulder to shoulder, hold hands, and walk forward at a steady pace. They walk on pavement, but then three students at the left hit mud and slow down. The other students keep going forward and the line bends toward the muddy field. What wave-propagation effect does this imitate, and why?

4. Waves passing from a slow medium into a faster medium undergo refraction.
    a. Does the frequency of the waves increase, decrease, or remain the same?
    b. Does the wavelength of the waves increase, decrease, or remain the same?

5. The horizontal distance from one crest to the next crest of a water wave is the wavelength of the wave. The *vertical* distance from crest to trough of a water wave is equal to
    a. the amplitude.
    b. half the wavelength.
    c. twice the amplitude .
    d. twice the wavelength.

6. This satellite TV dish reflects radio signals that arrive in parallel wave fronts from a great distance. To provide good reception, it must bring those wavefronts to a focus at the position of the arrowed antenna. Is the side of the dish seen here convex, or is it concave?

Anyone who lives near the shore knows that the *tides* cause the ocean level to rise and fall roughly twice a day. In Galveston Bay near Houston, the water level rises and falls about 35 cm between high tide and low tide. In Nova Scotia's Bay of Fundy the tides are enormous, *changing the water level by as much as 15 m twice a day!* That is about the height of a five-story building. The tidal flows are so large that powerful whirlpools form where water flowing out from the previous tide meets the incoming tide rushing in. Why are the tides so great at this place? The answer is *resonance* and the explanation comes later in this section!

## Multiple waves

**Can there be more than one wave?**

If you watch the ocean, you see ripples on top of waves: These are really small waves on the surface of larger waves. In most situations waves of many different amplitudes, wavelengths, and frequencies are present at the same time. The **superposition principle** explains how individual single-frequency waves add up to a complex many-frequency wave.

**An example of superposition**

The superposition principle says that the total amplitude at any point in space and time is equal to the sum of the amplitudes of all waves that occur then and there. For example, wave 1 in the figure below has a single wavelength and frequency, as does wave 2. By adding up the independent amplitudes from wave 1 and wave 2 at the same point in space and time, we create their combined waveform, wave 1 + 2, which contains two different wavelengths and frequencies *at the same time*. This example shows the superposition of two waves, but in a real-world situation there may be *many* waves present at the same time and in the same place.

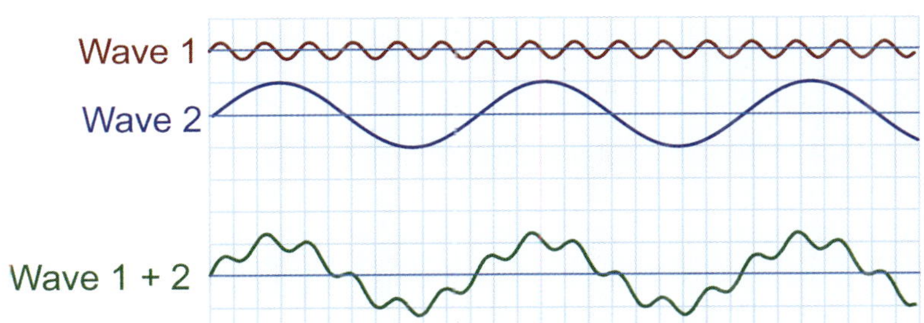


**Superposition principle**

The total amplitude at any point is the sum of the amplitudes of each individual wave at that same point.

**What is a "simple wave"?**

The simplest wave can be completely described with a single wavelength, a single amplitude, and a single frequency. A single-frequency wave is often referred to as a *sine wave* because the graphs of both amplitude versus position and amplitude versus time can be described by the *sine* function. As an example, consider simple sine waves as a function of position $x$ (in meters) and of time $t$ (in seconds). The sine function repeats every $2\pi$; therefore, $\sin(x)$ has a wavelength of $2\pi$ m. Likewise, $\sin(t)$ has a period of $2\pi$ s and a frequency of $(1/2\pi)$ Hz.

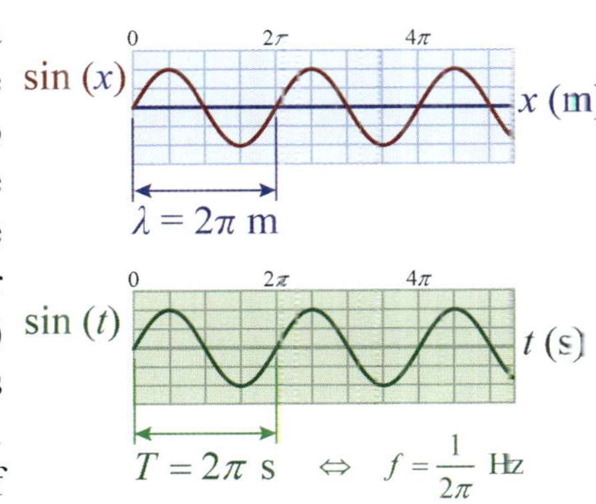

**Constructive interference**

When more than one wave is present they can add to make a larger amplitude or a smaller amplitude wave. If the sum of two waves has a larger amplitude we say **constructive interference** has occurred. The diagram on the right shows constructive interference. When two waves interfere constructively the amplitude of the total wave is the sum of the amplitudes of the smaller waves. Both the wavelength and frequency stay the same.

Constructive interference

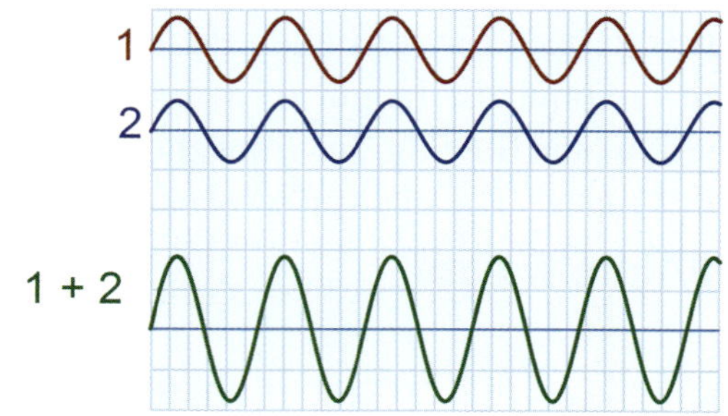


**Destructive interference**

Two waves can also combine to make a smaller wave. Two waves that are identical in frequency and amplitude, but offset in *time,* can add up to make a wave of *zero* amplitude. A situation where one or more waves add up to make a smaller amplitude wave is called **destructive interference**. The diagram shows 100%, or *total*, destructive interference. In many cases, however, there is only partial destructive interference.

Destructive interference

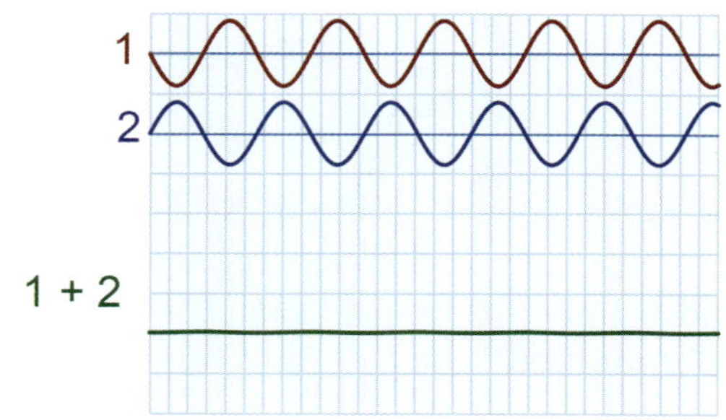


**Interference can be temporary**

If two wave pulses—one with positive amplitude and the other with negative amplitude—start on opposite sides of a snaky spring, something truly interesting happens in the middle. When the pulses meet they cancel each other out. For a moment the string is flat and both pulses vanish! The next moment the pulses reappear after passing through each other, and then they race away from each other. This is another example of destructive interference, and it shows how the energy of a wave may still be present even if it is momentarily in a form that cannot be seen.

Destructive interference between two wave pulses

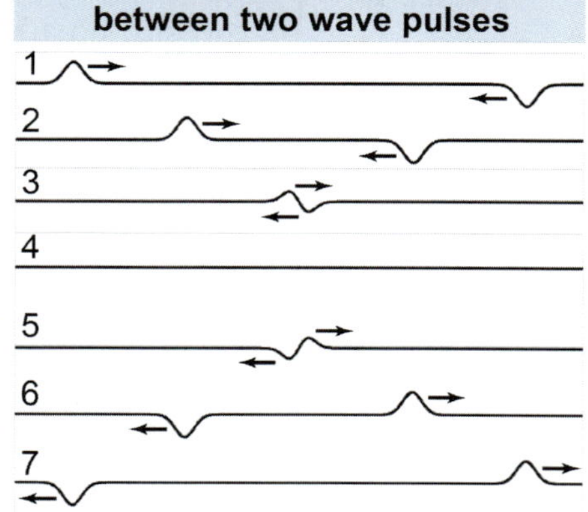


**Interference and superposition**

Interference is a special case of the superposition principle. In most realistic situations many waves will be present. Some waves will interfere constructively at some places and times. Other waves will interfere destructively at different places and times. The interaction of echoes in a concert hall is a good example of many sound waves interfering with each other. Certain music halls are known for *good* sound. This means interference between echoes from the walls and sound heard directly from the stage does not change the balance of frequencies heard by the listener in an unpleasant way.

# Investigation 15C Interference

**Essential questions** *How does the amplitude of a wave change when it encounters another wave?*

When multiple waves exist in the same place at the same time, *interference* occurs. Wave interference can be either *constructive* or *destructive*, causing the amplitude of the resultant wave to be different than the individual waves. In this investigation you will observe the interaction of wave pulses on a spring and draw a conclusion about when constructive and destructive interference occur in waves.

## Constructive and destructive interference

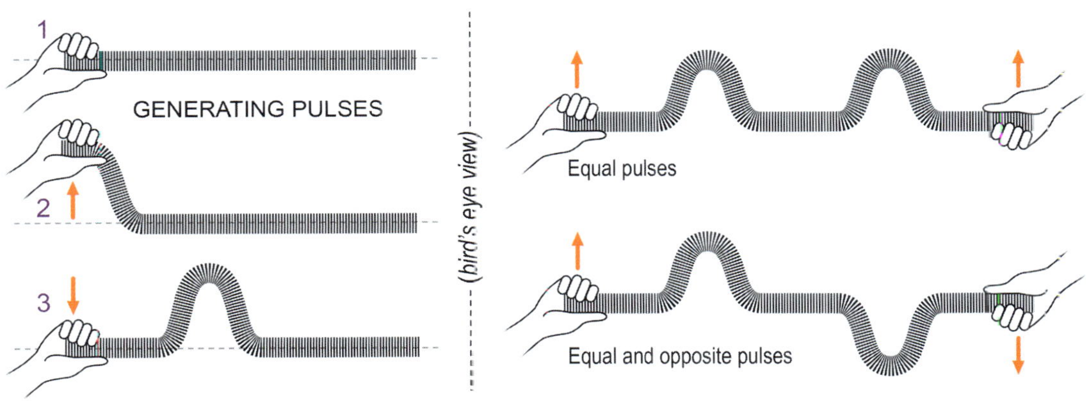


1. Stretch the slinky out flat on a table or the floor – one group member holding each end. Do NOT overstretch the slinky!
2. Send pulses down the slinky from each end and observe how they interact when they meet in the middle. Use your observations to answer the questions.
3. Test the following scenarios:
   - Equal pulses from each end
   - Big pulse from one end – small pulse from the other end
   - Equal and opposite pulses from each end
   - Big pulse from one end – opposite small pulse from the other end

a. *When the equal pulses met in the middle, how was the amplitude of the combined wave different from the individual pulses?*

b. *When the big and small pulses met in the middle, how was the amplitude of the combined wave different from the individual pulses?*

c. *When the equal and opposite pulses met in the middle, how was the amplitude of the combined wave different from the individual pulses?*

d. *Which scenarios experienced constructive interference? How do you know?*

e. *Which scenarios experienced destructive interference? How do you know?*

f. *For each scenario, describe what happened to the pulses after they met.*

## Resonance and standing waves

**Can reflected waves interfere?**

When a wave reflects from a boundary the reflection is another wave that interferes with the original wave. Boundaries often create situations where the initial wave and one or more reflections produce interference effects that can be surprising. The acoustical phenomenon of reverberation is one such effect. The extraordinary tides in the Bay of Fundy are another.

**Standing waves**

A **standing wave** occurs when there is constructive interference between a wave and its reflections from a boundary. You can create a standing wave by shaking one end of a spring so that wave pulses are launched at just the right rate. If the rhythm is correct, each new pulse adds constructively to the reflection of the previous pulse. This constructive interference is what causes the amplitude to grow into a standing wave.

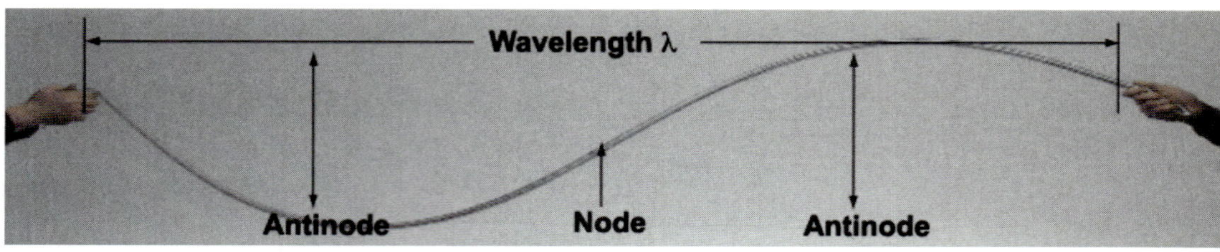


**Boundaries, nodes, and antinodes**

Standing waves are a form of *resonance.* Resonance in waves occurs when boundaries reflect a wave back on itself. The resonance for a vibrating spring (or string) occurs when the length of the spring is any multiple of half a wavelength. The diagram shows a standing wave where the length of the spring equals one full wavelength. One full wavelength is a complete "S" pattern, which has three *nodes* and two *antinodes.* A **node** is a stationary point where the amplitude is zero. An **antinode** is a point of maximum amplitude. In general, resonance occurs when the size of a system is proportionally related to the wavelength of a wave.

**Natural frequencies**

At different frequencies, different standing wave patterns appear on a vibrating string. The standing wave with the lowest frequency and longest wavelength is the *fundamental,* which is also called the *first harmonic*. The next higher frequency wave is the *second harmonic* at twice the frequency of the fundamental. The third harmonic has three times the frequency of the fundamental, and the pattern continues. Each harmonic is a vibrational *mode* of the string. A **mode** is a characteristic pattern of vibration that occurs at a resonance frequency of a system. The frequencies at which resonant modes occur are the *natural frequencies* of the system. Like the vibrating string, most systems have many modes and corresponding natural frequencies.

**Modes of a vibrating string**

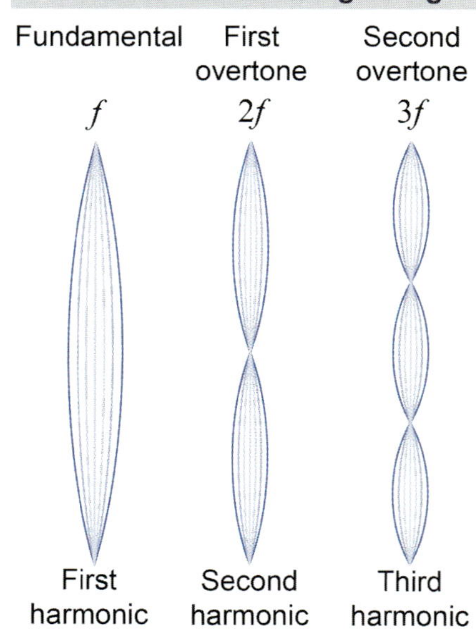

### How is resonance important?

Waves are a fundamental way through which energy moves. *Resonance* dramatically affects the energy content of waves. Resonance preserves the energy of waves of certain frequencies while all other frequencies rapidly damp out to zero amplitude. Every wave in your daily environment, including light waves, sound waves, vibrations of the ground, and radio waves, is shaped by resonance. Many technologies, such as microwaves, musical instruments, and magnetic resonance imaging (MRI) scanners explicitly use resonance.

### How does a guitar string vibrate at the right frequencies?

Consider what happens when you pluck the string of a guitar. A 67 cm E string at a tension of 72 N vibrates at 330 Hz. Why? You do not pluck the string 330 times per second, yet somehow the string turns your single pluck into vibrations at this precise frequency (and its harmonics). The first part of the explanation is that many waves are present in the initial oscillation, at many frequencies, because a string is capable of vibrating at an infinite number of frequencies.

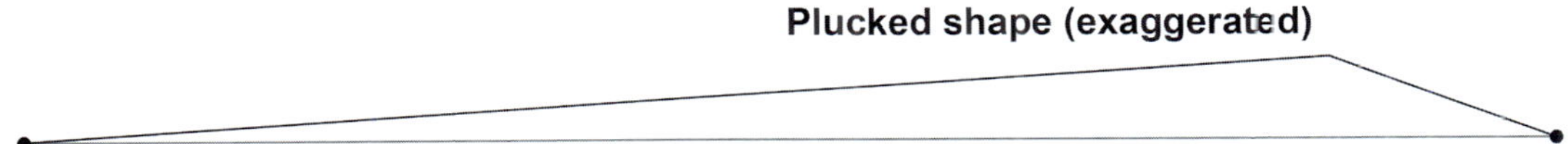


### Resonance and damping

Within a few oscillations, friction damps nonresonant frequencies to zero amplitude. Resonant frequencies, however, lose energy much more slowly. After a very short time *only the resonant frequencies are left with appreciable amplitude!* Resonance preserves only the energy of standing waves at the string's natural frequencies, so the string vibrates at these frequencies *no matter how it is plucked!*

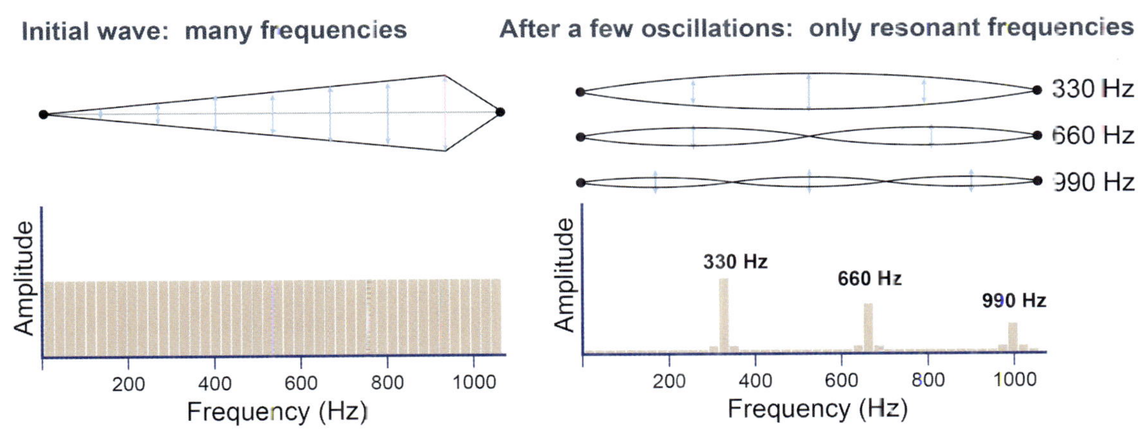


### Displaying resonant frequencies

The frequencies of oscillation are illustrated above in a form of bar chart called a *spectrogram,* in which the amplitude of each frequency is plotted as a vertical bar. You can easily see which resonant frequencies are present in the right-hand chart.

### Controlling frequency through resonance

Resonance in standing waves is created by constructive interference between a wave and its reflections from the boundaries of the system. Technologies such as musical instruments and microwave ovens control the frequency of waves by creating systems with boundaries that create resonance at the desired frequencies. MRI uses resonance in the reverse way. Systems also *absorb* energy very efficiently at their resonant frequencies and not at other frequencies. Atoms act like microscopic standing waves and atoms of different elements have different resonant frequencies.

# Investigation 15D

## Wavelength and standing waves

**Essential questions**

*What do we mean by the wavelength of a wave?*
*How can a wave have a length?*

Interactive simulation

A standing wave is a wave that has been trapped between two boundaries, such as the fixed ends of an elastic string. Because standing waves persist for long periods of time they make an excellent example for investigating wavelength and frequency.

### Part 1: Setting up a standing wave

1. Launch the interactive simulation of a vibrating string driven by an oscillator.
2. Change the oscillation frequency until you create a standing wave.
3. Change the frequency to see different standing wave patterns with different numbers of nodes and antinodes.

a. *Do the standing waves appear at all frequencies or just specific ones? Explain this using the idea of resonance.*
b. *Describe the role of reflection and interference in creating standing waves.*
c. *Sketch two or three different standing wave patterns and indicate the wavelength of each.*

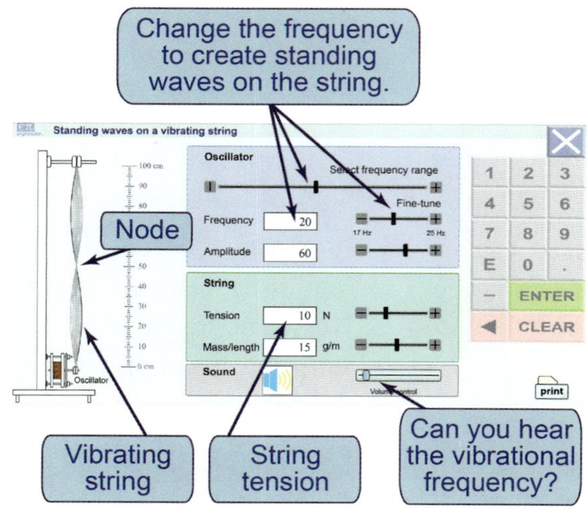


**What are the frequencies of the first through eighth harmonics?**

### Part 2: Relationship between wavelength and frequency

1. Adjust the frequency until you get the largest amplitude standing wave you can that has a single node between its fixed ends. This is the *second harmonic* because there are two antinodes.
2. Using the measuring scale, estimate the wavelength of the standing wave and record its wavelength and frequency.
3. Adjust the frequency to find the standing waves for the first through the eighth harmonics and record the wavelength and frequency of each.

a. *What is the relationship between frequency and wavelength? State it with a mathematical formula.*
b. *What units does the product of frequency × wavelength have? What is the interpretation of this quantity?*
c. *How could you control the frequency of a vibrating object through its wavelength? Describe at least two applications of this principle.*
d. *Going further: What happens to the fundamental frequency of the standing wave if you quadruple the tension in the string? What happens to the fundamental frequency if you quadruple the mass/length of the string?*

**How are waves useful?**

Waves and resonance are found throughout human technology. One specific area in which wave technology has had a profound impact is in medicine. A hundred years ago doctors had no way to sense directly what went on inside the body except to cut and look. Of course "cutting and looking" is painful, dangerous, and may cause more harm than help.

**Have you ever had an x-ray?**

Today if you fall hard on your hand you get an x-ray, like the one in the picture. Can you see where a bone has fractured? X-rays revolutionized the treatment of injuries because a doctor could *know* whether a bone was broken, where the break was, and how serious it was.

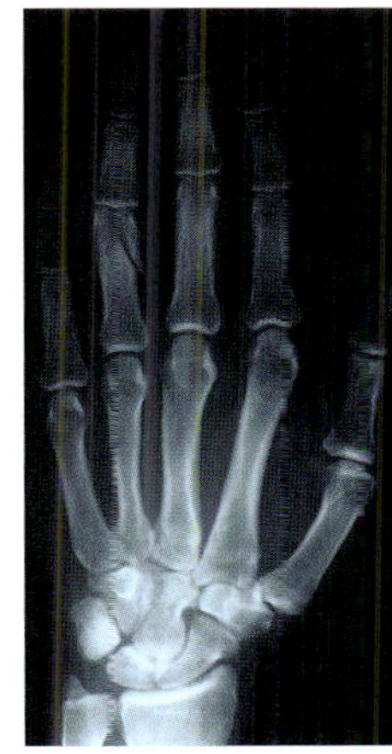

**How do x-rays work?**

X-rays work on the principle of absorption. The element calcium is a strong absorber of x-rays. Skin and muscle tissue are only weak absorbers of x-rays. When the x-rays went through the hand in the picture, they exposed and darkened the film directly behind any skin or muscle. Where there was bone, however, the x-ray energy was absorbed and therefore the film is less exposed.

**Are MRI scans different from x-rays?**

Magnetic resonance imaging scanners work on the resonant absorption of radio-frequency waves by hydrogen atoms. Hydrogen is in water molecules so it is found throughout the body, giving MRI the advantage of "seeing" different kinds of soft tissues. X-rays essentially see only calcium, which is dense only in bones. High-energy X-rays can also damage living cells, so doctors usually try to limit a patient's exposure.

**Why are magnets needed?**

Video

MRI makes detailed images of internal organs with no harm to the body. To make an MRI scan the patient is placed inside a very precise electromagnet. The nuclei of hydrogen atoms act like tiny magnets. When the magnetic poles of the hydrogen nucleus are aligned with the MRI magnetic field the atom has slightly less energy than when the poles are not aligned. This energy difference is *very* sensitive to the exact strength of the magnetic field.

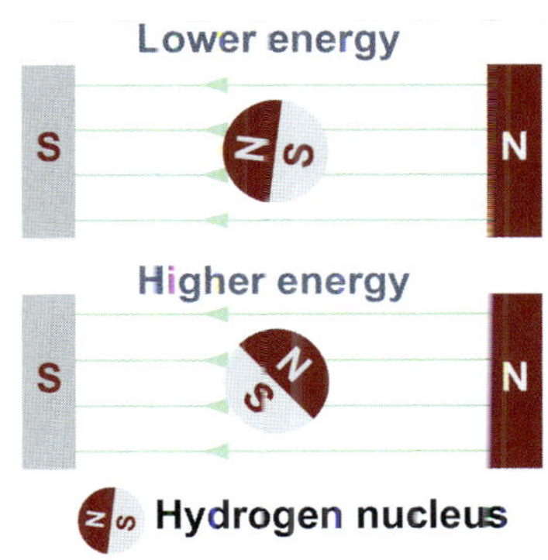


**How does MRI work?**

Hydrogen nuclei do not statically align to the MRI field but rather oscillate around the field with a particular frequency. If a radio wave has a frequency that is resonant with the oscillation of the hydrogen nuclei, it causes the nuclei to flip. The flip releases characteristic energy that is detected and used to create the image. By tuning the radio wave precisely, MRI technology can reveal whether the hydrogen is in fatty tissue, water, or other tissue types.

**Laser metrology**

Laser metrology is an application of light waves to measure distances with high precision. One method uses pulses of laser light and times how long it takes for each pulse to be reflected off a mirror and return to the instrument. This has been used to measure the distance to the Moon by reflecting laser light off mirrors installed on the Moon by the *Apollo* astronauts. Another method uses the interference between laser light and its reflection off the surface of an object. This method can be used in laser versions of coordinate measuring machines to measure the surface dimensions of manufactured parts.

Chapter 15

# Section 3 review

The *superposition principle* describes what happens when multiple waves overlap. It states that the wave amplitude at any place and time is the sum of the amplitudes from each individual wave. *Interference* is one possible result of superposition. When the combined amplitude of two waves is larger than either of the individual waves, the interference is *constructive*. If the combined amplitude decreases, the interference is *destructive*. A *standing wave* occurs when a wave interferes with its own reflection; it is characterized by *nodes* (places where the amplitude remains zero) and *antinodes* (where the amplitude is largest). In wave physics, *resonance* refers to the fact that certain physical systems will only sustain standing waves with certain frequencies. Resonance allows objects such as guitar strings and organ pipes to produce distinct musical notes.

| Vocabulary words | superposition principle, constructive interference, destructive interference, node, antinode, mode, standing wave |
|---|---|

## Review problems and questions

1. Is it possible for two waves to have individual amplitudes that are not zero while the *sum* of the two waves has an amplitude of zero? Explain.

2. Give three examples in which wave behaviors such as reflection, refraction, and absorption help doctors diagnose their patients' medical conditions.

3. Two friends stand on either end of a rope of length 2 m. One friend repeatedly shakes her end of the rope up and down by a few centimeters. By shaking the rope at different rates (frequencies), she generates all three of the standing-wave patterns shown here.

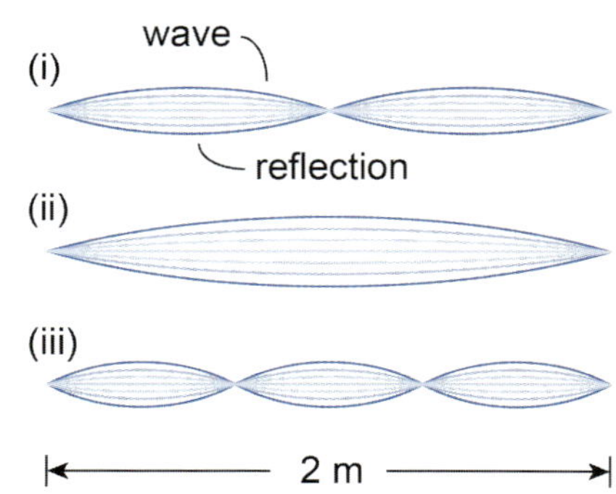


   a. In Pattern (i), what is the wavelength of the wave?
   b. Suppose that she completes one up-and-down hand motion in half a second (0.5 s) to generate Pattern (i). What are the period, frequency, and wave speed?
   c. Next, she generates Pattern (ii). What has doubled, the wavelength or the frequency?
   d. In Pattern (iii), how many wavelengths fit between the two ends of the rope?
   e. Which pattern represents the rope's fundamental mode of vibration?

4. For each standing wave pattern shown in the figure above, how many nodes are present? How many antinodes?

5. For a standing wave on a stretched string, what is the relationship between the number of nodes and the number of antinodes?

# Chapter review

### Vocabulary

Match each word to the sentence where it best fits.

### Section 15.1

| | |
|---|---|
| longitudinal | polarization |
| transverse | wave |
| wavelength | |

1. The distance measured from the top of one wave to the top of the next wave is called the _______.
2. A wave that oscillates up and down has a different _______ from a wave that oscillates left and right.
3. Sound, light, and the motion of water can all be thought of as a _______.
4. When a long rope is oscillated sideways, the wave created along it is a _______ wave.
5. Oscillations in ______ waves, such as sound waves, are parallel to the direction the wave travels.

### Section 15.2

| | |
|---|---|
| absorption | crest |
| diffraction | reflection |
| refraction | trough |
| wavefront | |

6. The lowest part of a wave is called its _______.
7. Waves undergoing _______ through a narrow opening will change their direction and the shape of their wavefronts.
8. A wave that is passing into a medium, thereby decreasing the wave's amplitude (and energy), is undergoing _______.
9. The top of a wave is also called its _______.
10. Waves bounce off of an obstruction in _______.
11. When a stone is dropped in a pond, the waves that travel outward from it have a circular _______.
12. A wave that changes speed and direction as it passes through a boundary between two regions is undergoing __________ .

### Section 15.3

| | |
|---|---|
| antinode | constructive interference |
| destructive interference | mode |
| node | phase |
| standing wave | superposition principle |

13. Noise-canceling headphones take advantage of the _______ of waves to reduce or eliminate repetitious sounds from the surrounding area.
14. For a standing wave on a string, the location of no movement is the _______.
15. An "A" string can vibrate in one _______ at 440 Hz or another _______ at 880 Hz.
16. Two water waves that add together to make a larger wave is an example of _______.
17. The amplitudes of waves add together according to the _______.
18. Two waves of the same frequency, wavelength, and amplitude can interfere with each other destructively if they have a different _______.
19. Two identical waves traveling in opposite directions on a string can create a/an _______.
20. On a standing wave, a/an _______ is a point where the oscillations of the medium have their greatest possible amplitude.

### Conceptual questions

### Section 15.1

21. Draw two waves, one with a longer wavelength than the other.
22. Draw two waves, one with a larger amplitude than the other.
23. Draw two waves, one with a longer period than the other.
24. If the frequency of a wave is tripled, what happens to the period of the wave?
25. A sound wave with a frequency of 343 Hz has a wavelength of 1.0 m. If the frequency is doubled to 686 Hz, what is the new wavelength?

# Chapter 15

# Chapter review

## Section 15.1

26. Habin took a wave motion rope and waved it sideways repeatedly to create waves. What kind of waves did he create, transverse, longitudinal, or a pulse?

27. Varujian took a wave motion rope and waved it sideways once. What kind of wave did he create, transverse, longitudinal, or a pulse?

28. Chakra took a Slinky® spring and repeatedly compressed and extended one end to create wave pulses. Did he create transverse or longitudinal waves?

29. Sven measured the distance along a wave for three complete oscillations. If he wants to know the wavelength, should he multiply or divide his number by 3?

30. Describe how to create (a) circular waves and (b) plane waves in a bathtub full of water.

31. Draw a wave on a position–time graph and label the amplitude and period.

32. Is a wave motion rope or a Slinky® spring the best choice of equipment to create *both* a transverse and a longitudinal wave?

33. Which sentence best expresses in words the meaning of the equation below?

$$v = \lambda f$$

a. The velocity of a wave equals its amplitude multiplied by its wavelength.
b. The velocity of a wave is equal to the ratio of its wavelength to its frequency.
c. Amplitude multiplied by frequency equals the wavelength.
d. The velocity of a wave equals its wavelength times its frequency.

34. You are given a long piece of rope with one end tied to the wall. Describe how you can create (a) a wave pulse and (b) a standing wave with the rope.

35. Imagine you are standing 100 m above the bottom of a valley on a rope bridge that is oscillating with a standing wave. Would you prefer to stand in a location that is a node or an antinode? Why?

36. Imagine you are designing an amusement park ride using a rope bridge that is oscillating with a standing wave. Do you think that customers would prefer to stand in a location that is a node or an antinode? Why?

## Section 15.2

37. Explain how the diffraction of waves plays a role in radio transmission and cellphone service.

38. Describe how refraction is used in optical imaging and ultrasound imaging.

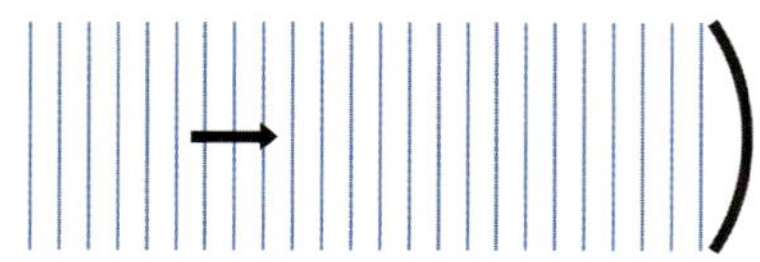

39. A student was conducting an investigation with a wave tank and a curved surface. Draw how plane waves reflect off of the curved surface in the illustration shown.

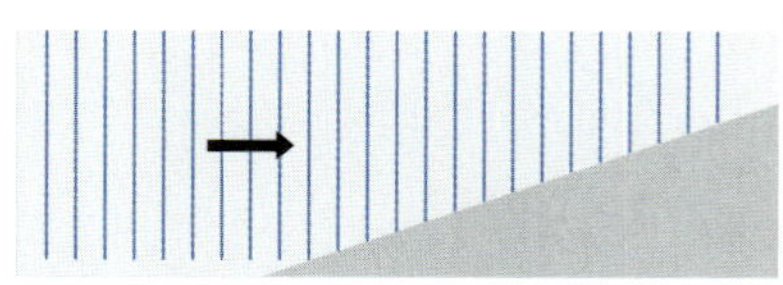


40. A student was conducting an investigation with a wave tank in which the waves refract from one medium into another in which the wave velocity is lower. Draw how plane waves refract from the boundary between two materials as shown in the illustration.

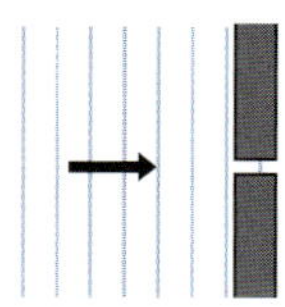

41. A student was conducting an investigation with a wave tank in which the waves diffract after passing through a small opening. Draw how plane waves diffract in the illustration shown.

42. What wave behavior explains why you can hear the radio playing in another room even though you cannot see it?

43. A person who shouts while hiking in a canyon is likely to hear an echo. What wave behavior explains this phenomenon?

44. Ocean waves entering the shallow water surrounding an island will bend so that the incoming crests are roughly parallel to the shore rather than at right angles to it. What wave behavior explains this phenomena?

45. Jack measured the wavelength of a wave using successive crests. Jill measured the wavelength using successive troughs. Who will measure the wavelength correctly?

46. Does a wave propagate parallel or perpendicular to its wavefront?

# Chapter review

## Section 15.2

47. Blue whales are able to communicate over very large distances using low-frequency sound waves. Their communication range is limited by background noise from shipping and by absorption. When a wave is absorbed, what wave property changes?

    a. wave velocity
    b. wavelength
    c. amplitude
    d. frequency

48. Light refracts as it passes the boundary between air and glass in a camera lens. Which of these wave properties change when the wave refracts, and which remain the same?

    a. wave speed
    b. frequency
    c. wavelength
    d. direction

49. When you are conducting an investigation using a ripple tank and a wave reflects off a surface with the same amplitude, where does its energy go?

50. When a light wave refracts through a glass lens, where does its energy go?

51. When a wave passes through a hole and undergoes diffraction on the other side, where does its energy go?

52. When a light wave is partially absorbed by passing through a glass window, where does its energy go?

## Section 15.3

53. If you want to make a light source brighter, would you use constructive or destructive interference?

54. Your assignment is to stretch an elastic band on which you will create a standing wave. Why is it better to mount the elastic band with a ring stand and clamps than to have your partner hold it between her hands?

55. Draw two waves with the same wavelength and amplitude that will interfere constructively and two that will interfere destructively.

56. A standing wave in a stretched string of length $L$ is vibrating at its fundamental frequency $f$. The length of the string is doubled, and it is again caused to vibrate at its fundamental frequency. What is the new wavelength? What is the new fundamental frequency?

### Quantitative problems

## Section 15.1

57. Shazi was using a ripple tank and stopwatch. She timed 4.0 s as the time it took for 10 wave crests to pass a certain location in the tank.

    a. What is the period of the waves from her data? Use the correct number of significant figures in your answer.
    b. What is the frequency of the waves?

58. How far does a wave travel during a time span corresponding to five periods of the wave's oscillation?

59. Two waves are traveling at the same speed in the same medium. Wave A has twice the wavelength of wave B. Which wave has a longer period?

60. A water wave has a wavelength of 204 m and a frequency of 0.5 Hz. How far does it travel in 1 s?

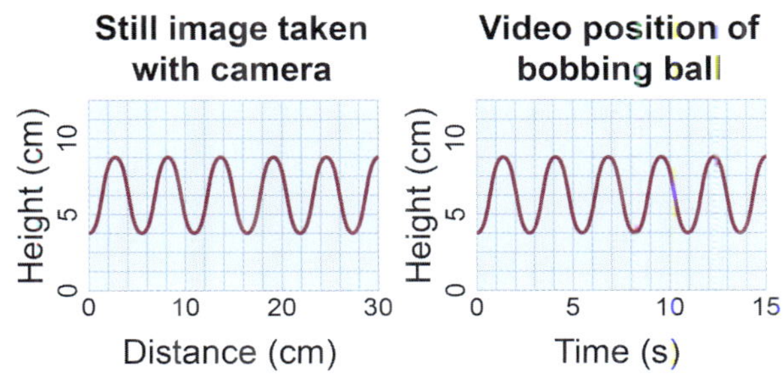


61. A student was conducting an investigation with water waves in a large wave tank. He took a still image with a digital camera (shown above left) to measure the water position at an instant in time. He also took a video of a ball bobbing up and down in the water (shown above right).

    a. What is the amplitude of the waves?
    b. What is the wavelength of the waves?
    c. What is the frequency of the waves?
    d. What is the velocity of the waves?

62. Infrasound is sound that is too low a frequency for our ears to detect. These sounds may cause nausea or feelings of discomfort. What is the wavelength of an inaudible 15 Hz infrasonic sound wave? The speed of sound in air is 343 m/s.

63. Sound speed in the ocean increases with depth. A submarine close to the surface, where the speed of sound is 1,480 m/s, produces a 10 kHz sound that is detected by a sonobuoy at a depth of 5.0 km, where the sound speed is 1,540 m/s.

    a. As sound speed increases, does the wavelength increase or decrease?
    b. What is the *change* in the wavelength of the sound between these two points?

# Chapter review

## Section 15.1

64. Water waves with a frequency of 4.5 Hz and a wavelength of 2.0 m are traveling across a small harbor that is 200 m wide. How long does it take the waves to travel from one side of the harbor to the other?

65. Blue light has a wavelength around two-thirds of the wavelength of red light. What is the frequency of blue light relative to the frequency of red light?

66. Red light has a wavelength of $6.56\times10^{-7}$ m and a speed of $3\times10^{8}$ m/s. What is the frequency of red light?

67. A wave has a frequency of $1.5\times10^{13}$ Hz and a speed of $3\times10^{8}$ m/s. What is its wavelength?

68. A radio station broadcasts at 99.5.
    a. What physical quantity is represented by the number 99.5?
    b. What are the units of the number 99.5?
    c. What does this number mean in terms of harmonic motion?
    d. If the speed of the broadcast waves is $3\times10^{8}$ m/s, what is its wavelength?

69. Radio waves broadcast at 95.3 MHz travel at a speed of $3\times10^{8}$ m/s. What is the wavelength of the radio waves? (Remember that one MHz is $10^{6}$ Hz.)

70. Radio waves with a wavelength of 2.95 m travel at a speed of $3\times10^{8}$ m/s. In megahertz, what is the frequency of the radio waves?

71. One AM radio station broadcasts at 1,350 kHz, while an FM radio station in the same city broadcasts at 89.7 MHz. All the radio waves travel at the same speed in air.
    a. Which one broadcasts at a higher frequency?
    b. Which one broadcasts at a longer wavelength?

72. A wave has a frequency of $5\times10^{14}$ Hz and a wavelength of $6\times10^{-7}$ m.
    a. What is the wave's speed?
    b. How far does the wave travel in 8 minutes and 18 seconds?

73. An enormous rogue wave has an amplitude of 17.0 m, a period of 14.0 s, and a wavelength of 340 m. It strikes an ocean liner, and at the instant of impact the skipper blasts a siren to alert a nearby vessel.
    a. What is the speed of the rogue wave?
    b. The speed of sound is 343 m/s. How long do sailors on the vessel 1,000 m away have to prepare for the wave once they hear the siren?

## Section 15.2

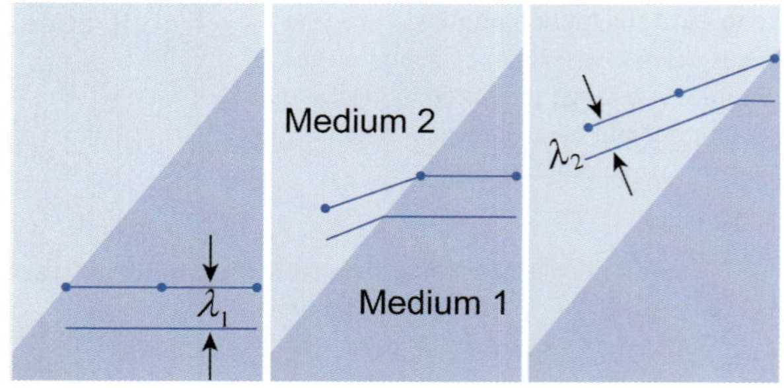


74. Two parallel wavefronts are moving upward through medium 1, separated by a wavelength $\lambda_1$. They impinge upon a boundary at an acute angle and refract into medium 2. There the wavefronts are separated by a smaller distance $\lambda_2$. The frequency is unchanged because each wavefront must immediately pass through the boundary.
    a. In which medium are these waves faster?
    b. Measurements indicate that $\lambda_2/\lambda_1 = 2/3$. What is the ratio of the wave speeds, $v_2/v_1$?
    c. What is the ratio of the frequencies, $f_2/f_1$?

## Section 15.3

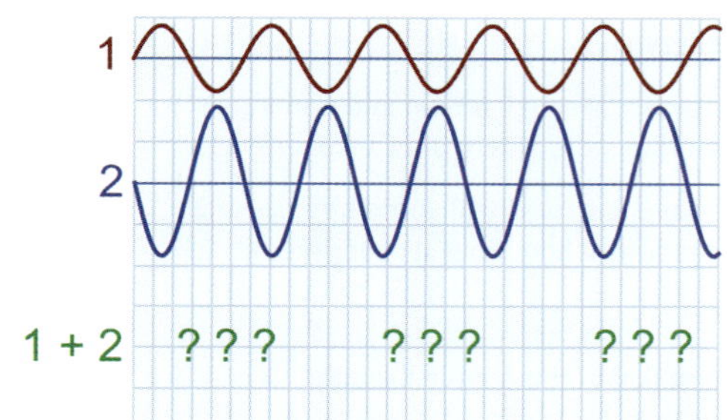


75. Wave 1 and wave 2, shown here, are superposed. How will the amplitude of the sum of the two waves compare to those of the original waves? The wavelength? The phase?

76. A large wave with an amplitude of 40 cm and a wavelength of 2.0 m undergoes interference with a second wave with an amplitude of 15 cm and a wavelength of 2.0 m.
    a. If the waves are 180° out of phase, what is the amplitude of the combined wave?
    b. If the waves are in phase, what is the amplitude of the combined wave?

77. The "E" string on a guitar has a length of 64 cm and vibrates with a fundamental frequency of 330 Hz.
    a. What is the wavelength of the wave in the string?
    b. What is the speed of the wave in the string?
    c. The speed of sound in air is 343 m/s. What is the wavelength of sound produced by the vibrating string?

# Chapter review

## Standardized test practice

78. Which waves have the most energy?

    A. low frequency, small amplitude
    B. high frequency, small amplitude
    C. low frequency, large amplitude
    D. high frequency, large amplitude

79. You and a friend create a standing wave with a rope. You both hold your hands still. In total, there are 5 nodes in this wave. How many antinodes are there?

    A. 3
    B. 4
    C. 5
    D. 6

80. A sound wave has a frequency of 500 Hz. What is the period of this wave?

    A. $2\times10^{-3}$ s
    B. 1 s
    C. 5 s
    D. $5\times10^{2}$ s

81. Which of the following terms are used to describe waves?

    I. wavelength
    II. mass
    III. velocity
    IV. force
    V. amplitude

    A. I and II only
    B. III, IV, and V only
    C. I and III only
    D. I, III, and V only
    E. all of the above

82. Which of the following is an example of a longitudinal wave?

    A. a plucked guitar string
    B. sound from a tractor
    C. a long spring when one end is moved side to side
    D. an airplane traveling north to south through Greenwich, England

83. An ocean buoy moves up and down once every 4 s because of passing water waves. What is the wavelength of the waves if their speed is 3 m/s?

    A. 0.75 m
    B. 1.33 m
    C. 7 m
    D. 12 m

84. When two waves overlap each other such that the combined amplitude is smaller than the amplitude of either wave, this is an example of

    A. constructive interference.
    B. destructive interference.
    C. diffractive interference.
    D. rarefactive interference.

85. What light–boundary interaction do eyeglasses use to correct vision?

    A. reflection
    B. refraction
    C. absorption
    D. diffraction

86. Two sound waves combine to create a wave with greater amplitude than either of them individually. What is this an example of?

    A. damping
    B. destructive interference
    C. wave–boundary interaction
    D. constructive interference

87. Which is *not* an example of wave phenomena at a boundary?

    A. Light is reflected off the surface of a mirror.
    B. Light is refracted into a rainbow by passing through a glass prism.
    C. Water waves are diffracted by passing through a small hole.
    D. Sound waves travel through the air.

88. A sound wave has an amplitude of 10 mm and a speed of 343 m/s. Which of the following could be the period of this wave?

    A. 500 Hz
    B. 10 m
    C. 0.1 $m^{-1}$
    D. 15 s

89. An electromagnetic wave traveling at $3\times10^{8}$ m/s has a 3 m wavelength. What is the frequency of this wave?

    A. 100 MHz
    B. 10 GHz
    C. 10,000,000 Hz
    D. 100 μHz

90. If a wave in the ocean has a frequency of 143 Hz and a wavelength of 2 m, what is the speed of this wave?

    A. 71.5 m/s
    B. 145 m/s
    C. 286 m/s
    D. 1,430 m/s

# Chapter 16 Sound

Musical instruments are among the very earliest technologies invented by humans. From the brass pots of the traditional Indonesian *gamelan* to the most ragged-edged electric guitar, all music is created from sounds using patterns of rhythm and pitch. Rhythm is created easily by tapping or plucking in time with the beat. Musical instruments create different pitches using objects that are shaped to *vibrate* at specific frequencies of sound. The fundamental frequency determines the pitch and the mix of overtones determines the quality or *timbre* of the sound.

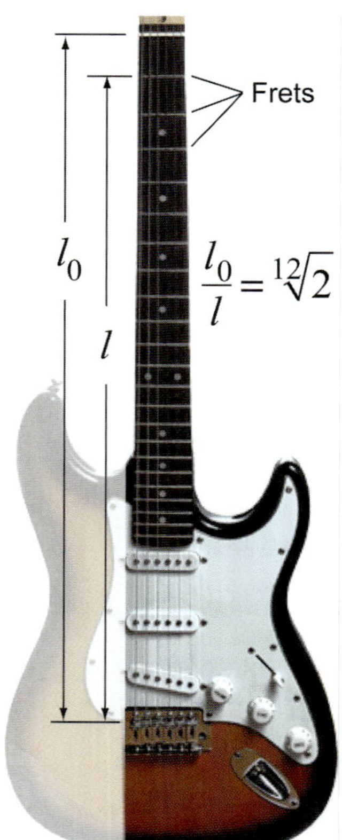


Stringed instruments such as the guitar control the pitch in three ways. The neck has a series of raised metal *frets* that provide a precise way to control the vibrating length of each string. By placing a finger just behind different frets the player changes the string's vibrating length, thereby changing the pitch. The shorter the vibrating length, the higher the pitch. Each successive fret in a modern guitar decreases the length of the vibrating string by the 12th root of two (approximately 1.059). The number 12 comes from the 12 half-steps in the western musical scale—A, A#, B, C, C#, D, D#, E, F, F#, G, and G#. The number 2 arises because the pitch of a note has twice the frequency of a note one octave below it.

The second method of changing pitch is to change the mass of the strings. The six strings of a guitar each have a different thickness. The low-E string has the greatest mass, and it vibrates 82 times per second (a frequency of 82 Hz) for its lowest note. The high-E string is the thinnest, and its highest playable note (22nd fret) has a frequency of 1,174 Hz.

The third method of controlling pitch is by changing the string tension, such as when a guitar is tuned. The string's frequency is proportional to the square root of its tension. Tightening a string's tension by a factor of 4 will raise its frequency by a factor of 2.

# Chapter study guide

## Chapter summary

Hearing is one of our five senses, which makes sound an important part of how we interact with the world. Sound is a longitudinal wave that travels to our ears as variations in pressure in the air. Pitch and loudness are commonly used to describe the sounds we hear, and both are related to physical properties of the sound waves themselves. We can distinguish different voices and musical instruments because the sound produced by each has a different *timbre*—a unique set of higher frequency overtones that make up the sounds. In this chapter, you will learn about the basic properties of sound and apply those concepts to understand why an approaching train has a higher pitch than a receding one, why jets produce a supersonic boom, and how a musical instrument works.

## Learning objectives

By the end of this chapter you should be able to

- describe how sound propagates;
- describe pitch and loudness for sound and how they relate to amplitude, frequency, and wavelength;
- solve problems involving frequency, wavelength, and the speed of sound;
- describe subsonic and supersonic motion;
- describe the Doppler effect, provide examples, and calculate the frequency shift;
- describe how sounds can include more than one frequency, interpret spectrograms, and explain how to distinguish different musical instruments;
- describe echoes and calculate the distance to a reflecting surface; and
- describe the concepts of reverberation, beats, and resonance in a pipe and provide examples of each.

## Investigations

16A: Sound waves
16B: Doppler effect
16C: How sound carries information
16D: Resonance and sound
Design a musical instrument

## Chapter index



## Important relationships

$$v = f\lambda$$

$$f = f_0 \frac{v_s}{v_s - v}$$

## Vocabulary

| | | |
|---|---|---|
| **pitch** | **speed of sound** | **decibel (dB)** |
| **supersonic** | **Doppler effect** | **microphone** |
| **frequency spectrum** | **Fourier's theorem** | **spectrogram** |
| **echo** | **phase** | **beats** |
| **harmonic** | | |

## 16.1 - Sound

Along with light, *sound* is one of the most important ways in which we experience the world. For most people, sound is a fundamental part of every moment. You might not hear the terms *frequency* and *amplitude* in everyday conversation but almost everyone knows these same properties by the names *low* and *high pitch* and *loudness*. For example, musical notes are different frequencies of sound. This section describes the basic properties of sound and how we perceive and understand voices and music.

### Basic properties of sound

**What are sound waves?**

Sound is a longitudinal wave, like the compression wave on a Slinky™ spring. Sound waves are similar except that they are much higher frequency and it is air that is being alternately compressed and expanded rather than the coils of a spring.

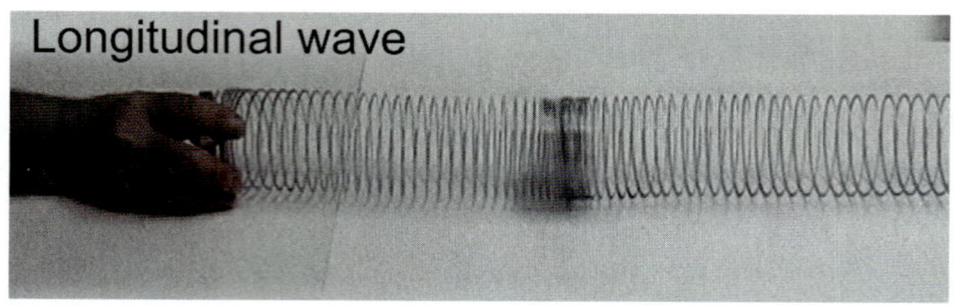


**The range of sounds we can hear**

We perceive different frequencies as having different **pitch**. The lowest frequency humans can ordinarily hear is a deep hum at a frequency around 20 Hz. Even this frequency is so fast that you cannot see the vibration with your eye; instead, you can only see a slight blur. The highest frequency that a young, healthy human ear can ordinarily perceive is a high-pitched whine at a frequency near 20,000 Hz.

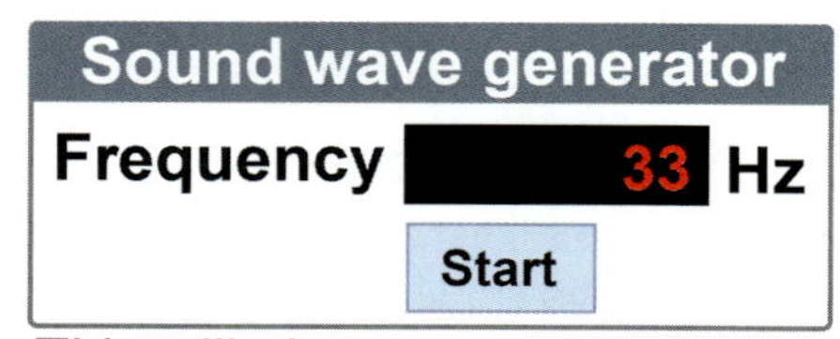


This will play a sound that increases in frequency from 20 to 20,000 Hz over 20 s.

**Loudness and pitch**

We also perceive sound to have a *loudness*. The loudness of a sound depends on the amplitude of the wave. A loud sound has a larger amplitude than a soft sound of the same frequency. A stereo's speaker moves back and forth a greater distance when producing a loud sound than when producing a soft sound. The larger amplitude of the speaker's motion causes larger amplitude pressure variations in the air.

**Multiple frequencies**

Almost all sound you hear contains many simultaneous frequencies at once. Even a "clean" musical instrument sound contains a dominant frequency, called the fundamental, and many *overtones*, which are additional frequencies that give the sound its characteristic "piano-ness" or "guitar-ness." The fundamental frequency is also called the first harmonic; the higher frequency overtones are called the second harmonic, third harmonic, and so on. The graph on the right is a *frequency spectrum*, which shows a range of frequencies up to around 2,500 Hz with the loudest peaks at 400, 800, 1,600, and 2,000 Hz.

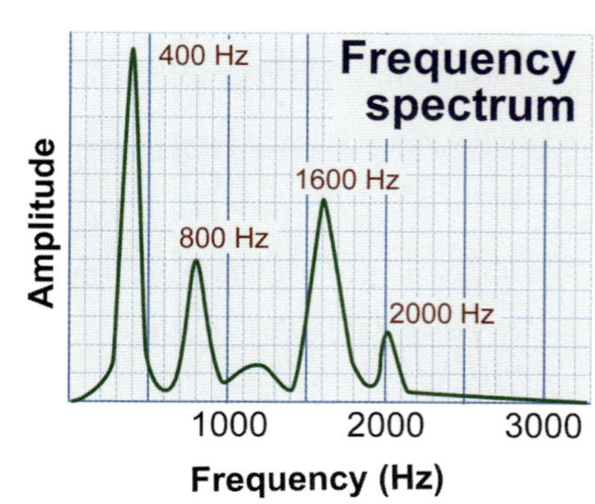

# Investigation 16A Sound waves

**Essential questions**

*What is a sound wave?*
*What are the properties of sound waves?*

Interactive simulation

Sound waves are traveling oscillations of air pressure. This interactive simulation allows you to determine the amplitude, wavelength, and frequency of some typical musical notes by matching the waveform in both time and space.

## Part 1: Matching the parameters of a sound wave

The red wave is the sound you are hearing, which you select with the piano keyboard keys. The black wave corresponds to the frequency, wavelength, and amplitude you set in the data boxes. The score is a percentage similarity between the red sound wave and the black wave with which you are trying to match it. The volume slider adjusts the volume of the sound (red wave). You will have to change the horizontal axis between time and length to determine both the frequency and the wavelength.

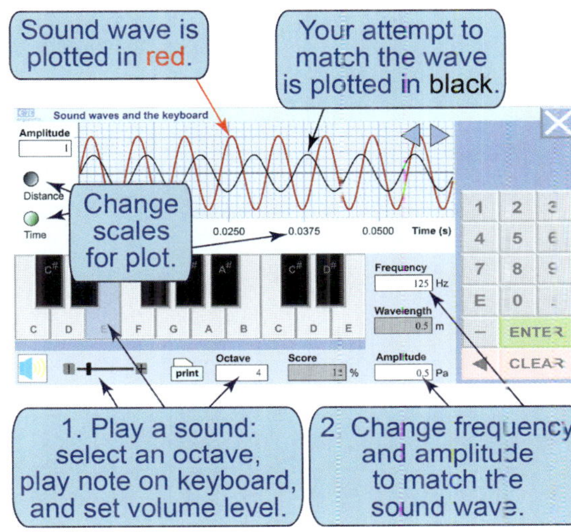


1. Choose a note and adjust your computer speakers so you can hear it.
2. Set the time and amplitude values on the graph until you can see at least a few cycles of the sound wave.
3. While time $t$ is plotted on the horizontal axis, see whether you can make the black wave match the red sound wave by adjusting the frequency and amplitude.
4. Switch to distance for the horizontal axis. Match your black wave to the red sound wave by adjusting the wavelength and amplitude. A good match has a score of greater than 95%. *You must match both frequency and wavelength to score 100%.*

*a. What is the frequency and wavelength of the note "C" on the left-hand side of the keyboard?*
*b. Describe how the observed wave varies with loudness.*
*c. Determine the frequency and wavelength for at least four different sounds.*
*d. From your data, discuss possible relationships between frequency and wavelength with your lab partners. Propose and test an equation that expresses your hypothetical relationship.*
*e. Using authoritative print or digital resources, research the value for the frequency of the note "A" in orchestral tuning. What is its accepted value and variation? Explain the origin of any variations.*

## Part 2: Going further with octaves

1. Devise and conduct an experiment to determine what happens to the frequency and wavelength of sound when you set the octave to different values. Note that the octave can only be set to positive integers.

*a. What do you conclude occurs by setting different octaves?*
*b. How does this fit with your prior knowledge of music?*

## The nature of sound waves

**What is sound?**

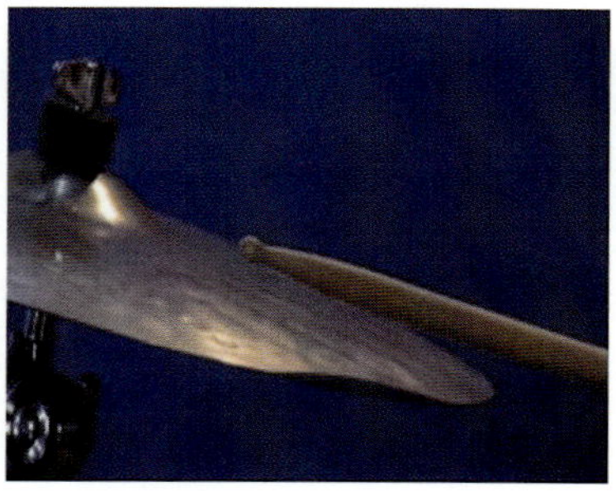

Sound is a very tiny oscillation of *pressure*. Imagine moving a metal cymbal up and down. When the surface moves upward, the air above is slightly compressed, which means the pressure is raised a little. When the surface moves downward, the air is drawn out, slightly lowering the pressure. Tapping the cymbal with a drumstick creates a much more *rapid* up-and-down oscillation of the metal surface. The result is a traveling oscillation of air pressure—a sound wave. Anything that vibrates in contact with matter makes sound waves.

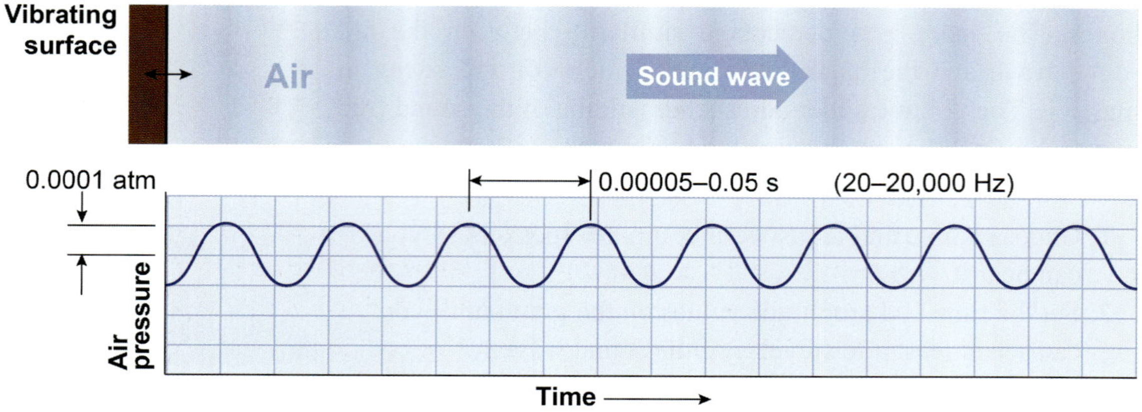


**How are sound waves different from other waves?**

Sound waves are rapid oscillations compared to waves in springs or in water. The period of oscillations in sounds that humans can hear is less than 0.05 s. This corresponds to a frequency of 20 Hz, which is the low-frequency limit to an average human ear. The high-frequency limit is about 20,000 Hz for a young person, but this declines to around 12,000 Hz by middle age.

**Can we feel the pressure of sound waves?**

Sound waves have a wide range of amplitudes. Typically the variation in pressure is around one part in 10,000 or 0.0001 atmospheres. This kind of pressure variation is far below the detection threshold of our sense of *touch*. Nonetheless, sound is such a rich environmental factor that virtually all higher animals have evolved a sense of hearing that is well adapted to detecting sound. The human ear is extremely sensitive and can easily detect pressure oscillations of less than one part in a million.

**How fast do sound waves travel?**

Sound waves travel faster than familiar objects in your everyday environment. For example, if someone across a 5 m room talks to you, the sound wave reaches your ear in 0.015 s. In air at room temperature (21°C) and one atmosphere of pressure, the **speed of sound** is 343 m/s (767 mph). The speed of sound is not constant but varies with temperature and pressure. The highest speed people normally attain, about 500 mph on a passenger jet, never exceeds the speed of sound.

**Frequencies of sound**

The frequencies of sound that the average human ear can perceive range from a low of around 20 Hz to a high of around 20,000 Hz. Most of the information contained in the human voice, however, is limited to the range between about 100 and 2,000 Hz The physical range of frequencies of sound is much greater than what humans can hear. Whales can sense sounds in water at frequencies below 10 Hz. Bats can sense sound frequencies in air higher than 100,000 Hz.

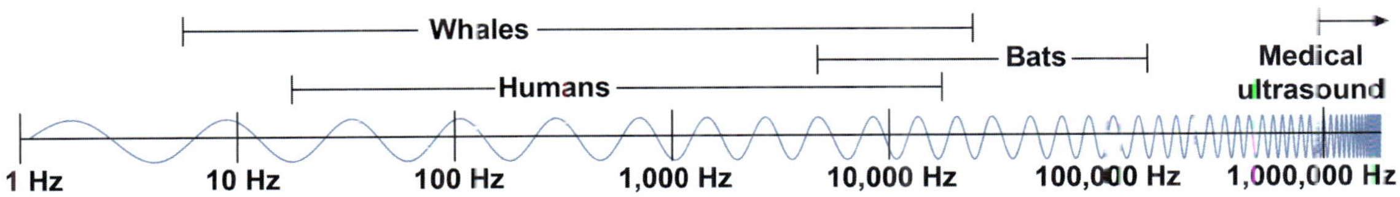


Frequencies of sound and average range of hearing

**What is ultrasound?**

Medical *ultrasound* technology uses sound waves at frequencies of $10^6$ Hz and higher. These frequencies are inaudible to the ear but pass readily through living tissue. Differences in tissue density reflect ultrasound waves back to a detector and allow sophisticated imaging without harm to the patient. The figure on the right is an ultrasound image of a 22-week old baby still in the womb.

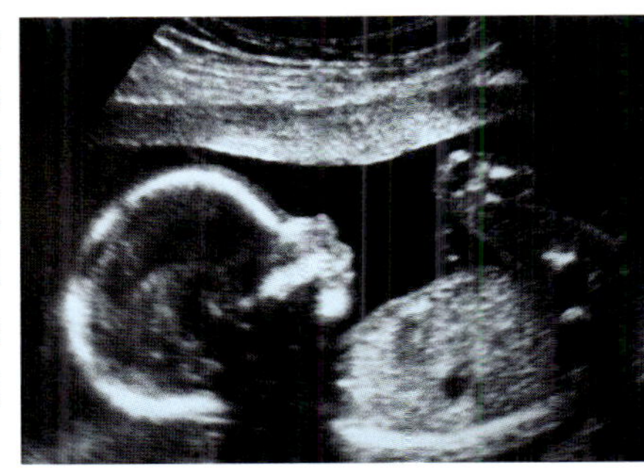

**Wavelengths of sound**

The wavelength of sound in air is comparable to the size of everyday objects. For example, a 1,000 Hz sound wave in air has a wavelength of 34 cm, or about the length of your forearm. The wavelength of sound is important in many technologies, including musical instruments. To design a vibrating object to make a certain frequency of sound, the size of the object must be comparable to the corresponding wavelength of the sound. A trombone is a good example. Pulling in the slide on a trombone results in a shorter wavelength vibration and a higher frequency sound.

**Wavelength and frequency of sound in air (at standard temperature and pressure)**

| $\lambda$ (cm) | $f$ (Hz) | Typical source |
|---|---|---|
| 840 | 41 | Low E bass guitar |
| 420 | 82 | Low E male bass voice |
| 70 | 500 | Average voice tone |
| 34 | 1,000 | Female soprano voice |
| 17 | 2,000 | Siren |
| 7 | 5,000 | Highest note on a piano |
| 3.4 | 10,000 | Whine of a turbine |
| 1.7 | 20,000 | Limit of human hearing |

**Wavelength varies with material**

The wavelength of sound varies depending on the properties of the material through which the sound is traveling. A 1 kHz (1,000 Hz) sound in air at 20ºC and 1 atmosphere has a wavelength of 34 cm. A 1 kHz sound wave in water has a wavelength of 150 cm, almost five times longer than in air. In general, the more resistant to compression a medium is, the longer the wavelength for a given frequency of sound. Steel is even more resistant to being compressed than water and the wavelength of a 1 kHz sound in steel is 5 m!

**Loudness and amplitude**

Loudness describes the perception of sound by your ear and brain. The loudness of sound is mostly determined by the amplitude of a sound wave. We say *mostly* because, to a human ear, the frequency also matters. A high-amplitude sound at a frequency of 40,000 Hz is *silent* to a human ear but quite loud to a bat! We use the **decibel (dB)** scale to measure noise levels because the human ear can respond to such a wide range of pressure changes. Zero decibels corresponds to the smallest pressure change that a healthy young listener can detect. Most sounds fall between 0 and 100 on the decibel scale, making it a very convenient set of numbers to understand and use.

**Loudness and the decibel scale**

| | |
|---|---|
| 10 dB | A whisper, 1 m away |
| 30 dB | Background sound, country |
| 40 dB | Background sound, city |
| 50 dB | Noise, average restaurant |
| 65 dB | Conversation, 1 m away |
| 70 dB | City traffic |
| 90 dB | Jackhammer, 3 m away |
| 120 dB | Physical pain |

**Decibel scale**

The decibel scale is *logarithmic*. In a logarithmic scale, equal intervals correspond to multiplying by 10 instead of adding equal amounts. For sound, every increase of 20 decibels (dB) means that the wave has 10 times greater amplitude. This is different from the linear scales you are familiar with. On a linear scale, going from 100 to 120 means that the amplitude increases by 20%. On a logarithmic scale, going from 100 to 120 dB means that the amplitude increases by a factor of 10—for example, from 100 to 1,000! Logarithmic scales are useful because they allow us to represent large ranges with convenient numbers. For example, sound waves with amplitudes from 0.00002 $N/m^2$ to 20 $N/m^2$ (also called *pascals* or Pa) can be represented by values from 0 to 120 on the decibel scale. The Richter scale for earthquake magnitude is also a logarithmic scale.

**Loudness and frequency**

The perception of loudness depends on frequency as well as amplitude. Sounds below 20 Hz or above 20,000 Hz are not perceived at all by the average human ear. A large amplitude sound wave at a frequency of 30,000 Hz is totally inaudible to a person. Of course, other animals have different sensitivities and this sound would be quite loud to a bat, which can hear frequencies past 100,000 Hz.

**Equal loudness curve**

Even within the range of hearing, the perceived loudness is affected by frequency. The equal loudness curve shows how sounds of different frequencies compare in perceived loudness to an average human ear. Sounds near 2,000 Hz seem louder than sounds of other frequencies, even at the same decibel level. For example, the equal loudness curve shows that a 40 dB sound at 2,000 Hz sounds just as loud as an 80 dB sound at 50 Hz. The human ear is most sensitive to sounds between 300 and 3,000 Hz. The ear is less sensitive to sounds outside this range. Not coincidentally, most of the frequencies that make up speech lie between 300 and 3,000 Hz.

**Equal loudness curve**

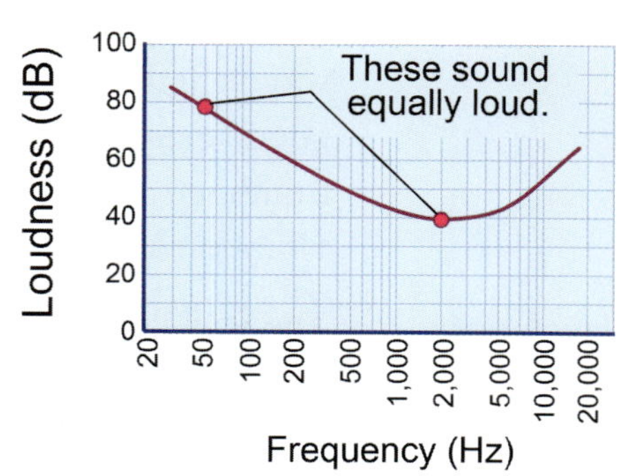

**What is the speed of sound?**

Sound travels at 343 m/s (767 mph) in air at 20°C and atmospheric pressure. Sound travels faster than most water waves but far slower than light waves. By comparison, sound is faster than all forms of human transportation except a few high-performance aircraft and rockets. Passenger jets typically travel at speeds of 550 mph or less.

**Speed, frequency, and wavelength**

The speed of sound is the product of frequency and wavelength, similar to other waves. With sound, this equations is most often applied to calculate frequency or wavelength of sound in air by assuming that the speed is 343 m/s.

Interactive equation

**Speed of sound**

$$(16.1) \quad v = f\lambda$$

$v$ = speed of sound = 343 m/s at sea level and 21°C
$f$ = frequency (Hz)
$\lambda$ = wavelength (m)

What is the frequency of sound that has a wavelength of 0.5 m?

| | |
|---|---|
| *Asked:* | frequency $f$ |
| *Given:* | $\lambda = 0.5$ m; assume $v = 343$ m/s |
| *Relationships:* | $v = f\lambda$ |
| *Solution:* | $v = f\lambda \rightarrow f = v/\lambda = (343 \text{ m/s}) \div (0.5 \text{ m})$ <br> $= 686$ Hz |
| *Answer:* | The frequency is 686 Hz. |

**Supersonic flight**

**Supersonic** describes motion at speeds greater than the speed of sound. Many military jets are capable of supersonic flight, as are most rockets. No passenger jets currently flying, however, are capable of supersonic flight. This is because the aerodynamics required for stable supersonic flight is different from that required for subsonic flight. The reason has to do with the *shock wave* that forms at the nose and leading edges of a supersonic aircraft.

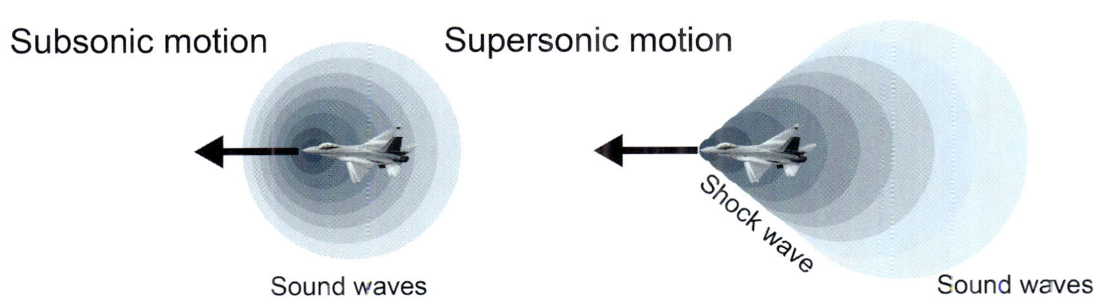


**The last supersonic passenger jet**

The pressure change across the shock wave causes a very loud sound known as a sonic boom. The shock wave can create severe turbulence that may make an aircraft unstable. It also takes a great deal of power to push through the atmosphere at supersonic speeds. Air travelers have been unwilling to pay the price, and the last supersonic passenger jet, the *Concorde*, was retired in 2003.

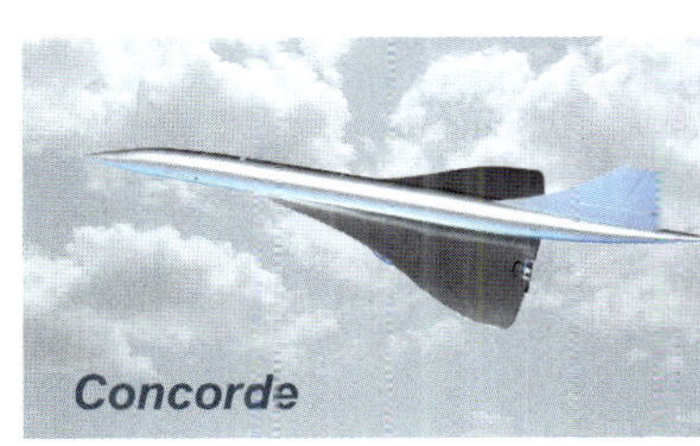

## The Doppler effect

**What is the Doppler effect?**

Consider a speaker that is moving forward. People in front of the moving speaker hear a higher frequency sound. People behind the moving speaker hear a lower frequency sound. The shift in frequency caused by motion is called the **Doppler effect** and it occurs when a source of emitted or reflected sound is moving at speeds less than the speed of sound.

**The Doppler effect**

The frequency of sound heard from a moving source depends on the observer's position.

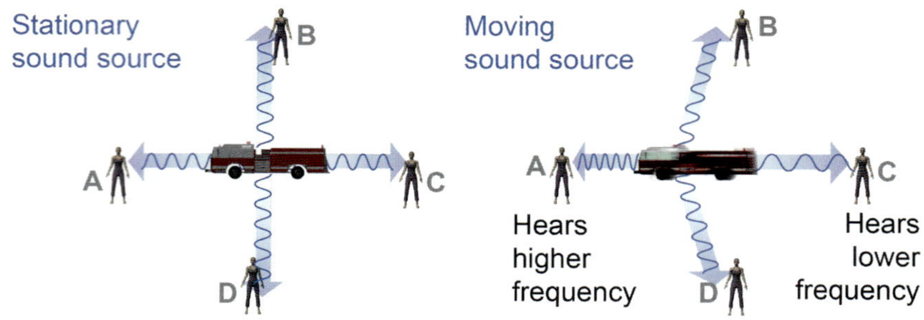


**Why does the frequency change?**

The Doppler effect is caused by relative motion along the line of sight between a source of sound and an observer. Consider yourself as the observer of a sound source moving toward you. Your ear "hears" the frequency at which wavefronts reach you. Between one wave and the next the source gets closer, so the second wave reaches you sooner. You hear a *higher* frequency. The opposite happens if the source is moving *away*. Successive waves reaching your ear are farther apart because the source is farther away with each wave. You hear a *lower* frequency.

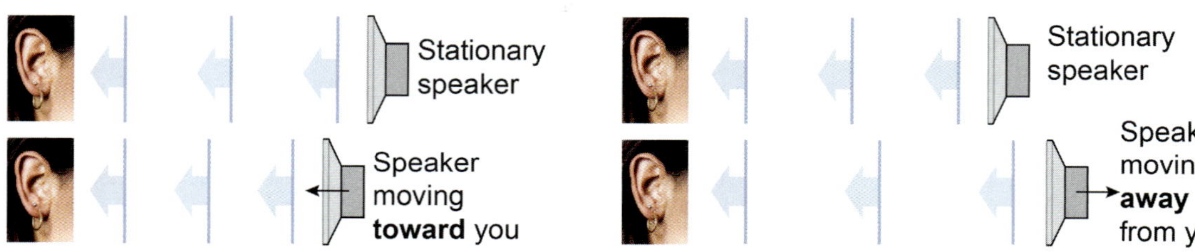


You hear **more** wavefronts per second. You hear **fewer** wavefronts per second.

**How much does the frequency change?**

Equation (16.2) gives the observed frequency $f$ when the source is emitting the sound at frequency $f_0$. In the equation it is assumed that the observer is at rest, and the source is moving *toward* the observer with velocity $v$. If the source is moving *away* from the observer the velocity should be *negative*. Equation (16.2) only holds for speeds below the speed of sound.

**(16.2)** $$f = f_0 \frac{v_s}{v_s - v}$$

$f$ = observed frequency (Hz)
$f_0$ = frequency of source (Hz)
$v$ = relative velocity of source to observer (m/s) [positive toward observer]
$v_s$ = speed of sound (m/s)

**Doppler effect**

**How is the Doppler effect used?**

The Doppler effect occurs for waves reflected from moving objects. The acronym RADAR originally stood for RAdio Detection And Ranging, in which reflected radio waves are analyzed for Doppler shift (speed) and time delay (range). Modern highway patrols use a laser form of Doppler radar to measure the speed of a car from a distance. The amount of the frequency shift is proportional to the speed of the car.

# Investigation 16B Doppler effect

**Essential questions** *What is the Doppler effect and how is it used in technology?*

Interactive simulation

The Doppler effect is the observed shift in frequency caused by relative motion between a source of sound and an observer of the sound. The frequency shift depends on the relative speed of the source and observer and on the speed of sound. By measuring the frequency shift of a reflected sound wave, it is possible to determine the velocity of the source. This is how police radar detects the speed of a moving car from a distance.

## Simulating the Doppler effect

You can set the frequency of the sound source and the velocity of the source and/or the observer. In either case the simulation will play the sound you would hear if you were the observer as the speaker moved toward you, away from you, or both sequentially.

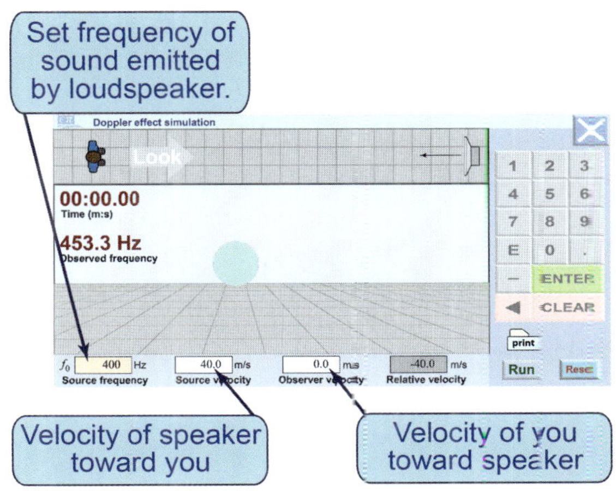


**Listen to how the observed frequency changes as you change the source and observer velocities.**

1. Choose a frequency of 400 Hz and a source velocity of 2 m/s (4.5 mph) and observe and record the frequencies as the speaker moves toward you and then away from you.
2. Choose a source velocity of 30 m/s (67 mph) and make the same observations.
3. Choose an *observer* velocity that moves you toward the speaker at 2 m/s and then at 30 m/s.
4. Choose a source velocity of 15 m/s and an observer velocity of 15 m/s. What happens to the frequencies?
5. Choose a source velocity of −15 m/s and an observer velocity of 15 m/s. What happens to the frequencies? Compare these values to your previous results.
6. Choose an observer velocity of −15 m/s and a source velocity of +15 m/s. What happens to the frequencies?

a. *Describe three different situations in which an observer hears a sound that is Doppler-shifted to a higher frequency.*

b. *Explain the difference between the frequency shifts you observed at velocities of 2 and 30 m/s. Express your answer in terms of the ratio of the velocity to the speed of sound (343 m/s).*

c. *Describe the difference between moving the source and moving the observer with respect to the Doppler shift. Does it matter which is moving? To what extent?*

d. *Can there be a situation in which both source and observer are moving and there is no Doppler shift? Explain why this is the case using the concept of relative velocity.*

e. *Can there be a situation where the source is moving at 15 m/s yet the Doppler shift corresponds to a velocity of 30 m/s? Explain how this can be using the concept of relative velocities.*

Chapter 16

# Section 1 review

Sound is a longitudinal pressure wave that can travel through many substances, though humans normally hear it through air. The human sense of hearing responds to sound waves with frequencies that range from about 20 Hz up to 20,000 Hz. High frequencies are perceived as high *pitch* and large-amplitude sound waves are perceived as *loud.* The amplitude of a sound wave is measured using the logarithmic *decibel* scale: An increase of 20 dB means that the amplitude of a sound wave has been multiplied by a factor of 10. The Doppler effect describes how a sound wave's pitch is altered when its source moves toward or away from the listener. Supersonic (faster-than-sound) motion through a substance creates a *shock wave* in that substance.

| | |
|---|---|
| **Vocabulary words** | **pitch, speed of sound, decibel (dB), supersonic, Doppler effect** |
| **Key equations** | $v = f\lambda$ $\quad\quad$ $f = f_0 \dfrac{v_s}{v_s - v}$ |

## Review problems and questions

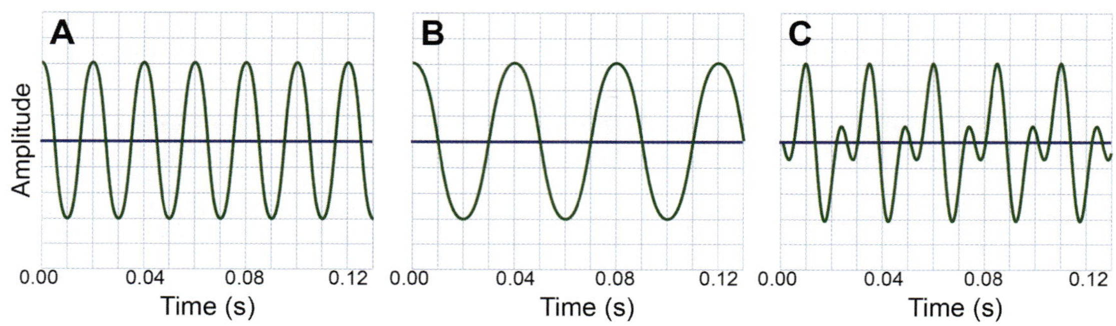


1. These three graphs show the relative amplitudes of three different sound waves, each as a function of time.
   a. Which of the three sound waves has the lowest pitch?
   b. Is that pitch high enough for the typical human to hear?

2. One valuable assistive technology is the *hearing aid*, an electronic device that contains a small amplifier to boost sound strength.
   a. Suppose that one hearing aid multiplies the *amplitude* of sound waves by a factor of 10. How many decibels will the hearing aid add to the sounds it detects?
   b. Another hearing aid adds 40 dB to the sounds it detects. By what factor does *it* multiply sound amplitude?

3. Return to the graphs near the top of this page. Suppose that Sound A and Sound B both came from the same source and had the same original frequency $f_0$. Now consider that one of the sounds was heard as the source approached you, while the other was heard after the source passed you and receded into the distance.
   a. Which sound (A or B) was from the source when *approaching*?
   b. The source was moving at a speed of 114 m/s. What was the source's actual sound frequency, $f_0$? (Assume a sound speed $v_s$ of 343 m/s.)

The true richness of sound and the ability to carry information lie in the fact that many frequencies are present simultaneously. The human ear can "listen" to more than 15,000 different frequencies of sound, forming a *sonic image* many times per second. The incredible information density in sound comes from the changing patterns of frequency and amplitude for thousands of different frequencies.

## Visualizing sound waves

**Seeing sound waves on a graph**

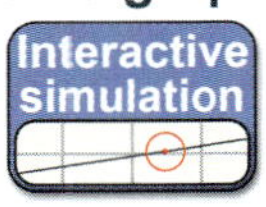


Sound waves oscillate much faster than water waves, and air is transparent, so variations in pressure are invisible. To visualize a sound wave scientists and engineers use several different kinds of graphs. The first shows the *waveform,* which describes how the pressure changes with time. To see the individual oscillations of a sound wave, the time axis must be enlarged to show a value much less than one second. The graph below shows 12.5 cycles in 0.03 s, so the frequency of the wave is

$$f = \frac{12.5 \text{ cycles}}{0.03 \text{ s}} = 416.7 \text{ Hz}$$

This is the musical note G# near the middle of the piano keyboard.

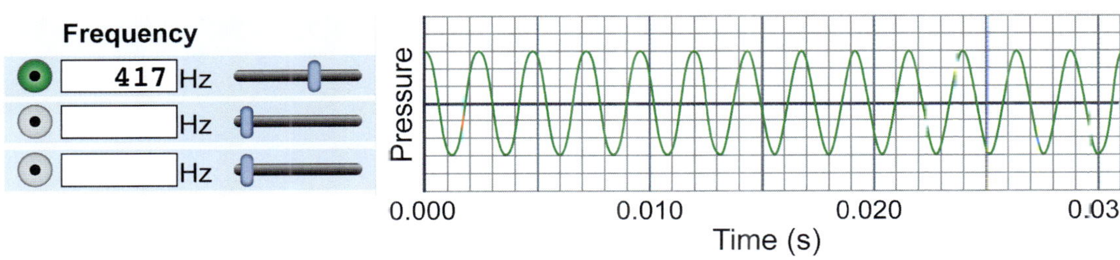


**Interpreting a soundtrack**

You may have seen a computer represent a sound as in the diagram on the right. This shows the pressure variation over real time in seconds and minutes. This graph shows the loud and soft variations of amplitude, but to see the individual oscillations of the sound waves, you have to zoom in on the time axis. The lower diagram shows part of the same sound magnified by a factor of 140 in time. The oscillations are now visible on the magnified time scale of 0.01 s per major division of the horizontal axis.

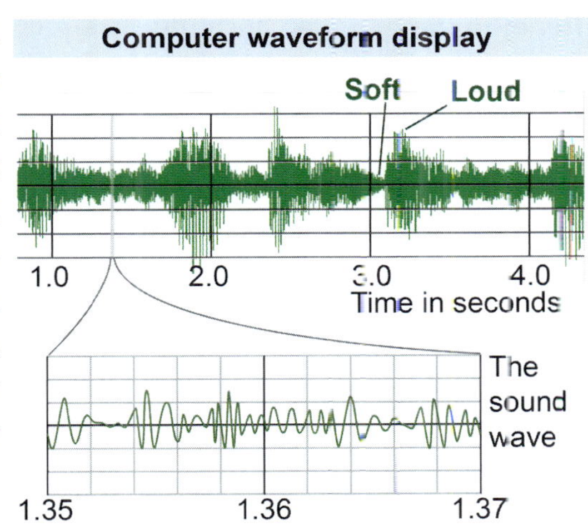


**Microphones and sound recording**

To record a sound wave, the pressure variation in the air is first converted into an electrical signal with a **microphone**. A microphone transforms a sound wave into an electrical signal with the same complex pattern of oscillation. In *digital* systems, such as CDs and computers, a sensitive circuit called an *analog-to-digital converter* (ADC) measures the amplitude of the electrical signal 44,100 times per second. The numbers are scaled and graphed to make the waveform display. The same string of numbers is recorded as data on a CD or as compressed data in other digital formats, such as MP3.

## The frequency spectrum

**Waveform and spectrum**

A single, pure frequency has the form of a sine wave. Every successive cycle of a sine wave is exactly the same as the last cycle. On a waveform graph a sine wave has uniform oscillations. On a frequency spectrum, a sine wave shows up as a single peak, like a bar graph. The diagram shows a 100 Hz sine wave with an amplitude of three units on both graphs. From the waveform, one cycle takes 0.01 s, so the frequency is 1 ÷ 0.01 s = 100 Hz. The **frequency spectrum** shows amplitude versus frequency. The same 100 Hz wave appears as a single bar at 100 Hz with an amplitude of three.

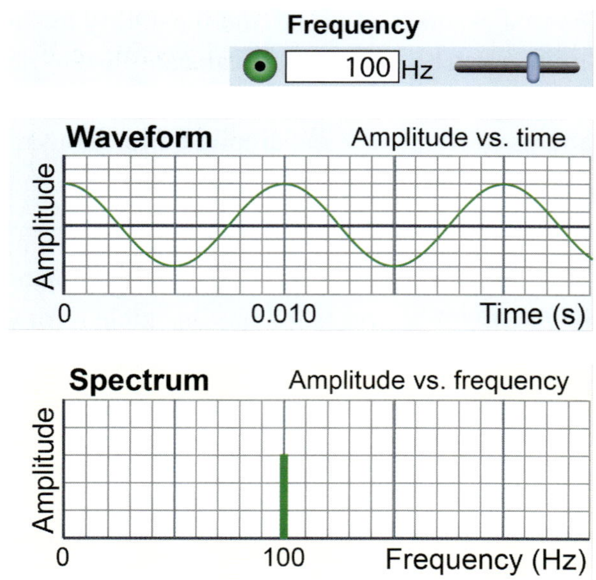


**Multi-frequency sound**

When more than one frequency is present, the wave oscillates in a more complicated pattern. The diagram on the right shows the addition of a 300 Hz wave, a 400 Hz wave, and a 450 Hz wave of the same amplitude. The resulting waveform does *not* have a simple repeating cycle like a sine wave, and therefore *you know it contains more than one frequency.*

**Reading the spectrum**

The spectrum has much more information because it tells you immediately that the complicated waveform is the addition of three sine waves. The three peaks are at 300, 400, and 450 Hz, and they have equal amplitude. The frequency spectrum breaks a complex wave down into its component frequencies and shows you the relative amplitude of each.

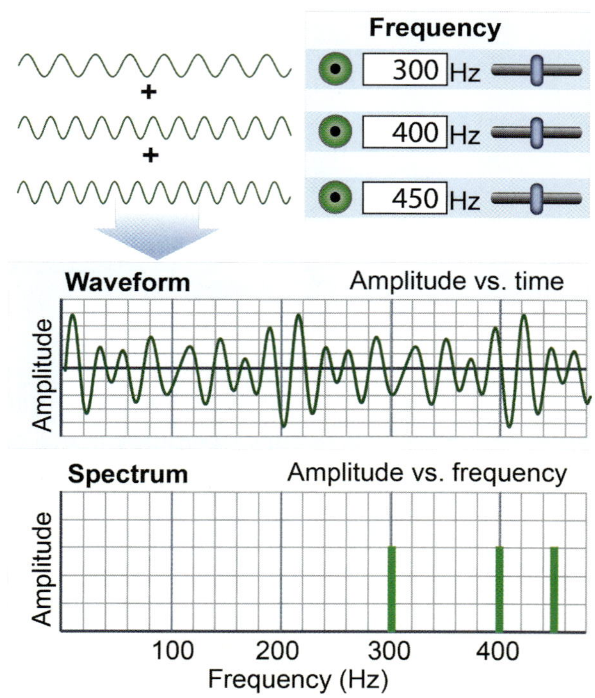


**Audio compression**

Accurate reproduction of sound in the frequency range of 20 to 20,000 Hz requires an enormous amount of data. One second of stereo, CD-quality sound includes 88,200 16-bit numbers or 1.41 million bits, corresponding to a *data rate* of 1.41 million bits per second (bps). Most sound is recorded, transmitted, or stored at lower data rates, but this is a tradeoff between file size and sound quality. For example, human speech includes mainly frequencies from 200 to 2,000 Hz. Telephones sound "tinny" because only this limited frequency range is included in a telephone signal. Digital audio files, such as the MP3 format, use *compression algorithms* to process digitally recorded sound to reduce the amount of data required per second. A standard MP3 file transmits only 128,000 bps, a compression of about 11:1 compared to CD-quality sound.

### Fourier's theorem

Our interpretation of complex waves as being made of single-frequency components is based on **Fourier's theorem**. This theorem says that any repetitive wave can be reproduced exactly by a *Fourier series* of single-frequency waves with different amplitudes. Fourier's theorem also includes a mathematical formula for finding the amplitude of each different frequency wave in the series. That is how a frequency spectrum is determined.

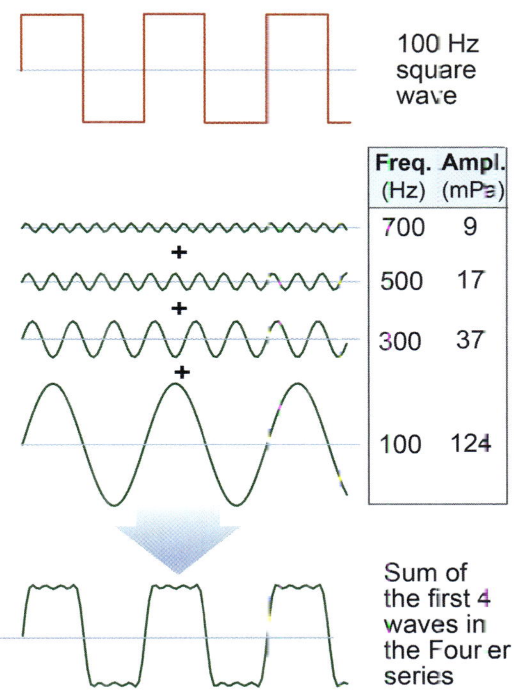


### Approximating a square wave

The top diagram shows the waveform of a 100 Hz square wave that has an amplitude of approximately 120 mPa. The first four sine waves in the Fourier series have frequencies of 100, 300, 500, and 700 Hz and amplitudes of 124, 37, 17, and 9 mPa. These four waves add up to a fairly good approximation of the original square wave. Adding more waves in the series would make the approximation even better.

### The frequency spectrum

The spectrum of a square wave shows the amplitude of each component frequency that makes up the total. The first four largest waves are at frequencies of 100, 300, 500, and 700 Hz. The spectrum is similar to a bar chart, with one vertical bar per frequency. In this chart, the height of the bar represents the amplitude of that frequency.

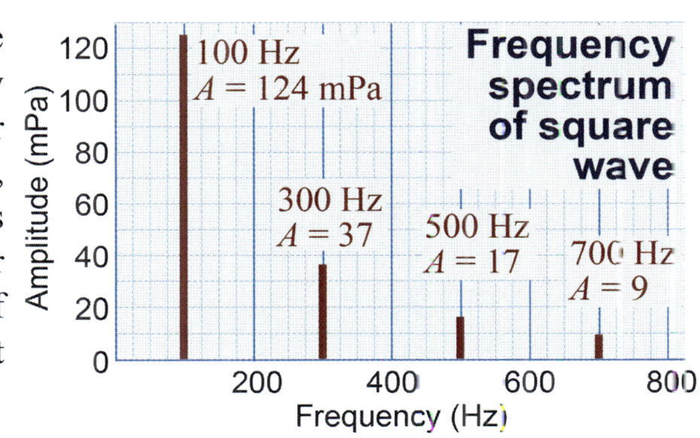


The relative heights of the different frequencies indicate the relationship between the amplitudes of the sound at each frequency. In the spectrum (at right), the fundamental frequency at 100 Hz has the largest amplitude.

### Real spectra

Sound ordinarily contains thousands of different frequencies, each with its own amplitude and phase. The spectrum on the right shows the sound wave from an acoustic guitar playing the note E. The frequency spectrum shows that the complex sound of the guitar is made up of many frequencies—similar to the square wave—only with a far more complex mix of different frequencies.

**Spectrum of "clean" guitar sound**

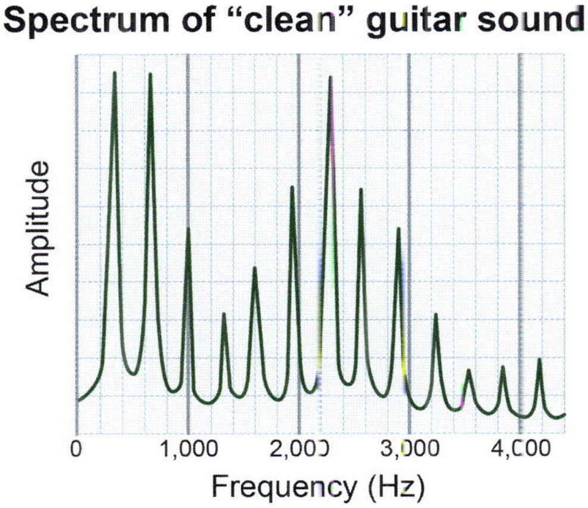

**Information content in sound**

How does sound carry information? Think about reading one word of a story. You recognize the word, but it does not tell you much about the story. When you read the whole story you put all the words together to get the meaning. The brain does a similar thing with different frequencies of sound. A single frequency by itself does not have much meaning. The meaning comes from patterns of changing amplitude in many frequencies together.

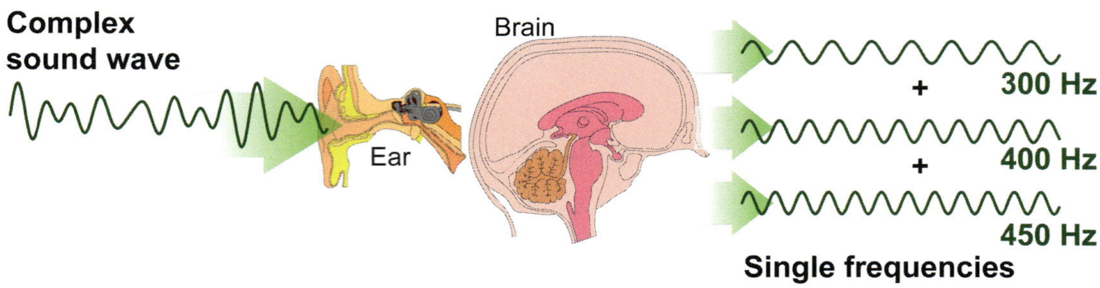


**Hearing sound**

When we hear, the nerves in the ear respond separately to each different frequency. The brain interprets the signals from the ear and creates a response to the individual frequencies contained in the sound wave. You hear the individual frequencies, which you can demonstrate by listening to multiple frequencies together.

**The spectrogram**

A **spectrogram** is a visual representation of sound that includes frequency, loudness, and time. The vertical axis represents frequency and the horizontal axis represents time. Color is usually used to represent loudness. The brighter the color, the higher the amplitude at that frequency. The example on the right shows a sequence of sounds from 200 to 500 Hz over a period of 10 s:

a. a loud sound at 200 Hz that lasts from 1 to 3 s;
b. a softer sound at 300 Hz that lasts from 3 to 5 s; and
c. a sound at 500 Hz that starts soft, gets loudest at 3 s, and gets softer again.

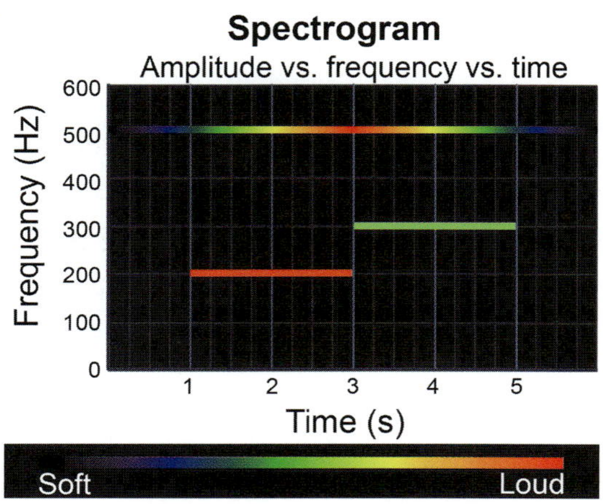


**How information is encoded in sound**

Fundamentally, information is *encoded* in patterns that are *decoded* by the ear and brain. In writing, the patterns are sequences of letters and spaces. In sound, the patterns are the changes in amplitude for about 15,000 different frequencies. The spectrogram to the right is for a male voice saying, "Hello." The word lasts from 0.1 s to about 0.6 s. You can see lots of sound below 1,500 Hz and two bands of sound between 2,000 and 4,000 Hz. Every person's spectrogram is different, even when they are saying the same word. You might infer that the chart represents an adult female voice if the frequencies were somewhat higher than those of a man.

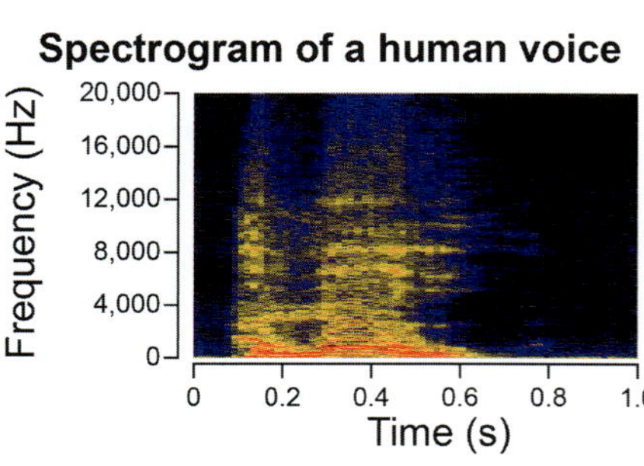

# Investigation 16C How sound carries information

**Essential questions** *How does sound carry information, such as words?*

Interactive simulation

The identity of a song, the meaning of words, and even a person's identity are information that we immediately decode from sound. Sound carries an extraordinary amount of information through the interference of many thousands of frequencies, each of which varies in phase and amplitude over time.

## Part 1: Multifrequency sound and the spectrum

1. The electronic resource shows a waveform graph (by default). Set a frequency of 300 Hz and adjust the volume. Set the time axis to display 0.02 s.
2. Add a 400 Hz sound, then a 450 Hz sound. Listen to the frequencies separately and together and observe the waveform.
3. Switch the graph to display a spectrum—a bar chart that shows the frequencies of the sound. Set the same three frequencies as before and observe the spectrum as you change the frequency and volume.
4. Starting with 300 Hz, use three frequencies in the ratios 1:3:5 to create the best approximation to a square wave.

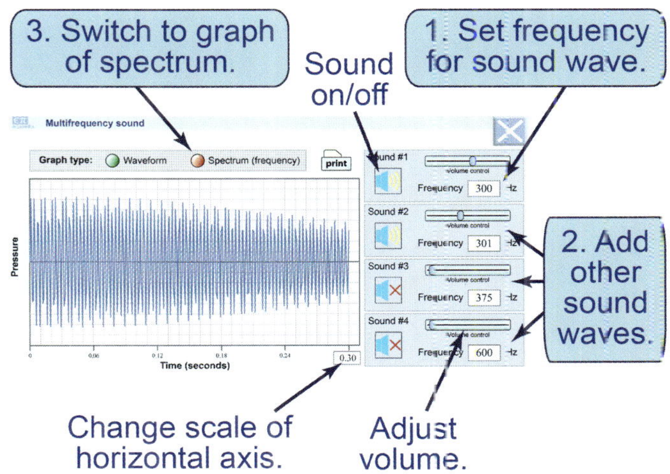


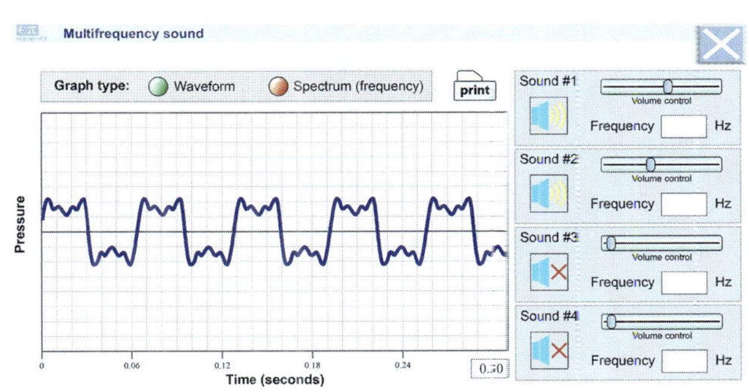


a. *Can you hear individual frequencies or do you only hear a mixture? What evidence supports your conclusion?*

b. *Compare the time and frequency graphs. Which best describes the frequency content of sound? Which describes the motion of your eardrum?*

c. *Research how many different and distinct frequencies the human ear can simultaneously perceive.*

# Section 2 review

Sound waves are invisible. Sensors and electronic devices are useful for displaying how the pressure of a sound wave varies with time on a *waveform* graph. Ordinary sound contains many simultaneous frequencies. The *frequency spectrum* shows amplitude versus frequency. The information content in sound is encoded in the patterns of how amplitude changes over time across many frequencies.

| Vocabulary words | microphone, frequency spectrum, Fourier's theorem, spectrogram |
|---|---|

## Review problems and questions

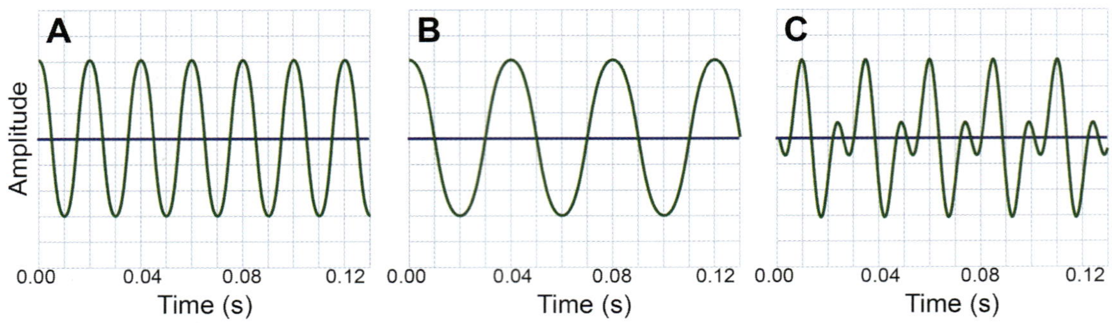


1. One of the three graphs shows a sound that contains two different frequencies.
   a. Which graph is it and how do you know?
   b. What is the lower frequency in this sound?
   c. What is the higher frequency in the sound?

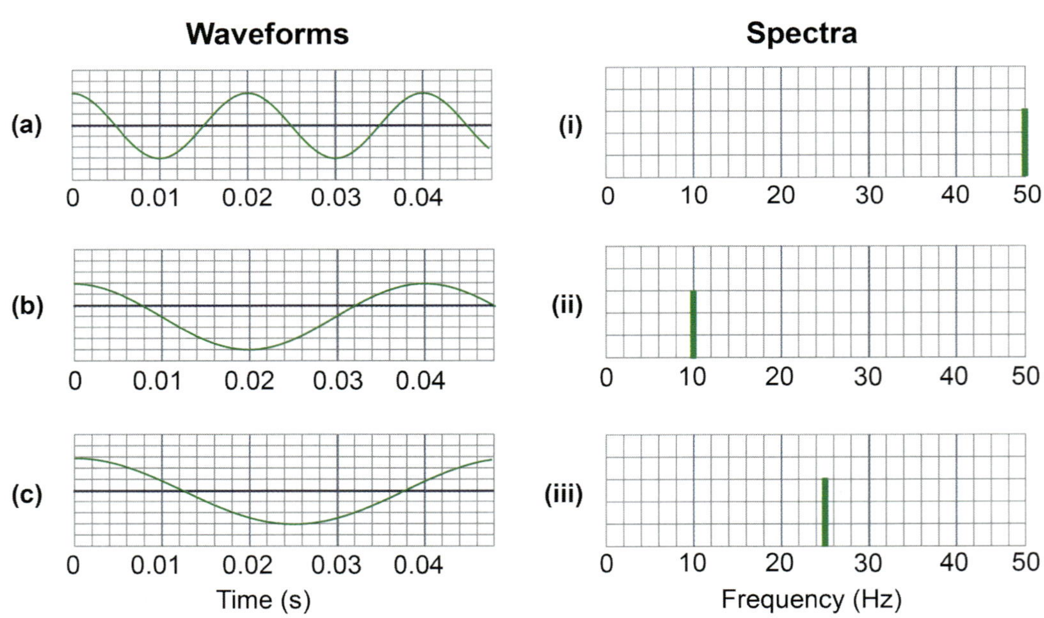


2. Match each waveform from the left-hand column with one spectrum from the right-hand column. Assume that the $y$-axis of each graph is amplitude.

3. Music can be recorded, transmitted, and stored digitally. Describe the tradeoffs between sound quality and file size in digital audio formats.

Virtually every sound you hear is affected by interference and resonance. In this context, the term *interference* does not mean anything *bad*. Interference just means that more than one frequency is present in the sound and that the overall wave form is the sum of multiple frequencies. Resonance in sound maximizes certain frequencies and eliminates or weakens others. Resonance occurs when you blow air across the top of a bottle, strike a tuning fork, or run a bow across a violin string. The distinctive frequency balance that makes a person's voice unique and recognizable comes from complex resonances involving your vocal cords, larynx, mouth, tongue, and even your nasal passages!

## Echoes and reverberation

**What causes echoes?**

If you yell "hello!" 170 m away from a large wall, the sound will reflect back to you one second later as a distinct *echo*. An **echo** is a reflected sound wave. One second is the round-trip time for the sound wave traveling at 340 m/s. But what will happen when the wall is much closer, say 10 m away? The round-trip echo time is 0.06 s. What will you hear?

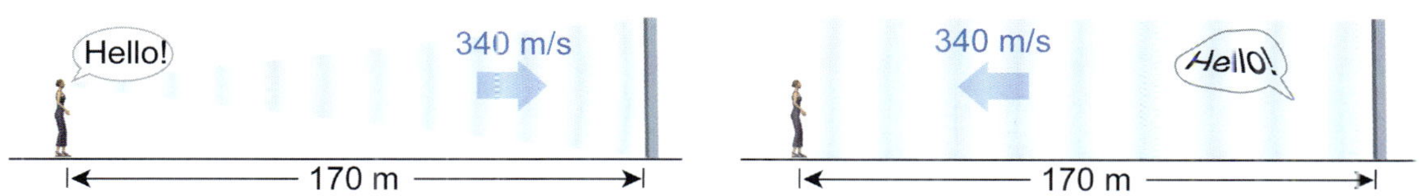


**What is reverberation?**

In a good concert hall you hear music directly from the stage but you also hear a multiple echo called *reverberation*. Reverberation is the addition of multiple reflections of a sound to the original sound. Each reflection is slightly delayed but not so much that the echo is distinctly recognizable. Reverberation adds liveliness, depth, and richness to sound and concert halls are specifically designed to created the right amount of "reverb." Many musicians use reverb effects that electronically add layers of echoes to their sound.

**Acoustical engineering**

Think about sitting in the middle of an auditorium. You hear sound directly along path A but also reflected sound (along path B, C, and others). The shape of the room and the wall surfaces are designed so that there is enough reverberation to be pleasant but not too much, which would "muddy" the sound. Furthermore, a good design avoids creating locations where indirect, reflected sounds interfere destructively with direct sound waves; in those "dead spots," certain frequency ranges will lose clarity or volume. Many auditoriums have movable ceiling panels ("clouds") that can be rearranged to balance the reverberation for different performances and instruments.

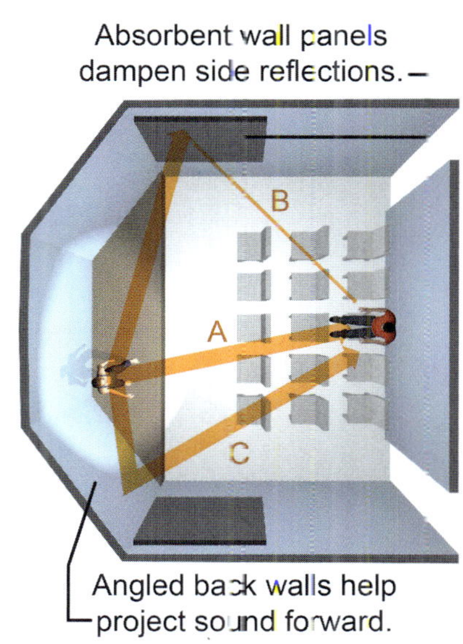

### Constructive and destructive interference

Interference can result from a *phase* difference between two or more sound waves. The **phase** of a wave describes a place in the wave's cycle with respect to the full cycle. A full wave is usually assigned a phase of 360°. That means a phase of 180° is one-half of a cycle. Two waves are *in phase* when both begin at the same point in their cycle. Two waves are 180° *out of phase* when one wave begins one-half cycle ahead of or behind the other wave. Sound waves that are the same frequency add up constructively when they are in phase, or destructively when they are out of phase.

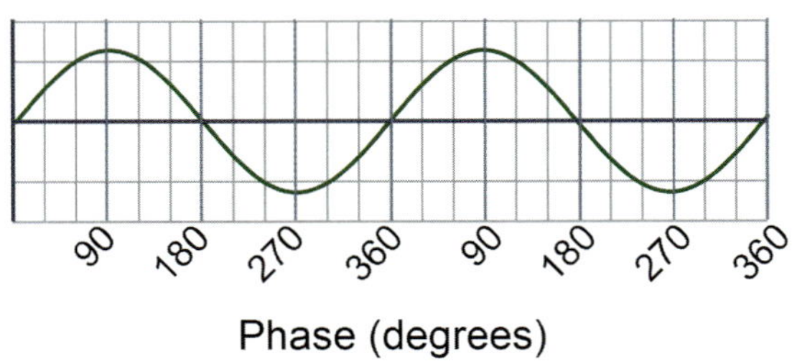


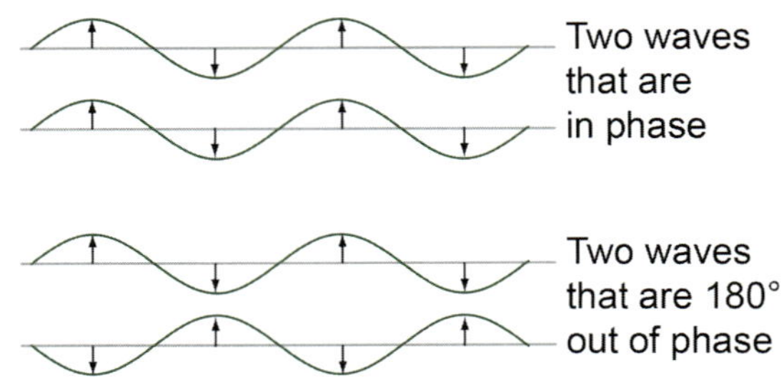


### Beats

Interactive simulation

When two frequencies of sound are close, but not identical, the sound waves drift in and out of phase and make **beats**. Sometimes the two waves are in phase, and the total is louder than either wave separately. Other times the waves are out of phase and they cancel each other out, making the sound quieter. The term *beats* refers to the rapid alternation in amplitude caused by this interference. The alternation in loudness occurs at the *beat frequency*, which is the difference between the two single frequencies. For example, a 120 Hz sound and a 140 Hz sound interfere to make beats at 20 Hz. The overall loudness of the combined sound would go up and down 20 times per second. In the e-Book, use the interactive tool on the left to investigate beats!

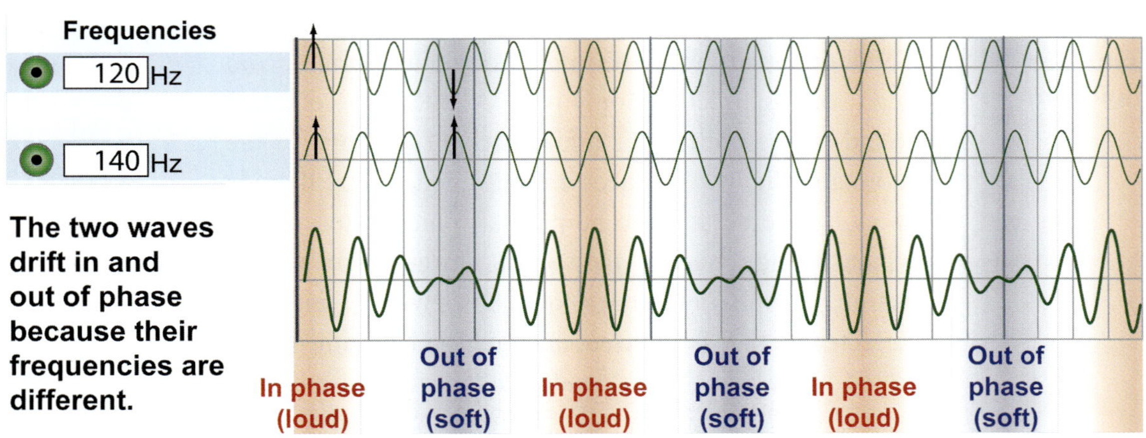


### What do beats sound like?

Instruments that are out of tune make beats. Many people find beat frequencies between 1 Hz and about 30 Hz unpleasant to listen to. Sounds that create beats are *dissonant* and may evoke tension and unrest. The frequencies in the musical scale are specifically chosen to reduce the occurrence of beats. The word *harmony* technically means a combination of sounds that do not make unpleasant beats.

**Resonance amplifies certain frequencies**

Musical instruments produce sound through *resonance*. Woodwind and brass instruments use a volume of air as a resonant *cavity*. In acoustics, a cavity is a volume of space enclosed by boundaries that contain and reflect sound. When many frequencies are present in a sound, frequencies that are resonant absorb and hold energy far more effectively than other frequencies. Within a few dozen oscillations—fractions of a second!—the energy in noise at all frequencies has been channeled into the amplitudes of only the resonant frequencies, which then dominate the sound.

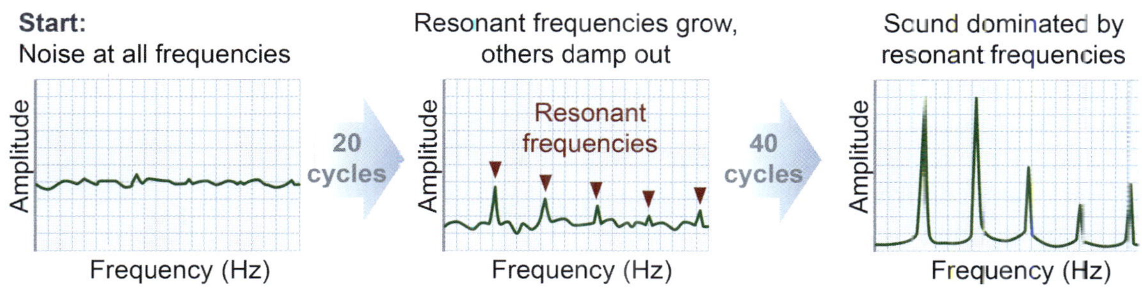


**Resonant modes of pipes**

The resonant *modes* of a cavity correspond to standing waves of air pressure that occur at specific frequencies. For a pipe that is closed at both ends, the resonant modes have *nodes* at the boundaries. For an open pipe, the resonant modes have *antinodes* at the ends. In both cases, the resonant frequencies are integer multiples of the fundamental. For a pipe that is closed at one end and open at the other the boundaries are a node and an antinode, respectively. An open/closed pipe has resonances that are *only odd-integer multiples* of the fundamental. The frequencies and wavelengths of the first four harmonics are shown below.

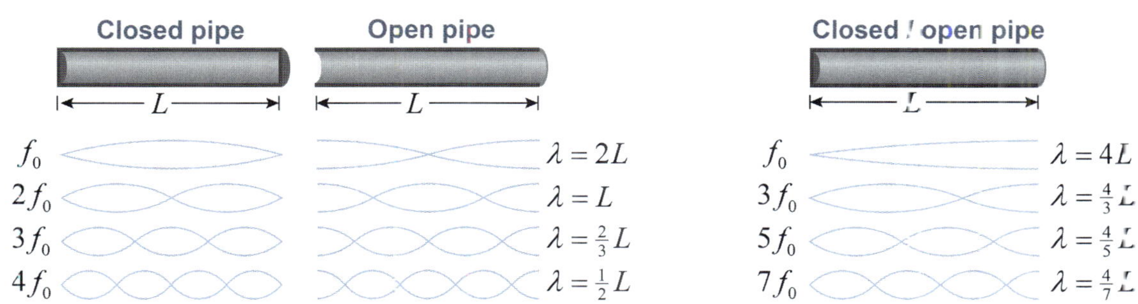


**Pipe organs**

The size of a vibrating object affects its natural frequency through resonance. For example, the pipes in a pipe organ are made in all different sizes, each designed to produce a specific wavelength and frequency of sound.

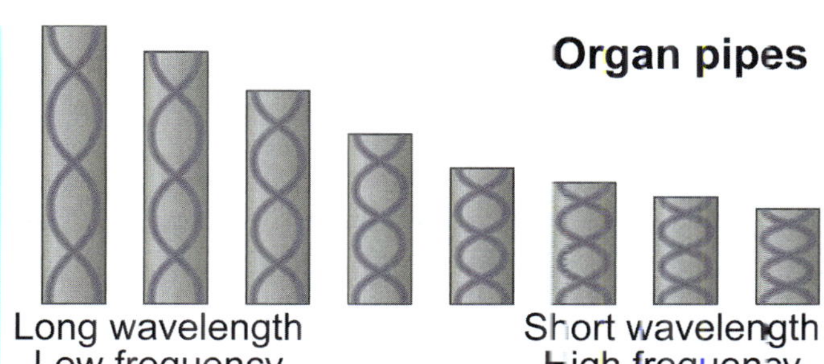

# Investigation 16D Resonance and sound

**Essential questions** *How do we create specific frequencies of sound, such as in music?*

A guitar string vibrates at its natural frequencies. Other objects, such as wine glasses and tuning forks, also vibrate at their natural frequencies. The frequencies are controlled by properties such as size, mass, and tension.

## Part 1: Measuring the natural frequency

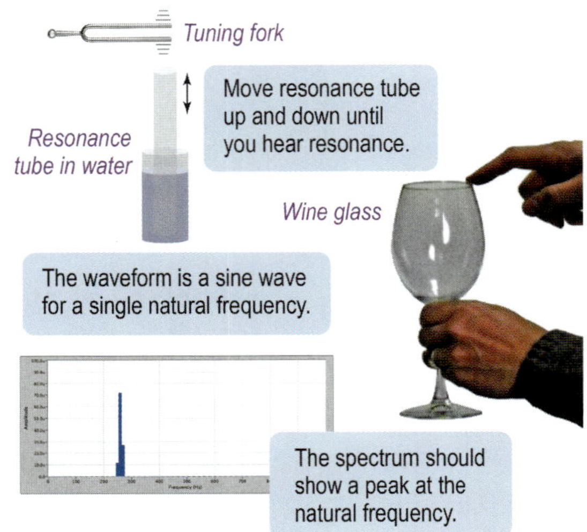


1. Open the **16D_ResonanceAndSound** experiment file.
2. Strike a tuning fork with a rubber mallet and hold it next to the microphone on your computer. Using the FFT in your software, confirm that its natural frequency matches the marked value.
3. Strike the tuning fork again and hold it over the resonance tube, partially immersed in water.
4. Raise and lower the tube until you hear the loudest amplification of the tuning fork note. Measure and record the height of the tube above the waterline. Repeat for two other tuning forks.
5. Firmly hold the stem of the wine glass while running a wet finger lightly around its rim. The glass should ring in a clear tone. Measure its natural frequency.
6. Add water to the glass in small increments and repeat the measurements.

a. *Why did the tube resonate at a particular position? What characteristics of a resonance tube determine its natural frequency?*

b. *Calculate the wavelength and then the product of wavelength and frequency for each tuning fork you tested, and record the data in a table.*

c. *Does the product of wavelength and frequency vary among tuning forks. Why or why not?*

d. *Propose a hypothesis that explains the variation in resonance frequency with the height of water in the wine glass. How is your hypothesis supported by your observations?*

## Part 2: Controlling the natural frequency through resonance

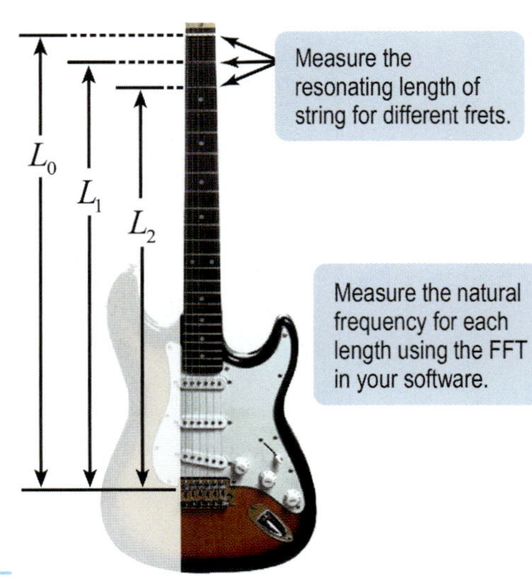


1. Measure the length of string for your instrument at each of its different frets, or fingerboard positions.
2. Use the FFT in your software to determine the natural frequency for each different string length.
3. Repeat the procedure for at least two additional strings.

a. *Graph frequency versus length of a single string. Explain why the graph has the shape that it does.*

b. *On the same graph, plot frequency versus length for a different string. Explain the difference between the two curves using Newton's laws.*

c. *Explain how resonance controls which frequencies persist in a vibrating system.*

**Why do instruments sound different?**

Different instruments have characteristic sounds just as different people have characteristic voices. As an example, the note E-330 Hz played on a guitar is recognizable as the same note when played on a piano. This note, however, sounds quite different when played on the two instruments. What is different and what is the same? The answer involves the properties of resonant objects and cavities.

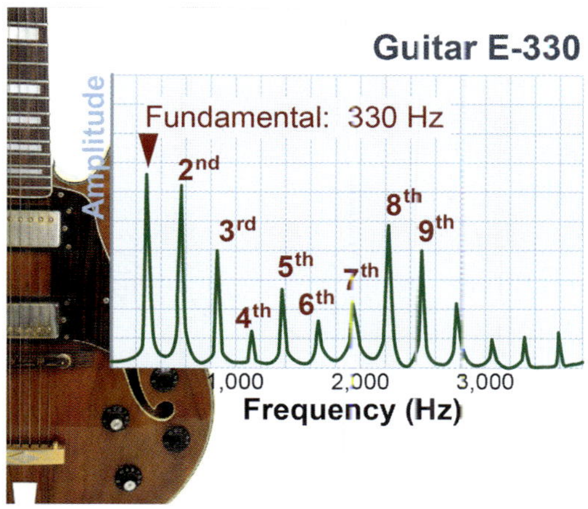


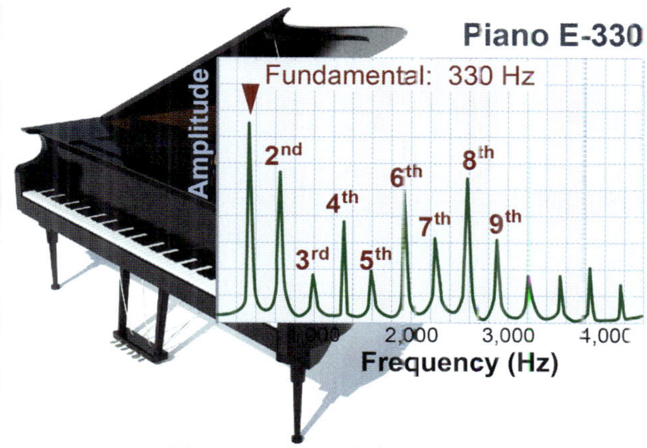


**Are notes a single frequency?**

The sound from a musical instrument is not a single pure frequency but contains many frequencies. The most important are *harmonics*. A **harmonic** is a frequency that is an integer multiple of the fundamental. Strings and air columns can vibrate at many harmonics. A real resonant oscillation in an instrument contains many harmonics at once. There are more than 40 distinct harmonics present in the note from a piano! The distinctive sound of an instrument results from:

1. the relative amplitude of each harmonic compared to the fundamental and
2. how quickly each harmonic grows and decays after a note is struck.

**Explaining the spectra**

In the guitar and piano spectra above, the fundamental frequency is the same but the third and fourth harmonics have different amplitudes. The guitar has a strong third and weak 4th and the piano has a weak third and strong fourth. The same 330 Hz fundamental is the reason why both notes sound like "E," and the difference in harmonics explains why the guitar sounds like a guitar and the piano sounds like a piano.

**Attack and decay times**

The *rate* at which the loudness of each harmonic rises and falls is another characteristic that makes an instrument's sound distinctive. The *attack* time is the time it takes to reach maximum loudness. The *decay* time is the time over which the sound dies away. Attack and decay times differ for each harmonic, and higher harmonics tend to attack and decay faster. Attack and decay depend on both resonance and damping, and they are very different for each instrument.

Each harmonic may have different attack and decay times.

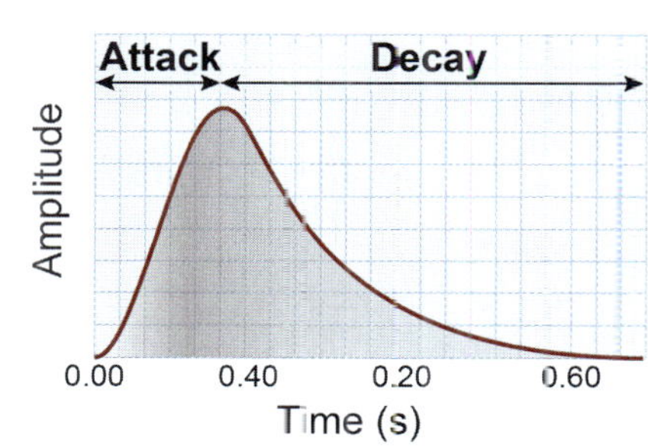

# Design a musical instrument

**Musical instruments**

A musical instrument creates frequencies of sound that match a musical scale. The beauty and richness of the sound come from the harmonics that are produced along with the fundamental frequencies.

**Design challenge**

Create a musical instrument that

1. plays the frequencies (within 5%) of all eight notes of a major scale,
2. has working parts consisting of a metal bar (or tube) with lengths less than 1 m, and
3. has a total mass of less than 5 kg.

Based on the above instructions, identify the *design criteria* and *design constraints* and include them in your written report.

**Modeling the system**

1. Measure the resonance frequencies of different lengths of your tube or bar material using the **DC_MusicalSound** experiment file.
2. Use your data to create a graphical and/or algebraic model that allows you to predict what length of tubing you need to produce each frequency in the scale.
3. Identify and record the variables that should be controlled. Describe how you designed your testing procedure to control for these variables.
4. Determine what changes can make small frequency corrections.
5. Explore how different methods of mounting the chime affect its sound. Consult additional resources (e.g., the library, a textbook, or the Internet) to obtain design ideas.

**Chimes of different length and/or mass**

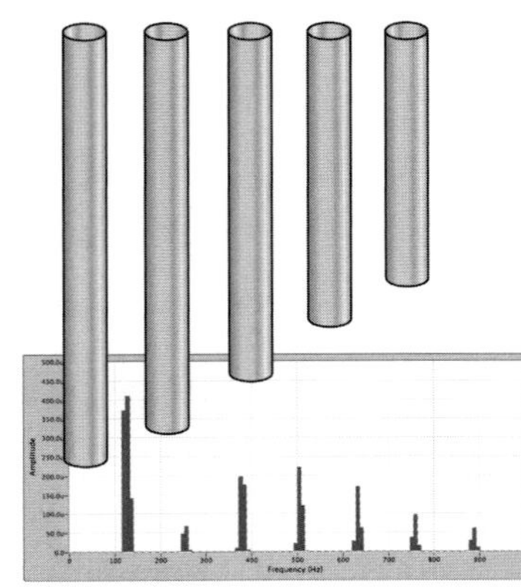

The spectrum shows a fundamental and several harmonics.

**Design**

Use your model to determine the design lengths you need to make your instrument. You will also need to design a way to support your resonant elements without dampening their vibration too much.

**Prototype**

Use appropriate tools, such as a pipe cutter, to create the resonant elements of your instrument. Assemble your prototype.

**Test**

Test your prototype by using the **MusicalSound** experiment file to measure the resonance frequency of each chime. Also look at the frequency spectrum for the harmonic content of each chime.

**Evaluate**

Compare the actual frequencies to the design criteria. Identify the strengths and weaknesses of your prototype's performance and document them in your written report. Determine any frequency adjustments that need to be made to the chimes and any changes that need to be made to your model.

**Revise**

Use your model to implement the design changes, such as removing material to raise a frequency. Implement the changes and test your revised design again. When you are ready, demonstrate the performance of your musical instrument by measuring the frequency of each chime.

**Pitch and rhythm**

The most basic elements of music are *pitch* and *rhythm*. The *pitch* of a sound is how high or low we hear its frequency. Pitch corresponds closely to frequency, but pitch is also a matter of *perception*. The way you hear a pitch can be affected by the sounds you hear immediately preceding, along with, and just after a note. Rhythm is the repeating time pattern in a sound. Rhythm can be loud or soft: tap-tap-TAP-tap-tap-TAP-tap-tap-TAP. Rhythm can be made with sound and silence or with different pitches.

**What is a musical scale?**

Although styles vary, all music is created from carefully chosen patterns of frequencies of sound called *musical scales*. Each frequency in a scale is called a *note*. The diagram below shows some of the notes on a piano along with their frequencies. There are eight primary notes in the Western musical scale, which correspond to the white keys on the piano.

| **C major scale** | C | D | E | F | G | A | B | C |
|---|---|---|---|---|---|---|---|---|
| **Frequency (Hz) (just tempered)** | 264 | 297 | 330 | 352 | 396 | 440 | 495 | 528 |
| Frequency ratio $f/f_0$ | 1 | $\frac{9}{8}$ | $\frac{5}{4}$ | $\frac{4}{3}$ | $\frac{3}{2}$ | $\frac{5}{3}$ | $\frac{15}{8}$ | 2 |
| Example (C-264) | $\frac{264}{264}$ | $\frac{297}{264}$ | $\frac{330}{264}$ | $\frac{352}{264}$ | $\frac{396}{264}$ | $\frac{440}{264}$ | $\frac{495}{264}$ | $\frac{528}{264}$ |
| Wavelength ratio $\lambda/\lambda_0$ | 1 | $\frac{8}{9}$ | $\frac{4}{5}$ | $\frac{3}{4}$ | $\frac{2}{3}$ | $\frac{3}{5}$ | $\frac{8}{15}$ | $\frac{1}{2}$ |
| **Frequency (Hz) (equal tempered)** | 261.6 | 293.7 | 329.6 | 349.2 | 392.0 | 440.0 | 493.9 | 523.2 |

**Why do note names repeat?**

The range between a frequency and twice that frequency is called an *octave*. Notes that are an octave apart in frequency share the same name because they sound similar. An 88-key piano can play notes across a little more than seven full octaves, corresponding to a frequency range from A-27.5 Hz to high C-4186 Hz.

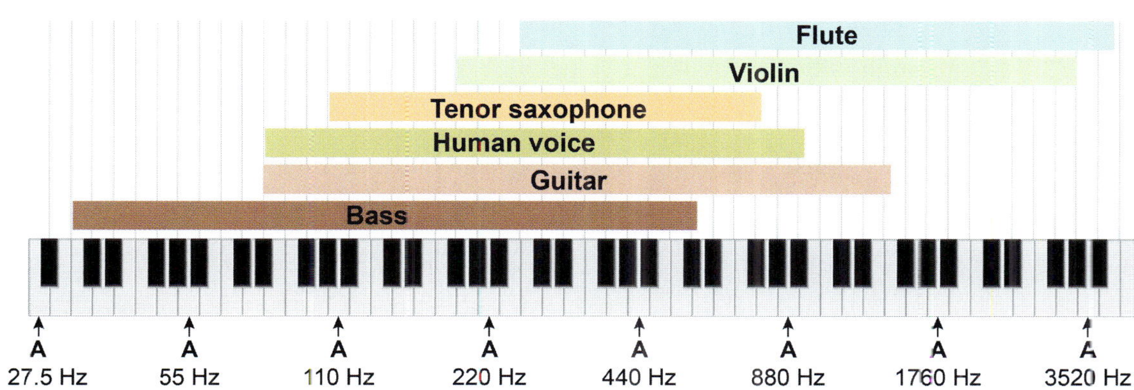


**Where does this scale come from?**

The ratios of the scale were discovered in the sixth century BC by Pythagoras, whose mathematical analysis of sound forms the basis of Western music. Look at the frequencies in the top figure: All of the frequencies are spaced apart such that none make unpleasant beats with another and *none of their harmonics make beats with the harmonics of other notes!* Since the mid-18th century, however, Western music has adopted the *equal tempered* scale, containing frequencies that deviate slightly from these perfect ratios.

## Noise cancellation

**Need for noise reduction**

Have you ever been in a place that was so noisy that ordinary conversation was impossible? Imagine being the pilot of a plane or the engineer on a locomotive where loud engine noise is a constant, irritating presence. Over the past few decades advances in electronics and miniaturization have made it possible to eliminate constant background noise *while allowing voices and other sounds to pass through!* This technology is called *active noise cancellation* and it is based on the destructive interference of sound waves.

**Everyday sounds**

The sounds of everyday life are made up of hundreds or thousands of different frequencies all superposed on top of each other. Although the human ear can separate out different frequencies, sometimes background noises become so loud that it is impossible to hear the desired sounds—such as someone's voice or music.

**Which sounds are canceled?**

In many cases the information you want to hear, such as a voice, has the following characteristics:

1. The amplitude changes relatively quickly, on time scales of a second or less and
2. most of the *information* is carried in frequencies higher than a few hundred hertz.

A background noise, such as a loud hum from an engine, has different characteristics:

1. The amplitude is relatively constant over time scales of 10 s or so and
2. the frequencies are low, typically less than a few hundred hertz.

Active noise cancellation uses the differences between these two kinds of sounds to interfere destructively with one while leaving the other unchanged.

**Applications of active noise cancellation**

Active noise cancellation technology is now used to improve safety for pilots, train conductors, and workers in noisy areas, but it has also appeared in cellphones, cars, and home stereo headphones. You may not even know that your cellphone uses antinoise to cancel out the sound of wind or other low-frequency noise from your phone calls. Many high-end cars include a "black noise" system, which uses internal microphones and special speakers to actively cancel tire and road noise from the interior of the car.

**Noise-canceling headphones**

Any noise-canceling technology includes a microphone that samples the external noises that you want to cancel out. In noise-canceling headphones, each earphone contains one or more small microphones located on the outside. The microphones are connected to a powerful computer chip that drives a miniature amplifier. The amplifier drives a speaker *inside* the headphones. This internal speaker can send sound directly into your ear.

**Noise-canceling headphones**

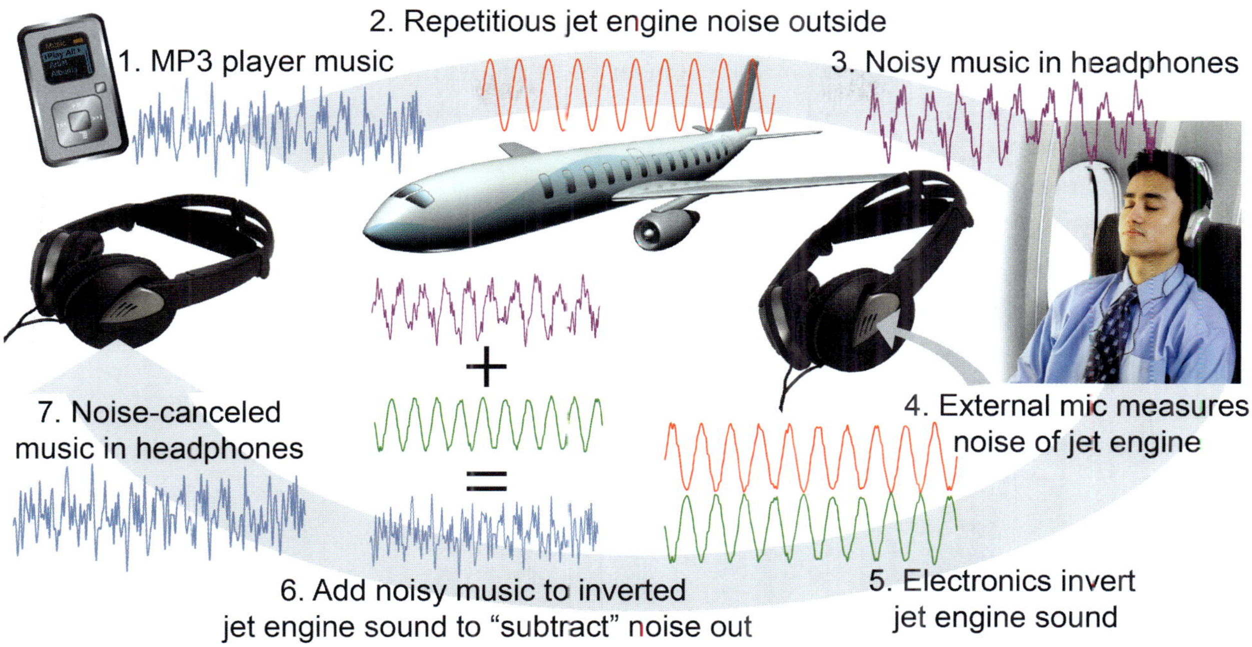


**Listening to music on an airplane**

Do you want to listen to music on headphones while on an airplane trip? If so, you might have to turn the headphone volume very high to hear the music over the regular whirring sound of the engines and air circulation system. Inside the headphones, your ear is hearing both the music and the jet noises superposed on top of each other. Noise-canceling headphones can reduce those external sounds so that you hear the music more clearly—and at a lower volume level!

**How the headphones work**

On the outside of noise-canceling headphones is a microphone that samples the sounds of the air and engine. Those signals are sent to a specialized computer chip called a *digital signal processor* that separates out the low frequencies. This chip then inverts the low-frequency sound—changes the sign from positive to negative or vice versa—and sends it to the amplifier. The amplifier plays this *antinoise* back into you ear at just the right amplitude to cancel with the low-frequency part of the sounds from the jet, but without affecting the sounds from your music. Thus

noise + antinoise = quiet!

**Voices can still be heard**

Even though the headphones cancel the jet noises, you can still hear the voice of the person next to you. Why? The antinoise signal does not cancel voices because these are mostly composed of higher frequencies and they do not repeat over periods of a few seconds. Repeated, low-frequency noises are canceled; nonrepeated, high-frequency sounds are not.

**Technology trade-offs**

Like all technologies, active noise cancellation has trade-offs. The headphones require additional power for both the chip and the internal amplifiers. There is also a degradation of audio quality, particularly for low-frequency (bass) sounds, and sometimes also an audible, high-frequency hiss.

# Section 3 review

Wave phenomena such as reflection, interference, and superposition are responsible for many acoustical and musical phenomena. *Echoes* and *reverberation* occur when sound waves are reflected. Although each musical note on a piano keyboard is associated with a particular fundamental frequency, most musical instruments create a wide spectrum of *harmonics*, which are integer multiples of the fundamental frequency. Musical intervals, such as the octave, correspond to simple integer ratios of frequencies. Interference of waves with two different frequencies produces *beats* and determines whether two or more notes sound pleasant when played together.

| Vocabulary words | echo, phase, beats, harmonic |
|---|---|

## Review problems and questions

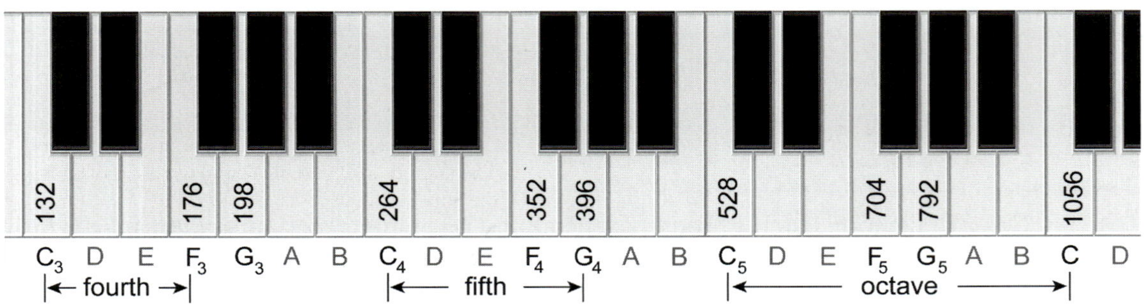


1. This depiction of a piano keyboard gives the letter names for notes in the C-major scale, along with their frequencies in hertz. *Intervals* such as the fourth, the fifth, and the octave refer to *ratios* of frequencies. For example, to go one *octave* to the right means *doubling* the frequency.
   a. What is the frequency ratio corresponding to the interval known as a *fifth*? (Express your answer as a ratio of integers and in decimal format.)
   b. What is the frequency ratio corresponding to the interval known as a *fourth*? (Express your answer as a ratio of integers and in decimal format.)
   c. What is the product of the two ratios you just computed?
   d. State a general relationship between the fourth, the fifth, and the octave.

2. *Beats* are heard when two tones of different frequencies occur. The beat frequency equals the difference of the two original frequencies.
   a. What is the beat frequency if $C_4$ and $C_5$ are played at the same time?
   b. Does that beat frequency correspond to one of the marked notes? If so, which?
   c. What is the beat frequency if $C_4$ and $G_4$ are played simultaneously?
   d. Does that beat frequency correspond to one of the marked notes? If so, which?
   e. What beat frequency results if $C_5$ and $F_5$ are played together?
   f. Does that beat frequency correspond to one of the marked notes? If so, which?
   g. The fourth, the fifth, and the octave are among the most pleasant-sounding and important intervals in Western music. Can you speculate why?

# Chapter review

## Vocabulary

Match each word to the sentence where it best fits.

### Section 16.1

| Doppler effect | decibel (dB) |
|---|---|
| pitch | speed of sound |
| supersonic | |

1. A sound that is 90 _______ is louder than an 80 _______ sound.
2. An airplane that is moving at 600 m/s is undergoing _______ motion.
3. When the sound of an approaching train is higher in pitch than the sound of a receding train, this is an example of the _______.
4. A musical note "B" differs in _______ from the note "C."

### Section 16.2

| Fourier's theorem | frequency spectrum |
|---|---|
| microphone | spectrogram |

5. A device for converting sound waves in air into electrical signals is called a _______.
6. A graphical tool for showing what oscillations are present in a wave is called a _______.
7. A graphical representation of sound and its various frequency contributions is called a _______.
8. When applying _______ to analyze the same note played by a piano and a flute, you can see that each has different contributions from higher frequency sounds.

### Section 16.3

| beats | echo |
|---|---|
| harmonic | phase |

9. When two sounds of nearly the same frequency are played at the same time, the sound waves drift in and out of _____, causing beats.
10. A piano and guitar playing the same fundamental tone will have different relative contributions of each _______ in their sound.
11. Sound reflected off the other side of the canyon, allowing Rosabella to hear a/an _______.
12. When you hear two sound waves at the same time, but they have slightly different frequencies, you might hear a slow pulsation of sound called _______.

## Conceptual questions

### Section 16.1

13. Describe the difference between loudness and pitch in musical sounds.
14. When conducting Investigation 16B on page 449, two students disagreed over a point. One of them said that the frequency of the sound would be higher if the *source* producing the sound were moving toward the observer The other students said that the frequency of the sound would be higher if the *observer* were moving toward the source producing the sound. Which student is correct?
15. How does the amplitude of a sound wave affect how you hear a sound wave?
16. Which of the following *two* statements are *not* true of sound?
    a. Sound is a transverse wave.
    b. Sound is a small traveling oscillation of pressure.
    c. The amplitude of audible sound waves may be less than one millionth of an atmosphere.
    d. Ordinary sound contains at most one or two different frequencies at a time.
17. ❮ When we talk of the "highest" or "lowest" notes on a piano, they are highest and lowest in what physical property of sound?
18. ❮ Is a high-frequency sound higher or lower in pitch? What is the relationship between pitch and frequency?
19. ❮ Which has a longer period, a musical note with a high pitch or one with a low pitch?
20. ❮ A violinist presses her finger down in the middle of a string she is playing. How does this change the wavelength of the sound produced by the string?
21. ❮❮ Why do sound waves typically oscillate at higher frequencies than water waves?
22. ❮❮ Estimate the loudness in decibels for the sound at a concert of (a) classical music with a solo piano, (b) a full orchestra, (c) an African drumming ensemble, and (d) a rock band.
23. ❮❮ There is a pipe in a church pipe organ that you want to sound a higher pitch. What can you do to the pipe to raise its pitch?

# Chapter review

## Section 16.1

24. A woman on a train hears a very loud siren from a long way away. She hears the frequency of the siren gradually decreasing but knows that the frequency of the siren should be constant. Which explanation is most likely correct?
    a. The train is moving toward the siren and slowing down.
    b. The train is moving toward the siren and speeding up.
    c. The train is moving at constant speed in a direction away from the siren.
    d. The train is moving at constant speed in a direction toward the siren.

25. A guitarist presses his finger down in the middle of a string he is playing. How does this change the frequency of the sound produced by the string?

26. Which equation best matches the meaning of this sentence?
"The wavelength of a sound wave is the speed of sound divided by the frequency of the sound."
    a. $\lambda = v \div f$
    b. $\lambda = fv$
    c. $v = \lambda f$
    d. $f = \lambda \div v$

27. Which has a higher frequency, a short-wavelength sound or a long-wavelength sound?

28. Do you think the equal-loudness curve is the same for bats and humans? Why?

29. Small animals typically have ears that are sensitive to higher frequencies than larger animals. Why might this be so?

## Section 16.2

30. A microphone converts sound waves into electrical waves. What part of a stereo system converts electrical waves into sound waves?

31. The identity of a single word is physically encoded in a sound wave as
    a. a different frequency.
    b. a pattern of different loudnesses.
    c. a pattern of changing amplitudes across many frequencies.
    d. a particular combination of certain frequencies heard together.

## Section 16.3

32. A student conducts an investigation into the resonance of sound by blowing air across the top of an empty soda bottle and measuring the frequency of the sound made by the bottle. The student then fills the bottle halfway with water. How should the frequency of the sound made by the half-full bottle compare to the frequency when the bottle was empty?

33. Which is the best choice of materials to conduct an investigation into the resonance of sound?
    a. a ripple tank
    b. a plucked, taut string
    c. a pendulum
    d. a dropped wine glass

34. A student conducts the resonance and sound investigation on page 460.
    a. The student is unsure whether the resonance tube will resonate for any tube length when the tuning fork is held over it. Write a well-defined question for her to ask.
    b. Write down a testable hypothesis corresponding to this question.
    c. When she tests this hypothesis, what is the variable that she will change?
    d. Knowing what you do about resonance tubes, what will she observe?
    e. The student measured that the length of tube above the water line that resonated was 3 mm. Is this a reasonable measurement?
    f. The student wants to modify how she implements the investigative procedure by instead using a tube that is closed at the top. Is this an appropriate choice of equipment for studying resonance with this procedure?
    g. The student wants to see whether using a loud sound at the tube's resonant frequency would cause the glass to break. (Do not try this yourself!) What would be an appropriate technology to choose to do this?

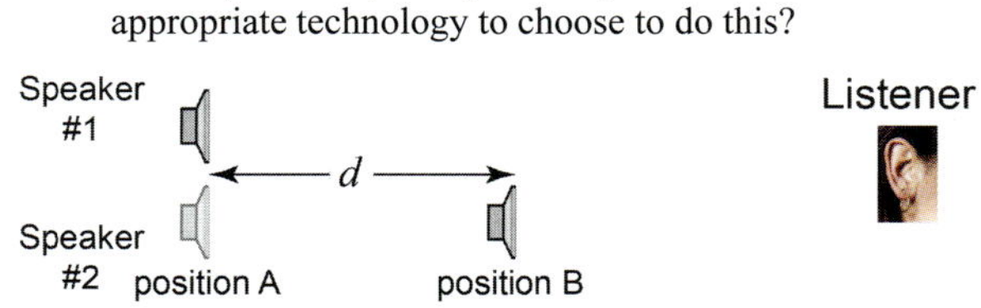


35. Sabrina was investigating the sound produced by two small speakers connected to her laptop computer. She first placed the speakers side by side ("position A") and listened to a tone at 400 Hz from several meters away, as illustrated in the drawing.
    a. As her lab partner Eko slowly moved one speaker closer to her, at "position B" Sabrina could barely hear the sound from the speakers. Why?
    b. Calculate the distance $d$ between positions A and B. (The speed of sound is 343 m/s.)

# Chapter review

### Section 16.3

36. Which is the best choice for reliably producing sound with a standardized frequency, a guitar string, a tuning fork, a bottle half-filled with water, or a resonance tube?

37. Why must you adjust the length of a resonance tube in order for it to come into resonance with an external sound, such as from a tuning fork?

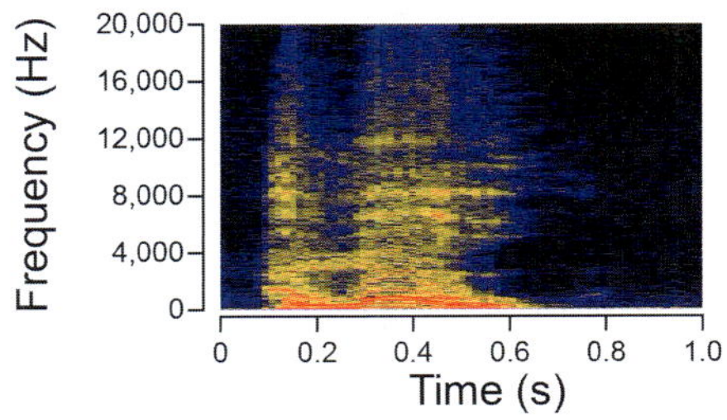


38. The chart of the frequency spectrum for a human voice is shown here. How would the chart change if the person spoke more quietly?

39. Sasha and Alicia are swinging on the playground. Describe visually what you would see if the two girls were swinging in phase, 90° out of phase, and 180° out of phase.

40. Giovanni is blind but his hearing has perfect pitch. How can he tell quickly if a marching band is marching toward or away from him?

41. A stationary observer hears the frequency of a horn of an approaching train as 333 Hz rather than the actual 300 Hz. If the train approached twice as fast, would the observed frequency decrease, increase, or stay the same?

42. A piano tuner listens to the beats between two strings played simultaneously to tell whether they are matched in frequency. What property of the beats tells the tuner that the two strings are perfectly in tune?

43. Define the term *harmonic* in the context of frequencies of sound. Describe how the characteristic sounds of different instruments, such as a guitar and a piano, come from harmonics.

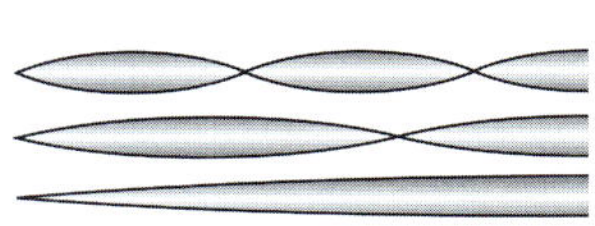

44. The standing wave patterns in the diagram represent resonances of which kind of system?

    a. a pipe with two open ends
    b. a pipe with two closed ends
    c. a pipe with one open and one closed end

## Quantitative problems

### Section 16.1

45. A sound wave has a wavelength of 0.96 m. How many times does this wave cause your eardrum to oscillate back and forth in 1 s?

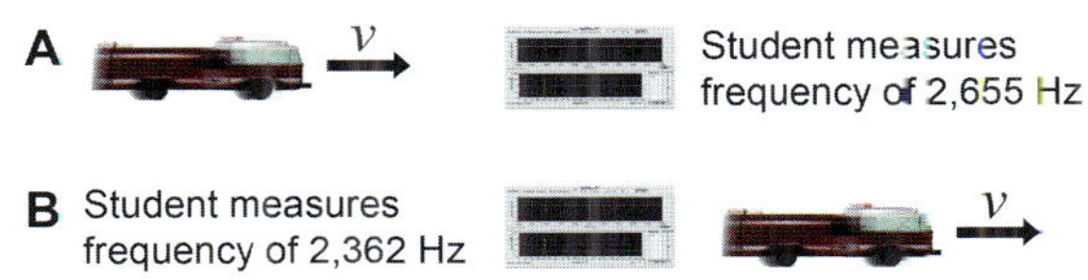


46. Two students use a spectrum analyzer to measure the frequency of the siren of a fire truck as it moves toward then away from them. They measured the frequency as 2,655 Hz when the fire truck was approaching (measurement "A") and 2,362 Hz as the fire truck we receding ("B").

    a. Why were the two frequency measurements different?
    b. If you were a firefighter riding on the truck, what would you observe to be the frequency of the siren?
    c. Calculate the speed of the fire truck.

47. An orchestral oboist sounds the note "A" at 440 Hz. If the speed of sound in air is 343 m/s, what is the wavelength of the note?

48. What is the wavelength of (a) a 100 Hz, (b) a 200 Hz, and (c) a 400 Hz sound wave in air?

49. Joseph wears ear protection at work because machinery in his work environment creates sounds up to 110 decibels (110 dB). Safety regulations do not allow regular exposure to sounds above 90 dB.

    a. By how many decibels must Joseph's ear protection reduce the sound of the machinery?
    b. By what factor is the amplitude of the sound wave reduced?

50. You are standing on the sidewalk. A car approaches while playing a musical note from a loudspeaker on its roof. The note being played is middle C ($f = 264$ Hz), but you hear the note as C#—the next-highest note on a piano, with a frequency of 280 Hz. How fast is the car moving? (Assume the sound speed $v_s$ is 343 m/s.)

51. Sound travels at a speed of 1,490 m/s in sea water at 10°C, compared to traveling at 343 m/s in air.

    a. Calculate the wavelength of a 440 Hz sound in water and air. How do they compare?
    b. Consider the sound made by a meteor hitting the surface of the ocean. How far away is the impact point if the sound is detected in the water 112 s before it is detected in the air?

## Section 16.1

52. Astronomers on Earth observe that the frequency of the light emitted by stars on one side of the Milky Way Galaxy is higher than the frequency emitted on the other side. What might cause this frequency difference?

53. Erin is conducting an investigation by setting up two stereo speakers facing each other and 2 m apart. She sets the stereo to play a note with a wavelength of 1 m. The speakers play in phase with each other.

    a. If she sits halfway between the two speakers, what does she hear?
    b. If she moves somewhat closer to one speaker than the other, what will she hear?
    c. What will Erin hear if she changes one speaker to play a note with a wavelength of 1.1 m and then sits halfway between the speakers?

## Section 16.2

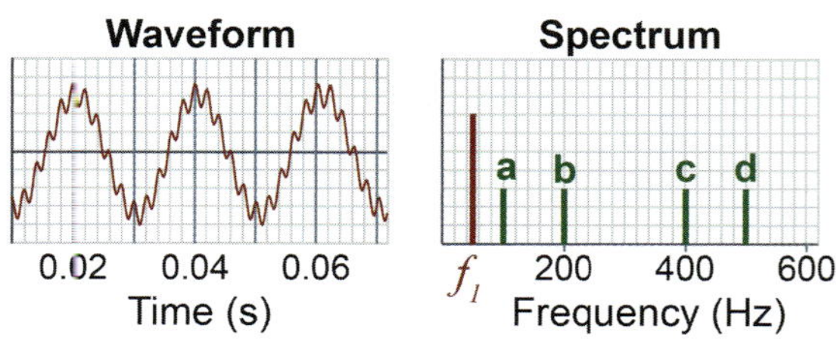


54. The waveform on the left contains two component frequencies. One is labeled $f_1$ on the spectrum. Which peak on the spectrum is the other frequency, a, b, c, or d?

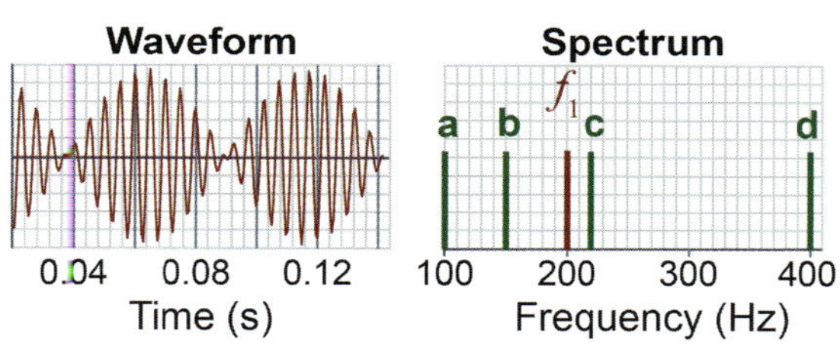


55. The waveform on the left contains two component frequencies. One is labeled $f_1$ on the spectrum. Which peak on the spectrum is the other frequency, a, b, c, or d?

## Section 16.3

56. A hiker in the middle of a canyon yells "Echo!" and then 4 s later hears his echo off of the walls of the canyon. How far away are the walls of the canyon?

57. A guitar string plays a fundamental note with a frequency of 335 Hz.

    a. What are the frequencies of the first, second, and third overtones?
    b. What are the frequencies of the first, second, and third harmonics?

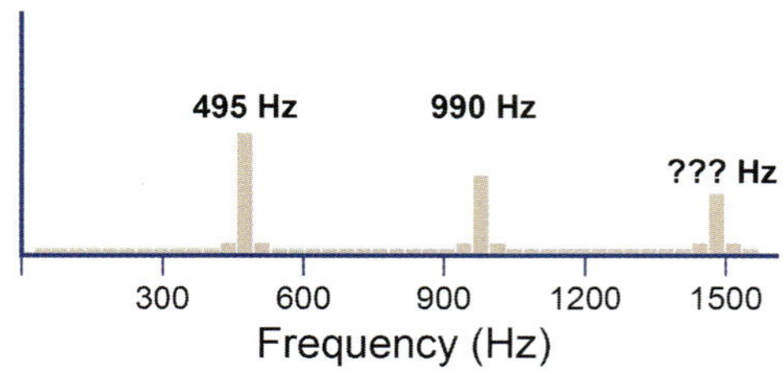


58. The spectrum above shows frequencies of a vibrating string one second after it was plucked.

    a. What is the quantity plotted on the vertical axis?
    b. Why are only some frequencies present and not others?
    c. What is the fundamental frequency for the string?
    d. What is the value of the missing resonant frequency?
    e. How did you *infer* the missing frequency?

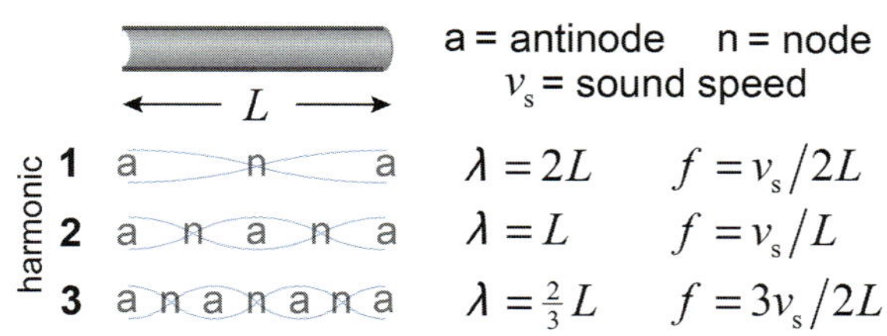


59. When blowing on a pipe with both ends open, a musician creates standing waves that have antinodes at both ends (because air molecules at both ends are free to oscillate back and forth). The diagram shows the nodes, antinodes, wavelength, and frequency of the first three harmonics.

    a. Write a formula that calculates *wavelength* ($\lambda$) when one knows the distance $d$ between any node and the nearest antinode.
    b. Write a formula that expresses wavelength in terms of pipe length $L$ and harmonic number $n_h$.
    c. Write a formula that expresses frequency in terms of pipe length, harmonic number, and sound speed $v_s$.
    d. Given a sound speed of $v_s = 343$ m/s and a 35-cm-long pipe with two open ends, what will be the frequency of the *fourth* harmonic?

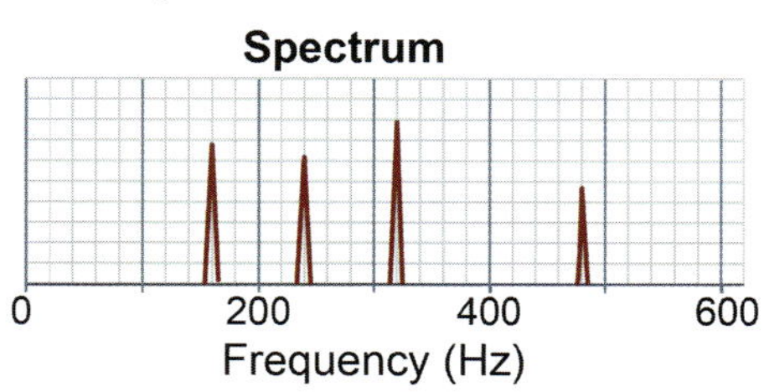


60. The spectrum shown here represents a musical sound, but parts of the spectrum are missing.

    a. What is the highest possible fundamental frequency?
    b. There are three missing frequencies in the range from 0 to 600 Hz. What are they?

# Chapter review

## Standardized test practice

61. Which is the best choice to amplify a sound?

    A. soundproofing insulation
    B. underwater sonic substratum echo mechanism
    C. tuning fork
    D. resonance tube

62. The sound from a train approaching you is higher in pitch than the sound from a train moving away from you. This is an example of

    A. supersonic motion.
    B. the Fourier theorem.
    C. the law of superposition.
    D. the Doppler effect.

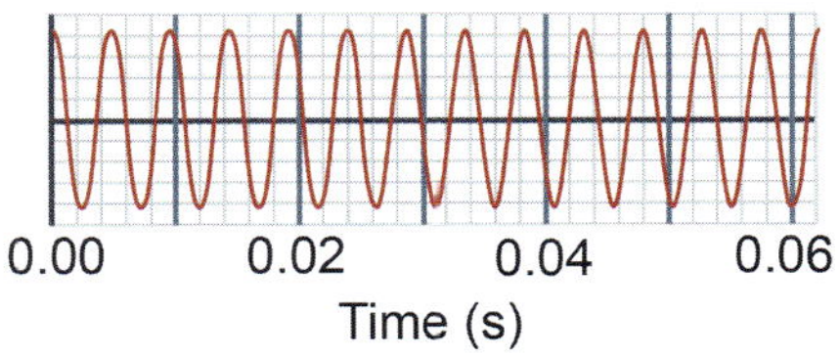


63. The graph is a representation of a sound wave traveling at 343 m/s. What is its wavelength?

    A. 0.61 m
    B. 1.00 m
    C. 1.65 m
    D. 20.6 m

64. Two loudspeakers play different notes, one at 440 Hz and the other at 430 Hz. What will a listener hear when these notes interfere?

    A. refraction
    B. beats
    C. reverberation
    D. diffraction

65. The lowest string on a violin has a fundamental frequency of G at 196 Hz. Which frequency represents the second harmonic of the string?

    A. 98 Hz
    B. 196 Hz
    C. 392 Hz
    D. 588 Hz

66. Which is *not* an example of resonance of sound?

    A. a tuning fork
    B. sounding a note by blowing air over a bottle
    C. playing a trombone
    D. a laser

67. A hiker in a canyon calls out "Hello!" and hears her echo back in 2 s. If the speed of sound is 343 m/s and her voice is at a frequency of 350 Hz, how far away is the canyon wall?

    A. 343 m
    B. 350 m
    C. 672 m
    D. 686 m

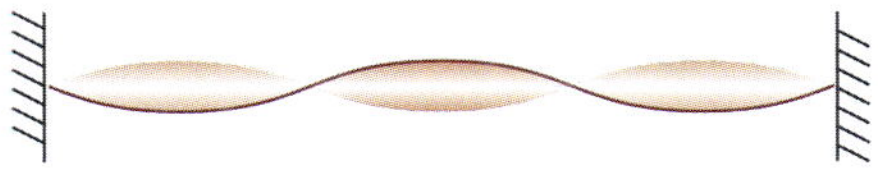

68. How many nodes and antinodes are there in the resonating string above?

    A. three nodes and two antinodes
    B. three nodes and three antinodes
    C. four nodes and three antinodes
    D. four nodes and six antinodes

69. A speaker plays a note at 30 dB. Nine other speakers are added and play the same note *in phase* at the same amplitude. What is their collective volume?

    A. 40 dB
    B. 270 dB
    C. 3.3 dB
    D. 50 dB

70. A sound column vibrates in an organ pipe of length 75 cm and with two open ends. If sound travels at 343 m/s in air, what is the frequency of the first harmonic in this pipe?

    A. 115 Hz
    B. 686 Hz
    C. 229 Hz
    D. 0.219 Hz

71. A 3 m sound wave is traveling in air at 343 m/s when it enters water, where the speed of sound is 1,482 m/s. If you assume that the wave loses no energy entering the water, what is its wavelength in water?

    A. 13 m
    B. 494 m
    C. 4.3 m
    D. 114 m

72. A sound wave of the note "A" has a frequency of 440 Hz in air. Which of these is closest to the wavelength of this sound? (The speed of sound in air is 343 m/s.)

    A. the length of a bug
    B. the height of a desk
    C. the length of a car
    D. the width of a highway

# Chapter 17 Electricity and Circuits

A lightning storm is a mesmerizing visual experience—from afar. The electricity in lightning or power lines carries such vast quantities of energy that coming in contact with either can send you to the hospital or worse. What is beautiful can also be very dangerous.

In 1752, American scientist and statesman Benjamin Franklin is said to have demonstrated that lightning was electrical in nature by carefully flying a kite just before a storm. *Don't try this!* Others who were not so careful were electrocuted repeating Franklin's experiment. Franklin's early studies of electricity were followed by the Italian physicist Alessandro Volta's discovery that electricity could be produced by alternating silver and zinc plates separated by cloth soaked in salt water (the first battery).

By 1950 virtually every American household had electric lights, TV, radio, and appliances—a consequence of the lasting societal impact of these and other scientists. You probably carry electronic devices around with you today, including cellphones, tablets, and even car door openers. While we take electricity for granted, electricity is one of many areas in which physics has fundamentally changed our way of life. Electric motors, generators, and transformers are fundamental technologies that allow electricity to be generated far away, transported to your home, and used in your personal life in a wide array of technologies.

Re-engineering the *power grid* that transmits and distributes electricity is one of the most pressing technical issues in the future of energy. Today, an unstable patchwork of old and new transmission lines at different voltages connect power plants with homes and businesses. Many foresee a future in which distributed wind, solar, and biomass energy sources feed the grid, possibly from each individual home. The existing power grid was never designed to accept multiple power inputs that fluctuate with local wind speed or cloud cover. Designing and building a "smart grid," able to reliably accept and distribute electricity, poses a challenge that will have to be met with solid physics and clever engineering.

Where will the science and technology of electricity take us in the next century?

# Chapter study guide

## Chapter summary

Electricity is all around us, being present in many of the devices we use every day and in the lightning that periodically accompanies strong storms. How does electricity work? In this chapter, you will learn the basics of electricity and circuits. Electricity and electric current are the flow of electric charges through a material such as a conductor. Electrical insulators, such as plastic or glass, prevent the flow of electricity. Simple electrical circuits contain wires, batteries, bulbs, resistors, and switches. Devices in circuits can be connected in series or in parallel. The current through a circuit is proportional to the voltage across it and inversely proportional to its resistance. Household circuits are generally connected in parallel, carry alternating current, and contain fuses or circuit breakers.

## Learning objectives

By the end of this chapter you should be able to

- distinguish between conductors and insulators based on their electrical properties;
- read and draw simple circuit diagrams;
- design and construct circuits with elements connected in series and parallel;
- calculate the current through, potential difference across, resistance of, and power used by circuit elements connected in series and parallel;
- use a digital multimeter or its equivalent to measure current, voltage, and resistance; and
- describe basic properties and components of the wiring system in a house.

## Investigations

17A: Electricity and circuits
17B: Voltage and batteries
Design a lemon battery
17C: Resistors and Ohm's law
17D: Series and parallel circuits
17E: Electrical power
17F: Compound circuits

## Chapter index



## Important relationships

$I = \frac{V}{R}$  $R = R_1 + R_2 + R_3 + \cdots$  $\frac{1}{R} = \frac{1}{R_1} + \frac{1}{R_2} + \frac{1}{R_3} + \cdots$  $P = IV$

## Vocabulary

| | | |
|---|---|---|
| electricity | electric current | ampere (A) |
| electric circuit | open circuit | closed circuit |
| electrical symbol | short circuit | volt (V) |
| potential difference | voltage | voltmeter |
| battery | Ohm's law | ohm (Ω) |
| resistance | resistor | digital multimeter |
| electrical conductor | electrical insulator | series circuit |
| equivalent resistance | parallel circuit | Kirchhoff's current law |
| Kirchhoff's voltage law | circuit breaker | kilowatt-hour (kWh) |

Devices that use electricity are all around us: for example, light bulbs, toasters, computers, cellphones, and televisions. Electricity is also found throughout the natural world, such as in lightning and the electric eel. You even have electrical signals traveling through the central nervous system, heart, and other muscles of your body. Electrical energy powers our modern lives, but like all technologies there are risks as well as benefits. Electricity is extraordinarily useful but it can also be *dangerous*. This section introduces electricity and electric current.

**Electricity all around us**

## Electric current

**What is electricity?**

The word **electricity** usually refers to the flow of *electric current,* typically through wires, conductors, and other electrical devices. We use electrical energy carried by electric current to power many devices, such as mobile phones, light bulbs, and electric stoves. **Electric current** is the flow of charged particles, usually electrons, inside a material that conducts electricity.

**Electric current is analogous to water in pipes**

Although electric current is invisible, you can visualize the concept by imagining the flow of water through a pipe. Just as the flow of water molecules carries mechanical energy that can power a hydroelectric generator, the flow of charged particles in a wire carries electrical energy that can do work, such as powering an electric motor. Just as pipes carry water through a house, *wires* carry electricity through a circuit.

**Electric current:**
**Charged particles flow through wire.**

**Water current:**
**Water particles flow through pipe.**

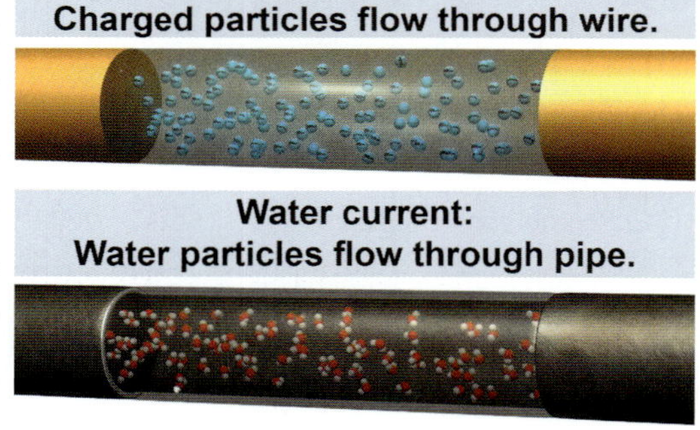

**What is current and how is it measured?**

**Digital multimeter**

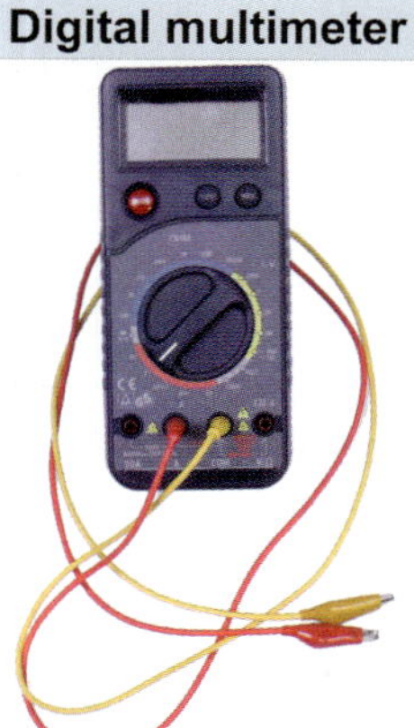

Electric current is measured in units called the **ampere (A)**, or amp. If a wire carries 1 A of current, this means that $6.2 \times 10^{18}$ electrons move across any cross section of the wire each second. One amp is a substantial amount of current, enough to power four household light bulbs. The electrical signals your body uses to control muscles are more typically in fractions of a *milliampere* (mA): 1 mA = 0.001 A. Many small electrical devices also use current in milliamperes. Electric current can be measured using an *ammeter* or the "amps" setting on a digital multimeter.

**What is a circuit?**

An **electric circuit** is a conducting path through which electric current can flow. You make a circuit whenever you plug in an electric appliance or light. The need to create a complete circuit is why plugs have at least two wires—one for the current to flow out of the wall socket and one for the current to flow back in again. Naturally occurring electric circuits are found in nerves in your body and the pathways of lightning.

**Electrical symbols**

**Electrical symbols in circuits**

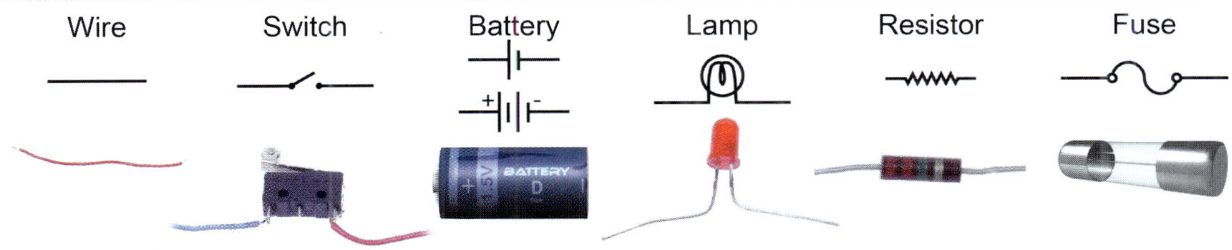


**How are circuits represented?**

A circuit diagram is a compact graphical tool that shows all the elements and their connections in an electrical circuit. Each circuit element—such as a wire, switch, battery, lamp, resistor, or fuse—has a unique symbol. The **electrical symbol** for each component has been standardized so that an electrical engineer can design a *schematic* drawing for a circuit without using pictures of the actual devices.

**What is the difference between an open and closed circuit?**

**Open circuit: No current flows.**

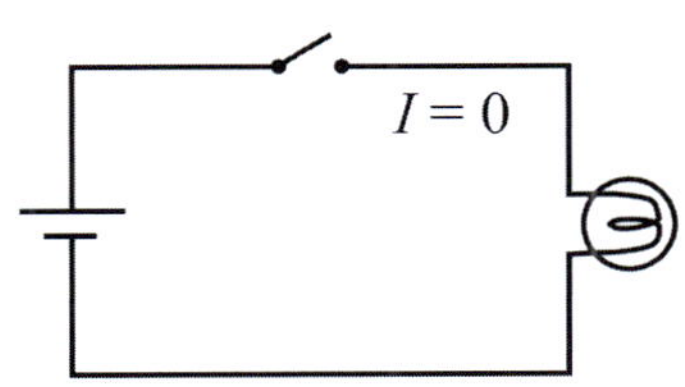


**Closed circuit: Current flows.**

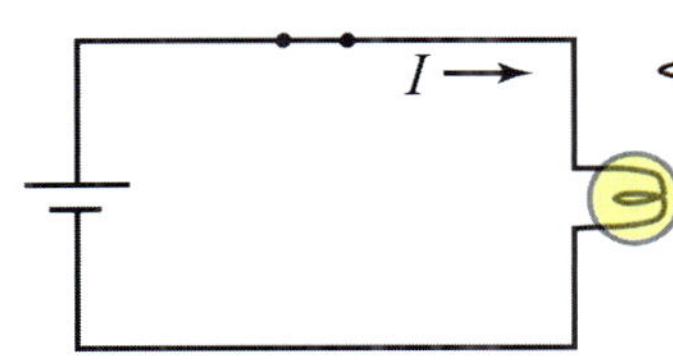


**Knife-blade switches can open and close circuits.**

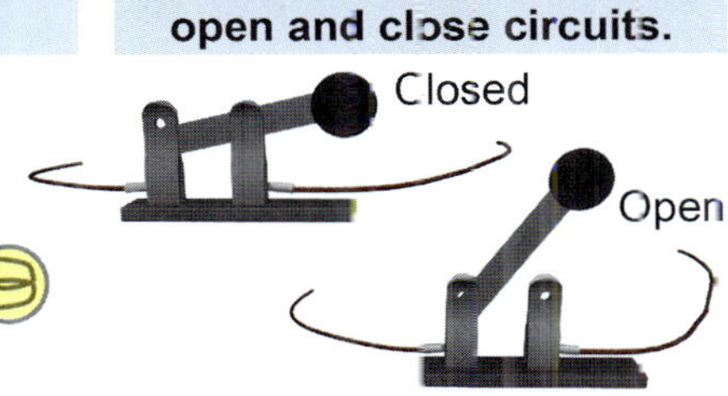


A **closed circuit** includes at least one continuous closed path of a conductor, such as a wire, allowing electric current to flow. An **open circuit** is a path in which the connection is broken in at least one place in the loop, preventing the flow of current. In the circuits illustrated above, the switch opens or closes the circuit. A switch is a useful device because it allows you to easily and safely control the flow of current. A *knife-blade switch* opens or closes a circuit when you raise or lower the handle, respectively.

**What is a short circuit?**

Closed circuits always include some device, such as a light bulb, that uses electrical energy. Accidentally connecting a wire directly between the terminals of a battery creates a *short circuit.* Without any device to dissipate electrical energy, a short circuit may cause a runaway flow of electric current. Short circuits should be avoided.

**Short circuit: Electric current follows the path of lowest resistance.**

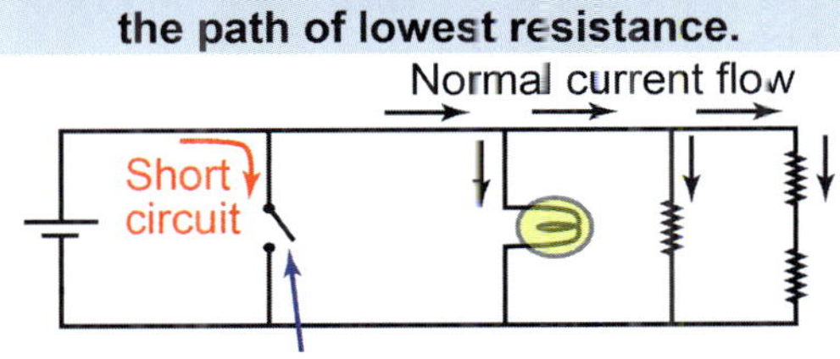


If this switch is closed, then it will cause a short circuit. Can you see why?

# Investigation 17A Electricity and Circuits

**Essential questions** *How do you build a circuit?*

The Modular Circuits kit provides an easy and quick way to prototype and design electronic circuits. This investigation explores how to build simple circuits using the kit, and how to relate the circuit to a comparable equivalent circuit diagram.

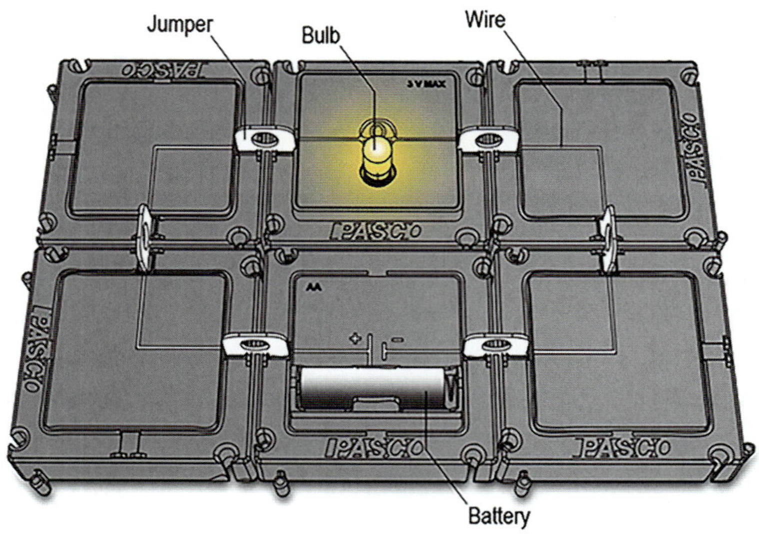


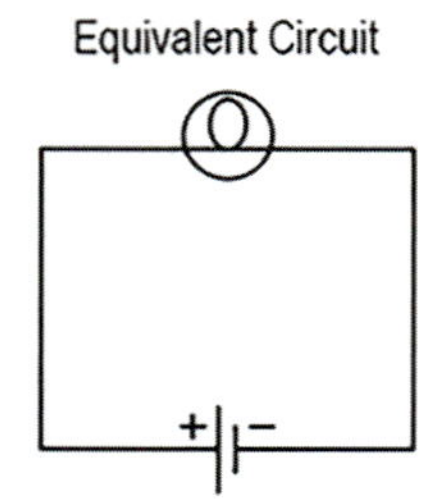


## Building a Simple Circuit

1. Connect a battery, bulb, and four wire corners together to create the circuit in the diagram.
2. Add six jumpers to your circuit to connect the components together and make the bulb light up.

a. *Remove a jumper. Why does the bulb not light up when the jumper is removed?*

b. *Remove the battery module from the circuit, turn it around so it is reversed, and reconnect it. Does the bulb still light up? What do you think reversing the battery changes in the circuit?*

c. *Connect the two jumpers on the bulb module with a bare wire to create a short circuit. What happens to the bulb when you create a short circuit? Try connecting jumpers on the wire and battery modules. Does connecting any two jumpers create a short circuit?*

d. *Add a switch and any necessary wire modules to your circuit so that you can turn on the light by closing the switch (closed circuit), and turn off the light by opening the switch (open circuit). Draw a circuit diagram of your new circuit, using the circuit symbols.*

e. *Describe another way you can create an open circuit without opening the switch.*

**What is a volt?**

The *volt* is named in honor of Alessandro Volta, who invented the first chemical battery in 1800. Volta discovered how to convert chemical energy into electrical energy by alternating plates of two different metals in a dilute acid or salt solution. In practical terms, a **volt (V)** tells you the amount of electrical power carried by one ampere of current. For example, each ampere of current flowing out of a 12 V battery carries 12 W of electrical power. The same ampere of current flowing out of a 120 V wall socket carries 120 W of electrical power. Voltage in a circuit is measured with a **voltmeter**.

**Water flows because of a difference in gravitational potential energy.**

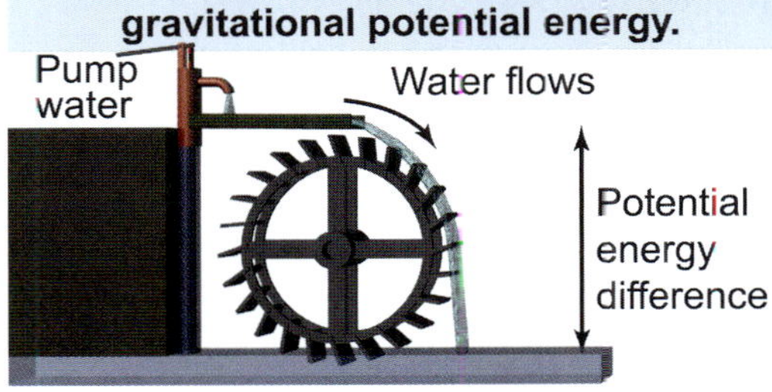


**Electric current flows because of a voltage (or electric potential difference).**

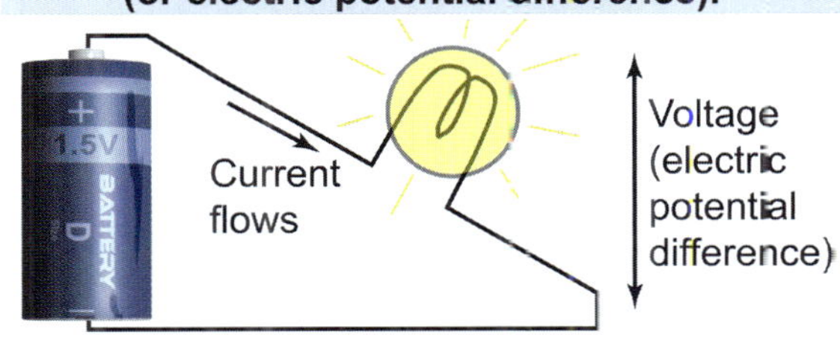


**How are voltage and current related?**

Current is what flows and does work. **Voltage**, in contrast, is what *causes* current to flow. The idea of *water pressure* in pipes makes a good analogy for voltage. A pressure difference causes water to flow in a pipe, and voltage causes electric current to flow in a wire. If there is no voltage, then no current flows. You can connect a loop of wire to a bulb in a closed circuit but current will *not* flow unless there is a battery or some other voltage source in the circuit. Current only flows in response to a voltage difference.

**Potential difference**

The term **potential difference** is another way to describe voltage. A 1.5 V battery has a potential difference of 1.5 V between its positive and negative terminals. Electrical potential difference is analogous to the difference in potential energy for water at two different heights.

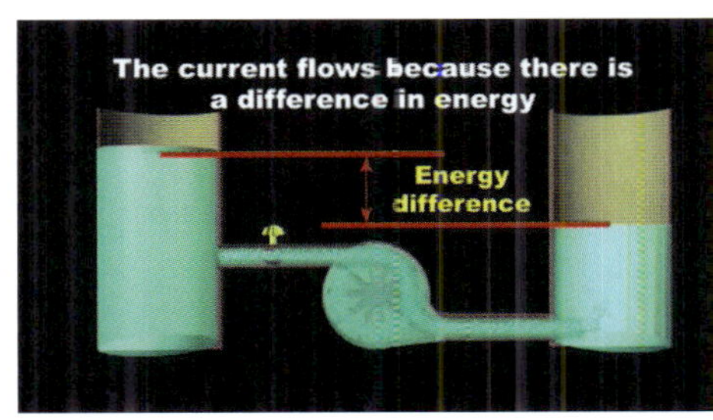


**What does the sign of voltage mean?**

Voltage can only be measured *between* two points, such as the two terminals of a battery. If we measure the voltage of an ordinary 1.5 V battery from the flat (−) end to the slightly protruding (+) end, we would record a voltage gain of +1.5 V, indicating an increase in electric potential. If we swapped the leads on our voltmeter so that we measured from the + to the − end of the battery, we would measure a voltage drop of −1.5 V. When we say that a battery is "1.5 V" we are implicitly choosing the negative end of the battery to be our reference of 0 V.

**Which direction does current flow?**

By convention, electric current flows from high (more positive) potential to low (more negative) potential. In circuit diagrams, an arrow is often used to indicate the direction of current flow. "Conventional" current flows out of the positive terminal of a battery and returns to the negative terminal. In most conductors it is actually negative electrons that move to carry electric current. Because electrons are negatively charged, the *electron flow* is opposite from the direction of conventional current.

# Investigation 17B Voltage and batteries

**Essential questions** *How do you connect batteries to increase their total voltage?*

When Alessandro Volta invented the first electric battery, he connected several individual battery cells together to create a more powerful composite battery. How did he connect them? In this investigation you will connect two batteries together and determine how to make the largest combined voltage.

There are two basic ways to connect electrical components together. In a *series* connection, the two components are connected one after another. In a *parallel* connection, the components are connected side-by-side—with the "tops" connected together and the "bottoms" connected together.

## Part 1: Connecting batteries in parallel

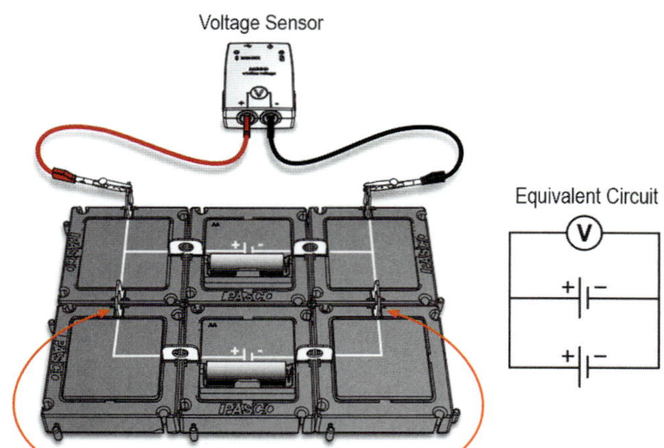

Connect positive terminals and negative terminals.

1. Open the experiment file **17B_VoltageAndBatteries**, and then power-on the Voltage Sensor and connect it wirelessly to your software.
2. Connect the two batteries in parallel: connect their positive terminals to each other and connect their negative terminals to each other.
3. Use the voltage sensor to measure the voltage across the two batteries connected in parallel.

a. *Read the voltage rating on the side of each battery. How does the printed voltage compare with what you measured?*

b. *How does the voltage of the two batteries connected in parallel compare with the voltage of either battery measured separately? Explain your answer.*

## Part 2: Connecting batteries in series

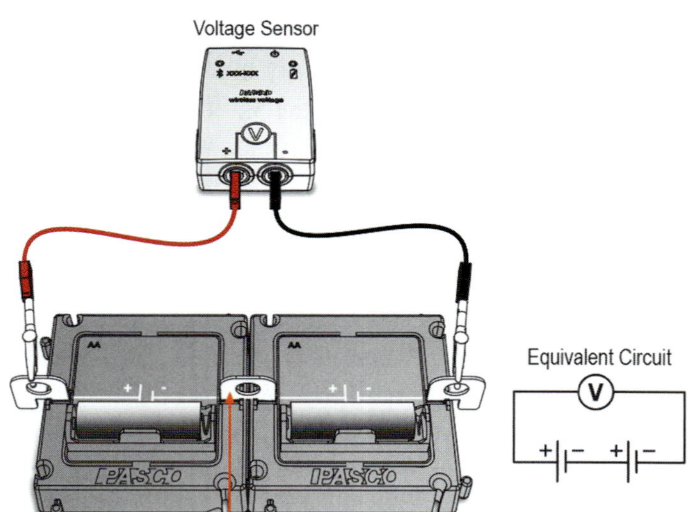

Connect negative terminal of one battery to positive terminal of other battery.

1. Connect the two batteries in series: connect the negative terminal of one battery to the positive terminal of the other battery.
2. Using the voltage sensor, measure the voltage across the two batteries connected in series.

a. *How does the voltage of the two batteries connected in series compare with the voltage of either battery measured separately? Explain your answer.*

b. *Compare the series and parallel circuits. Which would you use to create the largest voltage?*

c. *Which circuit design would you expect to find in a flashlight or other electrical device?*

d. *Open up the battery compartment of a flashlight and look at any visible wires to see how it is connected. Can you tell if the batteries are connected in series or parallel?*

**Power and voltage**

A **battery** converts stored chemical energy into electrical power. The amount of power per ampere of current depends on the voltage of the battery. One volt means one watt of power for each ampere of current. For example, 1 A of current flowing from a 1.5 V battery delivers 1.5 W of electrical power.

**Why does a battery "die"?**

A battery stores a fixed amount of energy, which is depleted as current flows. For example, an AA size (1.5 V) alkaline battery stores 10,000 J of energy. A current of 1 A will deliver 1.5 J/s (1.5 W). At this rate the battery's energy is depleted in under 2 hours. A lower current depletes the energy more slowly and the battery lasts longer. A larger "D" battery has the same voltage (1.5 V) but contains more chemicals and therefore has more stored energy. An alkaline "D" battery might contain 70,000 J of stored energy compared to only 5,000 J for a "AAA" size.

**Batteries in series and parallel**

Batteries can be connected in two ways. When they are connected in *parallel*—positive terminals together and negative terminals together—the combination has the same voltage as a single battery but can produce twice the current. When the batteries are connected end to end, or in *series*, the combination has twice the voltage but provides the same current as a single battery.

**Connecting batteries in series and parallel**

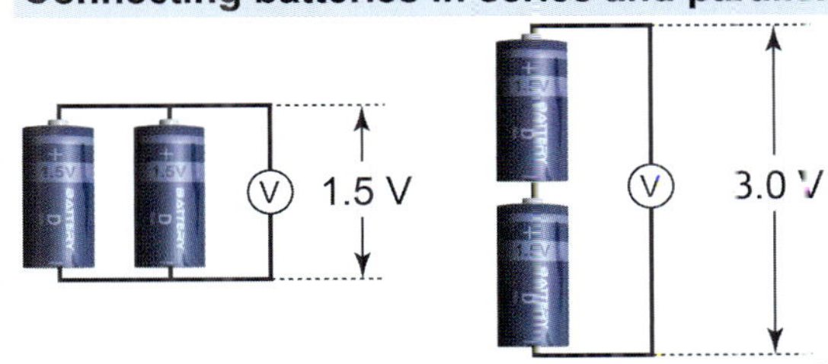


Batteries connected in *parallel* do not add their voltages.

Batteries connected in *series* add their voltages together.

**How does an alkaline battery work?**

Inside a typical battery, the positive terminal is one kind of metal and the negative terminal is a different kind of metal. Both metals are immersed in an *electrolyte,* a material that undergoes a chemical reaction. The reaction releases electrons that are repelled from the negative terminal, forming the electric current.

**Diagram of an alkaline battery**

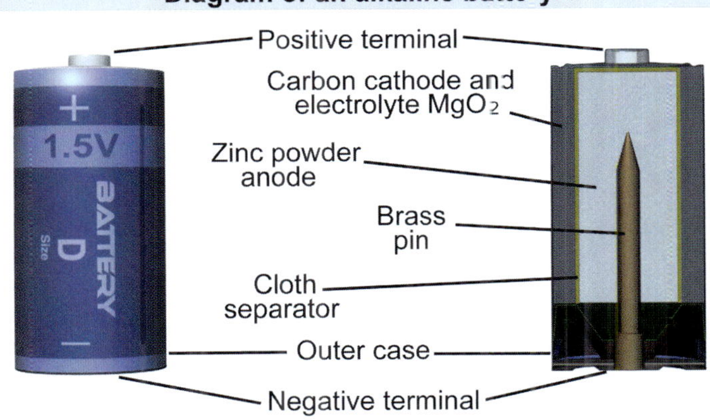


**Modern batteries**

Alkaline batteries use $MgO_2$ (manganese dioxide) mixed with carbon for the cathode, zinc (in powder form) for the anode, and potassium hydroxide as the electrolyte. A porous membrane separates the cathode and anode. A fresh alkaline battery produces a voltage of about 1.62 V and is guaranteed to maintain a voltage of at least 1.5 V for a certain amount of current and time. For example, a 1.5 V AA-size alkaline battery can supply a current of about 1 A for about 50 minutes before its voltage drops below 1.5 V. Rechargeable nickel–cadmium (NiCd) batteries have a voltage of 1.2 V. Lead acid batteries are used in car engines and usually have a voltage of 12 V.

**Disposal of batteries**

Dispose of batteries properly. Most alkaline batteries today no longer contain mercury and can be put in the trash. Rechargeable batteries—such as NiCd or car batteries—must be recycled at a waste management center or through an auto dealer or hardware store.

## Design a lemon battery

**Design challenge**

*Challenge:* Create a lemon battery with a voltage exceeding 2 V.
*Constraints:* Use up to five lemons, as many nails and wires as you need, and any metal objects you have.
*Performance evaluation:* The measured voltage must be $V \geq 2$ V, as measured by a voltage sensor or digital multimeter.

**How to build a lemon battery cell**

A simple lemon and two metal objects can create a single-cell battery. Both a store-bought alkaline battery and a lemon battery rely on chemical reactions that cause positive charges to move toward one metal and negative charges to move toward the other. If you insert two appropriate metal objects (called *electrodes*) into the lemon, the chemical reaction creates a voltage between the two metals.

**Modeling the system**

What parameters are important in creating a lemon cell and battery?

1. Try inserting two identical nails and measure the voltage between them. Then try two nails made from different metals. Record the voltage in each case.
2. Insert two nails close to each other and then far apart. Measure the voltage for each case.
3. Try to create a *two*-cell battery from your lemon. Try different kinds of nails and locations for the nails.

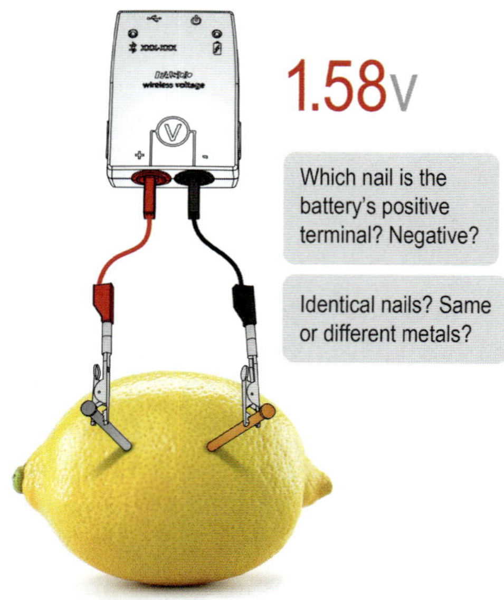


**Design**

Based on what you discover, design a lemon battery that consists of more than one cell and that will create a voltage of at least 2 V. Write down your design using a combination of text and diagrams. Make a prediction for your design's voltage, justify your prediction, and write down a procedure for testing and evaluating it.

**Prototype**

Construct the prototype of your lemon battery design using the equipment.

**Test**

Use a voltage sensor and the **DC_LemonBattery** experiment file, or a multimeter, to measure the voltage across your prototype lemon battery. Also measure the voltage across each individual cell in the battery. What were the variables you tested?

**Evaluate**

How well did the performance of your prototype match your prediction? Evaluate and document the strengths and weaknesses of your prototype's performance compared to the 2 V design criterion.

**Revise**

Evaluate the individual elements in your design and determine how to improve it. Create a revised design, prototype it, and test it. If you succeed in creating a 2 V battery, then try to create even higher voltage from the materials!

# Section 1 review

Electric current is the flow of electric charge through a material such as a conducting wire. Electric current flows in response to differences in electrical potential, or *voltage*. A battery uses chemical potential energy to generate a voltage, which can cause electric current to flow in a conductor. A closed circuit provides at least one complete conducting path for current to travel in a closed loop. An open circuit, such as a flashlight that has been turned off, contains a break in the conducting path and so does not enable current to flow. Circuits must contain a voltage source and can also contain components such as light bulbs, resistors, and switches.

| Vocabulary words | electricity, electric current, ampere (A), electric circuit, open circuit, closed circuit, electrical symbol, short circuit, volt (V), potential difference, voltage, voltmeter, battery |
|---|---|

## Review problems and questions

1. One *mole* of electrons is approximately $6.0\times 10^{23}$ electrons (that's six hundred billion trillion electrons). The ampere is the SI unit of electric current, but we could also measure current in moles of electrons per second. How many amperes equal one mole of electrons per second?

2. Portable electronic devices such as cellphones and media players have rechargeable batteries. When a battery is fully *charged* this means that it contains its maximum energy, which it can then supply to the electric charges that pass through it. The energy a battery provides is often stated in units of milliampere-hours, or mAh (based on its voltage). Suppose that your media player is rated at 100 mAh.
   a. How many electrons does 100 mAh represent?
   b. Suppose that you can listen to 5 hr of continuous music on this device after fully charging its battery. What is the average current in milliamperes (mA) that flows through the device while you listen to it?

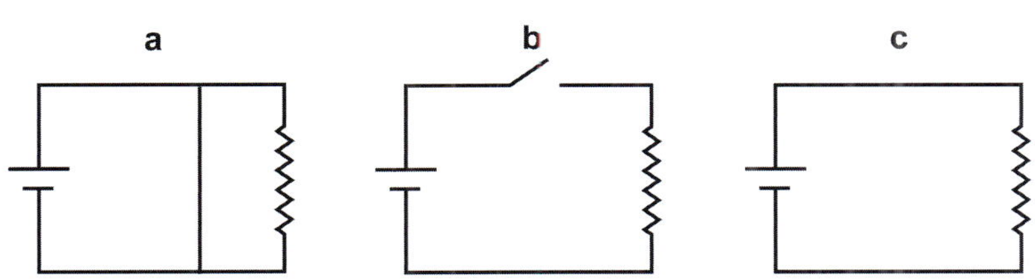


3. Identify each of the circuits shown above as a *closed circuit*, an *open circuit*, or a *short circuit*. Each term is used only once.

4. This circuit diagram for a simple flashlight shows the direction of current flow, as conventionally defined. It also shows the positive and negative terminals of the battery. What is the relationship between electron flow and the arrowed current shown here?

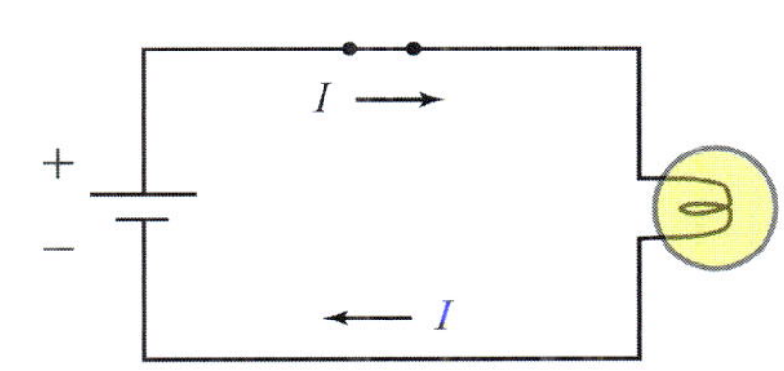

## 17.2 - Resistance

If you connect a 1.5 V battery across a circuit, how much current will flow through the circuit? It depends on the *resistance* of the circuit. Resistance is a measure of how easily current can flow through a circuit. Too much resistance permits only a small amount of current to flow. The connection among voltage, current, and resistance is an important part of understanding electric circuits.

### Ohm's law

**How much current flows?**

An ordinary lamp plugged into a 120 V receptacle draws a current of about 0.2 A when it is switched on. A hair dryer connected to the same voltage draws a current of 11 A. What property of the lamp and the hair dryer determine how much current they draw? The answer is a fundamental relationship of electricity.

**Ohm's law**

Electrical **resistance** measures how easily current can flow. At any given voltage, a higher resistance means less current flows. A lower resistance means more current flows. The relationship among current, voltage, and resistance is called **Ohm's law**, after the German physicist Georg Simon Ohm, who discovered it.

Interactive equation

**Ohm's law**

$$I = \frac{V}{R} \quad (17.1)$$

$V$ = voltage or potential difference (V)
$I$ = electric current (A)
$R$ = resistance (Ω)

**Units of resistance**

Resistance is measured in *ohms*. The symbol for an ohm is the Greek capital letter Omega or Ω. One **ohm (Ω)** is one volt per ampere. A resistance of 1 Ω means 1 A of current will flow when 1 V is applied. The word *resistance* describes how to think about Ohm's law. The current $I$ that flows is equal to the applied voltage $V$ divided by the resistance $R$ of the circuit.

**What is resistance?**

An upside-down water bottle makes a useful analogy for how resistance relates to current. A small opening has a large resistance and therefore only a small current of water dribbles out. A large opening has a low resistance and the current of water is large. Mathematically, resistance is in the denominator of the fraction (under voltage), so an increase in resistance makes the current smaller and a decrease in resistance makes the current larger. Doubling the resistance of a circuit results in only half as much current flowing.

**Upside-down water bottles are similar to resistance.**

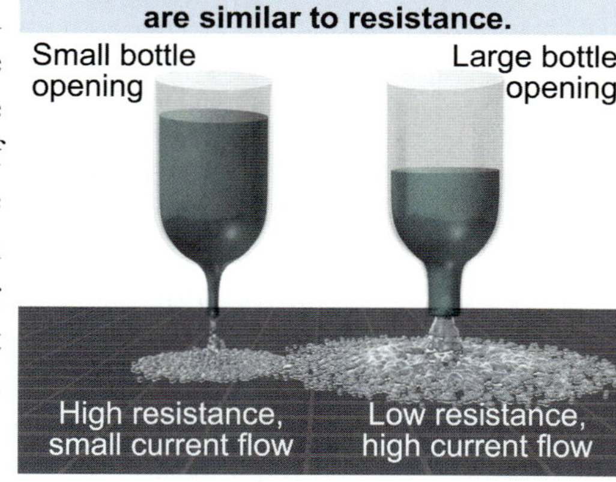


**Exceptions to Ohm's law**

Ohm's law is not a fundamental law of nature like Newton's three laws of motion. Instead, it is an observed relationship among voltage, current, and resistance that most materials obey in electric circuits most of the time. Exceptions to Ohm's law include devices such as *transistors* and *diodes*, for which the current flows through tiny structures of different materials with different electrical characteristics.

### Electrical devices and resistance

An ordinary light bulb has a resistance of 600 Ω. When placed in a circuit at 120 V, the light bulb draws 0.2 A and lights up as it was designed to do. A certain hairdryer has a resistance of 10.9 Ω. When connected to the same 120 V circuit, the hairdryer draws 11 A. Engineers design the resistance of electrical devices to draw the right current when connected to the right voltage. Every electrical device has a resistance and its resistance determines how much current the device draws in a circuit at the proper voltage.

**Using Ohm's law**

$$I = \frac{V}{R}$$

**High resistance**

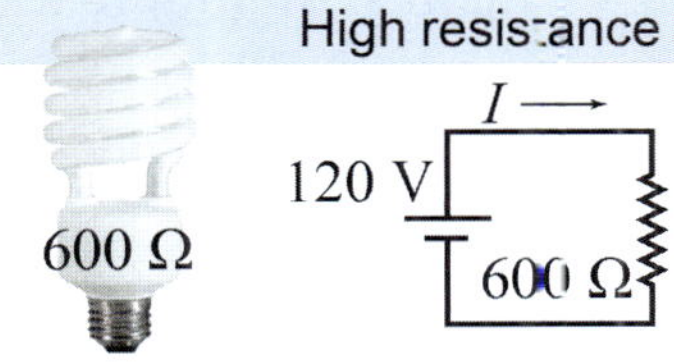


$$I = \frac{120\text{ V}}{600\ \Omega} = 0.2\text{ A}$$

**Low resistance**

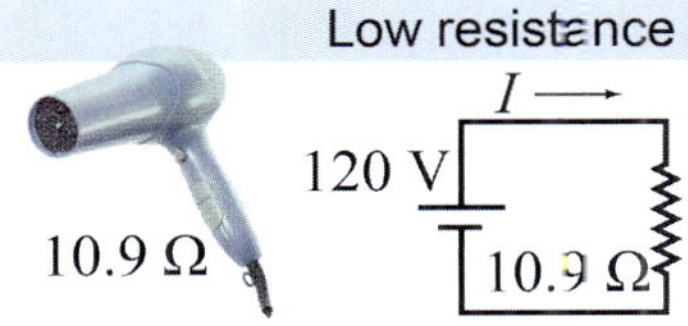


$$I = \frac{120\text{ V}}{10.9\ \Omega} = 11\text{ A}$$

### Resistors

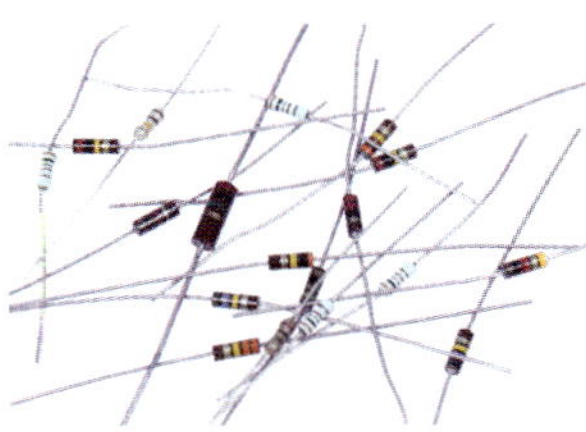

In electronic circuits resistance is created by *resistors,* such as the ones in the photo on the left. A **resistor** comes in many standard values of resistance. Resistors are used to control current, voltage, or both in electronic circuits.

### Resistance in circuits

The symbol for any resistance in a circuit diagram is a zigzag line. The example circuits for the light bulb and hair dryer show each as a zigzag line with the resistance labeled in Ω to the side. Potentiometers are variable resistors and have a knob that is turned to change the resistance.

**Both resistors and lamps have resistance.**

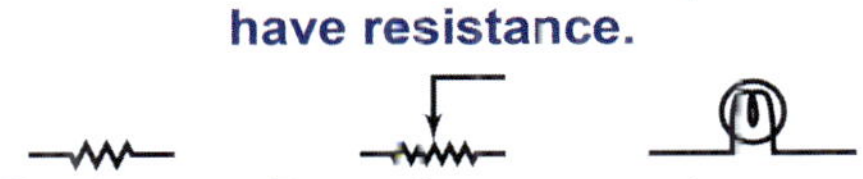


### What is the resistance of a short circuit?

Consider what happens when you connect the positive and negative terminals of a 1.5 V battery directly with a piece of wire, which has only a tiny resistance (such as 0.03 Ω). Ohm's law says that the current should be extremely large—50 A in this case! Such a high current is more than enough current to melt the wire and more than the battery can deliver. *That is why short circuits are dangerous!* A short circuit presents a low-resistance path for current to flow, allowing too much current. Batteries can safely deliver currents up to 2 or 3 A, and the thin wires in a typical circuit used in the laboratory are chosen to handle this amount of current safely.

### Calculating current using Ohm's Law

A simple circuit contains only wires, a light bulb with a resistance of 2 Ω, and a 1.5 V battery. How much current flows through the light bulb?

*Asked:* electric current $I$

*Given:* resistance $R = 2\ \Omega$, voltage $V = 1.5$ V

*Relationships:* Ohm's law for current: $I = V/R$

*Solution:*

$$I = \frac{V}{R} = \frac{1.5\text{ V}}{2\ \Omega} = 0.75\text{ A}$$

*Answer:* $I = 0.75$ A.

## Using a digital multimeter

**Settings of a multimeter**

The **digital multimeter** is a useful instrument that measures voltage, current, and resistance. The digital multimeter in the diagram below has a rotating central knob to select the type of measurement: voltage, current, or resistance. Within each measurement type there may be a number of different range settings. For example, there might be three voltage ranges: 0–2 V, 0–20 V, or 0–200 V. To make the most accurate voltage measurement of a 1.5 V battery you would select the 0–2 V range.

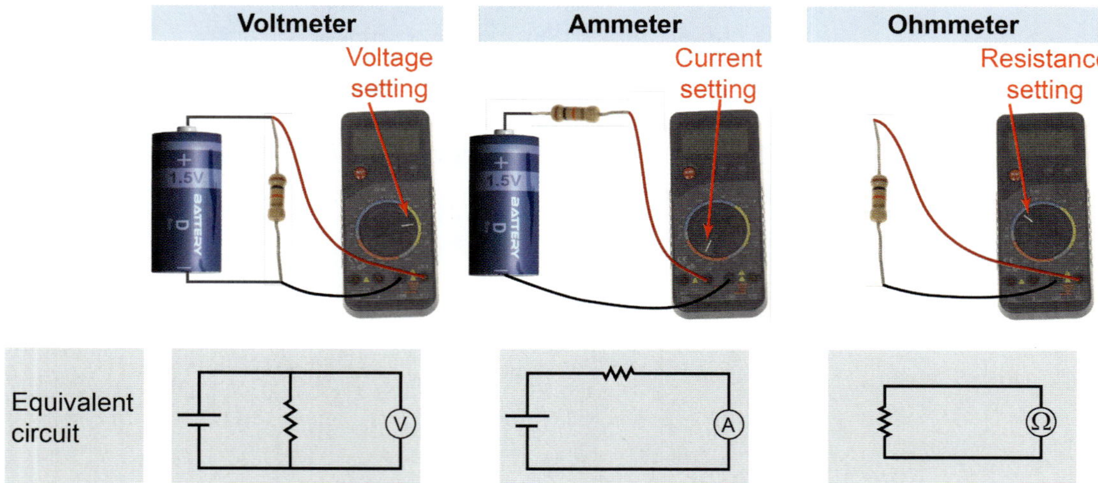


Connect the voltmeter in *parallel* with the circuit. Connect the ammeter in *series* with the circuit. Break the circuit to measure the resistive elements alone.

**Measuring with a multimeter**

Multimeters typically have red and black probe wires, or *leads*. The probe leads are connected to the matching (+) red and (−) black terminals of the meter. *The multimeter must be connected to the circuit in different ways to make different types of measurements.*

**Measuring voltage**

To measure voltage, the probe leads are touched to two points in a circuit. The meter reads the voltage difference between the two points. If the red lead is at a higher voltage than the black lead, the meter will read a positive value. If the black lead is at the higher voltage, the meter will read a negative value. Note that voltage cannot be measured *at one point* but only *between two points*. When the voltage function is selected the multimeter acts as a *voltmeter* and is represented in a circuit diagram as a circle containing the letter "V."

**Measuring current**

*To measure current, the meter must be inserted into the circuit, forcing the current to flow through the meter.* Current must flow through the multimeter to be measured. When the current function is selected the multimeter acts as an *ammeter* and is represented in a circuit diagram as a circle containing the letter "A."

**Measuring resistance**

To measure the resistance of a single electrical device, the device is isolated from the circuit and connected between the red and black leads of the meter. For most resistance measurements it does not matter which (red or black) lead connects to which terminal of the device being measured. When the resistance (Ω) function is selected the multimeter acts as an *ohmmeter* and is represented in a circuit diagram as a circle containing the Greek symbol Omega, "Ω." Most multimeters have resistance ranges from a few ohms to "20 M" (20 MΩ).

**Voltage and current sensors**

Voltage and current sensors can also be used as voltmeters and ammeters, and are ideal for exploring electricity and circuit concepts where voltage and current are changing. Using these sensors, voltage and current data can be monitored or recorded over time and displayed in a table and graph. Observing continuous data from the sensors helps reveal the dynamic properties of more complex circuit components and how they react to changing voltage and current.

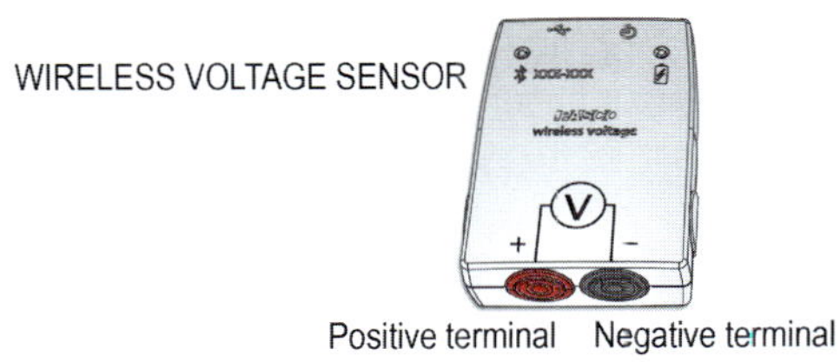


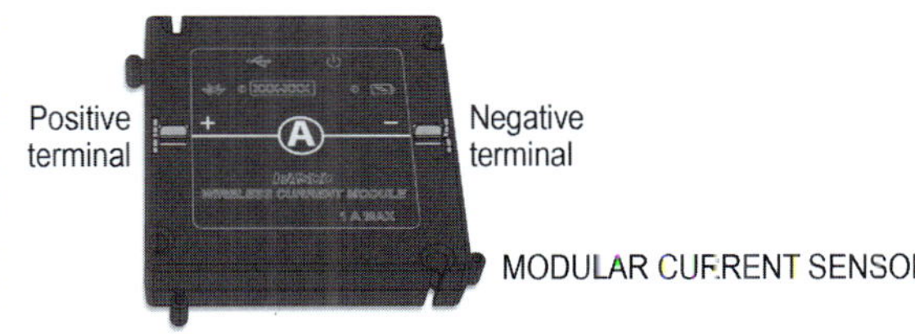


**Measuring voltage with a voltage sensor**

Most voltage sensors have both positive (red) and negative (black) terminals to connect to circuits, similar to a multimeter. To measure voltage, the probe leads from the sensor are touched to two points in a circuit where the voltage difference is measured between the two points. This *parallel* connection allows the user to measure the voltage difference *across* a circuit component, or set of components. In this configuration the sensor acts as a *voltmeter* and is represented in a circuit diagram as a circle containing the letter "V."

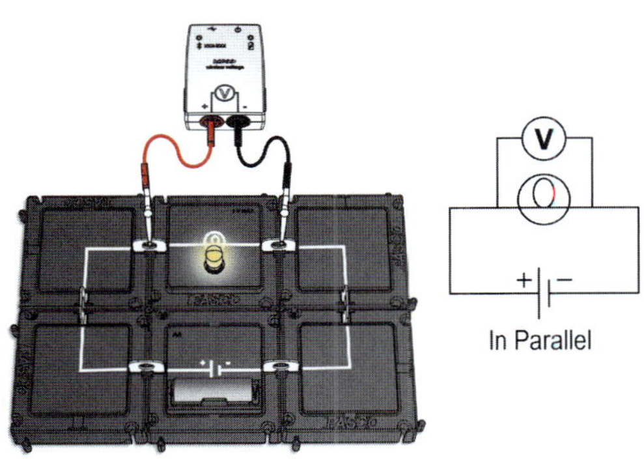


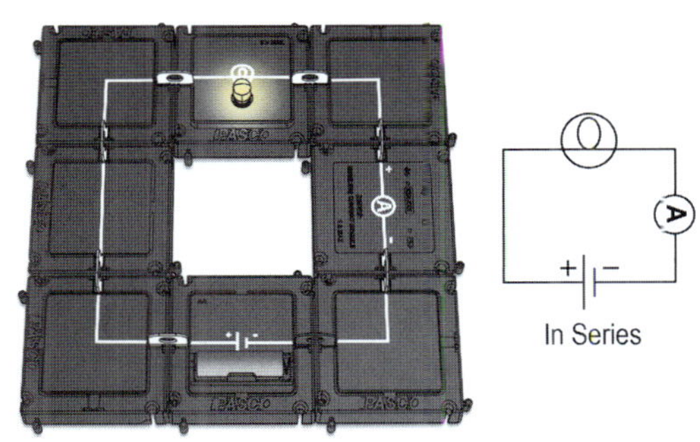


**Measuring current with a current sensor**

Current sensors also have positive and negative terminals like voltage sensors, but the terminals are not always color-coded red and black. To measure current, the sensor must be connected to the circuit in-line (*in series*) with circuit components to measure the current flowing through them. In this configuration the sensor acts as an *ammeter* and is represented in a circuit diagram as a circle containing the letter "A."

**Observing Polarity**

Connecting the positive and negative terminals on current sensors improperly may produce measurements with opposite sign, or *polarity*. Current sensors must be connected in such a way that the current flows in the direction from the positive terminal to the negative terminal. Connecting the sensor in the opposite orientation results in negative current measurements.

## Using resistors and wires

**Resistors**

The resistance of a resistor, in ohms, is often coded on its outside using a system of colored bands. Most common resistors have four color bands: three for the value of the resistance and the last band for its tolerance. The tolerance tells you how far off the actual resistance can be compared to the color code value.

**Reading resistors**

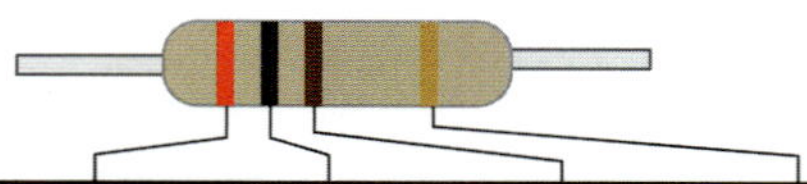

| Color | 1st Band | 2nd Band | Multiplier | Tolerance |
|---|---|---|---|---|
| Black | 0 | 0 | 1 Ω | |
| Brown | 1 | 1 | 10 Ω | ±1% |
| Red | 2 | 2 | 100 Ω | ±2% |
| Orange | 3 | 3 | 1 kΩ | |
| Yellow | 4 | 4 | 10 kΩ | |
| Green | 5 | 5 | 100 kΩ | ±0.5% |
| Blue | 6 | 6 | 1 MΩ | ±0.25% |
| Violet | 7 | 7 | 10 MΩ | ±0.1% |
| Gray | 8 | 8 | | ±0.05% |
| White | 9 | 9 | | |
| Gold | | | | ±5% |
| Silver | | | | ±10% |

**Examples**

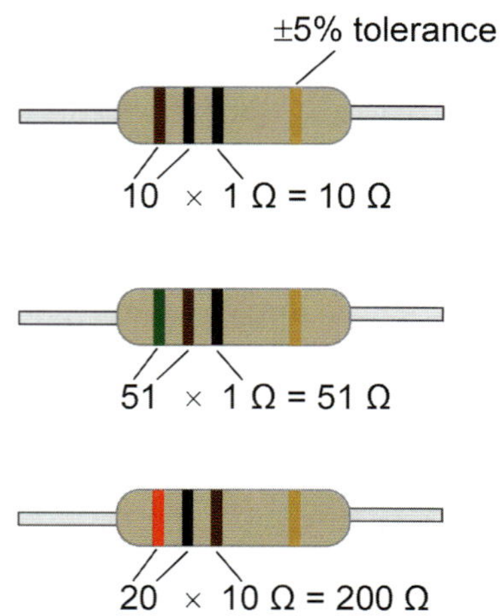


**Reading resistor values**

The first two color bands represent the first two digits of the resistor's value. In the example above the table, these are red and black, corresponding to 2 and 0, or 20. The third band represents a power of 10 multiplier. The brown third band corresponds to a multiplier of $10^1 = 10$. This resistor therefore has a resistance of 20 × 10 = 200 Ω. The fourth and final band provides the tolerance, which in the example is gold, or 5%. The actual resistance could therefore be anything within ±5%, or 190 Ω to 210 Ω.

**Stripping the insulation off a wire**

When constructing circuits, particularly with a circuit breadboard, you will typically need wires that have the insulation removed from both ends. Wires can be purchased with the ends already "stripped," but you can strip a wire yourself, too. A wire-cutter tool is used for this purpose. It has a sharp blade (like a scissors blade) for cutting off a length of wire. It also has a wire-stripping notch that, when you squeeze it around the wire and pull, will remove the outer insulation while leaving the inner conducting wire in place. Different wire-stripping tools have differently sized notches so that they work on wires with different diameters.

**How to strip a wire**

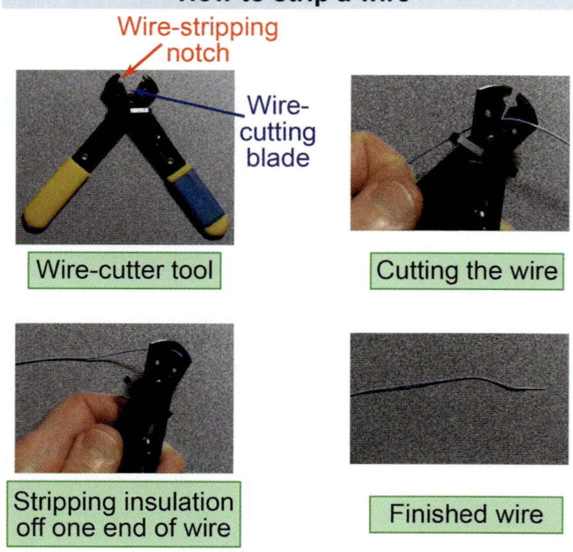

# Investigation 17C Resistors and Ohm's law

**Essential questions** *How is resistance measured?*

Ohm's law, $I = V/R$, is the fundamental relationship among current, voltage, and resistance in a circuit. Devices that measure resistance are based on Ohm's law. These devices apply a known voltage and/or current, then determine the resistance. In this investigation you will use a similar experimental technique to measure the resistance of a bulb.

## Part 1: Current through different resistors

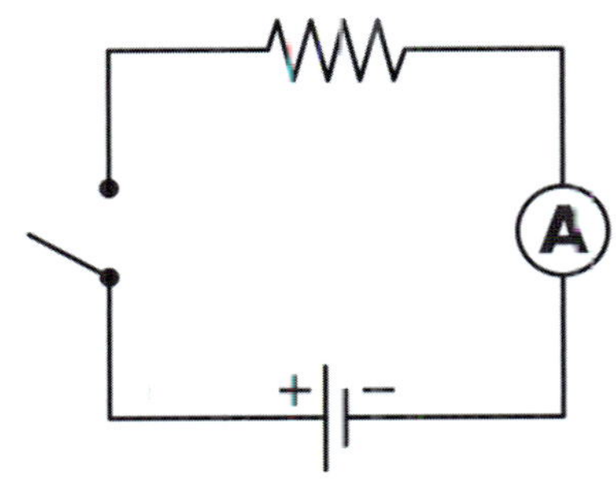


1. Open the experiment file **17C_ResistanceAndOhmsLaw**, and then power-on the current sensor and connect it to your software.
2. Construct the circuit shown using one battery, a switch, a current sensor, a 10-Ω resistor, and any necessary wire modules.
3. Start data collection, close the switch, and record the measured current in the data table.
4. Repeat the experiment for the 33-Ω and 100-Ω resistors. Tabulate your results for $R$ and $I$.

a. *Graph I on the vertical axis and R on the horizontal axis. What type of curve best fits this data?*
b. *Calculate 1/R for each resistor. Make a second graph with I on the vertical axis and 1/R on the horizontal axis to create a linear graph. Fit a line to determine the slope of the data. What is the value of the slope?*
c. *What does the slope of this graph represent? Why?*

## Part 2: Resistance of a light bulb

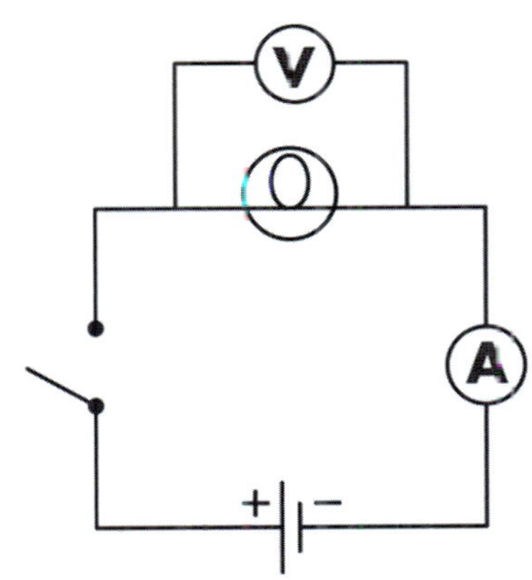


1. Power-on the voltage sensor and connect it to your software.
2. Replace the resistor with the bulb and attach the voltage sensor across the bulb.
3. Measure the current and voltage for the illuminated bulb.
4. Calculate the resistance of the bulb using Ohm's law, $R = V/I$.

a. *What is the resistance of the bulb? Show your work, including units.*
b. *Imagine you have a complicated circuit containing many resistors. Describe in words how you can use Ohm's law to find the effective resistance of the entire circuit.*

## Conductors and insulators

### Electrical conductors and insulators

Based on their electrical properties, materials can be divided roughly into *conductors* and *insulators.* A metal such as copper is commonly used as an **electrical conductor** because it offers little resistance. For example, 1 m of 20-gauge copper wire has a resistance of 0.03 Ω. At the other end of the scale, a material with high resistance is classified as an **electrical insulator**. Even a thin layer of plastic can have a resistance of 1,000,000 Ω or more. Electrical devices and wires are typically covered by an insulator, such as plastic. Other good insulating materials include glass and wood.

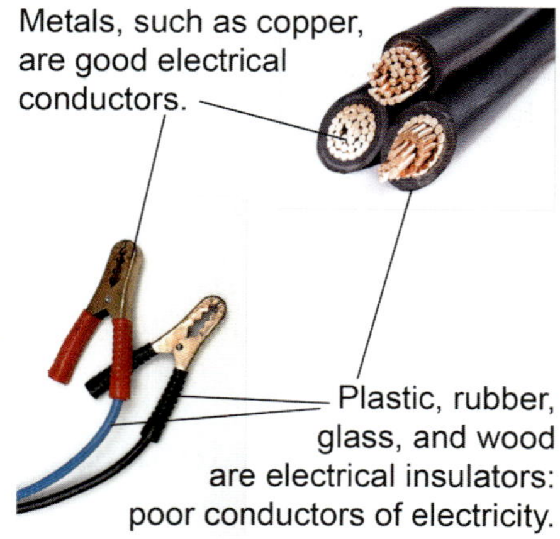


### Nature of electric current

In metals such as copper, the outermost electrons are not bound to a single atom, so they can flow and make *electric current.* These free electrons are what make copper a good electrical conductor. When there is a voltage across a conductor, electrons are attracted to the positive terminal and repelled from the negative terminal. The voltage provides a force that moves the electrons to create the electric current. The higher the voltage, the faster the electrons move, and the more power is transferred by the moving electrons to devices that can use the power, such as electric motors.

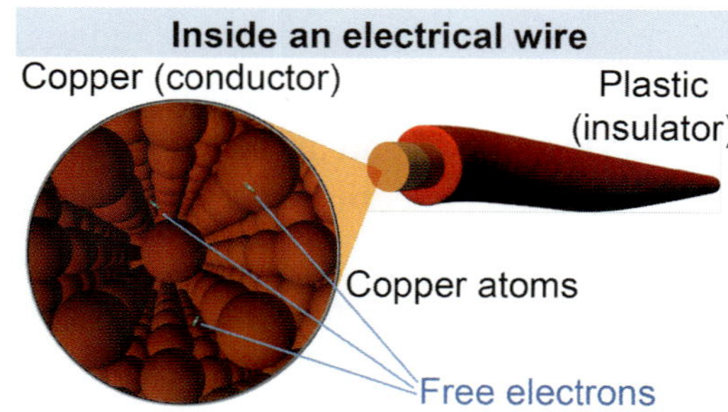


### Resistance and temperature

The resistance of metals increases somewhat as their temperature increases. As a result, some electrical circuits might only function properly over a limited range in temperature. Other circuit parts, particularly transistors, also can fail at extreme temperatures. Many electrical devices, such as computers, have cooling fans built into their housing to prevent overheating of the components.

### Resistance and the human body

Touching a 120 V wall outlet is dangerous. Electric current flowing through the body causes electric shock that can kill a person. A typical range for the body's resistance is 300–1,000 Ω—although it can be as high as 100,000 Ω if your skin is dry or only 100 Ω if you are soaked in saltwater. Currents through the body of 1 mA can be felt, while currents of 10 mA and higher can cause pain and muscle spasms.

### Wire gauge and resistance

Although the resistance of conductors is small, over a very long wire the resistance can add up. Wires with a large diameter—or small *gauge*—have less resistance than narrow wires. Home wiring typically uses a 12-gauge wire, which has a diameter of 2.05 mm and a resistance of 0.005 Ω per meter of wire. Electronic circuits often use 22-gauge wire, which has a diameter of 0.64 mm and a resistance of 0.053 Ω per meter of wire. High-voltage power transmission lines might have a diameter of between 4 and 25 mm.

### Resistivity

The *resistivity* $\rho$ of a material is an indicator of its resistance. Metals with higher values of resistivity have a higher resistance per meter of wire. Copper has a resistivity of $\rho = 1.68 \times 10^{-8}\ \Omega$ m, respectively. Semiconductors have resistivities of around $10^{-4}$ to $10^{-1}\ \Omega$ m, whereas insulators have resistivities of around $10^{9}$ to $10^{15}\ \Omega$ m.

# Section 2 review

Ohm's law states that the current $I$ through a conductor is the voltage drop $V$ divided by resistance $R$ (or, in other words, $I = V/R$). The SI unit for resistance is the ohm, abbreviated Ω. Conductors are materials that offer little resistance to current flow when subjected to a voltage. Insulators are materials that strongly resist current flow. Resistors are used in electric circuits to control current flow. Engineers design the resistance of electrical devices to draw the right current when connected to the right voltage.

| | |
|---|---|
| **Vocabulary words** | **Ohm's law, ohm (Ω), resistance, resistor, digital multimeter, electrical conductor, electrical insulator** |
| **Key equations** | $I = \frac{V}{R}$ |

## Review problems and questions

1. A resistor and voltage source are connected in a simple closed circuit, as shown here. The resistor obeys Ohm's law.

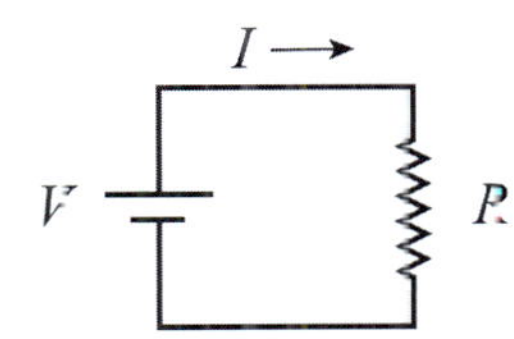


   a. If the voltage $V$ doubles but resistance stays the same, what happens to the current $I$?
   b. If the resistance $R$ doubles but the voltage remains unchanged, what happens to the current?

2. Two simple closed circuits are shown. Which carries a larger amount of current?

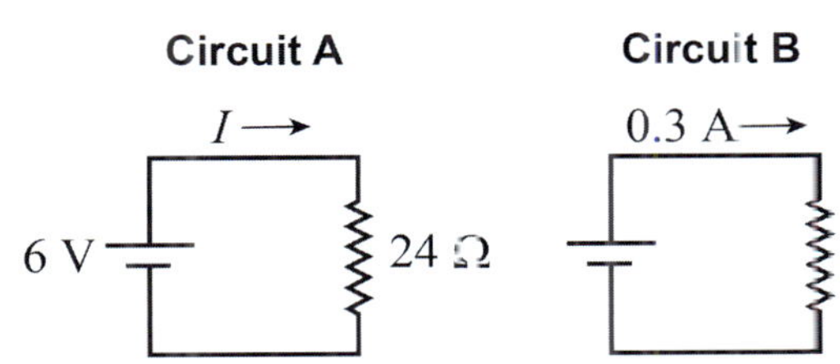


   i. A carries a *larger* current than B.
   ii. A carries a *smaller* current than B.
   iii. A and B carry *equal* amounts of current.
   iv. You cannot tell without more information.

3. Which of these is the correct equivalent to an ohm (Ω) ?
   a. V/A
   b. A/V
   c. V A
   d. V $A^2$
   e. $A^2$/V

4. A child coats the "top" (+ end) of a flashlight battery with thick shiny paint. He allows the paint to dry. He then inserts the battery into his flashlight and slides the switch to the "on" position. The bulb immediately gives off light. What can we conclude about the paint?
   a. The paint is an insulator.
   b. The paint contains free (mobile) electrons.
   c. The paint contains flakes of copper.

## 17.3 - Series and parallel circuits

You probably have many electrical devices at home that are plugged into wall sockets. You may use many of those devices—such as a refrigerator, lights, and stereo—at the same time. For each appliance to draw the correct amount of current, the circuits in your house must provide the same voltage to every device regardless of whether other devices are connected. In this section you will learn about the differences between series and parallel circuits and how these circuit types apply to the electrical wiring of a house.

### Series circuits and voltage drop

**How are electric circuits connected?**

A **series circuit** has only one path that electric current can follow. A series circuit connects the output of the first device to the input of the second device, the output of the second device to the input of the third device, and so on. In a series circuit all the current flows *in series* through each device, one after another.

**Series resistance**

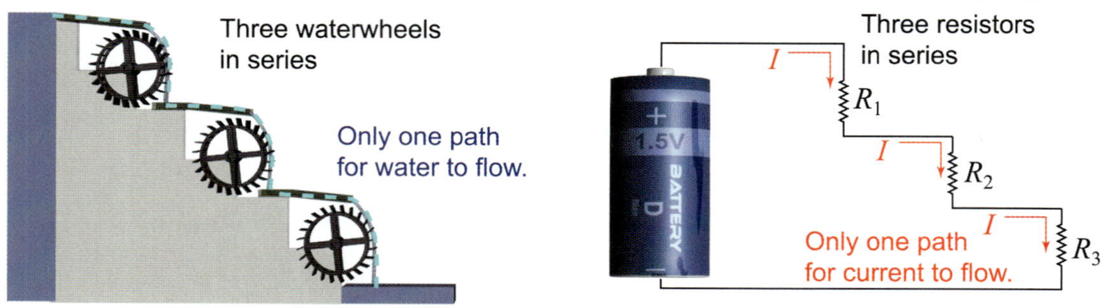


**How does voltage change in a series circuit?**

To understand how voltage changes across a resistor, think about three water wheels, each spanning one-third of the height of a waterfall. All the water flows onto each wheel, one after another, just as all the electric current flows through all three of the resistors. *The waterfall's total potential energy, however, only drops by one-third across each of the water wheels.* A series circuit of resistors is similar: The total voltage (or potential difference) only *partially* reduces (or drops) across each individual resistor. If the three resistors are equal, the voltage is reduced (or drops) by ⅓ across each resistor.

**Voltage drop**

Ohm's law is often written in the form of equation (17.2) to reflect the voltage drop across a resistor. When 1 A of current flows through 1 Ω resistor, there is a 1 V drop across that resistor. In a circuit with resistors in series there is a voltage drop across each resistor. The total voltage drop over a set of resistors in series equals the sum of the individual voltage drops.

Interactive equation

| (17.2) | $V = IR$ | $V$ = voltage drop (V)<br>$I$ = electric current (A)<br>$R$ = resistance (Ω) | **Ohm's law**<br>voltage-drop form |
|---|---|---|---|

**Power and voltage drops**

Voltage drops are evidence that electrical energy is being dissipated into other forms of energy, often heat. A drop in voltage is always accompanied by the dissipation of power. (You first learned about electrical power on page 266 in Chapter 9.) Any resistance dissipates electrical power when current flows through it. We will return to the topic of electrical power later in this section.

**How is Ohm's law applied to many resistances?**

Consider the circuit in the diagram that has three resistances and a battery connected in series.

1. How much current flows in this circuit?
2. How is Ohm's law applied?

The answer is that the current in the circuit adjusts itself so that the sum of all the voltage drops equals the voltage provided by the battery. Since it is a series circuit, the current $I$ is the same everywhere. In mathematics, the battery voltage $V$ equals the sum of the voltage drops across each of the three resistances, $R_1$, $R_2$, and $R_3$:

$$V = IR_1 + IR_2 + IR_3$$

**What is the resistance of this circuit?**

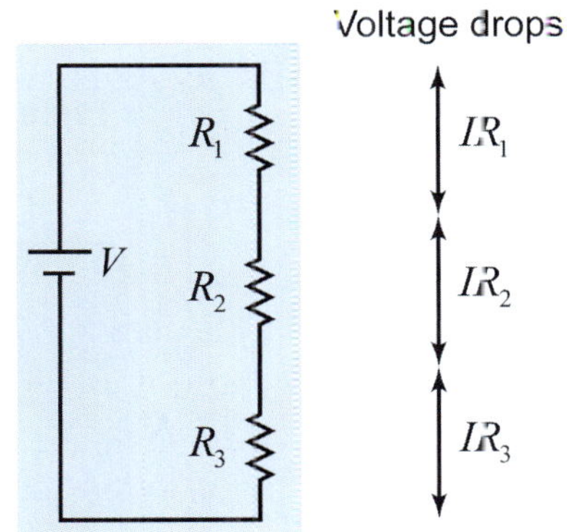


**Total resistance for a series circuit**

The current $I$ in the circuit is the same through each resistor; therefore, we can rewrite this equation by grouping the resistances together as

$$V = I(R_1 + R_2 + R_3)$$

This equation looks like Ohm's law if we make the substitution that the *total resistance R* of the circuit is equal to the sum of all the individual resistances:

$$R = R_1 + R_2 + R_3$$

This allows us to write the following for the total voltage and the total current:

$$V = IR, \qquad I = \frac{V}{R}$$

When resistors are connected in series, their total combined resistance is given by equation (17.3). The total resistance is the easiest way to calculate current in a series circuit. The current is the total voltage divided by the total resistance.

Interactive equation

**Equivalent resistance**
series resistors

$$(17.3) \quad R = R_1 + R_2 + R_3 + \cdots$$

$R$ = equivalent resistance (Ω)
$R_1$ = resistance 1 (Ω)
$R_2$ = resistance 2 (Ω)
$R_3$ = resistance 3 (Ω)

**Current through two resistors connected in series**

Two resistors with resistances of 10 Ω and 20 Ω are connected in series with a battery of voltage 10 V. What is the current through the circuit?

*Asked:* total current $I$ through the circuit

*Given:* voltage $V$ = 10 V; resistances $R_1$ = 10 Ω and $R_2$ = 20 Ω

*Relationships:* Ohm's law $I = V/R$;
equivalent resistance for series resistors: $R_{eq} = R_1 + R_2$

*Solution:* The equivalent resistance for the series circuit is

$$R_{eq} = R_1 + R_2 = 10\ \Omega + 20\ \Omega = 30\ \Omega$$

The current is then

$$I = \frac{V}{R_{eq}} = \frac{10\ \text{V}}{30\ \Omega} = 0.33\ \text{A}$$

*Answer:* $I$ = 0.33 A.

## Parallel circuits

**What is a parallel circuit?**

A **parallel circuit** contains *branches* that provide more than one path for electric current to follow. In a parallel circuit each branch can have a *different* current flowing through it. The total current in the circuit is the sum of the individual currents flowing in each branch.

**Comparing series and parallel circuits**

**Series resistance: one current path; lamps split the voltage drop**

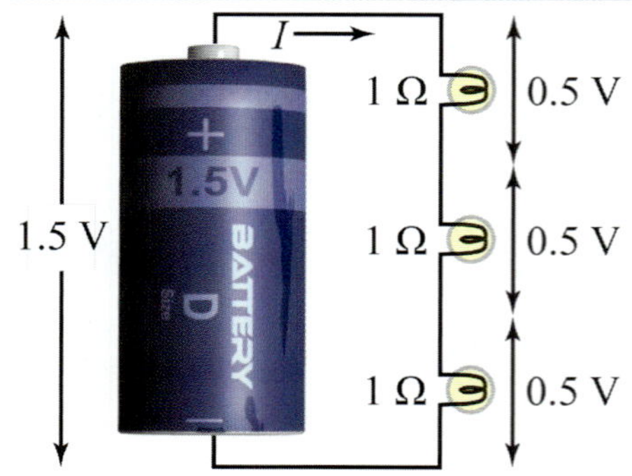


$$I_{total} = \frac{V}{R} = \frac{0.5\text{ V}}{1\,\Omega} = 0.5\text{ A}$$

**Parallel resistance: more than one current path; each lamp has the full voltage drop**

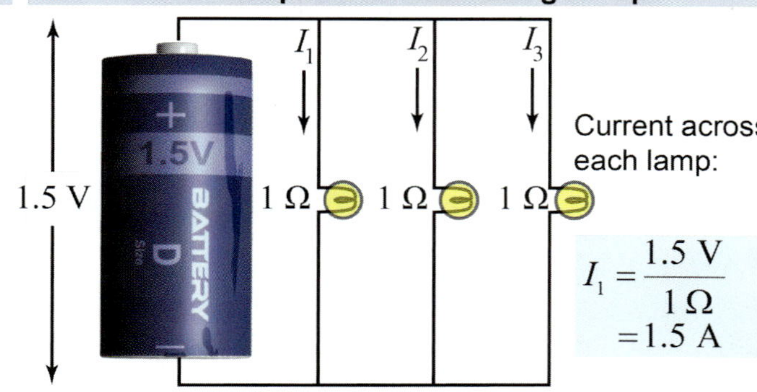


$$I_{total} = I_1 + I_2 + I_3 = 4.5\text{ A}$$

**Current and voltage in a series circuit**

Compare the series and parallel circuits in the diagram above. The current across a single bulb by itself would be $I = (1.5\text{ V})/(1\,\Omega) = 1.5\text{ A}$. In the series circuit, however, the voltage across each individual resistor (0.5 V) is a fraction of the voltage across all three resistors (1.5 V). Since the same current flows through all three resistors, we can calculate the current of the circuit by calculating the current through only one resistor, or $I = (0.5\text{ V})/(1\,\Omega) = 0.5\text{ A}$. The current in the series circuit is *less* than it would be if there were just one bulb. This happens because the overall resistance *seen by the battery* has increased: The resistances of each lamp add in series so that the total resistance is three times the resistance of a single lamp.

**Current and voltage in a parallel circuit**

In the parallel circuit, in contrast, each lamp has a direct low-resistance connection to both battery terminals. Because wires have negligible resistance, there are no voltage drops across the wires. Each lamp "sees" the full 1.5 V from the battery and therefore draws 1.5 A of current, as if it were the only lamp in the circuit. The total current from the battery is 4.5 A, which is the sum of the current in each of the three branches. The current in the three-bulb parallel circuit is *greater* than it would be in a circuit with a single bulb. In essence, the parallel circuit acts like three single-bulb circuits.

**Parallel circuit rules**

Keep in mind the following two rules to understand parallel circuits:

1. *The voltage is the same everywhere along a wire.*
   Voltage only changes across a circuit element, such as a resistor. Although not exactly true, this rule is a good approximation because wires have very low resistance.
2. *The current flowing out of any branch point, or junction, in the circuit must equal the current flowing into the branch point.*
   This rule is known as Kirchhoff's current law. In the parallel circuit shown there are three branches, and Kirchhoff's current law tells us that the current flowing out of the battery must equal the sum of the currents in each of the three branches.

**What is the current?**

The voltage in the two circuits below is the same, but the two-bulb circuit draws more total current. Why? To answer the question, consider that current flowing out of the battery obeys Ohm's law. The current is the battery voltage divided by the total resistance of the circuit. What resistance does the battery "see?"

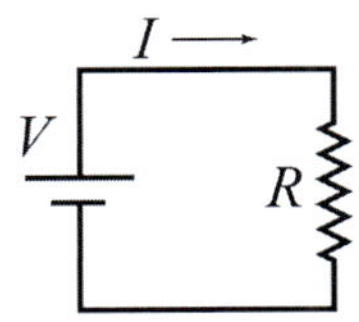

$$I = \frac{V}{R}$$

Single bulb circuit

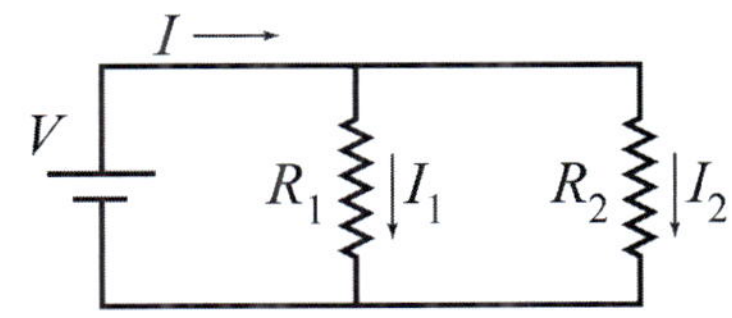

$$I = I_1 + I_2$$

$$I = \frac{V}{R_1} + \frac{V}{R_2}$$

Two-bulb parallel circuit

**Analyzing a parallel circuit**

According to Ohm's law, the current in each branch of the circuit is the battery voltage $V$ divided by the resistance of that branch ($R_1$ or $R_2$). Next, note that the current flowing out of the battery is the *sum* of the currents in the two branches (from Kirchhoff's current law).

Compare

$$I = \frac{V}{R}$$

Ohm's law

$$I = V\left(\frac{1}{R_1} + \frac{1}{R_2}\right)$$

Two-bulb parallel circuit

$$\frac{1}{R} = \frac{1}{R_1} + \frac{1}{R_2}$$

Total resistance

If you compare this with Ohm's law, the *inverse of the total resistance* as seen by the battery is equal to the sum of the inverses of the individual branch resistances. This leads directly to equation (17.4), which gives the total resistance $R$ for a parallel circuit containing three individual resistances:

Interactive equation

**Equivalent resistance**
parallel resistors

$$\textbf{(17.4)}\quad \frac{1}{R} = \frac{1}{R_1} + \frac{1}{R_2} + \frac{1}{R_3} + \cdots$$

$R$ = equivalent resistance (Ω)
$R_1$ = resistance 1 (Ω)
$R_2$ = resistance 2 (Ω)
$R_3$ = resistance 3 (Ω)

**Total resistance decreases**

Adding more branches to a *parallel* circuit always increases the total current in the circuit. From the perspective of the battery, the total resistance of the circuit must *decrease*, because more total current flows while the voltage stays the same (equation 17.4).

**Adding resistances in parallel**

What is the equivalent resistance of two 10 Ω resistors connected in parallel?

*Asked:* equivalent resistance $R_{eq}$

*Given:* individual resistances $R_1 = 10\ \Omega$ and $R_2 = 10\ \Omega$

*Relationships:*

$$\frac{1}{R} = \frac{1}{R_1} + \frac{1}{R_2} + \frac{1}{R_3} + \cdots$$

*Solution:* Add the resistances by finding a common denominator:

$$\frac{1}{R_{eq}} = \frac{1}{R_1} + \frac{1}{R_2} = \frac{1}{10\ \Omega} + \frac{1}{10\ \Omega} = \frac{1+1}{10\ \Omega} = \frac{1}{5\ \Omega}$$

If $(1/R_{eq}) = 1/(5\ \Omega)$, then $R_{eq} = 5\ \Omega$.

*Answer:* $R_{eq} = 5\ \Omega$.

# Investigation 17D

## Series and parallel circuits

**Essential questions** *What are the advantages and disadvantages of series versus parallel circuits?*

Have you ever had a string of holiday lights where one bulb is burned out, preventing all the other bulbs from lighting? Was it easy to find the burned out bulb? This investigation explores series and parallel circuits by connecting bulbs and observing their brightness. By comparing the two circuit types, you will learn why the wiring of *some* strings of lights allows one bad bulb to disconnect all the other bulbs.

### Part 1: Connecting bulbs in series

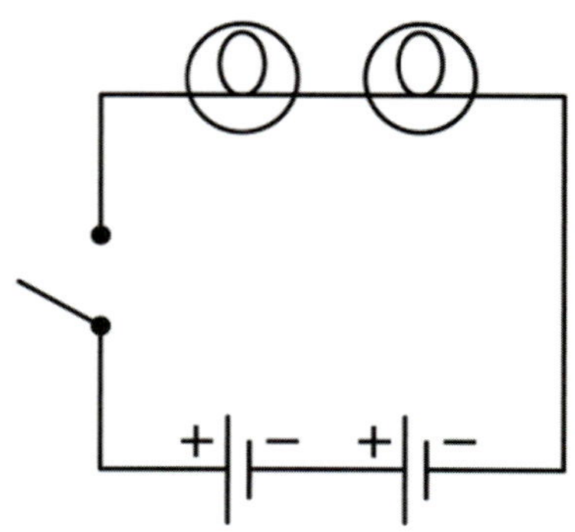


1. Create a circuit using two batteries, one bulb, a switch, and any necessary wire modules.
2. Close the switch and observe the brightness of the bulb.
3. Create a circuit with two bulbs in series, as shown in the circuit diagram. Compare the brightness of the two lamps to the previous circuit with one lamp.

*a. What property makes this a series circuit?*

*b. How bright are the two bulbs in series compared to a single bulb circuit? Why?*

*c. Unscrew one bulb from the series circuit. What happens to the other bulb? Why?*

### Part 2: Connecting bulbs in parallel

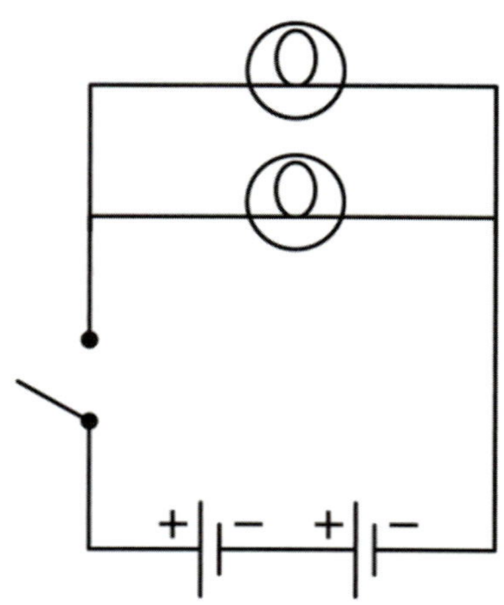


1. Create a circuit with two bulbs in parallel, as shown in the circuit diagram.
2. Compare the brightness of the bulbs in this circuit to the prior circuit with two bulbs in series.

*a. What property makes this a parallel circuit?*

*b. How bright are the parallel bulbs compared to the series bulbs? Compared to the single bulb? Why?*

*c. Unscrew one bulb from the parallel circuit. What happens to the brightness of the other bulb? Why?*

*d. Is a series or parallel circuit better for connecting a string of lights? Why?*

*e. Design a circuit of three bulbs that combines series and parallel arrangements, and sketch the circuit diagram. Predict the relative bulb brightness based on the previous experiments. Build the circuit and test your predictions. Were you correct?*

**Kirchhoff's current law**

The total amount of electric current entering any junction of a circuit must equal the total amount of current leaving that junction. This is known as **Kirchhoff's current law**, also known as Kirchhoff's first law, named after the German physicist Gustav Robert Kirchhoff (1824–1887). In the example on the right 40 mA enters a junction. If 10 mA leaves by one branch, then 30 mA must leave by the second branch to satisfy Kirchhoff's current law.

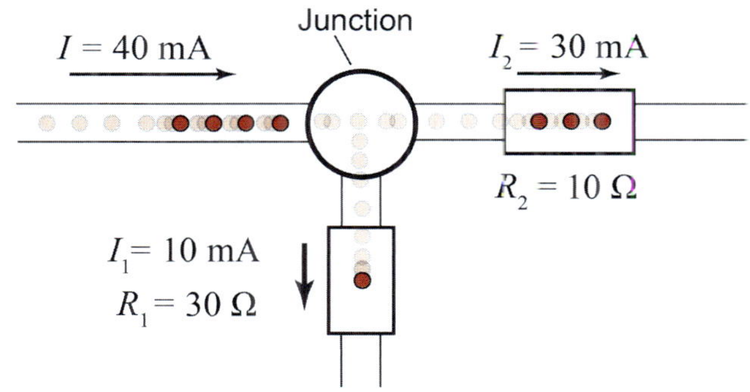


Kirchhoff's first law: *The sum of all currents entering a junction equals the sum of all currents leaving that junction.*

**Conservation of charge**

Kirchhoff's current law reflects a fundamental conservation law in physics: *the law of conservation of charge.* The law of conservation of charge says that the total electric charge of the universe is constant. If two positive charges go in, then two positive charges must also come out, much as every car that enters an intersection must be matched by one that leaves. We will explore electric charge more deeply in Chapter 18.

**Kirchhoff's first law and parallel circuits**

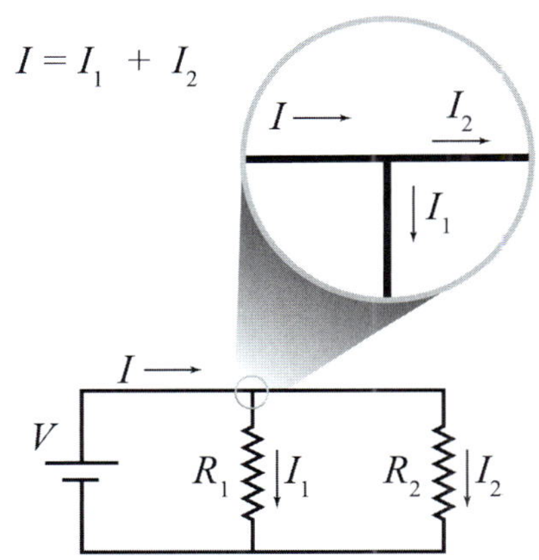


Kirchhoff's first law is usually brought up when discussing *parallel* circuits, because the *total* current (here labeled $I$) splits at one junction and merges at another. Current can only change at junctions, and it does not change when passing through a battery or resistor. In applying Kirchhoff's law to this parallel circuit, we first apply Ohm's law to each resistor individually. Next, we add the currents through each resistor to get the total current. Finally, we apply Ohm's law to the circuit as a whole to obtain the total resistance.

**Kirchhoff's first law and series circuits**

But Kirchhoff's first law is equally important in considering *series* circuits. Such circuits have no forks in the road, from the perspective of the moving charges—particles do not split off from one another and go in separate directions as current flows through a series circuit. Nevertheless, Kirchhoff's first law helps us see that the current through each resistor is the same in a series circuit. That insight is essential to solving for the total resistance of such a circuit.

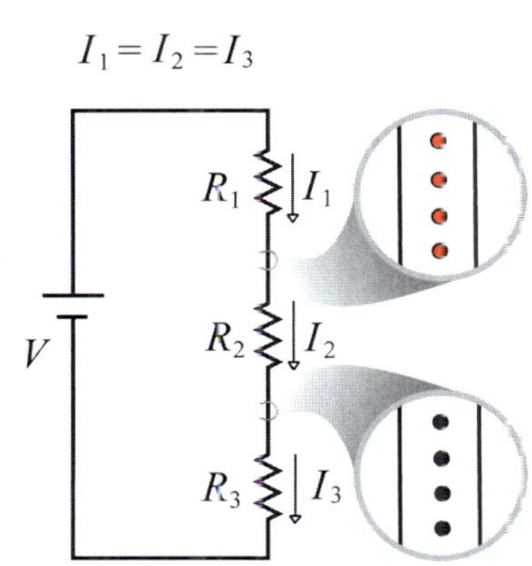

**Another conservation principle**

Kirchhoff's first law provides relationships between the currents flowing into or out of a junction. By contrast, **Kirchhoff's voltage law**, also known as Kirchhoff's second law, relates *voltages* to one another. It states that *the sum of the voltage differences (gains and drops) around a closed loop equals zero.* This is much like saying that you add energy to an object when you lift it and get it back as you put the object back down.

**Case study: a parallel circuit**

How can we apply Kirchhoff's second law to real-world design problems? Consider the now-familiar case of a parallel circuit with one battery and two resistors, shown at right. On page 493 we learned to apply Ohm's law to each resistor separately so we could calculate each resistor's current. Intuition then guided us to add the currents to get $I$.

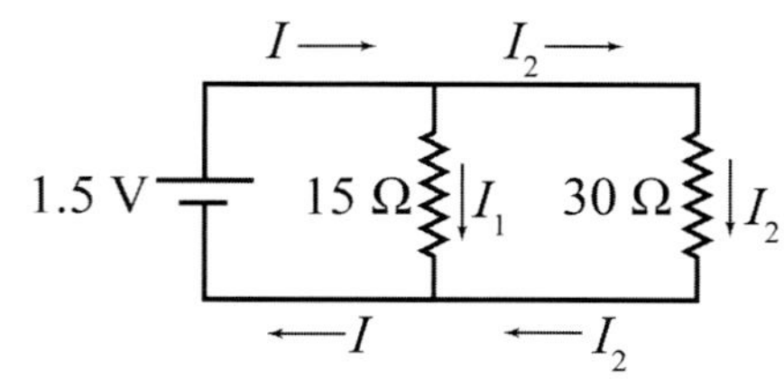


**Voltage drops around a closed loop**

Another approach is to look at the voltages across each element in a closed loop. One such loop, shown here, connects the battery to $R_1$. The battery provides a voltage increase of 1.5 V, which must equal the voltage drop in $R_1$, according to Kirchhoff's second law. By applying Ohm's law, that voltage drop must equal $I_1R_1$. Now we can solve for the current!

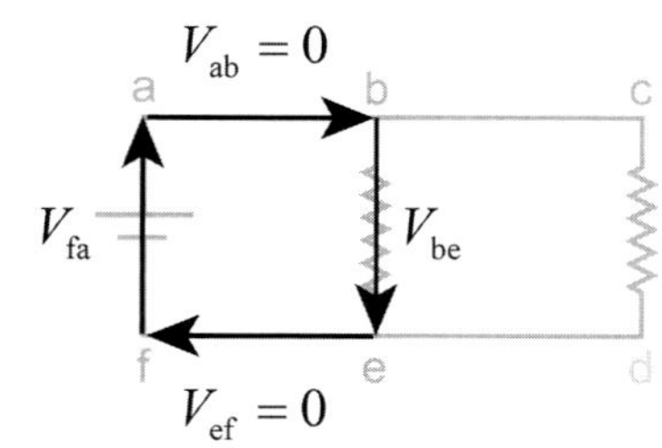


**One circuit may contain many loops**

Now take a look at the right-hand loop containing just the battery and $R_2$. Here, too, the voltage drop $I_2R_2$ must equal the battery voltage of 1.5 V, according to Kirchhoff's second law. Just as in the previous loop, the individual voltages add up to zero around *any* closed path that starts and ends at one point (such as $f$ at right).

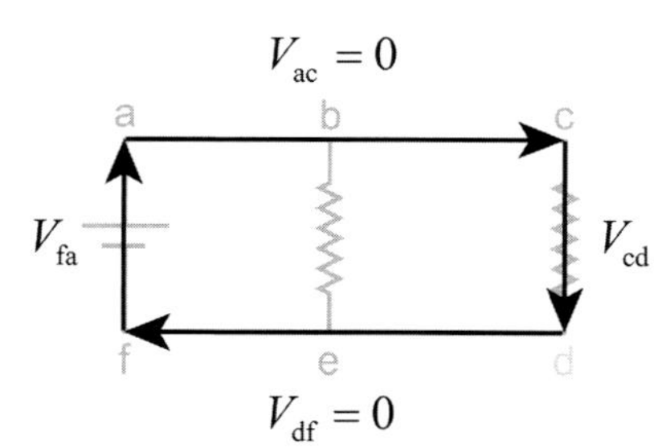


**Loops need not include batteries**

Finally, Kirchhoff's second law allows us to analyze closed loops that do not include any batteries. This enables us to quickly compare the *relative* amounts of current flowing through two resistors. In the example at right, the 15 Ω resistor will carry twice as much current as the 30 Ω one. Why? Both have the same voltage across them. Current is inversely proportional to resistance, so the resistor with half the resistance draws twice the current!

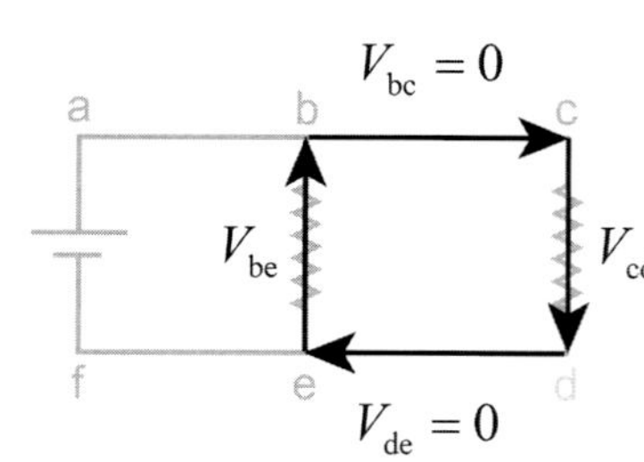

**Electrical power**

Have you ever accidentally touched a light bulb that was on for a long time? You may have burned your finger, because the bulb had heated up. Any resistive device—bulb or resistor—consumes power when a current flows through it. The power dissipated by a resistor is the product of the voltage drop $V$ across it and the current $I$ through it.

This light bulb uses 100 W of power.

Interactive equation

**Electrical power** in a circuit

$$P = IV \tag{17.5}$$

$P$ = electrical power (W)
$I$ = electric current (A)
$V$ = voltage or potential difference (V)

**Units of power**

Power and electrical power are measured in watts (W). In forces and motion, we thought of one watt as one joule of energy per second. In electricity, one watt can also be considered to be one ampere of current across a voltage drop of one volt.

**Power and the home**

In order to deliver 1,200 W of power for your microwave, the wires could deliver 1,200 A of current across 1 V or 10 A across 120 V. Since 1,200 A is a lot of current and would cause most wires to melt, home wiring systems typically use higher voltages (120 V) and carry smaller currents (10 A).

**Calculating electrical power**

In many electrical circuits, you will connect a battery or voltage source across a resistor or lamp. You know the voltage $V$ and the resistance $R$, but what is the power dissipated by the resistor? If you start with the power equation (17.5), you might realize that you don't know the current $I$. Therefore you can use Ohm's law $I = V/R$ from equation (17.1) to substitute for $I$ in the power equation:

**Electrical power** from voltage and resistance

$$P = IV = \frac{V}{R}V = V^2/R$$

When you know the current $I$ and resistance $R$, substitute for $V$ in the power equation (17.5) using the voltage drop version of Ohm's law $V = IR$ in equation (17.2):

**Electrical power** from current and resistance

$$P = IV = I(IR) = I^2R$$

**Current used by an appliance**

A 1,000 W immersion heater runs off of the 120 V wall outlet. Will the current it draws trip a 20 A circuit breaker?

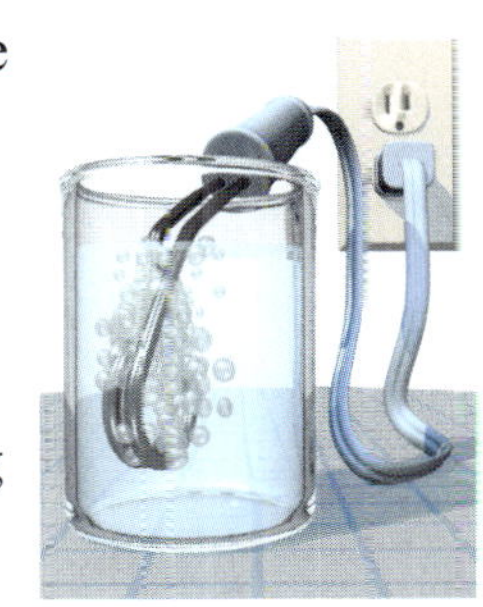

*Asked:* current $I$ drawn by heater compared to 20 A

*Given:* power $P$ = 1,000 W; voltage $V$ = 120 V

*Relationships:* $P = IV$

*Solution:* Solve the equation for the unknown current $I$ by dividing both sides by $V$:

$$P\frac{1}{V} = I\cancel{V}\frac{1}{\cancel{V}} \quad \Rightarrow \quad I = \frac{P}{V} = \frac{1{,}000\ \text{W}}{120\ \text{V}} = 8.33\ \text{A}$$

*Answer:* The current is less than the 20 A maximum for the circuit breaker. The circuit breaker will not trip.

# Investigation 17E Electrical power

**Essential questions** *How is power related to energy?*
*What determines the power rating of a device?*

Electrical devices are commonly rated in units of power, such as a 100-W light bulb or a 1200-W microwave. The electric company bills homes in units of energy. Why the difference? In this investigation, you will build circuits containing two different resistors in series and parallel (representing two different electrical devices) and measure the power and energy output of each device for each circuit.

## Part 1: Resistors in series

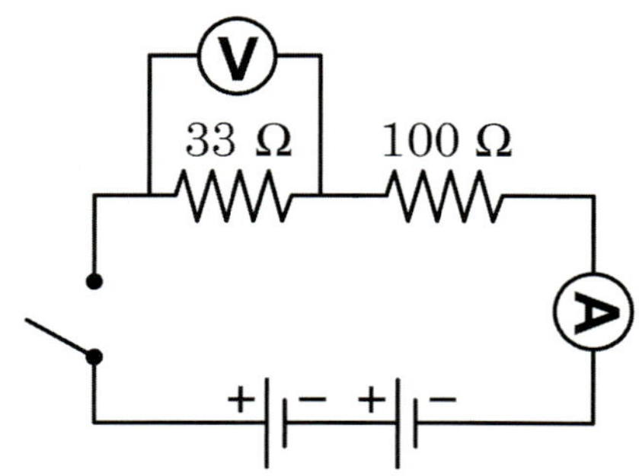


1. Open the experiment file **17E_ElectricalPower** and then power-on the Current and Voltage sensors and connect them to your software.
2. Construct the circuit shown. Connect the voltage sensor across the 33-Ω resistor.
3. Begin recording data and close the switch. Continue recording data until the energy has reached 1 joule.
4. Repeat the experiment for the 100-Ω resistor.

*a. What happens to the power output for each resistor when the switch is closed? The energy measurement?*
*b. Which resistor recorded a higher power output? Why?*
*c. Which resistor consumed 1 joule of energy the fastest? How is this related to the power output?*
*d. Fit a line to each energy versus time run. How does the value of the slope compare to the power output? What does the slope represent?*

## Part 2: Resistors in parallel

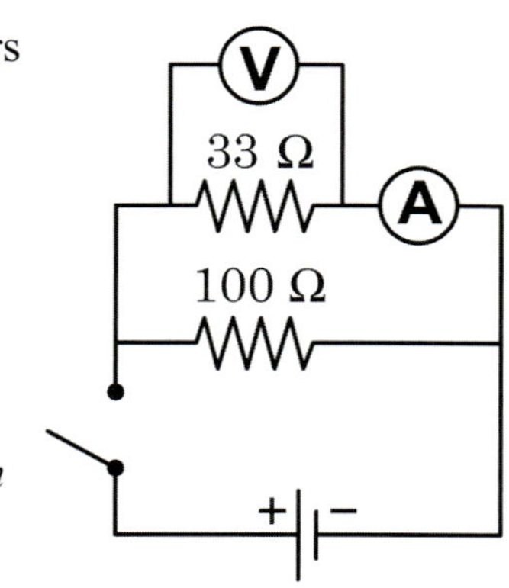


1. Construct a circuit using the same components as Part 1, but wire the resistors in parallel. Place the current sensor in series with the 33-Ω resistor and connect the voltage sensor across the 33-Ω resistor.
2. Begin recording data and close the switch. Continue recording data until the energy has reached 1 joule.
3. Repeat the experiment for the 100-Ω resistor.

*a. Does each resistor have the same power output as Part 1? Why or why not?*
*b. Which resistor has the higher power output? Is this the same as the results in Part 1? Why or why not?*
*c. Do the resistors consume energy faster in series or in parallel? Why?*
*d. Does the power output of a device depend on the properties of the device, how it is wired in the circuit, or both? Explain.*

**Channeling electrical energy**

A key circuit-design goal is to make electrical power appear *where* you want, in the *form* you need, and at the *rate* you desire. You may want three lamps lighting your bedroom as brightly as possible, or two electrical heaters warming your basement without starting a fire. Should you connect them in series or parallel? To answer this, we need to understand power in series and parallel circuits.

**Resistors in series**

In a series circuit, which element draws the most power? In the example at right, resistor B has twice the resistance of resistor A and both are connected to a 6 V battery. For each resistor, the power is $P = IV$. The voltage drop across each element may vary, so we eliminate $V$ by using Ohm's law. Substituting $IR$ for $V$ in the power equation results in $P = I^2R$. Since the current across each resistor is the same in a series circuit, the power is therefore proportional to the resistance—so the highest resistance element consumes the most power, 1.6 W versus 0.8 W!

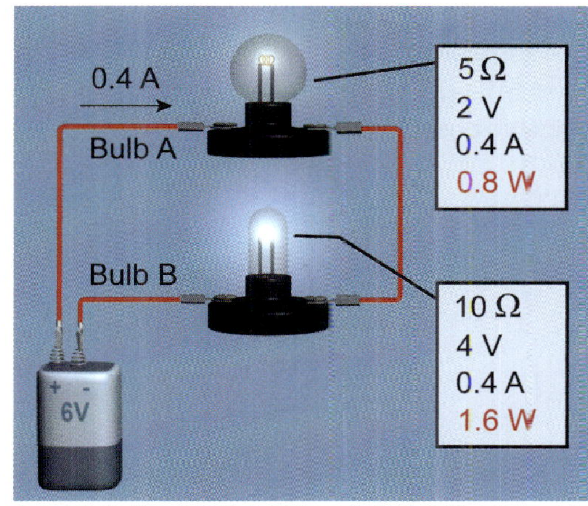


To understand this property, think of a stream of traffic approaching a narrow bridge. The traffic loses the most energy where it has to fight hardest to get through!

**Resistors in parallel**

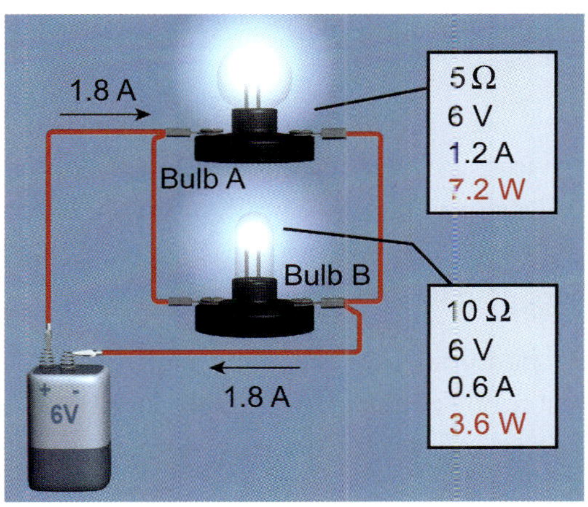


Things change dramatically if we wire the resistors in parallel. Now each is exposed to the full voltage provided by the battery. This means that the voltage drop $V$ is the same for each resistor, while current $I$ may differ from one to the next. We eliminate $I$ from $P = IV$ by inserting $I = V/R$. This yields $P = V^2/R$. In a parallel circuit, the power consumed by a resistor is *inversely proportional* to the resistance, so the smaller resistance (resistor A) operates at twice the power of the other! *Furthermore, even resistor B has a greater power output than it did in series!* Now the power dissipated by the 5 Ω resistor increases to 7.2 W and the power dissipated by the 10 Ω resistor is 3.6 W, for a total of 10.8 W (compared to only 2.4 W for the series circuit).

**Higher power in parallel circuits**

The same resistors will always produce more power when connected in parallel than when connected in series. The effect is similar to that of opening another bridge across a river that many drivers are eager to cross—many more vehicles can cross the river each minute.

**Practical applications**

Series circuits have several advantages. For example, putting a fuse or circuit breaker in series with an electrical heater ensures that the heater will be cut off if it draws an unsafe amount of current. In contrast, the heater and your computer should be in parallel, so that if one is turned off the other can stay on.

**Power in electrical circuits**

Interactive equation

You probably use at least 25 different electrical devices before eating lunch every day. Each converts electrical power into other forms of energy, such as heat, light, and motion. On page 519, we used the equation for electrical power ($P = IV$) as a practical definition of the volt: One volt means one watt per ampere. A current of one ampere flowing across a potential difference of one volt dissipates one watt of power.

**How much current do appliances use?**

Electrical devices are designed to draw the correct amount of current to transfer the power they need to operate when connected to the proper voltage. The power equation can be used to calculate the current drawn by a given appliance from its power requirement and the operating voltage. For most household appliances the operating voltage is 120 V. The current can therefore be found by dividing the power by the voltage.

**Calculate current.**

$$I = \frac{P}{V}$$

Power outlet
$V = 120$ V

**Home wiring uses parallel circuits.**

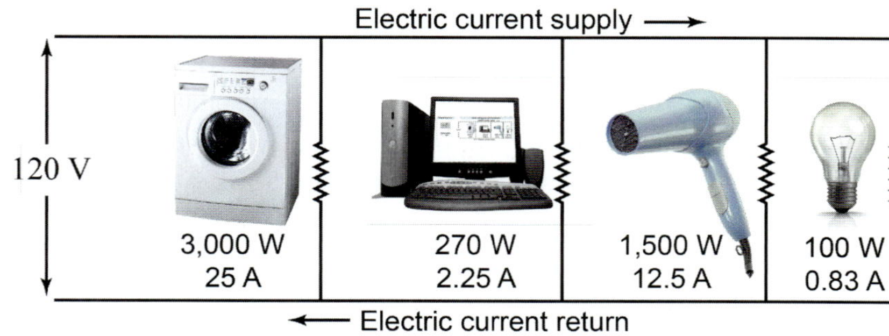


**What kind of circuits are in houses?**

Household outlets are typically wired in parallel circuits. Each outlet provides the full voltage from the power line regardless of what other appliances are connected because each outlet is on its own branch of the circuit. This means that some outlets can have appliances plugged in and running while others do not.

**Lights and light switches**

Lights are typically also wired in parallel. That way you can turn off lights in one room while leaving the lights in another room unaffected. A light *switch*, however, is placed *in series* with the lights on that branch (i.e., within a room) because you want the switch to turn off all the lights connected to it. The light switch changes the circuit for that room from open to closed.

**Energy consumption measured in kWh**

Electric power companies use the **kilowatt-hour (kWh)** to measure and collect payment for the total electrical energy you use. One kWh is 1,000 W (1 kW) used for 1 hr, or 3,600 s. A kilowatt-hour is a measure of *energy*, not power. Why? Remember that $P = E/t$, so $E = Pt$. A 100 W bulb, used for 10 hr continuously, will use 100 W × 10 hr = 1.0 kWh of energy.

**Circuit breakers**

What happens if you run a hair dryer, toaster oven, and microwave from the same parallel circuit? Each appliance draws about 10 A of current. The 12-gauge wire used to connect kitchen outlets can safely carry only 20 A before overheating. To protect against fire the **circuit breaker** would trip and disconnect the entire parallel circuit. When a circuit breaker trips you must turn off and disconnect some of the appliances overloading the circuit before resetting the circuit breaker.

**More complicated circuits**

How do you determine the equivalent resistance for circuits that have more than two resistances? The answer is that you break the circuit apart into individual series or parallel circuits and solve them individually. Then combine larger and larger elements of the circuit as series and parallel circuits.

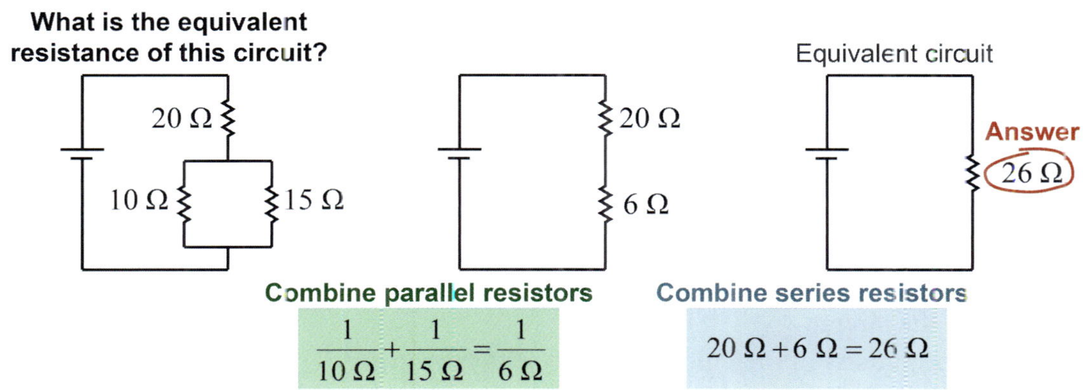


**An example**

How do you calculate the equivalent resistance of the three-resistor circuit above? Start by looking for a small portion of the circuit that, by itself, looks like a series or parallel circuit. In this example, the 10 Ω and 15 Ω resistors are in parallel, so we use equation (17.4) to determine that their equivalent resistance is 6 Ω. We replace the 10 Ω and 15 Ω resistors with a single 6 Ω resistor and look at the circuit again. Now we see that this 6 Ω resistor is in series with the 20 Ω resistor, so we use equation (17.3) to determine that their equivalent resistance is 26 Ω. Since we have now reduced the three-resistor circuit to a single, equivalent resistor, this is the answer.

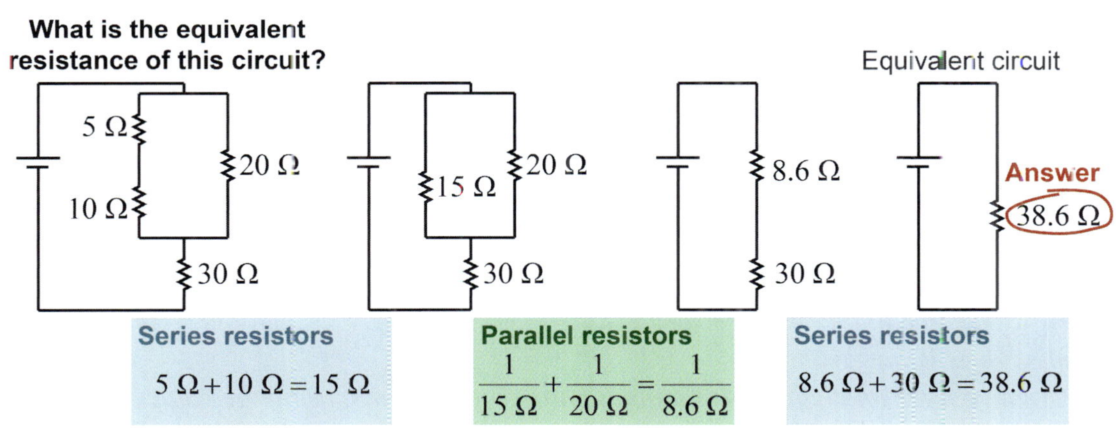


**A more complicated example**

The circuit above contains four resistors but we still use the same method for finding its equivalent resistance. In this case, the 5 Ω and 10 Ω resistors are in series so they add to 15 Ω using equation (17.3). This 15 Ω resistance is in parallel with the 20 Ω resistor, so we use equation (17.4) to calculate their equivalent resistance of 8.6 Ω. Finally, this 8.6 Ω resistance is in series with the 30 Ω resistor, which combine to make 38.6 Ω. It may take a few steps to calculate the equivalent resistance of 38.6 Ω, but all we needed were the equations for series and parallel resistance!

# Investigation 17F Compound circuits

**Essential questions** *How can voltage, current, and power be used to predict the behavior of electric circuits?*

In the first part of this investigation, you will predict the brightness of bulbs in a circuit. Bulb brightness is a measure of the dissipation of power. In the second part, you will explore the properties of a commonly used kind of circuit called a voltage divider.

## Part 1: Power dissipated by elements in a compound circuit

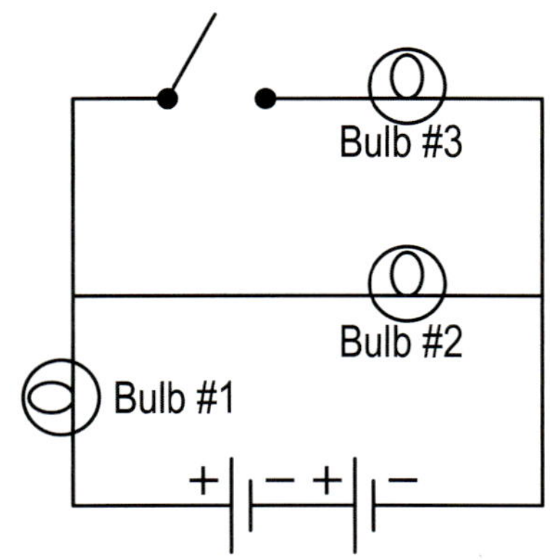


1. Open the experiment file **17F_CompoundCircuits**, and then power-on the Current and Voltage sensors and connect them to your software.
2. For the circuit at right, predict the relative brightness of the bulbs when the switch is closed and when it is open.
3. Construct the circuit.
4. Note the brightness of the bulbs when the switch is open.
5. Close the switch and note how each bulb's brightness changes.

a. *Using Kirchhoff's laws or an equation for power, justify your prediction for the relative brightness of the bulbs when the switch is closed.*
b. *Did the brightness of bulb #1 change when the switch was closed? Provide an explanation.*
c. *Add a voltage and current sensor to your circuit. Measure the voltage across and current through bulb #1 with the switch open and closed then calculate the power. How do the measurements justify your explanation?*
d. *Did the brightness of bulb #2 change when the switch was closed? Provide an explanation.*
e. *Measure the voltage across and current through bulb #2 with the switch open and closed. Calculate the power. How do the measurements justify your explanation?*

## Part 2: Voltage divider

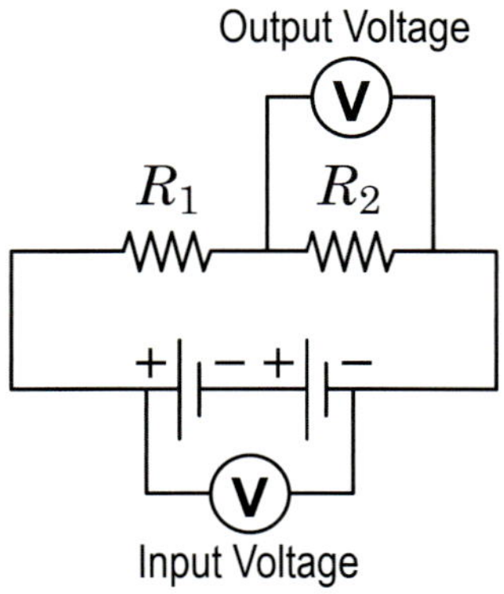


1. Construct the circuit shown using the resistor values in the table.
2. Measure the input voltage and the output voltage for each resistor combination in the table.
3. Record your measurements in the table.

a. *How does the output voltage vary with $R_1$?*
b. *How does it vary with $R_2$?*
c. *Create a mathematical model to predict the output voltage using the values of the input voltage and the two resistances.*
d. *Why is this circuit called a voltage divider?*

| Resistor $R_1$ | Resistor $R_2$ | Input voltage (V) | Output voltage (V) |
|---|---|---|---|
| 100 Ω | 10 Ω | | |
| 33 Ω | 10 Ω | | |
| 10 Ω | 10 Ω | | |
| 10 Ω | 33 Ω | | |
| 10 Ω | 100 Ω | | |

**History of DC and AC electricity**

The letters "DC" stand for *direct current,* the type of electricity produced by batteries. DC electricity was investigated by Italian physicists Luigi Galvani and Alessandro Volta around 1800. All of the circuits discussed so far in this chapter, and which you have built in the lab, have been DC circuits. Thomas Edison created the first electric light bulbs to operate on DC current.

**AC voltage**

The electricity in your home is *alternating current* or AC. With AC electricity the *polarity,* meaning which wire is positive, changes many times per second. Imagine a battery on a rotating wheel with contacts that alternately connected the wires to opposite poles of the battery. The voltage on one wire would alternate back and forth between positive and negative.

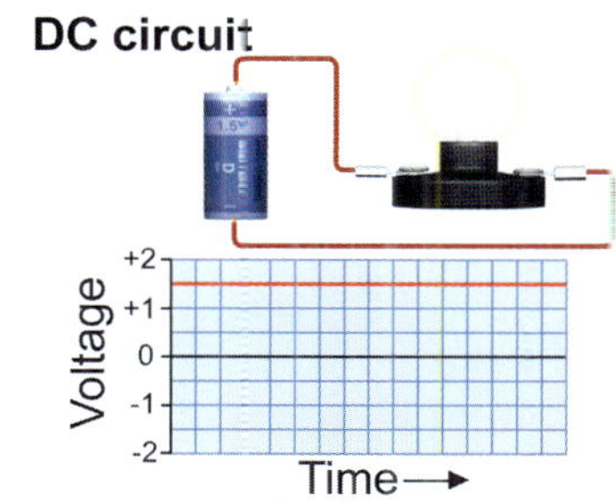


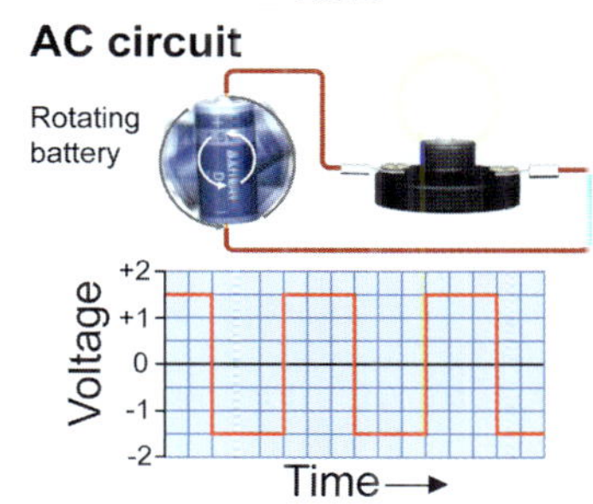


**AC current**

If the voltage alternates, so does the current. When the voltage is positive the current in the circuit is clockwise (see the graph). When the voltage is negative the current flows in the opposite direction. In actual AC circuits, the *hot* wire carries an alternating voltage that alternates from positive to negative. The other wire is held at a constant zero volts.

**Advantages of AC**

AC electricity is used for high-power applications because it is easier to generate and transmit over long distances. In Chapter 19 you will see that electric generators naturally make AC power. All the power lines you see overhead carry AC current. The plugs in the walls of your apartment or house or classroom also carry AC.

**What is the voltage in a wall receptacle?**

The 120 volt AC (VAC) electricity used in homes and businesses alternates between peak values of +170 V and −170 V. This kind of electricity is called 120 VAC because +120 V is the average positive voltage and −120 V is the average negative voltage. AC electricity is usually identified by the average voltage, not the peak voltage.

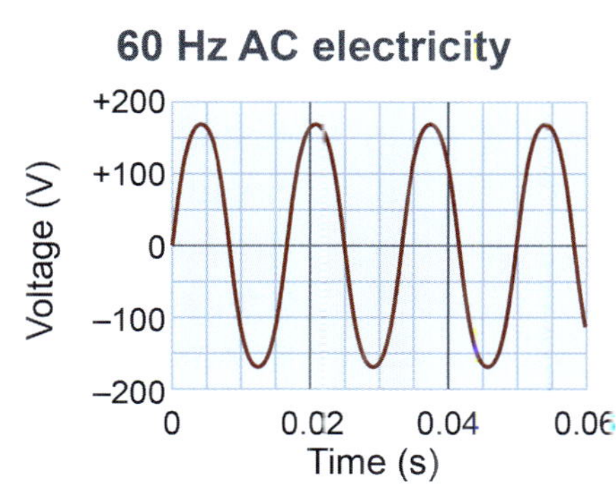


**Frequency is an important characteristic of AC power**

The frequency of AC electricity is an important characteristic. In the USA, AC power alternates at a frequency of 60 Hz. This means the voltage on the same wire switches back and forth between positive and negative 60 times each second. European and some South American countries operate their power grids at 50 Hz and 220 V. The higher voltage and different frequency mean electrical devices designed for 120 V, 60 Hz power do not work, or are destroyed, by lower frequency, higher voltage European power.

**Wires and circuit breakers**

The 120 VAC electricity comes into a typical home or building through a circuit breaker panel. The circuit breakers protect against wires overheating and causing fires. A 20 A breaker opens the circuit if the current exceeds 20 A for more than about a second. A typical circuit breaker panel has 15 A, 20 A, and 30 A breakers because the wires in a house are different sizes to safely carry different amounts of current.

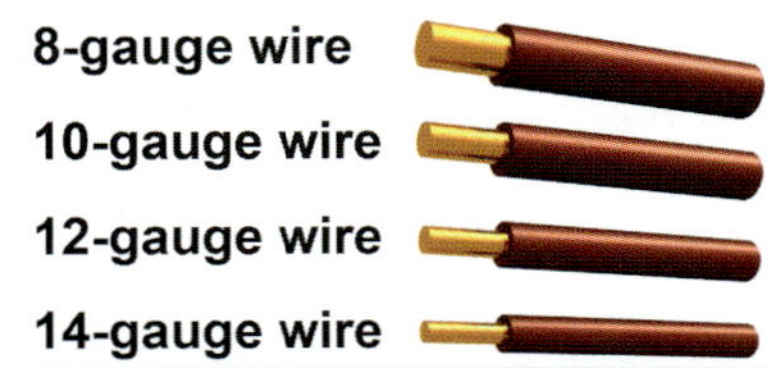


| Wire gauge | Current (A) |
|---|---|
| 8 | 50 |
| 10 | 30 |
| 12 | 20 |
| 14 | 15 |

**Different types of circuits**

Electrical appliances such as a saw or hair dryer may draw up to 20 A of current. Outlet circuits are typically wired with 12-gauge wire that can carry 20 A and they are connected through a 20 A circuit breaker. Light bulbs use less than 1 A each. Light circuits are typically wired with 14-gauge wire that can carry 15 A and they are connected through a 15 A circuit breaker. High-power devices such as an electric stove (50 A) or electric dryer (30 A) use special shaped plugs connected with 8- or 10-gauge wire.

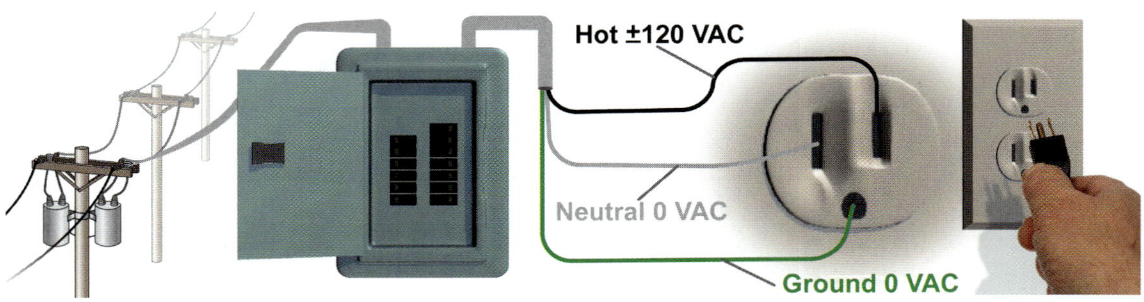


**Typical circuits**

Electric power in homes and buildings is distributed through parallel circuits. The diagram shows a typical outlet circuit. Each outlet, or receptacle, has three wires: hot (black), neutral (white), and ground (green). The hot wire is connected to 120 VAC through the circuit breaker. The neutral wire is connected to zero volts. When you plug something in, current flows in and out of the hot wire, through your appliance (doing work) and back through the neutral wire. The ground wire is for safety and is connected to the ground (0 V) near your house. If there is a short circuit in your appliance, the current flows through the ground wire rather than through you.

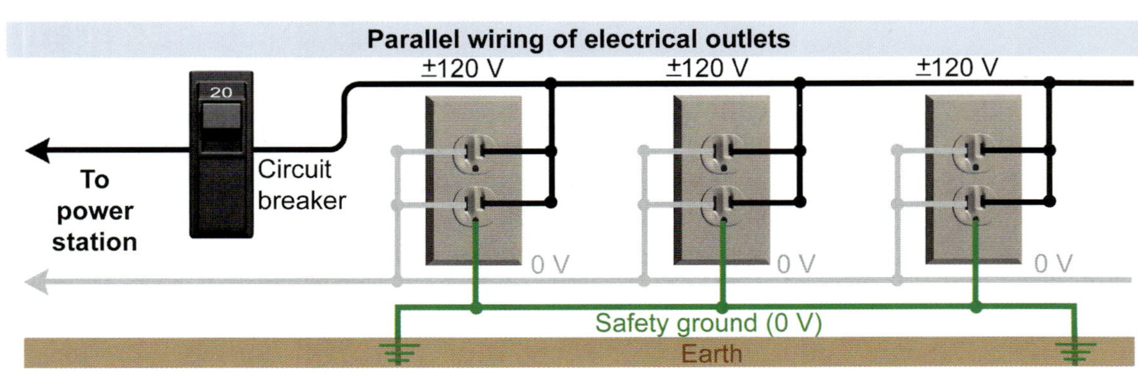

# Section 3 review

Most electrical circuits have more than one resistive component: Homes have many appliances, cars have two headlights, and so on. Resistors can be connected in series or in parallel. In a series circuit, each resistor carries the same amount of current but gets only a fraction of the total battery voltage. In a parallel circuit, each resistor "feels" the battery's entire voltage but only carries a fraction of the total current. A parallel circuit will exhibit higher wattage (power) than a series circuit built from the same resistors. The total current in a series or parallel circuit can be calculated with Ohm's law by using the equivalent resistance for that circuit.

| | |
|---|---|
| **Vocabulary words** | **series circuit, equivalent resistance, parallel circuit, Kirchhoff's current law, Kirchhoff's voltage law, circuit breaker, kilowatt-hour (kWh)** |

| | | | | |
|---|---|---|---|---|
| **Key equations** | $V = IR$ | $R = R_1 + R_2 + R_3 + \cdots$ | $\frac{1}{R} = \frac{1}{R_1} + \frac{1}{R_2} + \frac{1}{R_3} + \cdots$ | $P = IV$ |

## Review problems and questions

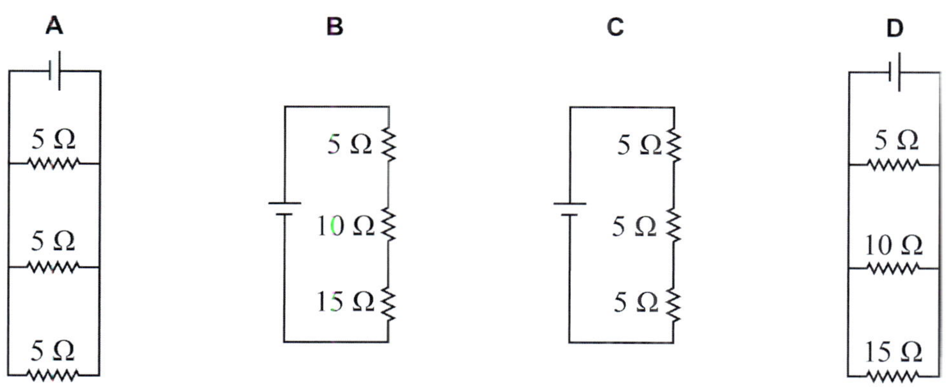


The following five questions pertain to the four circuits (A, B, C, and D) shown above.

1. In which circuit do the three resistors have equal voltage drops but carry different current?

2. In which circuit do the three resistors carry equal current but have different voltage drops?

3. Which of the circuits has the *lowest* equivalent resistance?

4. Which of the circuits has the *highest* equivalent resistance?

5. Assume that Circuits A, B, C, and D have identical batteries and convert all of their power to thermal energy. Which of the four circuits will produce the most heat?

6. Compile a list of the major appliances in your home and the power that each consumes.
   a. How much total power does your list of appliances use if they are all running at once?
   b. How much total current do they use?
   c. Which appliances use the most power?

# Chapter review

## Vocabulary

Match each word to the sentence where it best fits.

### Section 17.1

| | |
|---|---|
| **ampere (A)** | **battery** |
| **closed circuit** | **electric circuit** |
| **electric current** | **electrical symbol** |
| **electricity** | **open circuit** |
| **potential difference** | **short circuit** |
| **volt (V)** | **voltage** |
| **voltmeter** | |

1. Another term for voltage is _______.
2. The "1.5 V" label on an alkaline battery indicates its _______.
3. The voltage across a circuit component can be measured using an instrument called a/an _______.
4. When electricity can flow through a continuous loop it is called a/an _______.
5. When a break in a wire prevents the flow of electricity, it is called a/an _______.
6. The unit of measure for voltage is the _______.
7. Electric current is measured using the unit of _______.
8. A/An _______ converts chemical energy into electrical energy.
9. The electronics in a mobile phone is an example of a/an _______.
10. A drawing of an electrical circuit may contain one or more _______ to show how various circuit components are connected together.
11. If a wire is connected to bypass elements of a circuit, this is called a/an _______.
12. _______ is the flow of electric charges through a conductor.

### Section 17.2

| | |
|---|---|
| **Ohm's law** | **digital multimeter** |
| **electrical conductor** | **electrical insulator** |
| **ohm (Ω)** | **resistance** |
| **resistor** | |

13. The unit of electrical resistance is the _______.
14. Any electrical device that resists the flow of electric current is a/an _______.
15. A copper wire is a good example of a/an _______.
16. Wood and glass are each an example of a/an _______.
17. An electrical insulator has a very high _______ against the flow of electric current.
18. The _______ has different settings to measure voltage, current, and resistance.
19. A statement of _______ is that the current through a resistor is equal to the ratio of the voltage across it and its resistance.

### Section 17.3

| | |
|---|---|
| **Kirchhoff's current law** | **Kirchhoff's voltage law** |
| **circuit breaker** | **equivalent resistance** |
| **kilowatt-hour (kWh)** | **parallel circuit** |
| **series circuit** | |

20. When two resistors each have one end connected to the negative terminal of a battery and the other end connected to the positive terminal, then the resistors are connected in a/an _______.
21. In an electric circuit, you can draw many closed loops and analyze the voltage around each loop by applying _______.
22. If you connected too many electrical appliances to one outlet, you might trip the _______ in your house.
23. The power company typically charges by calculating electricity usage using units of the _______.
24. _______ states that all the current entering a circuit junction must equal the total current leaving that junction.
25. You can calculate the _______ using the equations for adding resistances in series or parallel.
26. When two resistors are connected such that the same current must flow through one resistor then the other, they are connected in a/an _______.

# Chapter review

## Conceptual questions

### Section 17.1

27. What conditions are required for electric current to flow?

28. Two tanks are each half full of water. What condition is necessary for the water to flow from one to the other? What is the analogous necessary condition for electric current in a conductor?

29. What flows through a circuit?

30. Imagine you have only a wire. Do electric charges flow through it? Why or why not?

31. Is voltage or current measured *across* a circuit?

32. Yuri was trying to use his 5 V power supply (a voltage source) for a light bulb circuit. He connected the red plug to his bulb, but it did not light. How was he using the power supply incorrectly?

33. Does voltage or charge *flow through* a circuit?

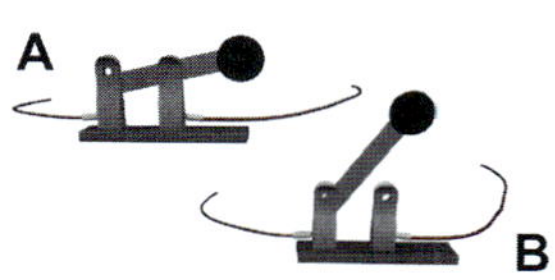


34. Which of these knife blade switches creates an electrical connection across it?

35. Why should you not insert *both* leads of a lamp into the same row of a circuit breadboard?

36. If you need a variable voltage, is it better to use a battery or a power supply?

37. When you are using copper wire to prototype a circuit, why is it better to use plastic-coated wire than bare wire?

38. Raymundo tried to illuminate a mini-lamp by touching its two leads to the positive terminal at the top of a battery. Did he succeed?

39. Describe how the energy of water behind a dam is analogous to the energy in a battery.

40. You are given three 1.5 V batteries, one resistor, and wires to connect them. Draw a circuit diagram to show how you can create a voltage of 4.5 V across the resistor.

41. Why do birds sitting on high-voltage power lines not get electrocuted?

### Section 17.2

42. What is the purpose of a wire cutter tool?

43. If the resistance of a resistor is unchanged but the current through it doubles, how does the voltage across the resistor change?

44. Which materials listed below are good conductors?

    a. aluminum
    b. rubber
    c. copper
    d. gold
    e. diamond

45. How do you connect the data acquisition probes for a digital multimeter to measure the voltage in a circuit?

46. Which of the following circuit elements typically has a measurable resistance: a wire, a switch, a battery, a lamp, and/or a resistor?

47. To measure the voltage across a resistor, do you connect the voltmeter in series or parallel with the resistor?

48. If the resistance of a resistor is doubled while the voltage across it is held constant, how does the current through the resistor change?

49. How are voltage, current, and resistance related to each other?

50. How would you characterize the resistance of an open circuit? How would you characterize the resistance of a short circuit?

51. To measure the current through a resistor, do you connect the ammeter in series or parallel with the resistor?

52. Describe how water flowing downward through the open neck of a bottle of water is analogous to the resistance of a resistor in an electric circuit.

53. Why are electrical wires usually covered by a layer of plastic?

54. Why shouldn't you use an electrical device while in the shower or bathtub?

55. If the resistance of a resistor is halved and the voltage across it is also halved, how does the current through the resistor change?

56. Should the part of a light switch that you touch be a good conductor or insulator? Why?

57. Air is not a good conductor of electricity. How might our lives be different if air were a good conductor?

# Chapter review

## Section 17.2

58. Your instructor makes the following statement:

"A short circuit is a wire or other conducting path that can have a large current when connected directly from the positive to the negative terminal of a battery. The current is large because the battery voltage is applied to a circuit with almost zero resistance."

Which of the following best summarizes the instructor's statement?

a. A wire or conducting path can have a large current.
b. A battery has essentially zero resistance.
c. A short circuit has essentially zero resistance and can result in large currents.
d. A short circuit is a circuit that connects the positive and negative terminals of a battery.

## Section 17.3

59. ❮ Do electric utility companies charge their customers using units of *energy* or *power*? Why not charge using the other quantity?

60. ❮ Ned turned on his home stereo, computer, washer, electric dryer, and the lights in every room. The circuit breaker then tripped. Is there anything he should do before flipping the circuit breaker switch?

61. ❮ How does a fuse differ from a circuit breaker?

62. ❮ How are electric current, voltage, and power related to each other?

63. ❮ When connecting a mini-lamp and socket to a circuit breadboard, why *must* you connect the leads to different rows of the circuit breadboard?

64. ❮ Four resistors are connected in series. The individual resistances are 10 Ω, 20 Ω, 99 Ω, and 23 Ω. If the current through the first resistor is 0.1 A, what is the current through each of the other three resistors?

65. ❮ Draw a circuit diagram with a battery, resistor, and lamp, where the lamp and resistor are connected in series.

66. ❮ Draw a circuit diagram with a battery, resistor, and two lamps, where the lamps and resistor are connected in parallel.

67. ❮❮ If two resistors are connected in series, is their effective resistance larger or smaller than either of their individual resistances? Why?

68. Design a circuit with a battery and two resistors all connected in series.

69. Design a circuit with two resistors connected in parallel with each other and both connected to a battery.

70. ❮❮ A sales person in a store tells you that all light bulbs labeled 100 W will radiate 100 W of power when connected to a 120 V outlet. As evidence to support his claim, he plugs two different 100 W bulbs into a socket and points out that they are equally bright.

a. In your analysis, did the scientific evidence support the sales person's claim?
b. If the sales person's claim were true, what should happen if you connected two 100 W bulbs *in series* to the power outlet?
c. Based on what you know about circuits, what will *actually* happen to the brightness of two 100 W bulbs connected in series with the power outlet? Would this support or contradict the sales person's assertion?
d. Based on what you know about circuits, propose an alternate hypothesis for how much power a 100 W bulb will radiate.

71. ❮❮ If two resistors are connected in parallel, is their effective resistance larger or smaller than either of their individual resistances? Why?

72. ❮❮ Do three lamps shine more brightly if they are connected in series or in parallel? Defend your answer.

73. ❮❮ Lamps are rated in terms of maximum power output based on the amount of heat they can withstand. If you connect a bulb to a lamp that draws more power than the lamp's rating, you can start a fire. Based on this information, are bulbs with higher resistive elements safer or more dangerous?

74. ❮❮ Estrella had one resistor connected to a battery, resulting in a current of 0.01 A. She wanted more resistance in her circuit to reduce the total current. So she took two identical resistors and connected their ends as shown above. Will she succeed?

75. ❮❮ Disassemble a flashlight that uses more than one battery, but without damaging its components or operation. Inspect the wires inside. How are the batteries connected together, in series or in parallel?

76. ❮❮❮ There are two resistors in a simple circuit, $R_1$ and $R_2$. Resistor $R_1$ has a voltage drop across it that is 1/3 the voltage drop across $R_2$. Which element has more current flowing through it?

# Chapter review

## Section 17.3

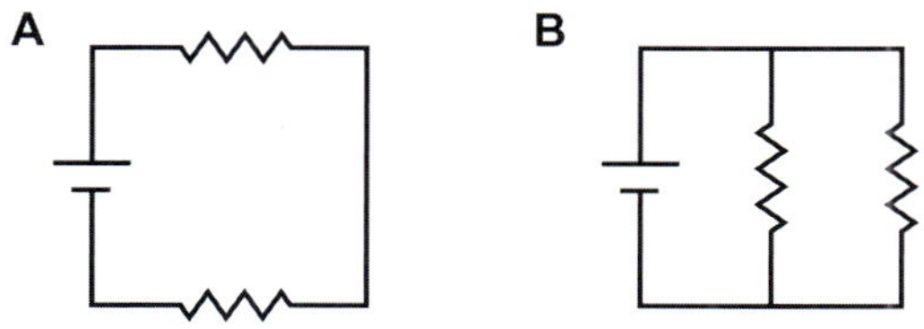


77. Consider the two circuit diagrams.
    a. Which illustration corresponds to a correctly constructed circuit of two resistors connected in series?
    b. Which illustration corresponds to a correctly constructed circuit of two resistors connected in parallel?

78. Why is it a terrible idea to try to connect a mini-lamp bulb to a wall socket?

79. ❨ Use the parts in the illustration above to draw connections for the following two circuits.
    a. Connect the two batteries in series with each other. Connect the two resistors in parallel with each other. Connect the two resistors together across the maximum voltage from the batteries.
    b. Connect the two batteries in parallel with each other. Connect the two resistors in series with each other. Connect the two resistors together across the maximum voltage from the batteries.
    c. What is the potential difference across each circuit?
    d. By interpreting the color bands, what is the resistance of each resistor? (*Hint:* The color bands read brown-black-red-gold.)
    e. What is the value of the effective resistance for each circuit?

## Quantitative problems

## Section 17.1

80. Three batteries, each having a voltage of 2.0 V, are connected in series. What is the total voltage across them?

81. ❨ Three batteries, each having a voltage of 2.0 V, are connected in parallel. What is the total voltage across them?

## Section 17.2

82. ❨ What is the voltage drop across a 50 Ω resistor when a current of 0.1 A flows through it?

83. ❨ What is the current through a 10 Ω resistor when a voltage of 5 V is applied across it?

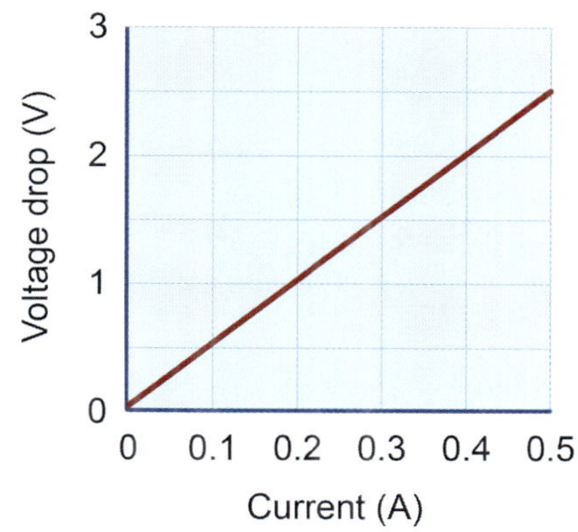


84. ❨❨ A resistor is subject to a variable voltage drop $V$, and it conducts varying amounts of current $I$ as a result. A student measures these voltages and currents and then graphs the measurements, as shown above.
    a. Calculate the slope of the plotted line by using the two endpoints. What is the slope's numerical value, and what is its unit?
    b. Does this resistor obey Ohm's law? Justify your response.
    c. Which of the quantities in Ohm's law corresponds to the slope of the plotted line?

85. ❨❨ A light bulb produces a light intensity according to the equation

$$Int = 3.5\sqrt{V}$$

where $V$ is the voltage applied to the bulb.
The resistance $R$ of the filament in the light bulb is a function of the intensity $Int$ of light and is related to it by the formula

$$R = 5.0 + 2.35(Int)^{1/3}\ \Omega$$

If the optimum intensity of the bulb is 12.5 lumens, what are the operating current and voltage?

Chapter 17

# Chapter review

## Section 17.3

86. Three resistors, each having a resistance of 25 Ω, are connected in series. What is their effective resistance?

87. A hair dryer and a curling iron have resistances of 15 Ω and 25 Ω, respectively, and are connected in series. They are connected to a 60 V battery. Calculate the

a. current through the circuit.
b. power used by the hair dryer.
c. power used by the curling iron.

88. A 100 W light bulb is plugged into a 120 V socket.

a. How much current is drawn by the light bulb?
b. Using your value for the current, determine the value of the bulb's resistance.

89. Three resistors, each having a resistance of 30 Ω, are connected in parallel with each other. What is the value of their effective resistance?

90. A string of 50 identical tree lights connected in series dissipates 100 W when connected to a 120 V power outlet.

a. What is the equivalent resistance of the string?
b. What is the resistance of each individual light?
c. How much power is dissipated by each light?

91. Suppose that you are experimenting with a 15 V source and two resistors: $R_1 = 2500\ \Omega$ and $R_2 = 25\ \Omega$. Find the current for a, b, c, and d below. What do you notice?

a. $R_1$ in series with $R_2$
b. $R_1$ in a circuit alone
c. $R_1$ in parallel with $R_2$
d. $R_2$ in a circuit alone

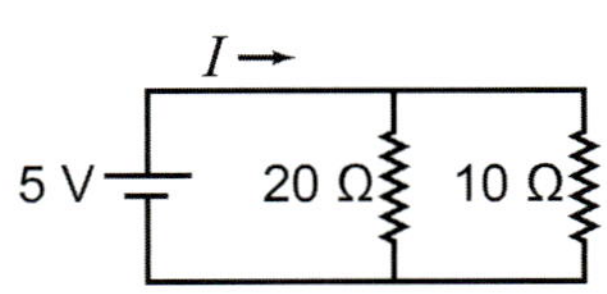


92. Consider the circuit shown.

a. Calculate the effective resistance.
b. Calculate the current $I$.

93. A 20 Ω lamp and a 30 Ω lamp are connected in series with a 10 V battery. Calculate the following:

a. the equivalent resistance
b. the current through the circuit
c. the voltage drop across the 20 Ω lamp
d. the voltage drop across the 30 Ω lamp
e. the power dissipated by the 20 Ω lamp
f. the power dissipated by the 30 Ω lamp

94. A hair dryer and a curling iron have resistances of 15 Ω and 25 Ω, respectively, and are connected in *parallel*. They are connected to a 60 V battery.

a. Calculate the current through the circuit.
b. Calculate the power used by the hair dryer.
c. Calculate the power used by the curling iron.

95. A 20 Ω lamp and a 30 Ω lamp are connected in parallel with a 10 V battery. Calculate the following:

a. the equivalent resistance
b. the current through the circuit
c. the voltage drop across the 20 Ω lamp
d. the voltage drop across the 30 Ω lamp
e. the power dissipated by the 20 Ω lamp
f. the power dissipated by the 30 Ω lamp

96. A 1,000 W immersion heater, a 700 W microwave oven, and a 1,200 W toaster oven are all plugged into a single 120 V outlet. The outlet is connected to a 15 A circuit breaker. Will the circuit breaker trip?

97. Use Ohm's law and the equation $P = IV$ to find the power of a simple circuit with a 5 V voltage source and a 2 kΩ resistor.

98. You have a 30 V source and two resistors: $R_1 = 10\ \Omega$ and $R_2 = 25\ \Omega$. Find the current in a circuit with the resistors connected in parallel. What is the current through each resistor if they are connected to the power supply separately? Do you notice anything?

99. A voltage divider is a series circuit that is used to produce a specific voltage across one of its resistors. Find a symbolic expression for the voltage $V_2$ across resistor $R_2$ using circuit elements $V$, $R_1$, and $R_2$.

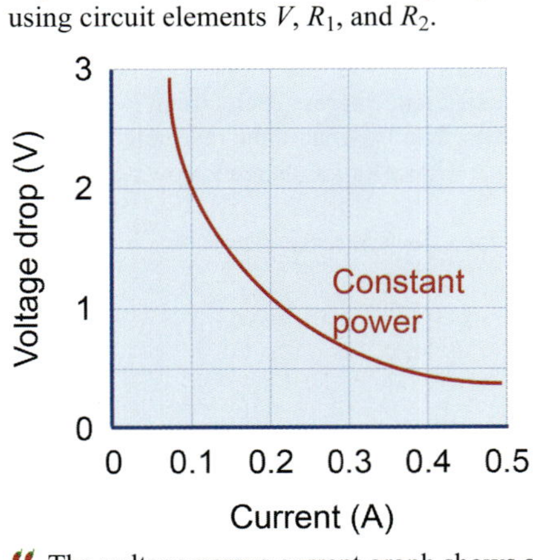


100. The voltage versus current graph shows a constant power curve of a resistor $R = 20\ \Omega$.

a. What is the operating voltage?
b. What is the operating current?

# Chapter review

## Standardized test practice

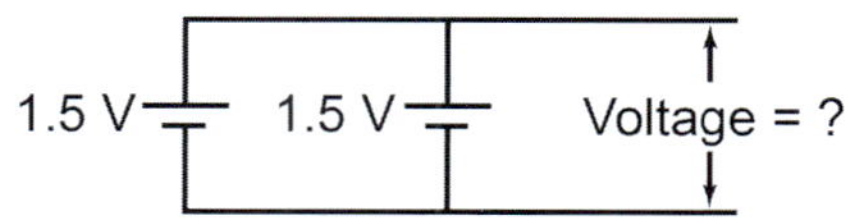


101. Two batteries are connected together as shown in the figure. What is the value of the voltage across them?

   A. 0 V
   B. 0.75 V
   C. 1.5 V
   D. 3 V

102. Which statement or statements below provide a correct practical definition of a voltage source?

   A. A voltage source creates an electric potential difference between two points in a circuit, such as the two ends of a battery.
   B. A voltage source provides the electrical potential energy needed for a circuit to operate.
   C. A voltage source is analogous to the pump in a system of circulating water.
   D. All of the above

103. Which of the following is not a good electrical insulator?

   A. wood
   B. air
   C. water
   D. plastic

104. What is the correct relationship among current, resistance, and voltage?

   A. Current is directly proportional to resistance and voltage.
   B. Current is directly proportional to resistance and inversely proportional to voltage.
   C. Current is inversely proportional to resistance and directly proportional to voltage.
   D. Current is inversely proportional to both resistance and voltage.

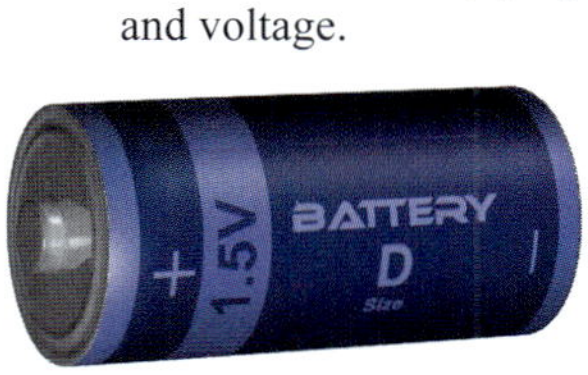


105. What does the label "1.5 V" on the battery mean?

   A. The voltage of the positive terminal is 1.5 V.
   B. The voltage of the negative terminal is 0 V.
   C. The potential difference between the two terminals is 1.5 V.
   D. All of the above

106. A circuit consists of a battery plus two resistors connected in parallel with each other. How do you connect the data acquisition probes for a digital multimeter to measure the current through one of the resistors?

   A. in parallel across the resistor
   B. in series with the resistor
   C. in parallel across the battery
   D. in series with the battery

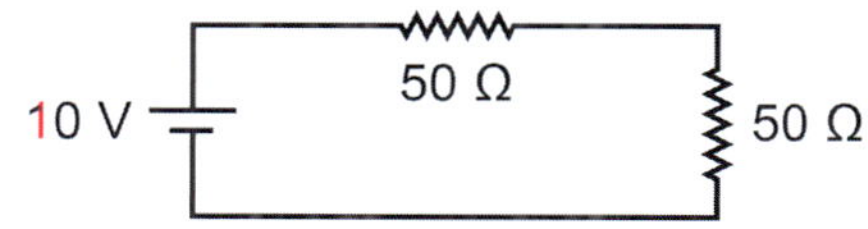


107. What is the current through the circuit shown?

   A. 0.1 A
   B. 0.4 A
   C. 1 A
   D. 4 A

108. A mobile phone charger draws 0.2 A of current from the 120 V wall outlet. How much energy does it use if it were plugged in for 24 hours?

   A. 0.024 kWh
   B. 0.58 kWh
   C. 35 kWh
   D. 2,070 kWh

109. A 50 Ω lamp is connected to a 120 V electrical outlet. How much power is consumed by the lamp?

   A. 2.4 W
   B. 20.8 W
   C. 288 W
   D. 6,000 W

110. Four 1 Ω resistors are connected in parallel across a 10 V battery. What is the equivalent resistance of the circuit?

   A. 0.1 Ω
   B. 0.25 Ω
   C. 4 Ω
   D. 10 Ω

111. Which of the following can be done with data acquisition probes and a digital multimeter?

   I. Voltage can be measured.
   II. Power can be measured.
   III. Continuity can be tested.

   A. I only
   B. I and II only
   C. I and III only
   D. I, II, and III

Chapter

# 18 Electric and Magnetic Fields

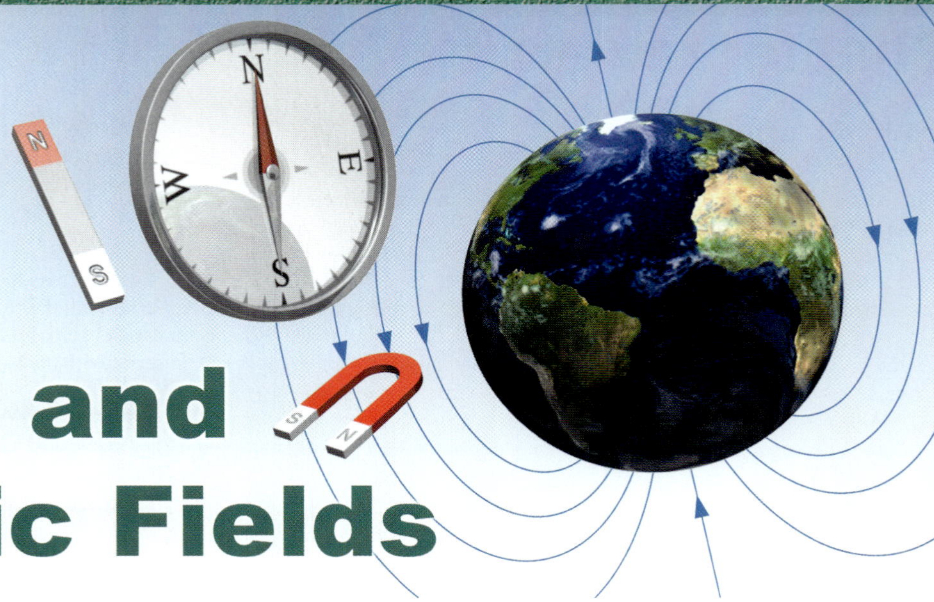


When you hear the word "weather" you probably don't think about streams of red-hot plasma swirling about in violent storms big enough to swallow a planet. In outer space, however, more than 99% of the mass of the ordinary matter in the universe exists as a red-hot ionized state of matter called a *plasma.* The body of plasma most important to our relatively cool, rocky Earth is the Sun. *Space weather* is created by the interplay of electrically conductive plasma and electrical and magnetic fields driven by the prodigious energy of the Sun.

The electricity we take for granted is distributed by enormous currents flowing in several-hundred-thousand miles of interconnected high-voltage wire that make up the *power grid.* Electricity in power plants is generated by rotating car-sized coils of wire in a magnetic field. The power grid includes huge loops of wire in the magnetic field of Earth and under certain circumstances the grid can act as its own generator. The magnetic field of Earth is relatively weak, but the area of a loop in the grid can be the size of a state or more. Anything that creates changes in the Earth's magnetic field causes powerful surge currents to flow in the power grid. These surge currents can cause large-scale losses of electricity, or *blackouts.*

A spinning ball of conducting plasma, the Sun acts like a gigantic electromagnet. Underneath the Sun's "surface," boiling motions tangle its magnetic fields and build up as much energy as trillions of atomic bombs in a single magnetic "knot." When a "short circuit" releases the energy in one of these knots, gigatons of charged particles get blasted outward, such as in the video at right from NASA's *Solar Dynamics Observatory.* Occasionally, a *coronal mass ejection* happens to point at Earth and some of those particles slam into our planet.

*Image credit: NASA and Solar Dynamics Observatory*

In 1989, a record-setting solar storm crippled the electrical grid in eastern Canada, leaving millions without power. To get advance warning of severe solar storms, NASA and the European Space Agency have several research spacecraft that constantly monitor the Sun. Someday this research will enable us to predict "space weather" just as we do ordinary weather.

# Chapter study guide

## Chapter summary

How does a compass work? The answer lies in the magnetic force exerted on the needle by Earth's magnetic field. How are computer cables shielded from stray electricity? The answer comes from the properties of electric fields and conductors. Magnetic and electric forces lie behind many phenomena and everyday items that we use, from the refrigerator magnet to the antistatic sheet in a clothes dryer. In this chapter, you will learn about static electricity and magnetism. In the process, you will learn why your hair stands on end when you touch a Van de Graaff generator!

## Learning objectives

By the end of this chapter you should be able to

- describe basic properties of magnetic and electric forces and identify everyday examples of each;
- calculate the electric force between two charges;
- draw electric field diagrams around point charges;
- draw magnetic field diagrams around a bar magnet;
- describe electrostatic induction and how an electroscope works;
- explain how electric shielding works and give practical examples;
- calculate electric potential and electric potential energy;
- infer the properties of an electric field based on equipotential lines; and
- describe how a capacitor works and calculate its stored charge and energy.

## Investigations

18A: Magnetic force between magnets
18B: Magnetic field around a magnetic
18C: Static electricity on transparent tape
18D: Electric field and electric force

## Chapter index



## Important relationships

$$F_e = \frac{k_e q_1 q_2}{r^2} \qquad F_e = qE \quad \text{or} \quad E = \frac{F_e}{q} \qquad E = \frac{k_e q}{r^2} \qquad V = \frac{E_p}{q} \qquad V = \frac{k_e q}{r}$$

$$E = \frac{k_e q}{r^2} \qquad E_p = \frac{k_e q_1 q_2}{r} \qquad q = CV \qquad E_p = \frac{1}{2}CV^2 \qquad C = \frac{\varepsilon A}{d}$$

## Vocabulary

| | | | |
|---|---|---|---|
| **magnetic** | **magnetism** | **permanent magnet** | **magnetize** |
| **magnetic force** | **magnetic poles** | **polarity** | **force field** |
| **magnetic field** | **magnetic field lines** | **ferromagnetic** | **magnetic domains** |
| **static electricity** | **electrostatics** | **electric charge** | **electrically neutral** |
| **electric force** | **electroscope** | **electrostatic induction** | **Coulomb's law** |
| **coulomb (C)** | **negative charge** | **positive charge** | **electric field** |
| **electric field lines** | **electric potential** | **equipotential** | **electric potential energy** |
| **capacitor** | **capacitance** | | |

## 18.1 - Magnetism

What does the word "magnet" mean to you? Do you think of a bar magnet, horseshoe-shaped magnet, or a refrigerator magnet? Do you think of how a compass responds to the magnetic field of Earth? **Magnetism** is the property of a material or object to attract other objects made out of iron, magnetite, or steel. This fascinating phenomenon has been a foundation of technology since ancient times. This section introduces the fundamental ideas underlying magnetism.

### Magnets

**Where are magnets used?**

Magnets abound in technology. Computer hard disk drives use magnets to read and write data. Magnets are used in speakers in home stereos and headphones. Magnetic clasps are on many handbags. Credit cards and ATM cards store data on magnetized stripes that are read by swiping the card through a reader with additional magnetic devices. Magnets are in electric generators that convert mechanical energy to electrical energy. Junkyards use large magnets to lift scrap metal and cars. Magnetic switches are part of many security alarm systems.

**Everyday objects containing magnets**

**What kinds of magnetism are there?**

**Magnetic objects**

Permanent magnets

Objects that can be magnetized

**Magnetic** objects are generally grouped into two categories: permanent magnets and objects that can be magnetized. A **permanent magnet** retains its magnetic properties at all times, regardless of whether or not it is near another magnetic object. A steel refrigerator door, however, is not a permanent magnet. The steel in the door has magnetic properties only when another magnetic object, such as a refrigerator magnet, is brought near enough to **magnetize** it. Iron, steel, nickel, and cobalt, as well as some naturally occurring minerals such as lodestone, can be magnetized.

**Are all materials affected by magnets?**

Most materials do not have noticeable magnetic properties. Magnetic influences pass right through them. Paper is a good example. Many people use magnets to hold pieces of paper on steel refrigerator doors. The magnetic force between a magnet and the steel passes right through the paper. Wood, plastic, and glass are other materials that do not block the influence of magnets. Iron and some other metals that are good conductors, however, can block or change the magnetic activity of nearby objects.

**Can a magnet be liquid?**

There are no liquid magnets. All magnets are solids. This is because the kind of magnetic effects we are discussing come from the small-scale organization of atoms within a material. Anything that disrupts the orderly arrangement of atoms, such as being *melted*, destroys ordinary magnetic effects. This is why heating up a permanent magnet past a certain temperature causes it to lose its magnetism.

**Magnetic poles**

A basic property of magnets is that they contain two **magnetic poles** of opposite **polarity**, referred to as the magnetic north pole and magnetic south pole. All magnets must have both a north and a south pole. If you cut a permanent magnet in half, each of the halves will have a north and a south pole! This is a fundamental difference between magnetism and electrostatics: You can separate positive charges from negative charges, but you cannot separate a magnetic north pole from its magnetic south pole. You will learn more about electric charges in section 2 of this chapter.

**Cutting bar magnet in half**

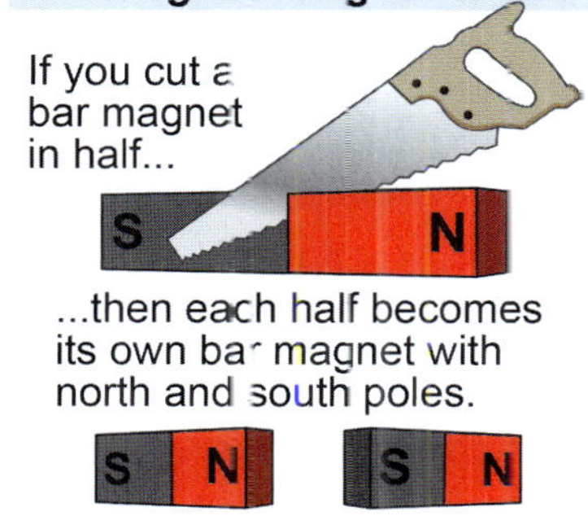


**Attraction and repulsion**

When two magnets are close to each other they exert attractive or repulsive **magnetic force** on each other depending on how their poles are aligned. When opposite magnetic poles—north and south—face each other, the magnetic force is attractive. When similar magnetic poles face each other, north–north or south–south, the magnetic force is repulsive.

**Attraction and repulsion of magnets**

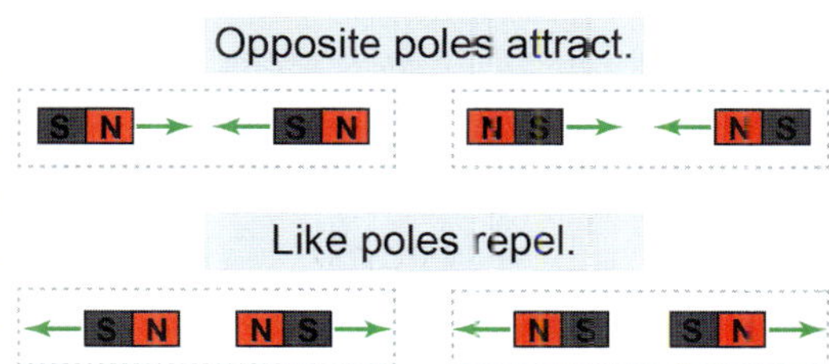


**Why magnets get stuck together**

Have you ever handled two very strong magnets that, when you joined them together, were stuck so strongly that you could barely separate them? Why do magnets that are so close together stick together so strongly? The answer is that the force between two magnets falls off *more rapidly* than an inverse-square-law relationship. This also means that the force at shorter distances *increases more rapidly* than an inverse-square-law relationship. So the forces become extremely strong at short distances, resulting in magnets that are stuck together and very hard to separate from each other!

**Magnetic forces**

Because magnets are *dipoles* the strength of the magnetic force decreases very quickly as they are separated. Think about the attraction between the north pole of one magnet and the south pole of a second magnet. The *south* pole of the first magnet is *repelled* by the south pole of the second magnet. The force between two magnets is an example of a dipole–dipole interaction because it includes four poles: two north and two south. Opposite magnetic forces on the two ends of a bar magnet can create a *torque* that causes the magnet to twist or rotate. Later in the chapter, we will see how magnetic torque can be used in a simple electric motor.

**Torque caused by magnetic force**

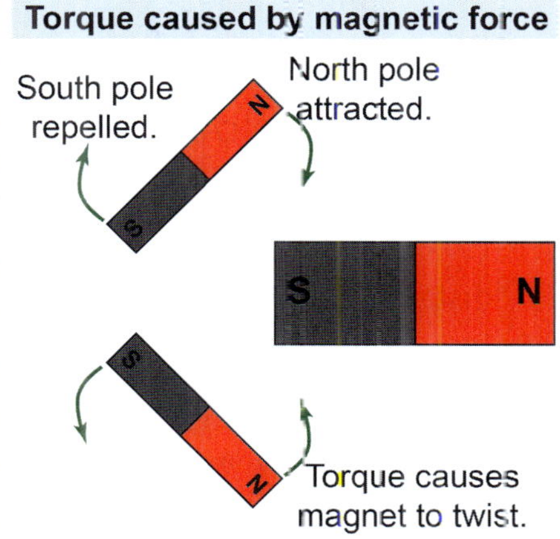

# Investigation 18A Magnetic force between magnets

**Essential questions** *At what distance does the magnetic force operate?*
*Can magnets be combined to make a stronger magnet?*

The magnetic force between two magnets will act to repel or attract them to each other, depending on the relative orientation of their magnetic poles. In this investigation, you will characterize the magnetic force between bar magnets. You will observe how the magnetic force varies with separation between magnets, whether combining magnets creates a more powerful magnet, and to what extent the magnetic force is affected by an intervening material.

## Part 1: Magnetic force between bar magnets

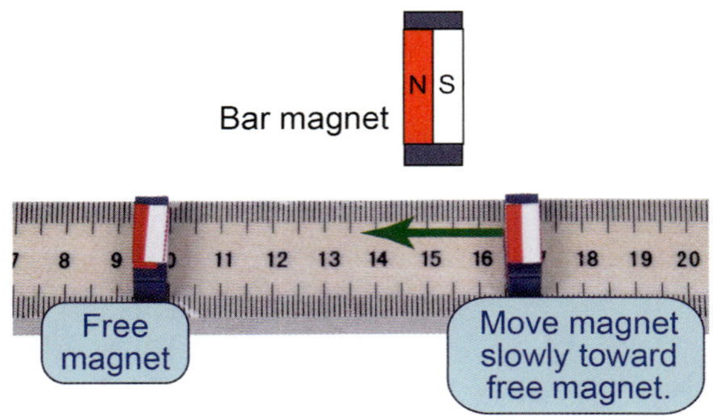

1. Place a free magnet and a test magnet on a ruler on the table.
2. Slowly move the test magnet closer to the free magnet until the free magnet begins to move. Measure the distance at which the free magnet begins its movement.
3. Repeat for different combinations of poles: north–north; south–south; and north–south.
4. Create a composite bar magnet by connecting five small magnets together. Use it as a test magnet and record the distance at which the free magnet begins its movement.

a. *What is the distance over which the magnetic force between these magnets acts?*

b. *Does the combination of five magnets create a stronger magnetic force?*

c. *Why does the free magnet only start to move suddenly? (Hint: Does the free magnet have to overcome a different force?)*

d. *In which configuration was more energy available, the individual bar magnet or the five magnet combination? Why?*

## Part 2: What blocks a magnetic force?

1. Hold a textbook in place between the magnets. Measure the distance at which the free magnet begins to move.
2. Repeat for various materials: glass, a block of wood, a block of metal, and a pile of nails.

a. *Do any of these materials block the magnetic force between bar magnets? Justify your answer.*

**How does force act at a distance?**

Consider two magnets attracting each other from some distance apart. If one magnet moves up, the other magnet is also pulled up. The question is *how?* How does the force get from one magnet to the other? If one magnet were to vanish, would the force on the other magnet instantly vanish? Suppose the two magnets were a kilometer apart. The force would be much smaller but the question of time remains. Would one magnet instantly respond when the other moved or would there be some time delay?

**How does gravity act at a distance?**

Consider a similar situation with gravity. Suppose the Sun were to instantly vanish; would the Sun's gravitational pull on the Earth stop immediately? The answer is *no!* Earth would continue to "feel" the Sun's gravity for 8.3 min. *Why?* Why does it take 8.3 min for Earth to "notice" that the Sun has vanished?

**Gravitational field**

The reason is that the Sun does not act on the Earth directly. Instead the Sun creates a gravitational field in space. The *field* acts on the Earth to keep the planet in a circular orbit. The Earth would still be affected by the Sun's gravity because the Sun's gravitational field is already there, in space, at Earth's orbit. Any change in the field travels at the speed of light, so the time interval of 8.3 min is also the time it takes the *lack of a gravitational field* to reach Earth's orbit.

**What is a force field?**

A **force field** is an organization of energy in space that creates a force on any receptive matter that passes within its influence. Although force fields are popular in science fiction, some force fields are nonetheless real. Gravity is one example. The Sun's gravitational field creates forces on each planet in the Solar System. The gravitational field travels at the speed of light, and therefore changes to the field also travel at the speed of light. The force between magnets also acts through a field, the *magnetic field*. The presence of magnetic poles creates a magnetic field that extends outward in the surrounding space. Other magnets "feel" forces through their interaction with the magnetic field. It is the *field* that creates forces, not the magnets interacting directly.

**Fields and energy**

It takes work to move the north poles of two magnets close to each other because they repel each other. If you then release the two magnets, they move apart. Their magnetic field can be thought of as storing potential energy that is converted into kinetic energy when the two magnets are released. In a similar way, potential energy is stored in gravitational fields and electric fields.

**Contact and noncontact forces**

When you rub your two hands together, you feel the force of friction between them. This is called a *contact force*. Contact forces involve direct interaction of matter. Gravitation and magnetism, by contrast, are capable of acting through empty space without any contact between objects. Both these *noncontact* forces act through fields. The electric force between two charged objects is a third example of a noncontact force that acts through a field. In fact, the interaction of two particles can be thought of as occurring in two steps: 1. the creation of the field and 2. the field interacting with matter.

**How does the force get from one magnet to another?**

The **magnetic field** describes the strength and direction of magnetic forces around a magnet. A compass needle is a magnet that is mounted so that it can spin freely. Compasses are sensitive detectors for magnetic fields. The diagram below shows how the needle of a compass changes at different points around a bar magnet.

**Using a compass to trace the magnetic field of a bar magnet**

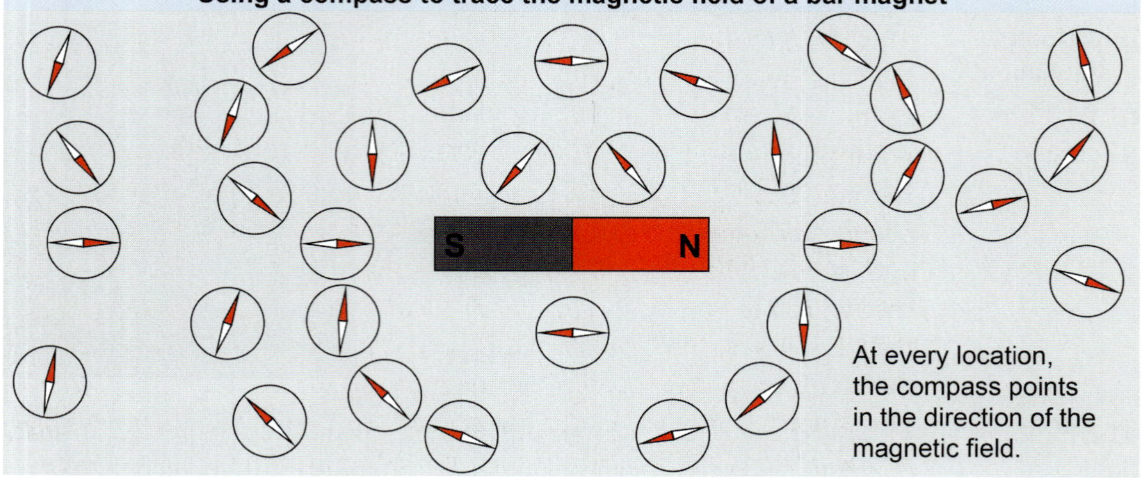


**How does a compass show magnetic fields?**

At every location we place a compass, its orientation tells us the direction of the magnetic force. The red end of a compass needle is typically the needle's north magnetic pole. The compasses in the diagram above show that the compass needle's north pole is attracted near the south pole of the bar magnet and repelled near the north pole of the bar magnet.

**What does a field diagram show?**

A magnetic field drawing uses lines and arrows to represent the force exerted on the *north magnetic pole* of a test magnet. In a drawing of the field around a bar magnet, the arrows point toward the south pole of the bar magnet and away from the north pole. These four rules apply to interpreting or making magnetic field diagrams:

**Field lines around a bar magnet**

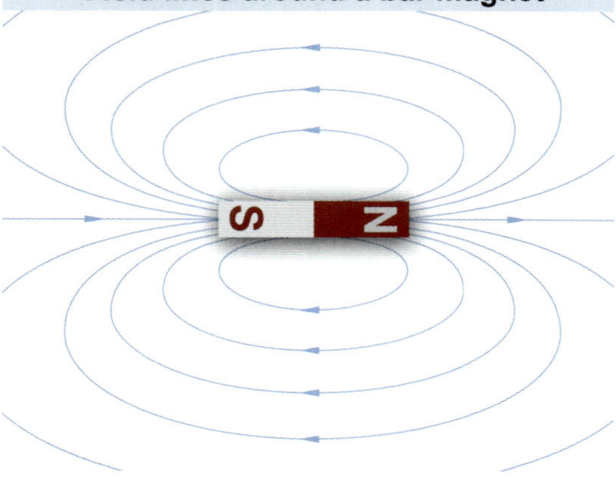


1. **Magnetic field lines** point into south poles and out of north poles.
2. The closer the field lines are, the stronger will be the magnetic force.
3. Field lines never cross.
4. Field lines always make closed loops (but the portion of the loop inside the magnetic source may not be pictured).

**Units for magnetic field**

The SI unit for magnetic field $B$ is the tesla (T). A field of 1 T is very strong! As you will learn on the next two pages, the Earth's magnetic field near the equator has a strength of around $3\times10^{-5}$ T or 30 μT. The powerful magnets in a magnetic resonance imaging machine can have a field strength of 1 T (or so).

**Earth's magnetic field**

Our planet has a magnetic field that is approximately aligned with its north–south axis. Humans cannot sense magnetism directly but some animals can, including birds, insects, and even sharks. Migratory birds can fly thousands of miles over open oceans and not lose their way because they sense their direction (north, south, east, or west) from the Earth's magnetic field. Humans do the same thing with a *compass*. The north magnetic pole of a compass points toward Earth's north geographic pole.

**Why does a compass point north, not south?**

*This should cause you to stop and think a minute!* Isn't the north pole of a magnet attracted to the *south pole* of another magnet? Why does the north pole of a compass magnet point *north* on Earth? The answer is that Earth's north *geographic* pole is a south *magnetic* pole. Long ago, people labeled the north magnetic pole of a magnet as the pole that points north! It was more than 1,000 years later when we figured out *why* the north magnetic pole of a compass needle points north. Although Earth's magnetic field can be *modeled* as a large bar magnet, in reality the Earth's magnetic field is likely created by electric currents within the Earth's core.

**Earth's magnetic field**

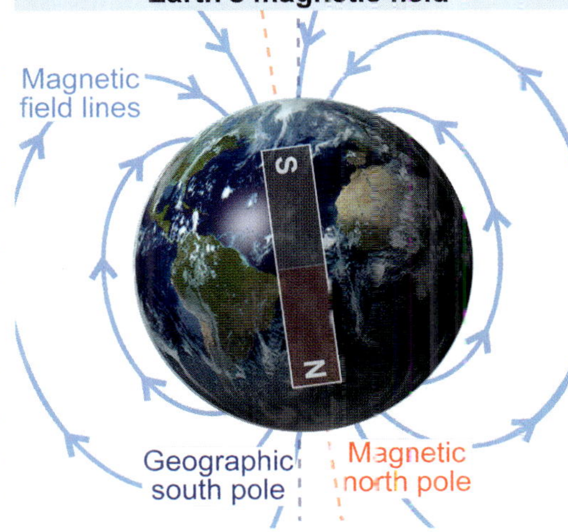


**Do compasses really point north?**

**Compass orienteering across the USA: Magnetic north differs from geographic north.**

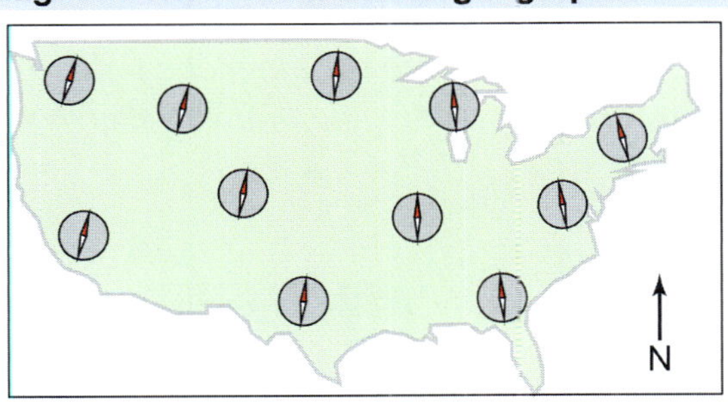


The direction of magnetic north is actually a few degrees off true north. The deviation between geographic north and magnetic north is called *magnetic declination* and it is important for navigation. A correction of up to 15° must be made to compass headings to align them to true geographic north. Magnetic declination differs around the world and even varies considerably from state to state within the USA.

**Changes in Earth's magnetic field**

Earth's magnetic field is slowly changing! A century ago, magnetic north was located at around 70°N latitude and is now around 83°N latitude. The *strength* of Earth's magnetic field has also decreased by about 6% in the past century. In fact, we know from the geological record that Earth's magnetic field completely reverses itself every million years or so. The last time the Earth's magnetic poles reversed was around 780,000 years ago. The animated map from the U.S. Geological Survey shows how magnetic declination has changed over the past 400 years.

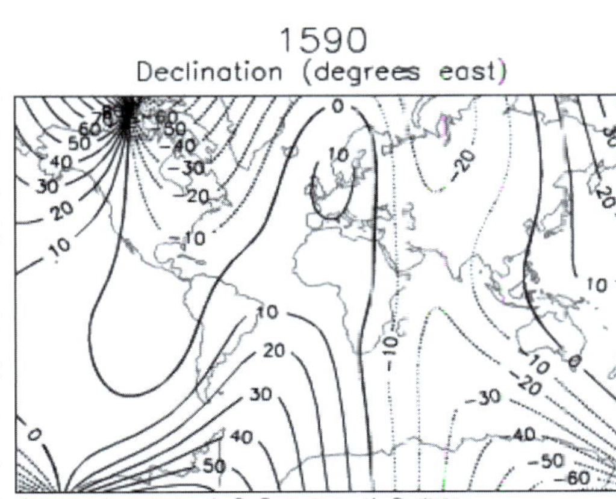


Model by A. Jackson, A. R. T. Jonkers, M. R. Walker. Phil. Trans. R. Soc. London A (2000), 358, 957–990.

# Investigation 18B

## Magnetic field around a magnet

**Essential questions** *How does a force field indicate the forces acting on an object?*
*What is the shape of the magnetic field around a bar magnet?*

Interactive simulation

We use magnets and magnetized objects every day, often without realizing it. All around us is a magnetic field—the Earth's magnetic field. How do these magnetic forces and fields work? In this investigation, you will explore the magnetic field surrounding a bar magnet and compare it to the Earth's magnetic field. Magnetic fields are measured in units of teslas (T).

### Part 1: Tracing the magnetic field of a bar magnet

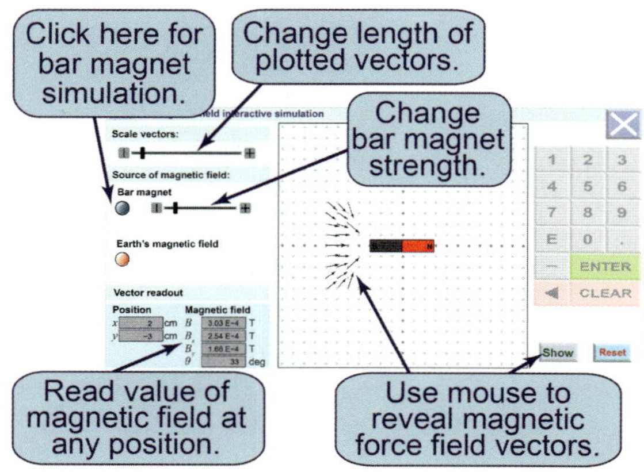


1. Launch the magnetic field interactive simulation and select the bar magnet option.
2. Pass your mouse over the diagram to reveal the magnetic field line vectors.
3. Find a few positions close to the north magnetic pole of the magnet from which you can follow the field arrows all the way to the south magnetic pole. Sketch the paths you followed.

a. *What does the direction and length of each field arrow indicate?*
b. *Determine the strength and direction of the magnetic field 5 cm to the right of the bar magnet's middle, 5 cm to the left the bar magnet's middle, and 5 cm above the bar magnet's middle.*
c. *If you increase the strength of the bar magnet, what happens to the magnetic field lines?*
d. *What is the meaning of the paths from north to south magnetic pole that you sketched out?*

### Part 2: Earth's magnetic field

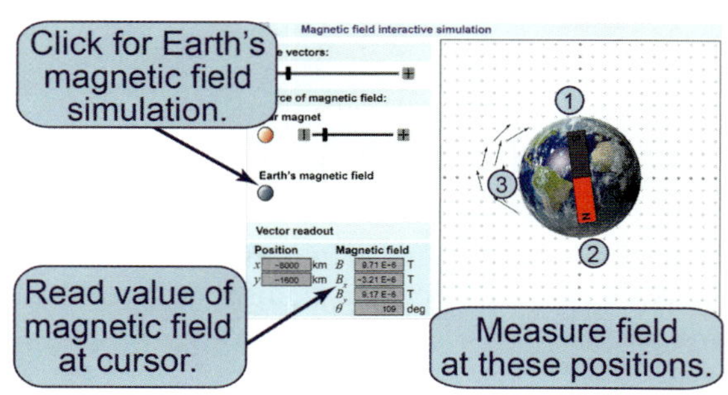


1. In the interactive simulation, click on the button to select the Earth's magnetic field option.
2. Pass your mouse over the diagram to reveal the magnetic field line vectors.
3. Find Earth's north and south magnetic poles.

a. *What is the strength of Earth's magnetic field near one pole (location #1 in the figure)? Near the other pole (#2)? Near the equator (#3)?*
b. *Where are Earth's north and south magnetic poles? Explain using magnetic field lines.*
c. *How does the strength of Earth's magnetic field near its surface compare to the field 5 cm from the bar magnet that you measured in the previous part?*
d. *Research online the strength of Earth's magnetic field and compare to your results in Question a above.*

### Part 3: Magnetic field of a bar magnet using iron filings

Have your partner place a bar magnet under a piece of poster board. Now sprinkle iron filings on the board. *How can you use iron filings to locate the poles of the bar magnet?*

### Diamagnetism and paramagnetism

Most materials have a small response to magnetic fields. Silver, lead, and copper slightly *repel* magnets of either polarity. This effect is called *diamagnetism*. Aluminum, magnesium, and tungsten slightly attract magnets of either polarity. These materials are called *paramagnetic*. In both cases the forces are so small that it takes sensitive instruments to detect them.

### Ferromagnetism

Some materials such as iron, however, are strongly attracted to magnets of either polarity. Your fingers can easily feel the attraction of a steel paper clip to a magnet. Iron, cobalt, and nickel belong to a special class of *ferromagnetic* materials. **Ferromagnetic** substances become magnetized and experience strong magnetic forces in the presence of external magnets.

### Magnetic properties of atoms in materials

The magnetic properties of matter come from the magnetic properties of atoms. Each atom acts as a tiny magnet. In most materials the orientation of the atoms is random, so the atomic magnetic fields cancel each other out. In permanent magnets, however, the atoms are, on average, slightly more aligned in one direction than all others. This partial alignment explains why there are permanent magnets.

Each atom in a material acts as a small magnet.

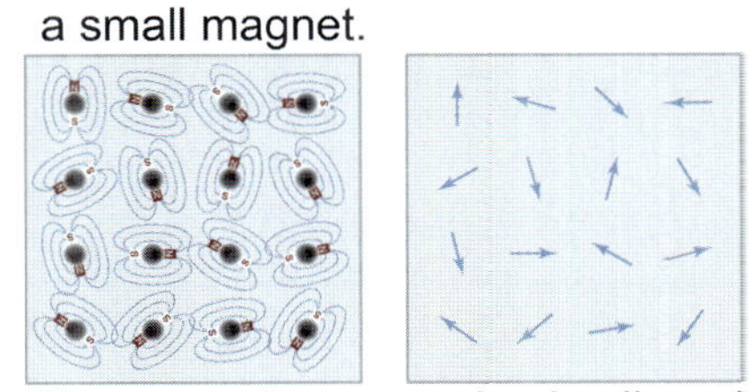

The magnets are randomly aligned, so there is no net magnetic field.

### Magnetic domains

The property that makes iron ferromagnetic is that it only takes a small amount of energy for an iron atom to flip its magnetic axis. Groups of iron atoms form **magnetic domains**. The atoms within a single domain have a similar magnetic alignment. Magnetic domains are small, from $10^{-6}$ to $10^{-4}$ m. In unmagnetized iron, adjacent domains are randomly scrambled, leaving an overall average magnetic field of zero.

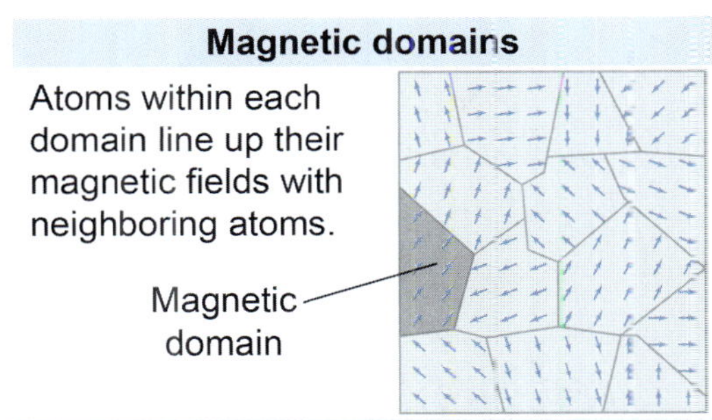


Magnetic fields cancel for these randomly oriented domains.

### Ferromagnetic materials

When an external magnet is brought near iron, magnetic domains *attracted* to the external magnet gain atoms and grow because iron atoms can easily change their magnetic orientation. Domains *repelled* by the external magnet lose atoms and shrink. The migration of atoms into *attractive* domains quickly *magnetizes* a volume of iron to attract either pole of the external magnet. Ferromagnetic materials are crucial to many technologies such as motors and generators.

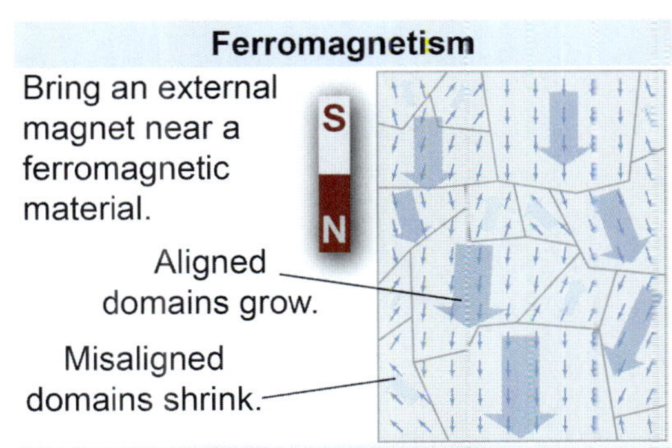


Result: The material has a net magnetic field.

# Section 1 review

Magnets are found in many objects we use every day, such as motors. Permanent magnets maintain their magnetic activity even when not in the presence of other magnets. Some magnetic materials can be temporarily magnetized by the presence of an external magnet. Magnets always have two different magnetic poles, which we call north and south. The force between magnetic poles is attractive between unlike poles and repulsive between like poles. The magnetic force becomes very strong as magnets are brought close together. Magnetic field lines are a tool for visualizing how magnetic fields vary in the space around a magnet. The planet Earth has a magnetic field. The south magnetic pole of Earth's magnetic field is close to, but not exactly aligned with, the planet's north geographic pole. The difference between geographic and magnetic north is called declination. Electricity and magnetism are connected with each other because electric current creates a magnetic field.

| Vocabulary words | magnetic, magnetism, permanent magnet, magnetize, magnetic force, magnetic poles, polarity, force field, magnetic field, magnetic field lines, ferromagnetic, magnetic domains |
|---|---|

## Review problems and questions

1. A teacher needs to store two strong permanent magnets in a box. The problem is that they attract each other, stick together, and are very hard to separate. Describe a way to store the two magnets in the same box so that they can be easily separated.

2. The north magnetic pole of a bar magnet is brought near the middle of a second bar magnet, i.e., halfway between its north and south magnetic poles. The second magnet is observed to spin by 90°. In what direction does it spin? Why? Draw a diagram to illustrate your answer.

3. What is the cause of Earth's magnetic field?

4. Describe the outcome of an experiment in which a student tries to create an isolated north magnetic pole by breaking off the north pole of a permanent magnet.

5. Which pole of a bar magnet will attract a paper clip? Why?

6. A typical bar magnet is placed on a table, approximately 10 cm away from a pile of iron filings.
   a. Does the bar magnet or the Earth produce a greater contribution to the magnetic field in the vicinity of the iron filings?
   b. How can you test your answer?

7. Which of these actions would be more likely to weaken the magnetic field of a bar magnet: placing it in a warm oven or placing it in the freezer? Why?

8. We know that like magnetic poles repel each other. Yet the north pole of a compass needle points roughly toward the Earth's North Pole. Why?

9. How can you use a compass to determine direction?

Have you ever been zapped by a metal doorknob after walking across a carpet on a dry day? Have you ever rubbed a balloon against your sweater and watched your hair stick out? These experiences are examples of **static electricity**. Static electricity is a reminder that all matter contains electric charge. We don't normally sense this because positive and negative charge are rarely separated. In extreme cases such as lightning, however, electric charge is revealed for the powerful phenomenon that it is. This section introduces the fundamental concepts of electric charges and forces.

## Static electricity

**Static electricity**

Static electricity is caused by a very tiny excess of positive or negative electric charge. Until the 18th century, however, the cause of static electricity remained unknown. In 1750, Benjamin Franklin proposed an experiment to demonstrate his hypothesis that static electricity and lightning were related. Within the next few years Franklin and others proved that "sparks" could be extracted from clouds. Lightning is nothing more than static electricity on a very large scale. The physics of lightning and charges at rest (static) is called **electrostatics** and is the subject of this section. Moving electric charges make electric *current*, which was discussed in the previous chapter.

**Why does the balloon attract your hair?**

When you rub a balloon against a wool sweater the balloon will subsequently attract your hair. Why? While rubbing, the sweater transfers some negative charges to the balloon. The balloon now has a net negative charge and the sweater has an equal net positive charge. A similar process takes place when you shuffle your feet along a rug. When you touch a metal conducting surface, such as a door knob, the built-up charge you have collected is attracted to its opposite partner through the metal conductor of the doorknob and, briefly, a tiny electric current flows. The current is what makes the zap!

**Static electricity**

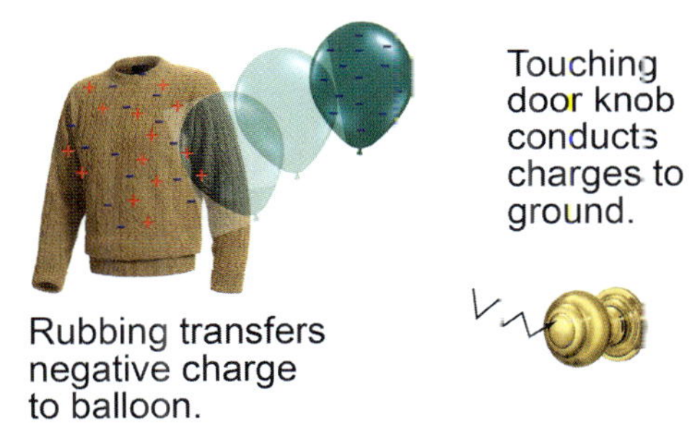


**Van de Graaff generator**

The amount of static electricity you can create with a balloon is relatively small. To produce a larger static charge you might use a *Van de Graaff* generator. Inside a Van de Graaff generator, a high-voltage power supply puts electric charge onto a moving insulating belt. When the belt reaches the top, a metal contact conducts the charge out onto the surface of a conducting metal sphere. The sphere of a Van de Graaff generator can reach thousands or even millions of volts. Fortunately, the high-voltage shock from a table-top-sized Van de Graaff does not contain enough electric *current* to seriously hurt you.

**How a Van de Graaff generator works**

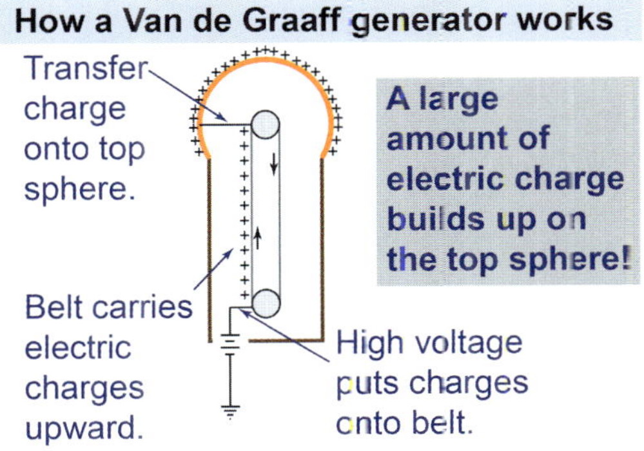

## Charge as a property of matter

### Where does charge come from?

Matter is made of atoms and electric charge is a property of the particles inside every atom. That means that on a very fundamental level, similar to *mass,* electric charge is a basic property of matter. Mass can only be zero or positive. (Even antimatter has positive mass!) Electric charge, however, can be positive, negative, or zero (neutral).

### What is inside an atom?

The nucleus of the atom contains positive charged particles called protons and neutral particles called neutrons. Outside the nucleus are tiny, fast-moving negative charged particles called electrons. The negative electrons are attracted to the positive protons in the nucleus and that is what holds atoms together. The electrons farthest from the nucleus of one atom may also be attracted to the nucleus of another atom, and the resultant sharing of electrons between two atoms is the underlying explanation for chemical bonds that group atoms into molecules and compounds. In fact, most of the behavior of matter—including static electricity and most of chemistry—comes from the activity of electrons.

**Structure of an atom**

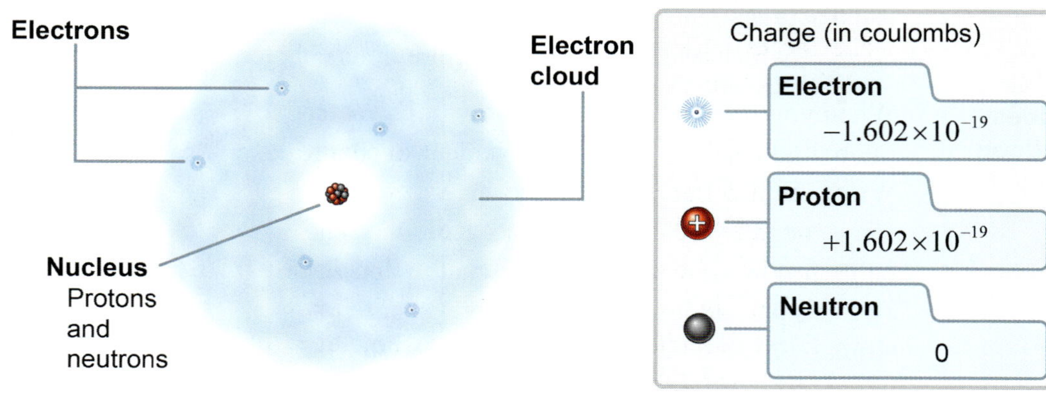


### What is the charge of an atom?

The charge on the proton and electron are exactly equal and opposite. A complete atom has an equal number of protons and electrons; therefore, a whole atom has zero net electric charge. Chapter 26 covers the atom in more detail; to understand electricity in this chapter, it is important to know that matter is made of atoms and that electric charge is a property of the protons and electrons in all atoms.

### What is the unit of electric charge?

The unit of electric charge is the coulomb (C) in honor of Charles-Augustin de Coulomb (1736–1806). A French physicist, Coulomb made the first accurate measurements of the electrical force between charges. *A coulomb is a very large amount of charge!* One coulomb of charge is equal to the charge of $6\times10^{18}$ electrons. Ordinary static electricity results from an excess charge of less than one-millionth of a coulomb.

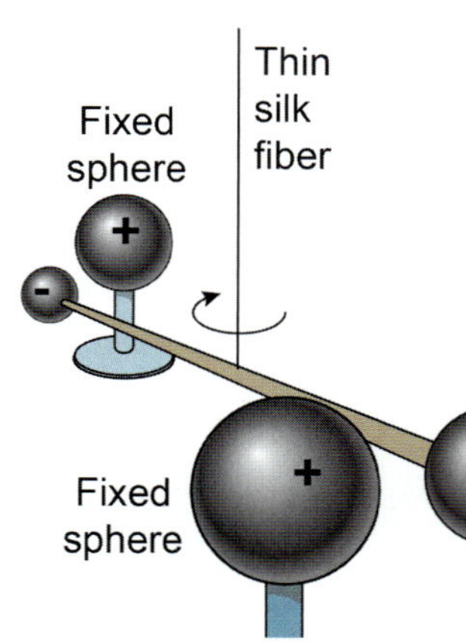


**Coulomb's experiment to measure the electric force**

**Positive and negative electric charge**

Just as there are two different magnetic poles in magnetism, there are two different kinds of **electric charge**: positive and negative. As you might expect, diagrams often use the plus sign (+) and negative sign (−) to tell them apart.

**Net charge**

Your body is full of electric charges: not just a few, but many—around $10^{29}$! You are not a walking lightning factory because every positive charge in every atom of your body is exactly balanced by a negative charge in the same atom. At macroscopic distances the electrical effects of positive and negative charges inside atoms cancel each other out. Your body is **electrically neutral**, which means that it has no *excess* of either positive or negative charge. In physics terms, your body has no *net charge.* Most objects around us are similarly electrically neutral. If this weren't the case, lightning would be everywhere, all the time!

**Net charge**

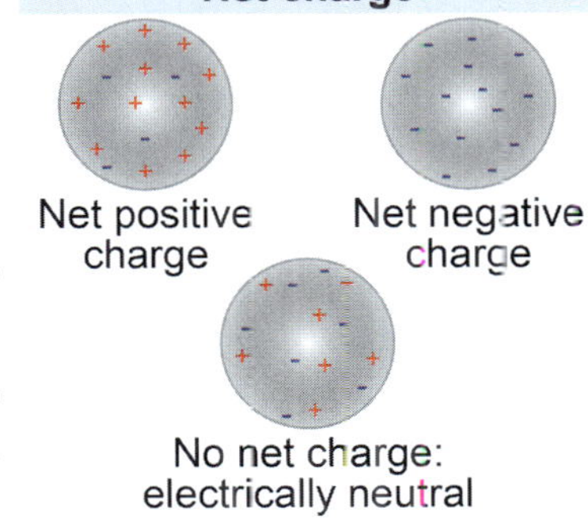


**Electric attraction and repulsion**

The forces between positive and negative charge can be attractive or repulsive, similar to the forces between north and south magnetic poles.

- The **electric force** between electric charges is enormously strong.
- The electric force is a universal force that exists between any two objects that contain net positive or negative charge.
- Electric forces hold atoms and molecules together by providing half of the underlying "glue" that maintains the structure of matter.
- Electric forces are so strong that, outside of atoms, positive and negative charges are rarely separated, *so the forces are almost never directly evident!*

**Do charges attract and repel?**

Similar to magnetic poles, opposite electric charges attract while similar electric charges repel. A positive charge repels other positive charges but attracts negative charges. A negative charge repels other negative charges but attracts positive charges. A neutral object, with zero charge, feels no electric force from either positive or negative charge.

**Attraction and repulsion of electric charges**

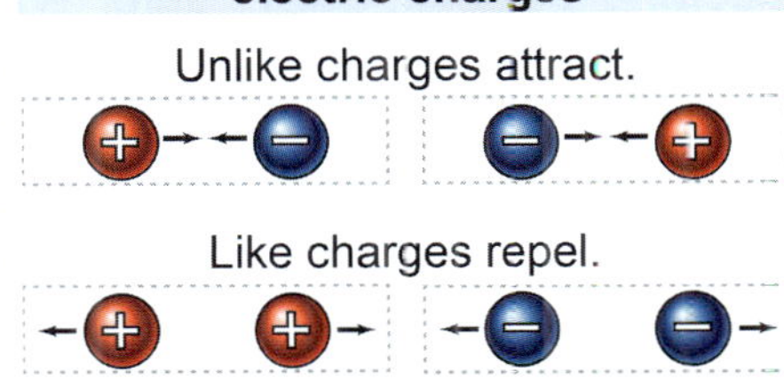


**Distribution of electric charges on an object**

The student's hair in the diagram is springing away from her head. Why? The answer is that the Van de Graaff generator she is touching has charged her body (and hair) uniformly negative. Like charges repel, so every negative charge repels every other negative charge as far as possible. The charges on the student's head spread outward from each other into her hair. This student has light, fine hair and the charge on each strand of hair repels each neighboring strand to create the truly *electrified* hair style you see in the picture.

**Why is her hair standing on end?**

The answer lies in how the electric charges are distributed on her!

# Investigation 18C

## Static electricity on transparent tape

**Essential questions** *How can you show that there are two different kinds of electric charge?*

There are many ways to create static electricity. This investigation shows how you can separate static charge on two pieces of ordinary transparent tape. Whether you observe attraction, repulsion, or both tells you about the polarity of the charge.

### Part 1: Charging the transparent tape

1. Take two pieces of transparent tape about 3–4 in long. Fold over a small part on one end of each tape strip to act as a handle.
2. Put both pieces of tape, sticky-side down, on the table (but not touching each other). Grab the "handles" and lift both tape strips up quickly.

*a. What happens when you bring the two tape strips close to each other (but not touching)?*
*b. What can you infer about the similarities or differences in the electric charges on the two strips?*
*c. What happens when you bring one strip near your hand? Why?*

### Part 2: Charging the tape in another way

1. Put one piece of tape on the table, sticky-side down. Place the other piece on top of it, sticky-side down.
2. Use the lower handle to pull both strips off of the table together. Rub the tape with your fingers to remove any charge.
3. Pull the tape strips apart by holding the handles.

*a. What happens when you bring these two strips near each other?*
*b. What happens with each strip when you bring it close to your hand?*
*c. What difference(s) between this and the previous experiment might have caused different behavior?*
*d. Find objects in the room that attract or repel both strips of tape.*
*e. Which experiment put like charges on both strips of tape? Which experiment put unlike charges on the two strips of tape?*
*f. How many different kinds of electric charges must there be? Why?*
*g. When the strips of tape attracted or repelled each other, one (or both) of them was moved slightly higher off the table—providing it with additional gravitational potential energy. Where did this energy come from?*

**Attracting positive and negative charges**

Electrons in insulating materials may be transferred by friction. For example, rubbing a glass rod with fur transfers some electrons from the fur to the rod. In a similar way, separating the two pieces of tape in Investigation 18C caused some negative charge to transfer from one tape to the other. The tape losing the negative charge became slightly positive and the one gaining the negative charge became slightly negative. With the tapes, you observed that uncharged objects in the room attracted *both* pieces of tape!

*How does an object attract both positively and negatively charged pieces of tape?*

**Electrostatic induction**

To answer the question, assume you have a negatively charged glass rod and you bring it near an uncharged metal sphere. Electrons nearest the rod are repelled—creating a slight excess of positive charge in the metal near the rod. The excess positive charge causes the rod to be attracted to the metal! With a glass rod the force is too small to feel but the attraction is strong enough to visibly bend a charged piece of tape. This is an example of **electrostatic induction**, which occurs when a charged object causes charges to rearrange on a different object.

Electrostatic induction

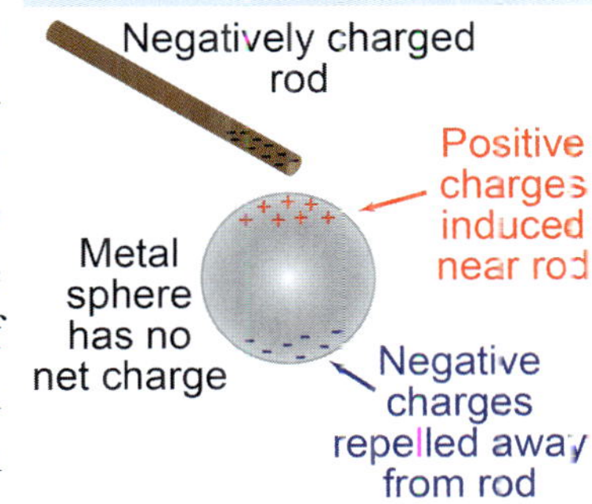


**Detecting electric charge with an electroscope**

An **electroscope** is a sensitive instrument that uses electrostatic induction to detect electric charge. In Figure B below, a positively charged rod is brought near a neutral electroscope. Negative charges are attracted toward the rod by induction, leaving a net positive charge on the leaves. The lightweight leaves spread apart because they have similar charges.

Charging an electroscope by induction and using a charged electroscope

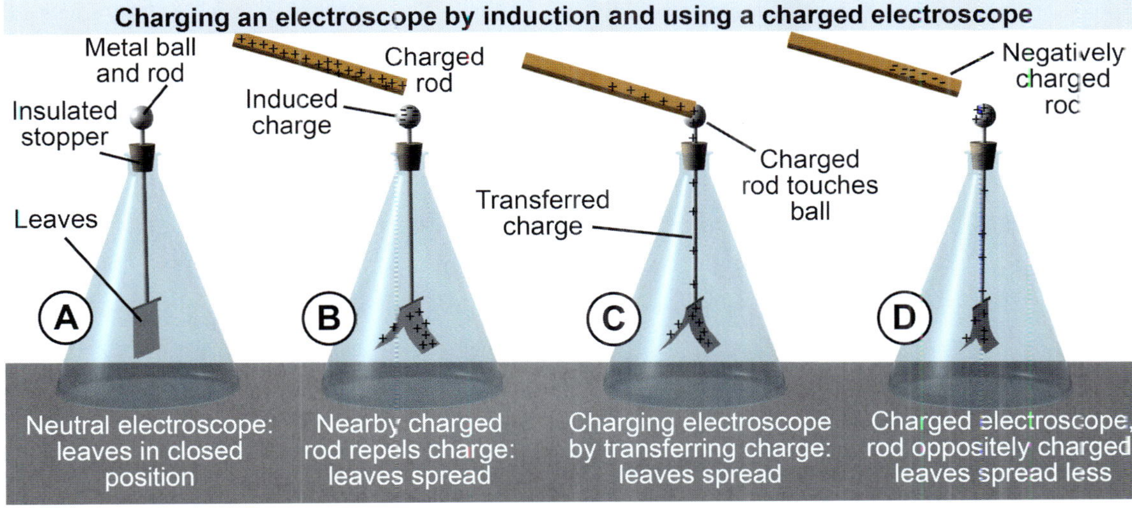


**Charging an electroscope**

If the rod *touches the electroscope* briefly (C) some of the charge is transferred to the electroscope. The electroscope becomes *charged* and its leaves stay spread apart. The leaves of a positively charged electroscope come closer together when a negative charge is near the electroscope (D). The leaves spread farther apart when a *positive* charge comes near. The charged electroscope can be used to determine whether an external charge is the same or opposite to the object that originally charged the electroscope.

**Electric force between two charged objects**

The electrostatic force exists between any two objects that have a nonzero net charge (positive or negative). The quantitative description of the electrostatic force is called **Coulomb's law** and is given in equation (18.1). Coulomb's law says the force between two charges is proportional to the product of the charges divided by the square of the distance between them. The unit of electric charge in Coulomb's law is the **coulomb (C)**.

Interactive equation

**Coulomb's law**

(18.1) $$F_e = \frac{k_e q_1 q_2}{r^2}$$

$F_e$ = electrostatic force (N)
$k_e$ = Coulomb constant = $9.0\times10^9$ N m$^2$ / C$^2$
$q_1$ = electric charge of object 1 (C)
$q_2$ = electric charge of object 2 (C)
$r$ = distance between the two objects (m)

**One coulomb is a lot of charge!**

The Coulomb constant is $k_e = 9\times10^9$ N m$^2$/C$^2$, which means that the electric force is *very* strong. There are about 2,000 C of positive and negative charges in one drop of water. If you could hold these charges one meter apart the attractive force between them would be 40,000,000,000,000,000 N ($4\times10^{16}$ N)! Because the coulomb force is so strong, on the rare occasions that separated charges do occur, such as static electricity, the amount of excess positive or negative charge is extremely small. Most applications in physics involve charges of *microcoulombs.* One microcoulomb (μC) is one-millionth of a coulomb, or $10^{-6}$ C.

**How does distance affect force?**

The Coulomb force follows an inverse square law, similar to the gravitational force. If two charged bodies are moved further apart by twice their distance, their mutual electric force is reduced by a factor of $1/(2^2) = ¼$.

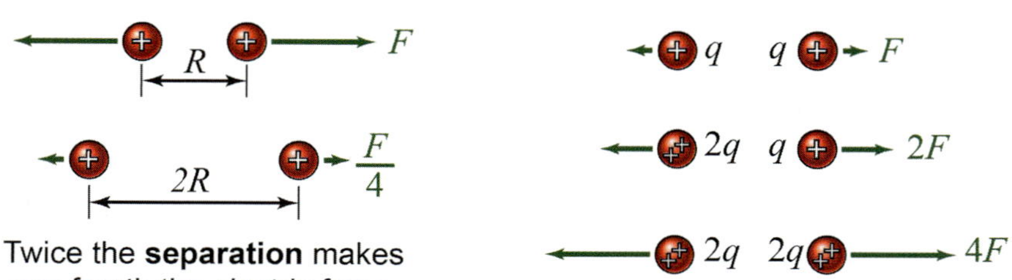


Twice the **separation** makes one-fourth the electric force.

Doubling either charge doubles the electric force.

Doubling both charges multiplies the force by 4.

**How does charge affect the force?**

The electric force between two charged objects depends on the product of the two charges. If charge $q_1$ doubles then the electric force also doubles. If charge $q_2$ also doubles then the force doubles again and is four times greater. If one of the two objects has zero net charge, then either $q_1$ or $q_2$ is zero and the force in equation (18.1) is zero.

**Example problem**

Two electric charges of +1 μC and −1 μC are separated by 1 cm. What is the strength and direction of the force between them?

*Asked:* the force $F_e$

*Given:* $q_1 = q_2 = 1\times10^{-6}$ C; $r = 1$ cm = 0.01 m

*Relationships:*

$$F_e = k_e \frac{q_1 q_2}{r^2}$$

*Solution:*

$$F_e = (9\times10^9 \text{ N m}^2/\text{C}^2)\frac{(10^{-6}\text{ C})(10^{-6}\text{ C})}{(0.01\text{ m})^2} = 90 \text{ N}$$

*Answer:* The force is 90 N and is attractive since the charges have opposite polarity.

# Section 2 review

There are two kinds of electric charges, positive and negative. Electrostatics is the study of electric charge at rest and electric current is electric charge in motion. Matter contains electric charge inside every atom. Most objects around us, however, have little or no net charge because each atom has exactly the same amount of positive and negative charge, which cancel each other. Electric charge is measured in units of coulombs (C). Electric charges exert forces on each other that can be calculated using Coulomb's law. Charges of similar sign (+/+ or −/−) repel each other whereas charges of opposite sign (+/−) attract each other.

| | |
|---|---|
| **Vocabulary words** | **static electricity, electrostatics, electric charge, electrically neutral, electric force, electroscope, electrostatic induction, Coulomb's law, coulomb (C), negative charge, positive charge** |
| **Key equations** | $F_e = \frac{k_e q_1 q_2}{r^2}$ |

## Review problems and questions

1. Describe the similarities and differences between the interactions of magnetic poles and the interactions of electric charges.

2. The leaves of an electroscope are initially apart from each other. When a charged rod is brought near the electroscope, the leaves get closer together. Which of the following is true?
   a. The electroscope is initially neutral and becomes charged.
   b. The electroscope is initially charged with the *same polarity* as the rod.
   c. The electroscope is initially charged with the *opposite polarity* of the rod.

3. Two charged objects are located 1 m apart. Calculate the magnitude and describe the direction for the electric force between them if the two charges are
   a. +1 C and +1 C,
   b. +1 C and −1 C, or
   c. −1 C and −1 C.

4. If Benjamin Franklin had actually flown a metal key on a kite in the rain during an electrical storm, many people think that he would have died on the spot. Describe why he might have been killed.

5. How is it possible for a charged object to attract a neutral object?

6. The atoms that make up the Earth and the Moon contain enormous amounts of charge as well as mass. The Earth and Moon exert strong gravitational forces on each other, and yet they exert no electrical force on each other. Explain how this is possible.

7. Two charged spheres placed a distance $D$ apart exert a repulsive force of 90 N on each other. How far apart should they be placed to reduce this repulsive force to 10 N?

## 18.3 - Electric fields

If you are in a car during an electrical storm, what should you do? Assuming the car has a metal body, the safest place to be is inside the car! The reason the inside of a metal car is safe lies in the nature of electric fields and conductors. The shell of a metal car is a good conductor that shields the inside from external electric fields, including fields induced by lightning.

What should you do if you are in a car during an electrical storm?

### Force field diagrams

**How do I draw the electric field?**

The **electric field** is the organization of energy that carries the electric force between charges. The electric field is represented with diagrams similar to those used for the magnetic field. **Electric field lines** represent the force exerted on a *positive test charge* at all points in space.

1. The electric force points *toward* negative charges.
2. The electric force points *away* from positive charges.
3. Lines are closer together where the force is stronger and farther apart where the force is weaker.

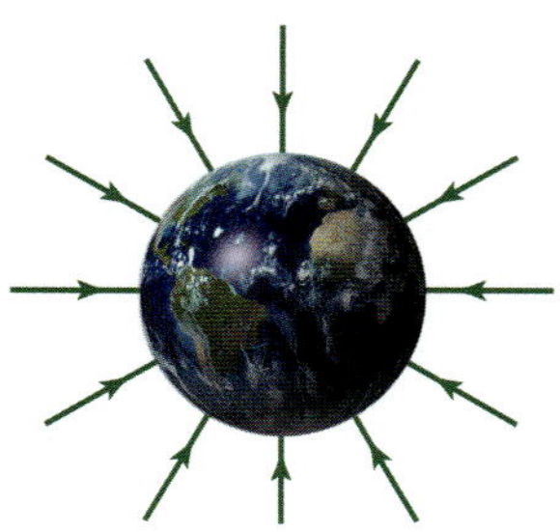

Gravitational force near Earth

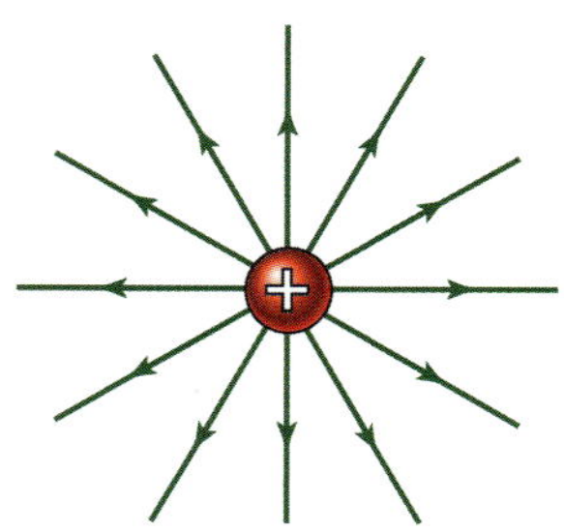


Electric force near a positive charge

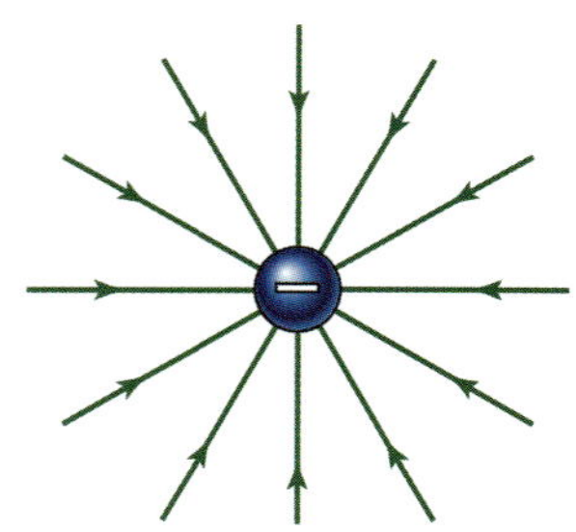


Electric force near a negative charge

**Different ways to draw force fields**

A *field-line diagram* showing gravitational forces near Earth is a set of radial lines with arrows pointing toward Earth's center. The arrows indicate the direction of the force on a test mass. The force is stronger where the lines are close together and weaker where the lines are far apart. Another way to represent a force field is with a *vector diagram.* In a vector field diagram, the *length* of each vector is proportional to the strength of the force at that location.

**Force field lines**

Field lines show the direction of force felt by an object.

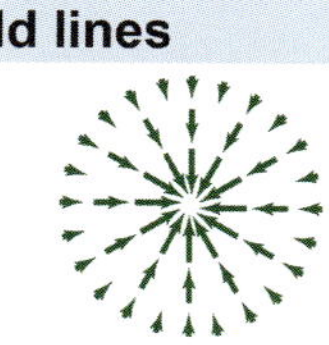

Vector force field shows the strength of the forces.

**How is the electric field different from the gravitational field?**

Earth's gravitational field is relatively easy to draw because gravity is only attractive, so all objects are attracted toward the planet's center. The electric field is more complex because field lines must represent both attractive and repulsive forces. For a single charge, the field lines are radial, similar to gravitational fields. Furthermore, the direction of the electric field arrows depends on the sign of the charge. The force points *outward,* away from a positive charge and *inward* toward a negative charge.

**The electric field *E***

The electric field is defined by equation (18.2). Equation (18.2) says that the force $F$ on a positive test charge $q$ is equal to the charge multiplied by the electric field strength $E$ at that point in space. The equation also defines the electric field $E$ as the force $F$ on the charged object divided by the amount of charge $q$.

**Electric field** and electric force

$$(18.2) \quad F_e = qE \quad \text{or} \quad E = \frac{F_e}{q}$$

$E$ = electric field (N/C)
$F_e$ = electric force (N)
$q$ = electric charge (C)

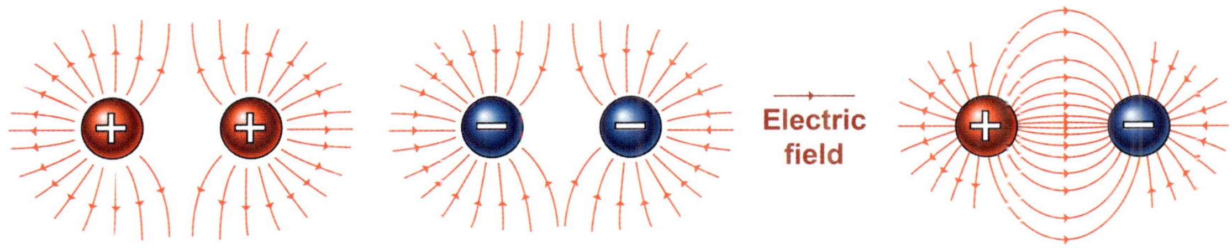


Two positive charges    Two negative charges    Two opposite charges

**Electric field between two like charges**

Consider the electric field around two point charges. At any position in space, the electric field $E$ is the force on an imaginary positive test charge at that position. A positive test charge is *repelled* by both positive charges, so the field lines point outward from each positive charge. A positive test charge is *attracted* to both negative charges, so the electric field lines point toward each negative charge.

**Electric field between positive and negative charges**

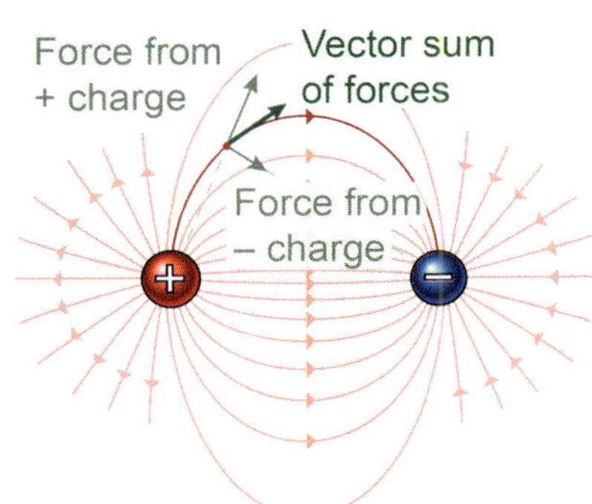


*The net force on a test charge at any position is the sum of the individual forces exerted by each charge in the system.* The electric field between a positive and negative charge generally points away from the positive charge and toward the negative charge. Between the charges the field points directly from positive to negative. At any point in space, the total electric field is the vector sum of the field from the positive charge and the field from the negative charge as if each were independent.

**The field between parallel plates**

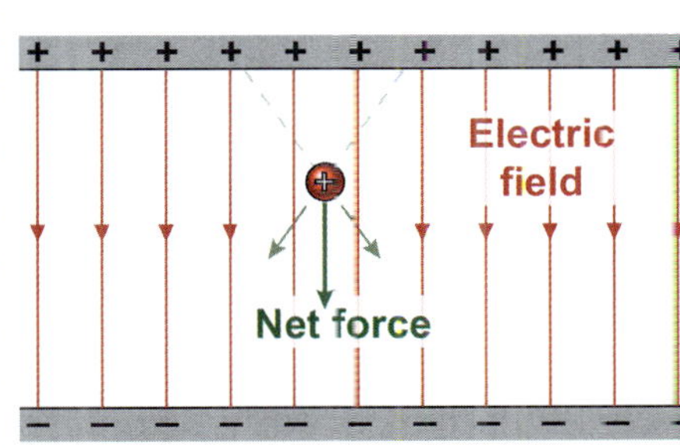


**The electric field between charged parallel plates**

What is the force on a positive test charge located between two oppositely charged parallel plates? The horizontal components of force cancel, leaving only a vertical force. Therefore, the electric field between parallel plates is purely vertical and points from the positive plate to the negative plate. The electric field is also constant in strength in the space between the plates. Why is it constant? Consider a positive test charge between the plates. When it moves away from the positive plate the repulsive force diminishes, but it simultaneously moves closer to the negative plate, causing the attractive force to increase. The two effects combine to create a constant electric field. Parallel plates are often used in experiments because they create a constant electric field and are found inside the capacitor, a commonly used device found in many electrical circuits.

# Investigation 18D Electric field and electric force

**Essential questions** *How does the electric force change with distance from a charged object? How does a force field indicate the electric forces acting on an object?*

Interactive simulation

Charged particles exert electric forces on each other. These forces are described by Coulomb's law. The strength and direction of the electric force on a test charge at any point in space is described by the electric field. This interactive simulation draws a vector diagram of the electric field as you change the electric charges that create it.

## Part 1: Electric force and Coulomb's law

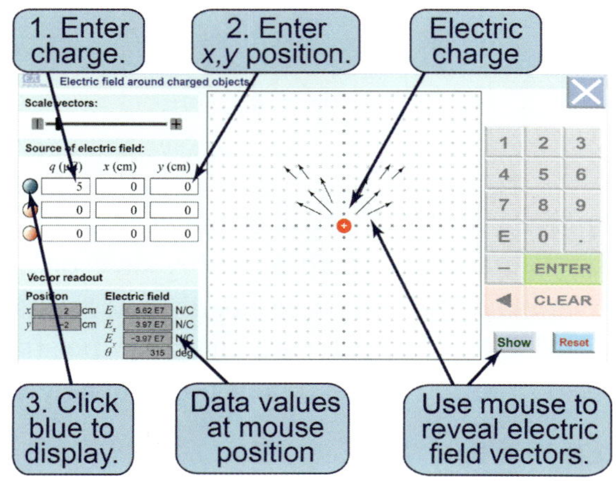


1. Place a $q = +10$ μC charge at the position $x = 0$ cm and $y = 0$ cm. Press the red button next to it; the button will turn green when it displays the charged object.
2. Pass the mouse over the plotting window to reveal the electric field vectors.
3. Measure the electric field at positions of 1, 2, 3, 4, and 5 cm away from the charged object.
4. Graph the electric field strength against distance away from the charged object.

a. *By what factor does the electric field change when you double the distance from the object? How about when you triple it?*

b. *What does the shape of your graph tell you about how the electric force varies with distance? Test your conclusion by creating a new, linear graph based on the same data.*

c. *What is the electric field of a neutral object? Enter an object with charge $q = 0$ μC and compare the field with your prediction.*

## Part 2: Electric field of multiple charges

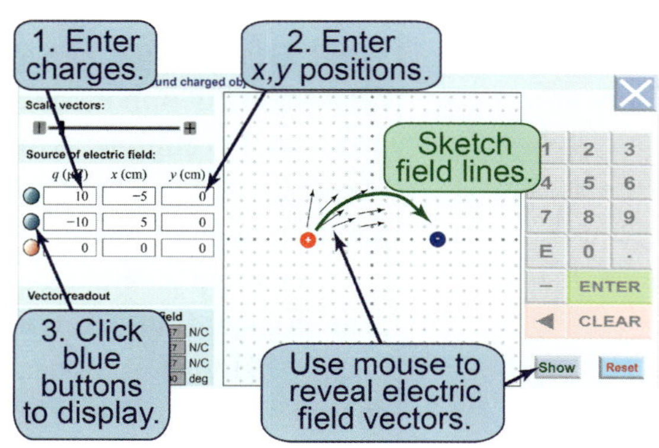


1. Create a new plot with two charged objects:
   a. a $q = +10$ μC charge at position $(x, y) = (-5 \text{ cm}, 0 \text{ cm})$ and
   b. a $q = -10$ μC charge at position $(x, y) = (+5 \text{ cm}, 0 \text{ cm})$.
2. Select several positions close to the positive charge and follow the arrows as they cross the graph. Sketch the paths in each case.

a. *How is your sketch related to the forces felt by charged particles?*

b. *How does your sketch compare with a figure somewhere else in this chapter?*

**Electric fields are all around us**

Low-intensity electric fields are almost always present in our environment. Every wire in your house generates electric fields, as does every atmospheric lightning bolt. Cellphones, wireless networks, and radios are sensitive to these fields. Dropped cellphone calls and static are partially caused by interference from stray electric fields. Fortunately, there is a way to *shield* against stray electromagnetic fields.

**Shielding effect of a conductor**

An electrical conductor, such as a metal, is a material in which charges are free to move. Consider what happens to a conductor placed in an electric field. Because they can move, charges feel the electric force and respond by moving almost instantly. The charges rearrange themselves so that *the electric field inside the conductor is zero!*

*Why is the field inside zero?*

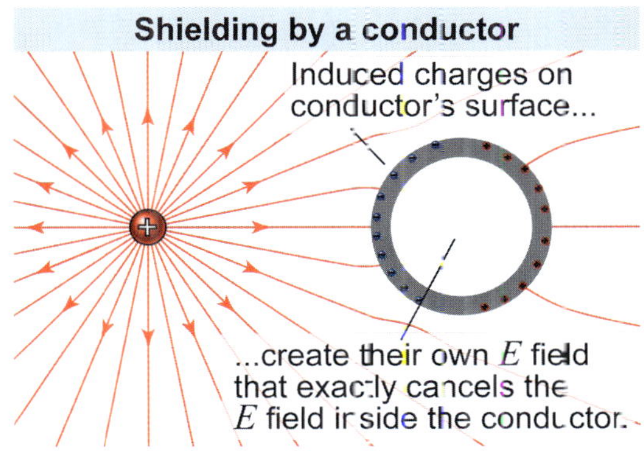


**How shielding occurs**

The field is zero because, if it were *not* zero, the charges in the conductor would continue to feel electric forces and continue to move in response. The charges in a conductor only stop moving when there is no longer any electrical force acting on them. When the electric force is zero inside the conductor, the rearranged charges have created an additional electric field that *exactly cancels out* the external electric field. This effect is called *shielding*.

**Shielding from electric fields**

Computer network signals can be corrupted by stray electric fields. To prevent this from happening, all computer network cables have a conducting metal layer surrounding the signal wires inside. The metal layer shields the wires from stray electric fields.

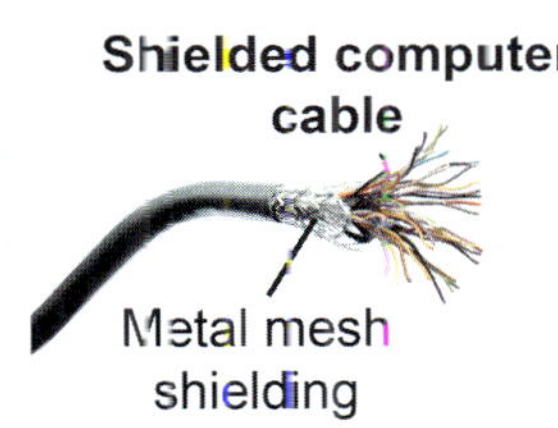


**Faraday cage**

The English scientist Michael Faraday in 1836 invented the *Faraday cage*. A Faraday cage is made of a conductor and uses the shielding effect to protect whatever is inside from external electricity and electric fields. A Faraday cage can be struck directly by lightning, yet a person inside is completely safe.

**Cars and lightning**

If you are driving through a lightning storm, the safest place to be is inside a metal car. The reason is *not* that the car's rubber tires are insulators. The correct reason is that the car's metal shell is a Faraday cage. If lightning were to strike the car, electric charge would redistribute itself around the outside of the car's metal shell to cancel out the electric field inside the car, leaving the interior safe.

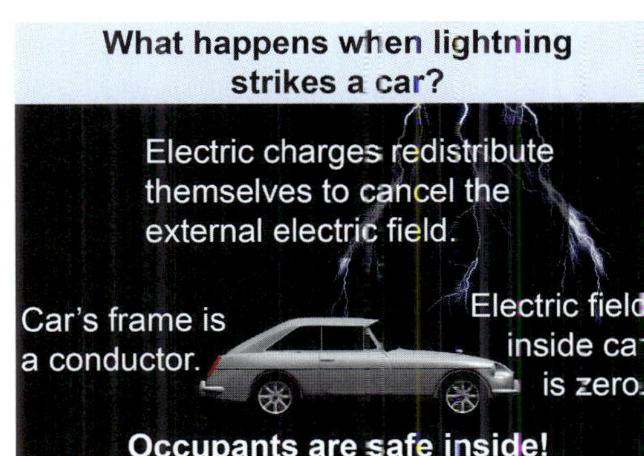

## Units of electric field

**Calculating electric field**

The electric field is a vector field that defines the strength and direction of the electric force in space. The magnitude of the electric field is given by equation (18.2). The units of electric field are newtons per coulomb (N/C). For example, if the electric field strength at a certain point in space equals 1 N/C, this means a 1 C charge placed at this point would feel a 1 N force. A 0.5 C charge would feel a 0.5 N force.

**Comparison with gravity**

The force of gravity near the Earth's surface is $F_g = mg$, whereas the electric force on a charged body is $F_e = qE$. The acceleration of gravity, $g$, can be viewed as a field in a similar way to the electric field $E$. The units of $g$ are N/kg and the units of electric field $E$ are N/C.

**How are charge and voltage related?**

We introduced the *volt* as a unit of electrical potential energy on page 477. We can now extend our earlier definition. A charge of 1 C gains 1 J of energy when it moves across 1 V of potential difference, so 1 V is 1 J/C.

Electric field: $\frac{\text{newton}}{\text{coulomb}} = \frac{\text{volt}}{\text{meter}}$

$$\frac{1\text{ N}}{1\text{ C}} = \frac{1\text{ N V}}{1\text{ J}} = \frac{1\text{ N V}}{1\text{ N m}} = \frac{1\text{ V}}{1\text{ m}}$$

volt: $1\text{ V} = \frac{1\text{ J}}{1\text{ C}}$ joule: $1\text{ J} = 1\text{ N m}$

**How are voltage and electric field related?**

The volt provides a much more practical way to define the electric field. One newton per coulomb is the same as one *volt per meter.* Consider the electric field between two parallel plates. If the plates are separated by 1 cm (0.01 m) and connected to a 1.5 volt battery, the electric field between them is 150 V/m. A 1 C charge placed between the plates feels a force of 150 N because 150 V/m = 150 N/C.

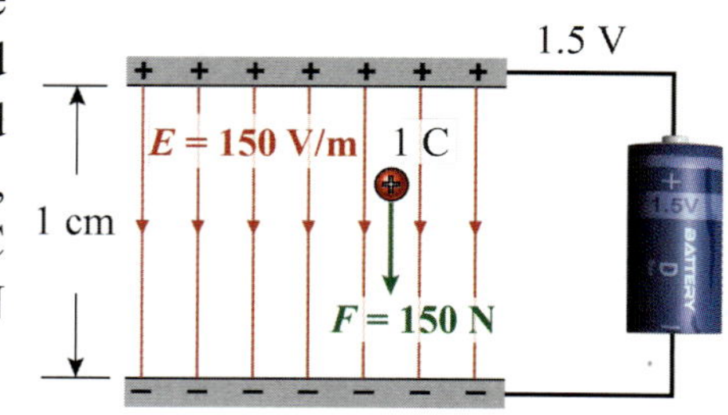


**Electric field of a point charge**

How can we calculate the electric field at a location near a point charge $q$? Assume we have a small, imaginary "test charge" $q_2$ located a distance $r$ away. Dividing Coulomb's law by $q_2$ gives the field of a point charge (equation 18.3):

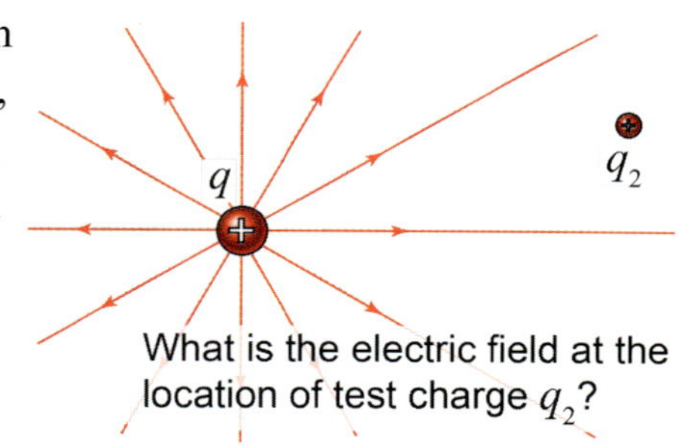

What is the electric field at the location of test charge $q_2$?

$$E = \frac{F_e}{q_2} = \frac{k_e q q_2/r^2}{q_2} = \frac{k_e q}{r^2}$$

The value of the test charge $q_2$ cancels out of the equation for electric field—which it should, because this is a *test* charge!

**Electric field for a point source**

$$(18.3)\quad E = \frac{k_e q}{r^2}$$

$E$ = electric field (N/C)
$k_e$ = Coulomb constant = $9.0\times10^9$ N m$^2$/C$^2$
$q$ = electric charge (C)
$r$ = distance between the two objects (m)

**Electric force and electric field**

Do you see the similarity between the electric *field* of a point charge in equation (18.3) and the electric *force* between two point charges in Coulomb's law (equation 18.1)? You simply multiply the electric field by charge $q_2$ to determine the electric force. Electric field is the *electric force per unit charge*, as you can also see in equation (18.2). The electric field of a point charge falls off as $1/r^2$, in the same way that the Coulomb force falls off with the distance squared.

**Discovering the charge on the electron**

By 1908 it was known that the atom contained positive and negative charges. One of the big unknowns was what the charges were. How could one measure the charge $e$ on a *single* electron? The value of the elementary charge $e$ is a fundamental quantity of physics, and its measurement was a challenge for experimenters of the day. The crucial experiments were done between 1908 and 1913 by the American physicist Robert Millikan at the University of Chicago. Working with his graduate student Harvey Fletcher, Millikan undertook a grueling series of very sensitive experiments that determined the charge on the electron to be approximately $1.59\times10^{-19}$ C. This is within 0.6% of its modern value of $1.602\times10^{-19}$ C.

**Millikan's apparatus**

In Millikan's experiment very tiny oil mist droplets are allowed to fall between two charged metal plates. Friction from an atomizer nozzle leaves each droplet with a small static charge. The voltage between the plates can be adjusted from 0 to >1,000 V, creating a uniform electric field between the plates. The electric field is adjusted until a droplet is suspended in equilibrium by balancing its weight against the electric force.

**Millikan's oil drop experiment**

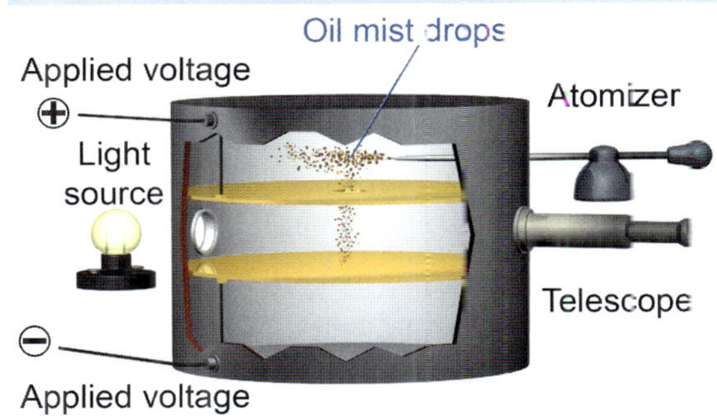


**Analyzing the data**

Using a telescope, Millikan adjusted the voltage until only a single drop was in sight. By carefully raising and lowering the voltage, and measuring the velocity of the oil drop, Millikan was able to determine the charge on one drop to a very precise value. His data showed that the *charge on an oil drop always occurred in multiples of $1.59\times10^{-19}$ C!*

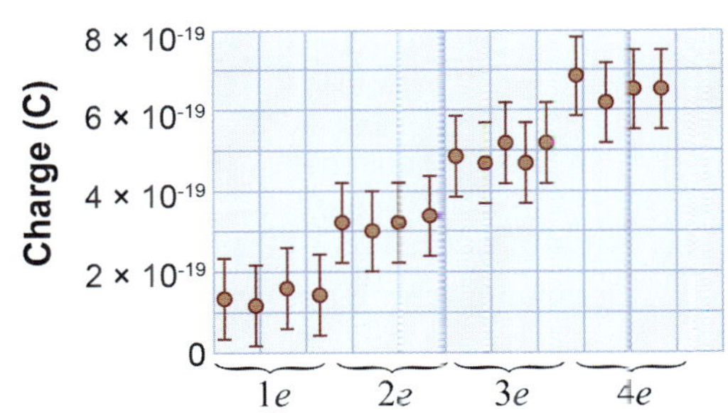


**The physics of Millikan's experiment**

The plate voltage provided Millikan with a very accurate measure of the electric field. To determine the charge, however, he also needed to know the mass of the oil droplet. To determine the mass, Millikan measured the *terminal velocity* as each droplet fell at constant speed. When falling at its terminal velocity, the weight of the oil drop is balanced against the air resistance, buoyancy, and electric forces combined. In a more advanced course on fluid dynamics, you may learn that the drag force $F$ on a sphere of radius $r$ moving through a liquid is $6\pi r\eta v$, where $v$ is the velocity and $\eta$ is the viscosity of the fluid (or air in this case). Once he had the radius, Millikan could work out the mass using the density of the oil. Millikan's early value for $e$ is lower than the modern value mostly because he underestimated the viscosity of air.

**Forces on a falling oil droplet**

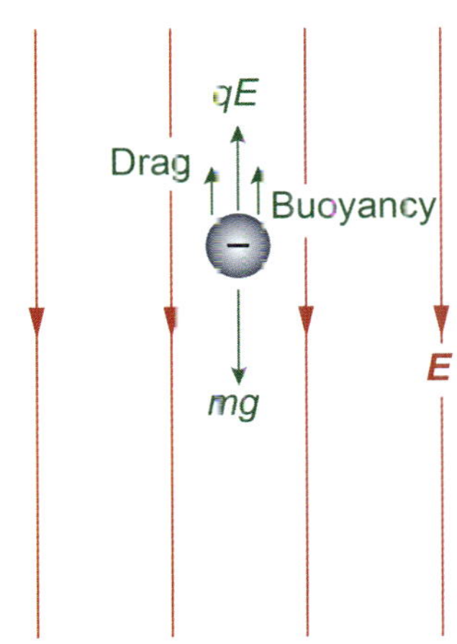

Chapter 18

# Section 3 review

A force field diagram is a tool for visualizing the electric forces around charged objects. The electric field is a vector that represents the force on a positive test charge. By convention, the *electric field* points in the direction of the electrostatic force on a *positive* test charge. Because of this, electric field lines point away from positive source charges and toward negative ones. The strength of the electric field (in newtons per coulomb, or N/C) is the force felt by a charged particle divided by the particle's electric charge. The closer the field lines are to each other, the stronger is the force.

| | |
|---|---|
| **Vocabulary words** | **electric field, electric field lines** |
| **Key equations** | $F_e = qE$ or $E = \frac{F_e}{q}$      $E = \frac{k_e q}{r^2}$ |

## Review problems and questions

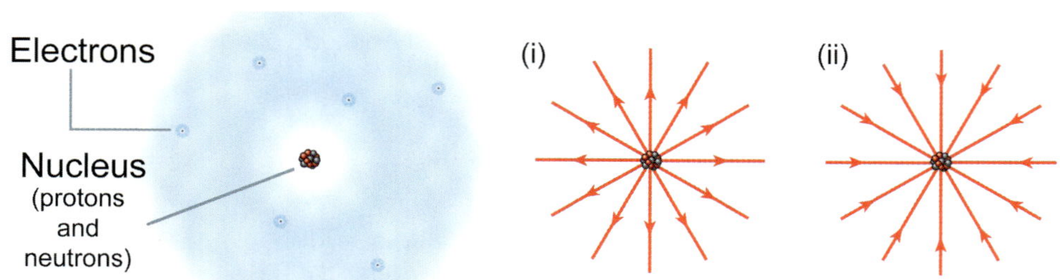


1. An atom consists of a positively charged nucleus surrounded by one or more negatively charged electrons, which occupy a region of space known as the electron cloud.
   a. Which of the diagrams depicts the *electric field* generated by the nucleus?
   b. Which diagram correctly shows the directions of electric forces *on the electrons*?
   c. Did you choose the same diagram for the previous two questions? Why or why not?

2. A commonly encountered unit of *charge* in consumer electronics is the milliamp-hour, or mAh, and in the USA the pound, or lb, is a commonly used unit of *force*. How many newtons per coulomb (N/C) would be equivalent to 1 lb/mAh?

3. Rebecca inflates a rubber party balloon, ties it to a string, and hangs it from the ceiling. Then she rubs the balloon with a wool sweater. Suppose that she transfers one trillion "extra" electrons ($10^{12}$ $e^-$) to the balloon. Assume, too, that you can pretend that all of this excess charge is located at the center of the balloon. What is the electric field strength 1 m from the center of the balloon?

4. Declare each of the following statements as *true* or *false*, based upon the accompanying electric field diagram.
   a. Object A is positively charged.
   b. Objects A and B have opposite charges.
   c. The electric field at point $x$ is stronger than that at point $y$.

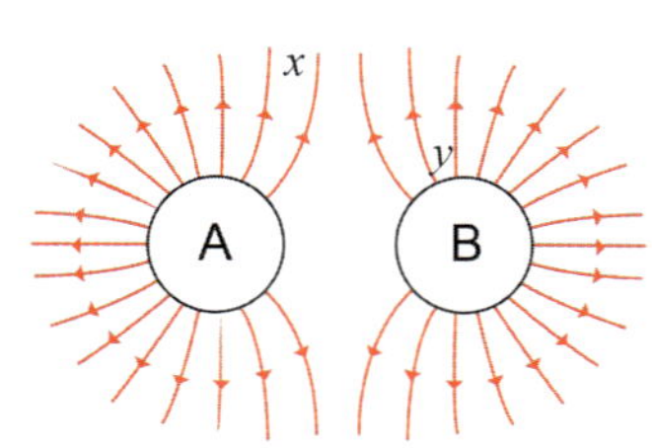

Stray electric fields create interference with many electronic devices, which can be addressed through shielding. Although you might think that electric fields and electronic devices are always best kept apart, you would be wrong! Electric fields inside conducting wires are what causes currents to flow. Furthermore, electric fields are used for a key component found on most circuit boards: the capacitor. Capacitors are used to store charge and electrical energy. In this section, you will learn more about electric fields and how two charged, parallel plates are used to create a capacitor.

## Electric potential

**Electric potential in circuits**

In Chapter 17 on page 492, you learned that any resistor connected across the terminals of a battery has the same voltage drop. An alternate way to describe this is that the positive terminal of the battery is at a high **electric potential**, and the negative end of the battery is at a low electric potential. The voltage *across* a battery—the difference between its two terminals—is also called the *potential difference*. In some textbooks you will find the "voltage" across a resistor, say, referred to as the "potential difference" across the resistor. The potential difference is simply the difference in electric potential from one side of the device to the other!

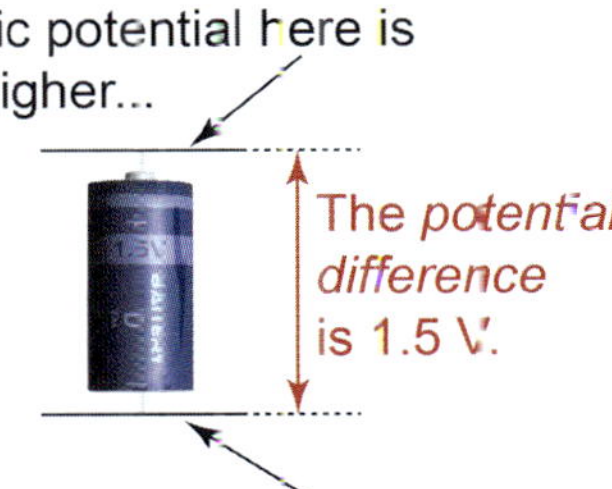


**Electric potential and electric fields**

While this is a useful description of electric potential for electric circuits, for electric fields it is more useful to define electric potential as the *potential energy per unit charge*, as in equation (18.4). Electric potential has units of joules per coulomb—although it can also be expressed in units of volts.

**Electric potential**

$$(18.4) \qquad V = \frac{E_p}{q}$$

$V$ = electric potential (V or J/C)
$E_p$ = electric potential energy (J)
$q$ = electric charge (C)

**Reference potential**

You can define the zero or reference electric potential at any location you want, just as you did for gravitational potential energy! The reference potential ought to be chosen in a way that is most simple or convenient. For electric circuits, the reference potential is usually chosen at the battery's negative terminal or at ground (which are often the same).

**Point source**

The electric potential for a point charge gets smaller the further away you are from the charge. The electric potential is also higher for higher electric charge. These relationships are part of equation (18.5), which describes the electric potential around a point source of charge $q$.

**Electric potential for a point source**

$$(18.5) \qquad V = \frac{k_e q}{r}$$

$V$ = electric potential (V)
$k_e$ = Coulomb constant = $9.0 \times 10^9$ N m$^2$/C$^2$
$q$ = electric charge (C)
$r$ = distance between the two objects (m)

**Properties of electric potential**

Inspection of equation (18.5) shows that the electric potential can be positive near a positive charge $q$ while it can be negative near a negative charge. (It is zero, of course, at its reference point.) This should not be surprising, since the electric (Coulomb) force can be both positive and negative, depending on whether the charges have the same or opposite sign.

## Equipotential lines

**Drawing electric potential**

The most useful visual representation of the electric potential is formed by mapping *lines of constant electric potential* called **equipotential** lines. For a point charge, the electric potential decreases as $1/r$ from equation (18.5), so the equipotential lines are concentric circles, as shown in the illustration below. Notice that, while the electric field direction may vary around the point charge, the electric field always points *perpendicular* to the equipotential lines.

**Equipotential lines**

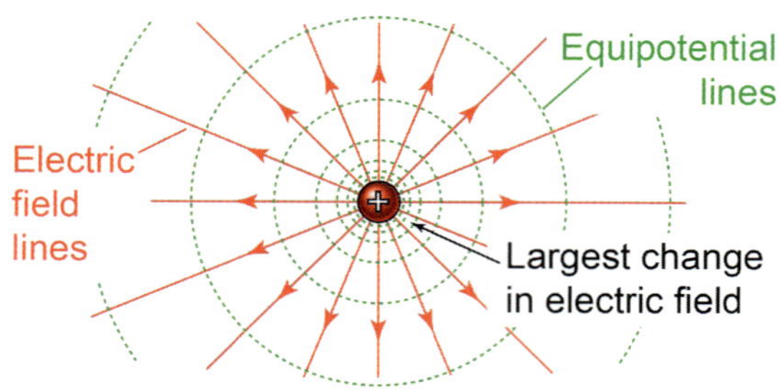


If you remain on an equipotential line, then the electric potential stays the same.

**Topographical map**

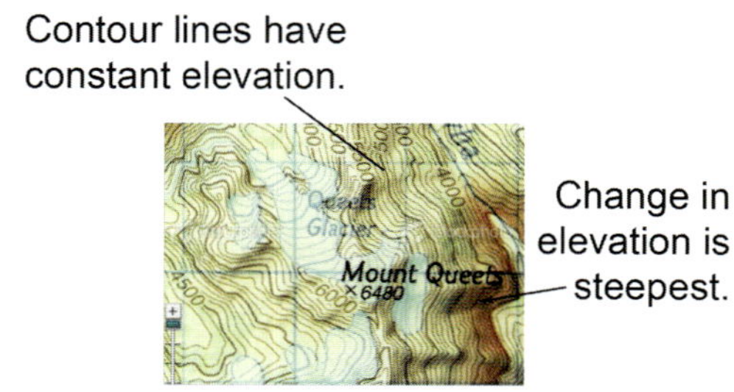


If you walk along a contour line, then your elevation remains the same.

**Rapidly changing electric field**

In a topographical map, the location where the contour lines are most "bunched up" is where the elevation is changing most rapidly. In a similar way, the density of equipotential lines indicates how rapidly the electric field is changing at that location. In the illustration above, the electric field is changing most rapidly near to the electric charge, as is indicated by how the electric potential lines bunch up there the most.

**Electric potential and conductors**

The electric field is zero everywhere inside a conductor, as you learned earlier on page 533. If the electric field is the same (zero!) everywhere inside a conductor, then all of the conductor is at the same potential—i.e., it is an equipotential. This is a general property of conductors in equilibrium, that they have the same electric potential everywhere inside.

**Reading equipotential lines**

If you are given a diagram with equipotential lines, you can infer the properties of the electric potential and electric field using some simple rules:

- The direction of the electric field is always *perpendicular* to equipotential lines,
- Electric field lines point from positive to negative electric potential,
- Locations where equipotential lines are "bunched up" correspond to a strong electric field.

**Reading equipotential diagrams**

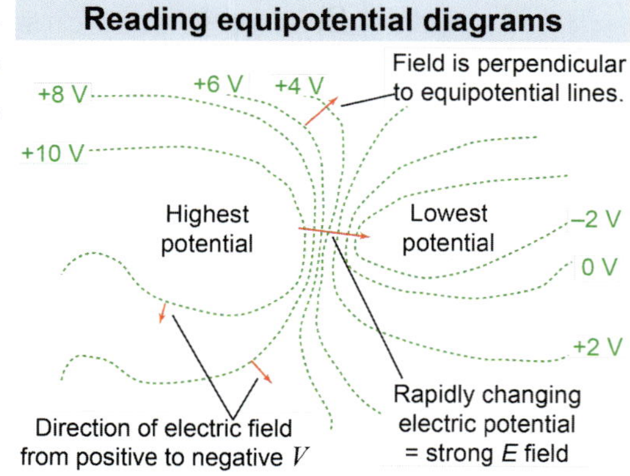

**Work and electrical energy**

Electric potential is the electrical potential energy per unit charge, as you learned on page 537. But what is the electrical potential energy for a given set of electric charges? The answer can be found by considering how much work is required to put those charges into their positions. This connection between work and energy should remind you of the work–energy theorem!

**Work done in an electric field**

What is the electric potential energy of an electric charge $q_2$ in the vicinity of another charge $q_1$? We will answer this question using work and energy by imagining *how much work has to be done to move $q_2$ to its location*. From equation (18.4), the energy of a charge in an electric potential is $E_p = qV$.

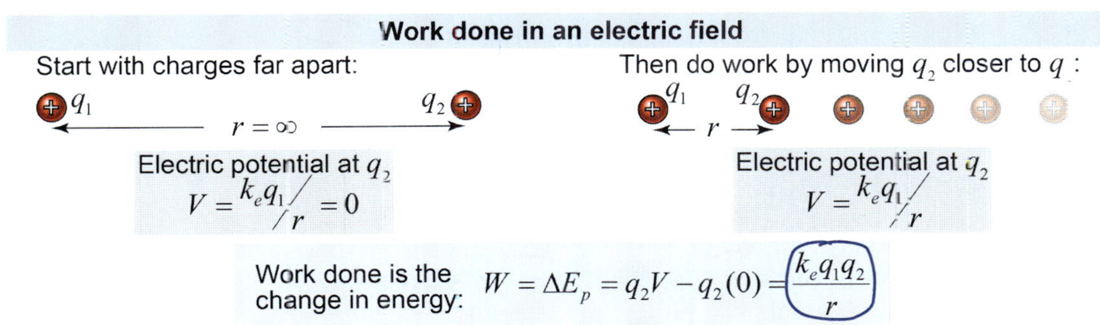


When the two charges are far away from each other, $r$ becomes very large and the potential $V$ goes to zero from equation (18.5). The energy likewise goes to zero when the charges are far apart. When the charges are close, however, the energy is $E_p = qV = k_e q_1 q_2 / r$.

**Electric potential energy**

The difference between the final energy and the initial energy is the work done to assemble the system—and is the electrical potential energy of the assembled system. The work done to move the charge into place provides an equation for the **electric potential energy** of a pair of point charges. It takes work to move two positive charges together, so the electric potential energy for two positive charges is positive in equation (18.6). It takes work to separate a positive and a negative charge, so their electric potential energy is negative. Electric potential energy can be thought of as being stored in their mutual electric field.

**Electric potential energy** for two charges

**(18.6)** $$E_p = \frac{k_e q_1 q_2}{r}$$

$E_p$ = electric potential energy (J)
$k_e$ = Coulomb constant = $9.0 \times 10^9$ N m$^2$/C$^2$
$q_1$ = electric charge #1 (C)
$q_2$ = electric charge #2 (C)
$r$ = separation (m)

**Parallel and perpendicular to electric field**

Moving two charges together or apart is equivalent to saying that they are being moved *along (or parallel to) the electric field lines*. It takes work to move a charged particle along an electric field line. How about if you moved a charged particle *along an equipotential line*? Since the electric potential does not change along an equipotential line, the electric potential energy also does not change from equation (18.4). Since equipotential lines are *perpendicular* to the electric field, no work is done against the electric force to move perpendicular to the electric field.

**Capacitors**

Capacitors are found in many circuits and electronics devices. A **capacitor** is a device for storing electric charge and hence electrical energy. Inside the capacitor are two parallel metal plates separated by a material called the *dielectric*. To make capacitors that store more charge, they are often wound into a spiral along with an insulating layer. The electrical symbol for a capacitor consists of two parallel lines that represent the parallel metal plates inside the device.

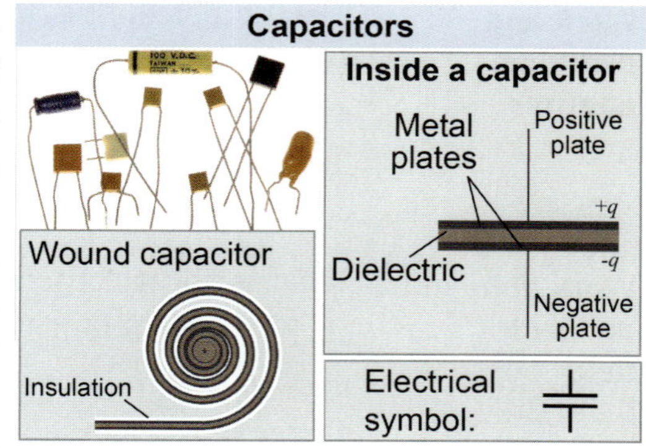


**Charging a capacitor**

A capacitor is uncharged until it is connected to a voltage or potential difference across its two metal plates. In circuits, a capacitor is usually connected in series with a resistor and battery. When the circuit is completed (or the switch closed), the capacitor begins to charge. As positive charge accumulates on one plate, an equal amount of negative charge accumulates on the other plate. The difference in charge creates a potential difference or voltage between the two plates.

Charging a capacitor

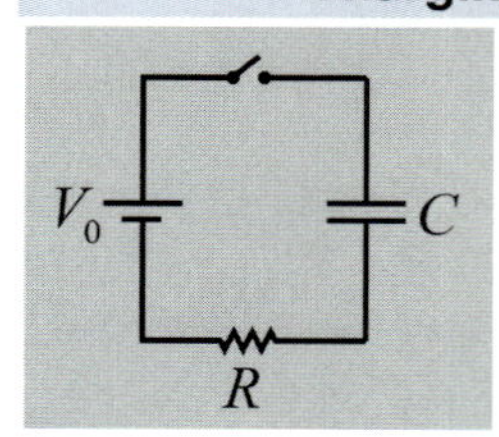

Capacitors are usually connected in series with a resistor.

When the switch is closed, the capacitor begins to charge.

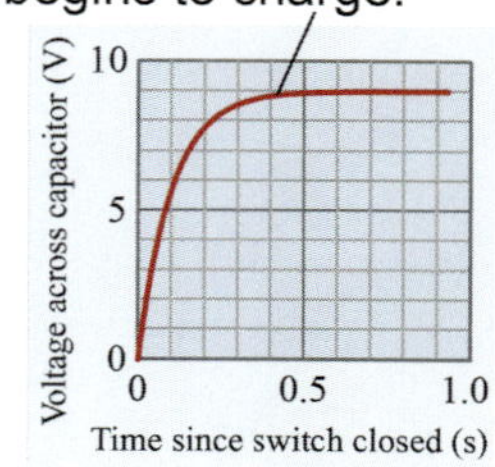


**Charge stored in a capacitor**

The charging time for most capacitors in circuits is very short—milliseconds or less! Once fully charged, the charge on each of its plates is given by equation (18.7).

| | | | |
|---|---|---|---|
| **(18.7)** | $q = CV$ | $q$ = electric charge (C)<br>$C$ = capacitance (C/V)<br>$V$ = electric potential (V) | **Charge stored in a capacitor** |

**Capacitance**

The new quantity in equation (18.7) is the **capacitance**, which is a measure of how effectively the capacitor can store charge. The units of capacitance are the farad (F), named after Michael Faraday. Just as one coulomb is a very large amount of charge, one farad is also a very large capacitance. Typical values for capacitance range from a picofarad (1 pF = $10^{-12}$ F) to a microfarad (1 μF = $10^{-6}$ F). Since $q = CV$, capacitance can also be expressed as coulombs per volt (C/V).

**Energy stored in a capacitor**

Since one plate of a capacitor stores positive charge while the other stores negative charge, a capacitor can also be considered as a store of electrical potential energy. The stored electrical energy in a capacitor is given by equation (18.8). As a larger voltage is applied across a capacitor, the stored energy increases—as $V^2$!

| | | | |
|---|---|---|---|
| **(18.8)** | $E_p = \frac{1}{2}CV^2$ | $E_p$ = stored energy (J)<br>$C$ = capacitance (F)<br>$V$ = voltage or potential difference (V) | **Energy stored in a capacitor** |

**Electric field between charged metal plates**

Inside a capacitor are two metal plates that are separated by a small gap. When positive charge builds up on one plate, an equal but opposite charge builds up on the other plate. The charged plates create an electric field between them. In the center, the electric field is nearly uniform, but the field curves a bit near the edges of the plates. Charged parallel plates are used to create uniform electric fields, such as when focusing the electron beam in a cathode ray tube.

**Electric field between two charged metal plates**

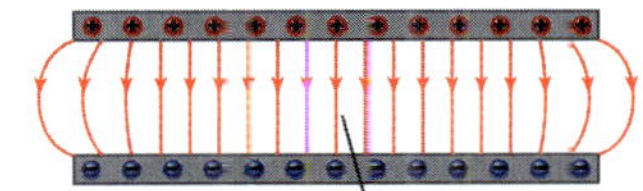

Electric field is approximately uniform throughout the center.

**Speed of an electric charge between two plates**

The electric field created between charged parallel plates can also be used to accelerate charged particles parallel to the field direction. If a positively charged particle is set free close to the positively charged plate, how fast will it be moving when it reaches the negatively charged plate? This problem can be solved using energy conservation. Initially, the charged particle has positive electrical potential energy $E_p = qV$ but no kinetic energy (because it starts at rest). At the end, it has kinetic energy $E_k = \frac{1}{2}mv^2$ but no electrical potential energy (because we set the reference potential as zero for the "negative" plate). Equating initial and final energies leads to the final velocity for the charged particle:

$$E = qV = \tfrac{1}{2}mv^2 \quad \Rightarrow \quad v = \sqrt{\frac{2qV}{m}}$$

**Velocity of a charged particle between two parallel plates**

Initial: $E_k = 0$ $\quad E_p = qV$

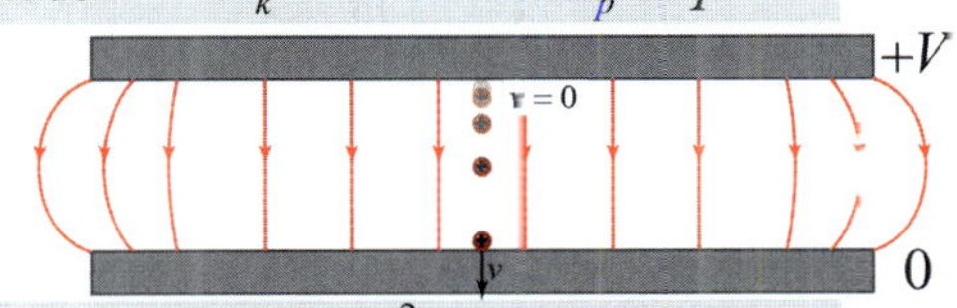


Final: $E_k = \frac{1}{2}mv^2$ $\quad E_p = 0$

Energy conservation $E_{tot} = qV = \frac{1}{2}mv^2$ $\Rightarrow$ Final velocity $v = \sqrt{\frac{2qV}{m}}$

**Capacitance of parallel plates**

How can you calculate the capacitance for two parallel metal plates? The capacitance increases as the area of the plates is increased. The capacitance also increases as the separation between the plates is *decreased*; the closer the plates are to each other, the more charge the capacitor can store per volt. The capacitance for a parallel plate capacitor is given in equation (18.9).

**Capacitance of parallel plates**

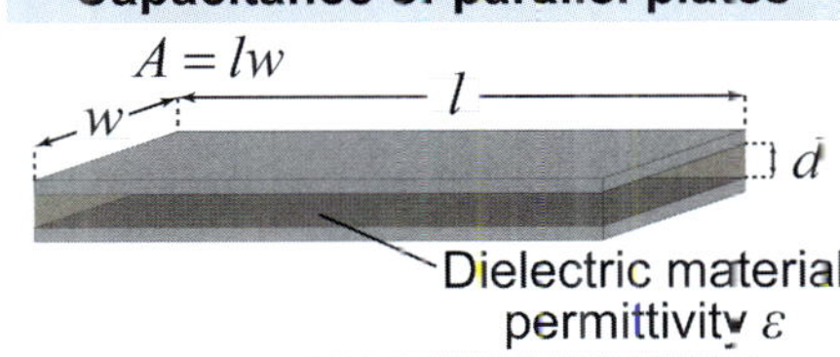


Capacitance $\quad C = \dfrac{\varepsilon A}{d}$

$$\textbf{(18.9)} \quad C = \frac{\varepsilon A}{d}$$

$C$ = capacitance (F)
$\varepsilon$ = permittivity ($C^2\ N^{-1}\ m^{-2}$)
$A$ = area of the plates ($m^2$)
$d$ = separation between plates (m)

**Capacitance**
parallel plate capacitor

**Dielectrics**

In equation (18.9), a new quantity was introduced: the *permittivity* $\varepsilon$ of the material between the plates. The permittivity is a measure of how much electric field is created in a material per unit charge. The permittivity of a vacuum is $\varepsilon_0 = 8.9\times10^{-12}\ C^2/(N\ m^2)$. Materials with higher values of $\varepsilon$ are used between the plates to improve the strength of the electric field and hence increase capacitance. Typical dielectrics improve the performance of capacitors by factors of 2–8.

### Stored charge and voltage

The most common use of parallel plate capacitors is in small devices in electric circuits or electronics boards. In these electrical applications, the stored charge on the capacitor can be calculated using equation (18.7), and the stored electrical potential energy can be calculated using equation (18.8).

### Capacitors in parallel

When more than one resistor is connected, the properties of the circuit can be modeled using an *equivalent resistance*, as you learned on page 491. Similarly, two or more capacitors in a circuit can be modeled with an *equivalent capacitance*. When connected in parallel, the total charge on each capacitor is $q = CV$, so the combined charge is

**Capacitors connected in parallel**

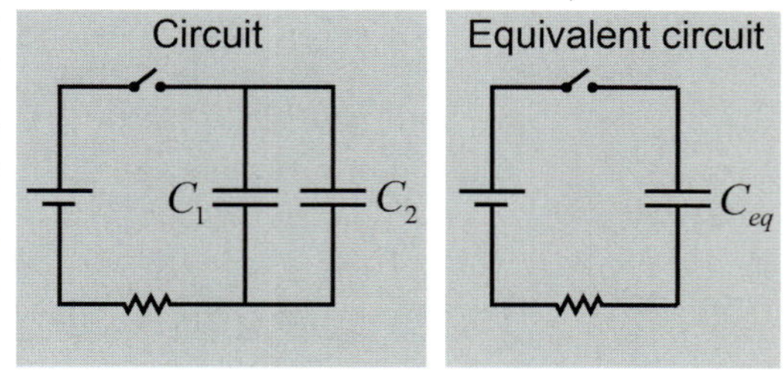


$$C_{eq} = C_1 + C_2$$

$$\begin{aligned} q_{eq} &= q_1 + q_2 = C_1V + C_2V \\ &= (C_1 + C_2)V \end{aligned}$$

This means that the equivalent capacitance is the direct sum of the individual capacitances $(C_1 + C_2)$ when they are connected in parallel.

**(18.10)** $C_{eq} = C_1 + C_2$

$C_{eq}$ = equivalent capacitance (F)
$C_1$ = capacitance #1 (F)
$C_2$ = capacitance #2 (F)

**Equivalent capacitance**
connected in parallel

### Capacitors in series

When two capacitors are connected in series, the charge collected on each capacitor must be the same because of conservation of charge. In other words, the negative charges that build up on the plate of $C_1$ must be equal but opposite to the charges that build up on the positive plate of $C_2$ because the electrical wire between the capacitors is isolated and electrically neutral. The total voltage across the two capacitors is the *sum* of the voltages across each capacitor, or $V = V_1 + V_2$. The individual capacitors have charges $q = C_1V_1$ and $q = C_2V_2$. The ratio of charge to effective capacitance is equal to this combined voltage:

**Capacitors connected in series**

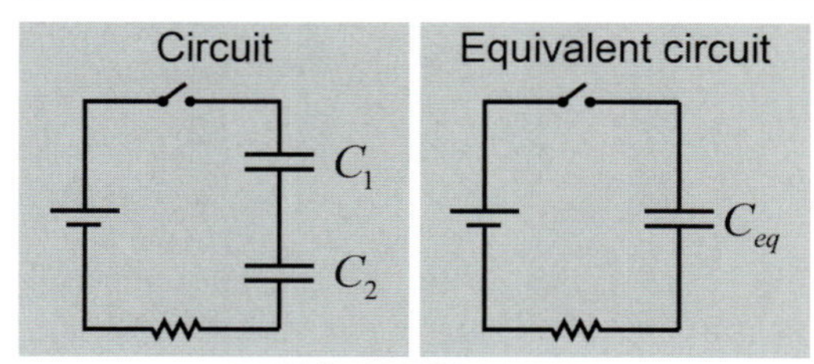


$$\frac{1}{C_{eq}} = \frac{1}{C_1} + \frac{1}{C_2}$$

$$\frac{q}{C_{eq}} = V = V_1 + V_2 = \frac{q}{C_1} + \frac{q}{C_2} \Rightarrow \frac{1}{C_{eq}} = \frac{1}{C_1} + \frac{1}{C_2}$$

**(18.11)** $\dfrac{1}{C_{eq}} = \dfrac{1}{C_1} + \dfrac{1}{C_2}$

$C_{eq}$ = equivalent capacitance (F)
$C_1$ = capacitance #1 (F)
$C_2$ = capacitance #2 (F)

**Equivalent capacitance**
connected in series

### Comparison to resistors

Resistances add directly when connected in series, but they add as reciprocals (inverses) when connected in parallel. It is the opposite for capacitors! Capacitances add directly when connected in *parallel*, but they add as reciprocals when connected in *series*.

# Section 4 review

*Electric potential energy* refers to the energy stored in a charged particle when you move that particle into a particular location. *Electric potential* is the electric potential energy per unit charge, and it has units of *volts* (one volt being one joule per coulomb). *Equipotential lines* trace regions of equal electric potential, much as contour lines trace regions of equal elevation on topographic maps. A *capacitor* is an electrical component that can store a certain amount of electric charge per volt applied to it; the amount of stored charge is proportional to the applied voltage and the capacitance.

| | |
|---|---|
| **Vocabulary words** | **electric potential, equipotential, electric potential energy, capacitor, capacitance** |

| | | | | |
|---|---|---|---|---|
| **Key equations** | $V = \frac{E_p}{q}$ | $V = \frac{k_e q}{r}$ | $E_p = \frac{k_e q_1 q_2}{r}$ | $q = CV$ |
| | $E_p = \frac{1}{2}CV^2$ | $C = \frac{\varepsilon A}{d}$ | $C_{eq} = C_1 + C_2$ | $\frac{1}{C_{eq}} = \frac{1}{C_1} + \frac{1}{C_2}$ |

## Review problems and questions

1. Which of the following diagrams could be applied to a fully charged 1.5 V battery? (More than one selection may be correct.)

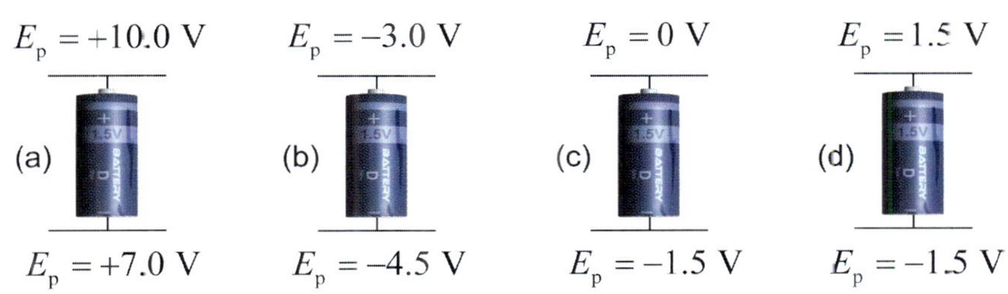


2. Charge 1 (with $q_1$ coulombs) generates the electric field shown here. Charge 2 (with $q_2$ coulombs) is brought to a distance $r$ from Charge 1.
   a. How would doubling $q_1$ affect the electric potential energy $E_p$?
   b. What about doubling $q_2$?
   c. How about doubling *both* charge values?
   d. Finally, what about doubling the distance, $r$?

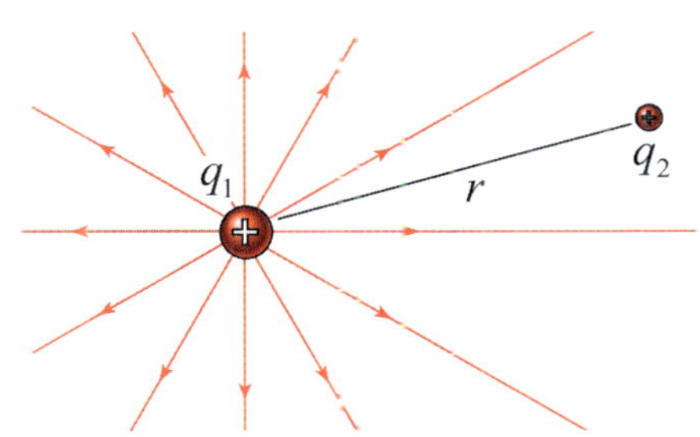


3. Two circuits are built with identical components, but one has the pair of microfarad capacitors in parallel, while the other has them in series. Which will be able to build up a greater charge?

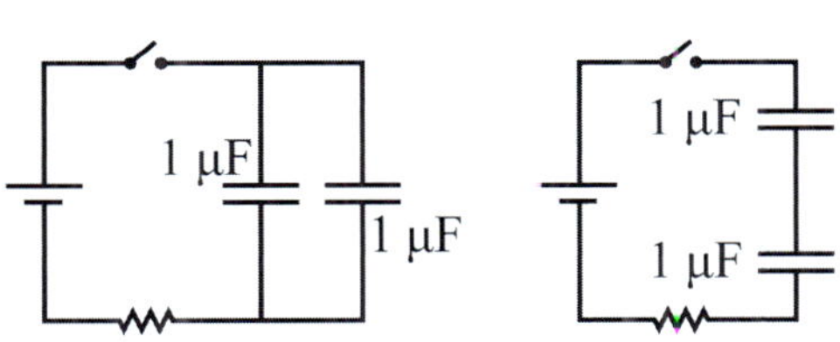


4. The positively (*top*) and negatively (*bottom*) charged plates of a parallel plate capacitor are shown here, along with the electric field between them. Draw several equipotential lines for the region between the plates. Indicate which has the highest electric potential.

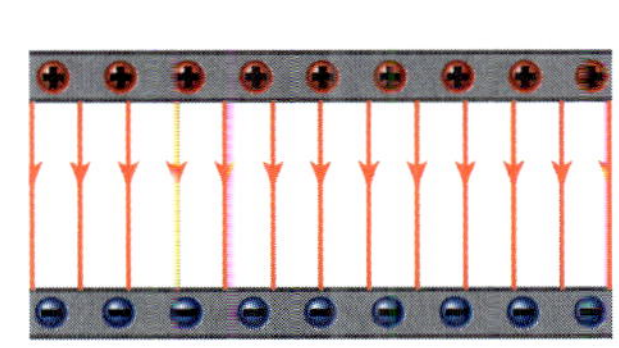

Chapter 18

# Chapter review

## Vocabulary

Match each word to the sentence where it best fits.

### Section 18.1

| | |
|---|---|
| **ferromagnetic** | **force field** |
| **magnetic** | **magnetic domain** |
| **magnetic field** | **magnetic field lines** |
| **magnetic force** | **magnetic poles** |
| **magnetism** | **magnetize** |
| **permanent magnet** | **polarity** |

1. Iron is an example of a _______, a nonmagnetic material that can be magnetized by a nearby magnet.
2. You can _______ paper clips by bringing them close to a magnet.
3. The Earth has a _______ that causes compass needles to point north.
4. When visualizing the magnetic force in a diagram, the _______ are drawn in the direction in which the north pole of a test magnet would feel the magnetic force.
5. A _______ material has both a north and south pole.
6. The Earth's _______ are misaligned with geographic north and south.
7. Labeling the positive end of an electrically charged object, or the north end of a bar magnet, are ways to indicate the _______ of each.
8. The direction and magnitude of the gravitational force or magnetic force around an object can be visualized with a _______ diagram.
9. The like poles of two bar magnets exert a _______ of repulsion.
10. A _______ is found in some paramagnetic materials and consists of a region of atoms that have mutually aligned magnetic field directions.
11. A bar magnet is an example of a _______.

### Section 18.2

| | |
|---|---|
| **Coulomb's law** | **coulomb (C)** |
| **electric charge** | **electric force** |
| **electrically neutral** | **electroscope** |
| **electrostatic induction** | **electrostatics** |
| **negative charge** | **positive charge** |
| **static electricity** | |

12. Electric charge is measured in units of the _______.
13. The proton has a/an _______.
14. An object that contains 200 C of positive charges and −200 C of negative charges is _______.
15. Electrons and protons have the same __________, but with opposite sign.
16. The study of electrical charges at rest is called _______.
17. The _______ is an instrument used to detect charged objects.
18. The equation for calculating the electric force between two electric charges is called _______.
19. _______ is illustrated by bringing an electrically charged object near an electroscope and seeing the foil leaves separate.
20. The buildup of electric charge on the surface of a balloon is an example of _______.
21. The electron has a/an _______.

### Section 18.3

| | |
|---|---|
| **electric field** | **electric field lines** |
| **electricity** | |

22. The curved lines with arrows in a diagram visualizing the electric field are called _______.
23. A positive charge will be surrounded by an _______.

### Section 18.4

| | |
|---|---|
| **capacitance** | **capacitor** |
| **electric potential** | **electric potential energy** |
| **equipotential** | |

24. In a field diagram, two locations that are both at +7.3 V are on the same _______ line.
25. The unit of _______ is named after Michael Faraday.
26. Two charged, parallel plates that are separated by a small distance form a/an _______.
27. Measuring the work required to move two charged objects near each other is equivalent to measuring their _______.
28. The _______ is the electrical potential energy divided by the charge.

# Chapter review



## Conceptual questions

### Section 18.1

29. Identify five different kinds of objects in your home that use magnets.
30. Is a refrigerator door a permanent magnet?
31. Does a compass needle show the magnitude or direction of the Earth's magnetic field?
32. ❬ Describe a procedure to use a magnetic compass to find the magnetic north pole of a bar magnet.
33. ❬ A sheet of paper is held in place by two bar magnets on either side of it. Is the paper now magnetized? Why or why not?
34. ❬ The north magnetic pole of a bar magnet is observed to demagnetize a credit card's magnetic stripe. Will the south magnetic pole of the same bar magnet also demagnetize the stripe?
35. ❬ Describe how a compass can be used to trace out the magnetic field around a bar magnet.
36. ❬ Will a compass needle point in the same or opposite direction as a magnetic field line?
37. ❬ One end of a bar magnet is labeled with a red "N" for north magnetic pole, but the other end is not labeled. Is the unlabeled end of the magnet a north magnetic pole, a south magnetic pole, or neither?
38. ❬ Sketch the magnetic field lines around a bar magnet.
39. ❬ Is Earth's magnetic north pole located near Earth's geographic north pole? Why or why not?
40. ❬ Why don't most magnets align themselves with the Earth's magnetic field?
41. ❬ When you want to trace out the magnetic field of a bar magnet, why do you place a piece of paper or poster board on top of a bar magnet before pouring iron filings out of a container onto it?
42. ❬ As a compass moves around a bar magnet, the compass needle is repeatedly caused to move. What is the source of the needle's kinetic energy?
43. ❬❬ Cynthia is standing at the location that corresponds to Earth's north magnetic pole. What direction will her compass point? How about if she were standing at Earth's south magnetic pole?
44. ❬❬ Why don't compasses work when they are held upside-down or sideways?
45. ❬❬ A physics teacher opened a box of bar magnets and found that none of them had north and south ends labeled. How could he figure out their magnetic poles? What other equipment would he need?
46. ❬❬ Explain the differences between how you would find the direction of geographic north if you had a compass but were located in (a) California, (b) Missouri, or (c) Maine.
47. ❬❬ When you use iron filings to trace out the magnetic field around a bar magnet, how can you distinguish the magnet's north and south magnetic poles from each other?
48. ❬❬ Two bar magnets are connected together, end to end, with the north end of one connected to the south end of the other. Sketch the magnetic field lines around this composite bar magnet.
49. ❬❬ Imagine two identical bar magnets placed side by side, with their poles aligned in the same direction. Does the magnetic force cause a torque on either bar magnet? How about if their poles were aligned in opposite directions?
50. ❬❬ Imagine two identical bar magnets placed next to each other such that the north magnetic pole of one is near the middle of the other bar magnet. Does the magnetic force cause a torque on either bar magnet?
51. ❬❬ What is the evidence that Earth's magnetic field can be represented by a large bar magnet?
52. ❬❬ Which has more energy available, the magnetic field created when the north poles of two bar magnets are moved to a separation of 1 cm or 10 cm?
53. ❬❬❬ What might cause the Earth's magnetic field to *change* over time?

### Section 18.2

54. Why does the gravitational force between the Sun and Earth dominate over the electrical force?
55. Do negatively charged particles attract or repel each other? What about positively charged particles?
56. Why don't the electrical charges in your body attract those in the wall?
57. Why is it a bad idea to discharge static electricity on the circuit board inside a computer?
58. Provide three examples of electric force in everyday life.
59. ❬ Why does a comb sometimes attract your hair toward it?
60. Why do you sometimes get shocked when you touch a doorknob on a dry day?

# Chapter review

## Section 18.2

61. Which equation below correctly matches the meaning of this sentence: "The electrostatic force between two charged particles equals the product of their charges divided by the square of the distance between them, all multiplied by a constant."
    a. $F = k_e \frac{q_1 q_2}{r}$
    b. $F = k_e \left(\frac{q_1 q_2}{r}\right)^2$
    c. $F = k_e \frac{q_1 q_2}{r^2}$
    d. $F = k_e \frac{{q_1}^2 {q_2}^2}{r}$
62. Lightning carries enormous electrical energy. What are the electrical characteristics of lightning bolts? How do they form? Use advanced searches to gather information from *multiple* authoritative sources. Integrate the information into a cohesive scientific narrative. Include an assessment of the strengths and limitations of each source.
63. When the leaves of an electroscope repel one another, do they have like or unlike charges?
64. ❮ Why do charged objects seem to hold more static charge in the winter?
65. ❮ What causes clothes to cling together when removed from a dryer?
66. ❮ Is it required for a charged object to touch the ball of an electroscope for its leaves to spread?
67. ❮ Describe three differences between the electric force and the gravitational force.
68. ❮ What is the relationship among electric charge, distance, and electric force?
69. ❮ If two charged objects are moved to twice the separation, how does the mutual electric force between them change?
70. ❮ If the distance between two charged objects is halved, how does the mutual electric force between them change?
71. ❮ What is happening when we ground an object (or connect it to ground)?
72. ❮ Gratia rubs two balloons against her sweater and discovers that the balloons attract her hair. Will the balloons attract or repel each other or neither?
73. ❮ What weather conditions are conducive to static electricity buildup?
74. ❮ How can you use an electroscope to determine whether an object were electrically charged or neutral?
75. ❮❮ How can you charge an object with positive charges only using another object that has negative charge?
76. ❮❮ Two objects experience a mutual electrical force equal to their mutual gravitational force. If they are moved to twice the distance apart, what is the new relationship between their mutual electrical and gravitational forces?
77. ❮❮ How can you use an electroscope to determine whether two electrically charged objects have the same *polarization*, i.e., the same sign for their net charge?
78. ❮❮ Why are electric charges distributed on the *surface* of a metal sphere rather than throughout its interior?
79. ❮❮ A small Van de Graaff generator might have a voltage of a thousand volts or more. Why don't people die when they touch it in a lab?

## Section 18.3

80. An isolated, positively charged sphere creates an electric field in the surrounding region of space. The electric field has a value of $E_1$ at a distance of 1 m from the charge.
    a. If you double the distance to 2 m, by what factor does $E$ change?
    b. If you triple the distance to 3 m, by what factor does $E$ change?
    c. At what distance from the charge is the field equal to zero?
81. ❮ List three kinds of force fields.
82. ❮ List three examples of contact forces, and three examples of noncontact forces.
83. ❮ Sketch the electric field around two identical, positively charged particles separated by a small distance.
84. ❮ An electric force field diagram is composed entirely of arrows pointing from left to right. In what direction will a negatively charged object feel an electric force?
85. ❮ The electric field around a charged object has field lines that, everywhere on its surface, point toward its center. What is the sign of its electric charge?
86. ❮ A 1 C charge placed at a given point in space feels a force of 100 N.
    a. What force would be felt by a 2 C charge placed at this same location?
    b. What force would be felt by an 0.5 C charge placed at this same location?
    c. What is the magnitude of the electric field at this location?

# Chapter review

Chapter 18

## Section 18.3

87. Can an electrically neutral object be polarized?

88. Sketch the electric field around three identical, negatively charged particles, located at the corners of an equilateral triangle.

89. Sketch the electric field between two particles, one with charge +2 μC and the other with charge +1 μC.

90. Xiu and Yves were preparing to make a sensitive measurement of electric charge using an electroscope. They noticed, however, that every time Yves brought his tablet computer close to the electroscope to take notes, the leaves of the electroscope would spread apart. What was going on?

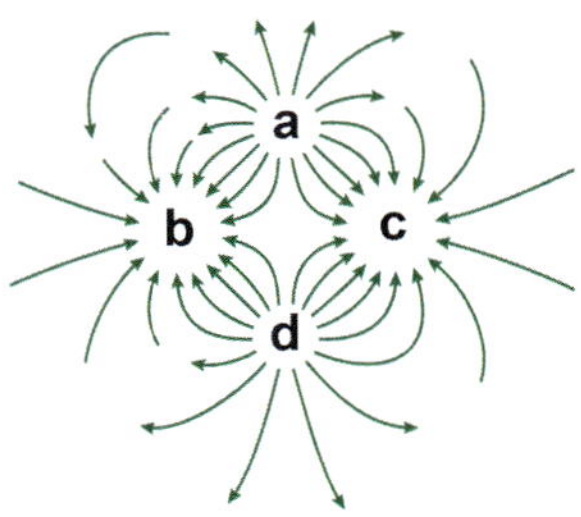


91. What four electric charges create the electric field in the diagram shown? Identify which are positive and which are negative.

## Section 18.4

92. *Electric potential* states how much energy is available per unit _______.

    a. mass
    b. charge
    c. current
    d. capacitance

93. Two capacitors are each charged by a 5 V battery. If the capacitance of Capacitor A is twice that of Capacitor B, what is the ratio of the charge that each holds?

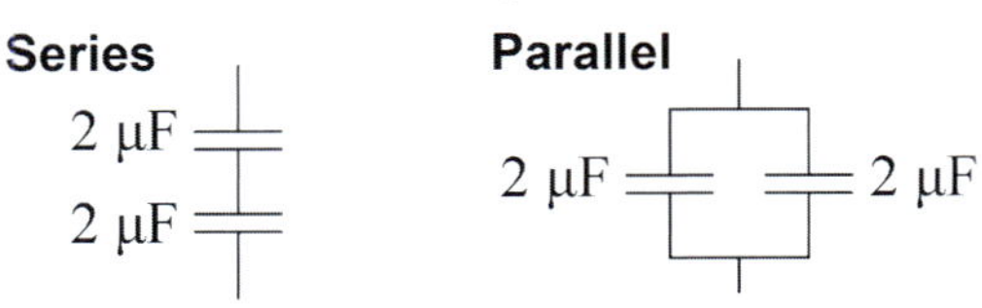


94. Eva has two capacitors, each with a capacitance of 2 μF. She would like to install the equivalent of one 4 μF capacitor into a circuit she is building. Should she put the two capacitors in series or parallel?

## Quantitative problems

## Section 18.1

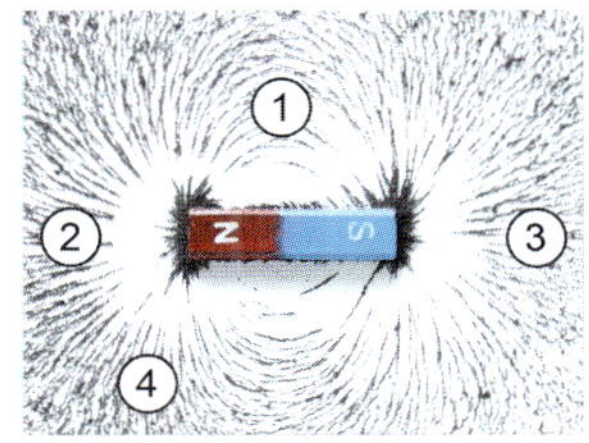


95. For the bar magnet in the figure, draw the direction the compass needle will point for each of the four locations identified in the figure.

96. A large bar magnet (mass of 0.4 kg) exerts a 5 N force on a small bar magnet (mass of 0.1 kg) located 20 cm away. Calculate the force exerted by the small bar magnet on the large one.

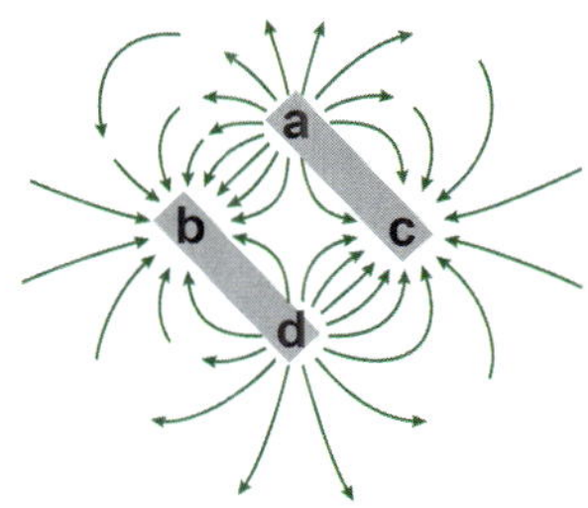


97. Two bar magnets create the magnetic field in the diagram. Which are north poles and which are south poles?

98. Two hikers consult a map and determine that, at their location, the magnetic declination is 10° west. If they hike 1 km north according to their compass needle, then how far off will they be from 1 km true north?

99. The magnetic force $F_m$ between two bar magnets was measured for various separations $d$ between them and tabulated below. Does the magnetic force follow an inverse-square law? Why or why not? Draw a graph to justify your conclusion.

**Magnetic force between two bar magnets**

| Distance $d$ (cm) | Force $F_m$ (N) |
|---|---|
| 5.0 | 3.0 |
| 10.0 | 0.38 |
| 15.0 | 0.11 |
| 20.0 | 0.047 |
| 25.0 | 0.024 |

## Section 18.2

100. Two charged objects are located near each other. How does their mutual electric force change if their:

    a. separation is doubled?
    b. separation is halved?
    c. individual charges are each doubled?

101. One coulomb of charge passes through a point in an electric circuit every ten seconds. What is the current in the circuit?

102. Two identical charges are located 1 m apart and feel a 1 N repulsive electric force. What is the charge of each particle?

103. A positive and a negative charge are located 25 cm apart and feel an attractive electric force of 5 N. What is the direction and magnitude of the electric force they feel when separated by 50 cm?

104. Compare the electric force experienced by an electron in the hydrogen atom to the gravitational force experienced by the electron.

105. What is the electric force experienced by a negatively charged particle $Q_1 = -10\ \mu C$ with respect to a positively charged particle $Q_2 = 20\ \mu C$ located a distance $r = 10$ cm away?

106. Two 10 g objects with charges of +5 μC and −10 μC are separated by 1 cm.

    a. What is the direction of the electric force?
    b. What is the magnitude of the force between them?
    c. What is the acceleration on the positive charge?
    d. What is the acceleration on the negative charge?

107. Two objects with charges +20 μC and −20 μC are located 5 cm apart.

    a. What is the strength of the electric force between them?
    b. What mass would experience an equivalent force due to gravity near the Earth's surface?

108. A particle with a charge of −100μC is located halfway between two other particles: One is located at $x = -5$ cm and has charge −50μC; the other is located at $x = +5$ cm and has charge +50μC. What are the direction and magnitude of the force on the middle particle?

## Section 18.3

109. A uniform electric field can be created between two parallel plates that bear equal but opposite electric charges. Suppose that two such plates have a potential difference of 6 V between them. How far apart do they need to be to create an electric field of 50 N/C? (The plates are in a vacuum.)

110. An electron has a mass of $9.1 \times 10^{-31}$ kg and a charge of $-1.6 \times 10^{-19}$ C. Suppose you could isolate one electron in a perfect vacuum and then create an electric field to pull upward on the electron. How strong would the field have to be to counteract the electron's weight? (In other words, how strong would the field have to be to put the electron in a state of force equilibrium?)

111. Three charged particles, having equal positive charges, are equidistant at the vertices of an equilateral triangle. Sketch their combined electric field.

## Section 18.4

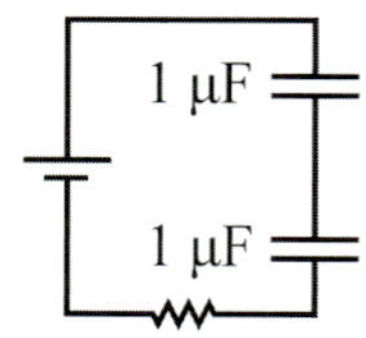


112. Two 1 μF capacitors are connected in series as shown on the schematic. What is the total capacitance of the circuit?

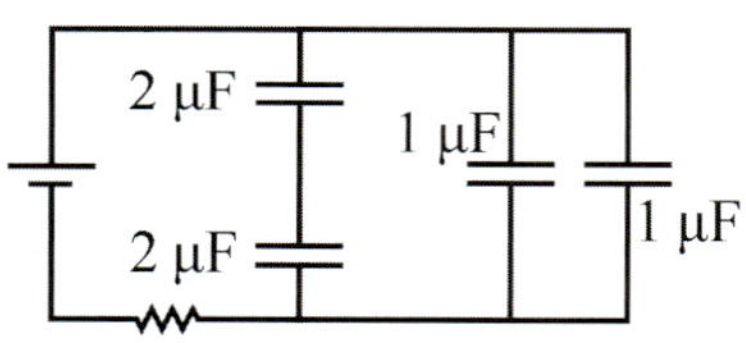


113. Four capacitors are connected in a circuit as shown on the schematic. What is the equivalent capacitance of the circuit?

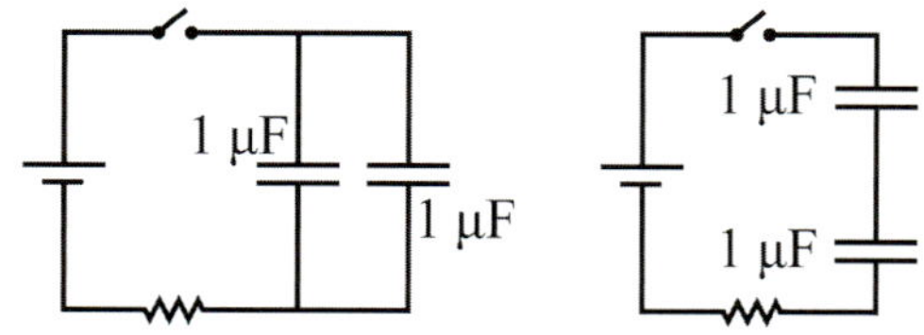


114. Two 1 μF capacitors can be wired in parallel (left panel) or series (right panel). Suppose that either circuit can be charged with a 12 V battery by closing the switch. How much charge would each circuit store in its capacitor network?

# Chapter review

## Standardized test practice

115. Which is true of Earth's north magnetic pole?

A. It is located near the geographic north pole.
B. It is located at the geographic north pole.
C. It is located near the geographic south pole.
D. It is located at the geographic south pole.

116. What is the correct application for an electroscope?

A. to trace a magnetic field
B. to detect the presence of electric charge
C. to measure electric current
D. to take an image of electricity

117. A student rubbed two nonmetal rods, each of a different material, with two different cloths. He found that the two rods then attracted each other. What is the most likely explanation?

A. The combination of rods and cloths put positive charges onto one rod and negative charges onto the other.
B. The combination of rods and cloths put positive charges onto both of the rods.
C. The combination of rods and cloths put negative charges onto both of the rods.
D. Both rods were neutral.

118. What is the strength of the electric force between two charges, each of charge +1 μC, separated by 1 cm?

A. 0.009 N
B. 90 N
C. $9\times10^{9}$ N
D. $9\times10^{25}$ N

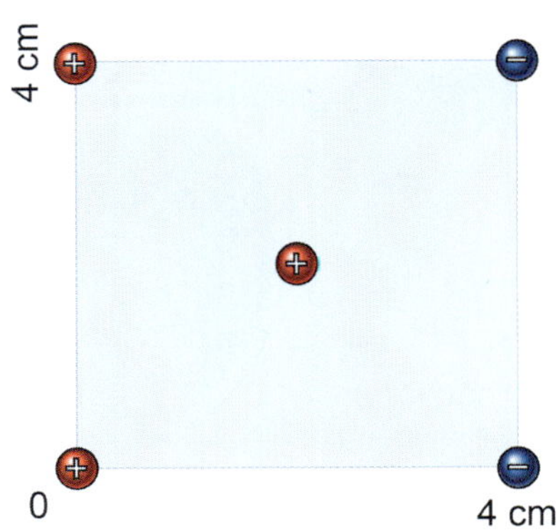


119. Four charges are placed on the corners of a rectangle as shown. If a positive charge is placed exactly at the center of the rectangle, what would happen to it?

A. It will move upward.
B. It will not move.
C. It will move diagonally to the lower right corner.
D. It will move to the right.

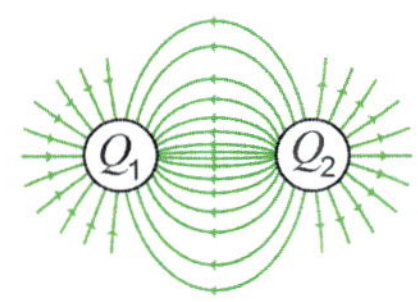


120. In the figure, what is the sign of the charges $Q_1$ and $Q_2$, respectively?

A. positive and positive
B. positive and negative
C. negative and positive
D. negative and negative

121. Which of the following does not create a force field?

A. gravitational force
B. centrifugal force
C. electric force
D. magnetic force

122. Which of the following can be used to detect the presence of a magnetic field?

I. a compass needle
II. iron filings
III. a bar magnet

A. I only
B. II only
C. I and II only
D. I, II, and III

123. The force between similar magnetic poles is repulsive.

In this sentence, the word "similar" means:

A. looks the same.
B. is about the same size.
C. has the same magnetic polarity.
D. is equidistant from all other magnets.

124. Which of the following is *not* true?

A. A resistor dissipates energy.
B. A capacitor generates energy.
C. The electric field is related to electric charge.
D. A battery stores chemical energy.

125. If the distance between two positive charges increases by a factor of 2, what happens to the force between those charges?

A. It decreases by a factor of 2.
B. It does not change.
C. It increases by a factor of 4.
D. It decreases by a factor of 4.

Chapter

# 19 Electromagnetism

According to the U.S. Department of Energy, in 2009 American drivers used 24 trillion megajoules of energy to drive cars and trucks. This enormous amount of energy is equivalent to the total output of 760 major power plants. Yet, by the laws of physics, all of the energy is ultimately *wasted!*

Think about driving a car. You run the engine to convert chemical energy in gasoline into kinetic energy of the moving car. When you stop, the brakes convert all that kinetic energy into waste heat and wear. Of course, you also get where you are going, so that counts for something! Can some of this wasted energy be recovered?

One of the key features of electric and hybrid vehicles is that they can recover *some* of the car's kinetic energy while the car is slowing down or braking. During the *regenerative braking* process, some of the car's kinetic energy is converted into electrical energy and then stored in batteries for later use. A car can be driven an additional distance using this stored electrical energy, thereby delivering more driving miles from the same gallon of gasoline.

**How regenerative braking works**

Moving car and wheels turn driveshaft inside motor/generator.

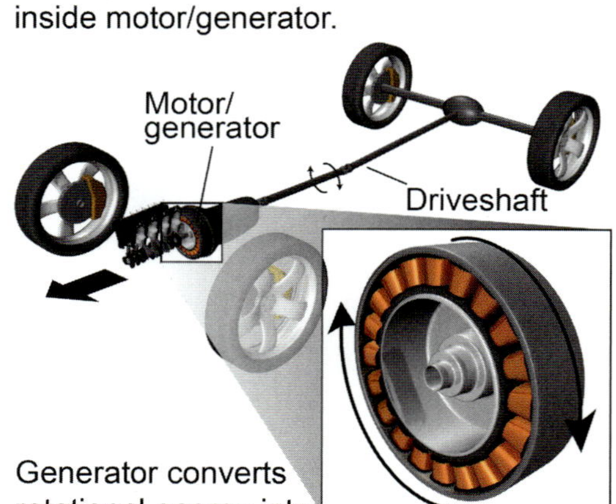


Generator converts rotational energy into electrical energy.

Electric and hybrid cars have an electric motor that can drive the car by converting electricity into rotational motion of the driveshaft. When the car is slowing down, the electric motor runs in reverse and becomes an electric generator. The driveshaft continues to turn as a result of the car's forward motion, thereby acting as a turbine inside the generator. The generator converts this rotational motion into electricity and partially recharges the battery.

The electric motor and generator are based on the connection between electricity and magnetism—called *electromagnetism*!—which you will learn about in this chapter. The magnetic force between two magnets and the electric force between two charged objects—described in Coulomb's law—are two aspects of a single electromagnetic force. Electromagnetism is the physics behind many technologies all around you, such as the loudspeaker, microphone, computer memory, and the cathode ray tube (or old-style television).

# Chapter study guide

## Chapter summary

Electricity and magnetism were considered two different phenomena until 1820, when the Danish physicist Hans Christian Ørsted discovered the connection between them. Moving charged particles, such as electric current in a wire, create a magnetic field around them. Changing magnetic flux, such as through a loop of wire, generates electric current in the wire. Electromagnetism lies at the heart of many technologies, such as the electromagnet, electric motor, generator, and transformer.

## Learning objectives

By the end of this chapter you should be able to

- describe and investigate the connection between electricity and magnetism;
- describe key scientists and their contributions to the field of electromagnetism;
- describe how an electromagnet works and determine its polarity;
- list technologies based on electromagnetism;
- define magnetic flux and describe magnetic induction;
- describe the components and operation of an electric motor and generator;
- calculate the magnetic force on a moving charged particle and its cyclotron radius;
- calculate the forces felt by two current-carrying wires and a wire loop rotating in a magnetic field; and
- describe the principles behind the mass spectrometer, galvanometer, and cathode ray tube.

## Investigations

19A: Build an electromagnet
19B: Electric motor
Design a Rube Goldberg machine
19C: Build a paperclip motor

## Chapter index



## Important relationships

$$\Phi_B = BA\cos\theta$$

$$\text{emf} = -N\frac{\Delta\Phi_B}{\Delta t}$$

$$\frac{V_s}{V_p} = \frac{N_s}{N_p}$$

$$F_B = qvB\sin\theta$$

$$B = \frac{\mu_0 I}{2\pi r}$$

$$F_B = ILB$$

## Vocabulary

**electromagnetism**
**solenoid**
**armature**
**electromagnetic induction**
**transformer**
**permeability of free space**
**cathode ray tube (CRT)**
**electromagnets**
**electric motor**
**magnetic flux**
**Lenz's law**
**mass spectrometer**
**voice coil**
**polarity**
**commutator**
**Faraday's law of induction**
**electrical generator**
**cyclotron radius**
**galvanometer**

## 19.1 - Magnetic fields and the electric motor

During a class lecture in 1820, Danish physicist Hans Christian Ørsted placed a compass near a wire that was carrying electric current. The compass needle moved as if the current-carrying wire were a magnet. Switching the direction of voltage in the wire caused the current to flow in the opposite direction in the wire; the compass needle also switched direction. Prior to Ørsted's discovery, physicists had treated electricity and magnetism as if they were two separate subjects. The English chemist and physicist Michael Faraday showed the connection between changes in magnetic fields and induced electric current in 1822. By 1861, the Scottish physicist James Clerk Maxwell tied together the fundamental properties of electromagnetism in his four equations, possibly the most elegant equations in all of science. Taken together, these scientists showed that magnetism and electric currents were two different manifestations of one underlying physical phenomenon: **electromagnetism**.

### Magnetic field around a current-carrying wire

**Magnetic field around a current-carrying wire**

Ørsted's discovery was that electric current creates a magnetic field. What direction is that magnetic field? If the current flows in a straight wire, the magnetic field lines form concentric circles around the wire. What does a circular magnetic field line mean in practice? Imagine placing a compass needle (a test magnet) some distance away from the wire. Draw an imaginary circle centered on the wire that passes through the compass needle. The compass needle's north magnetic pole will point in a direction that is *tangent* to the circle surrounding the wire. In other words, the magnetic field will point in a direction that is *perpendicular* to the radius of the circle at that location along the circle. If the compass is moved in a circle around the wire, its direction will rotate so that it always points tangential to a circle centered on the wire.

**Magnetic field around a current-carrying wire**

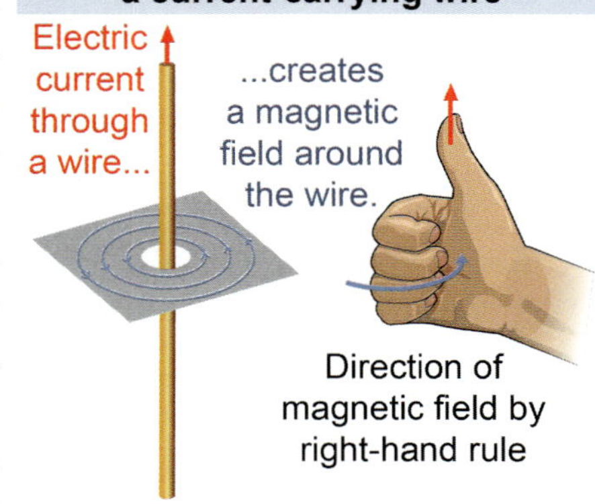


**The right-hand rule**

The right-hand rule is a way to visualize the direction of the magnetic field caused by an electric current. Point the thumb of your right hand in the direction of the electric current and your fingers curl in the direction of the magnetic field lines.

**Tangent of a circle**

**Tangent of a circle**

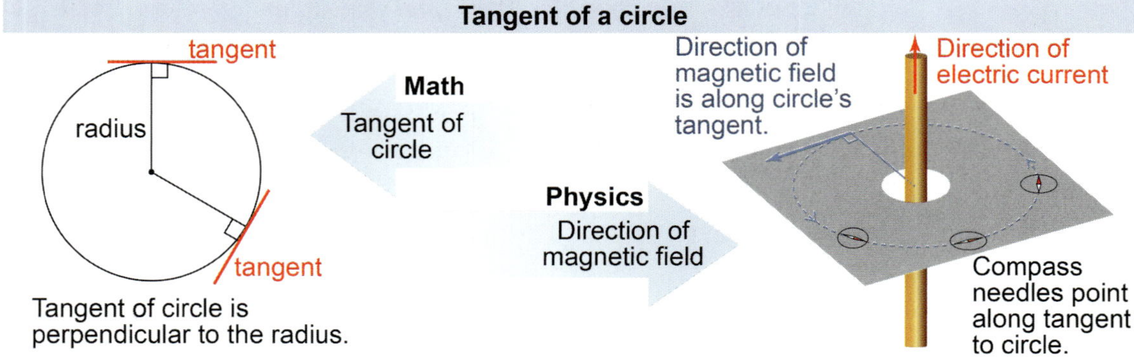


The magnetic field lines around a current-carrying wire trace out concentric circles. That means that, at any particular location, the direction of the magnetic force is *tangential* to a circle. To find the tangent to a point on a circle, draw a radial line connecting the point with the center. The tangent is *perpendicular* to this radial line.

**Using electricity to create a *strong* magnetic field**

Although a straight, current-carrying wire does create a magnetic field around it, the field usually isn't very strong. We could make the magnetic field stronger by passing more current through it, but if we use too much current the wire will overheat and melt! How can we use a current-carrying wire to create a *powerful* magnetic field?

**Magnetic field of a straight wire** | **Magnetic field of a single loop** | **Magnetic field of a wire coil** | **Magnetic field of a solenoid**

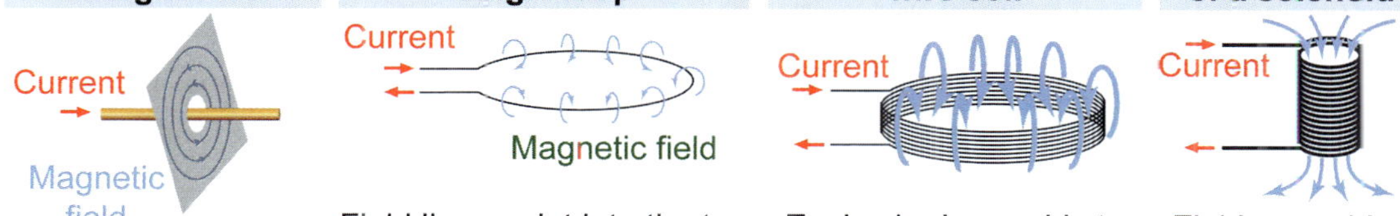


Field lines point into the top and exit the bottom of loop.

Each wire loop adds to create a stronger field

Field resembles a magnet!

**Magnetic field of a wire loop**

The answer is to use many wires wrapped in a coil (or a series of loops)! When a single piece of wire is bent into a wire loop, the resulting magnetic field everywhere along the curved wire follows the same right-hand rule. In the figure above, all the magnetic field lines *point into the top* of the wire loop and *exit the bottom*. When the same wire is wrapped into many coils, the magnetic field produced by each coil is in the same direction—so they add! Any device that uses electric current through multiple loops of wire to create a magnetic field is a **solenoid**. Many solenoids are created in a cylindrical shape that is significantly longer than the diameter of each wire loop.

**Electromagnets**

**Electromagnets** are magnets that are created by the flow of electric current. Most electromagnets are created by wrapping the wire coil around a metal core, which increases its strength. The strength also increases by wrapping more coils. You can make a simple electromagnet with wire wrapped around an iron nail and connected to a battery.

**Right-hand rule for coils and electromagnets**

The orientation of the north and south poles of the electromagnet—also called its **polarity**—depends on the direction of the current in the wire. If the direction of current flow is switched, the polarity of the electromagnet is also switched and the north and south poles reverse. The right-hand rule is used to determine the polarity of the magnetic field. When you wrap your fingers around the wire in the direction the current flows (from the positive to the negative terminal of the battery), your thumb points in the direction of magnetic north.

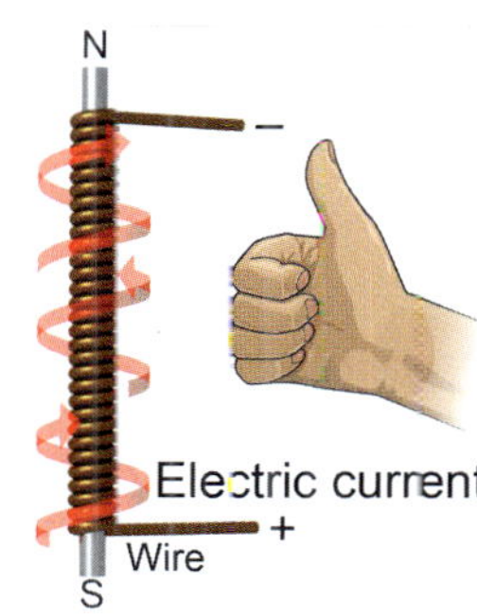


**Applications of electromagnets**

Video

The strength of an electromagnet can be controlled by the amount of current that flows in the coil. Using electronic devices, engineers can control the amount *and the direction* of current and are so able to design electromagnets for all types of applications from security systems to motors. A powerful electromagnet is used by cranes in junkyards to pick up and move heavy metal objects.

# Investigation 19A Build an electromagnet

**Essential questions**
*How are electricity and magnetism connected to each other?*
*How can you predict the polarity of an electromagnet?*

An electromagnet uses electric current to create a magnetic field. An electromagnet is used in an electric motor to attract and repel magnets on a spinning disk. By switching the direction of the electric current, the magnetic field also changes its polarity. In the electric motor the electric current direction is switched back and forth rapidly to spin the motor at high speed. In this investigation you will create a simple electromagnet using a metal nail, ordinary wire, and a battery.

## Part 1: Build the electromagnet

1. Cut a long piece of wire, approximately 1 m in length. Strip the insulation off of the two ends of the wire.
2. Starting about 10 cm from one end of the wire, wrap the wire tightly around the nail. Start from the head of the nail. As you wrap the coil, make sure it is tight and that the wire does not cross itself.
3. Stop wrapping at a point 0.5–1.0 cm before you get to the tip of the nail. You should have at least 10 cm of free wire on this end of the nail.
4. Attach the wire from the tip of the nail to the positive terminal of the battery. Attach the other end of the wire to the negative terminal.

*a. Does your electromagnet deflect a compass needle?*
*b. How many paper clips can it lift?*
*c. Propose additional investigative questions: What variables could you change that might affect the strength of your electromagnet?*

**Electromagnet unconnected**

**Electromagnet connected**

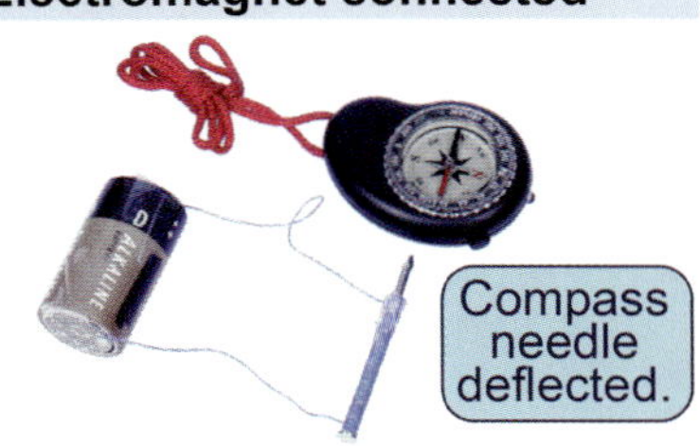


## Part 2: Determine its polarity

The polarity of a magnet indicates which end is north or south.

*a. Using a model including the right-hand rule, predict the polarity —which end is north and which is south—of your electromagnet. Justify your answer. (Hint: Inspect carefully how you wound the wire around the nail and include that information in your model.)*
*b. Using the compass, check your prediction. Were you right?*

**What is the polarity of each electromagnet?**

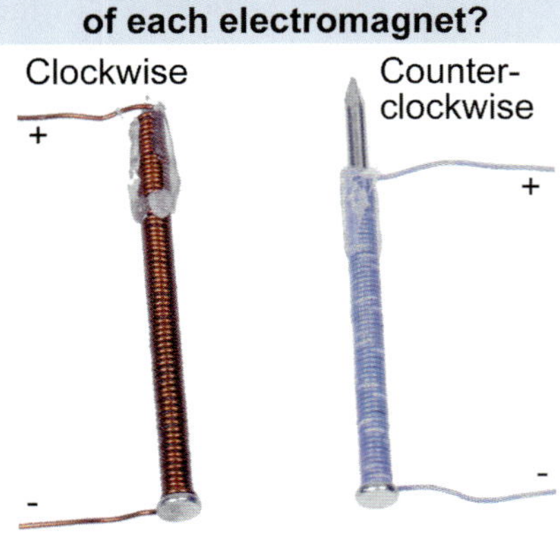

**Devices that use electric motors**

The **electric motor** is a device that uses electricity and magnets to transform electrical energy into mechanical energy. This transformation of energy helped revolutionize human technology. Today, electric motors are everywhere and they range in size from the fingernail-sized motor that makes your cellphone vibrate to the room-sized motors used in factories and locomotives. Recently, scientists have developed motors that are as thin as a human hair! These tiny motors are called nanomotors and will some day find medical and other applications.

**Electric motors work through magnetic forces**

Electric motors operate through the generation of attractive and repulsive forces between magnets. Those forces are controlled so that they always point in the direction of rotation. As a rotating magnet approaches the stationary magnet, the two magnets must have opposite poles to create an *attractive* magnetic force. As the magnets pass each other, they now must exert a *repulsive* magnetic force from similar poles to push them apart. The only way to accomplish this attraction and repulsion is if one of the two magnets—either on the rotor or the stationary magnet—periodically flips its polarity between north to south! At least one set of magnets must be *electromagnets* so that it can be electrically controlled to flip its polarity.

**Basic operation of an electric motor**

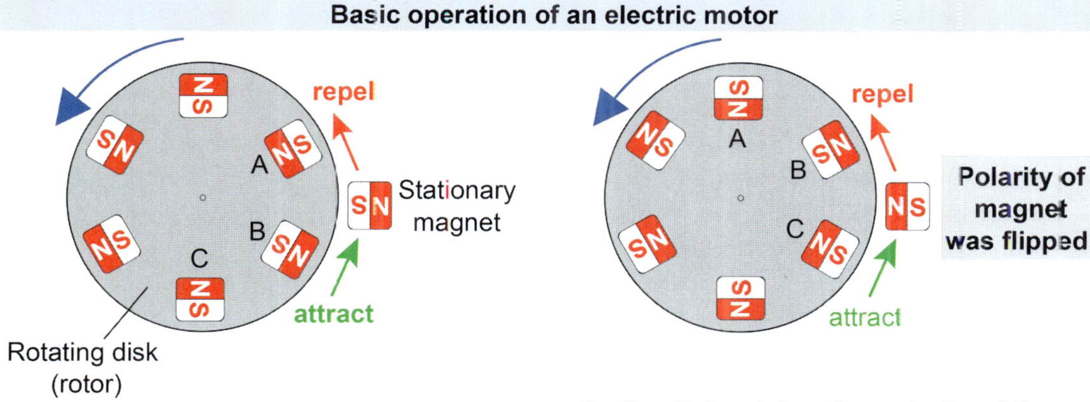


**Magnet A is repelled and magnet B is attracted.**

**As the disk rotates, the polarity of the stationary magnet flips, keeping the magnetic force in the same direction.**

**Synchronization with polarity reversal**

The polarity of the stationary magnet must be reversed when the rotating magnets are in the right position. If the reversal of the polarity does not occur at the right time, the motor is not very efficient in converting electrical energy to mechanical energy. An efficient motor requires precise synchronization between the switching of the magnet polarity and the passage of each rotating magnet.

# Investigation 19B Electric motor

**Essential questions** *What is the connection between electricity and magnetism in a motor?*

Interactive simulation

The electric motor is found in many appliances and technologies, including the turntable of a microwave or CD/DVD player, the spinning blades of an electric shaver, and the electric motor in a hybrid electric vehicle. The physics behind the electric motor is the connection between electricity and magnetism. But how does this connection work? In this investigation, you will explore how to use *changing* electric current in an electromagnet to spin the rotor of a simulated electric motor. How fast can you spin the rotor in this motor?

## Part 1: Reversing the electromagnet's polarity to spin the rotor

Open the interactive simulation of an electric motor.

1. Press [Run] to start the rotor spinning.
2. Press the [N]/[S] button to switch the polarity of the electromagnet at an appropriate time so that the rotor will spin faster.
3. See how fast you can get the rotor to spin!
4. Change the strength of the electromagnet using the slider and see how fast you can get the rotor to spin.

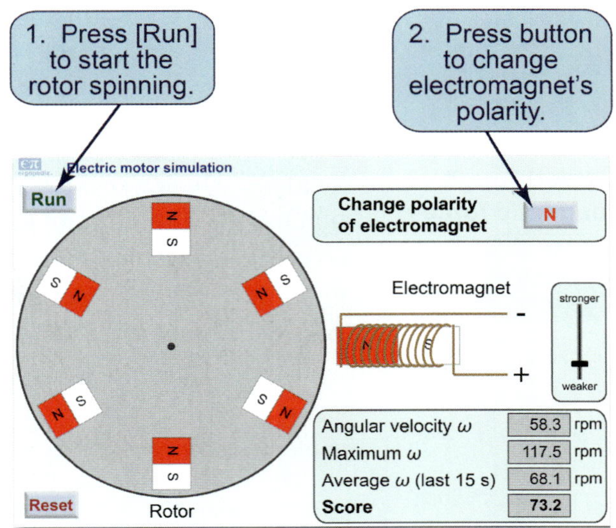


**If you repeatedly change the electromagnet's polarity at the "right" time, then the rotor will spin faster and faster.**

***How fast can you make it spin?***

*a. What happens when you switch the electromagnet's polarity when the electromagnet is halfway between two magnets on the rotor?*
*Does the rotor spin faster or slower?*

*b. What happens when you switch the electromagnet's polarity when the electromagnet is closest to a magnet on the rotor?*
*Does the rotor spin faster or slower?*

*c. When is the ideal time to switch the polarity of the electromagnet? Draw a sketch.*

*d. Write down the angular velocity, maximum angular velocity, average angular velocity, and score for your best run in rotating the motor. Why are the three values of angular velocity different?*

*e. Can you spin the rotor faster when the electromagnet is strong or weak?*
*Why do you think this is the case?*

## Part 2: Changing magnetic fields and electric current

1. Create a coil of wire with at least 10 loops.
2. Connect the two wire ends to the lowest current setting of a digital multimeter.
3. Quickly insert a stack of magnets into the center of the coil while watching the current reading.
4. Hold the magnets in place and watch the current reading.
5. Quickly remove the magnets from the coil and watch the current reading.

*a. What happens to the current through the wire as the magnets move into the coil?*
*b. What happens to the current through the wire as the magnets are held fixed inside the coil?*
*c. What happens to the current through the wire as the magnets are removed from the coil?*
*d. Describe in words the relationship you observe between magnetic fields and electric current.*

**How does a *real* electric motor work?**

On the previous two pages, you learned the principles behind the operation of the electric motor. What does the inside of a *real* electric motor inside a power tool or toy look like? How does a real electric motor work?

**Electromagnets on the rotor**

In the electric motor described up until now, *the polarity of the stationary magnet changes* to keep the rotor spinning, while the magnets on the rotor have fixed polarity. In many electric motors, however, *the rotor magnets are electromagnets and alternate their polarity*, while the stationary magnets have fixed polarity. In the image at right, the coil of electric wires, wound around a metallic core, are visible for each of the electromagnets on the rotor. The rotating set of electromagnets in a motor is called the **armature**.

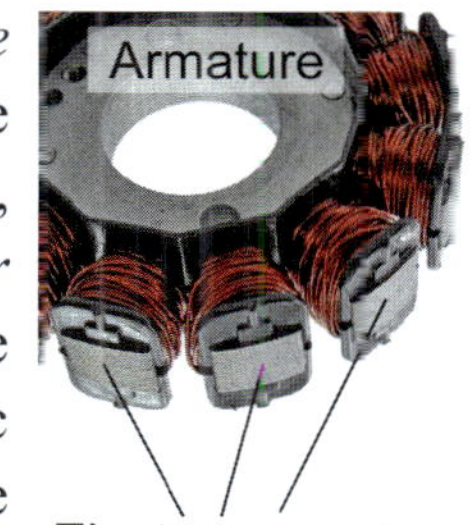


**Components of a typical motor**

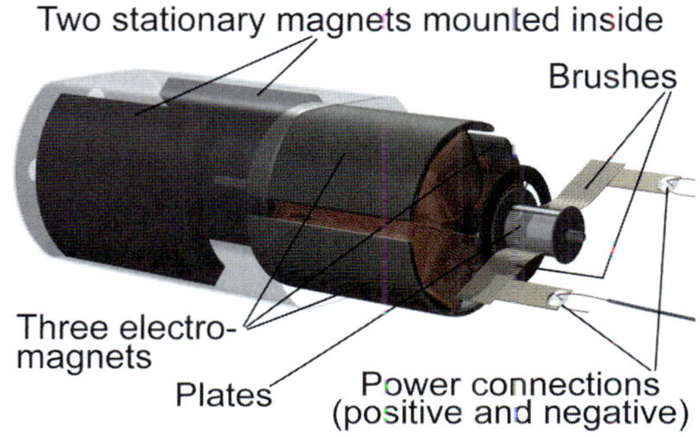


A typical electric motor found in an electric device is quite small, only 3–10 cm, but it can deliver plenty of power for basic applications. If you take apart an electric motor (and always disconnect it from power before doing so!), you can easily see the working parts inside.

**How is electrical power supplied to a rotating motor?**

In the motor, rotating electromagnets are electrically connected to the power supply through *brush contacts* that automatically switch the polarity at the right time. There are two brushes, one connected to the positive terminal of the power supply and the other to the negative terminal. At any given time, the brushes make contact with two of three metal plates, each called a **commutator**, mounted on the axle of the motor. The commutators in turn connect to the electromagnets on the armature. The brushes and commutators work together to switch positive and negative power to the electromagnets.

**Connections within the motor**

If you took a snapshot of the electric motor in action, what would the connections inside look like? In the figure at right, the brushes are in contact with two of the commutator plates, causing current to flow from top to bottom of the electromagnet at the top of the armature. Using the right-hand rule, we know that this electromagnet has its south magnetic pole pointing up. The magnetic force between this electromagnet and the two stationary magnets creates a counter-clockwise torque on the armature, causing the armature to rotate.

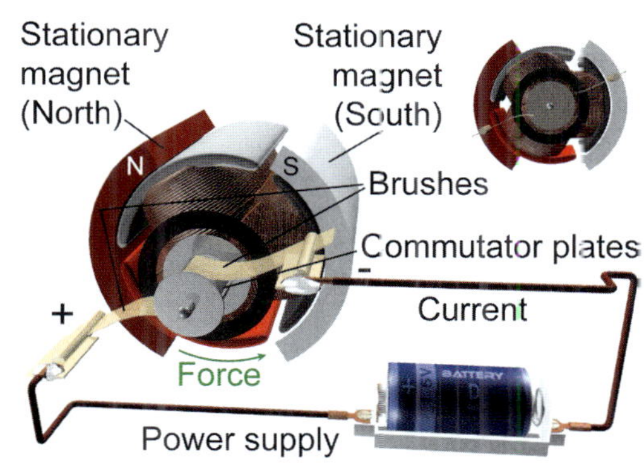

## Design a Rube Goldberg machine

**Design challenge**

Use the materials provided by your teacher to design, build, and refine a Rube Goldberg machine within the following constraints:

a. The initial step releases a steel ball. After release the device must operate without assistance.
b. It must use at least four different kinds of energy.
c. It must use the electromagnet you constructed earlier in this chapter.
d. The final transformation extends a spring or rubber band.

When operating your device, make suitable measurements to calculate the ratio of the output energy to the input energy.

**Assessment**

You will be assessed on the number of energy transformations and/or transfers in your device and the number of unique simple machines used in the design.

**What is a Rube Goldberg machine?**

Rube Goldberg (1883–1970) was a cartoonist who was best known for drawing fanciful and complicated machines that performed what was ultimately a simple task. In the example at right, the subject's act of moving food to his mouth triggered a series of operations that eventually wiped his mouth with a napkin. You already saw a simple example in Prince Ludwig's contraption on page 335. In a Rube Goldberg machine, humor is *de rigueur*!

Self-Operating Napkin

**Energy transformations**

At the heart of any Rube Goldberg machine is a series of transformations of energy, such as gravitational potential energy, elastic potential energy, linear kinetic energy, rotational energy, magnetic energy, or electrical energy. Many Rube Goldberg devices use a variety of the six simple machines, too.

**Design**

Sketch out a basic design of your machine where you break it into individual subsystems and their functions. For each functional element, note where there might be more than one design that you will want to explore. Write down the energy transformations or simple machines involved in each subsystem.

**Prototype**

Construct a prototype of your design. Identify where human understanding, ergonomics, or economics influenced how you constructed the prototype.

**Test**

Test each individual subsystem and, once they work individually, progress to testing the entire machine. Document the performance through data, which might include taking video of the machine's elements in operation.

**Evaluate**

Evaluate each subsystem as well as the overall performance. Are there parts of your machine that can be improved?

**Revise**

Write down design revisions and implement them.

**Final evaluation**

Your teacher will evaluate how well you accomplished the task and score your device.

# Section 1 review

A magnetic field is created whenever current flows, and a device that turns electrical energy into magnetic energy is called an *electromagnet*. If current flows through a straight length of wire, the magnetic field at any point is tangential to a circle that is centered upon the wire. A *solenoid* is a long coil of current-carrying wire, and it generates a magnetic field that resembles a bar magnet's. Electric motors rely on the attractive and repulsive forces between permanent magnets and electromagnets. For an electric motor to efficiently operate, the electromagnets' poles must periodically alternate between N and S. The motor then can use electric current to generate torque and perform mechanical work.

For more information to research the development of the concept of the electromagnetic force, see

*Hidden Attraction: The History and Mystery of Magnetism* by G. L. Verschuur,
*A History of Electricity and Magnetism* by H. W. Meyer, and
*The Electric Life of Michael Faraday* by A. W. Hirshfeld.

| Vocabulary words | **electromagnetism, electromagnets, polarity, solenoid, electric motor, commutator, armature** |
|---|---|

## Review problems and questions

1. Research and describe the historical development of the concept of the electromagnetic force.

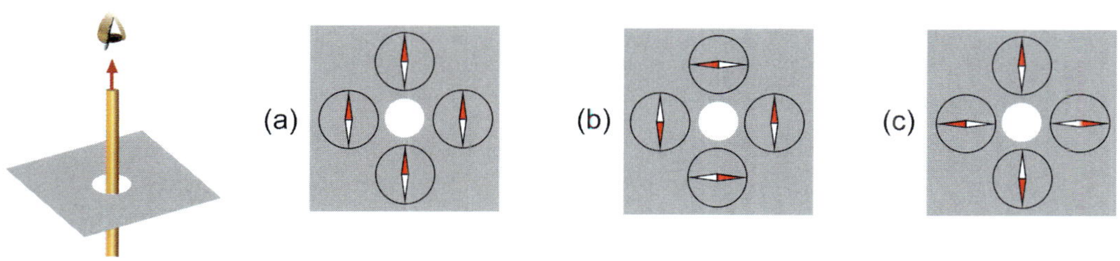


2. You look down upon a current-carrying wire with a cardboard square around it. (Positive current is approaching you.) You place four magnetic compasses on the cardboard stage. Which pattern will you see? (Recall that each red tip has a N polarity.)

3. You have been given an unusually colored magnetic compass. As shown in the top diagram, you place it next to a properly labeled bar magnet to determine what the compass's white and blue tips represent. Next (bottom diagram), you bring the compass up toward one end (marked "?") of a solenoid, which is carrying current as shown here. Which way will the blue tip of your compass needle point?

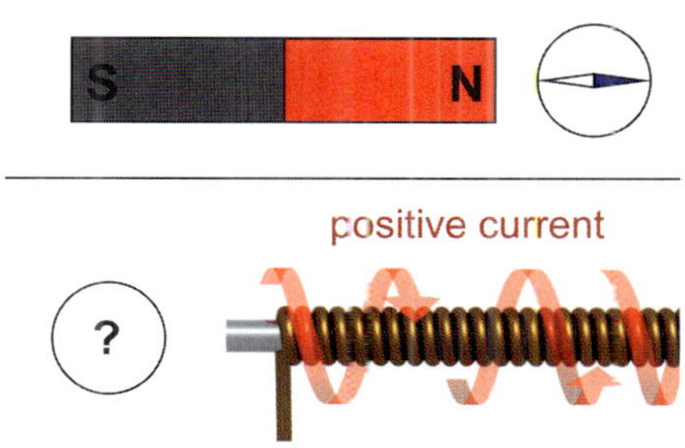

## 19.2 - Induction and the generator

Electricity and magnetism are related to each other: Moving electric charges, or electric current, generates a magnetic field. Can you reverse this process, using a magnet to create an electric current? If you placed a magnet inside a solenoid, would a current be produced in the wire? The phenomenon at the heart of this question is *electromagnetic induction*, which is the principle behind electric generators in power plants—as well as transformers that convert electricity to high voltages for long-distance transmission.

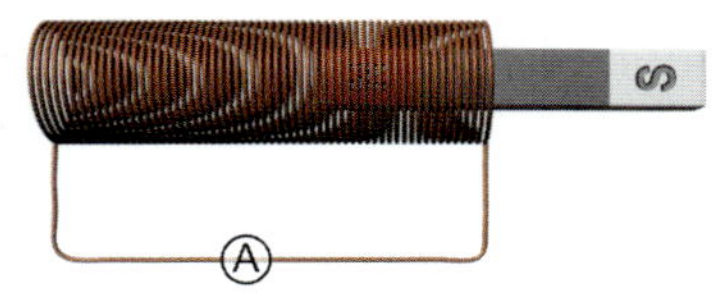


**Will a bar magnet inside a solenoid produce an electric current through the wire?**

### Magnetic flux through a loop of wire

**Magnetic flux**

The direction and strength of a bar magnet's magnetic field can be visualized by using field lines, as you learned on page 518. The strength of the magnetic field passing through a wire coil determines the **magnetic flux**. When the imaginary surface formed by the coil is perpendicular to the magnetic field, many field lines pass through the coil and there is a large magnetic flux through it. When the coil's surface is parallel to the field, no lines pass through it and it has zero magnetic flux. Magnetic flux depends on the strength of the field, the area of the coil, and the relative orientation between the field and the coil, as shown in equation (19.1).

**Magnetic field lines pass through coil.** | **No field lines pass through coil.**

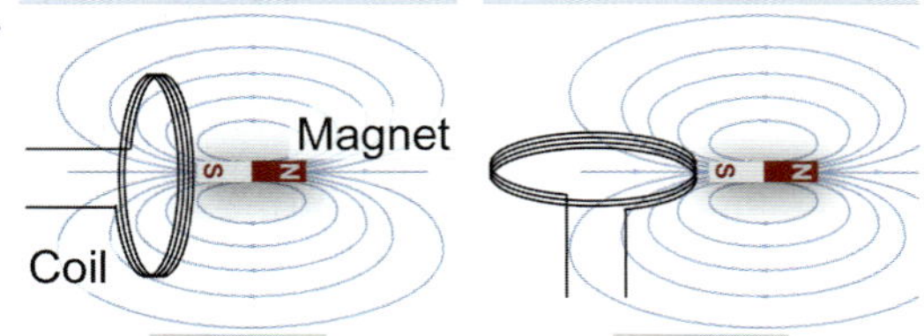


$\Phi_B > 0$ — Positive magnetic flux through coil

$\Phi_B = 0$ — No magnetic flux through coil

**Magnetic flux** through a wire loop

$$\text{(19.1)}\quad \Phi_B = BA\cos\theta$$

$\Phi_B$ = magnetic flux (T m$^2$)
$B$ = magnetic field (T)
$A$ = area of loop (m$^2$)
$\theta$ = angle between field and normal (deg)

**Using the equation**

Since $\Phi_B$ is the product of magnetic field (measured in tesla or T) and area (measured in square meters), the units of $\Phi_B$ are T m$^2$, which is also called the weber (Wb). Be careful in determining the angle $\theta$ between the magnetic field and the *normal to the surface of the coil*—as you can see both in the illustration at right and in the solved problem below.

Angle between $B$ and normal to coil

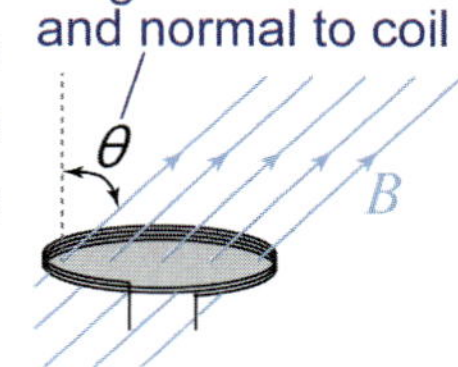


**Calculating magnetic flux through a wire loop**

A 0.5 cm square wire loop inside a generator lies between the poles of two magnets, as shown in the figure at right. If the magnetic field strength is 0.1 T, what is the magnetic flux through the loop?

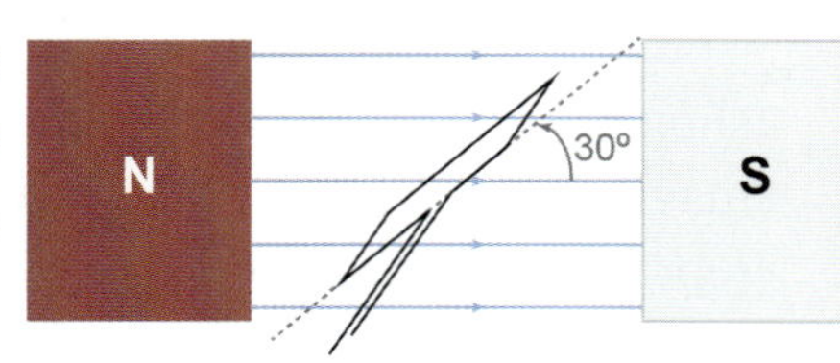


| | |
|---|---|
| *Asked:* | magnetic flux $\Phi_B$ |
| *Given:* | loop dimensions $l = 0.5$ cm by $w = 0.5$ cm; magnetic field $B = 0.1$ T |
| *Relationships:* | $\Phi_B = BA\cos\theta$ area $= A = l \times w$ <br> angle between magnetic field and normal to wire loop $\theta = 90°–30° = 60°$ |
| *Solution:* | $\Phi_B = BA\cos\theta = Blw\cos\theta = (0.1\text{ T})(0.005\text{ m})(0.005\text{ m})\cos 60°$ |
| *Answer:* | $\Phi_B = 1.25 \times 10^{-6}\text{ T m}^2 = 1.25 \times 10^{-6}\text{ Wb}$ |

**Induced current**

Does a magnet inside a solenoid cause electricity to flow in the solenoid's wire? It depends! When the magnet is moving into the solenoid, an ammeter connected to the circuit shows electric current called an *induced current*. When the magnet stops moving, the current stops, too. When the magnet is moved backward out again, current starts up again—but flows in the opposite direction! The magnet induces a current in the wire, but *only when the magnet is moving* relative to the wire. The phenomenon of a moving magnetic field inducing an electric current is called **electromagnetic induction**.

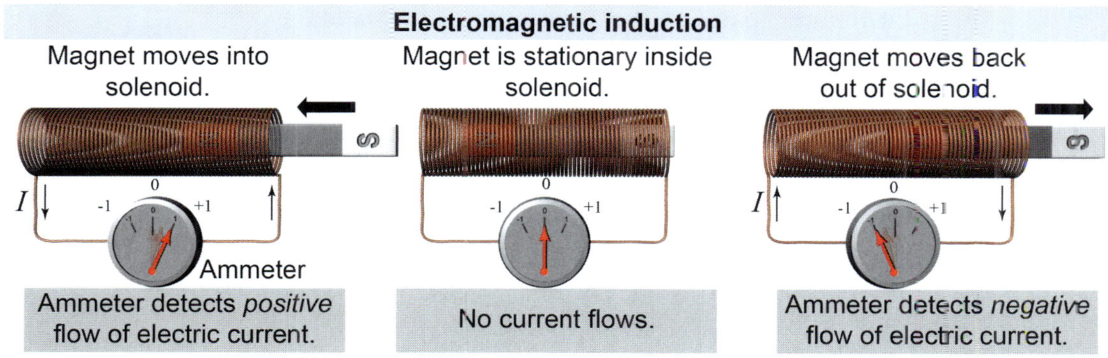


**Changing magnetic flux induces electric current**

What was happening when the magnet moved in and out of the solenoid? The movement of the magnetic field caused a change in the magnetic flux through every loop of wire in the solenoid. *Changing magnetic flux induces electric current in a conductor.* Since electric current will flow only when there is an electric field inside the conducting wire, we could also say that *changing magnetic flux induces an electric field in the wire.*

**Induced emf**

In magnetic induction, work is being done to move the electric charges around the coil—i.e., to create the electric current. The work done per unit charge is called the *electromotive force* (emf), although it is not really a force. The units of emf are joules per coulomb or volts. In magnetic induction, a changing magnetic flux creates an *induced emf*.

**Faraday's law**

In 1831, Michael Faraday conducted a series of experiments similar to the illustration above to discover electromagnetic induction. **Faraday's law of induction** (equation 19.2) states that the *induced emf* in a circuit is proportional to both the rate of change of the magnetic flux and the number of coils in the wire.

**Faraday's law**

$$\text{(19.2)} \quad \text{emf} = -N\frac{\Delta\Phi_B}{\Delta t}$$

emf = induced emf (V or J/C)
$N$ = number of turns of coil
$\Delta\Phi_B$ = change in magnetic flux (T m$^2$)
$\Delta t$ = change in time (s)

**Lenz's law**

When a current is induced, in what direction will it flow? **Lenz's law** states that the direction of the induced current is such that the current's magnetic field acts to oppose the original change in magnetic field. Lenz's law describes how the induced current acts to resist changes in magnetic flux—similar to how the inertia of an object causes it to resist changes in its velocity. Lenz's law therefore provides the minus sign in equation (19.2)!

**What is the direction of the induced current in the loop?**

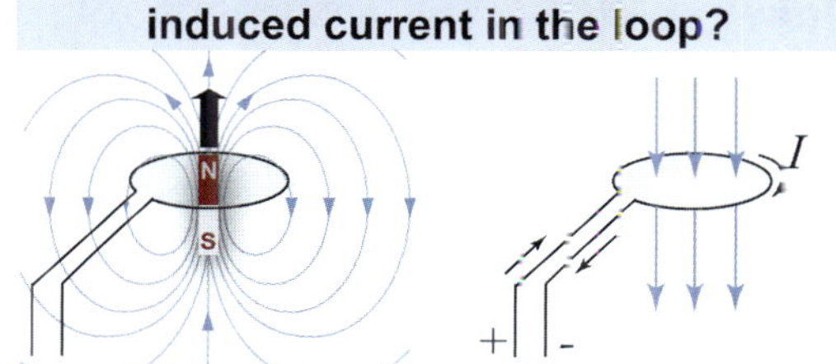


*Increasing* magnetic flux through loop

Induced current in direction to create *downward* field

**What happens if you crank a motor shaft by hand?**

Look back at the diagram of how the electric motor works on page 557. What happens if, instead of powering the motor with electricity, you now *crank the axle of the motor by hand*? What happens when you attach the leads to a voltmeter, rather than driving the device with voltage from a battery?

**Hand cranking a motor generates voltage**

The result of hand cranking a motor is shown in the figure at right. As the coil on the armature passes the stationary magnets, electromagnetic induction causes a voltage to be induced in the wire. As a given coil approaches the *north* magnetic pole stationary magnet, the magnetic flux through the coil *increases*—which induces a *negative* voltage according to Faraday's law. If the hand were to crank the generator in the opposite direction, the same coil would instead approach the *south* stationary magnet, causing magnetic flux through the coil to *decrease* and inducing a *positive* voltage. The electric motor has become an **electrical generator!**

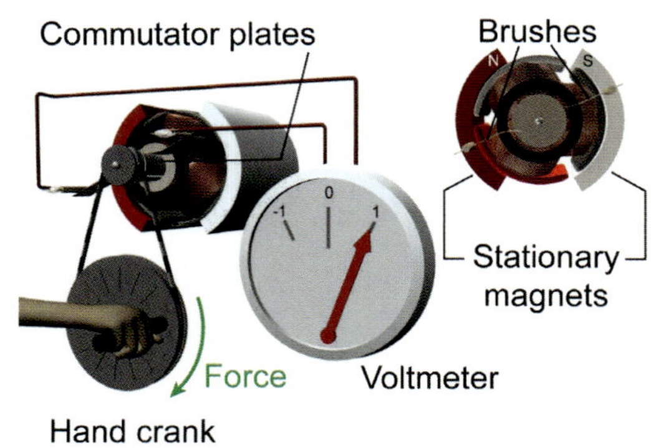


**Rotating wire coil in a magnetic field**

The heart of the generator consists of a wire coil that is rotating in a magnetic field, as shown in the figure at left. As the coil rotates, the magnetic flux through the plane of the coil changes, which induces an electric current in the coil. The magnetic flux increases and decreases as the coil rotates, causing a negative and positive induced voltage, respectively.

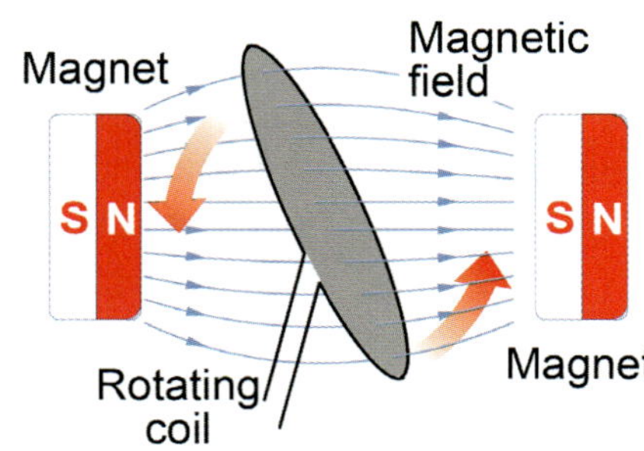


**Symmetry between electricity and magnetism**

Many laws of physics display *symmetry*. The science of electromagnetism shows us a symmetry between electricity and magnetism. The flow of electric current creates a magnetic field, and a changing magnetic field results in the flow of electric current. An electric motor uses electricity to generate mechanical energy; an electric generator uses mechanical energy to generate electricity. The motor and generator have the same basic physics behind them, except that one uses the reverse process of the other!

**Power plants use generators**

The generation of electricity by wind generators, hydroelectric dams, nuclear plants, the engine in a car, or gas- and coal-fired plants is accomplished with electrical generators. Power plants have rotating machines called turbines that generate the rotational motion and power the electrical generator. Generators create alternating current, which is the kind of electricity used by houses and businesses. Most of the electricity you use every day was created by an electric generator.

**Power plants use electric generators.**

**What voltage should be used to transmit electricity?**

Power plants are often located far away from the houses they power. What voltage should be used to transmit the electricity? Suppose you wanted to deliver 120 kW of power (enough to supply around 100 houses) through a low-resistance (0.1 Ω) power line. Should you transmit it at 120 V or at 2,400 V? The low-voltage line carries a lot of current, which results in 83% of the power being lost in the transmission lines. This is not good! The high-voltage line performs much better, only losing 0.2% of the power during transmission.

**How should you transmit 120 kW?**

| | Equation | 120 V line | 2,400 V line |
|---|---|---|---|
| Current in 0.1 Ω transmission line | $I = \frac{P}{V}$ | 1,000 A | 50 A |
| Power losses in transmission line | $P = I^2R$ | 100 kW | 250 W |
| Lost power | $\frac{P}{120\text{ kW}}$ | 83% | 0.2% |

**High-voltage lines reduce power losses**

This is why high-voltage lines are used for long distance transmission of electricity: to reduce power losses in the lines. But how do we change the voltage of the AC transmission line? To do this requires a *transformer*.

**How a transformer works**

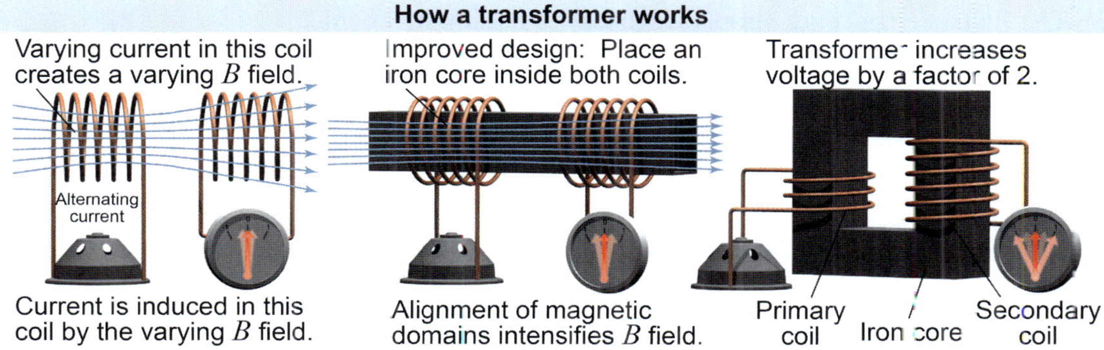


**Transformers**

A **transformer** uses *mutual induction* to transfer electrical power from one transmission line to another. The changing magnetic field produced by the primary coil induces a changing current in the secondary coil. The efficiency of a transformer is increased by winding the coils around a ferromagnetic core, because the magnetic domains in the core align to the magnetic field. A transformer exploits induction by introducing a twist: using a *different number of loops in the primary and secondary coils*. The induced voltage increases with each additional loop in the secondary coil. But energy (and power) must be conserved! When the induced voltage in the secondary is larger than the primary, the induced current must be smaller.

**Transformer voltage**

$$\frac{V_s}{V_p} = \frac{N_s}{N_p} \quad (19.3)$$

$V_p$ = primary coil voltage (V)
$V_s$ = secondary coil voltage (V)
$N_p$ = number of primary coils
$N_s$ = number of secondary coils

**Using transformers**

The ratio of the voltages of a transformer is given by the ratio of the number of coils in the primary and secondary, as shown in equation (19.3). A "step-up" transformer increases the voltage, whereas a "step-down" transformer decreases the voltage. Transformers only work with *varying* input voltages and currents—such as AC circuits—because only a changing magnetic field will induce a current in the secondary.

Chapter 19

# Section 2 review

Electric current can be created by a process called *electromagnetic induction*—by changing the *magnetic flux* through a conducting loop. The flux is the product of the loop's area, magnetic field strength, and orientation of the loop relative to the field. *Faraday's law of induction* tells us how much voltage will be generated. A *generator* uses induction to change mechanical work into electrical energy. A *transformer* uses the changing magnetic flux caused by changing voltage in one wire to induce a changing voltage in another wire.

| Vocabulary words | **magnetic flux, Faraday's law of induction, electromagnetic induction, Lenz's law, electrical generator, transformer** |
|---|---|

| Key equations | $\Phi_B = BA\cos\theta$ | $\text{emf} = -N\dfrac{\Delta\Phi_B}{\Delta t}$ | $\dfrac{V_s}{V_p} = \dfrac{N_s}{N_p}$ |
|---|---|---|---|

## Review problems and questions

1. The figure represents a simple electrical generator. Mechanical forces (such as wind) cause a wire coil to rotate in the presence of a magnetic field. Suppose that the rotation rate is beyond your control but that you can modify the properties of the generator itself. List three ways you could increase the voltage it produces.

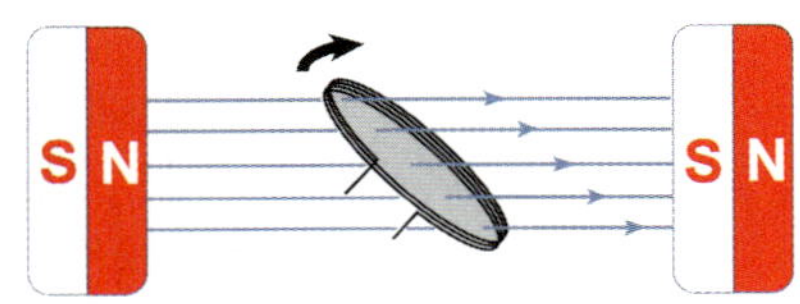


2. George's electrical toy requires two 1.5 V AA cells (in series, for a total of 3 V). He only has one AA cell, but he has a transformer with a 2:1 turn ratio. Can he use his toy?

3. Marcia inserts the N pole of a bar magnet into the right side of a wire coil and induces a temporary positive current. Would she induce (+) or (−) current if she
   a. inserted the S pole into the right side of the coil?
   b. inserted the N pole into the *left* side of the coil?
   c. inserted the S end into the left side?

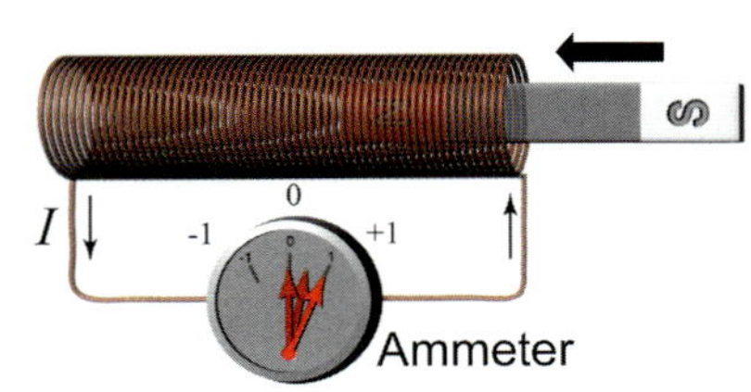


4. A loop of wire rests at right angles to a uniform, upward-pointing magnetic field. The loop is 20 cm wide and 60 cm long. The coil begins to turn and then stops after rotating 90°. The magnetic field has a strength of 0.1 T. What is the magnetic flux through the loop in each of the states shown here?

(a) (b)

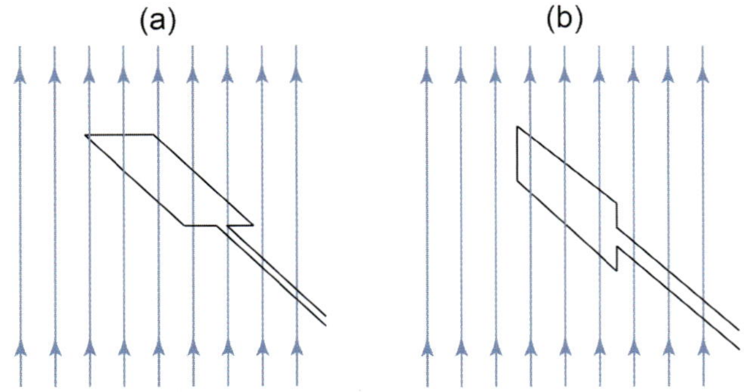

(a) $\theta = 0°,\ \cos\theta = 1$ (b) $\theta = 90°,\ \cos\theta = 0$

Particle physicists use "bubble chambers" (or more modern spark chambers) to track the paths of colliding *elementary particles*—protons, electrons, and other subatomic particles. In the collision illustrated at right, two of the particles that were created follow spiral trajectories. Why do these particles travel along spiral paths in a bubble chamber? The answer has to do with electromagnetism. The same phenomenon is used in older television sets to create the picture on the screen!

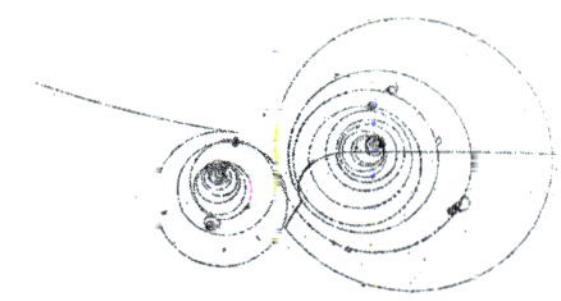

Why do these particles travel in spirals?

## Magnetic force on a moving electric charge

**Magnetic force on a moving electric charge**

At the beginning of this chapter, you learned that moving electric charges—also called electric current—create a magnetic field around them. It is also the case that an electric charge will feel force when moving in a magnetic field. This magnetic force depends not only on the strength of the magnetic field $B$ but also on the charge $q$ of the particle. Moving positive charges feel a magnetic force in one direction, whereas negative charges with the same velocity will feel a force in the opposite direction. Neutral particles, such as neutrons, feel no magnetic force at all! Furthermore, the strength of the magnetic force depends on *how fast the charged particle is moving*. Slow particles feel only a weak force, whereas high-speed particles feel a strong force. Equation (19.4) gives the force on a charged particle in a magnetic field.

Charged particles are deflected when they move through a magnetic field.

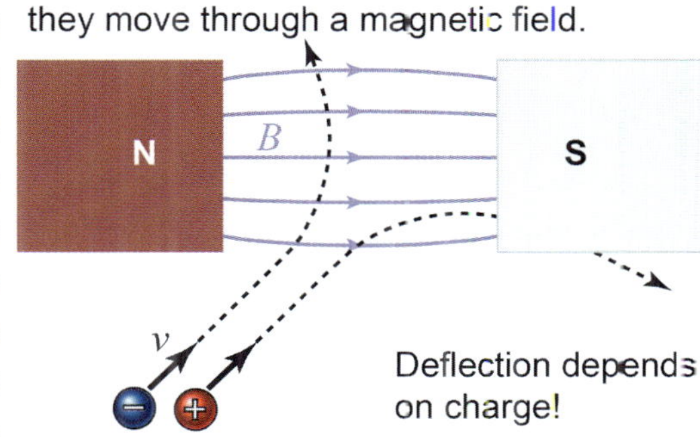


**Magnetic force** on a moving charge

$$(19.4) \quad F_B = qvB\sin\theta$$

$F_B$ = magnetic force (N)
$q$ = electric charge (C)
$v$ = speed (m/s)
$B$ = magnetic field (T)
$\theta$ = angle between velocity and field (deg)

**Right-hand rule for magnetic force on a moving charge**

The direction of the magnetic force is determined by the right-hand rule. You have seen other right-hand rules before for the magnetic field around a current-carrying wire and the magnetic field of an electromagnet! In this case, your thumb represents the direction of the velocity, your index finger the direction of the magnetic field, and your middle finger the direction of the resulting magnetic force *on a positively charged particle*. Negative charges will feel a force in the opposite direction.

**Right-hand rule**

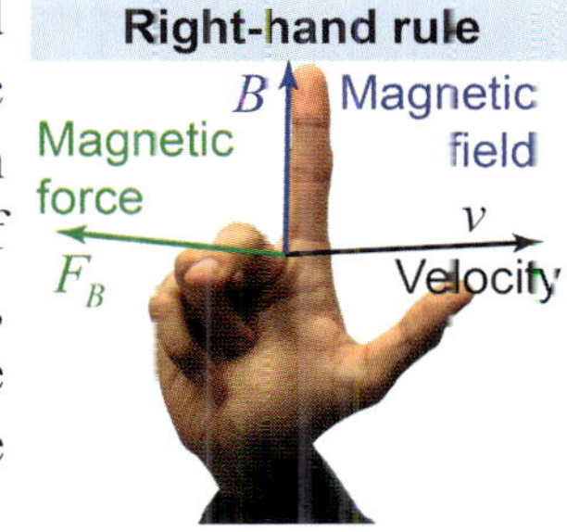


**Comparing gravitational, electric, and magnetic forces**

The magnetic force has a property that is very different from gravity or the electric force. Two masses that are separated by 1 m feel the same mutual gravitational attraction, regardless of the direction or magnitude of their velocity vectors. Likewise, two electric charges exert the same electric force on each other (from Coulomb's law), regardless of their velocities. But the magnitude and direction of the magnetic force felt by a moving charged particle depends on the magnitude and direction of its velocity.

## Mass spectrometer

**Identifying elements and isotopes in a gas**

The **mass spectrometer** is a scientific instrument used to identify the elements and isotopes present in a gas. Inside the mass spectrometer, magnetic forces on moving charged particles are used to measure the masses of individual atoms. How does it work?

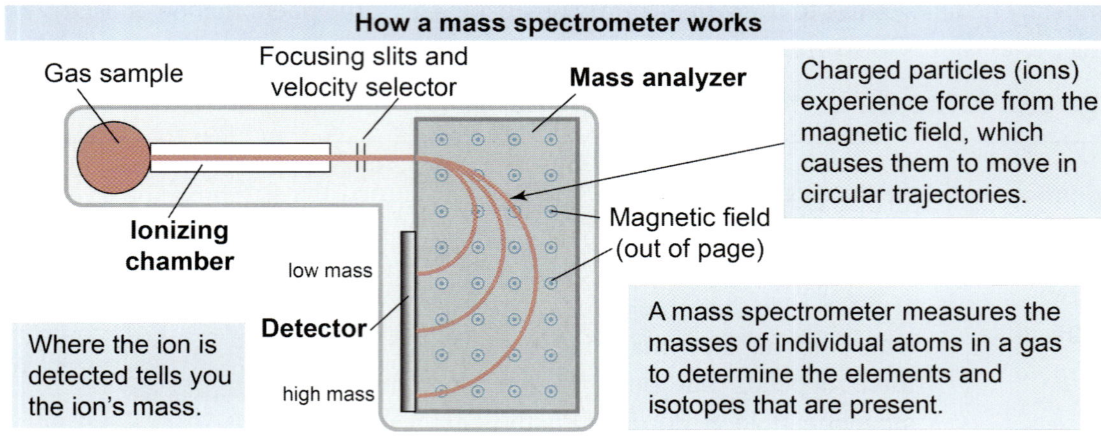


**How a mass spectrometer works**

In the ionizing chamber, atoms are ionized (into positive ions) by removing one (or more) of their outermost electrons. Ions of a particular velocity are isolated in the velocity selector and then emerge in a narrow beam as a result of focusing slits. In the mass analyzer, these moving charges pass through a strong magnetic field. The atoms experience a magnetic force that causes them to move in circles; the *radius* of the circle depends on the *mass of the ion*. Ions of different masses will strike different parts of the detector—allowing you to determine their masses.

**Why do the ions move in circles?**

According to the right-hand rule, the magnetic field exerts a force on the ions that is always perpendicular to their velocity. As you learned on page 211, a constant force applied perpendicular to the object's velocity will result in circular motion! The radius of circular motion for a charged particle in a uniform magnetic field is referred to as the **cyclotron radius** (or gyroradius). This is the explanation behind the circular tracks in the bubble chamber on the previous page.

**Cyclotron radius**

Magnetic force on particle is perpendicular to its velocity:

$$F_B = qvB\sin\theta = qvB$$

($\sin\theta$ crossed out and replaced by 1)

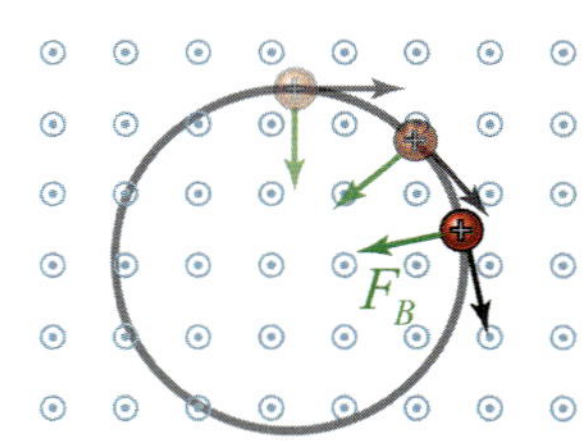


Equate this force with the centripetal force:

$$F_c = \frac{mv^2}{r}$$

**Circular motion:** $r = \dfrac{mv}{qB}$

**Mass is determined from the cyclotron radius**

The cyclotron radius can be calculated by using equation (19.4) and the equation for the centripetal force. Set the centripetal force $F_c = mv^2/r$ equal to the magnetic force on the ions, $F_B = qvB$ (where $\sin\theta = 1$, because the field is perpendicular to the velocity), then solve for the mass of the ions to obtain equation (19.5).

**(19.5)** $$m = \frac{rqB}{v}$$

$r$ = radius of trajectory (m)
$q$ = charge of particle (C)
$B$ = magnetic field strength (T)
$v$ = particle speed (m/s)
$m$ = particle mass (kg)

**Mass of an ion** given its gyroradius

The mass of the ion is therefore proportional to the radius of curvature of its trajectory through the mass analyzer. That's how the mass spectrometer works!

**Direction of magnetic force around a current-carrying wire**

A straight wire carrying an electric current creates a magnetic field around the wire in concentric circles. This magnetic *field* can exert a magnetic *force* on nearby moving charged particles. Using the right-hand rule in the illustration at right, you can see that a positively charged particle moving in the same direction as the electric current will feel a magnetic force *toward the wire*. Likewise, a positively charged particle moving in the opposite direction as the electric current will feel a magnetic force *away from the wire*.

**Direction of magnetic force around a current-carrying wire**

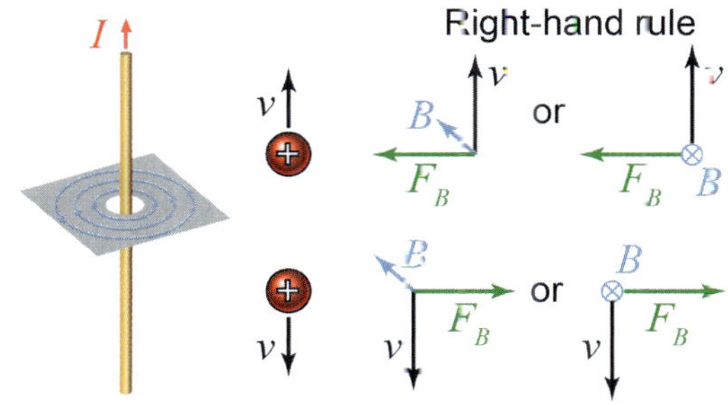


**Strength of magnetic *field* around a wire**

If the direction of the magnetic force is given by the right-hand rule, what is the *magnitude* or strength of the magnetic field and force? The magnitude of the magnetic *field* around a straight, current-carrying wire is given by equation (19.6). The magnetic field $B$ is stronger for a larger electric current $I$ and becomes weaker as the distance $r$ from the wire increases.

**Strength of magnetic force around a current-carrying wire**

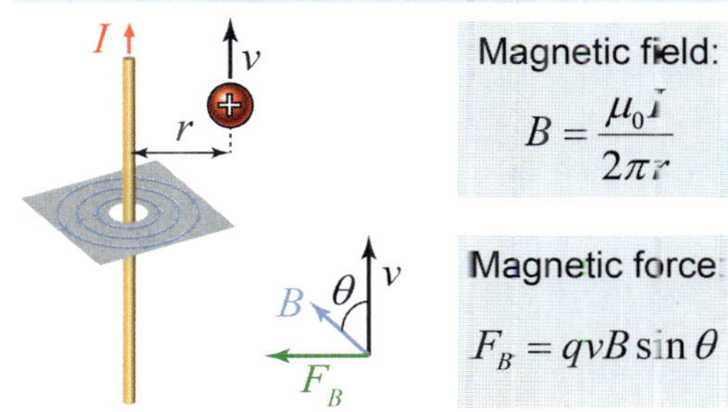


**Magnetic field** around a current-carrying wire

$$(19.6) \quad B = \frac{\mu_0 I}{2\pi r}$$

$B$ = magnetic field (T)
$\mu_0$ = permeability constant = $4\pi \times 10^{-7}$ T m/A
$I$ = current (A)
$r$ = distance from wire (m)

**Magnetic force around a wire**

Once the magnitude of the magnetic field is known, the magnetic force can be calculated by using equation (19.4), or $F_B = qvB \sin\theta$, where $\theta$ is the angle between the magnetic field and velocity vectors.

**Using the equation**

In equation (19.6), most of the quantities are familiar: magnetic field $B$, measured in teslas (T); electric current $I$, in amperes; and distance $r$ from the wire , in meters. What is $\mu_0$? It is called the **permeability of free space**, which is a physical constant describing how effectively a magnetic field can be established in a vacuum. (It is sometimes called the *vacuum permeability*.) But all you really need to know to use equation (19.6) is that $\mu_0$ is a physical constant equal to $4\pi \times 10^{-7}$ T m/A!

**Calculate the magnetic field near a wire**

Calculate the magnetic field 10 cm away from a straight wire carrying 2 A of current.

*Asked:* magnetic field $B$

*Given:* distance $r = 0.1$ m; electric current $I = 2$ A

*Relationships:* $B = \mu_0 I/(2\pi r)$

*Solution:*

$$B = \frac{\mu_0 I}{2\pi r} = \frac{(4\pi \times 10^{-7}\ \text{T m/A})(2\ \text{A})}{2\pi(0.1\ \text{m})} = 4 \times 10^{-6}\ \text{T}$$

*Answer:* $B = 4 \times 10^{-6}\ \text{T} = 4\ \mu\text{T}$

### Magnetic fields from two current-carrying wires

Inside a current-carrying wire, charged particles (electrons!) are moving. When one such wire is near another, each wire creates a magnetic field that exerts a magnetic force on the moving charges *in the other wire*. What happens?

**Magnetic force between two current-carrying wires**

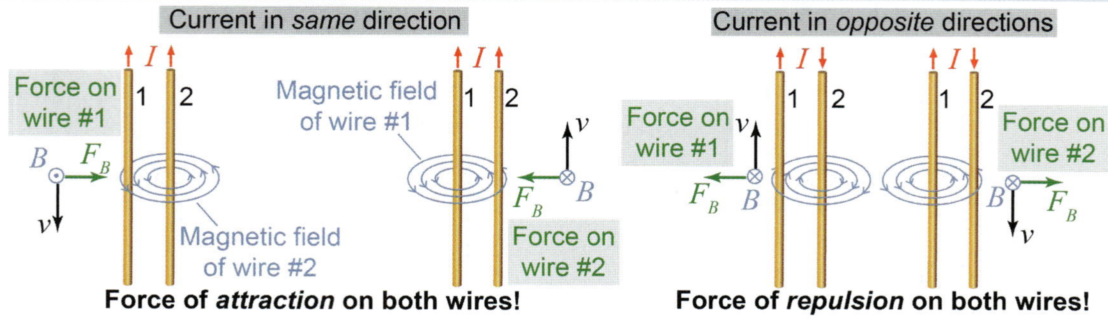


### Magnetic force between two current-carrying wires

The direction of the magnetic force on either wire depends on the direction current is flowing in both. If the current in the two wires is in the same direction, then applying the right-hand rule shows that the magnetic force *attracts* the two wires to each other, as shown in the illustration above. If the currents are in opposite directions, then the magnetic force causes the wires to *repel* each other.

### Force on a wire in a uniform magnetic field

The top illustration depicts the *direction* of the magnetic force on each wire that is in a magnetic field, as determined by using the right-hand rule. But what is the *magnitude* of the force? Intuitively, we should expect the magnetic force to be stronger if the length of the wire is longer, or if the current flowing through it is larger, or if the magnetic field around it is stronger—which is the relationship in equation (19.7)!

**Force on a current-carrying wire in a uniform magnetic field**

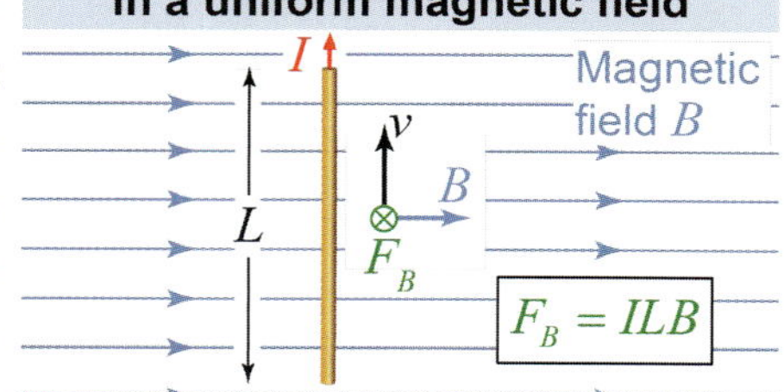


**Magnetic force** on a current-carrying wire

$$(19.7) \qquad F_B = ILB$$

$F_B$ = magnetic force (N)
$I$ = current (A)
$L$ = length of wire (m)
$B$ = magnetic field (T)

### Loudspeakers

The operation of the loudspeaker uses the magnetic force on a current-carrying wire (the **voice coil**) in a fixed magnetic field. As the signal current through the coil varies, the resulting force on the voice coil moves it up and down. The direction of current in the illustration causes the speaker cone to move upward; when the current reverses, the cone moves downward. The voice coil is attached to a speaker cone, which is often made of foam. When the cone moves up and down, it creates pressure waves—or sound!—in the air.

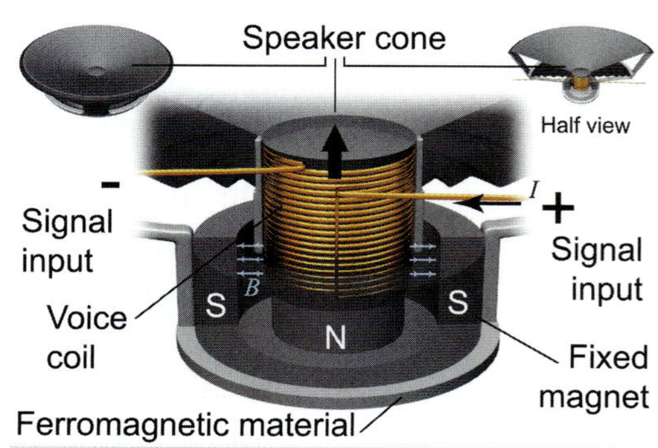


Varying signal voltages cause the voice coil to move up and down.

**What is the force on a *loop* of wire?**

A straight, current-carrying wire in a uniform magnetic field can experience a magnetic force, as we learned on the previous page. What happens if the wire is twisted into a loop or coil? (This is often called a *current loop*.) This is the principle behind a number of useful devices, such as the generator and the galvanometer.

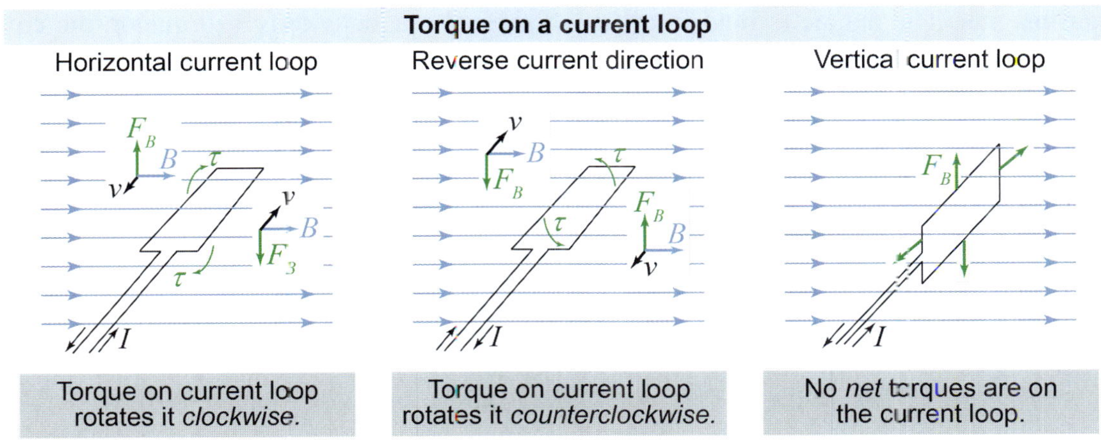


**Torque on a current loop in a magnetic field**

Let's start with the situation where the plane formed by a rectangular current loop is *parallel* to the direction of the magnetic field, as shown in the left-hand panel of the figure above. The current through the loop, shown in the figure, causes the right side of the loop to experience a downward magnetic force—by the right-hand rule on page 565. The left side of the loop, however, experiences an upward force because the current (and electric charge) flows in the opposite direction. These two forces create *torques* that will rotate the current loop clockwise! If the current is reversed, as in the middle figure, then the forces are reversed and the torque rotates the loop in the opposite, counterclockwise direction.

**No net torque when forces cancel**

A current loop in a uniform magnetic field *may* feel a torque—but not always! In the right-hand figure above, the plane of the current loop is *perpendicular* to the direction of the magnetic field. Now each segment of the rectangular loop experiences a force—but in outward directions. Each of the opposite forces cancel, resulting in no net force or torque. The loop is in equilibrium!

**Galvanometer**

The current loop in a magnetic field rotates in a manner that depends on the *direction of the current flowing through the loop*. This is the principle behind a scientific instrument called the **galvanometer**. A galvanometer detects and measures the current flowing through a wire using the rotation of the wire loop in a uniform magnetic field. A small spring is often mounted on the loop to keep it at the zero position when no current is flowing. A galvanometer can be used as an *ammeter* and can be a sensitive detector of small currents.

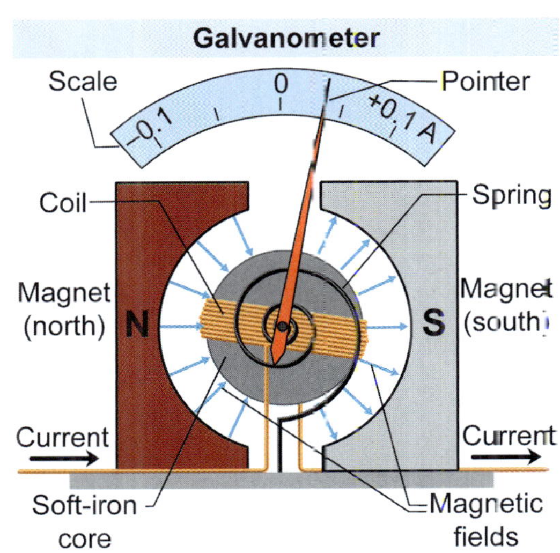

# Investigation 19C  Build a paper clip motor

**Essential questions** *How do magnetic forces cause an electric motor to spin?*

An electric motor uses electric current to create rotational motion, which can then be used to spin a fan or propel an electric car. In this investigation you will create a simple electric motor using a current-carrying coil and a magnet. The magnetic field from the permanent magnet exerts a torque on the coil, causing it to spin. The principle behind the electric motor is the same as for a galvanometer.

## How it works

When current flows through the coil shown on the left, the magnetic field exerts a *clockwise* torque that causes it to rotate. This rotation brings it to the position shown on the right, where the torque acts *counterclockwise*. Under these conditions, the coil can never spin continuously. To ensure that the torque acts in only one direction, the current (and therefore the torque) must shut off during half of each rotation.

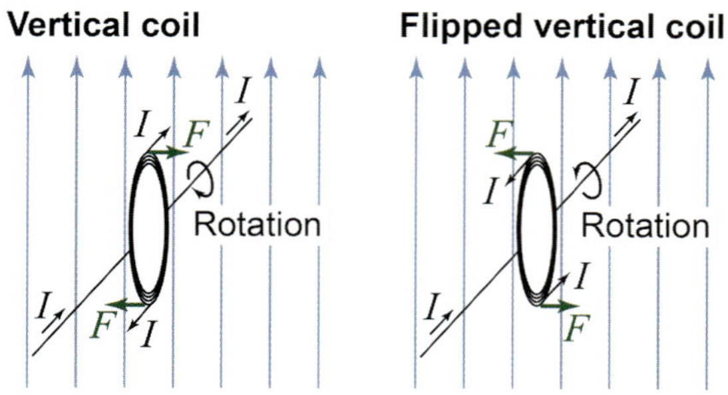


Forces cause *clockwise* rotation.

Forces cause *counterclockwise* rotation.

## Build the motor

1. Wrap approximately 80 cm of magnet wire tightly around a "D" cell battery, leaving a few centimeters free at each end.
2. Remove the coil. Create axles by wrapping each free end once around the coil to hold it tightly together.
3. Create a stand from two large paperclips and make a loop in each to hold the axles. Tape one support to the tabletop.
4. Test that your coil spins freely by inserting an axle into the support, setting the second support in place, and flicking the coil.
5. Remove the coil and use a knife to remove the insulation from the upward-facing sides of both axles. Do *not* remove insulation from the back side of the axles.
6. Place the coil into the supports. Connect the supports to the battery so that current will flow when the coil is in the face-down position.
7. Bring the magnet close to the coil. You may need to start it spinning but the magnet will keep it spinning.

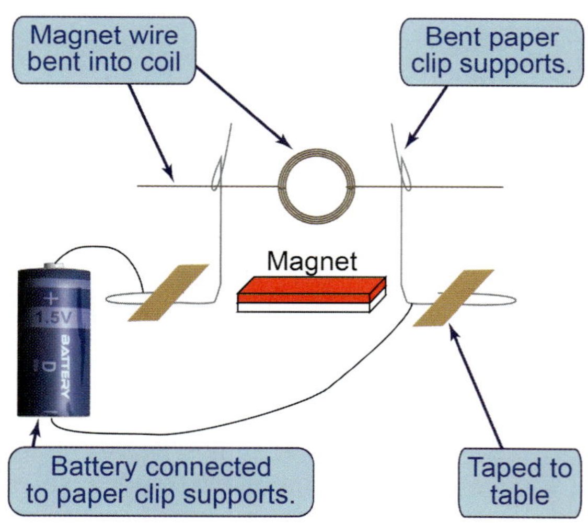


a. *Predict the effect that changing the number of batteries or magnets will have on the operation of the motor. Explain your prediction using the equation $F = ILB$.*

b. *Test your predictions. Were you correct?*

c. *Propose additional investigative questions: What variables could you change that might affect the operation of your motor?*

**Cathode ray tube**

For many years television sets and monitors used a **cathode ray tube (CRT)** to project the image. Three beams of electrons (for red, green, and blue) are created and focused in the rear of the CRT. An incoming electrical signal passes through electromagnets that create a varying magnetic field that deflects and steers the electron beams to different locations on the screen. Three masks separate the three electron beams and then each causes a phosphor layer to fluoresce. The steering magnets vary rapidly to create 24 images per second on the screen.

**How a cathode ray tube works**

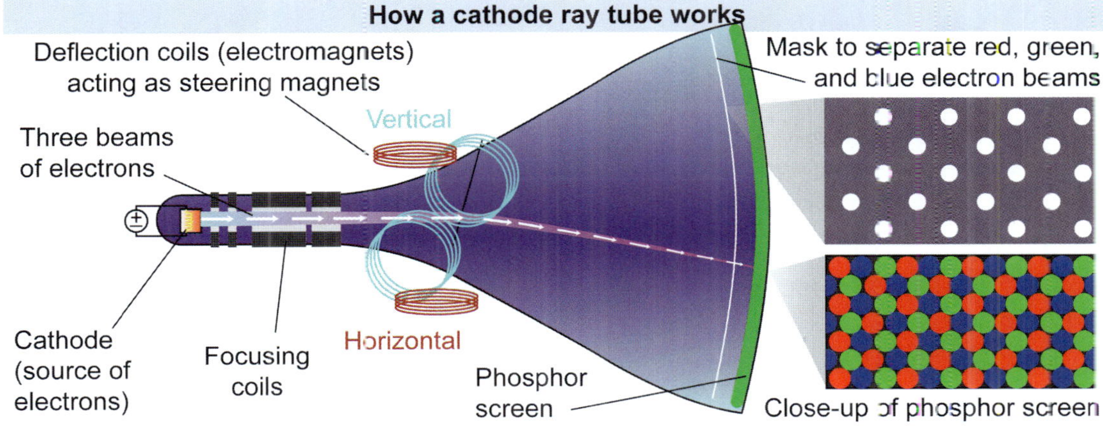


**Fun with a CRT!**

You can see the connection between electricity and magnetism by using a strong magnet to deflect the electron beam inside a CRT. Watch the screen change while you move a horseshoe magnet around the screen of the CRT. Why does this happen?

**Superconducting magnets**

Electromagnets are typically used when creating a strong magnetic field, but even copper wire has enough resistance to cause it to lose a lot of energy as heat. Some materials called *superconductors* have been found that exhibit *no resistance* at very low temperatures—whether at a few or tens of degrees above absolute zero. *Superconducting magnets* are electromagnets constructed from superconducting materials and they lose no energy as heat! Superconducting magnets are used to produce the strong magnetic fields in magnetic resonance imaging machines in hospitals, mass spectrometers, and particle accelerators, such as the Large Hadron Collider along the border of France and Switzerland.

**Magnetically levitating trains**

One application of superconducting magnets is to magnetically levitate high-speed trains. In the design illustrated at right, there are two sets of magnets. Magnetic attraction between the superconducting magnets of the undercarriage and the guideway lifts the "maglev" train above the track. Attraction and repulsion between magnets mounted on the side of the track maintain the horizontal alignment of the train. Magnetically levitating trains are currently in commercial use in China and Japan.

**Magnetic levitation using superconducting magnets**

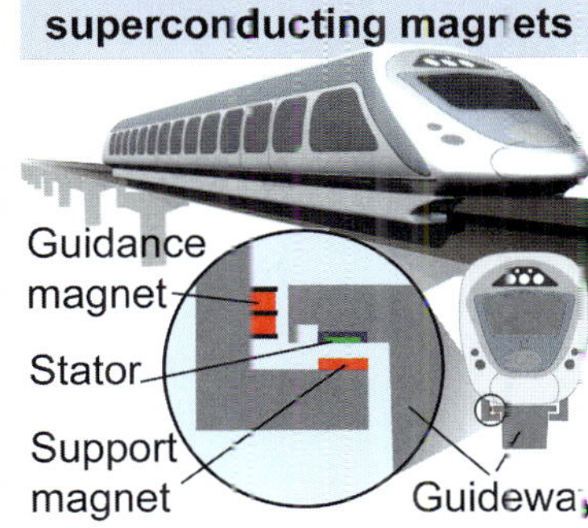

## Hybrid cars

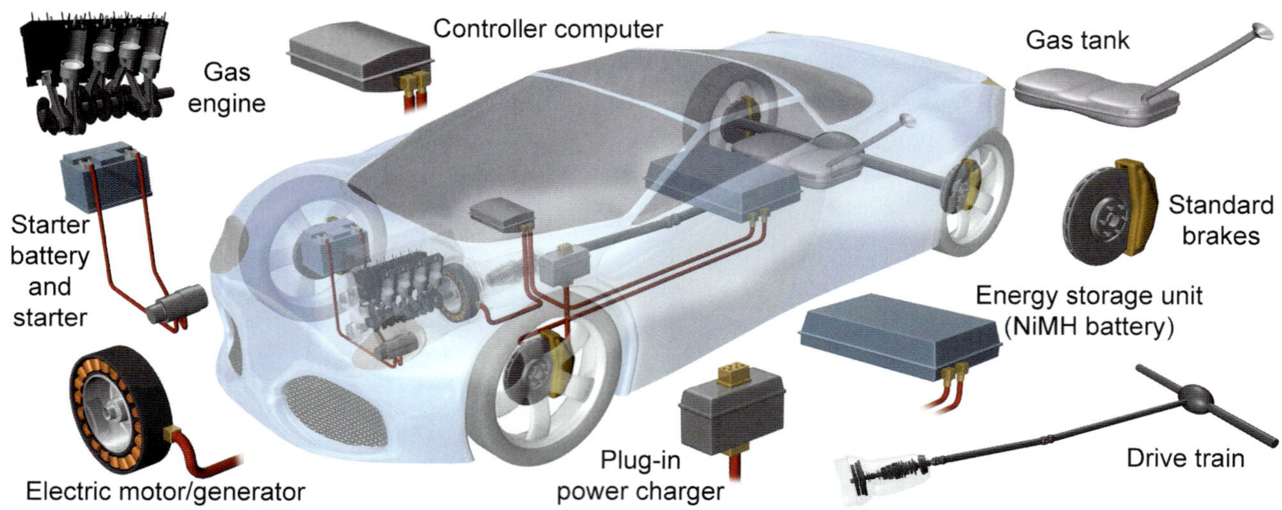


**What makes a car a *hybrid* car?** Although the first widely available hybrid was commercially produced beginning in 1997, working prototypes existed as early 1900! A hybrid electric vehicle is a car that has both a gas engine and an electric motor. It is therefore powered both by gasoline and by electricity stored in a battery.

**Electric motor** At the heart of a hybrid (or electric!) car lies an electric motor that transforms electrical energy from a battery into rotational motion of the drive shaft. As you learned on page 555, the electric motor is based on the link between electricity and magnetism.

**Gas engine** Hybrid cars have a standard gas engine under their hoods, but the engines are typically small and would be considered underpowered if they were the sole power source for the car. (Fully electric cars, in contrast, have no gas engine at all!)

**Storage batteries** All cars, including hybrid cars, have a car battery under the hood that is used to start the engine. But a hybrid car also contains a large number of nickel metal hydride (NiMH) batteries—usually located underneath the rear seats or trunk—that are dedicated to powering its electric motor.

**Generator** One major difference with the hybrid cars of a century ago is that today's hybrid cars usually contain a *generator* for recharging the battery. When the gas engine is running, but delivers more power than needed to move the car, the excess power is used to drive the generator. The generator is also used during *regenerative braking*.

**Regenerative brakes** When braking, modern hybrid and electric cars transform some of their kinetic energy into electrical energy using regenerative braking, which you learned about at the beginning of the chapter. The regenerative braking system in hybrid cars is *in addition* to the standard brakes—using friction on brake pads—that all cars have.

**Plug-in hybrid** With improved batteries, the electric storage capacity of some hybrid cars has been increased to the point that the car could operate on the electric motor *alone* for a few or a few dozen miles. It is more efficient to recharge these high-capacity batteries between trips using a power outlet, which is why some hybrid cars are called "plug-in hybrids."

**Operation of subsystems**

When a hybrid car is being driven, the different subsystems—gas engine, electric motor, generator, and regenerative brakes—are not always in operation at any given moment. Depending on the driving conditions, the engine might be off and the brakes might be regenerating the battery, for example.

**Accelerating slowly**

When the car is stopped and then starts up *slowly*, the low power requirements can be met by the electric motor alone. In this mode, the gas engine is off, and the car runs highly efficiently off the batteries. The car also runs on the motor alone when cruising at moderate speeds, such as under 40 mph.

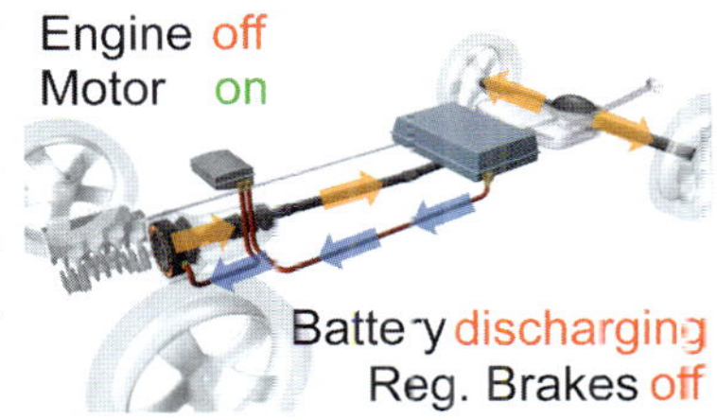


**Accelerating rapidly**

If the driver presses the accelerator pedal down a lot to accelerate rapidly, more power is needed than can be delivered by the motor alone. In rapidly accelerating operation, the engine and motor both are running to provide maximum power. The battery's charge is being used to power the motor.

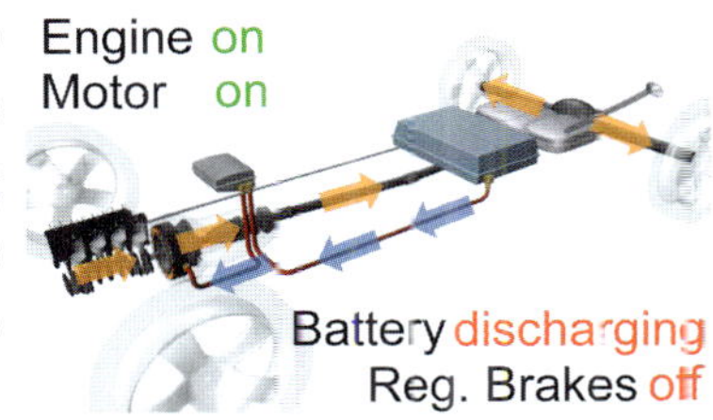


**Cruising at highway speeds**

When driving at highway speeds, the car needs enough power to overcome substantial air resistance. The engine is on, but often it delivers more energy than needed just to move the car forward. In this operational mode, the excess power from the gas engine powers the electric generator, thereby recharging the battery.

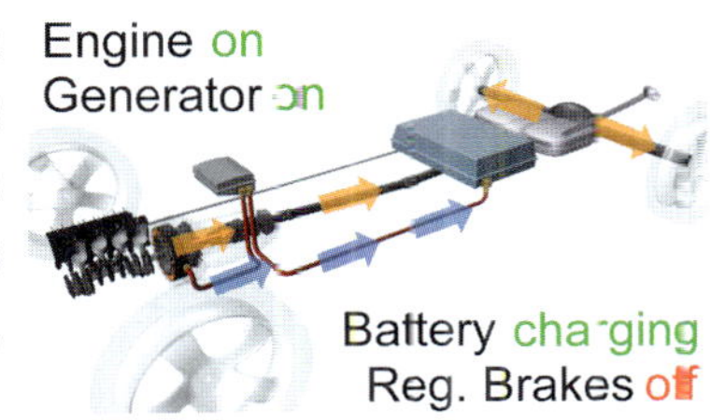


**Coasting**

If the driver sees a slowdown in traffic—or a red traffic light—far ahead, she may switch to coasting mode. The gas engine turns off and the electric motor is not needed, so no energy is expended. This is equivalent to putting a standard car's engine into neutral, except that the hybrid car's gas engine has shut off temporarily to save energy!

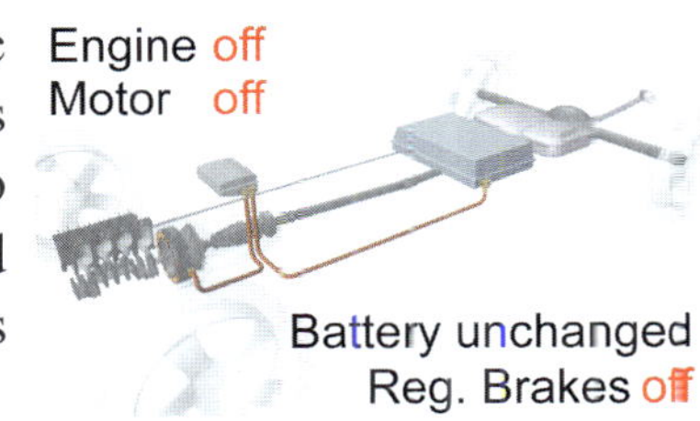


**Braking slowly**

If the driver wants to slow down somewhat more rapidly, she releases the accelerator pedal altogether and the regenerative braking system turns on. The car now slows down, because the car's kinetic energy is being transformed into electrical energy for storage in the battery. The engine and motor are both turned off, so no energy is consumed.

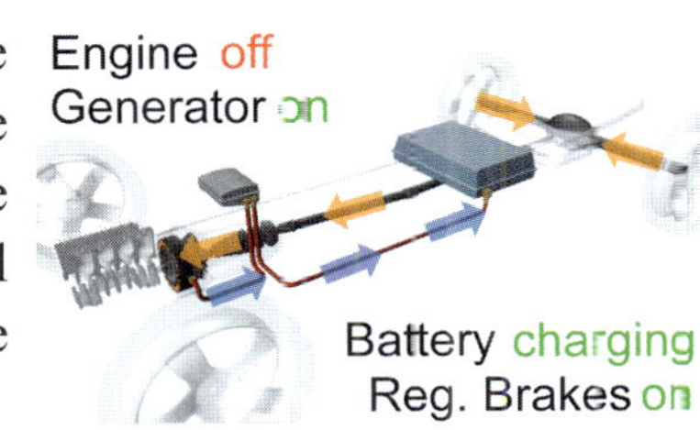


**Braking rapidly**

If the driver needs to brake more rapidly, she presses the brake pedal. In this mode, both the standard brakes and the regenerative braking systems are in operation. Once again, the engine and motor are both turned off, so no energy is consumed.

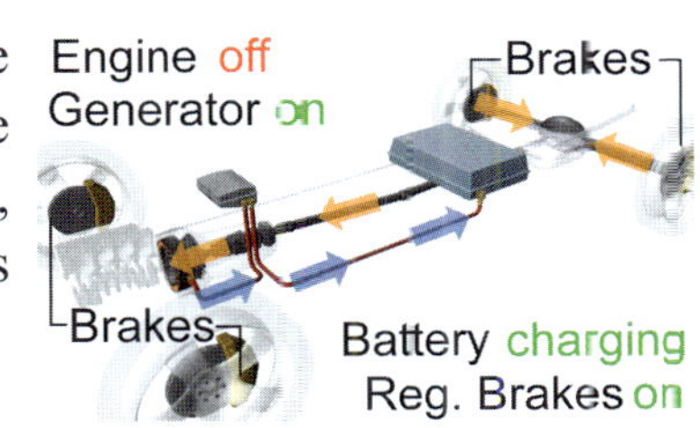

# Section 3 review

When a charged particle moves within a magnetic field, it feels a force. The force vector is perpendicular to the magnetic field *and* to the particle's velocity vector; this means that the force often causes charged particles to follow curved paths. The strength of the force is proportional to the magnetic field strength, the particle's charge, and the particle's speed. This force has many practical applications because it can separate particles with differing masses or electrical charges.

| Vocabulary words | **mass spectrometer, cyclotron radius, permeability of free space, voice coil, galvanometer, cathode ray tube (CRT)** |
|---|---|

| Key equations | $F_B = qvB\sin\theta$ | $m = \dfrac{rqB}{v}$ | $B = \dfrac{\mu_0 I}{2\pi r}$ | $F_B = ILB$ |
|---|---|---|---|---|

## Review problems and questions

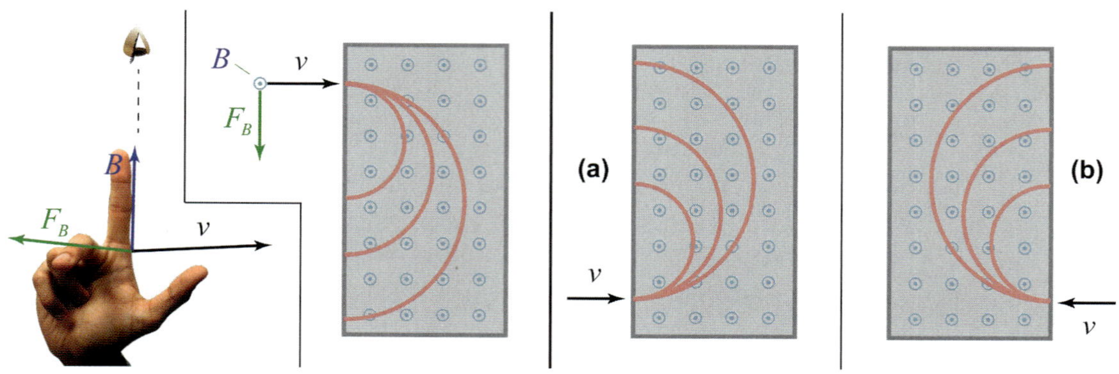


1. As shown here, the right-hand rule tells us the direction of the force on a *positively* charged particle that moves with velocity $v$ through a magnetic field $B$. The first magnetic spectrometer shows the trajectories of positively charged particles entering from the left. What is the sign (+ or −) of the particle charge in Case (a) and Case (b)? The magnetic field emerges from the page toward the viewer in all three spectrometers.

2. In each spectrometer shown on this page, three charged particles enter with identical velocities, but soon they follow different paths. Which of the following could explain why their paths diverge? (Recall that each of the three particles is in the same magnetic field.)
   a. The particles have differing masses.
   b. The particles have differing amounts of electric charge.
   c. Either a or b.
   d. Neither a nor b.

3. Calculate the strength of the electromagnetic force upon the positive charge in each of these three cases. (Note that the force will point toward you, at right angles to the page, since the charge is positive and the magnetic field and velocity vectors both lie in the plane of the page.) The field strength, speed, and charge remain constant; only the angle between the two vectors is varied.

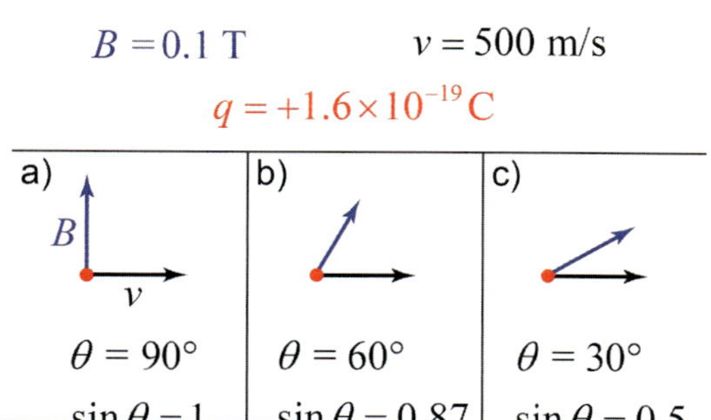

# Chapter review

## Vocabulary

Match each word to the sentence where it best fits.

### Section 19.1

| | |
|---|---|
| armature | commutator |
| electric motor | electromagnetism |
| electromagnets | polarity |
| solenoid | |

1. When you identify which end of a magnet is north you have determined its _______.
2. As an electrical device rotates about its axis, the curved plates that create alternating voltages are called the _______.
3. The subject of _______ connects the flow of electricity to the magnetic force.
4. _______ can be created by wrapping wires carrying electricity around a metal nail.
5. When Ravi wrapped a long wire around a cardboard tube he was creating a/an _______.
6. The part of an electrical device that has rotating electromagnets is the _______.
7. The _______ converts electrical energy into rotational motion of its axle.

### Section 19.2

| | |
|---|---|
| Faraday's law of induction | Lenz's law |
| electrical generator | electromagnetic induction |
| magnetic flux | transformer |

8. A hydroelectric power plant has a/an _______ that converts the rotational motion of the turbines into electricity.
9. Electricity from a hydroelectric power plant is usually converted to high voltage using a/an _______.
10. The _______ is a measure of the amount of magnetic field that passes through a closed loop of wire.
11. Voltage can be created in a conductor by nearby, changing magnetic flux in a process called _______.
12. That a changing magnetic field through a loop of 10 wire coils will induce more current than the same changing magnetic field through a loop of 3 wire coils is a consequence of _______.
13. That an induced current flows in the direction to oppose a changing magnetic flux follows from _______.

### Section 19.3

| | |
|---|---|
| cathode ray tube (CRT) | cyclotron radius |
| galvanometer | mass spectrometer |
| permeability of free space | voice coil |

14. The _______ is a scientific instrument that can determine what elements are present in a gas.
15. An indication of how well a magnetic field can be established in a vacuum is called the _______.
16. Older kinds of television sets and computer monitors use a _______ to form the picture.
17. In a bubble chamber image, the _______ can be used to determine the mass and charge of a particle.
18. A scientific instrument that is a sensitive detector of small electric currents is the _______.
19. At the heart of a loudspeaker is the _______, which moves up and down in response to changing electric current through it.

## Conceptual questions

### Section 19.1

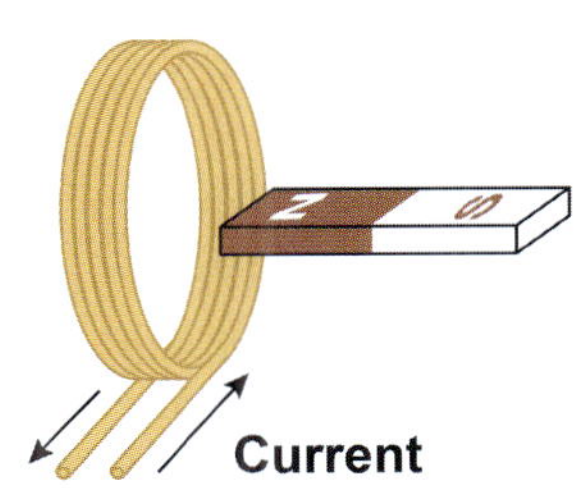


20. A coil of wire carries a current as shown in the diagram. Will the bar magnet be attracted or repelled by the coil?

# Chapter review

## Section 19.1

21. Research the work of James Clerk Maxwell. What physical phenomena do his four equations describe? Why was his work so important? Use a standard format for citations in your paper.

22. A stationary electric charge is surrounded by an electric field. What kind of field surrounds a moving charge?

23. Describe how changes in the understanding of the science of electricity and magnetism led to the development of three different technologies in use today.

24. Define what the armature in a motor is. Use the words rotor and electromagnet in your definition.

25. An electromagnet is placed a few centimeters west of a compass. The compass needle deflects so that it points due west. How strong is the electromagnet's magnetic field, compared to the Earth's magnetic field, at the location of the compass needle?

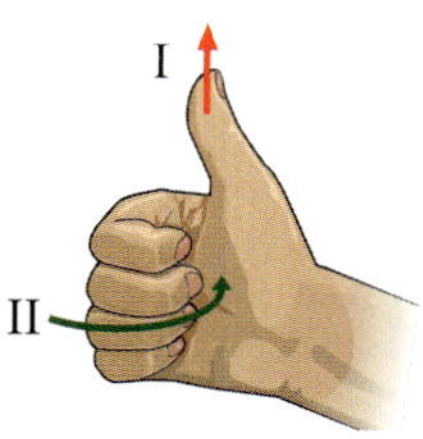


26. Which of the following pairs of quantities can be represented by the straight arrow I and the curved arrow II shown here?

    a. I = electric current and II = magnetic field
    b. I = magnetic field and II = electric current
    c. Both a and b are valid representations.
    d. Neither a nor b is a valid representation.

27. A physics teacher has placed a horizontal magnetic compass next to a vertical current-carrying wire.

    a. Describe or draw how the compass points (toward the wire?, away from it?, etc.).
    b. What will the compass needle do if the teacher reverses the direction of the current?

28. When inducing a voltage in a wire coil, to *maximize* the induced voltage you should ______ the number of loops and ______ the rate of flux change.

    a. decrease; decrease
    b. decrease; increase
    c. increase; decrease
    d. increase; increase

29. In doing research for a paper, a student came across the following text: "English chemist and physicist Michael Faraday showed the connection between changes in magnetic fields and induced electric current in 1822." Which of the following are appropriate ways to include this information in the paper without plagiarizing?

    a. Include the sentence word for word, without quotations, followed by a superscript footnote number such as [10].
    b. Include the sentence word for word, enclose the words with quotation marks, and follow the text with a citation to the book (enclosed in parentheses).
    c. Include the sentence word for word, enclose the words with quotation marks, follow the text with a superscript number for footnoting, and include a citation to the book at the end of the paper.

30. Draw the magnetic field lines around two wires that are parallel to one another and have current flowing in the same direction. Do the two wires attract or repel, provided they are close enough together?

31. Explain how a *tangent* is used to determine the direction of a magnetic field near a current-carrying wire.

32. A horizontal wire entering a hole in a wall carries electric current. Using a diagram, show the direction of the magnetic field around the wire if the current travels into the wall. Use a second diagram to show the magnetic field direction if the current flows away from the wall.

33. Research and describe the contributions of three scientists to the understanding of the connection between electricity and magnetism.

34. Joe and Sarah are experimenting with electricity and magnetism. Both are carrying handheld magnetic field detectors; Joe is wearing a sweater that has become electrically charged. When Joe runs past Sarah, she detects a slight magnetic field, but he doesn't. What's going on?

## Section 19.2

35. Which of the following statements is the best description of the following equation?

$$\text{emf} = -N\frac{\Delta\Phi_B}{\Delta t}$$

    a. Changing emf will cause the magnetic flux to change in a coil.
    b. Changing magnetic flux will induce an emf in a coil.
    c. Changing magnetic field will cause an electromotive current to flow.
    d. Induced emf will change the magnetic flux in a coil.

# Chapter review

## Section 19.2

36. Which of the following equations best represents the following statement?
"In a transformer, the ratio of the secondary to primary voltage equals the ratio of the number of secondary to primary coils."
    a. $V_{s/p} = N_{s/p}$
    b. $S_V / P_V = S_N / P_N$
    c. $V[S:P] = N[S:P]$
    d. $V_s / V_p = N_s / N_p$

37. Siegfried placed a $10^{-5}$ T bar magnet inside a 56-coil solenoid of length 4 cm and diameter 1 cm. His lab partner, Rob, then measured the current through the solenoid's wire.
    a. What is the value of the current Rob measures?
    b. What can Siegfried do to change the value of the current Rob will observe?

38. Lehi makes his own transformer by winding 10 loops of wire around one side of an iron rod and 20 loops of wire on the other side. Then he connects a DC (constant voltage) battery to the first coil. After he connects the battery, he checks the voltage coming out of the second coil, but there is no voltage. What did he do wrong?

39. A constant magnetic field points upward from the floor to the ceiling everywhere in the room. There are three loops of wire: A, B, and C. Loop A is flat on a table, B is perpendicular to the table, and C is at a 45° angle to the table.
    a. List the loops in order of least to greatest magnetic flux.
    b. In which, if any, of the loops will an electric current be induced?

40. A magnetic field points down and is decreasing in strength. The induced current in a wire loop on the top of a table will flow ___________ (clockwise, counterclockwise) when viewed from the top.

41. A magnetic field points into the page and is increasing in strength. For a wire loop in the plane of the page, will the induced current flow clockwise or counterclockwise?

42. When power plants generate electricity, they use a transformer to transmit the electricity through power lines at a low current to reduce power loss. Then a transformer at your house converts it back to the voltage that your appliances use. Which transformer is "step up" and which is "step down?" How do you know?

## Section 19.3

43. Which statement best expresses the following equation for the magnetic force on a moving, charged particle?

$$F_B = qvB\sin\theta$$

    a. The magnetic force is the product of charge, speed, magnetic field, and the sine of the angle between its velocity and the magnetic field.
    b. The magnetic field is the product of charge, speed, magnetic force, and the sine of the angle between its velocity and the magnetic field.
    c. The magnetic field is the sum of charge, speed, magnetic force, and the cosine of the angle between its velocity and the magnetic field.
    d. The magnetic force is the product of charge, speed, magnetic field, and the sine of the angle between its charge and its velocity.

44. In a particle accelerator experiment, a particle produced by a collision was observed to move in a spiraling pattern. What property of the particle can you infer from this observation?

45. In a particle accelerator experiment, two particles observed after a collision were moving in spirals—but rotating around their spirals in the opposite direction. What intrinsic property can you infer about the two particles?

46. Why is an image on the screen of a cathode ray tube affected by moving a horseshoe magnet near to it?

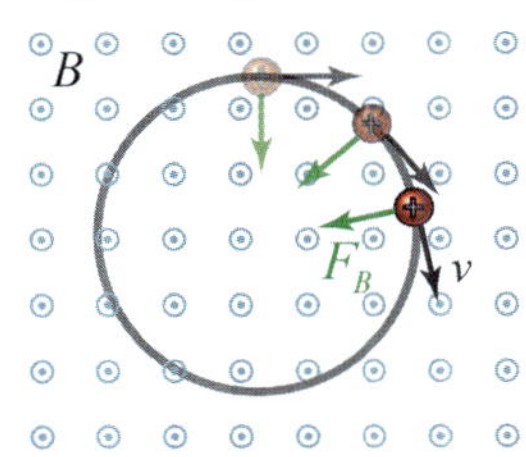


47. This diagram shows a uniform magnetic field $B$ emerging toward you from the plane of the page. It also shows a positively charged particle moving with velocity $v$. The charged particle experiences a force $F_B$ due to its motion within the field.
    a. If you were using the right-hand rule for magnetic force on a moving charge, which of the labeled vectors would correspond to your thumb?
    b. Which of the labeled vectors would correspond to your index finger?
    c. Suppose the particle became negatively charged, but nothing else in the diagram changed. Which way would $F_B$ point?
    d. Suppose $B$ reversed direction, but nothing else in the diagram changed. Which way would $F_B$ point?

## Quantitative problems

### Section 19.1

48. An electromagnet located 10 cm east of a compass deflects the compass needle 45° east. If the Earth's magnetic field is $5\times10^{-5}$ T, what is the strength of the electromagnet's magnetic field at the location of the compass needle?

### Section 19.2

49. Joaquim's household electric service delivers alternating current at 120 V, but his homebuilt robot requires 30 VAC. The transformer in his power converter has 1,200 turns on the side that receives the 120 V current. How many turns are required on the 30 V side of the transformer?

50. A typical installation of power lines on a street, carried by wooden poles, carries the electricity at 12,000 V.

    a. By what factor must this voltage be changed before running it through a wire to a household?
    b. Somewhere along the street is a transformer that changes the voltage for household use. Is this a step-up or step-down transformer?
    c. Assuming the wires entering and exiting this transformer have the same resistance per meter, what is the ratio of the dissipated power per meter for those wires?

51. Colin and Suzette have heard that you can generate electricity by spinning an extension cord like a jump rope. Suppose you could spin a circular wire loop once per second in the presence of Earths magnetic field ($B = 3\times10^{-5}$ T near the equator).

    a. What would be the maximum magnetic flux $\Phi_B$ through the loop if the loop had a radius of 1 m?
    b. If you went from zero flux to this maximum value in a quarter of a second ($\Delta t = 0.25$ s), what would be the magnitude of the average induced voltage?

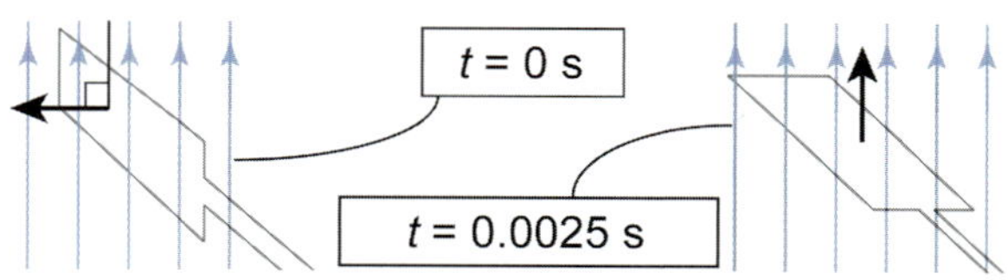


52. The generator illustrated here consists of one rectangular wire coil, 20 cm by 60 cm, in a uniform, constant 0.1 T magnetic field. An external force rotates the loop a quarter turn (from $\theta = 90°$ to $\theta = 0°$) in 0.0025 s. Which of the following is approximately equal to the average voltage delivered by the generator during this quarter turn? (Ignore the sign, or polarity, of the voltage.)

    a. 5 mV
    b. 5 V
    c. 5,000 V

53. In a physics experiment, a rectangular coil with an area of 1 $cm^2$ is placed in a 0.0001 T magnetic field.

    a. If the plane of the coil is perpendicular to the magnetic field, what is the magnetic flux through the coil?
    b. If the plane of the coil is parallel to the magnetic field, what is the magnetic flux through the coil?

### Section 19.3

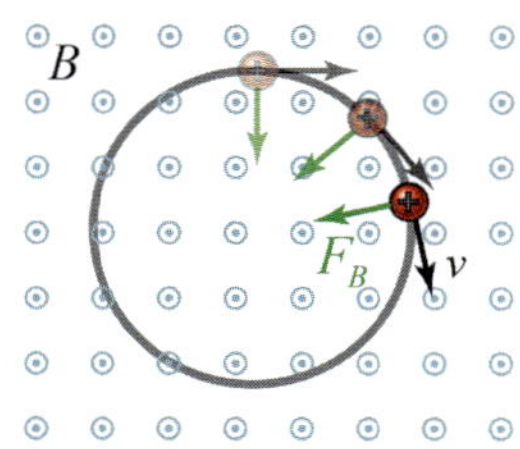


54. When a charged particle moves at right angles to a uniform magnetic field, the magnetic force acts like a centripetal force. That force keeps the particle on a circular "orbit" until the particle loses energy. The circle's radius is known as the cyclotron radius (see page 566).

    a. Suppose you double the particle charge but leave its speed and mass unchanged. Will this alter the gyroradius? If so, by what factor?
    b. Suppose you double the particle mass but leave its speed and charge unchanged. Will this alter the gyroradius? If so, by what factor?
    c. Suppose you double the strength of the magnetic field but leave the speed, mass, and charge of the particle unchanged. Will this alter the gyroradius? If so, by what factor?

55. Positively charged protons are the nuclei of hydrogen atoms, and they travel through space after being launched by *coronal mass ejections.* Some of these protons later collide with the Earth's magnetic field. Suppose that a proton reached the stratosphere at a speed of 300 km/s and encountered a magnetic field with strength $B = 10$ nT $= 10^{-8}$ T.

    a. If the proton's motion were parallel to the magnetic field, what magnetic force (in newtons) would it feel?
    b. If the proton were to move at right angles to the magnetic field, what force would it feel?
    c. The proton's mass is $m_p = 1.67\times10^{-27}$ kg, and its electrical charge is $q_p = +1.60\times10^{-19}$ C. If the proton were to be trapped by the magnetic field, what would its *cyclotron radius* be? (The cyclotron radius is defined on page 566.)

# Chapter review

## Standardized test practice

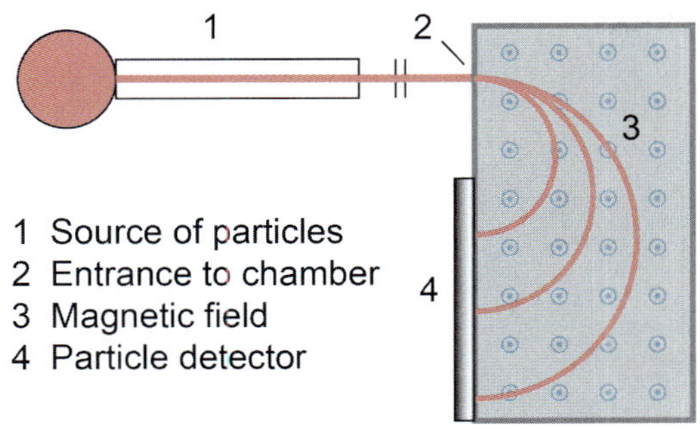


56. The device shown fires charged particles into a chamber that has a uniform magnetic field. There, the particles are deflected by magnetic forces. The magnetic field comes toward you at right angles to the plane of the page. What is the sign of the charge for the particles whose paths are shown here?

A. positive only
B. negative only
C. positive and negative
D. There is not enough information provided.

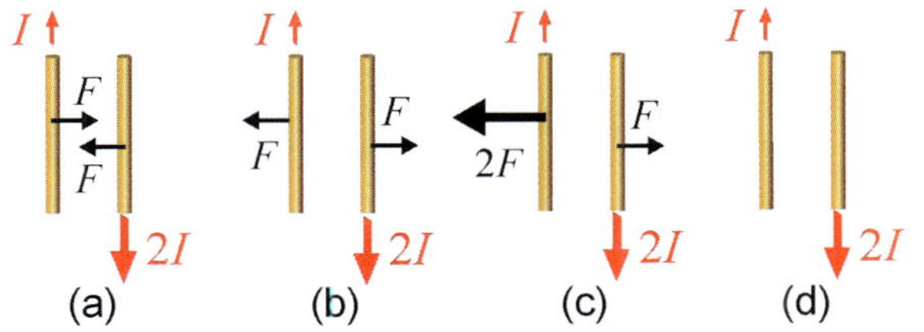


57. Two equally long parallel wires carry currents that are flowing in opposite directions, with Wire 2 carrying twice as much current as Wire 1. Which of the following statements can correctly describe the forces that each wire exerts upon the other?

A. Wires 1 and 2 attract each other with equal strength.
B. Wires 1 and 2 repel each other with equal strength.
C. Wire 1 repels Wire 2 with a force $F$, and Wire 2 repels Wire 1 with twice as much force.
D. Neither wire attracts or repels the other.

58. One of the following wire loops would *not* induce any electric current. Which is it?

A. A wire loop at rest in a constant magnetic field.
B. A wire loop at rest in a changing magnetic field.
C. A wire loop that rotates around an axis at right angles to a constant magnetic field.
D. A wire loop with a permanent magnet continuously moving into and out of it.

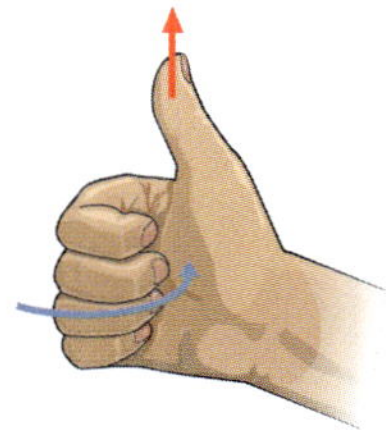

59. The right-hand rule is shown for the magnetic field produced by the current in a straight wire. Which of the following statements best expresses this rule?

A. If you point your fingers in the direction of the magnetic field, then your thumb will indicate the direction of the current.
B. If you point your thumb in the direction of the current, then your curled fingers show the direction of the magnetic field.
C. If you curl your fingers in the direction of the current, then your thumb will point in the direction of the magnetic field.
D. If you move your thumb in the direction of the magnetic field, then your fingers will point in the direction of the current.

60. A motor spins because the magnetic field of an electromagnet in the motor changes direction, alternately attracting and repelling a permanent magnet. What causes the magnetic field to change direction?

A. The current flow increases and decreases.
B. The distance between magnets changes.
C. The current direction is reversed.
D. The current is switched on and off.

61. One part of an iron core is wound with three loops of wire. Another part of the same iron core is wound with a six-loop coil. Electric current is passed through the first coil and a light bulb, motor, or other load is attached to the second. What type of device is this?

A. an electric motor
B. an electric generator
C. a step-up transformer
D. a step-down transformer

62. Which of the following is an example of a *magnetic* force within an electric motor?

A. the force between the armature and the stator, driving the motor
B. the force of friction against the armature, generating heat
C. the force that the motor can apply to another object
D. the force that pushes electrons through the electromagnet

# Chapter 20 Light and Reflection

Near noon on a sunny day, more than 500 W of radiant power from the Sun falls on every square meter of the ground. Sunlight is great for warming *you* up, but the power of sunlight is too diffuse to cook food. The purpose of a *solar oven* is to gather the power of sunlight over a large area and focus it, creating enough concentrated power to cook food. In a solar-thermal power plant the same principle is used to boil water into high-pressure steam, which then turns an electric generator to make electricity.

You can build a powerful solar oven using little more than basic physics, geometry, cardboard, and aluminum foil. The principle is to use curved or angled reflecting surfaces to direct sunlight from many areas onto the same area. Sunlight reaches Earth in nearly parallel rays. A *parabola* has the property that parallel rays reflect to a central point called the *focus*. A one-square-meter parabolic mirror can easily achieve temperatures of 200ºC or higher in full sun. Near the focus, a cup of water will boil within minutes. In fact, the radiant power near the focus can be so concentrated that one must be careful to avoid getting seriously burned.

Building a solar oven might be a nifty extra-credit lab or after-school project for a high school student. For millions of the world's poorest people, however, this application of physics is a matter of life or death. In Darfur and eastern Chad, myriad refugees from Sudan's civil war are living in refugee camps. The work of gathering fuel and preparing food falls primarily to the women living there. It is a dangerous area; gathering firewood is both risky and can take hours. Even if wood could be safely gathered, smoke from burning wood in a densely populated camp is a health hazard and the pursuit of firewood can quickly strip an area of trees.

The solar oven is a life-saving solution for thousands of refugee families across Africa. Many people now cook most of their meals with sunlight, creating no smoke and cutting down no trees. Solar ovens also are used in many places to sterilize drinking water. This application of basic physics has saved countless lives and also helped preserve fragile ecological areas.

# Chapter study guide

## Chapter summary

Light carries energy and information and varies in intensity and color. Light has both wave and particle properties and can change direction when interacting with matter. Many physical phenomena, such as eclipses and shadows, can be explained by using the concept of light rays traveling in straight lines. The ray model for light is useful for showing how light is reflected and refracted by optical devices. Optical systems may create *images*, which are illusions created by reflecting or refracting light rays. Images may be larger (magnified), smaller, or inverted compared to the *objects* they represent.

## Learning objectives

By the end of this chapter you should be able to

- describe seven properties of light;
- explain how shadows, eclipses, and day and night occur on Earth;
- define and calculate intensity;
- define the normal and the angles of incidence and reflection;
- describe reflection and refraction;
- characterize the images formed by different optical devices;
- predict the location of images for a flat mirror using a ray diagram;
- determine magnification; and
- characterize the properties of images formed by spherical mirrors.

## Investigations

Design a pinhole camera
20A: Magnification of mirrors and lenses
20B: Reflection in a plane mirror
20C: Image formation for curved mirrors

## Chapter index



## Important relationships

$$I = \frac{P}{A} \qquad m = \frac{h_i}{h_o} \qquad \theta_i = \theta_r$$

## Vocabulary

| | | |
|---|---|---|
| ray diagram | inverse square law | object distance |
| image distance | mirror | lens |
| prism | reflection | refraction |
| specular reflection | diffuse reflection | magnification |
| transparent | opaque | translucent |
| incident ray | reflected ray | normal |
| law of reflection | angle of incidence | angle of reflection |
| image | convex | concave |
| optical axis | focal point | focal length |

## 20.1 - Properties of light

Virtually all people and animals see the world through light. Televisions, candles flames, light bulbs, and the Sun emit light. Trees, cars, people, the sky, and almost everything else reflect light. In either case, "seeing" is the reception and interpretation of the energy in light by the eye and brain. How objects *emit* light was not understood until the early 20th century and is discussed in Chapter 26. How light is reflected, refracted, and dispersed were all understood (to some extent) since the time of Galileo.

### Seven properties of light

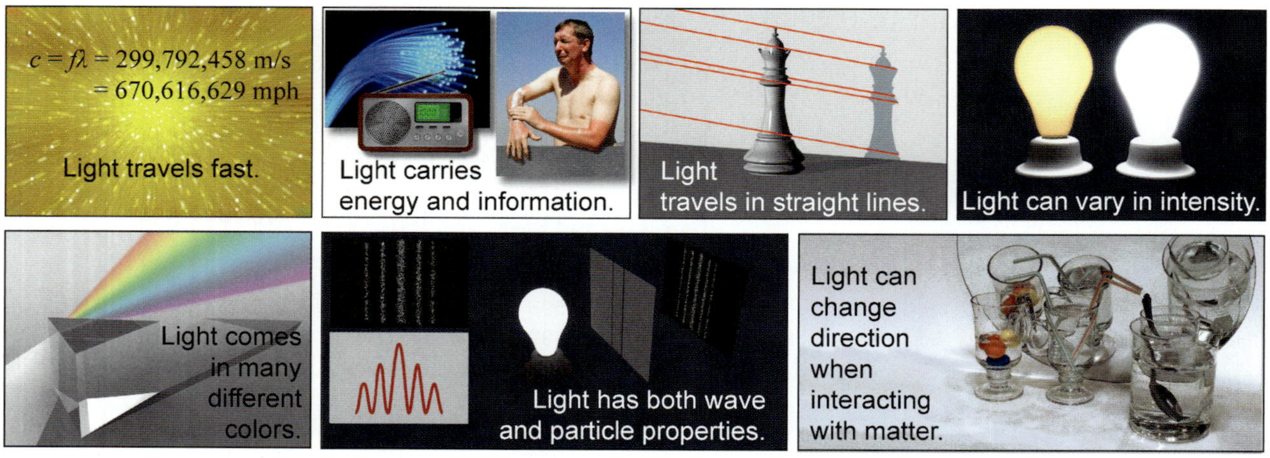


**Travels fast** Light travels faster than anything else, moving 300 million meters per second ($3\times10^8$ m/s). This is fast enough to circle the Earth 7.5 times in one second. It takes just over eight minutes for light to travel 150 million kilometers from the Sun to the Earth.

**Carries energy and information** Light carries radiant energy and information. A lizard warms itself in the sun by absorbing energy from sunlight. Optical fibers use light to transmit information or communicate. Internet traffic and telephone calls travel through optical fiber cables.

**Travels in straight lines** When unimpeded by matter or objects, light travels in straight lines. This is the explanation for why shadows form.

**Has different intensities** Light can vary in intensity. The difference between dim and bright light bulbs is an example. The energy of light depends on both its intensity and its frequency.

**Has different colors** White light can be dispersed by a prism into the rainbow of colors. Light is also called electromagnetic radiation, a more general term spanning a broader spectrum of "colors" that includes x-rays, ultraviolet light, infrared radiation, and radio waves.

**Has wave and particle behavior** For many applications light behaves as a wave, having frequency, wavelength, phase, and polarization. In the quantum world, however, light also acts like a particle called a photon. A photon is a tiny, discrete bundle of light energy.

**Direction can be diverted** When light interacts with matter its direction can be changed. Light is reflected by mirrors, refracted by glass or water, or diffracted when passing through a narrow slit. Microscopes, telescopes, and the eye use optics to focus light and form images.

### Shadows

A shadow is direct evidence that light travels in straight lines. If light were to curve around objects, then shadows would not form as they do. The sharpness of a shadow is determined both by the size of the light source and the relative separation between the light source, object, and the surface (screen) on which the shadow appears. Dispersed light sources, such as overhead fluorescent tubes, create blurry shadows. If an object is close to a light source, then the edges of its shadow also become blurred.

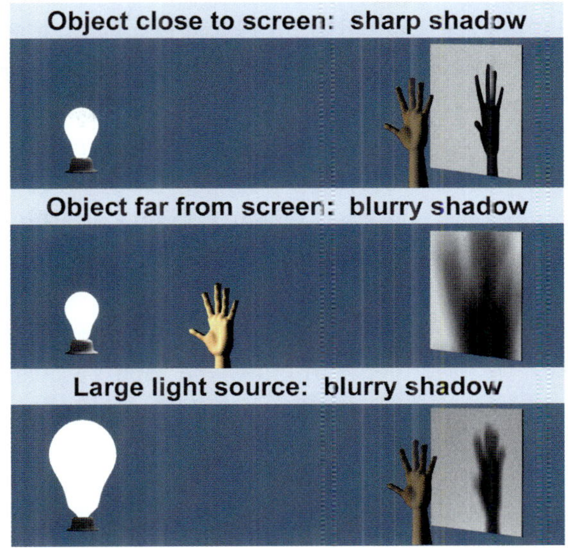


### What are light rays?

The straight line paths followed by light are shown in diagrams in the form of *light rays*. Light rays use arrows to show the direction in which the light travels. Think of a light ray is an idealized, infinitely narrow beam of light, like a laser beam.

### Ray diagrams

How a shadow forms is explained with a light **ray diagram**. Areas are illuminated where light rays can freely travel to the ground. Areas are shadowed where light rays are blocked. Although every point on an illuminated surface receives light, a ray diagram typically includes only a few of those light rays that best illustrate what we are trying to explain. For a shadow, the important rays are those that just graze the outline of the object casting the shadow.

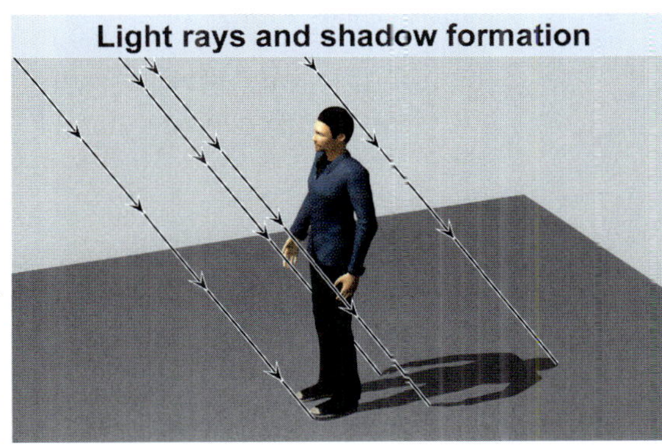


### Three different models of light

This representation of light is called the *ray model* and it is useful for describing shadows, reflection, and refraction. Other models of light, such as the wave model and the photon model, are better for explaining other phenomena. The wave model describes polarization, interference, and diffraction. The photon model describes how light interacts with atoms.

### What causes day and night?

Light from the Sun can only illuminate areas of Earth with a direct line of sight to the Sun. At any moment all areas of Earth facing the Sun are in daylight. Areas facing away from the Sun are in shadow and experience night. In essence, night is caused by Earth's own shadow! The 24-hour cycle of day and night occurs because Earth's 24-hour rotation continually changes the part of the planet facing the Sun.

Light rays from the Sun only illuminate one side of the Earth, creating a light and a dark side.

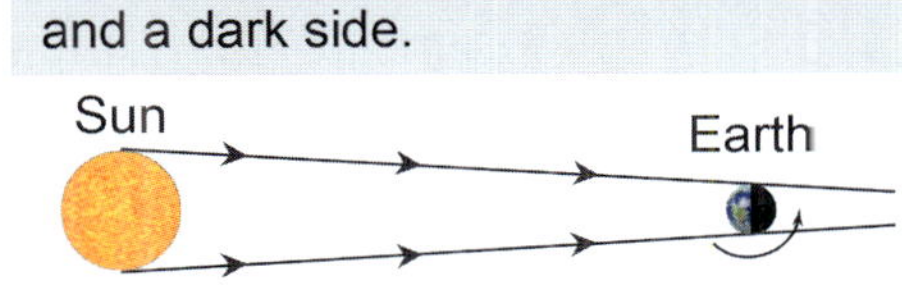


The Earth's rotation causes any place to experience both night and day.

## Eclipses and the phases of the Moon

**What causes eclipses?**

If you have ever seen a solar or lunar eclipse, you may have asked this question:

*Why do eclipses occur?*

Ancient people certainly asked the question and invented many contradictory answers, usually involving supernatural powers. To answer the question using physics, you need to know the relative positions of the Sun, the Moon, and the Earth, as well as the fact that light travels in straight lines.

**Solar eclipse**

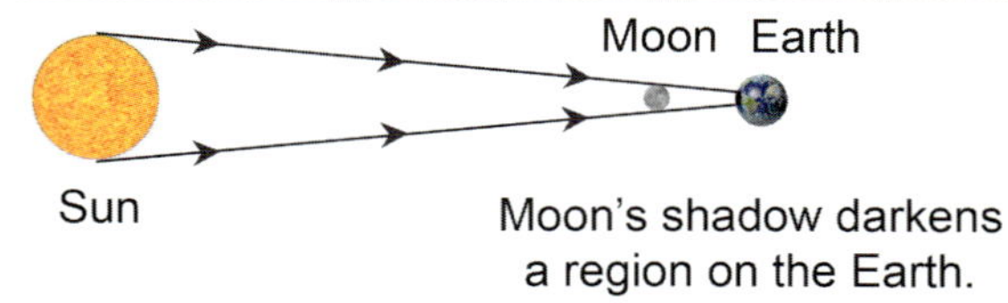


**Lunar eclipse**

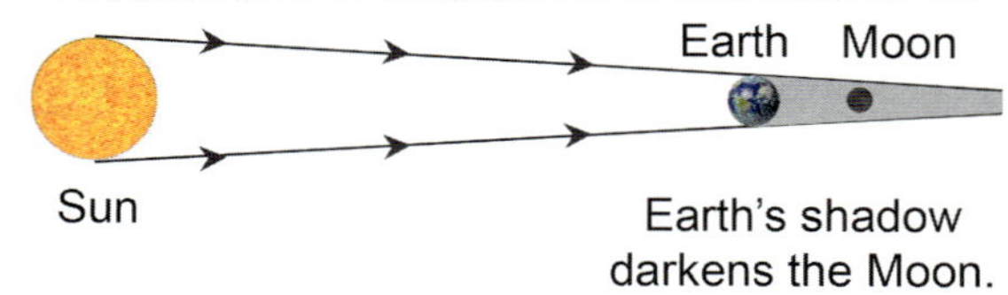


**Solar and lunar eclipses**

Eclipses occur when one celestial body blocks the light coming from another body. In other words, one of them is casting a shadow! When the Moon lines up between the Sun and the Earth, the Moon casts a shadow onto the Earth, blocking the Sun for a small region on the Earth. This creates a solar eclipse in which some or all of the Sun appears to vanish. When the Earth lies between the Sun and the Moon, the Earth's shadow—which is larger than the Moon's shadow—engulfs the Moon, blocking sunlight from reaching the Moon. This creates a lunar eclipse in which the Moon appears to vanish. Solar and lunar eclipses do not occur very often because the plane of the Moon's orbit around the Earth is inclined a little bit relative to the plane of the Earth's orbit around the Sun.

**What causes the phases of the Moon?**

Have you ever wondered whether Earth's shadow creates the crescent Moon? How does that explain the full Moon? Why is the dividing line between light and dark straight instead of curved in a half Moon? Why does the Moon go through these *phases*? The answer is that a shadow *does* create the observed phases of the Moon, except that the shadow is not the shadow of Earth but the shadow of the Moon itself.

**Explaining the phases of the Moon**

Video

Sunlight travels in straight lines toward the Moon, which illuminates the side of the Moon facing the Sun. The Moon orbits around the Earth, which causes the illuminated side of the Moon to change its orientation relative to the Earth. Sometimes, we can see the fully illuminated side of the Moon, which is called a "full Moon." Other times, the illuminated side is facing completely away from the Earth, which is called a "new Moon." In between are the other phases of the Moon: crescent, quarter, and gibbous.

**Phases of the Moon**

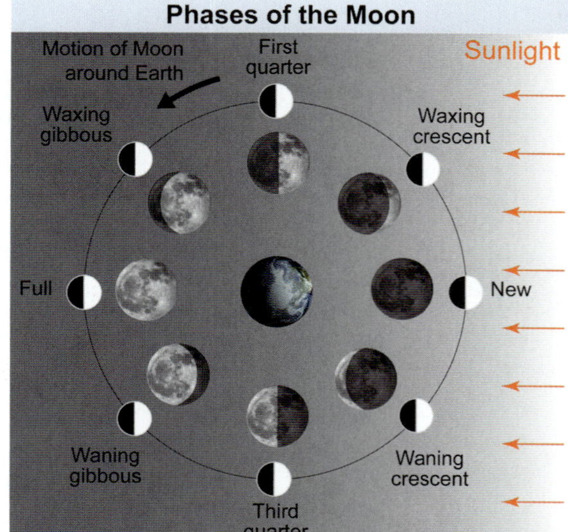

**Light intensity**

The brightness of light is described by intensity $I$, which has units of watts per square meter (W/m$^2$). Thus intensity is a measure of the power per unit area, as shown in equation (20.1). The Sun, a bright light source, has an intensity around 500 W/m$^2$. The light of a flashlight spot on the floor in a darkened room might be 1 W/m$^2$.

Interactive equation

**Light intensity**

$$I = \frac{P}{A} \tag{20.1}$$

$I$ = light intensity (W/m$^2$)
$P$ = power (W)
$A$ = area (m$^2$)

**Intensity and the inverse square law**

As you move farther from a light source, the intensity of light decreases as the inverse square of the distance. Going twice as far decreases the intensity by $2^2 = 4$. This is not a law of physics but a result of geometry. Consider the light from a point source that emits a total power $P$ equally in all directions. At any distance $r$, the power is spread out over a sphere of area $4\pi r^2$. The intensity of the light is $I = P/(4\pi r^2)$. A relationship in which a quantity decreases as $1/r^2$ is called an **inverse square law**. The intensity of light from a point source obeys an inverse square law.

**Light intensity and the inverse square law**

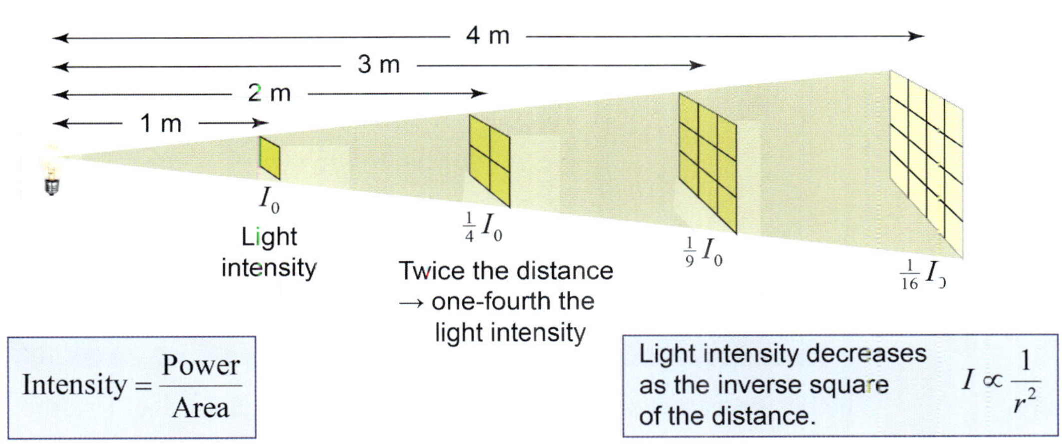


**Luminous intensity**

*Luminous intensity* is a physical quantity related to intensity, but it is defined in a way that factors in the response of the human eye to visible light. The unit of luminous intensity is the candela (Cd), which is approximately the amount of visual light produced by a typical candle. Luminous intensity is one of the seven fundamental physical quantities of science.

**The intensity of sunlight**

The energy output of the Sun is called the solar *luminosity* $L$, which has a value of $3.8\times10^{26}$ W. The Earth is $1.5\times10^{11}$ m away from the Sun. At Earth's orbit the intensity of sunlight is equal to the Sun's luminosity divided by the area of a sphere the size of Earth's orbit. This works out to 1,365 W m$^{-2}$ at the top of Earth's atmosphere.

**Gravitational force**

The inverse square law appears in other aspects of physics as well as light. As you learned on page 216, the gravitational force follows an inverse square law. In Chapter 18, you learned that the electric force (Coulomb's law) is also an inverse square law.

# Design a pinhole camera

**Design challenge**

In this field investigation you will design a pinhole camera that produces an image of the Sun at least 5 mm in diameter. You will use the pinhole camera to observe features on the Sun. Note that your eyes can be damaged by looking directly at the Sun. *At no time should you directly look at the Sun.*

**How a pinhole camera works**

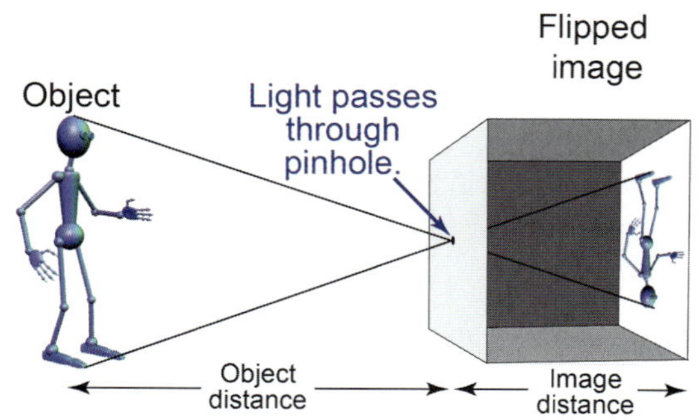


$$\frac{\text{Image distance}}{\text{Object distance}} = -\frac{\text{Image height}}{\text{Object height}}$$

**Geometry of the pinhole camera**

In this project, you will adjust **image distance** and pinhole size to obtain a sharp image of the Sun. As shown in the figure at right, the image distance is related to the image height, object distance, and object height. In this investigation, the **object distance** is the distance from the Earth to the Sun, which is $1.50 \times 10^8$ km. The *object height* is the Sun's diameter, $1.39 \times 10^6$ km.

**Equipment**

**Materials for pinhole camera**

**Assembled camera**

**Design**

The design parameters are image distance and pinhole size. Use the equation to predict the image distance (or length) of your camera, and choose an appropriately sized box. How will you view the screen inside the box? How can you vary the image distance? What is the optimal size for the pinhole? Address these issues in your design.

**Prototype**

Construct the prototype of your pinhole camera. Can you view the screen easily?

**Test**

Test your design using a bright light bulb in the classroom. Can you image the filament of the bulb? Test a few different pinhole sizes.

**Evaluate**

How does the image size and quality vary when the pinhole size varies? What is the optimal pinhole size? How does the image size vary as the screen position (or image distance) is varied? What are the tradeoffs? Design and describe an investigative procedure to address each question. What are the independent, dependent, and controlled variables in each case? How much data do you need to collect to evaluate each parameter?

**Revise**

Use your evaluation to improve your design. Which pinhole size and image distance work best? Go outside and test your design on the Sun.

# Section 1 review

Light travels in straight lines at $3 \times 10^8$ m/s. Light carries energy and information and can vary in intensity. The intensity of light has units of power divided by area, or watts per square meter. Light has both wave and particle properties, and its energy determines its color. Light rays can be diverted when they interact with matter. The properties of light explain the formation of shadows, why Earth has cycles of night and day, why there are eclipses of the Sun and Moon, and why there are phases of the Moon. Light intensity decreases with distance following an inverse square law. Gravitational and electrical forces are other examples of inverse square laws.

| | |
|---|---|
| **Vocabulary words** | **ray diagram, inverse square law, object distance, image distance** |
| **Key equations** | $I = \frac{P}{A}$ |

## Review problems and questions

1. The distance between the Earth and the Sun is one astronomical unit (AU), which is $1.50 \times 10^8$ km. A light year (ly) is the distance light travels in one year.
   a. How many minutes does it take light to travel from the Sun to the Earth?
   b. How many kilometers is a light year?
   c. How many AU is a light year?

2. a. What is the difference between a solar and a lunar eclipse?
   b. Which body's shadow projects onto another in a solar eclipse? In a lunar eclipse?
   c. Which shadow is larger? Why?

3. The dwarf planet Makemake has an average orbital distance of 45.8 AU from the Sun. If sunlight reaching the Earth has an intensity of 1,365 W/m$^2$, what is the intensity of sunlight reaching Makemake?

4. Where is the lens of a pinhole camera located?

5. The light intensity at a spot 1.0 m away from a bright lightbulb is 30 W/m$^2$.
   a. What is the intensity at a distance of 2.0 m?
   b. What is the intensity at a distance of 0.5 m?
   c. At what distance from the bulb will the intensity be 15 W/m$^2$?

6. Sometimes near midday on a sunny day, you can see many images of the Sun projected onto the ground underneath a leafy tree. What creates the images?

## 20.2 - Optical devices

Light travels in straight lines in a vacuum but light rays may bounce (reflect) or bend (refract) upon interacting with matter. Many technologies and objects exhibit the reflection and/or refraction of light, including eyeglasses and contact lenses, mirrors in dressing rooms, rear-view mirrors on cars, a glass of water, and even the cut and polished diamond in a ring. The science and technology of manipulating light is the field of *optics*.

### Examples of optical devices

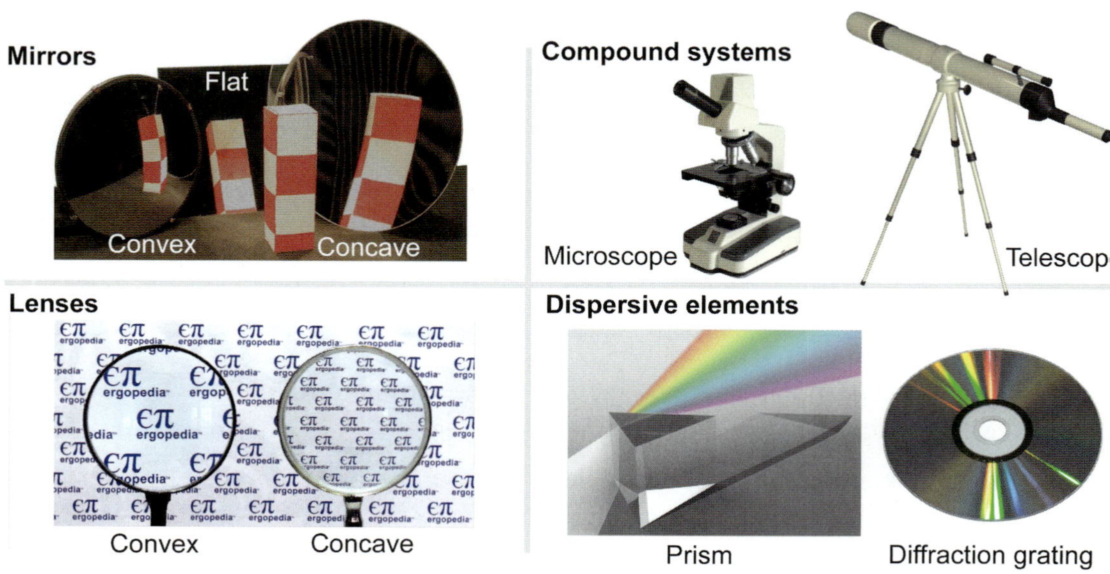


**Mirrors**

A **mirror** is a basic optical device that diverts light using *reflection*. Most large mirrors are flat, such as the mirror above the sink in a bathroom. Mirrors can also have curved surfaces, which create unusual reflections—especially the wavy mirrors found in a funhouse that make you look skinny or short! Most mirrors are made from clear glass with a thin layer of aluminum or silver sprayed on the back side to reflect light. Many metals can also be polished smooth and shiny to act as a mirror.

**Lenses**

A **lens** has a curved surface that bends light by *refraction*. Lenses are found in many optical technologies including magnifying glasses, microscopes, and all types of cameras including video cameras and the camera on the back of a mobile phone. Lenses are usually made from transparent polished glass or plastic.

**Prisms**

A **prism** is an optical device with flat surfaces that uses refraction to divert the path of light. Because different colors of light refract at slightly different angles, a prism may *disperse* white light into its constituent colors. Dispersion explains how a triangular prism separates a beam of white light into the colors of the rainbow. Prisms in technology are made from transparent polished glass or plastic. In nature, droplets of water act as tiny prisms to create rainbows.

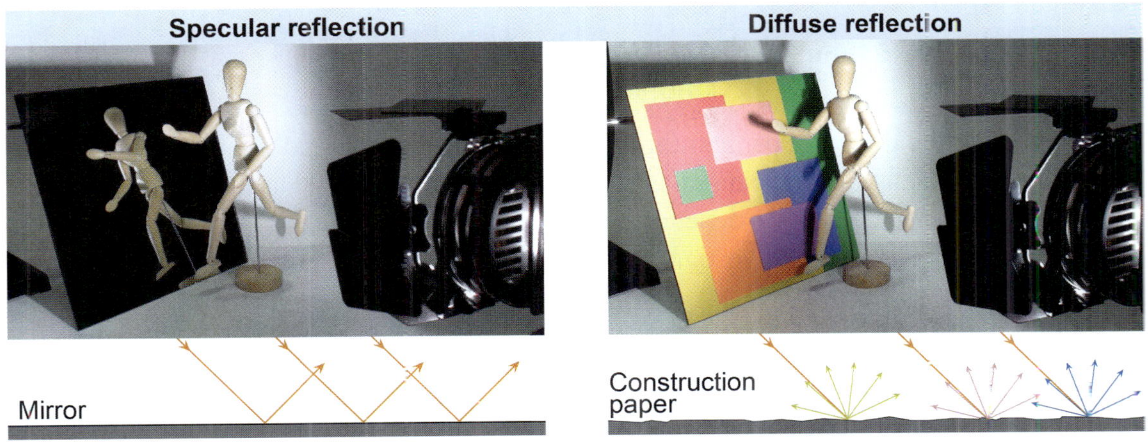


**Reflection**

**Reflection** occurs when light strikes a surface and some or all of the light "bounces" back. The diagram above illustrates two types of reflection. When light is reflected from a mirror you don't see the mirror itself. Instead, you see the reflected image of the mannikin. When light is reflected from the painted board you see the surface of the board itself.

**Specular reflection**

A mirror or shiny metal surface creates *specular reflection.* **Specular reflection** forms images because each ray that falls on the surface results in a single reflected ray at the same angle. You see an image because multiple light rays striking a mirror retain their original angles with respect to each other after reflection.

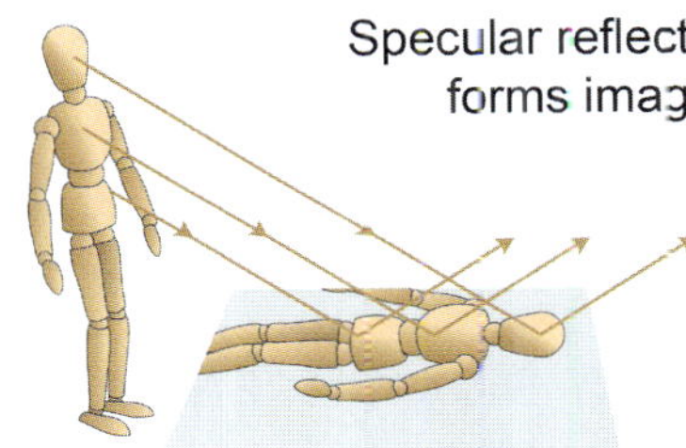

Specular reflection forms images.

**Diffuse reflection**

When light strikes a surface that is not shiny, such as a wall or a page in a book, the reflection is *diffuse.* In **diffuse reflection**, each light ray is absorbed and re-emitted (scattered) in a random direction. The scattered light rays come from the surface itself and that explains why you can see the color and texture of a diffuse reflecting surface.

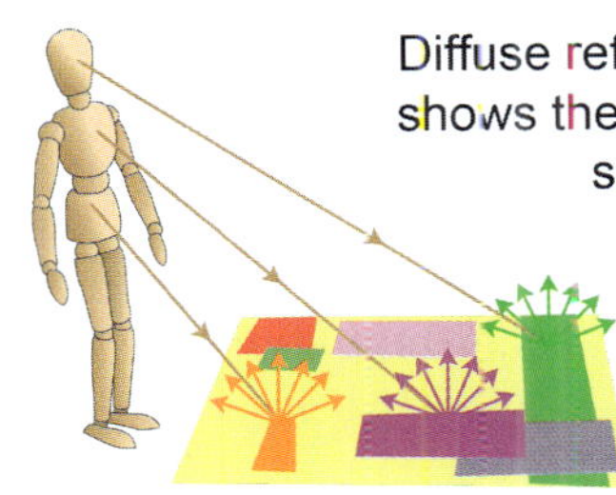

Diffuse reflection shows the actual surface.

**Refraction**

Light rays may *bend* when they cross a boundary between different materials. The bending of light rays by surfaces is called **refraction**. Refraction is why the wrench in the photograph appears to "break" at the water's edge. It's not really broken; the light refracting through the water causes the lower image of the wrench to be offset slightly. Different materials, such as glass, water, or diamond, refract light by different amounts.

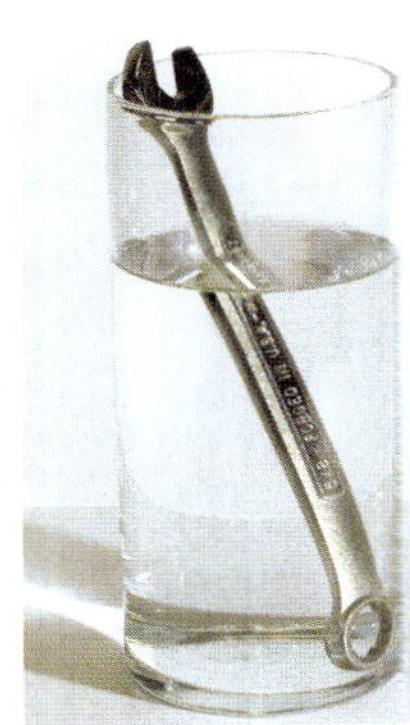

Refraction bends light rays making objects seem to be in a different place.

# Investigation 20A

## Magnification of mirrors and lenses

**Essential questions** *What kinds of images can be made by lenses and mirrors?*

Mirrors change the direction of light through reflection. Lenses bend the direction of light through refraction. Both mirrors and lenses can create images. The images may be upside-down, right-side-up, larger, or smaller. How the image appears usually depends on the geometry of the lens or mirror and also on the location of the object and the observer. This investigation explores images formed by single lenses and mirrors.

### Part 1: Which mirrors or lenses magnify an object?

You will investigate the optical effects of three basic kinds of mirrors (flat, convex, and concave), two kinds of lenses (convex and concave), and a prism.

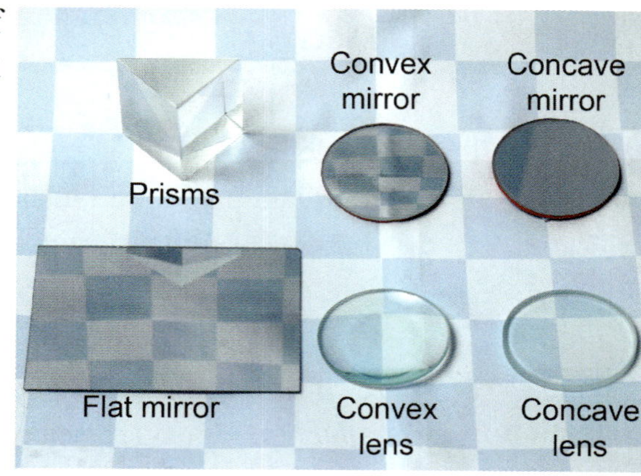


1. Look into the mirrors from different distances.
2. Hold the lenses and prism over the table at different heights and look at the table through them.

a. *Which optical devices create magnified images (either larger or smaller)?*

b. *Are there any optical devices that create both enlarged and reduced size images? Under what conditions?*

c. *Do any of the devices create* inverted *(upside-down) images?*

### Part 2: Magnification of a convex lens

A magnifying glass is a commonly available optical device that is constructed using a convex lens.

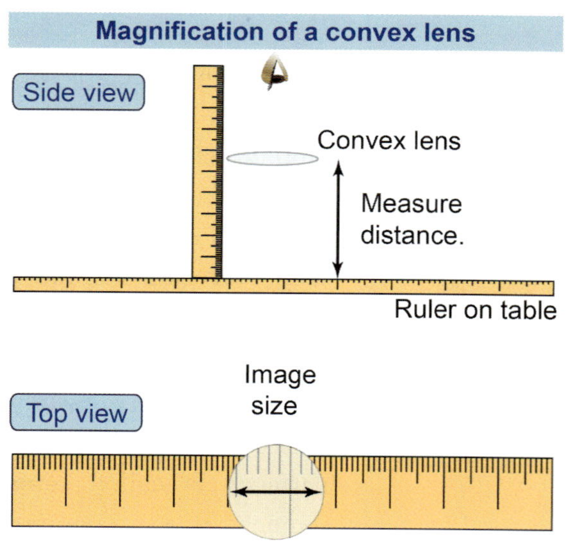

**Magnification of a convex lens**

1. Place a ruler flat on the table and look down at it from around 0.5 m.
2. Place a *convex lens* directly onto the ruler (distance $d_o = 0$) and record the diameter of the lens. Since the image fills the entire lens, this is the image size, $h_i$.
3. Use the markings on the ruler to measure *how much* of the object (the actual ruler) is visible through the lens. This is the object size, $h_o$.
4. Repeat these measurements with the lens held at three different distances above the ruler. For each trial, record the image size $h_i$, object size $h_o$, and the distance $d$ between the ruler and the lens.

a. *Find the magnification of the lens for each trial by calculating* $m = h_i/h_o$.

b. *How does the magnification vary with the distance from the object (the ruler) to the lens?*

c. *Does the appearance of the image change as the distance between the ruler and the lens increases?*

**Magnifying glass**

If you look in a flat mirror in a dressing room, your image is the same height as you are. A magnifying glass held above this page, however, produces an image of the page that is larger than the page itself. Curved optical elements—lenses and curved mirrors—can be used to create magnified images. In some cases, the magnified image is larger than the object; in others, the image is smaller than the object.

**Different mirrors have different magnifications.**

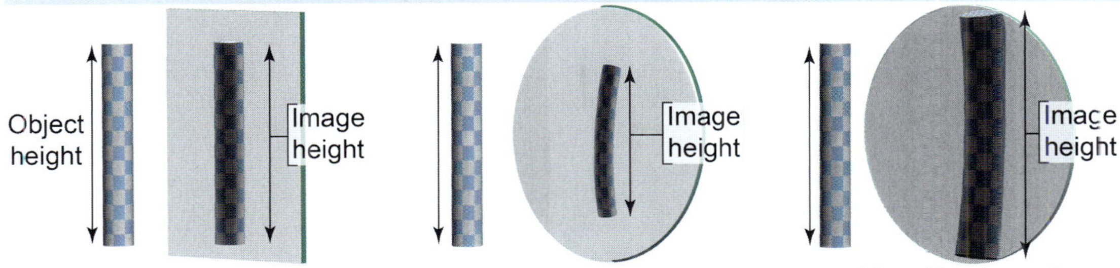


Magnification = 1
Image same size as object

Magnification < 1
Image smaller than object

Magnification > 1
Image larger than object

**Magnification**

The **magnification** of a mirror or lens is the ratio of the image and object heights.

Interactive equation

**Magnification**

$$m = \frac{h_i}{h_o} \quad (20.2)$$

$m$ = magnification
$h_i$ = image size (m)
$h_o$ = object size (m)

**Positive and negative magnification**

Can you hold a concave mirror such that your image is *inverted*, i.e., upside-down? The values of the heights used in equation (20.2) can be either positive *or negative* depending on whether the image is upright (positive height) or inverted (negative height). Look at equation (20.2) again: A negative value for $h_i$ will produce a negative value for the magnification $m$. A negative magnification represents an inverted image, while a positive magnification is an upright image.

**Example problem**

What is the magnification of the magnifying glass?

*Asked:* magnification $m$

*Given:* Object height: 17.5 squares (5 mm/square) are visible through the lens so the object "height" is 87.5 mm. Image height: The image fills the lens so the image "height" is the diameter of the lens 135 mm (27 squares).

*Relationships:* $m = h_i / h_o$

*Solution:*

$$m = \frac{h_i}{h_o} = \frac{135 \text{ mm}}{87.5 \text{ mm}} = 1.54$$

*Answer:* The magnification is 1.54.

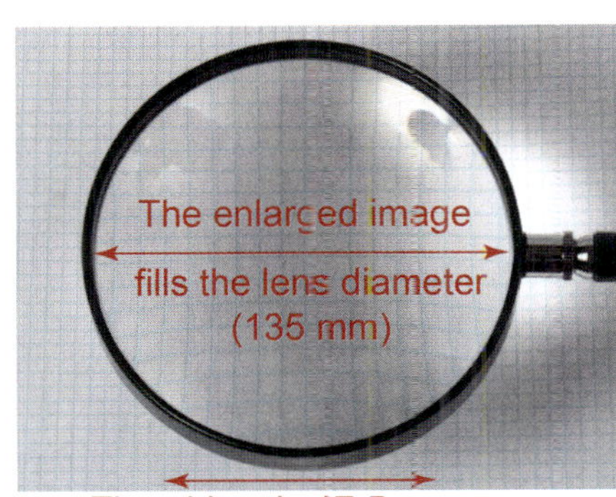


The object is 17.5 squares of graph paper (87.5 mm).

## Transparency

### Transparency and opacity

Window glass is *transparent* whereas wood is *opaque*. In the context of physics, a perfectly **transparent** object allows light rays to pass through with no loss of energy. A perfectly **opaque** object absorbs 100% of any light energy and completely stops light rays. Transparency and opacity depend on the material of the medium and also on its shape and surface characteristics. Even glass can be opaque when ground up as a powder.

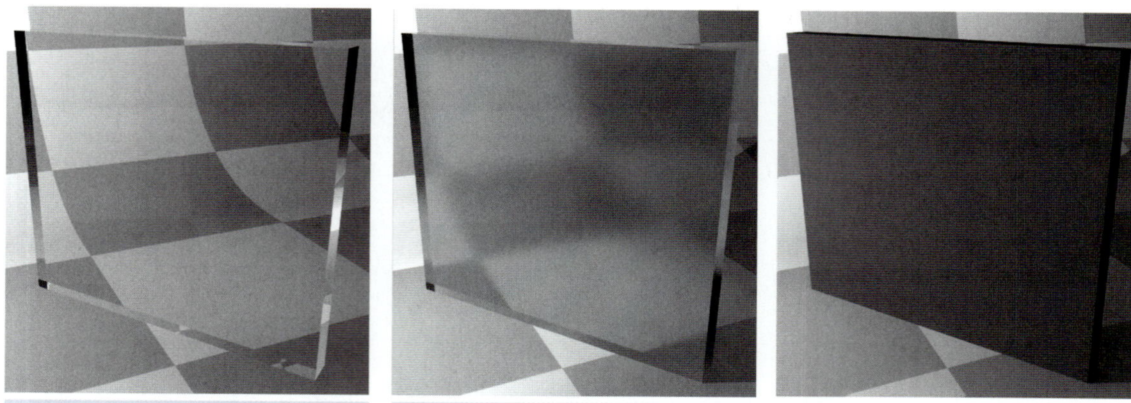

Transparent | Translucent | Opaque

### Translucence

Many bathroom windows let in light but you cannot see an image through them. Frosted windows are **translucent**, which means that light rays pass through but their direction is scattered so the information (images) gets blurred or is completely eliminated. Transparent materials can be made translucent by changing the surface texture.

### Transparency, translucency, and opacity

Many objects may display all three properties—transparency, translucency, and opacity—to various degrees. The diagram at right shows an example. The optical properties depend on the application. Optical glass for eyeglass lenses must be nearly 100% transparent with 0% opacity (although a darkened coating may be applied to the surface to turn them into sunglasses).

A screen between you and the ball can be transparent, opaque, or translucent.

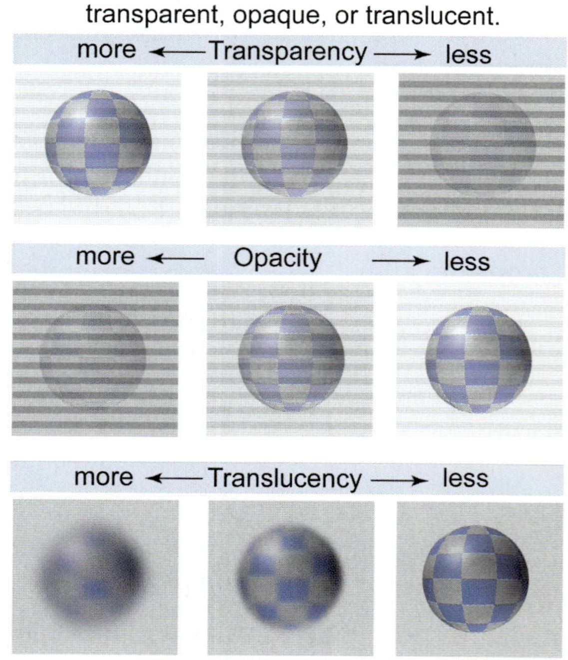


### Transparency may vary with wavelength

A substance that is transparent to visible light may not be transparent to other wavelengths of light. For example, sunglasses may be 50% transparent to visible light yet only be 0.1% transparent to ultraviolet light (99.9% opaque). Earth's atmosphere is transparent to visible light but partially opaque to infrared light. This results in the "greenhouse effect" in which solar energy arrives as visible light but heat energy cannot escape as infrared light.

# Section 2 review

Optical devices may change the path of light rays through reflection and refraction. Reflection bounces light off of a surface, such as a mirror. Specular reflection occurs with shiny surfaces and diffuse reflection occurs with nonshiny surfaces, such as paper. Refraction bends the path of light as the light passes from one medium into another, such as from glass into air and vice versa. Refraction is the principle behind prisms and lenses. Both reflection and refraction can be used to form images. Magnification occurs when an image has a different apparent size than an object. Dispersion occurs when an optical element separates white light into its constituent colors, such as a prism creating a rainbow. Optical media may have varying degrees of opacity, transparency, or translucency.

| Vocabulary words | **mirror, lens, prism, reflection, refraction, specular reflection, diffuse reflection, magnification, transparent, opaque, translucent** |
|---|---|
| **Key equations** | $m = \frac{h_i}{h_o}$ |

## Review problems and questions

1. Match each diagram (A, B, and C) to the appropriate name (i–iii) and function (a–c) from the list.

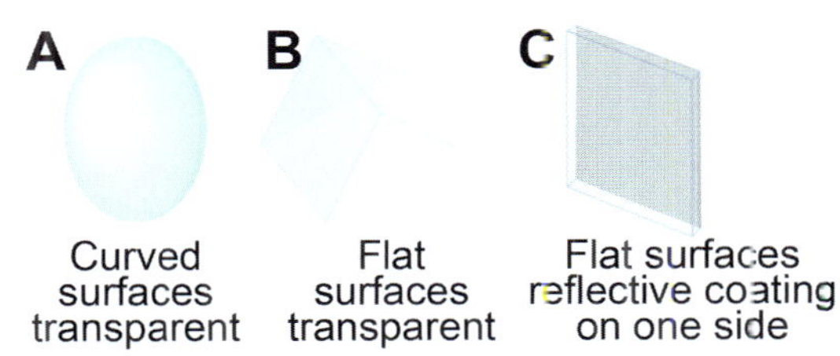


| Name | Function |
|---|---|
| i. Mirror | a. Refracts light and may magnify. |
| ii. Lens | b. Reflects light and may show images. |
| iii. Prism | c. Refracts light without magnification. |

2. A ruler of length 30 cm is submerged in a tank of water. The image seen outside the tank is measured to be 36 cm long.
   a. What is the magnification of this system?
   b. In centimeters, how long would a meter stick in the tank of water appear to be?

3. A piece of sandpaper may turn a transparent plastic window into a translucent plastic window. Propose a hypothesis to explain why sanding the surface makes the window translucent. Use the concept of light rays in your explanation.

4. How large does the image of a 10-m-tall tree appear when using an optical device with a magnification of 0.3?

5. In Investigation 20A on page 590, which optical device would you choose to
   a. create a magnified (enlarged) image?
   b. create an upside-down or inverted image?
   c. create an image the same size as the object?

## 20.3 - Reflection and images

Intuitively, we know the spoon in the water glass in the picture is not broken at the water surface and that there is only one real glass and one reflected image. In this section we answer the question of how the illusion in the mirror is created. In the next chapter, on refraction, we will address the illusion of the broken spoon.

Consider the spoon as the *object.* The object is the real thing made of matter that reflects light from the room into your eye. The *image* in the mirror is an illusion created by changing the direction of light rays so that they appear to come from behind the mirror. To understand how this happens we use ray diagrams. Ray diagrams describe how light gets from the object to your eye and how images are formed.

### Light rays and ray tracing

**Representing light as waves or rays**

Light can be represented in different ways including wavefronts and light rays. The wavefronts of visible light are so close together, typically $5\times10^{-7}$ m, that wave-front diagrams are only useful when the size of something in the system is similarly small. The diffraction and interference of light are phenomena that require this approach.

**Wave representation**

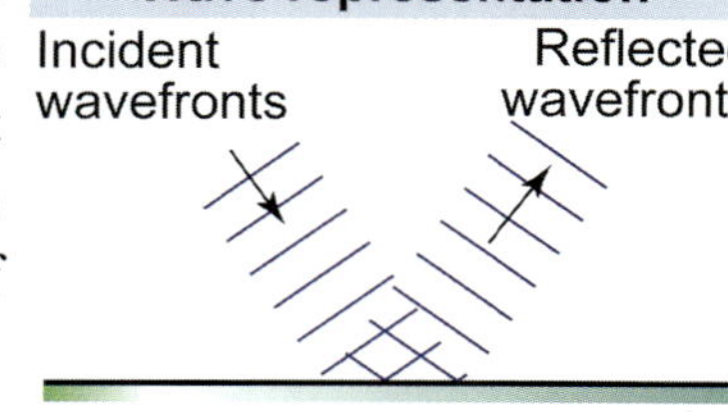


For the purposes of optics and images it is more instructive to represent light with rays. Light rays travel in straight lines through space and represent the direction of energy flow. You may imagine a single light ray as a laser beam that has a specific color and comes from a specific place.

**Ray representation**

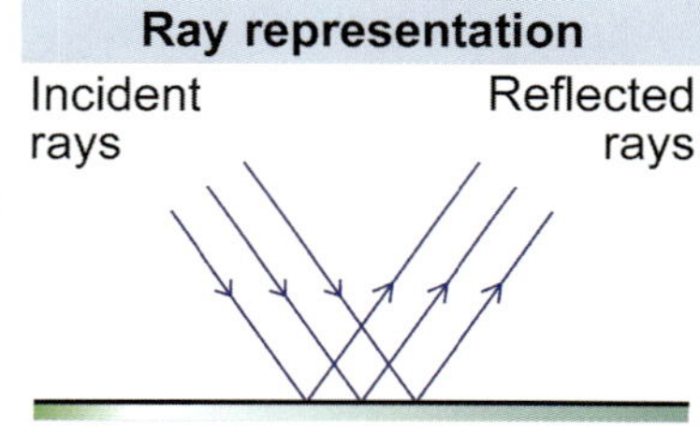


**What are the incident and reflected rays?**

When light strikes a mirror, light rays show the relationship between the *incident* light and the *reflected* light. The **incident ray** represents light approaching the mirror. The **reflected ray** represents light reflected from the mirror surface. Incident and reflected rays usually have arrows to denote the direction the light travels.

**What is ray tracing?**

*Ray tracing* is the process of determining the path of one or more representative light rays through an optical system. The diagram on the right shows a simple ray trace of three rays passing through a lens, reflecting off a mirror, and being observed. Note that the observer looks *through* the system yet the ray diagram is drawn *from the side.* Drawing from the side is necessary because the sizes and positions of objects and images are best represented from the side. As we will see later, the intersection of multiple light rays (drawn from the side) indicates where an image may be formed.

**Example of a ray diagram**

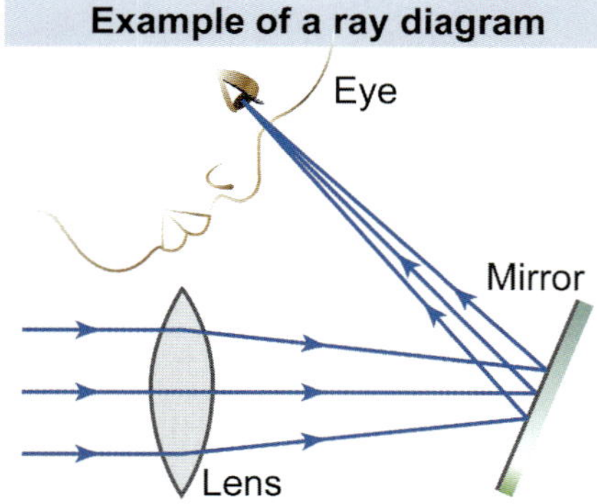

# Investigation 20B Reflection in a plane mirror

**Essential questions** *How and where does an image form in a mirror?*

If you look at yourself in a mirror, where is your *image* located? In this investigation, you will use three different techniques for locating the image produced by a flat mirror.

## Part 1: Locate the image using parallax

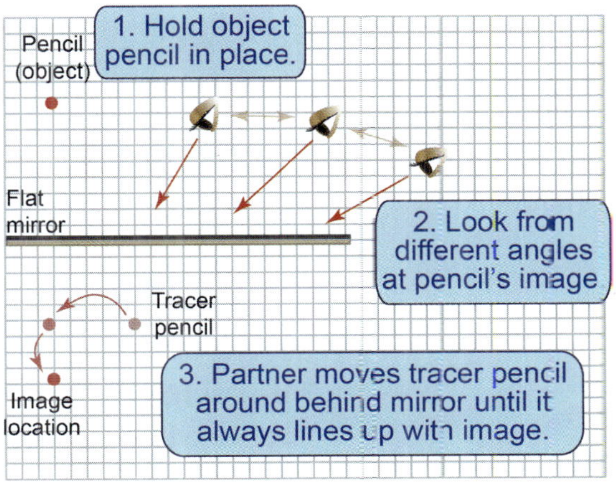


1. Mount a flat mirror vertically on top of a piece of graph paper and trace its location.
2. Hold a tall "object" pencil near the shiny side of the mirror. Mark its location.
3. Move a second "tracer" pencil behind the mirror until it lines up with the reflection of the "object" pencil for every angle you look in the mirror.

a. What does the location of the tracer pencil signify?
b. Measure the distance from the mirror to the object and from the mirror to the image. Which is larger, or are these distances the same?

## Part 2: Locate the image using pins

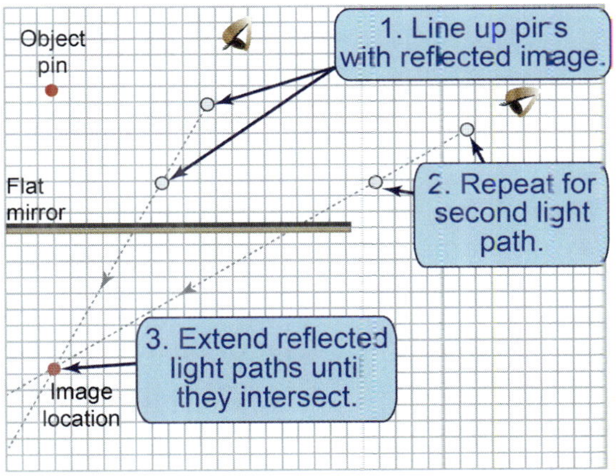


1. Replace the "object" pencil with a pin.
2. Look at the image of the pin and place two pins in the graph paper in line with this image.
3. Move sideways and repeat with two more pins.
4. Draw lines connecting each pair of pins to each other; extend these lines until they intersect.

a. What does each of the two lines represent?
b. Predict where the image will be located and why.
c. Does your image location agree with the location from Part 1? Was your prediction correct?

## Part 3: Trace the rays using a laser pointer and index card

*Caution: never look directly at a laser.*

1. Mount a laser pointer horizontally using a ring stand and a 90° rod clamp. Align the laser to point through the object pin and strike the mirror at an angle.
2. Use the edge of an index card to locate and trace the path of the laser light (incident and reflected) directly onto the graph paper.
3. Position the laser at a different angle to the mirror and trace the new path.

a. Label the incident and reflected rays. Do the rays obey the law of reflection?
b. How can you find the image using the light paths you made?
c. Compare the laser pointer method to the other two. Which is better? Why?

## Law of reflection

**How are incident and reflected rays related?**

The diagram on the right shows how a flashlight beam reflects from a flat mirror. If the beam approaches at a low angle the reflected beam also leaves the mirror at a low angle. If the flashlight beam approaches nearly perpendicular to the mirror the reflected beam is also nearly perpendicular to the mirror. Reflection shows *symmetry* of angles between the incident and reflected beams.

**Symmetry of incident and reflected rays**

**How do we describe the geometry of reflection?**

Reflection is symmetric in angle. Therefore, to describe reflection we start by defining angles. The **normal** is an imaginary line drawn *perpendicular* to the reflecting surface at the point of intersection with the incident ray. All angles are defined relative to the normal.

$\theta_i$: The **angle of incidence** is the angle between the incident ray and the normal.

$\theta_r$: The **angle of reflection** is the angle between the reflected ray of light and the normal.

The **law of reflection** states that the angle of incidence equals the angle of reflection (equation 20.3).

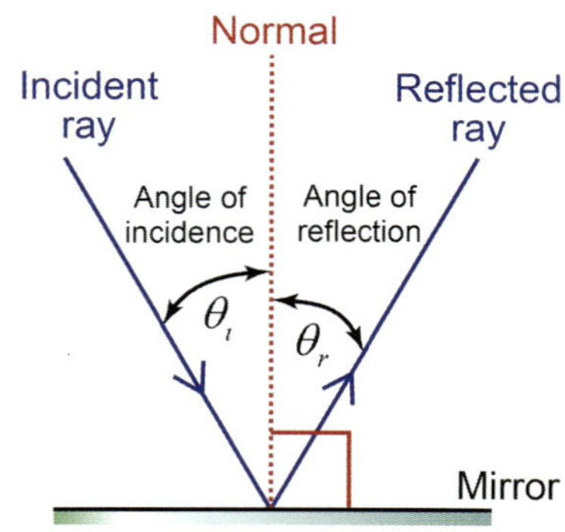


Interactive equation

**(20.3)** $\theta_i = \theta_r$

$\theta_i$ = angle of incidence (degrees)
$\theta_r$ = angle of reflection (degrees)

**Law of reflection**

**Drawing a ray diagram for reflection of light**

1. Draw the incident light ray,
2. Draw the normal to the surface at the point of contact,
3. Determine the angle of incidence,
4. Draw the reflected ray on the other side of the normal line, such that the angle of reflection equals the angle of incidence.

**Example problem**

A light ray strikes a flat mirror at an angle of 50° to the mirror surface. Draw a ray diagram for the incident and reflected light rays. Label the angles of incidence and reflection and the normal.

| | |
|---|---|
| *Asked:* | a ray diagram |
| *Given:* | angle between the incident ray and the (flat) mirror, 50° |
| *Relationships:* | $\theta_i = \theta_r$ |
| *Solution:* | From geometry, the angle of incidence is $\theta_i = 90° - 50° = 40°$ From the law of reflection, the angle of reflection is also 40°. |

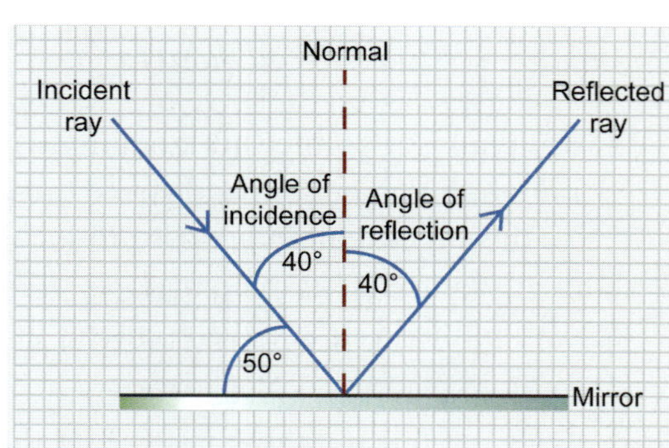

**What is an image?**

An **image** is an illusion created with light. To understand images, ask yourself why you can see the same point on a ball from many angles? The answer is that light from *each point* on the ball spreads out in all directions. Your eye captures those rays that happen to enter your pupil and forms the image of that point of the ball on your retina. An *image* is an organization of light rays that has the same spatial pattern as the original light rays coming from the actual object.

**Where is the image of the ball located?**

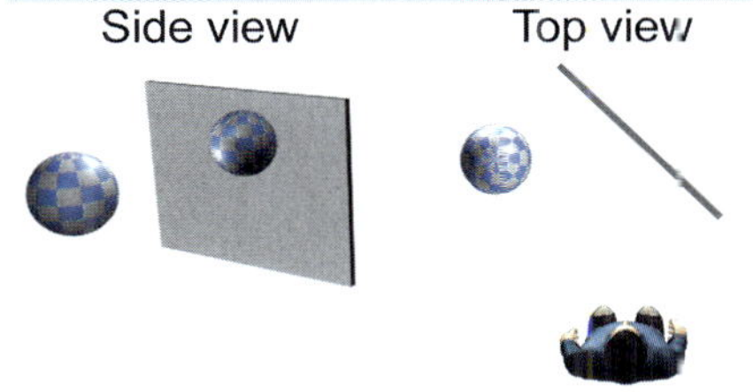


**Using a ray diagram to locate the image**

Consider the reflected image of the ball in a flat mirror. Three light rays that leave point A on the ball are reflected from the mirror. For each of the three rays, the law of reflection gives the direction of the reflected ray (angle of reflection equals angle of incidence). *The reflected rays appear as if they came from point B behind the mirror!* Point B is where you see the *image* of point A.

**Image formation for a flat mirror**

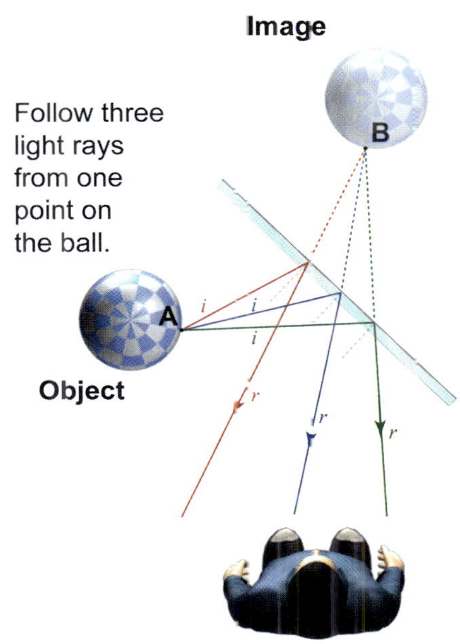


**The images in a flat mirror**

The image in a mirror forms because multiple rays from the *same* point on an object appear to come from a single point somewhere else. Your eye sees the exact same organization of light in the reflected rays as it would if the ball were actually at the image location. Your visual system does not sense "mirror" and perceive things as a reflection. Instead, your visual system receives just light rays. If the light rays appear to come from behind the mirror, then you "see" the ball at that point behind the mirror (the image).

**Image and object distances are equal**

Look carefully at the illustration above: The image of the ball appears behind the mirror at a distance equal to the actual ball's location in front of the image. The distance between the mirror and the ball is the *object distance*, while the distance between the mirror and the ball's image is the *image distance*. For reflection in a flat mirror, the image and object distances are equal.

**A reflected image reverses left and right**

The next time you look in a mirror notice that your image appears to have left and right sides reversed, as if you were standing facing yourself. Try raising your left hand. Your image will raise the hand on *your* left, which is the image's *right* hand. To understand why this happens, consider the reflection of the letter "F" (illustrated at right). Point A on the right-hand side of the object appears at point B on the left-hand side of the image.

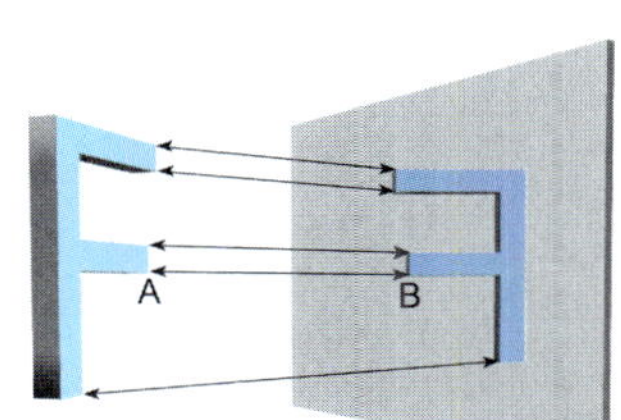


Image reflected in a flat mirror reverses left and right.

## Locating an image with a ray diagram

**How to locate an image formed by a mirror**

Ray diagrams are a good way to analyze an optical system to find the position, size, and orientation of any images. The procedure is the same for all types of optical systems. To start, several rays are traced leaving at different angles from a single point on an object. If these rays come together again at a point then a *real* image forms at that point. If the rays leave the optical system and *appear* to come from some point in space, the point they appear to come from is a *virtual image*. The following steps describe how to find the image of an arrow in a flat mirror.

**Step 1: Draw two incident rays**

An arrow makes a convenient object for ray tracing. Draw two incident rays from the top of the arrow to the mirror. You can choose any two rays but it is easiest to choose one ray perpendicular to the mirror and the second at an angle.

Draw two light rays incident upon the mirror. (One ray is drawn straight toward the mirror.)

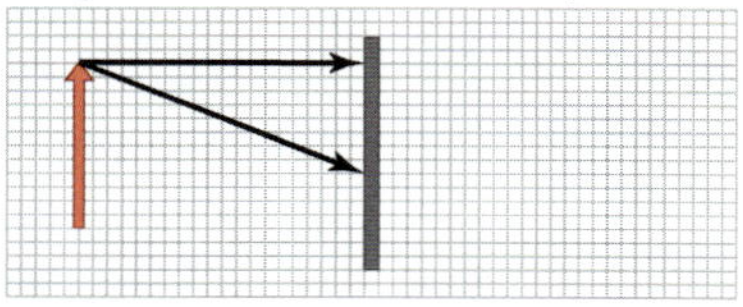

**Step 2: Draw the normals**

Draw the normals to the surface of the mirror where each incident ray contacts the mirror. The normal is a line perpendicular to the mirror passing through the point of contact. For the top light ray in the figure at right, the normal is along the incident light ray itself—that was the reason for choosing to draw this ray.

Draw the normal to the mirror's surface at the location each ray strikes it.

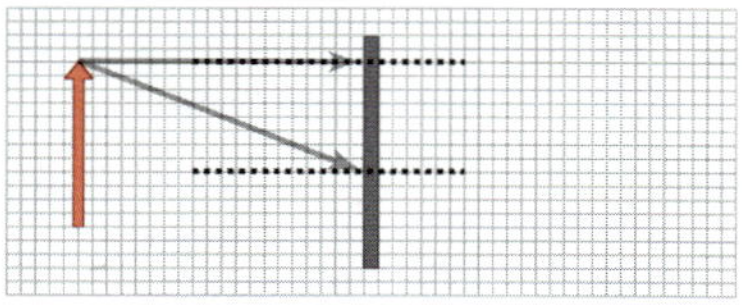

**Step 3: Draw the reflected rays**

Apply the law of reflection to draw the reflected rays. The top ray makes an angle of 0° with the normal, so the angle of reflection is also 0°. This ray reflects back upon itself. For the bottom ray in the example, the angle of incidence is 21.5° and therefore the angle of reflection is 21.5°.

Draw the reflected rays using the law of reflection. The top ray reflects back upon itself!

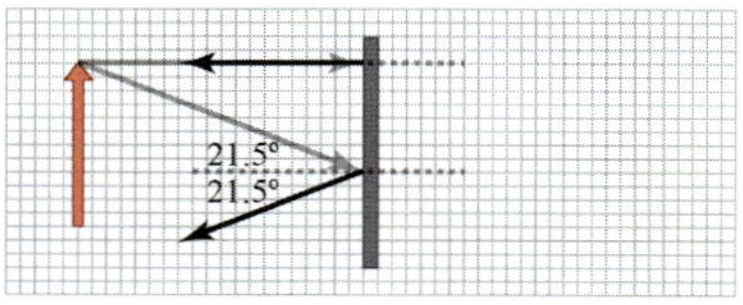


**Step 4: Extend the reflected rays**

The reflected rays diverge and will not meet in front of the mirror. Nonetheless, those divergent rays appear to come from a point *behind* the mirror. Extend both reflected rays behind the surface of the mirror. The two extended lines meet at the location of the image of the tip of the arrow.

Extend the reflected rays until they intersect behind the mirror at the location of the image.

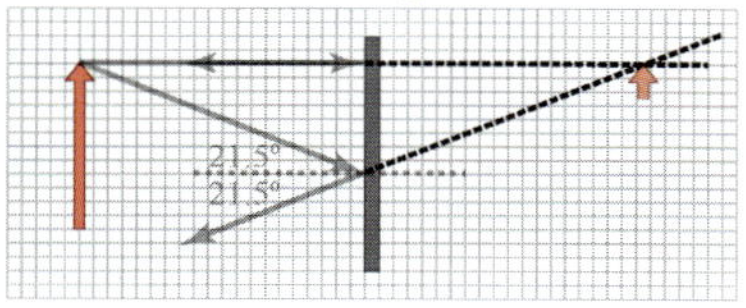


**Step 5: Find the image of the bottom of the arrow**

A full image of the arrow includes every point on the arrow. Fortunately, an image's size and orientation can be found by tracing rays from only *two* points. Choose the bottom of the arrow as the second point and follow the same steps to find the normals and reflected rays. Extend both reflected rays until they intersect at the image location for the bottom of the arrow.

Repeat to find the image location for the bottom of the arrow.

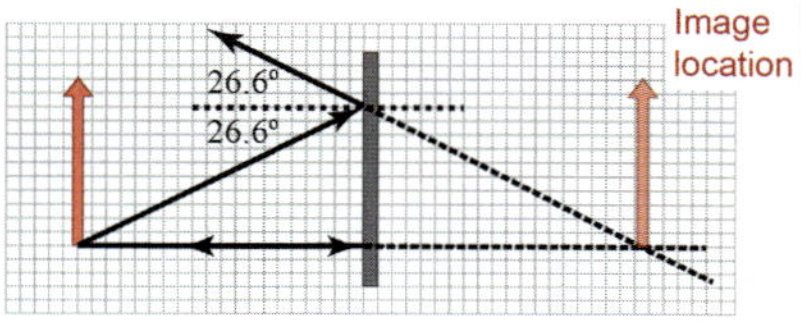

# Section 3 review

Ray diagrams are used to trace the path of light rays through an optical system. The angle of incidence and angle of reflection are measured with respect to the normal to the reflecting surface. The law of reflection states that the angle of incidence equals the angle of reflection. When light is reflected from a flat mirror, an image forms behind the mirror at the point from which the reflected rays appear to diverge. The image distance for a flat mirror equals the object distance, which is the distance between the object and the reflecting surface.

| | |
|---|---|
| **Vocabulary words** | **incident ray, reflected ray, normal, law of reflection, angle of incidence, angle of reflection, image** |
| **Key equations** | $\theta_i = \theta_r$ |

## Review problems and questions

1. Does the law of reflection apply to specular reflection, diffuse reflection, or both types of reflection?

2. In the Diego Velasquez work *Venus and Cupid*, the artist depicted the subject looking into a mirror. Is she looking at herself in the mirror? Why or why not?

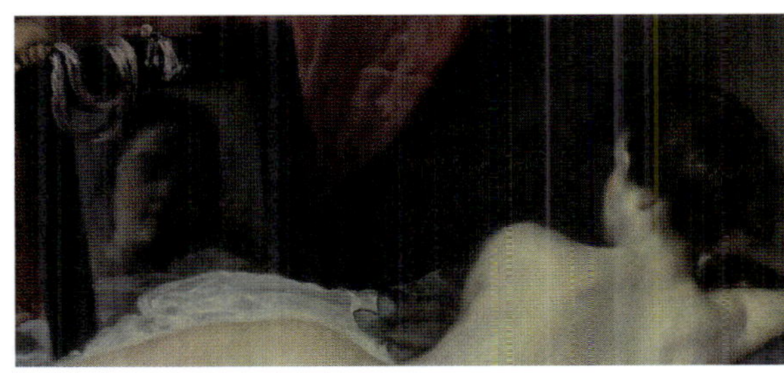

3. Some interior designers place a large, flat mirror on one wall of a room to make the room feel bigger. Explain their argument using the physics of mirrors and image formation.

4. A laser beam hits a mirror at an angle of incidence of 65°.
   a. What is the angle of reflection of the laser beam?

   The laser then hits a mirror perpendicular to the first.

   b. What is the angle of incidence of the laser to this mirror?
   c. What is the angle of reflection from the second mirror?

5. A young man is standing 87.3 cm in front of a flat mirror. He sees the image of a light bulb at an angle of reflection of 57.9°. The light bulb is physically located 175 cm in front of the mirror. What is the image distance for the image of the bulb in the mirror (i.e., the distance behind the mirror at which the image of the bulb appears)?

6. In drawing a ray diagram, a student drew an incident ray that passed along a normal to the surface of the mirror. What is the value of the angle of reflection?

## 20.4 - Spherical mirrors

The image in a flat mirror is a life-size, undistorted "picture" of the original object. The image in a *curved* mirror may be upside-down, magnified, or distorted in other ways. The simplest curved mirror to analyze has the shape of a section of a sphere. Think of a shiny, hollow ball—and then cut off a piece of it. A **convex** mirror has a reflecting surface that bulges outward. A **concave** mirror has a reflecting surface that cups inward, like a bowl.

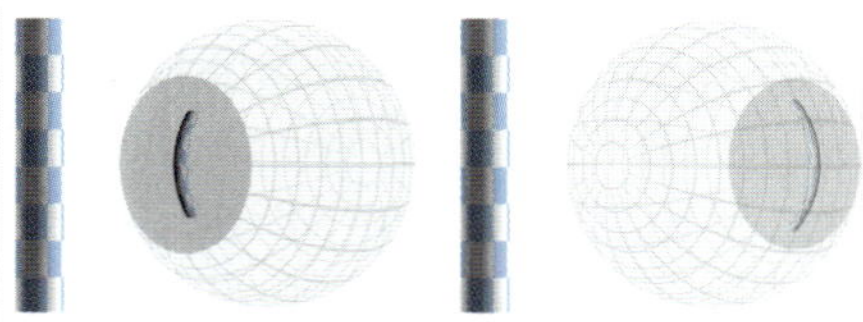

Convex and concave mirrors are segments of spherical mirror surfaces.

### Reflection in spherical mirrors

**Center of curvature**

The reflecting surface of a spherical mirror is a segment of the surface of a sphere. The *center of curvature* of the mirror is the center of the sphere. The *radius of curvature* of the mirror is the radius of the sphere.

**Reflection from a curved mirror's surface**

The law of reflection applies to reflection from a curved surface. The normal, however, has a different direction at each location on the surface. The normal to any point on the surface of a spherical mirror is a radial line that goes from the center of curvature through that point. In the figure at right, an incident ray striking the mirror at an angle of 25° to the normal reflects at an angle of 25° on the opposite side of the normal.

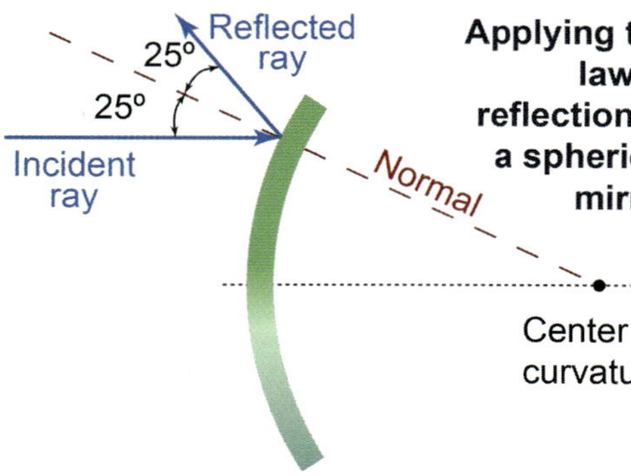


**Applying the law of reflection to a spherical mirror**

**Diverging mirror (convex)**

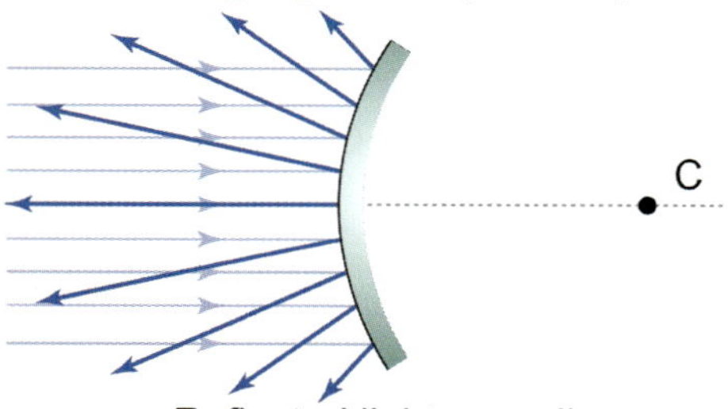


Reflected light rays *diverge* away from each other.

**Converging mirror (concave)**

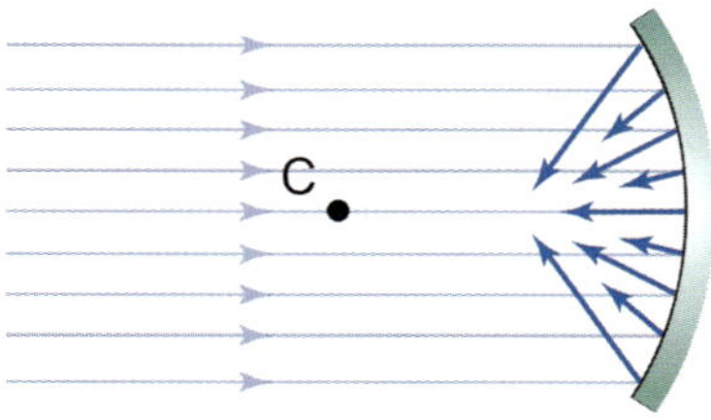


Reflected light rays *converge* toward each other.

**Diverging mirrors**

Consider the same series of parallel light rays incident upon two different spherical mirrors. The reflected rays from the *convex* mirror diverge from each other. A convex mirror is therefore called a *diverging mirror*. The side rear-view mirrors on a car are slightly convex. The divergence provides a wider field of view compared to a flat mirror, causing objects to look farther away than they actually are.

**Converging mirrors**

The reflected rays from the *concave* mirror come together or converge with each other. A concave mirror is a *converging mirror*. The focusing mirror of a solar oven is a converging mirror. Large astronomical telescopes use concave mirrors. The ability to see distant objects depends on collecting as much light as possible. It is far easier to create a large-diameter, optically perfect mirror than it is to make an optically perfect lens of equal size.

**What kinds of images occur in a curved mirror?**

The image in a spherical mirror may be larger or smaller than life size (magnified) and may be upside-down (inverted) or upright. These characteristics also change with different object distances. For example, a concave mirror creates a right-side-up, magnified image at close object distances and an upside-down, reduced-size image at far object distances. Because of their high reflectivity and ability to magnify image sizes, curved mirrors are the central element in all large telescopes.

**Diagram of a convex mirror**

The first step in ray tracing for a spherical mirror is to draw the *optical axis*. The **optical axis** (or *principal axis*) is the path of a light ray traveling down the center of the optical system. For a spherical mirror, the optical axis is a line perpendicular to the mirror's surface at its center and passing through the center of curvature. Halfway between the center of curvature and the mirror is the *focal point*. The **focal point** is the point through which all incident rays parallel to the optical axis pass (or appear to pass) after they are reflected. The focal point is a very important feature for understanding how a curved mirror reflects light.

**Properties of a convex mirror**

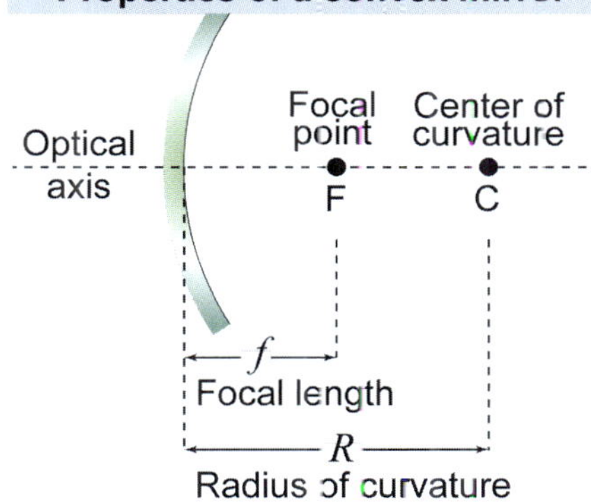


**Focal length and radius of curvature**

The **focal length** of a curved mirror (or a lens!) is the distance between the mirror and its focal point. A simple rule to remember is that the focal length of a spherical mirror is equal to one-half of the radius of curvature. Equivalently, the radius of curvature is twice the focal length.

$$R = 2f$$

$R$ = radius of curvature (m)
$f$ = focal length (m)

**Radius of curvature** of a spherical mirror

A mirror with a short focal length (or radius of curvature) has a highly curved surface. A mirror with a very long focal length has a surface that is only slightly curved. As the focal length gets longer and longer the curved surface more closely resembles a flat mirror.

**Ray tracing rules for spherical mirrors**

1. Incident rays through the center of curvature reflect back upon themselves.

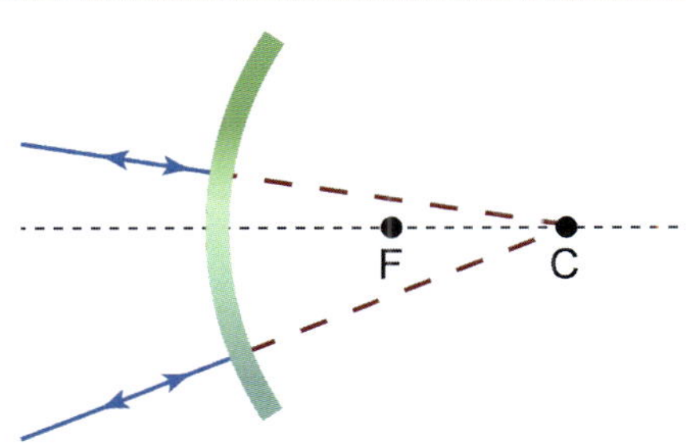


2. Incident rays parallel to the principal axis reflect through the focal point.

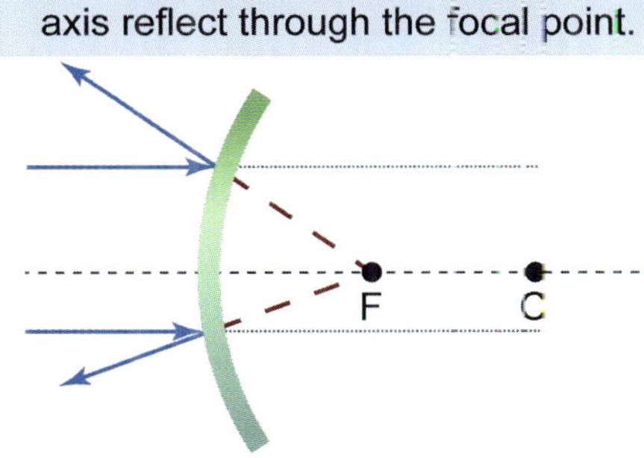


When tracing rays for a curved mirror, the two most useful rays to trace follow these basic rules. In the next investigation on page 602, these two rules will be used to predict the location and magnification of images formed by curved mirrors.

1. Any incident ray that would pass through the center of curvature is incident normal to the mirror surface and therefore reflects back on itself.
2. Any incident ray parallel to the optical axis reflects along a line that passes through the focal point.

# Investigation 20C Image formation for curved mirrors

**Essential questions** *What are the magnifications of convex and concave mirrors? Where are the images located for each?*

Interactive simulation

*Flat* mirrors produce images that are the same size as the object and located an equal distance behind the mirror. What is the magnification for a *curved* mirror surface? Where are its images located? In this investigation, you will use an interactive simulation to study the properties of image formation for curved mirrors. You will then use the mirrors in the equipment kit to check the conclusions you drew from the simulations.

## Part 1: Image location and magnification for a convex mirror

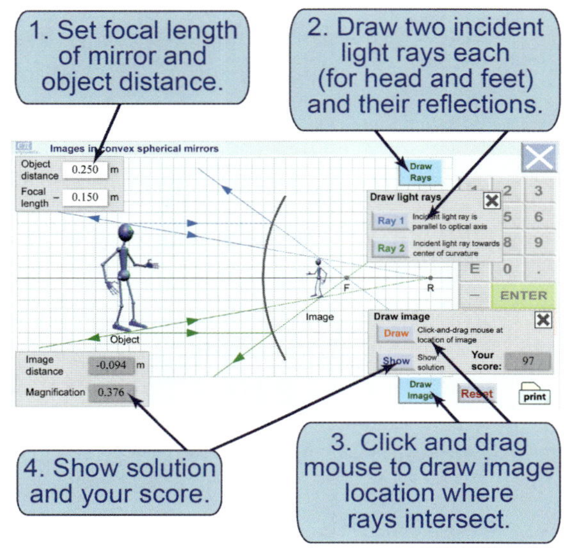


1. Set the convex mirror's focal length to 20 cm and the object distance to 30 cm.
2. Press the buttons to show how the two incident light rays reflect from the mirror's surface.
3. Press the [Draw image] button and then click and drag the mouse at the location of the image.
4. Show the solution and record your score. Tabulate the object distance, image magnification, image orientation (upright or inverted), and image location (in front of or behind the mirror).
5. Repeat for object distances of 10 and 50 cm.

*a. Describe how you found the image location.*

*b. Based on the simulation, what kinds of images does a convex mirror create?*

*c. Pick up a real convex mirror and look at your image from different object distances. Do your simulation results agree with what you see?*

## Part 2: Image location and magnification for a concave mirror

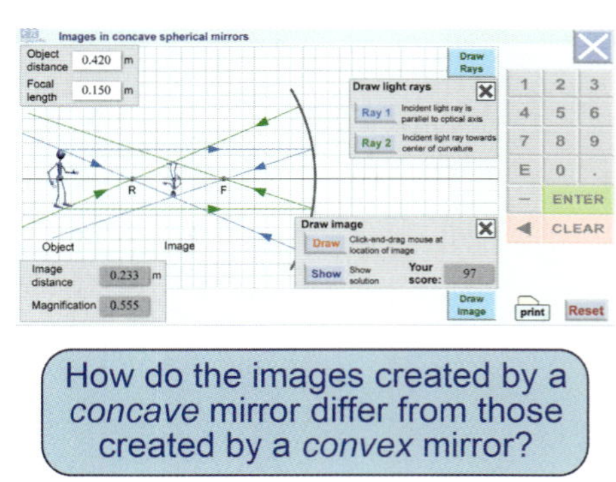


1. Set the concave mirror's focal length to 20 cm and the object distance to 50 cm.
2. Repeat the procedure from Part 1, identifying the location of the image formed by the concave mirror.
3. Repeat for object distances of 10 and 30 cm. Tabulate the results as before.

*a. Based on the simulation, what kinds of images does a concave mirror create? Under what conditions?*

*b. How do the images created by a concave mirror differ from those created by a convex mirror?*

*c. Pick up a real concave mirror and look at your image from different object distances. Do your simulation results agree with what you see?*

# Section 4 review

Spherical mirrors can produce images that are larger or smaller than life size and right-side up or upside-down. These mirrors can be converging (concave) or diverging (convex). The center of curvature is the center of the sphere that defines the mirror's reflecting surface. The optical axis is the path of a ray traveling through the center of an optical system. The optical axis of a spherical mirror is a line perpendicular to the mirror surface that passes through the center of curvature of the mirror. The focal point of a spherical mirror is the point through which rays parallel to the optical axis are reflected. The focal length of a spherical mirror is equal to half the radius of curvature.

| Vocabulary words | convex, concave, optical axis, focal point, focal length |
|---|---|

## Review problems and questions

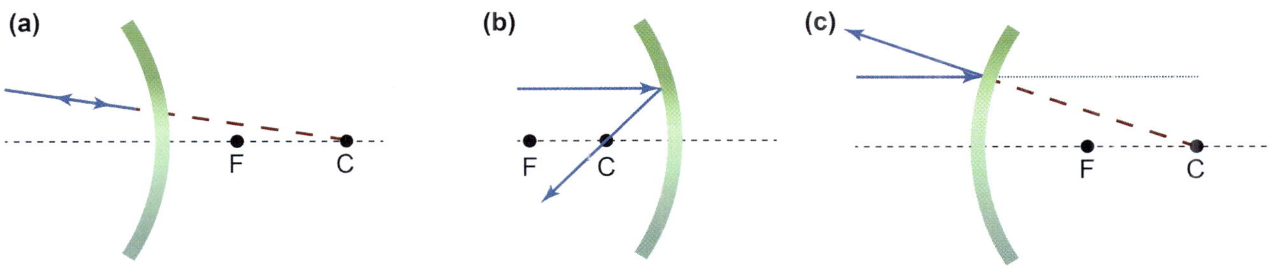


1. A student drew the ray diagrams (above) to describe the reflection of light from different spherical mirrors. Each drawing has an error in it. For each illustration (a–c), describe what the student did wrong in drawing and/or labeling each ray diagram.

2. Which of the following statements is *not* correct for drawing a ray diagram for a spherical mirror?

   a. Incident rays parallel to the optical axis reflect along a line passing through the focal point.
   b. The image is drawn where the two incident rays intersect.
   c. You can draw as many incident rays as you want in drawing a ray diagram.
   d. The image formed by a convex mirror is located behind the mirror.

3. Which of the following statements is *not* correct for drawing a ray diagram for a spherical *concave* mirror?

   a. The object arrow can be placed at the center of curvature.
   b. The object arrow can be placed at the focal point.
   c. The image is always located between the center of curvature and the mirror surface.
   d. The object and the image arrows can be located at the same distance from the mirror.

Chapter 20

# Chapter review

## Vocabulary

Match each word to the sentence where it best fits.

### Section 20.1

| | |
|---|---|
| **image distance** | **inverse square law** |
| **object distance** | **ray diagram** |

1. The intensity of light decreases by a factor of 9 when the distance increases by a factor of 3 because light follows a/an _______.
2. Jean-Claude is taking a picture with his camera of his friend located 2.5 m away. The _______ is 2.5 m.
3. Inside a camera, the separation between the lens and the detector (film or a CCD) is the _______.
4. A/An _______ is useful to represent how light changes direction with mirrors, lenses, and prisms.

### Section 20.2

| | |
|---|---|
| **diffuse reflection** | **lens** |
| **magnification** | **mirror** |
| **opaque** | **prism** |
| **reflection** | **specular reflection** |
| **translucent** | **transparent** |

5. _______ is when light waves can bounce off a material in any direction.
6. A mirror image off a surface is caused by _______.
7. A/An _______ window provides privacy while still letting in light.
8. Most glass used in windows is _______ and does not blur or darken images.
9. The curved surface of a/an _______ may magnify images.
10. A camera lens uses _______ to change the apparent size of the subject in the photograph.
11. Light, water, and sound waves can exhibit _______ when they bounce off a boundary.
12. Specular reflection is produced by a/an _______.
13. A/An _______ disperses light into its individual colors.

### Section 20.3

| | |
|---|---|
| **angle of incidence** | **angle of reflection** |
| **image** | **incident ray** |
| **law of reflection** | **normal** |
| **reflected ray** | |

14. The angle of incidence is measured between the _______ and the incident ray.
15. When you look in a mirror, a light ray from the image you are observing is a/an _______.
16. You can look at a/an _______ of yourself in the mirror.
17. The relationship between the angle of incidence and the angle of reflection is related by the _______.
18. The _______ is the angle formed by the incident ray.
19. The ________ is always equal to the angle of incidence.

### Section 20.4

| | |
|---|---|
| **concave** | **convex** |
| **focal length** | **focal point** |
| **optical axis** | |

20. The imaginary line drawn through the center of a spherical mirror is its _______.
21. A diverging, or _______, mirror reflects rays of light away from each other.
22. A/An _______ mirror is depressed inward.
23. A light ray that is incident upon a spherical mirror and travels parallel to the principal axis will be reflected through the _______.

## Conceptual questions

### Section 20.1

24. In an eclipse on Earth, the Sun or Moon is blocked from view.
    a. In a solar eclipse, what order are the Earth, Moon, and Sun in?
    b. What about in a lunar eclipse?
25. In designing a pinhole camera, such as in the design project on page 586, what quantity will you vary to project a 3-mm-high image of the Sun onto the back screen of the camera?

# Chapter review



## Section 20.1

26. Will a lunar eclipse occur at the time of a new Moon, full Moon, or quarter-Moon?

27. Will a solar eclipse occur at the time of a new Moon, full Moon, or quarter Moon?

28. ❮ The radius of Jupiter's orbit is approximately five times the radius of Earth's orbit around the Sun. How much different is the intensity of the Sun's radiation at Jupiter compared to that at Earth?

## Section 20.2

29. What optical device is used in eyeglasses? Why?

30. ❮ What does a negative magnification indicate?

31. ❮ Distinguish between specular reflection and diffuse reflection and give an example of each.

32. ❮❮ Describe the concept of parallax and its application for locating the image formed by a plane mirror.

33. ❮❮ After waxing a car, Daniel can see his own face in it. Is this an example of specular or diffuse reflection? Explain.

34. ❮❮ Brian is spear fishing and having some trouble. Even though he aims his spear directly at the fish he sees in the water, he misses every time. What property of light causes this? Explain.

35. ❮❮ Do any surfaces exhibit neither specular nor diffuse reflection?

## Section 20.3

36. At what angle is the normal line drawn to the boundary?

37. You receive a coded message. The phrase is reversed left to right, and all the letters are backward. What should you do to decode the message? Why?

38. What is the most important safety consideration for using a laser pointer?

39. What is an advantage of using a ring stand to mount a flashlight for a laboratory investigation, rather than having your partner hold the flashlight?

40. ❮ A small figurine is located in front of a plane mirror. Describe a procedure for determining the location of the *image* of the figurine in the mirror.

41. ❮ Explain the term "mirror image."

42. ❮ Look at the detail from *Candle in Mirror*, a painting by Georges de la Tour. Why does the image of the candle appear to be lower than the candle itself? Provide at least two possible explanations.

43. ❮❮ On a sheet of graph paper, draw a straight line in the middle to represent a plane mirror. Draw a small, filled circle located 7.3 cm in front of the mirror. Draw two light rays from the filled circle that are incident upon the mirror's surface. Use a protractor and the law of reflection to draw the reflected ray for each incident ray. Extend the reflected rays and locate the image for the filled circle. Using your drawings on the graph paper and a metric ruler, determine the image distance.

44. ❮❮ How can ray diagrams address both the ray and wave models for light?

45. ❮❮ Do any surfaces not follow the law of reflection? How does light behave differently at these surfaces?

46. ❮❮ How does the law of reflection affect drawing ray diagrams?

## Section 20.4

47. ❮ What property of a spherical mirror defines its center of curvature?

48. ❮❮ In the 1434 Jan Van Eyck work *The Arnolfini Portrait*, the artist painted a mirror behind his subjects. By looking at the nature of the images in the reflection, what kind of mirror is it?

## Section 20.4

49. What is the size of an image in a convex mirror compared to the size of the object itself?

50. If you double the radius of curvature of a spherical mirror, how does its focal length change?

51. What is the size of an image in a concave mirror compared to the size of the object itself?

## Quantitative problems

## Section 20.1

52. Suppose that a reading lamp holds a 15 W light bulb 1 m above the book that lies open in your hands. You want to double the intensity of the light that falls upon the page. Which of the following will work? (Assume that all light bulbs convert the same percentage of their electric power into light.)
    a. Double the wattage of the light bulb.
    b. Move the lamp until it is half as far from the book (that is, to a distance of 50 cm).
    c. Either a or b.
    d. Neither a nor b.

53. The Sun emits radiant energy (visible light, ultraviolet light, x-rays, etc.) at a rate of $3.8\times10^{26}$ W. The Earth is $1.5\times10^{11}$ m from the Sun.
    a. What is the intensity of this light, in watts per square meter, at the Earth's distance from the Sun? (State your answer to two significant figures.)
    b. By what factor would you divide this intensity, if you wanted to calculate the value at Pluto's distance from the Sun? (On average, Pluto is roughly 40 times as far from the Sun as the Earth is.)
    c. Suppose you are on a spaceship at Pluto's distance from the Sun. How large (in square meters) must your solar panels be, if they are to produce 1,300 W of electric power for your ship? (Assume that the panels can capture *all* the sunlight that strikes them and turn it into electricity.)

## Section 20.2

54. A lens has a magnification of 0.75. How long would the image of a 16-cm-long pencil be through the lens?

55. Harley is at an amusement park. She is 5 ft, 10 in tall. When she stands in front of a curved mirror in the funhouse, her reflection is 7 ft tall. What is the magnification of the mirror?

56. A scanning electron microscope magnifies objects by 2,000 times. A biologist is examining an Archean cell by using this microscope. The microscope produces an image of the cell that is 5 cm long. How long is the cell?

57. A pencil 7 in in length is held up to a curved mirror with a magnification of −½. How long is the image of the pencil that this mirror produces? Will the image be right-side-up or upside-down (inverted)?

## Section 20.3

58. A ray of light hits a mirror at an angle of incidence of 43°. At what angle does it reflect?

59. Magdalena is standing 60 cm from her bathroom mirror, looking at her reflection in it. How far away from her is her reflected image?

60. A light beam reflects off a mirror at a reflection angle of 75°. What was its angle of incidence?

61. At what angle does a ray of light need to hit a mirror for the angle between the incident ray and reflected ray to be a right angle?

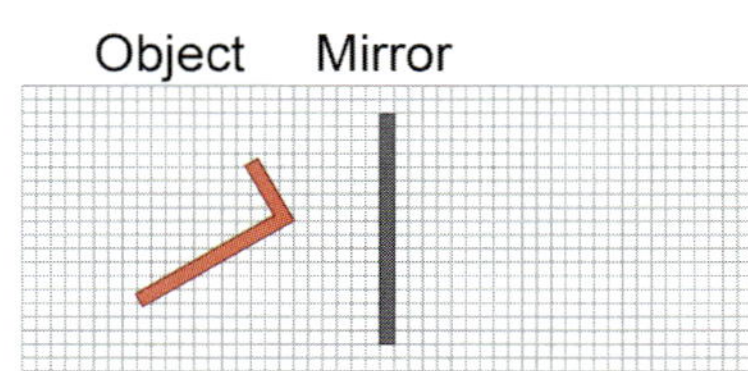


62. The drawing shows the positions of an L-shaped object and a flat mirror.
    a. Draw the reflected image of the object at its correct location. (Suggestion: Use the rules of locating an image with a ray diagram.)
    b. What is the magnification of the image?
    c. What is the orientation of the image?

63. If a ray of light hits a mirror at a 30° angle to the glass, what will be the angle of reflection?

64. A ray of light hits a mirror and is reflected. The ray comes in at a 30° angle *with respect to the mirror*. What is the angle between the incident ray and the reflected ray?

## Section 20.4

65. An image produced by a spherical mirror is 1/3 as large as the original object and has the same orientation. What is the magnification of the mirror?

66. A 1.6-m-tall woman is standing 75 cm in front of a spherical mirror that has a radius of curvature of 60 cm. If her image is 47 cm tall, what is the focal length of the mirror?

# Chapter review



## Standardized test practice

67. A ray of light hits a mirror at an angle of 72° to the normal. What is the angle of reflection?

    A. 28°
    B. 72°
    C. 144°
    D. 180°

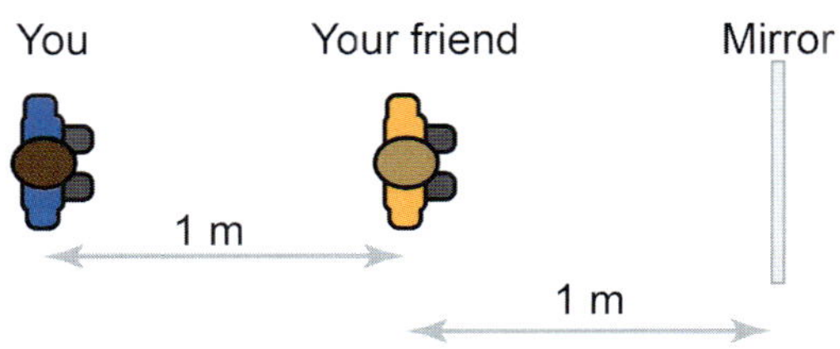


68. A friend and you are looking in a large, flat mirror against a wall. The friend is 1 m away from the mirror and you are 1 m directly behind her. How far are you from the image of your friend in the mirror?

    A. 0 m
    B. 1 m
    C. 2 m
    D. 3 m

69. What property of light explains the formation of shadows?

    A. Light travels in straight lines.
    B. Light has different colors.
    C. Light has wave and particle properties.
    D. Light can change direction.

70. Which of the following are properties of the images formed by a plane mirror?

    I. The image is located behind the mirror.
    II. The image distance is equal to the object distance (although it may be negative).
    III. The image reverses left and right.
    IV. The image is smaller than the object.

    A. I and II only
    B. II and III only
    C. I, II, and III only
    D. I, II, III, and IV

71. Which of the following optical devices uses refraction?

    I. mirror
    II. lens
    III. prism

    A. I only
    B. II only
    C. I and II only
    D. II and III only

72. What is the best choice in the following list for mounting a laser pointer to a ring stand?

    A. a ring clamp
    B. a 90° rod clamp
    C. wood glue
    D. nylon string

73. A 20-cm-wide image is seen through a compound system with a magnification of 2.5. What is the width of the subject?

    A. 8 cm
    B. 20 cm
    C. 30 cm
    D. 50 cm

74. A man is standing 2 m in front of a flat mirror and facing the mirror. How far away from him is his image located?

    A. His image is located at his location, i.e., 0 m away.
    B. His image is located 2 m in front of him.
    C. His image is located 4 m in front of him.
    D. His image is located 2 m behind him.

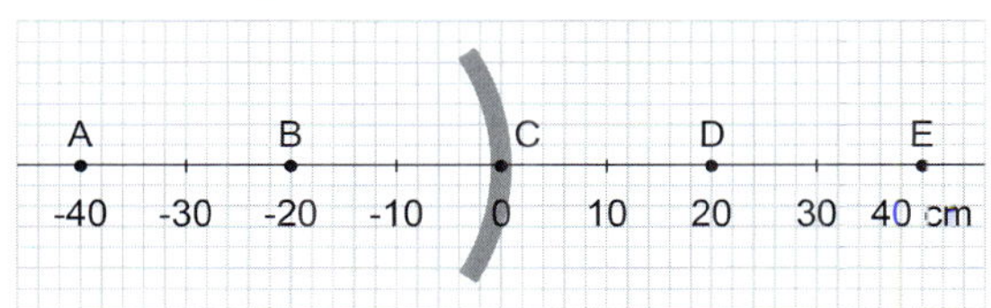


75. The curved mirror depicted has a focal length of 10 cm. Which point on the number line is the center of curvature?

    A. A
    B. B
    C. C
    D. D
    E. E

76. For which of the following would a ray diagram be useful?

    I. predicting where a shadow will appear for an object held under a light source
    II. representing the electronic circuitry of a lamp
    III. displaying the path of light in a periscope

    A. I only
    B. I and II only
    C. I and III only
    D. II and III only

77. Which kind of mirror can create an image that is larger than the object?

    A. a flat mirror
    B. a convex mirror
    C. a concave mirror
    D. None of the above is correct.

# Chapter 21 Refraction and Lenses

The eyes of all higher animals, including yourself, contain natural lenses. In technology, lenses are ubiquitous, being essential to every camera, microscope, eyeglass, projector, microscope, and optical fiber system. Lenses have become so much a part of technology that for the price of a few movie tickets you can buy a small telescope much more powerful than the one Galileo used to change our worldview in 1610.

Although spectacle lenses were made in Italy as early as the 13th century, optics remained a wealthy person's novelty for the next 300 years. A lens maker in the Netherlands named Hans Lippershey is credited with the first design of a telescope in 1608 that used *two* lenses at the opposite ends of a tube. His prototypes had magnifications of only three, but news of his invention spread remarkably fast across Europe and Southeast Asia. By 1609 news of Lippershey's telescope reached Galileo Galilei, then a mathematician at the University of Padua.

Telescope image credit: NASA

Galileo ground his own lenses and varied parameters of the telescope design to improve it. Galileo's early instrument used a convex lens at the entrance to the telescope and a concave lens as an eyepiece. Most people, including Galileo, thought the device would have applications in the military, such as for viewing enemy troops from a distance, or in navigation on the seas.

By 1610, Galileo pointed his new telescope at the heavens. Galileo drew the craters and shadows he saw on the Moon. He observed four bright points of light near the planet Jupiter, saw that they moved from night to night in orbits around the planet, and deduced that they were moons of Jupiter. He saw the phases of Venus and made the connection to the phases of the Moon. Even though his telescope had low magnification by modern standards, Galileo could see that the nebulous Milky Way resolved into hundreds and thousands of individual stars. Galileo's discoveries put our own world into its true context—that the Earth orbits the Sun and moons orbit their planets, a model called *heliocentrism*. Astronomy as a science was born.

# Chapter study guide

## Chapter summary

When light encounters a boundary between two materials, such as between air and glass, the light can be reflected, refracted, or both. In this chapter, you will learn how refraction *bends* light rays as they pass from one medium to another. How much the light is bent is determined by the *index of refraction* for each of the two materials. The physics of refraction explains how lenses create images and how prisms disperse light. Refraction varies with color and this variation is the explanation for the dispersion of light by a prism (or water droplets) into a rainbow. Lenses appear in a wide variety of devices, from telescopes and microscopes to cameras, and occur naturally in the eyes of living creatures.

## Learning objectives

By the end of this chapter you should be able to

- describe refraction and provide examples of it in real life;
- define the index of refraction;
- calculate the angle of refraction at a boundary;
- describe why internal reflection occurs;
- calculate the critical angle at a boundary;
- characterize different lenses based on their image properties;
- use ray tracing to locate the image created by a convex lens;
- calculate image distance and magnification for a convex lens; and
- explain how compound optical devices, such as the telescope and microscope, work.

## Investigations

21A: Refraction of light
21B: Creating real and virtual images with lenses
21C: Image formation for a convex lens
21D: Build a microscope and a telescope

## Chapter index



## Important relationships

$$n_i \sin\theta_i = n_r \sin\theta_r$$

$$\sin\theta_c = \frac{n_r}{n_i}$$

$$\frac{1}{d_o} + \frac{1}{d_i} = \frac{1}{f}$$

$$m = -\frac{d_i}{d_o}$$

## Vocabulary

| | | |
|---|---|---|
| refraction | index of refraction | Snell's law of refraction |
| critical angle | total internal reflection | focal length |
| focal point | convex lens | concave lens |
| real image | virtual image | rods |
| cones | | |

# 21.1 - Refraction

Imagine observing a distant lion in a telescope. Through the lens the lion appears close enough to bite you, yet the actual lion is safely 50 m distant. The apparent closeness of the image in the telescope is an *illusion* created by bending light rays. The illusion works because we "see" objects through the *light* reflected from objects. Anything that changes the light between the object and our eyes, such as refraction, can make objects appear larger, smaller, closer, inverted, or otherwise distorted.

## Refraction

**What is refraction?**

**Refraction** is the bending of light rays, usually upon crossing a boundary between different materials such as air and glass or water. The refraction is caused by a change in the speed of light as it passes from one material into another. A ray of light that approaches an interface is *refracted* when it changes direction upon crossing the interface. A spoon in a glass of water is an excellent example of refraction. The spoon is not actually broken by the water surface, but it appears that way because the air–water boundary refracts light rays. The part of the spoon that is under water appears in a different place because light reflected from the spoon is refracted as it passes from the water to glass to air.

**How do we understand refraction?**

To describe refraction we consider a single *incident ray* approaching a boundary between two transparent materials. The normal is an imaginary line perpendicular to the boundary and passing through the point where the light ray crosses. When light passes from air into water we observe that the light rays bend *toward* the normal. When light passes from water into air we observe that the light rays bend away from the normal.

**From air into glass, refraction bends a light ray *toward* the normal.**

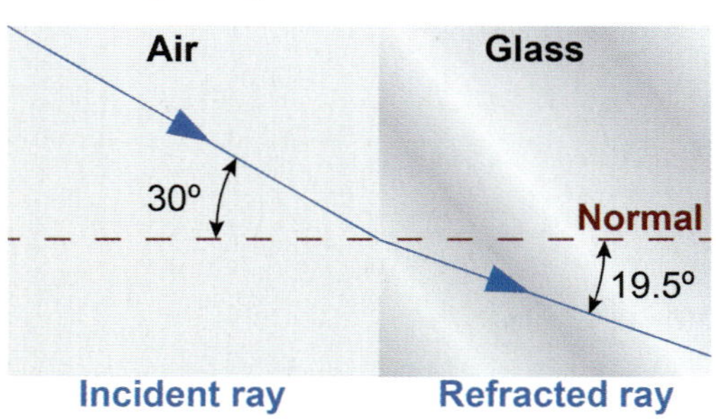


**From glass into air, refraction bends a light ray *away* from the normal.**

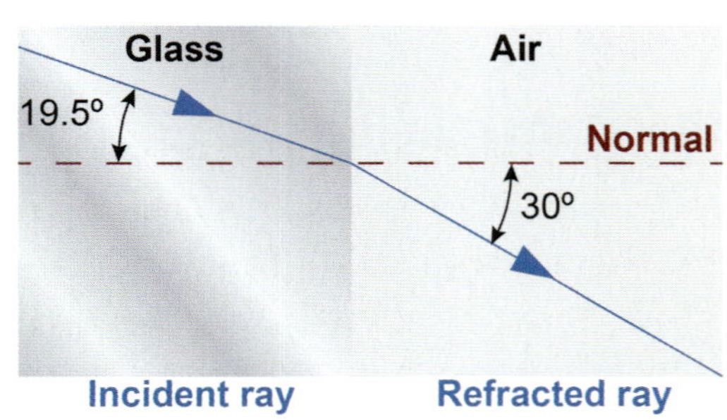


**Do all materials refract equally?**

Looking at different materials we observe that incident light rays at the same angle are refracted by different amounts. A light ray with an angle of incidence of 30° might be deflected by 10.5° when passing from air into glass. The same incident ray is deflected by only 7.9° when passing from air into water. The ability to refract light is determined by a material's **index of refraction**. The index of refraction is greater for glass (1.5) than for water (1.33); therefore, light is refracted more strongly in glass compared to water.

# Investigation 21A Refraction of light

**Essential questions** *How does light refract at a boundary?*
*What is the index of refraction of water?*

Interactive simulation

Refraction may change the direction of light rays passing from one medium to another. The differences in index of refraction between the two media determine how much refraction occurs. In this investigation, you will analyze light rays passing through air and water and determine the index of refraction of water.

## Part 1: Trace the path of light through air and water

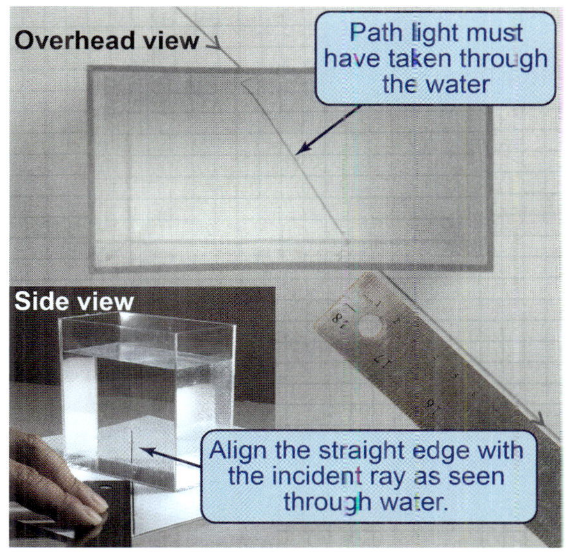


1. Place the container in the middle of a piece of graph paper and trace its outline. Remove the container.
2. Use a straight edge to draw an incident ray that intersects the left side of the container outline at an incident angle of about 45° to 50°.
3. Fill the container with water and replace it.
4. On the far side of the container, look through the water to view the incident ray. Align a straight edge with the incident ray as seen through the water. Use the straight edge to draw the refracted ray.
5. Empty and replace the container. Align a straight edge with the incident ray as seen through the empty container. Draw and label the "refracted" ray.
6. Remove the container. Connect the path of the light rays through both the full and empty containers.

a. Did the light bend through the empty container? Through the water? Why or why not?
b. When the light bent, in how many places did it bend?
c. Construct the normals for both boundaries using dashed lines. With a protractor, measure and record the angles of incidence and refraction.
d. Use the Snell's law calculator to calculate the index of refraction of water for each boundary.

## Part 2: Investigate the effect of angle on refraction

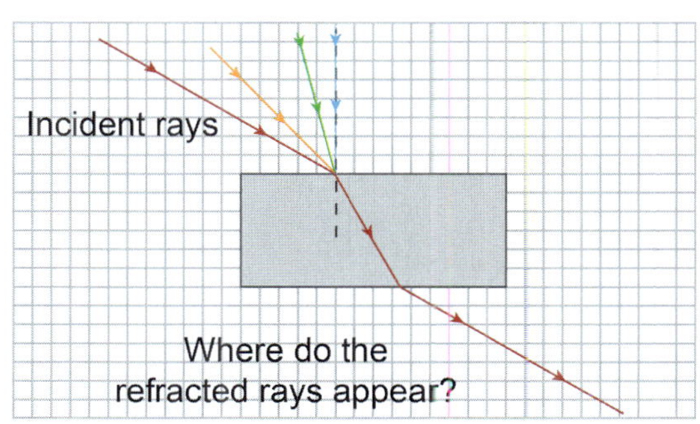


1. Construct three more incident rays at angles smaller than 45°, including one ray with an angle of incidence of 0°.
2. Refill and replace the container of water.
3. Look through the container to view the incident rays. Draw the refracted rays with a straight edge.
4. Remove the container and connect the path of the light rays. Then replace the container and view the complete set of rays.

a. Did all rays through the water refract? Which rays refract the most? The least?
b. What two factors have you found to determine how much a light ray will bend?
c. Reverse position and look through the water to view the refracted rays. What do you notice? Why?

**Do materials bend light differently?**

The angle by which a light ray is refracted toward or away from the normal depends on the index of refraction on both sides of the boundary. The higher the difference between the two indices of refraction on either side of the boundary, the greater the refraction. Among common materials, the index of refraction varies from 1 (air) to 2.42 (diamond). The table on the right lists some typical values. The greatest difference we would normally see occurs between air and diamond. In part, this explains the characteristic "sparkle" of a polished diamond.

| Material | Index of refraction $n$ |
|---|---|
| vacuum | 1.0000 |
| air | 1.0003 |
| water | 1.33 |
| ice | 1.31 |
| acrylic | 1.49 |
| window glass | 1.52 |
| flint glass | 1.62 |
| leaded glass | 1.7 |
| diamond | 2.42 |

**How can I determine the angle of refraction?**

**Snell's law of refraction** relates the angles of the incident and refracted rays to the index of refraction on both sides of the refracting boundary. The formula says that the product of the index of refraction multiplied by the sine of the angle is the same on both the incident and refracting sides of the boundary. In equation (21.1) for Snell's law of refraction, , $n_i$ and $\theta_i$ refer to the medium containing the incident ray. Similarly, $n_r$ and $\theta_r$ refer to the medium containing the refracted ray.

Interactive equation

**Snell's law of refraction**

$$n_i \sin\theta_i = n_r \sin\theta_r \quad (21.1)$$

$n_i$ = index of refraction, incident medium
$n_r$ = index of refraction, refracted medium
$\theta_i$ = angle of incidence (degrees)
$\theta_r$ = angle of refraction (degrees)

**How do I use Snell's law?**

When using Snell's law, always remember to reference the angles of incidence and refraction to the normal, *not to the boundary interface between the materials!* The solved problem below illustrates what we mean. The *sine* of an angle is a mathematical function that varies from zero to one as the angle changes from 0° to 90°. In standard mathematical notation, sin(0°) = 0 and sin(90°) = 1.

**Refraction of light from air to water**

A light ray traveling from air ($n_{air}$ = 1.0) into the ocean is deflected, according to the figure at right. What is the index of refraction of the water?

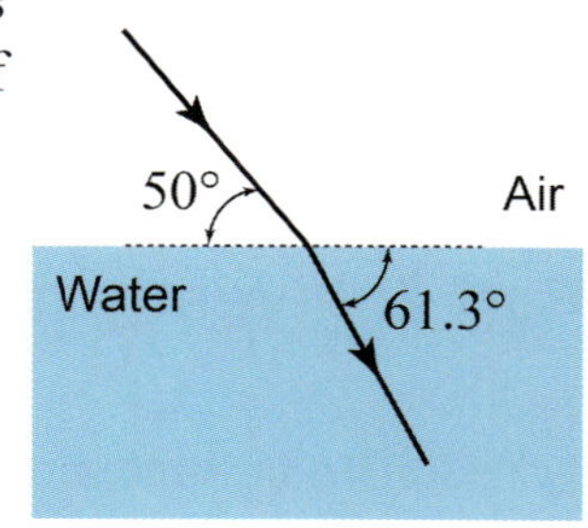


*Asked:* index of refraction of water, $n_w$

*Given:* index of refraction of air, $n_a = 1.0$; figure showing incident and refracted angles

*Relationships:* Snell's law: $n_a \sin\theta_i = n_w \sin\theta_r$

*Solution:* From the figure, recognize that the incident angle is

$$\theta_i = 90° - 50° = 40°$$

and the refracted angle is

$$\theta_r = 90° - 61.3° = 28.7°$$

Solve for the index of refraction of water:

$$n_w = \frac{n_a \sin\theta_i}{\sin\theta_r} = \frac{(1.0)\sin 40°}{\sin 28.7°} = 1.34$$

*Answer:* $n_w = 1.34$.
(Note that this is the value for *saltwater*!)

**Internal reflection**

When light passes from air into glass, there is usually some reflection and some transmission. The transmitted light is the refracted ray and it bends toward the normal according to Snell's law. When light passes from glass into air, however, something different may happen. When the angle of incidence is 41° the refracted ray bends 80° away from the normal. When the angle of incidence is 43° *there is no refracted ray at all!* The incident ray is reflected back into the glass and *no light is transmitted into the air!* This phenomenon is known as **total internal reflection**.

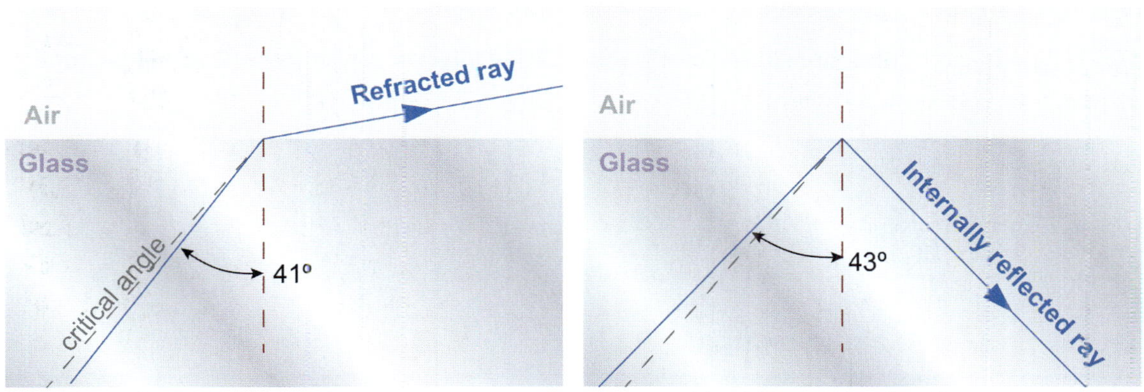


Ordinary refraction — Total internal reflection

**Critical angle for internal reflection**

The **critical angle** is the angle at which a refracted ray bends 90° away from the normal. When the angle of incidence is greater than the critical angle an incident ray is totally reflected and there is no refracted ray. The critical angle depends on the ratio of the indices of refraction on either side of the boundary. Equation (21.2) tells us that the sine of the critical angle is the lower refractive index divided by the higher refractive index.

Interactive equation

**Critical angle for internal reflection**

$$\textbf{(21.2)}\quad \sin\theta_c = \frac{n_r}{n_i}$$

$\theta_c$ = critical angle (degrees)
$n_r$ = index of refraction for refracted medium
$n_i$ = index of refraction for incident medium

**How do optical fibers work?**

An *optical fiber* is a pipe for light rays that works on the principle of total internal reflection. Consider sending light through a glass rod at an angle of incidence greater than the critical angle (with respect to the wall of the rod, not the end). At every point, light is totally internally reflected off the boundary between glass and air and bounces back into the glass. If the glass rod is made very thin, it becomes flexible, but still traps light by total internal reflection. Optical fiber cables form the backbone of all high-speed telecommunications lines, spanning continents and oceans carrying voice and network data. In another application, a bundle of optical fibers is the core of a flexible *image pipe* in which each fiber transmits one dot of the image.

**Calculating the critical angle**

What is the critical angle for light passing from glass ($n$ = 1.50) to air ($n$ = 1.00)?

*Asked:* critical angle $\theta_c$ going from glass to air

*Given:* incident material (glass) $n_i$ = 1.50, refracted material (air) $n_r$ = 1.00

*Relationships:* $\sin\theta_c = n_r / n_i$

*Solution:*

$$\sin\theta_c = \frac{n_r}{n_i} = \frac{1.00}{1.50} = 0.667 \quad\Rightarrow\quad \theta_c = \sin^{-1}(0.667) = 41.8°$$

*Answer:* $\theta_c$ = 41.8°. Compare this angle to the diagram at the top of the page!

# Section 1 review

Refraction of light occurs at the boundary between two different media, such as between glass and air. In refraction, a light ray is bent toward the normal if it passes into a material with a higher index of refraction and away from the normal if the second material has a lower index. A greater difference between the indices of refraction of the two materials causes light to bend more. When light passes from a higher index material (such as water) to a lower index material (such as air), there is a *critical angle* of incidence. At angles of incidence greater than the critical angle, a light ray is totally reflected back into the incident medium and there is no refracted ray. This is called *total internal reflection.*

| **Vocabulary words** | **refraction, index of refraction, Snell's law of refraction, critical angle, total internal reflection** |
|---|---|

| **Key equations** | $n_i \sin\theta_i = n_r \sin\theta_r$ | $\sin\theta_c = \frac{n_r}{n_i}$ |
|---|---|---|

## Review problems and questions

1. Light from air enters a piece of amber ($n = 1.55$) on a necklace at an angle of 45°. What is the angle of refraction of the light inside the amber?

2. Light passing through glass is completely internally reflected when it hits the glass–air barrier at an angle of incidence of 40.0°. What kind(s) of glass could this possibly be? Use the table of indices of refraction to identify different types of glass.

3. Eddie goes into a jewelry store to sell a large diamond cube that has been in the family for generations. The jeweler needs to test the cube to confirm that it is really diamond, but she cannot hit it with a hammer because she might break it. So she tests the block's ability to refract light. She shines a laser into the block at various angles and measures the refracted angle. Her data are in the table to the right.

| $\theta_i$ | $\theta_r$ |
|---|---|
| 0° | 0° |
| 30.0° | 19.6° |
| 45.0° | 28.3° |
| 60.0° | 35.5° |

   a. What is the index of refraction for this block?
   b. What material is this block made out of?
   c. Using optics, is there a different way to test whether the cube is diamond?

4. A physicist has three materials (*A*, *B*, and *C*) with indices of refraction 1, 2, and 3, respectively.
   a. Does a light ray shining from *A* into *B* deflect away from or toward the normal?
   b. Does a light ray shining from *C* into *B* deflect away from or toward the normal?
   c. Does a light ray shining from *A* into *C* deflect away from or toward the normal?
   d. If the angle of incidence is always 30°, does the light ray deflect more in a or in c?

Optics is the science and technology of light. One of the most important inventions in optics is the *lens*. There is a lens in each of your eyes, and lenses are found in cameras, microscopes, telescopes, and projectors. Lenses use curved surfaces to refract light in specific ways.

## Lenses

**Deflection increases with slope**

Consider a light ray encountering the three shapes of glass in the diagram below. A flat-sided shape does not deflect the incident ray at all. A slope-sided shape, however, deflects the incident ray and the deflection increases with the slope of the surfaces.

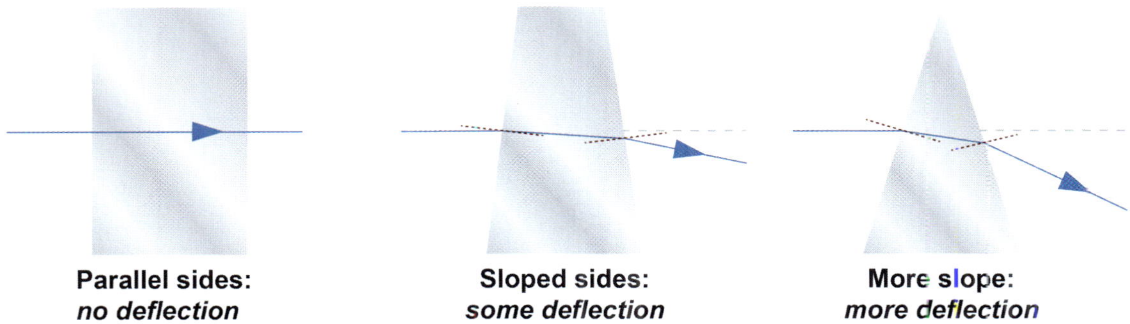


**How do lenses work?**

In a lens, the surfaces are continuously curved. At the center, on the *optical axis,* a light ray is not deflected. Light rays that enter the lens parallel to the optical axis are refracted more and more as their distance from the optical axis increases. In a perfect lens, the surface is curved in just the right way so as to bend all the light rays—no matter what their distance from the optical axis—to meet at the **focal point**.

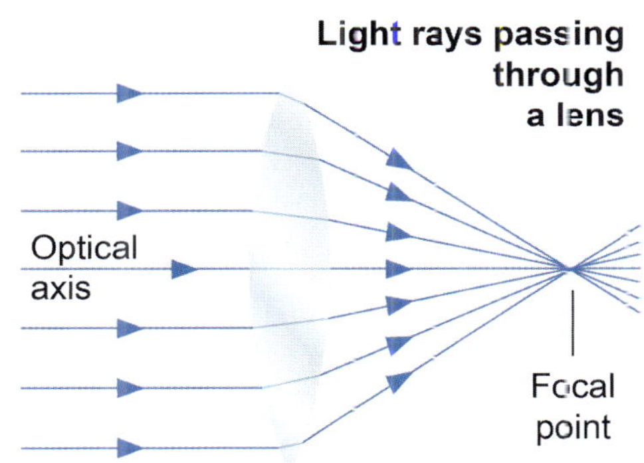


**How are lenses designed?**

The **focal length** of a lens is the distance from the center of the lens to the focal point. The focal length depends on the curvature of the surfaces and the index of refraction of the lens material. A thick lens with strongly curved surfaces deflects light rays more and therefore has a shorter focal length than a thin lens with more gradual curves. A lens made of material with a high $n$ has a shorter focal length than a lens made of a low-$n$ material.

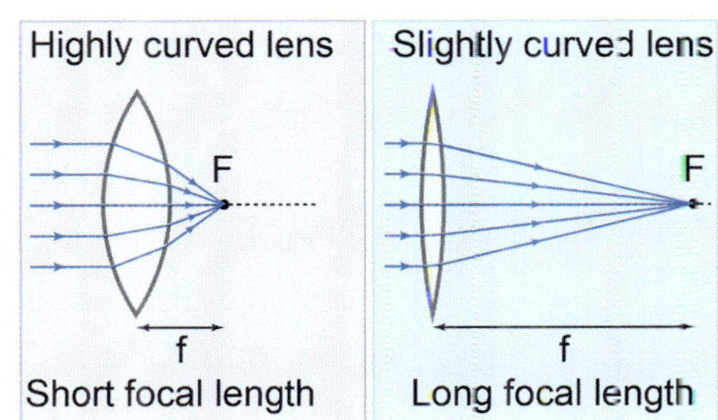


**The benefit of a larger camera lens**

The diameter of a lens determines how much light is collected at the focal point. A large-diameter lens collects more light than a small-diameter lens. That is why expensive cameras have larger lenses. It takes a minimum amount of light to register an image on the electronic detector in a digital camera. A large lens collects enough light for the camera to be able to take a good picture even in dim light, such as indoors. Cellphone cameras can only take good pictures in bright light because they have much smaller lenses.

## Converging and diverging lenses

**Converging lenses**

There are two basic kinds of lenses: converging and diverging. They differ in how they bend parallel light rays that are incident upon them. A **convex lens** is a converging lens, which means that it bends parallel light rays *toward* the optical axis. When a light ray strikes the curved surface of a converging lens, it is bent toward the normal through Snell's law—just as it bends for a sloped side of a glass prism. These bends cause incident parallel light rays to converge to an image location beyond the lens. A converging lens is typically convex, that is, thicker in its middle and thinner around its edges. A magnifying glass is a commonly used converging lens.

**Converging lens**

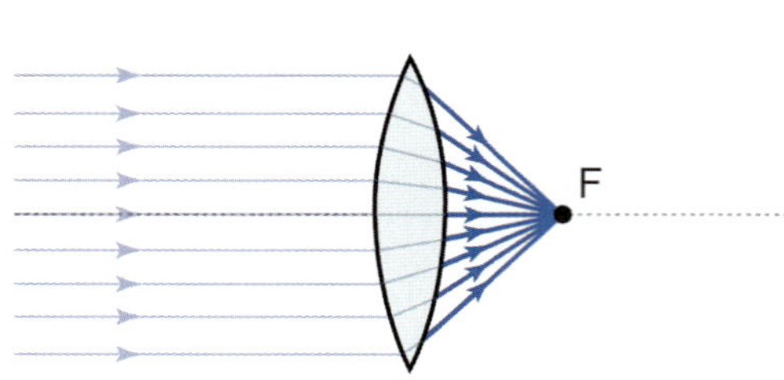


Incident rays parallel to the optical axis *converge* toward the far focal point.

**Diverging lens**

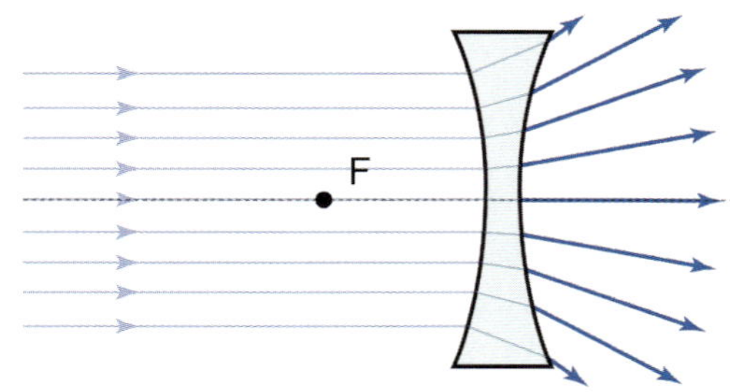


Incident rays parallel to the optical axis *diverge* away from the near focal point.

**Diverging lenses**

A **concave lens** is a diverging lens, which means that the lens bends parallel light rays *away* from the optical axis. When incident light strikes the surface of a diverging lens, light is still refracted toward the normal—but in this case the normal to the surface points away from the optical axis! A diverging lens bends light in a way that *appears to focus it at a point on the front side of the lens.* A diverging lens is thinner in its middle and thicker around its edges—its surface is concave. A diverging lens can be used as a peep hole in a door to allow you to see a wide field of view outside the door.

**Different kinds of converging and diverging lenses**

The illustration above shows only one example of each kind of lens. A *biconvex* lens has an outwardly curved surface on both sides. There are also other kinds of convex lenses, such as when one side is flat (a *planoconvex* lens). Similarly, there are diverging lenses other than the biconcave lens shown above. In physics class you may focus only on symmetrical biconvex and biconcave lenses, but when optics are designed for technology or scientific instruments, a wider variety of lenses are typically used to correct for aberrations and distortions.

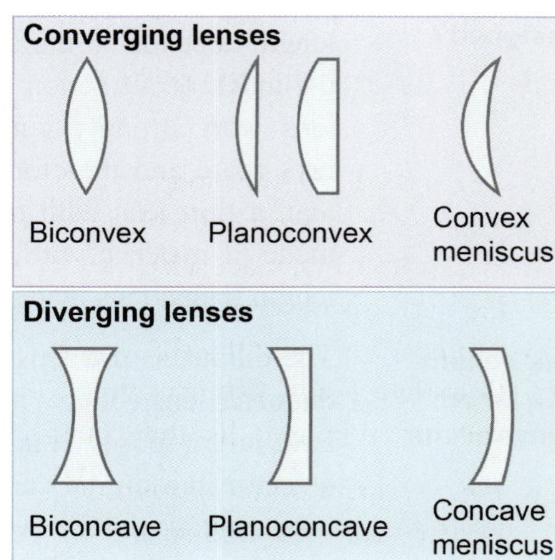

# Investigation 21B — Creating real and virtual images with lenses

| Essential questions | *What is the difference between real and virtual images?* *How can you create real and virtual images with a convex lens?* |
|---|---|

Optical devices can create either *real* or *virtual* images. What's the difference between the two? A real image is formed at a location where light rays *actually converge*, while a virtual image is formed at a location where light rays *only appear to converge*. In practical terms, a real image can be projected onto a piece of paper but a virtual image cannot. This morning you saw a *virtual* image of yourself in the mirror. Why was it virtual? Because you cannot put a sheet of paper behind the mirror and project your image onto it! A flat mirror produces a virtual image. How about a convex lens?

## Part 1: Create real and virtual images with a convex lens

*Caution: Never look directly at the LED light through the lens.*

1. Mount the light source (the "object") on the track near one end. Create the *object* by illuminating LEDs of different colors on the light source.
2. Mount a convex lens at a distance of 60 cm from the LED light source. Mount the screen on the opposite side of the lens from the object.
3. Adjust the screen location until you produce the sharpest possible image. Record whether the lens can produce a real image for this object distance.
4. Move the lens closer to the object in steps of 5 cm and repeat the experiment, determining in each case whether a real image is formed.

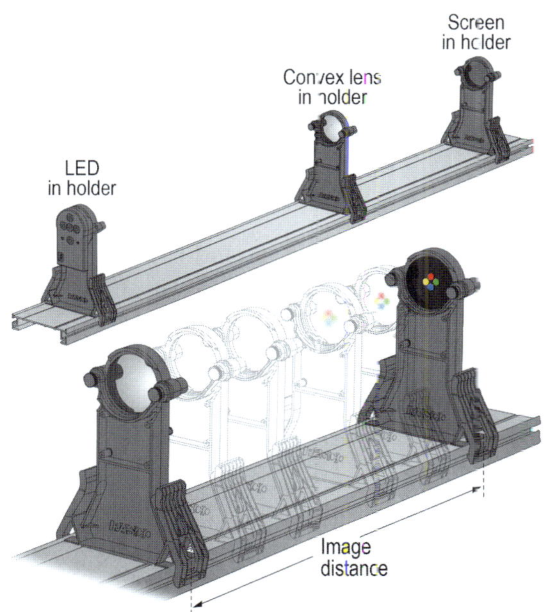

Move screen until sharp image forms.

*a. If you cannot project a focused image onto the screen no matter what distance, then what kind of image is being produced by the lens? Explain.*

*b. At what object distances is the image real? At what distances is it virtual? How do you know?*

## Part 2: Image formation for other optical devices

1. Leave the LED light source (the "object") in place.
2. Repeat the experiment with a convex lens, a concave mirror, and a convex mirror.
3. For the mirrors, place a half screen in front of the lens (where the light rays are reflected!) so only half the mirror is covered. This allows you to look for the image without blocking all the light.
4. With each of these three devices, determine whether it produces a real or a virtual image.
5. Complete a table listing each optical device and the kind of image it produces. If it produces both real and virtual images, list the conditions under which it produces each kind of image.

*a. Which optical devices produce real images? Which produce virtual images? Which optical devices produce both real and virtual images?*

*b. Under what conditions do these optical devices produce each kind of image?*

## Real and virtual images

**Objects and images**

Optical devices take light from an object and produce an image. Objects are physical things, such as the Sun, the light from a computer monitor, or a blue wall. Images are pictures of those things that are formed when light rays meet. Images can be produced by mirrors, lenses, prisms, or combinations of them.

**Real images**

There are two kinds of images that you can create with optics: real and virtual images. A **real image** is a place where light rays physically meet and can be projected onto a piece of paper. You can hold a magnifying glass in the sunlight and focus an image of the Sun onto a piece of paper. The light rays from the Sun that pass through the glass come together to form an image on the paper, which can cause it to catch fire!

A magnifying glass produces a *real* image of the Sun...

...that can be projected onto a piece of paper.

**Virtual images**

A **virtual image** occurs when the light rays *appear* to come together, but they don't *actually* do so. This might sound strange. Virtual images are formed behind a flat mirror or in front of a convex lens. Remember when we found that light rays appeared to come together to form an image *behind* a flat mirror? The light rays aren't actually coming together behind the mirror—they only appear that way!

A flat mirror produces a *virtual* image behind the mirror...

...that cannot be projected onto a piece of paper.

**Testing for real or virtual images**

The practical test for whether an image is real or virtual is to try to project the image onto a piece of paper. Look at the image of yourself in a mirror. Have a friend hold a piece of paper at the location of your image—behind the mirror!—and tell you whether he can see your image on the paper. Not a chance! The light bounced off the *front* of the mirror; it did not travel behind the mirror. It's a *virtual* image!

**Convex lenses can create different kinds of images**

In Investigation 21B on page 617, you experimented with convex lenses to create different kinds of images. A convex lens, such as a magnifying glass, can create many different kinds of images—depending on where the object is located relative to the lens. When the object is located close to the lens (within the focal length), the image is magnified, virtual, and upright; this is the typical mode for using a magnifying glass. When the object is located between one and two focal lengths from the lens, the image is magnified, *real*, and *inverted*. When the object is located farther away, the image is *reduced in size*, real, and inverted.

**Which optical devices produce real or virtual images?**

**Image properties for some optical devices**

| Optical device | Object location | Image type | Magnification | Image orientation |
|---|---|---|---|---|
| Flat mirror | Anywhere | Virtual | Same size | Upright |
| Convex mirror | Anywhere | Virtual | Reduced | Upright |
| Concave lens | Anywhere | Virtual | Reduced | Upright |
| Convex lens | $d_o > 2f$ | Real | Reduced | Inverted |
| Convex lens | $f < d_o < 2f$ | Real | Magnified | Inverted |
| Convex lens | $d_o < f$ | Virtual | Magnified | Upright |

### Rules for ray tracing for converging lenses

A ray diagram can be used to find the location of the image for a converging lens using these rules:

1. Incident rays parallel to the optical axis refract through the *far* focal point.
2. Incident rays passing through the center of the lens refract straight through the lens *undeflected*.
3. Incident rays passing through the *near* focal point refract parallel to the optical axis.

The image is formed where all three refracted lines intersect. In the example at right, the image is inverted, real, and smaller than the object (magnification less than one). It is real because it is located on the far side of the lens from the object—and thus is composed of real light that could be projected onto a piece of paper.

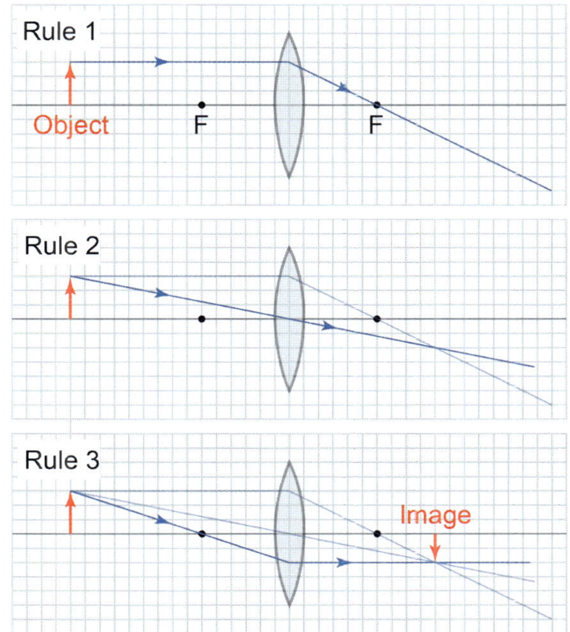


The image is formed where all three refracted lines intersect.

### Rules for ray tracing for diverging lenses

In the same way, a ray diagram can be used to locate the image for a diverging lens. Rule #2 is the same as for a converging lens, but the other two rules are slightly different:

1. Incident rays parallel to the optical axis refract along a line passing through the *near* focal point.
2. Incident rays passing through the center of the lens refract straight through the lens *undeflected*.
3. Incident rays directed towards the *far* focal point refract parallel to the optical axis.

In the example at right, the image is upright, virtual, and smaller than the object. The image is virtual because it is on the same side as the object, and hence the image cannot be projected onto a piece of paper.

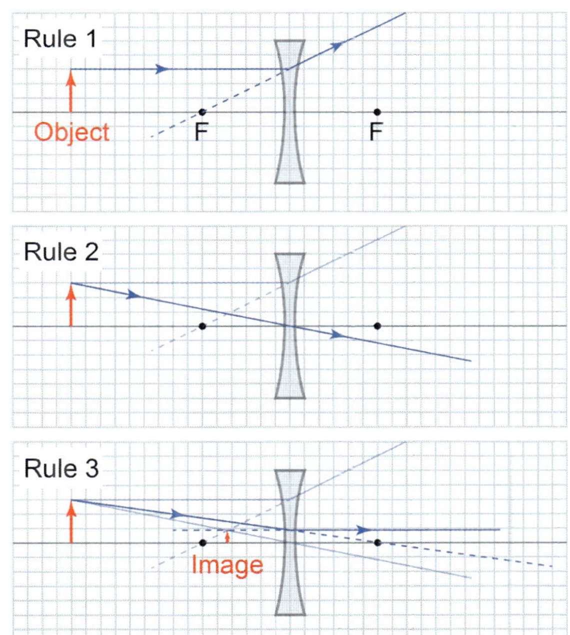


The image is formed where all three refracted lines (or extensions of those lines) intersect.

### Extending lines in ray diagrams

Note that there are a number of cases where you have to extend the light rays in a ray diagram *backward*—such as in Rules 2 and 3 in the diverging lens example. These virtual light rays are usually drawn as *dashed lines*. If the image is formed at the intersection of these dashed lines, then it is a virtual image because no actual light is being focused at that location.

# Investigation 21C Image formation for a convex lens

**Essential questions**
*How does a convex lens form an image?*
*How do you measure the focal length of a lens?*

Interactive simulation

Light travels in straight lines unless its path is diverted by an optical device—such as a mirror or lens. A convex lens is used in cameras and refracting telescopes to redirect light rays in order to focus light and form an image. In this investigation, you will use an interactive simulation and an actual lens to explore the physics behind image formation by a convex lens.

## Part 1: Use ray tracing to locate images

1. Enter a focal length of 20 cm and an object distance of 60 cm.
2. Press [Draw rays].
3. Click and drag the mouse to draw each incident and refracted ray.
4. Press [Draw image]. Click and drag the mouse to draw the image at the location where the refracted rays intersect.
5. Record the image distance obtained from the simulation and print out a copy to compare your solution with the simulation's solution. Describe the image properties (upright or inverted, real or virtual, and magnified or reduced in size).
6. Repeat for object distances of 40, 20, and 8 cm. Tabulate your results.

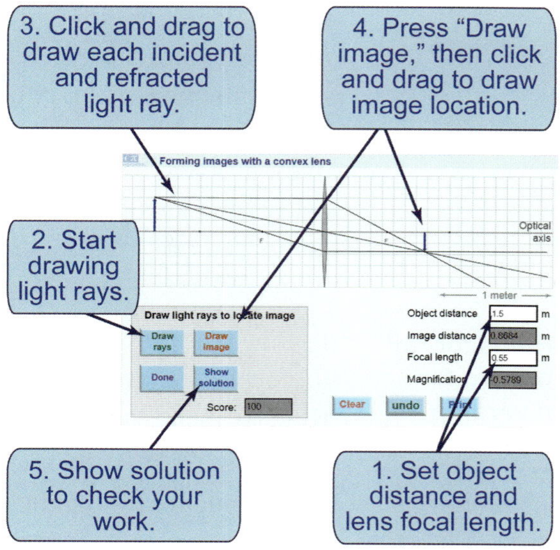


*a. What type of image is produced when the object distance is greater than the focal length? Equal to the focal length? Less than the focal length?*

## Part 2: Test the predicted image locations

1. Set up the optical bench as shown. Place the LED light source (the "object") at an object distance of 60 cm from a convex lens with a 20 cm focal length.
2. Slide the screen along the optical bench on the far side of the lens until the sharpest possible image is projected onto the screen.
3. Measure the image distance and compare it to your prediction.
4. Repeat for object distances of 40 and 20 cm.

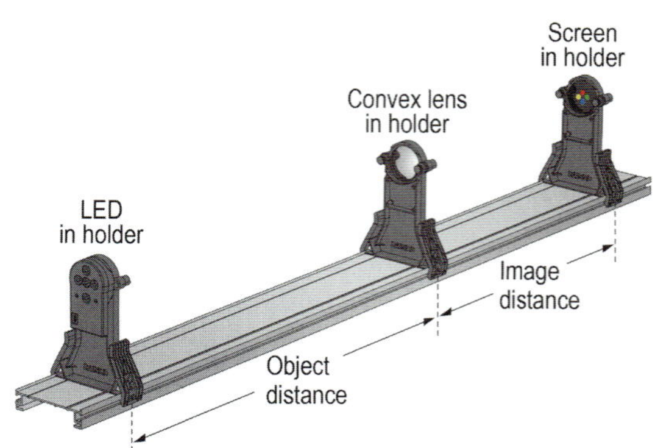


*a. Do your actual measured image distances match your predictions?*

*b. What will the image distance be if you place the LED light source far away (such as at a distance of several meters)? Check your prediction by making the measurement.*

*c. Why can't you use an object distance of 10 cm with this technique?*

**Locating the image produced by a convex lens**

Different kinds of lenses produce different image properties: real or virtual, upright or inverted, and magnified or reduced. Each configuration also produces an image at a particular location. How can you predict the location of an image? The relationship between the object distance $d_o$, the image distance $d_o$, and the focal length $f$ of the lens is called the *thin lens formula*. If you know two of these quantities, you can use the thin lens formula to calculate the third quantity. Equation (21.3) is also called the *Gaussian lens formula*.

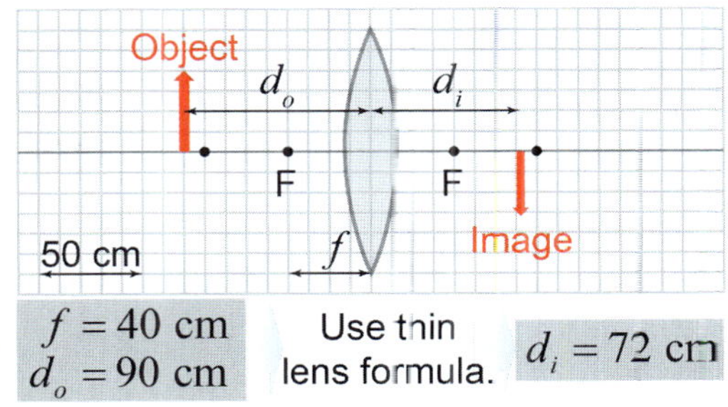


Interactive equation

**Thin lens formula**

**(21.3)** $$\frac{1}{d_o} + \frac{1}{d_i} = \frac{1}{f}$$

$d_o$ = object distance (m)
$d_i$ = image distance (m)
$f$ = focal length (m)

**Distant objects produce images near the focal point**

What happens to the location of the image when the object is moved further and further away from the lens? As you can see in the figure at right, when the object is moved to $d_o$ = 260 cm—which is 6.5 times larger than the focal length of the lens—the image is located slightly beyond the far focal length. As the object is placed further and further away from the lens—as the object distance "goes to infinity"—its images will be produced closer and closer to the focal point.

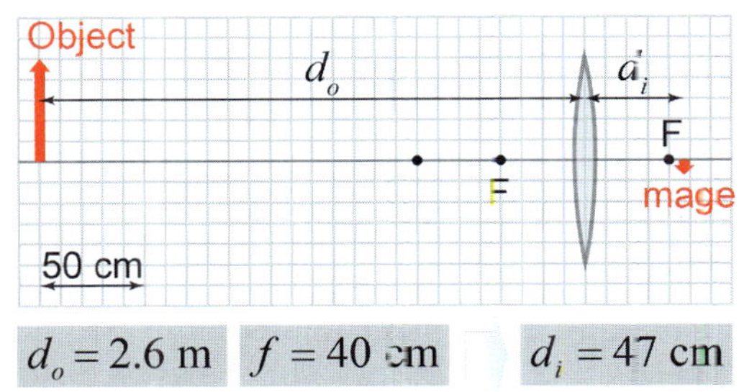


**Calculating the focal length of a lens**

A student has a convex lens but does not know its focal length. She sets up a light source 75 cm in front of the lens and then uses an index card to determine that it produces an image on the other side of the lens 37.5 cm away from the lens. What is the focal length of the lens?

*Asked:* focal length $f$ of the lens

*Given:* Object distance $d_o$ = 75 cm; image distance $d_i$ = 37.5 cm

*Relationships:*

$$\frac{1}{d_o} + \frac{1}{d_i} = \frac{1}{f}$$

*Solution:* Insert the values into the thin lens equation:

$$\frac{1}{f} = \frac{1}{d_o} + \frac{1}{d_i} = \frac{1}{75\text{ cm}} + \frac{1}{37.5\text{ cm}} = 0.0133\text{ cm}^{-1} + 0.0267\text{ cm}^{-1} = 0.04\text{ cm}^{-1}$$

Now take the inverse of both sides of equation to solve for focal length:

$$f = \frac{1}{0.04\text{ cm}^{-1}} = 25\text{ cm}$$

*Answer:* The focal length is 25 cm.

## Using the thin lens formula

**Real and virtual images in the equation**

When applying the thin lens formula (equation 21.3), a real image has a positive image distance, $d_i < 0$. A virtual image has a negative image distance, $d_i < 0$. A negative image distance means the image is on the same side of the lens as the object.

**Sign conventions for lenses**

| Quantity | Positive | Negative |
|---|---|---|
| Focal length $f$ | Converging lens | Diverging lens |
| Image distance $d_i$ | Real image | Virtual image |
| Magnification $m$ | Upright image | Inverted image |

**Magnification and the thin lens formula**

In the previous chapter, we learned that the magnification of an image is the ratio of the image height to the object height. An alternate equation for the magnification is the *negative* of the ratio of the image distance to the object distance.

Interactive equation

**Magnification** alternate definition

$$(21.4) \qquad m = -\frac{d_i}{d_o}$$

$m$ = magnification
$d_i$ = image distance (m)
$d_o$ = object distance (m)

**Inverted images have negative magnification**

Why is there a minus sign in equation (21.4)? Look at the ray diagram for a convex lens on page 1226. *The image is inverted*, which corresponds to the minus sign in equation (21.4). A negative magnification means that the image is inverted, whereas a positive magnification is an upright image.

**Virtual images and the thin lens formula**

A magnifying glass with a focal length of 20 cm is held 15 cm above the page of a book. (a) How far from the lens is the image located? (b) Is the image real or virtual? (c) What is the magnification of this image? (d) Is the image upright or inverted?

*Asked:* (a) image distance $d_i$; (b) whether image is real ($d_i > 0$) or virtual ($d_i < 0$); (c) magnification $m$ of the image; (d) whether the image is upright ($m > 0$) or inverted ($m < 0$)

*Given:* object distance $d_o = 15$ cm, focal length $f = 20$ cm

*Relationships:*

$$\frac{1}{d_i} + \frac{1}{d_o} = \frac{1}{f}, \quad m = -\frac{d_i}{d_o}$$

*Solution:* (a) Solve for $1/d_i$ in the thin lens formula and calculate:

$$\frac{1}{d_i} = \frac{1}{f} - \frac{1}{d_o} = \frac{1}{20\text{ cm}} - \frac{1}{15\text{ cm}} = (0.05 - 0.0667)\text{ cm}^{-1} = -0.0167\text{ cm}^{-1}$$

Invert both sides of the equation to get $d_i = 1/(-0.0167\text{ cm}^{-1}) = -60$ cm. (b) The image distance is *negative*, so this is a virtual image and it is therefore located on the *same side as the object*.

(c) Use the magnification equation:

$$m = -\frac{d_i}{d_o} = -\frac{-60\text{ cm}}{15\text{ cm}} = +4$$

(d) The magnification is *positive*, so the image is upright.

*Answer:* (a) $d_i = -60$ cm. (b) It is a virtual image. (c) Magnification $m = +4$. (d) It is an upright image.

# Section 2 review

Lenses use curved surfaces to focus light rays into an image. The focal length of a lens depends on the curvature of the surface of the lens and the index of refraction of the lens material. Converging lenses can create real or virtual images, depending on the object distance relative to the focal length of the lens. Real images occur when light rays physically meet at a point; virtual images occur when the light rays only appear to meet at a point. The thin lens formula is used to calculate object distance, image distance, or focal length when the other two quantities are known.

| | |
|---|---|
| **Vocabulary words** | **focal length, focal point, convex lens, concave lens, real image, virtual image** |

| | | |
|---|---|---|
| **Key equations** | $\frac{1}{d_o} + \frac{1}{d_i} = \frac{1}{f}$ | $m = -\frac{d_i}{d_o}$ |

## Review problems and questions

1. Does a flat mirror produce real or virtual images?

2. Does a concave lens produce real images, virtual images, or both?

3. A thin convex lens produces an image that is magnified and inverted. Where is the object located?

4. Cassandra has a lens with a focal length of 20 cm. Where should she place an object to get an image at a distance of 60 cm?

5. Damien places an object at a distance of 90 cm from a lens, and the image distance is 30 cm.
   a. What is the focal length of his lens?
   b. What is the magnification?

6. A one-dollar bill is 6.5 cm high by 15.5 cm wide. When the bill is placed 26 cm away from a thin convex lens, an inverted image of the dollar bill is projected onto the wall 1.0 m away from the bill on the far side of the lens.
   a. Is the image real or virtual? How do you know?
   b. What is the image distance $d_i$?
   c. What are the dimensions of the image?

7. A simple projector shines light through a slide to project an enlarged image of the slide on the wall. Slides with a width of 6 cm are placed 11 cm, behind a lens with a focal length of 10 cm.
   a. Where does the image appear? How far should the projector be from the wall?
   b. What is the magnification?
   c. How large is the image?
   d. What does the orientation of the image mean?

## 21.3 - Compound optics

How do binoculars, telescopes, microscopes, and cameras differ from a magnifying glass? A magnifying glass uses only one convex lens. Most optical instruments, however, use more than one optical element—lens, mirror, or disperser (e.g., prism)—and are called *compound optics*.

### Telescopes

**Refracting telescopes**

Galileo's refracting telescope of 1609 used a convex *objective lens* (the lens at the entrance of the telescope) and a concave *eyepiece lens* (the lens in front of the observer's eye). Within two years, the German mathematician and astronomer Johannes Kepler had invented a modified design that used two convex lenses.

**Keplerian refracting telescope**

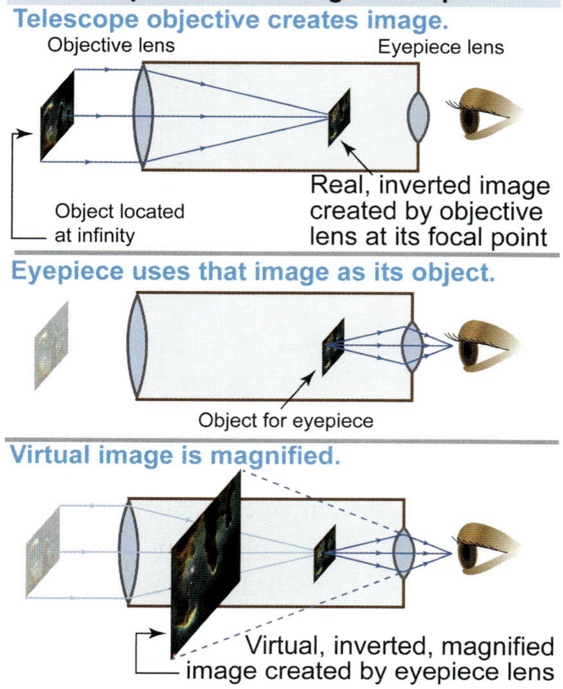


**How Kepler's telescope works**

In Kepler's telescope, the objective lens takes the light from a distant object and forms an image at the focal point of the lens. This is not a very useful image by itself—try using a magnifying lens to look at something on the far wall! Instead, this image is *used as the object* for the eyepiece lens, which produces a *magnified* image. This then is the useful feature of a telescope: to magnify the image of a distant object, whether it is a far-away mountain peak or a galaxy.

**Disadvantages of refracting telescopes**

The largest refracting telescope today is 1 m in diameter, which is the maximum lens size that can be supported around its edge. Refracting telescopes also suffer from *chromatic aberration*: Light of different colors focuses at different points because the refractive index of glass varies with wavelength (as you will learn in Chapter 22).

**Reflecting telescopes**

Isaac Newton built the first reflecting telescope to avoid the chromatic aberration problem in refractors. Modern telescopes also use curved mirrors instead of lenses because the large mirrors can be supported easily from the back. In a reflecting telescope, the light comes into the telescope and is redirected by a concave primary mirror. It then reflects off of a convex secondary mirror and is brought to a focus through a hole in the primary mirror.

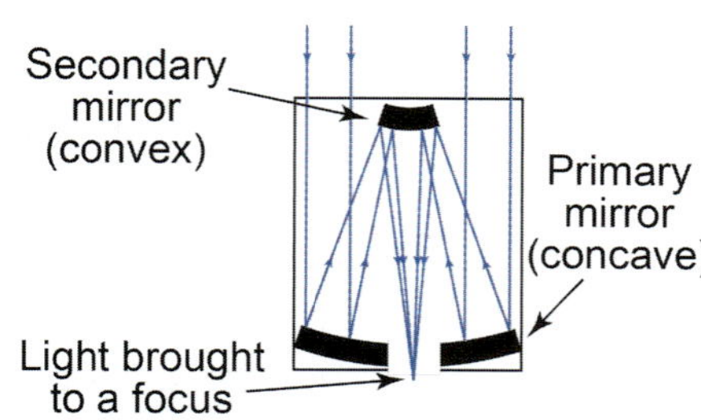

**Reflecting telescope**

**Other telescopes**

There are many other kinds of telescopes, such as the Very Large Array of radio telescopes in Socorro, New Mexico. Each of these radio telescopes reflects radio waves—light with very long wavelengths—off a primary and a secondary mirror onto the scientific instruments. A radio telescope is a *reflecting* telescope!

**Binoculars**

Galileo's contemporaries immediately began using his telescope for military purposes, e.g., in order to see enemy troop positions and ships. Today people will instead use binoculars for field work. If binoculars only used objective and eyepiece lenses, then they would produce inverted images, just as Galileo's telescope did. *Porro prism binoculars*, however, deliver an *upright* image by passing the light through internal reflections in two prisms. One prism reflects the light twice horizontally, while the other reflects the light twice vertically. Binoculars are also popular with amateur astronomers because they are easy to use and can image a wide field.

**How Porro prism binoculars work**

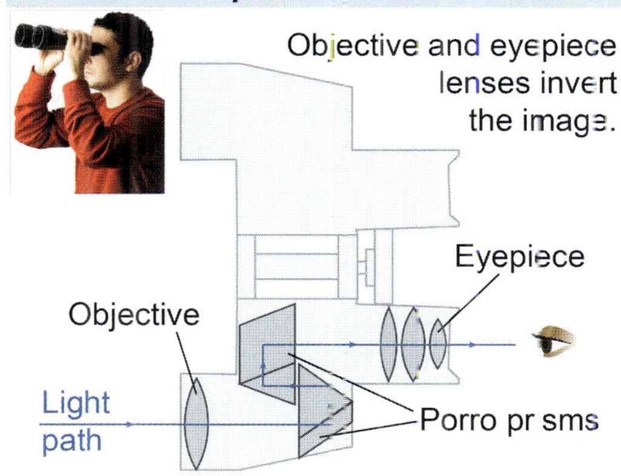


Internal reflections in the Porro prisms make the image upright.

**Compound microscope**

The compound microscope uses the same principles as the refracting telescope: An objective lens forms an image of the sample, and the eyepiece lens uses that image as its object to form a magnified image of the sample. A prism is often used to create two internal reflections to divert the light to a comfortable viewing angle. Most microscopes have a lamp under the sample to illuminate it. Typical microscopes can achieve magnifications of 50× or more.

**Compound microscope**

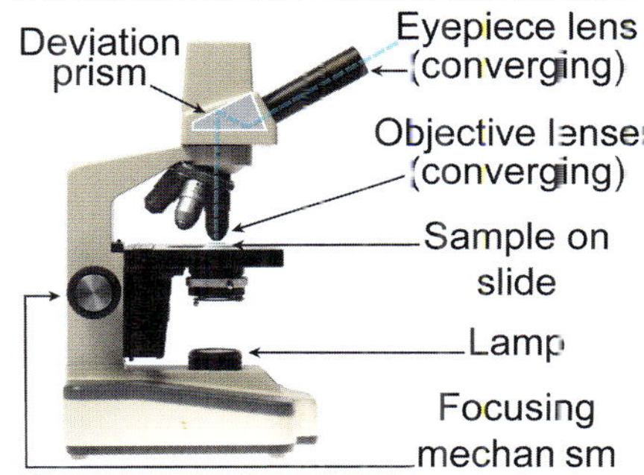


**SLR camera**

Both digital and film cameras use multiple lenses to focus the light into an image. The magnification of the camera can be changed by moving the position of the front mount lenses. Older cameras recorded images onto photographic film, but digital cameras use a CCD as a detector. Professional photographers mostly use the *single-lens reflex (SLR) camera*, which allows them to see the image directly through the lens—not through an electronic finder screen. The *reflex mirror* is mounted at a 45° angle to the incoming light and diverts the light to the eyepiece. The problem with using only a reflex mirror is that the camera's lenses invert the image. So the light passes through internal reflections in a *roof pentaprism*, which make the image upright again. When the photographer pushes the button to take a picture, the reflex mirror flips out of the way briefly to allow light to reach the detector.

**Single-lens reflex (SLR) camera**

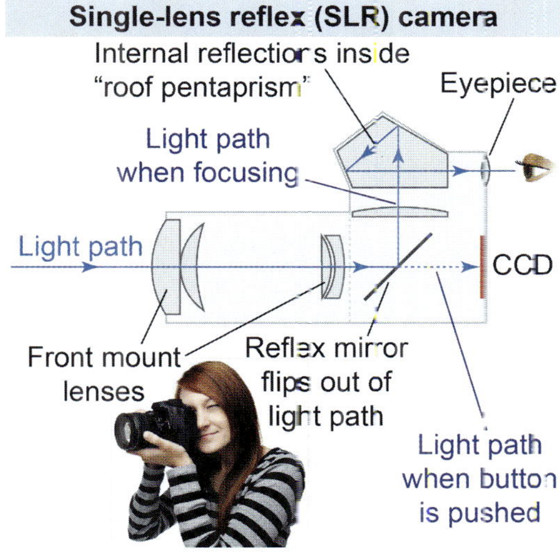

# Investigation 21D Build a microscope and a telescope

**Essential questions** *How do a microscope and telescope work?*

Interactive simulation

The basic microscope and the telescope are each a compound optical device that uses two lenses. The object is imaged by the first (or objective) lens; this image is used as the object by the second (or eyepiece) lens. By choosing particular types of objective and eyepiece lenses, and by separating the two lenses by a suitable distance, both the microscope and the telescope can produce magnified images.
*Caution: Never look directly at the Sun with either a microscope or a telescope!*

## Part 1: Build a simple microscope

1. Use three lenses (illustrated at right).
2. Using two lenses at a time, determine which two lenses create a *microscope* by generating a magnified image of a nearby object.
3. Construct a microscope on the optics track. Adjust the lens positions to improve the microscope magnification without significant distortion.

*a. What kinds of lenses are used to make a microscope?*
*b. Which lens (if any) has a shorter focal length, the objective or eyepiece?*

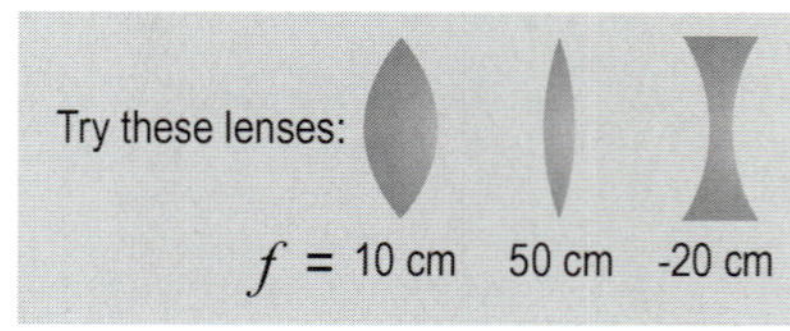


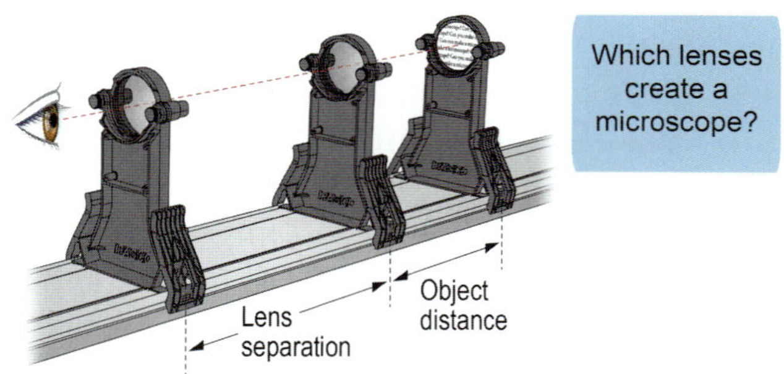


## Part 2: Model the simple microscope

1. Enter your microscope's parameters into the compound optics simulation.
2. Make small modifications to the positions of the lenses to see what range in positions produces a large magnification.

*a. Where is the object for the objective lens located relative to its focal point?*
*b. Where is the eyepiece's object located relative to its focal point?*
*c. How is the separation of the lenses related to the focal lengths of the two lenses?*

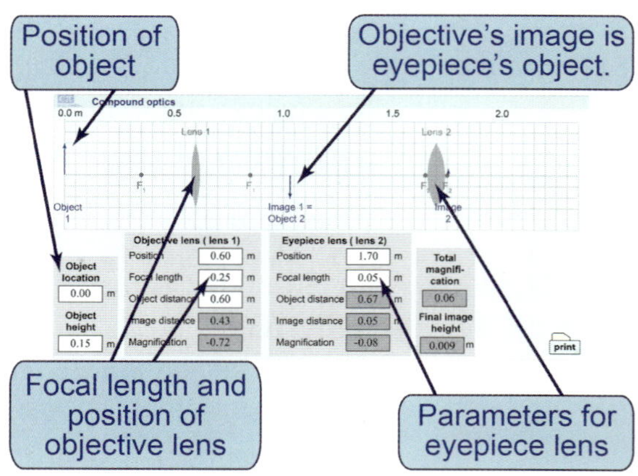


## Part 3: Build and model a simple telescope

1. Use the same lenses as in Part 1. Holding two lenses at a time—one at arms length and the other close to your eye—determine which lenses create a telescope that magnifies a distant object.
2. Construct your telescope on the optical track.
3. Devise a method to measure its magnification for a particular eye position.

*a. What kinds of lenses are used to make a Galilean telescope? A Keplerian telescope?*
*b. Which lens (if any) has a shorter focal length, the objective or the eyepiece?*

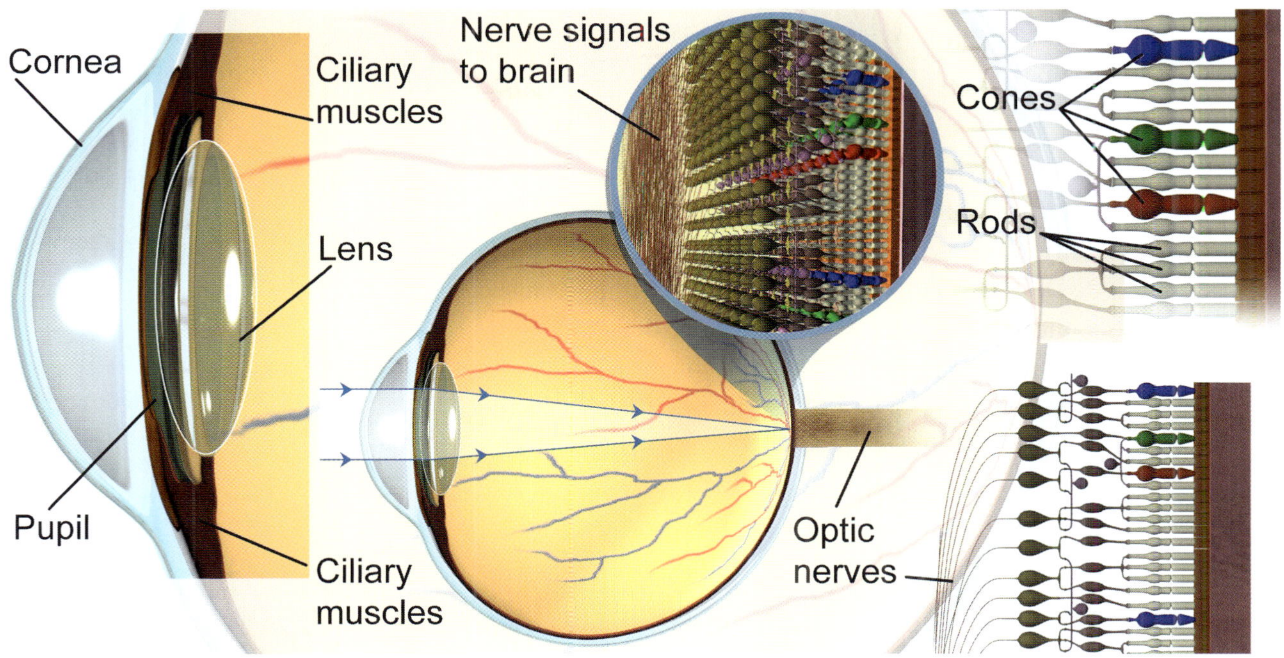


**Lens forms image on the retina**

You probably take your eyes for granted. But how do they work? Your eye is very similar to a magnifying glass: A converging lens at the entrance to the eye focuses light into an image on the back of the eye (called the retina).

**Iris dilates to let in more light**

How does your eye manage to see both in bright sunlight and dark nighttime? The pupil is an entrance hole that can vary its size to change the amount of light entering the eye. The iris *dilates*—opens the pupil wider—in dim conditions to let in more light. It contracts to block out some light in bright conditions.

**Lens can change its shape and focal length**

One amazing feature of the eye is that the lens can vary its shape! The purpose of the lens is to form an image on the retina. If you want to look at a nearby book—or distant trees—the ciliary muscles in your eye will change the shape of its lens to *change its focal length*. The eye thus maintains the image location on the retina. *Accommodation* is the eye's process of changing the focal length of its lens.

**How is light detected by the eye?**

At the retina, the eye has two kinds of *photoreceptors* called *rods* and *cones* that detect light. When light hits a photoreceptor cell, it initiates a chemical process that in turn creates an electrical signal that is sent to the brain. Most of the eye's photoreceptor cells are **rods**, which detect the *intensity* of light but not its color. The rods tell the brain how bright the light is at every point across your retina.

**Cones detect RGB colors**

But your eyes don't see in black and white—they see in color! There are three different kinds of photoreceptor cells called **cones**: red, green, and blue. The cones in your eye are similar to the RGB color emission process in a computer monitor! When red light strikes your retina, only the red cones are stimulated; with cyan light, the blue and green cones are stimulated. A typical human eye has approximately 125 million rod cells but only 6 million cones. Since there are relatively few cone cells, the eye cannot distinguish among colors at low light levels.

## Correcting eyesight

**Corrective lenses**

Do you wear glasses or contact lenses? If so, then it is likely because your eyes do not always properly *accommodate* or change their focal length to focus images onto your retinas. The purpose of corrective lenses is to bend the path of incoming light so that the eye better focuses the light onto the retina.

**Nearsighted vision and its correction**

Interactive simulation

A person whose eyes form an image in front of the retina has *nearsightedness* or myopia and can only focus well on objects that are located close to the eyeball. This condition can be caused either by the eyeball being longer than normal or by a cornea that has too much curvature. Nearsightedness can be corrected with *diverging* lenses. These eyeglasses spread apart light rays so that the eye will focus them further back and onto the retina.

**Nearsightedness and its correction**

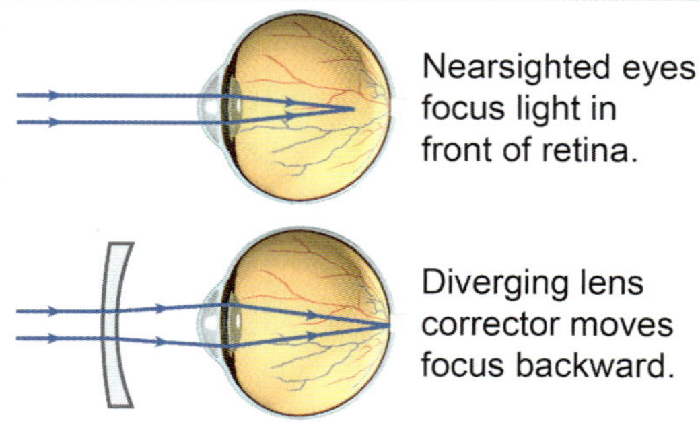


**Farsighted vision and its correction**

A person whose eyes form an image behind the retina has *farsightedness* or hyperopia and can only focus well on more distant objects. This condition can be caused either by the eyeball being shorter than normal or by a cornea that has too little curvature. Farsightedness can be corrected with *converging* lenses. These eyeglasses bring light rays together so that the eye will focus them further forward and onto the retina.

**Farsightedness and its correction**

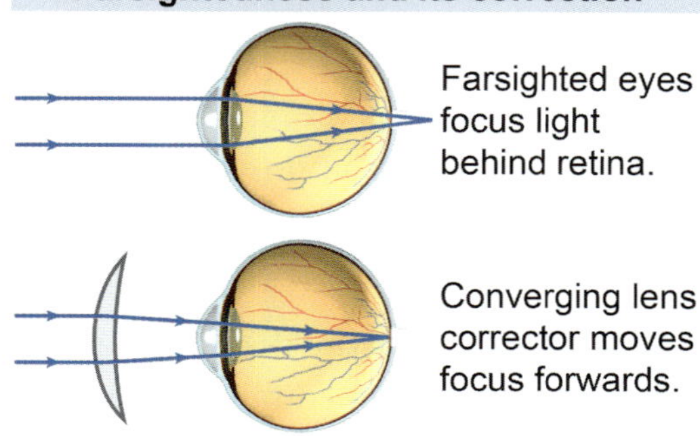


**Astigmatism**

In some people the cornea is not round, but is instead flattened more like a cylindrical lens. A person with *astigmatism* will see images that are extended more in one direction than another because the light is focusing at different distances from the retina. This causes lines in one direction, such as horizontally, to be in focus while vertical lines are blurred. Astigmatism can be corrected with a cylindrical lens that compensates for the shape of the cornea.

**Color blindness**

The eye has three types of cones that are sensitive to red, green, or blue light. Some people have an impairment or loss of function for one or more of these types of cones, with red or green cones being the most commonly deficient. This *color blindness* or color vision deficiency prevents affected people from distinguishing between different colors. A person with red–green color blindness cannot see the number in the illustration at right. No vision corrector is available, but there are applications for mobile devices that help affected people to distinguish color.

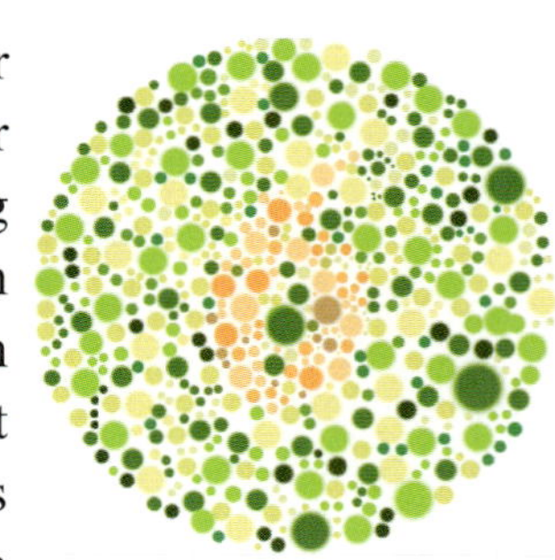

# Section 3 review

A single optical element might be useful for a bathroom mirror or magnifying glass, but most technologies that use optics include more than one optical device and are called compound optics. The refracting telescope and microscope are historically important optical instruments that produce inverted and magnified images. In their basic design, both use two lenses; the image formed by the first lens acts as the object for the second lens. Modern binoculars use multiple lenses, but they also take advantage of internal reflections within a pair of prisms so that the image is upright. The human eye uses a lens with variable shape (or focal length) to focus on objects either nearby or far away. Within the human eye, rods and cones detect the intensity and colors of light, respectively.

| Vocabulary words | rods, cones |
|---|---|

## Review problems and questions

1. Do the lenses in a camera produce a real or a virtual image?

2. Why is your vision sharpest when the pupils of your eyes are contracted or very narrow?

3. Identify the part of the human eye that causes each of the following vision problems and what is used to correct for it.
   a. farsightedness
   b. nearsightedness
   c. color blindness
   d. astigmatism

4. Imagine using the optical telescope you constructed in Investigation 21D on page 626 to create an image on a screen of the letter "F" printed on a piece of paper. Now cover up the top half of the entrance (or objective) lens for the telescope.
   a. Will you continue to see the entire letter "F" or only part of it?
   b. Will the image magnification be changed?
   c. Will the brightness of the image change?
   d. Explain your answers for parts a through c.

5. A microscope usually has a rotating set of objective lenses of different focal lengths. A student was using the objective lens with a 10 mm focal length and found that the images were magnified 40 times. What will happen to the magnification if the student switches to an objective lens with a shorter, 4 mm focal length?

# Chapter review

## Vocabulary

Match each word to the sentence where it best fits.

### Section 21.1

| | |
|---|---|
| **Snell's law of refraction** | **critical angle** |
| **index of refraction** | **refraction** |
| **total internal reflection** | |

1. _______ is the relationship among the indices of refraction, the angle of incidence, and the angle of refraction.
2. Light passing into a new material may exhibit _______ and change direction.
3. _______ occurs in glass at angles of incidence greater than 41°.
4. The _______ of air is 1.0003.
5. A high index of refraction results in a small _______.

### Section 21.2

| | |
|---|---|
| **concave lens** | **convex lens** |
| **focal length** | **focal point** |
| **real image** | **virtual image** |

6. A _______ produces a virtual image that is upright and reduced in size.
7. A _______ bends light rays towards the optical axis.
8. The distance from the surface of a lens to the focal point is the _______
9. A _______ can be projected on a screen.
10. When an image appears at a certain location, such as behind a plane mirror, but the light rays do not actually converge there, it is a _______.
11. Parallel light rays that are incident upon a converging lens will be focused at the _______.

### Section 21.3

| | |
|---|---|
| **cones** | **rods** |

12. The brain is told how bright light is by _______ in the eye.
13. The _______ in an eye perceive color.

## Conceptual questions

### Section 21.1

14. Explain why a straightedge (or two pins) are needed to trace a light ray in Investigation 21A on page 611.
15. A ray of light shines through an interface between glass and water and bends toward the normal. Is the ray passing from the glass into the water or the other way around?
16. For each of the following, indicate whether the phenomenon described involves refraction or reflection.
    a. You see yourself in a mirror.
    b. Objects look larger when viewed through a magnifying glass.
    c. Objects look broken when semi-submerged in water.
    d. A room looks unusual when viewed through a glass.
    e. You can see the room in which you are standing in somebody else's sunglasses.
    f. People who cannot see well can often see better with glasses.
    g. At night when you look out of your house through a window, you can still see the room you are standing in.
17. Amelie shines a light ray from air into a material with $n = 1$. If she changes the second material to have $n = 2$, what happens to the sine of the angle of refraction?
18. If light rays going from window glass to leaded glass are deflected toward the normal, in which direction are light rays deflected when passing from leaded glass to window glass?
19. A light ray shines from acrylic ($n = 1.49$) into ice ($n = 1.31$).
    a. Would the ray deflect toward the normal or away from it?
    b. If the acrylic was replaced with diamond ($n = 2.4$), would the ray deflect more or less?
20. Ahmad has materials *A*, *B*, *C*, and *D*, with indices of refraction 1, 2, 3, and 4, respectively. He shines a light ray from *A* into *B* with angle of incidence 40° and angle of refraction 18.7°. What other pair of materials could he use to get the exact same angles of incidence and refraction?
21. Aisha has materials *A*, *B*, and *C*. Material *A* and *B* have indices of refraction of 1 and 3, respectively. The critical angle for a ray traveling from *B* to *A* is the same as the critical angle for a ray traveling from *C* to *B*. What is the index of refraction of material *C*?
22. What is the difference between a reflected ray and a refracted ray? How does each relate to the incident ray?

# Chapter review

## Section 21.1

23. What piece of equipment would you use to measure an angle: a compass, a magnetic compass, a metric rule, a protractor, or a refractor?

24. ⟪ Light refracts from one material into another with some angle of incidence and angle of refraction. If you change the angle of incidence so that the sine of the angle doubles, what happens to the sine of the angle of refraction?

25. ⟪ A ray of light enters a glass window at a certain angle of incidence and passes through. How will the angle of refraction compare when that ray of light exits the other side of the window?

26. ⟪ What is the range of possible values of the critical angle when light is passing into a material with a higher index of refraction (for example, from air to water)?

27. ⟪ Typically, a mirror is a reflective surface covered in glass. Does the refraction of light through glass affect the angle at which light reflects? Explain.

28. ⟪⟨ Ilana shines a ray of light through water into two mystery materials, *A* and *B*, with an angle of incidence of 30°. The angle of refraction for material *A* is 15°. The angle of refraction for material *B* is 20°.

    a. Does the ray bend more or less in material *B* than in material *A*?
    b. Which material has the higher index of refraction?

29. ⟪⟨ Suppose a light ray shines through water, refracts into a thin straight layer of glass, and then refracts from the glass into air. Explain why the angle of refraction is the same into the air even if you remove the glass. (*Hint*: Write down Snell's law for each time the ray refracts.)

## Section 21.2

30. Describe the procedure for ray tracing to locate the image formed by a convex lens.

31. Kaleb has a lens, an object, and a screen. No matter where he puts the object, he can't project it onto the screen. What kind of lens does he have?

32. Claudius is looking through a lens in a door that allows him to see a very large field of view. Is his lens a diverging or converging lens?

33. Name a common use for a convex lens.

34. An object is placed in front of a thin, convex lens and its image has a negative magnification. What does this imply about the image?

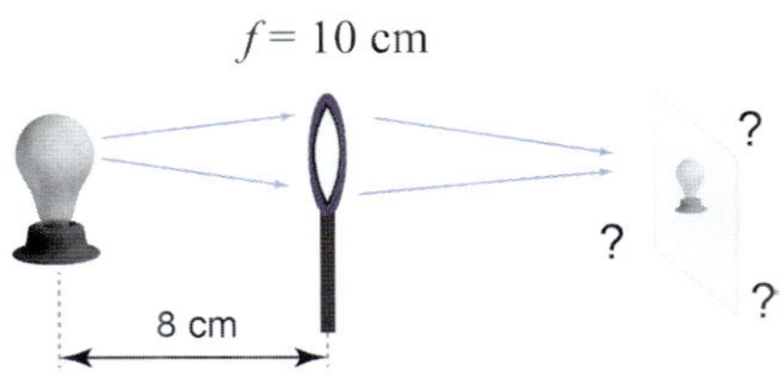


35. A student has a magnifying lens with a focal length of 10 cm and holds it 8 cm from a light source. Will she be able to project an image of the light source onto a piece of paper?

36. ⟨ What material discussed in this chapter would make the most powerful lens, i.e., bend light the most? Why?

37. ⟨ Linda wants to project an image onto a screen using a lens. Should she use a concave or convex lens?

38. ⟪ Where should an object be placed in relation to a convex lens to produce a virtual image?

39. ⟪ An object is placed several focal lengths away from a convex lens. Describe the resulting image.

40. ⟪⟨ Use a piece of graph paper to draw the following and find the location and size of the image *graphically*. Draw a bi-convex lens and an optical axis passing through it. The convex lens has a focal length of 20 cm, so set a scale on your graph and draw both focal points. Place an object at an object distance of 30 cm from the lens and with a height of 5 cm. Draw light rays passing through the convex lens to locate the image.

    a. From your drawing, what is the image distance?
    b. From your drawing, what is the height of the image?
    c. Is the image inverted or upright?
    d. Is the image real or virtual?
    e. From your drawing, what is the magnification of the image?

## Section 21.3

41. Which of the following compound optical devices usually has a prism inside?

    a. a refracting telescope
    b. a reflecting telescope
    c. Porro binoculars
    d. a compound microscope
    e. an SLR camera

42. How do binoculars deliver an upright image of the object although refracting telescopes produce an inverted image?

# Chapter review

### Section 21.3

43. What combination of lenses would you use to create a telescope similar to Kepler's?

44. What combination of lenses would you use to recreate Galileo's telescope?

## Quantitative problems

### Section 21.1

45. Light enters a rhinestone at an angle of incidence of 40.0° and refracts to an angle of 23.4°. What is the index of refraction of the rhinestone?

46. Light from air enters a diamond ($n = 2.42$) on a ring at an angle of 45°. What is the angle of refraction of the light when inside the diamond?

47. Light in a window ($n = 1.52$) hits the boundary with air at an angle of 55°. Will this light reflect off the boundary?

48. Imagine that at the end of Earth's atmosphere there is a clear boundary between air ($n = 1.0003$) and the vacuum of space ($n = 1$). (This is not actually the case.)

    a. If sunlight enters the atmosphere from space at an angle of incidence of 30°, what is the angle of refraction?
    b. What is the critical angle for light leaving Earth's atmosphere?

49. A cube of ice ($n = 1.31$) sits in an unknown liquid. Light in the liquid hits the ice at an angle of 35.0° and refracts to an angle of 41.1°. What is the index of refraction of this liquid?

50. A ray of sunlight in the air ($n = 1.00$) hits a calm pool of water ($n = 1.33$) at an angle of 40.0° and is refracted as it passes the boundary.

    a. What is the angle of refraction after the ray has entered the water?
    b. Later, it is cold, and a layer of ice ($n = 1.31$) has formed on top of the water. If the ray enters the ice at the same angle, travels through the ice, and enters the water, what is its angle of refraction in the water?
    c. Create an algebraic relationship for a light ray that passes through three materials ($n_1$, $n_2$, $n_3$) successively that relates the initial angle ($\theta_1$) to the final angle ($\theta_3$).

### Section 21.2

51. Melinda places an object 50 cm in front of a thin convex lens with a focal length of 30 cm. What is the magnification of the image?

52. Shireen has a magnifying glass and wants to use it to focus the Sun's light onto a sheet of paper. When she holds the magnifying glass 6 cm away from an object, the object is magnified by 4. How close should she hold the paper to the magnifying glass to catch the focused sunlight?

53. A 3-cm-wide stamp is held 4 cm in front of a magnifying glass that has a focal length of 12 cm.

    a. Where does the image appear?
    b. How large is the image of the stamp?

54. An object is placed 40 cm away from a lens and an image is projected 120 cm away on the other side. What is the focal length of this lens?

55. Saul places an object 40 cm in front of a thin convex lens. The resulting image has a magnification of 2.5. What is the focal length of the lens?

56. Maria places an object 30 cm in front of a thin convex lens with a focal length of 50 cm.

    a. What is the magnification of the image?
    b. Is the image inverted or upright?
    c. Is the image real or virtual?

57. Vladamir wants to project a piece of artwork that is 0.5 m tall onto a screen that is 3.0 m tall. He has a convex lens with a focal length of 2.0 m.

    a. At what distance from the lens should he place his artwork to completely fill the screen?
    b. How far should the lens be from the screen?
    c. What orientation should his artwork have to produce an upright image?

### Section 21.3

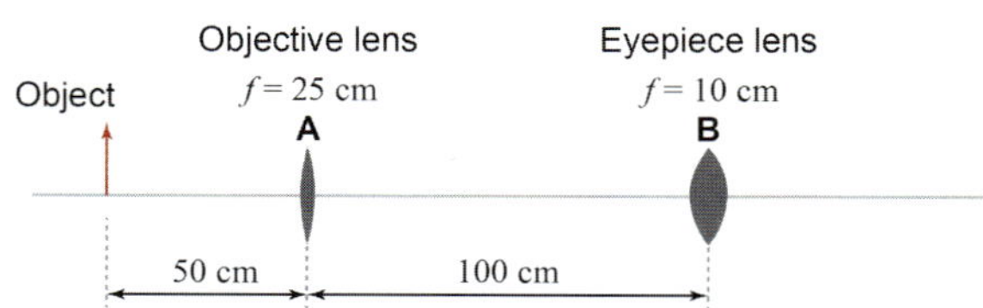


58. The two lenses shown are separated by 100 cm to form a compound optical device for an object placed 50 cm in front of the objective.

    a. What is the image distance for the image formed by lens A?
    b. What is the magnification for lens A?
    c. Is this image real or virtual?
    d. Is this image upright or inverted?
    e. What is the object distance for lens B?
    f. What is the image distance for the image created by lens B?
    g. Is this image upright or inverted?
    h. Is this image real or virtual?
    i. Are these lenses properly set up to act as a telescope?

# Chapter review

## Standardized test practice

59 Which of the following is an example of refraction?

A. light bouncing off a mirror
B. light passing through a small hole
C. light passing through a window pane
D. light heating up a lizard

60. A material has an index of refraction of 1.82. What is the critical angle of this material when it is surrounded by air?

A. 1.82°
B. 24.4°
C. 33.3°
D. 45°

61. A ray of light in air ($n = 1.00$) enters water ($n = 1.33$) at an angle of 50°. What is the angle of refraction?

A. 29°
B. 35°
C. 45°
D. 50°

62. A light ray shines from a diamond ($n = 2.42$) into water ($n = 1.33$), and no refraction occurs. What is the angle of incidence?

A. 0°
B. 30°
C. 45°
D. 60°

63. A physicist has three mystery materials, *A*, *B*, and *C*. He shines light from one material into another and records which direction the light bends, obtaining the following results:

- When shining light from *A* into *C*, the ray bends away from the normal.
- When shining light from *B* into *C*, the ray bends toward the normal.

Which of the following correctly orders the three materials from least index of refraction to greatest?

A. *A, B, C*
B. *A, C, B*
C. *B, C, A*
D. *C, A, B*

64. Yvette shines a light through diamond ($n = 2.42$) so that it strikes an unknown material at an angle of incidence of 39°. All she can observe is whether total internal reflection occurs. Which of the following two materials can she differentiate?

A. air ($n = 1$) and acrylic ($n = 1.49$)
B. acrylic ($n = 1.49$) and glass ($n = 1.5$)
C. glass ($n = 1.5$) and amber ($n = 1.55$)
D. amber ($n = 1.55$) and diamond ($n = 2.42$)

65. A light ray refracts from glass ($n = 1.5$) into two different materials with an angle of incidence of 45°. The first material has $n = 1.33$ and an angle of refraction of 52.9°. The second material has an angle of refraction of 41.5°. What is its index of refraction?

A. 1.6
B. 1.4
C. 1.2
D. 1.0

66. A light ray refracts from glass ($n = 1.50$) into two different materials with an angle of incidence of 45°. The first material has $n = 1.33$ and an angle of refraction of 52.9°. The second material has an angle of refraction of 49.3°. What is its index of refraction?

A. 1.6
B. 1.4
C. 1.2
D. 1.0

67. Which of the following statements about a light ray striking an interface from air to water is the most accurate?

A. All of the light ray always refracts.
B. All of the light ray sometimes refracts.
C. Some of the light ray always refracts.
D. Some of the light ray sometimes refracts

68. Light traveling through an unknown material strikes amber ($n = 1.55$) at an angle of incidence of 58° and undergoes total internal reflection. Which of the following could be that material?

A. ice ($n = 1.31$)
B. acrylic ($n = 1.49$)
C. leaded glass ($n = 1.7$)
D. cubic zirconia ($n = 2.15$)

69. An object is placed 50 cm away from a lens with a focal length of 30 cm. What are the properties of the resulting image?

A. real, inverted, and magnified
B. real, upright, and magnified
C. virtual, upright, and reduced
D. virtual, inverted, and magnified

70. Ami wants to project a real, magnified image of an object onto a screen using a convex lens with a focal length of 20 cm. How far should she place the object from the lens?

A. 5 cm
B. 25 cm
C. 45 cm
D. 65 cm

# Chapter 22 Electromagnetic Radiation

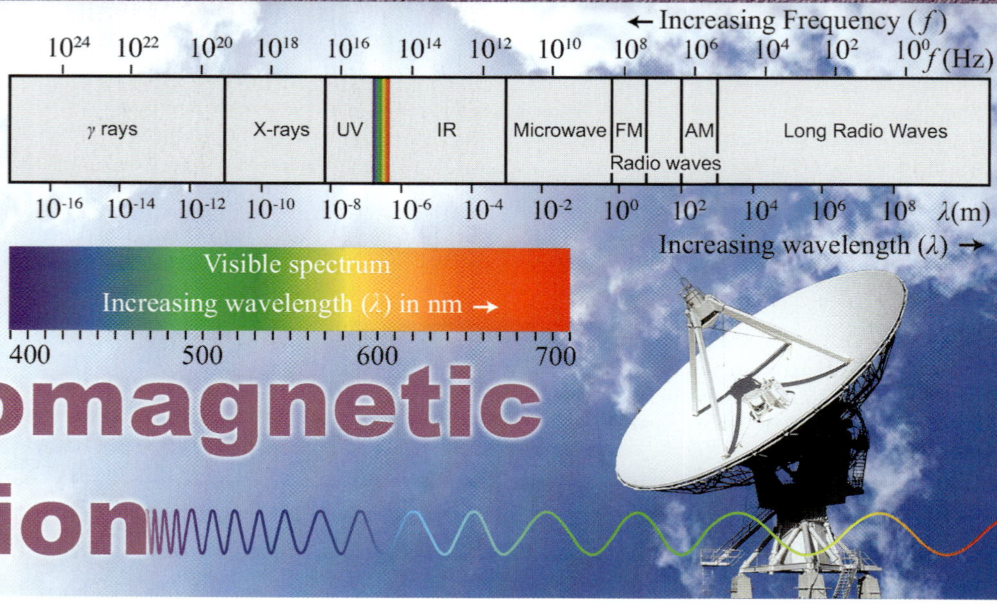


Few technologies have woven themselves so deeply into our daily routine as the mobile phone. In 1975 a mobile phone was a 2 lb (1 kg) "brick." Today, cellphones are smaller and lighter than most wallets and more powerful than a room-sized mainframe computer from 1975. Smartphones enable billions of people to communicate and also to access the Internet, watch movies, take pictures, play music, study, and even navigate! Cellphones receive and transmit data using electromagnetic waves, similar to the waves that heat food in a microwave oven.

Electromagnetic waves are a form of *radiation*, which in physics simply means the movement of energy through space. The most familiar electromagnetic (EM) radiation is *visible light.* Visible light includes a range of energy that is too low to break chemical bonds but high enough to stimulate rhodopsin molecules in your retina to send signals to your brain.

Electromagnetic radiation is pure energy in the form of oscillations of the electromagnetic field traveling through space at the enormous velocity of 300,000 km/s. The electromagnetic spectrum comprises waves with energies both higher and lower than visible light. Radio waves, microwaves, and infrared radiation have lower energy than visible light, while x-rays and gamma rays have higher energy.

Have you ever noticed that cellphone reception gets noticeably worse when it rains or is foggy? That is because different substances are transparent to different energies of EM radiation. For example, x-rays pass readily through animal tissue, but they are absorbed by bone. The microwaves used by cellphone transmissions easily pass through *dry* air but, like the microwaves in an oven, they are absorbed by *water.* A small but significant percentage of the microwave signal from a cellphone is absorbed or scattered by atmospheric water vapor, and that explains the poor reception.

Although rain hinders communications, the variations in microwave signal strength can be useful in constructing detailed, real-time weather maps. One research team recently used data from 2,500 microwave ground stations in the Netherlands to track individual rainstorms as they moved across the country. These maps made from cellphone signal strength can be constructed more frequently and with better resolution than many traditional weather maps.

# Chapter study guide

## Chapter summary

Light is all around us, and not just the kind of light we can see with our eyes. Visible light is one small part of the entire electromagnetic spectrum, which ranges from gamma rays and x-rays to infrared radiation and radio waves. Electromagnetic waves are oscillations in electric and magnetic fields, can travel through a vacuum, and propagate at the speed of light (300,000 km/s). But although light often behaves as a *wave*, one of the amazing insights from a century ago is that sometimes light will instead behave as a *particle*. This dual nature of light is one of the first indications of the richness and strangeness of *quantum* phenomena at the microscopic size scales of the atom.

## Learning objectives

By the end of this chapter you should be able to

- describe the connection between electromagnetic fields and light;
- calculate wave and refraction properties related to the propagation of light;
- describe characteristics and behaviors of electromagnetic waves;
- describe the electromagnetic spectrum and provide examples of phenomena and technologies in the x-ray, ultraviolet, visible, infrared, and radio regions;
- describe and explain the photoelectric effect; and
- list evidence for both the wave and particle properties of light.

## Investigations

22A: Color, frequency, and wavelength of light
22B: Detect infrared light with a prism
22C: Photoelectric effect

## Chapter index



## Important relationships

$c = f\lambda$ $\quad$ $n = \dfrac{c}{v}$ $\quad$ $E = hf$ $\quad$ $E_k = hf - W_0$

## Vocabulary

light, vacuum, index of refraction, spectrum, ultraviolet light, radio waves, spectrograph, quantum physics, work function, optical fibers,
electromagnetic wave, scattering, electromagnetic spectrum, gamma rays, infrared radiation, polarization, photon, photoelectric effect, detector,
speed of light, dispersion, radiation, x-rays, microwaves, diffraction pattern, Planck's constant, threshold frequency, pixel

## 22.1 - Light and electromagnetism

When you look at sunlight streaming in through the window you probably don't associate the light with mental images of electricity or magnets. You probably don't think of *atoms* either. The nature of light was a mystery for thousands of years until physicists of the late 19th century discovered the connection among light, electricity, and magnetism. In the early 20th century physicists went further, connecting the emission and absorption lines of light to the properties of the atom (Chapter 26).

### Light and the oscillating electric field

**What is light?**

In 1865, physicist James Clerk Maxwell developed a theory of electricity and magnetism that predicted the existence of traveling waves. He calculated the speed of these waves and found that it matched the speed of light. Based on this evidence, Maxwell made a strong argument that light was a form of electricity and magnetism. Just 24 years later, in 1889, Heinrich Hertz proved experimentally the existence of Maxwell's electromagnetic waves and demonstrated conclusively that these waves had all the properties of light.

**How does the electric force get from one charge to another?**

Consider a stationary positive charge. The charge creates an electric field satisfying Coulomb's law that reaches out in every direction and affects other charges. A negative charge nearby feels the electric force of attraction from the positive charge. Maxwell asked the following question: *How fast does the force get from one charge to the other?*

**Oscillating electric and magnetic field creates an electromagnetic wave (light)**

| Electric field | Oscillating electric field | Oscillating electric field and magnetic field |
|---|---|---|

| A charge creates an electric field that expands outward and exerts forces on other charges. | An oscillating charge creates a wiggle in the electric field that expands outward and exerts oscillating forces on other charges. | Light is a traveling, oscillating electromagnetic wave. The electric and magnetic fields are perpendicular. |
|---|---|---|

**What happens when charges move?**

In Chapter 18 we considered the electric field of static charges. Now, consider a positive charge that oscillates up and down. The charge's motion creates a corresponding wiggle, or *wave,* in the electric field! That wave travels at the speed of light and causes distant charges to respond. *An oscillating charge creates a traveling oscillation in the electromagnetic field that travels outward at the speed of light.*

**Light is an electromagnetic wave**

**Light** is a traveling oscillation of the electromagnetic field. In the previous chapter we learned that a *moving electric charge* (current) creates a magnetic field. Therefore, the oscillating positive electric charge in the above example creates an oscillating *magnetic field* as well as an electric field. Once started, the electric field generates a magnetic field as it changes. The magnetic field in turn generates an electric field as *it* changes. The electric and magnetic fields propagate with each other in an **electromagnetic wave**.

**Light can propagate in a vacuum**

Light waves can travel through space as well as many gases, liquids, and solids. The ability to propagate in a vacuum (space) makes light waves different from sound waves, which require matter (such as air) to propagate. Another big difference is speed. The speed of sound can be estimated by timing an echo with a stopwatch. Light is about a million times faster than sound and it takes sensitive experiments to measure the speed of light.

**Light waves can travel through a medium or a vacuum.**

**Comparing light and sound waves**

If you compare light and sound, you notice that light travels far faster than sound. The wavelength of visible light is the size of atoms whereas the wavelength of sound is the size of macroscopic objects. The frequency of light is trillions of hertz compared to audible sound, which ranges from 20 to 20,000 Hz.

| | Visible light | Audible sound |
|---|---|---|
| Speed | $3\times10^{8}$ m/s | 343 m/s |
| Typical wavelength | $6\times10^{-7}$ m | 0.78 m |
| Typical frequency | $5\times10^{14}$ Hz | 440 Hz |

**Speed of light**

The frequency, wavelength, and speed of light are related by equation (22.1), which is similar to the relationship for other waves. The **speed of light** (or $c$) is the product of frequency and wavelength.

Interactive equation

**Speed of light**

$$(22.1)\quad c = f\lambda$$

$c$ = speed of light (m/s) = $3.00\times10^{8}$ m/s in a vacuum
$f$ = frequency (Hz)
$\lambda$ = wavelength (m)

**Speed of light through different media**

The speed of light is usually assumed to be *in a vacuum*, e.g., in outer space where there is no air or other gas. Just as the speed of sound changes depending on the material it is traveling through (such as air or water in the ocean), the speed of light differs when it travels through different materials. Light travels slightly slower when it passes through air and significantly slower in water. But the *frequency* of the light stays the same, even when it travels from one material to another.

**Frequency of visible light**

Calculate the frequency of visible light (in a vacuum) if the wavelength is 500 nm.

*Asked:* frequency $f$ of the light

*Given:* wavelength of the light, $\lambda$ = 500 nm; speed of light, $c = 3\times10^{8}$ m/s

*Relationships:* $c = f\lambda$

*Solution:* First convert from nanometers to meters

$$500 \text{ nanometers} = 5\times10^{-7}\text{ m}$$

Then, solve for frequency and substitute values.

$$f = \frac{c}{\lambda} = \frac{3\times10^{8}\text{ m/s}}{5\times10^{-7}\text{ m}} = 6\times10^{14}\text{Hz}$$

# Investigation 22A Color, frequency, and wavelength of light

**Essential questions**
*How is color related to the frequency and wavelength of light?*
*How are changes in frequency related to changes in wavelength?*

Interactive simulation

What physical property or properties of light make red light different from blue or green light? In this short interactive simulation, you will compare a visible spectrum of light—ranging from violet to red light—with the frequencies and wavelengths of the light.

## Part 1: Relating the color of light to its frequency and wavelength

In the interactive simulation, there is a horizontal spectrum of light ranging from violet to red. You can change the light in any of three different ways:

1. by moving the slider from left to right;
2. by entering a value in the box for the frequency (e.g., using "6.0 e14" to represent $6.0 \times 10^{14}$ Hz); or
3. by entering a value in the box for the wavelength (in nanometers).

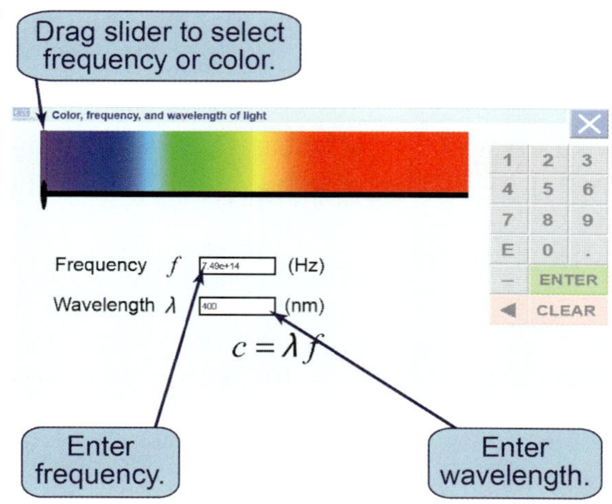


a. *What is the wavelength of blue light? Cyan? Green? Record the wavelengths using scientific notation and correct SI units.*

b. *What color of light corresponds to a wavelength of $\lambda = 580$ nm?*

c. *What color of light corresponds to a frequency of $f = 6.9 \times 10^{14}$ Hz?*

d. *What color(s) of the visible spectrum have the highest frequencies? Longest wavelengths?*

e. *Put the following colors in order of increasing frequency: blue, green, indigo, orange, red, violet, and yellow. Can you think of a mnemonic to remember this order of the colors?*

f. *How are frequency and wavelength related to each other? In other words, if you change the light to have a longer wavelength, how does the frequency of the light change?*

## Part 2: Compare to RGB color combinations

1. On page 24, you looked for the best RGB match for each of the colors on the right.
2. Now using the interactive simulation of the visible light spectrum, look for the best match in the visible light spectrum for each of the colors on the right.
3. Tabulate the wavelengths you found and describe whether you consider the color to be a good match.

a. *For which colors could you make a good match with the visible light spectrum?*

b. *For which colors could you find no suitable match? Explain why this is the case.*

**What is the wavelength for the best match to each of these colors?**

**Which ones cannot be matched? Why?**

**Scattering of light by particles**

If air is clear, why is the sky blue? If you look up on a clear day, and don't look directly at the Sun, you see a blue sky instead of the black of space. The explanation of the blue sky involves the effects of reflection, refraction, and dispersion together. Every cubic centimeter of air contains small particles of dust and water droplets. When light strikes these particles some fraction of the light is absorbed and subsequently re-emitted—usually in a different direction. **Scattering** is the process of small particles absorbing and re-emitting light.

**Scattering of light by particles**

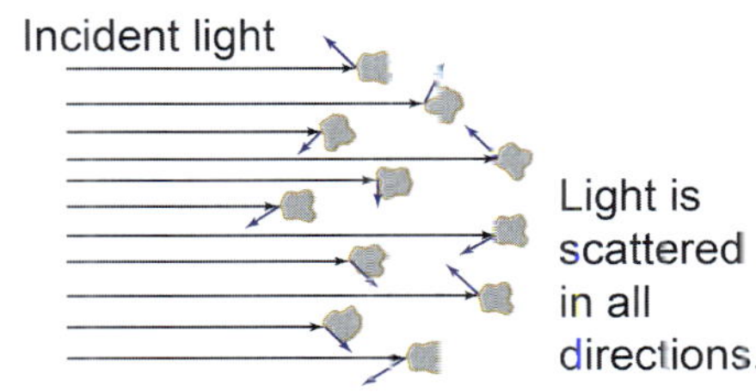


**Scattering depends on wavelength of light**

Scattering depends on the sizes of the particles and the wavelength of the incident light. Just as long-wavelength water waves can easily pass around small stones, long-wavelength light waves can pass by small particles in the atmosphere. Short wavelength-light—blue or violet in color—can be scattered by both small and large particles in the atmosphere.

**Why is the night sky black?**

On page 23 we learned that colors of light on a digital device can be represented as combinations of the additive primary colors: red, green, and blue. When all three colors appear in equal amounts, the result is white light; the absence of all three colors is black. At night there is no sunlight, so the night sky appears black.

**Why is the daytime sky blue?**

When you look away from the Sun in the *daytime* sky, you are seeing sunlight *scattered* by the particles of the atmosphere. The small particles in the atmosphere scatter much more blue light than they scatter red light, because blue light has shorter wavelengths that are similar in size to the atmospheric particles. The sky appears blue because blue light is preferentially scattered into your eyes by atmospheric particulates. Red light is not scattered as much and mostly travels straight from the Sun to the ground.

**Why the daytime sky appears blue**

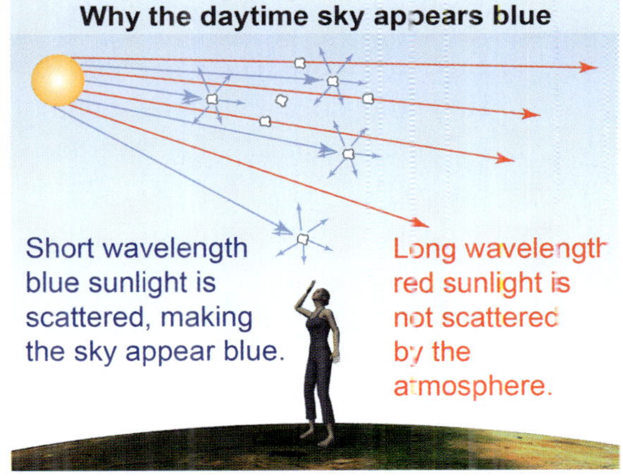


**Why are sunsets red?**

At sunset, the Sun is very near the horizon. This means that the path of the Sun's light must pass through much more air to reach our eyes near sunset than it does near noontime. All that extra air results in more and more scattering of the Sun's blue wavelengths —until there is little or no blue light left. Sunsets appear reddish-orange because all the blue light from the Sun has been scattered by the time the light reaches your eyes because of its long path through air.

**Why sunsets appear red**

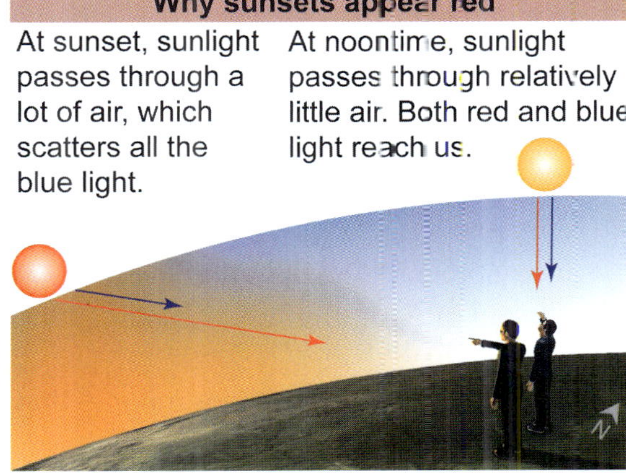

# Section 1 review

Light is a traveling transverse wave in an oscillating electric and magnetic field. The speed of light in a vacuum is a constant at $3\times10^8$ m/s. The colors of visible light are directly related to the wavelength of the light. Light can be scattered by small particles, but scattering of light waves depends on the relative sizes of the particles and the wavelength of the light. This is similar to the idea that long-wavelength water waves can easily pass around small boulders but short-wavelength water waves cannot. Scattering of light is the explanation for why the daytime sky is blue and sunsets are red.

| Vocabulary words | light, electromagnetic wave, speed of light, vacuum, scattering |
| --- | --- |

| Key equations | $c = f\lambda$ |
| --- | --- |

## Review problems and questions

1. Use a diagram and describe with words the connection between the field of an electric charge and light waves.

2. Lightning strikes a tree one mile (1.609 km) away from you. Light from the lightning bolt travels at $3\times10^8$ m/s while sound waves from the accompanying thunderclap travel at about 343 m/s.
   a. How long does it take the light to reach you?
   b. How long does it take the sound of the thunderclap to reach you?
   c. Formulate a simple rule for estimating the distance to a lightning strike.

3. How are electromagnetic waves similar to sound waves? How are they different?

4. How are the electric and magnetic fields in an electromagnetic wave related to each other?

5. Which has higher frequency, cyan light or yellow light?

6. a. What is the frequency of light with a wavelength of 560 nm?
   b. An angstrom (Å) is $10^{-10}$ m. What wavelength, in nanometers, corresponds to 6,563 Å?
   c. What is the wavelength of light corresponding to 121.5 MHz?
   d. Which has a higher frequency, light with a wavelength of 1.7 μm or light with a frequency of $7.3\times10^{13}$ Hz?

7. Imagine looking up at the night sky. If you look in any given direction at a tiny patch of sky, you will eventually come across a star or galaxy. That little patch of sky would then appear white. The German astronomer Heinrich Wilhelm Olbers in 1823 described this as a paradox: Why is the night sky black instead of white? Offer an explanation to resolve this paradox.

8. If we were on a planet—or moon—that had no atmosphere, what color would the daytime sky be?

What do a medical PET scanner, a mobile phone, an ear thermometer, and a microwave oven have in common? The answer is that all three technologies use regions of the *electromagnetic spectrum*. The PET scanner uses high energy gamma rays, the ear thermometer uses infrared or thermal radiation, and the oven and mobile phone both use microwaves. Gamma rays, infrared radiation, and microwaves are types of "light," differing from the light we see (and from each other) in frequency, wavelength, and energy.

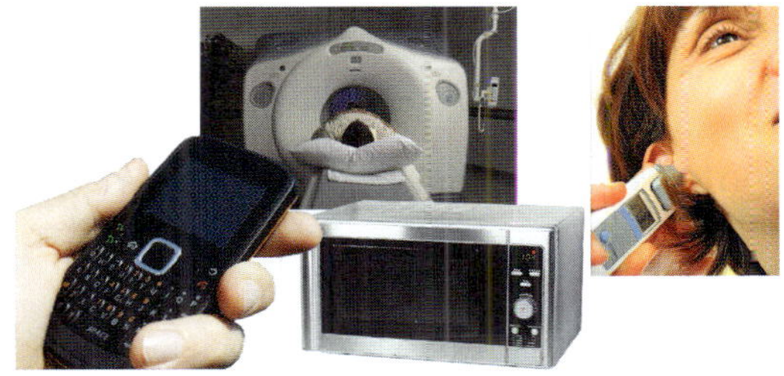

**What do these have in common?**

## Dispersion of light

**What is dispersion?**

When you look at a rainbow, you see sunlight separated into red, orange, yellow, and so on. **Dispersion** is the separation of light into its constituent colors or wavelengths. Dispersion in a rainbow is caused by refraction. The water droplets that cause a rainbow refract blue light slightly more than red light, so your eyes see blue and red in a slightly different place.

**Refractive index varies with color**

The diagram on the right shows how different colors diffract different amounts through an imaginary prism. The index of refraction for crown glass is $n = 1.513$ for red light, but it is slightly larger, $n = 1.532$, for violet light. The index varies between 1.513 and 1.532 as the color changes from red through violet. The higher value of $n$ for violet light means that violet light is refracted more than red light. In the illustration at right, the violet light rays are bent the most, emerging from the prism at the bottom. Red rays are bent the least. Intermediate colors are bent at angles in between red and violet.

**Dispersion caused when refraction varies with the color of light**

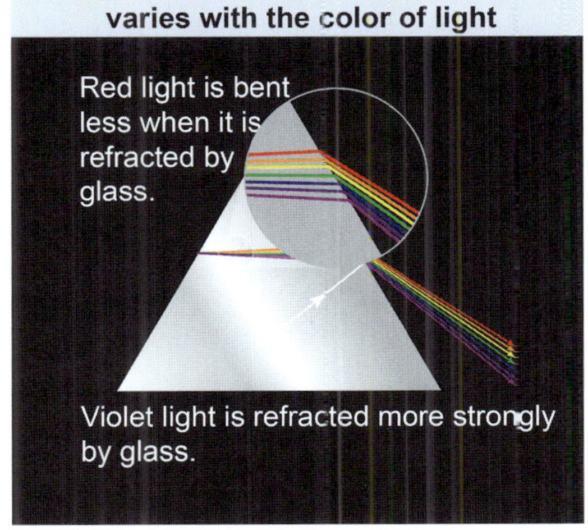


**Examples of dispersion**

There are many examples of the dispersion of light all around you. Four are shown in the illustration on the right. Later in this chapter you will learn the physics behind each one of these examples. The prism and rainbow are both based on the refraction of light as it passes from one medium to another—into glass for the prism or water for the rainbow. The compact disc is a reflection *diffraction grating*, while the novelty Rainbow Peepholes™ contain a transmission diffraction grating. Diffraction gratings have thousands of tiny, repeating grooves or rulings that break light up into many independent beams that interfere with each other to create dispersion.

**Examples of the dispersion of light**

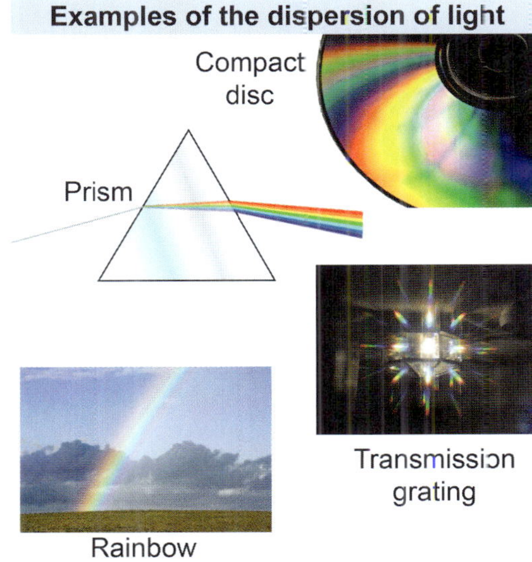

## The speed of light in different materials

**Speed of light changes in different materials**

You can ride a bicycle fast on hard pavement, but when you switch surfaces to sand, gravel, or mud you must slow down. Your speed on a bicycle depends on the *medium* you are traveling on. In the same way, the observed speed of light depends on the properties of the medium in which the light travels.

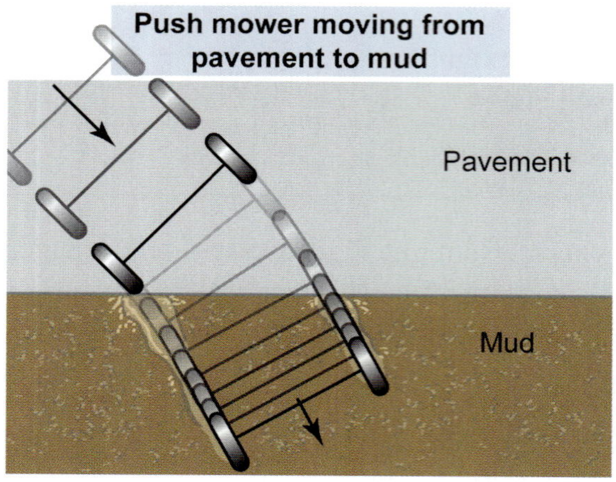


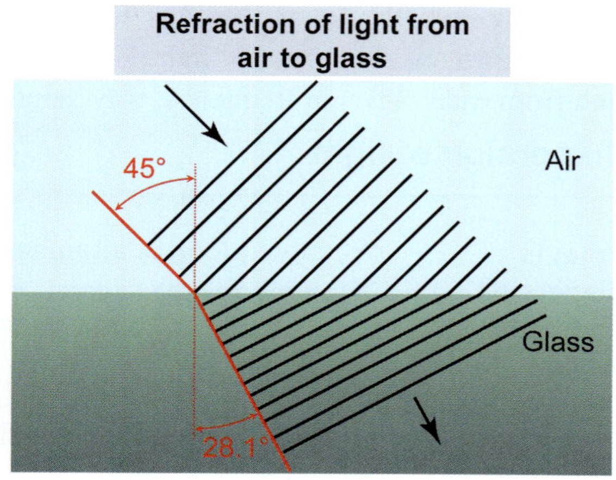


**Refraction and changes in the speed of light**

The physical phenomenon of refraction—where light bends as it propagates from one medium to another—is a consequence of the *change* in the speed of light between those two media. The **index of refraction** for a medium, such as glass or water, represents how the speed of light has changed relative to a vacuum and is described by equation (22.2).

Interactive equation

**Index of refraction**

$$(22.2) \quad n = \frac{c}{v}$$

$n$ = index of refraction
$c$ = speed of light in a vacuum (m/s)
$v$ = speed of light in medium (m/s)

**Light changes wavelength, not frequency**

When the speed of light $v$ changes as it moves from one medium to another, how are the frequency $f$ and/or wavelength $\lambda$ affected? Frequency cannot change because the number of waves per second that arrive must also depart. Instead, the *wavelength* changes from one medium to another ($c = f\lambda$). Look at the illustration above. Notice how the spacing between waves is smaller in glass than in air. The wavelength of light is shorter in glass than in air because the speed of light is lower in glass than in air.

**Index of refraction for saltwater**

What is the speed of light through saltwater if its index of refraction is 1.34?

*Asked:* speed of light $v$ in saltwater

*Given:* index of refraction of saltwater, $n_w = 1.34$;
speed of light in air (nearly same as a vacuum), $c = 3.0 \times 10^8$ m/s

*Relationships:* index of refraction for water, $n_w = c/v$

*Solution:* Solve for the speed of light in water by multiplying by $v$:

$$v \times n_w = \cancel{v} \times (c/\cancel{v})$$

then dividing by $n_w$:

$$(v\cancel{n_w})/\cancel{n_w} = c/n_w \quad \Rightarrow \quad v = c/n_w$$

$$v = (3.0 \times 10^8 \text{ m/s})/1.34 = 2.2 \times 10^8 \text{ m/s}$$

*Answer:* $v = 2.2 \times 10^8$ m/s

### What causes a rainbow?

A rainbow is caused by the same physical phenomenon that is behind the prism: the dispersion of light. In dispersion, white light is separated into its constituent colors, from red to green to violet.

**How rainbows form**

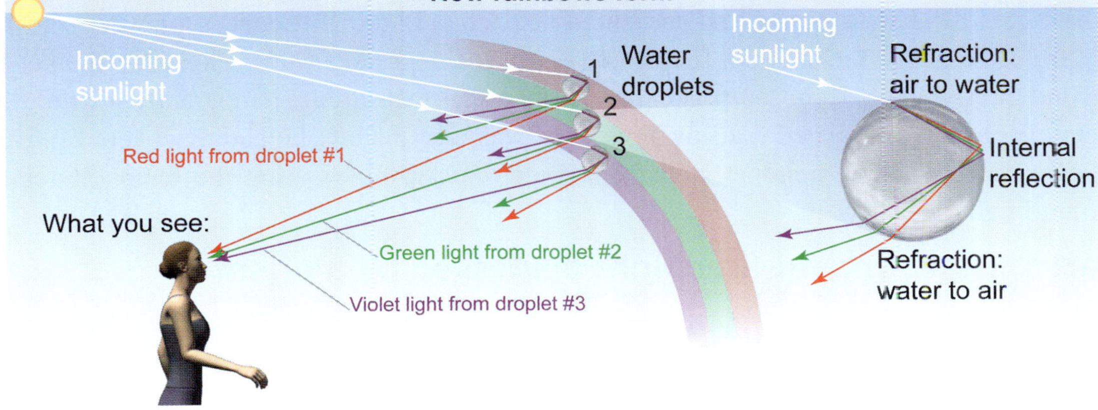


### Water droplets disperse sunlight

You can only see a rainbow when you are facing away from the Sun and looking relatively low toward the horizon at air that is humid—full of water droplets. But what causes the effect of the rainbow? The water droplets disperse the white sunlight, refracting the red light rays less than violet light rays. The refraction occurs at the interface between air ($n = 1$) and water ($n = 1.33$). Inside the water droplets, the light rays also reflect internally, because the rays strike the inside back surface of the droplet at an incidence angle greater than the critical angle. The violet light rays emerging from the water droplet travel closer to parallel with the ground; the emerging red light rays travel more directly toward the ground.

### How does refraction cause a rainbow?

The rainbow is created because your eyes are seeing light rays of different colors coming from different water droplets—either higher or lower in the sky. The red light in a rainbow comes to your eyes from the highest water droplets, because the droplets have refracted the red light by the least angle. Violet light in the rainbow comes from the lowest water droplets, because the violet light rays were bent the most by refraction. That is why red is always on top of a rainbow; violet is at the bottom.

### Double rainbows

The same water droplets that produce a rainbow can also produce a second rainbow—higher in the sky than the primary rainbow—called a *double rainbow*. A double rainbow is produced when sunlight inside the water droplets undergoes *two internal reflections*, not one. Because there are two internal reflections, the relative directions of the colors in a double rainbow are inverted with respect to the primary rainbow; i.e., the double rainbow has red at the bottom while the primary rainbow has red at the top. A double rainbow is much fainter than the primary rainbow because of the added losses resulting from the additional reflection.

**Double rainbow**

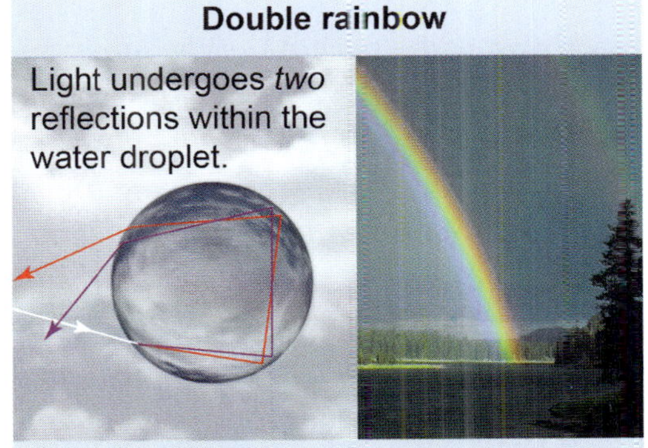


Double rainbow's colors are inverted compared to the primary rainbow.

# Investigation 22B Detect infrared radiation with a prism

**Essential questions** *How can we detect infrared light if it is invisible?*

William Herschel discovered infrared light in 1800. Herschel conducted an experiment in which he passed white light through a prism and measured the temperature differences between the colors of the visible spectrum. To gauge the ambient temperature near his experiment, Herschel placed an additional thermometer beyond the red light. To his surprise, this thermometer measured a temperature greater than that of any of the others. Herschel speculated that invisible light was responsible; he had discovered infrared radiation. In this investigation, you will conduct Herschel's experiment by measuring the temperatures of blue, yellow, and infrared light.

## Part 1: Detecting infrared light dispersed by a prism

1. Attach a triangular prism to a cardboard box to disperse sunlight onto the bottom of the box.
2. Cover the bulbs of three alcohol thermometers with black paint or black permanent marker.
3. Experiment with the slit width and tilt of the prism and box to obtain a sharp, bright spectrum.
4. Attach two thermometers to the bottom surface of the box so that the bulbs are in the blue and yellow light. Attach the third several millimeters *beyond* the red part of the spectrum.
5. Let the system come to equilibrium for 5–10 min, then measure the temperature differences between the thermometers.
6. Switch the places of the thermometers twice, each time letting the temperatures equilibrate and measuring the temperature differences.
7. Record each set of results and average them.

**Detecting infrared light**

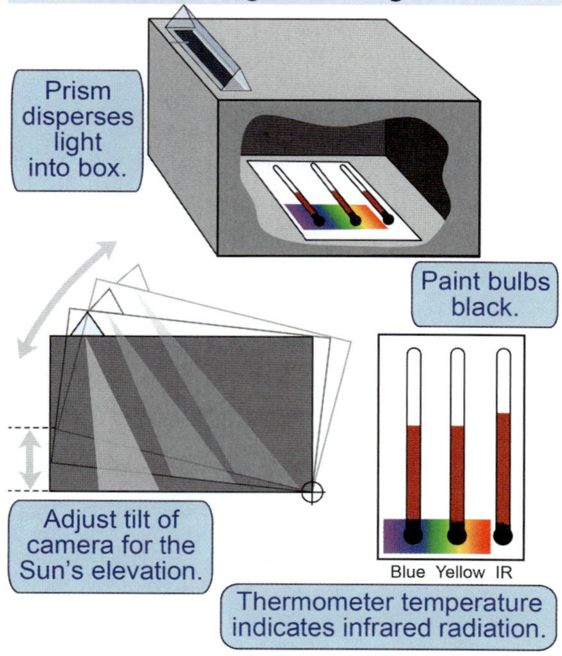


*a. What is the purpose of the black paint or permanent marker?*
*b. Do you need to cover the box while the thermometers are heating?*
*c. Why should you switch the positions of the thermometers for each set of measurements?*
*d. Which thermometer gives the highest reading each time?*
*e. How well do you think this method would work for detecting ultraviolet light?*
*f. Predict how the experimental results might differ if you used a different light source.*
*g. Explain whether the prism works better when illuminated by a diffuse or a narrow beam.*
*h. Analyze your observations: Do they require infrared light to explain the solar spectrum?*
*i. Meet with another lab group and present to each other your designs and performance. Critique each other's observational data. Evaluate whether or not the other group's observations require infrared light to be part of the solar spectrum.*
*j. As a result of this discussion, are there ways that you could revise your experiment to produce a more definitive result? Implement these changes and collect a new set of observations of the Sun.*

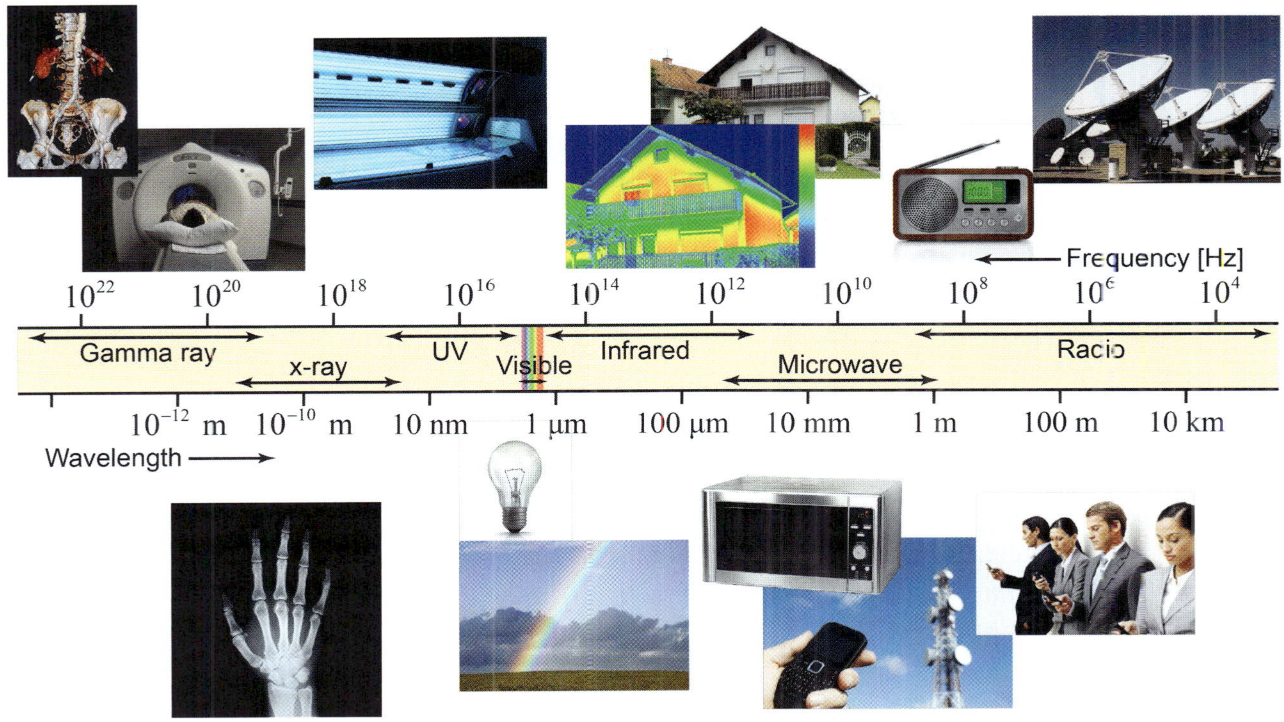


Interactive equation

There is a lot more to electromagnetic **radiation** than meets the eye! While our eye detects visible light, there are many other wavelengths of light—such as **gamma rays**, infrared radiation, and radio waves—that together comprise a *spectrum* of EM waves. The broad variety of EM waves are referred to as a **spectrum** because they span a wide range—many orders of magnitude—in frequency, wavelength, and energy.

**Orders of magnitude**

The term *orders of magnitude* refers to powers of 10. If a quantity spans one order of magnitude, then it varies by a factor of 10. Two orders of magnitude are a factor of $10^2 = 100$, three orders of magnitudes are $10^3 = 1{,}000$, and so on. Mathematically, powers of 10 are expressed as the exponent $n$ in $10^n$.

**Frequency range of the electromagnetic spectrum**

The **electromagnetic spectrum** spans many orders of magnitude in the frequency of EM waves. In the illustration above, you can read off the frequencies and wavelengths associated with each region of the EM spectrum. From the lowest frequency radio waves at $10^3$ Hz to the highest frequency gamma rays at $10^{23}$ Hz (and higher), the EM spectrum spans *over 20 orders of magnitude* in frequency!

**Comparison with sound waves**

Electromagnetic waves are transverse waves that can propagate through a vacuum, whereas sound waves are longitudinal pressure waves that must travel through a medium. Our eyes can only detect a narrow sliver of the EM spectrum—less than one order of magnitude—but our ears can detect several orders of magnitude of frequency for sound. Modern instrumentation can detect a broader range of the EM spectrum than the sound spectrum. Our eyes see colors within the EM spectrum that are composites of three primary colors (red, green, and blue), whereas our ears can sense many different frequencies of sound simultaneously.

## Regions of the electromagnetic spectrum

**High-energy radiation**

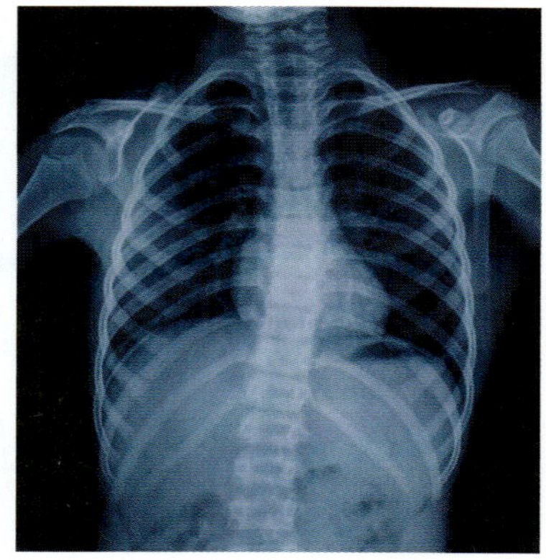

The highest energy forms in the electromagnetic spectrum are gamma rays (or $\gamma$ rays) and **x-rays**, which are dangerous to humans in large quantities. X-rays are energetic enough to travel through the human skin but not quite powerful enough to pass through bone, so they can provide an image of the condition of a patient's skeleton without the need for invasive surgery. Medical imaging often uses a gamma ray PET scan (which tracks an ingested, radioactive substance) alongside an x-ray CAT scan to provide a three-dimensional view of the patient's internal physiological systems.

**Ultraviolet light**

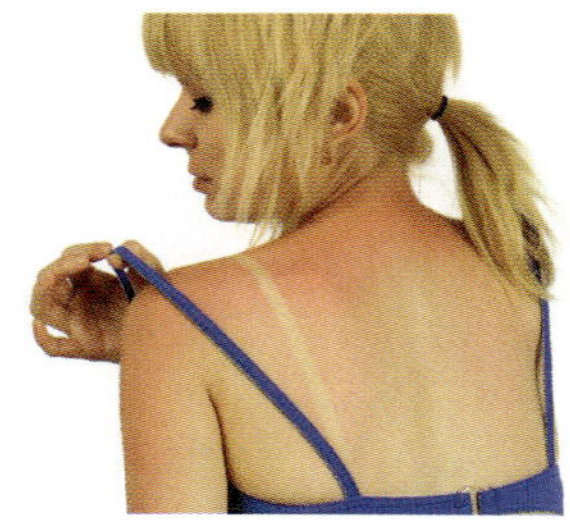

**Ultraviolet light** is more energetic than visible light but is mostly, although not entirely, blocked by the Earth's atmosphere. The higher energies of ultraviolet light make it more damaging to human skin cells. That is why it is important to wear sunblock on a sunny day to prevent damage to your skin.

**Visible light**

When most people talk about light they usually refer to optical or visible light, because that is what our eyes can perceive. The colors of visible light appears to us as combinations of red, green, and blue. Light bulbs, the Sun, and other stars radiate much of their light energy as visible light.

**Infrared or thermal radiation**

**Infrared radiation** (or thermal radiation) has wavelengths longer than, and energies lower than, visible light. Every body in the universe emits radiation because of its temperature and mostly at infrared wavelengths. Infrared imaging cameras have become useful devices for checking for thermal leaks in a house in order to reduce energy consumption. Infrared goggles are also used on the battlefield at night to detect the thermal radiation from enemy forces. Infrared weather satellites are sensitive detectors of water vapor in clouds—even at night!

**Radio waves and microwaves**

When Heinrich Hertz first showed the existence of electromagnetic waves, he did so using *radio* waves. **Microwaves** are used in ovens and mobile phone technology. Radio waves are used for communication with artificial satellites and AM and FM radio broadcasting. The longest wavelengths of radio waves are used to communicate with submarines.

# Section 2 review

When white light passes through a glass prism, blue light is refracted more strongly than red light, resulting in dispersion; glass has a higher index of refraction for blue light than for red light. While all colors of light travel at the same speed in a vacuum, light of different colors travels at different speeds through a medium such as glass. The refractive index of a material represents the ratio of the speed of light in a vacuum to the speed of light in the material. Electromagnetic waves or light do not only come in the *visible* colors that you can see in a rainbow or in light dispersed by a prism. Electromagnetic radiation encompass a broad *spectrum* of waves, from low-frequency radio waves and microwaves to high-frequency x-rays and gamma rays. The electromagnetic spectrum is used in many different modern technologies.

| | |
|---|---|
| **Vocabulary words** | **dispersion, index of refraction, electromagnetic spectrum, radiation, spectrum, gamma rays, x-rays, ultraviolet light, infrared radiation, microwaves, radio waves** |
| **Key equations** | $n = \frac{c}{v}$ |

## Review problems and questions

1. Describe one piece of technology that is based on each of the following wavelength regimes of the electromagnetic spectrum: x-ray, visible, infrared, microwave, and radio.

2. Compare photons detected by a CCD digital camera and a thermal imaging camera. Which photons have a longer wavelength? Which have a higher frequency?

3. Astronomical observations at 1 THz can only be made at high elevations where there is little water vapor. What wavelength is this light? In what part of the electromagnetic spectrum does it fall?

4. How might a doctor use visible radiation to diagnose a patient?

5. Dog ears can hear frequencies of sound that are higher than a human can hear. If an animal could see frequencies of light higher than humans can, what spectrum would they be able to see?

6. Many people have radios in their car. When they listen to them, are they listening to radio waves? Explain.

7. Calculate the speed of light in crown glass for the following:
   a. violet light ($n = 1.532$)
   b. red light ($n = 1.513$)
   
   Which travels faster through crown glass, violet light or red light?

8. During a talk show segment a medical guest recommended the use of "safe" UVB (ultraviolet B) tanning beds for getting vitamin D. Vitamin D is important for human health, yet many Americans are deficient in the vitamin. The World Health Organization recommends against using a tanning bed for cosmetic purposes and classifies the beds as "carcinogenic to humans." Research this issue and evaluate the validity of this doctor's claims that you should not "believe all the negative hype on tanning beds." Cite specific evidence for your conclusion.

## 22.3 - Dual nature of light

Earlier in this chapter, light was described as an electromagnetic *wave* that propagates through oscillating electric and magnetic fields. In this section, you will learn that light can also be described as a *photon*, a tiny and indivisible packet of energy resembling a particle. So which is it, a wave or a particle? This perplexing conundrum lies at the heart of the microscopic world of the atom and quantum physics itself.

### Polarization

**Oscillating string**

Try oscillating a string vertically through a slit cut into a piece of cardboard. If the slit is cut in the same direction as the string's oscillations, then the oscillations can pass through the cardboard. If the slit is perpendicular, then the oscillations are stopped altogether. This oscillating string serves as an analogy for the **polarization** of light, where the oscillations of light can be passed or blocked by a polarizing filter.

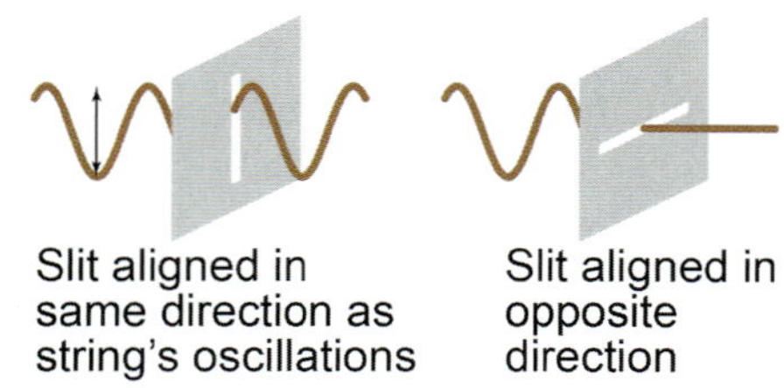


**Polarization of light**

Light is made of oscillations in the electric and magnetic field. If those oscillations are randomly oriented, the light is *unpolarized*. If the oscillations occur more in one direction than another, the light is *polarized*. Polarization is a property of the oscillation of a transverse wave in a direction perpendicular to the direction of propagation.

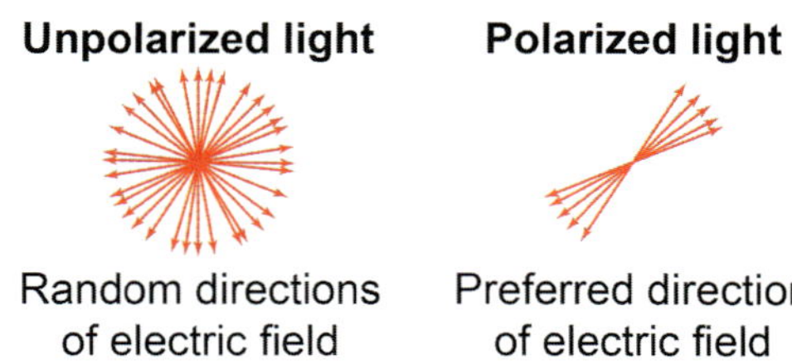


**Reflection polarizes light**

It can be difficult to see into a lake or swimming pool because of the glare of overhead light reflected off the surface of the water. Light that reflects off of *nonmetallic* surfaces, such as water or snow, becomes polarized in a direction parallel to the surface, i.e., in a horizontal direction. By viewing this reflected light through a vertically polarized filter—perpendicular to the direction of the horizontally polarized reflected light—the glare is reduced and you can see underwater more easily. Many sunglasses are sold with vertically polarized filters to reduce glare from reflected light. Polarized filters for cameras can increase contrast of the clouds in the sky. Look at the sky through a polarized filter or sunglasses. Does it sharpen the clouds?

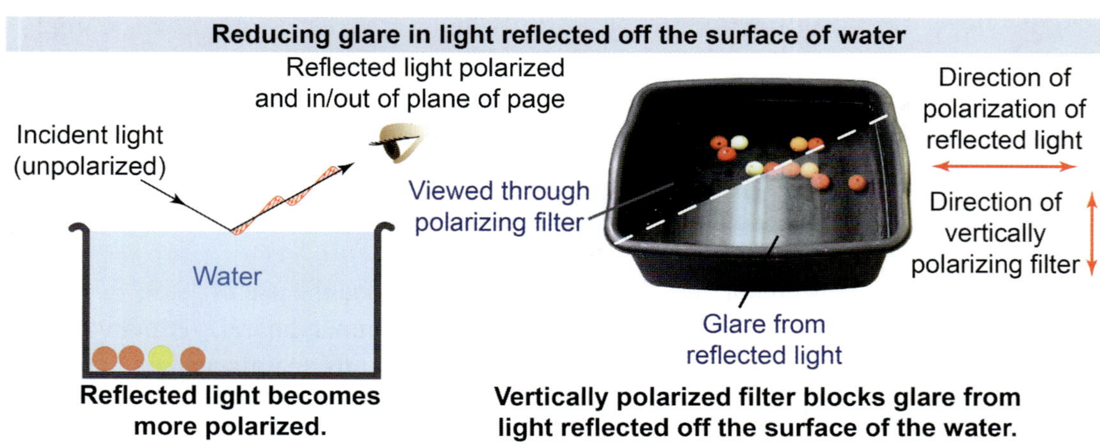

### Diffraction of light

If you shine light at a small object, such as the paper clip in the figure at right, you'll see something peculiar: a series of bright and dark rings (or bands) surrounding the shadow of the object! This **diffraction pattern** of bright and dark bands is caused by constructive and destructive interference when light passes around a barrier or through an opening.

**What causes the light and dark rings surrounding the shadow of the paper clip?**

### Wave diffraction

How do diffraction and interference create this pattern? On page 424, you learned how waves *diffract* around the edge of an obstacle. Sound waves can diffract around the wall of a room, but light waves cannot, because the wavelength of light is much smaller than the size of the barrier. What happens if you use a smaller barrier, such as a paper clip? Does light diffract around it?

### Diffraction of light around a barrier

When light waves pass a barrier, a little bit of the light's intensity diffracts around the barrier's corner. These waves can interfere with light waves diffracting around the *other* side of the barrier, which creates a pattern of interference. If the diffracted light is projected onto a screen, such as in the illustration at the top of this page, you will see a pattern of bright and dark bands around the edge of the barrier. The bright bands are constructive interference, whereas the dark bands are destructive interference. The brightness of the interference bands falls off away from the edge of the barrier.

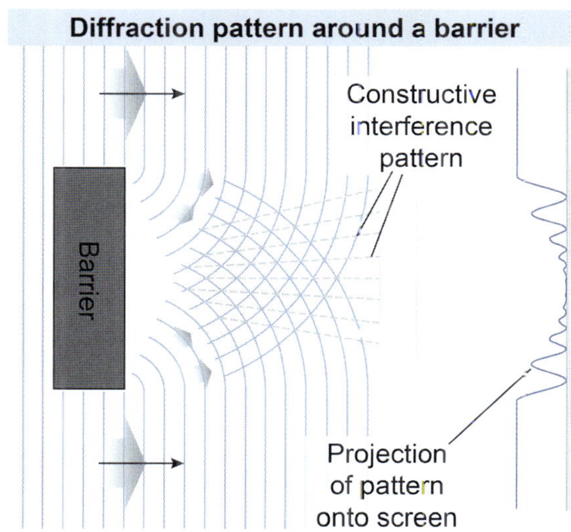


### Wave behavior of light

The illustration at the top of this page shows this diffraction pattern for light passing around a paper clip. The diffraction pattern of light is an example of how *light can behave like a wave.*

### Monochromatic light

The diffraction pattern is difficult to see when you use ordinary, white light. Diffraction is a wave phenomenon, so it is best demonstrated *when all the light has the same wavelength*. Light waves are called *monochromatic* if they have the same wavelength, although they may differ in amplitude or phase. Consider the green light color of the illustration at the top of the page. A monochromatic, green light source was used to create it. Furthermore, the light must be *coherent* or in phase to see the interference pattern from diffraction around a barrier. Lasers are often used as sources of coherent light.

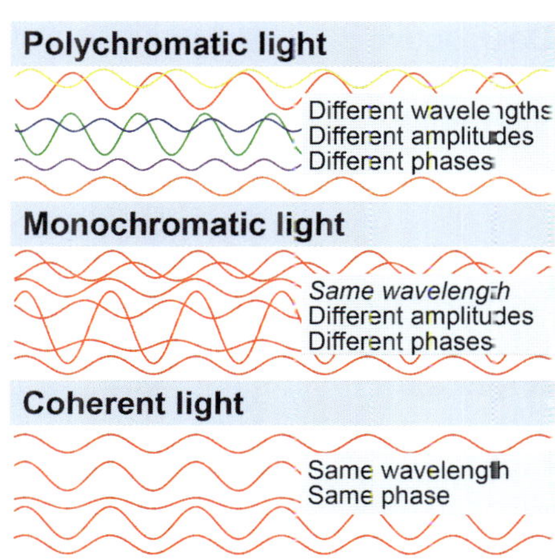

**Diffraction through a tiny slit**

If light diffracts *around* an object, does it also diffract *through* a narrow opening? In Chapter 15, you learned that waves diffract through a narrow slit with a pattern similar to that shown in the illustration at right. As long as the slit or opening is very narrow compared to the wavelength of the waves, then the diffracted waves form nearly circular arcs. Does light behave in the same way?

Diffraction through an infinitely narrow slit

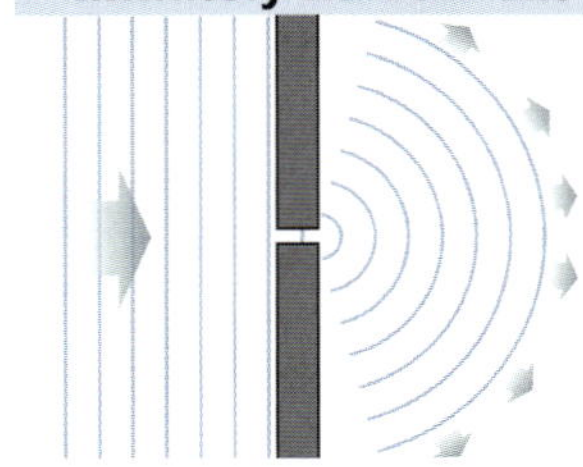

**Real slits**

The diffraction pattern shown above represents diffraction through a slit *of infinitely small width*. How many real slits are that narrow? None! In real physics situations, the slit has a measurable width, which leads to a more complicated diffraction pattern. The reason for this pattern is that light through one side of the slit can interfere with light passing through the other side of the slit.

Diffraction of light through a single slit

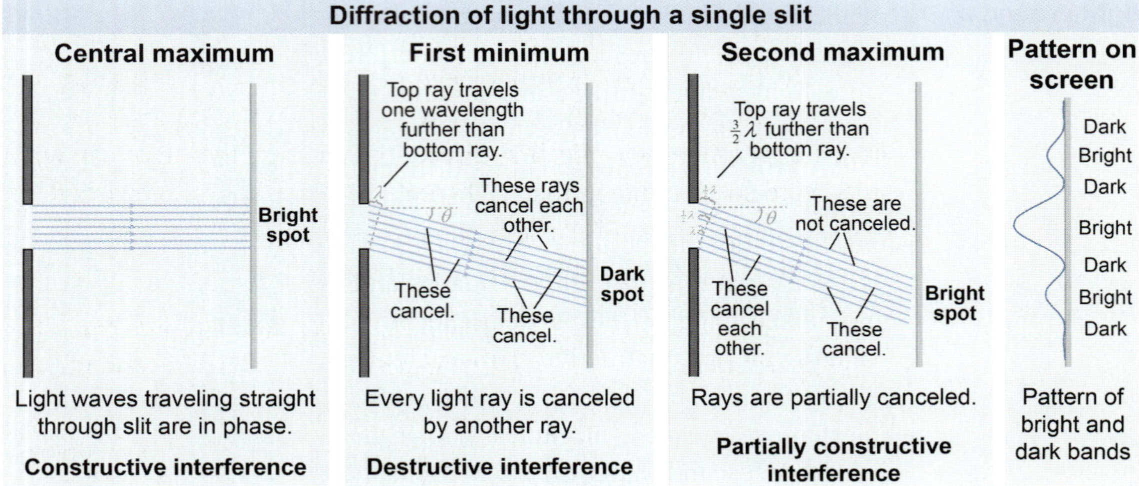


**Single slit diffraction pattern**

The illustration above shows light passing through a realistic slit and the diffraction pattern produced on a screen behind the slit. Light passing straight through the slit adds constructively to create a bright, central spot on the screen.

**Destructive interference dark spot**

At larger projection angles $\theta$ onto the screen, light from different parts of the slit travels different distances to reach the screen. A ray from the top of the slit has to travel one-half a wavelength further than a ray from the middle of the slit, resulting in destructive interference. In fact, every ray through the slit at that angle will be canceled by another ray! This causes a dark spot on the screen at that location.

**Constructive interference bright spot**

At a larger angle, light from one side of the slit will travel $(3/2)\lambda$ further than light from the other side. At this angle, some of the light interferes destructively, while one-third of the light interferes constructively, leading to a bright spot on the screen.

**Light diffracts like a wave**

This pattern of bright and dark spots repeats at larger angles, although the brightness of the spots gets fainter. This diffraction pattern shows that light is interfering with itself, which means that light is exhibiting wave behavior.

**Young's double slit experiment**

In 1801, the English physicist Thomas Young devised an experiment to determine whether light behaved as a wave or a particle. Up until then, physicists disagreed on the nature of light. Young's insight was to pass *monochromatic* and *coherent* light through two narrow and closely spaced slits and then observe the pattern of the light on a screen. (Coherent light means that the waves are in phase with each other.) If light were composed of particles, then the particles would pass straight through the two slits and produce two bright lines of light on the screen. If light acted as a wave, then a distinctive *interference pattern* would be created by light waves passing through the slits.

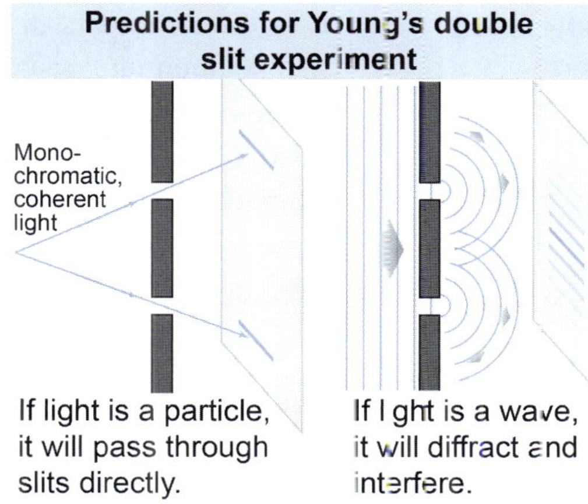


**Light interferes like a wave**

Young found that light interfered with itself to produce a set of *interference fringes* on the screen. The brightest fringe was located *between* the slits, a location predicted by the wave model but where no particle could have reached. Interference is a behavior of waves, so Young's experiment was the first direct confirmation that light can act as a wave. Young's results were so convincing that 19th-century physicists assumed that light was a wave and thoroughly rejected the particle model until Albert Einstein proposed a theory that successfully explained the photoelectric effect.

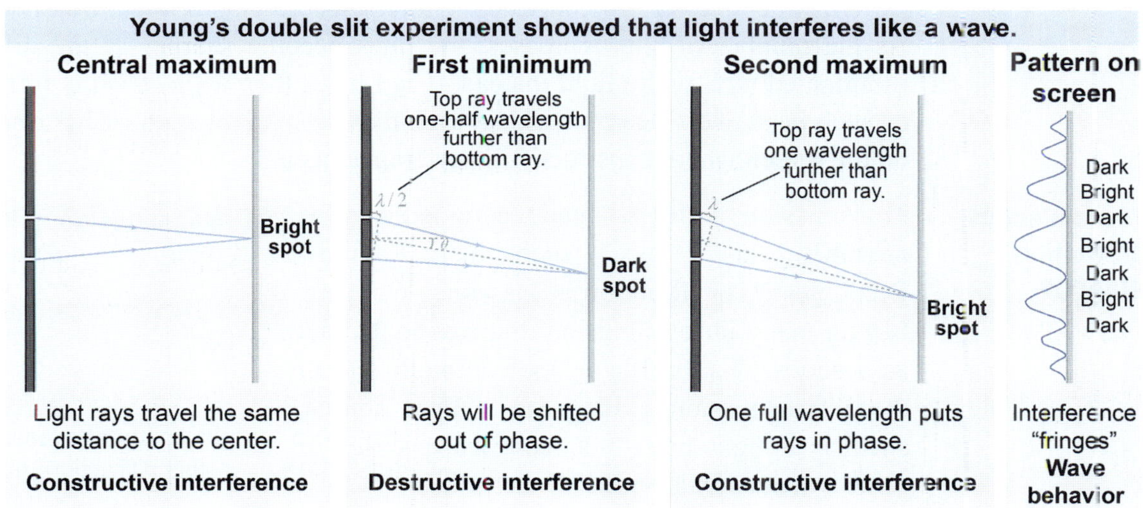


**Explanation of the interference pattern**

What causes the interference pattern? Light through both slits will travel an equal distance to a location on the screen halfway between the two slits, producing a central maximum. At a small angle $\theta$ from the center, light from one slit will have to travel an additional distance of $\lambda/2$, or one-half of a wavelength. That causes *destructive* interference, because the two waves are out of phase. At a larger angle, light from one slit will have to travel an additional distance of $\lambda$, or one full wavelength. That causes *constructive* interference, because the two waves are fully in phase. The pattern repeats as the angle increases, resulting in an interference pattern.

## Diffraction gratings and spectrographs

**Diffraction and dispersion**

Look at a compact disc illuminated by a room light. Do you see different colors or even a rainbow? What causes the different colors of light to bounce off the CD? The compact disc has many tiny grooves or rulings that encode the electronic files or music. When light strikes the surface, these grooves act similarly to a series of tiny slits that diffract and disperse the incident light.

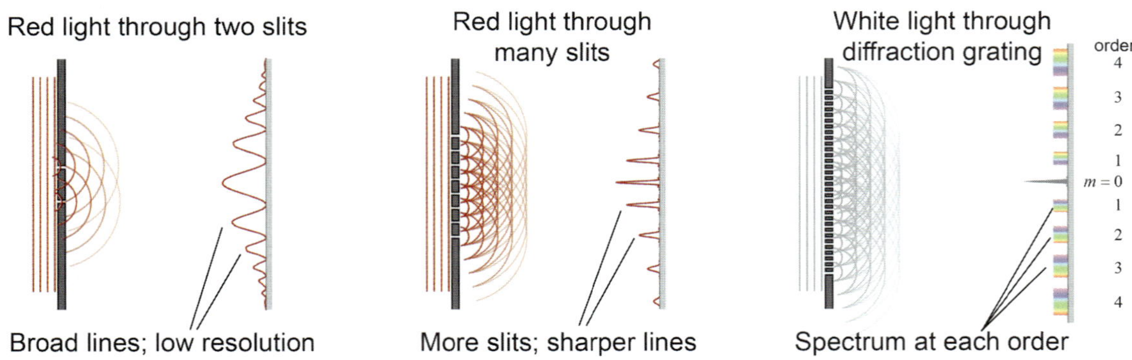


**Diffraction with many slits**

To understand how the CD dispersion works, imagine shining red light through two slits to produce the interference pattern on the previous page. As the number of slits increases, the individual peaks in the interference pattern become sharper. The reason for this sharper pattern is that, with many more slits, there are fewer places at which all the diffraction patterns line up to create constructive interference. This is a feature of *diffraction gratings*: As the number of slits increases, the lines become sharper. If the diffraction grating is illuminated with white light instead of red light, then a *spectrum* is formed at the location of each peak. This occurs because each different wavelength of light has its own location on the screen where constructive interference occurs.

**Transmission diffraction gratings**

Transmission gratings—made commercially as novelties and toys, such as Rainbow Peepholes™ or Rainbow Glasses®—work by having many little rulings on thin plastic that act as a series of slits. If you look through one of these transmission gratings at a light source, such as a light bulb, you can see a series of rainbow spectra dispersed by the rulings.

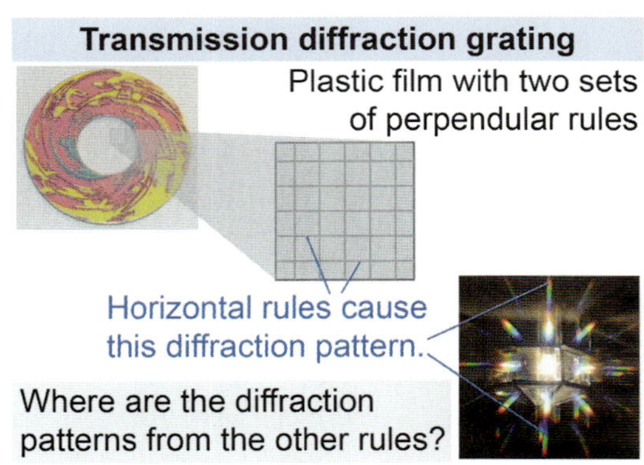


**Reflection gratings and spectrographs**

The CD acts as a *reflection* grating that disperses light by reflection instead of by allowing the light to pass through it. Reflection gratings are more effective than transmission gratings, so they are more commonly used for scientific research and commercial applications. A **spectrograph** is a scientific instrument that uses a diffraction grating to disperse light into its spectrum. Spectrographs can be small, hand-held devices, or they can be the size of a large truck when built for the largest telescopes on Earth.

**Light is radiant energy**

Radiant energy from the Sun's travels to the Earth as electromagnetic waves over a wide band of frequencies. The majority of sunlight is infrared, visible, and ultraviolet light. The total power of sunlight is 1368 W per square meter on the side of the Earth facing the Sun. The total solar power received by Earth is $1.2\times10^{17}$ W or 120,000 TW. By comparison, the total power consumed by humans is about 15 TW.

**Sun radiates energy that travels to us as electromagnetic waves.**

**Energy of light**

How much energy is in *just one* of the frequencies present in the Sun's radiation? The relation between frequency and energy is given by the Planck relation, equation (22.3).

Interactive equation

$$(22.3)\quad E = hf$$

$E$ = photon energy (J)
$h$ = Planck constant $= 6.63\times10^{-34}$ J s
$f$ = frequency (Hz)

**Planck relation for the energy of light**

The constant of proportionality is called **Planck's constant**, named after the German physicist Max Planck, who inferred its properties in 1900. Planck's constant connects the energy of electromagnetic radiation to its frequency. Its value is $6.63\times10^{-34}$ J s.

**Photons**

Equation (22.3) might appear straightforward, but it actually represents a fundamental change in scientific thought! Planck's relation describes a universe where light's energy only comes in discrete packets called the **photon**. There is no bundle of light energy smaller than $hf$—all larger quantities of light energy are made up of whole number multiples of $hf$. When Planck—and later Albert Einstein—thought about light as being composed of little bundles of energy, they changed the way we think of light.

**Quantum physics**

You may have noticed that Planck's constant is of the order of $10^{-34}$! The Planck equation affects matter and energy on the microscopic scale. A photon of visible light energy is a tiny quantity, appropriate to the size and energy of a single atom. The photon is an example of **quantum physics**. The word *quantum* means a physical quantity that occurs only in discrete units, or bundles. A photon is a quantum of light energy. *A quantum cannot be divided.* A single photon is the smallest possible quantity of light. You cannot have half of a photon or 1.5 photons. Light occurs in integer multiples of whole photons. You will learn about other quanta in Chapter 26!

**Number of photons in 1 J of energy**

How many visible light photons of frequency $f = 6\times10^{14}$ Hz are there in 1 J of energy?

*Asked:* number of photons in 1 J of energy

*Given:* frequency $f = 6 \times 10^{14}$ Hz; Planck's constant $h = 6.63 \times 10^{-34}$ J s

*Relationships:* Planck relation $E = hf$

*Solution:* The energy of one photon of light (i.e., joules per photon) is

$$E = hf = \left(6.63 \times 10^{-34}\text{J s}\right) \times \left(6 \times 10^{14}\text{Hz}\right) = 4.0 \times 10^{-19}\text{ J}$$

The number of photons in 1 J (i.e., photons per joule) is the inverse:

$$1/\left(4.0 \times 10^{-19}\text{ J/photon}\right) = 2.5 \times 10^{18}\text{ photons/J}$$

*Answer:* There are $2.5\times10^{18}$ photons of this frequency in 1 J of energy.

**What is the photoelectric effect?**

Hertz in 1887 discovered that an ultraviolet light source shining on a metal's surface would increase electrical sparks—later recognized to be electrons ejected from the metal—across the gap between two metal plates. The **photoelectric effect** occurs when electrons are ejected by illuminating a metal surface with light.

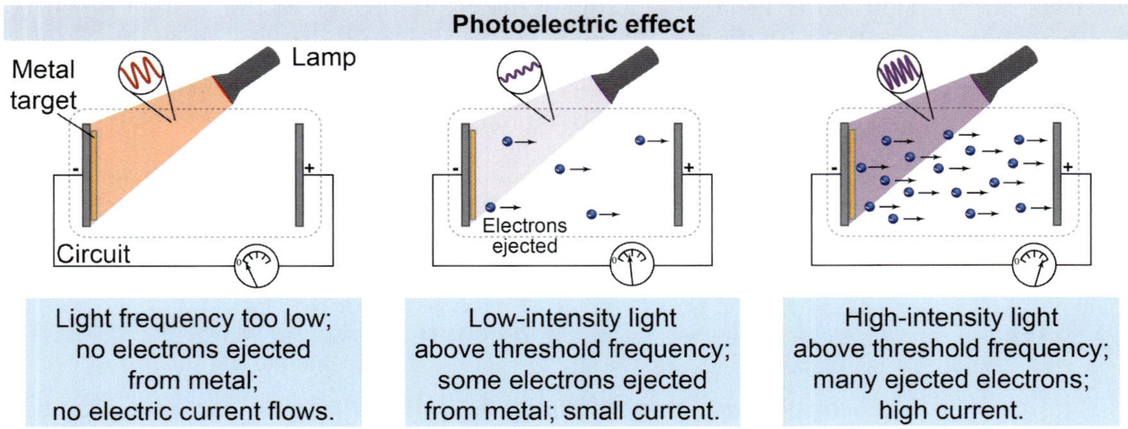


**Puzzles of the photoelectric effect**

Subsequent researchers showed several curious results. Lower frequency light, such as red light, never liberated electrons no matter how bright the light source. Higher frequency light, such as ultraviolet light, would always eject electrons no matter how faint the light source—although the number of electrons varied with the intensity of the light. In classical physics it was assumed that light was an electromagnetic *wave*, which meant that the energy of the ejected electrons should be a function of the intensity and frequency of the light, not just frequency. What was wrong with this picture?

**Threshold frequency explains the photoelectric effect**

Albert Einstein in 1905 explained the photoelectric effect with two insights. First, he assumed that light was composed of quantized *photons*, each with its own, indivisible bundle of energy. Second, the atoms in the metal held onto their electrons with a binding energy called the work function. Only if the photon's energy was high enough to exceed the work function would an electron be ejected. Since Planck had already proposed a photon's energy to be related to its frequency, the minimum energy of the work function corresponded to a **threshold frequency** for electrons to be ejected.

**Work function and the electron's energy**

How much energy do the ejected electrons have? Part of the energy of the incident photon liberates the electron from the metal; the remainder of the photon's energy is converted into kinetic energy of the electron. The photon energy required to liberate the electron is called the **work function**. The kinetic energy of the electron is therefore the difference between the incident photon energy and the work function of the metal, as given by equation (22.4). The work function is a property of the metal; different metals have different values for their work function.

Interactive equation

**Photoelectric effect**

$$E_k = hf - W_0 \quad (22.4)$$

$E_k$ = maximum energy of ejected electron (J)
$h$ = Planck's constant = $6.63\times10^{-34}$ J s
$f$ = frequency of incident light (Hz)
$W_0$ = work function for metal (J)

**Electron volt**

The value for the work function of a material is usually very small, on the order of $10^{-19}$ J. Because of this, it is convenient to use a different unit of energy to express the work function. The electron volt is a unit of energy that is used to express changes in energy on an atomic scale. One electron volt is equal to $1.602\times10^{-19}$ J.

# Investigation 22C Photoelectric effect

**Essential questions** *How does the photoelectric effect depend on the frequency and intensity of the incident light? Why does it require a quantum physics explanation?*

Interactive simulation

Nineteenth-century physicists uncovered a most peculiar interaction between light and matter that they called the *photoelectric effect*. Light shining on a metal will liberate electrons, creating electric current. Lower frequency light, however, does not liberate electrons—no matter how bright the source of the light. Why would a faint but blue light source create electric current while an intense red light source would not? In this investigation you will recreate the photoelectric effect in an interactive simulation.

## Part 1: Threshold frequency

1. Launch the interactive simulation. Select calcium.
2. Set the light color to violet using the frequency slider.
3. Increase the intensity of the light source to around 50% so that electrons are ejected from the metal surface.
4. Change the light to different wavelengths—such as red light or ultraviolet—and observe the changes to the ejected electrons.
5. Repeat this test for three other metals.

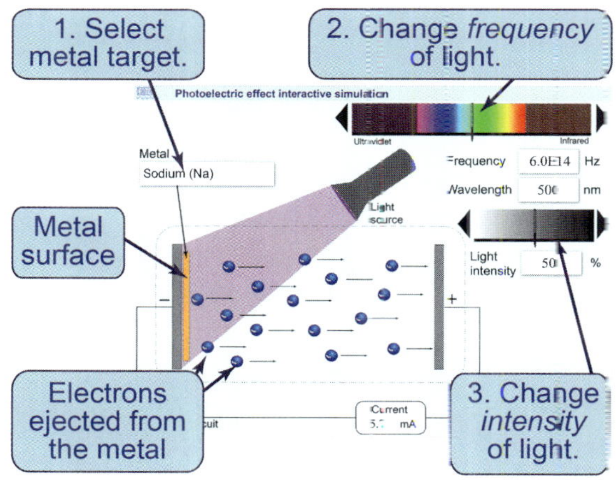


How do the number of electrons ejected from the metal depend on the *intensity* of the light? On the *frequency* of the light?

a. *Does UV light eject electrons? Green light? Red?*
b. *What is the threshold frequency? Are electrons ejected at lower or higher frequencies than this threshold?*
c. *Above the threshold frequency, how does the electron speed change with frequency?*
d. *Above the threshold frequency, how does electric current change with frequency? Collect and tabulate current data every 5 s for 30 s for two different frequencies to support your answer.*
e. *Research the work function, threshold frequency, and/or threshold wavelength for the metals that you analyze, such as by looking them up in the* CRC Handbook of Chemistry and Physics. *Do your values agree with the experimentally measured values?*

## Part 2: Intensity of the light

1. Set the frequency *above* the threshold. Observe the ejected electrons as you vary the light intensity.
2. Set the frequency *below* the threshold. Observe the ejected electrons as you vary the light intensity.

a. *How does the number of ejected electrons vary with light intensity above the threshold frequency?*
b. *How does the number of ejected electrons vary with light intensity below the threshold frequency?*
c. *Above the threshold frequency, how does the electron speed change with the light intensity?*
d. *Why does the photoelectric effect require a quantum explanation? In your answer, refer to the threshold frequency and the effect of changing the intensity of the light.*

## Digital cameras

**Film and digital cameras**

The invention of photographic emulsion in the 1850s revolutionized how we document history and society. The invention of the charge-coupled device (CCD) by Willard Boyle and George Smith in 1969 led to many advances in astronomy and other fields. Digital cameras based on CCDs have made photography easy to do and ubiquitous in society. How does a digital camera work?

**How a digital camera and a CCD work**

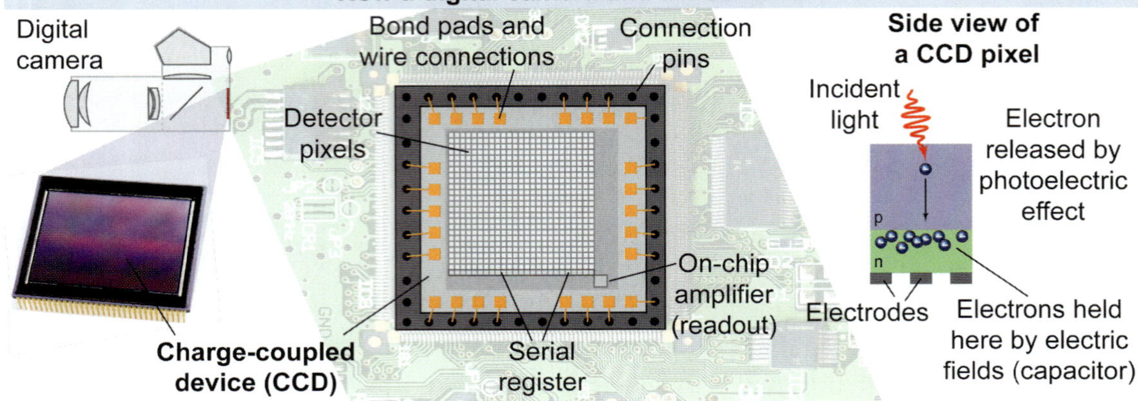


**Detectors use the photoelectric effect**

The light **detector** in a digital camera is the CCD. A detector is an instrument used to detect and measure radiation. Photons of visible light enter the camera and strike the silicon surface of the CCD, liberating electrons through the photoelectric effect. A CCD typically has millions of tiny *pixels*, each of which acts as both a detector and a storage bin. Each pixel is a *p-n* junction semiconductor device, which acts very similar to photovoltaic cells used in solar arrays. In a CCD, each **pixel** acts as a capacitor, where electric fields keep the electrons in place until it is time to read out the picture.

**Reading out each pixel**

The accumulated charge in a pixel corresponds to the number of photons that were incident upon the pixel and liberated an electron. An amplifier converts the charge of the stored electrons into a voltage that is read and stored. If more photons strike a pixel, then it stores more electrons and results in a higher voltage when the pixel is read out. The CCD shifts the charge for the first row's pixels one by one across the serial register to the amplifier where the voltage is read out. The CCD then shifts the remaining pixels down by a row and reads out the second row's pixels one by one. In this way the CCD creates an image of the number of photons detected in each pixel.

**Read pixels across a row.** **Shift and then read the next row.**

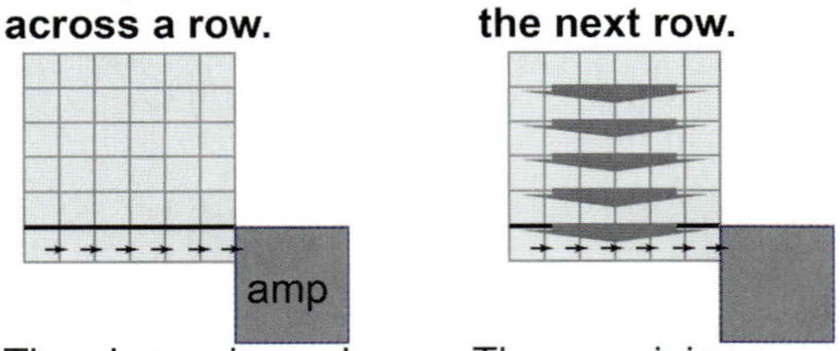


The charge in each pixel is shifted in the serial register to be read out by the amplifier one by one.

The remaining rows are shifted down by one row and then the bottom pixels are read one by one.

**Colors with digital cameras**

To create a *color* image with a CCD, alternating pixels have a red, green, or blue filter mounted in front of them, called a *Bayer mask*. In the digital image, adjacent RGB pixels are combined to create a color that you can see on a computer screen.

### Compton effect

Einstein's explanation for the photoelectric effect required light to act as a *particle*, not a wave. Is there any other evidence for the particle nature of light? In 1916, Einstein predicted that *photons have momentum*. Arthur Compton tested the prediction by scattering x-rays off of electrons. The scattered x-rays decreased in energy, while the recoiling electrons increased in energy *and momentum*. To conserve total momentum in the collision, the photons must have momentum—a property of particles! The Compton effect further demonstrated that light can act as a particle.

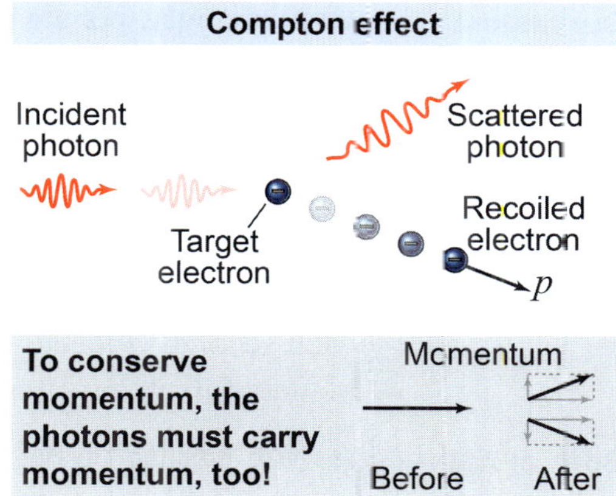


### Light is a wave and a particle

As you have learned in this section, light can exhibit properties of waves *and* particles. This is called the *dual nature of light*, and it lies at the heart of modern physics. The traditional physics or *classical* view of light is that it is a wave, but the atomic view of physics is that light is a particle.

### Light as a wave

Both the wave and particle models for light can explain that light reflects and refracts when it strikes a boundary between two media. But only the wave model can explain the phenomena of polarization, diffraction, and interference exhibited by light. Young's double slit experiment in particular provides strong evidence for the wave nature of light.

**Evidence for the wave and particle models of light**

| Property | Wave model | Particle model |
|---|---|---|
| Reflection | ✓ | ✓ |
| Refraction | ✓ | ✓ |
| Polarization | ✓ | ✗ |
| Diffraction | ✓ | ✗ |
| Interference | ✓ | ✗ |
| Photoelectric effect | ✗ | ✓ |
| Compton effect | ✗ | ✓ |

### Light as a particle

The wave model fails, however, to explain either the photoelectric effect or the Compton effect. To explain the photoelectric effect, Einstein had to assume that light was made up of particles called photons and that their energies were equal to $hf$. The explanation for the Compton effect further required photons of light to carry momentum, which is a particle phenomenon.

### Birth of quantum physics

The new physics of these two experiments demonstrated that the energy of a photon is *indivisible*; i.e., it cannot be split up into pieces. Photon energies of a given frequency are *quantized* in that they can only take on a discrete energy known as a *quantum*. If you shine light of a particular frequency on an object, then your beam of light cannot contain just any amount of energy: It can only contain energy that has *integer multiples* of the quantum energy. This was the first indication of the new physics of the quantum world. Quantum physics was born!

**Our fastest messenger**

Until recently, people could only hear each other as far as they could shout—across a canyon or village, perhaps. Now, though, you can talk with a friend halfway around the globe as if she were next door. We owe this amazing state of affairs to electromagnetic radiation, which travels through air at 300,000 kilometers per second—roughly 10 million times as fast as highway traffic. At this speed, light, radio waves, and other forms of EM radiation can circle the globe 7½ times each second. (Indeed, they can reach the Moon in a heartbeat and the planet Mars in a matter of minutes.)

**Birth of AM radio**

As you learned on page 654, EM radiation has a dual personality: It can act like a wiggling wave or like a particle smashing abruptly into matter. Ocean-crossing conversation originally focused on the wave nature of EM radiation. In 1902, inventor Guglielmo Marconi transmitted a simple coded message from North America to Europe. Marconi had built a machine that pushed electrons up and down along a metal wire. These electrons were accelerated tens or hundreds of thousands of times per second (tens or hundreds of kilohertz, or kHz), and this generated radio waves a kilometer or more in length. Shortly thereafter, speech and music were broadcast using radio waves, which has impacted society ever since.

**How AM radio works**

How did this process work? First, a radio station's microphone "heard" an audible tone and turned it into an electrical signal. (It could do this because sound waves pushed back and forth on a magnet within the microphone, and this generated a vibrating current within a coil of thin wire.) Next, a special circuit mixed this audio-frequency signal with the higher frequency radio waves that the station used to reach its listeners. This constantly changed the amplitude of those radio waves, making them strong when sound was loud and weak when it was quiet. *Amplitude modulation* (AM)—a technology that remains in widespread use—was born.

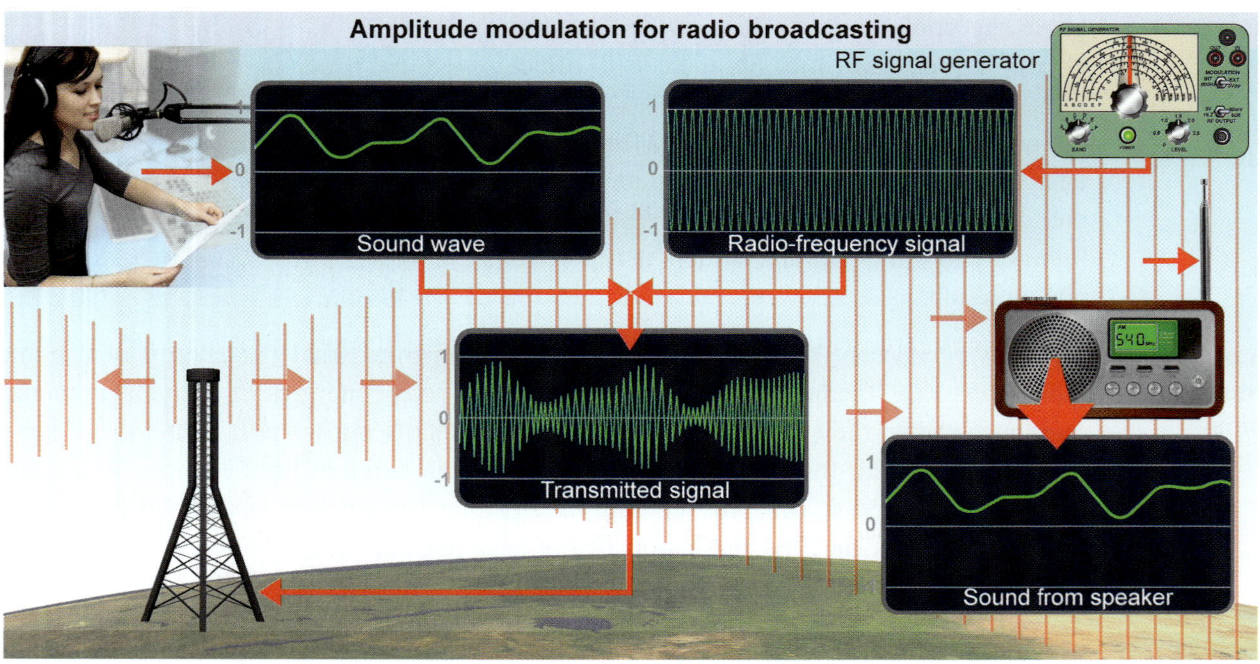

**The Internet era**

While bringing a world of news and entertainment to millions, AM radio and similar technologies have limitations: Interference can add an annoying hiss and reception quality can vary. Today's technology, however, allows us to reliably digitize, store, and exchange nearly perfect copies of speech, music, and even high-definition video with people across the globe. Many forms of modern communication use infrared (IR) and visible light, which are EM waves that vibrate much more rapidly than radio waves. With frequencies of many trillions of hertz (terahertz, or THz), such waves can easily convey billions of bytes (or gigabytes) of data each second—making it possible to create the impression of smooth, realistic motion across a large screen.

**Particulate photons**

Whereas radio transmissions are usually described as *waves*, visible and infrared forms of light are best thought of as *particles*, since their wavelengths are microscopically small. (The near-infrared photons most commonly used in telecommunications have wavelengths of 1,550 nm or $1.55 \times 10^{-6}$ m, roughly 1/100th the width of a human hair.) Furthermore, visible and IR radiation usually are detected by devices that rely on the photoelectric effect, which you read about on page 654. When a near-infrared or visible photon smacks into such a device, it knocks an electron loose, thereby creating electric current. This collision can most easily be understood by thinking about the light as a stream of individual photons.

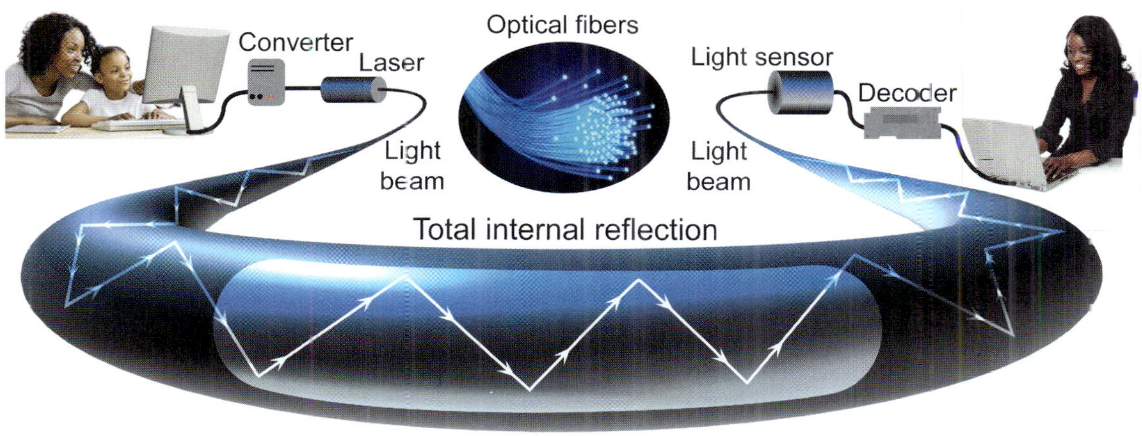


**Channeling light**

Visible and near-infrared light can't wiggle through clouds and fog the way radio waves can. Yet such light has fantastically high frequencies and information-packing power. How can we use light to transmit data from city to city and country to country? By guiding it from place to place along as smooth a road as possible. And that road is a bundle of **optical fibers**, which are spaghetti-like strands of glass. (Optical fibers are so ubiquitous that you can now find them in novelty stores.) Light enters an optical fiber much like it enters the lens in your eye: by *refraction* from air into the fiber material. Once inside, the light is prevented from escaping from the fiber by a series of total internal *reflections*, until it reaches the far end of the fiber. The internal reflections off the inside surface of the fiber can only happen if the light enters the fiber at a suitably small angle of incidence!

# Section 3 review

Is light a wave or a particle? This question confounded physicists for many centuries. In the 19th century, physicists were fairly certain that light was a *wave*, because they knew from Young's double slit experiment that light can diffract and then interfere with itself. But the photoelectric effect, discovered at the end of the century, could not be explained by the wave model. Einstein soon resolved the problem of the photoelectric effect by explaining that light acts as little *particles* called photons, based on Planck's equation for their energy, $E = hf$. While light can exhibit wave properties, at the atomic level light often exhibits properties of a particle. Quantum physics is the study of phenomena at the atomic and nuclear level where many physical properties are quantized, or only come in discrete values. An early success of the quantum theory was in explaining the photoelectric effect through quantized energy of photons.

| | |
|---|---|
| **Vocabulary words** | **polarization, diffraction pattern, spectrograph, photon, Planck's constant, quantum physics, photoelectric effect, threshold frequency, work function, detector, pixel, optical fibers** |
| **Key equations** | $E = hf$ $\qquad$ $E_k = hf - W_0$ |

## Review problems and questions

1. Describe evidence that light behaves like a particle and evidence that light behaves like a wave.

2. a. What is the energy of a photon of light that has a wavelength of $\lambda = 2$ μm?
   b. What is the energy of a photon of light that has $\lambda = 0.5$ μm?
   c. What part of the electromagnetic spectrum does each photon correspond to?

3. Explain how to find the maximum kinetic energy of an electron ejected from a material via the photoelectric effect.

4. A researcher was designing an experiment to demonstrate the photoelectric effect using the metal nickel, which has a work function of 5.15 eV. The electron volt, or eV, is a more convenient way to express photon energies than using joules. $1 \text{ eV} = 1.602 \times 10^{-19}$ J.
   a. Explain what is meant physically by the work function.
   b. Calculate the threshold frequency for nickel. To what region of the electromagnetic spectrum does this correspond?
   c. Will light at an energy of 7.35 eV result in the production of photoelectrons from nickel? In what region of the electromagnetic spectrum is this light?
   d. Will infrared light at a wavelength of 1 μm produce photoelectrons?
   e. Will violet light (400 nm) produce photoelectrons?
   f. Will ultraviolet light at 200 nm produce photoelectrons?

5. Many forms of modern communication use wave pulses of infrared and visible light to transfer digital information. Describe why infrared (IR) or visible light, rather than radio frequencies, is used for transmitting the signals needed for high-definition video.

# Chapter review

## Vocabulary

Match each word to the sentence where it best fits.

### Section 22.1

| | |
|---|---|
| electromagnetic wave | light |
| scattering | speed of light |
| vacuum | |

1. The daytime sky appears blue because of the _______ of blue light by small particles in the Earth's atmosphere.
2. The _______ is $3\times10^8$ m/s.
3. A/An _______ is an oscillating electric and magnetic field.
4. The volume inside a bell jar, when it has been evacuated of all matter, is a/an _______.
5. Electromagnetic waves are also called _______.

### Section 22.2

| | |
|---|---|
| dispersion | electromagnetic spectrum |
| gamma rays | index of refraction |
| infrared radiation | microwaves |
| radiation | radio waves |
| spectrum | ultraviolet light |

6. The mobile phone network transmits information at the wavelength region corresponding to _______.
7. Sunscreen is important to protect your skin from the damaging effects of _______.
8. The Sun's light energy travels to us as electromagnetic _______.
9. Radio waves, infrared light, ultraviolet light, and x-rays are all part of the _______.
10. The most energetic form of electromagnetic waves is _______.
11. The _______ of a material represents how the speed of light within the material differs from the speed of light in a vacuum.
12. Thermal leaks from houses can be detected using a camera that is sensitive to _______.
13. The longest wavelength electromagnetic waves are _______.
14. A triangular glass prism separates white light into its constituent colors through _______.
15. The _______ of visible light are the colors, wavelengths, and frequencies from red to green to violet.

### Section 22.3

| | |
|---|---|
| Planck's constant | detector |
| diffraction pattern | optical fibers |
| photoelectric effect | photon |
| pixel | polarization |
| quantum physics | spectrograph |
| threshold frequency | work function |

16. _______ is/are a technology that transmits light and information using a process of total internal reflection.
17. The _______ is the photon energy needed to liberate an electron from a metal.
18. The _______ is an experiment that is explained using wave–particle duality.
19. Light shining on a metal will eject no electrons unless the light exceeds the _______.
20. The _______ is a scientific instrument that disperses light.
21. The _______ projected onto a screen shows waves alternately adding and canceling each other.
22. That light can be considered as an indivisible bundle of energy is a phenomenon of the field of _______.
23. The image taken by a digital camera might have thousands or millions of individual _______(s).
24. The value of _______ is $6.63\times10^{-34}$ J s.
25. That light comes in discrete packets is the essential concept of the _______ model of light.

## Conceptual questions

### Section 22.1

26. Which sentence corresponds best to the equation $c = f\lambda$?
    a. Current equals footage times length.
    b. Velocity of sound equals frequency times amplitude.
    c. Speed of light equals frequency times wavelength.
    d. Speed of light equals frequency divided by wavelength.
27. In a vacuum, which waves travel fastest, radio, infrared, or ultraviolet?

Chapter 22

# Chapter review

## Section 22.1

28. Describe the relationship between the index of refraction of a material and the speed light travels in that material.

29. Sound waves cannot travel through a vacuum, but electromagnetic waves can. Can you think of an example of an electromagnetic wave traveling through a vacuum?

30. When electromagnetic waves enter a new medium, such as water, what happens to their speed? Is this similar to what happens to sound waves?

31. What happens when you slightly change the frequency of light in the visible spectrum? How does this compare to changing the frequency of a sound wave?

32. ❮ Which color of visible light has the lowest frequency? Which color has the highest frequency?

33. ❮ Why was the equation $c = f\lambda$ used in this chapter instead of $v = f\lambda$?

34. ❮❮ What is the range of wavelengths (in nanometers) that the human eye can see?

35. ❮❮ How are frequency and wavelength related for light?

36. ❮❮❮ In general, the speed of light decreases in denser materials whereas the speed of sound increases. For example, light travels at $3\times10^8$ m/s in air, $2.26\times10^8$ m/s in water, and $1.24\times10^8$ m/s in diamond. Sound travels travels at 343 m/s in air, 1,483 m/s in water, and 12,000 m/s in diamond. Why do you think this is? What is different about the way light and sound propagate?

## Section 22.2

37. The index of refraction of glass is larger for blue light than for green light. Which kind of light, blue or green, travels faster through glass?

38. The index of refraction of glass is larger for blue light than for green light. Which kind of light, blue or green, will have its path bent more when dispersed through a triangular glass prism?

39. Cell phones expose users to radio frequency (RF) radiation. Possible health effects of this exposure are the subject of current study and controversy. Research this issue by citing and quoting from two studies or articles: one providing evidence of a medical risk and the other indicating that any risk is negligible.

40. What characteristic of x-rays allows them to be used to image the bone structure of patients in medical facilities?

41. How do doctors use gamma rays to help treat their patients?

42. Inspect a chart of the electromagnetic spectrum and identify what type of electromagnetic radiation corresponds to each of the following.

    a. wavelength of 10 μm
    b. frequency of 1,050 kHz
    c. wavelength of 656 nm
    d. frequency of 0.1 THz

43. ❮ While skiing, Daniela looked at her friend Mauro through a thin piece of plastic but found it difficult to see him because of the glare off the snow. Why was the glare from the snow reduced when she rotated the plastic by 90°?

44. ❮❮ How does a microwave oven cook food? Research how this technology works and write a short essay describing how the microwaves are generated, what wavelengths or frequencies are involved, how the waves heat the food, and how the machine prevents the microwaves from escaping.

45. ❮❮ *The Boston Globe* reported in 2012 that a private company was taking pictures of area houses using cars with specially designed, roof-mounted cameras. "The cameras were designed to take pictures of homes on either side of the car as we roamed the streets of Belmont —pictures of heat seeping out from windows, eaves, and inadequately insulated walls... [This] was what you might call HeatView."

    a. At what portion of the electromagnetic spectrum do you think the cameras operate? Why?
    b. If it is legal to take pictures of the houses from the street using ordinary cameras operating at visible-light wavelengths, do you think it would be legal to take pictures at the wavelengths that these heat-sensing cameras are operating? Explain.
    c. The *Globe* article concludes, "Some architects, homeowners, and commercial landlords are already hunting for ways to cut energy consumption. But the key to success for this crop of software start-ups—and the key to curbing our need for new generating capacity—will be getting the mainstream to care, and spend money." Write a short essay describing how a company might use the information gathered by these heat-sensitive cameras to sell customers energy-saving upgrades to their homes. In your essay, communicate clearly what scientific information you have taken from the article and what inferences you made from the article.

46. ❮❮ Compare photons detected by a CAT scanning machine to those emitted by a sunning lamp. Which photons have longer wavelength? Higher frequency? Higher energy?

# Chapter review

## Section 22.2

47. When constructing the enclosure to detect infrared light in Investigation 22B on page 644, why is it helpful to cover the inside of the box with black paper?

48. Many sound waves can diffract around buildings—which means that a sound made on one side of the building can be heard on the other side. Light in the visible part of the spectrum, however, cannot diffract around a building—you can't see what is on the other side of a building. What is the main difference between them? Does this suggest a part of the electromagnetic spectrum that would be able to diffract around a building?

49. ‹‹ When using a triangular glass prism to disperse white light, is it more effective to use a diffuse light source or a narrow beam of light?

50. ‹‹‹ Why do night goggles detect infrared radiation? Why not gamma or ultraviolet radiation?

## Section 22.3

51. Why shouldn't you look directly at the Sun when using a hand-held visual spectroscope outdoors during a field investigation?

52. Describe the photoelectric effect and how it relates to the dual nature of light.

53. Polarized glasses block some of the light and reduce glare. Why is it still dangerous to look at the Sun when wearing polarized glasses?

54. ‹ How is the vibrational frequency of an electromagnetic wave related to its energy?

55. ‹ Explain what is meant by the number of "megapixels" in a digital camera.

56. ‹ For the signal of an AM radio station at 1,050 kHz, what does the frequency 1,050 kHz represent? Is it the frequency of the musical sounds broadcast by the station?

57. ‹‹ One student proposed a scientific explanation that light is a particle, based on the photoelectric effect. A second student tested the first student's hypothesis by conducting the double slit experiment, which showed that light interferes as a wave. The second student therefore proposed the scientific explanation that light is a wave. In your evaluation of the two scientific explanations and the evidence, who is right?

58. ‹‹ In the double slit experiment for light, how will the spacing of the interference bands (or fringes) vary with the wavelength of the light?

59. ‹‹‹ Sometimes a picture taken with a CCD-based camera shows a long streak trailing from a very bright source, such as a lamp, bright star, or the Sun. What property of the CCD do you think causes this?

### Quantitative problems

## Section 22.1

60. How much faster do radio waves travel in a vacuum than infrared light?

61. How many times faster is light than sound?

62. Light takes approximately 8.3 min to travel from the Sun to the Earth. About how far is the Sun from Earth in kilometers?

63. Astronomers often state the distances to other stars in terms of *light-years*. Using scientific notation, calculate the number of kilometers in one light-year. State your answer with two significant figures, using $c = 3.0 \times 10^8$ m/s for the speed of light in a vacuum.

64. You are directing a science-fiction movie in which Earthlings use light to communicate with the inhabitants of another planetary system. An Earth scientist sends a greeting and receives an answer a few seconds later. Is this possible given what you know about the speed of light? (The nearest star system to our own is roughly $4 \times 10^{16}$ m away in the far-southern constellation of Centaurus.)

65. ‹ The *Voyager 1* spacecraft is currently more than 19 billion km from the Earth. How long will it take a signal sent by the spacecraft to reach the Earth?

## Section 22.2

66. Sort the following kinds of electromagnetic radiation in order of increasing wavelength: yellow light, microwaves, x-rays, ultraviolet light, and violet light.

67. Sort the following kinds of electromagnetic radiation in order of increasing frequency: red light, radio waves, gamma rays, infrared light, and green light.

68. An electromagnetic wave with the same frequency as the A string on a violin (440 Hz) would lie in what part of the electromagnetic spectrum? How about one with the same wavelength (78.4 cm)? Why are these different?

69. ‹ Calculate the frequency of electromagnetic radiation that has a wavelength of 5 μm. What kind of electromagnetic radiation is it?

70. ‹ The Very Large Array radio telescope often observes at a wavelength of 3.6 cm. What is the frequency of these radio waves in gigahertz (GHz)?

# Chapter review

## Section 22.2

71. ❮ Compare 980 on the AM radio dial with 98 on the FM dial.
    a. Which frequency is higher?
    b. By what factor is it higher?
    c. What is the wavelength corresponding to each frequency?
    d. Which has a longer wavelength?

72. ❮❮ Television broadcasts on frequencies between 54 MHz (channel 2) and 806 MHz (channel 69).
    a. What are the wavelengths corresponding to these frequencies?
    b. How do these frequencies compare to AM/FM radio broadcast frequencies?

73. ❮❮❮ Communication with submarines typically uses frequencies below 1 kHz.
    a. What wavelengths of electromagnetic radiation does that correspond to?
    b. Name an object in your everyday environment that is similar in size to those wavelengths.
    c. How can such wavelengths be broadcast?
    d. Can the submarines broadcast at these wavelengths?

## Section 22.3

74. Calculate the energy of a 500 nm photon of light.

75. ❮ Approximately how many photons of infrared light ($\lambda$ = 6 μm) are there in 1 eV of energy?

76. ❮ A typical radio wave has a wavelength of 6 cm.
    a. Approximately how many photons of this type of radio "wave" are there in 1 eV of energy?
    b. Why might electromagnetic radiation at radio wavelengths often be treated as waves, not particles?

77. ❮ Sort the following kinds of electromagnetic radiation in order of increasing energy: ultraviolet light, visible light, infrared light, radio waves, gamma rays, and x-rays.

78. ❮❮ One electron volt or eV is an energy of $1.602\times10^{-19}$ J. The kiloelectron volt (keV) is often used to express the energy of photons in a particular region of the electromagnetic spectrum. Using an electromagnetic spectrum chart, determine what kind of electromagnetic radiation corresponds to an energy of 1 keV.

79. ❮❮ Chromium has a work function of $7.2\times10^{-19}$ J. What is the threshold frequency for this material?

80. ❮❮ A metal only ejects electrons by the photoelectric effect when a light with a wavelength of 286 nm or smaller shines on the metal. What is the work function of the metal?

81. ❮❮ Some kinds of electromagnetic radiation are typically described using units of energy, such as the electron volt (eV, where 1 eV = $1.602\times10^{-19}$ J).
    a. What frequency corresponds to electromagnetic radiation having an energy of 2.0 keV?
    b. What kind of electromagnetic radiation is it?

82. ❮❮ How many photons of orange light ($\lambda$ = 600 nm) does it take to make 1 J of energy?

83. ❮❮ A student who was investigating the photoelectric effect measured the threshold frequency for sodium to be $4.4\times10^{14}$ Hz. The student subsequently increased the frequency of the incident radiation to $7.0\times10^{14}$ Hz.
    a. What kind of electromagnetic radiation corresponds to the frequencies of $4.4\times10^{14}$ and $7.0\times10^{14}$ Hz?
    b. Based on his measurement, how much work is required to free an electron from the surface of the metal sodium?
    c. How much kinetic energy did the electrons ejected from the surface have when the frequency of the incident light was $4.4\times10^{14}$ Hz?
    d. How much kinetic energy did the electrons ejected from the surface have when the frequency of the incident light was $7.0\times10^{14}$ Hz?

84. ❮❮❮ A student investigates the photoelectric effect using calcium. For each frequency of the incident light, she measures the stopping voltage that she needs to apply between the electrodes to stop the flow of electrons. Graph the data and use your graph to find the threshold frequency for calcium.

| **Frequency *f* of incident light (Hz)** | **Stopping voltage *V* (V)** |
|---|---|
| $0.75\times10^{15}$ | 0.1 |
| $0.80\times10^{15}$ | 0.3 |
| $0.95\times10^{15}$ | 0.7 |
| $1.20\times10^{15}$ | 1.3 |

# Chapter review

## Standardized test practice

85. The highest energy electromagnetic waves are

A. gamma radiation.
B. ultraviolet radiation.
C. infrared radiation.
D. radio waves.

86. The constant of proportionality between the energy of a photon and its frequency is known as the

A. speed of light.
B. Rydberg energy.
C. Einstein constant.
D. Planck constant.

87. A photon with a frequency of $6\times10^{14}$ Hz has an energy of

A. $4.0\times10^{-19}$ J.
B. $5.0\times10^{-7}$ J.
C. $1.8\times10^{23}$ J.
D. $5.4\times10^{31}$ J.

88. The Moon is typically about 384,000 km from Earth. Very roughly, how long would it take light to travel from Earth to the Moon?

A. a second
B. a minute
C. an hour
D. a day

89. Which of the following statements is true about sound waves and light (electromagnetic waves)?

A. They cannot travel through a perfect vacuum.
B. They are transverse waves.
C. They are compressional (longitudinal) waves.
D. The shorter their wavelength, the higher is their frequency.

90. Which of the following best explains how a raindrop separates white light into different colors?

A. Water absorbs red light more than blue light.
B. Red and blue photons have different speeds when traveling through water.
C. Blue and red photons have different energies.
D. Rainbows only occur shortly before sunset.

91. Which of the following provides evidence that light can behave like a wave?

A. the Compton effect
B. the photoelectric effect
C. both A and B
D. neither A nor B

92. Which of the following provides evidence that light can behave like a wave?

A. diffraction
B. polarization
C. interference
D. all of the above

93. Which of the following is closest to the speed of light?

A. 2.99 m/s
B. 343 m/s
C. $3\times10^{8}$ m/s
D. $10^{10}$ m/s

94. One astronomer said that light is a wave. Another astronomer rejected this scientific explanation. In his critique, he said that light is a particle because individual photons are observed by detecting them with a CCD camera. What is the best explanation of why his critique is incorrect?

A. Light can exhibit both wave and particle properties.
B. Light does not create an image on a CCD detector; electrons do.
C. Some of the photons striking the CCD do not liberate any electrons.
D. Photons have nothing to do with light.

95. One Calorie (food calorie) is equivalent to 4,184 J. Suppose you eat a 300 Calorie sandwich. How many infrared photons would it take to match that amount of energy? (Assume that each photon has a wavelength of 10 μm, or $10^{-5}$ m, which is typical of the electromagnetic radiation emitted by a human body.)

A. $7.0\times10^{8}$
B. $6.3\times10^{15}$
C. $2.1\times10^{17}$
D. $6.3\times10^{25}$

96. The index of refraction in water is 1.330 for red light ($\lambda = 700$ nm), and 1.345 for violet light ($\lambda = 400$ nm). How much longer (in nanoseconds) will it take violet light than red to cross 1 m of water?

A. $3.3\times10^{-9}$ ns
B. $2.2\times10^{-7}$ ns
C. 220 ns
D. 330 ns

97. Both x-ray machines and PET scan devices work as medical imaging instruments using electromagnetic radiation that has what properties?

A. low energy and short wavelength
B. low energy and long wavelength
C. high energy and short wavelength
D. high energy and long wavelength

# Chapter 23 Properties of Matter

Circling the planet Saturn is one of the most enigmatic worlds in our Solar System. Technically a moon, Titan has a diameter of 5,150 km and is slightly larger than the planet Mercury. Unique among moons, Titan has a solid surface, a dense atmosphere, oceans, rivers, and even weather! Out of a hundred or so planets and moons in our Solar System, only four have solid surfaces and substantial atmospheres: Earth, Mars, Venus, and *Titan.* Like Earth, Titan has rainstorms, seasons, river valleys, mountains, and oceans. Unlike Earth, however, the temperature on Titan is so cold that the liquid in Earth's rivers and oceans is the same substance that makes solid mountains on Titan. The *gases* that might be in Earth's atmosphere fall as liquid rain on Titan and flow in its mysterious dark rivers.

Credit: NASA, ESA, JPL, and the University of Arizona

The mean surface temperature on Titan is −190°C (−270°F). At this extremely low temperature the phases of ordinary matter become very different from what we experience on balmy Earth. For example, water is as hard as granite. Titan's rocks and surface are largely water ice with hydrocarbons mixed in. Liquid water on Titan is more analogous to volcanic magma here on Earth. Instead of $H_2O$, Titan's "water cycle" consists of liquid methane and ethane. The boiling point of methane ($CH_4$)—the *natural gas* used for cooking and heating on Earth—is −161°C, about 30° above Titan's surface temperature. Flammable natural gas and ethane flow in vast rivers and collect in deep oceans on Titan. Titan's lakes and oceans contain hundreds of times more hydrocarbon fuels than all the known reserves of fossil fuels on Earth. Unfortunately, these vast resources are also 1.3 billion kilometers away, nearly 10 times the distance from Earth to the Sun.

On January 14, 2005, the *Huygens* probe, a joint effort between NASA and the European Space Agency (ESA), landed on Titan. *Huygens* is the first and only human technology to land in the outer Solar System. The car-sized probe took more than eight years to reach Titan and carried enough battery power to collect data for 90 minutes on Titan's surface.

# Chapter study guide

Chapter 23

## Chapter summary

Temperature measures the kinetic energy in random motion of atoms or molecules. Units of temperature are degrees Celsius (°C), degrees Fahrenheit (°F), and kelvins (K). Heat is the total thermal energy in a quantity of matter. The specific heat of a substance describes its thermal energy per unit mass per degree. Forces in fluids act through pressure. Pressure has units of pascals ($N/m^2$) and pounds per square inch (psi or $lb/in^2$). Weight creates hydrostatic pressure both within the atmosphere and beneath the surface of water. Bernoulli's equation relates pressure, volume, and height in a moving fluid. The pressure, volume, and temperature of a gas are related by the ideal gas law. The kinetic theory of matter relates the microscopic interactions of atoms to macroscopic properties such as pressure and temperature. The ideal gas law and the concept of specific heat are explained by the kinetic theory.

## Learning objectives

By the end of this chapter you should be able to

- distinguish between temperature and heat;
- convert between Fahrenheit and Celsius temperatures;
- describe how the particle nature of matter explains temperature;
- explain why the phases of matter occur at different temperatures;
- use the ideal gas law to calculate pressure, volume, or temperature;
- calculate thermal energy of a substance using its specific heat;
- calculate pressure underwater from depth;
- use Bernoulli's equation to explain fluid effects; and
- explain key concepts of the kinetic theory of matter.

## Investigations

23A: Phases of matter
23B: Specific heat of water and steel
23C: Pressure of an ideal gas

## Chapter index



## Important relationships

$T_C = \frac{5}{9}(T_F - 32)$ $N_A = 6.022 \times 10^{23}$ $E = \frac{3}{2}k_BT$ $Q = mc_p\Delta T$ $P = \frac{F}{A}$

$T_F = \frac{9}{5}T_C + 32$ $PV = nRT$ $PV = Nk_BT$ $\rho = \frac{m}{V}$ $P = \rho gd$

$T_K = T_C + 273.15$ $\rho gh + \frac{1}{2}\rho v^2 + P = \text{constant}$ $v_{th} = \sqrt{\frac{3k_BT}{m}}$ gas: $c_v = \frac{3}{2}\frac{N_Ak_B}{m_{mol}}$ solid: $c_p = \frac{3N_Ak_B}{m_{mol}}$

## Vocabulary

| | | | |
|---|---|---|---|
| temperature | Brownian motion | thermometer | Celsius scale |
| Fahrenheit scale | absolute zero | Kelvin scale | Avogadro's number |
| mole | kinetic theory | Boltzmann's constant | phases of matter |
| gas | liquid | solid | heat |
| thermal energy | calorie | Calorie | specific heat |
| fluid | pressure | compressible | incompressible |
| Boyle's law | Charles's law | ideal gas | ideal gas law |
| density | Bernoulli's equation | streamline | statistical mechanics |
| Maxwellian distribution | | | |

## 23.1 - Temperature and heat

In physics, the concepts of *hot* and *cold* do not refer to *heat,* but instead they describe *temperature.* Hot means higher temperature and cold means lower temperature. The cause of temperature and the relationship between temperature and heat remained mysteries for thousands of years. In 1905, Albert Einstein derived the compelling connection between atoms and temperature—one of his many discoveries that year—proving the existence of atoms. This section explains the meaning of temperature and its connection to *thermal energy* and *heat.*

### Temperature and thermometers

**Measuring temperature**

Humans and most other animals have some sense of temperature as part of the sense of touch. Our temperature sense is not very accurate, however, so for more quantitative purposes **temperature** is measured with a **thermometer**. Many properties of matter change with temperature and many different kinds of thermometers have been invented, such as alcohol oral thermometers and infrared ear thermometers. Many technologies, such as *thermostats*, include a temperature sensor, which is the measuring part of an electronic or digital thermometer.

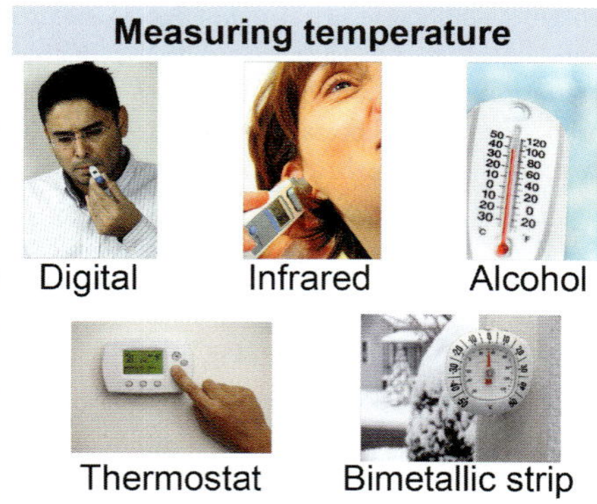


**Brownian motion**

If you use a powerful microscope to observe a tiny speck of pollen floating in water, then you observe the pollen speck to move constantly in a jerky, irregular way. This motion mystified Irish botanist Robert Brown in 1827, and it remained unexplained for the next 85 years. Several questions defied conventional explanation:

Why doesn't a pollen speck remain still in calm water?
Why does it jump about?
Where does the energy of motion come from?

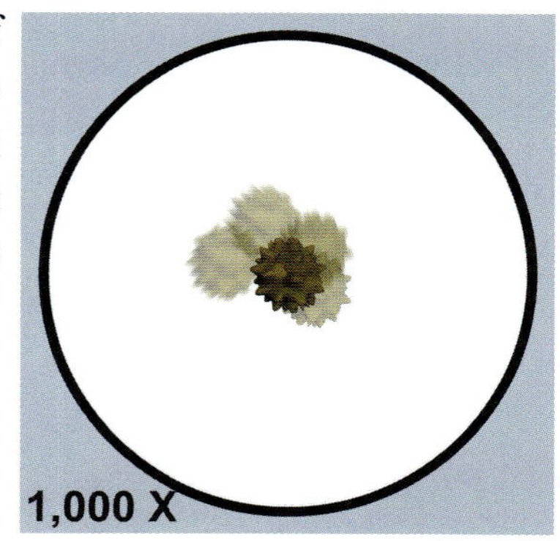


**What is temperature?**

**Brownian motion** was explained by Einstein in 1905 when he proposed that matter on the atomic level *is constantly in agitated motion.* Brownian motion occurs because a pollen speck is so small that individual collisions with water molecules are evident.

1. Individual atoms and molecules are *never* at rest but are constantly agitating about and bumping into each other trillions of times per second.
2. There is a kinetic energy associated with the random microscopic motion of atoms and molecules, and this kinetic energy is what temperature measures. Higher temperature means that the individual atoms in matter are moving about on the microscopic scale with more kinetic energy.

**The meaning of temperature**

Temperature measures the average kinetic energy in random motion of atoms and molecules.

**What are the units of temperature?**

The Fahrenheit and Celsius scales are commonly used to measure temperature. Both are based on the properties of water. The **Celsius scale** defines zero degrees as the freezing point of water and 100 degrees as the boiling point of water. The **Fahrenheit scale** defines 32 degrees as the freezing point of water and 212 degrees as the boiling point of water.

**Converting from Fahrenheit to Celsius**

The Fahrenheit scale has 180 degrees between the boiling and freezing points of water. The Celsius scale, however, has 100 degrees between the same two points. Therefore, a change of 180°F is equivalent to a change of 100°C. To convert from degrees Celsius to degrees Fahrenheit, you multiply by a factor of 9/5 then add 32°, because 0°C is the same as 32°F. Equations (23.1) and (23.2) are used to convert temperatures between the two scales.

**Relative temperature scales**

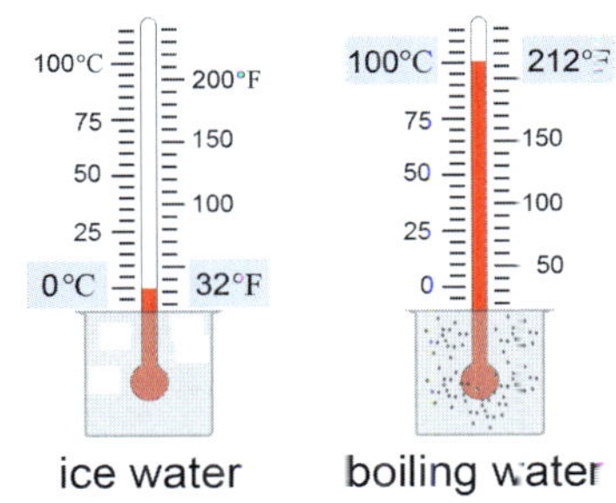


Interactive equation

$$\text{(23.1)} \quad T_C = \frac{5}{9}(T_F - 32)$$

$T_C$ = temperature in degrees Celsius (°C)
$T_F$ = temperature in degrees Fahrenheit (°F)

**Celsius** temperature scale

$$\text{(23.2)} \quad T_F = \frac{9}{5}T_C + 32$$

$T_F$ = temperature in degrees Fahrenheit (°F)
$T_C$ = temperature in degrees Celsius (°C)

**Fahrenheit** temperature scale

**Absolute zero**

The choice of 0°C as the freezing point of water is quite arbitrary in that water molecules do *not* have zero kinetic energy at 0°C. Instead, the kinetic energy in molecular motion is effectively zero at a much lower temperature called *absolute zero*. **Absolute zero** is −273.15°C (−459.7°F) and is the minimum possible temperature that matter can have. *Temperature cannot go lower than absolute zero.*

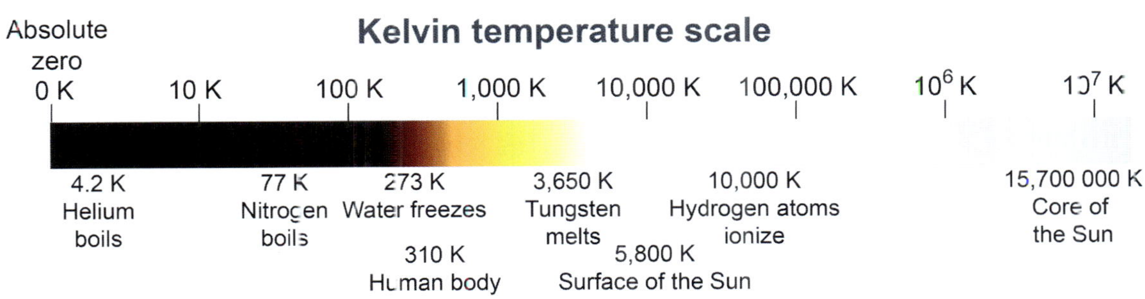


**The Kelvin scale**

The **Kelvin scale** of temperature starts at absolute zero. There are no "degrees" in the Kelvin scale, even though 1 K represents the same temperature difference as 1°C. Since temperatures in kelvins are referenced to absolute zero, a temperature on the Kelvin scale is called an *absolute temperature*. Most equations relating to heat, such as the gas laws, use absolute temperatures in kelvins. You can convert between degrees Celsius and kelvins by using equation (23.3).

Interactive equation

$$\text{(23.3)} \quad T_K = T_C + 273.15$$

$T_K$ = temperature in kelvins (K)
$T_C$ = temperature in degrees Celsius (°C)

**Kelvin** temperature scale

**How many atoms are in matter?**

Using very clever experiments, in 1895 French physicist Jean Perrin determined that there were about $6 \times 10^{23}$ atoms in 1 g of hydrogen, the lightest element. Subsequent experiments have refined **Avogadro's number** to its modern value of $N_A = 6.022 \times 10^{23}$. The number is named in honor of the Italian scientist and teacher Amadeo Avogadro. One **mole** is defined as Avogadro's number. For example, one mole of hydrogen atoms is $6.022 \times 10^{23}$ atoms. The mole is one of the seven fundamental quantities of the SI system.

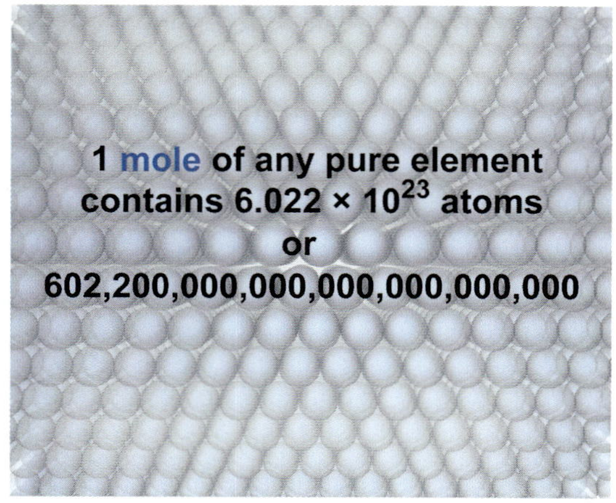


| | | |
|---|---|---|
| **(23.4)** $N_A = 6.022 \times 10^{23}$ | $N_A$ = number of atoms in one mole | **Avogadro's number** |

**How do you get from atoms to properties like temperature?**

You use statistics to calculate an average of a set of numbers, such as the average rent for an apartment in a certain city. The **kinetic theory** of matter uses statistics to analyze the average behavior of trillions of individual atoms. Kinetic theory shows how the average behavior of $10^{23}$ particles produces observable properties such as temperature and pressure. A fundamental result of kinetic theory is that the average energy of a single atom resulting from random thermal motion in any particular direction is given by equation (23.5), where $k_B$ is **Boltzmann's constant** and $T$ is the absolute temperature in kelvins. Equation (23.5) relates the microscopic property of an atom's thermal energy $U$ to its macroscopic property of temperature $T$.

Interactive equation

| | | |
|---|---|---|
| **(23.5)** $E = \frac{3}{2} k_B T$ | $E$ = energy (J)<br>$k_B$ = Boltzmann's constant = $1.38 \times 10^{-23}$ J/K<br>$T$ = absolute temperature (K) | **Thermal energy per atom** |

**Understanding Boltzmann's constant**

If we did not have the historical measure of temperature in "degrees," then we might instead use equation (23.5) to *define* temperature in units of *energy*. A temperature of 20°C (293 K) is equivalent to $6.07 \times 10^{-21}$ joules per particle, for example. Think of Boltzmann's constant and Avogradro's number as numerical "bridges" between the microscopic world of atoms and molecules and macroscopic properties such as density, temperature, and pressure. Boltzmann's constant has a very small value at $1.38 \times 10^{-23}$ J/K, which indicates that individual atoms have very small energies.

**Mass of one atom**

One mole of carbon has a mass of 12 g. What is the mass of a single atom?

| | |
|---|---|
| *Asked:* | mass of a single atom |
| *Given:* | 12 g = 1 mole of carbon atoms |
| *Relationships:* | 1 mole = $6.022 \times 10^{23}$ particles |
| *Solution:* | $\frac{12\text{ g}}{\text{mol}} \times \frac{1\text{ mol}}{6.022 \times 10^{23}\text{ atoms}} = 1.99 \times 10^{-23}$ g/atom |
| *Answer:* | $1.99 \times 10^{-23}$ g/atom |

**The phases of matter**

Matter can be a solid, liquid, gas, or plasma—four of the **phases of matter**—depending on its temperature. In the kinetic theory, the phase of matter for a particular substance is related to the average thermal energy of its atoms or molecules. At any given temperature, the phase of a material depends on the strength of the attractive forces that pull atoms and molecules together compared to the chaotic agitation of thermal motion.

**The gas phase**

In a **gas** the thermal motion of atoms completely overcomes attractive forces between molecules. Molecules in a gas are typically far apart from each other. Water becomes a gas above its *boiling point* of 100°C. Of the three phases, gas has the highest kinetic energy per molecule and occurs at the highest temperature.

**The liquid phase**

Between 0°C and 100°C water is a **liquid**. Molecules in a liquid have enough thermal energy to break away temporarily from their neighbors and change places with one another, but liquid molecules do not have enough thermal energy to separate completely and become a gas. Liquids flow because molecules can change places with their neighbors.

**The solid phase**

Below 0°C water is a **solid**. Matter becomes solid when thermal energy is too low to overcome intermolecular forces. Molecules in a solid are still vibrating, but an average molecule does not have enough energy to break away from its immediate neighbors or to exchange places with another molecule. That is why solids hold their shape.

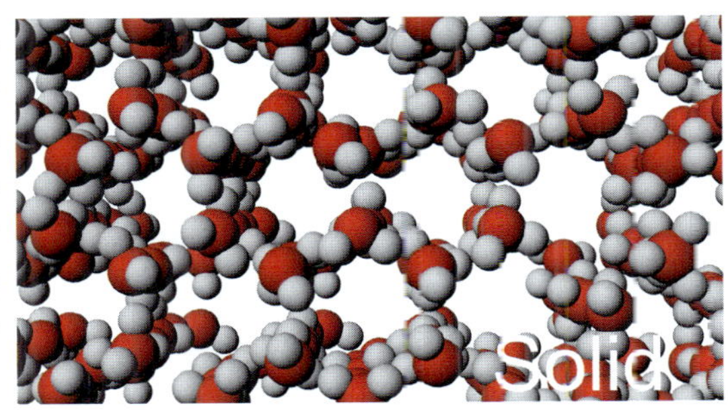


**The plasma phase**

Plasma is a less well know phase of matter. We don't regularly come in contact with plasma, due to the high temperatures required (10,000 K for a hydrogen atom and even higher temperatures for other atoms!). The plasma phase occurs when a gas is at a temperature high enough that some (or all) of the electrons break free form their atoms or molecules. Speeding around in a plasma are both (negatively charged) free electrons and the rest of the atoms (the positively-charged ions). Plasma is a good conductor of electricity, can block some wavelengths of light, and, because it is composed of charged particles, can be manipulated by both electric and magnetic fields.

**Energy and phase changes**

It takes energy to be absorbed by a solid for it to *melt* into a liquid. Similarly, energy is absorbed by a liquid to *evaporate* it into a gas. This is known as an endothermic process. Conversely, an exothermic process occurs when energy is released, such as when a gas *condenses* back into a liquid or when a liquid *freezes* back into a solid.

# Investigation 23A  Phases of matter

**Essential questions** *How do we explain how the same substance can be a solid, liquid, or gas at different temperatures?*

Interactive simulation

Temperature measures the kinetic energy that comes from random motion of individual atoms (or molecules). In matter, atoms are attracted to neighboring atoms by *intermolecular forces.* These attractive forces create potential energy, much like springs. Whether matter is solid, liquid, or gas depends on the balance between the attractive forces between neighboring atoms and the agitation of thermal energy that tends to separate atoms from each other.

## Simulating the phases of matter

The interactive model simulates a collection of magnified atoms in slow motion. You may choose different elements and set the temperature (in kelvins) using numbers or the arrow buttons.

**How can you tell when the atoms are in a solid, liquid, or gaseous state?**

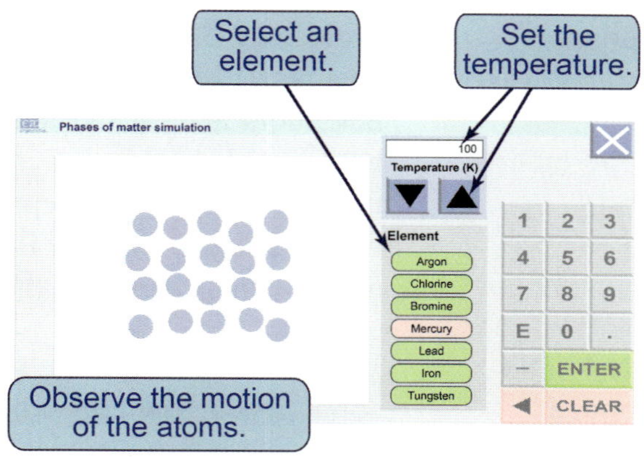


**What is the melting point of mercury? What is the boiling point?**

1. Choose mercury as an element.
2. Watch the simulation for temperatures of 200 K (−73°C), 300 K (27°C), and 1,000 K (727°C). You may have to watch for a few minutes as the atoms reach equilibrium at the chosen temperature.
3. Try the simulation for different elements.

a. What different behaviors of the atoms in the model do you associate with the phases of solid, liquid, and gas?

b. How can you use the visual behavior of the atoms in the simulation to determine the melting point of an element? The boiling point?

c. Determine the approximate melting and boiling points for the elements chlorine, bromine, mercury, iron, lead, and tungsten. Tabulate your results.

d. Research the melting and boiling points for each of these elements, as well as for argon. Convert the data to kelvins and compare them to what you observed in the simulation.

e. Why is it difficult to determine the melting and boiling point for argon using this simulation?

f. By referring to what you learned from this simulation, describe the connection between energy as viewed at the macroscopic scale and the motions of the particles.

## Understanding this simulation

This simulation uses 20 atoms in two dimensions. Real matter contains $10^{20}$ or more atoms in three dimensions and the atoms move *much* faster. Like all models, the approximations included in the simulation approximate behavior that is similar to, but not exactly like, the behavior of real atoms in matter.

**Is heat different from temperature?**

The words *heat* and *temperature* are often used interchangeably in conversation, but in physics they are different.

1. *Temperature* describes the *average* kinetic energy in random thermal motion per atom or molecule.
2. **Heat** and **thermal energy** describe the *total amount of energy* resulting from temperature in a quantity of matter containing many atoms or molecules.

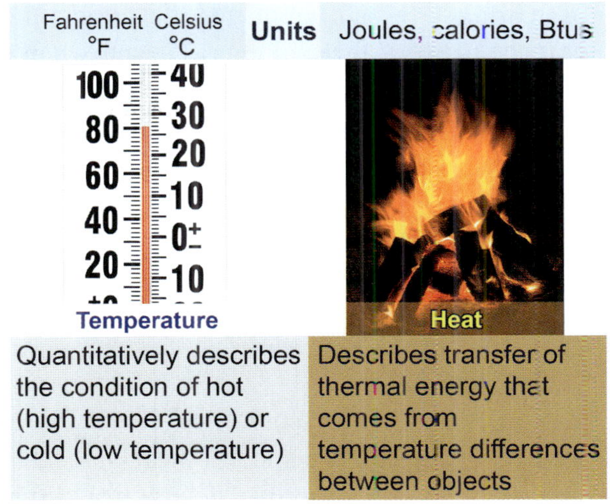


More specifically, *thermal energy* is the energy associated with temperature, whereas *heat* is thermal energy that can be transferred from one object to another owing to their difference in temperature. For example, heat flows from a hotter object to a colder object.

**Units of thermal energy and heat**

Heat is a form of energy and is therefore measured in joules. For historical reasons, however, there are many other units for heat.

- One **calorie** is 4.184 J.
- One food **Calorie** is 1 kilocalorie (kcal) or 4,184 J.
- One British thermal unit (Btu) is 1,055 J.

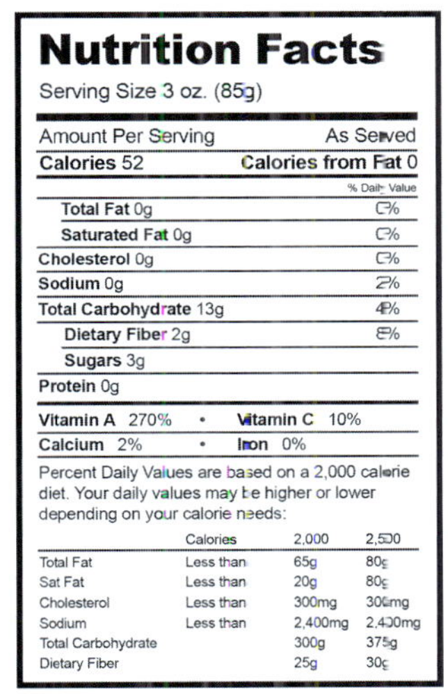

**Nutrition Facts**

Serving Size 3 oz. (85g)

| Amount Per Serving | As Served |
|---|---|
| **Calories** 52 | **Calories from Fat** 0 |
| | % Daily Value |
| **Total Fat** 0g | 0% |
| **Saturated Fat** 0g | 0% |
| **Cholesterol** 0g | 0% |
| **Sodium** 0g | 2% |
| **Total Carbohydrate** 13g | 4% |
| **Dietary Fiber** 2g | 8% |
| **Sugars** 3g | |
| **Protein** 0g | |

**Vitamin A** 270% • **Vitamin C** 10%

**Calcium** 2% • **Iron** 0%

Percent Daily Values are based on a 2,000 calorie diet. Your daily values may be higher or lower depending on your calorie needs:

| | Calories | 2,000 | 2,500 |
|---|---|---|---|
| Total Fat | Less than | 65g | 80g |
| Sat Fat | Less than | 20g | 80g |
| Cholesterol | Less than | 300mg | 300mg |
| Sodium | Less than | 2,400mg | 2,400mg |
| Total Carbohydrate | | 300g | 375g |
| Dietary Fiber | | 25g | 30g |

Nutrition labels list the energy content of foods in kilocalories: Calories with an uppercase "C"! Home heating systems and fuels are rated in Btus. For example, a typical home heating furnaces might deliver 100,000 Btus of heat per hour.

**How many joules in a Calorie?**

Nutritional labels use Calories as the unit of energy, so how many joules of energy are there in 1 Calorie? Start with 1 Calorie and use the unit conversions between Calories/calories and calories/joules to convert a Calorie into joules:

$$1 \text{ Calorie} = (1 \text{ Calorie})\left(\frac{1{,}000 \text{ calories}}{1 \text{ Calorie}}\right)\left(\frac{4.184 \text{ J}}{1 \text{ calorie}}\right) = 4{,}184 \text{ J}$$

**Heat from friction**

All forms of energy can be transformed into heat. A common example is heat generated through *friction*. Work done against friction is partially transformed into heat. For example, a friction force of 1 N acting over 1 m converts 1 J of work into heat. Moving parts get warm and can even melt from the heat generated through friction.

**Heat from fuels**

A liter of gasoline releases 35,000,000 J of chemical energy when completely burned with oxygen. This chemical energy is almost entirely released from the burning reaction in the form of heat. Most of the heat (65%) flows out of the radiator and tailpipe. On average, only about 13% of the heat released from the fuel goes toward physically moving the car.

## Specific heat

**Does equal temperature mean equal thermal energy?**

Different materials contain different amounts of thermal energy, even when they are at the same temperature. For example, it takes 4,184 J of energy to raise the temperature of 1 kg of water by 1°C. By comparison, it only takes 900 J to raise the temperature of 1 kg of aluminum by 1°C. The quantity of matter (1 kg) is the same, but the change in thermal energy per degree of temperature is different.

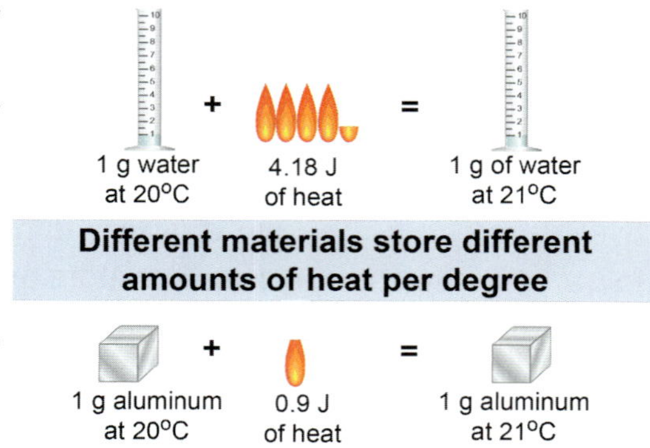

Different materials store different amounts of heat per degree

**Specific heat**

The amount of thermal energy per degree per kilogram is called the **specific heat**. Water has a specific heat of 4,184 J/(kg °C), which means that water contains 4,184 J of thermal energy per degree per kilogram. Aluminum has a lower specific heat of 900 J/(kg °C) and contains 900 J of thermal energy per degree per kilogram. Thermal energy (as well as heat) depends on temperature, mass, and specific heat.

**Equation for specific heat**

The change in heat $Q$ for an object depends on its mass $m$ and its change in temperature from $T_1$ and $T_2$, as given by equation (23.6). The specific heat $c_p$ is determined by the material and can vary widely. The capital letter $Q$ is traditionally used to represent thermal energy or heat, even though the letter $E$ is used to represent other forms of energy.

Interactive equation

**Specific heat**

$$Q = mc_p(T_2 - T_1) = mc_p\Delta T \qquad (23.6)$$

$Q$ = thermal energy or heat (J)
$m$ = mass (kg)
$c_p$ = specific heat (J kg$^{-1}$ °C$^{-1}$)
$T_1$, $T_2$ = temperature (K or °C)
$\Delta T$ = change in temperature (K or °C) = $T_2 - T_1$

**About specific heat**

The specific heat of ordinary matter varies from about a hundred to a few thousand joules per kilogram and per degree Celsius. The table on the right gives some examples for common solids, liquids, and gases. Solid metals have the lowest specific heat. Heavier atoms (or denser materials) typically have lower specific heats.

**Specific Heats**

| Material | $c_p$ [J/(kg °C)] |
|---|---|
| Water | 4,184 |
| Wax | 3,430 |
| Pine (wood) | 2,500 |
| Gasoline | 2,220 |
| Water ice | 2,000 |
| Oak (wood) | 2,000 |
| Water vapor | 1,970 |
| Oil | 1,800 |
| Plastic | 1,600 |
| Air | 1,000 |
| Asphalt | 920 |
| Aluminum | 900 |
| Glass | 840 |
| Granite | 790 |
| Steel | 470 |
| Lead | 130 |
| Gold | 130 |

Solid | Liquid | Gas

**Water**

Liquid water has the highest specific heat of common substances, which means that it takes a lot of energy to change its temperature. This property makes liquid water a good moderator of temperature on Earth, creating a suitable environment for life. The $H_2O$ molecule is fairly light, and intermolecular forces between water molecules are unusually strong because the water molecule is *polar*. The polar arrangement of the atoms contribute to the cohesive behavior of water and also makes water a versatile solvent. Notice that the specific heat of water is quite different depending on whether the phase is solid, liquid, or gas. Specific heat depends on the strength of interactions between molecules. Since water expands upon freezing, ice molecules are farther apart than liquid water molecules, reducing the strength of the intermolecular forces.

# Investigation 23B

## Specific heat of water and steel

**Essential questions** *What is the difference between temperature and heat?*
*How does the specific heat differ between water and steel?*

In everyday conversation we often use the words heat, thermal energy, and temperature interchangeably. In physics, however, they are different (albeit related) quantities. Temperature is the measure of the average kinetic energy per atom or molecule. Thermal energy is the total energy a substance contains. Heat is the flow of thermal energy from a hotter to a colder material. In this investigation you will see that water and steel contain very different amounts of thermal energy for the same mass—even though they are at the same temperature.

### Specific heat of steel

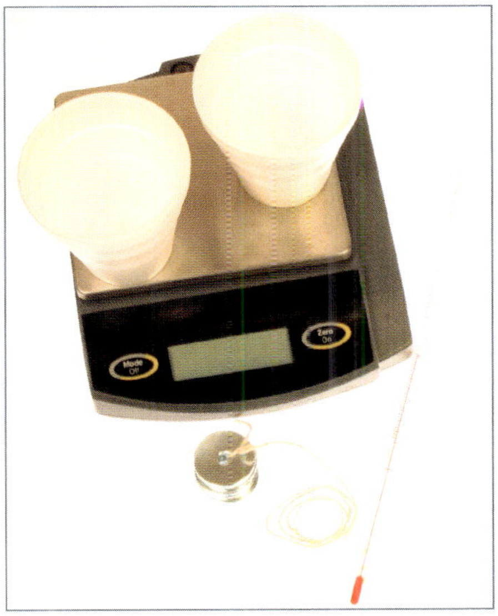

**Materials:** You will need two foam cups (8 oz or larger), water, a hot plate (or very hot water from tap), ice, a mass scale, 10 or more steel washers ($m \geq 100$ g), string, a Celsius thermometer (or other temperature probe), a ring stand, and a clamp.

1. Tie the steel washers together with the string.
2. Tare (or zero) the scale with one of the empty foam cups. Measure the mass of the steel washers. Choose a number of steel washers that will have a total mass close to 100 g.
3. Cover the washers with ice and cold water. Allow to equilibrate close to the freezing point. Measure the temperature of the cold water.
4. Mount the foam cup in the ring stand and ring clamp.
5. Heat water in an appropriate container using a hot plate.
6. Measure 100 g of hot water into the other foam cup. Measure the temperature of the water.
7. Quickly move the steel washers from the ice water into the hot water. Stir them for a minute or two until the temperature equilibrates. Measure the final temperature of the water.
8. Repeat the experiment by adding 100 g of ice water to the cup containing 100 g of hot water.

a. *Which caused a larger change in the temperature of the hot water, the 100 g of 0°C steel or the 100 g of 0°C ice water?*
b. *Which material do you think has a higher specific heat, steel or water? Why?*
c. *Calculate the specific heat of steel from your data given the specific heat of water:* $c_{water}$ *= 4.18 J/(g °C) = 4,180 J/(kg °C). Ask your teacher for the correct value—or look it up in an authoritative reference book, such as the CRC Handbook of Chemistry and Physics. Was your calculation correct?*
d. *Approximately equal masses of hot water and cold steel were mixed, yet the final temperature is not halfway between the two temperatures. Why not?*
e. *Now imagine you are given an unknown metal. Explain very specifically and in detail how you would design an experiment to determine whether the unknown metal is steel.*
f. *How did the distribution of energy between the washers and water change during the experiment?*
g. *What would happen to the final temperature if you put twice as many (i.e., 200 g) cold steel washers into the water? How does the final temperature depend on the relative mass of the steel and water?*

# Section 1 review

In everyday life we might use the words temperature and heat interchangeably. In physics, however, they are different concepts. Temperature measures the average kinetic energy in random thermal motion per atom or molecule. Temperature is measured on the Celsius, Fahrenheit, and Kelvin scales. Both Celsius and Fahrenheit are based on the phase changes of water but the Kelvin scale is based on absolute zero. Matter takes different phases (solid, liquid, or gas) depending on its temperature. Heat, or thermal energy, is the total energy in random thermal motion for a whole *collection* of atoms or molecules. The strength of intermolecular bonds varies among materials. For this reason, equal amounts of different substances usually contain different amounts of heat even at the same temperature. The *specific heat* is the thermal energy (or heat) per unit mass for a substance, per degree Celsius.

| | |
|---|---|
| **Vocabulary words** | **temperature, Brownian motion, thermometer, Celsius scale, Fahrenheit scale, absolute zero, Kelvin scale, Avogadro's number, mole, kinetic theory, Boltzmann's constant, phases of matter, gas, liquid, solid, heat, thermal energy, calorie, Calorie, specific heat** |

| | | | |
|---|---|---|---|
| **Key equations** | $T_C = \frac{5}{9}(T_F - 32)$ | $T_F = \frac{9}{5}T_C + 32$ | $T_K = T_C + 273.15$ |
| | $N_A = 6.022 \times 10^{23}$ | $E = \frac{3}{2}k_B T$ | $Q = mc_p(T_2 - T_1) = mc_p\Delta T$ |

## Review problems and questions

1. An Italian exchange student living in the United States wants to set his apartment's thermostat to 20°C. At what temperature in degrees Fahrenheit should he set the thermostat?

2. An American tourist in Paris, Goldie Locks, set her hotel room thermostat to 68°—not realizing that the thermostat works in degrees Celsius instead of degrees Fahrenheit.
   a. In an hour or two, will her hotel room feel hot, cold, or just right?
   b. At what temperature (in degrees Fahrenheit) did she set the thermostat?
   c. Would you expect there to be some property of the thermostat that would prevent her from setting it at 68°? Explain.

3. A racing cyclist burns 1,260 Calories in an hour, which go into powering his bicycle.
   a. How many joules does he burn in an hour?
   b. How many watts of power does he generate?
   c. How many incandescent light bulbs rated at 100 W could he theoretically power with this energy?
   d. How many fluorescent bulbs rated at 100 W could he power?

4. Three 30 g metal balls, one each of aluminum, copper, and lead, are placed into a large beaker of hot water for a few minutes. [The specific heats of aluminum, copper, and lead are 903, 385, and 130 J/(kg °C), respectively.]
   a. Which, if any, of the balls will reach the highest temperature? Explain.
   b. Which, if any, of the balls will have the most thermal energy? Explain.

5. What is the energy of a mole of carbon atoms at a room temperature of 20°C?

Both liquids and gases are *fluids*. A **fluid** is a form of matter that easily flows and changes shape in response to an applied force. Fluids conform to the shape of their container under the force of gravity. The motion of fluids is important in all areas of science, including the flow of water and air in the environment, aerodynamics, and even blood flow within your body.

## Pressure

**What are the units of pressure?**

The concept of *pressure* is important because forces in fluids act differently from forces on rigid solids. **Pressure** is force per unit area as described by equation (23.7); in the SI system, pressure has units of newtons per square meter ($N/m^2$). In the English system, pressure has units of pounds per square inch ($lb/in^2$), often abbreviated *psi*. Our atmosphere at sea level has an average pressure of 101,325 $N/m^2$, or 14.7 psi.

Interactive equation

**Pressure**

$$P = \frac{F}{A} \qquad (23.7)$$

$P$ = pressure (Pa or $N/m^2$)
$F$ = force (N)
$A$ = area ($m^2$)

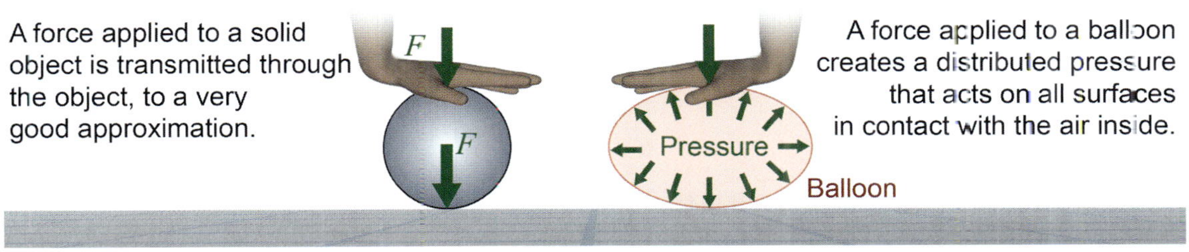


**How is pressure related to force?**

Pressure creates distributed forces on all surfaces in contact with a fluid. Think about pushing on a bowling ball compared to pushing on an inflated balloon. When you squeeze a balloon the fluid (air) inside exerts a uniform outward pressure everywhere within the balloon. The pressure in the balloon increases until the pressure multiplied by the area of your hand balances the downward applied force. Another way to state equation (23.7) is that *force = pressure × area.*

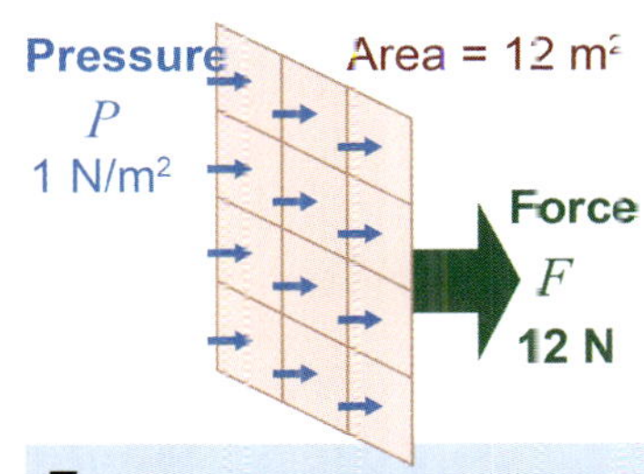


Force = pressure × area

$$F = P \times A$$

**Gauge pressure**

Tire gauges read the *gauge pressure*, which is pressure *in addition to that of the atmosphere*. A gauge pressure of 5.3 psi means the *absolute pressure* is 14.7 + 5.3 = 20.0 psi.

**What causes pressure?**

The cause of pressure is found in the microscopic behavior of atoms, similar to the case for temperature. Think about the air molecules near the inner surface of the balloon. Any molecule that hits the surface bounces back. The act of "bouncing back" is a change in motion that is caused by the balloon surface exerting a force on each air molecule. The reaction force is the molecule exerting a force back against the balloon. Pressure is the result of adding up the reaction forces from trillions of molecules colliding with every square centimeter each second.

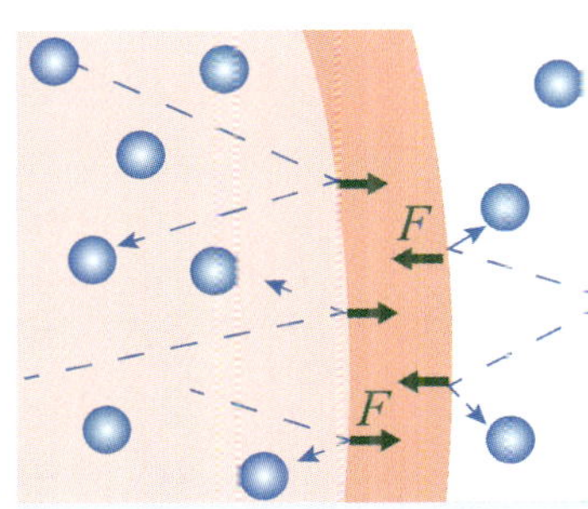


Pressure comes from trillions of molecular impacts.

**How are liquids and gases different?**

A fluid is **incompressible** when its volume stays roughly constant despite changes in pressure. Most liquids are incompressible. A fluid is **compressible** when it changes volume proportional to a change in pressure. Gases are compressible. A gas will expand its volume to fill any size container, usually changing both pressure and temperature in the process.

**Pressure and volume**

Imagine a volume of gas confined in a cylinder by a sliding piston. As the piston moves up and down, the amount of gas (mass) stays the same, but the volume, pressure, and temperature can change. If the volume is *decreased* then the pressure *increases*. This is mathematically stated by *Boyle's law*, which says that the change in pressure is inversely proportional to the change in volume, or $P \propto 1/V$, when temperature is held constant. The converse is also true: If the volume goes up, then the pressure goes down.

**Boyle's law**

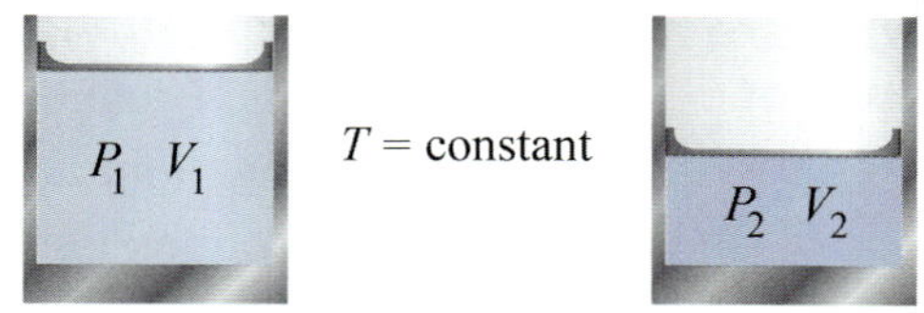


If the temperature stays the same then pressure is inversely related to volume.

$$\frac{P_1}{P_2} = \frac{V_2}{V_1} \quad \text{or} \quad P_1V_1 = P_2V_2$$

Boyle's law is often written in the form $P_1V_1 = P_2V_2$, where $P_1V_1$ are the initial pressure and volume and $P_2V_2$ are the final values. With this form, the final pressure and volume are inversely proportional as $P_2 \propto 1/V_2$.

**Pressure, temperature, and volume**

When pressure is kept constant, gases expand and contract proportional to their absolute temperature, or $V \propto T$. This rule is known as *Charles's law* after French scientist and balloonist Jacques Charles.

If instead the volume is kept fixed, then the pressure changes proportional to the change in absolute temperature, or $P \propto T$. This last rule is known as *Gay-Lussac's law* after another French scientist, Joseph Gay-Lussac, who discovered it experimentally in 1809.

**Charles's law**

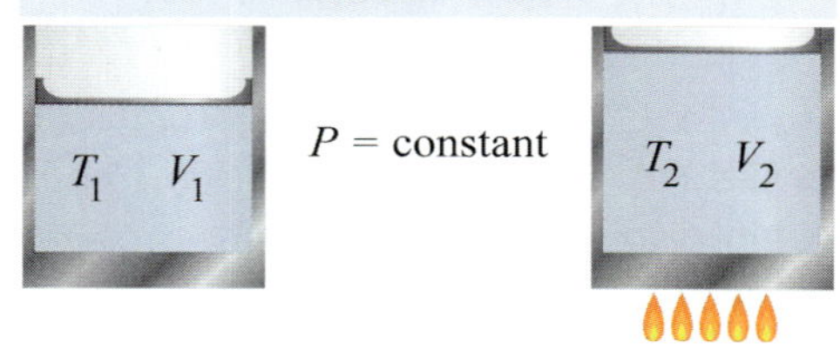


At constant pressure, the volume of a gas is proportional to its temperature.

$$\frac{V_1}{V_2} = \frac{T_1}{T_2} \quad \text{or} \quad \frac{V_1}{T_1} = \frac{V_2}{T_2}$$

**Use absolute temperature!**

The temperature and pressure that appear in these equations are *absolute* values. For temperature that means $T$ must be in kelvins, which are referenced to absolute zero. To convert to kelvins add 273 to the Celsius temperature. For example, 10°C is 283 K. The absolute pressure must include the pressure of the atmosphere if appropriate. An ordinary tire pressure gauge reads pressure *above atmospheric*. A *gauge pressure* of 100,000 Pa therefore represents an absolute pressure of 201,325 Pa after including 101,325 Pa for the atmosphere.

**Gay-Lussac's law**

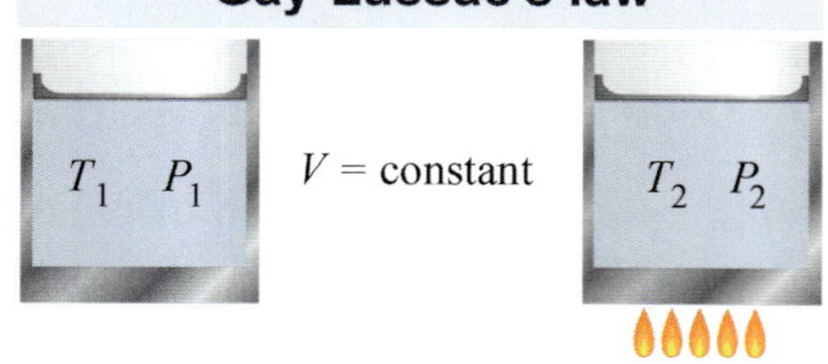


At constant volume, the pressure of a gas is proportional to its temperature.

$$\frac{P_1}{T_1} = \frac{P_2}{T_2} \quad \text{or} \quad P_1T_2 = P_2T_1$$

### The ideal gas law

Boyle's law, Charles's law, and Gay-Lussac's law were discovered experimentally. Our modern understanding of matter combines all three equations into the **ideal gas law** in equation (23.8). An **ideal gas** is an imaginary form of matter consisting of randomly moving particles that have mass but zero volume. These particles interact with each other only by perfectly elastic collisions that exchange energy and momentum among the particles and any walls confining the gas. Practically speaking, atoms are so small that real gases behave like ideal gases at low pressures (such as at one atmosphere) and at temperatures significantly above the material's boiling point.

Interactive equation

**Ideal gas law** for particles

$$(23.8) \qquad PV = Nk_BT$$

$P$ = pressure (Pa or N/m$^2$)
$V$ = volume (m$^3$)
$N$ = number of particles (molecules)
$k_B$ = Boltzmann's constant = $1.38\times10^{-23}$ J/K
$T$ = absolute temperature (K)

### How many particles are in a gas?

The quantity $N$ in the ideal gas law is the number of particles. This is typically a very large number! At atmospheric pressure of 101,325 Pa, a cubic meter of air at room temperature of 293 K contains $2.5\times10^{25}$ or 25 trillion trillion molecules!

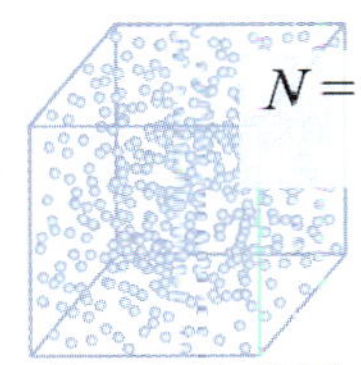


$P$ = 101,325 Pa (1 atm)
$V$ = 1 m$^3$
$T$ = 293 K (20°C)

$$N = \frac{PV}{k_BT} = \frac{(101{,}325 \text{ Pa})(1 \text{ m}^3)}{(1.38 \times 10^{-23} \text{ J/K})(293 \text{ K})} = 2.5 \times 10^{25}$$

### A more useful form of the ideal gas law

The number of particles is so large that equation (23.8) is inconvenient to work with, even though the equation accurately relates particle motion to temperature and pressure. Equation (23.9) is more useful because it is stated in terms of *moles* of gas $n$. The number of moles is calculated using Avogadro's number: $n = N/N_A$. Boltzmann's constant is replaced in equation (23.9) by the *ideal gas constant*, which has the value $R$ = 8.31 J/(mol K).

Interactive equation

**Ideal gas law** for moles

$$(23.9) \qquad PV = nRT$$

$P$ = pressure (Pa or N/m$^2$)
$V$ = volume (m$^3$)
$n$ = number of moles of gas
$R$ = gas constant = 8.31 J/(K mol)
$T$ = absolute temperature (K)

### Temperature of ideal gas

A pump compresses a volume of gas as shown in the diagram. What is the final temperature of the gas if the gas starts at 20°C (293 K)?

*Asked:* final temperature $T_2$

*Given:* $P_1, P_2, V_1, V_2, T_1$ [at right]

*Relationships:* $PV = nRT$

*Solution:* The quantity of gas stays the same:

$$\frac{P_1V_1}{T_1} = \frac{P_2V_2}{T_2} \quad \rightarrow \quad T_2 = \frac{T_1P_2V_2}{P_1V_1}$$

$$= \frac{(293 \text{ K})(1{,}100{,}000 \text{ Pa})(0.00008 \text{ m}^3)}{(101{,}325 \text{ Pa})(0.0008 \text{ m}^3)}$$

$$= 318 \text{ K } (45°\text{C})$$

*Answer:* 318 K or 45°C

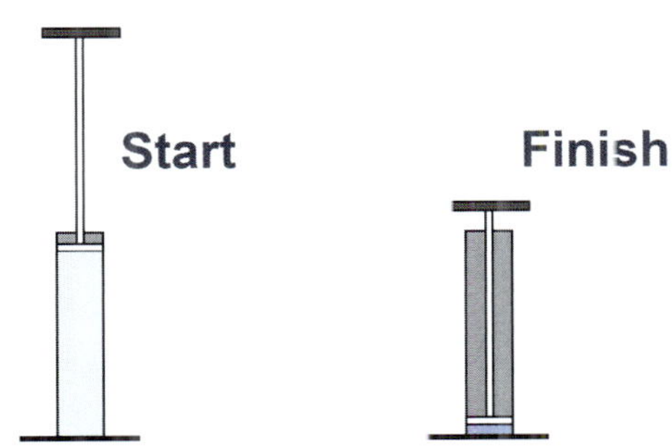


| Start | Finish |
|---|---|
| $V_1 = 0.0008$ m$^3$ | $V_2 = 0.00008$ m$^3$ |
| $P_1 = 101{,}325$ Pa (14.7 psi) | $P_2 = 1{,}100{,}000$ Pa (160 psi) |
| $T_1 = 293$ K (20°C) | $T_2 = ?$ |

# Investigation 23C Pressure of an ideal gas

**Essential questions** *What is the relationship among pressure, volume, and temperature for a gas?*

Interactive simulation

In this simulation you will explore the behavior of ideal gases under different values of pressure, temperature, and volume. The simulation will change volume, temperature, and pressure according to the quantity and type of gas and the conditions you specify.

## Properties of various ideal gases

1. Enter values in any of the text boxes to run the simulation.
2. The temperature, volume, and pressure are limited by realistic properties of materials.

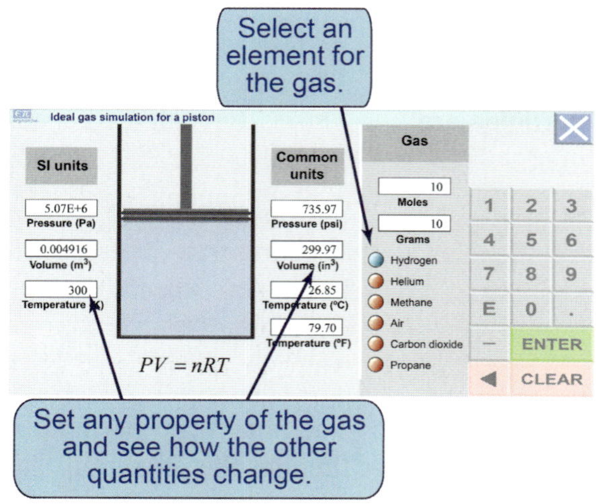


a. *What volume does 1 mol of air occupy at 101,325 Pa and 0°C (273 K)? Compare this to the volume of other gases at the same conditions.*

b. *Tire air pressure changes when temperatures change. Suppose a tire is inflated to a gauge pressure of 32 psi (absolute pressure of 46.7 psi) at 20°C. What is the pressure if the temperature drops to −20°C?*

c. *If the same tire were to be driven on a road across the Sahara Desert, how would the pressure change? Estimate its gauge pressure and justify your value.*

d. *What happens to the pressure when you double the quantity of gas while keeping volume and temperature constant?*

e. *How much does the pressure change when you increase the volume by a factor of 10 while keeping temperature constant?*

f. *If the temperature doubles, by how much does the volume have to change if pressure is held constant?*

g. *What is the volume of 1 g of helium at atmospheric pressure and 0°C?*

h. *What is the volume of 1 g of carbon dioxide at atmospheric pressure and 0°C?*

i. *A propane tank for a barbeque contains 9,000 g of propane (20 lb). What volume does this occupy as a gas? Compare this to the actual volume of the propane tank. What do you conclude?*

j. *At room temperature and for the same volume and gas mass, which of the gases in the simulation (hydrogen, helium, methane, air, carbon dioxide, or propane) has the highest pressure? Why do you think that is the case?*

**Density of fluids**

The **density** of a fluid is the ratio of the mass divided by the volume, as shown in equation (23.10). Air has a density of 1.29 kg/m$^3$ at standard temperature and pressure (STP, which means atmospheric pressure and a temperature of 0°C). The density of gases changes rapidly with temperature and pressure. Water has a density of 1,000 kg/m$^3$. Mercury (metal) is liquid at room temperature and has a density of 13,580 kg/m$^3$!

**Densities of fluids**

| Fluid | $\rho$ (kg/m$^3$) |
|---|---|
| Helium | 0.18 |
| Air | 1.29 |
| Carbon dioxide | 1.98 |
| Gasoline | 740 |
| Oil | 900 |
| Water | 1000 |
| Mercury | 13,500 |

Liquid | Gas (at STP)

Interactive equation

**Density**

$$\rho = \frac{m}{V} \qquad (23.10)$$

$\rho$ = density (kg/m$^3$)
$m$ = mass (kg)
$V$ = volume (m$^3$)

**Buoyancy**

Because fluids can flow, an object or another fluid may either sink or float when resting on a second fluid. An object of lower average density than a fluid is *buoyant* and floats on that fluid. An object of higher average density sinks. Oil has a lower density than water; therefore, oil floats on water. A ping-pong ball has a lower *average* density than either water or oil because its interior is filled with air. A ping-pong ball therefore floats above both oil and water.

**Pressure from weight**

The weight of a fluid creates *hydrostatic pressure*. Hydrostatic pressure increases with depth under water. At any given depth $d$, the pressure $P$ in newtons per square meter equals the weight of a square meter column reaching all the way back up to the surface. This weight depends on the density of the fluid $\rho$ and the acceleration due to gravity $g$, as seen in equation (23.11). Denser fluids (such as water) create more pressure at a given depth when compared to lighter fluids such as air. Even air at its low density of $\rho$ = 1.29 kg/m$^3$ creates a pressure of more than 100,000 Pa because of the nearly 100 km height of the Earth's atmosphere.

Interactive equation

**Hydrostatic pressure**

$$P = \rho g d \qquad (23.11)$$

$P$ = pressure (Pa or N/m$^2$)
$\rho$ = density (kg/m$^3$)
$g$ = acceleration of gravity
= 9.8 N/kg at Earth's surface
$d$ = depth (m)

**Pressure from the atmosphere**

Atmospheric pressure increases with depth and decreases with altitude.

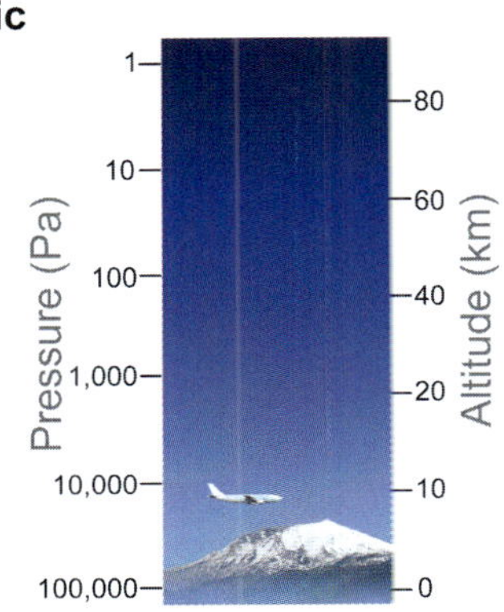


**Pressure under water**

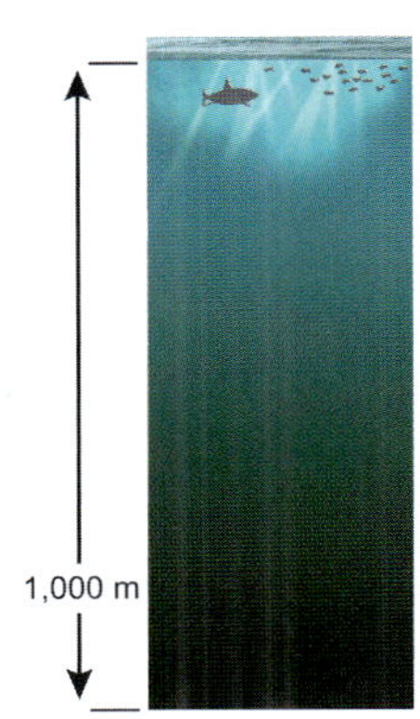


Mass
$m = \rho V$
$= 1{,}000{,}000$ kg

Weight
$F_w = mg$
$= 9{,}800{,}000$ N

Pressure
$P = \dfrac{F}{A} = 9{,}800{,}000 \text{ N/m}^2$

The pressure under the ocean can be enormous, reaching almost 10 million N/m$^2$ at a depth of 1000 m.

**How is pressure related to energy?**

Energy can be thought of as the ability to exert force and do work. Since force is pressure times area, energy and pressure are therefore closely related. One joule of work is done when a pressure of one pascal pushes a surface of one square meter a distance of one meter. The loss of energy by the fluid may lead to a decrease in temperature, pressure, or both.

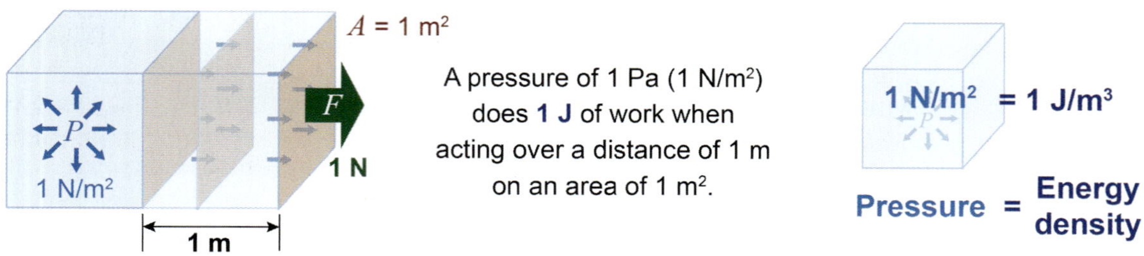


**Pressure is energy density**

A pressure of one newton per square meter represents a *potential energy density* of one joule per cubic meter. Energy density has units of $J/m^3$, analogous to mass density, which has units of $kg/m^3$. The potential energy in a volume of fluid is therefore equal to its volume multiplied by its pressure.

**Fluid energy**

To calculate the energy of a fluid, imagine a tank with a hole in the side through which a stream of water squirts. Pressure energy inside the tank is converted to kinetic energy when the water squirts out. The energy of a small mass $m$ of water is the sum of the gravitational potential energy $mgh$ (relative to the top of the tank), kinetic energy $\frac{1}{2}mv^2$, and potential energy from pressure, $PV$.

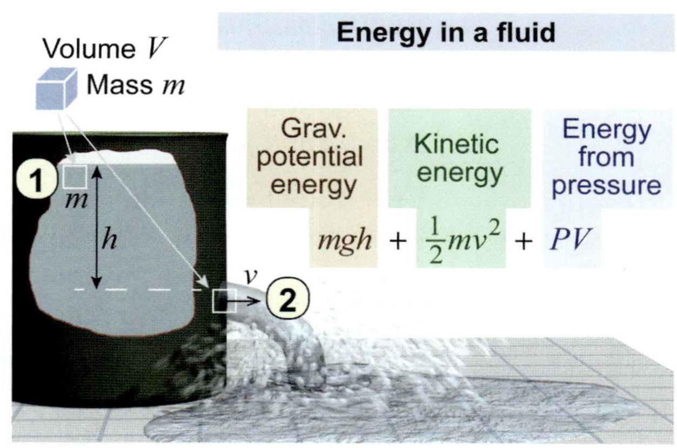


**Bernoulli's equation**

For fluids it is usually more convenient to use density $\rho$ instead of mass. Divide the energy equation by volume $V$ and the result is *Bernoulli's equation,* which is given in equation (23.12). **Bernoulli's equation** is the law of conservation of energy applied to the motion of a specific element of mass within the fluid that is moving along a *streamline.* A **streamline** is an imaginary line that follows the motion of a particular element of fluid as it moves.

$$\left(\frac{1}{V}\right)mgh+\left(\frac{1}{V}\right)\frac{1}{2}mv^2+\left(\frac{1}{V}\right)PV=\left(\frac{1}{V}\right)E$$ **Energy equation**

**(23.12)** $$\rho gh + \frac{1}{2}\rho v^2 + P = \text{constant}$$

$\rho$ = density ($kg/m^3$)
$g$ = gravitational acceleration = 9.8 N/kg
$h$ = height (m)
$v$ = speed (m/s)
$P$ = pressure (Pa or $N/m^2$)

**Bernoulli's equation** along a streamline

**What are streamlines?**

Streamlines are drawn parallel to the direction of flow and show how fluids flow around the surfaces of objects such as vehicles and aircraft. Where streamlines are close together, speed is higher. Where streamlines are farther apart, speed is lower. Energy is conserved only along the same streamline, because a streamline tracks the motion of the same specific mass of fluid as it flows.

Streamlines trace the direction of air flow around a moving vehicle.

**What does Bernoulli's equation mean?**

Bernoulli's equation tells us that *along a streamline* the variables of height, pressure, and speed are related by energy conservation. If any of these three increases, at least one of the other two must decrease. For example, if the speed of a fluid goes up, the pressure may go down to compensate.

**How does a plane fly?**

*Flight* is an important application of fluid flow in which the fluid is air. In order to fly, a plane in motion must generate a vertical lift force at least equal to its weight. The purpose of a *wing* is to manipulate airflow in a specific way to create and control lift forces. The cross section of a wing has the shape of an *airfoil.* As shown in the illustration below, the shape of an airfoil and the angle of attack force air along streamline (A) to be divided into two paths: air flowing across the top of the wing (B) and air flowing under the wing (C). During the same time interval, air flowing over the wing must travel a further distance than air flowing under the wing, so the speed of air over the top surface of the wing (B) is higher than the speed below the wing (C). This difference in the speed of the airflow is the key to flying.

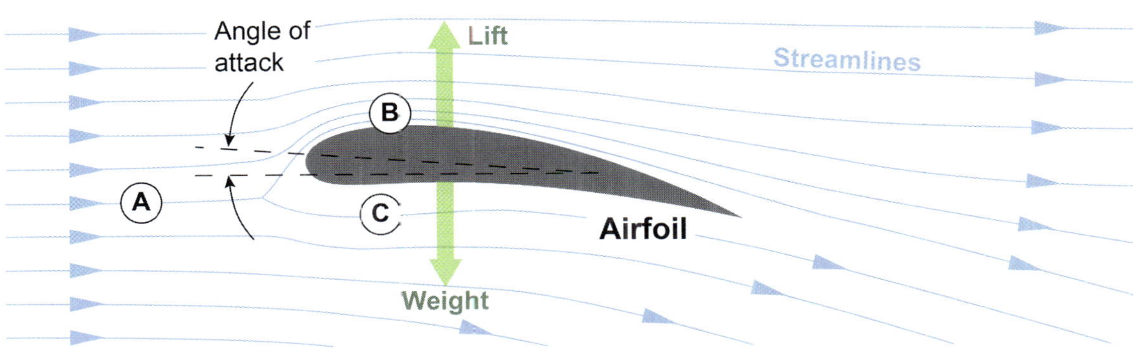


**What creates the lift force?**

According to Bernoulli's equation, the sum of energy densities must be equal for both paths that originate along the same streamline at (A). Increasing the speed above the wing therefore means that the *pressure* on the top surface of the wing is lower than the pressure on the lower surface—since above and below the wing all the other parameters in Bernoulli's equation ($\rho$, $g$, and $h$) are equal. The difference between the pressures above and below the wing, when multiplied by the area of the wing, generates the lift force. Because the lift force is created by the motion of air over the wing, the force increases as the square of the plane's speed. That is why planes need to accelerate along a runway until they reach take-off speed. A plane cannot get off the ground until it reaches a speed at which the lift force exceeds the plane's weight.

# Section 2 review

A fluid is a form of matter that easily flows and changes shape in response to an applied force. Forces in fluids (gases and liquids) act through pressure. Pressure is force per unit area and has units of pascals ($N/m^2$) and pounds per square inch (psi or $lbs/in^2$). Weight creates hydrostatic pressure, including atmospheric pressure and pressure underwater. Energy conservation along a streamline leads to Bernoulli's equation, which relates pressure, volume, and height in a moving fluid.

| Vocabulary words | fluid, pressure, compressible, incompressible, Boyle's law, Charles's law, ideal gas, ideal gas law, density, Bernoulli's equation, streamline |
|---|---|

| Key equations | | | |
|---|---|---|---|
| | $P = \frac{F}{A}$ | $PV = Nk_BT$ | $PV = nRT$ |
| | $\rho = \frac{m}{V}$ | $P = \rho gd$ | $\rho gh + \frac{1}{2}\rho v^2 + P = \text{constant}$ |

## Review problems and questions

1. A cubic box contains gas at a gauge pressure of 400,000 $N/m^2$ (400,000 Pa) and measures 0.5 m × 0.5 m × 0.5 m. What force does the gas inside the box exert on any single side of the box?

2. Put the following pressures in order from lowest to highest.
   a. 1 atmosphere
   b. 1 psi
   c. 1 $N/m^2$
   d. 100 psi
   e. 10,000 Pa

3. A submarine must be able to dive to a depth of 800 m. What is the absolute pressure at this depth? How does this compare to atmospheric pressure of 14.7 psi?

4. A cylindrical container full of liquid mercury is 80 cm high and has a radius of 10 cm. The density of mercury is 13,580 $kg/m^3$.
   a. What is the volume of the container?
   b. What is the mass of the mercury?
   c. What pressure does the container exert on the floor? Assume that the mass of the container is negligibly small compared to the mass of the mercury.

5. Calculate the net force on a 3 m square window when a wind outside blows parallel to the window at 20 m/s. Assume that the quantity $\rho gh + ½\rho v^2 + P$ is the same inside and outside but that a sudden wind gust allows the outside pressure to drop compared to the inside pressure where $v = 0$. Express your answer in newtons and pounds. Do you think this is a significant problem for building engineers? Why? (Assume the density of air to be 1.29 $kg/m^3$.)

6. A certain quantity of gas has an absolute pressure of 250,000 Pa at 20°C. What would the pressure be if the temperature increased to 1,000°C while the volume remained constant?

In the first section of this chapter, we described how matter is made of particles—atoms and molecules—that are in constant thermal motion. The ideas of specific heat and thermal energy were mostly described at the *macroscopic* level that you experience in daily life. In this section, we dig into the *microscopic* physics of matter on the molecular scale. Starting with ideal gases we will see how properties such as specific heat are derived from the motions and interactions of atoms. In any ordinary quantity of matter, such as the gas in a balloon, there are so many atoms that their average behavior must be studied *statistically*. The kinetic theory uses statistical techniques to relate the macroscopic properties of matter to the physics of matter on the microscopic scale.

## Statistics and macroscopic properties

**Statistical mechanics**

Atoms are so small that we do not experience individual particles. Instead, we experience the average behavior of trillions and trillions of particles. *Statistics* is the branch of mathematics that deals with the properties of large groups of numbers. When you calculate an *average* you are using statistics. **Statistical mechanics** is the branch of physics that explains how the average behavior of trillions of microscopic particles creates properties such as temperature, density, and pressure.

**Ordered and random energy**

The kinetic energy of a single particle is ½$mv^2$, similar to the kinetic energy of a baseball. A whole collection of particles can actually have *two* kinds of kinetic energy: *ordered* and *random*. To appreciate the difference, consider a handful of ping-pong balls. If you throw the whole handful, the group of balls has ordered kinetic energy because there is an average velocity. But suppose you put them in a jar and then vigorously shake the jar up and down. The balls will bounce madly around as individuals *but the average velocity of the whole group together is zero*.

**Ordered kinetic energy**

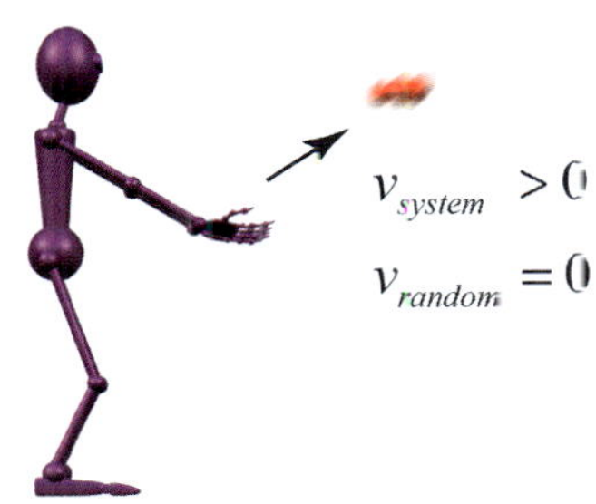


**Random kinetic energy**

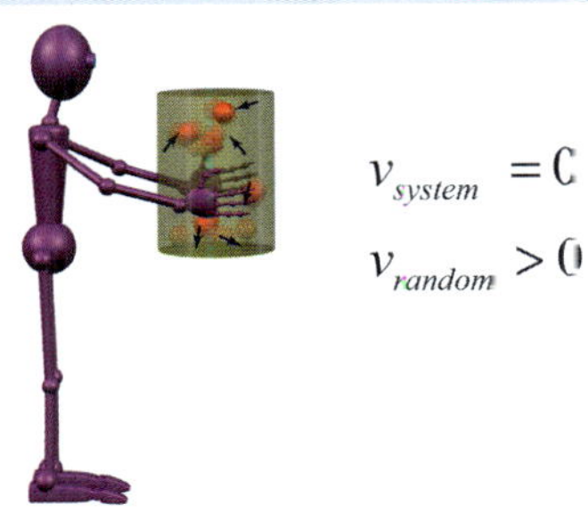


Random motion—such as for the ping-pong balls in a jar—has an average velocity of zero, but the *average speed is not zero*. That is why the bouncing balls inside the jar have nonzero kinetic energy. Normal matter may have $10^{23}$ atoms bumping into each other a trillion times a second. The constant bumping creates exchanges of energy and momentum between particles, and this keeps the motion truly random.

**The kinetic theory of matter**

Avogadro's number is such an enormously large value than only statistical methods can deal with atoms in any observable quantity of matter. The kinetic theory of matter applies the principles and tools of statistical mechanics to create a comprehensive explanation for how matter behaves. For example, temperature is the average kinetic energy of molecules resulting from random motion. Boltzmann's formula for the average thermal energy per degree of freedom of a single atom ($E$ = ½$k_BT$) comes from kinetic theory. Kinetic theory provides the fundamental explanation for virtually all the properties of matter we observe, including pressure, specific heat, electrical conductivity, and viscosity.

**A thought experiment**

Consider a thought experiment in which 10,000 moving particles are placed in a box. At the start, half the particles are moving 100 m/s to the right and half are moving at the same speed to the left. The total kinetic energy is *E*. Collisions between particles are perfectly elastic, which means that any energy lost by one particle in a collision is gained by the other. The total kinetic energy of all the particles therefore stays constant over time. What happens to the particle velocities over time?

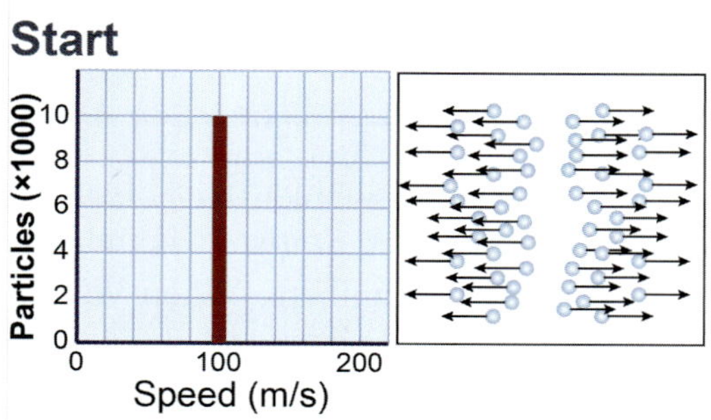


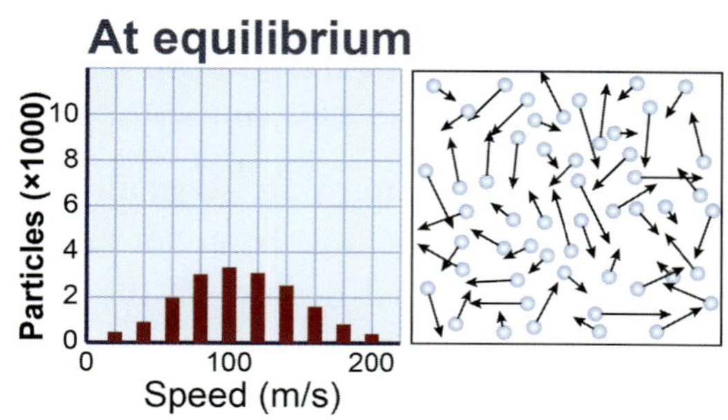


**What happens over time?**

Over time, particle velocities become scattered in all directions by collisions. Eventually, an equilibrium is reached in which, on average, equal numbers of particles are moving in all directions with *different speeds*. A bar graph of particle speeds shows a characteristic shape with few particles at very high or very low speeds and a single maximum. This graph is called a *distribution function* because it shows how particle speeds are distributed over the whole population of particles.

**Do all particles have the same speed?**

In 1860, the Scottish physicist James Clerk Maxwell used statistical mechanics to calculate the speeds in a gas of interacting particles. The **Maxwellian distribution** predicts the percentage of particles at each speed as a function of the temperature. The graph on the right shows the Maxwellian distribution of speeds at 100 K, 1000 K, and 5,000 K. The total area under each curve is the same—100%. That means that 100% of the particles have speeds that fit the curve for any given temperature.

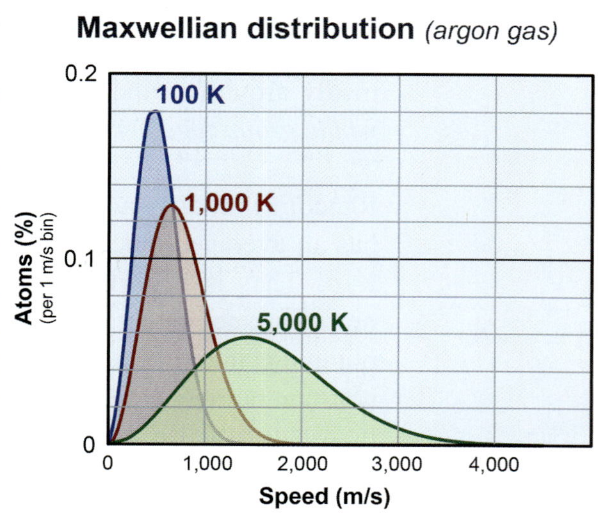


**What does the graph mean?**

1. At any temperature above absolute zero there will be a range of particle speeds. There are some particles with very low speeds and some with very high speeds at all temperatures.
2. As the temperature increases, the average speed increases.
3. The width of the distribution of speeds also increases with temperature.

**How is energy related to temperature?**

A gas for which pressure, temperature, and volume are steady and constant has a Maxwellian distribution of particle speeds. Knowing how many particles there are at any speed allows us to find the *average thermal speed*, which is the speed of a particle with average kinetic energy given by equation (23.5). This equation (repeated below) is derived directly from the Maxwellian distribution.

Interactive equation

**Thermal energy per atom**

$$E = \frac{3}{2}k_B T$$

$E$ = energy (J)
$k_B$ = Boltzmann's constant = $1.38 \times 10^{-23}$ J/K
$T$ = absolute temperature (K)

**Thermal speed**

We determine the average thermal speed $v_{th}$ for a particle in a gas by setting the kinetic energy equal to the average thermal energy from equation (23.5):

$$\frac{1}{2}mv_{th}^2 = \frac{3}{2}k_B T$$

Solving this for the average thermal speed leads to the following equation.

**Thermal speed**

$$\textbf{(23.13)} \quad v_{th} = \sqrt{\frac{3k_B T}{m}}$$

$v_{th}$ = thermal speed of particles (m/s)
$k_B$ = Boltzmann's constant = $1.38 \times 10^{-23}$ J/K
$T$ = temperature (K)
$m$ = mass (kg)

**How fast do gas particles move?**

Argon is a monatomic gas that makes up about 1% of Earth's atmosphere. Slightly heavier than oxygen or nitrogen, argon has an atomic mass of 40 g/mol. For argon gas at room temperature of 293 K (20ºC) the average thermal speed is 427 m/s, or 956 mph! As argon heats up, the average speed of the particles becomes even greater, reaching 1,765 m/s when the temperature is 5,000 kelvin!

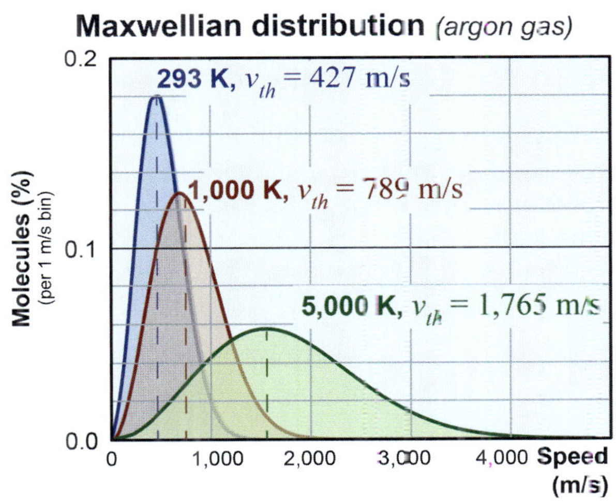


**Thermal speed for air molecules**

The thermal speed of air is *higher* than that of argon because the thermal speed varies as the *inverse* square root of the particle mass. The average molecular mass of air (29 g/mol) is *smaller* than that for argon; therefore, the thermal speed of air is *higher*. At 293 K the average thermal speed of air is 502 m/s or 1,120 mph. The *speed of sound* is related to the thermal speed because sound waves move through collisions between air molecules. In air, the speed of sound at 20ºC is 343 m/s and increases as the square root of the temperature, in agreement with the formula for the thermal speed.

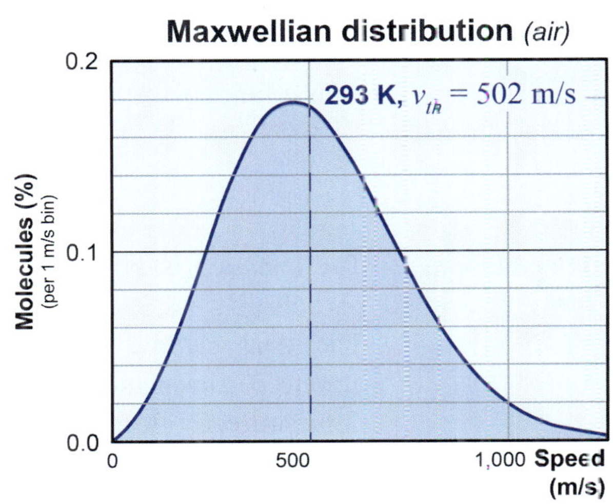

**Deriving the ideal gas law**

The ideal gas law connects the microscopic motion of particles in a gas to their macroscopic quantities of temperature and pressure. To see how this connection is made, we calculate the average pressure on the sides of a container resulting from collisions of gas particles.

**Single gas particle in a box**

Consider a particle traveling with velocity $v_x$ in the $x$-direction in a box of length $L$. The particle hits the side of the box, which has area $A$, and rebounds with velocity $-v_x$.

Pressure is force divided by area. Force is the rate of change of momentum, $F = \Delta P/\Delta t$. After colliding with the wall, the particle's change in momentum is $-2mv_x$.

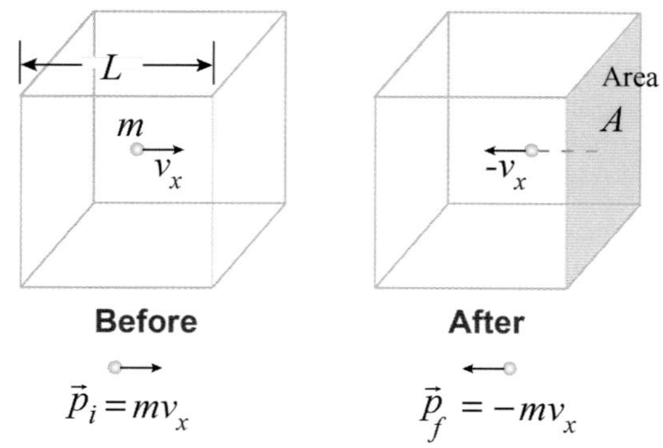


$$\Delta\vec{p} = -mv_x - mv_x = -2mv_x$$

**What is the average force on the particle?**

The particle travels back and forth a distance $2L$ between collisions with the right-hand wall. The time it takes to traverse a distance of $2L$ is $\Delta t = 2L/v_x$. The average force is therefore given by

$$F = \frac{\Delta P}{\Delta t} = \frac{-2mv_x}{\left(\frac{2L}{v_x}\right)} = -\frac{mv_x^2}{L}$$

**What is the average pressure on the wall?**

The sign is negative because this is the force *on the particle* that changed its momentum. To calculate the pressure *on the wall*, we need the positive reaction force on the wall. The average pressure comes from this positive force divided by the area $A$ of the wall:

$$P = \frac{F}{A} = \frac{mv_x^2/L}{A} = \frac{mv_x^2}{A \times L} = \frac{mv_x^2}{V} \Rightarrow PV = mv_x^2$$

where we have used volume $V = A \times L$. Note that $PV$ is the kinetic energy associated with motion in the $x$-direction.

**How is temperature related to particle speed?**

Next we need to relate $v_x$ to the average thermal speed $v_{th}$. No one direction is favored over another in our imaginary box, so the average velocities in the $x$-, $y$-, and $z$-directions are equal. We can express this mathematically with $v_x^2 = v_y^2 = v_z^2$. The thermal speed $v_{th}$, when broken into components, satisfies the relationship $v_{th}^2 = v_x^2 + v_y^2 + v_z^2 = 3v_x^2$, so we derive $v_x$ as a function of temperature $T$:

$$v_x^2 = \frac{1}{3}v_{th}^2 = \frac{1}{3}\left(\frac{3k_BT}{m}\right) = \frac{k_BT}{m}$$

**The ideal gas law**

The last step is to substitute this relationship for $v_x$. The result tells us that for a single particle $PV = k_BT$. If the box contains $N$ particles we have the ideal gas law, $PV = Nk_BT$. The ideal gas law comes directly from the addition of trillions of microscopic collisions among particles and their container. Thus the ideal gas law for a collection of $N$ particles comes directly from the properties of a single particle.

| $PV = k_BT$ | $PV = Nk_BT$ |
|---|---|
| single particle | $N$ particles |

**How does a gas store thermal energy?**

When heat is added to a gas, the particles move faster and the temperature increases. One way to look at this is to envision the gas as *storing* thermal energy in the motion of its constituent atoms. In fact, all matter stores thermal energy in a similar way. *Specific heat* is another macroscopic property that can be derived from statistical mechanics and particle motion.

**Constant pressure vs. constant volume**

For a gas there is a difference between adding heat at constant *pressure* and adding heat at constant *volume*. If pressure is constant, then volume changes and some of the heat energy becomes work since $P\Delta V \neq 0$. The specific heat at constant pressure has the symbol $c_p$.

In contrast, if heat is added at constant *volume,* then no work is done and all the energy goes to changing pressure. The specific heat at constant volume has the symbol $c_v$.

Constant pressure | Constant volume

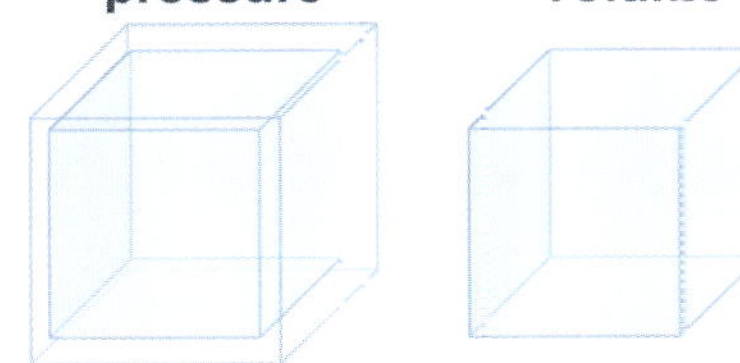

Adding heat causes expansion, so $P\Delta V > 0$. | Adding heat increases pressure.

**Deriving the specific heat**

Consider one mole of gas, which contains $6.022 \times 10^{23}$ ($N_A$) particles. The mass is $m_{mol}$, the mass of one mole. The internal energy $U$ of a monatomic gas of $N_A$ particles is $(3/2)N_Ak_BT$. Set this equal to the heat $Q$ it takes to raise the temperature of this mass by an amount $\Delta T$:

| heat | internal energy | | |
|---|---|---|---|
| $Q = m_{mol}c_v\Delta T$ | $U = \frac{3}{2}N_Ak_BT$ | $\rightarrow$ | $m_{mol}c_v\Delta T = \frac{3}{2}N_Ak_BT$ |

**The result**

To be correct in setting these equal, the heat $Q$ must be the amount to raise the temperature from absolute zero to $T$. In kelvins the temperature change is $\Delta T = T$. Solve this for $c_v$, and the result is the specific heat at constant volume:

$$m_{mol}c_vT = \frac{3}{2}N_Ak_BT \quad \rightarrow \quad c_v = \frac{3}{2}\frac{N_Ak_B}{m_{mol}}$$

**Helium ($m_{mol} = 4$ g)**

$$c_v = \frac{3}{2}\frac{(6.022\times10^{23})(1.38\times10^{-23}\text{ J/K})}{(0.004\text{ kg})}$$
$$= \mathbf{3{,}120\ J/(kg\ K)}$$

**Argon ($m_{mol} = 40$ g)**

$$c_v = \frac{3}{2}\frac{(6.022\times10^{23})(1.38\times10^{-23}\text{ J/K})}{(0.040\text{ kg})}$$
$$= \mathbf{312\ J/(kg\ K)}$$

**Comparing theory and experiment**

These theoretical values are in excellent agreement with experimental measurements for helium and argon gases. The agreement between measured and predicted specific heat is strong evidence supporting the kinetic theory of matter. For air, however, the coefficient in front of the equations is 5/2 instead of 3/2. This is because air is 99% *diatomic* gases ($O_2$ and $N_2$). Each molecule can have kinetic energy of rotation around two axes as well as kinetic energy of translation along the three coordinate directions, which leads to the five degrees of freedom in the numerator of the coefficient 5/2.

**How are solids different from gases?**

In a solid, the atoms (or molecules) are close together and interact strongly. Imagine a 3-D array of masses connected by springs that represent the bonds between neighboring atoms. Like actual springs, interatomic bonds store potential energy when extended or compressed away from equilibrium. Because atoms are so close together, the concept of thermal *speed* is not strictly accurate because thermal energy is shared between kinetic energy and potential energy in bonds.

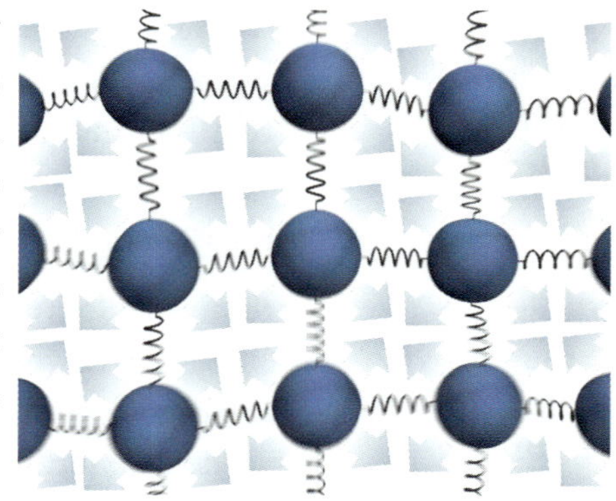

**Does kinetic theory work with solids?**

Consider a monatomic solid such as a metal. Kinetic theory still holds that each independent degree of freedom has an energy of $\frac{1}{2}k_BT$ as long as the temperature is not too low. In addition to the three degrees of freedom from motion in $x$-, $y$-, or $z$-directions, there are three additional degrees of freedom associated with potential energy stored in the bonds (or "springs") that connect atoms to each other.

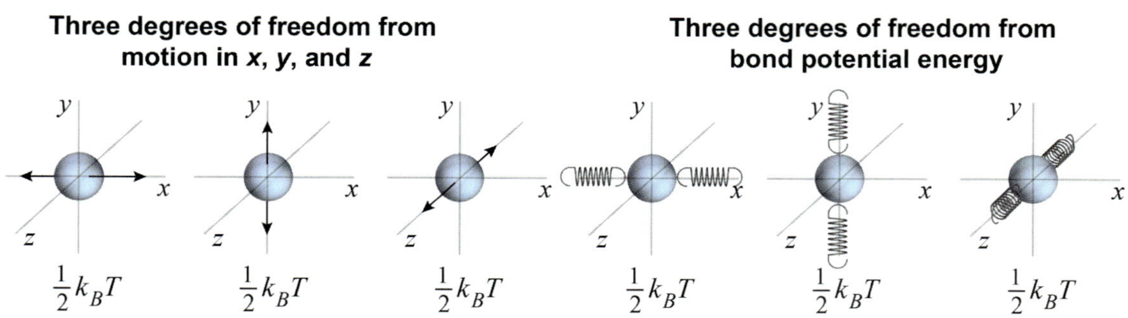


**The specific heat of a metal**

The total thermal energy is therefore $3k_BT$ per particle. If we let the thermal energy for one mole of particles be equal to the heat $Q$ to raise the temperature from absolute zero to temperature $T$, we get the following result for the specific heat of a monatomic solid:

$$\underbrace{Q = m_{mol}c_pT}_{\text{heat}} = \underbrace{U = 3N_Ak_BT}_{\text{internal energy}} \Rightarrow \underbrace{c_p = \frac{3N_Ak_B}{m_{mol}}}_{\text{specific heat}}$$

**Does the specific heat vary?**

This model explains the *law of Dulong and Petit*, which says that the specific heat of a monatomic solid decreases as the inverse of the atomic mass. The predicted specific heats of aluminum, copper, and lead are 898, 385, and 130 J/(kg K), respectively. These compare well with the actual measured values of 910, 390, and 130 J/(kg K), as shown in the illustration at right.

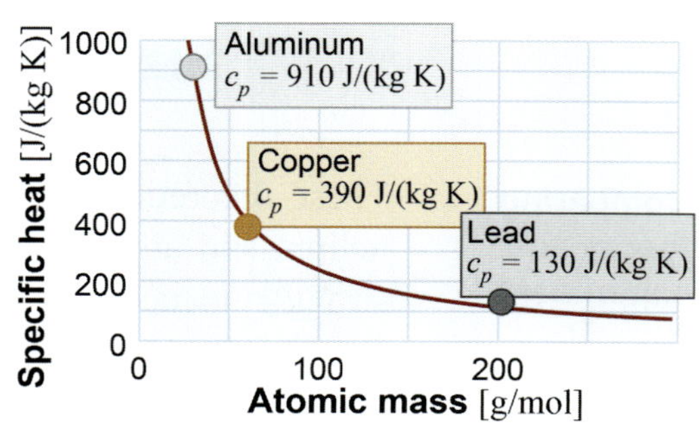

**Electronic devices are built with semiconductors**

Modern life would not be the same without electronics, which rely on the special properties of *semiconductors*. A *semiconductor* is a material in which a very small number of electrons are free to carry electric current. Semiconductors have electrical resistivities in between the values for insulators and conductors. Silicon is the most widely used semiconductor, being used in 99.9% of all computer chips and electronic devices.

**What makes semiconductors useful?**

Semiconductors can be used to construct electronic devices that can be changed from an insulator to a conductor by applying a tiny voltage, like setting an electronic switch. Atoms bond with other atoms through *valence electrons*, which are the outermost electrons in an atom. Silicon has four valence electrons, and silicon atoms form a *crystal* in which each atom shares its four valence electrons with four neighbors.

***n*-type semiconductors**

Adding an impurity of 1 phosphorus atom per 10 million silicon atoms increases the conductivity by a factor of 20,000. Phosphorus atoms have five outer electrons compared with silicon's four. When a phosphorus atom bonds with a lattice of silicon atoms, four of the five valence electrons of the phosphorus atom pair up with the neighboring silicon atoms. The extra electron does not pair up and is free to carry current. Adding a phosphorus impurity to silicon makes an *n*-type semiconductor in which current is carried by electrons.

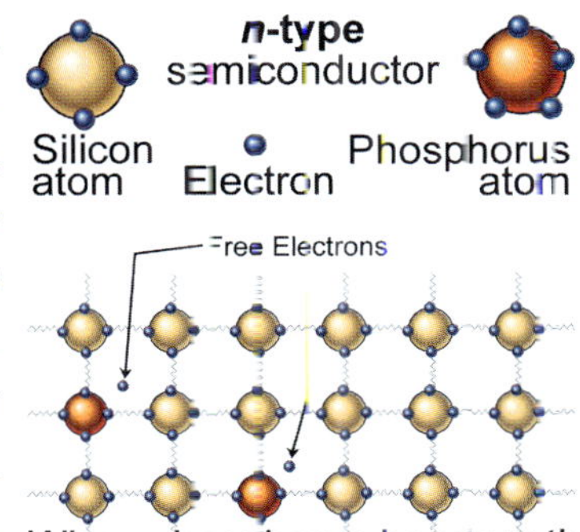


When phosphorus bonds with a silicon lattice, one phosphorus electron is left free.

***p*-type semiconductors**

A boron atom has three valence electrons, one fewer than silicon. When a boron atom bonds with lattice silicon atoms, the boron atom captures an electron from a neighboring silicon atom. The silicon atom is left with a missing electron and a net positive charge called a *hole*. The positive silicon atom attracts an electron from one of its neighbors, and the hole moves. The new hole takes an electron from its neighbor and the hole moves again. As electrons jump from atom to atom in one direction, the positive hole moves in the opposite direction and can carry current. Silicon with a boron impurity is a *p*-type semiconductor in which current is carried by holes.

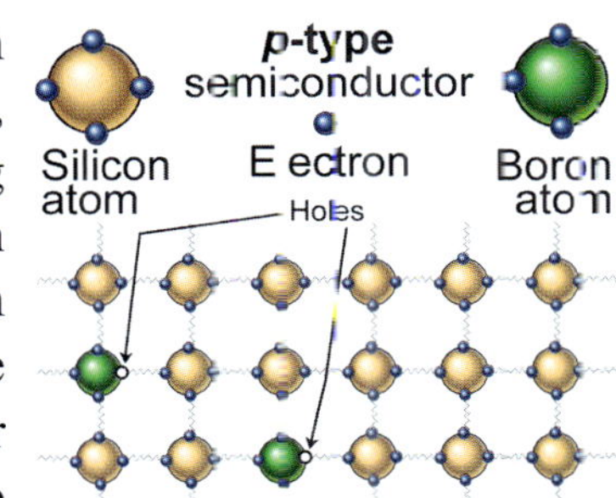


When a boron atom bonds with a silicon lattice, the boron atom takes an electron from a silicon atom.

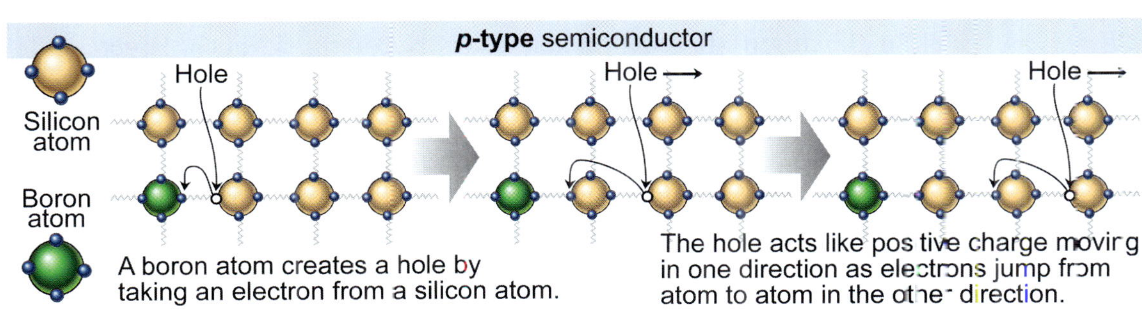


A boron atom creates a hole by taking an electron from a silicon atom.

The hole acts like positive charge moving in one direction as electrons jump from atom to atom in the other direction.

# Diodes and transistors

**The *p-n* junction**

Diodes and transistors are electronic devices built by combining *n*-type and *p*-type semiconductors. Consider what happens at a *p-n* junction where *p*-type and *n*-type semiconductors meet. Initially the *n* side has free electrons and the *p* side has holes. Negative electrons from the *n* side flow over to the *p* side and combine with positive holes. The *n* side becomes positively charged and the *p* side becomes negatively charged. In silicon, electrons move until the charge difference is 0.6 V, enough to keep any additional electrons from crossing over.

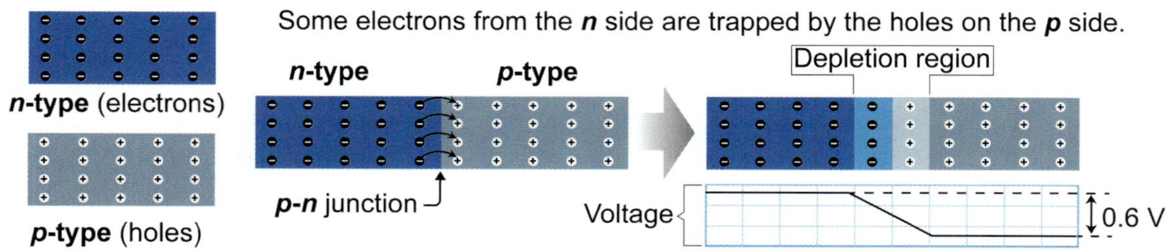


**How does the *p-n* junction work?**

Right at the *p-n* junction itself, a *depletion region* forms that has neither electrons nor holes. The depletion region is an insulator because it has no movable charges that can carry current. Nonetheless, the depletion can grow and become a stronger insulator—or disappear and allow current to flow—depending on the external voltage applied to the material.

**Diodes**

In short, a *p-n* junction is a diode, allowing electric current to flow one way and blocking it the other way. To see how this works, consider an applied external voltage that attracts electrons on the *n* side and holes on the *p* side. The depletion region gets larger and current is blocked because increasing the voltage just makes the depletion region an even better insulator. Now suppose the opposite voltage is applied. Both electrons and holes are repelled toward the depletion region. The depletion region shrinks and the *p-n* junction becomes a conductor for applied voltages above 0.6 V.

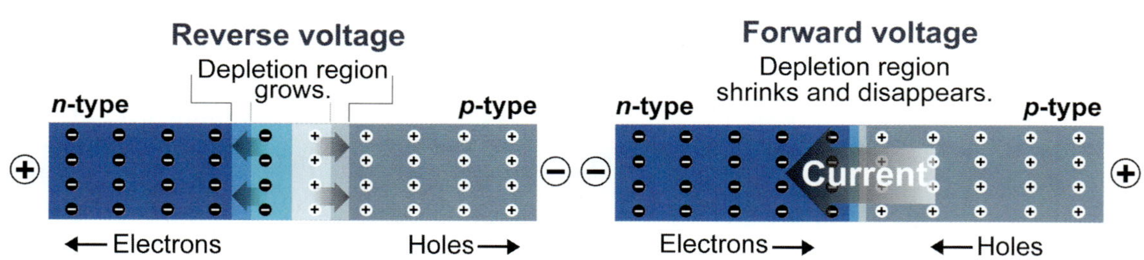


**Transistors**

A transistor is constructed from *three* semiconducting materials called the collector, base, and emitter. A transistor allows a very small current flowing in the base to control a much larger current from collector to emitter. A simple transistor has two *p-n* junctions back to back. The base terminal of an *npn* transistor is connected to the central *p*-type region. This layer is much thinner than the *n*-type layers on either side. A *pnp* transistor is the inverse, with an *n*-type semiconductor sandwiched between two layers of *p*-type. Transistors are the fundamental building blocks of electronic amplifiers, such as in your music player, and logic circuits in computers.

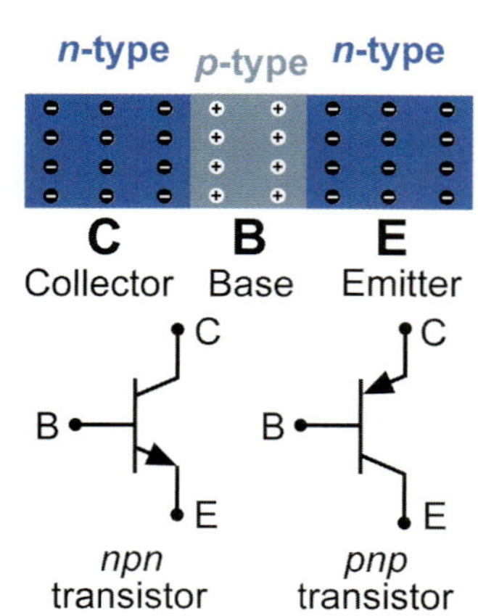

# Section 3 review

*Statistical mechanics* explains how the average behavior of trillions of microscopic particles creates macroscopic properties such as temperature, density, and pressure. This explanation is known as the *kinetic theory* of matter. One important prediction of kinetic theory is that, in thermal equilibrium, each degree of freedom has a mean thermal energy of $\frac{1}{2}k_BT$. The kinetic theory also shows that particles in a gas assume an equilibrium range of speeds described by the *Maxwellian distribution*. The average thermal speed of a gas particle can be predicted by using the Maxwellian distribution and depends on temperature, mass, and Boltzmann's constant.

Another result of kinetic theory is the *ideal gas law*, which relates pressure, volume, temperature, and quantity for a gas. The ideal gas law can be derived by calculating the average force exerted on the walls of a container by the impact of *N* particles per second. The specific heat of an ideal gas is derived by summing the kinetic energies of all constituent atoms. In contrast to atoms in a gas, atoms in a solid interact strongly with each other. The bonds between neighboring atoms constitute additional degrees of freedom because they can store potential energy (akin to "springs"). Kinetic theory predicts that the specific heat of a monatomic solid is twice that of an ideal gas.

| | |
|---|---|
| **Vocabulary words** | **statistical mechanics, Maxwellian distribution** |

| | | |
|---|---|---|
| **Key equations** | $E = \frac{3}{2}k_BT$ | $PV = Nk_BT$ |
| | gas: $c_v = \frac{3}{2}\frac{N_Ak_B}{m_{mol}}$ | solid: $c_p = \frac{3N_Ak_B}{m_{mol}}$ |

## Review problems and questions

1. Calculate the mean thermal speed of an atom in xenon gas at room temperature (21°C). The molecular mass of xenon is 131.2 g/mol.

2. How many particles are in one cubic meter of air at a pressure of one atmosphere (101,325 Pa) and a temperature of 20°C?

3. The specific heat of an unknown gas is measured to be 620 J/(kg °C). The gas is most likely to be which of the following?
   a. helium (4 g/mol)
   b. neon (20.1 g/mol
   c. argon (40 g/mol)
   d. xenon (137 g/mol)

4. Calculate the theoretical specific heat of silver. The atomic mass of silver is 107.9 g/mol. How does this compare with the actual measured value? Research possible explanations for any discrepancy.

# Chapter review

## Vocabulary

Match each word to the sentence where it best fits.

### Section 23.1

| | |
|---|---|
| Avogadro's number | Boltzmann's constant |
| Brownian motion | Calorie |
| Celsius scale | Fahrenheit scale |
| Kelvin scale | absolute zero |
| calorie | gas |
| heat | kinetic theory |
| liquid | mole |
| phases of matter | solid |
| specific heat | temperature |
| thermal energy | thermometer |

1. The ________ is a temperature scale on which the boiling point of water is 212°.
2. The wood in a table is an example of matter in its ________ phase.
3. The __________ is a unit of energy that is often found on food nutritional labels.
4. A/An __________ is a unit of thermal energy, commonly used in chemistry, that is equal to the amount of energy required to heat one gram of water by one degree Celsius.
5. ___________ is a form of energy associated with the temperature of a single object.
6. The ________ is a temperature scale on which the freezing point of water is at 0°.
7. The qualities of hot or cold are described by ________.
8. It takes 4.18 J to raise one gram of water by one degree Celsius because the ________ of water is 4.18 J/(g °C).
9. A mole of nitrogen atoms contains a number of individual atoms defined by _______.
10. _______ provides the conversion between microscopic thermal energy per particle and macroscopic temperature.
11. Temperature is measured with a/an _______.
12. The helium inside a balloon is an example of matter in its ________ phase.
13. Gas, liquid, and solid are the three most commonly encountered __________.
14. One _______ of an element contains $6.022 \times 10^{23}$ individual atoms.
15. The _______ is a connection between the microscopic behavior of collections of particles and their macroscopic properties.
16. The _______ is an absolute scale for temperature.
17. The lowest possible temperature is called _______.
18. A pollen speck floating is water is observed to jiggle around when seen through a microscope. This chaotic behavior is known as ________.
19. Apple juice is an example of matter in its ________ phase.

### Section 23.2

| | |
|---|---|
| Bernoulli's equation | Boyle's law |
| Charles's law | compressible |
| density | fluid |
| ideal gas | ideal gas law |
| incompressible | pressure |
| streamline | |

20. The equation $PV = nKT$ is an expression of the _______.
21. An imaginary line through a fluid tracing the path of motion of a fluid particle is called the _______.
22. Energy conservation along a streamline is often expressed as $\rho gh + ½mv^2 + P =$ constant, which is also called _______.
23. A/An _______ is a substance that can change shape and flow readily in response to applied forces.
24. A/An _______ fluid will change its volume in proportion to changes in pressure.
25. For a gas at constant pressure, the change in volume is proportional to the change in absolute temperature according to _______.
26. _______ is force per unit area.
27. A/An _______ fluid does not change its volume in proportion to changes in pressure.
28. In a/an _______, randomly moving point particles only collide elastically with each other and the walls of the container.
29. _______ is measured in kilograms per cubic meter in SI units.
30. _______ is often expressed as $P_1V_1 = P_2V_2$.

# Chapter review

## Section 23.3

**Maxwellian distribution** **quantum physics**
**statistical mechanics**

31. The mathematical function describing the distribution of velocities in an ideal gas is called the ________.

32. The branch of physics in which large ensembles of particles are analyzed is called ________.

## Conceptual questions

## Section 23.1

33. Examine the nutrition label on a can of soda. How many Calories does it contain? How many joules is this? If this energy is used to raise the temperature of 1 kg of water at room temperature (21°C), will the water boil?

34. A student wanted the equipment to be warmed up and ready for his investigation the next day, so he left the hot plate on overnight. Why is this a bad idea?

35. The amount of energy that must be used to heat 1 kg of a substance by 1°C is called the

    a. heat.
    b. specific heat.
    c. temperature.
    d. thermal energy.

36. What is the purpose of using a ring clamp to hold a styrofoam cup full of hot water?

37. Would you use a ring clamp or a 90° rod clamp to hold a glass test tube in place?

38. At 25°C, oxygen is a gas, but aluminum is a solid. Why are they in different phases even though they are at the same temperature?

39. When you heat a flask of water, how are you changing the *ordered* kinetic energy of the water molecules (if at all)? Likewise, how are you changing the *random* kinetic energy of the molecules (if at all)?

40. You are about to pour cold water into one cup and hot water into a second cup. How can you can put the same amount of thermal energy into each cup?

41. Which is a larger change in temperature, 1°F or 1°C?

42. In the phases of matter simulation, set the element to argon and the temperature to 200 K. Describe in your own words the behavior of the argon atoms, and use their behavior to determine whether they form a solid, a liquid, or a gas.

43. For each of the following, state whether it is a correct report of a temperature (physically possible and using correct notation). If it is not correct, explain why.

    a. −722.1°F
    b. −25 K
    c. 230.0°C
    d. 293°K

44. If the molecules in the desk in front of you are moving, then why is the desk standing still?

45. Rose fills a bathtub with hot water and takes her laundry out of the dryer. Then she goes to do something else. When she gets back the water in the tub is still hot but her laundry has cooled off. Which probably has a higher specific heat, water or her clothes?

46. A large pitcher contains 1 L of water at 10°C. How would the total heat (or thermal energy) of the water change if you:

    a. double the amount of water at the same temperature?
    b. change the temperature from 10°C to 20°C?
    c. switch the liquid from water to gasoline, but for the same mass and temperature?

47. More than a century ago, people often used hot water bottles to heat their bed sheets before climbing in. Why didn't they use hot metal objects of similar mass instead?

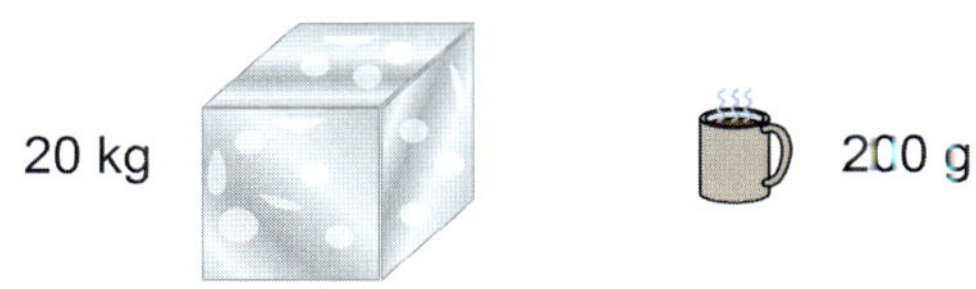


48. Compare a 20 kg block of ice to a 200 g cup of hot tea.

    a. Which one has more thermal energy? Why?
    b. If you poured the hot tea into an indentation on the top of the ice block, would you expect the ice to melt or the tea to cool to the freezing point? Why?

49. Areas of the world that are close to oceans or other large bodies of water usually have climates that have a narrower range in temperature. Why is this so?

50. Your friend is having trouble understanding what the average kinetic energy of atoms has to do with overall temperature of a substance. Help him understand by giving an example in everyday life where the average of individuals in a system tells you something about the overall system.

# Chapter 23

# Chapter review

## Section 23.1

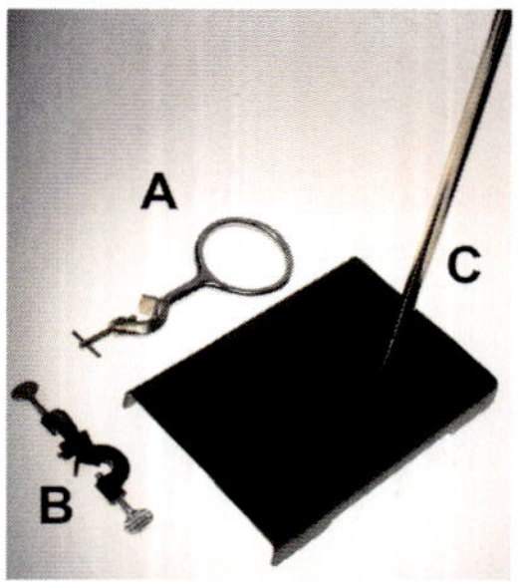


51. Identify the three pieces of equipment shown: ring stand, ring clamp, and 90° clamp.

52. Why shouldn't you place a styrofoam cup directly on a hot plate?

53. Some of the older Celsius thermometers contain mercury. What precautions should be taken when using them?

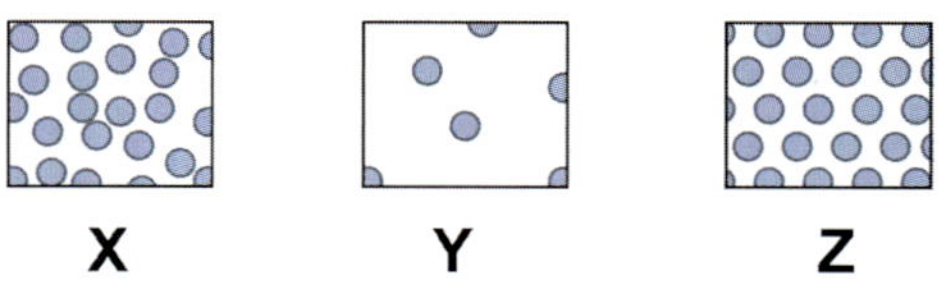


54. The three diagrams describe the phases of matter in which order?
    a. X: solid, Y: liquid, Z: gas
    b. X: gas, Y: liquid, Z: solid
    c. X: liquid, Y: solid, Z: gas
    d. X: liquid, Y: gas, Z: solid

55. Which of the following is the best description of the concept of specific heat?
    a. Specific heat is the number of Celsius degrees of temperature change per kilogram of matter.
    b. Specific heat is the energy needed to raise one kilogram by one degree Celsius.
    c. Specific heat is the mass in kilograms of one cubic meter of matter at 0°C.
    d. Specific heat is the temperature change in degrees Celsius when one kilogram of matter absorbs one joule of heat.

56. A Fahrenheit degree represents how much temperature change compared to a Celsius degree?
    a. less
    b. more
    c. the same amount
    d. one degree

57. Describe a situation in which two objects have the same temperature but contain different amounts of heat.

58. Object A has twice the specific heat of Object B. A candle flame applies the same amount of heat to both objects. Which of the following is true and why?
    a. Object A is warmer than Object B.
    b. Object B is warmer than Object A.
    c. Object A and Object B are at the same temperature.
    d. It is impossible to say without knowing the masses of the objects.

59. The caloric theory explains temperature and heat with the hypothesis that temperature measures the quantity of a substance called *caloric* that is present in objects proportional to their heat content and temperature. In this theory, warmer objects contain more caloric than colder objects. *The theory must account for the following pieces of empirical evidence:* (1) Heat flows from hot to cold but not from cold to hot. (2) An object has the same mass when hot as it does when cold.
    a. Evaluate the caloric explanation of temperature, using logical reasoning to argue why one of the pieces of evidence is consistent with the caloric theory and one is not.
    b. Suppose experimental testing could only measure mass to the nearest 0.1 g. How might that affect whether the observations confirm or refute the caloric theory?
    c. A student places a beaker of hot water on a sensitive balance. The balance shows that the mass decreases as the water cools off. The student argues that this evidence supports the caloric theory because the lost mass represents the flow of caloric out of the cooling water. Analyze and critique the student's reasoning. (*Is there another possible explanation for the reduction in mass?*)
    d. Propose an investigative procedure to evaluate the caloric explanation with a better experimental test in which the alternative explanation in Part c is not an important factor. What is your testable hypothesis?

## Section 23.2

60. If the temperature of a fixed volume of gas increases, and the quantity (mass) of the gas stays the same, what must happen?
    a. The pressure must decrease.
    b. The pressure must increase.
    c. The gas must condense into a liquid.
    d. The gas molecules break apart into separate atoms

# Chapter review

### Section 23.2

61. In this chapter, you learned the last of the seven *fundamental quantities* in physics. Name all seven and provide the SI unit for each.

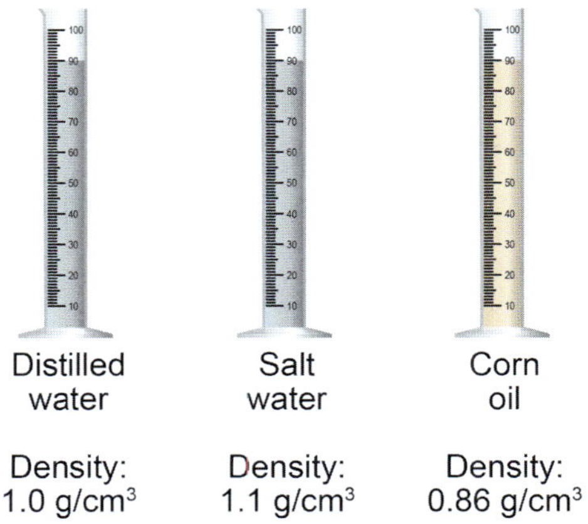


62. The three graduated cylinders in the diagram above have the same depth of liquid. Which one has the highest pressure at the 20 mL mark? Explain your reasoning in one sentence.

63. The units of pressure are newtons per meter squared and the units of volume are cubic meters. What are the units of pressure × volume? What other very important physical quantity has these units?

### Section 23.3

64. In the kinetic theory, what contributes to the internal energy of a solid other than the molecular kinetic energy of the individual particles?

65. When you take in a deep breath, does the air inside your lungs have more ordered or random kinetic energy?

66. Which equation would you use to calculate the thermal energy per atom for a gas?

    a. $v_{th} = \sqrt{3k_BT/m}$
    b. $E = \frac{1}{2}kx^2$
    c. $E = \frac{3}{2}k_BT$

67. The air in Luigi's classroom has a temperature of approximately 20°C, so he calculated $v_{th}$ = 502 m/s. What is the speed of the air molecules in his classroom?

## Quantitative problems

### Section 23.1

68. Which feels hotter, 100°F or 40°C?

69. During the summer an air conditioner is set to cool the room air down to 76°F. What temperature does this correspond to in degrees Celsius?

| **SuperZboiler** | **Model 1234** | | |
|---|---|---|---|
| **Serial #:** | 1234A3454 | **Fuel type:** | #2 |
| **Type:** | Oil-fired boiler | **Pump:** | 150 psi |
| **Voltage:** | 120/240 V | **Draft:** | –0.01 to –0.02 |
| **Amps:** | 3.4/1.8 A | **Max output:** | 120,000 btu/hr |
| **Frequency:** | 60 Hz | **Altitude:** | 0–5000 ft |
| **Short Circuit:** | 10 kA | **Purge time:** | 90 s |
| | | **Nozzle:** | 3.5 gal/45° |

70. The heating boiler or furnace in your home has a label similar to the example shown here. Which of the numbers describes the heat output? What units are used? Express the heat output of your own furnace or boiler in watts.

71. The hottest temperature ever recorded in the continental United States was 56.7°C. What does this correspond to in degrees Fahrenheit?

72. How much energy is required to raise the temperature of 200 g of water from 10°C to 20°C? (The specific heat of water is 4.18 J $g^{-1}$ $°C^{-1}$.)

73. Your lab supervisor tells you to heat a chemical until it is 165°F. What should your thermometer (which measures in degrees Celsius) read when it reaches the correct temperature?

74. In order to use a certain equation, you need to plug in the absolute temperature (K). The temperature is 25°C. What number should you use?

75. A mole is a furry animal with an average mass of 0.8 kg. How much mass would a mole of moles (the animal) have?

76. What is the mass of one krypton atom? (The atomic mass of krypton is 84 g.)

77. Rory eats a candy bar and then runs for a long time. He wonders how much energy he got from the candy bar to do all that running, so he looks at the label, which says "Serving Size 1 bar, Calories Per Serving: 260." How many joules of food energy did the candy bar provide?

78. If you add 100 g of water at 80°C to 500 g of water at 20°C, and there is no heat loss to the surroundings, what is the final temperature of the mixture of water?

79. A 0.5 kg ball of an unknown metal absorbs 5,775 J of energy when it heats up from 20°C to 50°C.

    a. Calculate the specific heat capacity of the material.
    b. Look up the specific heat capacities of various metals. What metal is the ball made of?

# Chapter review

## Section 23.1

80. 《 Yesterday it was 20ºC outside. Today, it is 25ºC.
    a. How much higher, in kelvins, is the temperature today than it was yesterday?
    b. How much higher is the temperature in degrees Fahrenheit?

81. 《 In a balloon at room temperature (25ºC), how much thermal energy does the average helium atom have?

82. 《 Alonzo puts 2 kg of 20ºC water [specific heat $c_p$ = 4,184 J/(kg ºC)] on the stove. The stove provides 502,080 J of thermal energy to the water. What is the maximum temperature that the water can be heated to by this energy?

83. 《 Hugh heats 10 kg of aluminum (ask your teacher for the specific heat if you do not remember it) from 15ºC to 55ºC. What is the least amount of work he could have done to heat it?

84. 《《 A hot, 100-g glass prism is placed in an insulated 300-mL sample of water at room temperature (22°C), causing the temperature of the water to come to equilibrium at 25°C. What was the initial temperature of the hot glass prism? [The specific heat of glass, $c_{p,g}$, is 664 J/(kg °C).]

85. 《《 Amelia performs an experiment to determine the type of an unknown metal. She places a 0.10 kg block of the metal at 25ºC into 1.0 kg of 75ºC water. After stirring for a few minutes she measures the temperature of both the water and metal to be 74.54ºC.
    a. The specific heat of water is 4,184 J/(kg ºC). How much thermal energy did the water lose in going from 75ºC to 74.54ºC?
    b. How much thermal energy did the metal gain in going from 25ºC to 74.54ºC?
    c. What is the specific heat of the unknown metal?
    d. The specific heats of aluminum, copper, and lead are 900, 386, and 128 J/(kg ºC), respectively. Which of these three metals is the unknown metal?
    e. Explain in detail what Amelia should do if her experimental value for the specific heat doesn't correspond with any of the given values.

## Section 23.2

86. ‹ Suppose that a typical compact car's gas tank has a capacity of 13 gallons. When the tank is full, how much mass and weight does fuel add to the car? (Express the fuel weight in both newtons and pounds). The density of gasoline is about 740 kg/m$^3$. Use the conversion factors 1 m$^3$ = 1,000 liters (L), 1 gal = 3.78 L, and 1 lb = 4.45 N.

87. ‹ For the following questions, consider an ideal gas that is currently at a temperature of 27°C.
    a. If the gas is heated by 1°C while the pressure is held constant, by what fraction will its volume change?
    b. If the volume is held fixed while the gas is cooled by 1°C from 27°C, by what fraction will its pressure change?
    c. If the temperature of the gas is held constant while its volume is expanded by 1%, by how much will its pressure change?

88. 《 Car tires are typically inflated to a pressure of roughly 35 psi (pounds per square inch). Perform the following calculations with a precision of three significant figures.
    a. Convert this pressure into a number of pascals by using the following conversion factors and definitions: 1 in = 2.54 cm, 1 cm = 0.01 m, 1 lb = 4.45 N, and 1 Pa = 1 N/m$^2$.
    b. Assuming that a compact car's mass is 1,200 kg, calculate the weight supported by *each* of its four tires (in newtons).
    c. Divide the force from Part b by the pressure from Part a to get an area in square meters. Also express this area in square inches.
    d. What is the physical significance of this area?

## Section 23.3

89. What is the thermal energy per atom for argon gas at room temperature (68°C)?

90. A car tire has a volume of 10 L and is inflated to a gauge pressure of 30 psi (207,000 Pa) at 20°C. How many air molecules are there inside the tire?

91. 《 A midsize airship (blimp) contains roughly 200,000 ft$^3$ of helium gas at an approximate pressure of 1.1 atmospheres. 1 ft$^3$ = 0.0283 m$^3$ and one atmosphere of pressure = $1.01\times10^5$ N/m$^2$.
    a. Suppose a blimp's helium is at 12°C (around 54°F). How many moles of helium does the blimp contain?
    b. What is the mass of this quantity of helium? (Helium has a molar mass of 16.0 g/mol.) Compare this to the mass of a compact car (about 1,200 kg).
    c. Imagine that this helium could cool by 1°C at constant volume. How much energy would this liberate? Compare this quantity to one kilowatt-hour. (1 kWh = $3.6\times10^6$ J.)
    d. What is the average thermal speed of a helium atom at 12°C? Compare this to a typical highway speed (30 m/s, or about 66 mph).

# Chapter review

## Standardized test practice

92. Two different hot metal objects, both at the same temperature, are each added to the same quantity of water that is at room temperature. Which metal will raise the water to a higher temperature?

A. the metal with the higher specific heat
B. the metal with the higher mass
C. neither, because both metals are at the same temperature
D. It will depend on both the mass and the specific heat of each metal.

93. You have a cup of 0.5 kg of water at 98°C and a cup of 0.3 kg of water at 56°C. If you mix them together, at what temperature will they reach equilibrium?

A. 82°C
B. 84°C
C. 77°C
D. 80°C

94. Which of the following temperatures is the *coldest*?

A. 100°C
B. 210°F
C. 370 K
D. 120°C

95. If you placed a large beach ball in water at room temperature, is it likely that you would see Brownian motion?

A. Yes, because the ball and the water are both above absolute zero.
B. Yes, because matter is continually in motion at a molecular level.
C. No, because the temperature is not high enough.
D. No, because the ball is too large for water molecule collisions to be noticeable.

96. Which of the following would you predict to have the highest specific heat?

A. light, polar molecules in the gas phase
B. light, polar molecules in the liquid phase
C. a heavy metal in the solid phase
D. a heavy metal in the liquid phase

97. A 10 kg zinc cannonball has been sitting in the desert sun all day. Its temperature is 40°C. Night falls on the desert and the temperature falls to a chilling 5°C. How much heat will the cannonball release as it cools to air temperature? (The specific heat capacity of zinc is 390 J $kg^{-1}$ $°C^{-1}$.)

A. 95 J
B. 19,500 J
C. 136,500 J
D. 175,500 J

98. Manuel heats a rigid container of an ideal gas, raising its temperature from 50°C to 100°C. Which statement accurately describes its change in pressure?

A. The gas pressure is doubled.
B. The gas pressure is halved.
C. The gas pressure is more than doubled.
D. The gas pressure is increased by a factor that lies between 1 and 2.

99. Nephi adds 502,080 J of energy to some water (specific heat of 4,184 J $kg^{-1}$ $K^{-1}$) by heating it from 20°C to 60°C. What is the mass of the water?

A. 0.3 kg
B. 2 kg
C. 3 kg
D. 4,800 kg

100. Sample 1 is an ideal gas with a volume of 0.25 $m^3$, a temperature of 300 K, and a pressure of 100,000 $N/m^2$. Sample 2 has the same composition, but a volume of 0.5 $m^3$, a temperature of 400 K, and a pressure of 50,000 $N/m^2$. Which statement correctly compares the number of particles in the two samples?

A. Sample 1 contains more particles than Sample 2.
B. Sample 2 contains more particles than Sample 1.
C. Samples 1 and 2 hold equal numbers of particles.
D. You cannot tell without more information

101. Ten grams (10 g) of helium are at a temperature of 300 K. What is the average thermal energy per atom?

A. $4.98\times10^{-23}$ J/atom
B. $6.21\times10^{-21}$ J/atom
C. $3.74\times10^{4}$ J/atom
D. $2.71\times10^{26}$ J/atom

102. The intermolecular forces in a certain substance are such that at room temperature atoms can switch neighbors, but they are not completely separated. At room temperature it must be in the ____ phase. If it is then cooled sufficiently to change phase, it will most likely be in the _____ phase.

A. liquid; solid
B. liquid; gas
C. solid; liquid
D. solid; gas

103. You are cooling an unknown gas. When the substance reaches a temperature 10°C, the gas starts to condense and liquid droplets form. When the substance reaches a temperature of −10°C, the liquid becomes solid. What is the boiling point of this substance?

A. −10°C
B. 0°C
C. 10°C
D. More information is required.

# Chapter 24 Heat Transfer

Heat flows from the Sun to Earth through the vacuum of space in the form of electromagnetic radiation. Less well known is that Earth must reradiate the same amount of energy back into space again! For our planet's temperature to remain constant (on average), the energy input from the Sun must equal the energy output from Earth.

While the 5,500°C Sun's radiant energy comes chiefly in the form of visible light, the much cooler Earth (14°C on average) glows in infrared light. Furthermore, Earth's atmosphere is not equally transparent to all forms of electromagnetic radiation. Much of the Earth's infrared glow is blocked by gases such as carbon dioxide, methane, and water vapor. By itself this is no cause for alarm; indeed, without these *greenhouse gases* Earth would be too cold to sustain life.

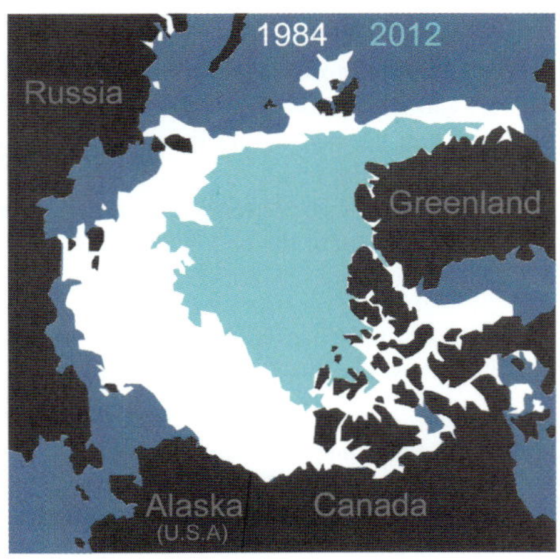


But Earth's temperature is set by a very delicate balance of energy flow, and when its atmosphere becomes more opaque to infrared light, our planet responds by heating up. Atmospheric carbon dioxide has increased by 20% since the 1950s, in large part because of our burning fossil fuels such as coal, oil, and natural gas. Most climate scientists conclude that this has caused the record-breaking high temperatures of recent years. Another global consequence of heat transfer can be seen in satellite images of the Arctic Ocean, where the summertime ice cap has recently shrunk by more than 25% over 10 years. This threatens the survival of iconic species such as the polar bear.

Even bigger changes may be in store. Since water absorbs much more sunlight than does ice, the Arctic is now retaining much more of the Sun's radiant energy than it did in the past century. This may set off a positive feedback loop, where temperatures rise, more ice melts, oceans absorb more sunlight, and temperatures rise again. Some scientists have proposed stabilizing measures that would make the Arctic "shiny" once more, including seeding the atmosphere with tiny particles to reflect sunlight or help form clouds. The science of Earth's changing thermal energy balance has led to a society-wide discussion of how to address its consequences.

# Chapter study guide

## Chapter summary

Heat flows at a rate proportional to the difference in temperature. Thermal equilibrium occurs when temperatures are equal. Heat is exchanged through *conduction, convection,* and *radiation.* In *free convection,* fluid moves owing to differences in density caused by temperature changes. In *forced convection* fluid flow is driven, such as by wind, pumps, or fans. Convection is much more effective than conduction in heat transfer involving fluids, such as air or water. Radiation travels at the speed of light and can pass through empty space. Conduction and convection are far slower and occur only through matter. All objects at temperatures above absolute zero emit and absorb thermal radiation. The average temperature of Earth and the other planets depends on the balance between emitted thermal radiation and radiation absorbed from the Sun.

## Chapter index



## Learning objectives

By the end of this chapter you should be able to

- explain the concept of thermal equilibrium;
- explain why hot objects cool at a decreasing rate when left in a constant-temperature environment;
- explain the similarities and differences among conduction, convection, and radiation;
- calculate the power of heat conduction through simple solid shapes such as rods and walls;
- explain the meaning of the wind chill index in terms of heat transfer; and
- relate the blackbody curve to the color temperature of a light source such as a compact fluorescent bulb.

## Investigations

24A: Heat transfer
24B: Visualizing convection

## Important relationships

$$P = \frac{\kappa A}{L}\Delta T$$

$$P = hA\Delta T$$

$$P = \varepsilon\sigma AT^4$$

$$P = \varepsilon\sigma A(T_\infty^4 - T^4)$$

## Vocabulary

**thermal equilibrium**
**thermal conductivity**
**emissivity**
**Newton's law of cooling**
**convection**
**blackbody**
**heat transfer**
**heat transfer coefficient**

## 24.1 - Thermal equilibrium and heat flow

Can you think of a way to cook soup on a campfire with only a wooden or wicker pot? The secret is to use fire-heated rocks to transfer heat. In fact, the children's story *Stone Soup* was inspired by the real traditional practice of using hot stones to boil water. Using this technique, Native Americans and other ancient peoples cooked soups and stews in water-tight wicker baskets! This is quite clever because a wicker basket would never survive being heated directly by a fire. Think about the flow of thermal energy, or *heat,* that occurs when you drop a hot rock in a container of cold water.

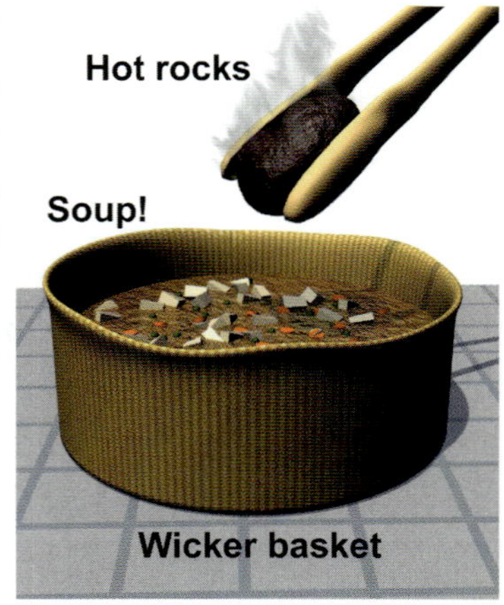


- How does the temperature of the rock change?
- How does the temperature of the water change?
- How do the two temperatures compare after some time has passed?

### Thermal equilibrium

**When does heat flow?**

Whenever two objects are at different temperatures, heat will flow from the higher temperature object to the lower temperature object. The same is true for different areas of the *same* object. Heat naturally flows from regions of higher temperature to regions of lower temperature. In the example of boiling water with hot stones, thermal energy flows *out* of a hot stone and *into* the cold water instead of the other way around. The fact that heat flows from hot to cold is the most common consequence of the *second law of thermodynamics*, the law that governs the flow of energy in real systems.

**Whenever there is a temperature difference, heat flows *from* the higher temperature object *to* the lower temperature object.**

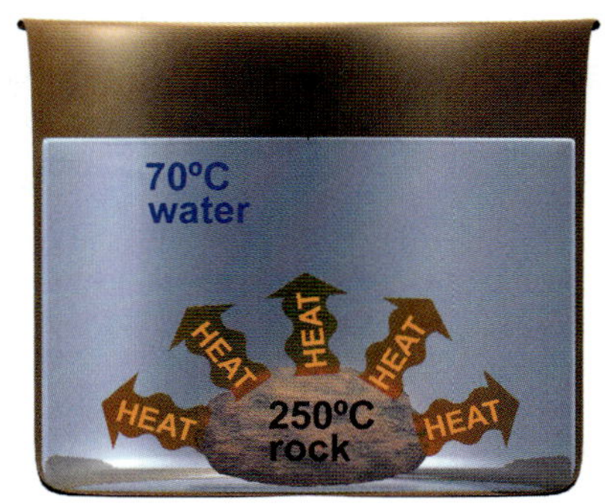


**When does heat stop flowing?**

Heat continues to flow until all objects are at the same temperature. The condition in which everything is at the same temperature is called **thermal equilibrium**. In thermal equilibrium, no heat flows because the temperatures are the same. For example, if hot rocks are immersed in cold water, thermal equilibrium occurs once both the rocks and the water reach the same temperature. Of course, unless the air is at the same temperature, heat will still flow between the pot and the *air!* This heat flow is why hot objects cool down.

**Thermal equilibrium occurs when temperatures are equal.**

No heat flows in thermal equilibrium.

**Is equilibrium reached instantaneously?**

When the hot rock is placed in the cold water, does the water heat up *instantaneously*? Of course not! It usually takes some time for two objects to reach thermal equilibrium. The *rate* that two objects come into thermal equilibrium is discussed on the next page.

**A cooling curve**

Consider a cup of 85°C coffee on a table in a room where the air temperature is 21°C. Over time, the hot coffee cools down until it eventually reaches the same temperature as the air. The graph of temperature versus time, illustrated at right, is called a *cooling curve.* Notice that the temperature drops quickly at the start and then falls more gradually over time. This is typical of how heat is exchanged. The largest changes in temperature occur when the differences in temperature are greatest.

Cooling curve

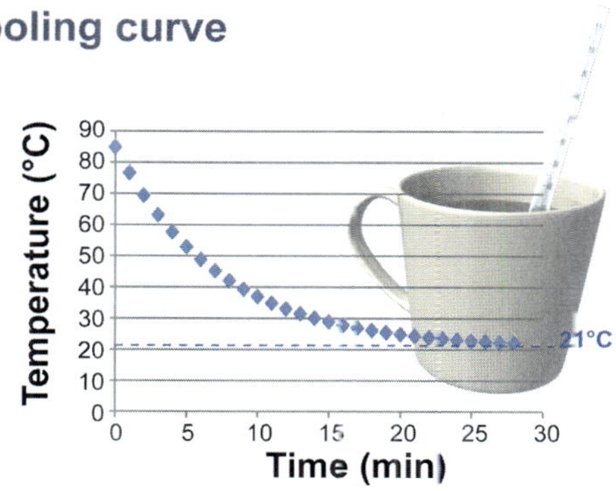


**Heat transfer**

**Heat transfer** is the study of how thermal energy, or heat, moves. The coffee cools because heat transfer moves thermal energy from the coffee and cup into the room air. In this chapter we will consider *how* heat moves through the mechanisms of conduction, convection, and radiation. We will also consider *how quickly* heat moves, which is the *rate* of heat transfer. Since heat is a form of energy, the heat transfer rate is energy per unit time, which has units of joules per second (otherwise known as *watts*).

**Why does the cooling curve change slope?**

The *rate* of heat transfer between the coffee cup and the room depends on the difference in temperature. For example, when the temperature difference is high—such as at the start—the temperature changes by 8°C in the first minute. This represents an average heat transfer rate of 144 J/s or 144 W. Later, when the coffee has cooled to 30°C, the temperature changes by only 1.2°C per minute because the rate of heat transfer has dropped to 21 W.

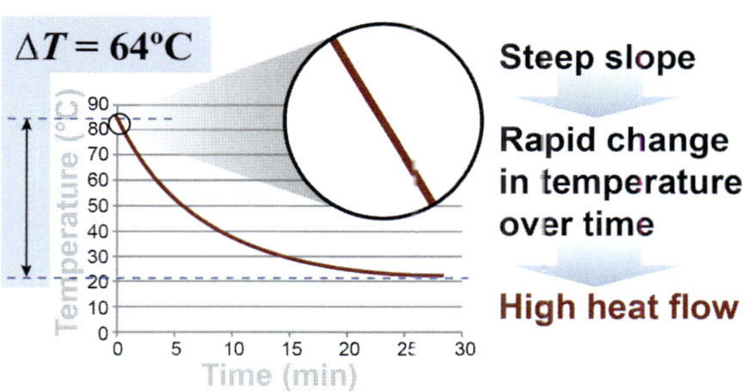


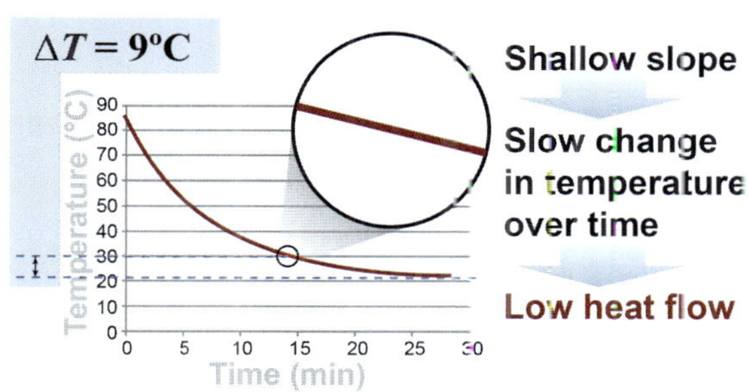


**Newton's law of cooling**

The observation that the rate of heat transfer depends on the difference in temperature is called **Newton's law of cooling**. This "law" is not really a law but a rule of thumb that applies approximately to many situations. Newton's law of cooling makes common sense. When there is a big temperature difference, heat flows quickly and causes the temperature to change quickly, too. When there is a small temperature difference, heat flows slowly and temperature also changes slowly.

**Environmental temperature**

When solving problems like the cooling coffee cup, we often assume that the temperature of the room stays constant even though energy is absorbed by the room's air. This is usually an acceptable approximation because the thermal energy from the coffee cup cannot appreciably change the large room's temperature. The room may also maintain a constant temperature because it exchanges heat with the environment around it, such as via the its central heating (and cooling) system or through the walls with the outside air.

**Do materials conduct heat differently?**

In heat transfer, *conduction* is the transfer of heat through or between materials by direct contact. Imagine a bar of copper and an identical size piece of wood. One end of each is immersed in boiling water and the other end is in your hand. If you did the experiment you could comfortably hold the wood but the copper would quickly become too hot to hold. The reason is that heat flows through copper 4,000 times more readily than heat flows through wood. This is why cooking spoons are typically made out of wood or plastic—or at least have wood or plastic handles—rather than being made entirely of metal.

**Explaining conduction**

In heat conduction, energy moves but not the matter itself. On the molecular level, hotter molecules have more kinetic energy and therefore more random motion. As neighboring molecules bump into each other they transfer kinetic energy. This transfer of intermolecular kinetic energy is the fundamental explanation for heat conduction.

**Thermal conductors and insulators**

Different materials have very different abilities to conduct heat. The table (at right) compares the **thermal conductivity** of different materials. A thermal conductivity of one watt per meter per degree, or 1 W/(m °C), means a 1°C temperature difference across 1 m of length causes 1 W of heat to flow through 1 m$^2$ of cross-sectional area. You can evaluate whether a material is a thermal conductor or insulator based on the value of its thermal conductivity $\kappa$ listed in the table (at right). Materials with high thermal conductivity, such as copper, are classified as thermal conductors. Materials with low thermal conductivity, such as foam or air, are classified as thermal insulators. Other materials, such as glass, are intermediate heat conductors.

Thermal conductivity

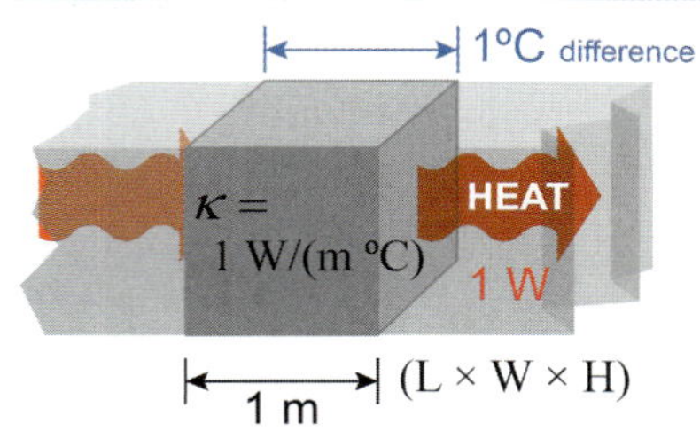


| Material | $\kappa$ W/(m °C) |
|---|---|
| Diamond (IIa) | 2,650 |
| Copper | 401 |
| Aluminum | 226 |
| Steel | 43 |
| Rock | 3 |
| Glass | 2.2 |
| Ice | 2.2 |
| Water (liquid) | 0.58 |
| Wood | 0.11 |
| Wool | 0.038 |
| Fiberglass insulation | 0.038 |
| Air | 0.026 |
| Styrene foam | 0.025 |

Thermal conductors

Thermal insulators

**Can you feel thermal conductivity?**

Think about touching a foam cup and a metal cup that have both been sitting on a table long enough to be at the *same temperature* as the room. The metal cup feels cold but the foam cup feels warm even though both are at the same temperature! What you feel is the difference in thermal conductivity. Metal conducts heat away from your skin much faster than foam. The skin touching the metal cup gets cooler, closer to room temperature, than the skin touching the foam cup.

### What is convection?

A hair dryer heats a surface by blowing hot air. This is an example of **convection**. Convection is the transfer of heat by the motion of liquids and gases. You can experience convection yourself with a candle flame. Your hand gets hot even a half meter above the flame while at the same distance to the side of the flame produces little sensation of heat at all. You feel heat directly above the flame because the hot air rises and carries the heat with it, an example of convection. To the side, there is some heat transfer through conduction and radiation, but air is an insulator so the amount of heat is barely detectable.

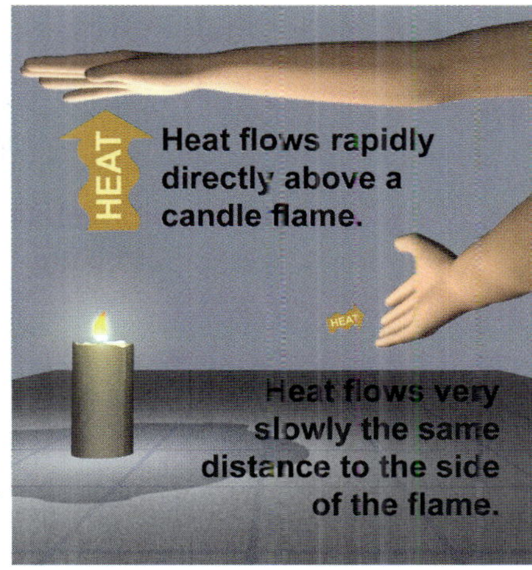


### Forced and free convection

*Free convection* occurs when the fluid moves on its own because of differences in density caused by differences in temperature. The convection above the candle flame is an example of free convection. *Forced convection* occurs when fluid flow is driven by wind, pumps, fans, or other means. The blower in a hair dryer is an example of forced convection. Convection is the dominant means of large-scale heat transfer within Earth's atmosphere and its oceans. Vast wind patterns move heat through the atmosphere. Enormous currents circulate warm ocean water from equatorial areas to higher latitudes and cold water from the poles to warmer latitudes.

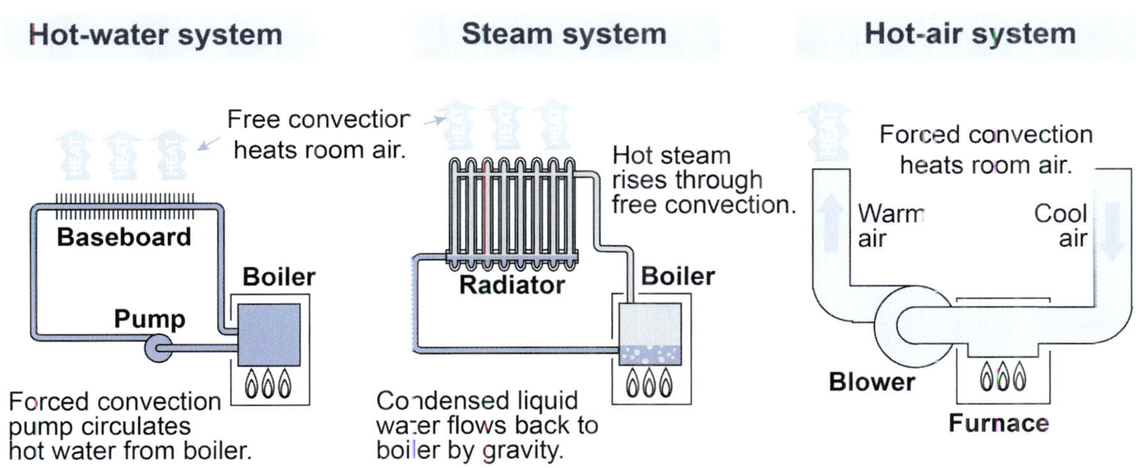


### How do heating systems work?

In many situations convection transfers heat between a surface and water or air. For example, many homes are heated by steam or hot water in baseboards or radiators. Free convection heats the air, which flows around the room. In a hot-water system, finned-tube *baseboards* are heated by forced convection from a pump that circulates hot water from the boiler throughout the house. In a steam system, hot steam rises through natural convection through *radiators*, carrying heat to each room. In a forced-air system the furnace heats air that is circulated by a blower to each room in the house. The air blower creates forced convection—moving heat by moving hot air.

### What is thermal radiation?

The physics definition of *radiation* means the dispersal of energy through space. Both conduction and convection occur only through matter—neither can transfer energy through the vacuum of space. Radiation is different. Radiant energy can travel through space without the presence of matter. *Thermal radiation* consists of electromagnetic waves (including light) produced by objects because of their temperature. All objects with a temperature above absolute zero give off thermal radiation, not just the Sun.

### Can we see thermal radiation?

We do not normally see thermal radiation from room-temperature objects because the energy is too low to be detected by the human eye. An *infrared camera*, however, sees infrared radiation very well and is the basis for "night vision" goggles. Night vision systems amplify low-energy thermal radiation to create images in higher energy visible light.

### When does thermal radiation become visible?

The power versus wavelength graph shows how light from thermal radiation is spread over a range of energy. An object at room temperature does not "glow" because the curve for 20°C does not extend into visible wavelengths. An electric stove element at 600ºC glows a dull red, and you can see that the graph for 600ºC just extends into the visible range. The Sun's photosphere has a temperature of 5,500ºC and radiates across the entire spectrum of visible light, providing the "white" sunlight we see.

**Power vs. wavelength for thermal radiation**

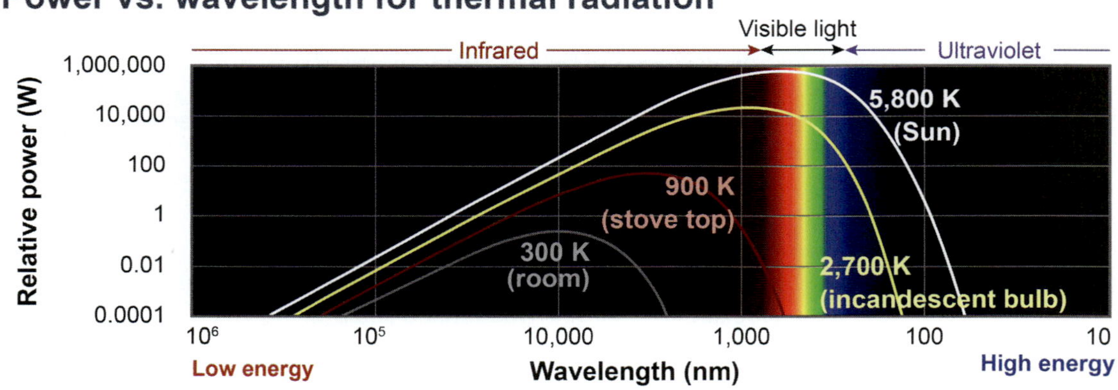


### Radiation balance

The transfer of thermal radiation goes both ways because all objects constantly receive thermal radiation from everything else in their environment. Otherwise, all objects would eventually cool down to absolute zero by radiating their energy away. The temperature of an object fluctuates depending on the object's overall energy balance. If radiation is the only means of heat transfer, such as in space, the temperature rises if more radiation is absorbed than emitted. The temperature falls if more radiation is emitted than absorbed. Space objects such as asteroids have a temperature that represents a balance between radiation absorbed and radiation emitted.

**Conduction and convection**

The thermal energy carried by conduction and convection is fundamentally the energy in random motion of atoms and molecules in matter. This is why conduction and convection only occur in matter. It also explains why heat moves relatively *slowly* in both conduction and convection. Energy must be ultimately transferred by trillions of tiny, random collisions between individual molecules. Conduction occurs in all phases of matter: solid, liquid, and gas. Convection occurs only in the liquid and gas phases in which entire collections of molecules can move as a whole in addition to their individual thermal motion.

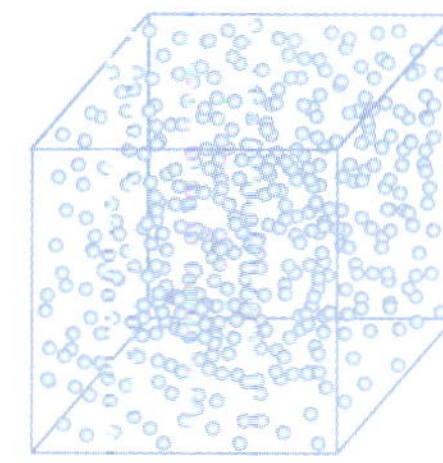

**Conduction and convection both occur through the contact between atoms and molecules in matter.**

**Radiation**

Radiation is fundamentally different from conduction and convection. Electromagnetic waves are pure energy independent of matter. This is why radiation can transfer heat through the vacuum of space. Radiation also moves at the speed of light, $3 \times 10^8$ m/s, far faster than other types of heat transfer. When radiation encounters matter, some of the electromagnetic energy is transformed into molecular thermal motion—ordinary heat.

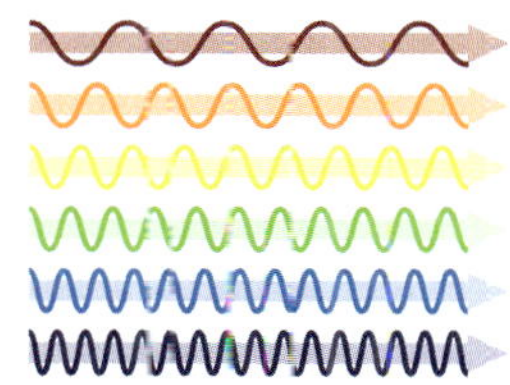

**Radiation occurs through electromagnetic waves and travels at the speed of light.**

**Realistic heat transfer**

Our world of matter and energy includes solids, liquids, and gases at temperatures far above absolute zero. All three forms of heat transfer act together to determine the rate at which thermal energy is exchanged. Which of the three is most important depends on the temperature and on the circumstances. Although it is uncommon for all three modes of heat transfer to be equally important in a single transformation of energy, heat transfer usually occurs in many steps. In both nature and technology, a different form of heat transfer may be most important at each different step. For example, in a solar-thermal power station, radiant energy heats a metal tube. The heat travels through the walls of the tube by conduction, where it heats water into steam, which in turn carries the heat to a turbine by convection.

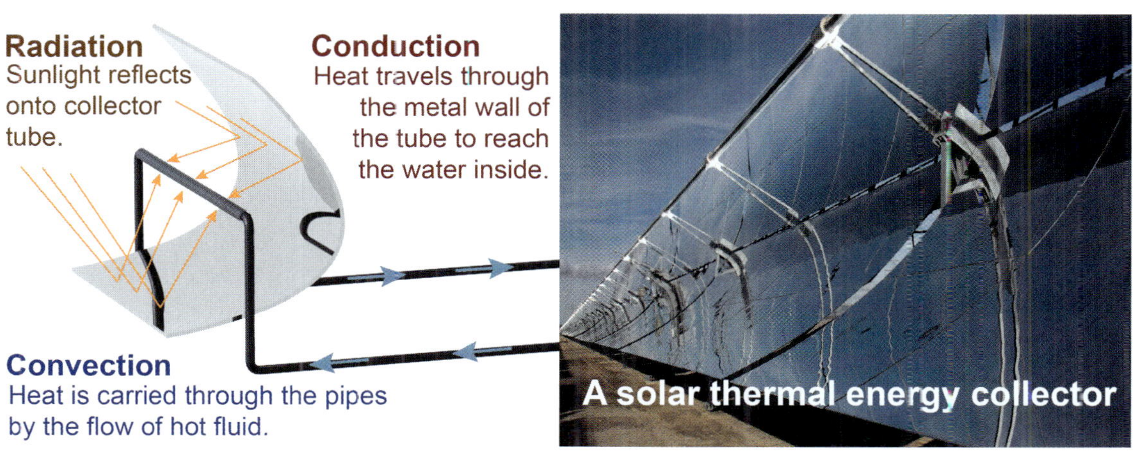

# Investigation 24A Heat transfer

**Essential questions** *How quickly does heat move?*

Interactive simulation

No object in our environment stays at a constant temperature for very long. As temperature changes everything, including our bodies, constantly exchanges heat with the environment and other objects. In this investigation you will model the heat flow through a wall of a typical home.

## Part 1: How heat transfer is balanced for a house

The interactive model shows a wall with windows that represent the heat transfer properties for a typical house.

1. Start with a wall area of 90 $m^2$ with R4 insulation (uninsulated) and window area of 10 $m^2$ single-pane glass.
2. Set the inside temperature to 21°C, outside to 0°C, and wind speed to 1 m/s, and choose a sunny day.

a. Describe the heat that flows and its distribution among convection, conduction, and radiation.

b. How does the heat transfer change when the wind speed is increased to 15 m/s?

c. How does the heat transfer change when the outside conditions change from sunny to cloudy to night?

d. How high can you set the inside temperature before a 15,000 W furnace cannot keep up?

e. What might you want to install inside to reduce a large heat transfer *into* the house.

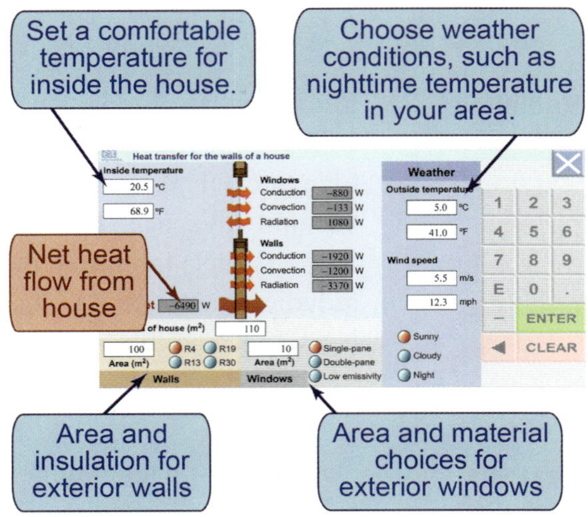


**How can you design the windows and walls of a house so you can use a lower power furnace to heat the house?**

## Part 2: House conditions appropriate to your area

1. Set the parameters to be typical for houses in your area and use R13 wall insulation.
2. Set the outside weather conditions to nighttime values typical for winter in your area.
3. Set the inside temperature to a value you consider to be comfortable, such as 21°C (70°F).

a. How much heat must the furnace produce to balance the heat transfer through the walls and glass?

## Part 3: Going further

1. Design and carry out a virtual experiment to determine the effect of "R-value" in terms of heat transfer through the wall.
2. Design and carry out a virtual experiment that determines the effect of different window construction options in sunny, cloudy, or night conditions.

a. What are the effects of "R-value" on heat transfer through the wall?

b. Describe the parameters of the

control *used throughout this investigation. How did you use the control in the investigation? Why is it useful to have a control?*

# Section 1 review

The rate of heat flow is proportional to the difference in temperature. Thermal equilibrium occurs when temperatures are equal. Heat flow stops in thermal equilibrium. The rate of heat transfer between a warmer and a colder body slows down as the temperature difference between them gets smaller. Heat *conduction* is the transfer of heat through or between materials by direct contact. *Thermal conductivity* describes the rate at which heat is conducted through a substance. *Convection* is the transfer of heat by the motion of liquids and gases, such as hot air rising above a candle flame. *Free convection* occurs when the fluid moves on its own owing to differences in density. *Forced convection* occurs when fluid flow is driven by wind, pumps, fans, or other means. *Thermal radiation* is the transfer of heat energy through electromagnetic waves, including light. Radiation can travel through space without the presence of matter whereas both conduction and convection occur only through matter. All three mechanisms of heat transfer (conduction, convection, and radiation) usually act at the same time, but not all are equally effective at any given temperature or in any given situation.

| Vocabulary words | thermal equilibrium, Newton's law of cooling, heat transfer, thermal conductivity, convection |
|---|---|

## Review problems and questions

1. Describe a situation you experienced in the last 24 hours in which heat flows through
   a. conduction.
   b. convection.
   c. radiation.

2. Some people place mirrored reflectors in their car windows on hot, sunny days to reduce the heat gain by
   a. conduction.
   b. convection.
   c. radiation.

3. If you drop some confetti above a warm radiator, the flakes of paper will rise as warmed air above the radiator rises. This is a good example of heat transfer through
   a. conduction.
   b. convection.
   c. radiation.

4. The end of a silver spoon gets warm a few moments after the bowl of the spoon is immersed in hot soup. The end of the spoon gets warm mainly because of
   a. conduction.
   b. convection.
   c. radiation.

## 24.2 - Conduction and convection

The models in this section are useful approximations for heat transfer by conduction and convection. The heat conduction equation is the closest model to a true law of physics because it accurately describes most situations. Models of convection, however, are far more complex and empirical in nature. The simple form of the equation $P = hA\Delta t$ buries enormous complexity in the heat transfer coefficient $h$. The value of $h$ can vary by more than a million depending on the shape of a surface, the flow rate, the specific fluid, and other factors. Realistic calculations of convection must come directly from experimental data and a model that is 10% accurate for one situation often gives unrealistic predictions for a different geometry or flow speed. Models of heat transfer should always be assumed to be approximations subject to test by experiment for anything other than the simplest geometry.

### The heat conduction equation

**The heat conduction equation**

The heat conduction equation (24.1) describes the thermal energy that flows through a material because of a difference in temperature between one part of the material and another. Since the flow of energy is *power*, the equation gives the power $P$ in watts (W) for a temperature difference $\Delta T$. The material may be a solid, liquid, or a gas. In equation (24.1) the following geometry is assumed:

- Heat flows across area $A$ through length $L$.
- The temperature difference $\Delta T = (T_2 - T_1)$ occurs across the same distance $L$.
- The material has a uniform thermal conductivity $\kappa$.

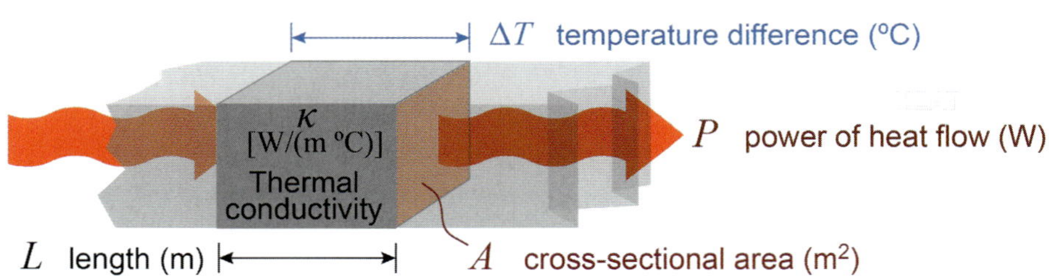


**Heat conduction**

$$P = \frac{\kappa A}{L}\Delta T \quad (24.1)$$

$P$ = power (W)
$\kappa$ = thermal conductivity (W m$^{-1}$ °C$^{-1}$)
$A$ = cross-sectional area (m$^2$)
$L$ = length (m)
$\Delta T$ = temperature difference (°C)

**How do I use the conduction equation?**

Equation (24.1) is most useful when applied to solids because in most realistic situations, convection moves heat more rapidly than conduction in liquids and gases. To use the equation, you need to know something about the shape of the object the heat flows through. Two geometric factors are important:

1. Heat flow is directly proportional to the *area* the heat flows through. Double the area, and the heat flow also doubles for the same temperature difference.
2. Heat flow is inversely proportional to *length*. If you double the length the heat has to flow through, the power of heat flow is reduced by half.

**Heat conduction through a rod**

Consider our previous example. A solid rod is held at one end by a person's hand, fixing the temperature at about 30ºC. The other end is immersed in boiling water at 100ºC. The heat flows through a length of 20 cm, or 0.2 m, between the water surface and the hand. The rod has a diameter of 1.9 cm. What is the rate of heat flow if the material is copper compared to wood?

**Heat conduction**

$$P = \frac{\kappa A}{L}\Delta T$$

$P$ = power (W)
$\kappa$ = thermal conductivity (W m$^{-1}$ °C$^{-1}$)
$A$ = cross-sectional area (m$^2$)
$L$ = length (m)
$\Delta T$ = temperature difference (°C)

**The geometry of heat flow**

When analyzing any conduction problem the first step is to identify the geometry so that you can determine the appropriate area $A$ and length $L$.

1. Heat flows in the direction from higher temperature to lower temperature.
2. The length $L$ in equation (24.1) is measured *along* the direction the heat flows.
3. The area $A$ in equation (24.1) is *perpendicular* to the direction heat flows.

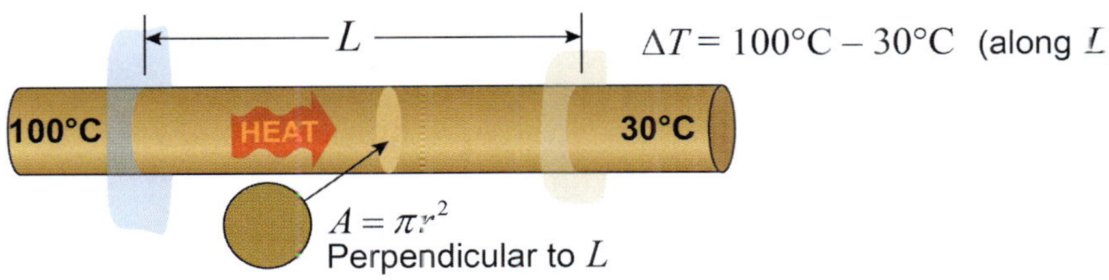


**Geometry for heat flow along a rod**

**Choosing $A$ and $L$**

In this situation the two ends of the rod are at different temperatures, so heat flows along the length of the rod from the hotter end to the cooler end. The value of $L$ is therefore measured along the rod's length. The heat crosses through the cross-sectional area ($\pi r^2$) of the rod, and therefore $A = \pi r^2$.

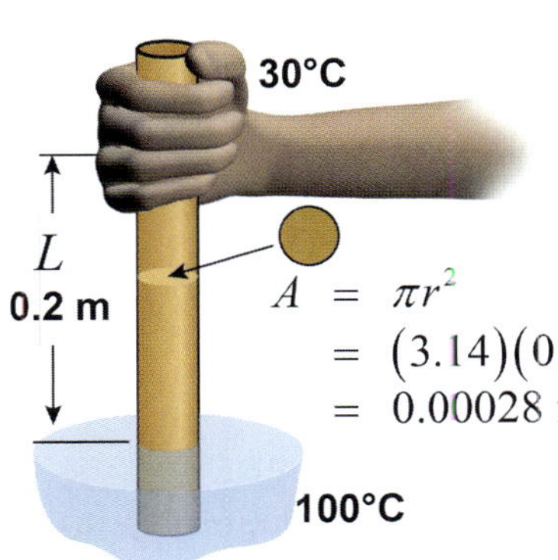


$$A = \pi r^2 = (3.14)(0.0095\text{ m})^2 = 0.00028\text{ m}^2$$

$$P = \frac{\kappa A}{L}\Delta T$$

**For copper** $\kappa = 401$ W m$^{-1}$ °C$^{-1}$

$$P = \frac{(401\text{ W m}^{-1}\text{ °C}^{-1})(0.00028\text{ m}^2)}{0.2\text{ m}}(70\text{°C}) = \mathbf{39.3\ W}$$

**For wood** $\kappa = 0.11$ W m$^{-1}$ °C$^{-1}$

$$P = \frac{(0.11\text{ W m}^{-1}\text{ °C}^{-1})(0.00028\text{ m}^2)}{0.2\text{ m}}(70\text{°C}) = \mathbf{0.011\ W}$$

**What the difference in heat means**

Notice that 40 J of heat per second (40 W) flows up the copper rod while only 0.011 J of heat per second (0.011 W) flows up the wood rod. This is a good reason to use wood for spoon handles instead of copper! A heat flow of 40 W can raise the temperature of your skin by about 5°F per second—reaching the boiling point of water in 25 s.

## Heat conduction through a wall

**How do we calculate heat flow through a wall?**

Many situations involve the heat flowing through a *wall.* The width and height of a wall are large compared to its thickness. The geometry for calculating heat conduction through a wall is shown in the diagram on the right. If $T_2 > T_1$ then heat flows from left to right through the thickness of the wall. The length $L$ appearing in equation (24.1) is the wall thickness since that is the dimension in the direction of heat flow. The area $A$ is the wall area, which is perpendicular to the length direction and is the area *through* which the heat flows.

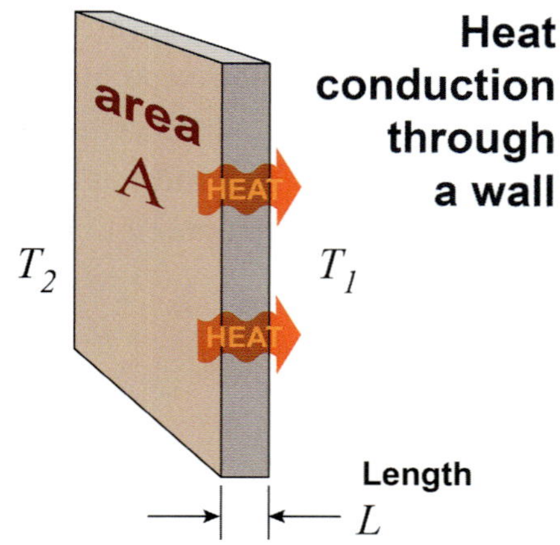


**Heat conduction**

$$P = \frac{\kappa A}{L}\Delta T$$

$P$ = power (W)
$\kappa$ = thermal conductivity (W m$^{-1}$ °C$^{-1}$)
$A$ = cross-sectional area (m$^2$)
$L$ = length (m)
$\Delta T$ = temperature difference (°C)

**An example problem**

Compare an aluminum cup and a foam cup of boiling water immersed in ice water at 0°C. Both cups have a radius of 5 cm and are immersed 10 cm into the cold water. How fast does heat flow through the walls of each cup if the wall thickness is 5 mm? In this case the *area* through which the heat flows is the area of the cup contacting the water. The *length* through which the heat flows is the thickness of the cup wall.

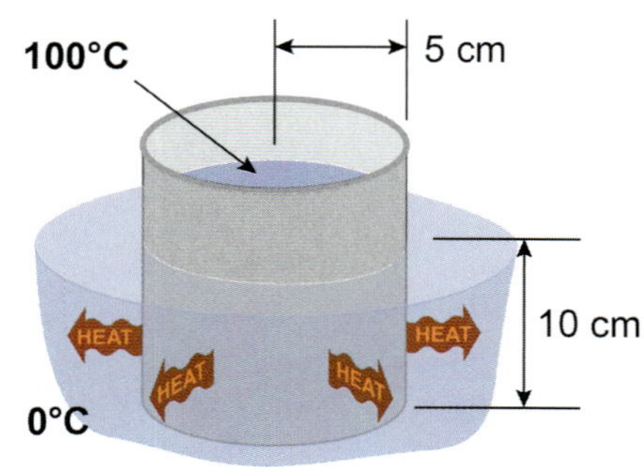


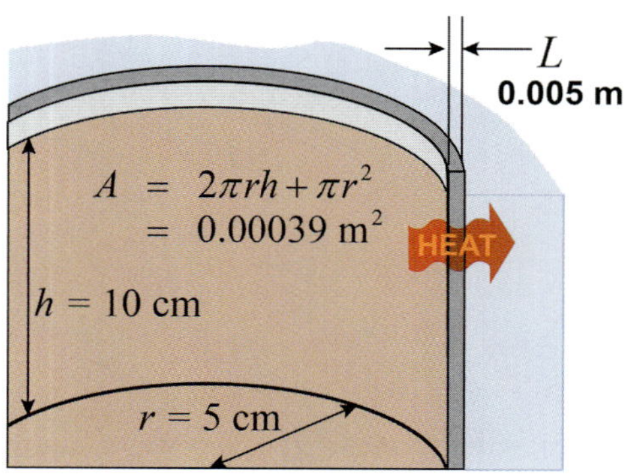


$$P = \frac{\kappa A}{L}\Delta T$$

**For aluminum** $\kappa = 226\ \text{W m}^{-1}\ \text{°C}^{-1}$

$$P = \frac{(226\ \text{W m}^{-1}\ \text{°C}^{-1})(0.039\ \text{m}^2)}{0.005\ \text{m}}(100\text{°C}) = \mathbf{176{,}000\ W!}$$

**For foam** $\kappa = 0.025\ \text{W m}^{-1}\ \text{°C}^{-1}$

$$P = \frac{(0.025\ \text{W m}^{-1}\ \text{°C}^{-1})(0.039\ \text{m}^2)}{0.005\ \text{m}}(100\text{°C}) = \mathbf{19.5\ W}$$

**Why are cups not often made of metal?**

The rate of heat transfer for the aluminum cup is 176,000 W! This would initially cool the water by 112ºC per second. This heat flow is so rapid that the water near the inner surface of the cup would cool almost instantly to 0ºC. The remaining water in the middle of the cup would cool a bit more slowly. For the foam cup the heat flow is only 19.5 W, and the hot water loses only 0.01ºC per second. Both cups have the same wall thickness and contact area, but the thermal conductivity of aluminum is 226 W/(m ºC) compared to 0.025 W/(m ºC) for foam.

**What is convection?**

Convection is the movement of heat through the motion of matter, in the form of flowing gas or liquid. *Convection can transfer heat very rapidly.* As an example, consider a glass window in still air separating an inside temperature of 20°C from an outside temperature of 0°C.

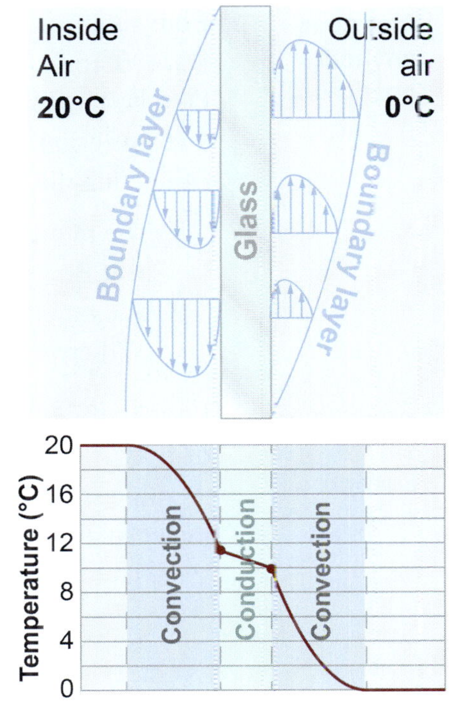


**Convection at a window**

Near the inner surface of the glass the air is cooled and sinks. The farther the air is from the top of the glass, the faster it moves. A *boundary layer* of downward moving cool air flows down the window. Put your hand near the bottom of a window on a cold day and you can feel the *draft* created by this cold layer of air falling from the glass surface. The opposite happens on the cold outside of the glass. Air near the bottom warms up and rises. An upward moving boundary layer forms on the outside carrying heat upward.

**How do convection and conduction compare?**

To compare the heat transferred by conduction and convection, consider heat moving across the 20 cm of air next to the window. Air is a good insulator and the heat conduction is small, less than 3 W across a square meter. In comparison, the heat flow by free convection is more than 100 W! This is thirty times greater. The important conclusion is that *convection is far more effective at moving heat in liquids and gases than conduction.* In almost all cases in which heat moves through a fluid (liquid or gas) convection plays a major role, usually exceeding conduction by at least an order of magnitude.

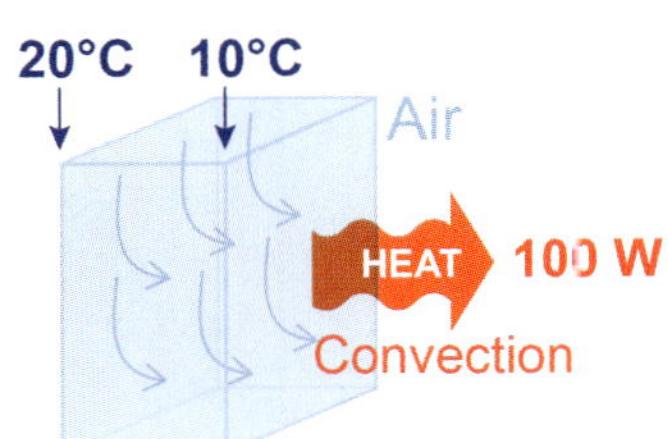


Convection moves heat *much* faster through fluids than conduction.

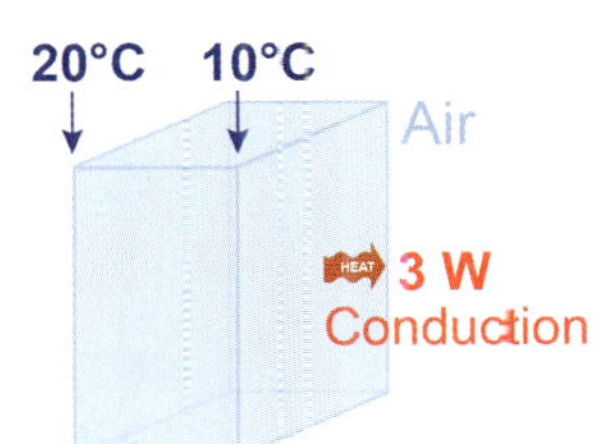


**How does wind affect convection?**

Consider the same window in a stiff wind. Wind allows cold air to replace warm air right at the window surface far more rapidly. The heat transfer rate typically increases by a factor of 10–100. Forced convection, in which fluid is moved by an external source (such as wind), is an extremely effective way to move heat.

**Convection in water**

When water boils in a pot, the heat is typically applied at the bottom. If you watch carefully, you can see convection occur. Heated water rises in multiple convection cells, which mix cooler water near the top with heated water near the bottom.

**What does convection depend on?**

Convection is a complex phenomenon. An accurate description involves fluid flow, heat flow, friction, viscosity, surface roughness, buoyancy, and about a dozen other factors. Despite the complexity, a couple of general principles can be used to create a useful model of convection. This model applies to convective heat transfer between a surface and a fluid contacting the surface.

1. The rate of heat transfer is approximately proportional to the area $A$ through which the heat flows.
2. The rate of heat transfer is approximately proportional to the temperature difference $\Delta T$ between the surface and the fluid a distance away from the surface.

**Convective heat transfer equation**

Equation (24.2) is widely used to describe convective heat transfer. The complicated effects of flow speed and surface conditions are grouped together in the **heat transfer coefficient** $h$. A value of $h = 1$ W/(m$^2$ °C) means that one watt of heat is transferred from each square meter of area when the temperature difference is one degree Celsius. Practical heat transfer systems typically have a value of $h$ between 10 and 10,000 W/(m$^2$ °C).

**Heat convection**

$$\textbf{(24.2)} \quad P = hA\Delta T$$

$P$ = power (W)
$h$ = heat transfer coefficient (W m$^{-2}$ °C$^{-1}$)
$A$ = cross-sectional area (m$^2$)
$\Delta T$ = temperature difference (°C)

**Optimizing for convection**

Equation (24.2) suggests two common ways to increase heat transfer by convection:

1. Increase the area $A$ contacting the fluid and/or
2. increase the heat transfer coefficient $h$ by raising the flow rate.

**What are the fins for?**

Most convective heat transfer devices have *fins*. Fins are extensions of heat-conducting metal that can dramatically increase the surface area available for heat transfer by convection. For example, a computer chip is cemented to the finned aluminum block in the illustration. The finned surface is 20 times larger than the surface of the chip itself. The finned aluminum structure is an example of a *heat exchanger.* A heat exchanger is a device specifically designed to enhance heat transfer by convection.

**Fan**
creates forced convection

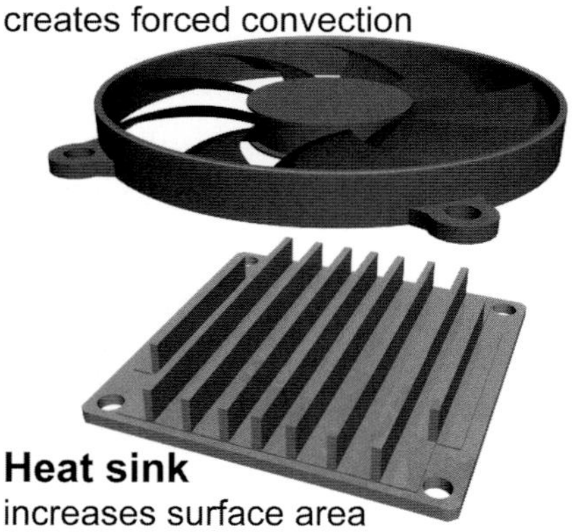

**Heat sink**
increases surface area

**Why is there a fan?**

The second part of the cooling mechanism is the fan that snaps on top of the heat exchanger. The fan forces air to flow across the surface of the fins, greatly increasing the value of the heat transfer coefficient. In practice, the use of a fan often raises the heat transfer coefficient by a factor of 10–20 over free convection.

**Temperature difference**

Cooling fins and a fan can be insufficient to cool some devices, such as a room packed full of computer network servers. Such rooms are typically air conditioned, which acts to increase the temperature difference in equation (24.2) and thus to increase the convective heat transfer.

**Why do windy days feel colder?**

We have all experienced the difference in comfort between a cold day and a cold, *windy* day. Wind dramatically increases the heat transfer coefficient, increasing the effectiveness of heat flow out of your body. To quantify this effect, in 1941, explorers Paul Siple and Charles Passel, of the U.S. Antarctic Service Expedition, performed a very simple experiment during the frigid Antarctic winter. Siple and Passel measured how long it took for a gallon of water to freeze at different air temperatures and wind speeds. For example, they found a temperature of 0°F with a 30 mph wind froze the water in the same time as a temperature of −26°F with no wind.

**Siple and Passel's experiment**

1 gallon of water freezes solid in equal time.

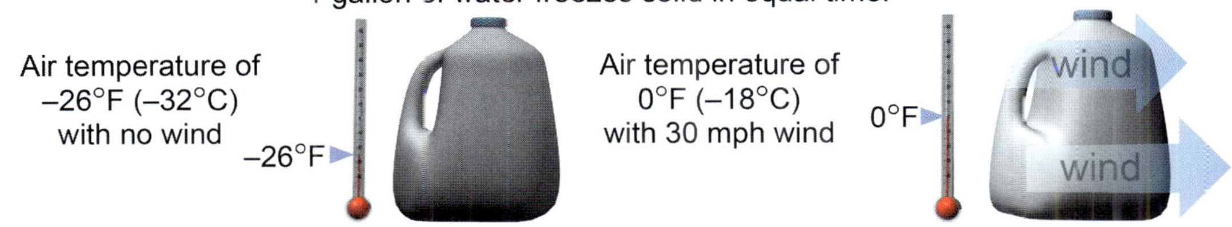


**What is the wind chill?**

To illustrate their findings, Siple and Passel created the *wind chill index*. A wind chill of −26°F corresponds to an air temperature of 0°F with a 30 mph wind. This means that a body loses heat at the same rate as if the air temperature were actually −26°F without any wind. The chart below shows the equivalent wind chill temperatures for different temperatures and wind speeds.

**Wind chill equivalent temperatures (°F / °C)**

| Wind speed | Equivalent temperatures with wind chill | | | | |
|---|---|---|---|---|---|
| 0 | −20°F −29°C | −10°F −23°C | 0°F −18°C | 10°F −12°C | 20°F −7°C |
| 10 mph (4.5 m/s) | −41°F −41°C | −28°F −33°C | −16°F −27°C | −4°F −20°C | 9°F −13°C |
| 20 mph (8.9 m/s) | −48°F −44°C | −35°F −37°C | −22°F −30°C | −9°F −23°C | 4°F −16°C |
| 30 mph (13.4 m/s) | −53°F −47°C | −39°F −39°C | −26°F −32°C | −12°F −24°C | 1°F −17°C |
| 40 mph (17.9 m/s) | −57°F −49°C | −43°F −42°C | −29°F −34°C | −15°F −26°C | −1°F −18°C |

**How do I interpret wind chill temperatures?**

The wind chill temperature represents an imaginary temperature at which there is an equivalent heat loss rate between free convection with no wind and forced convection at the actual temperature and wind speed. Wind chill does *not* mean that the actual temperature goes down with the wind speed! Many people make this mistake when interpreting the meaning of wind chill. For example, the antifreeze in a car radiator will freeze at −30ºF. Will the radiator freeze if the air temperature is −10ºF with a 30 mph wind? The answer is *no!* The radiator will lose heat as if the temperature was −39ºF according to the wind chill index but the lowest temperature the radiator will reach is the same as the air temperature, or −10ºF, so it will not freeze.

## Convection on the Earth

**What causes ocean currents?**

Convection drives much of the Earth's climate and much of the behavior of the ocean as well. On a very large scale, enormous currents circulate heat from the equator to the poles. Dense, cold water from both polar regions sinks to the ocean floor and flows toward the equator. Warmer surface water from the equator circulates back toward the poles.

**Surface currents in the world's oceans**

**What is El Niño?**

This movement of water transfers a huge amount of thermal energy and is responsible for the periodic weather pattern known as El Niño. El Niño, also known as the *southern oscillation*, is a localized warming (about 0.5ºC) of the surface water in the southern Pacific ocean that occurs about every 3–7 years and causes heavier-than-average storms. El Niño is caused by an oscillation in the flow of the convection currents in the Pacific.

**What causes weather?**

Convection in the atmosphere causes weather. Air passing over warm ocean water rises as it warms and carries water vapor with it. As the rising air cools in the cold upper atmosphere the water vapor condenses to form clouds. Cool air circulates back toward the ground (a process called *subsidence*) in large convection cells that circulate both heat and water between the atmosphere and the ground. When the heat flow is large enough, the clouds may develop into thunderstorms, hurricanes, or tornadoes. Modeling atmospheric heat transfer in an effort to understand climate is an active area of research involving physics, meteorology, computer science, and mathematics.

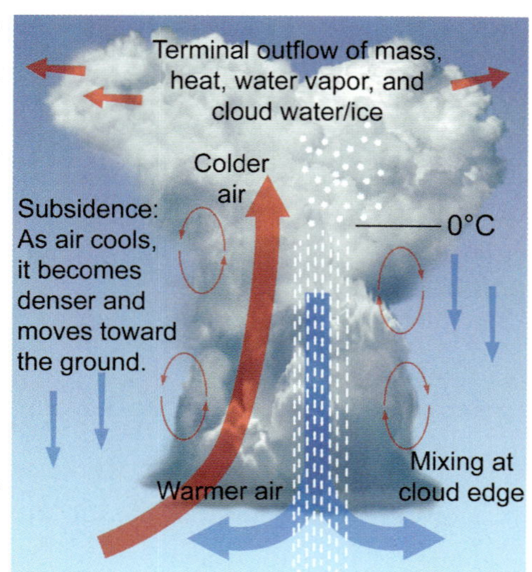

# Investigation 24B Visualizing convection

**Essential questions** *How does convection transfer heat?*

Convection transfers heat through the movement of matter. When matter such as air or water moves, the thermal energy contained in that matter also moves. Convection is the dominant mode of heat transfer in fluids such as air and water.

## Part 1: Observing convection

1. Fill a pan with cold water and set it on a hot plate at high heat.
2. Drop one tablet of egg dye into the pot and observe the motion of the colored water as the tablet dissolves.
3. Turn off the heat and repeat the experiment with another tablet.

a. *What differences do you observe when the heat is on compared to when the heat is off?*

b. *Calculate the equivalent power in heat transferred when 1 kg of water at 20°C replaces 1 kg of water at 10°C in 1 s.*

c. *Calculate the power in heat transferred by conduction through a cubic container filled with 1 kg of water. One wall of the container is at 20°C and the opposite wall is at 10°C.*

d. *How do the two heat transfer power values compare with each other? Which is greater? Are they much different or only a little different?*

## Part 2: Comparing convection and conduction

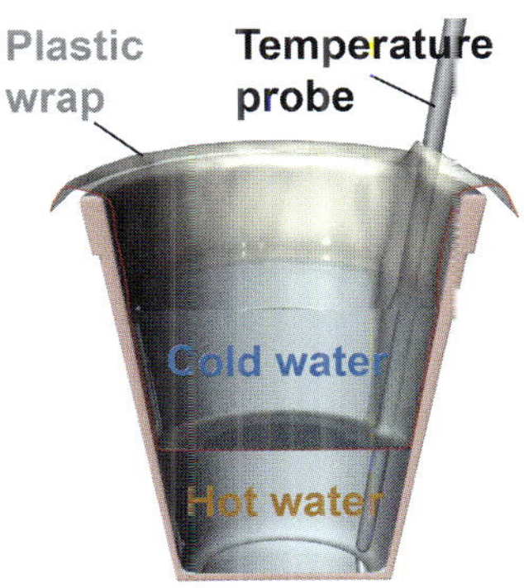


1. Fill half a 20 oz foam cup with hot water and insert a temperature probe against one side.
2. Loosely set a sheet of thin plastic wrap over the cup and start recording the temperature at a rate of once every 5 s.
3. Pour some ice water into the plastic wrap so it can exchange heat with the hot water through the plastic wrap but is not allowed to mix. Take temperature data for 3 min.
4. Repeat the experiment without the plastic wrap, allowing the hot and cold water to mix. Keep the data collection at the same rate so you can compare the two experiments.
5. Create graphs of these two cooling curves using the same horizontal and vertical scales.

a. *Describe the difference in the rate of temperature change you observe when the water is allowed to mix compared to when it is prevented from mixing by the plastic wrap.*

b. *Estimate the rate of heat transfer through the plastic wrap. Is the plastic wrap a factor in limiting the overall heat transfer by conduction? (The specific heat of water is 4,184 J/[kg °C].)*

c. *Discuss the relative contributions of conduction and convection in this experiment.*

d. *How did the distribution of energy between the hot and cold water change when you used the plastic wrap? How did it change when you removed the plastic wrap?*

# Section 2 review

The power of heat flowing through conduction varies directly as the thermal conductivity, the area through which the heat flows, and the temperature difference and inversely as the length the heat flows through. Examples are developed for calculating heat conduction through rods and walls. Convection is more complex than conduction, but convection can transfer much more heat in circumstances where fluid such as air or water is free to flow. For example, the heat flow next to a window via conduction may be a few watts per square meter, whereas the heat flow through convection may be 100 W or more. The equation for convection is a model in which many complexities are wrapped up in the heat transfer coefficient $h$, which can vary from less than 10 to more than a million depending on the exact situation. For this reason, convection calculations are best treated as approximations subject to experimental verification.

| | | |
|---|---|---|
| **Vocabulary words** | **heat transfer coefficient** | |
| **Key equations** | $P = \frac{\kappa A}{L}\Delta T$ | $P = hA\Delta T$ |

## Review problems and questions

1. How much heat can flow through a solid, round aluminum rod that is 0.50 m in length with a diameter of 1.0 cm? The temperature difference is 50°C between the two ends of the rod.

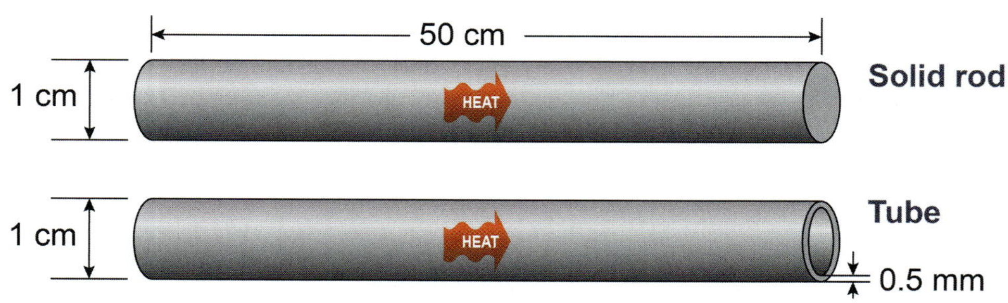


2. Repeat the calculation of the first problem if the rod is replaced by a hollow tube with a wall thickness of 0.50 mm. The outside diameter and length are the same and the temperatures are the same.

3. An insulated wall has an average thermal conductivity of 0.080 W/(m °C). How much heat passes through the wall if the area is 20 $m^2$ and the thickness is 0.2 m? The temperature difference is 20°C between the inside surface of the wall and the outside surface.

4. A cooling surface for a computer chip has a heat transfer coefficient of 50 W/($m^2$ °C). The outside air temperature is 25°C, the chip temperature is 100°C, and when the chip is operating it must dissipate 100 W or else it will overheat. How much surface area does the heat sink need?

Thermal radiation is a source of electromagnetic waves, including sunlight and ultraviolet and infrared light. Thermal radiation is emitted by all objects with temperatures above absolute zero and thermal radiation is also absorbed by all objects.

## The Stefan–Boltzman equation

**What model describes radiative heat transfer?**

Unlike convection and conduction, the power emitted as thermal radiation increases as the *fourth power* of the temperature. This means that doubling the temperature increases the amount of radiated power by a factor of $2^4 = 16$! Like other forms of heat transfer, the power is also proportional to the area; bodies with a larger surface area emit more radiative power than bodies of smaller surface area at equal temperature. Equation (24.3) gives the power of thermal radiation as a function of temperature $T$, area $A$, and emissivity $\varepsilon$.

**Stefan–Boltzmann equation**

$$P = \varepsilon \sigma A T^4 \qquad (24.3)$$

$P$ = power (W)
$\varepsilon$ = emissivity (dimensionless)
$\sigma$ = Stefan–Boltzmann constant = $5.67\times10^{-8}$ W m$^{-2}$ K$^{-4}$
$A$ = emitting surface area (m$^2$)
$T$ = absolute temperature (K)

**What does the model tell me?**

The temperature in equation (24.3) has units of *kelvins* because thermal radiation depends on the absolute temperature. The Stefan–Boltzmann constant $\sigma$ has the value $5.67\times10^{-8}$ W/(m$^2$ K$^4$). At a temperature of 1 K (−272°C), a square meter of a perfect emitter would radiate $5.67\times10^{-8}$ W of power. This is a very small quantity, but radiative heat transfer increases quickly with temperature. At temperatures over 500°C, radiation almost always transfers more heat than convection or conduction. For example, the diagram below compares the different modes of heat transfer between two parallel plates separated by 20 cm of air.

**Heat transfer between parallel plates**

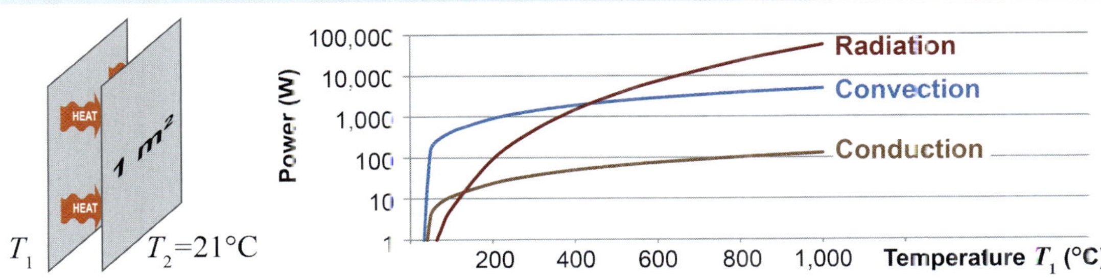


**What is the emissivity $\varepsilon$?**

The factor $\varepsilon$ in equation (24.3) is the *emissivity*. The **emissivity** describes how much power is radiated by a surface compared to a perfect *blackbody* at the same temperature. In physics, a **blackbody** is a surface that appears completely black—meaning it absorbs 100% of the radiation falling on the surface and reflects nothing. A perfect blackbody absorber is also a perfect blackbody emitter and therefore has an emissivity $\varepsilon = 1$. Most real surfaces reflect some light and have values of emissivity that range from 0.1 to 0.95. Very shiny surfaces, such as chromed steel, have low emissivities, as low as 0.05 or less.

**What is a blackbody?**

An object appears perfectly black when it absorbs all light that falls on its surface, reflecting nothing. If all radiation is absorbed, then any radiation coming off the surface can only be thermal radiation. In physics, a perfect blackbody is a surface that reflects nothing and emits pure thermal radiation. That does not mean an object looks black! The Sun is a pretty good blackbody and so is the white-hot filament of an incandescent light bulb. For both the Sun and the filament, all light coming from the surface is thermal radiation, and none is reflected from other sources.

**The blackbody spectrum**

Thermal radiation always includes a range of different wavelengths from low-energy infrared to high-energy ultraviolet. In 1900, German physicist Max Planck deduced the spectrum of blackbody radiation, which many feel is the first experimental success of the new *quantum theory*. The diagram below shows the blackbody spectrum for several different temperatures.

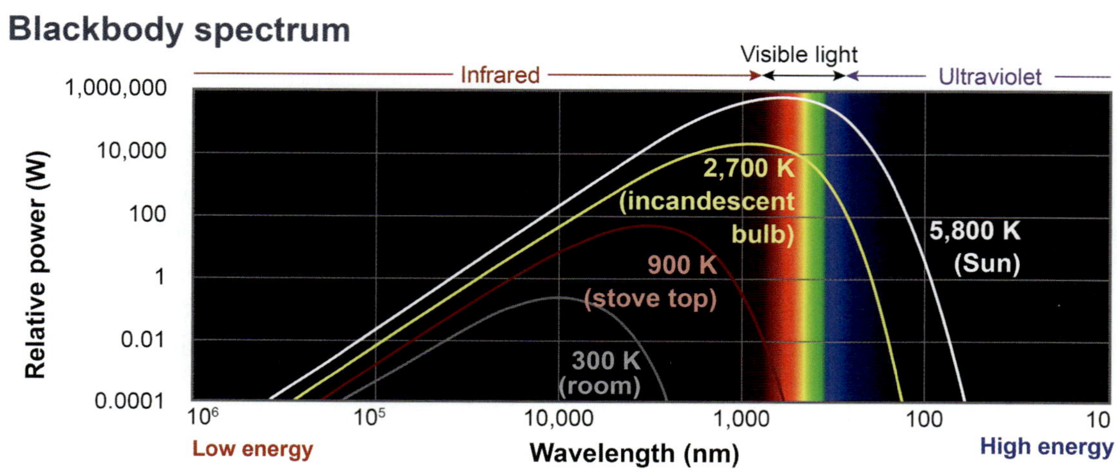


**Sunlight and light from bulbs**

An incandescent bulb filament, at 2,700 K, includes the whole range of visible light. Most of the power emitted by an incandescent bulb, however, is at infrared wavelengths, transferring heat instead of providing visible light. Incandescent light bulbs are quite inefficient, emitting less than 10% of their energy in visible light. The Sun has a surface temperature of 5,800 K. From the 5,800 K blackbody curve you can see that the solar spectrum produces visible light, infrared light, and quite a lot of potentially harmful, high-energy, ultraviolet light as well.

**Color temperature**

Compact-fluorescent bulbs come in soft white, cool white, and daylight *color temperatures*. Each type corresponds to a temperature in kelvins. The bulb has been designed to give off light in which the relative intensities of red, green, and blue light are balanced according to the blackbody spectrum for the bulb's color temperature. A cool white bulb at 4,100 K has more blue in its light than a warm white at 2,900 K.

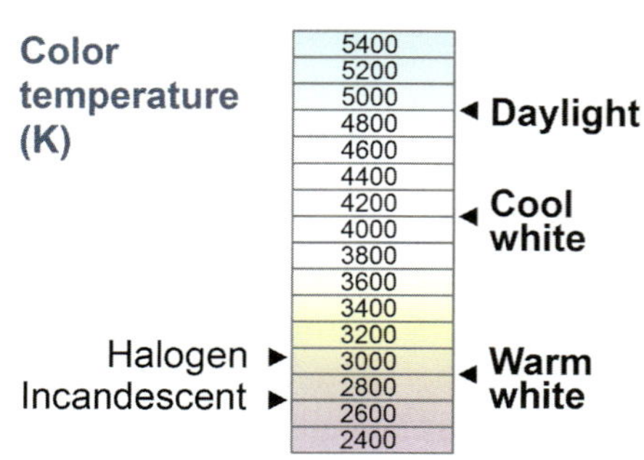

**How is radiant heat transfer determined?**

The Stefan–Boltzmann relation tells half the story by giving the *emitted* power in radiation. In reality, every object that emits radiation also *absorbs* radiation emitted by other objects in the environment. Radiative heat transfer depends on the difference between energy absorbed and energy emitted. In general, this is very complicated because exact shapes and positions matter in determining how much radiation emitted from one object is absorbed by another. Nevertheless, a useful simple case occurs when an object is surrounded by an environment at a constant temperature. The net power exchanged by radiation between a body at one temperature and the environment at a different temperature is given by equation (24.4).

**Radiant heat transfer**

$$P = \varepsilon\sigma A(T_\infty^4 - T^4) \qquad (24.4)$$

$P$ = power (W)
$\varepsilon$ = emissivity (dimensionless)
$\sigma$ = Stefan–Boltzmann constant = $5.67\times10^{-8}$ W m$^{-2}$ K$^{-4}$
$A$ = emitting surface area (m$^2$)
$T_\infty$ = environment temperature (K)
$T$ = object temperature (K)

**Radiation by living bodies**

The radiative heat transfer from a human body makes a good example. The surface area of an adult is around 2 m$^2$. Assume skin has an emissivity of 1 since only transparent, metallic, or reflective surfaces have emissivities much lower than 0.9. When the room is at 20ºC (68ºF) the average surface temperature of a person's clothing is about 28ºC or 301 K, as shown by the thermal infrared camera image on the right. An infrared camera "sees" in infrared then electronically converts the infrared to colors to represent temperature.

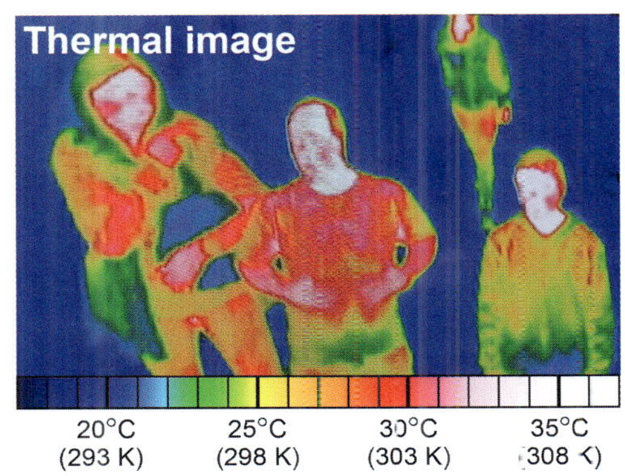


**The net radiated power**

The radiation emitted from 2 m$^2$ at 301 K is 931 W. The same surface absorbs 836 W from the room radiating back at 293 K. The *net* heat transfer is just under 100 W. Humans radiate about 100 W *constantly* just by being alive! If you evaluate this heat loss over 24 hours it comes out to 8.2 MJ of energy, or 1,960 calories, the approximate energy equivalent of three meals.

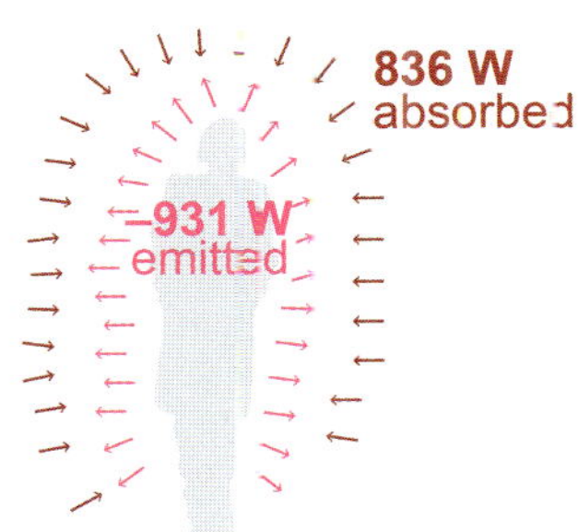


**Net heat transfer**

$$P = \varepsilon\sigma A(T_\infty^4 - T^4)$$
$$= -95 \text{ W}$$

**How objects "select" their temperature**

The temperature of any object reflects the balance of heat gained and heat lost. If more heat is gained, temperature increases. Increased temperature, in turn, means heat is lost more rapidly. In a steady state situation, the temperature comes to equilibrium at just the point where the heat lost through all mechanisms (conduction, convection, and radiation) equals the heat absorbed.

**What determines the temperature of Earth?**

Earth is isolated in space, far from any source of significant thermal radiation except for the Sun. Earth's temperature is a balance between radiant energy absorbed from the Sun and energy radiated away into the cold of space. For a simple and surprisingly accurate model for the temperature of our planet, we can assume that the solar energy absorbed by Earth's disk, with an area of $\pi R_e^2$, is radiated back into space over the entire surface area of the planet, $4\pi R_e^2$.

Power absorbed by Earth = Power radiated by Earth

$$\left(\frac{P_{sun}}{4\pi R_{orbit}^2}\right)\pi R_e^2 = \sigma(4\pi R_e^2)T_e^4$$

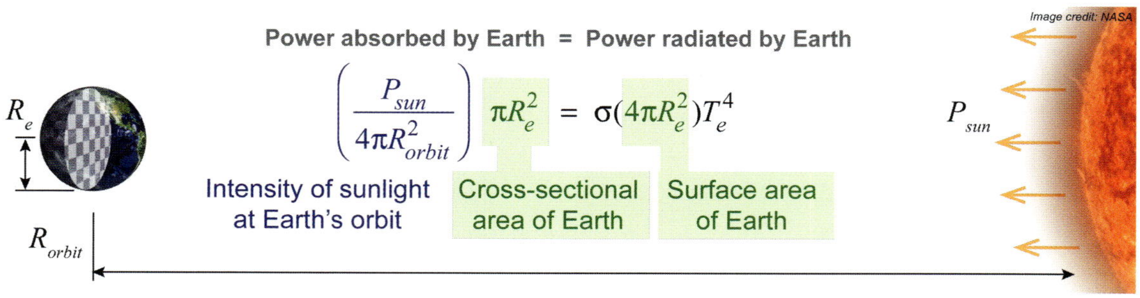


**A model for the temperature of a planet**

The Sun's power $P_{sun}$ is spread out over a sphere of radius $R_{orbit}$. The amount of power intercepted by Earth is on the left of the equation above. On the right is the radiated power of Earth with an average temperature $T_e$ and emissivity $\varepsilon = 1$ assumed. Notice that Earth's radius $R_e$ cancels from both sides. After taking the fourth root to solve for the temperature, we have a single equation model that predicts the average temperature of a planet given only the power of the Sun and the planet's distance from the Sun.

$$T_e = \left(\frac{P_{sun}}{16\pi\sigma}\right)^{\frac{1}{4}} \frac{1}{\sqrt{R_{orbit}}}$$

**Model**

$$T_e = \left(\frac{\overbrace{3.94\times10^{26}\text{ W}}^{P_{sun}}}{16\pi\,\underbrace{(5.67\times10^{-8}\text{ W/[m}^2\text{K}^4])}_{\sigma}}\right)^{\frac{1}{4}} \frac{1}{\underbrace{\sqrt{1.5\times10^{11}\text{ m}}}_{R_{orbit}}} = 280\text{ K } (7°\text{C})$$

**Prediction**

**How accurate is this model?**

The average *surface* temperature of Earth is 287 K (13ºC). The average temperature of the atmosphere and clouds is slightly lower, in excellent agreement with this model. The model is accurate because a planet such as Earth has been in existence long enough to come to *steady-state equilibrium* with regard to energy balance. The same is true for other celestial bodies that are not stars, including comets, asteroids, and moons. The temperature of each is set by the balance of radiative heat transfer. One *astronomical unit* (AU) is equal to Earth's orbital radius. When the radiative balance model is written in astronomical units the result is a simple but very accurate way to estimate the average temperature of any body in the Solar System!

$$T = \frac{280}{\sqrt{R}}$$

$T$ = absolute temperature (K)
$R$ = orbit radius (AU)

**Average temperature of an object in the Solar System**

**Earth's heat balance**

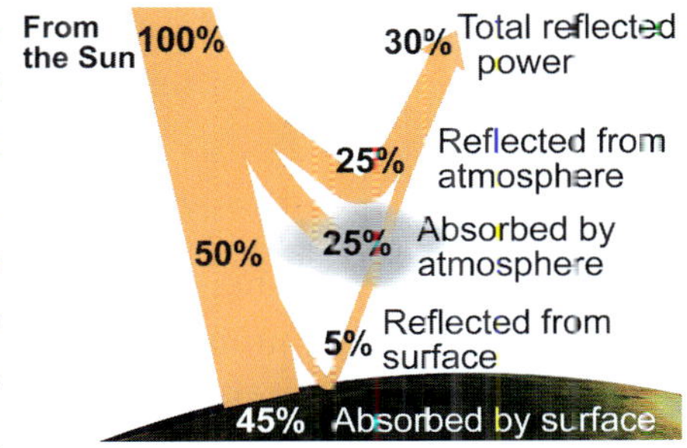


The Sun radiates 1,350 W of energy on every square meter of Earth that faces it. About 25% of that radiant energy is reflected from the Earth's atmosphere, and another 25% is absorbed by the atmosphere. Meanwhile, 45% of the sunlight is absorbed by the Earth's surface, while 5% is reflected from the surface back into outer space. The Earth maintains its average temperature because it is in thermal equilibrium. But what would happen if the atmosphere suddenly started *reflecting* more light back into space or if the atmosphere began to *absorb* more of the Sun's energy? The temperature on the Earth would change to bring itself back to thermal equilibrium. This is the basic problem posed by *climate change*.

**Climate change**

The scientific field of climate change studies broadly involves the following topics:

1. *Global warming* deals with direct evidence that the average, global temperature of the Earth has increased over the past century or more.
2. *Climate change* is based on evidence of the *effects* of global warming, such as rise in sea levels and decrease in sea ice.
3. The rise in atmospheric *greenhouse gases*, particularly carbon dioxide ($CO_2$), provides evidence as to the cause of global warming.
4. That human activity has caused much, if not most, of the rise in greenhouse gases is offered as an *anthropogenic explanation*.

**Global warming**

What is the evidence for global warming? Many scientists have been assembling global averages of surface air temperatures taken on land over the past 150 years. The data show a significant rise in average temperature over the past 50 years. Could this be the result of flawed methodology? One indication that it is not is that independent studies, using different methodologies, have reached similar results, known as a *consensus*.

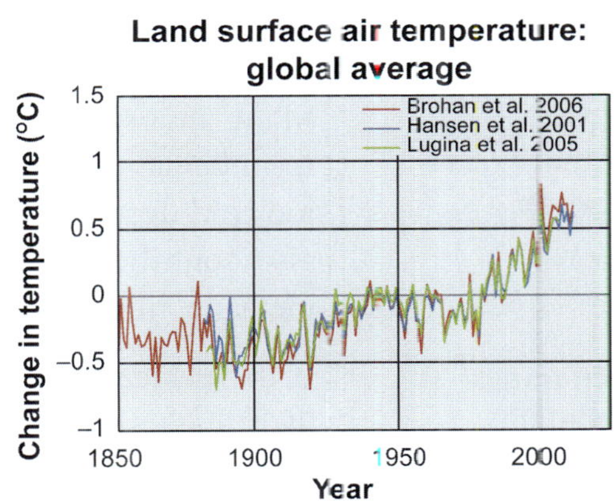


**Other temperature measurements**

Let's dig deeper into analyzing the quality of the scientific evidence. Could the land surface air temperature measurements themselves be fundamentally flawed? Other climate scientists have assembled equally detailed sets of data of the marine air temperature, the sea surface temperature, and the temperature of the atmosphere in the lower troposphere. All these separate sets of measurements show similar trends in the change of temperature over the past half century.

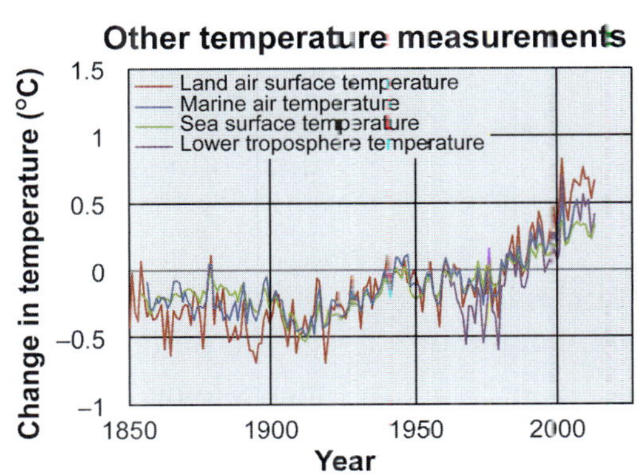

## Explanations of climate change

**Anthropogenic origin**

What could be causing this global warming? The consensus explanation of the scientific community, as reflected in reports from the Intergovernmental Panel on Climate Change, is that human activity is likely to be the main driver. Is this *anthropogenic* explanation reasonable, based on the scientific evidence?

**Carbon dioxide in the atmosphere**

One main piece of evidence that points to human activity is that the composition of the atmosphere has changed subtly but significantly over the same time period. The atmospheric concentration of carbon dioxide ($CO_2$), a greenhouse gas, has risen steadily, increasing by more than 25% since 1880. Carbon dioxide is a product of many human activities, particularly the burning of fossil fuels for transportation and electricity generation.

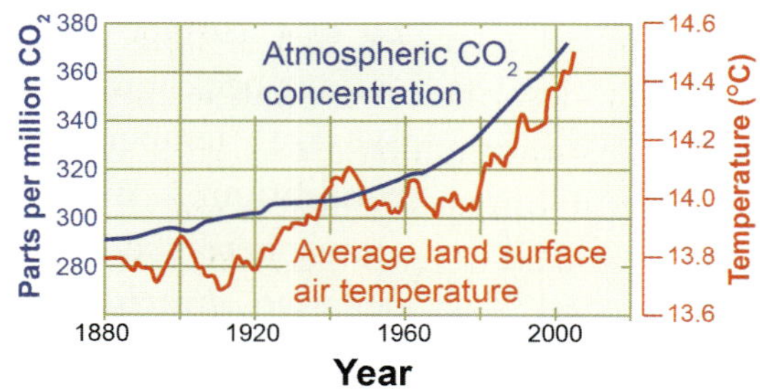


**Historical data on $CO_2$**

Could $CO_2$ concentrations simply fluctuate up and down over time, such that we are in a natural cycle? By using ice cores drilled through antarctic ice, scientists can study the concentration in $CO_2$ over hundreds of thousands of years. Antarctic ice far below the surface trapped a record of the atmosphere's concentration at the time the ice formed. The data show that current $CO_2$ levels are unique in the past 650,000 years—they are not the result of a periodic fluctuation.

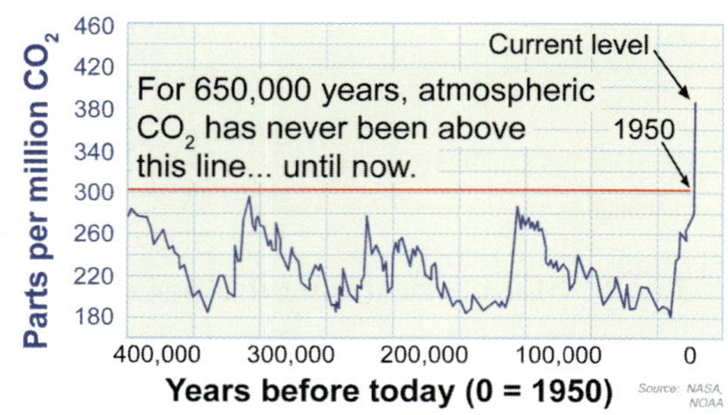


**Other greenhouse gases**

Many atmospheric gases act as *greenhouse gases* by trapping some of the Earth's energy from being radiated into space. If the Earth had no greenhouse gases, it would actually be too cold to support life as we know it! The major greenhouse gas is water vapor, but $CO_2$ also contributes significantly to the greenhouse effect. Other greenhouse gases include methane, nitrous oxides, and chlorofluorocarbons.

**Effects of climate change**

If the evidence of global warming is to be believed, then there should be other global effects caused by warming. This process is one way to analyze, critique, and evaluate scientific evidence and explanations. The effects of climate change have been documented in melting Arctic sea ice, rising sea levels, rising specific humidity, decreasing snow cover, and shrinking glaciers.

**Rising sea level measurements**

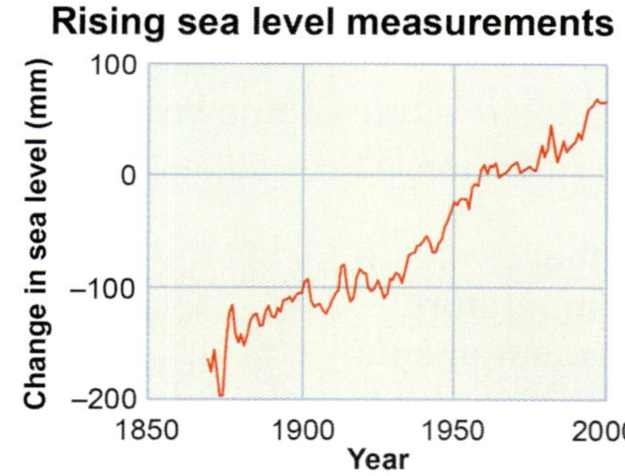


**Societal implications**

Progress in science sometimes leads to a broad discussion across society as to what should be done. If human activity is implicated as the primary cause of climate change, then should human activity be modified? If so, how? One common proposal is to change electricity generation to renewable energy sources that do not generate carbon dioxide. Do you think anything should be done to address climate change?

# Section 3 review

Thermal radiation comprises electromagnetic waves, including sunlight and ultraviolet and infrared light, that are emitted as a result of thermal motion of atoms and molecules. Thermal radiation is emitted by all objects with temperatures above absolute zero and thermal radiation is also absorbed by all objects. The power in thermal radiation over all wavelengths increases as the absolute temperature to the fourth power. A perfect *blackbody* is a surface that absorbs all incident radiation and emits only thermal radiation. The *blackbody spectrum* describes the relationship between power and wavelength for thermal radiation. An object in thermal equilibrium balances heat gained with heat lost. In space, conduction and convection cannot occur so radiation dominates heat transfer. Radiative heat balance determines the average temperature of Earth, the planets, and all other Solar System objects except the Sun itself.

| **Vocabulary words** | **emissivity, blackbody** | |
|---|---|---|
| **Key equations** | $P = \varepsilon\sigma AT^4$ | $P = \varepsilon\sigma A(T_\infty^4 - T^4)$ |

## Review problems and questions

1. The Stefan–Boltzmann equation describes how much energy every object in the universe radiates away. But in the universe, some objects are warming up, while others are cooling down. How can this be?

2. If you heat steel from room temperature (20°C) to its melting point at 1,510°C, by what increased factor will it radiate thermal energy?

3. Telescope mirrors are coated with a thin layer of a shiny metal so that light striking them will mostly reflect, rather than be absorbed. For *infrared* astronomy, however, there is another important reason: As the mirror becomes less reflective, it radiates more energy at infrared wavelengths—which can overwhelm the images from distant stars or galaxies. Why does a less-reflective mirror radiate more infrared radiation?

4. The dwarf planet Eris has an average orbital distance from the Sun of 68 AU. Estimate the average temperature of Eris based on a radiation balance between the thermal energy it receives from the Sun and the energy Eris radiates as a blackbody. Express your answer in degrees Celsius.

5. While the Earth has a wide range of temperatures across its surface, its average temperature is approximately 15°C or so. How much power is radiated by the Earth at that temperature, assuming the Earth is 100% emissive? (The radius of the Earth is 6,370 km.) Compare this power to the Hoover Dam, which can generate up to 2,080 MW.

Chapter 24

# Chapter review

## Vocabulary

Match each word to the sentence where it best fits.

### Section 24.1

| | |
|---|---|
| **Newton's law of cooling** | **convection** |
| **heat transfer** | **thermal conductivity** |
| **thermal equilibrium** | |

1. The study of how thermal energy moves is called _______.
2. When two objects are in _______, no heat will flow between them.
3. As the temperature difference between two objects becomes large, the heat flow between them increases according to _______.
4. _______ is heat transfer that occurs through the motion of matter, such as forced air in a building's heating system.
5. A material with a higher _______ can more readily conduct heat.

### Section 24.3

| | |
|---|---|
| **blackbody** | **emissivity** |
| **heat transfer coefficient** | |

6. An object with a/an _______ of zero emits no thermal radiation from its surface.
7. A/An _______ absorbs all of the thermal radiation that is incident upon it.
8. The measure of the rate of heat flow between a surface and a fluid is called the _______.

## Conceptual questions

### Section 24.1

9. How many different ways can heat be transferred? Name and describe each one.
10. A hot cup of coffee is cooling to room temperature over the course of 20 minutes following a standard cooling curve. Which statement best represents its cooling behavior?
    a. The coffee's temperature changes fastest when the temperature difference between the coffee and the room air is smallest.
    b. The coffee's temperature changes fastest when the temperature difference between the coffee and the room air is largest.
    c. The coffee's temperature changes at a constant rate.
11. Using the term "thermal conductivity," explain why a copper ladle in a pot of boiling water feels hotter than a wood ladle.
12. When investigating the heat transfer properties of a house, such as in Investigation 24A on page 1407, you want to maintain an inside temperature of 20°C using a 2,500 W heater during the winter.
    a. What are the variables related to the design of the house that can be varied to change the heat transfer properties of the house?
    b. How would you pose a well-defined question to investigate what insulation is required?
    c. When determining the optimal size for the surface area of the walls of a house, one student obtained a value of 1.7 $m^2$. Is this a reasonable answer?
    d. When another student was setting the parameters for a house, she entered an area for the windows that was 1% of the area of the walls. Is this a reasonable design?
    e. Jeremiah wanted to use the simulation to model the heat flow for his own house. What measurements and observations would he need to implement his idea for an investigation? What equipment does he need to make those measurements?
    f. Jeremiah poses the question for his investigation, "Is my house radiating energy?" Is this a well-defined question for his investigation with the simulation?
    g. Is the simulation an appropriate kind of technology for Jeremiah to determine how much relative energy is flowing out of the north side of his house as compared to the south side of his house at noontime?

### Section 24.2

13. Identify each of the following actions with *conduction* or *convection.*
    a. blowing on a spoonful of hot soup to cool it
    b. shoving a warm can of soda into a bucket of ice to cool the soda
    c. replacing thin, single-pane windows with thick, double-pane ones to better insulate your home
14. Decide whether each of the following statements correctly describes the process of *conduction*, *convection*, or *both*.
    a. It requires a temperature difference to transfer thermal energy from one place to another.
    b. The larger the temperature difference between two places, the faster it transfers thermal energy.
    c. It operates efficiently in fluids but not in solids.
    d. It will burn your skin if you grab a red-hot piece of metal with bare hands.

# Chapter review

## Section 24.2

15. To cool a miniature circuit, an engineer glues it to one end of a metal rod. She then dips the other end of the rod in water. Which of the following would *speed up* the rate at which the circuit is cooled? (More than one choice may be correct.)

    a. replacing the first rod with a shorter one
    b. immersing the rod in warmer water than before
    c. replacing the metal rod with a wooden one of the same shape and size

16. ❮ Once again, an engineer glues a small circuit to one end of a metal rod to help keep it from getting too hot. She then dips the other end of the rod in water. Each of the following actions will change one variable in the right-hand side of the conduction formula, $P = (\kappa A/L)\Delta T$. Which variable applies to each of the following actions?

    a. replacing the first rod with a shorter one
    b. immersing the rod in warmer water than before
    c. replacing the metal rod with a wooden one of the same shape and size

17. ❮ Heat conducts across a wall according to the equation

$$P = \frac{\kappa A}{L}\Delta T$$

Which of the following choices best completes the sentence?
The power conducted decreases as

    a. the height of the wall increases.
    b. the width of the wall increases.
    c. the thickness of the wall increases.
    d. the conductivity of the insulating material increases.

18. ❮❮ In the heat *conduction* equation, the energy-transfer rate ($P$) depends on a material's *thermal conductivity* ($\kappa$), which has SI units of watts per meter per degree Celsius. By contrast, in the heat *convection* formula, the *heat transfer coefficient* has SI units of watts per *squared* meter per degree Celsius. Explain this difference.

## Section 24.3

19. Which of the following statements correctly compares radiation with conduction and convection?

    a. Only radiation requires a temperature difference in order to transfer thermal energy from one place or object to another.
    b. Radiation can travel through a perfect vacuum, whereas conduction and convection both require the presence of matter.
    c. In all three cases, the rate of energy transfer (or power $P$) is proportional to the difference in temperature.

20. Which of the following astronomical objects will have the peak of its blackbody radiation at the *shortest* wavelength?

    a. a molecular cloud at a temperature of 10 K
    b. a high-mass star with surface temperature of 10,000 K
    c. diffuse, ionized gas at a temperature of $10^6$ K

21. ❮ Most stars behave more or less like blackbodies, with their surfaces approximately obeying the Stefan–Boltzman law. Use this fact to explain how some relatively cool stars (*red giants*) can emit much more radiant energy than some relatively hot stars (*white dwarfs*)—even though the dwarfs have much higher temperatures.

22. ❮❮ Sea level has been measured for centuries using tide gauges, but more recently it has been measured using satellite data. Research these two sets of data to answer the following questions.

    a. Analyze each approach. How does each measurement technique work? Provide a written explanation of the technical details for each of the two processes.
    b. Critique each method and compare each method to the other. Would you trust one set of measurements more than another? Why?
    c. Does either data set or both show a trend over time in the change of the average sea level? Does the trend appear significant?
    d. Compare each independent data set to the other and evaluate them. Do they agree with each other?

23. ❮❮ One objection to scientific evidence for global warming is that the temperature measurements are unreliable, possibly because some temperature gauges are situated too close to buildings. Research the topic of temperature measurements and global warming.

    a. Analyze the measurement techniques, particularly the standard method of how temperature gauges should be mounted for weather measurements
    b. Critique whether or not siting issues create systematic errors in the global warming temperature data.
    c. Evaluate whether the warming trends are compromised by thermometer gauge siting issues.

24. ❮❮❮ While the Antarctic land ice sheet has been shown to be losing mass by melting, a consequence of climate change, the amount of Antarctic *sea* ice has been increasing. How can this be? (*Hint*: Think about the freezing point of fresh and salt water.)

## Quantitative problems

### Section 24.1

25. A small house is in thermal equilibrium at noontime on a cold, winter day. The furnace generates 1,500 W of heat, while the walls conduct and convect 1,800 W away from the house. How much thermal radiation from the Sun is helping to warm the house?

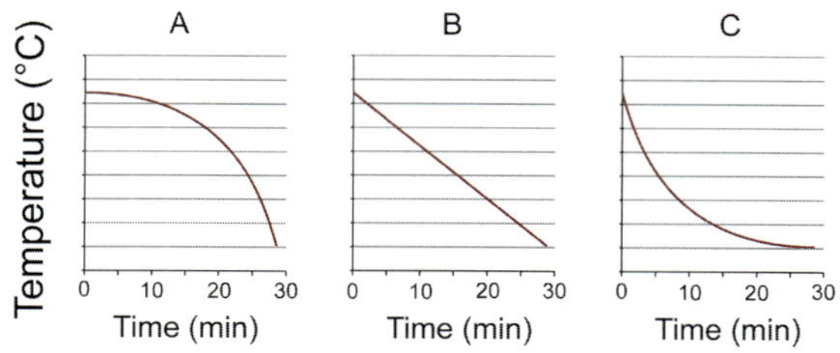


26. According to Newton's cooling law, the *rate* at which a substance cools is proportional to the *difference* in temperature between that substance and its surroundings. If that is true, which one of these three curves best represents a cup of coffee that cools from brewing temperature to room temperature?

### Section 24.2

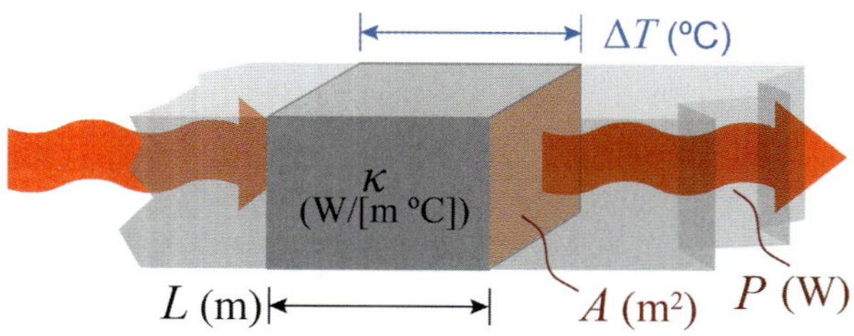


27. A temperature difference of $\Delta T$ exists between the inside (left) and outside (right) of a building. The wall has a thickness $L$. Thermal energy is conducted at a rate $P$ through a section with area $A$. The wall material has a thermal conductivity $\kappa$.

   a. If $\Delta T = 20°C$, $L = 40$ cm, and $\kappa = 0.3$ W/(m °C), how many watts of thermal energy will flow through each square meter of the wall?
   b. Now suppose that $\Delta T = 20°C$ and $L = 40$ cm, as before, but 30 W of thermal energy flows through each square meter of the wall. What is the new value for $\kappa$?
   c. Which would better insulate this building, a 10-cm-thick wall of brick ($\kappa = 1.3$) or a 2-cm-thick sheet of gypsum board ($\kappa = 0.2$)?

28. The exterior walls of a small wooden cabin have a total cross-sectional area of 36 m$^2$ and a thermal conductivity of 0.15 W/(m °C). Joe has discovered that he can keep the cabin 10°C warmer than the outdoors at night. To do so, however, he has to run a 500 W space heater. How thick are the cabin walls?

### Section 24.3

29. The Stefan–Boltzman law tells us how much energy will be emitted *in the form of radiation* by every square meter of an object's surface. It reads $P = \varepsilon\sigma AT^4$, where
   $P$ is the emitted power in watts;
   $\varepsilon$ is the emissivity, which ranges from 0 to 1;
   $\sigma$ is a constant that equals $5.67\times10^{-8}$ W m$^{-2}$ K$^{-4}$;
   $A$ is the surface area in square meters; and
   $T$ is the temperature of the surface *in kelvins*.

   a. You and your baby sister have the same body temperature and emissivity, but your surface area is four times as great as hers. What is the ratio of your thermal power to hers?
   b. Two space probes are in outer space, far from the Sun, but still operating thanks to their radioactive power packs. Probe A is maintaining a surface temperature of 5 K, and Probe B is at 10 K. If they are otherwise identical, what is the ratio of Probe B's thermal power to that of Probe A?

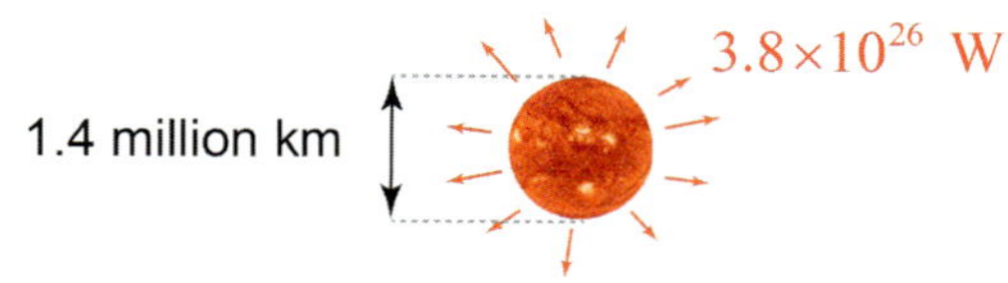


30. Although it certainly isn't black, the Sun is nearly a *blackbody*: Most of the light it gives off is actually thermal radiation. Use the Stefan–Boltzman law (with $\varepsilon = 1$) to estimate the Sun's *effective temperature* in kelvins. The Sun's diameter is 1.4 million km, and its *luminosity* is $3.8\times10^{26}$ W.

31. Once again, consider two spacecraft that are far from our Sun. They are identical except for their surface temperatures: Probe A is at 5 K and Probe B is at 10 K.

   a. Outer space itself has a blackbody temperature! It emits thermal radiation at a temperature of 2.7 K, chiefly in the form of microwaves. How many watts would one square meter emit if it were a perfect blackbody ($\varepsilon = 1$) at 2.7 K?
   b. The *net* thermal energy loss from each spacecraft takes into account the fact that the surroundings give some energy back! The formula for radiant heat transfer can be written as $P = \varepsilon\sigma A(T^4_\infty - T^4)$, where $T_\infty$ is the temperature of the environment (outer space) and $T$ is the temperature of the object (the probe). What is the *net* thermal energy transfer from each probe to outer space in watts? Each probe has an area of 4.0 m$^2$ and an emissivity of $\varepsilon = 0.5$.

# Chapter review

## Standardized test practice

32. Two objects in thermal contact with each other can be said to be in *thermal equilibrium* if

A. they contain identical amounts of thermal energy.
B. they have identical temperatures.
C. they have identical masses.
D. they have identical specific heats.

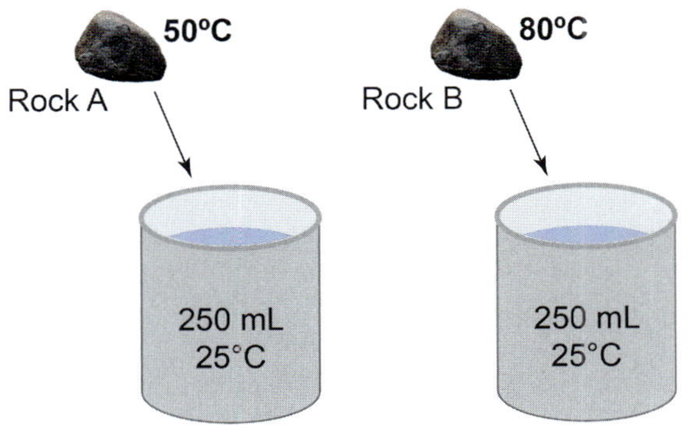


33. You have two identical containers, each holding 250 mL of water at 25°C. You also have two identical rocks, except that Rock A is at 50°C and Rock B is at 80°C. (The rocks have the same masses, compositions, sizes, and shapes.) You put one rock into each of the water cups at the same time. Which one of the following statements is true?

A. Rocks A and B will lose thermal energy at the same rate.
B. Rock A will lose thermal energy faster than Rock B.
C. Rock B will lose thermal energy faster than Rock A.
D. You cannot choose among A, B, or C without more information.

34. Which of the following processes requires a physical medium (a solid, liquid, or gas) to transfer thermal energy from one object to another?

A. conduction, convection, and radiation
B. conduction and convection, but not radiation
C. conduction only
D. None of these requires a physical medium.

35. A 300 g rock at 20°C is completely immersed in 800 g of water at 40°C. Which one of the following statements is true right after the rock goes in?

A. The rock and the water give each other thermal energy, but the rock gives more than it gets.
B. The rock gives the water thermal energy and gets no thermal energy in return.
C. The rock and the water give each other thermal energy, but the water gives more than it gets.
D. The water gives the rock thermal energy and gets no thermal energy in return.

36. A planet in a stable orbit maintains a balance between the sunlight energy it absorbs and the infrared energy it emits. Suppose that our Earth were replaced with a new Earth that had *twice the surface area* but was otherwise identical. Which one of the following statements would be true?

A. The big Earth would capture twice as much sunlight as the small Earth, so it would be hotter.
B. The big Earth would give off twice as much thermal radiation as the small Earth, so it would be cooler.
C. The big Earth would capture twice as much sunlight but also give off twice as much infrared light, so its temperature would be the same.
D. The big Earth would capture *twice* as much sunlight but give off *four times* as much infrared light, so it would be cooler.

37. For a student's written report on Investigation 24A, which of the following statements is best in using a formal style and objective tone?

A. I checked out what happened when the windows were big, and the results weren't so good.
B. The building lost tons of heat when I made the windows too large.
C. The heat losses from the building were largest when the windows were very large.
D. It was stupid of me to make the windows so large.

38. Adult humans emit about 1,000 W of thermal energy in the form of radiation (mostly as infrared light). But they only have to generate one-tenth of that amount (that is, roughly 100 W) by digesting the food they eat. Why?

A. Their clothes keep the rest of their body heat in.
B. They get the rest from sleeping and exercising.
C. Physics doesn't apply to human beings.
D. Their environment transfers roughly 900 W of thermal energy back to them (in the form of ever-so-slightly cooler infrared light).

39. One implication of global warming is the hypothesis that many glaciers should, at least in part, melt. In your analysis, which observation would *not* be evidence in support of this hypothesis?

A. The surface area of a glacier reduced over time.
B. The thickness of a glacier reduced over time.
C. The salinity of a glacier reduced over time.
D. The mass of a glacier reduced over time.

Chapter

# 25

# Thermodynamics

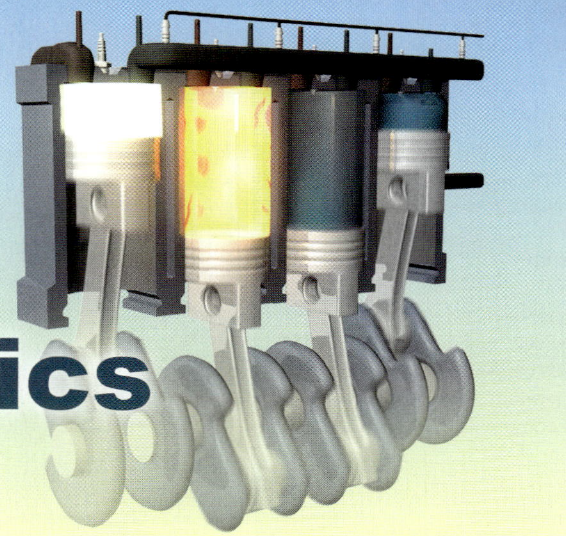

Both Leonardo da Vinci and Isaac Newton drew up designs for self-propelled vehicles, but neither actually built one. The first self-propelled "car" was built in 1769 by a French engineer and mechanic, Nicolas Joseph Cugnot. Really a three-wheeled tractor, Cugnot's invention used a steam engine for power and barely managed a speed of 2½ miles per hour.

A practical car requires a compact, lightweight, and reliable *engine* that can produce the equivalent power of a team of horses. This crucial technology came in 1876 when German engineer Nikolaus Otto invented the first practical *internal combustion engine.* True four-wheel automobiles arrived 10 years later when, in 1886, Gottlieb Daimler and Karl Benz independently created improved engines, cars, and companies to build them. The two inventor's companies merged in 1924 to create Daimler-Benz AG, the forerunner of today's Mercedes-Benz.

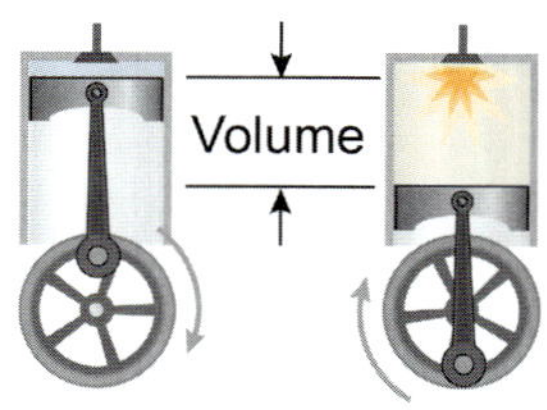


An internal combustion engine operates on a volume of gas/air mixture not much larger than a one liter soda bottle. The exploding mixture forces a piston down, turning the crank and spinning the engine. The larger the volume displaced by the piston as it moves, the more air and fuel can be burned per cycle, raising the engine's power output. Otto's first engine displaced 6 L but produced only 4 hp.

The first high-performance V8 car engine was created by the Ford Motor Company in 1932. Called the *flathead* for its odd-shaped valve covers, Ford's 1932 engine displaced 3.6 L and initially produced 65 hp. The engine was so successful that its variations appeared in Ford cars and trucks for the next 21 years. Flathead Ford engines are still highly sought-after by hot-rod enthusiasts today. Of course, the essence of all good technology is continuing innovation, and the internal combustion engine is no exception. The gasoline engine in the Ford Fusion hybrid generates 165 hp from only 2.5 L of displacement, making it 350% more efficient and able to drive 47 miles on one gallon of gas. The modern supercharged 5.8 L performance engine generates 650 hp, or 112 hp/L, a whopping 625% higher efficiency than was achieved in the 1932 Flathead.

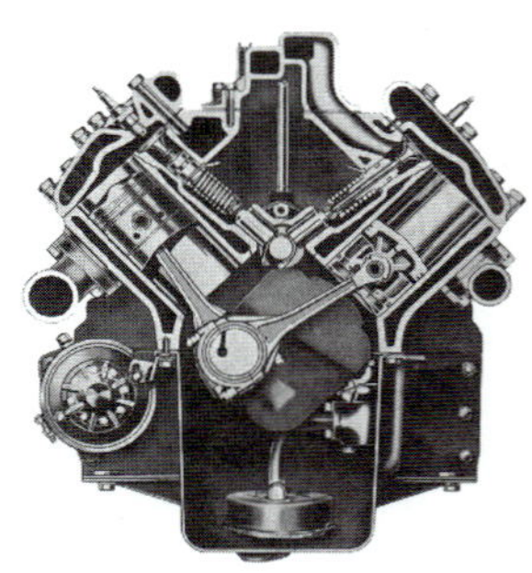

Cross section of the 1932 Flathead Ford V8

# Chapter study guide

Chapter 25

## Chapter summary

Thermodynamics is the study of heat, work, and how both transform to and from other forms of energy, particularly mechanical energy. At the core of this subject lie the laws of thermodynamics, which define how temperature, energy, and entropy work on the macroscopic level. The four laws—numbered zero through three!—define thermodynamic equilibrium, state energy conservation, describe how heat flows from hot to cold objects, and describe reversible and irreversible processes, respectively. In this chapter, you will learn the basic principles of the laws of thermodynamics and apply them to understand how steam engines in old locomotives, gasoline engines in cars, and refrigerators work.

## Learning objectives

By the end of this chapter you should be able to

- state the laws of thermodynamics and provide examples of each;
- define entropy and provide examples of reversible and irreversible systems;
- explain the concept of absolute zero and its relation to thermodynamics;
- describe how a heat engine and refrigerator work;
- draw and interpret *PV* diagrams, including calculating the work done by a system;
- describe isothermal and adiabatic processes and provide examples of each; and
- calculate the Carnot efficiency of a heat engine and explain why it can never be 100%.

## Investigations

25A: Heat engines

## Chapter index



## Important relationships

$$\Delta E = Q + W$$

$$\Delta S \geq 0$$

$$PV^{\gamma} = \text{constant}$$

$$\text{efficiency} = \left(1 - \frac{T_1}{T_2}\right) \times 100\%$$

## Vocabulary

**thermodynamics**
**entropy**
**third law of thermodynamics**
**isothermal process**
**Carnot efficiency**
**zeroth law of thermodynamics**
**second law of thermodynamics**
**heat engine**
**adiabatic process**
**refrigerator**
**first law of thermodynamics**
**reversibility**
***PV* diagram**
**Carnot cycle**

## 25.1 - Thermodynamics

**Thermodynamics** was developed to study systems in which heat energy is converted into other forms of energy such as mechanical or electrical energy. The prefix *thermo* refers to temperature and *dynamic* refers to motion. Thermodynamic systems are not "steady state" but instead involve temperature, pressure, or other variables changing dynamically over time. A good example is the internal combustion engine in a car. Designing an engine to get better efficiency—more miles per gallon—requires a thorough understanding of thermodynamics, the physics of thermal energy in motion.

### The first real steam engine

**Where did thermodynamics begin?**

You could argue that the transition from the agricultural world of horse-drawn carriages to today's mechanized world of automobiles and electric power stations began in 1712. That is when an English ironworker named Thomas Newcomen built the first working steam engine. Newcomen lived in an area where coal mining was a major occupation. Mines are essentially holes in the ground, which means that there is always a need to prevent flooding caused by the seepage of water into the mine. Newcomen's engine operated the first commercially successful pump that used heat instead of animals for motive force.

Newcomen steam engine

**How did Newcomen's steam engine work?**

To understand how Newcomen's engine works, consider that a liter of liquid water expands to occupy 1,654 L of gas when it boils into steam. That same 1,654 L volume of steam contracts back to 1 L if it condenses back into liquid water again.

**The engine in operation**

A cycle starts with the piston pulled up by the heavy mass attached to the end of a rocking beam. The steam valve opens and steam replaces air in the cylinder. The steam valve now closes and a jet of cold water sprays into the cylinder, condensing the steam back into liquid. This creates a partial vacuum in the cylinder and air pressure forces the piston down, lifting the heavy mass. When the piston gets to the bottom, the air valve opens and now the heavy mass pulls the rocking beam back down, lifting the piston back up and operating the water pump. This cycle repeats up to 12 times per minute. Newcomen and his partners built more than 700 of these machines at mines across England.

**Where does the force come from?**

The force that operates the steam engine comes from the outside air! When the steam condenses back into liquid water it reduces the pressure under the piston. The pressure of the air above the piston pushes the piston down with great force. The descending piston lifts the pumping rod.

### System and surroundings

Thermodynamics is usually applied to analyze the behavior of a *system* as it exchanges matter or energy with the *surroundings*. The system includes matter we are interested in that may be taking in or giving up energy. The surroundings are everything else. For example, a piston engine traps a volume of gas in a cylinder. The gas is the system. The gas may absorb heat from a fire, or it may exchange work with the surroundings when the piston moves. To simplify the analysis we often assume the surroundings are an infinite reservoir of heat and therefore may exchange work and energy with the system without a change in temperature or pressure. This is valid as long as the surroundings are "large" compared to the system.

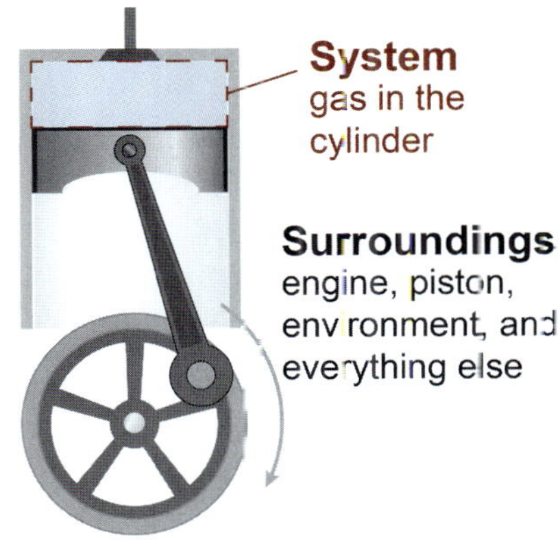


The system may exchange mass, work, and energy with the surroundings.

### Processes

In thermodynamics, any change in a system occurs through a *process*. We distinguish "process" from "change" because there are many different ways in which a particular change can occur. For example, you can add heat to a gas and keep the volume fixed. Both temperature and pressure increase in this process. You can also add the same amount of heat while letting the gas freely expand. This is a different process and the results are different.

**Compression**

Any process that ***decreases*** the volume of a given mass of a gas. The pressure and temperature may increase.

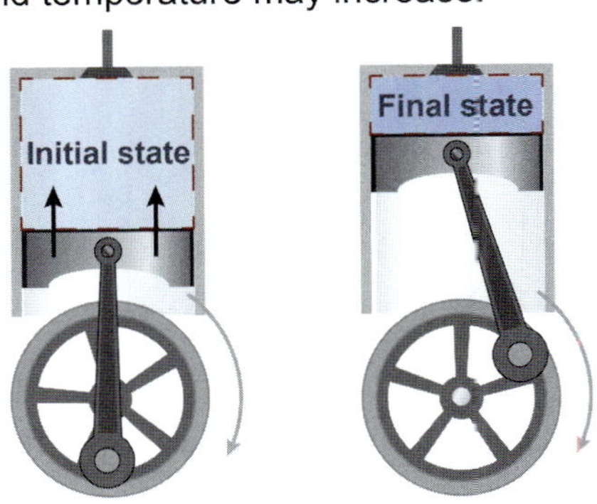


**Expansion**

Any process that ***increases*** the volume of a given mass of a gas. The pressure and temperature may decrease.

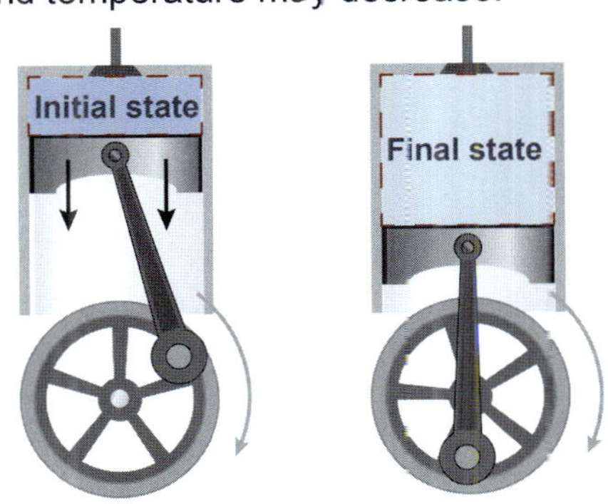


### What kinds of processes occur?

Two common processes involve *expansion* and *compression*. In expansion, the volume of a system increases. In the diagram above, expansion occurs when the piston moves down, making the volume of gas in the cylinder larger. Compression is the opposite process. In compression, the volume of the system decreases. In the diagram, compression occurs when the piston moves up, reducing the volume of the trapped gas in the cylinder. The actual quantity of matter may remain the same during expansion or compression! Gases expand and contract to fill their container.

## The zeroth and first laws of thermodynamics

### The zeroth law

Two objects in thermodynamic equilibrium with a third object are also in thermal equilibrium with each other. For example, suppose the air temperature is 20°C surrounding a stone placed next to a bowl of water. The stone and water eventually come to thermal equilibrium with the air at 20°C. Suppose you now drop the 20°C stone into the 20°C water. The **zeroth law of thermodynamics** says that no heat will flow between them *because* both are at the same temperature. That is the real meaning of the zeroth law—that heat cannot naturally flow between any two objects at the same temperature.

**The zeroth law of thermodynamics**

Two objects in thermodynamic equilibrium with a third object are also in thermodynamic equilibrium with each other.

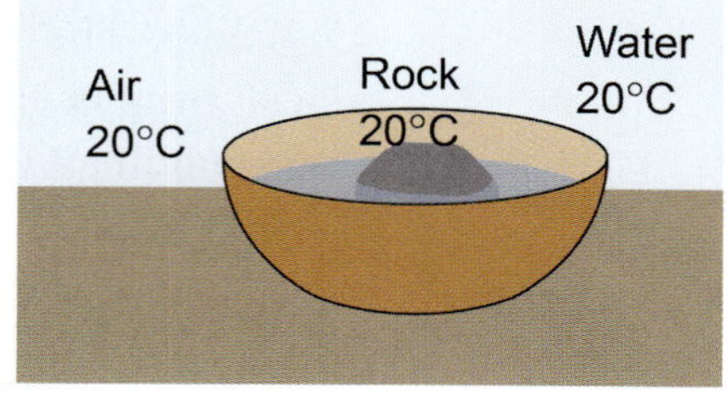


### The first law

The **first law of thermodynamics** is the law of energy conservation applied to processes that involve heat energy and at least one other form of energy, such as work. The work done by a system can be no more than the heat lost by the system. If a system cools and loses 100 J of thermal energy *by any process,* the most you can ever expect is 100 J of mechanical work. Another way to interpret the first law of thermodynamics is to think of heat as simply another form of energy subject to the same conservation law as potential energy, kinetic energy, and work.

| | | |
|---|---|---|
| **(25.1)** $\Delta E = Q + W$ | **The change in energy is the heat added to the system plus the work done on the system.** | **The first law of thermodynamics** |

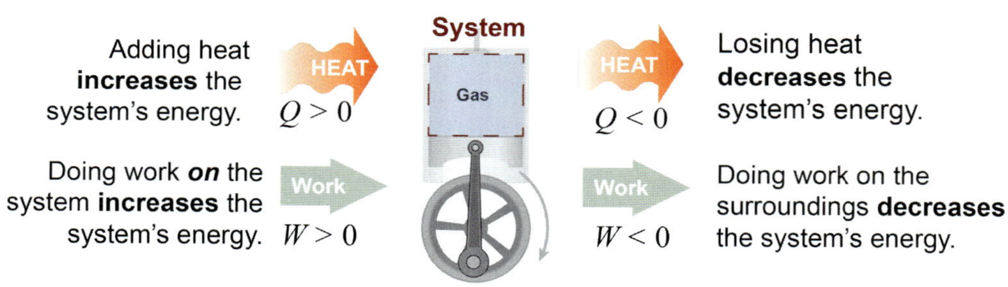


### The first law and a gas

The engine in a car and the steam turbine in a power station involve converting heat into mechanical work through the expansion of a gas. To apply thermodynamics to these systems it is useful to express the work done, $W$, in terms of pressure $P$ and volume $V$. Recall that work is the product of force and distance, which is equivalent to the product of pressure and volume. If a gas experiences a volume change $\Delta V$, then the work done by the system is the product of the pressure and the change in volume:

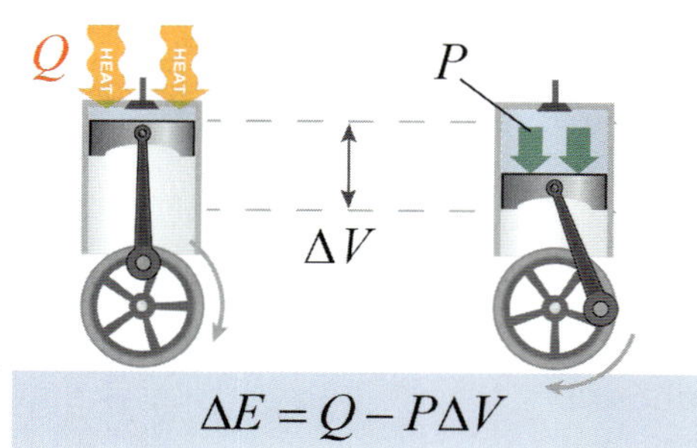


$$W = -P\Delta V.$$

The negative sign means that this is work done *by* the system.

**The flow of heat**

We observe that heat naturally flows from higher temperature to lower temperature and never the other way around. If a cup of coffee at 80ºC is left in a 20ºC room, heat always flows *from* the coffee *to* the room. We do not observe the opposite behavior. We never see heat flow from the cooler air into the warmer coffee. Why? Newton's laws of motion allow things to go forward or backward equally. Why does heat only flow in one direction, from higher temperature to lower temperature?

### The second law of thermodynamics

Heat only flows spontaneously from higher temperature to lower temperature and not the other way.

**Reversibility**

The ability to go equally forward or backward is called *reversibility.* A reversible process may proceed identically in the other direction when the inputs are reversed. If cooling were reversible, it would be possible for a 20ºC coffee to warm back up to 80ºC by exchanging heat with room air. The room air certainly has sufficient thermal energy, so energy conservation is *not* the reason. We observe, however, that *any process in which heat flows is irreversible!* Why? Why is heat flow irreversible?

**Entropy**

Suppose the cup loses $Q$ = 1,000 J of heat at a temperature $T$ = 80ºC. Divide the heat by the temperature and you get a new property, $S = Q/T$, called **entropy** and having units of energy per degree. The coffee cup loses 2.8 J/K of entropy while the room air gains 3.4 J/K for a net change of +0.6 J/K. The **second law of thermodynamics** is really about entropy; *entropy* determines whether a process is reversible. If the cooling were to reverse, then the net change in entropy would be −0.6 J/K, but *processes that decrease total entropy do not occur spontaneously!*

Coffee cup

$$\Delta S = \frac{Q}{T} = \frac{1{,}000\text{ J}}{353\text{ K}} = -2.83\text{ J/K}$$

Room air

$$\Delta S = \frac{Q}{T} = \frac{1{,}000\text{ J}}{293\text{ K}} = 3.41\text{ J/K}$$

**Net change** **+0.6 J/K**

**(25.2)**

$$\Delta S \geq 0$$

**Total entropy can only increase or remain the same.**

**The second law of thermodynamics**

**What does $\Delta S \geq 0$ mean?**

1. A process is reversible *only* if entropy remains constant, i.e., $\Delta S = 0$.
2. The total entropy of a closed system always increases if any energy is exchanged as heat, i.e., $\Delta S > 0$.
3. *All naturally occurring processes only proceed in the direction that increases total entropy.*

**Why the net change in $Q/T$ is always positive**

Heat flows only when there is a temperature difference, $T_{high} > T_{low}$. When heat is exchanged between two temperatures, the quantity $S = Q/T$ summed for both sides of the exchange will *always* increase because the ratio $Q/T_{high}$ is always less than the ratio $Q/T_{low}$. The numerator $Q$ is the same and the denominator is different, with $T_{low}$ less than $T_{high}$.

**Why is entropy important?**

Since the ratio $Q/T$ always results in a net increase, you might reason that we have not proved *anything* by inventing this new property called entropy. You might be right if that were the whole story! But entropy involves a much deeper concept that lies at the very heart of our modern understanding of physics. While we used the coffee cup argument to introduce the concept, we have not yet explained *why* entropy must increase and what entropy represents.

**Reversibility and time**

The concept of **reversibility** is fundamental to the physics of *time.* Consider two equal masses connected by a string passing over a frictionless pulley. Completely elastic springs on either side give each mass a bounce back upward when it touches down. Now imagine starting the masses moving and recording a video. In a reversible, frictionless world the masses would bounce up and down, exchanging potential and kinetic energy *forever.* The video would look exactly the same running forward *or running in reverse*. In a frictionless, reversible world forward or backward in time are equivalent and equally possible.

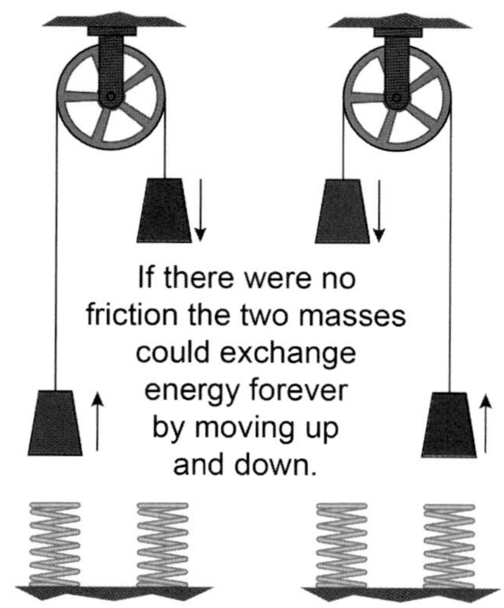


**Why friction changes things**

Now consider the impact of friction. Once heat flows, $\Delta S > 0$ and some of the energy of motion is irreversibly transformed into heat with each bounce. The second bounce is a little slower than the first. The third is a little slower than the second. On a videotape there is a measurable difference between forward and backward. *Irreversibility causes a difference between time moving forward and backward.* In the statement that total entropy must always increase, the second law of thermodynamics fixes the direction of time as forward, in the direction of increasing entropy.

**Where entropy comes from**

In Chapter 23, we talked about temperature and the particle nature of matter. When particles interact on the atomic scale energy and momentum are perfectly conserved and therefore each interaction is reversible. Time *does not* have a forward preference in the frictionless quantum world. The irreversibility arises when we consider $10^{23}$ particles on the macroscopic scale. Although there is no fundamental law against reversibility, the chance of a process involving so many particles randomly reversing itself is so low it will never occur in a thousand times the age of the universe. Entropy is fundamentally related to *order* and the second law is fundamentally a statistical argument that systems that are disordered do not spontaneously go back into highly ordered states.

### What is entropy?

Entropy describes the *disorder* in a system, which is the number of different ways the system can be arranged internally and yet still have the same macroscopic properties, such as temperature and density. To illustrate the idea, consider a system of nine boxes containing three particles. This system has an average "density" of 0.33 particles per box. There are 504 ways to arrange those three particles in the nine boxes, but many of those arrangements repeat—such as #6 and #16 in the figure below—because the three particles are identical. After removing duplications, there are 84 *unique* ways to arrange the three particles in the nine boxes. Each of the 84 different ways has the same "density"; therefore, the "entropy" is 84.

28 of the 504 ways to arrange 3 particles in 9 boxes

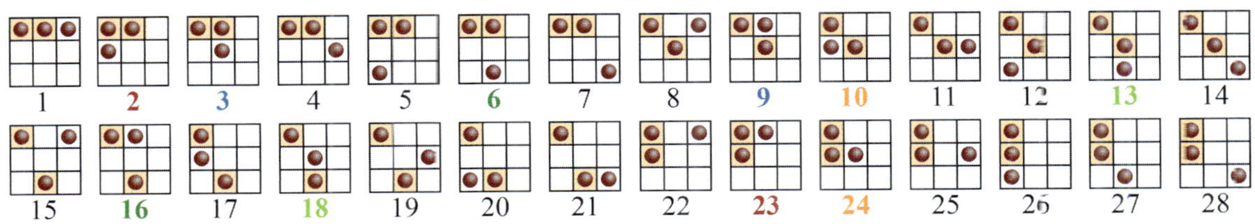


There are 84 *different* ways to arrange the 3 particles in 9 boxes.

### Why does entropy increase?

This system exchanges "density" by mixing with another system that has one particle in nine boxes. The other system has an "entropy" of nine because there are nine ways to put one particle in nine boxes. The entropy of the new, 18-box system is 3,060 because there are 3,060 different ways to organize four particles in 18 boxes and all have the same "density" of (4/18). The total entropy, or disorder, has increased from 93 in the original state of the system to 3,060 in the new state.

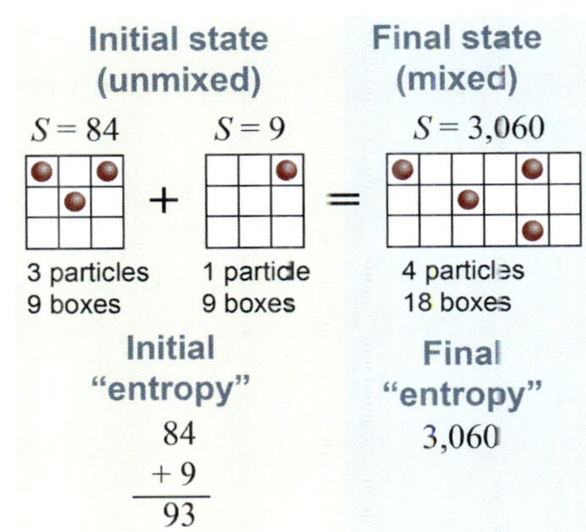


### What does reversibility have to do with entropy?

To explain irreversibility, let the particles constantly collide and redistribute themselves randomly among the 3,060 different arrangements of the 18-box system. *What is the chance of going back to the original state of three particles in the first nine boxes and one particle in the other nine boxes?* The answer is that there are 84 × 9 = 756 ways for this to happen out of the 3,060 different possibilities, so the chance is 756/3,060, or about 25% that the mixing will spontaneously reverse.

### Reversibility and entropy in real systems

One gram of air in an "empty" 1 L soda bottle contains $2 \times 10^{22}$ molecules. Entropy measures the number of different ways, or *degrees of freedom*, in which the individual positions and velocities of each molecule can be switched around while keeping the same macroscopic properties (such as temperature) for the whole collection. Similar to mixing density in the particle/box example, heat moving from hot to cold always increases the number of degrees of freedom in a system. With 4 particles in 18 boxes there was a 25% chance that a mix would randomly reverse. With, say, $10^{23}$ particles in $10^{29}$ boxes, the chance of cold and hot randomly "unmixing" is so vanishingly small it will never occur in a trillion times the age of the universe. That is why increasing entropy makes a process irreversible.

## The third law and absolute zero

### What is the entropy at absolute zero?

Entropy gives us a new insight into the meaning of absolute zero. There is precisely *one* way to organize all the atoms to have minimum energy: A substance at absolute zero is perfectly ordered. The **third law of thermodynamics** says that *absolute zero* is the minimum possible temperature; this is the temperature at which the entropy of a system is zero. That means that the system has no ability to transfer heat to any other system by any means. Technically, quantum physics tells us that the real energy can never be exactly zero, but for all practical purposes a system at absolute zero has effectively zero thermal energy. Absolute zero is 0 K, or −273°C (−459°F). This is the lower limit of temperature, and no temperature can be lower than absolute zero.

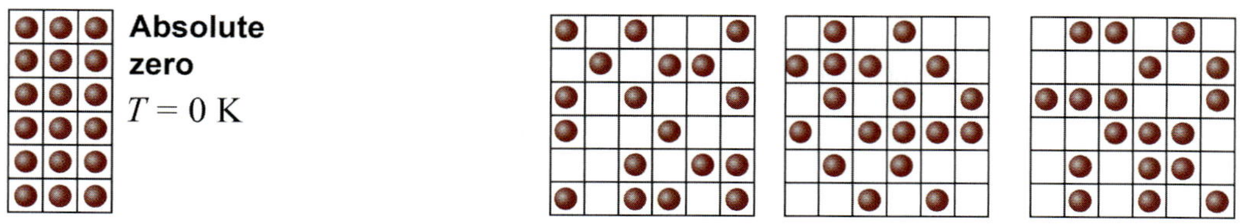


**Entropy = 0**
A perfectly ordered system has only one possible configuration.

**Entropy > 0** $T > 0$ K
System has some **disorder.** "Disorder" means the system has a number of possible configurations.

### Why does entropy increase with temperature?

Entropy increases extremely rapidly as temperature increases above absolute zero. Entropy also increases with volume. This occurs because each particle can move in three dimensions $(x, y, z)$, each with three possible momentum components $(p_x, p_y, p_z)$. Each possible variation of position and momentum represents a different "box" that can be occupied by a particle. The more thermal energy there is per particle, the more "boxes" or *microstates* each particle can potentially occupy. The increase in the number of microstates is what fundamentally creates the increase in entropy.

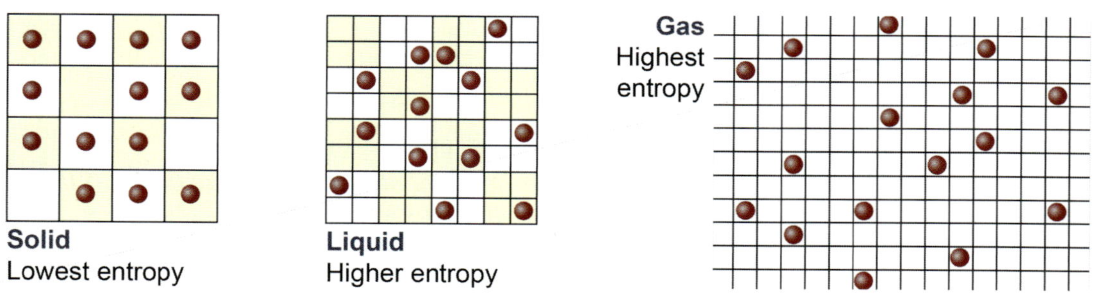


**Solid**
Lowest entropy

**Liquid**
Higher entropy

### Do solids, liquids, and gases have the same entropy?

At equal pressure a substance has lower entropy as a solid, higher entropy as a liquid, and even higher entropy as a gas. The entropy difference occurs because molecules in a solid have less energy and therefore fewer degrees of freedom to move around. Although the average spacing of molecules in a liquid is similar to a solid, molecules in a liquid have more thermal energy. More thermal energy means more variation of momentum, increased degrees of freedom, and higher entropy. As a gas, molecules have both more space and more variation in momentum. Gas has the highest entropy of the three ordinary phases of matter. Since entropy changes involve heat, changing phase also requires heat—even if the temperature remains constant. For example, it takes 334 J of heat to turn 1 g of ice at 0°C into 1 g of liquid water at 0°C.

# Section 1 review

Thermodynamics is the field of physics studying how thermal energy can be converted to or from other forms of energy, usually mechanical or electrical energy. The steam engine is an example of a thermodynamic system, where heat drives a shaft to pump water or power a train. The laws of thermodynamics describe the physics behind thermodynamic processes such as the steam engine. The zeroth law implies that no heat will flow between two objects at the same temperature, which means that the two are in thermal equilibrium with each other. The first law is a thermodynamic restatement of energy conservation, where the change in energy of a system equals the heat added to it and the work done on it. The second law states that entropy cannot decrease for any naturally occurring process. This law implies that, when heat is transferred in a thermodynamic process, the process is not reversible. Heat spontaneously flows from hotter objects to colder ones, but the reverse never occurs. The third law states that absolute zero is the minimum possible temperature and the temperature at which the entropy of a system also drops to its minimum.

| | |
|---|---|
| **Vocabulary words** | **thermodynamics, zeroth law of thermodynamics, first law of thermodynamics, entropy, second law of thermodynamics, reversibility, third law of thermodynamics** |

| | | |
|---|---|---|
| **Key equations** | $\Delta E = Q + W$ | $\Delta S \geq 0$ |

## Review problems and questions

1. Jean wants to build a machine that has a wheel that will never stop turning. She claims that it will work because energy is conserved. Is Jean right? Use the first law of thermodynamics to explain.

2. When a container of gas expands, is positive work done on the gas or is the gas doing positive work on the surroundings?

3. Kaleb fills a glass with half lemon-lime soda and half root beer. The two drinks mix together. Explain why Kaleb's drinks will not naturally separate themselves from one another after being mixed.

4. After learning Newton's first law, Natasha figures that if she gets her car up to the speed she wants, it will continue moving at that speed until she applies the brakes. When she performs an experiment to test this, however, her car slows down and eventually stops without her pressing the brakes. When she gets out to examine her car, she notices that the tires are warm. What happened?

5. Match each statement below to the corresponding law of thermodynamics.
    a. No engine can be 100% efficient.
    b. A system can only have zero entropy if it is cooled to absolute zero (0 K).
    c. Energy is conserved in thermodynamic systems.
    d. If A and B are at the same temperature, and B and C are at the same temperature, then A and C must also be at the same temperature.

6. Simon received several balloons filled with pure nitrogen gas for his birthday. He left them in his room for a couple days, and they all deflated. Will the nitrogen gas that escaped ever separate itself from the other gas in his room?

A *heat engine* is a system that converts thermal energy, or heat, into other forms of energy, usually mechanical work. This includes the gasoline or diesel engine in any modern car and also the enormous turbines that spin the generators in every nuclear-, gas-, or coal-powered electric generating station. In this section you will learn the fundamental physics behind all heat engines, including why a 100% efficient heat engine is impossible.

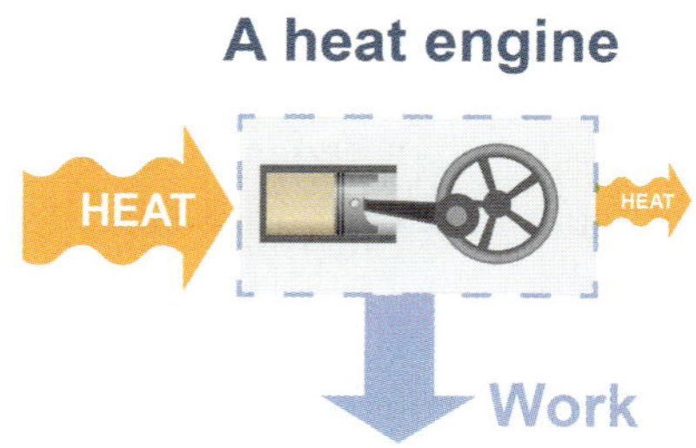


## Heat engines

**The first steam engine**

Newcomen's 1712 steam engine was the first industrial heat engine. Its efficiency, however, was very low and the engine burned a lot of coal. In 1763, Scottish instrument maker James Watt was called upon to repair one of Newcomen's engines. Instead, Watt devised a much-improved engine. Over the next decade Watt's improvements to the steam engine were so successful that the unit of power (the watt) is named after him. "Inventor of the steam engine" and "James Watt" are nearly synonymous today. Modern heat engines, including Watt's design, use pressure rather than vacuum to provide the mechanical force that converts thermal energy into work.

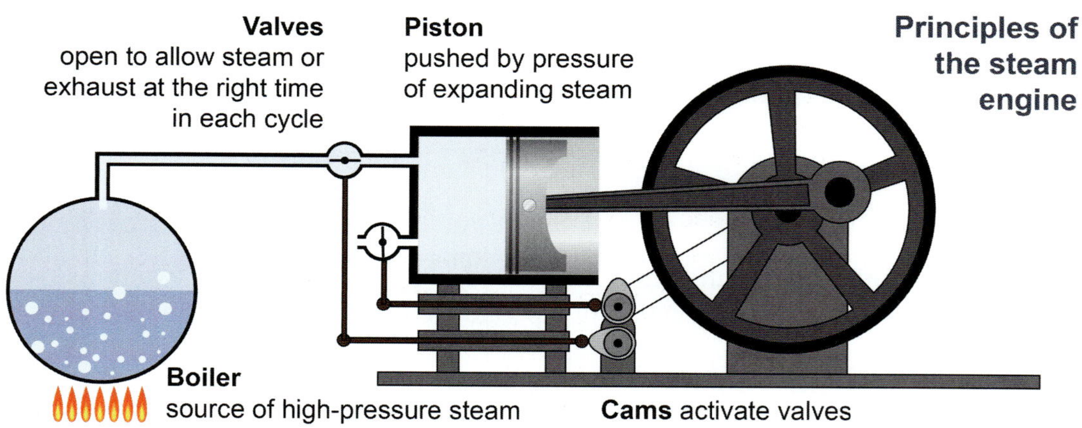


**How does a heat engine work?**

Conceptually, all heat engines extract energy from the flow of heat between a higher temperature and a lower temperature. It looks simple on a diagram. In practice, however, this is a very complex process involving moving pistons or spinning turbines. Real heat engines convert heat to mechanical energy by manipulating the pressure, volume, temperature, and/or phase of a *working fluid*. In a car engine the working fluid is mostly air. Heat comes from the ignition of a gasoline/air mixture. Energy is extracted by the work done as the hot (air/gas) working fluid pushes against a piston, expanding its volume. In an electric generating station the working fluid is steam. Heat may come from nuclear reactions, gas, or coal. In all three cases the heat is used to boil water into high-pressure steam, which expands by spinning a turbine, which turns an electric generator.

### How does a heat engine operate?

Virtually all heat engines convert heat into mechanical work through the expansion of a gas, referred to as the *working fluid.* The engine in most cars uses a *piston* moving in a *cylinder.* The trapped volume of gas inside the cylinder is the working fluid.

a. When heat is added, the gas temperature and pressure increase.
b. The increase in pressure pushes the piston and does work on the surroundings. As the gas expands, it also cools.
c. When the piston reaches the end of its travel, at maximum volume the pressure is released by opening a valve.
d. The piston returns to its starting position to begin a new cycle in step a.

The piston is connected to a rotating *flywheel* that moves the piston back to its starting point.

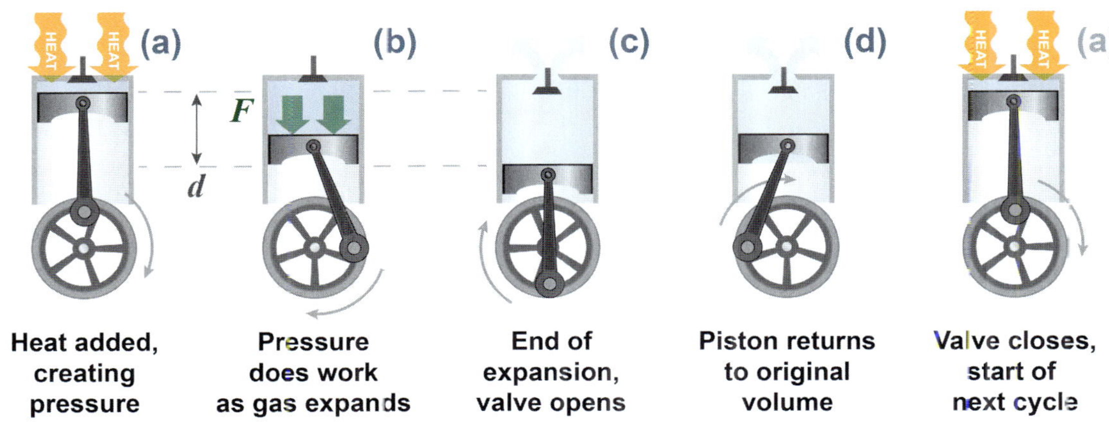


### Calculating the work done

The work done by the engine is the force exerted by the gas pressure multiplied by the distance the piston moves. Assume that the piston moves a small distance $\Delta d$ over which the pressure is approximately constant. The work done is therefore

$$W = F\Delta d$$

Force is pressure × area. The area × distance is the change in *volume* $\Delta V$. Therefore another way to write the work done by an expanding gas is as the pressure multiplied by the change in volume:

$$W = P\Delta V$$

### Efficiency

According to the first law of thermodynamics, the change in energy, $\Delta E$, is equal to the total heat $Q$ added *to* the system minus the work $W$ done *by* the system. (Alternatively, you could *add* the work done *on* the system.) You might think that it could be theoretically possible to invent an engine that would take in 100 J of heat and do 100 J of output work. Over the next few pages we will see that this is *impossible!*

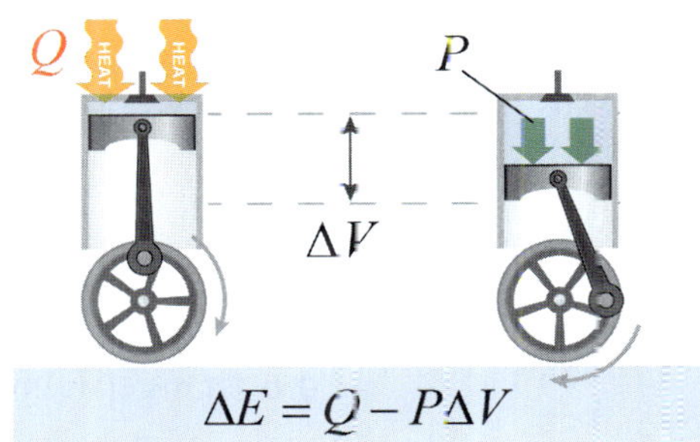

**How are changes in pressure and volume represented?**

Since pressure and volume change constantly, we use a graph to determine the work done by the expanding gas pushing on the piston. The pressure versus volume graph, or ***PV* diagram**, has pressure on the vertical axis and volume on the horizontal axis. On this graph, *area equals work.* If the pressure scale is in newtons per meter squared and the volume scale is in cubic meters, then area has units of joules.

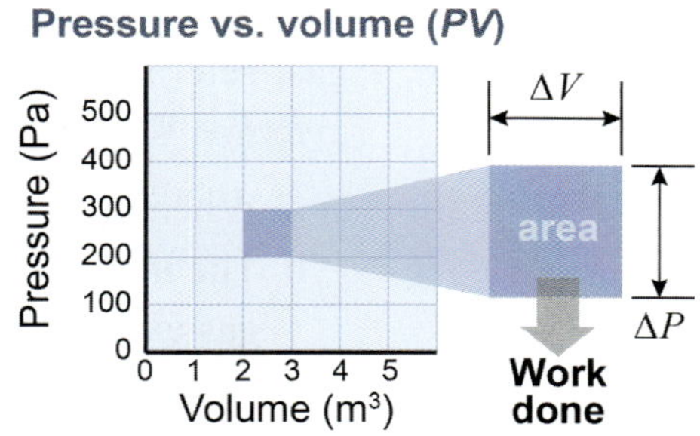


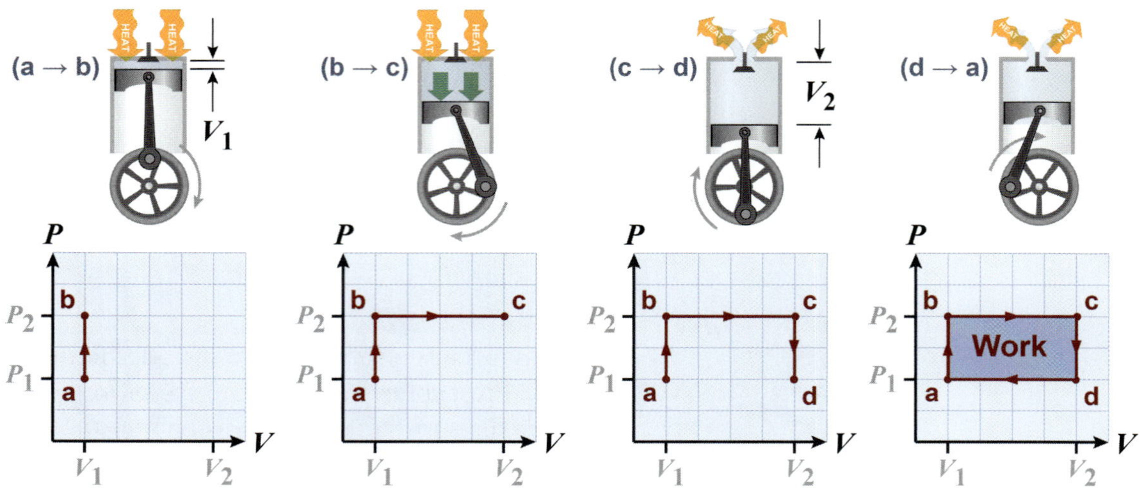


**An illustrative but impractical example**

Suppose we could create a heat engine in which the pressure and volume change like the diagram above. How much work is done? How much heat is added? How efficient is the engine at converting heat into work?

- At point (a) the volume is at a minimum. From (a) to (b) heat is applied to the engine and the pressure of the gas rises.
- From (b) to (c) the gas expands, doing work on the piston. In this (unrealistic) example we continue to add enough heat to keep pressure constant as the gas expands.
- From (c) to (d) the valve opens and reduces the pressure back to its starting point. Heat flows out of the engine into the surroundings.
- From (d) to (a) the piston moves back up.

**How much work is done?**

The work done by the expanding gas on the piston in one cycle is the area enclosed by the curve *abcd.* In terms of pressure and volume, the area enclosed by one cycle on the graph is $(P_2 - P_1)(V_2 - V_1)$. The important result here is that the *area enclosed by the cycle on a PV diagram equals the work done by the gas in one cycle.*

**How does the energy change in a cycle?**

The gas in a thermodynamic cycle returns to its original pressure and volume at the end of a complete cycle. This means that *the net energy change of the gas over a full cycle is zero!* If the change in energy is zero, then by the first law of thermodynamics the work done by the gas must equal the heat input, or $Q = P\Delta V$.

**Problem-solving strategy**

*PV* diagrams are useful for understanding thermodynamic processes, such as the cycle of a heat engine. It takes some practice to interpret a *PV* diagram. Start by looking at how pressure and volume change for each individual segment on the *PV* diagram. Then use any additional information from the problem, such as whether the quantity of gas or heat is constant.

**A process example**

The diagram shows a process in which a gas goes from A to B and no heat is added. Which of the following is true and why?

a. The energy of the gas remains constant.
b. The gas does work on its surroundings and the gas loses energy.
c. The surroundings do work on the gas and the gas gains energy.

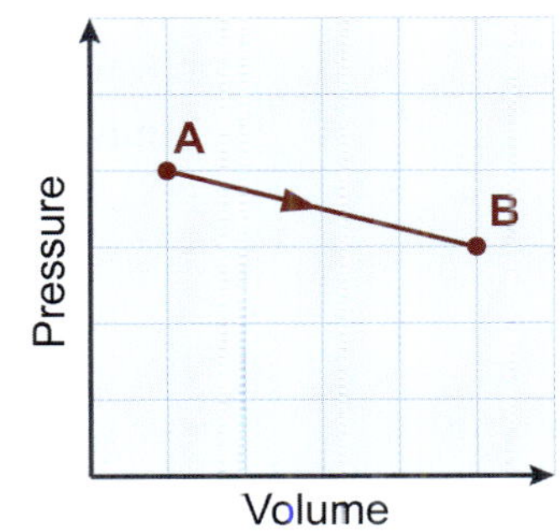


*Asked:* whether the work done is positive, negative, or zero

*Given:* *PV* diagram showing pressure decreasing and volume increasing

*Relationships:* $\Delta E = Q - P\Delta V$

*Solution:* $\Delta V$ is positive because the diagram shows a volume at B greater than the volume at A. Pressure is also positive; therefore, $P\Delta V > 0$ and the gas does work on the surroundings. The energy of the gas must decrease because no heat is added that might offset the work done (i.e., $Q = 0$).

*Answer:* Choice b is correct; the gas does work on the surroundings and loses energy in the process.

**Thermodynamic cycle example**

The diagram on the right shows a cycle in which pressure and volume change for a fixed quantity of gas.

a. Which part of the process compresses the gas? How do you know?
b. Does the complete cycle absorb heat energy or give off heat energy? How do you know?

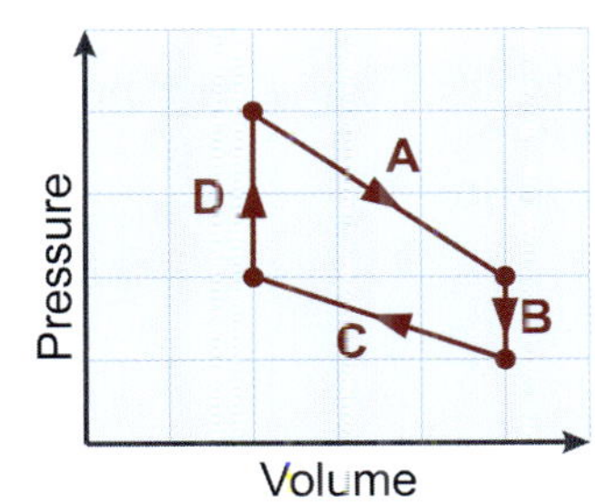


*Relationships:* $\Delta E = Q - P\Delta V$

*Solution:*

a. Compression is a reduction in volume and only process C shows a reduction in volume.
b. In any complete cycle, the net energy change of the gas is zero. If the net change in work is positive, then heat must be added.
Process A expands the gas and therefore does work on the surroundings, so $P\Delta V$ is positive.
Process C compresses the gas, so the surroundings do work on the gas and $P\Delta V$ is negative.
Since the pressure is higher for A than C and the volume change is the same, the net change in work is positive and *heat must be added.*

## Efficiency of a heat engine

**How much heat is required?**

The efficiency of a heat engine is the ratio of work output divided by heat input. The process from a to b occurs at constant volume; the heat added, $Q_{ab}$, therefore depends on the moles of gas, the molar specific heat at constant volume, $C_V$, and the change in temperature. The process from b to c occurs at constant pressure, so the heat added, $Q_{bc}$, instead depends on the molar specific heat at constant pressure, $C_P$:

**Example of a heat engine**

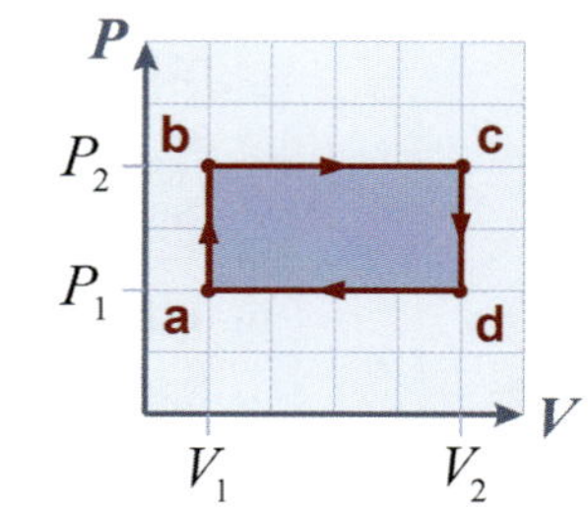


$$Q_{ab} = nC_V(T_b - T_a), \qquad Q_{bc} = nC_P(T_c - T_b)$$

**Molar specific heats**

The equations above and below use the *molar specific heat*—which you learned about on page 674. Typical values for air are $C_V = 20.8$ J/(mol K) and $C_P = 29.1$ J/(mol K).

**Temperature at each point in the cycle**

Assume an ideal gas for which $PV = nRT$. The temperatures at points a, b, c, and d are therefore determined by the pressures and volumes:

$$T_a = \frac{P_1V_1}{nR}, \qquad T_b = \frac{P_2V_1}{nR}, \qquad T_c = \frac{P_2V_2}{nR}, \qquad T_d = \frac{P_1V_2}{nR}.$$

**Calculating input heat**

After substituting for the temperatures we find the heat inputs $Q_{ab}$ and $Q_{bc}$:

$$Q_{ab} = \frac{C_v}{R}V_1(P_2 - P_1) \qquad Q_{bc} = \frac{C_p}{R}P_2(V_2 - V_1)$$

**How would this theoretical engine perform?**

Consider a large car engine, in which each piston has a 10 cm diameter, stroke of 10 cm, and clearance of 2 cm and operates between 1 and 5 atmospheres (100,000 to 500,000 Pa). Apply the above equations to calculate the *input heat* as

$$Q_{ab} + Q_{bc} = 157 \text{ J} + 1{,}374 \text{ J} = 1{,}531 \text{ J}.$$

The work done in the thermodynamic process is the area inside the $PV$ diagram, or

$$(P_2 - P_1)(V_2 - V_1) = 314 \text{ J}.$$

The input heat is almost five times the work output of the engine! Furthermore, to keep the pressure constant from b to c requires that the temperature be $T_c$ = 15,363°F! A real engine could not sustain this high a temperature.

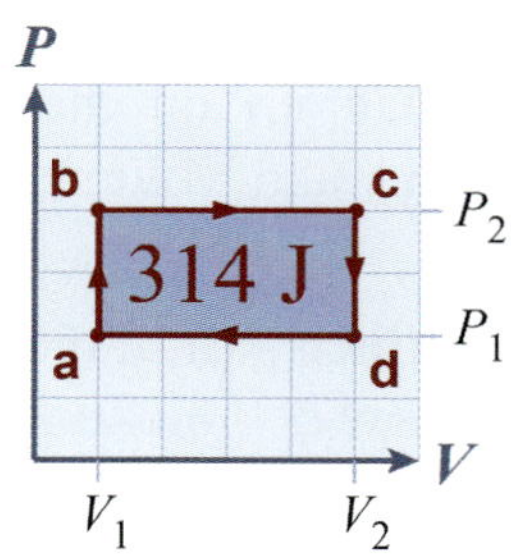


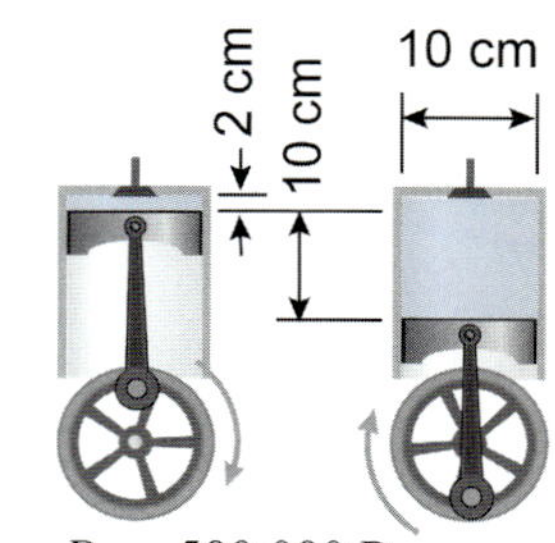


$P_2$ = 500,000 Pa
$P_1$ = 100,000 Pa

**Calculations**

$T_a$ = 293 K (68°F)
$T_b$ = 1,465 K (2,178°F)
$T_c$ = 8,790 K (15,363°F)
$T_d$ = 1,758 K (2,705°F)
$Q_{ab}$ = **157 J**
$Q_{bc}$ = **1,374 J**

**Efficiency**

$$\eta = \frac{314 \text{ J}}{157 \text{ J} + 1{,}374 \text{ J}}$$

**= 21%**

**Why is the efficiency so low?**

This thermodynamic cycle has a terrible efficiency of 21%. Heat not converted to work is rejected to the surroundings during steps c → d and d → a. According to the first law, the heat rejected is the heat added minus the work done, or 1,271 J (79%). Soon we will see how to get better efficiency, but, according to basic thermodynamic principles, no heat engine operating above absolute zero can convert 100% of heat into output work.

**Isothermal processes**

An **isothermal process** is a change in pressure and volume in which temperature is kept constant. In this case, pressure and volume are inversely related to each other. If pressure increases, volume decreases and vice versa. The diagram shows isothermal processes at different temperatures for an ideal gas. Although not quite achievable in a real engine, isothermal heat transfer has the highest possible theoretical efficiency.

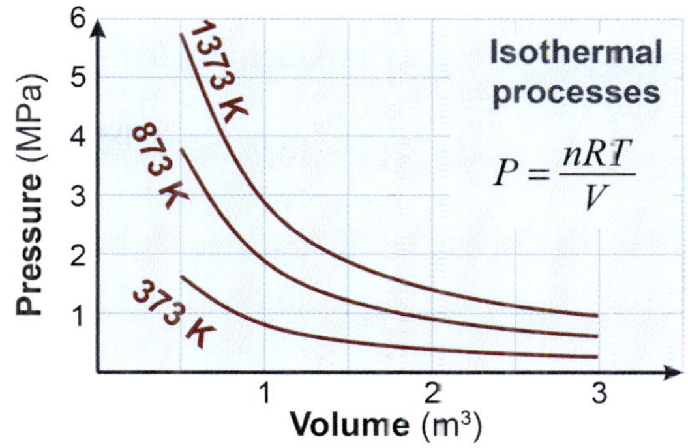


**Adiabatic processes**

Another important ideal is an **adiabatic process** in which *zero* heat is exchanged. A process that occurs very quickly, such as a gas expanding rapidly, is approximately adiabatic because there is not enough time for much heat to flow in or out. A process that takes place in a perfectly insulated container is also adiabatic. The diagram shows that adiabatic processes change pressure more rapidly than isothermal processes.

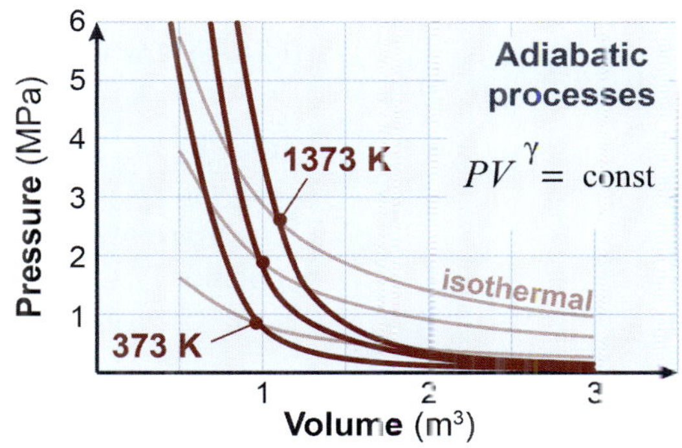


**Adiabatic pressure–volume relationship**

On a pressure–volume graph, an adiabatic process is a curve with a shape different from that of the isothermal curve. The adiabatic relationship between pressure and volume is given by equation (25.3).

**Adiabatic process**

**(25.3)** $PV^{\gamma} = \text{constant}$

$P$ = pressure (Pa)
$V$ = volume ($m^3$)
$\gamma$ = ratio of specific heats = $c_p/c_v$

**Ratio of specific heats**

In equation (25.3), the quantity $\gamma$ is the ratio of specific heats, $\gamma = c_p/c_v$. As you learned on page 689, $c_p$ is the specific heat at constant pressure and $c_v$ is the specific heat at constant volume.

**Why are adiabatic processes important?**

In the last section, we learned that adiabatic processes are *reversible* because no heat flows, making the change in entropy ($Q/T$) equal to zero. Reversibility is the "zero friction" limit of efficiency. The ideal heat engine should therefore use reversible processes as much as possible.

**Ideal heat engines**

Adiabatic and isothermal processes are important because an ideal heat engine can be modeled by two isothermal curves and two adiabatic curves. This thermodynamic cycle, called the *Carnot cycle*, has the highest possible theoretical efficiency of any heat engine that operates by changing the pressure, volume, and temperature of a gas. All real heat engines have efficiencies less than that of the ideal Carnot cycle. The cycle was discovered by the French engineer Sadi Carnot in 1824. For this work, Carnot is recognized as the "father" of the science of thermodynamics.

## Carnot cycle

### What is the Carnot cycle?

Consider an ideal-gas heat engine that uses a piston that starts at position a.

1. Heat is added and the gas is allowed to expand isothermally by converting some of the heat into work following the isothermal curve (a → b).
2. The gas is then allowed to expand further adiabatically (b → c), thereby extracting more work by lowering the temperature. This is a frictionless, reversible change that extracts the most possible work from the expanding gas.
3. Once the piston reaches its lowest point c the gas is compressed again but held to constant temperature along an isothermal curve (c → d). The heat generated by compression is transferred to the surroundings at the lower temperature $T_c$.
4. The last step in the cycle is to compress the gas back to its starting point reversibly (d → a), for which $Q = 0$.

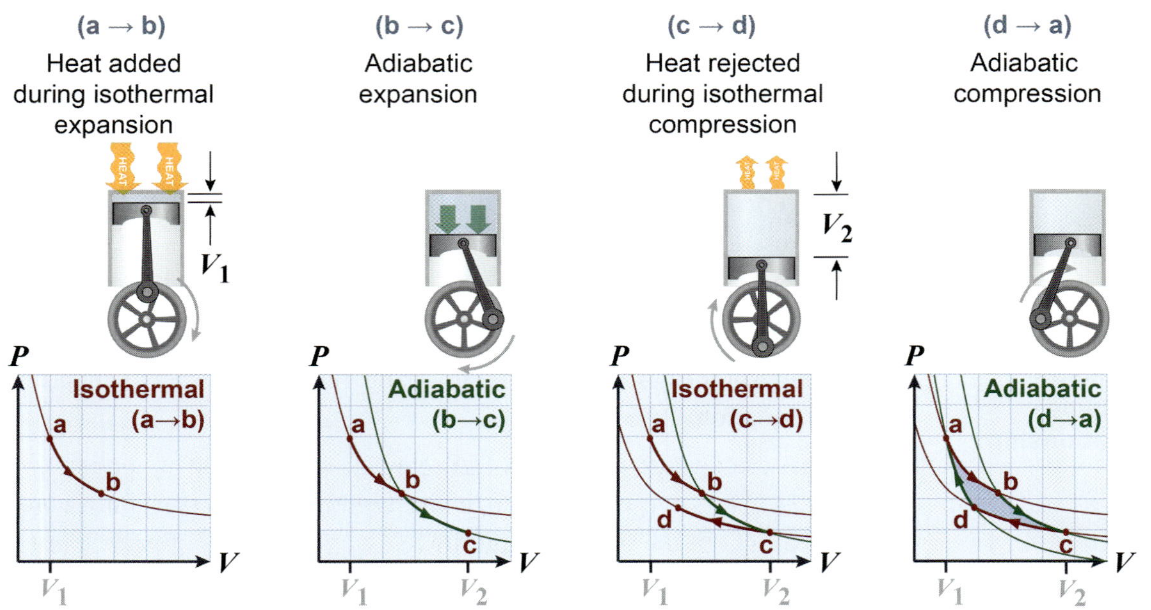


### How is the Carnot cycle analyzed?

The work done during the **Carnot cycle** is the shaded area on the *PV* graph, but this is difficult to calculate because of its curved shape. Instead, Carnot used the graph of temperature versus entropy (*TS*) to analyze this cycle. On the *TS* graph, area is equal to *heat.* An isothermal process is represented by a horizontal line (a → b). Because no heat flows ($Q = 0$), an adiabatic process is a vertical line (b → c) representing no change in entropy. The entire cycle is a rectangle on the *TS* graph defined by the two isothermal processes (a → b and c → d) and the two adiabatic processes (b → c and d → a).

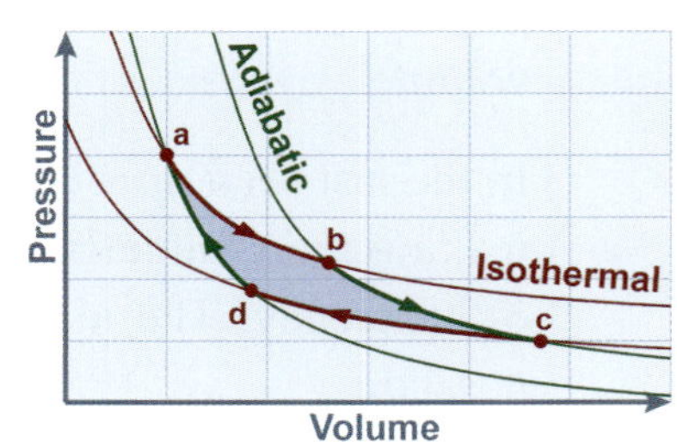

**The Carnot cycle on a *PV* graph**

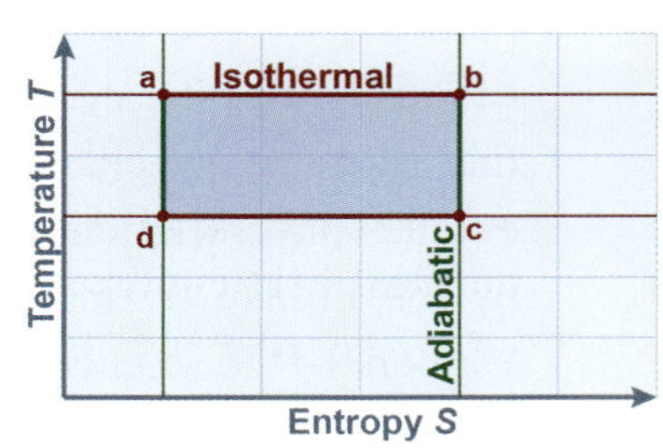

**The Carnot cycle on a *TS* graph**

**Analysis of the Carnot cycle**

The temperature–entropy (*TS*) graph allows us to analyze the efficiency of the Carnot cycle. Consider the graph drawn with absolute zero as the lower limit of the temperature axis. With this choice of scale, the heat added to the engine is equal to the pink shaded area on the left-hand graph. The heat rejected by the engine is the pink shaded area in the middle graph. The heat that is converted to work is the blue shaded area on the right-hand graph, which is the intersection of the other two.

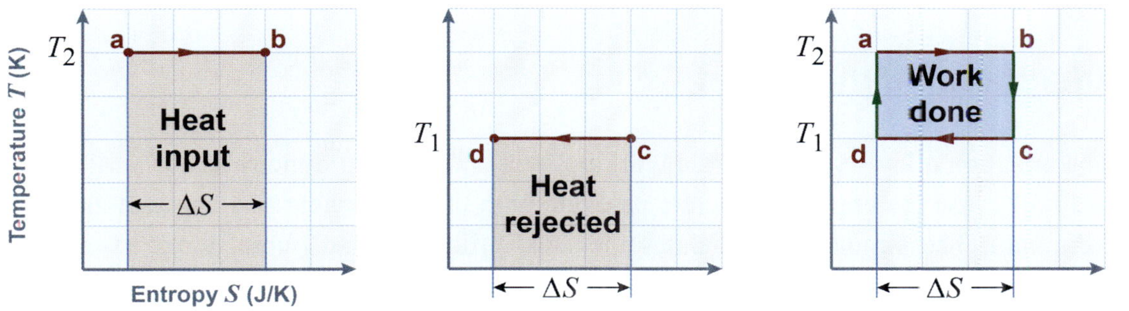


$Q_{in} = T_2\Delta S$ $\quad$ $Q_{rej} = T_1\Delta S$ $\quad$ $W = Q_{in} - Q_{rej} = (T_2 - T_1)\Delta S$

**What is the efficiency of the Carnot cycle?**

The efficiency of the Carnot cycle is the work output divided by the heat input:

$$\text{efficiency} = \frac{W}{Q_{in}} = \frac{(T_2 - T_1)\Delta S}{T_2\Delta S}$$

Notice that the change in entropy, $\Delta S$, cancels out because it appears in both the numerator and denominator. We can rearrange the remaining equation into a very useful form that depends only on the highest and lowest temperatures. For a Carnot engine operating between temperatures $T_1$ and $T_2$, the **Carnot efficiency** is given by equation (25.4) below. This is the highest theoretically possible efficiency for any heat engine that converts heat into work.

**(25.4)** $$\text{efficiency} = \left(1 - \frac{T_1}{T_2}\right) \times 100\%$$

$T_1$ = lower temperature (K)
$T_2$ = higher temperature (K)

**Carnot efficiency**

**What the equation tells us**

No heat engine that operates above absolute zero can be 100% efficient because some heat is rejected along path c → d in the thermodynamic cycle. Equation (25.4) tells us that the highest efficiency occurs when the engine operates between the largest difference in temperatures. A typical car engine reaches a maximum combustion temperature of around 2,500 K (4,000°F). The exhaust temperature is 1,100 K (1,500°F). The Carnot efficiency for an ideal heat engine operating between these two temperatures is 56%. A modern internal combustion engine has an operating thermal efficiency that ranges from about 20% for a production car up to 34% for a highly tuned race engine.

# Investigation 25A Heat engines

**Essential questions** *How does a heat engine work?*

Interactive simulation

A heat engine converts thermal energy (heat) into mechanical energy (work). Most heat engines use a gas as a working fluid. The gas does work by expanding. In this investigation, a simulated heat engine illustrates the *PV* diagram and models the efficiency of a heat engine at different temperatures.

## Understanding the model

This simulation allows you to set the working temperatures and dimensions of an ideal heat engine. Pressing [Run] will cause the engine to operate its cycle with the parameters you have set and plot the corresponding *PV* diagram. The simulation calculates the ideal efficiency and output work by assuming that the engine operates on air. The engine will allow you to make the cylinder as large as 0.5 m in either dimension.

## Construct your model of a Carnot heat engine

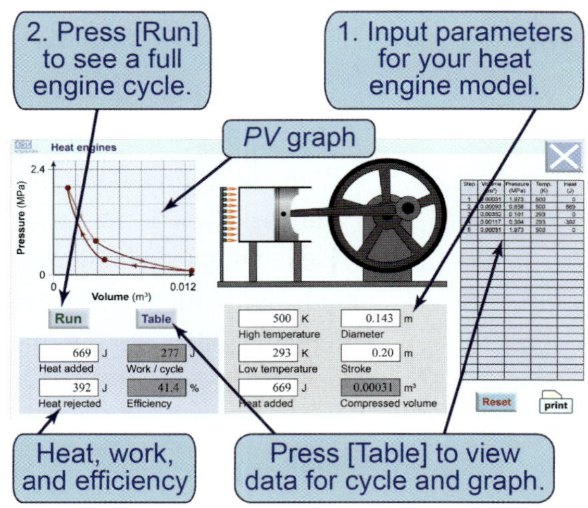


1. Set the initial operating temperatures to 550 and 1,293 K. Set the piston diameter to 20 cm (0.2 m) and the stroke to 20 cm (0.2 m).
2. Press [Run] to see a cycle. Use the data table to look at the pressure and volume data.
3. Alternately, you may set the heat input and the simulation will calculate the maximum temperature. Set a heat input of 1,000 J and see what happens.

a. *What is the operating efficiency at the parameters above?*

b. *What is the highest pressure reached? What is the lowest pressure?*

c. *What could you do to improve the efficiency of the engine? Propose two possible changes and then design an experiment to test each of your changes. Your experiments should have procedures for how you will change variables, and which variables you will change, to determine the effect of your changes.*

d. *What conclusions do you draw from your experiments?*

e. *What is the maximum possible theoretical efficiency given the limit of 293 K as the lowest operating temperature and 6,000 K as the highest operating temperature? What might happen to the piston at this high temperature?*

**What is a refrigerator?**

The invention of the refrigerator is the reason you can buy fresh fruits and vegetables in February in New England, at a time when no crops are growing outside. Food in a refrigerator cools or freezes because a refrigerator extracts heat from the cold food and rejects the heat into the warmer room. In the refrigerator, heat flows from cold to hot, in the direction opposite to that prescribed by the second law of thermodynamics! This is done through a *reversed* thermodynamic cycle.

**How can a refrigerator work?**

A **refrigerator** does not violate the laws of thermodynamics because it takes energy input to cause heat to flow from cold to hot. If you *unplug* your refrigerator, then the situation reverts to normal and heat flows from the warmer room to the colder refrigerator interior. In the reversed thermodynamic cycle work is done *on* the working fluid instead of by the fluid.

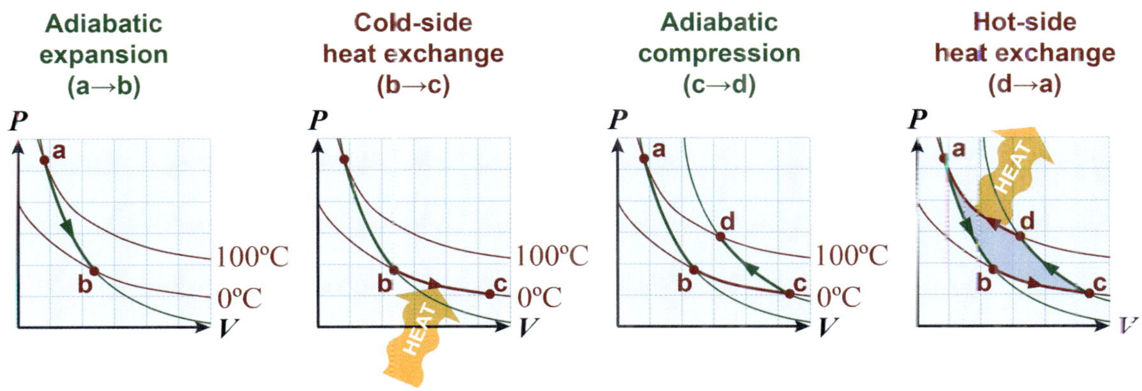


**The reversed cycle**

When a fluid is compressed, its temperature increases *even if no heat is exchanged*. When a fluid is allowed to freely expand, its temperature decreases even if no heat is exchanged. This change in temperature is what allows refrigeration to work.

1. Consider a reversed gas cycle that starts at 100°C at point a. The gas is allowed to expand and cool during process a → b.
2. During step b → c the 0°C cold gas absorbs heat because it is colder than the inside of the refrigerator.
3. During step c → d the gas is adiabatically compressed and its temperature is raised back to 100°C.
4. During step d → a the gas is now *warmer* than the room, so it rejects the heat it absorbed during step b → c into the room.
5. The cycle is reversed, so the shaded area represents work done *on* the gas to move the heat from a colder temperature to a hotter temperature.

**Two-phase cycles**

In this example we used a gas. Modern refrigerators, however, operate using a *two-phase* cycle. The working fluid is liquid during the high-pressure, hot part of the cycle during which heat is rejected to the ambient environment. The expansion that cools the fluid finishes at a low pressure so that the fluid becomes a gas. The cold gas may be −20°C or lower, which is colder than inside the freezer and therefore able to absorb heat from chilled food.

Chapter 25

# Section 2 review

The heat engine is a machine that uses heat to do mechanical work. The operation of a heat engine uses some kind of gas as a *working fluid*; when that working fluid is heated, its pressure increases and drives the piston in the engine. The thermodynamic processes in a heat engine can be plotted on a pressure–volume or *PV* diagram. The area enclosed by a thermodynamic process on a *PV* diagram is the work done. An isothermal process is one in which the temperature is kept constant, whereas an adiabatic process is a typically fast process that has no net exchange of heat. Both processes appear as curves in a *PV* diagram. The efficiency of all heat engines is less than 100%. The Carnot cycle describes an idealized heat engine that has maximum efficiency, although this efficiency is still less than 100% because the heat is always rejected to a medium at a temperature that is above absolute zero. Refrigerators seem to defy the second law of thermodynamics by taking heat from a cold location and rejecting it into the warmer surroundings. But refrigerators use input energy to do work on the gas or fluid inside them.

| Vocabulary words | **heat engine, *PV* diagram, isothermal process, adiabatic process, Carnot cycle, Carnot efficiency, refrigerator** |
|---|---|
| **Key equations** | $PV^{\gamma} = \text{constant}$    $\text{efficiency} = \left(1 - \frac{T_1}{T_2}\right) \times 100\%$ |

## Review problems and questions

1. In the first section of this chapter the first law of thermodynamics was described as energy conservation with the equation $\Delta E = Q + \Delta W$. But in this section, the operation of a heat engine was described by using an equation with the *opposite sign* for the last term: $\Delta E = Q - P\Delta V$. Which is correct? Should it be a positive or a negative sign?

2. In the illustration below, four *PV* graphs are plotted (I–IV).
   a. Which plot(s) corresponds to a Carnot engine cycle? Which corresponds to a refrigerator?
   b. In which plot(s) is the work done *by* the gas? *On* the gas?

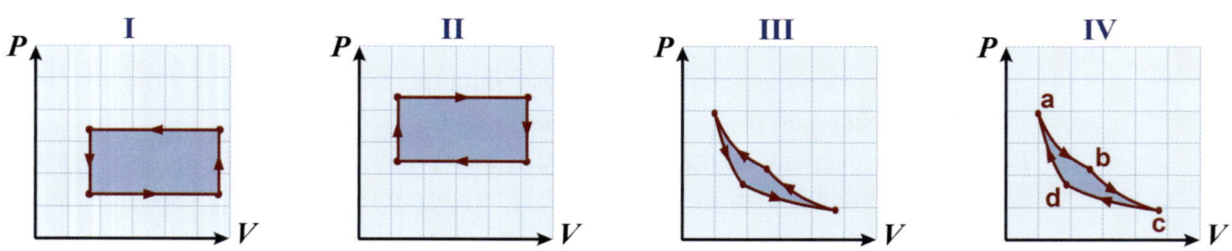


3. Refer to the four curved segments (a–b, b–c, c–d, and d–a) in the right-hand illustration (IV) above to answer the following questions.
   a. Which curved segment(s) correspond to *adiabatic* processes? Adiabatic *compression*?
   b. Which curved segment(s) correspond to *isothermal* processes? One where heat is *rejected*?
   c. Which curved segment(s) are *reversible* processes? Explain.
   d. Is the machine or technological process represented by a–b–c–d overall a reversible process?

# Chapter review

## Vocabulary

Match each word to the sentence where it best fits.

### Section 25.1

| | |
|---|---|
| **entropy** | **first law of thermodynamics** |
| **reversibility** | **second law of thermodynamics** |
| **thermodynamics** | **third law of thermodynamics** |
| **zeroth law of thermodynamics** | |

1. No temperature exists below absolute zero, which is a consequence of the _______.
2. That no heat will flow between two objects at the same temperature is a consequence of the _______.
3. Heat and its transformation to and from mechanical energy is the subject of _______.
4. If a thermodynamic process has the property of _______, then it can run forward and backward to return to its original state.
5. The _______ of a system is a measure of the thermal energy that is not available to do work.
6. A consequence of the _______ is that no heat can flow from a colder object to a warmer object unless work is done on them.
7. The change in energy of a system equals the heat added to it plus the work done on it as defined by the _______.

### Section 25.2

| | |
|---|---|
| ***PV* diagram** | **Carnot cycle** |
| **Carnot efficiency** | **adiabatic process** |
| **heat engine** | **isothermal process** |
| **refrigerator** | |

8. A/An _______ is a four-stage process of an idealized heat engine.
9. The area subtended by a thermodynamical process on a/an _______ is equal to the work done on the system.
10. A system that can convert thermal energy into mechanical energy is a/an _______.
11. A/An _______ uses a reverse Carnot cycle to extract heat from a colder object by doing work.
12. The _______ illustrates that no real-world heat engine can convert all of the input heat into output work.
13. Changes in pressure are inversely proportional to changes in volume at constant temperature in a/an _______.
14. No heat is added or taken away from a system during a/an _______.

## Conceptual questions

### Section 25.1

15. Which equation best represents the first law of thermodynamics?
    a. $\Delta S \geq 0$
    b. $PV = nRT$
    c. $W = -P\Delta V$
    d. $\Delta E = Q + W$
16. Why don't coffee cups ever reheat themselves?
17. Selena has created a machine that converts heat into mechanical energy. She claims that her machine is so efficient that she can create 120 J of mechanical energy using only 90 J of heat. Do you believe Selena's claim?
18. Firnen claims that he can defy the law of entropy, as he can sort his bag of candies by color and separate the colors from one another. Has Firnen broken the second law of thermodynamics?
19. You own a machine that becomes more efficient as you add more heat to it as a power source. Yet no matter how much heat you add to the machine, it never is more than 100% efficient. Explain this using the laws of thermodynamics.
20. Describe how the first law of thermodynamics is related to the notion of energy conservation.
21. If the volume of a gas changes by $\Delta V$, choose the sentence that best describes what is meant by the negative sign in the following equation:

    $$W = -P\Delta V$$

    a. Pressure is directly proportional to the change in volume.
    b. Work is done on the gas.
    c. Work is done by the gas.
    d. The product of pressure and volume is given by the second law of thermodynamics.
22. What quantity is used in physics to measure *disorder*?

Chapter 25

# Chapter review

## Section 25.1

23. What happens to the internal energy of a system if it does work on something else?

24. The hot coffee in a mug cools from 80°C to room temperature over the course of 15 minutes. An equal amount of hot coffee, also at 80°C, in an insulated travel mug cools to room temperature over the course of 4 hours. In which mug did the coffee experience a greater change in entropy?

## Section 25.2

25. Describe what an *adiabatic* process is for a gas.

26. Describe what an *isothermal* process is for a gas.

27. What is the difference between a *heat engine* and a *heat pump*? Give one example of each.

28. Imagine that the gas within a piston expands, pushing against an outside pressure. Which of the following statements are correct? (There may be more than one right answer.)
    a. The gas within the piston does positive work on its surroundings.
    b. The gas within the piston does negative work on its surroundings.
    c. The surroundings do positive work on the gas within the piston.
    d. The surroundings do negative work on the gas within the piston.

29. Which statement is consistent with the following equation for a Carnot cycle?

$$\text{efficiency} = \left(1 - \frac{T_1}{T_2}\right) \times 100\%$$

    a. The efficiency of a Carnot engine is equal to the ratio of the two temperatures it is operating between.
    b. No Carnot engine with $T_1$ above absolute zero can be 100% efficient.
    c. When the upper and lower temperatures of a Carnot engine are very close to each other, then it will have nearly 100% efficiency.
    d. By inverting the operating temperatures, a Carnot engine can operate at >100% efficiency.

30. Mirette put together a business model for a company that would build *ideal* Carnot engines that convert 100% of the input heat into energy. Would it be a good idea to invest your life savings in her company?

## Quantitative problems

## Section 25.1

31. The gas within a piston expands from an initial volume of 25 mL to a larger volume of 125 mL, against a constant pressure of 100,000 Pa (approximate atmospheric pressure at sea level). How much work does the gas in the piston perform?

32. A bearded dragon lizard basks in the desert Sun during the day and reaches a body temperature of 40°C. At night the air temperature quickly becomes 15°C. During the course of the night the lizard loses 100 J of heat while his body temperature equilibrates to the nighttime air temperature. What is his change in entropy?

33. A hot rock is plucked from a fire pit and dropped into a bucket of cold water. The rock has a temperature of 80°C, and it transfers an initial 500 J of thermal energy to the water. Assume that the temperatures of the rock and water remain unchanged during this first instant.
    a. Does the entropy of the rock increase or decrease? By how much?
    b. Does the entropy of the water increase or decrease? By how much?
    c. Does the total entropy of the rock+water system increase or decrease? By how much?

## Section 25.2

34. Suppose that an internal combustion engine operates between an internal temperature of 2,000 K and an exhaust temperature of 1,000 K. What is the engine's maximum theoretical efficiency expressed as a percentage?

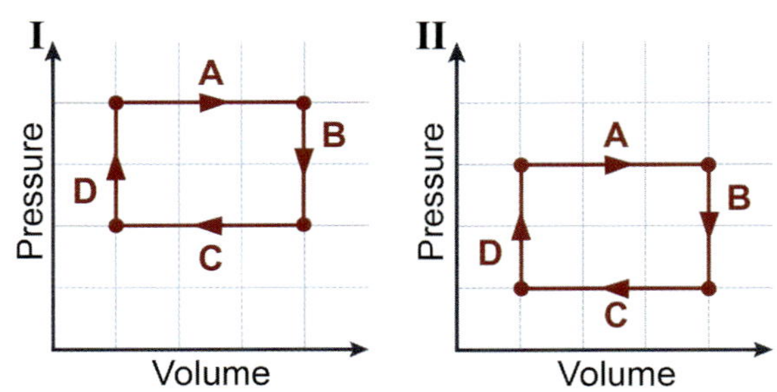


35. These two *PV* diagrams each depict one thermal cycle of a piston containing an ideal gas. The pistons are physically identical, but Piston I operates at higher pressure, on average, than Piston II. The lower-left corner of the coordinate grid represents the origin (zero pressure and volume).
    a. Which piston does more (positive) work during the expansion phase (I, II, or a tie)?
    b. Which piston does more (negative) work during the contraction phase (I, II, or a tie)?
    c. Which piston does more (positive) work during one complete cycle (I, II, or a tie)?

# Chapter review

## Standardized test practice

36. An ideal Carnot heat engine is operating between temperatures $T_{high}$ and $T_{low}$, where $T_{high}$ is the highest temperature that occurs within the piston during one cycle, and $T_{low}$ is the lowest. Which of the following formulas correctly gives the efficiency that such an engine has for converting input *heat* to output *work*? (Both temperatures must be stated in kelvins.)

   A. efficiency = $(1 + T_{high}/T_{low}) \times 100\%$
   B. efficiency = $(1 + T_{low}/T_{high}) \times 100\%$
   C. efficiency = $(1 - T_{high}/T_{low}) \times 100\%$
   D. efficiency = $(1 - T_{low}/T_{high}) \times 100\%$

37. Helium gas can be found in a balloon, but at low temperatures it can be in other phases. Which of the following phases of matter for helium will have the highest entropy?

   A. solid
   B. liquid
   C. semi-liquid
   D. gas

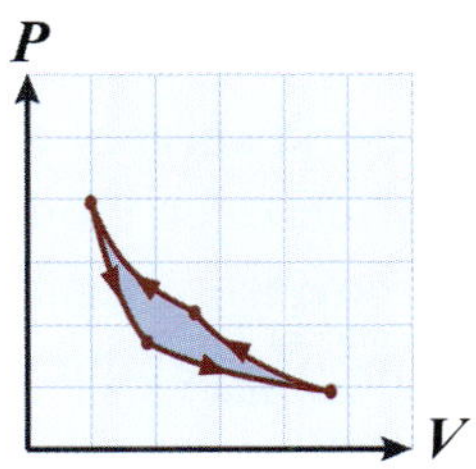


38. In the pressure–volume diagram shown, what kind of thermodynamic process is depicted?

   A. a refrigerator
   B. a heat engine
   C. a Carnot cycle
   D. an adiabatic isotherm

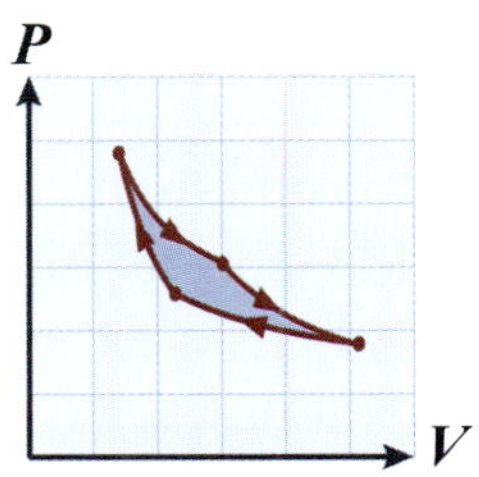


39. For one complete cycle plotted in the *PV* diagram shown, which of the following statements is correct?

   A. No net work is done during the cycle.
   B. Net work is done on the gas during the cycle.
   C. Net work is done by the gas during the cycle.
   D. The gas increases its temperature.

40. Which statement best describes how a refrigerator works?

   A. It uses energy to cause heat to flow from a colder location to a hotter location.
   B. It does work by causing heat to flow from a hotter location to a colder location.
   C. It cools through reverse osmosis.
   D. It rejects heat from the environment into the enclosure.

41. Which one of the following SI units (labels) could properly describe the product of pressure and volume?

   A. newton (N)
   B. joule (J)
   C. watt (W)
   D. horsepower (hp)

42. Which sentence best describes the following thermodynamics equation?

$$\Delta S \geq 0$$

   A. The change in speed is always greater than or equal to zero.
   B. Entropy does not change.
   C. The change in static energy for a heat engine is greater than or equal to zero.
   D. Total entropy only increases or stays the same.

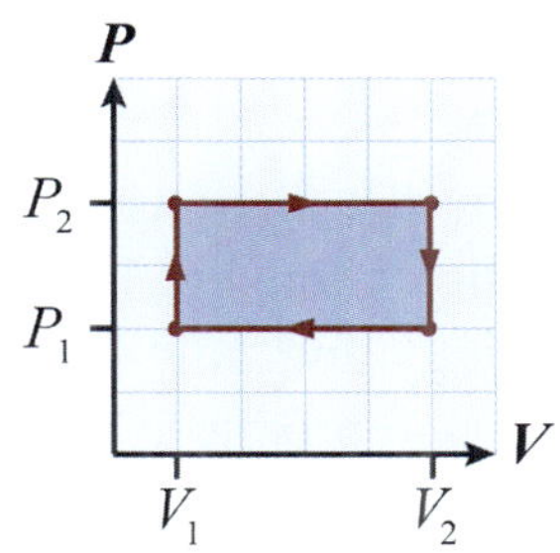


43. For one complete cycle of the thermodynamic process shown in the illustration, which is the equation for the work done *by* the gas?

   A. $(P_1 - P_2) / (V_1 - V_2)$
   B. $(P_2 - P_1) / (V_2 - V_1)$
   C. $(P_1 - P_2) \times (V_1 - V_2)$
   D. $(P_2 - P_1) \times (V_2 - V_1)$

44. Imagine that 100 J of thermal energy spontaneously moved *from* cool air (at 20°C) *to* a warm cup of tea (at 60°C). Which of the laws of thermodynamics, if any, would this process violate?

   A. the first law of thermodynamics only
   B. the second law of thermodynamics only
   C. both the first and second laws of thermodynamics
   D. none of the laws of thermodynamics

Chapter

# 26

# Quantum Physics and the Atom

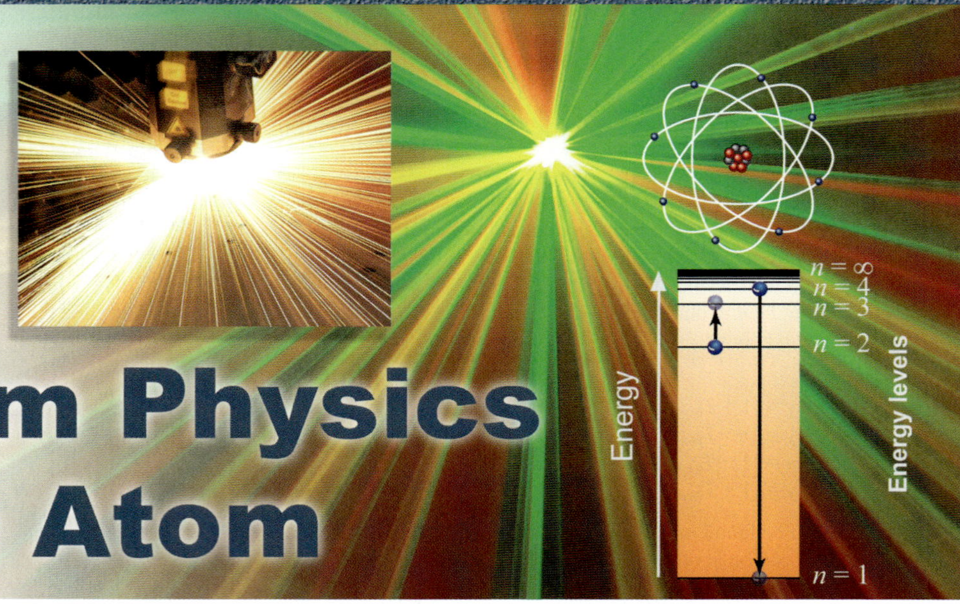


In 1942 science-fiction writer Isaac Asimov published the first of the Foundation series stories in which he described a blaster that projected a beam so powerful it could slice steel as if it were butter. Not only does such a machine exist today, but there are thousands of them that cut steel plates into parts for other machines every hour of every day.

The *laser* took 50 years to make it from science fiction to science fact and today there are lasers that read movies from discs in your DVD player, lasers in supermarket price scanners, lasers used in surgery to correct eye problems, and lasers that transmit millions of gigabytes of Internet data worldwide over thousands of miles of optical fiber cables. Lasers are used to align machines, as rangefinders to measure distances, as cutting guides for power tools, and in computer printers. Lasers are one of thousands of examples of how physics touches your life.

Today, *laser* is a word found in the dictionary, but it was originally an acronym for Light Amplification by Stimulated Emission of Radiation coined in 1957. Lasers rely on *quantum* properties of certain types of atoms to absorb light of specific energy and hold it for a short time. By trapping these energized atoms between parallel mirrors, the light *emission* produced by one atom *stimulates* a matching emission from another atom, *amplifying* the original light. Bouncing back and forth between mirrors a trillion times per second, the pure light of a laser builds up almost instantly. A small hole in one mirror lets a fraction of the light out to make the beam.

The first lasers used crystals of pure ruby to make light. Today's red, blue, and green visible-light lasers use semiconductors, similar to computer chips, to do the same thing at a very low cost. A low-power laser pointer can be bought at any office supply store for $10.

Image credit: J. Stein / Keck Observatory

Researchers today might use a laser to align optical elements, to measure distances with fine precision, or to search for gravitational waves. Researchers at an observatory can use lasers to trace the Earth's atmosphere, allowing astronomers to produce sharp images of stars and galaxies by removing distortions caused by the atmosphere. All of this has become possible because we have a deep understanding of the structure of the atom.

Chapter 26

# Chapter study guide

## Chapter summary

By the beginning of the 20th century, scientists knew there were elements or different kinds of atoms, but they did not know the structure of the atom. In a few short decades, the main properties of the atom and its elementary particles were discovered. Fundamental to these discoveries was learning about the connection between the changes in atomic energy levels and the emission or absorption of photons by the atom. These scientific advances indicated, however, that there were very strange properties in the quantum theory underlying the atom: the probabilistic nature of the atom, the uncertainty principle, wave–particle duality of matter, and particles that carry momentum but no mass. These ideas were so unexpected and weird that even Albert Einstein resisted them at first. In this chapter you will learn about our modern understanding of the atom and the quantum theory.

## Learning objectives

By the end of this chapter you should be able to

- describe the structure of the atom and solve isotope problems;
- describe evidence for a dense and positively charged nucleus;
- describe evidence for the Bohr model of the hydrogen atom;
- describe emission and absorption of a photon by an atom and calculate the photon's properties;
- describe how to identify elements using spectroscopy; and
- summarize the main features of the quantum theory, including quantization, uncertainty, probability, and wave–particle duality.

## Investigations

26A: Rutherford scattering experiment
26B: Phosphorescence
26C: Energy levels of the hydrogen atom
26D: Identifying elements using spectroscopy
Design an infrared pulse monitor

## Chapter index



## Important relationships

$A = Z + N$  $E = hf$  $\lambda = \frac{h}{p}$  $\Delta p \Delta x \geq \frac{h}{4\pi}$  $\Delta E \Delta t \geq \frac{h}{4\pi}$

## Vocabulary

| | | | |
|---|---|---|---|
| phosphorescence | atom | electron | proton |
| neutron | atomic number | elementary charge | nucleus |
| isotopes | mass number | orbit | Bohr model |
| quantum number | quantized | energy levels | ground state |
| electron volt (eV) | excited state | absorption | emission |
| scattering | spectral line | line spectrum | continuum spectrum |
| spectrograph | Pauli exclusion principle | uncertainty principle | wave function |
| stimulated emission | laser | spontaneous emission | |

## 26.1 - Structure of the atom

Greek philosophers first proposed the idea of atoms more than 2,500 years ago. It seemed logical that if you kept grinding up matter into smaller bits you should eventually reach a limit as to the smallest *possible* bit of matter. This they named *atomo* (ἁτομο), meaning indivisible. Leucippus's and Democritus's intellectual deduction of the atom's existence was far beyond their ability to prove by experiment. For the next 2,300 years few believed in atoms and the idea remained a minor curiosity. Today the fact that matter *is* made of atoms is a central theme in our understanding of the universe. We also know that atoms are not indivisible but are themselves composed of even smaller particles—protons, neutrons, and electrons.

### The discovery of atomic structure

**Historical concept of atoms**

The modern theory of the **atom** dates to John Dalton's 1805 hypothesis that each element is a different kind of atom and that atoms combine together to form *compounds*. Dalton suggested that atoms of different elements have different masses and sizes. By 1867, the Russian chemist Dmitri Mendeleev had created the first version of the periodic table, which organized the known data about elemental masses and grouped elements by similar properties. Mendeleev's work was a clue that atoms have an internal *structure* that is responsible for the different observable properties of the elements.

**Mendeleev's first periodic table**

ОПЫТЪ СИСТЕМЫ ЭЛЕМЕНТОВЪ.

ОСНОВАННОЙ НА ИХЪ АТОМНОМЪ ВѢСѢ И ХИМИЧЕСКОМЪ СХОДСТВѢ.

| | | | | | |
|---|---|---|---|---|---|
| | | | Ti=50 | Zr=90 | ?=180. |
| | | | V=51 | Nb=94 | Ta=182. |
| | | | Cr=52 | Mo=96 | W=186. |
| | | | Mn=55 | Rh=104,4 | Pt=197,1. |
| | | | Fe=56 | Rn=104,4 | Ir=198. |
| | | | Ni=Co=59 | Pl=106,6 | O-=199. |
| H=1 | | | Cu=63,4 | Ag=108 | Hg=200. |
| | Be=9,4 | Mg=24 | Zn=65,2 | Cd=112 | |
| | B=11 | Al=27,4 | ?=68 | Ur=116 | Au=197? |
| | C=12 | Si=28 | ?=70 | Sn=118 | |
| | N=14 | P=31 | As=75 | Sb=122 | Bi=210? |
| | O=16 | S=32 | Se=79,4 | Te=128? | |
| | F=19 | Cl=35,5 | Br=80 | I=127 | |
| Li=7 | Na=23 | K=39 | Rb=85,4 | Cs=133 | Tl=204. |
| | | Ca=40 | Sr=87,6 | Ba=137 | Pb=207. |
| | | ?=45 | Ce=92 | | |
| | | ?Er=56 | La=94 | | |
| | | ?Yt=60 | Di=95 | | |
| | | ?In=75,6 | Th=118? | | |

Д. Менделѣевъ

**Building blocks of atoms**

**J. J. Thomson**

The first confirmation of structure in atoms came in 1897 when British physicist J. J. Thomson demonstrated that cathode rays were made of negatively charged particles emitted from matter and were much smaller than atoms. He hypothesized that *electrons* came from inside the atom. By 1909, Ernest Rutherford of England and New Zealand showed experimentally that nearly all of the mass of the atom is concentrated in a small, dense, central nucleus containing positively charged particles, which he called protons.

**Electrons, protons, and neutrons**

Today, we have a fairly good understanding of the structure of atoms. Atoms contain three particles called electrons, protons, and neutrons. The **electron** is the lightest of the three and is a small, fast-moving particle with negative electric charge. The **proton** has positive electric charge and 1,836 times more mass than the electron. The **neutron** has slightly more mass than the proton but has zero electric charge. The heavy protons and neutrons are located in the nucleus of the atom, while the light electrons can be thought of as whizzing around the outer part of the atom. The nucleus of the atom is so small that, if the nucleus were the size of a grain of rice, the whole atom would be the size of two football fields. That is a small nucleus, indeed!

**Atom is composed of a small, dense nucleus surrounded by electrons**

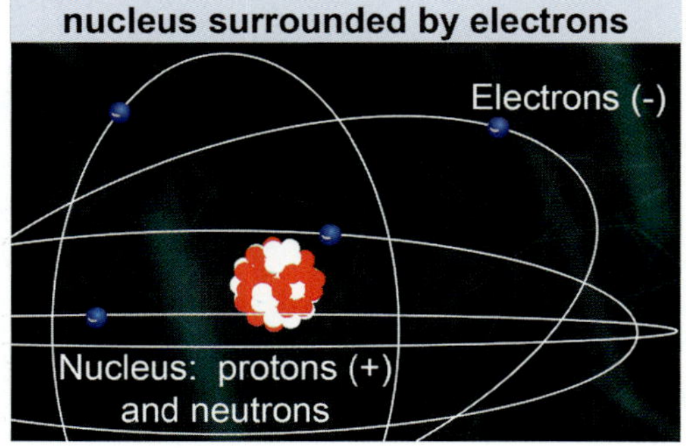

### Elementary particles

The building blocks of atoms are protons, electrons, and neutrons. The electron is the smallest and least massive, being 1/1,836 the mass of the proton. The proton and neutron have nearly the same mass.

**Elementary particles**

| | Mass | Charge |
|---|---|---|
| Proton | $1.673\times10^{-27}$ kg | $+1.602\times10^{-19}$ C |
| Neutron | $1.675\times10^{-27}$ kg | 0 |
| Electron | $9.11\times10^{-31}$ kg | $-1.602\times10^{-19}$ C |

### The elementary charge

The electron and proton have exactly equal and opposite electric charges. The charge of the electron is $-1.602\times10^{-19}$ coulomb (C) and the charge on the proton is $+1.602\times10^{-19}$ C. The value $e = 1.602\times10^{-19}$ C is called the **elementary charge** and is represented by a lower-case $e$. *Electric charge only occurs in multiples of the elementary charge.* You can have a charge of $-e$, 0, $e$, $5e$, or $10^{20}e$, but nature does not allow charges of $1.5e$, $-3.3e$, or any noninteger multiple of the elementary charge. The neutron has zero electric charge but is important for another force, as we will see on page 761.

### How are atoms different?

*All atoms of the same element have the same number of protons.* Atoms of different elements contain different numbers of protons and electrons. For example, every atom of hydrogen has one proton and one electron. Every atom of helium has two protons and two electrons. The number of protons is called the **atomic number**. The atomic number of carbon is 6, telling you that every atom of carbon has six protons. The lightest element is hydrogen with 1 proton per atom and the heaviest naturally occurring element is uranium with 92 protons per atom.

| Element | Protons |
|---|---|
| Hydrogen | 1 |
| Helium | 2 |
| Lithium | 3 |
| Beryllium | 4 |
| Boron | 5 |
| Carbon | 6 |
| Nitrogen | 7 |
| Oxygen | 8 |

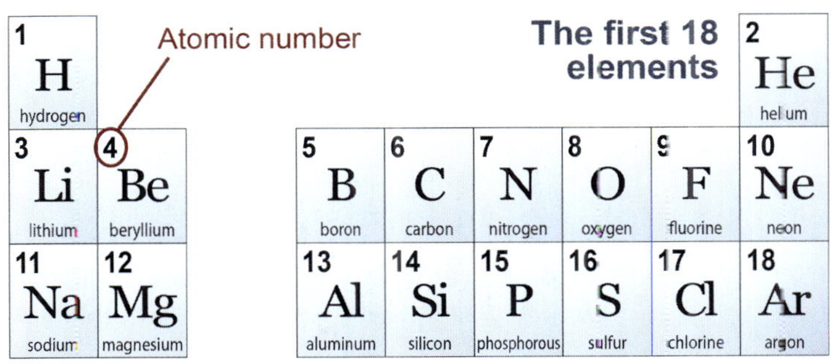


### Why don't we normally experience charge?

*Neutral atoms contain an equal number of protons and electrons.* For example, a carbon atom has six protons and six electrons. The elementary charge on the proton and electron are *exactly* the same, but opposite in sign, and so they cancel each other out and the *total charge of an atom is exactly zero.* Atoms are electrically *neutral*, which means that their net charge is zero. The reason we do not ordinarily feel electrical forces in the macroscopic world is not because matter does not contain charge, but because neutral atoms contain the same amount of positive and negative charges.

**Carbon atom has a net charge of zero.**

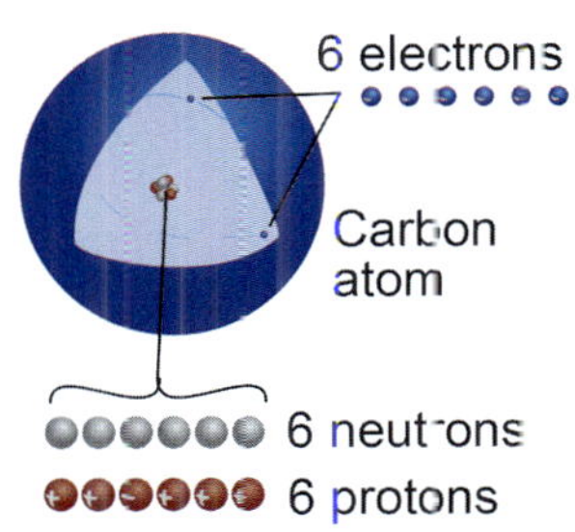

## Discovery of the nucleus

**How are atoms arranged inside?**

By 1900 it was known that atoms contained tiny negative electrons. Atoms were electrically neutral, and therefore they had to contain an equal amount of positive charge. But it was a complete mystery how the electrons and the positive charge were arranged inside the atom. The prevailing theory, nicknamed the "plum pudding model," supposed the electrons were sprinkled within a uniform positive substance, like raisins in an English plum pudding.

**"Plum pudding" model of the atom**

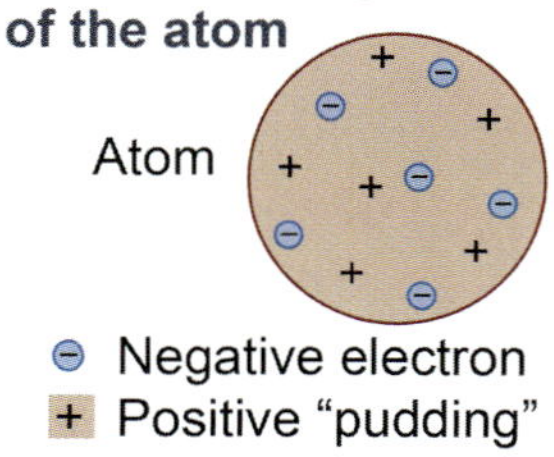


**The crucial experiment**

In 1909, Ernest Rutherford and Hans Geiger, along with undergraduate student Ernest Marsden, did the crucial experiment in which they shot millions of *alpha particles*—which were already known to have positive charge—at a thin gold foil. By observing how many alpha particles were scattered at different angles they expected to shed some light on how positive and negative charges were arranged inside the atom. Rutherford expected to find that most of the alpha particles were deflected by very small angles as they "plowed through" the positive "pudding" inside. Instead, they found something wholly unexpected!

1. Virtually all the alpha particles passed through completely unaffected, as if they had somehow completely *missed* every atom.
2. A very few alpha particles bounced off at large angles and some even bounced backward!

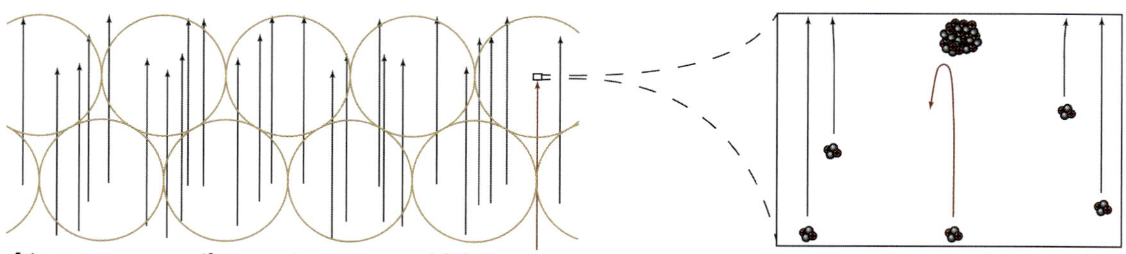

**Atoms are mostly empty space,** which is why virtually all the alpha particles hit nothing at all.

**Positive charge and mass are concentrated in a tiny dense nucleus.**

**What did the observations mean?**

From the first observation, the team deduced that most of the space inside an atom must be *empty.* The second observation could only be explained if nearly all an atom's mass, and all of the positive charge, were concentrated in a tiny volume, which they called the *nucleus.* Only a massive positive charge could have bounced a fast alpha particle straight backward. From the ratio of particles that "missed" to ones that "hit" the nucleus Rutherford was able to estimate that the size of the nucleus was 100,000 times smaller than that of the atom!

**The nucleus**

Inside an atom, neutrons and protons are grouped in a tiny **nucleus** at the center. The nucleus is extremely small. If the atom were the size of your classroom, the nucleus would be smaller than a grain of sand in the center. Because electrons are so light, 99.97% of an atom's mass is concentrated in the nucleus. Only 0.03% is in the electrons zipping around the mostly empty space outside the nucleus.

# Investigation 26A

## Rutherford scattering experiment

**Essential questions** *How do we know that the atom contains a small, positively charged nucleus containing nearly all the atom's mass?*

Interactive simulation

How did Ernest Rutherford determine that the nucleus of the atom was small, massive, and positively charged? In this interactive simulation, you will bombard a nucleus with alpha particles and watch their trajectories. You will have the opportunity to expand on Rutherford's experiment by using different masses of target nuclei.

### Background on Rutherford scattering

Rutherford, in his famous experiment, shot alpha (α) particles at a thin foil of gold atoms as a way of probing the *internal structure* of the atom. An alpha particle is the nucleus of a helium atom, consisting of two protons and two neutrons. Rutherford observed that most of the alpha particles passed through the foil unaffected. But some of the alpha particles were deflected by the gold atoms, which he interpreted as evidence for a dense, massive, and positively charged nucleus of the atom. How did his experiment work?

### Simulation of Rutherford scattering

1. Choose gold ($Z = 79$) as the target nucleus.
2. Press Run to shoot α particles at the target nuclei of the gold foil. Allow the simulation to run for some time to collect sufficient numbers of deflected particles.
3. Enter the number of undeflected and deflected particles (partially and strongly) in Table 1.
4. Determine the percent of particles which were deflected and undeflected and enter the data in Table 2.
5. Change the target nucleus to Iron ($Z = 26$) and repeat the experiment.

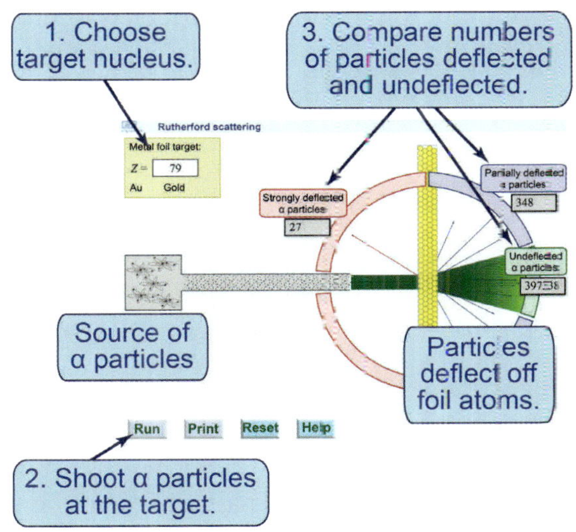


**Why are α particles deflected by the target nuclei?**

a. *What fraction of alpha particles are deflected by a gold nucleus? What evidence supports your conclusion?*

b. *What happens to alpha particles that pass near the nucleus? Those that pass far from the nucleus? Why?*

c. *Change the number of protons in the target nucleus. How do the trajectories of the alpha particles change? Why?*

d. *Use the evidence from this simulation to argue that the nucleus of the atom must be (i) small, (ii) massive, and (iii) positively charged.*

Table 1: Frequency of deflected vs. undeflected particles

| | Gold (Z = 79) | Iron (Z = 26) |
|---|---|---|
| Deflected | | |
| Undeflected | | |
| Total | | |

Table 2: Relative frequency of deflected vs. undeflected particles

| | Gold (Z = 79) | Iron (Z = 26) |
|---|---|---|
| Deflected | | |
| Undeflected | | |
| Total | | |

## Forces in the atom

**Structure and size**

All atoms have an internal structure of electrons loosely spread around a vast, mostly empty volume with a tiny dense nucleus at the center. The volume of the atom is defined by the *electron cloud,* which is roughly $10^{-10}$ m in diameter. The nucleus is much smaller, of the order of $10^{-15}$ m in diameter. Atoms of different elements vary somewhat in size but by less than you might think. More than 99.9% of the mass is in the nucleus. A uranium atom has 240 times the mass of a hydrogen atom and yet its size is only about three times larger.

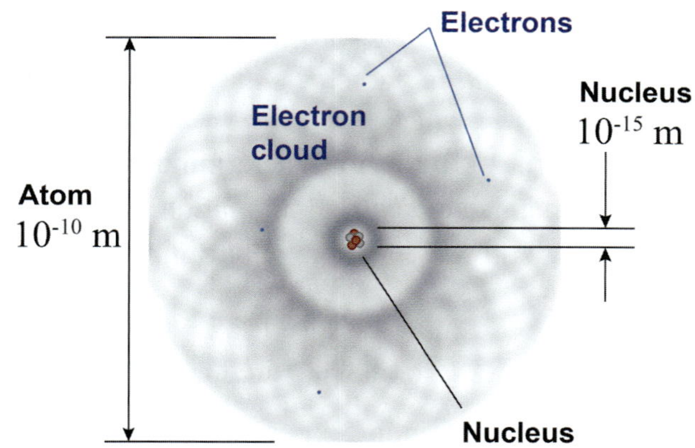


**What holds atoms together?**

The negatively charged electrons in the cloud are attracted to the positive charge in the nucleus. The force of attraction (from Coulomb's law) between the protons and electrons binds the electrons to the nucleus. The electrons do not "fall into" the nucleus because they have kinetic energy and momentum. A useful analogy is Earth orbiting the Sun. The Sun's gravitational force pulls Earth inward, but the planet's kinetic energy and momentum keeps the Earth on an orbit around the Sun. A similar balance between attraction and momentum keeps electrons out of the nucleus and determines the size of an atom.

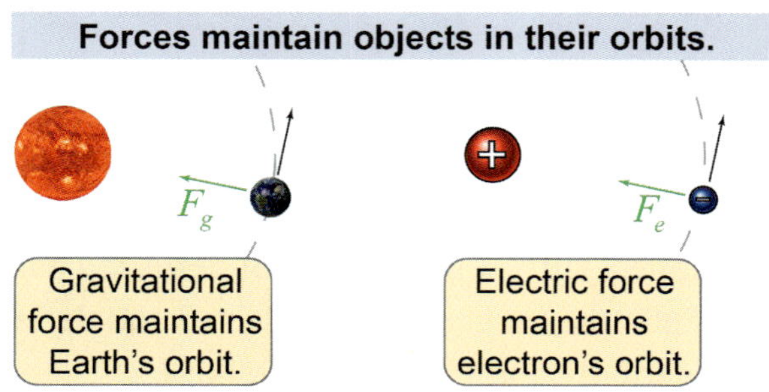


**How do atomic forces affect us?**

Gravity may be the most obvious force in everyday life, but it is the electric force that holds matter together beneath the surface. Electric forces also bond atoms to other atoms to create compounds, such as *salt.* Salt is a compound formed by sodium atoms and chlorine atoms bonded together. While electrostatic forces are not as "forcelike" compared to gravity on the macroscopic scale, these forces are what determine all the underlying properties of matter.

**What about gravity in atoms?**

Gravity is intrinsically the weakest of the four known forces of nature. It takes a planet-size large mass to create enough gravity to make a significant force. Electrons, protons, and neutrons have far too little mass for gravity to be significant. The ratio of the gravitational force between an electron and a proton compared to the force of electrostatic attraction is $10^{-40}$. This is a very small number. Gravity is so completely insignificant on the scale of atoms that there are no known atomic effects caused by gravity. As a result, we do not have a workable theory for gravity on small scales and this is a major unsolved mystery in physics.

**Why don't the protons in the nucleus repel each other apart?**

The nucleus contains both protons and neutrons. Since protons are positively charged and neutrons do not have any charge, why does the nucleus stay together? The electrostatic repulsion between two protons is enormously strong and should tear the nucleus apart instantly. The answer is that there is another type of force that is attractive and even larger than the electrostatic force. This attractive force is what keeps the nucleus together and it is called the strong nuclear force.

**Nucleus**

A nucleus may contain many protons, all with positive charge.

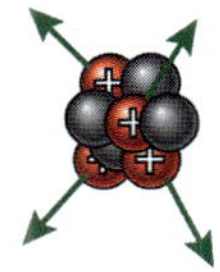

The protons all repel each other through electrostatic forces.

**What holds the nucleus together?**

**Helium nucleus**
2 protons
2 neutrons

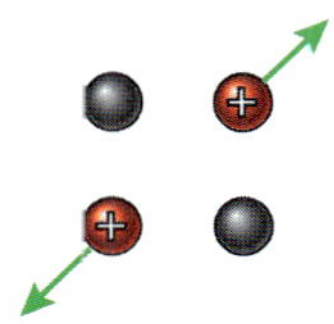

**Electrostatic forces**

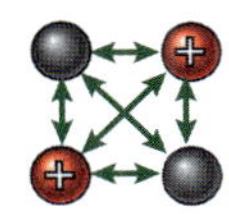

**Strong nuclear forces**

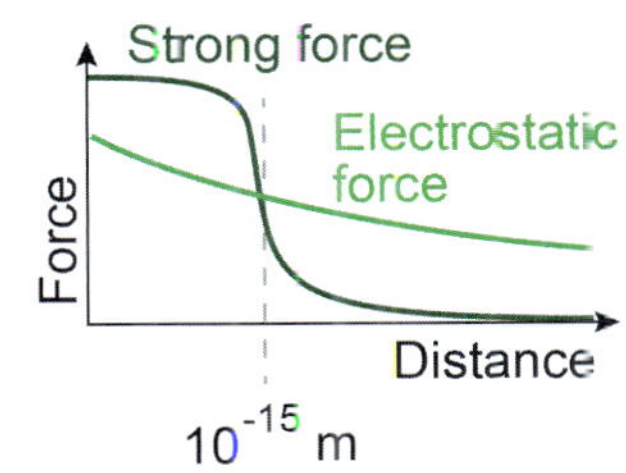


**Strong nuclear force**

The strong nuclear force attracts every proton or neutron to every other proton or neutron regardless of electric charge. The strong force binds together the protons and neutrons in the nucleus with such a great attractive force that enormous amounts of energy are involved in any process that changes the nucleus.

**Short-range force**

The strong force is peculiar in that it falls off in strength very quickly at distances greater than $10^{-15}$ m. For this reason the strong nuclear force is called a *short-range force*. While the strong nuclear force is very powerful within the nucleus, its short-range nature means it is much weaker than the electrostatic force over the greater volume of an atom outside the nucleus.

**Are there other forces?**

Physicists know of four fundamental forces in nature that govern the behavior of everything from the motion of planets to the motion of atoms. These forces are gravity, electromagnetism, the strong nuclear force, and the *weak nuclear force*. The strong force affects protons and neutrons but has no effect on electrons. The *weak nuclear force* affects electrons and other particles called *neutrinos*. Neutrinos are very light, almost massless particles similar in some ways to photons of light. The weak force causes a free neutron outside a nucleus to break up spontaneously into a proton, an electron, and a type of neutrino. This process, called *beta decay*, is one form of *radioactivity* discussed further in Chapter 27.

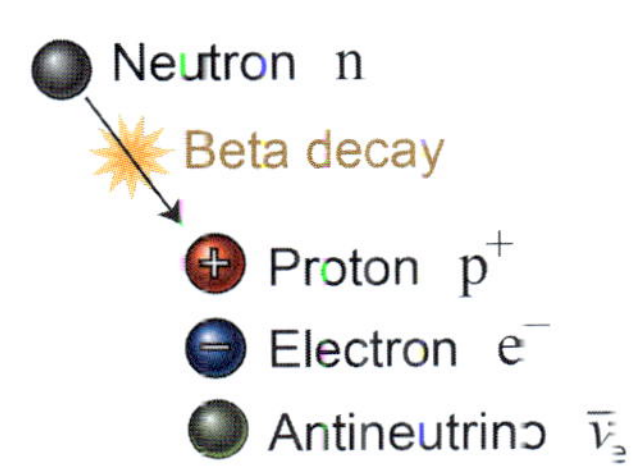


**Beta decay of a neutron**

$$n \rightarrow p^+ + e^- + \bar{\nu}_e$$

## Isotopes and the atomic mass

**The number of neutrons**

All atoms of the same element have the same number of protons, but they may have different numbers of *neutrons*. For example, a random sample of carbon atoms will contain several different *isotopes*. **Isotopes** are atoms that have the same number of protons but different numbers of neutrons. All the isotopes of carbon have six protons but one isotope has six neutrons, another has seven neutrons, and a third isotope has eight neutrons.

**Nuclei of 3 isotopes of carbon**

Carbon-12 Carbon-13 Carbon-14

| Isotope | Protons | Neutrons | Total |
|---|---|---|---|
| Carbon-12 | 6 | 6 | 12 |
| Carbon-13 | 6 | 7 | 13 |
| Carbon-14 | 6 | 8 | 14 |

Mass number

**Mass number**

The **mass number** is the total number of nucleons—protons plus neutrons—in the nucleus. Different isotopes are identified by their different mass numbers. For example, carbon-12 is the isotope of carbon with a mass number of 12, corresponding to six protons and six neutrons. Carbon-13 also has six protons but has seven neutrons for a mass number of 13. Equation (26.1) relates the atomic number $Z$ to the mass number $A$ and neutron number $N$.

Interactive equation

**Atomic mass number**

$$A = Z + N \qquad (26.1)$$

$A$ = atomic mass number (number of nucleons)
$Z$ = atomic number (number of protons or charge of nucleus)
$N$ = number of neutrons

**Writing isotopes**

Isotopes can be written with the full element name and mass number, such as carbon-13, or with the element symbol and a superscript preceding it to show the mass number, such as $^{13}C$. When physicists say "carbon twelve," they are talking about the isotope of carbon with six neutrons and a mass number $A = 12$.

**Isotopes in nature**

A random sample of most elements in nature will contain a mixture of isotopes. This is measured with a *mass spectrometer*, which you learned about on page 566. Its operation is based on the interaction of charge with electric and magnetic fields. For neon the detected signal has a large peak and a small peak. The large peak corresponds to $A = 20$ ($^{20}Ne$) and the small peak corresponds to $A = 22$ ($^{22}Ne$). The size of the peaks is directly related to the relative abundance of the isotopes neon-20 and neon-22, which is approximately 9:1.

**Neon mass spectrum**

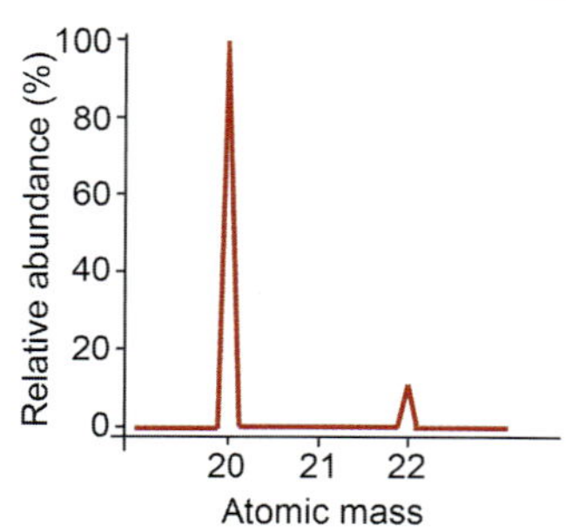


**Radioactive isotopes**

Carbon-12 is a *stable* isotope, which means that an atom of $^{12}C$ will remain $^{12}C$. Other isotopes may be *unstable* because they will spontaneously change into other isotopes. Carbon-14 is an example of an unstable isotope. Carbon-14 is *radioactive* and over time atoms of $^{14}C$ decay into atoms of nitrogen-14. Radioactivity and the element carbon are discussed on page 806.

# Section 1 review

All matter is made up of fundamental building blocks called atoms. The properties of matter are determined by the structure of the atoms within it. Within each atom are elementary particles: protons, neutrons, and electrons. Scattering experiments show that the nucleus is small, massive, and positively charged. The orbit of the electron around the nucleus is maintained by the electric force between the negatively charged electrons and the positively charged protons in the nucleus. All atoms of a particular element have the same atomic number (and hence have the same number of protons and electrons), but they may be different isotopes depending on the number of neutrons in the nucleus. The atomic mass number of an isotope is the sum of the number of protons and neutrons.

| | |
|---|---|
| **Vocabulary words** | **atom, electron, proton, neutron, atomic number, elementary charge, nucleus, isotopes, mass number** |

| | |
|---|---|
| **Key equations** | $A = Z + N$ |

## Review problems and questions

1. Describe the evidence that demonstrates why we know the nucleus of an atom is small, is massive, and contains positive charge.

2. Describe the history of at least three changes in our understanding of the properties of the atom since 1850.

3. What property of atoms determines the overall organization of the periodic table of the elements?

4. Compare the three primary elementary particles (protons, neutrons, and electrons) on the basis of their location within an atom, relative mass, and charge.

5. An element has a mass number $A = 84$ and a neutron number $N = 48$. Which element is this?

6. An atom is made up of 19 protons and 20 neutrons. What is the total mass of this atom?
   a. $6.528\times10^{-29}$ kg
   b. $6.528\times10^{-26}$ kg
   c. $6.528\times10^{-27}$ kg
   d. $3.350\times10^{-26}$ kg

7. The proton is approximately 1,836 times more massive than the electron. If we calculate the mass of an atom by considering only the mass of the protons and the neutrons, what is the percent error introduced in the calculation?

When electricity is passed through a low-pressure neon gas inside a glass tube, the gas glows. Neon signs are usually red, although they can also have green, yellow, or orange colors. But they are never blue! Why? Neon gas has unique *emission lines* in the green, yellow, orange, and red wavelengths of visible light but none at blue wavelengths. If you ever see a glowing blue sign at night you will know one thing: It's not a *neon* sign. Some other gas is inside the tube!

**Neon lights and spectrum**

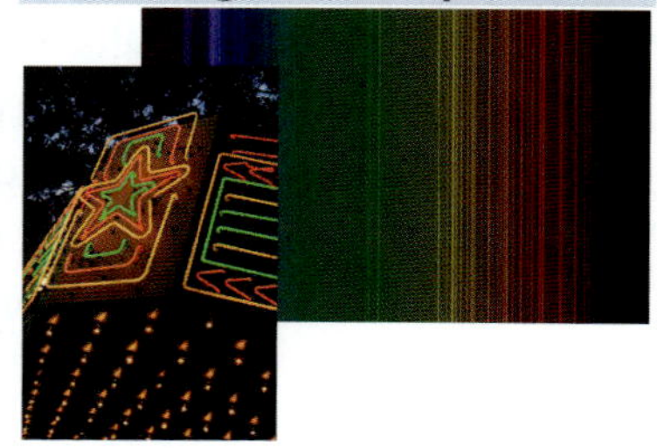

### Spectrum of hydrogen and the Bohr atom

**The hydrogen spectrum and Balmer's magic formula**

When electricity is passed through hydrogen gas, the gas glows and gives off light. The light, however, is *not* like the light given off by an incandescent light bulb. Instead of a smooth rainbow of colors, hydrogen gas gives off a few *very specific* colors and nothing in between. Other elements were observed to give off different patterns of colors. The pattern, called a *spectrum*, is unique for each element. Why this occurred was a mystery. In 1885, a Swiss school teacher named Johann Balmer discovered that the wavelengths of the light in the hydrogen spectrum obeyed a precise mathematical relationship, now known as Balmer's formula.

**Balmer's magic formula**

| $n$ | $\lambda$ (nm) | |
|---|---|---|
| 3 | 656 | red |
| 4 | 486 | blue-green |
| 5 | 434 | blue-violet |
| 6 | 410 | violet |
| 7 | 397 | ultraviolet |

$$\frac{1}{\lambda} = \left(1.097\times10^{7}\ \text{m}^{-1}\right)\left(\frac{1}{2^2}-\frac{1}{n^2}\right)$$

Rydberg constant

Hydrogen spectrum

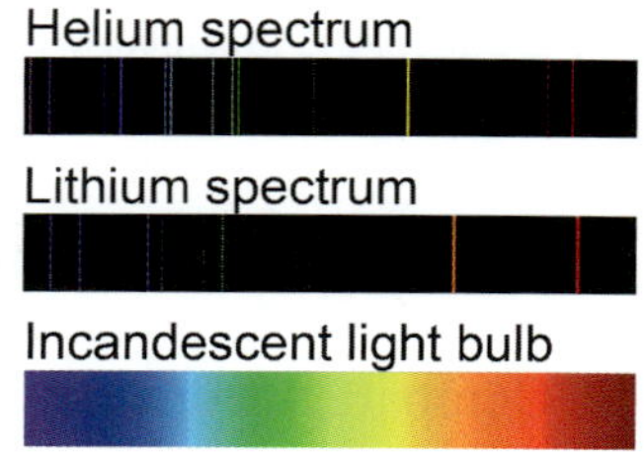


**The riddle of the formula**

Balmer had found the pattern but he could not explain why it occurred. In Balmer's formula, $n$ is an integer, such as 3, 4, or 5. When $n = 3$, the formula predicts a wavelength of 656 nm, which matches exactly the red line in the hydrogen spectrum. What the formula implied was that something inside hydrogen atoms acted like a series of switches. The atomic switches could be set to any integer, such as 3 or 4, but not any number in between, such as 2.5.

**Bohr and the birth of the quantum theory**

Danish physicist Niels Bohr deduced a brilliant explanation for Balmer's formula in 1913. Bohr proposed that the electron makes circular orbits around the nucleus. The electron's energy depends on the radius of the orbit. An electron changing orbits could emit light of specific wavelengths proportional to the energy difference between the orbits. The **Bohr model** marked the beginning of the quantum theory of the atom.

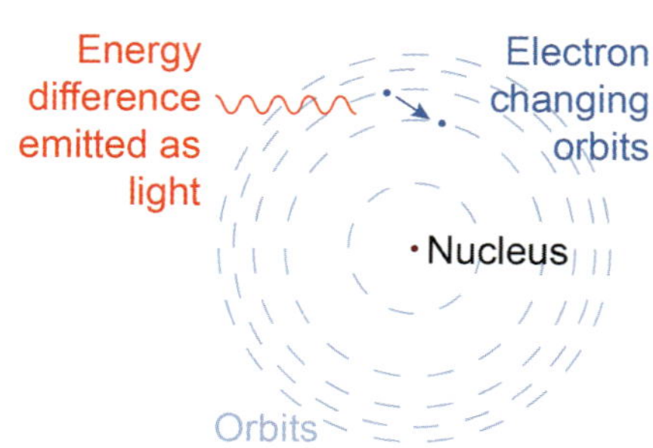

**Electrons act as waves in the atom**

How do electrons "know" which orbits to occupy in an atom? In his 1913 paper, Bohr showed that this behavior would explain the spectrum but he could not say *why* the electron could occupy only those orbits. Ten years later in 1924, French physicist Louis de Broglie made a most unusual proposal. He proposed that there was an intrinsic connection between waves and electrons. When electrons were confined to a small space, de Broglie suggested that they have a wavelength related to energy, like light.

**De Broglie's electron orbit resonances in the Bohr model**

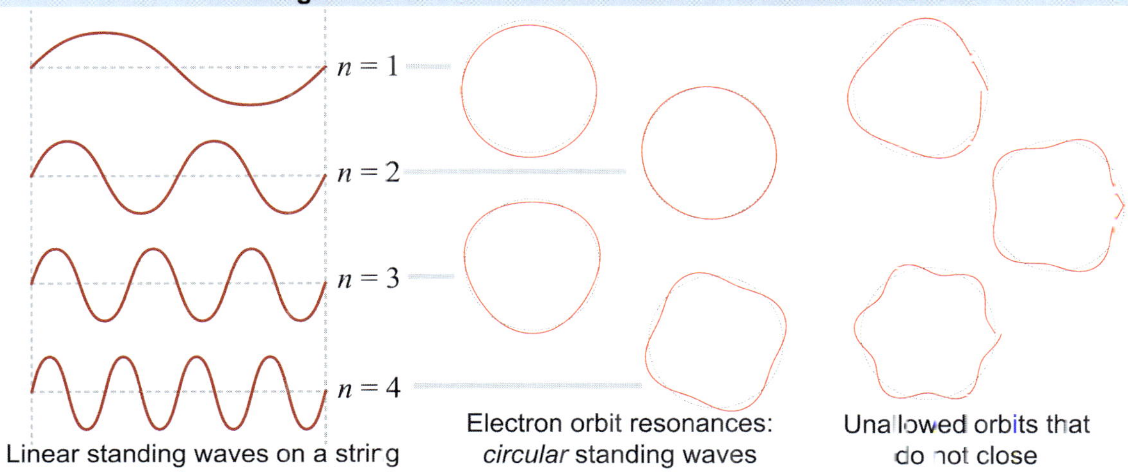


**Electron orbits are resonances**

If electrons had a wavelength then certain orbits would be *resonances* of the electron's "wavelength" as it traveled around the circumference of its orbit. This is similar to the harmonics on a vibrating string. Bohr's fixed orbits would correspond to *standing waves*, occurring when one complete electron wavelength exactly matched the circumference of a circular orbit. Since the circumference of a circular orbit depends on its radius, restricting the circumferences to integer multiples of wavelength forced the electron to be at specific distances from the nucleus. This explained why the electron orbits have discrete energies. The smallest orbit would be the fundamental, the next would be the second harmonic, and so on. Electrons would not be found at any other wavelengths because the electron "wave" would undergo destructive interference with itself.

**Quantum number**

The radius of the electron's orbit in the Bohr model is at integer multiples $n = 1, 2, 3, \ldots$ of the minimum radius. These integer values of $n$ are called the **quantum number** for the electron's orbit.

**Quantum theory**

One of the results of quantum theory is that, when a particle is confined to a small system, such as within an atom, the energy the particle can have is *quantized.* The term **quantized** means that the particle's energy can only assume discrete values determined by the system. An electron in a quantum system may have an energy corresponding to the first quantum level, or the second level, or the third. The electron *cannot*, however, have an energy in between the quantized values. A remarkable insight of Bohr's model of the hydrogen atom is how closely the quantum theories of light and the atom are linked. The atom can change its energy through emission and absorption of single photons of light.

# Investigation 26B Phosphorescence

**Essential questions** *What property of light is required to cause a material to phosphoresce?*

Glow-in-the-dark plastic may seem like a toy, but it has powerful physics embedded into it! The science behind it is called **phosphorescence**. When you expose a phosphorescent material to light it will glow for a while afterward. Phosphorescence is an accessible demonstration of some of the unusual properties of the physics of the atom. In this short investigation you will explore what kind of light is necessary to make the plastic phosphoresce. Why do some colors make it glow while others do not? Here's a hint: Think about what physical quantity is different for red versus green or blue light.

## Making the material phosphoresce

1. For this investigation you will need a dark room, since ambient light may cause the plastic to glow. Pull the window shades and turn off the overhead lights.
2. Turn on the *red* LED lamp and hold it over the plastic material for a few seconds. Then remove the lamp and see if the material phosphoresces—glows in the dark. Record your observations.
3. Repeat with the *green* LED lamp.
4. Repeat with the *blue* LED lamp.

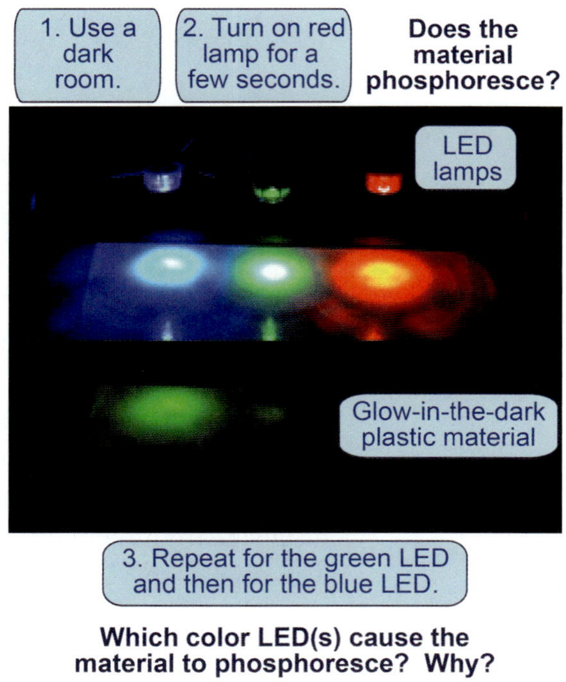


a. *Which color(s) of light caused the material to phosphoresce?*

b. *What is physically different about this color of light compared to a color that does not cause phosphorescence?*

c. *Explain what property of the light you think is required to cause a material to glow in the dark.*

d. *Try turning on the classroom lights. Then turn them off. Does the material glow? Explain why or why not.*

e. *Do you think that infrared light would cause this material to phosphoresce? Why or why not? Describe an experiment using sunlight and a triangular prism that could test your hypothesis.*

f. *Do you think that ultraviolet light would cause this material to phosphoresce? Why or why not? Describe an experiment that could test your hypothesis. What is the light source for your experiment?*

g. *How is phosphorescence different from* fluorescence*? Research the topic to provide justification for your explanation.*

**Electron orbits are energy levels**

Electrons in an atom can only have an energy that matches one of the atom's *energy levels*. **Energy levels** are the allowed energies within a quantum system, such as an atom. In the hydrogen atom, with one electron, each energy level has a different principal quantum number $n$, which can be any integer from one to infinity. The principal quantum number is like the floor number in a tall building. The energy level is analogous to the potential energy of the floor, except that the floors are not equally spaced! As the diagram shows, the spacing between energy levels gets smaller as $n$ gets larger.

**Energy levels of the hydrogen atom**

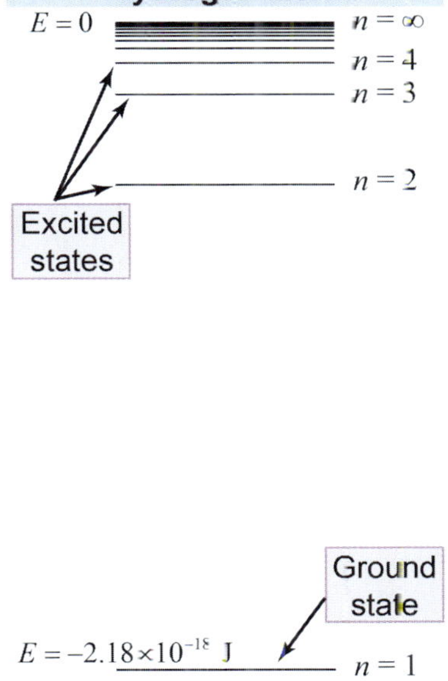


**Atom's energy changes when electron changes orbit**

Atoms change their energy when electrons move up or down between energy levels. For example, an electron at the $n = 1$ energy level might make a transition to the $n = 2$ level. Conservation of energy requires that the atom absorb a photon of light with precisely the same energy gained by the electron. Transitions may also skip levels. For example an electron might go from the 23rd energy level to the 5th energy level. The atom conserves energy by emitting a photon with an energy equal to the energy difference between the 23rd level and the 5th level. An electron can even go from $n = 1$ to $n = \infty$, which corresponds to the electron being removed altogether from the atom!

**Ground state and excited states of the hydrogen atom**

**Energy level transitions**

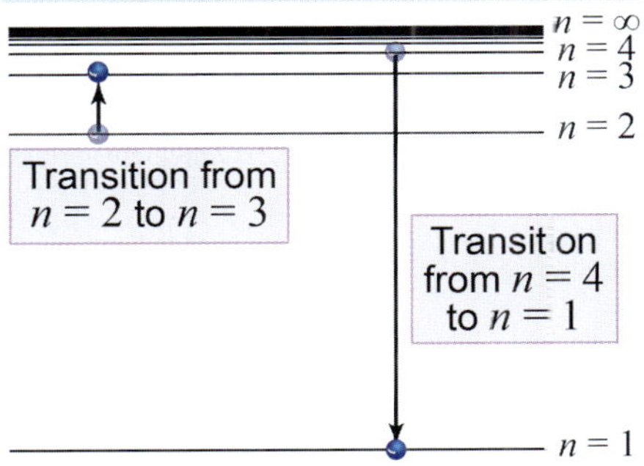


The $n = 1$ state is the *lowest* energy level, called the **ground state**. Atoms in their ground state have the lowest possible energy. The important thing to remember about the ground state is that the electron energy *cannot get any lower*. From the ground state, the electron can only change to a higher energy level. An **excited state** is a configuration of the atom in which one or more electrons are in higher energy levels than they could be. Excited states occur when an atom absorbs the energy of a photon.

**Electron volt as a unit of energy**

The ground state of the hydrogen atom is at an energy of $-2.18\times10^{-18}$ J. A more convenient unit for representing atomic energy levels is the **electron volt (eV)**, which is the amount of energy gained by an electron that moves across one volt. As a result, the electron volt (or eV) is equal to one volt times the charge of the electron, or $1\text{ eV} = 1.602\times10^{-19}$ J. The ground state of the hydrogen atom can be written as $(-2.18\times10^{-18}\text{ J})/(1.602\times10^{-19}\text{ J/eV}) = -13.6$ eV. Writing the ground state energy as $-13.6$ eV avoids having to keep track of all those powers of 10! Whenever working with the electron volt, the important point to remember is that it is a unit of *energy*.

## Emission and absorption spectra

### Light and atomic energy levels

When an atom makes a transition to a lower energy level, where does the excess energy go? Likewise, where does an atom get the energy needed to make a transition to a higher energy level? The answer to both questions involves *light.* Atoms change their internal energy by rearranging electrons among energy levels through emitting or absorbing light.

### Emission of light

If an atom lowers its energy by $\Delta E$, a photon is *emitted* with a frequency given by Planck's relation $\Delta E = hf$. Discrete spectra occur because the atom's energy is "stored" in the arrangement of its electrons. The atom can only lose an amount of energy corresponding to the difference between two energy levels. This restricts the frequency and wavelength of light emitted. For example, a hydrogen atom that drops an electron from the $n = 3$ energy level to $n = 2$ gives off red light with a wavelength of 656 nm, corresponding to the red line in the hydrogen spectrum.

**Emission line is produced when electron transitions to a *lower* energy level.**

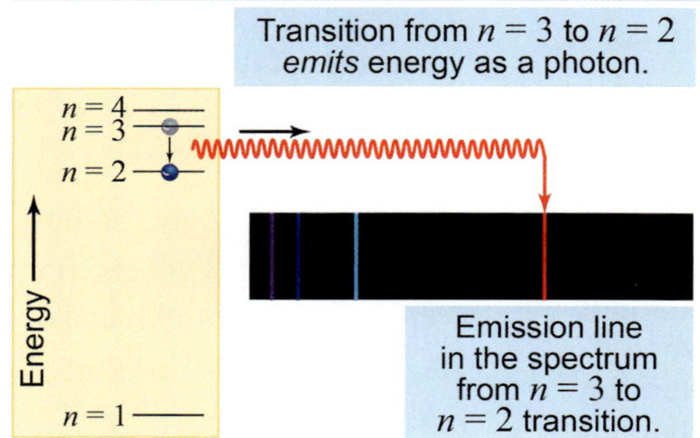


### Absorption of light

An atom gains energy when it absorbs a photon of light. Absorption is also quantized and an atom can *only* absorb light with a photon energy equal to the difference between two energy levels. For an atom to increase its energy by $\Delta E$ it must *absorb* a photon with a frequency given by $\Delta E = hf$. For example, a hydrogen atom absorbs a photon of red light with a wavelength of 656 nm by promoting one electron in the $n = 2$ energy level up to the $n = 3$ level.

**When electron absorbs energy from a photon, it transitions to a *higher* energy level.**

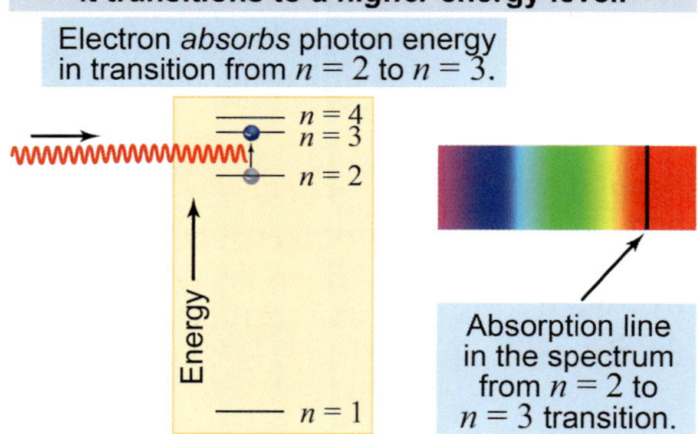


### Scattering of light

**Scattering of light**

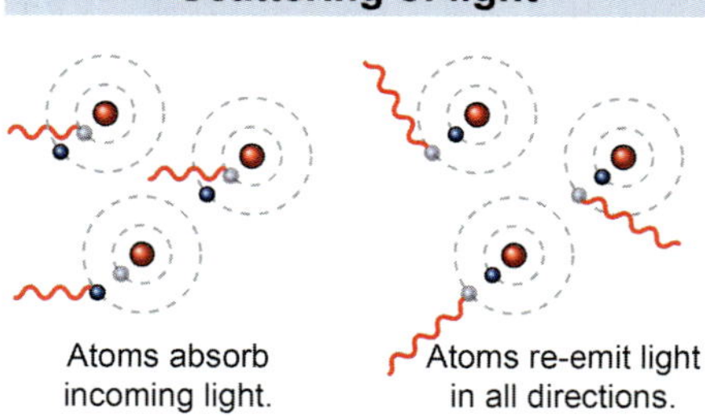


Imagine that an atom absorbs a photon of light to raise its energy level, but soon after it emits another photon of light to lower its energy level back again. In what direction does the newly emitted photon travel? The answer is that the photon may be emitted in *any direction.* When an atom emits a photon of light, there is no telling in advance what direction it will be traveling. This curious property of the absorption and re-emission of light by atoms is called **scattering**. You can see scattering of light every day in the blue sky and red sunsets.

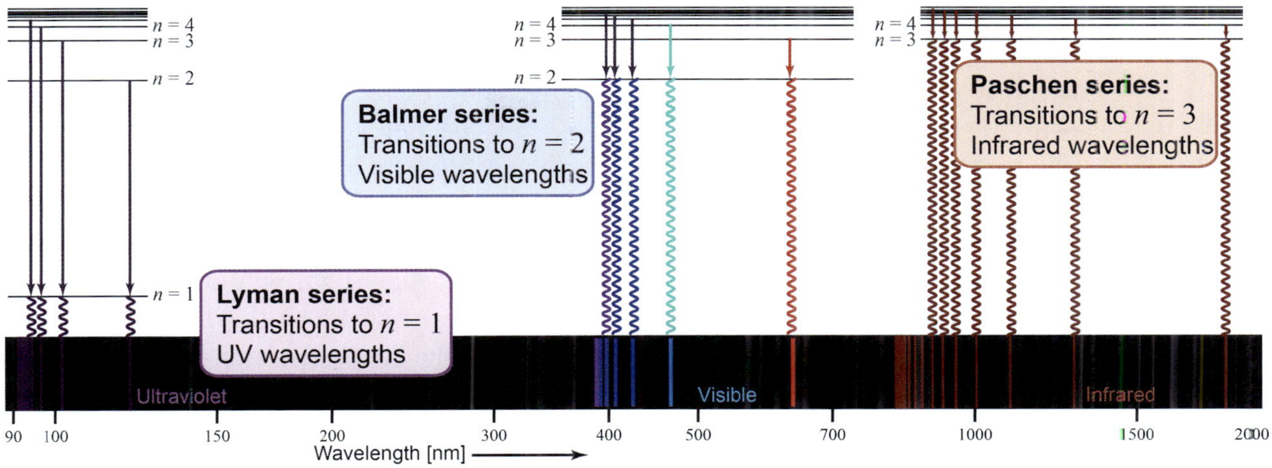


**Spectrum of hydrogen in Bohr's model**

The hydrogen atom offers the simplest example of emission and absorption because it has only one proton in the nucleus and one electron orbiting around it. Some of the unusual features of the quantum theory were accepted remarkably quickly because of the success of Bohr's model in matching the observed wavelengths of the line spectrum for hydrogen.

**Emission of light by the hydrogen atom**

Hydrogen gas emits spectral lines at ultraviolet (UV), visible, and infrared wavelengths. Atomic transitions to and from the $n = 1$ energy level correspond to wavelengths of light in the ultraviolet and are named the Lyman series. Atomic transitions between the $n = 2$ energy level and higher $n$ levels correspond to wavelengths of light in the visible and near-UV and are named the Balmer series. Atomic transitions between $n = 3$ and higher quantum numbers correspond to photon energies in the infrared. The lines corresponding to the Lyman series have the highest energy, followed by the lines of the Balmer series.

**Example problem**

A hydrogen atom has an electron at an energy level of $-5.45\times10^{-19}$ J. It suddenly makes a transition to the $-2.18\times10^{-18}$ J energy level.
(a) Did it emit or absorb a photon? (b) What is the energy of the photon?
(c) What is the photon's wavelength?
(d) What part of the electromagnetic spectrum does this wavelength correspond to?

*Asked:* (a) whether it emits or absorbs a photon; (b) energy of the photon; (c) wavelength of the photon; (d) region of the electromagnetic spectrum

*Given:* initial energy level of $-5.45\times10^{-19}$ J, final energy level of $-2.18\times10^{-18}$ J

*Relationships:* Planck equation $E = hf$, speed of light equation $c = f\lambda$

*Solution:* (a) The atom *lost* energy; therefore, it emitted a photon.
(b) The atom's change in energy equals the photon's energy:

$$E_{photon} = -5.45 \times 10^{-19}\text{ J} - (-2.18 \times 10^{-18}\text{ J}) = 1.64 \times 10^{-18}\text{ J}$$

(c) Substitute the speed of light equation into the Planck equation to get $\Delta E = hc/\lambda$. Solving for $\lambda$ gives

$$\lambda = \frac{hc}{\Delta E} = \frac{(6.63 \times 10^{-34}\text{ J s})(3 \times 10^{8}\text{ m/s})}{-5.45 \times 10^{-19}\text{ J} - (-2.18 \times 10^{-18}\text{ J})} = 121.7\text{ nm}$$

(d) This wavelength is in the *ultraviolet* on the figure on page 645.

*Answer:* (a) It *emitted* the photon; (b) $1.64\times10^{-18}$ J; (c) 121.7 nm; and (d) ultraviolet portion of the EM spectrum.

# Investigation 26C Energy levels of the hydrogen atom

**Essential questions** *What wavelengths of light are absorbed and/or emitted by a hydrogen atom? How are these wavelengths related to quantum phenomena?*

Interactive simulation

In Bohr's model of the hydrogen atom, the energy levels of the electron are *quantized*, meaning that they can only take certain, discrete values. For the electron to change energy levels, the atom must either absorb or emit light at specific wavelengths or specific energies. These energies correspond to the difference in the energy levels of the electron. In this interactive simulation, you will shine light onto a hydrogen atom and determine the wavelengths of light that can be absorbed—and then re-emitted—by this hydrogen atom. Only certain photon frequencies (or wavelengths) will be absorbed by the hydrogen atom—the rest will pass through the atom untouched!

## Bohr's model of the hydrogen atom

The Bohr model describes a hydrogen atom where only certain energy levels are allowed. To transition between energy levels, the atom must gain or lose the difference in energy between the two levels. If a photon of light with exactly one of these energy differences strikes the hydrogen atom, then the photon may be absorbed—causing a change in the atom's energy level. Bohr's model proved to be successful for calculating the absorption and emission line spectra of hydrogen for transitions between $n = 2$ and higher values of $n$, called the Balmer series.

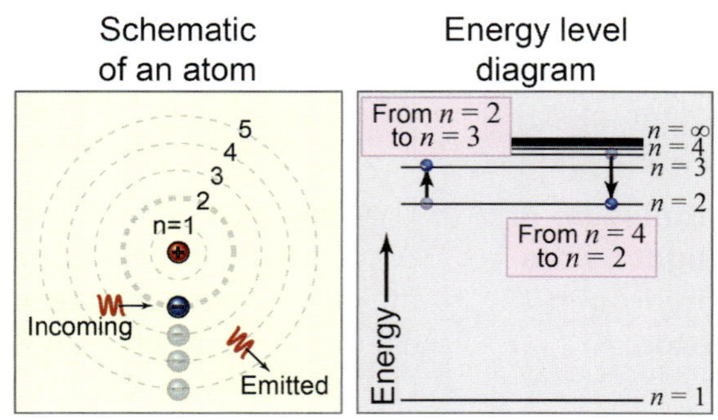


## What wavelengths are absorbed and emitted by the hydrogen atom?

1. Launch the interactive simulation and press the [Run] button.
2. Adjust the wavelength or frequency of the light. If the light has a wavelength corresponding to an energy level transition, then it will be absorbed—and the hydrogen atom will subsequently emit a photon at the same wavelength.
3. Find at least three of the wavelengths that are absorbed by the hydrogen atom.

*a. What happens when the hydrogen atom absorbs a photon? What happens afterward?*

*b. What are the wavelengths of light that the hydrogen atom could absorb? What energy level transitions do these correspond to?*

*c. Are the wavelengths evenly spread throughout the spectrum? Describe their distribution as a function of wavelength and/or color.*

*d. In this simulation, all the incident photons travel in one direction. What direction do the emitted photons travel? Explain why this happens.*

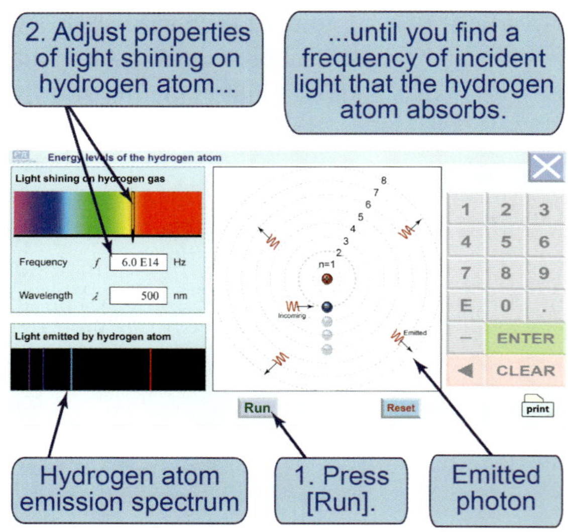


**How many emission lines can you produce with the hydrogen atom?**

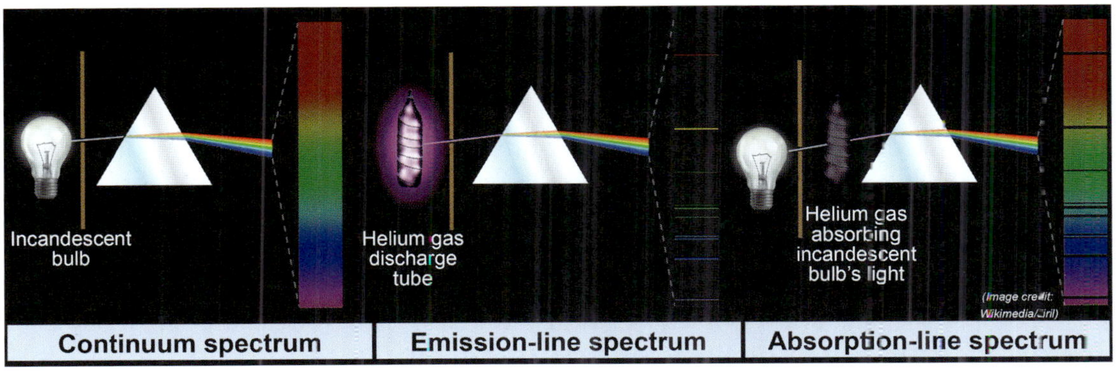


**Continuum spectrum**

When white light passes through a prism the light is dispersed into its individual wavelengths and a *spectrum of light* is created. For the rainbow produced by the prism, even though the light has different colors, there is a smooth or continuous spectrum of light called a **continuum spectrum**.

**Line spectra from atomic energy level transitions**

The light produced by energy level transitions of the atom, however, consists of *discrete*, not continuous, wavelengths. In a **line spectrum**, each individual wavelength of light corresponding to the emission or absorption of light by the atom is called a **spectral line**. When the atoms are emitting light, bright lines appear in this emission spectrum. A discharge tube contains a gas that produces an emission spectrum when a voltage is applied across it. When the atoms are along the line of sight toward a background light source they absorb the light to create an absorption spectrum. An absorption spectrum often appears as black lines on an otherwise continuous spectrum of light. These black lines are called absorption lines.

**Each element has a unique line spectrum**

Every element contains a different number of protons and electrons, so the electrical energy between its charged particles will be different. This results in every atom having a unique signature in its line spectrum. Hydrogen can be distinguished from other elements by comparing their line spectra.

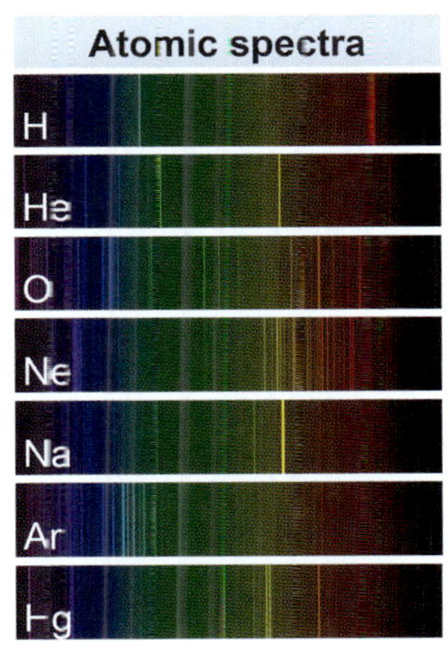


**Spectrograph**

An instrument that disperses light into its constituent wavelengths is called a **spectrograph**. The simplest kind of spectrograph is the prism. Light of any kind—not just white light!—can enter the prism through one side and is dispersed into a spectrum as it exits out the other side.

**Two different representations of emission-line spectra**

Emission-line spectra are often represented with a black background and bright, vertical emission lines, such as in the top panel of the figure on the right. Another method of representing the data is through a line plot, as shown in the bottom panel of the figure. Both of these representations show the same emission lines.

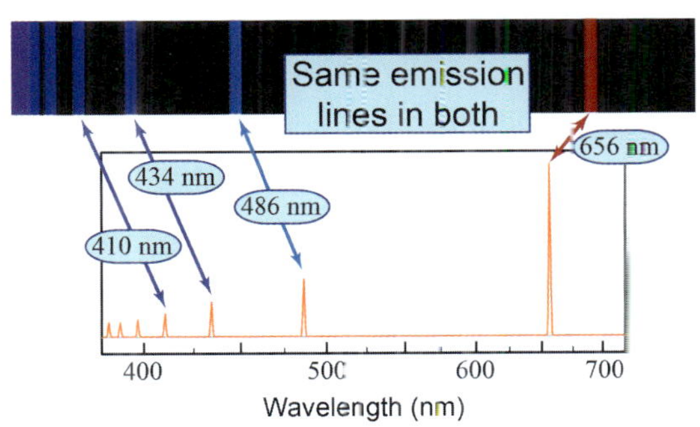

# Investigation 26D Identifying elements using spectroscopy

**Essential questions** *How can you use spectroscopy to identify elements?*

Every atomic element has a different number of protons and electrons, which means that every element has a unique set of energy levels. When an electron makes a transition between energy levels, a photon of light is either emitted or absorbed. Every element's unique set of energy levels has a corresponding unique set of emission and absorption lines—a signature. In this investigation you will identify unknown elements using their spectroscopic signature.

## Part 1: Hand-held visual spectrograph and discharge lamps

*Safety warning: Never point a spectrograph directly at the Sun or any other bright light source. Never touch discharge tubes because they can be very hot.*

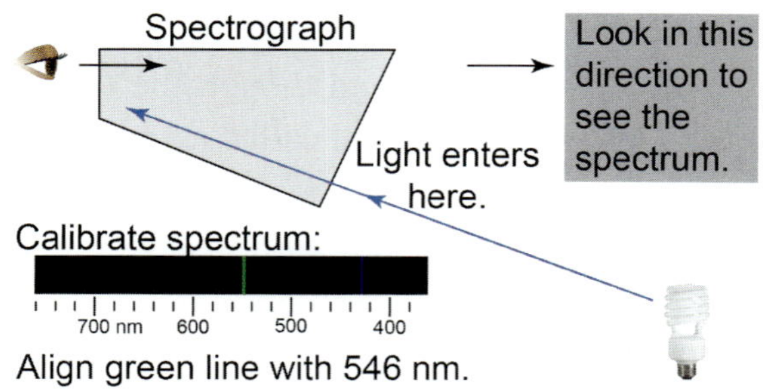


1. Calibrate the hand-held visual spectrograph by pointing it at a fluorescent lamp and adjusting the scale to align the green line with 546 nm.
2. Turn on the power for the discharge lamps.
3. Point the spectrograph at the fluorescent lamp and discharge lamps.

a. *What elements can you identify in the overhead fluorescent lamp? Why?*

b. *Based on the spectra you observe, which of the discharge lamps (H, He, Ne, or Ar) is which?*

## Part 2: Identifying mystery elements in spectra

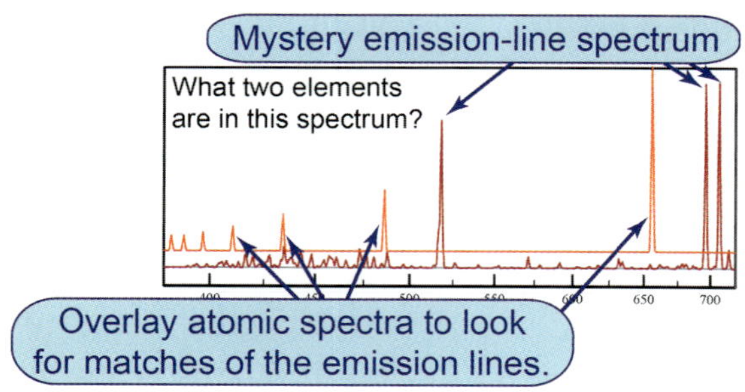


1. The first four sheets of emission-line spectra should be printed on clear transparencies. The other pages should be printed on white paper.
2. Slide the transparencies over the mystery spectra to identify the element(s) they contain.

a. *How did you use the known atomic spectra to identify the unknown ones?*

b. *Describe at least three physical properties that can be used to identify elements.*

## Part 3: Identifying elements in stars and the Sun

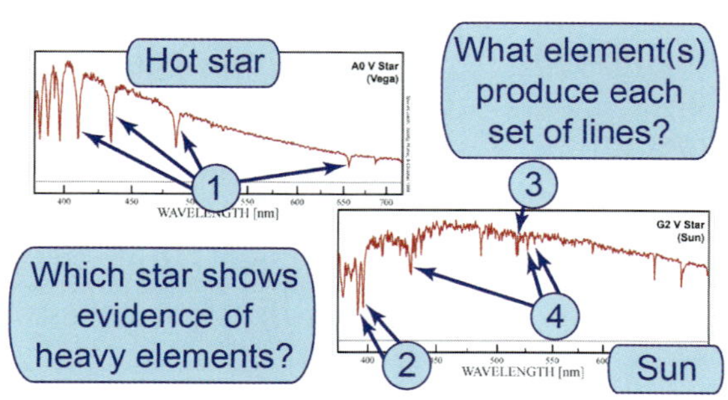


1. Use the atomic spectra to identify elements in the absorption lines of various stars.

a. *How does an absorption-line spectrum differ from an emission-line spectrum?*

b. *Which star(s) show evidence for heavier elements? Why might this happen?*

### What happens in atoms with many electrons?

How are electrons organized in an atom with many electrons? The fact that each element has a discrete spectra is evidence that all atoms have energy levels. The critical clue to understanding multielectron atoms is found in their *chemical behavior*. Atoms combine chemically with other atoms by sharing electrons. For example, water ($H_2O$) contains two atoms of hydrogen and one atom of oxygen chemically combined into a molecule.

### Mendeleev's discovery

Dmitri Mendeleev deduced a repeating or *periodic* pattern in how elements combine. Elements with 3 electrons (Li), 9 electrons (Na), 19 electrons (K), and 37 electrons (Rb) combine with oxygen in 2:1 ratios to make $Li_2O$, $Na_2O$, $K_2O$, and $Rb_2O$. Elements with 2, 10, 18, and 36 electrons are all noble gases (He, Ne, Ar, and Xe) that make no chemical bonds with other elements. Elements with 4 electrons (Be), 12 electrons (Mg), and 20 electrons (Ca) combine with oxygen in a 1:1 ratio to make BeO, MgO, and CaO. This evidence points to some internal structure in the atom that repeats at 2, 10, 18, and 36 electrons.

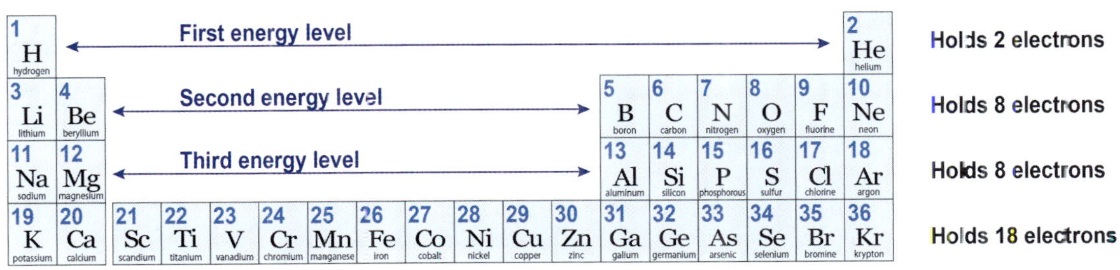


### Energy levels and electrons

Each energy level holds a fixed number of electrons. The lowest energy level can hold two electrons. A third electron must occupy an open state in the second energy level. The second energy level can hold eight electrons. That means an element with 10 electrons completely fills all the quantum states in the first and second energy levels. The 11th electron has to go into the third energy level. The *periodic table* is actually a diagram showing how many electrons fit in each energy level.

### Quantum states

Neils Bohr proposed that electrons in an atom must occupy stable *quantum states*. In physics, a "quantum state" refers to a particular allowed set of values for momentum and energy within a system. Think of a small theater with multiple levels of seating. The lowest level has only two seats—corresponding to two allowed quantum states. The second level has eight seats—corresponding to eight allowed quantum states. The third level has eight more seats, and so on.

### The Pauli exclusion principle

The **Pauli exclusion principle**, deduced by Austrian physicist Wolfgang Pauli in 1925, states that no two electrons can be in the same quantum state in the same atom. Once a quantum state is occupied by an electron, all other electrons are "excluded" from that state. The Pauli exclusion principle, combined with Bohr's quantum states provides the theoretical explanation for the elements. The periodic properties repeat when all the quantum states are filled at one energy level and the next electron starts the next higher energy level.

## Quantum numbers

### Schrödinger and the quantum numbers

Erwin Schrödinger proposed the fundamental equation of quantum mechanics in 1926. When Schrödinger's equation is applied to the atom, the result is a family of different solutions. Each solution corresponds to *one* allowable quantum state for an electron and is characterized by four *quantum numbers.* Each quantum state in the atom has a unique "quantum address" of the four quantum numbers, *n, l, m,* and *s.* The principal quantum number *n* is the one that appears in Balmer's formula.

### What are quantum numbers?

Think of an atom like a multilevel theater where each seat can hold up to one person. The best seats are closest to the stage but there are a limited number. The quantum numbers are the address code for each seat. The principal quantum number *n* tells you the row. The last number, *s,* tells you whether the seat is on the right or left side. The number *l* tells you what section and *m* is a specific seat in that section.

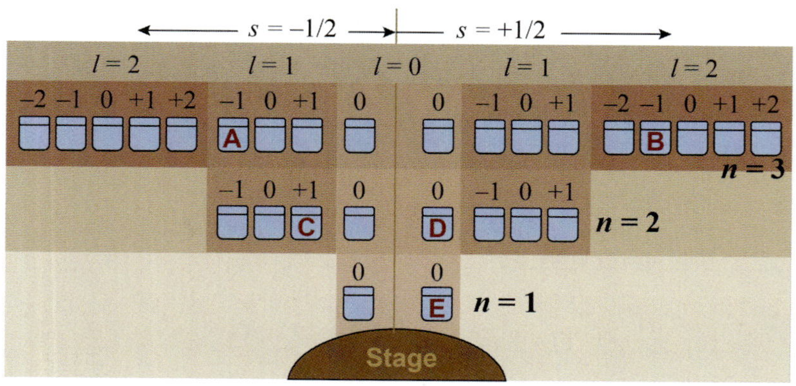


| | Row | Section | Seat | Side |
|---|---|---|---|---|
| | *n* | *l* | *m* | *s* |
| A | 3 | 1 | –1 | –1/2 |
| B | 3 | 2 | –1 | +1/2 |
| C | 2 | 1 | +1 | –1/2 |
| D | 2 | 0 | 0 | +1/2 |
| E | 1 | 0 | 0 | +1/2 |

### Angular momentum and magnetic quantum numbers

Like the seats in our imaginary auditorium, every quantum state in the atom is identified by a unique combination of the four quantum numbers. The orbital angular momentum quantum number, *l*, can only be a positive integer from zero to $n - 1$. For example, if $n = 1$, the only possibility is $l = 0$. If $n = 2$, then *l* can be 0 or 1. The magnetic quantum number, *m,* is an integer that can go from $-l$ to $+l$. For example, if $l = 2$, then *m* can have any of five values between −2 and +2: $m = -2, -1, 0, 1,$ or 2. The last quantum number, *s,* is the electron *spin* angular momentum. An electron can have only one of two different spin states, up (+½) or down (−½).

### Orbital shapes

All the electrons in the atom are attracted to the nucleus but repelled from each other. This dance of attraction and repulsion creates complex three-dimensional shapes for each electron "orbital." The orbital angular momentum and magnetic quantum numbers determine the spatial shape. When $l = 0$ the orbital is spherically symmetric. When $l = 1$ there are three perpendicular states along each of the coordinate axes. The $l = 0$, 1, 2, and 3 states correspond to the *s, p, d,* and *f* orbital notation in chemistry.

**Electron states (orbitals)**

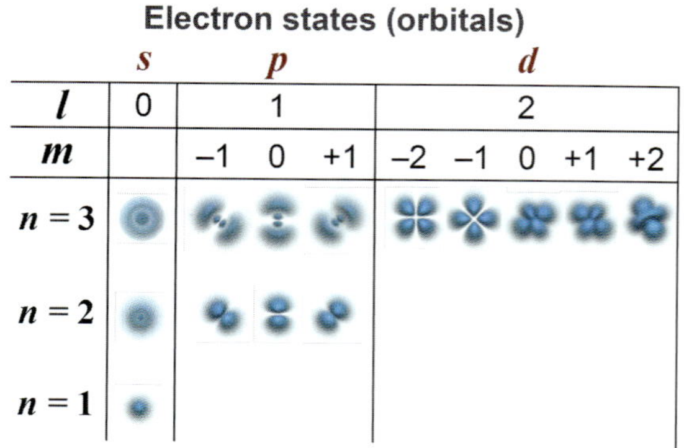

**Design challenge**

*Challenge:* Create an infrared (IR) monitor to detect the human pulse in a finger.
*Materials:* circuit components, breadboard, plastic tube, and foam pieces.
*Performance evaluation:* Measure student's heart rate.
*What is the need or problem that this product addresses? What are the design criteria?* Write your answers in your report.

**Infrared pulse monitors**

Human flesh is partially translucent at IR wavelengths. A pulse monitor has an IR LED that radiates light across the fingertip and a photodetector on the other side that detects light passing through the fingertip. As the heart beats, arterial blood vessels expand and contract, blocking varying amounts of the IR light.

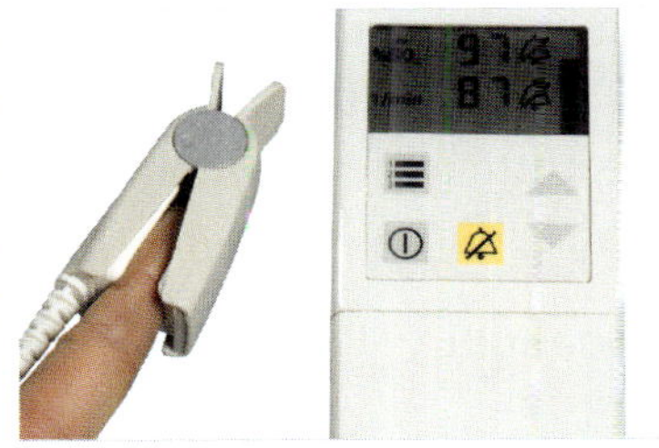

**Build an IR LED circuit**

1. Build the LED lamp and switch portion of the circuit. Test it by verifying that the LED goes on and off when the switch is pressed.
2. Build the QED (IR-emitting diode) portion of the circuit and the second LED lamp (and other 100 Ω resistor) portion of the circuit. The LED should be illuminated.
3. Build the QSD (IR-sensing diode) portion of the circuit. The QSD should be detecting light overhead, which will cause the LED to turn off.
4. Add series resistors and adjust the potentiometer so that the LED illuminates when you cover the sensor with your hand.
5. Point the QED and QSD at each other. Modify the resistance (series resistors and potentiometer) so that pressing the switch turns the sensor's lamp on and off.

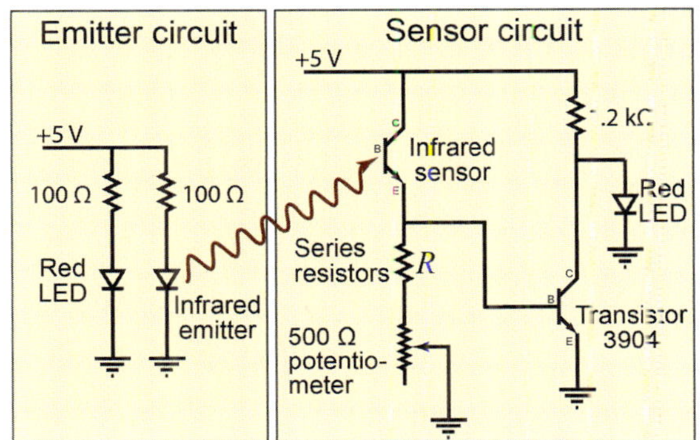


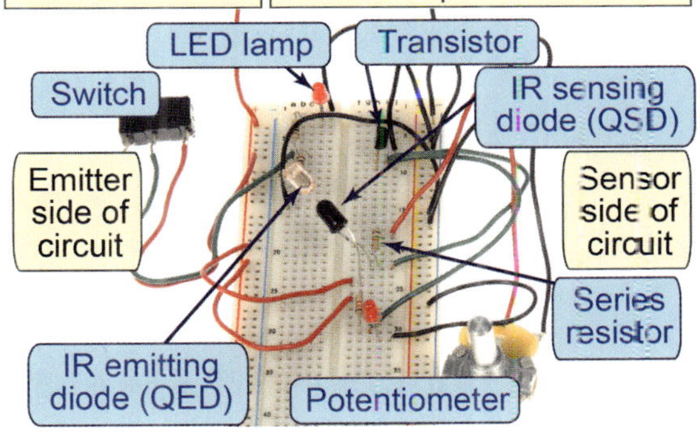


**Modeling the system**

Place your finger between the emitter and sensor and optimize the circuit performance to detect your pulse. (*Hint*: Add one or more resistors in series and adjust the potentiometer.)

**Design**

Create a design based on this circuit for an IR pulse monitor using the tube, foam, duct tape, and/or other common materials.

**Prototype**

Construct the prototype of your IR pulse monitor using the equipment.

**Test**

Measure your pulse. Is your measurement reasonable and easy to take?

**Evaluate**

How well did your design work? Evaluate its performance.

**Revise**

Based on your evaluation, improve your design. Prototype and test it.

# Section 2 review

The Bohr model for the hydrogen atom allowed electrons in atoms only to occupy particular, discrete orbits corresponding to quantized energy levels. Bohr could not explain *why* electrons favored these particular orbits; de Broglie proposed that the electron behaves like a *wave* at the atomic scale, so electron orbits correspond to circular standing waves or resonances. Quantized energy levels provide a direct explanation for absorption and emission of photons by atoms. Emission lines are produced by an atom when an electron makes a transition from a higher to a lower energy level. Absorption lines occur when the electron in an atom absorbs a photon with a specific energy, moving it from a lower to a higher energy level. An absorption-line spectrum appears when atoms absorb particular wavelengths of light from the spectrum of a background light source.

| Vocabulary words | phosphorescence, orbit, Bohr model, quantum number, quantized, energy levels, ground state, electron volt (eV), excited state, absorption, emission, scattering, spectral line, line spectrum, continuum spectrum, spectrograph, Pauli exclusion principle |
|---|---|

## Review problems and questions

1. By using the Bohr model of the atom, should the size of the helium atom be larger or smaller than that of the hydrogen atom? Why?

2. One hydrogen atom has an electron in its $n = 1$ energy level, while another atom has an electron in its $n = 2$ level. Which one requires more energy to move the electron to the $n = 3$ level? To the $n = \infty$ level? Why?

3. What property of atoms produces an emission-line spectrum?

4. The hydrogen atom produces three distinct sets of spectral lines known as the Lyman, Balmer, and Paschen series.

   a. Describe the differences in the light emitted in these three series.
   b. Explain how these differences are related to the energy level transitions of the atom.

5. The electron in an atom changes energy levels, and in the process it emits a photon at a wavelength of 656.3 nm.

   a. Is the electron energy level higher before or after it emits the photon?
   b. In joules, what is the difference in energy between the two energy levels?
   c. What is this energy difference in electron volts?

6. An atom absorbs a photon of light that has a frequency of $10^{14}$ Hz.

   a. What is the energy of this photon of light?
   b. What is the energy difference between the energy level of the atom before it absorbs the photon and the energy level after it absorbs the photon?
   c. What part of the electromagnetic spectrum does this photon correspond to?

Newton's laws along with concepts of force, velocity, and acceleration were adequate to describe everything humans knew up until around the turn of the 20th century. The experiments that revealed the structure of the atom cast classical physics on its head because we observed many things that classical physics could not explain. Between 1900 and 1925 a new branch of physics, *quantum physics*, took shape and our understanding of nature has never been the same. Quantum physics describes the physical laws at a small scale—the scale of the atom and the elementary particles. At the microscopic scale, particles such as electrons do not move according to Newton's laws. Light, which has no mass, is found to have *momentum*. Even though we do not directly perceive the quantum world, many technologies, from GPS satellites to lasers, are derived from quantum physics.

## Wave–particle duality

**Photons**

The quantum theory of light is very different from the wave theory. Wave theory says that you can reduce the energy of a light wave as much as you want by reducing the amplitude. According to quantum theory, however, you cannot split a photon. Light can be 1 photon, 10 photons, or 10 trillion photons—but never half a photon. As we learned on page 653, Planck discovered that light has a *particle* nature when observed on a very small scale.

**Photon energy**

(26.2) $$E = hf$$

$E$ = energy (J)
$h$ = Planck's constant = $6.63\times10^{-34}$ J s
$f$ = frequency (Hz)

**de Broglie and matter waves**

A classical *particle* is a like a tiny ball. It has a definite size, mass, and position. In 1924, Louis de Broglie proposed that this intuition is wrong when things are as small as an atom. In the quantum world, a particle is not like a tiny ball at all. Instead, its mass, size, and even its location is spread out into a wave. The wavelength of de Broglie's *matter wave* is given by equation (26.3).

**de Broglie wavelength**

(26.3) $$\lambda = \frac{h}{p}$$

$\lambda$ = wavelength (m)
$h$ = Planck's constant = $6.63\times10^{-34}$ J s
$p$ = momentum (kg m/s)

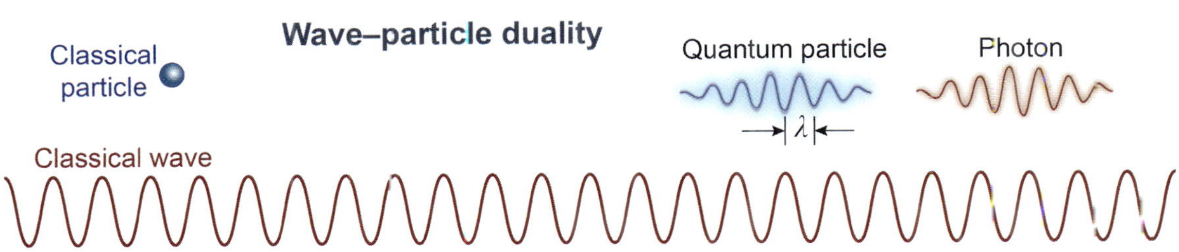


**What is duality?**

The fact that light has particle aspects and matter has wave aspects on the quantum level is called *wave–particle duality,* which we learned about on page 657. The particle nature of light becomes evident when the energy of a system gets close to the photon energy. The wave nature of matter becomes evident when the size of a system becomes comparable to the de Broglie wavelength. Both equations involve Planck's constant $h$, which is a characteristic of the quantum world.

## Double slit experiment for electrons

**Questions to think about**

The behavior of electrons in atoms is complex and *strange!* Why should there be energy levels? Why should the first level hold two electrons, the second hold eight electrons, and so on? What creates the structure reflected in the periodic table and consequentially makes the infinite variety of matter possible, including life? Presenting the mathematics of quantum theory is more than we can do in this book, but we want to provide a conceptual framework for this incredibly successful theory.

**Double slit experiment**

Consider a barrier with two small slits. Electrons are emitted and some pass through the slits and fall on a screen where they are detected. If electrons were classical particles they would be detected in two places, directly in line with each slit. In fact, when the experiment is done with very small slits, something very different happens. We do *not* see two maxima in front of the slits. Instead, we see an *interference pattern* that is characteristic of two waves.

**Electron double slit experiment**

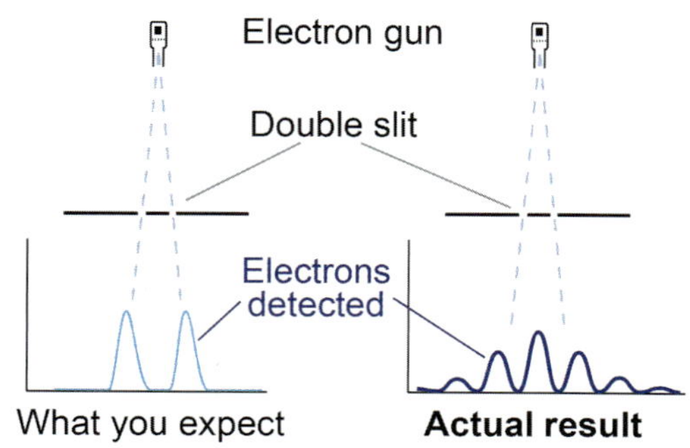


**Single electrons and the double slit**

The experiment gets even more interesting if we let only *one* electron at a time pass through the slits. A single electron is detected at one place on the screen each time the electron gun fires. The weird thing occurs when we record where each electron strikes the screen for 10,000 electrons *one at a time.* We accumulate the *same interference pattern* one electron at a time as we did when there were many electrons at once!

**Electron double slit experiment with a single electron at a time**

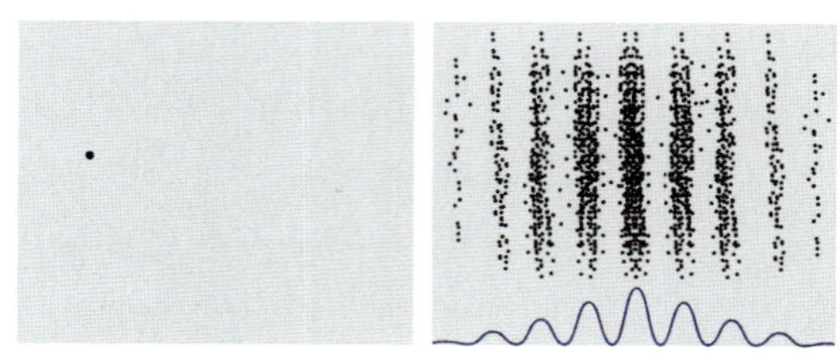

Each electron registers as one dot on the screen. **We still see an interference pattern build up!**

**The electron must be a wave**

The one-at-a-time result implies that a single electron somehow passes through *both slits at the same time to interfere with itself!* Our classical concept of an electron as a particle with a definite position is *not* adequate to explain the double slit experiment. If the electron is a wave, however, then we can explain the two-slit diffraction pattern by constructive and destructive interference.

**Why we don't usually see the electron wave**

Early experimentalists did not see the electron wave because the slits have to be on the order of the electron wavelength in size. Electrons are so light that a 1.5 V battery will accelerate them to a velocity of 726,000 m/s! At this velocity, the electron wavelength is 1 nanometer ($10^{-9}$ m). Quantum effects generally become evident when a system is of the order of the electron wavelength.

Energy of charge $q$ moving across voltage $V$

$$qV = \frac{1}{2}mv^2 \rightarrow v = \sqrt{\frac{2eV}{m_e}}$$

$$v = \sqrt{\frac{2(1.6\times10^{-19}\text{ C})(1.5\text{ V})}{9.11\times10^{-31}\text{ kg}}}$$

$$= 726{,}000\text{ m/s}$$

**Light has momentum!**

According to classical physics, light is pure energy and has no mass. No mass implies no momentum since the classical definition of momentum is mass times velocity, $p = mv$. But the wave-particle duality for light allows *photons to have no mass but still have momentum.*

**Compton's experiment**

In 1923, Arthur Compton conducted a series of experiments on light in which he directed x-rays to hit a graphite target. Compton measured the scattered x-rays at various angles. Although the incident x-rays had a single wavelength, at any given angle Compton found the reflected light to have two different wavelengths. One of the two wavelengths was the incident wavelength $\lambda_1$, while the other was a slightly longer wavelength $\lambda_2$. The longer wavelength radiation implied that the energy of this second reflection was lower. Compton explained the shift in energy by postulating that the incident radiation was not a wave but rather a collection of particles called photons, each with energy $E = hf$.

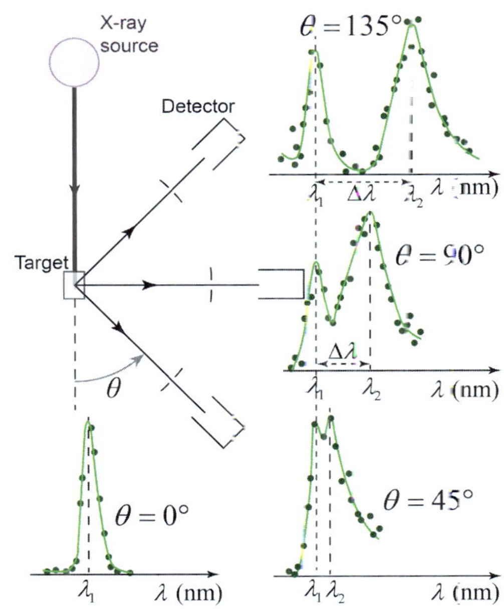


**Why did the frequency change?**

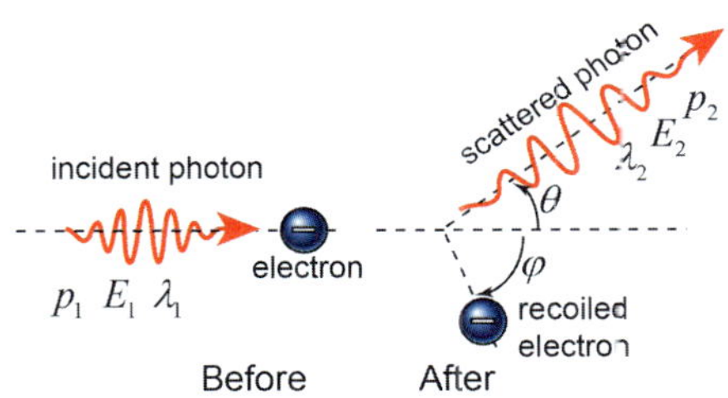


Compton deduced that the shift in frequency came from an elastic collision between a photon and an electron in a carbon atom in the graphite. Compton calculated the energy of the scattered photon by applying the laws of energy and momentum conservation:

**momentum before** = **momentum after**

*and*

**energy before** = **energy after**

**Compton's conclusions**

The difference $\Delta\lambda$ between the scattered wavelength $\lambda_2$ and the incident wavelength $\lambda_1$ increases as the detector angle increases. This result cannot be explained by the wave theory of light. Compton concluded that the photon "*carries with it directed momentum as well as energy.*" Photons of light conserve energy and momentum in their interactions, much like particles. A complete description of radiation has both wave aspects, to explain interference and diffraction, and particle aspects, to explain how photons interact with matter.

**Compton scattering data**

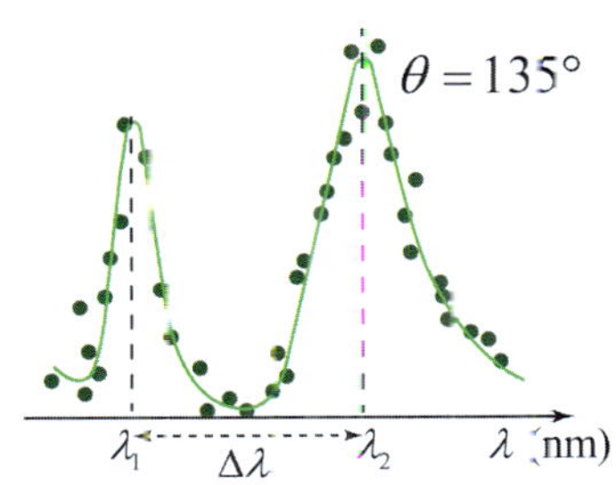

**When do objects become "quantum"?**

Classical physics, such as Newton's laws, accurately describes the macroscopic world. Quantum theory describes the microscopic world. How do these two very different theories match up? The answer is not just philosophical. There cannot be a sharp boundary at which things suddenly start behaving "quantumly." The *correspondence principle* states that quantum theory must transition smoothly into classical physics when the number of particles, energy, or size of a system becomes large.

**The de Broglie wavelength of a baseball**

The de Broglie wavelength gives an approximate length scale for the transition between classical physics and quantum physics. Quantum effects become increasingly important as the size of a system becomes closer to the de Broglie wavelength of the particle. As an example, calculate the de Broglie wavelength of a 0.1 kg ball moving with a speed of 100 km/hr.

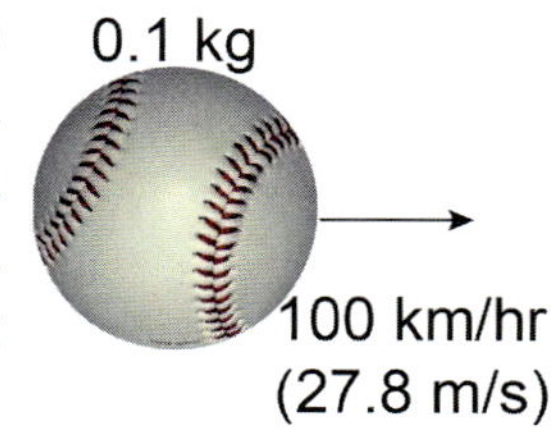


| | |
|---|---|
| *Asked:* | the de Broglie wavelength $\lambda$ |
| *Given:* | $m = 100$ g, $v = 100$ km/hr |
| *Relationships:* | $\lambda = h/p$, $h = 6.63\times10^{-34}$ J s, $p = mv$ |
| *Solution:* | First convert speed to meters per second: $v = (100{,}000 \text{ m})/(3{,}600 \text{ s}) = 27.8$ m/s<br>Calculate the momentum: $p = mv = (0.1 \text{ kg})(27.8 \text{ m/s}) = 2.78$ kg m/s<br>and so the wavelength is: $\lambda = (6.63\times10^{-34} \text{ J s})/(2.78 \text{ kg m/s}) = 2.38\times10^{-34}$ m |

The quantum wavelength of a baseball is $10^{-34}$ m! The typical size of a system including a baseball is at least the size of the ball, which is $10^{34}$ times larger than the de Broglie wavelength. The scale of the baseball's system is so much larger than the de Broglie wavelength that we can completely ignore quantum effects.

**How about electrons?**

Now consider the wavelength of an electron moving at the same velocity. What is the de Broglie wavelength of the electron? (The electron's mass is $9.11 \times 10^{-31}$ kg.)

| | |
|---|---|
| *Asked:* | the de Broglie wavelength $\lambda$ |
| *Given:* | $m = 9.11 \times 10^{-31}$ kg, $v = 27.8$ m/s |
| *Relationships:* | $\lambda = h/p$, $h = 6.63\times10^{-34}$ J s, $p = mv$ |
| *Solution:* | $p = mv = (9.11\times10^{-31} \text{ kg})(27.8 \text{ m/s}) = 2.53\times10^{-29}$ kg m/s<br>$\lambda = (6.63 \times10^{-34} \text{ J s})/(2.53\times10^{-29} \text{ kg m/s}) = 2.62\times10^{-5}$ m |

**Electron microscopy**

The electron's wavelength is around $10^{-5}$ m. This is larger than an atom and comparable to the size of a virus particle. The *electron microscope* is an instrument that uses electron waves instead of light waves to make very high magnification images, such as this image of a dust mite hiding in the fur of an animal (right). Optical microscopes cannot image anything smaller than the wavelength of visible light. The electron wavelength can be much shorter, allowing the electron microscope to image much smaller objects.

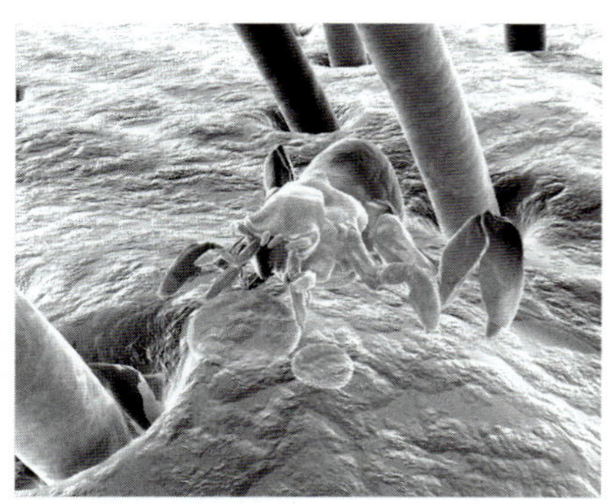

Electron micrograph of a mite on hair

**You cannot observe a quantum system without disturbing it**

When we observe a baseball, we see light that hits the baseball and reflects from it into our eyes. How do we "see" an electron? On the quantum level, *observing* is not so straightforward. If we detect an electron by scattering a photon from it, we disturb the electron, so it moves off in an unknown direction. In fact, any act of "observing" on the quantum level disturbs the very system we are trying to observe, changing it in random ways. In 1927, the German physicist Werner Heisenberg asked the following: *Is it possible to know simultaneously both the momentum and the position of an electron?*

**The uncertainty principle**

Heisenberg's answer is the **uncertainty principle**, which states that no possible means of measurement can simultaneously determine momentum and position to better than the limits of equation (26.4).

**Uncertainty principle**

$$(26.4)\quad \Delta p \Delta x \geq \frac{h}{4\pi}$$

$\Delta p$ = uncertainty of momentum measurement (kg m/s)
$\Delta x$ = uncertainty of position measurement (m)
$h$ = Planck's constant = $6.63\times10^{-34}$ J s

**What does the uncertainty principle mean?**

The quantities $\Delta x$ and $\Delta p$ are the *uncertainties* in position and momentum. By *uncertainty,* we mean that a measurement of position that is 100% accurate has zero uncertainty: $\Delta x = 0$. Equation (26.4) tells us that it is not possible to know the position and momentum of a quantum particle to any accuracy better than $\Delta p\Delta x \geq h/4\pi$. The more accurately we know a particle's momentum ($\Delta p$ = small), the less we can know about the particle's position ($\Delta x$ = large). This is not because we need better measurements! The uncertainty principle says that the quantum universe is *fundamentally* unpredictable when it comes to single particles.

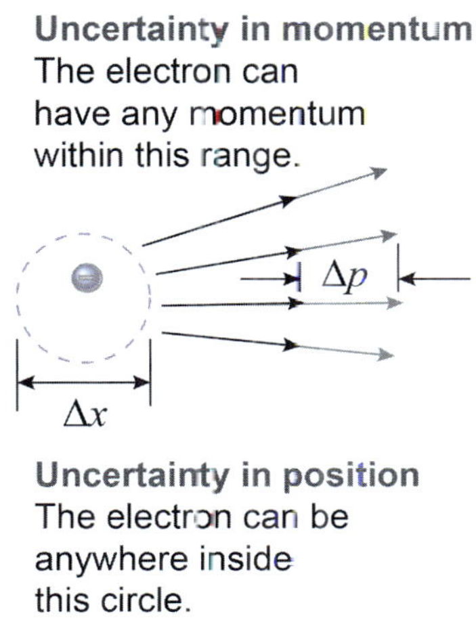


The uncertainty principle can also be written in terms of the energy measurement and the time that it takes to perform this measurement.

**Uncertainty in energy and time**

$$(26.5)\quad \Delta E \Delta t \geq \frac{h}{4\pi}$$

$\Delta E$ = uncertainty of energy measurement (J)
$\Delta t$ = uncertainty of time measurement (s)
$h$ = Planck's constant = $6.63\times10^{-34}$ J s

**Uncertainty and probability**

The uncertainty principle starts to make sense when you think about a particle as a wave. What does the idea of "position" of a wave mean? Heisenberg's principle pointed toward *probability* as the meaning of de Broglie's matter waves. Where the wave amplitude is largest, you have the highest probability of finding the particle *if* you did an experiment that looked for it. But "highest probability" does not mean 100%. Because the wave is distributed over space, there is a nonzero probability of finding the electron anywhere that its matter wave has significant amplitude.

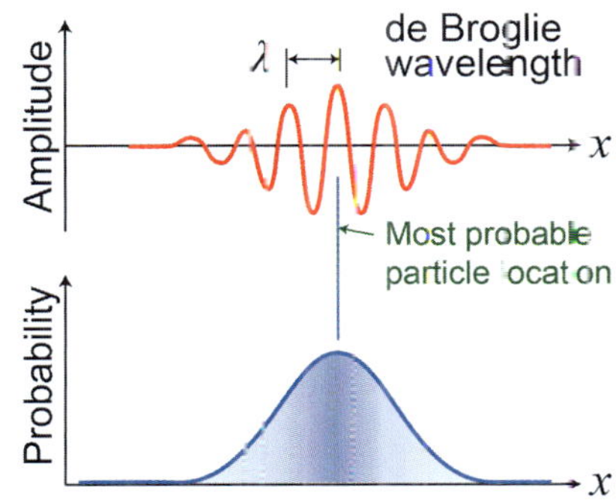

**Uncertainty and the quantum world**

In the quantum world, anything that can happen, *does happen.* In fact, unless something is specifically forbidden from happening, it *must* happen. As one example, consider the energy–time uncertainty relationship, which states that the total uncertainty in time and energy must be greater than $h/4\pi$. The uncertainty principle permits temporarily breaking the law of conservation of energy so long as the event reverses itself quickly enough. A particle might appear out of nothing, then disappear again *if* it happened so fast that it was within the energy and time limit of $h/4\pi$.

**Virtual particles and Feynman diagrams**

The quantum "vacuum" is not truly empty. There is considerable experimental evidence supporting the conclusion that *virtual* particles of matter and antimatter are continually popping into existence and disappearing again, out of pure nothing. Physicists use *Feynman diagrams* (named after theoretical physicist Richard Feynman) to represent quantum events such as virtual particle creation. In a Feynman diagram, time goes from left to right and space goes from bottom to top. The creation and annihilation of a virtual electron–positron pair is shown within the blue circle represented by Heisenberg's uncertainty principle.

Feynman diagram showing virtual electron–positron pair

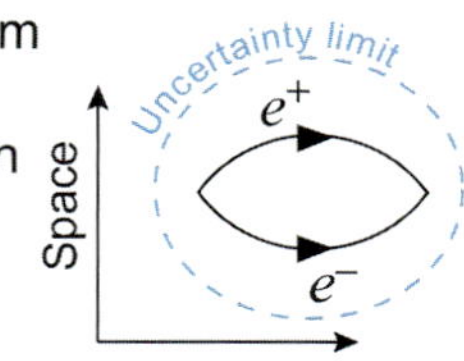


Feynman diagram showing two electrons colliding and exchanging a virtual photon

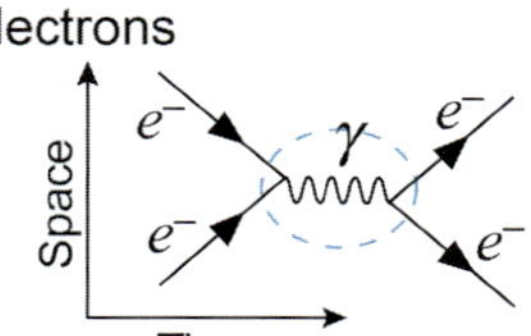


**How are quantum forces transmitted?**

The lower Feynman diagram shows the quantum interpretation of the collision between two electrons. Electrons repel each other through the exchange of a virtual photon. The photon is created and destroyed within the Heisenberg uncertainty limit so it can never be directly detected. The diagram illustrates that *forces are transmitted by virtual particles!* The electromagnetic force is transmitted through the exchange of virtual photons of light.

**The quantum world is in constant, random fluctuation**

Space and time on the quantum scale behave in very nonintuitive ways. The quantum world is in a constant state of random fluctuation. Quantum space itself is not a smooth, featureless void in which matter and energy exist. Instead the quantum vacuum is a continual chaotic storm of virtual particles flashing in and out of existence. This is one reason why it is impossible to predict the state of a quantum system. Even if you know the state of the system at one moment, random fluctuations make the next moment uncertain and the moment after that even more uncertain.

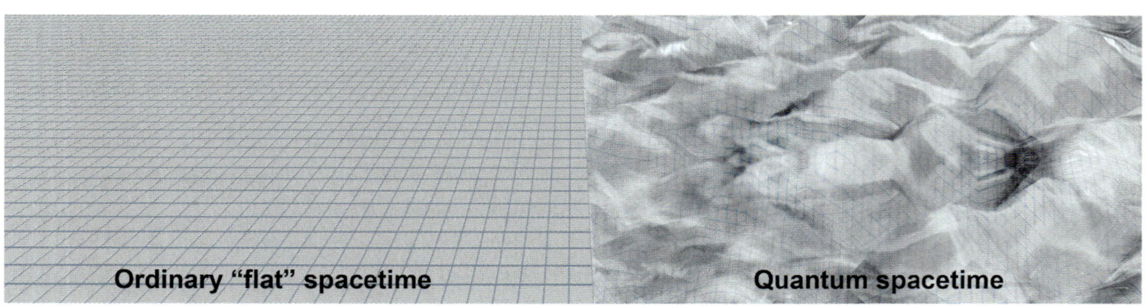

**Determinism**

The quantum behavior of a particle is very different from the motion of a stone tossed in the air. According to classical physics, such as Newton's laws, if you know the initial velocity and forces you can calculate exactly the path the stone will follow. This kind of universe is *deterministic*, meaning that the future can be determined from knowledge of the present.

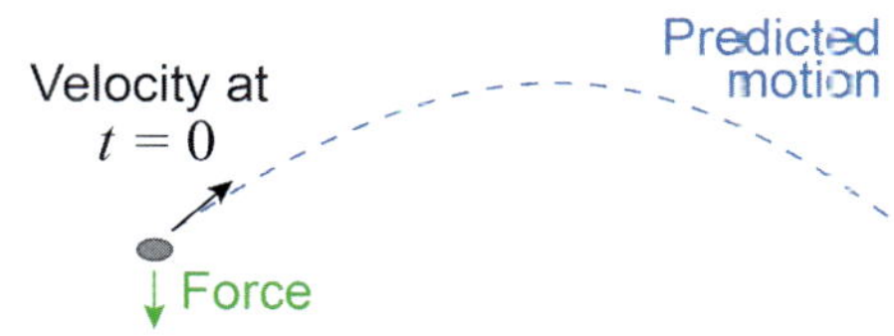


The motion of a stone is *predictable* given knowledge of its position, velocity, and the forces that act on it.

**The effect of quantum fluctuations**

You cannot make a similar prediction about the future state of a single electron. Even if you knew the electron's position and velocity at one moment, its future quickly becomes randomized by quantum fluctuations. There might be a 30% chance the electron lies in a certain range of velocities and positions and a 15% chance the electron has a different range of positions and speeds.

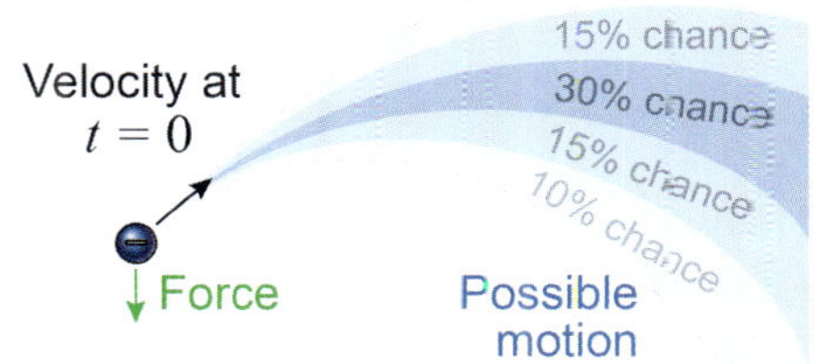


The motion of an electron cannot be predicted with certainty.

**Probability**

The core concept of quantum physics is *probability*. The outcome of a quantum calculation is the probability of a system being in a particular state of position, energy, or momentum. For example, we might calculate a 30% chance that any electron has a certain range of energy in a certain system. We cannot, however, predict whether any one *specific* electron will have that energy.

**The nature of quantum calculations**

Fortunately, we usually don't care about a single electron! Instead, we are interested in the average behavior of trillions of electrons. Statistics and probability allow very accurate predictions about the average behavior of a *collection* of particles even though we cannot predict the behavior of any single particle.

**An example of probability**

Consider tossing a coin. The coin may land heads-up or tails-up. The probability of landing heads-up is 50%. If you flip a coin *once* you cannot predict with any confidence whether it will land heads-up. Nonetheless, knowing the probability is 50% means that you *can* predict that if you flipped the coin 1,000 times the most likely outcome is 500 heads-up results and that there is an 80% chance that the number of heads-up results is between 480 and 520. The graph at right shows the chance of getting between 460 and 540 heads-up results out of 1,000 tosses. This graph is called a *probability distribution* because it shows you how the probability of a specific outcome is distributed over all possible outcomes. As with the coin tosses, probability *does* allow quantitative predictions for a collection of many particles.

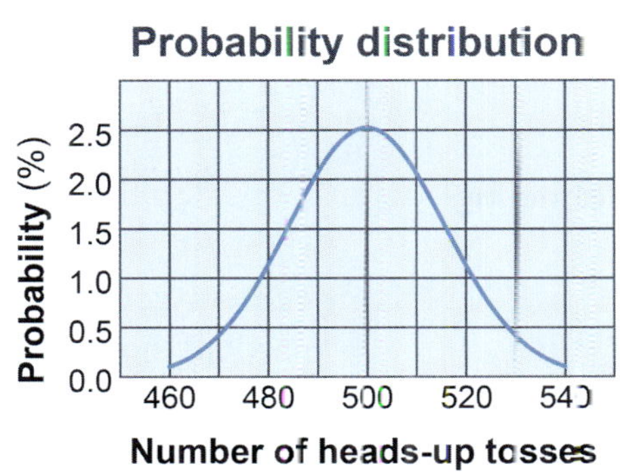

**The wave function**

In 1926, Austrian physicist Erwin Schrödinger proposed that a quantum electron can be described by a *wave function* Ψ*(R)*. The **wave function** is a three-dimensional description of the electron's amplitude over space. The wave function has positive and negative values and can interfere with itself or other wave functions. Every characteristic of the electron can be determined from its wave function, including its momentum, energy, and position.

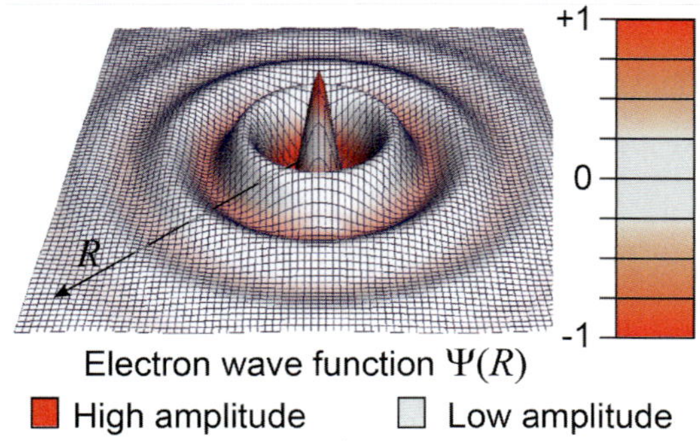

Electron wave function $\Psi(R)$

**How to calculate the probability from Ψ**

To determine the probability of finding the electron in a given region of space, you calculate the square of the wave function, $\Psi^2$, over that region. The time average of $\Psi^2$ is the probability. The blue-shaded region under the graph of $\Psi^2$ is the probability distribution of the electron in that region of space.

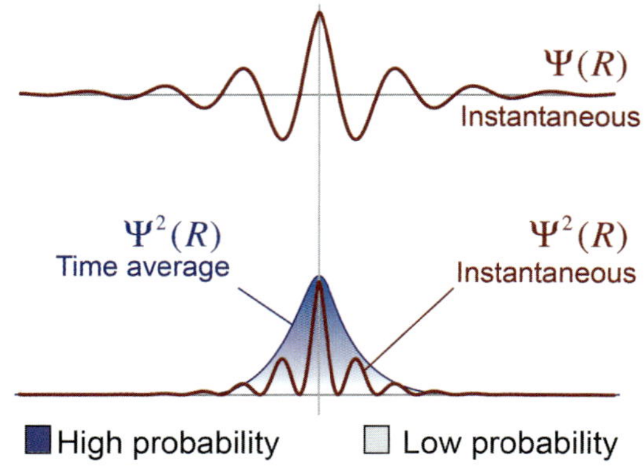

The probability distribution is the time average of $\Psi^2(R)$

**Schrödinger's equation**

Schrödinger's equation describes how the wave function interacts with matter and energy. The equation is essentially a quantum expression of the conservation of energy. We include it because it is beautiful in a very deep way once you learn to read the language in which it is written. You are *not* expected to solve it in this course! You will learn how to use it in a college course in quantum physics.

**Schrödinger's equation**

$$i\hbar\frac{\delta}{\delta t}\Psi = -\frac{\hbar^2}{2m}\nabla^2\Psi + V\Psi$$

- Rate at which Ψ changes over time: $i\hbar\frac{\delta}{\delta t}\Psi$
- Kinetic energy: $-\frac{\hbar^2}{2m}$
- How the wavefunction changes over space: $\nabla^2\Psi$
- Potential energy: $V\Psi$

**Tunneling**

Consider the position versus energy diagram on the right. In classical physics, a particle with energy $E_0$ cannot get from point A to point B. A quantum particle, however, can *tunnel* through the barrier even if it does not have enough energy! The wave function decreases in amplitude *but does not vanish* in areas of space where energy conservation is violated. Tunneling occurs when the energy barrier is "thin" enough that the wave function still has some amplitude on the other side. An electron with energy $E_0$ has a finite probability of being found at point B.

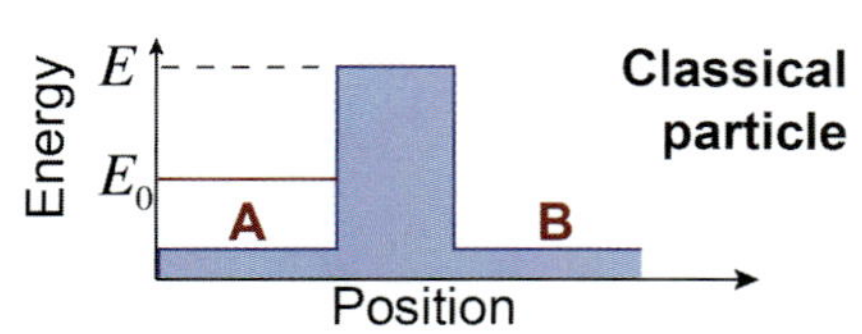


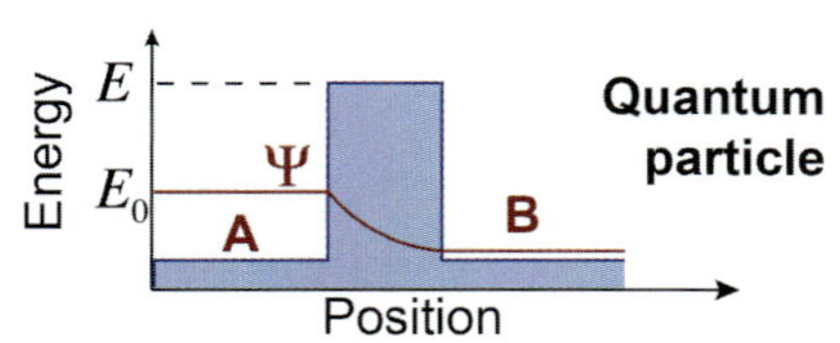

**Lasers and their applications**

Lasers can be found in many everyday technologies, such as barcode scanners at the grocery store, laser pointers, and laser printers. They are also used in metal cutting, communications, medical surgery, and astrophysics. Have you ever wondered how they work?

**Spontaneous emission and stimulated absorption**

The **laser** is based on atomic energy level transitions—but with a twist. When an electron transitions on its own from a higher to a lower energy level, this is called **spontaneous emission**, and the atom loses energy in the form of a photon. It is called spontaneous because it is not caused by an external force or energy. The emitted photon can travel in any direction. For the electron to transition from a lower energy to a higher energy level, however, it must gain energy from an external source—namely, by the *stimulated* absorption of a photon. Stimulated absorption is another name for the absorption of a photon. Spontaneous emission and stimulated absorption are the two kinds of energy level transitions that were covered earlier in this chapter. The change in energy between the two atomic levels corresponds to the energy of the photon, whether it was absorbed or emitted.

**Electron energy level transitions**

**How stimulated emission works in a laser**

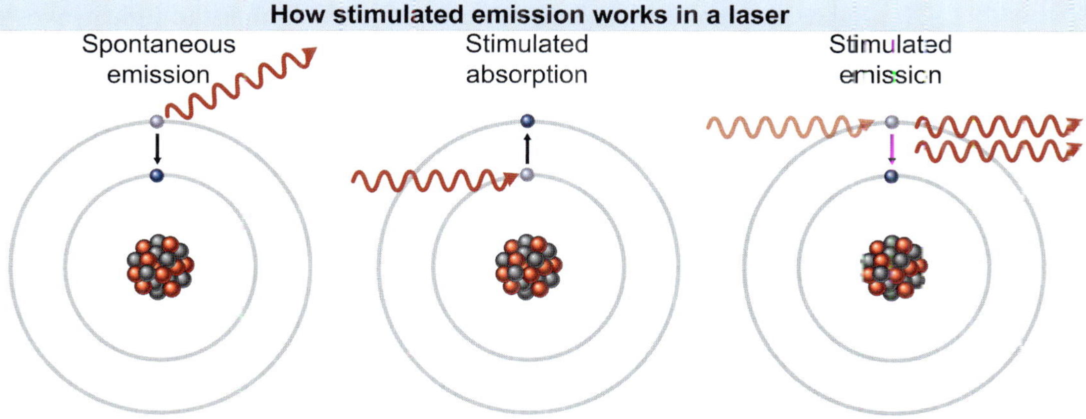


**Stimulated emission**

In the laser, a different kind of emission of a photon takes place. In **stimulated emission** an incoming photon induces the atom to change from a higher to a lower energy state. For this to happen, the incoming photon must have an energy exactly equal to the difference between the two energy levels for the atom. There are five key features of stimulated emission:

1. Energy is required initially to raise the electrons to a higher energy level.
2. Both the energy and the direction of the incoming photon are not affected.
3. The emitted photon travels in exactly the same direction as the incoming photon
4. The emitted photon is exactly in phase with the incoming photon.
5. The incoming light is amplified, because for every one input photon there are two output photons.

The laser is based on this physical process of stimulated emission. The term **laser** is an acronym for **l**ight **a**mplification by **s**timulated **e**mission of **r**adiation, coined in 1957 by physicist Gordon Gould of Columbia University.

**Coherent light**

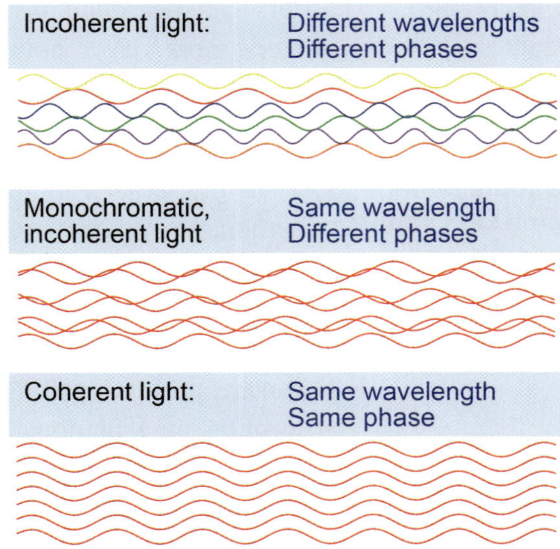


The laser uses stimulated emission to produce *coherent light*. Ordinary white light from a light bulb is incoherent: It consists of light of many different wavelengths and many different phases. When white light is passed through a filter (or colored glass), it becomes monochromatic light, because it consists of only one color of light (such as red). This monochromatic light is still incoherent, however, because it may have many different phases. The problem with incoherent light, even if it is monochromatic, is that the waves can add both constructively and destructively. The laser produces monochromatic light that is also *in phase*—which is called coherent light. The main advantage of coherent light is that the light waves add constructively, so that a laser beam can have a very high intensity (even if the beam is narrow). The high intensity of laser light is why you should never look directly at a laser.

**Designing a laser**

To design a laser, ask the following questions: What is the desired energy of the laser light? (For example, do we want a red or a blue laser?) What is the power required? The answer to the first question determines the type of material that can be used in the laser. The atoms that release the laser light are contained in this material. The answer to the last question determines the geometry of the device and the energy source that we use to power it. For example, a laser pointer device can be powered by a small battery but a metal cutting laser must be powered by a large power supply.

**Building a laser**

Constructing a laser requires three components:

1. a substance that contains atoms whose electrons can be excited (or energized) to the desired energy levels;
2. two parallel mirrors, one with perfect reflection and one that allows a small fraction of light to go through; and
3. an energy source that will be used to excite the electrons to the desired energy levels.

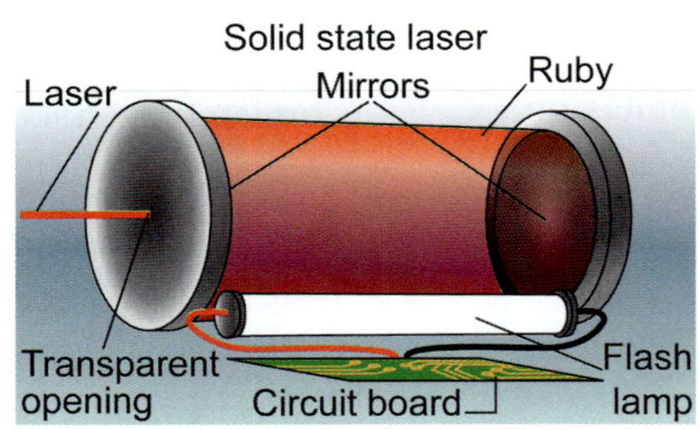


The atoms are located between two parallel mirror surfaces that reflect the emitted photons back and forth. One of these mirrors allows a small fraction—about 1%—of the photons to pass through it. These escaping photons form the laser beam. The atoms are excited by some energy source such as a strong light source or a voltage.

# Section 3 review

Physics at atomic scales contains many surprises. According to the quantum theory, light has momentum, just like matter. And matter (such as electrons) can exhibit both wave and particle behavior, just as you learned earlier about light. In quantum physics, the position and momentum of a particle cannot be known *simultaneously* to high precision. Instead, there is an inherent *uncertainty* in how well we can know position and momentum (or energy and time). But perhaps the strangest feature of the quantum world is that many of the properties of particles are *probabilistic* or statistical. You don't know where an electron is located: you only have *probabilities* for its being in various locations.

| | |
|---|---|
| **Vocabulary words** | **uncertainty principle, wave function, stimulated emission, laser, spontaneous emission** |

| | | |
|---|---|---|
| **Key equations** | $E = hf$ | $\lambda = \frac{h}{p}$ |
| | $\Delta p \Delta x \geq \frac{h}{4\pi}$ | $\Delta E \Delta t \geq \frac{h}{4\pi}$ |

## Review problems and questions

1. In the electron version of the double slit experiment, electrons are detected on a screen and show an interference pattern. Does each individual electron pass through one or the other slit?

2. Imagine a modification to the electron double slit experiment where you place a detector next to one slit to determine *at the location of the slit* whether or not each electron passed through that slit. Will the electrons still show an interference pattern on the screen?

3. One of the best-known paradoxes of quantum physics is called "Schrödinger's cat." Imagine a black box that contains a cat and a randomized device that, if triggered, kills the cat. If you wait a while, you won't know whether the device has triggered and killed the cat, or whether the cat is still alive. If you open up the box and look inside, then you can tell whether the cat is alive or dead. But before you open up the box, according to the concepts of quantum physics, is the cat alive or dead?

4. Mauro studied quantum physics and has what he thinks is a great idea for a new invention he calls the Earth Tunneler: a tunneling device that will allow him to pass through the solid Earth and show up on the far side. He insists that the concepts of quantum tunneling will allow him to tunnel straight through the Earth. Is his invention a solid idea?

5. Research the two kinds of laser eye surgery called LASIK and PRK. What wavelengths of lasers are used in them? What is the laser used for?

6. If you want to know the position of an electron to within the radius of the atom of around 0.5 angstrom ($0.5\times10^{-10}$ m), what is the best precision with which you can know its velocity?

Chapter 26

# Chapter review

## Vocabulary

Match each word to the sentence where it best fits.

### Section 26.1

| | |
|---|---|
| **atom** | **atomic number** |
| **electron** | **elementary charge** |
| **isotopes** | **mass number** |
| **neutron** | **nucleus** |
| **proton** | |

1. Atoms that have six protons and six neutrons or six protons and seven neutrons are both _______ of carbon.
2. The nuclear particle with zero charge is the _______.
3. The _______ is the lightest of the three elementary particles forming an atom.
4. Oxygen-16 has eight protons and eight neutrons in its _______.
5. An atom's _______ is usually represented by *Z*.
6. A/An _______ is the smallest particle of an element that can exist in isolation.
7. The proton and the electron both carry the same _______ but with opposite signs.
8. An atom that has three protons and four neutrons has a/an _______ of seven.

### Section 26.2

| | |
|---|---|
| **Bohr model** | **Pauli exclusion principle** |
| **absorption** | **continuum spectrum** |
| **electron volt (eV)** | **emission** |
| **energy levels** | **excited state** |
| **ground state** | **line spectrum** |
| **orbit** | **phosphorescence** |
| **quantized** | **quantum number** |
| **scattering** | **spectral line** |
| **spectrograph** | |

9. The wavelengths of the Balmer emission lines of hydrogen were successfully explained by the _______.
10. In the Bohr model of the hydrogen atom, electrons can only occupy a quantized _______.
11. The $n = 1$ energy level is called the _______.
12. _______ is the process in which an atom gains energy from a photon of light.
13. Many physical quantities at the atomic level can only take _______ values.
14. A/An _______ is an instrument that can be used to disperse light to study its emission lines.
15. One particular energy level transition in an atom corresponds to one particular wavelength of light called a/an _______.
16. Light from one direction that is absorbed by atoms and re-emitted in other directions is called _______.
17. After illumination by a light source, a material may emit light in other directions through a process called _______.
18. The energy level of an electron is described by its integer _______.
19. An incandescent light bulb generally does not emit light at distinct wavelengths but rather as a/an _______.
20. Although the SI unit of energy is the joule, the _______ is a more natural unit for calculating atomic energy levels.
21. When an electron in an excited state loses energy by producing a photon, this is called _______.
22. No two electrons can occupy the same orbit according to the _______.
23. A pattern of distinct emission or absorption wavelength features for an atom is an example of its _______.
24. An atom emits or absorbs light when its electron changes between two _______.

### Section 26.3

| | |
|---|---|
| **laser** | **spontaneous emission** |
| **stimulated emission** | **uncertainty principle** |
| **wave function** | |

25. The probability of finding a particle at a particular location and at a particular time is determined from its _______.
26. It is impossible to know simultaneously both the momentum and position of an electron because of the _______.
27. A scientific instrument that produces a beam of coherent light is called a/an _______.
28. A laser pointer works using the phenomenon of _______.

# Chapter review



## Conceptual questions

### Section 26.1

29. For the gold isotope $^{194}_{79}Au$, how many neutrons are there?

30. ❮ Use diagrams to show how our understanding of atoms has changed over the past several thousand years.

31. ❮ In Rutherford's model for the atom, why aren't the electrons ejected?

32. ❮ Explain how the Rutherford scattering experiment showed that the positive charge in an atom must be concentrated, rather than spread throughout the atom.

33. ❮❮ In the Rutherford scattering experiment, why do most alpha particles pass through the gold foil without deflection?

34. ❮❮ In the Rutherford scattering experiment, why are very few alpha particles deflected backward?

35. Thomson's experiment showed that cathode rays were deflected by charged plates. He concluded that

    a. cathode rays carried a negative charge.
    b. ordinary light behaved in the same manner.
    c. the gas present in the tube affected the deflection.
    d. different metals emitted different types of cathode rays.

### Section 26.2

36. A number of commercial products, such as Rainbow Peepholes™ or Rainbow Glasses® that you used in Investigation 26D on page 772, contain small transmission gratings etched onto clear plastic. What property of light produces the multiple rainbows that you see when you look through one at a small light source?

37. When matching laboratory emission-line spectra against that for a series of different elements, why is it useful to print one set onto clear, plastic transparencies?

38. ❮ If you wanted to obtain a spectrum of the Sun with a hand-held visual spectroscope while outdoors during a field investigation, how would you do so *safely* without pointing the spectroscope directly at the Sun?

39. ❮ Describe the connection between energy levels and orbits of the electron and whether or not an atom is a good conductor.

40. ❮ How can you use a hand-held visual spectroscope to distinguish the elements in two different gas discharge tubes?

41. Distinct energy levels are present in which model of the atom?

    a. Bohr's quantum atom
    b. Rutherford's alpha particle
    c. Schrödinger's cloud model
    d. Thomson's plum pudding model

42. Create a presentation on the historical theories and experiments leading to our present model of the atom. Use digital media, such as a slide presentation, and include both text and visual elements to enhance understanding and add interest.

43. ❮ How can emission-line spectra be used to distinguish different elements?

44. ❮❮ If the electron in a hydrogen atom is in the $n = 3$ energy level, how many different energies of photons can it emit?

45. ❮❮ In the de Broglie model of the electron orbit, the electron is treated like a *wave*. How does this address the question raised by Bohr's model of why electron orbits are quantized?

46. ❮❮ Why are energy levels usually written as negative energies?

47. ❮❮ When identifying elements by their spectra in Investigation 26D on page 772, you first matched spectra for individual gas discharge lamps and later matched spectra with the Sun and other stars. Why was it more useful to do the investigation in this order and not the other way around?

48. ❮ A physicist measures the wavelengths of light given off by hydrogen gas. Despite taking thousands of measurements, he only measures light at a discrete set of wavelengths, and nothing in between. Explain his results.

49. ❮❮❮ If the electron in a hydrogen atom is in the $n = 3$ energy level, how many different energies of photons can it absorb?

50. ❮❮❮ A discharge tube ionizes the gas inside—removing one or more electrons from its outer shell—when you apply a voltage across the tube. Why do you think that different discharge tubes require different voltages?

### Section 26.3

51. Explain how the operation of the laser in a laser pointer is related to quantum theory and the atom.

52. ❮ Describe three technological applications of quantum phenomena.

# Chapter review

## Quantitative problems

### Section 26.1

53. If a target nucleus in Rutherford's scattering experiment were suddenly to double its number of *neutrons*, how would the force felt by each alpha particle change?

54. If a target nucleus in Rutherford's scattering experiment were suddenly to double its number of *protons*, how would the force felt by each alpha particle change?

55. How many electrons does it take to create a charge of $-1\ \mu C$?

56. How many protons does it take to make a charge of 1 pC?

57. ❮ Calculate the ratio of the mass of the proton to that of the electron.

### Section 26.2

58. How many electron volts are there in one joule of energy?

59. ❮ Consider an electron in each of two energy levels at −2 eV and −1 eV. Which electron can be ionized by lower frequency light?

60. ❮ In the hydrogen atom, which transition emits a higher frequency photon of light, $n = 3$ to 2 or $n = 4$ to 2?

61. ❮❮ The ground state of the hydrogen atom is at −13.6 eV. What wavelength of light (in nanometers) is required to ionize a hydrogen atom, i.e., remove the electron from the ground state to $n = \infty$?

62. ❮❮ Consider three different energy levels of a particular atom: $E_1 = -10$ eV, $E_2 = -3$ eV, and $E_3 = -1$ eV. List all the different possible energies for electron transitions among these three energy levels. What other energy level transitions are there if you also include transitions with the $E_\infty = 0$ (ionization) level?

63. ❮ Niels Bohr's model of the atom described it as having fixed "positions" (represented by a combination of numbers and letters) that could only hold one electron each. What are these "positions" called?
    a. quantum states
    b. quantum orbitals
    c. quantum ions
    d. quantum equations

64. ❮❮ A hydrogen atom with its electron in the ground state of −13.6 eV absorbs a photon with an energy of 15.0 eV that ionizes the atom. What is the kinetic energy of the electron ejected from the atom?

65. ❮❮ The $n = 2$ state of the hydrogen atom is at −3.4 eV. What wavelength of light (in nanometers) is required in to ionize a hydrogen atom with its electron in the $n = 2$ energy level?

66. ❮❮❮ In the Bohr model of the hydrogen atom, the energy levels are given as

$$E_n = -13.6\frac{1}{n^2}\text{ eV.}$$

    a. Calculate the energy levels in electron volts for $n = 1, 2, 3, 4$, and 5.
    b. Calculate the energy lost (in electron volts) by the electron that makes a transition from $n = 2$ to 1.
    c. What is the frequency of the photon (in hertz) that would be emitted if the electron changes its energy level from $n = 2$ to 1?
    d. What is the frequency of the photon (in hertz) that would be emitted if the electron changes its energy level from $n = 3$ to 2?
    e. A photon with a wavelength of 434 nm is absorbed by the atom. What is the energy of the photon (in electron volts)? What energy level transition was caused by absorbing this photon?

### Section 26.3

67. ❮ Green light has a wavelength of $\lambda = 500$ nm ($5.0\times10^{-7}$ m) in a vacuum.
    a. What is the energy of one photon of green light? Express this both in joules and electron volts. ($1\text{ eV} = 1.6\times10^{-19}$ J.)
    b. What is the momentum (in kilogram-meters per second or newton-seconds) of an electron that has the same wavelength as a photon of green light?
    c. Divide your answer to Part b by the mass of an electron ($m_e = 9.11\times10^{-31}$ kg). What does this ratio correspond to?
    d. Compare your answer in Part c to the speed of light in a vacuum ($c = 3.0\times10^8$ m/s).

68. ❮❮ A photon of red light has a wavelength of 700 nm in a vacuum ($\lambda = 5.0\times10^{-7}$ m). Use this as a starting point to investigate the consequences of the uncertainty principle:
    a. Calculate the energy in joules of one photon of red light.
    b. Suppose that you have an electron whose kinetic energy equals the value you computed in Part a. Now imagine that you wish to measure the electron's energy with 10% precision. According to the uncertainty principle, for how long (in seconds) must you observe the electron?
    c. Compare the measurement time interval you calculated in Part b with the period of a photon of green light.

# Chapter review

## Standardized test practice

69. Which of the following instruments does *not* act as a spectrograph?

A. a glass prism
B. a gas discharge tube
C. a transmission diffraction grating
D. a hand-held visual spectroscope

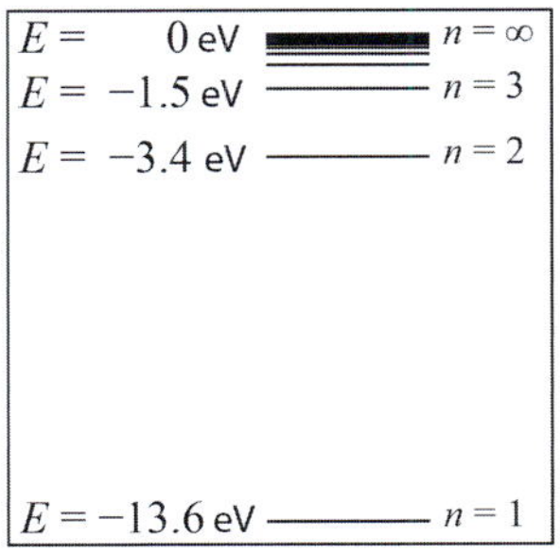


70. In the energy level diagram shown, how much energy will the atom emit in the form of a photon if its electron moves from the $n = 3$ to the $n = 1$ energy level?

A. 1.5 eV
B. 1.9 eV
C. 12.1 eV
D. 13.6 eV

71. Which phenomenon requires the quantization of light in order to explain it?

A. electromagnetic radiation
B. Rutherford scattering
C. the photoelectric effect
D. alpha decay

72. In a hydrogen atom with radius $2.5\times10^{-11}$ m, what is the electric force of attraction between the proton and the electron? (The Coulomb constant is $9.0\times10^{9}$ N m$^2$/C$^2$ and the elementary charge is $1.602\times10^{-19}$ C.)

A. $3.7\times10^{-7}$ N
B. $2.2\times10^{26}$ N
C. 58 N
D. $4.1\times10^{-17}$ N

73. An atom has an energy of −16 eV for its ground state, whereas its first excited state is at −4 eV. Photons with energy of 6 eV bombard the atom while it is in its ground state. What will happen to the atom?

A. Nothing will happen.
B. The atom's electron will absorb one photon and transition to the first excited state.
C. The atom's electron will absorb two photons and transition to the first excited state.
D. The atom's electron will absorb one photon and immediately re-emit a photon with the same energy.

74. Johann Balmer proposed a formula that could be used to predict quantitatively the wavelengths of the emission lines of hydrogen in the visible portion of the electromagnetic spectrum. What physical intuition was behind his formula?

A. Electrons act as both a particle and a wave.
B. The energy levels of an atom are quantized.
C. The speed of light is constant in all reference frames.
D. The position and velocity of an electron cannot be perfectly determined simultaneously.

75. The electron of a hydrogen atom changes its level from $n = 1$ to $n = 2$. What is true of the atom?

A. The atom emitted a photon of energy equal to the energy of the $n = 2$ level.
B. The atom absorbed a photon of energy equal to the energy of the $n = 1$ level.
C. The atom emitted a photon of energy equal to the difference between the two energy levels.
D. The atom absorbed a photon of energy equal to the difference in energy between the two energy levels.

76. An experimenter illuminates an incandescent light bulb and a hydrogen gas discharge tube and passes the light of each through a prism. More colors are observed from the incandescent bulb than from the hydrogen gas. Why?

A. The incandescent bulb is hotter.
B. The gas in the incandescent bulb is rarefied.
C. The hydrogen gas is hotter.
D. Only certain atomic transitions can be made by the hydrogen gas.

77. What historical experiment demonstrated that the nucleus is massive and occupies a tiny fraction of the volume of the atom?

A. Young's double slit experiment
B. Rutherford's scattering experiment
C. Millikan's oil drop experiment
D. Hertz's photoelectric effect experiment

78. The particles that orbit the nucleus are

A. protons.
B. neutrons.
C. electrons.
D. compounds.

79. The production of light by the laser in a laser pointer is based on the concept of

A. stimulated emission.
B. stimulated absorption.
C. spontaneous emission.
D. stimulated absorption.

# Chapter 27 Nuclear Physics

Image courtesy NASA

The ultimate source of energy in the Solar System is a chain of nuclear fusion reactions in the core of the Sun that combines four hydrogen nuclei into one helium nucleus. Many scientists envision the day when the electricity to power our civilization will come from harnessing the same kinds of reactions here on Earth. There is enough deuterium in seawater to provide limitless fuel for fusion energy, deliver clean power with no carbon emissions or pollution, and make us independent from fossil fuels. Of course, if it were *easy* it would have been done long ago! Creating nuclear power from fusion energy here on Earth requires us to tame the reactions that power the Sun. The brightest physicists have been working for almost 70 years to develop a practical fusion reactor, a challenge that has been described as the most difficult technological feat ever attempted by humans.

The easiest fusion reaction to produce on Earth is the fusion of deuterium and tritium. Deuterium ($^2_1$H) is an isotope of hydrogen with one neutron. Tritium ($^3_1$H) is an isotope of hydrogen with two neutrons. If deuterium and tritium nuclei can get close enough, they react to make a helium nucleus and a neutron. The reaction frees an enormous amount of energy. One kilogram of deuterium fuel, a soda-bottle full, releases more than 400 trillion joules of energy. This is equivalent to the energy content of 4,700 metric tons of gasoline, enough to fill 280 tanker trucks.

For fusion to occur, the nuclei must have enough energy to get close enough to react before their mutual repulsion pushes them away again. In the core of the Sun this occurs at about 15 million K. At that temperature, the reaction rate is about 1 in 2 trillion million ($2 \times 10^{18}$) atoms per second. The Sun's core is incredibly dense and huge, so a tiny reaction rate still produces a lot of total energy. On Earth, however, a practical reactor needs to produce many more reactions in a much smaller space. The core of a *tokamak* reactor, such as two of the authors of this book worked on, reaches a temperature five times hotter than the Sun's core, or roughly 85 million K. Confining and controlling matter at 85 million K is no small feat of engineering. The powerful electromagnets that confine the hot fusion plasma use magnetic fields 20,000 times as strong as the Earth's magnetic field. They are cooled with liquid nitrogen (at 77 K) to dissipate the heat from electric currents of 100,000 A. Nowhere else in the known universe is there as extreme a temperature difference—from 85,000,000 to 77 K—as is found in a fusion reactor across the space of barely one meter!

# Chapter study guide

## Chapter summary

Why can a relatively small power plant keep a submarine running for months or years at a time? The fuel at the heart of a nuclear power plant packs an immense amount of energy into a small mass. Nuclear physics is a description of the properties of the nucleus that can produce large amounts of energy through mass–energy equivalence. The strong nuclear force binds the nucleus together by overcoming the repulsion between the protons in the nucleus, thus providing the binding energy at the heart of nuclear power. Nuclear fission reactions provide the energy behind nuclear power, whereas nuclear fusion reactions power the core of the Sun. The weak nuclear force governs radioactive decay, which emits alpha, beta, or gamma radiation. Radioactive decay is at the heart of carbon-14 dating. The standard model brings together elementary particles and the forces that govern their interactions.

## Learning objectives

By the end of this chapter you should be able to

- describe the three types of radioactive decay and solve nuclear reaction equations involving them;
- solve radioactive half-life problems, including carbon-14 dating;
- describe the strong and weak forces and evidence for their existence;
- describe the historical development of the concepts of the strong and weak nuclear forces;
- describe mass–energy equivalence, explain its role in nuclear reactions, and calculate the energy produced by nuclear reactions;
- describe features in the design and operation of a nuclear power plant;
- describe applications of nuclear physics, including radiation therapy and diagnostic imaging; and
- summarize the standard model, including the four fundamental forces of nature.

## Investigations

27A: Controlling a nuclear fission reaction

## Chapter index



## Important relationships

$$E = mc^2$$

$$N = N_0\left(\frac{1}{2}\right)^{t/t_{1/2}}$$

$$\underbrace{{}^{1}_{0}\text{n} + {}^{14}_{7}\text{N}}_{\text{Reactants}} \rightarrow \underbrace{{}^{14}_{6}\text{C} + {}^{1}_{1}\text{p}}_{\text{Products}}$$

## Vocabulary

**strong nuclear force**
**rest energy**
**radioactive decay**
**gamma decay**
**carbon dating**
**chain reaction**
**control rods**
**dose**
**ionization**
**atomic mass unit (amu)**
**binding energy**
**alpha decay**
**weak nuclear force**
**nuclear reaction**
**fusion**
**nuclear waste**
**rem**
**radiotherapy**
**mass–energy equivalence**
**mass deficiency**
**beta decay**
**half-life**
**fission**
**nuclear energy**
**positron**
**Geiger counter**

The nucleus of the atom contains both protons and neutrons. Protons are positively charged and neutrons do not have any charge. The coulomb repulsion between protons as close together as they are in a nucleus is enormously strong and should tear the nucleus apart instantly. *Why does the nucleus stay together?* The fact that atomic nuclei are stable implies that another type of force must exist that is attractive and much stronger than the coulomb force.

If positive charges repel each other...

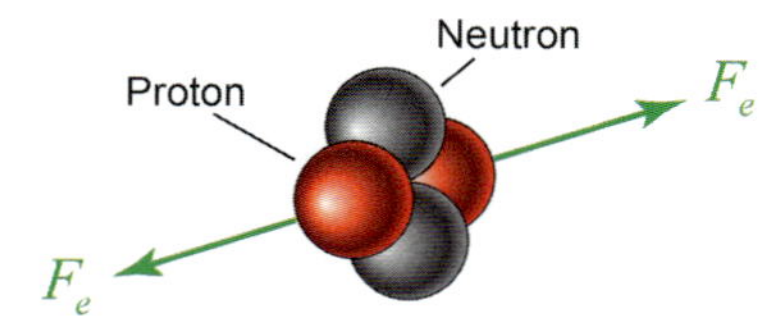


**...then what holds the protons and neutrons in the nucleus together?**

## Evidence for the strong nuclear force

**Discovery of the proton and neutron**

When Rutherford discovered in 1909 that the nucleus of the atom was tiny and had positive charge, the evidence shed no light on whether the nucleus had any structure or was one particle. In 1918, Rutherford discovered the proton by observing that hydrogen atoms are sometimes produced when nitrogen atoms are bombarded with alpha particles. A hydrogen nucleus contains a single proton. In 1932, James Chadwick discovered another electrically neutral elementary particle called the neutron and measured its mass to be nearly the same as the proton. Physicists then knew that there were *individual* protons and neutrons in the nucleus. They were then faced with the problem of what holds the nucleus together?

**Strong force holds nucleus together**

The primary evidence for the existence of a strong nuclear force is that otherwise the nucleus would immediately disintegrate from the powerful repulsion of positively charged protons. Since we observe that many nuclei do *not* fly apart, we deduce there must be a strong force attracting protons and neutrons. This force is called the **strong nuclear force** and it is a short-range attractive force acting between protons and neutrons (but not electrons). Modern particle physicists usually refer to the strong nuclear force as the *strong interaction*. Of course, knowing a force exists does not explain what causes the force! The next step in the development of nuclear physics was to develop a theory to explain the strong force.

**Forces between two protons in a nucleus**

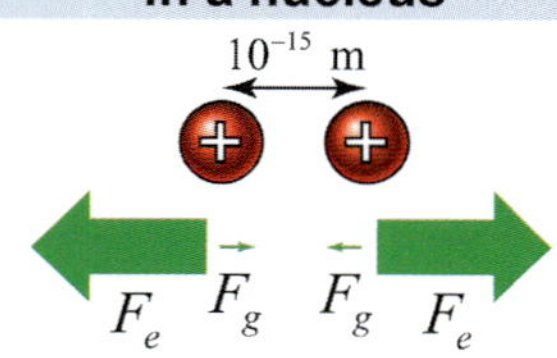


| Gravitational force | Electrical force |
|---|---|
| $F_g = 1.9\times10^{-34}$ N | $F_e = 230$ N |

Some other force must be present to balance the electrical force!

**Mesons carry the strong force**

In 1935, Japanese theorist Hideki Yukawa proposed a model in which the strong force would work through a hitherto unknown type of massive, elementary particle called the *meson*. Yukawa named his predicted particles after the Greek word *mesos* (μέσος)—which means "intermediate"—for the role mesons play in mediating the strong force. Experimental confirmation of Yukawa's theory came in 1947 when an international team led by English physicist Cecil Power discovered the first meson, a pi meson (or pion).

**Quarks**

The strong interaction not only binds the atomic nucleus together, it also governs the interactions among *quarks*, the elementary particles from which protons and neutrons are made. The existence of quarks was independently proposed by Murray Gell-Mann and George Zweig in 1962. The first quark was experimentally confirmed in 1968 at the Stanford Linear Accelerator Center, providing additional evidence for the strong interaction and the theory behind it.

**Summary of properties**

The strong nuclear force has a number of important properties that determine its effect on atomic particles. The strong nuclear force

- is only attractive, not repulsive;
- acts only on nuclear particles or the particles that make up nucleons, not electrons;
- acts on charged particles (protons) and neutral particles (neutrons); and
- is a short-range force.

**Strong force attracts protons and neutrons**

The strong nuclear force attracts every proton or neutron to every other proton or neutron regardless of electric charge. Both protons and neutrons are classed as *nucleons*, and the strong force attracts all nucleons to all nearby nucleons. The strong force does not affect electrons just as the electric force does not affect neutrons.

**Short- and long-range forces**

The strong force is peculiar in that it acts only when particles are very close to each other, such as within an atomic nucleus. For this reason the strong nuclear force is called a *short-range force*. Although the strong nuclear force is very powerful within the nucleus, its short range means that it is unimportant beyond the scale of the nucleus. By comparison, the electromagnetic force is a *long-range force*. Two charges can exert an electromagnetic force on each other at any distance. The electric force gets weaker as $1/r^2$ but the strong force effectively drops to *zero* at distances much larger than $10^{-15}$ m!

**What does short range mean?**

To get a sense of what "short range" means, consider two deuterium nuclei moving toward each other. A deuterium nucleus consists of a proton and a neutron. If the kinetic energy is low, the nuclei repel each other because the Coulomb force pushes them back apart before they get close enough for the strong nuclear force to take hold. At higher kinetic energies, if the separation becomes smaller than about $10^{-15}$ m, the strong nuclear force snaps them together and they become an alpha particle (helium nucleus).

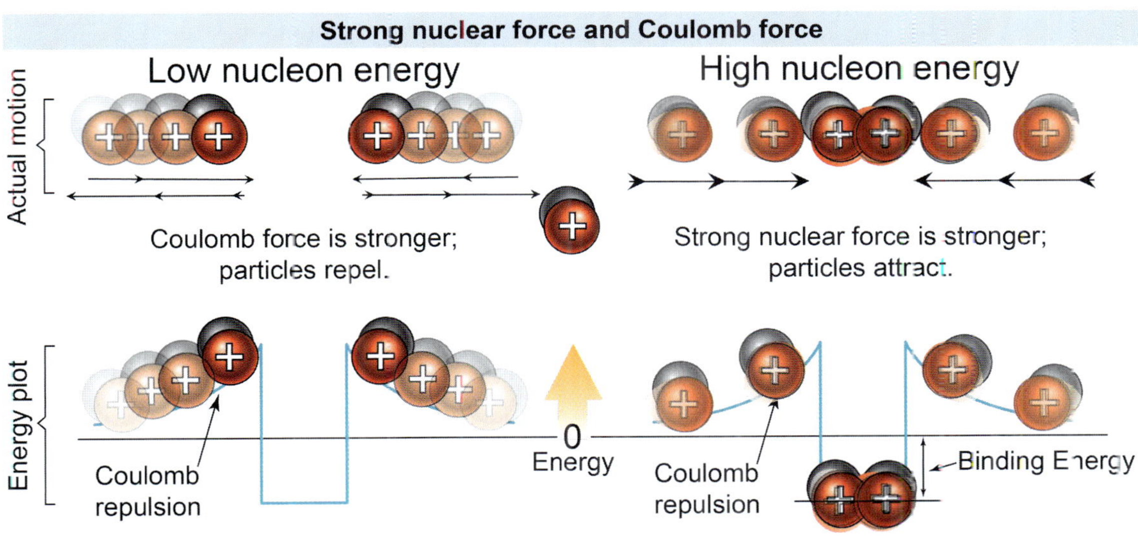


**Binding energy**

It takes energy to separate nucleons once they are bound together by the strong force. This energy is called the *binding energy*. Nuclear binding energy is enormous. For example, to separate the protons and neutrons in 1 kg of water (the amount in a 1 L bottle) would take $10^{13}$ J! This is the equivalent energy of 100,000 gallons of gasoline, all derived from the nuclear binding energy in just *one* liter of water.

## The atomic mass unit

**Atomic mass unit**

The masses of the proton and neutron are so small (about $1.67 \times 10^{-27}$ kg) that nuclear scientists often work in *atomic mass units*. One **atomic mass unit (amu)** (1 amu) is defined to be one-twelfth (1/12) the mass of the carbon-12 atom. Another way to state this is that the atomic mass of carbon-12 is defined as being exactly 12 amu. Why one-twelfth? Carbon-12 has six protons and six neutrons, so the atomic mass unit is about the average mass of a nucleon.

**Atomic mass unit (amu)**

$$1 \text{ amu} = 1.661 \times 10^{-27} \text{ kg}$$

One atomic mass unit is $1.661 \times 10^{-27}$ kg, which is slightly less than the mass of a single proton. The electron has a mass of 0.0005 amu, which means that the electron has around 0.05% of the mass of a nucleon.

**Information in the periodic table**

The periodic table of the elements contains information about nuclear structure as well as chemical properties. The atomic number is the number of protons in the nucleus and the atomic mass is the average mass of one atom in atomic mass units. Note that the atomic mass given in a periodic table is the *average* mass, which includes all isotopes that naturally occur on Earth. The atomic mass for carbon, 12.011 amu, is not precisely 12.000 because naturally occurring carbon contains a small amount of the isotopes carbon-13 and carbon-14. The atomic mass also includes the mass of the electrons, which is very small compared to the mass of the nucleus.

**Element information**

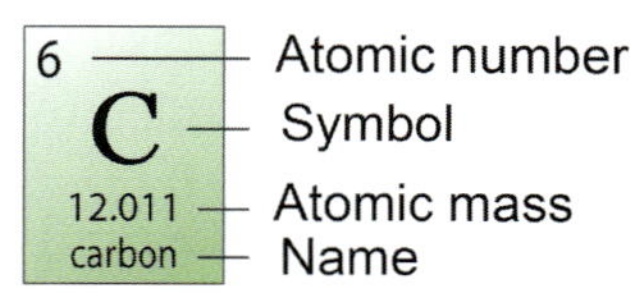


**Notation for nuclear isotopes**

An isotope is identified by providing both the atomic number $Z$ and the mass number $A$. The number of neutrons, $N$, in the isotope can be calculated by using the relationship $A = N + Z$. One way to write down an isotope uses the name of the element and the atomic mass number, such as "carbon-12." Alternatively, a special notation is used that includes the abbreviation for the element, the atomic number, and the atomic mass number. A *subscript* to the left of the element name represents the atomic number $Z$. A *superscript*, also to the left of the element, represents the atomic mass number $A$. The general form of this special notation for any isotope of an element X is ${}^{A}_{Z}\text{X}$.

**Notation for isotopes**

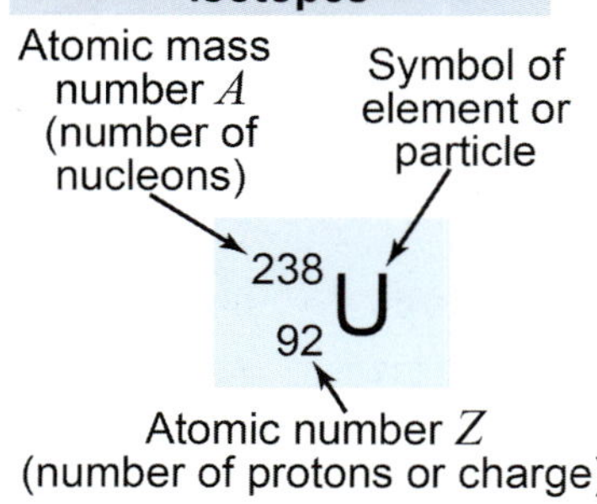


**Elementary particles**

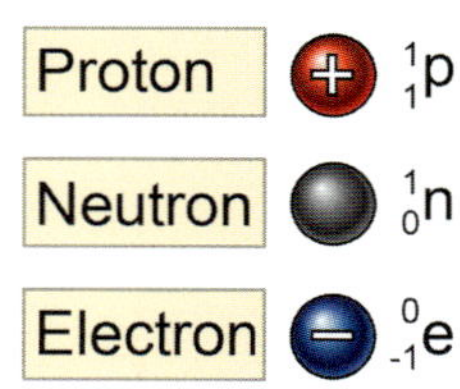


**Examples**

For example, carbon-12 has $A = 12$ and $Z = 6$ and is denoted as ${}^{12}_{6}\text{C}$. The isotope uranium-238 is written as ${}^{238}_{92}\text{U}$. To determine the number of neutrons, $N$, we solve the equation $A = N + Z$, or $238 = 92 + N$. The number of neutrons in uranium-238 is $N = 238 - 92 = 146$.

**Elementary particles**

The same notation is used to represent the elementary particles. The proton is ${}^{1}_{1}\text{p}$ because it has one nucleon and one positive elementary charge. The neutron is ${}^{1}_{0}\text{n}$ because it is neutral. The electron is ${}^{0}_{-1}\text{e}$ because it is not a nucleon and has negative charge.

**How is mass related to energy?**

You may have heard Einstein's equation many times, but did you ever stop to think about what it means? The equation, which relates energy, mass, and the speed of light, is a succinct statement of a powerful concept: that mass can be converted into pure energy and vice versa. This **mass–energy equivalence** is at the heart of nuclear physics and its applications, such as nuclear power generation. Equation (27.1) describes exactly how much mass is associated with a given amount of energy *and* how much energy it takes to create a given amount of mass.

Interactive equation

**Mass-energy equivalence**

$$(27.1) \quad E = mc^2$$

$E$ = energy (J)
$m$ = mass (kg)
$c$ = speed of light = $3\times10^8$ m/s

**Even at rest, matter has energy**

Einstein's equation tells us how much **rest energy** is contained in matter itself. All matter has rest energy even if the matter has no kinetic energy, potential energy, thermal energy, or any other type of energy we have discussed. Rest energy is intrinsic to matter itself. Whether the matter is stopped or moving, matter *itself* is energy and energy is matter.

**Matter contains vast quantities of rest energy**

The factor $c^2$ is an enormous number: $c^2 = 9 \times 10^{16}$ m$^2$/s$^2$. Matter contains vast quantities of rest energy through the mass–energy equation. Nuclear reactions convert some of the rest energy into other forms of energy, such as heat. When one kilogram of uranium reacts in a nuclear power plant, about 0.7% of its mass is converted to energy. This tiny fraction is more than a million times more energy than burning one kilogram of coal or oil.

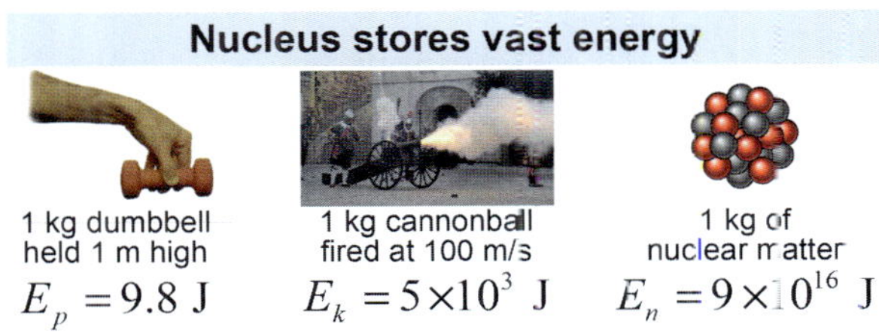


**Electron volt (eV)**

Nuclear scientists often express mass in energy units based on equation (27.1). The atomic mass unit (amu) corresponds to a mass of $1.661\times10^{-27}$ kg. The energy equivalent of 1 amu is calculated from Einstein's relationship:

$$E = mc^2 = (1.661 \times 10^{-27}\ \text{kg}) \times (2.998 \times 10^8\ \text{m/s})^2 = 1.493 \times 10^{-10}\ \text{J}$$

The energy unit used in nuclear physics is the electron volt (eV). One eV corresponds to $1.602 \times 10^{-19}$ J (page 767). Therefore, in energy units,

$$(1\ \text{amu})\ c^2 = 931.5\ \text{MeV},$$

which is 931.5 million eV or $9.315 \times 10^8$ eV.

**Calculating mass–energy**

What is the difference in rest energy between a neutron and a proton? (The masses of the neutron and proton are $m_n = 1.675 \times 10^{-27}$ kg and $m_p = 1.673 \times 10^{-27}$ kg, respectively.)

*Asked:* difference in rest energy

*Given:* mass of the neutron, $m_n = 1.675 \times 10^{-27}$ kg;
mass of the proton, $m_p = 1.673 \times 10^{-27}$ kg

*Relationships:* $E = mc^2$; speed of light, $c = 2.998 \times 10^8$ m/s

*Solution:* The difference in the two particle's rest energies is the difference between their two masses multiplied by $c^2$:

$$\Delta E = (\Delta m)c^2 = (1.675-1.673)\times(10^{-27}\ \text{kg})\times(2.998\times10^8\ \text{m/s})^2 = 1.8\times10^{-13}\ \text{J}$$

*Answer:* $1.8 \times 10^{-13}$ J

**Why is binding energy important?**

The energy $E_b$ that must be supplied to separate a nucleus into its constituent protons and neutrons is called the **binding energy**. Different elements, and different isotopes, have different amounts of binding energy. For example, if eight protons and eight neutrons make two helium-4 nuclei, then the amount of binding energy is different compared to the same particles arranged as a single oxygen-16 nucleus. *The difference in binding energy between isotopes provides the energy for nuclear reactions.*

**Nuclear mass deficiency**

When free nucleons come together in a nucleus, the binding energy is released. That means the nucleus has less rest energy *and therefore less mass* than the separate particles it is made from. The difference in mass, called the **mass deficiency**, is exactly equal to the binding energy required to hold the nucleus together. Binding energy and the mass difference are related through Einstein's mass–energy equivalence equation.

**An example: nucleus of helium-4**

As an example, consider the nucleus of helium-4, which has two protons and two neutrons. The mass of the individual protons and neutrons is 4.0319 amu.

$$\text{Two protons: } m_{2p} = 2 \times (1.6726 \times 10^{-27}\text{ kg}) = 2 \times 1.0073\text{ amu} = 2.0146\text{ amu}$$

$$\text{Two neutrons: } m_{2n} = 2 \times (1.6749 \times 10^{-27}\text{ kg}) = 2 \times 1.0087\text{ amu} = 2.0173\text{ amu}$$

$$\text{Total mass of nucleons: } m_{2p} + m_{2n} = 4.0319\text{ amu}$$

The mass of the helium-4 nucleus is 4.0015 amu. This is 0.0304 amu *less* than the sum of the masses of the individual protons and neutrons. The mass difference of 0.0304 amu is helium-4's mass deficiency, equal to 28.32 MeV in energy units:

$$E_b = (0.0304\text{amu}) \times (931.5\text{MeV/amu}) = 28.32\text{MeV}$$

**Binding energy per nucleon**

A useful quantity in nuclear physics is the *binding energy per nucleon*, or the binding energy $E_b$ divided by the number of nucleons, $A$. For the example of helium-4, the binding energy per nucleon ($E_b/A$) is (28.3 MeV)/4 = 7.08 MeV. The binding energy per nucleon increases with atomic number up to element 56, which is iron (Fe). The binding energy per nucleon decreases slowly with atomic number for elements heavier than iron. The shape of the binding energy graph reflects a balance between the repulsion of the electric force (among the protons) and the attraction of the strong force.

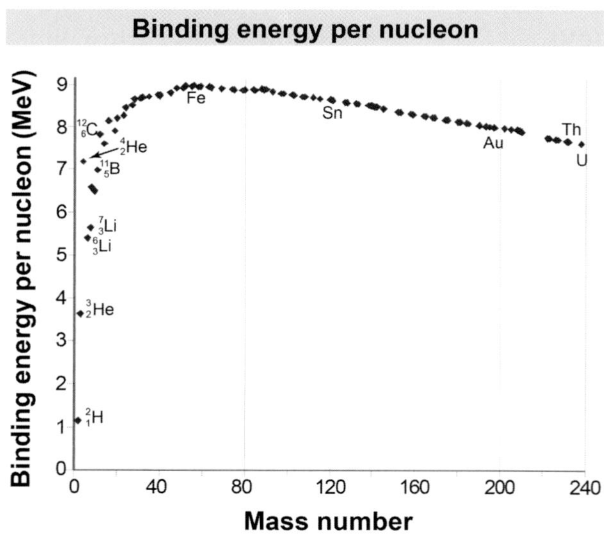


**What does the binding energy graph mean?**

The graph of binding energy per nucleon reflects the ultimate energy source of our universe. Stars release nuclear energy by combining light elements and moving from left to right, from hydrogen toward iron. Nuclear reactors release energy by splitting up heavy atoms and moving from right to left, from uranium toward iron.

**Forces in the nucleus**

The richness of nuclear phenomena from radioactivity to nuclear energy generation is derived from the balance between the attractive and repulsive forces among the nucleons. The attractive nuclear force is experienced by both protons and neutrons, whereas the repulsive Coulomb force is experienced only by the positively charged protons.

**Neutrons: the "glue" in the nucleus**

For all elements heavier than hydrogen, the balance of forces in the nucleus depends strongly on the number of neutrons. The "neutron glue" that holds the nucleus together comes from the action of the strong nuclear force. Without enough neutrons, the overall effect of the strong nuclear force is not sufficient to hold the nucleus together against the Coulomb force. If there are enough neutrons, then the attraction from the strong nuclear force overcomes the Coulomb force and the nucleus stays together. To achieve this force balance, every element heavier than helium has least one neutron for every proton.

**Chart of nuclides and nuclear stability**

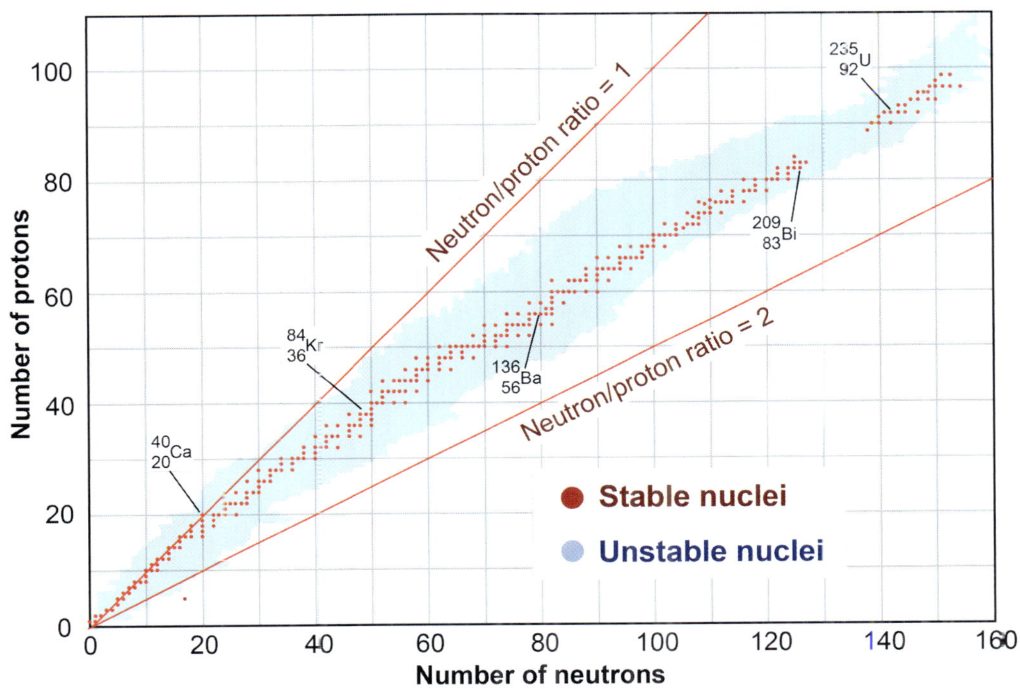


**The graph of protons vs. neutrons**

The balance of protons and neutrons is beautifully illustrated by graphing the number of protons versus the number of neutrons. The upper red line represents nuclei with equal numbers of protons and neutrons ($N = Z$). The red circles represent stable nuclei, such as carbon-12. The blue region on the graph shows nuclei, such as uranium-238, that may exist but are unstable (or radioactive). The white area represents combinations of neutrons and protons that cannot remain bound to form a nucleus, even for a nanosecond.

**Nuclear stability**

Notice that stable nuclei tend to lie close to the line $N = Z$ up to atomic number 20 (calcium). Past calcium, the stability line bends down, indicating more neutrons than protons. For example, uranium-235 has 92 protons and 143 neutrons, which is 51 more neutrons than protons. As the size of the nucleus increases, the short range of the strong force cannot bind the outermost nucleons as tightly as those near the center. To maintain equilibrium in large nuclei requires an excess of neutrons. The extra neutrons "dilute" the effect of the repulsive Coulomb force.

## Periodic table and electron structure

The periodic table is an indispensable tool in science. It is organized by the chemical behavior of the elements and it gives us a map for predicting the ability of elements to combine and form chemical compounds. The periodic table does not contain information about isotopes since isotopes of the same element have the same electron structure and therefore the same chemical properties. The only isotope-related information in the periodic table is contained in the atomic mass of the elements, which represents the average mass for all isotopes of that element.

**Periodic table colored by element groups**

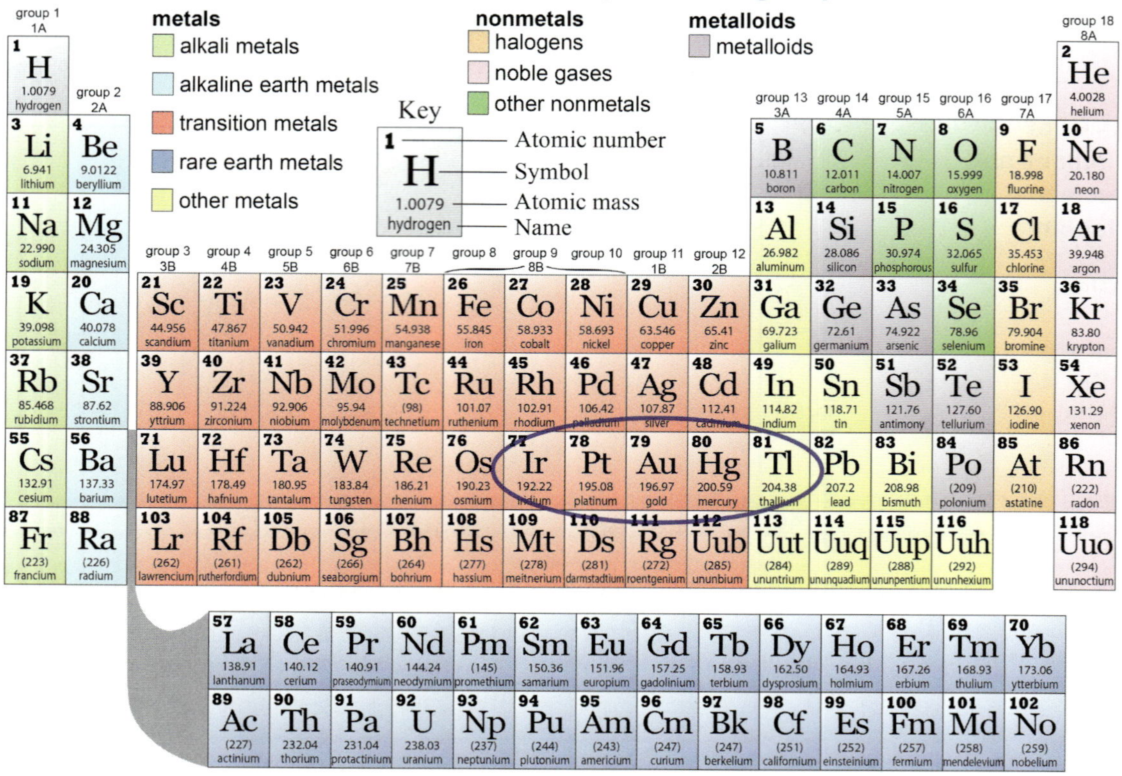


## Chart of nuclides and nuclear properties

In nuclear science, nuclear engineering, nuclear medicine, and any other nuclear discipline, nuclides need to be organized according to their isotopes. The chart of nuclides is a table that maps the atomic number $Z$ and the neutron number $N$ for each element. The chart of nuclides also contains information about the radioactive properties of the nuclides. Here is an example of a portion of the chart of nuclides centered at gold $^{197}_{79}Au$. Note the various isotopes for gold. The row above gold gives information about the isotopes of mercury (Hg), which is to the right of Au on the periodic table. The row below Au gives information about the isotopes of platinum (Pt). The green squares represent stable isotopes. The time shown on the white squares is the half-life of the particular isotope—which we will learn about on page 805.

**Portion of chart of nuclides**

| Atomic number $A$ | 116 | 117 | 118 | 119 | 120 |
|---|---|---|---|---|---|
| 81 | $^{197}$Tl 2.83 h | $^{198}$Tl 5.3 h | $^{199}$Tl 7.4 h | $^{200}$Tl 26.1 h | $^{201}$Tl 73.6 h |
| 80 | $^{196}$Hg | $^{197}$Hg 64.1 h | $^{198}$Hg | $^{199}$Hg | $^{200}$Hg |
| 79 | $^{195}$Au 183 d | $^{196}$Au 6.18 d | $^{197}$Au | $^{198}$Au 2.70 d | $^{199}$Au 3.14 d |
| 78 | $^{194}$Pt | $^{195}$Pt | $^{196}$Pt | $^{197}$Pt 18.3 h | $^{198}$Pt |
| 77 | $^{193}$Ir | $^{194}$Ir 19.2 h | $^{195}$Ir 2.5 h | $^{196}$Ir 52 s | $^{197}$Ir 9.9 min |

Neutron number $N$

# Section 1 review

We infer the presence of a strong nuclear force because positively charged protons in the nucleus would otherwise repel each other. The strong nuclear force is much stronger than the Coulomb force, but its range does not extend beyond the approximate size of a nucleus. The energy captured in the nucleus is called binding energy and is calculated by Einstein's equation of mass–energy equivalence. The mass of a nucleus of an element is less than the mass of its constituent nucleons. The difference is called mass deficiency. The mass deficiency and binding energy of a nucleus are related through Einstein's equation.

For more information to research the historical development of the concepts of the strong (and weak) forces, see

*Coming of Age in the Milky Way* by Timothy Ferris,
*Strange Beauty* by George Johnson, and
*Story of the W and Z* by Peter Watkins.

| Vocabulary words | **strong nuclear force, atomic mass unit (amu), mass–energy equivalence, rest energy, binding energy, mass deficiency** |
|---|---|
| **Key equations** | $E = mc^2$ |

## Review problems and questions

1. Research and describe the historical development of the concept of the strong nuclear force.

2. What is the rest energy of an electron (whose mass = $9.11 \times 10^{-31}$ kg)?
   a. $1.0 \times 10^{-47}$ J
   b. $3.0 \times 10^{-39}$ J
   c. $2.7 \times 10^{-22}$ J
   d. $8.2 \times 10^{-14}$ J

3. In your body, are there more protons than neutrons? More protons than electrons?

4. The isotope ${}^{16}_{8}O$ has a mass of 15.99491 amu. What is its binding energy?

5. Calculate the mass deficiency for the nucleus of tritium ${}^{3}_{1}H$, which is an isotope of hydrogen with a mass of 3.0160 amu. (The mass of a proton is 1.0073 amu, the mass of a neutron is 1.0087 amu, and the mass of an electron is 0.0005 amu.)

6. What is the binding energy of the nucleus of a tritium atom?

## 27.2 - Radioactivity

Radioactivity describes the property of certain nuclei to emit energy. The word *radioactivity* was coined by Polish-born French chemist and physicist Marie Curie, who was the first to systematically investigate the phenomenon. She was the first woman to win a Nobel prize and the only woman ever to win *two* Nobel prizes (one each in physics and chemistry). Radioactivity carries energy derived from the structure of the nucleus. The energy associated with radioactivity can be hazardous to biological systems since it can penetrate tissues and deposit energy, thereby damaging cells and vital organs. Radioactivity, however, may also be used as a diagnostic and treatment tool, again by virtue of its ability to penetrate matter.

### Radioactive decay

**Radioactive decay**

Radioactivity is released by nuclear processes known as *radioactive decay.* A **radioactive decay** is a process by which a nucleus spontaneously undergoes a change that emits energy. The three most common forms of radioactive decay are denoted by the first three letters of the Greek alphabet: alpha ($\alpha$), beta ($\beta$), and gamma ($\gamma$) decay. Alpha and beta decay result in the emission of particles from the nucleus; gamma decay is the release of a high-energy photon. As a result, alpha and beta decay change the atomic number of the decaying atom, but gamma decay does not.

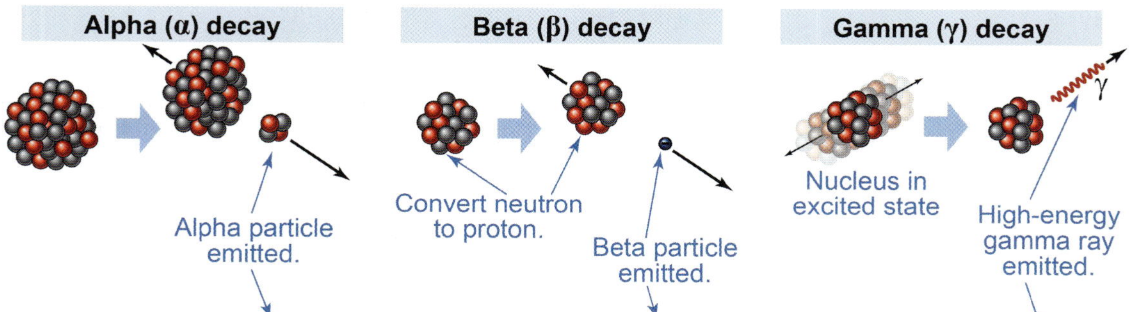


$^{238}_{92}U \rightarrow {}^{234}_{90}Th + {}^{4}_{2}He$ $\qquad$ $^{14}_{6}C \rightarrow {}^{14}_{7}N + {}^{0}_{-1}e$ $\qquad$ $^{12}_{6}C \rightarrow {}^{12}_{6}C + \gamma$

**Nuclear stability**

When a nucleus decays, it changes from a state with less binding energy to a state with more binding energy. The change in binding energy is both the cause of the change and the source of the emitted energy. For example, when uranium-238 decays via alpha decay it becomes thorium-234. The binding energy per nucleon in the $^{238}_{92}U$ nucleus is 7.58 MeV. For $^{234}_{90}Th$, the binding energy per nucleon is 7.61 MeV. The difference in the binding energy of the two nuclei (7.61 − 7.58) MeV = 0.03 MeV is the energy associated with the released alpha particle. Note that $^{234}_{90}Th$ is more stable because the binding energy is higher.

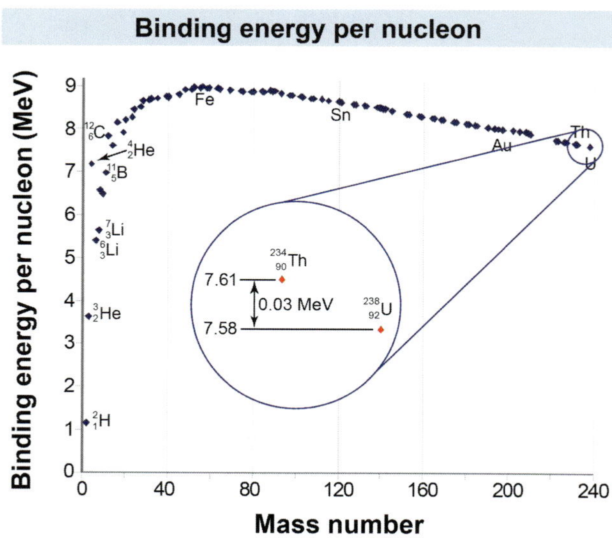

**Alpha decay** An alpha particle is the nucleus of a helium atom, consisting of two protons and two neutrons. Since the alpha particle has no electrons, it has a net *positive* charge, which makes it different from an ordinary, neutral helium atom. When **alpha decay** occurs, the nucleus emits an alpha particle that carries a modest amount of kinetic energy away with it. The energy of an alpha particle is low enough to be blocked by just a piece of paper!

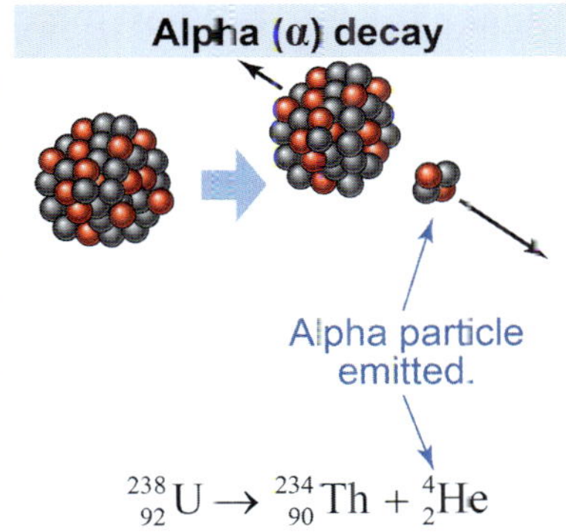


**Beta decay** A neutron may spontaneously transform into a proton and an electron: $^{1}_{0}n \rightarrow ^{1}_{1}p + ^{0}_{-1}e$. This type of transformation is done via a process called **beta decay** (β decay). When a nucleus undergoes beta decay, it emits an energetic electron or a positron. These energetic particles are called beta particles—denoted as $\beta^-$ or $\beta^+$. *Beta particle* is a historical term for an electron. Originally, it was Ernest Rutherford who coined the term beta radiation to distinguish it from the alpha radiation that he also discovered and used to study the nucleus.

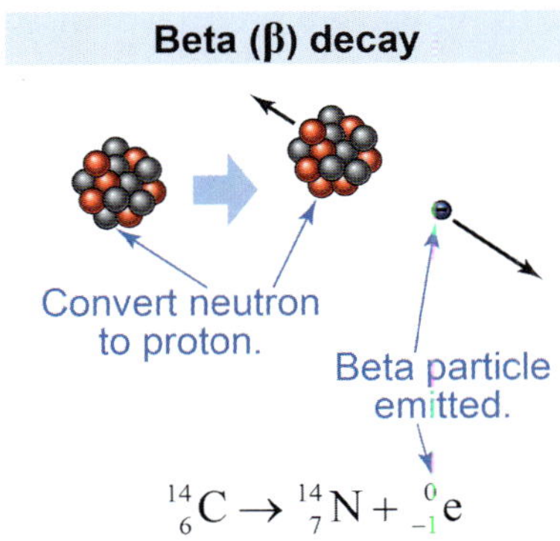


**Positrons** Similarly, a proton may transform into a neutron and a positively charged particle called a positron: $^{1}_{1}p \rightarrow ^{1}_{0}n + ^{0}_{1}e$. The positron is denoted by $^{0}_{1}e$ or $\beta^+$ and it is a positively charged electron.

**Gamma decay** In Chapter 26 we learned that quantum theory forces electrons confined to an atom to have energy levels. The same is true of nucleons confined to the nucleus, except that the energies are much higher because they involve the strong nuclear force. A nucleus is in an *excited state* when a proton or neutron occupies an energy level above the nuclear ground state. An excited nucleus releases energy by emitting a high-energy photon or γ ray via the process of **gamma decay** (γ decay).

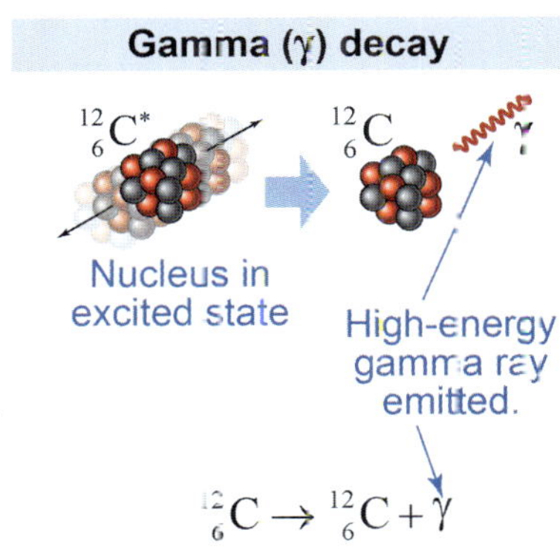


**What causes gamma decay?** Gamma decay is almost always the result of leftover energy from another decay or nuclear reaction. When a nucleus undergoes alpha or beta decay, some of the change in binding energy appears as kinetic energy of the emitted particle. Part of the remaining energy puts the nucleus in an excited state. The excited nucleus goes to a lower energy state by releasing a γ ray.

**Gamma ray energy** Gamma decay is analogous to the emission of visible light by an atom as an electron transitions from a higher to a lower energy level. The energy of a γ-ray photon, however, is much higher than the energy of a visible photon. Gamma rays are at the high-energy and high-frequency region of the electromagnetic spectrum. Gamma rays are very penetrating and to stop them requires material with high atomic mass and high density, such as lead.

## Weak nuclear force

**What causes beta decay?**

What force causes beta decay, in which a neutron is converted into a proton and an electron? The neutron has no electric charge, so its decay cannot result from the electromagnetic force. The electron is not a nucleon, so it is not affected by the strong nuclear force. The gravitational force is far too weak. Since none of these three forces can be responsible, the existence of beta decay is evidence for the presence of a fourth force of nature, the *weak nuclear force*.

**Another particle in beta decay?**

Radioactivity was discovered in 1896 by French scientist Henri Becquerel (Marie Curie's professor). In 1900, Becquerel showed that beta rays were identical to electrons, recently discovered by J. J. Thompson. Over the next few decades, physicists realized that whenever beta decay occurred, some energy and momentum were missing and not accounted for by the proton and electron. Based on this evidence, Austrian theorist Wolfgang Pauli argued in 1930 that there had to be another, unseen particle produced by beta decay that carried no charge and had little or no mass.

| Beta decay | n | $\longrightarrow$ | p | + | e | + | $\bar{\nu}_e$ |
|---|---|---|---|---|---|---|---|
| | Neutron | | Proton | | Electron | | Antineutrino |

**Neutrinos**

In 1934, Italian physicist Enrico Fermi proposed a model for the weak nuclear force governing beta decay that incorporated Pauli's particle, which Fermi called the *neutrino*. Because it is so hard to detect, the neutrino (actually the *antineutrino*!) was not experimentally discovered until 1956 by Americans Frederick Reines and Clyde Cowan. Neutrinos are not affected by the electromagnetic force because they have no charge. They are also not affected by the strong nuclear force, which allows them to pass right through the Earth! In fact, neutrinos are *only* affected by the weak nuclear force.

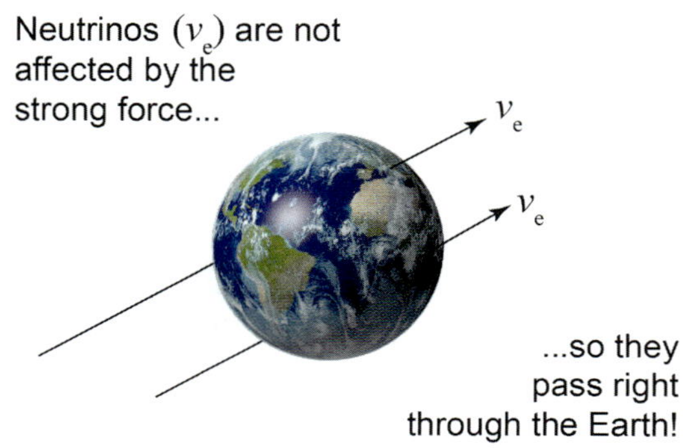


**Radioactivity and the weak nuclear force**

Beta decay occurs through the **weak nuclear force**, which is one of the four fundamental forces of nature. The weak force has its name because this force is about a million times weaker than the strong nuclear force and the electromagnetic force. The weak force also operates at shorter distances than even the strong force. All types of matter are affected by the weak force however, the effects of the weak force are typically only observed when strong or electromagnetic forces are absent or inactive.

**Effects of the weak force**

The effects of the weak nuclear force are important for carbon-14 dating, which you will learn about on the next two pages. Beta decay also occurs in the core of the Sun as part of the nuclear fusion reaction that converts two hydrogen atoms into a deuterium atom. Beta decay in the Sun's *proton–proton* reactions produces positrons instead of electrons.

**W and Z bosons**

In 1968, Americans Sheldon Glashow and Steven Weinberg and Pakistani Abdus Salam proposed a theory for the weak nuclear force that went beyond Fermi's earlier model. The new model argued that the weak force is mediated through three not-yet-seen and massive particles, the $W^+$, $W^-$, and $Z^0$ bosons. The particles were detected in 1983 by a large collaboration led by Italian physicist and inventor Carlo Rubbia and Dutch physicist Simon van der Meer.

### Decay of radioactive materials is statistical

Radioactive isotopes decay *spontaneously*; that is, the decay reaction occurs randomly without outside influences. If we look at the nucleus of a radioactive isotope, we do not know when it will undergo radioactive decay. It could be in the next second, a few minutes from now, or in days, years, centuries—or even in a few millennia. In the macroscopic world, the forces of gravity and electromagnetism are *deterministic*, meaning that we can write down the equations and can calculate exactly how every object moves and interacts with the environment. In the microscopic world of the nucleus, however, the weak force and the decay of radioactive atoms is *statistical*, not deterministic. This is one way in which our everyday world and the quantum world are fundamentally different.

### Half-life

The number of atoms that will decay radioactively in a fixed period of time is determined by the isotope's **half-life**. In the time period of one half-life, half of those radioactive atoms will undergo radioactive decay. The half-life is denoted by the symbol $t_{1/2}$.

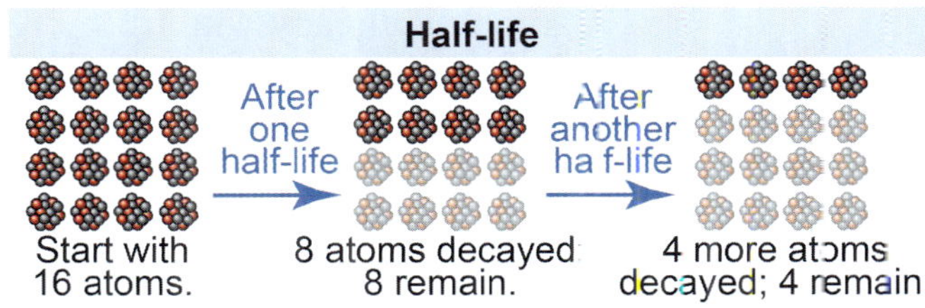


Imagine that you have 16 atoms of the same radioactive isotope and it has a half-life of one hour. After one hour, half of those atoms—eight of them—will have decayed, while the other eight are unchanged. Now wait another hour. Of the eight previously undecayed atoms, half of them—four more atoms—will decay in this second hour, while the other four atoms are still unchanged. Wait again for an hour. At the end of this third hour, half of the four undecayed atoms—two more atoms—will have decayed, while the other two still stay the same. That is how half-life works: After the time period of one half-life, half of the atoms will have decayed radioactively.

### Half-life for various nuclides

| Nuclide | Half-life | Decay type |
|---|---|---|
| $^{238}_{92}U$ | 4.5 × $10^9$ years | alpha |
| $^{131}_{53}I$ | 8.05 days | beta |
| $^{137}_{55}Cs$ | 30.2 years | beta |
| $^{220}_{86}Rn$ | 55.6 s | alpha |
| $^{14}_{6}C$ | 5,730 years | beta |

Some radioactive isotopes have a very short half-life, whereas others have a half-life of billions of years. For example, uranium-238 has a half-life of 4.5 billion years and so it has been around since the formation of our Solar System. Iodine-131 and cesium-137 are radioactive products of nuclear fission. Fluorine-18, which is a isotope used in positron emission tomography (PET), has a half-life of about 110 minutes and so it does not occur naturally. Isotopes with very short half-life are generated in the laboratory.

### Example of carbon-14 half-life

Every half-life, the number of radioactive nuclei decreases by half. The number of carbon-14 nuclei in a sample will decrease by half every 5,730 years. As carbon-14 decays via beta decay, nitrogen-14 is generated.

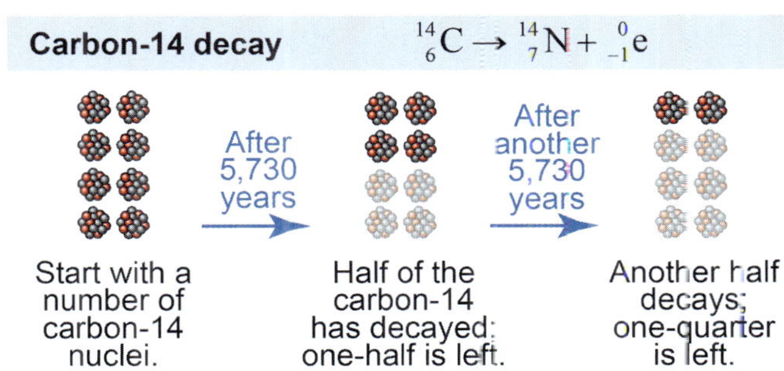

## Carbon dating

**How carbon dating works**

**Carbon dating** is the technique of comparing the concentrations of carbon-12 and carbon-14 to determine the age of a biological material. While the most common isotopes of carbon, ${}^{13}_{6}C$ and ${}^{12}_{6}C$, have stable nuclei, ${}^{14}_{6}C$ is radioactive. Carbon-14 is produced steadily in Earth's upper atmosphere, where ${}^{14}_{7}N$ nuclei interact with neutrons produced by cosmic rays entering the atmosphere. Living organisms constantly exchange carbon with the environment maintaining a constant ratio of about one atom of carbon-14 for every $10^{13}$ atoms of carbon-12. When an organism dies, active carbon exchange with the environment stops. Carbon-14 already within the organism decays radioactively with a half-life of 5,730 years. This decay provides a natural clock which starts ticking the moment an organism dies. For example, carbon dating can determine the age of ancient Egyptian papyrus samples and the charcoal from prehistoric campfires. Carbon dating works reliably up to about 10 times the half-life, or 57,300 years. After 10 half-lives there is not enough carbon-14 remaining to measure accurately.

**Carbon dating**

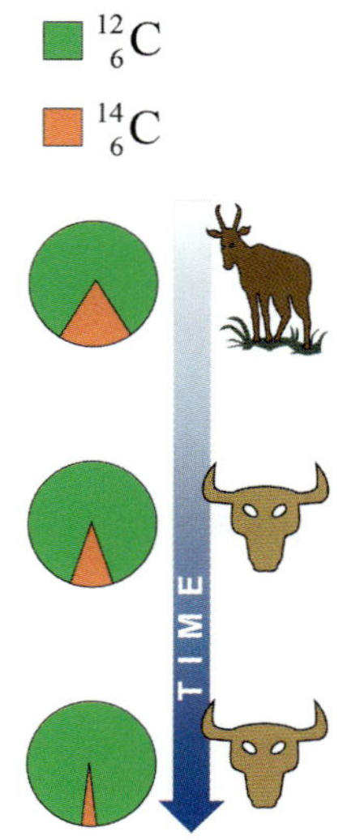


**Mathematics of carbon dating**

The rate of radioactive decay—how many nuclei decay per unit time—is related to the half-life $t_{½}$. If we have a radioactive sample that initially has $N_0$ nuclei, then after time $t$ the number of radioactive nuclei remaining in the sample is given by equation (27.2).

Interactive equation

**Decay rate equation**

$$\textbf{(27.2)} \quad N = N_0\left(\frac{1}{2}\right)^{t/t_{½}}$$

$N$ = amount of sample after time $t$
$N_0$ = initial amount of sample
$t$ = time elapsed (s)
$t_{½}$ = half-life (s)

**Radiometric dating**

The general technique for dating matter using radioactivity is called radiometric dating. Besides carbon-14, a number of other radioactive isotopes are used for dating. For example, climate scientists measure the concentration of oxygen-18 and oxygen-16 in ice core samples to understand the composition of the atmosphere over time. Geologists use this same technique to date rock samples by measuring the concentration ratio of uranium-238 to plutonium-239. This technique allows geologists to explore geologic times because the half-life of uranium-238 is 4.5 billion years.

**Calculating age using carbon-14 dating**

An ancient fern specimen has one-eighth the amount of carbon-14 per gram as compared to a living creature. How old is the specimen?

| | |
|---|---|
| *Asked:* | the number of years ago the fern died |
| *Given:* | The mass of carbon-14 has decayed to one-eighth its original value. |
| *Relationships:* | For the time period of one half-life, half of the radioactive atoms will decay radioactively.<br>The half-life of carbon-14 is 5,730 years. |
| *Solution:* | After one half-life, half of the carbon-14 atoms will be left. After the second half-life, one-half of the remaining one-half carbon-14 atoms will remain, or one-fourth. After the third half-life, one-eighth will remain. The fern therefore died three half-lives of carbon-14 ago: 3 × 5,730 years is 17,190 years. |
| *Answer:* | The fern died 17,190 years ago. |

# Section 2 review

Radioactive decay is the spontaneous emission of an energetic particle or photon from a nucleus. The three types of radioactive decay are called alpha, beta, and gamma. Radioactive materials are also identified by their half-life, the time in which half of the original atoms have decayed. Carbon dating is an effective method for dating the age of biological relics and is based on the 5,730 year half-life of carbon-14. (See page 1592 for further references to research the historical development of the concept of the weak nuclear force.)

| Vocabulary words | radioactive decay, alpha decay, beta decay, gamma decay, weak nuclear force, half-life, carbon dating |
|---|---|
| Key equations | $N = N_0\left(\frac{1}{2}\right)^{t/t_{1/2}}$ |

## Review problems and questions

1. Research and describe the historical development of the concept of the weak force.

2. Uranium-235 undergoes beta decay according to this nuclear reaction equation:

$$^{235}_{92}\text{U} \rightarrow {}^{A}_{Z}? + {}^{0}_{-1}\text{e}$$

What is the element and isotope produced?

3. The unstable uranium isotope 235 undergoes radioactive decay to produce thorium-231 following the equation

$$^{235}_{92}\text{U} \rightarrow {}^{231}_{90}\text{Th} + {}^{A}_{Z}?$$

What is the atomic and mass number of the emitted particle? What is the name for this particle and type of decay?

4. In the radioactive decay equations in questions #2 and 3 above, do you expect the binding energy of the resultant nucleus (or nuclei) to be larger or smaller than the binding energy of the original nucleus?

5. A two level decay process starting with americium-240 ends up with plutonium-236, as indicated by the arrows on this relevant portion of the chart of nuclides. These two levels are indicated by the arrows marked A and B on the chart of nuclides.

What are these two decay processes?

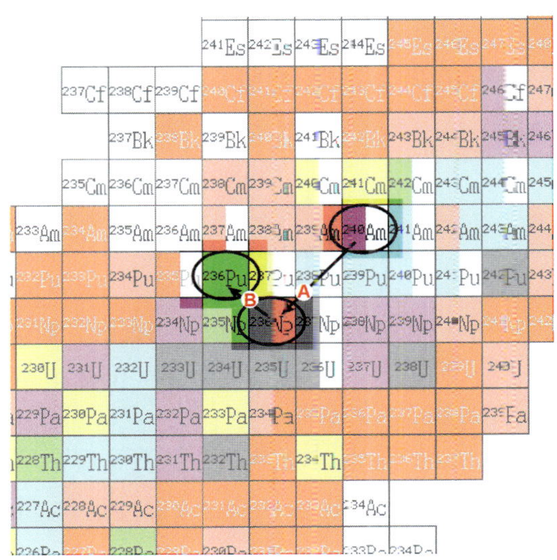

## 27.3 - Nuclear reactions

A *nuclear* reaction is any process that changes the nucleus of an atom. Nuclear reactions may change one element into another or change one isotope into a different isotope of the same element. Because nuclear reactions involve the strong nuclear force, the energy is typically a millions times greater compared to *chemical* reactions. Chemical reactions rearrange shared electrons but do not change the nucleus.

**Nuclear vs. chemical reactions**

Nuclear reactions create new elements.

Chemical reactions create new compounds from elements.

### Writing balanced nuclear reaction equations

**Representing nuclear reactions**

Nuclear reactions can change an element into a different element by altering the number of protons and neutrons. Nuclear reactions are represented by nuclear reaction equations, just as chemical reactions are represented by chemical reaction equations. Every nuclear reaction equation has two parts: reactants and products. Consider the nuclear reaction that produces carbon-14 in the atmosphere. In this nuclear reaction, a neutron interacts with nitrogen-14 to produce carbon-14 and a proton. The equation for this reaction is

$$\underbrace{{}^{1}_{0}\text{n} + {}^{14}_{7}\text{N}}_{\text{Reactants}} \rightarrow \underbrace{{}^{14}_{6}\text{C} + {}^{1}_{1}\text{p}}_{\text{Products}} \qquad (27.3)$$

**How to balance nuclear reactions**

A correct nuclear equation must show mass balance by having the mass numbers equal on both sides of the equation. First, this means that the total number of protons plus neutrons must be the same before the reaction and after the reaction. Second, the total charge before the reaction must equal the total charge afterward, which is the basic law of charge conservation. These two fundamental rules for balancing nuclear reaction equations can be used to determine an unknown element in a reaction.

**Balancing rules for nuclear reaction equations**

**Mass number balance:**
Total mass number of reactants
= total mass number of products.

$${}^{226}_{88}\text{Ra} \rightarrow {}^{222}_{86}\text{Rn} + {}^{4}_{2}\text{He}$$

**Charge balance:**
Total charge of reactants
= total charge of products.

**An example**

Consider the incomplete reaction equation

$${}^{1}_{0}\text{n} + {}^{235}_{92}\text{U} \rightarrow {}^{A}_{Z}\text{X} + {}^{94}_{38}\text{Sr} + {}^{1}_{0}\text{n} + {}^{1}_{0}\text{n}$$

To find the unknown element X we start by applying the charge conservation (proton balance) rule:

$$92 = Z + 38 \Rightarrow Z = 54$$

From the periodic table we see that the element with atomic number 54 is xenon. Next we identify the xenon isotope by applying the mass balance rule:

$$1 + 235 = A + 94 + 1 + 1 \Rightarrow A = 140$$

So the unknown nucleus is xenon-140 or ${}^{140}_{54}\text{Xe}$. The complete reaction is

$${}^{1}_{0}\text{n} + {}^{235}_{92}\text{U} \rightarrow {}^{140}_{54}\text{Xe} + {}^{94}_{38}\text{Sr} + {}^{1}_{0}\text{n} + {}^{1}_{0}\text{n}$$

**Three types of nuclear reactions**

Any change in the nucleus of an element is done via a nuclear reaction. The change may be due to radioactivity when a nucleus changes spontaneously or it may be the result of an interaction between two nuclei. In either case, we use a nuclear equation to represent the process. In general, there are three categories of nuclear reactions:

1. spontaneous reactions,
2. fission reactions, and
3. fusion reactions.

The rules for writing and balancing the equations for these reactions are the same.

**Decay and spontaneous reactions**

The decay of a nucleus happens spontaneously. It is characterized by the half-life of the process and it is described by a nuclear equation. An example of a spontaneous reaction is the decay of the isotope americium-235, which decays by emitting an alpha particle according to the equation

**Americium alpha decay**

$$ {}^{240}_{95}\mathrm{Am} \rightarrow {}^{236}_{93}\mathrm{Np} + {}^{4}_{2}\mathrm{He} $$

**Alpha decay in a smoke detector**

This reaction is used as the basis for smoke detector systems. In the presence of smoke, the alpha particle that is emitted by this decay interacts with the smoke particles, which it ionizes. The ionized particles in turn create a current that is detected by electronics, triggering the audible or visual indicators of the alarm.

**Fission reactions break apart a large nucleus**

In a fission reaction, the nucleus breaks apart not spontaneously but as a result of an interaction with another particle. Although the nucleus of uranium-235 is an alpha-emitter (with a half-life of 704 million years), when it is bombarded by neutrons with enough energy the nucleus can also be broken apart in a fission reaction. The equation that represents this reaction is

**Uranium-235 fission reaction**

$$ {}^{1}_{0}\mathrm{n} + {}^{235}_{92}\mathrm{U} \rightarrow {}^{140}_{54}\mathrm{Xe} + {}^{94}_{38}\mathrm{Sr} + {}^{1}_{0}\mathrm{n} + {}^{1}_{0}\mathrm{n} $$

Notice that there is one free neutron on the left-hand side but two neutrons on the right-hand side. This observation is the basis for nuclear energy generation and we will discuss it later in this chapter.

**Fusion reactions combine two nuclei**

The nuclear reaction by which two light nuclei *combine* to create a heavier nucleus is called a fusion reaction. An example of this process occurs in the core of the Sun, which produces energy by combining nuclei with low atomic numbers. A typical fusion reaction that combines two isotopes of hydrogen, deuterium (${}^{2}_{1}\mathrm{H}$) and tritium (${}^{3}_{1}\mathrm{H}$), is

**Deuterium–tritium fusion reaction**

$$ {}^{2}_{1}\mathrm{H} + {}^{3}_{1}\mathrm{H} \rightarrow {}^{4}_{2}\mathrm{He} + {}^{1}_{0}\mathrm{n} $$

This fusion reaction releases a large amount of energy, as can be seen from the plot of binding energy per nucleon. In general, the energy released from fusion reactions is much larger than the energy released from fission reactions. For this reason (and because of some other advantages of fusion over fission), fusion is considered a promising energy source for the future.

### Splitting a large nucleus releases energy

A **nuclear reaction** derives its energy from the binding energy of the nucleus. If a large nucleus such as uranium is broken apart, energy can be released. Splitting a nucleus is called a *nuclear fission reaction*. An example is the nuclear reaction that splits the $^{235}_{92}U$ nucleus into two products, strontium-94, and xenon-140. The energy released by this reaction can be estimated with an energy balance calculation of the binding energy.

**Binding energy per nucleon**

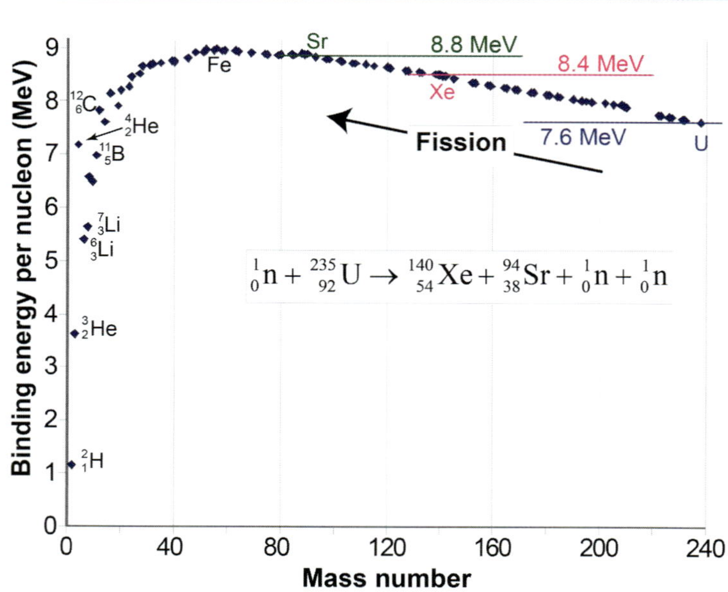


From the binding energy plot we see that the binding energy per nucleon for uranium, xenon, and strontium are approximately 7.6, 8.4, and 8.8 MeV, respectively. The energy released is then approximately equal to

$$\begin{aligned} E_{\text{released}} &= (8.4\text{ MeV}) \times A_{\text{Xe}} + (8.8\text{ MeV}) \times A_{\text{Sr}} - (7.6\text{ MeV}) \times A_{\text{U}} \\ &= (8.4\text{ MeV}) \times 140 + (8.8\text{ MeV}) \times 94 - (7.6\text{ MeV}) \times 235 = 217\text{ MeV} \end{aligned}$$

This is very close to the actual energy release, which is 208 MeV.

### Fission

**Nuclear fission reaction**

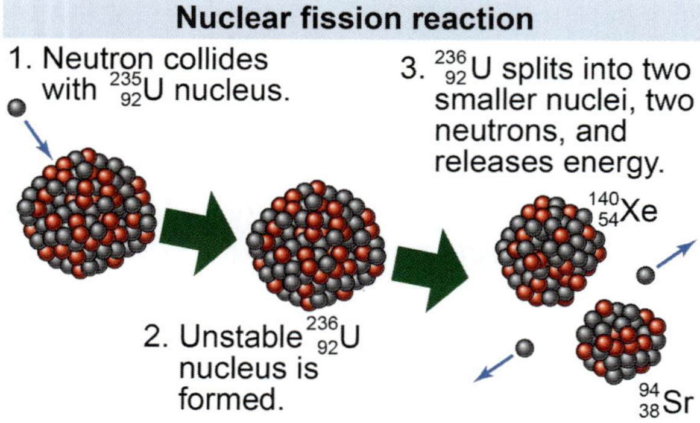


What initiates a **fission** reaction? In the late 1930s, German physicist Lise Meitner and chemists Otto Hahn and Fritz Strassmann discovered that bombarding uranium atoms with neutrons can initiate a fission reaction. The incoming neutron is initially absorbed into the $^{235}_{92}U$ nucleus, turning it into an unstable $^{236}_{92}U$ nucleus. The $^{236}_{92}U$ nucleus then undergoes decay into a pair of nuclei, ejecting two neutrons.

### Nuclear chain reaction

Bombarding a $^{235}_{92}U$ atom with one neutron can cause a nuclear fission reaction that ejects *two* neutrons. If there are many uranium atoms nearby, then those two ejected neutrons can initiate two new fission reactions that eject a total of four neutrons. Those four neutrons can start four new fission reactions that produce eight neutrons, and so on. This process is called a **chain reaction** and it is the key to continuous nuclear energy production.

**Nuclear chain reaction**

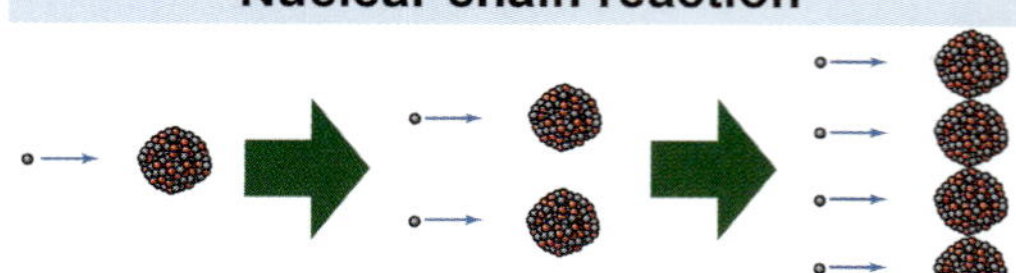

Each fission reaction creates two neutrons that start two more reactions.

**Nuclear fusion**

Whereas the nuclear fission reaction creates energy by splitting a heavy nucleus into two nuclei, the nuclear **fusion** reaction creates energy by *combining two light nuclei* into a larger nucleus. Fusion is the most important phenomenon in nature: It powers the Sun and so is ultimately the energy source for all biological and physical processes on Earth. A representative fusion reaction combines deuterium and tritium and yields helium. The energy released by this reaction is about 18 MeV. Previously, we calculated that the energy released by a fission reaction is about (200 MeV)/(235 nucleons) = 0.85 MeV/nucleon. As shown above, the energy released by fusion is (18 MeV)/(5 nucleons) = 3.6 MeV/nucleon. Per unit mass, the energy released by a fusion reaction is about four times larger than that released by a fission reaction.

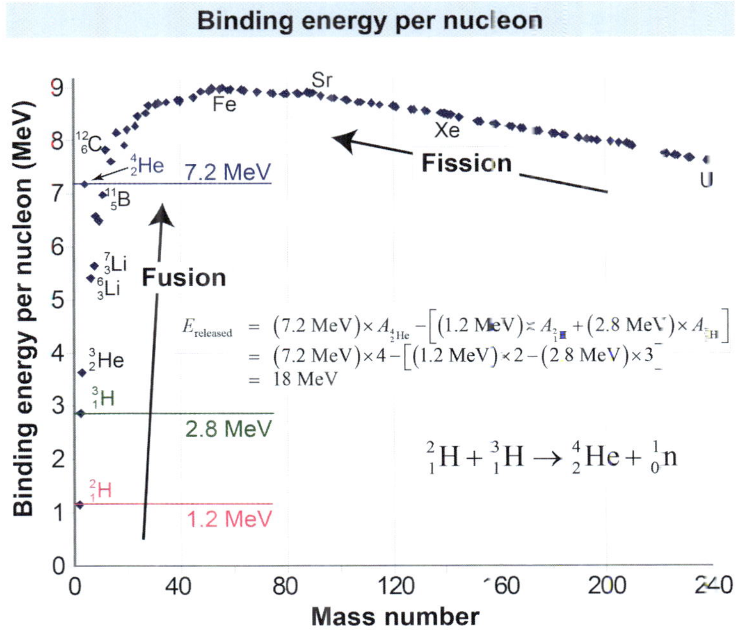


**Fusion powers the Sun**

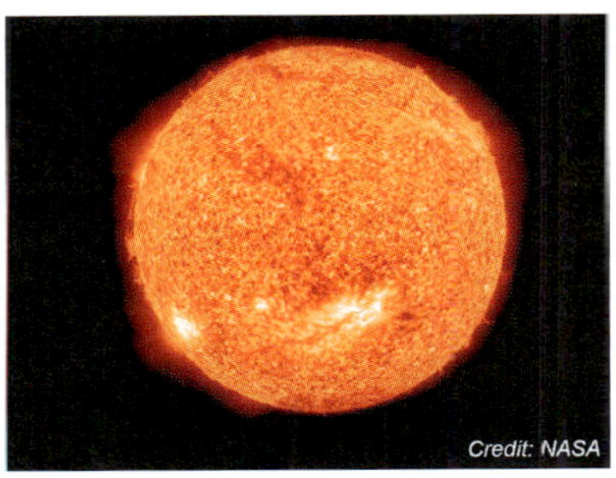


Most of the visible matter in the universe, including the matter inside of stars, is in the form of the element hydrogen. In the core of the Sun and other stars, the temperature and the pressure are so high that nuclear fusion reactions convert abundant hydrogen into helium, releasing energy in the process. The Sun produces nuclear fusion energy in its core that then turns into thermal energy of the matter in the Sun. The thermal energy is eventually transported outward to the surface of the Sun and then radiated into space.

**Elemental composition of the Sun**

The overall reaction that fuels the Sun is called the proton–proton cycle. The p–p cycle converts four hydrogen atoms (containing four protons) into a helium atom (containing two protons and two neutrons). The Sun is composed mostly of hydrogen and helium (91.2% hydrogen and 8.7% helium). Larger nuclei up to carbon and oxygen are present at a very small fraction. The energy required to fuse nuclei of larger atomic mass is much greater and it is found in stars that are in a different stage of their evolution. Mass–energy equivalence is behind fusion, which powers the Sun!

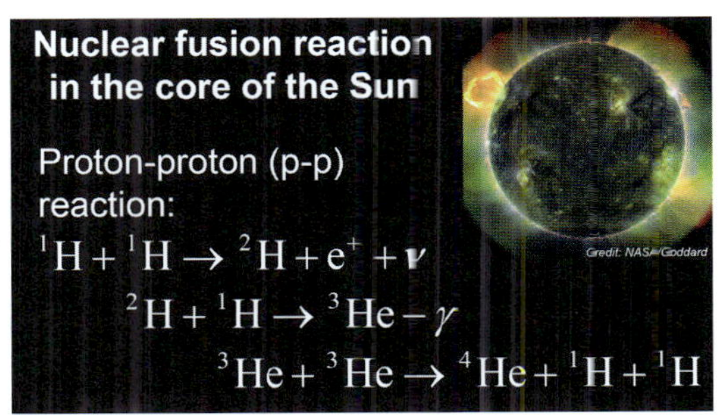

# Section 3 review

Nuclei exist because protons and neutrons have lower energy when bound together. The binding energy is stored in the nucleus and can be released if the nucleus changes, such as through a nuclear reaction. Nuclear reactions can be represented by nuclear reaction equations. Nuclear reactions may split nuclei (fission) or combine nuclei (fusion). Both types of reactions can release energy. Commercial nuclear reactors generate energy by nuclear fission whereas the Sun generates energy with nuclear fusion reactions.

| | |
|---|---|
| **Vocabulary words** | **nuclear reaction, fission, chain reaction, fusion** |
| **Key equations** | $\underbrace{{}^{1}_{0}\text{n} + {}^{14}_{7}\text{N}}_{\text{Reactants}} \rightarrow \underbrace{{}^{14}_{6}\text{C} + {}^{1}_{1}\text{p}}_{\text{Products}}$ |

## Review problems and questions

1. How is mass–energy equivalence related to nuclear fission and fusion?

2. Distinguish between *atomic* physics and *nuclear* physics.

3. Distinguish between *chemical* reactions and *nuclear* reactions.

4. Balance the following reaction equation by providing the missing product:

$$ {}^{2}_{1}\text{H} + {}^{3}_{2}\text{He} \rightarrow {}^{A}_{Z}\text{X} + {}^{1}_{1}\text{H} $$

5. For the reaction ${}^{A}_{Z}\text{X} \rightarrow {}^{22}_{10}\text{Ne} + {}^{0}_{1}\text{e}$ what is the unknown ${}^{A}_{Z}\text{X}$?
   a. ${}^{21}_{10}\text{Ne}$
   b. ${}^{22}_{11}\text{Na}$
   c. ${}^{23}_{10}\text{Ne}$
   d. ${}^{23}_{12}\text{Mg}$

6. Determine the name of the unknown element, the atomic number $Z$, and the atomic mass number $A$ in this nuclear reaction equation:

$$ {}^{1}_{0}\text{n} + {}^{235}_{92}\text{U} \rightarrow {}^{141}_{56}\text{Ba} + {}^{A}_{Z}? + 3{}^{1}_{0}\text{n} $$

7. In this nuclear reaction equation:

$$ {}^{1}_{0}\text{n} + {}^{235}_{92}\text{U} \rightarrow {}^{A}_{Z}? + {}^{87}_{35}\text{Br} + 3{}^{1}_{0}\text{n} $$

   determine the atomic mass number $A$, atomic number $Z$, and the name of the unknown element.

8. Smoke detectors contain radioactive isotope americium-235 and emit alpha particles, yet manufacturers claim that the products are safe. Evaluate the validity of this claim by researching natural exposure levels, medically-recommended limits, or exposure from medical imaging.

The nucleus stores and is capable of releasing large amounts of energy as it converts mass to energy via Einstein's mass–energy equivalence. The massive energies from nuclear phenomena have many applications in modern society, ranging from energy production to radiation therapy to medical and industrial noninvasive imaging. The decay properties of certain nuclei can also be used to target the deposition of their energy, such as in radiation therapy.

## Fusion energy

**Fusion in the Sun**

Since 1950 scientists and engineers have been working to design a fusion reactor that will make controlled production of fusion energy possible on Earth. In the Sun, the fusion reactions of the p–p cycle are driven by the pressure energy that is available in the extremely dense core of the star. The large mass of the Sun compresses the core to high densities, 150 times the density of ordinary water. On the Earth, however, it is not practical to achieve the same density as inside the Sun. As a result, if we want to initiate fusion reactions here on Earth, we must heat the fusion reactants to temperatures higher than the core of the Sun—higher than 16 million K!

**Fusion with deuterium**

Other fusion reactions are considered better candidates for nuclear power, but they generally require even higher temperatures than found in the Sun's core. The most promising reactions involve deuterium, an isotope of hydrogen that has an extra neutron in its nucleus. Although deuterium is somewhat rare in seawater —0.016% of hydrogen atoms are deuterium—Earth's massive oceans effectively provide a nearly inexhaustible supply of fuel. What is needed to create a practical fusion reactor on Earth that burns deuterium?

**Deuterium and tritium for nuclear fusion reactors**

$$^{2}H + ^{2}H \rightarrow ^{3}He + n$$

$$^{2}H + ^{2}H \rightarrow ^{3}H + ^{1}H$$

$$^{2}H + ^{3}H \rightarrow ^{4}He + n$$

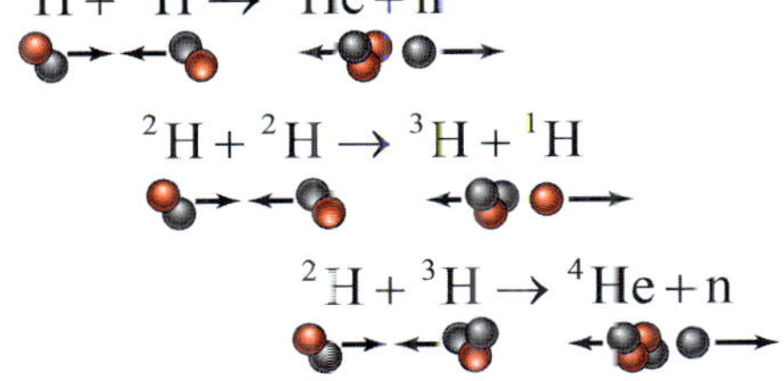

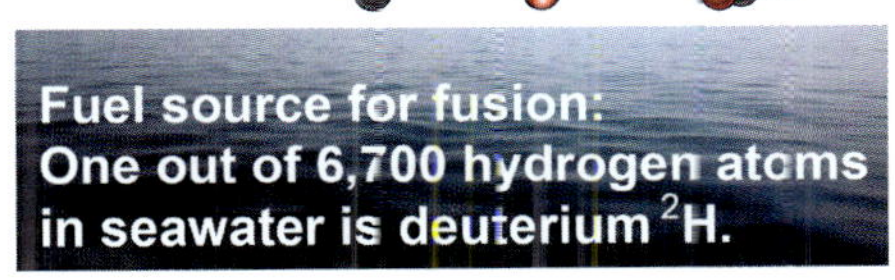


**Deuterium nuclear reactors**

**Tokamak fusion reactor**

Credit: Princeton Plasma Physics Laboratory

High temperatures of 400 million K are required for a fusion reactor and have been achieved in the laboratory. The current engineering challenge is how to *confine* the hot, ionized gas for long enough so that the fusion reactions can proceed and generate enough usable energy. The *tokamak* design uses strong magnetic fields to confine the hot gas in a doughnut-shaped chamber. The *inertial confinement* design, in contrast, zaps tiny fuel pellets with powerful lasers to create many tiny, thermonuclear hydrogen explosions. Future research and development will show whether either the tokamak or inertial confinement methods can be made practical to construct a large-scale, commercial fusion reactor.

## Nuclear power

**Nuclear energy**

The term **nuclear energy** is generally used to refer to nuclear *fission* reactions that are maintained through a chain reaction. Each fission reaction produces a significant amount of energy that can be transformed into other forms of energy—such as by heating water into steam. The steam can turn a steam turbine, creating mechanical energy, which then powers an electric generator to turn the energy into electricity. Nuclear power is typically used to generate electricity. In the USA, which is the world's largest producer of nuclear power, about 20% of the electricity is generated by nuclear reactors.

**Design of a nuclear reactor and power plant**

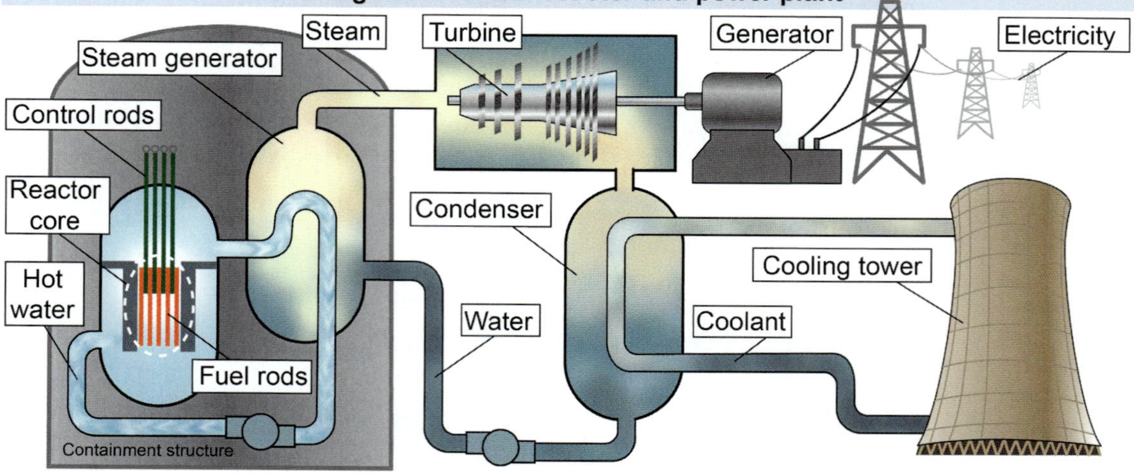


**Nuclear reactor**

At the heart of a nuclear power plant is the reactor core containing the nuclear fuel rods immersed in a circulating water system. It includes all the radioactive systems, and it is housed in a thick containment vessel. Water heated by the nuclear reactions exchanges energy with a secondary water system, which drives the steam turbine to generate electricity. Water from the secondary system exchanges heat with a tertiary system, which directs water into the large cooling tower. A key design principle of the nuclear power plant is to keep all the radioactive material (fuel rods and water) in the primary containment system, so that the other parts of the system do not become radioactive.

**Moderator and control rods**

One fundamental challenge in the design of a nuclear reactor is that the fission reaction requires a slow neutron to hit the uranium nucleus, yet *highly energetic neutrons* are produced by a fission reaction. In most reactors the fuel is immersed in water, which acts as a *moderator*, slowing down the speed of the neutrons enough that they can initiate new fission reactions to carry on the chain reaction. The second challenge is preventing the nuclear reaction from becoming uncontrolled. **Control rods** are used in a nuclear reactor to limit the rate of fission reactions to a safe level.

**Nuclear waste**

One technical problem with nuclear fission reactions is that the spent fuel, although not capable of producing enough reactions to drive the reactor, remains radioactive for many years. Nuclear reactors create **nuclear waste** that must be stored for a long time in shielded containers at a safe site. In the USA, no long-term storage site has been agreed upon. Consequently, nuclear engineers must design nuclear power plants to store most nuclear waste on-site.

# Investigation 27A Controlling a nuclear fission reaction

**Essential questions**

*How are chain reactions necessary for nuclear power?*
*How do control rods work in a nuclear reactor?*

Interactive simulation

Nuclear reactions produce the energy that is at the heart of nuclear power. Most nuclear reactors use uranium as the fuel for the reactions. When a uranium nucleus is struck by a neutron, the nucleus fissions. Each nuclear fission reaction releases two neutrons, which cause fission in two nuclei, releasing four neutrons, and so on starting a chain reaction. In this investigation you will investigate the nuclear chain reaction inside a reactor and control the reaction rate using control rods. Control rods are made out of a material such as cadmium that absorbs neutrons. The control rods move in and out of the reactor core to allow fewer or more reactions by regulating the number of fission-producing neutrons.

## Controlling the nuclear chain reaction inside a nuclear reactor

1. Launch the interactive simulation. Press [Run] to start fission reactions by allowing neutrons into the reactor.
2. Move the control rods up and down to control the nuclear chain reactions.
3. Monitor the temperature probe to keep the reactor within its safe operating range as you adjust the control rods.
4. Note the length of time you kept the reactor within the optimal temperature range.

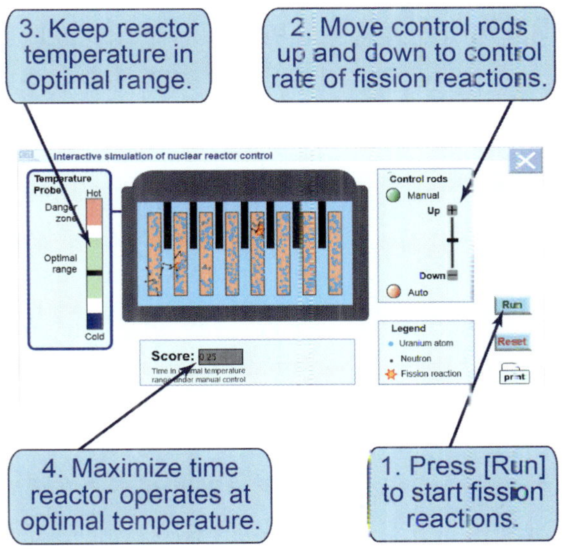


a. *What is required to start the first nuclear reaction? Write down the nuclear reaction equation to back up your answer.*

b. *What happens to the chain reaction when the control rods are fully inserted into the reactor? What happens when the rods are fully removed?*

c. *What happens to the temperature inside a nuclear reactor if the control rods are fully removed? How could this be a safety issue?*

d. *Determine the optimal position of the control rods to produce a steady power output for the nuclear reactor. Explain why this position makes sense to control a nuclear chain reaction.*

**Positrons and gamma rays**

The properties and physical location of radioactive nuclei can be observed by detecting the products of their decay with specific detectors. Many useful nuclear reactions release high-energy photons also called $\gamma$ rays (gamma rays). Detecting these $\gamma$ rays is a very well developed technology and it is used in many applications including medical diagnostic imaging. Radioactive nuclei such as ${}^{18}_{9}F$ (fluorine-18) decay by emitting a **positron**. (This process is a type of $\beta$ decay in which a positron is emitted instead of an electron.) When a positron and an electron come into contact, which happens almost instantaneously in matter, they immediately "disappear" or annihilate each other. This annihilation reaction generates two $\gamma$ rays, each with an energy of 0.511 MeV, which are very penetrating and can be detected with $\gamma$ ray detectors.

**Positron–electron annihilation**

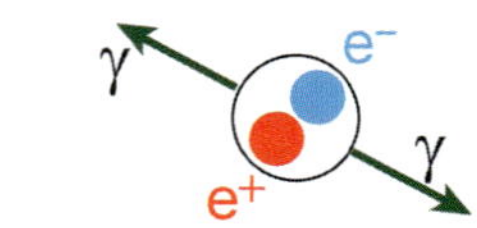


**Positron emission tomography (PET)**

Engineers have designed special systems that detect these rays and map their origin in space. This technique is called positron emission tomography, or PET for short, and is used for high-resolution medical diagnostic imaging. The science of PET is based on the decay of ${}^{18}_{9}F$, which has a half-life of 110 minutes. Since a half-life of 110 minutes is very short, ${}^{18}_{9}F$ does not occur naturally. Scientists have to produce these isotopes close to the place of use. To do so they use an accelerator to generate high-energy protons that interact with a target containing ${}^{18}_{8}O$, which produces ${}^{18}_{9}F$ according to the reaction:

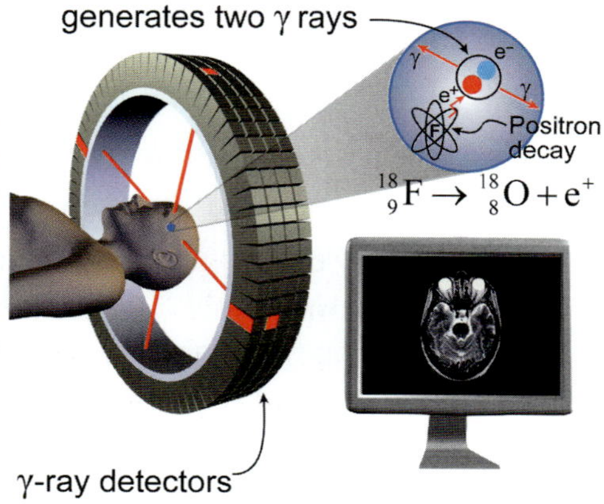


$$ {}^{1}_{1}H + {}^{18}_{8}O \rightarrow {}^{18}_{9}F + {}^{1}_{0}n $$

**PET for medical diagnostic imaging**

Once ${}^{18}_{9}F$ is produced it is ingested into the bloodstream. The ${}^{18}_{9}F$ isotopes concentrate in cells with high activity such as brain and tumor cells. Fluorine-18 then decays, releasing $\gamma$ rays, which are detected by the $\gamma$-ray detectors surrounding the patient. The $\gamma$ rays from the annihilation reaction are always produced in pairs that emit their energy in opposite directions. Special mathematical techniques and software have been developed that can construct an image of the region according to the concentration of ${}^{18}_{9}F$. Radiologists examine the resulting images in the same way that they look at an x-ray image of an arm to see whether a bone is broken. By precisely mapping organs and tissues, a medical doctor can target treatment to the correct part of the body.

**Magnetic resonance imaging**

Magnetic resonance imaging (MRI) takes advantage of a different nuclear property. The human body is full of water. Protons in the nucleus of hydrogen in the water molecule possess a quantum mechanical "spin" that will align itself with an outside magnetic field in a process called *nuclear magnetic resonance*. A magnetic resonance imaging machine has a powerful electromagnet that aligns the protons in the patient's body in this way. The machine varies the electromagnetic field (using radio waves) to cause the protons to rotate around. As the protons rotate, they give off a unique signal that the MRI machine detects to visualize the patient's internal tissues.

**Ionization**

With sufficient energy, an electron can be removed from its orbit, leaving the atom positively charged or *ionized*. As we learned in Chapter 26, the amount of energy required to completely remove an electron from its orbit varies but it roughly falls in the range between 5 and 20 eV. This is a small amount of energy, but the effects that it has on the chemical and biological properties of the elements and molecules are significant. **Ionization** is associated with the emission of radiation.

**Ionization depends on energy**

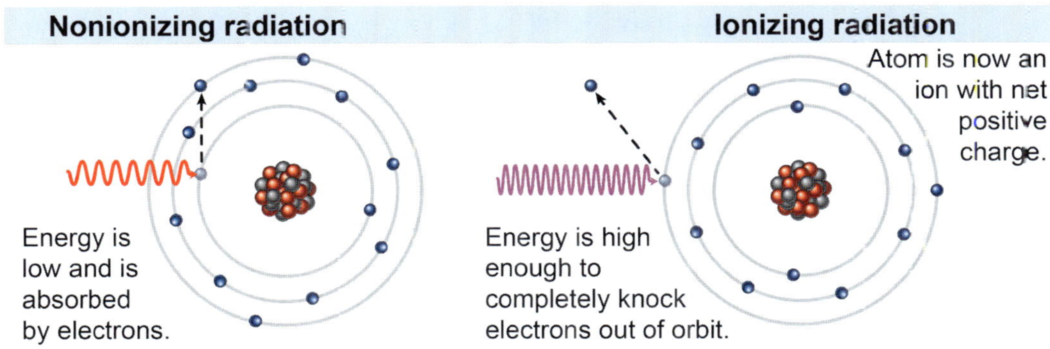


**Biological effects of radiation**

The normal functions of biological systems depend on the various chemical compounds that make up the system. Chemical bonds of molecules such as DNA depend on the exchange or sharing of electrons between and among the various elements that make up the molecules. When a biological system is exposed to radiation and the energy of the radiation is sufficient to result in ionization, the chemical bonds are affected. When a bond is broken or changed in any way, the properties and the function of the associated molecule may also change. Biological systems are able to correct the damage caused by small amounts of radiation. As the radiation increases, however, the damage to the molecules and proteins causes permanent damage such as genetic defects and cancer.

**How do we measure radiation?**

The energy associated with radiation is used to detect it. When radiation interacts with matter it deposits energy in it, resulting in a number of changes at the atomic and molecular level. Radiation detection devices such as the **Geiger counter** detect these changes and use electronic processing to present the result as an audible popping sound. Radiation detectors are used for maintaining a safe working environment around radiation sources and to search for a potential radiation leak.

**Radiation dose**

Humans who remain close to radioactive substances experience a total exposure to radiation over time called the **dose**. The unit of measurement for exposure to radioactive or ionizing radiation is called the **rem**. A rem is a large dose of radiation and so exposures are usually estimated in millirem, or one-thousandth of a rem.

**Radiotherapy**

Although large doses of radioactivity can be harmful to the human body, small targeted doses of radiation are also used as a medical therapy to kill cancer cells in the body. In nuclear medicine, radiation therapy (or radiotherapy) is a process of targeting radiation toward a part of the body that has many cancer cells in order to kill them. Some cancers, such as leukemia, respond well to radiotherapy, but others do not. The dose of radiation is chosen to balance the benefits of killing cancerous cells with the risks of damaging healthy cells.

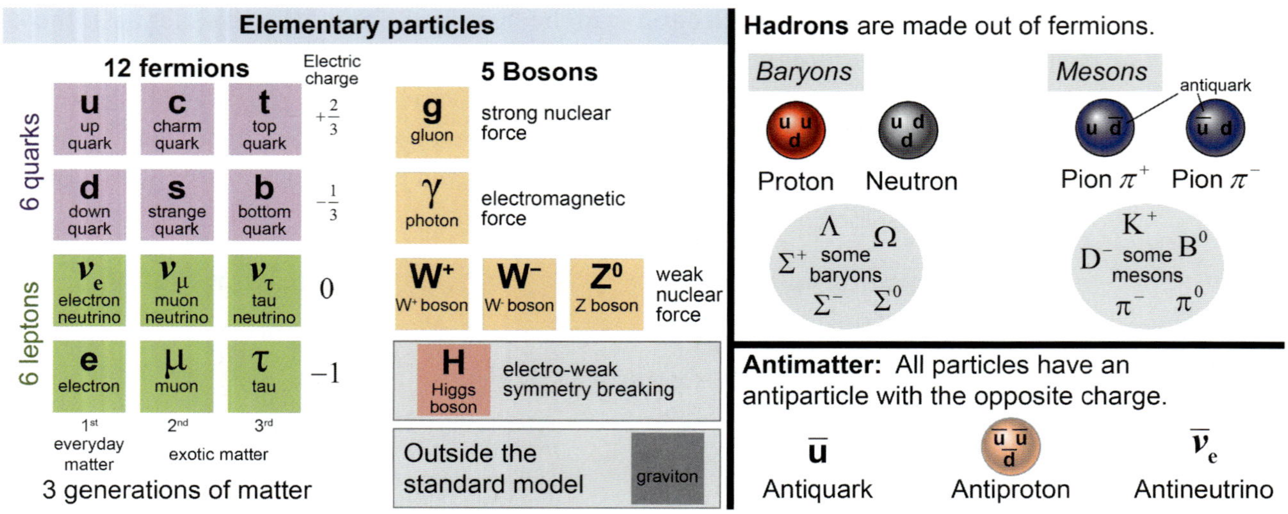


**Standard model** In the past few decades, physicists have used ever more powerful particle accelerators to collide particles at higher and higher energies. They have discovered many new particles and have created a *standard model* that explains the new physics. The model is a consensus theory of particles, matter and antimatter, particle interactions (or forces), and quantum physics. Particles are classified according to their electric charge, how they interact with each other, and family (or "generation").

**Elementary particles** In the standard model, there are two kinds of elementary particles: fermions and bosons. All particles in the universe ultimately are made up of fermions.

**Quarks** Quarks are particles that are smaller than protons or neutrons. There are six different quarks that come in three pairs: up and down, charm and strange, and top and bottom. Quarks are never seen directly or found in isolation but can only be found inside hadrons, such as protons or neutrons. Baryons, including neutrons and protons, are composed of three quarks. Mesons are composed of a quark and an antiquark. Quarks are affected by the strong nuclear force.

**Leptons** Leptons are the other category of fermion. Electrons and neutrinos are two examples of leptons. Leptons are not affected by the strong nuclear force but obey the Pauli exclusion principle.

**Three families** Quarks and leptons are grouped into three different families or "generations." Everyday matter is made up from the lowest mass group of particles, comprising the electron, the electron neutrino, and the up and down quarks. The two higher mass groups are called "exotic" because they are usually only seen in very high energy conditions, such as inside a particle accelerator.

**Matter and antimatter** In the standard model every particle also has an *antimatter* opposite, an antiparticle with the same mass and other properties but opposite charge. The positron is the antimatter twin of the electron and has positive charge. Neutral particles are their own antiparticle, because they are electrically neutral. When a particle and its antiparticle twin collide in *pair annihilation*, a large amount of energy is released.

**Bosons** Bosons are elementary particles that mediate (or cause) forces to happen. In the standard model, there is at least one boson for each of the four forces of nature.

**Four fundamental forces**

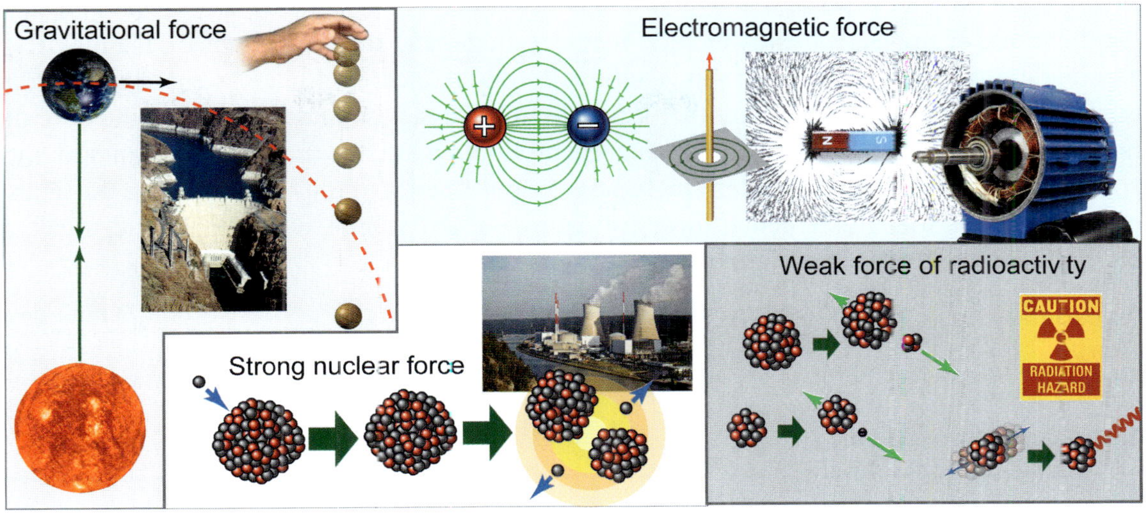


**Four interactions**

In the standard model, there are four fundamental *interactions* of nature: the gravitational, electromagnetic, strong, and weak forces. Forces are often called "interactions" to emphasize how elementary particles are acting on each other.

**Forces of classical physics**

The *gravitational force* is an attractive force that acts between any two objects or particles that have mass. Gravity prevents you from floating away from the surface of the Earth, maintains the Earth in orbit around the Sun, and governs the orbit of the Sun in the Galaxy. The *electromagnetic force* acts in two different ways: Electrically charged objects or magnetic objects attract or repel each other. The electric and magnetic forces are two manifestations of one underlying phenomenon: electromagnetism. The gravitational and electromagnetic forces are part of "classical" physics, which was fairly well understood by the beginning of the 20th century.

**Forces of nuclear physics**

At the dawn of the 20th century, new physics was necessary to explain the atomic world. From the new physics two new forces emerged. The *strong nuclear force* acts only over very short distances at the scales found in and around the nucleus of an atom. The strong force is required to hold the nucleus together, because otherwise the electric force of repulsion between the protons would force them apart. The *weak force* governing radioactive decay is the fourth and final member of the fundamental forces of nature. Nuclear particles periodically undergo radioactive decay (such as alpha or beta decay): the weak force provides a framework for understanding radioactivity. The strong and weak forces comprise the new physics that arose in the quantum world of atoms and atomic nuclei.

**Grand unified theories**

Together, these four forces provide a complete explanation for all the processes through which matter and light interact. In recent decades, theoretical physicists have tried to go one step further. Their theories unify two or more of these forces into a single underlying force, much in the same way that electricity and magnetism were unified. The new theories are called *Grand Unified Theories*, because they seek to unify different forces of nature under a common framework. Do you think that, in your lifetime, a model will successfully unify all four forces?

# Section 4 review

Applications of nuclear physics include medical diagnostic imaging (both PET and MRI), radiation therapy, and nuclear power. Fission reactions, continually maintained by a nuclear chain reaction, are at the heart of a nuclear power plant. Fusion is another kind of nuclear reaction and it occurs in the core of the Sun. Fusion is not yet a proven source for regular energy production on the Earth. The standard model is the modern consensus of elementary particles (including bosons such as light) and how the particles interact with each other. There are four fundamental forces of nature—gravitational, electromagnetic, weak, and strong—and all but gravity are encompassed in the standard model.

| Vocabulary words | nuclear energy, control rods, nuclear waste, positron, dose, rem, Geiger counter, ionization, radiotherapy |
|---|---|

## Review problems and questions

1. List three applications of nuclear physics.

2. Compare the advantages and disadvantages of electrical generation by nuclear power versus burning coal.

3. What are some of the challenges encountered by researchers who are trying to develop nuclear power using fusion?

4. Which of the two following reactions is a fission reaction and which is a fusion reaction?

   a. ${}^{2}_{1}H + {}^{3}_{1}H \rightarrow {}^{4}_{2}He + {}^{1}_{0}n$

   b. ${}^{1}_{0}n + {}^{235}_{92}U \rightarrow {}^{144}_{56}Ba + {}^{89}_{36}Kr + 3{}^{1}_{0}n$

5. Which of the two reactions above is most likely to be found inside a nuclear power plant, and which is most likely to be found in the core of the Sun?

6. a. In a *fission* reaction, will the combined binding energies of the nuclei in the products differ from the binding energy of the original nucleus? If so, explain the origin of the difference.

   b. In a *fusion reaction*, will the binding energy of the resulting nucleus differ from the combined binding energies of the original nuclei? If so, explain the origin of the difference.

7. Describe the role of control rods in a nuclear power plant.

8. Describe what happens if an electron collides with a positron.

9. Which scientific instrument is the best choice to detect the presence of radioactivity?

   a. a galvanometer
   b. a rem
   c. a magnetic resonance imager
   d. a Geiger counter

10. A quark is a type of particle that is never seen directly or in isolation.

    a. What are the six types of quarks?
    b. What common particles are composed of three quarks?

# Chapter review

## Vocabulary

Match each word to the sentence where it best fits.

### Section 27.1

| | |
|---|---|
| atomic mass unit (amu) | binding energy |
| mass deficiency | mass–energy equivalence |
| rest energy | strong nuclear force |

1. A particle that is not in motion still contains _______ that can be calculated from Einstein's famous equation.
2. The difference in _______ between the particles before and after a nuclear reaction is the energy source of nuclear power.
3. The electric force of repulsion would push the protons in the nucleus apart if not for the _______ that holds them together.
4. If you total the individual masses of the nucleons in a nucleus it will not equal the total mass of the nucleus. The difference is called the _______.
5. The number 14.007 represents the average mass of a nitrogen atom using the unit _______.
6. Einstein's famous equation is a statement of _______.

### Section 27.2

| | |
|---|---|
| alpha decay | beta decay |
| carbon dating | gamma decay |
| half-life | radioactive |
| radioactive decay | weak nuclear force |

7. Carbon-14 has a/an _______ of 5,730 years.
8. An atom that is unstable and may emit energetic particles is called _______.
9. The age of a fossil can be estimated by using a process called _______.
10. In _______, the total number of neutrons, protons and electrons in an atom remains the same.
11. _______ is when a nucleus emits a particle with two protons and two neutrons.
12. Alpha and beta particles can be emitted by a nucleus in a process called _______.
13. Radioactive decay is governed by the _______.
14. _______ is when a nucleus emits an electron and is accompanied by the transformation of a neutron into a proton.

### Section 27.3

| | |
|---|---|
| fission | fusion |
| nuclear chain reaction | nuclear reaction |

15. Through a _______ an element is changed into another.
16. Nuclei are split in a nuclear reaction called _______.
17. Nuclei are combined together in a nuclear reaction called _______.
18. An example of a _______ is when one neutron starts a fission reaction that produces two neutrons, and both neutrons start a new fission reaction, producing a total of four neutrons.

### Section 27.4

| | |
|---|---|
| Geiger counter | control rods |
| dose | ionization |
| nuclear energy | nuclear waste |
| positron | radiotherapy |
| rem | |

19. Cancer cells can be killed in the human body through a medical process called _______.
20. A/An _______ is a unit of measure of the impact of radioactivity on humans.
21. If you have been exposed to too much radioactivity, you have received a dangerous _______.
22. A nuclear power plant's spent fuel is called _______.
23. In a nuclear reactor, the rate of reactions can be slowed down by inserting the _______.
24. The antimatter particle twin of the electron with a charge of $+1.6 \times 10^{-19}$ C is the _______.
25. Reactions in the atomic nucleus produce _______.
26. After _______, the remaining atom has one fewer electron and a net positive charge.
27. An instrument that pops every time a radioactive particle passes through it is called a/an _______.

# Chapter review

## Conceptual questions

### Section 27.1

28. Prepare a short oral report on Albert Einstein and how his discovery of mass–energy equivalence influenced the world we live in today. Use at least three vocabulary words from this chapter.

29. Using the periodic table, determine the atomic number of lead.

30. Name two different nucleons.

31. Positive charges repel one another. Yet several positively charged particles are stuck very close to one another in all the atoms around you. What keeps them together?

32. How would the physical world change if the strong nuclear force no longer worked?

33. How does the rest energy of an object on a table compare to its potential energy?

34. What is the equation for mass–energy equivalence? What processes does it empower?

35. Using the periodic table, determine the atomic mass of copper.

36. Research Japanese theorist Hideki Yukawa and how mesons have influenced scientific thought today about the strong force. Prepare a two minute oral report on this scientist.

37. ❮ How is the periodic table organized?

38. ❮ What is the difference between atomic mass and atomic mass unit?

39. ❮ Why is the mass of an atom nearly exactly equal to the mass of its nucleus?

40. ❮ Isotope C has a more stable nucleus than isotope D. Which has a higher binding energy?

41. ❮ Imagine a nucleus the size of your classroom. Would you expect this nucleus to be stable? Why or why not? Use mass–energy equivalence in your explanation.

42. ❮ Describe evidence for, and the effects of, the strong nuclear force.

43. ❮ Describe the difference between *element* and *isotope.*

44. ❮❮ Which will have the higher binding energy: helium-4 (2 protons and 2 neutrons) or helium-3 (2 protons and 1 neutron)? Use your knowledge of the forces acting in the nucleus to predict the answer. Explain your prediction.

45. ❮❮ How can isotopes be found on the periodic table?

46. ❮❮ In what region of the periodic table would you expect to find an atom with $Z = 5$? An atom with $Z = 89$?

### Section 27.2

47. Research and write a short report on the contributions that Clyde Cowan, Sheldon Glashow, Frederick Reines, Mohammad Abdus Salam, or Steven Weinberg made to the scientific understanding of the weak force and the impact that his research had on society.

48. Describe evidence for, and the effects of, the weak force.

49. Carbon dating allows us to determine when a fossil was created by measuring the ratio of radioactive carbon to nonradioactive carbon. Which of the four fundamental forces allows carbon dating?

50. Describe two technological applications of radioactive decay.

51. Research the work of Henri Becquerel. Write a couple of sentences about his research and what radioactivity is. Compare this to how radioactivity is portrayed in movies, TV shows, and novels.

52. For a paper on the history of the discovery of the weak nuclear force, which of the following are acceptable ways to write citations?
    A. "Weak nuclear force." *Wikipedia.org.* Wikipedia, accessed May 23, 2012.
    B. From the Internet.
    C. Johnson, George. *Strange Beauty.* New York: Vintage Books, 1999.

53. ❮ Cesium-137 has a half-life of 30 years, whereas iodine-131 has a half-life of 8 days. Both were produced by the Chernobyl nuclear disaster in 1986. Which radioactive isotope will still remain in the environment in significant quantities decades after the disaster?

54. ❮❮ Why does alpha or beta decay change one element into another while gamma decay does *not* change the element?

55. ❮❮ Smoke detectors contain radioactive isotope americium-241 and emit alpha particles. According the Environmental Protection Agency (EPA), these products are safe. Assess the extent to which specific evidence provided on the EPA website supports this claim.

### Section 27.3

56. Create a histogram (bar graph) that visually displays the energy per nucleon released in a fission reaction versus a fusion reaction. (The information needed is in the text.)

# Chapter review

## Section 27.3

57. What is the difference between fission and fusion?

58. Which nuclear reaction, fission or fusion, normally involves low-mass nuclei? High-mass nuclei? Explain.

59. In a nuclear reaction, does the total mass stay the same? Does the total energy stay the same? Why?

## Section 27.4

60. How does nuclear fission influence your daily life?

61. How is nuclear physics used to diagnose patients?

62. Should a larger fraction of the country's electricity be produced using nuclear power? Synthesize relevant information about nuclear fission from multiple authoritative print and digital sources, craft a cohesive report supporting your opinion, and provide properly formatted citations.

63. A student is designing an investigative procedure using the interactive simulation of control rods in a nuclear reactor in Investigation 27A on page 815.
    a. What variable can the experimenter change?
    b. Write a well-defined question that can be answered using the simulation about the effects of leaving the control rods fully retracted from the reactor.
    c. Write down a testable hypothesis that can address that question.
    d. What observations should she include in the procedure to address that question and hypothesis?
    e. What limitations of the simulation might the student remedy by using an additional piece of equipment?
    f. Would it be a good idea to choose a different piece of technology, such as running the procedure on a real nuclear reactor instead of a simulation?
    g. The student wants the experimenter to measure the distance of the control rods from the top of the reactor and compare it to the size of a typical reactor. Is this a reasonable measurement to include in the procedure? Do you expect the experimenter to provide reasonable answers?

64. How is radiation therapy related to nuclear physics?

65. How is nuclear power related to properties of the nuclei of atoms?

66. How is magnetic resonance imaging related to quantum physics?

67. The gas in a ceiling fluorescent light has a temperature of 15,000 K. If the temperature is so high, why is it that we do not feel incredible heat energy coming from it?

## Quantitative problems

## Section 27.1

68. Identify the unknown particle in the nuclear reaction equation

$$^{2}_{1}\text{H} + ^{2}_{1}\text{H} \rightarrow ^{3}_{2}\text{He} + \text{X}$$

69. How many atomic mass units are there in one kilogram of matter?

70. If the nuclei of nitrogen-14 and lithium-6 combine, what isotope is the product?

71. If neon-20 spontaneously breaks in two and one of the products is the isotope carbon-11, what is the other isotope produced by the reaction?

72. What is the isotope X produced by the following reaction?

$$^{2}_{1}\text{H} + ^{18}_{8}\text{O} \rightarrow \text{X} + ^{1}_{1}\text{p}$$

73. Calculate the mass of a carbon-12 atom in kilograms.

74. Calculate the mass of a carbon-13 atom in kilograms. Make sure to look up the mass of carbon-13 in amu.

75. What is the charge of an atom of carbon-13 that is doubly ionized, i.e., it has had two of its electrons removed?

76. What is the difference in the nuclei of the isotopes of oxygen-16, oxygen-17, and oxygen-18? What is the difference in their numbers of electrons?

77. How much rest energy does 0.001 kg of matter have?

78. How many atoms are there in one kilogram of carbon-12?

79. Carbon-12 is a common isotope of carbon.
    a. How many protons and how many neutrons does this isotope have?
    b. What is the sum of the masses of these protons and neutrons? (The mass of a proton and neutron are 1.0073 and 1.0087 amu, respectively.)
    c. The mass of a carbon-12 nucleus is 11.9967 amu. What is the mass deficiency?
    d. Convert the mass deficiency to kilograms and use Einstein's equation to calculate the binding energy in joules. ($1.661 \times 10^{-27}$ kg = 1 amu.)

80. Calculate the rest energy of a proton, a neutron, and an electron in joules. (A proton has a mass of $1.673 \times 10^{-27}$ kg, a neutron has a mass of $1.675 \times 10^{-27}$ kg, and an electron has a mass of $9.11 \times 10^{-31}$ kg.)

# Chapter review

## Section 27.2

81. If you start with 1,280 atoms of an unstable isotope, how many half-lives will it take (on average) for you to end up with 10 or fewer atoms of the original isotope?

82. Through what decay process does uranium-238 decay to thorium-234?

83. ( Cesium-137 undergoes beta decay.

    a. Write down the nuclear reaction equation including the reactants and the products.
    b. This decay is evidence of what fundamental force?

84. ( An atom undergoes beta decay. If the product is nickel-60, what was the original isotope?

85. ( Seaborgium-271 undergoes alpha decay. Write down the nuclear reaction equation.

86. ( An atom undergoes alpha decay, producing bismuth-207. Write down the nuclear reaction equation.

87. ((( A laboratory technician starts with a mass of 2.79 kg of radioactive polonium-197 that has a half-life of 53.6 s. It is enclosed in a 23.9 kg lead case. How long will it take for the polonium-197 to decay radioactively to 0.78% of the original amount?

88. ( An archeologist measured the mass of carbon-14 in a fossil to be 1 g, and then calculated the age of the fossil to be 28,650 years old. What was the original mass of carbon-14 in the fossil 28,650 years ago?

89. (( Krypton-85 decays with a half-life of 10.8 years. If we start out with 50 mg of krypton-85, how much would remain after 30 years?

90. ((( A sample that contains iron-59 is measured with a radiation counter, which records about 11 decays per minute. After 20 days the same sample is measured to give about 8 decays per minute. Calculate the half-life of iron-59 in days.

91. ( The radioactive decay of polonium-210 is described by the following equation:

$$^{210}_{84}\text{Po} \rightarrow {}^{?}_{?}\text{Pb} + {}^{4}_{2}\text{X}$$

where X is the particle emitted upon decay. Identify the correct isotope of lead and type of decay.

    a. lead-206; alpha decay
    b. lead-208; alpha decay
    c. lead-206; beta decay
    d. lead-208; positron emission

## Section 27.3

92. ( The product of one fission reaction for a uranium-235 atom is given by the following reaction equation:

$$^{235}_{92}\text{U} + {}^{1}_{0}\text{n} \rightarrow {}^{143}_{60}\text{Nd} + {}^{90}_{40}\text{Zr} + 3{}^{1}_{0}\text{n} + ?\text{e}$$

How many electrons are produced by the reaction?

93. ( What is the type and the unknown product of the following nuclear reaction equation?

$$^{2}_{1}\text{H} + {}^{3}_{1}\text{H} \rightarrow {}^{4}_{2}\text{He} + \text{X?} + \text{energy}$$

    a. fusion; proton
    b. fusion; neutron
    c. fission; neutron
    d. fusion; two neutrons

94. (( When a neutron hits the nucleus of a plutonium-239 atom, the subsequent nuclear reaction creates cesium-148, a mystery isotope, and three neutrons. Write down the nuclear reaction equation and determine the mystery isotope.

95. ( For the reaction:

$$^{3}_{2}\text{He} + {}^{1}_{1}\text{H} \rightarrow {}^{A}_{Z}\text{X} + {}^{0}_{1}\text{e}$$

what is the unknown ${}^{A}_{Z}\text{X}$?

    a. ${}^{2}_{1}\text{H}$
    b. ${}^{3}_{2}\text{He}$
    c. ${}^{4}_{2}\text{He}$
    d. ${}^{5}_{3}\text{Li}$

## Section 27.4

96. (( A small nuclear power plant has a set of fuel rods that weigh 50.0 kg. After some time, the fuel rods weigh 49.9 kg. How much energy did the fuel rods release?

97. (( To produce energy in nuclear power plants, uranium-235 undergoes fission according to the equation

$$^{1}_{0}\text{n} + {}^{235}_{92}\text{U} \rightarrow {}^{99}_{42}\text{Mo} + {}^{?}_{50}\text{Sn} + 2{}^{1}_{0}\text{n} + \text{energy}$$

Determine the missing value.

98. (( The nuclear reaction

$$^{1}_{0}\text{n} + {}^{235}_{92}\text{U} \rightarrow {}^{144}_{56}\text{Ba} + {}^{89}_{36}\text{Kr} + 3{}^{1}_{0}\text{n}$$

releases around 200 MeV of energy. Convert this energy to joules.

# Chapter review



## Standardized test practice

99. In the nuclear reaction equation

$$^{236}_{92}U \rightarrow {}^{140}_{54}Xe + {}^{A}_{38}Sr + 2{}^{1}_{0}n$$

what is the value of $A$?

A. 86
B. 92
C. 94
D. 95

100. Two neutral isotopes of the same element have the same number of

A. protons and neutrons.
B. protons and electrons.
C. neutrons and electrons.
D. protons, neutrons, and electrons.

101. A particular atom has 29 neutrons and 55 nucleons. What is the atomic number of this atom?

A. 26
B. 52
C. 55
D. 81
E. 84

102. If the strong nuclear force did not act on the nucleons of an element, what would be the binding energy of that element?

A. zero
B. $1.67\times10^{-27}$ J
C. infinity

103. One of the difficulties of creating fusion reactions on Earth is that ______________ is required.

A. a low pressure
B. extremely rare fuel
C. a high temperature
D. a low temperature

104. Lead-192 has a half-life of 3.5 minutes. If you start with 8 kg of lead-192, how much will remain after 17.5 minutes?

A. 1.6 kg
B. 0.5 kg
C. 0.46 kg
D. 0.25 kg

105. When a radioactive isotope emits a beta particle, its atomic number will:

A. increase by one.
B. remain unchanged.
C. decrease by one.
D. decrease by two.

106. In a common fission reaction, the reactants have a mass of $3.920\times10^{-25}$ kg but the products have a mass of $3.917\times10^{-25}$ kg. How much binding energy is released in one of these reactions?

A. $3\times10^{-28}$ J
B. $9\times10^{-20}$ J
C. $2.7\times10^{-11}$ J
D. $4.2\times10^{-6}$ J

107. Which of the following is a significant application of Einstein's mass–energy equivalence equation?

A. mass spectrometers
B. astronomy
C. nuclear power
D. wind turbines

108. The owner of a nuclear power plant accidentally sets the control rods to go completely down into the reactor core. What will likely happen to the turbine?

A. It will slow down.
B. It will speed up.
C. It will change direction.
D. It will continue spinning at the same speed.

109. When a radioactive isotope emits an alpha particle, its atomic mass number will

A. increase by 4.
B. increase by 2.
C. decrease by 2.
D. decrease by 4.

110. A scientist manages to synthesize 200 g of seaborgium-266. After 90 s, 25 g of seaborgium-266 remain. What is the half-life of seaborgium-266?

A. 90 s
B. 270 s
C. 30 s
D. 8 s

111. Consider an isotope with a stable nucleus and one with an unstable nucleus. Which will have the highest mass deficiency and highest binding energy?

A. the isotope with a stable nucleus
B. the isotope with an unstable nucleus
C. neither, since all isotopes have the same mass deficiency and binding energy

112. Control rods are inserted or removed from a nuclear reactor to

A. slow down the speed of the neutrons.
B. initiate the nuclear reactions.
C. moderate the rate of nuclear reactions.
D. control the fusion reaction.

# Appendix

## Contents:

**Table 1: Fundamental constants**

| Quantity | Symbol | Value | Units |
|---|---|---|---|
| Speed of light in a vacuum | $c$ | $3.00\times10^{8}$ | m/s |
| Gravitational constant | $G$ | $6.67\times10^{-11}$ | N m$^2$/kg$^2$ |
| Boltzmann's constant | $k_B$ | $1.38\times10^{-23}$ | J/K |
| Stefan–Boltzmann constant | $\sigma$ | $5.67\times10^{-8}$ | W/(m$^2$ K$^4$) |
| Planck's constant | $h$ | $6.63\times10^{-34}$ | J s |
| Avogadro's number | $N_A$ | $6.022\times10^{23}$ | mol$^{-1}$ (or particles/mol) |
| Gas constant | $R$ | 8.315 | J/(mol K) |
| Permittivity of free space | $\varepsilon_0$ | $8.85\times10^{-12}$ | C$^2$/(N m$^2$) |
| Permeability of free space | $\mu_0$ | $4\pi\times10^{-7}$ | T m/A |
| Mass of the electron | $m_e$ | $9.11\times10^{-31}$ | kg |
| Mass of the proton | $m_p$ | $1.6726\times10^{-27}$ | kg |
| Mass of the neutron | $m_n$ | $1.6749\times10^{-27}$ | kg |
| Charge of the electron | $e$ | $-1.60\times10^{-19}$ | C |
| Coulomb's law constant | $k_e$ | $8.99\times10^{9}$ | N m$^2$/C$^2$ |
| Absolute zero | 0 K | −273.15 | °C |

**Table 2: Physical data**

| Quantity | Symbol | Value | Units |
|---|---|---|---|
| Strength of gravity at Earth's surface | $g$ | 9.81 | m/s$^2$ or N/kg |
| Speed of sound at 20°C and 1 atm | $c_s$ | 343 | m/s |
| Standard atmospheric pressure | atm | $1.01\times10^{5}$ | Pa or N/m$^2$ |

**Table 3: Astronomical data**

| Quantity | Symbol | Value | Units |
|---|---|---|---|
| Astronomical unit (average Earth–Sun distance) | AU | $1.50\times10^{11}$ | m |
| Average Earth–Moon distance | | $3.84\times10^{8}$ | m |
| Radius of the Sun | $r_\odot$ | $6.96\times10^{8}$ | m |
| Radius of the Earth | $r_\oplus$ | $6.37\times10^{6}$ | m |
| Radius of the Moon | | $1.74\times10^{6}$ | m |
| Mass of the Sun | $m_\odot$ | $1.99\times10^{30}$ | kg |
| Mass of the Earth | $m_\oplus$ | $5.98\times10^{24}$ | kg |
| Mass of the Moon | | $7.35\times10^{22}$ | kg |

**Table 4: Fundamental units and physical quantities in the International System of Units (SI)**

| Physical quantity | Symbol | Fundamental unit | Unit abbreviation |
|---|---|---|---|
| Mass | $m$ | kilogram | kg |
| Length | $x, d, l$ | meter | m |
| Time | $t$ | second | s |
| Electric current | $I$ | ampere | A |
| Temperature | $T$ | kelvin | K |
| Luminous intensity | $I_v$ | candela | cd |
| Amount of a substance | $n$ | mole | mol |

**Table 5: Derived units and physical quantities in the International System of Units (SI)**

| Derived quantity | Symbol | Derived unit | Unit | Equivalence in fundamental units |
|---|---|---|---|---|
| Area | $A$ | square meter | | $m^2$ |
| Volume | $V$ | cubic meter | | $m^3$ |
| Speed or velocity | $v$ | meter per second | | m/s |
| Acceleration | $a$ | meter per second squared | | $m/s^2$ |
| Density or mass density | $\rho$ | kilogram per cubic meter | | $kg/m^3$ |
| Force | $F$ | newton | N | $kg\ m/s^2$ |
| Energy and work | $E$ or $W$ | joule | J | $kg\ m^2\ s^{-2}$ |
| Power | $P$ | watt | W | $kg\ m^2\ s^{-3}$ |
| Pressure | $P$ | pascal | Pa | $N/m^2$ or $kg\ m^{-1}\ s^{-2}$ |
| Frequency | $f$ | hertz | Hz | $s^{-1}$ |
| Magnetic field | $B$ | tesla | T | $kg\ A^{-1}\ s^{-2}$ |
| Magnetic flux | $\Phi$ | weber | Wb | $kg\ m^2\ A^{-1}\ s^{-2}$ |
| Electric charge | $q$ | coulomb | C | A s |
| Electric potential, potential difference, or voltage | $V$ | volt | V | $kg\ m^2\ A^{-1}\ s^{-3}$ |
| Electric field | $E$ | volt per meter or newton per coulomb | V/m<br>N/C | $kg\ m\ A^{-1}\ s^{-3}$<br>$kg\ m\ s^{-2}\ C^{-1}$ |
| Resistance | $R$ | ohm | Ω | $kg\ m^2\ A^{-2}\ s^{-3}$ |
| Capacitance | $C$ | farad | F | $A^2\ s^4\ kg^{-1}\ m^{-2}$ |

Table 6: Numbers greater than or equal to one

| Description | Scientific notation | Equivalent value | SI prefix | Abbreviation |
|---|---|---|---|---|
| One quadrillion | $1\times10^{15}$ | 1,000,000,000,000,000 | peta | P |
| One trillion | $1\times10^{12}$ | 1,000,000,000,000 | tera | T |
| One billion | $1\times10^{9}$ | 1,000,000,000 | giga | G |
| One hundred million | $1\times10^{8}$ | 100,000,000 | | |
| Ten million | $1\times10^{7}$ | 10,000,000 | | |
| One million | $1\times10^{6}$ | 1,000,000 | mega | M |
| One hundred thousand | $1\times10^{5}$ | 100,000 | | |
| Ten thousand | $1\times10^{4}$ | 10,000 | | |
| One thousand | $1\times10^{3}$ | 1,000 | kilo | k |
| One hundred | $1\times10^{2}$ | 100 | hecto | h |
| Ten | $1\times10^{1}$ | 10 | deca (deka) | da |
| One | $1\times10^{0}$ | 1 | | |

Table 7: Numbers less than one

| Description | Scientific notation | Equivalent value | SI prefix | Abbreviation |
|---|---|---|---|---|
| One tenth | $1\times10^{-1}$ | 0.1 | deci | d |
| One hundredth | $1\times10^{-2}$ | 0.01 | centi | c |
| One thousandth | $1\times10^{-3}$ | 0.001 | milli | m |
| One ten thousandth | $1\times10^{-4}$ | 0.0001 | | |
| One hundred thousandth | $1\times10^{-5}$ | 0.00001 | | |
| One millionth | $1\times10^{-6}$ | 0.000001 | micro | µ |
| One ten millionth | $1\times10^{-7}$ | 0.0000001 | | |
| One hundred millionth | $1\times10^{-8}$ | 0.00000001 | | |
| One billionth | $1\times10^{-9}$ | 0.000000001 | nano | n |
| One trillionth | $1\times10^{-12}$ | 0.000000000001 | pico | p |
| One quadrillionth | $1\times10^{-15}$ | 0.000000000000001 | femto | f |

**Table 8: Conversion factors**

| Quantity | Some conversions | | |
|---|---|---|---|
| **Length** | 1 in = 2.54 cm | 1 cm = 0.3048 in | 1 m = 39.37 in |
| | 1 foot (ft) = 12 in | 1 yard (yd) = 3 ft | 1 mile (mi) = 5,280 ft |
| | 1 mi = 1.609 km | 1 km = 0.621 mi | 1 angstrom (Å) = $10^{-10}$ m |
| | 1 AU = $1.50\times10^{11}$ m | 1 light year (ly) = $9.46\times10^{15}$ m | |
| **Area** | 1 $m^2$ = 10.76 $ft^2$ | 1 $in^2$ = 6.452 $cm^2$ | 1 acre = 43,560 $ft^2$ = 4,048 $m^2$ |
| **Volume** | 1 $m^3$ = 1,000 liters (L) | 1 L = 1,000 $cm^3$ | 1 gallon = 3.785 L |
| **Time** | 1 year (yr) = 365.24 day | 1 day = 24 hr | 1 hr = 60 min |
| | 1 min = 60 s | 1 yr = $3.156\times10^{7}$ s | 1 day = 86,400 s |
| **Speed and velocity** | 1 m/s = 2.237 mi/hr (mph) | 1 mph = 1.609 km/hr | 1 m/s = 3.600 km/hr |
| **Force** | 1 N = 0.2248 pounds (lb) | 1 lb = 4.448 N | 1 ton = 2,000 lb |
| **Pressure** | 1 Pa = 1 N/$m^2$ | 1 Pa = $1.45\times10^{-4}$ lb/$in^2$ (psi) | 1 atm = $1.013\times10^{5}$ Pa |
| **Energy** | 1 J = $10^{7}$ erg | 1 calorie = 4.184 J | 1 Calorie = 1,000 calorie |
| | 1 eV = $1.602\times10^{-19}$ J | 1 kWh = $3.6\times10^{6}$ J | 1 Btu = 1,055 J |
| **Power** | 1 horsepower (hp) = 746 W | 1 hp = 550 ft lb/s | |
| **Mass** | 1 kg = 1,000 g | 1 amu = $1.6605\times10^{-27}$ kg | 1 kg has a weight of 2.205 lb |
| **Angular measures** | 1 radian (rad) = 57.3° | π rad = 180° | 1 rad/s = 9.55 rev/min (rpm) |

**Table 9: Mathematical symbols**

| Symbol | Description | Symbol | Description |
|---|---|---|---|
| $\propto$ | is proportional to | $=$ | is equal to |
| $\equiv$ | is equivalent to | $\approx$ | is approximately equal to |
| $\geq$ | is greater than or equal to | $\leq$ | is less than or equal to |
| $>$ | is greater than | $<$ | is less than |
| $\infty$ | infinity | $\Delta x$ | change in $x$ |
| $\pi$ | = 3.14159 | $e$ | = 2.7183 |

**Table 10: Greek alphabet**

| Name | Uppercase | Lowercase | Name | Uppercase | Lowercase | Name | Uppercase | Lowercase |
|---|---|---|---|---|---|---|---|---|
| Alpha | Α | α | Iota | Ι | ι | Rho | Ρ | ρ |
| Beta | Β | β | Kappa | Κ | κ | Sigma | Σ | σ |
| Gamma | Γ | γ | Lambda | Λ | λ | Tau | Τ | τ |
| Delta | Δ | δ | Mu | Μ | μ | Ypsilon | Υ | υ |
| Epsilon | Ε | ε | Nu | Ν | ν | Phi | Φ | φ |
| Zeta | Ζ | ζ | Xi | Ξ | ξ | Chi | Χ | χ |
| Eta | Η | η | Omicron | Ο | ο | Psi | Ψ | ψ |
| Theta | Θ | θ | Pi | Π | π | Omega | Ω | ω |

# Periodic table colored by element groups

**metals**
- alkali metals
- alkaline earth metals
- transition metals
- rare earth metals
- other metals

**nonmetals**
- halogens
- noble gases
- other nonmetals

**metalloids**
- metalloids

**Key**

| 1 H 1.0079 hydrogen |
|---|

1 — Atomic number
H — Symbol
1.0079 — Atomic mass
hydrogen — Name

| group 1 1A | group 2 2A | group 3 3B | group 4 4B | group 5 5B | group 6 6B | group 7 7B | group 8 8B | group 9 8B | group 10 8B | group 11 1B | group 12 2B | group 13 3A | group 14 4A | group 15 5A | group 16 6A | group 17 7A | group 18 8A |
|---|---|---|---|---|---|---|---|---|---|---|---|---|---|---|---|---|---|
| 1<br>H<br>1.0079<br>hydrogen | | | | | | | | | | | | | | | | | 2<br>He<br>4.0028<br>helium |
| 3<br>Li<br>6.941<br>lithium | 4<br>Be<br>9.0122<br>beryllium | | | | | | | | | | | 5<br>B<br>10.811<br>boron | 6<br>C<br>12.011<br>carbon | 7<br>N<br>14.007<br>nitrogen | 8<br>O<br>15.999<br>oxygen | 9<br>F<br>18.998<br>fluorine | 10<br>Ne<br>20.180<br>neon |
| 11<br>Na<br>22.990<br>sodium | 12<br>Mg<br>24.305<br>magnesium | | | | | | | | | | | 13<br>Al<br>26.982<br>aluminum | 14<br>Si<br>28.086<br>silicon | 15<br>P<br>30.974<br>phosphorous | 16<br>S<br>32.065<br>sulfur | 17<br>Cl<br>35.453<br>chlorine | 18<br>Ar<br>39.948<br>argon |
| 19<br>K<br>39.098<br>potassium | 20<br>Ca<br>40.078<br>calcium | 21<br>Sc<br>44.956<br>scandium | 22<br>Ti<br>47.867<br>titanium | 23<br>V<br>50.942<br>vanadium | 24<br>Cr<br>51.996<br>chromium | 25<br>Mn<br>54.938<br>manganese | 26<br>Fe<br>55.845<br>iron | 27<br>Co<br>58.933<br>cobalt | 28<br>Ni<br>58.693<br>nickel | 29<br>Cu<br>63.546<br>copper | 30<br>Zn<br>65.41<br>zinc | 31<br>Ga<br>69.723<br>galium | 32<br>Ge<br>72.61<br>germanium | 33<br>As<br>74.922<br>arsenic | 34<br>Se<br>78.96<br>selenium | 35<br>Br<br>79.904<br>bromine | 36<br>Kr<br>83.80<br>krypton |
| 37<br>Rb<br>85.468<br>rubidium | 38<br>Sr<br>87.62<br>strontium | 39<br>Y<br>88.906<br>yttrium | 40<br>Zr<br>91.224<br>zirconium | 41<br>Nb<br>92.906<br>niobium | 42<br>Mo<br>95.94<br>molybdenum | 43<br>Tc<br>(98)<br>technetium | 44<br>Ru<br>101.07<br>ruthenium | 45<br>Rh<br>102.91<br>rhodium | 46<br>Pd<br>106.42<br>palladium | 47<br>Ag<br>107.87<br>silver | 48<br>Cd<br>112.41<br>cadmium | 49<br>In<br>114.82<br>indium | 50<br>Sn<br>118.71<br>tin | 51<br>Sb<br>121.76<br>antimony | 52<br>Te<br>127.60<br>tellurium | 53<br>I<br>126.90<br>iodine | 54<br>Xe<br>131.29<br>xenon |
| 55<br>Cs<br>132.91<br>cesium | 56<br>Ba<br>137.33<br>barium | 71<br>Lu<br>174.97<br>lutetium | 72<br>Hf<br>178.49<br>hafnium | 73<br>Ta<br>180.95<br>tantalum | 74<br>W<br>183.84<br>tungsten | 75<br>Re<br>186.21<br>rhenium | 76<br>Os<br>190.23<br>osmium | 77<br>Ir<br>192.22<br>iridium | 78<br>Pt<br>195.08<br>platinum | 79<br>Au<br>196.97<br>gold | 80<br>Hg<br>200.59<br>mercury | 81<br>Tl<br>204.38<br>thallium | 82<br>Pb<br>207.2<br>lead | 83<br>Bi<br>208.98<br>bismuth | 84<br>Po<br>(209)<br>polonium | 85<br>At<br>(210)<br>astatine | 86<br>Rn<br>(222)<br>radon |
| 87<br>Fr<br>(223)<br>francium | 88<br>Ra<br>(226)<br>radium | 103<br>Lr<br>(262)<br>lawrencium | 104<br>Rf<br>(261)<br>rutherfordium | 105<br>Db<br>(262)<br>dubnium | 106<br>Sg<br>(266)<br>seaborgium | 107<br>Bh<br>(264)<br>bohrium | 108<br>Hs<br>(277)<br>hassium | 109<br>Mt<br>(278)<br>meitnerium | 110<br>Ds<br>(281)<br>darmstadtium | 111<br>Rg<br>(272)<br>roentgenium | 112<br>Uub<br>(285)<br>ununbium | 113<br>Uut<br>(284)<br>ununtrium | 114<br>Uuq<br>(289)<br>ununquadium | 115<br>Uup<br>(288)<br>ununpentium | 116<br>Uuh<br>(292)<br>ununhexium | | 118<br>Uuo<br>(294)<br>ununoctium |

| | | | | | | | | | | | | | |
|---|---|---|---|---|---|---|---|---|---|---|---|---|---|
| 57<br>La<br>138.91<br>lanthanum | 58<br>Ce<br>140.12<br>cerium | 59<br>Pr<br>140.91<br>praseodymium | 60<br>Nd<br>144.24<br>neodymium | 61<br>Pm<br>(145)<br>promethium | 62<br>Sm<br>150.36<br>samarium | 63<br>Eu<br>151.96<br>europium | 64<br>Gd<br>157.25<br>gadolinium | 65<br>Tb<br>158.93<br>terbium | 66<br>Dy<br>162.50<br>dysprosium | 67<br>Ho<br>164.93<br>holmium | 68<br>Er<br>167.26<br>erbium | 69<br>Tm<br>168.93<br>thulium | 70<br>Yb<br>173.06<br>ytterbium |
| 89<br>Ac<br>(227)<br>actinium | 90<br>Th<br>232.04<br>thorium | 91<br>Pa<br>231.04<br>protactinium | 92<br>U<br>238.03<br>uranium | 93<br>Np<br>(237)<br>neptunium | 94<br>Pu<br>(244)<br>plutonium | 95<br>Am<br>(243)<br>americium | 96<br>Cm<br>(247)<br>curium | 97<br>Bk<br>(247)<br>berkelium | 98<br>Cf<br>(251)<br>californium | 99<br>Es<br>(252)<br>einsteinium | 100<br>Fm<br>(257)<br>fermium | 101<br>Md<br>(258)<br>mendelevium | 102<br>No<br>(259)<br>nobelium |

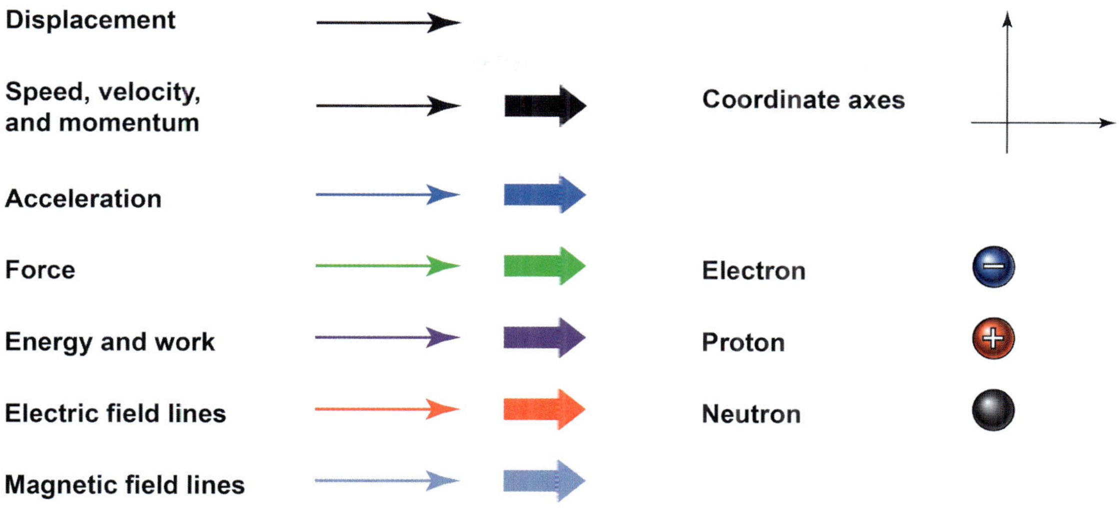
Displacement
Speed, velocity, and momentum
Acceleration
Force
Energy and work
Electric field lines
Magnetic field lines
Coordinate axes
Electron
Proton
Neutron

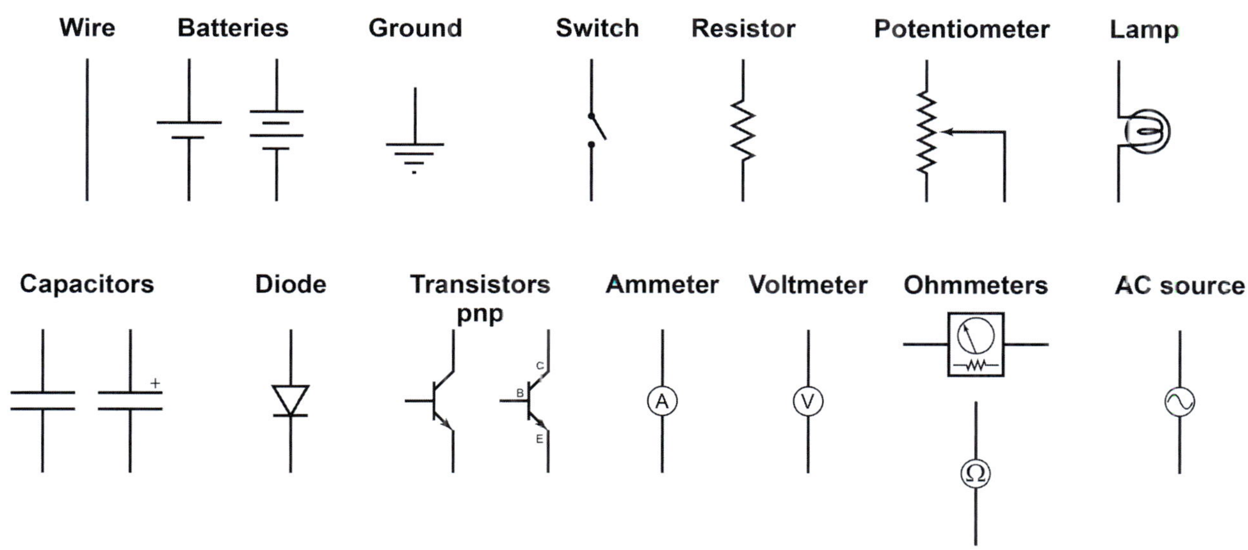
Wire
Batteries
Ground
Switch
Resistor
Potentiometer
Lamp
Capacitors
+
Diode
Transistors pnp
C
B
E
Ammeter
A
Voltmeter
V
Ohmmeters
Ω
AC source

Here are some rules that should be followed in *every* investigation.

1. Be responsible for your behavior. Broken glassware, spilled acid, and other hazards result from not paying attention. This means no running, pushing, practical jokes, or horseplay in the lab.
2. Know the location and methods of operation of all safety equipment, including the eye wash station, the fire extinguisher, and the safety shower.
3. Do not perform experiments that you have not been authorized to perform. This means using only the quantities your instructor provides. Ask if you don't understand the lab instructions.
4. Never remove chemicals from the lab.
5. Know the hazards of the chemicals you are using. You can find these on the material safety data sheets (MSDS), which lists physical characteristics (melting and boiling point, etc.), toxicity and health concerns, storage and reactivity, disposal instructions, and protective equipment (goggles, gloves, etc.).
6. Never work alone in a lab. An instructor must always be present when you are in the laboratory.
7. Wear approved eye protection, including safety goggles, at all times, except when explicitly excused from doing so by your instructor. Don't let your eyes be destroyed by a splash of acid!
8. If a chemical gets in your eyes, wash them immediately with flowing water from a sink or eye wash station for at least 15 minutes. However, do not point a high-pressure stream of water at your eye. Have someone alert the instructor and get medical attention immediately.
9. If a chemical spills on your body or clothing, alert your instructor immediately. Wash the affected area with running water. If the affected area is large, use a safety shower. Remove any clothing affected by chemicals to prevent further reactions with skin.
10. Notify your instructor immediately in case of cuts, burns, or other injuries.
11. Wear closed-toe shoes in the laboratory at all times. Bare feet and sandals are prohibited.
12. Clean up spills or broken glass immediately. Use one of the bins provided for broken glass. No paper, plastic, or other trash should go in the broken-glass bin.
13. Do not eat or drink anything in the laboratory. You never know which chemicals may have been used on a bench or table.
14. Wash your hands thoroughly before leaving the laboratory. Many chemicals, especially organic solvents, can be easily absorbed through the skin.
15. Avoid breathing fumes of any kind. If you are instructed to smell something, use your hand to gently waft the scent near your nose. Do not stick your nose into the end of a test tube or into any chemical reaction container.
16. Although this course will not use them, fume hoods are necessary when working with many hazardous chemicals, such as strong acids.

17. Never use suction from your mouth to pipette a chemical. Use a pipette bulb.
18. Long hair and loose clothing must be tied back or restrained in the lab. Hair and clothes burn!
19. In case of fire, let everyone know immediately and evacuate the laboratory. If the fire is small, you may engage a partner and use the fire extinguisher together to put the fire out. If the fire is larger, evacuate the building and pull the fire alarm.
20. If you are on fire, get under the safety shower. If a fellow classmate is on fire, it is also your responsibility to help him or her get under the safety shower.
21. Know the fire hazards of chemicals you are working with. Never put flammable liquids near an open flame.
22. Never point an open test tube containing a potential reaction at anyone. Outgassing, sputtering, and even flame can result from chemical reactions.
23. Keep your work area neat. You are less likely to knock things over if glassware and equipment are not cluttered with books or papers.
24. Do not force a stopper into a tube or a glass tube into a stopper. If you must put a tube through a stopper, use lubrication such as soap or glycerol and protect your hands with a glove or towel.
25. Be alert when handling glassware. A hot beaker looks just the same as a cold one, but it hurts a lot more when you touch it.
26. Know how to dispose of any chemicals or other waste you may generate in your laboratory activities. Waste containers are always provided for toxic or hazardous chemicals. There may be special containers or posters with useful disposal instructions.
27. Always replace the cap on chemicals when you have removed what you need.
28. Do not use dirty utensils to remove chemicals from containers. This may contaminate the supply.
29. Take only what you need, and never return excess chemicals to the container. This helps avoid waste and maintain the purity of the chemical supply for future students.
30. Clean up before leaving. Wash all used glassware and store it on a drying rack or other facility.
31. Read the lab investigation carefully before starting. It can be unsafe if you do not follow the written procedure.
32. If you do not understand a laboratory or safety rule, ask your teacher for clarification.
33. Before you begin an investigation, have all materials ready, including glassware, safety goggles, electronic balances, chemicals, probes, calculators, and computers.
34. Plan your laboratory procedure—be familiar with the questions you are being asked, and prepare testable hypotheses! Be ready to carry out your procedure!

## A

**absolute zero** — lowest possible temperature matter can have, corresponding to the minimum energy that its atoms or molecules can have; the zero point of the Kelvin temperature scale.

**absorption** — (1) process by which a wave loses energy to a material as it passes through it;
(2) process by which the electron of an atom takes in the energy of a photon of light, thereby increasing the electron's energy and moving it to a higher energy level.

**acceleration** — rate of change in velocity or the change in velocity divided by the change in time. Units of acceleration are units of speed divided by units of time, or $m/s^2$ (sometimes written as m/s/s).

**accuracy** — how well a measurement's value agrees with a commonly accepted value.

**additive primary colors (RGB)** — three colors (red, green, and blue) that, when added together in various combinations, can create any other color. Also called the additive color process.

**adiabatic process** — thermodynamic process in which no heat is added or removed from a system. Because no heat is exchanged, an adiabatic process is reversible.

**alpha decay** — radioactive decay in which a nucleus emits an alpha particle (consisting of two protons and two neutrons), thereby reducing the original atom's atomic number by two.

**ampere (A)** — unit of electric current, equivalent to one coulomb of charge per second.

**amplitude** — maximum displacement of an oscillation from its equilibrium, or average, value.

**angle of incidence** — angle between a light ray that strikes a surface and the normal to the surface at the point of contact.

**angle of reflection** — angle between a light ray that reflects off a surface and the normal to the surface at the point of reflection.

**angular momentum** — momentum of an object due to its rotation or spin. Angular momentum is the product of the radius of rotation and the momentum, or $L = r \times mv$. Angular momentum has units of kg $m^2/s$.

**angular velocity** — rate at which a body rotates about an axis or center. Units are radians per second (rad/s).

**antinode** — point of maximum amplitude on a standing wave.

**armature** — rotating part of an electric motor containing electromagnets.

**atom** — smallest particle of an element that can exist in isolation or in a combination with other atoms. Consists of a heavy nucleus surrounded by one or more electrons.

**atomic mass unit (amu)** — standard unit of measure for expressing the mass of an atom. One amu is defined as one-twelfth of the mass of a carbon-12 atom or $1.661\times10^{-27}$ kg.

**atomic number** — number associated with an atom that is equal to the number of protons in its nucleus.

**average** — value obtained from a set of measurements by adding up all their measurement values and dividing by the total number of measurements

**average velocity** — distance or displacement an object travels divided by the time taken, regardless of any variations in velocity during the time interval.

**Avogadro's number** — $6.022\times10^{23}$ particles, the number of particles in one mole.

**axis** — (1) imaginary straight line about which an object rotates;
(2) one of the reference lines used to depict a coordinate system.

## B

**bar graph** — a graph in which the height or length of columns or bars represent the values of data

**battery** — device that transforms or converts chemical energy into electrical energy or electric current. A battery's voltage, or potential difference, is measured in volts (V).

**beats** — pulsations of sound resulting from the superposition of two waves of different frequencies. The beat frequency is the difference in frequency between the two waves.

**Bernoulli's equation** — expression of energy conservation along a streamline for a frictionless fluid. Often expressed as
$\varrho gh + ½mv^2 + P =$ constant.

**beta decay** — radioactive decay in which a neutron in the nucleus of an atom spontaneously converts into a proton and an electron. The proton remains in the nucleus, increasing the atom's atomic number by one, while the energetic electron is emitted as a beta particle.

**binding energy** — net energy that must be supplied to overcome the strong nuclear force in order to split an atomic nucleus into its constituent particles.

**biomechanics** — study of the motion of various parts of the human body (or other living organisms).

**black hole** — gravitational singularity predicted by general relativity.

**blackbody** — absorbs 100% of the electromagnetic radiation (i.e., thermal radiation) incident upon its surface.

**block and tackle** — system of pulleys that changes the *magnitude* of the input force, not just its direction. Often used in sailing.

**Bohr model** — model of the hydrogen atom, proposed by Danish physicist Niels Bohr, where the electron moves in circular orbits only at discrete or quantized radii from the nucleus.

**Boltzmann's constant** — constant describing the conversion between microscopic thermal energy per particle and macroscopic temperature. The value of Boltzmann's constant is
$k_B = 1.38\times10^{-23}$ J/K.

**Boyle's law** — for a gas at constant temperature, the change in pressure is inversely proportional to the change in volume. Often written $P_1V_1 = P_2V_2$.

**brittle** — material property describing the tendency to break before deforming. Brittle is the opposite of *elastic*.

**Brownian motion** — random, jerky movement of extremely small particles floating in still water observed by Robert Brown in 1827.

## C

**calorie** — unit of heat energy corresponding to the quantity of heat required to increase the temperature of 1 g of water by 1°C, or 4.18 J. The calorie is written with a lowercase "c" and is not to be confused with the capital "C" unit of the Calorie, which is equal to 1,000 calories.

**Calorie** — unit of heat energy equal to 1,000 calories or 4,180 J. The Calorie, written with a capital "C" and typically used to list the energy content in foods, should not be confused with the calorie that is written with a lowercase "c."

**capacitance** — measure of the ability of a capacitor (or other conductor) to store electric charge. The unit of capacitance is the farad (F), although it is also measured in coulombs per volt (C/V).

**capacitor** — device with two parallel conducting plates separated by a narrow gap usually filled with an insulating material. A capacitor is a device that stores electric charge.

**carbon dating** — procedure for determining the time since death or age of an organic material by measuring the fraction of the radioactive isotope carbon-14.

**Carnot cycle** — four-stage process of an idealized heat engine, consisting of:
(1) adding heat during isothermal expansion;
(2) adiabatic expansion;
(3) rejecting heat during isothermal compression; and
(4) adiabatic compression.

**Carnot efficiency** — ideal maximum percentage of the input energy that can be converted into work by a heat engine.

**Cartesian coordinates** — coordinate system in which positions are specified with coordinate values along two or three perpendicular axes labeled $x$, $y$, and $z$.

**cathode ray tube (CRT)** — vacuum tube in which a cathode emits a beam of electrons that are deflected by magnetic fields before hitting a phosphorescent screen. By suitably controlling the signal in the electromagnets, an image can be created on the screen.

**Celsius scale** — relative temperature scale where 0°C is defined as the temperature that water freezes and 100°C is the temperature that water boils.

**center of mass** — average position of all the mass in an object. The center of mass is the point around which the object balances in any direction.

**centripetal acceleration** — any acceleration perpendicular to an object's velocity that causes it to move in a circle. The centripetal acceleration is directed towards the center of the circle.

**centripetal force** — any force perpendicular to an object's velocity that causes it to move in a circle. The centripetal force is directed towards the center of the circle.

**chain reaction** — chemical or nuclear reaction where the products of one reaction can create more than one additional reaction. A nuclear chain reaction is at the heart of nuclear fission.

**Charles's law** — for a gas at constant pressure, the change in volume is directly proportional to the change in absolute temperature. Often written as $V_1/T_1 = V_2/T_2$.

**circuit breaker** — safety device in a circuit that stops the flow of current when a maximum current is exceeded. A circuit breaker can be reset manually after the current is brought below the maximum allowed value.

**closed circuit** — electric circuit where the switch is connected (or closed), allowing electricity to flow through the circuit. The opposite of an open circuit.

**closed system** — isolated system that cannot exchange matter or energy with its surroundings.

**coefficient of kinetic friction** — ratio of the force of dry sliding friction between moving surfaces divided by the normal force acting between the surfaces.

**coefficient of rolling friction** — ratio of the force of dry sliding friction between moving surfaces divided by the normal force acting between the surfaces.

**coefficient of static friction** — ratio of the maximum force of static friction between two stationary surfaces divided by the normal force acting between the surfaces.

**collision** — interaction that causes one or more objects to change their velocity.

**commutator** — device attached to the armature of a motor (or generator) that provides an electrical connection and allows the direction of current to be changed periodically during the rotation of the motor. In an electric motor, the commutator plates typically comes in electrical contact with the brushes.

**compass** — instrument that indicates direction relative to the north–south and east–west axes, usually based on sensing Earth's magnetic field.

**component** — (1) part of a vector; (2) projection of a vector against one of the coordinate axes.

**component force** — force that lies along one of the three coordinate axes ($x$, $y$, or $z$).

**compound machine** — machine consisting of two or more simple machines.

**compressible** — fluid that significantly changes its volume in proportion to changes in pressure.

**concave** — curved form, usually for a mirror or lens, that is depressed inwards in its middle.

**concave lens** — diverging lens that is thinner in its middle than at its outer part. A concave lens produces a virtual and upright image that is reduced in size.

**cones** — light-sensitive cells in the eye that respond to color.

**constructive interference** — occurs when two or more waves, of the same phase, add together to create a larger wave.

**continuum spectrum** — spectrum apparently containing all wavelengths of light over a broad range in wavelength. Also called a continuous spectrum.

**control rods** — rods made of a material that absorbs neutrons. They are used to control the rate of nuclear reactions occurring in a nuclear reactor.

**convection** — transfer of heat through the motion of matter, such as blowing hot air or circulating a liquid coolant.

**conversion factor** — ratio used to transform a quantity expressed in one unit (or set of units) into another unit (or set of units). In a conversion factor, the same quantity appears in both the numerator and denominator but is expressed in two different units. A conversion factor has a value of exactly one. Example: the conversion factor from centimeters to meters is (1 m/100 cm).

**convex** — curved form, usually for a mirror or lens, that bulges out in the middle.

**convex lens** — converging lens that is thicker in its middle than at its outer part. If the object is located outside the focal point, then the image is inverted and real. If the object is located inside the focal point, then the lens becomes a magnifying glass producing a virtual and upright image.

**coordinates** — individual values that collectively specify a position in one-, two-, or three-dimensional space.

**correlation coefficient** — a number between +1 and -1 which describes the goodness of fit for a regression

**coulomb (C)** — SI unit for electric charge.

**Coulomb's law** — attraction or repulsion between two charged objects is inversely related to the square of the distance between them and directly related to the magnitude of each of the charges. Also expressed as $F_e = k_e q_1 q_2 / r^2$.

**crest** — highest point or maximum amplitude of a wave.

**critical angle** — angle of incidence at which light is totally reflected back into a material.

**cycle** — something that repeats in time at regular intervals, such as one full swing of a pendulum.

**cyclotron radius** — radius of circular (or helical) motion of a charged particle moving in a uniform magnetic field. Also called gyroradius.

## D

**damping** — gradual decrease in amplitude and energy of a wave due to friction or other energy-loss mechanisms.

**decibel (dB)** — unit of power or amplitude that scales logarithmically, or by powers of 10. For example, an increase of 20 dB represents a factor of 10 increase in amplitude. An increase of 10 dB represents a factor of 10 increase in power, while an increase of 20 dB represents a factor of 100 increase in power.

**decimal places** — number of digits in a numerical value that are to the right of the decimal point.

**density** — ratio of mass to volume or the mass per unit volume.

**dependent variable** — variable in an experiment that changes in response to changes made in the independent variable. The dependent variable is typically plotted as the vertical or *y*-axis (the ordinate) in a graph.

**destructive interference** — occurs when two or more waves, which are out of phase with respect to each other, combine to create a smaller wave.

**detector** — scientific instrument used to indicate the presence of radiation.

**diffraction** — property of waves that describes how waves bend around obstacles and spread out when passing through openings.

**diffraction pattern** — pattern of constructive and destructive interference, usually of light projected onto a screen, caused by diffraction around a barrier or through a narrow opening.

**diffuse reflection** — light that strikes a surface that is rough (or not shiny) will scatter in many directions. Light striking the paint on a wall or the pages of a book are examples of diffuse reflection.

**digital multimeter** — instrument with different settings that can measure current (ammeter), voltage (voltmeter), or resistance (ohmmeter).

**dispersion** — separation of white light into its constituent colors (or wavelengths), such as by a prism or a rainbow.

**displacement** — vector quantity that represents a change in position. Also used to describe the difference between an object's initial and final positions. Measured in meters (m).

**Doppler effect** — shift in frequency of a wave that is emitted from a source that is moving with respect to the observer.

**dose** — total amount of radiation received by a person. Measured in rems.

**drag coefficient** — dimensionless number that describes the effect of the shape of an object on fluid friction. An aerodynamic shape has a low drag coefficient and therefore lower air resistance compared to a shape with a higher drag coefficient.

## E

**echo** — reflected sound wave.

**efficiency** — ratio of the output energy or power divided by the input energy or power for any process that transforms energy. (Note that input and output can be defined in different ways.)

**elastic collision** — type of collision in which objects bounce off each other in a way that conserves total kinetic energy. (As in all collisions, momentum is also conserved for an elastic collision.)

**elastic potential energy** — potential energy stored in an object that is transformed into kinetic energy when the object itself is transformed in shape, configuration, or length.

**elasticity** — property of a material to bend or deform without breaking. Rubber is very elastic while glass is not.

**electric charge** — fundamental property of matter that comes in two varieties we call positive and negative.

**electric circuit** — electrical device that has a complete or continuous path that electricity can flow through.

**electric current** — flow of electric charges, usually in wires or other conductors. Measured in amperes (A).

**electric field** — force field created by electric charge. Electric field is a vector with the direction that a positive test charge would move, and whose magnitude is the force per unit charge. Measured in force per unit charge with units of newtons per coulomb, abbreviated N/C.

**electric field lines** — vector arrows that depict visually the direction and magnitude of an electric field.

**electric force** — fundamental force that charged objects exert on each other.

**electric motor** — machine that transforms or converts electrical energy into mechanical energy.

**electric potential** — electrical potential energy per unit charge for an electric field.

**electric potential energy** — change in potential energy per unit charge in an electric field.

**electrical conductor** — material, such as a metal, that offers low resistance to the flow of electric current.

**electrical generator** — machine that converts rotational motion to electricity

**electrical insulator** — material that is a poor conductor of electric current.

**electrical symbol** — symbol used to represent a component of a circuit in a circuit diagram.

**electrically neutral** — object that has a net charge of zero.

**electricity** — flow of electric current through a conductor, such as a wire. Electricity is sometimes also used to describe electrostatic charges.

**electromagnetic induction** — inducing (or creating) a voltage in a conductor by changing the magnetic field near the conductor.

**electromagnetic spectrum** — range of frequencies of electromagnetic radiation extending from low frequencies (radio waves and microwaves) to higher frequencies (infrared radiation, visible light, ultraviolet light, x-rays, and gamma rays).

**electromagnetic wave** — transverse wave resulting from oscillating and perpendicular electric and magnetic fields and traveling at the speed of light. Also called electromagnetic radiation. Examples of electromagnetic waves, in order of increasing wavelength, are gamma rays, x-rays, ultraviolet radiation, visible light, infrared radiation, microwaves, and radio waves.

**electromagnetism** — interaction of electric currents (or fields) and magnetic fields.

**electromagnets** — magnets that are created by the flow of electricity.

**electron** — negatively charged subatomic particle with a mass of $9.109\times10^{-31}$ kg and a charge of $-1.602\times10^{-19}$ C.

**electron volt (eV)** — unit of energy corresponding to the amount of energy gained by an electron that moves across a potential difference of one volt (V). One eV is equal to $1.602\times10^{-19}$ J, which corresponds to one volt multiplied by the charge of the electron.

**electroscope** — scientific instrument used to detect charged objects.

**electrostatic induction** — process of electrically charging an object by holding it close to another charged object.

**electrostatics** — study of electric charges at rest.

**elementary charge** — indivisible electric charge carried by a single proton (positive) or electron (negative). Usually denoted *e* and equal to $1.602\times10^{-19}$ coulomb (C).

**emissivity** — relative ability of a surface to emit thermal energy by radiation. The emissivity is the ratio of the actual power emitted as thermal radiation divided by the power emitted by a perfect blackbody at the same temperature. The value of emissivity $\varepsilon$ is between zero (no power radiated) and one (radiates as a perfect blackbody), or $0 \le \varepsilon \le 1$.

**energy** — quantity that causes matter to change and mediates how much change occurs.

**energy levels** — set of quantum states for the electron energies in an atom. The electron must absorb or release energy to change from one level to another.

**engineering** — creative application of science to design technologies that solve problems or meet needs.

**entropy** — thermodynamic quantity calculated by dividing heat by temperature. The entropy of a system measures the thermal energy per degree that is *not* available for converting to output work.

**equilibrium** — state of a system in which all forces are in balance. There is zero net force on a system in equilibrium.

**equipotential** — two or more positions in an electric field that have the same electric potential. In an electric field diagram, equipotential lines are lines of constant electric potential.

**equivalent resistance** — total or combined resistance of a circuit with more than one resistor Also called the effective resistance

**error** — the unavoidable difference between a real measurement and the unknown true value of the quantity being measured

**error bars** — vertical and horizontal lines added to the plotting of a data point to show the uncertainty in the x and y values

**escape velocity** — minimum initial velocity needed to completely break away from a gravitational field. Objects with less than the escape velocity eventually fall back towards the parent body. Objects with greater than the escape velocity may never come back if launched on the right trajectory.

**excited state** — energy level of an atom that has higher energy than the ground state. An electron will move from the ground state to an excited state by absorbing energy from light; if it returns to the ground state it will emit light.

**experiment** — controlled situation set up specifically to observe what happens under a well-defined set of circumstances.

**exponent** — value of $b$ in an expression $a^b$.

## F

**Fahrenheit scale** — relative temperature scale where 32°F is defined as the temperature that water freezes and 212°F is the temperature that water boils.

**Faraday's law of induction** — changing magnetic flux through a coil of wire will induce an emf in the wire that is proportional to the number of loops and the rate of change of the magnetic field.

**ferromagnetic** — a type of substance, such as iron, that experiences strong magnetic forces when in the presence of a magnet. Iron, nickel, and cobalt are ferromagnetic materials.

**first law of thermodynamics** — change in energy of a system is the heat added to the system plus the work done on the system. Referred to as the conservation of energy for thermodynamics.

**fission** — nuclear reaction in which a nucleus is split, thereby releasing large amounts of energy.

**fluid** — substance that can change shape and flow readily in response to any applied forces, such as a liquid or a gas.

**focal length** — distance from the center of an optical device to the focal point.

**focal point** — point at which light rays that are parallel to the optical axis of an optical device either meet (converge) or diverge after being reflected or refracted.

**force** — action on a body that causes change in motion. Measured in newtons (N).

**force field** — field that exerts a force on objects in it. Examples include magnetic, electric, and gravitational fields.

**Fourier's theorem** — any repetitive signal, such as a multi-frequency wave, can be replaced by a series of single-frequency sine waves of different amplitudes and phases.

**free fall** — motion of an object when the only force acting on it is the force of gravity.

**free-body diagram** — sketch that isolates a single object and uses arrows to represent the strength, location, and direction of all forces acting on the object.

**frequency** — rate at which a cyclic behavior repeats. Measured in hertz (Hz), where 1 Hz = 1 cycle per second or 1/s. Frequency is the inverse of the period.

**frequency spectrum** — graph showing amplitude on the vertical axis and frequency on the horizontal axis. The frequency spectrum tells you the relative amplitudes of the different frequencies in a signal.

**frequency table** — a table in which a large data set data has been sorted into a smaller number of categories with each category representing a range of values or a count of items - useful for seeing large scale patterns in data with high point-to-point variability

**friction** — resistive force caused by motion that always acts to oppose motion.

**fulcrum** — pivot point or center of rotation for a lever.

**fusion** — nuclear reaction in which two nuclei merge together to form a larger nucleus.

## G

**galvanometer** — instrument used to detect very small electric currents by detecting the magnetic force on a loop of wire.

**gamma decay** — radioactive process during which a nucleus goes to a lower energy state by emitting a high-energy photon called a gamma ray.

**gamma rays** — highest frequency form of electromagnetic radiation.

**gas** — phase of matter that flows to fill completely its container and contains particles that experience little or no attractive force to each other.

**gear** — toothed wheel that meshes together with another toothed wheel in order to transmit motion.

**gear ratio** — ratio between the rates that the last gear and the first gear rotate in a machine.

**Geiger counter** — scientific instrument used to detect and count the rate of ionizing radiation.

**gravitational potential energy** — form of potential energy due to gravity. Close to the Earth's surface, gravitational potential energy is equal to $mgh$. Measured in joules (J).

**ground state** — lowest energy level for electrons in an atom.

## H

**half-life** — average time for one-half of the atoms in a substance to decay radioactively.

**harmonic** — vibrating mode of oscillation with a frequency that is an integer multiple of the frequency of the fundamental vibrating mode.

**heat** — thermal energy that flows from one object to another due to differences in their temperatures or because one object is undergoing a phase change. Measured in joules (J).

**heat transfer** — study of how thermal energy moves, often through conduction, convection, and/or radiation.

**heat transfer coefficient** — measure of the rate of heat flow between a surface and a fluid per square meter of contact area per degree of temperature difference. The heat transfer coefficient $h$ is measured in power per unit area per degree of temperature difference. In SI units, the heat transfer coefficient is measured in watts per square meter per degree Celsius (W $m^{-2}$ $^{o}C^{-1}$).

**hertz (Hz)** — unit of frequency. One hertz = 1 cycle per second, or 1/s or $s^{-1}$.

**Hooke's law** — restoring force of a spring is proportional to the displacement (extension or compression) of the spring. Usually expressed as $F = -kx$, where $k$ is the spring constant and $x$ is the displacement.

**horsepower (hp)** — unit of power equal to 746 watts.

**hypothesis** — tentative explanation that may be tested and verified or revised based on the outcome of observations and experiments.

## I

**ideal gas** — theoretical gas of randomly moving point particles that interact only through elastic collisions with each other and the walls of their container.

**ideal gas law** — relationship between pressure, volume, temperature, and the number of particles for an ideal gas. Usually expressed as $PV = Nk_BT$.

**ideal mechanical advantage** — ratio of the distance moved by the input force to the distance moved by the output force. The ideal mechanical advantage corresponds to a machine with no efficiency losses.

**image** — picture of an object that is formed by optical devices through reflection and/or refraction.

**image distance** — distance between an optical element (such as a mirror or lens) and the image formed by it.

**impulse** — (1) change in momentum of an object; (2) product of force multiplied by the time over which the force acts. Units either kg m/s (kilogram meter per second) or N s (newton second).

**incident ray** — light ray from an object that strikes a surface or boundary between two materials.

**inclined plane** — flat and smooth surface that is tilted at an angle with respect to an acceleration or force. Also called a ramp.

**incompressible** — fluid that does not change its volume (or density) despite changes in pressure.

**independent variable** — variable in an experiment that is typically changed by the experimenter in order to cause a change in the dependent variable. The independent variable is typically plotted as the horizontal or $x$-axis (the abscissa) in a graph.

**index of refraction** — ratio of the speed of light in a vacuum to its speed in a material. A measure of how much a light ray bends when it passes from a vacuum into the material.

**inelastic collision** — type of collision where the objects stick together after impact, are deformed, and/or convert some kinetic energy to heat. Momentum is always conserved in collisions, but energy is not conserved in an inelastic collision. In a *perfectly inelastic* collision, the objects stick together after impact.

**inertia** — property of matter that resists changes in speed or direction. The property of inertia derives from an object's mass.

**inertial reference frame** — reference frame that has a zero or constant velocity (both speed and direction).

**information** — describes how matter and energy are arranged in time and space.

**infrared radiation** — form of electromagnetic radiation at somewhat lower frequencies than visible light. Often emitted as a form of radiant heat.

**input arm** — distance on a lever between the location the input force is applied and the fulcrum.

**input force** — force applied to a machine.

**instantaneous velocity** — velocity of an object measured at a particular time (not over a time interval).

**intensity** — power per unit area (or the ratio of power to area) for forms of traveling energy such as light and sound. Measured in watts per square meter ($W/m^2$).

**inverse square law** — any law in which a physical quantity varies inversely as the square of the distance from a location.

**ionization** — process by which an atom or molecule becomes charged by the removal or the addition of an electron.

**isothermal process** — thermodynamic process in which a system undergoes a change in pressure and volume while temperature is held constant. During an isothermal process, pressure and volume are inversely proportional to each other.

**isotopes** — atoms that have the same number of protons but different numbers of neutrons. Isotopes will have the same atomic number (because they have the same number of protons) but different atomic masses (because they have different numbers of neutrons).

## J

**joule (J)** — unit of work and energy in the SI. A force of one newton, acting for one meter, requires one joule of work.

## K

**Kelvin scale** — absolute scale for temperature that starts at absolute zero and measures the average energy of atoms or molecules. Measured in kelvins (K).

**kilowatt-hour (kWh)** — measure of consumed energy typically used by power companies. Equivalent to the energy transferred by one kilowatt of power in one hour.

**kinetic energy** — energy due to mass in motion. Measured in units of joules (J).

**kinetic theory** — theory of matter that uses statistical mechanics to derive observable macroscopic properties, such as temperature and pressure, from the microscopic behavior of collections of particles.

**Kirchhoff's current law** — sum of all currents entering a junction must equal the sum of all currents exiting that junction.

**Kirchhoff's voltage law** — sum of the voltage drops (voltage differences) around a closed loop equals zero.

## L

**laser** — scientific instrument that produces a beam of coherent light, i.e., that has the same frequency, direction, and phase. Acronym for light amplification by stimulated emission of radiation.

**law of conservation of energy** — total energy in a closed system does not change over time. While energy cannot be created or destroyed, it can be transformed from one form to another, just as long as the total energy remains constant.

**law of conservation of momentum** — total momentum of a system is constant in the absence of outside forces acting on the system.

**law of reflection** — when a light ray reflects off a surface or boundary between two materials, the angle of reflection is equal to the angle of incidence.

**law of universal gravitation** — force of attraction between two objects is directly proportional to each of their masses and inversely proportional to the distance between them.

**length** — separation between two points. A fundamental measure of space with common units of meters, feet, and miles.

**lens** — transparent optical device that refracts light and is shaped in such a way either to converge or diverge light rays to focus them into an image.

**Lenz's law** — direction of the current induced in a wire is such that the current's induced magnetic field opposes the change in magnetic field that originally induced the current.

**lever** — elongated, rigid object that may rotate around a point (the fulcrum) under the action of one or more forces. The lever is one of six kinds of simple machines.

**lever arm** — perpendicular distance from the center of rotation to the line of action of a force.

**light** — form of radiant energy consisting of electromagnetic waves traveling at the speed of light.

**line of action** — imaginary line that passes through a point and follows the direction a force acts.

**line spectrum** — distinct pattern of emission or absorption spectral lines, each line corresponding to a unique wavelength of light, coming from a material when viewed in a spectrograph.

**linear momentum** — momentum due to an object's motion in a straight line, which is the product of mass and velocity. Also called *momentum*. Linear momentum has units of kg m/s.

**linear regression** — a mathematical technique for calculating the best straight line fit to a set of x-y data

**liquid** — phase of matter that maintains its fixed volume, flows to take the shape of its container, and contains particles with intermolecular bonds that are continually being broken and reformed.

**longitudinal** — type of wave where the oscillations are along the direction of motion of the wave itself.

**lubrication** — technology of using fluids such as oil or other materials to reduce friction.

## M

**machine** — device that changes the magnitude and/or direction of a force. A machine is a mechanical system that is capable of doing work.

**macroscopic** — scale that is large enough to be directly sensed, such as the scale of things in the everyday world.

**magnetic** — property of a material that describes its ability to exert a force on a magnet or a magnetic material.

**magnetic domains** — region of a material comprised of many atoms with their individual magnetic fields aligned. By having their magnetic fields aligned they add together rather than cancel each other out.

**magnetic field** — force field surrounding magnetic objects. Measured in units of teslas (T).

**magnetic field lines** — the arrows that depict the direction of the magnetic force in a magnetic field. Magnetic field lines are drawn with arrows pointing away from magnetic north pole and towards magnetic south pole.

**magnetic flux** — measure of the strength of a magnetic field through an area. Magnetic flux is measured in tesla square meters (T $m^2$) or webers (Wb).

**magnetic force** — force exerted on an object in a magnetic field. Magnetic force can be either attractive or repulsive depending on both the alignment of the object in the field and the object's material properties.

**magnetic poles** — two regions in a magnetic field, called magnetic north and south poles, that produce the strongest magnetic force.

**magnetism** — (1) physical phenomenon produced by the motion of electric charge, resulting in attractive and repulsive forces between objects; (2) property of a material or object to attract other objects made out of iron, magnetite, or steel.

**magnetize** — to develop a magnetic field in an object in response to applying an external magnetic field.

**magnification** — change in the apparent size of an object by an optical device or system.

**magnitude** — single number that represents the length, strength, or "size" of a vector without reference to direction.

**mass** — measure of the quantity of matter. Measured in units of kilograms in the International System of Units (SI). English units are slugs (rarely used).

**mass deficiency** — difference between the sum of the mass of nucleons making up a nucleus and the total mass of the nucleus. The mass deficiency is related to binding energy through Einstein's mass–energy equivalence equation.

**mass number** — total number of protons and neutrons in a nucleus.

**mass spectrometer** — scientific instrument used to measure the masses of individual atoms in a gas using the force exerted on them in a uniform magnetic field. The instrument can be used to identify the elements and isotopes present in a gas.

**mass–energy equivalence** — principle that mass can be converted into energy and vice versa. Represented by the relationship $E = mc^2$.

**material strength** — ability of a material or object to sustain force without permanent deformation or breaking.

**matter** — physical substance that has mass and takes up space.

**Maxwellian distribution** — mathematical function that describes the distribution of velocities in an ideal gas in thermal equilibrium.

**measurement** — communication of quantity that includes both a value and a unit. The unit provides a commonly understood interpretation for the value. For example, *2.5 kilograms* is a measurement in which *2.5* is the value and *kilograms* is the unit.

**mechanical advantage** — ratio of the output force to the input force for a machine.

**mechanical energy** — energy that comes from position or motion. Gravitational potential energy, elastic potential energy, rotational energy, and kinetic energy are examples of mechanical energy.

**microphone** — device that transforms the pressure variations in a sound wave into electrical oscillations with the same pattern of variation with time.

**microscopic** — scale that is too small for direct observation with the senses. In physics, this is typically the size of atoms and molecules (or even smaller).

**microwaves** — form of electromagnetic radiation at somewhat higher frequencies than radio waves.

**mirror** — optical device that reflects light.

**mode** — characteristic pattern of resonant vibration of a system which occurs at one of the system's natural frequencies.

**model** — relationship that connects variables, such as an equation or graph. Models can be quantitative—giving numerical results—or qualitative—giving only a descriptive explanation.

**mole** — quantity of $6.022 \times 10^{23}$, which is the number of particles in one *mole* of matter. Often used when considering the number of atoms or molecules in a quantity of matter. The SI unit for the mole is the mol.

**moment of inertia** — quantity expressing an object's resistance to change its state of rotation about a particular axis, calculated as the product of mass times the square of the radius from that axis. For a distribution of masses, the moment of inertia is the sum of the individual products of mass times radius squared. Moment of inertia is measured in units of kg $m^2$.

**momentum** — vector equal to the product of velocity and mass. Momentum has units of kg m/s (kilogram meter per second) or N s (newton second).

## N

**net force** — combination of all the forces that act on a body.

**neutron** — neutral subatomic particle with zero electric charge that has a mass of $1.675 \times 10^{-27}$ kg.

**newton (N)** — SI unit of force that is equivalent to 1 kg $m/s^2$. A one newton net force applied to a one kilogram mass will accelerate the mass by 1 $m/s^2$.

**Newton's first law of motion** — an object at rest remains at rest, and an object in motion continues with constant velocity, unless acted on by a net force.

**Newton's law of cooling** — rate of heat flow is approximately proportional to the temperature difference.

**Newton's second law of motion** — acceleration of an object is proportional to the net force acting on it and inversely proportional to its mass.

**Newton's third law of motion** — whenever one object exerts a force on another object, the second object exerts an equal and opposite force on the first object.

**node** — stationary point on a standing wave. A point of minimum or zero amplitude.

**normal** — line that is perpendicular to another line or surface.

**normal force** — force perpendicular to a surface. Typically created where objects touch other objects or surfaces, such as floors or walls.

**nuclear energy** — type of energy created by reactions in the atomic nucleus.

**nuclear reaction** — process in which the nucleus of an element is changed into a different isotope of that element or a different element altogether.

**nuclear waste** — unwanted byproducts of nuclear reactions that are radioactive, have long half-lives, and must be disposed in safe, long-term storage.

**nucleus** — tiny dense core of an atom that contains all the protons and neutrons, measuring approximately 1/10,000 of the diameter of the atom.

## O

**object distance** — distance between the object and the optical element (such as a mirror or lens).

**objectivity** — description only of what has actually occurred without opinion, interpretation, or exaggeration.

**ohm (Ω)** — unit of electrical resistance. A resistance of one ohm (1 Ω) means a current of one amp will flow if a voltage of one volt is applied. Named in honor of Georg Simon Ohm.

**Ohm's law** — current in an electric circuit is proportional to the voltage applied and inversely proportional to the resistance, or $I = V/R$. Named in honor of Georg Simon Ohm.

**opaque** — material that allows no light to pass through it.

**open circuit** — electric circuit with a break that prevents electricity from flowing through it. The opposite of a closed circuit.

**open system** — system on which outside influences can act, such that matter or energy can be added or removed from the system.

**optical axis** — (1) for a mirror, a line perpendicular to the surface of a mirror that joins its focal point and center of curvature.
(2) for a lens, a line connecting the centers of curvature of the surfaces of the lens.

**optical fibers** — thin, transparent, and flexible tube, usually made of glass or plastic, that transmits light. Often used for communications technology.

**orbit** — regular, curved path one object travels around another object that is maintained by an attractive force such as gravity. Usually refers to the gravitationally determined path of a planet or satellite around another, larger body.

**orbital period** — time it takes a satellite to complete one orbit with respect to the fixed stars or celestial sphere. Earth's orbital period is 365.24 days.

**origin** — point in space where the position = 0 m in one dimension or (0,0,0) m in three dimensions. The origin is the fixed reference point for position data and is the location where all the coordinate axes intersect.

**oscillator** — system that exhibits harmonic motion.

**output arm** — distance on a lever between the location from which the output force is applied and the fulcrum.

**output force** — force produced by a machine.

## P

**parallel circuit** — electric circuit where the electric current can take more than one path.

**Pauli exclusion principle** — quantum rule that no two electrons (or fermions) may occupy the same quantum state in the same system.

**period** — time it takes to complete one full cycle of an oscillation. Equal to the inverse of the frequency of oscillation.

**periodic force** — force whose magnitude or direction has a repetitive or cyclic pattern.

**permanent magnet** — magnetic object that retains its magnetism even when it is not in an external magnetic field.

**permeability of free space** — physical constant describing how effectively a magnetic field can be established in a vacuum. Also called the *vacuum permeability*. The permeability of free space is used in the equation for the magnetic field surrounding a current-carrying wire.

**phase** — *Waves:* location of a wave with respect to its complete cycle. Usually given in degrees, where 360 degrees is one full cycle. A phase of 180 degrees is a half cycle.
*Matter:* state of solid, liquid, gas, or plasma.

**phases of matter** — matter can exist in only four different states: gas, liquid, solid, and plasma. The first three are regularly encountered as part of everyday life.

**phosphorescence** — type of light emission by a material where there is a delay between the absorption and re-emission of the light. The delay is caused by the electrons being excited to higher energy levels that do not decay rapidly back to lower energy levels.

**photoelectric effect** — ejection of electrons from a metal when light of certain frequencies is incident upon it.

**photon** — smallest, discrete packet of electromagnetic energy that makes up light in the quantum theory.

**photovoltaic cell** — device that converts energy from the Sun's light into electricity.

**pie graph** — a circular graph resembling slices of a pie in which each slice represents the relative size of one part of a group in relation to the whole group

**pitch** — perceived frequency of sound. High pitch is high frequency and low pitch is low frequency.

**pixel** — smallest element of an electronic image, such as the picture from a digital camera.

**Planck's constant** — fundamental constant that determines the scale of quantum phenomena. Planck's constant corresponds to the ratio of the energy of light to its frequency. The value of Planck's constant is $h = 6.63 \times 10^{-34}$ J s.

**polar coordinates** — coordinates that use radius (or magnitude) and angle to define points in space.

**polarity** — condition of a system that describes opposing physical characteristics. Examples include the poles of a magnet and electrical charges.

**polarization** — property of a transverse wave that describes the direction of the oscillation in a plane perpendicular to the wave's direction of motion.

**position** — vector quantity that locates a unique point in space relative to an origin.

**positron** — particle that has the same mass as an electron and electric charge of $+1.6 \times 10^{-19}$ C.

**potential difference** — difference in the electric potential between two locations. In electric circuits, potential difference and voltage are often used interchangeably.

**potential energy** — energy of position—often due to gravity, but also includes other forms such as elastic potential energy and pressure. Measured in joules (J).

**pound (lb)** — English unit of force equal to 4.448 N. The force due to the Earth's gravity is often expressed in pounds.

**power** — rate at which work is done or energy is transferred. The units of power are watts (W). One watt is one joule per second.

**precession** — wobbling of a spinning object, such that the object's axis of rotation traces out an imaginary cone.

**precision** — degree of mutual agreement among a series of measurements of the same quantity.

**pressure** — force per unit area. SI units are pascals (Pa) and 1 Pa = 1 N/m$^2$. English units are pounds per square inch (lb/in$^2$) or psi.

**prism** — optical device, usually a triangular block of transparent material such as glass, that separates white light into its constituent colors through dispersion.

**projectile** — moving body traveling only under the influence of gravity.

**proton** — positively charged subatomic particle with a mass of $1.673\times10^{-27}$ kg and charge of $+1.602\times10^{-19}$ C.

**pulley** — wheel that acts as a lever to change the direction of an input force. One of six types of simple machines.

***PV* diagram** — graph of pressure $P$ as the vertical axis and volume $V$ as the horizontal axis on which thermodynamic processes can be analyzed. The area enclosed by a process on a $PV$ diagram is the work done by the system.

## Q

**quadratic** — equation containing a dependence of one variable on another variable raised to the second power. For example, an equation is quadratic in $x$ when it includes terms involving $x^2$.

**qualitative** — description or explanation that does not involve or predict numerical values.

**quantitative** — description or explanation that involves or predicts numerical values.

**quantized** — concept applying to the atomic scale where physical quantities, such as energy, are permitted only to take discrete values. Examples: the energy of a photon of light is quantized as $E = hf$; and the orbits of an electron in Bohr's model of the hydrogen atom are quantized as $n = 1, 2, 3,...$

**quantum number** — dimensionless value that specifies a state in the quantum theory. The quantum number is an integer for the orbits of the electron, but can take half-integer values for other quantum states such as the spin of the electron.

**quantum physics** — field of science that explains physical phenomena at atomic scales. The quantum theory is notable in describing many quantities as having only discrete, not continuous, values called quanta. In quantum physics, light and elementary particles exhibit both particle and wave phenomena.

## R

**radian (rad)** — angle that has an arc length that is equal to its radius. One radian is equivalent to approximately 57.3°. One full circle is 360 degrees or $2\pi$ radians. It is not necessary to include "rad" when expressing an angle in radians because all angles are assumed to be in radians unless otherwise noted (e.g., when given in degrees).

**radiant energy** — energy transferred by electromagnetic waves. Also called electromagnetic energy.

**radiation** — (1) the process of emitting electromagnetic, or radiation, energy; (2) the particles and energies emitted from radioactive substances.

**radio waves** — longest wavelength region of the electromagnetic spectrum.

**radioactive decay** — spontaneous, partial disintegration of a nucleus that is accompanied by the release of an energetic particle or radiation.

**ramp** — inclined surface or plane used for elevating objects. One of six types of simple machines.

**ramp coordinates** — coordinate system where one axis is parallel to the slope of a ramp or inclined plane. Since motion down an inclined plane is parallel to the plane, ramp coordinates are the natural axes to use for the equations of motion.

**range** — horizontal distance a projectile travels between its launching point and the point where it hits the ground.

**ray diagram** — diagram showing how the direction of light rays changes as they are reflected and/or refracted by optical devices.

**reaction force** — one member of an action–reaction pair of forces that is equal in strength and opposite in direction to the action force that is its counterpart.

**real image** — image formed by light rays that converge at the location of the image. A real image can be viewed or projected on a screen.

**reference frame** — coordinate system assigning positions and times to events. Two different reference frames can be in motion with respect to each other.

**reflected ray** — light ray that bounces off a surface or boundary between two materials.

**reflection** — process by which waves (including light) bounce off or return by striking a surface or the boundary between two different materials.

**refraction** — process by which waves (including light) change direction by traveling into or through a surface or boundary between two materials.

**refrigerator** — device that uses a reverse Carnot cycle to do work to transfer thermal energy from a colder object to a warmer one.

**rem** — unit of measure for ionizing radiation that estimates its impact on a human. Acronym for roentgen equivalent man.

**renewable energy** — comes from natural sources of energy—such as solar, water from rainfall, wind, tides, or geothermal heat—that are replenished naturally and presumably can be used as an energy source indefinitely.

**reproducibility** — ability to obtain the same results for experiments or observations by doing the same experiment or making the same observation in the same way.

**resistance** — opposition that a conducting material or device offers to resist the flow of current. Measured in ohms (Ω).

**resistor** — electrical device that resists the flow of electric current.

**resolution of forces** — process of breaking a single force into the equivalent set of one or more component forces.

**resonance** — condition where the frequency of a periodic force matches a natural frequency of an oscillating system. At resonance, energy accumulates over many cycles of the periodic force and oscillations can become very large, even when the causative force is relatively small.

**rest energy** — energy equivalent to the mass of a particle or body at rest, i.e., with zero speed, given by the equation $E = mc^2$.

**resultant vector** — vector sum of two or more vectors.

**reversibility** — property of an ideal thermodynamic process that can return to its original state by running the same series of steps backwards.

**revolution** — spinning motion of an object about an axis outside of the object. The Earth, for example, revolves around the Sun.

**rods** — light-sensitive cells in the eye that respond to differences in brightness of the light.

**rolling friction** — friction force that occurs between two surfaces in rolling contact with each other.

**rotation** — spinning motion of an object about an axis within the object. The Earth, for example, rotates about its axis every 24 hours.

**rotational energy** — mechanical energy of rotation for an object, equal to one-half of the product of its moment of inertia and the square of angular velocity, or $E_r = \frac{1}{2}I\omega^2$. Rotational energy is measured in units of joules (J).

**rotational inertia** — apparent resistance of an object to change its state of rotation. The measure of rotational inertia about a particular axis is called the *moment of inertia*.

## S

**safety factor** — multiplier typically between 2 and 20 that compares calculated forces and actual design specifications. For example, a safety factor of 10 means a structure that is calculated to support a force of 5 N is designed to withstand 50 N.

**satellite** — natural or technological object that is gravitationally bound to, and travels in a repeating orbit around, a larger object such as a planet

**scalar** — quantity that can be completely described by a single value (without a direction). Mass, pressure, and temperature are scalars.

**scale** — (1) relative size of things, as in large scale versus small scale;
(2) association between representation on a drawing or model and the real parameter being represented, such as 1 cm = 1 N on a force diagram.

**scattering** — absorption of light by matter followed by emission of light in random directions. Scattering is done by atoms or materials with particle sizes that are small compared to the wavelength of the light.

**scientific method** — rigorous method of developing and testing knowledge by comparing hypotheses against objective observational evidence.

**scientific notation** — method of writing numbers that uses a coefficient multiplied by a power of 10. For example, the number 1,500 is written in scientific notation as $1.5 \times 10^3$.

**screw** — device consisting of a spirally grooved body and grooved head that converts rotational motion into translational motion. One of six types of simple machines. A screw is often used to fasten two materials together.

**second law of thermodynamics** — heat only flows spontaneously in nature from hotter to colder objects. Alternately stated that processes occur in nature in a manner that maintains or increases entropy. The implication of the second law is that heat cannot be transferred from a colder object to a hotter object without external work being done on the system.

**series circuit** — electric circuit where the electric current has only one path.

**significant figures** — those digits in a number that carry meaning as to the precision of the number or the measurement of it. Leading zeroes (the zeroes on the left of the first digit) are not significant because they are placeholders that indicate the scale of the number or measurement.

**simple machine** — mechanical device that uses only one kind of motion to change the magnitude or direction of a force. The six kinds of simple machines are lever, pulley, wheel and axle, ramp, wedge, and screw.

**sliding friction** — friction force between two surfaces that are moving relative to each other through sliding. Also called kinetic friction.

**slope** — ratio of the increase in the vertical or $y$-direction divided by the change in the horizontal or $x$-direction.

**Snell's law of refraction** — ratio of the sine of the angle of incidence to the sine of the angle of refraction is a constant. The value of the constant is the ratio of the indices of refraction for the two materials.

**solenoid** — coil of wire, usually cylindrical, that uses electric current to create a magnetic field.

**solid** — phase of matter that does not flow but instead holds its shape and volume, because its particle's energies are insufficient to break intermolecular bonds.

**specific heat** — amount of energy it takes to change the temperature of one unit of mass by one degree. Measured in joule per kilogram per degree Celsius ($J\ kg^{-1}\ °C^{-1}$) or joule per gram per degree Celsius ($J\ g^{-1}\ °C^{-1}$). Also called specific heat capacity.

**spectral line** — individual wavelength of light that is emitted or absorbed by a material as viewed in a spectrograph.

**spectrogram** — three-variable graphic representation of sound that plots time on the horizontal axis, frequency on the vertical axis, and amplitude as different colors.

**spectrograph** — scientific instrument that disperses light into its different wavelengths (or colors). A simple example of a spectrograph is a glass prism. Also called a spectrometer or spectroscope.

**spectrum** — characteristic wavelengths of light emitted or absorbed by an object.

**specular reflection** — reflection of light off smooth or shiny surfaces that produces a visible image of the object. Light striking a mirror is an example of specular reflection.

**speed** — ratio of distance traveled to the time taken, or distance divided by time. Speed is also the magnitude of the velocity vector and therefore cannot be negative.

**speed of light** — constant value of the speed that light (or electromagnetic radiation) travels in a vacuum. The speed of light has a value of $c = 3.0\times10^8$ m/s.

**speed of sound** — speed that a sound wave travels in a medium. The speed of sound is 343 m/s in air at 1 atmosphere and 21°C.

**spontaneous emission** — process by which an excited atom releases energy by producing a photon of light, causing an electron to transition from a higher to a lower energy state.

**spring** — elastic device specifically designed to produce a controllable restoring force when it is deformed.

**spring constant** — quantity representing the value of the restoring force exerted by a particular spring (or other elastic material) when it is stretched or compressed. Measured in units of force divided by length, or newtons per meter (N/m).

**standard deviation** — the average deviation of all the data points away from the average value of the whole set of data points

**standard error** — the error in the average (or mean) of a set of measurements - mathematically equal to the standard deviation divided by the square root of the number of measurements

**standing wave** — stationary wave pattern created by two identical waves traveling in opposite directions, such as on a vibrating string.

**state** — particular configuration of all the elements in a system, usually in reference to a defined energy of the system.

**static electricity** — buildup of electrical charge, either positive or negative, on the surface of an insulating material.

**static friction** — friction force that occurs between two surfaces that are at rest with respect to each other.

**statistical mechanics** — branch of physics that uses statistical analysis to relate the macroscopic properties of matter, such as temperature and pressure, to the microscopic behavior of collections of atoms or molecules.

**stimulated emission** — process in which an atomic electron in an excited state interacting with an electromagnetic wave of a specific frequency drops to a lower energy level, thereby transferring its energy to the wave.

**streamline** — imaginary line through a fluid tracing the path of motion followed by a particular element of mass within the fluid.

**stress** — force per unit of cross-sectional area. Stress has units of pressure, which are measured in pascals (Pa) or psi.

**strong nuclear force** — strongest of the fundamental forces of nature and the force that is responsible for holding the protons and neutrons in an atom's nucleus together.

**subtractive primary colors (CMYK)** — three colors of pigments (cyan, magenta, and yellow) that absorb light and, when mixed in various proportions, can reflect any color. The "K" in CMYK refers to a black pigment used to create darker shades.

**superposition principle** — when two or more waves overlap, the amplitude of the resulting wave is the sum of the amplitudes of the individual waves.

**supersonic** — moving faster than the speed of sound, which is 343 m/s (767 mph) in air at standard temperature and pressure.

**surface area** — number of square units that completely cover the exterior or exposed surface of an object.

**system** — group of related and interacting objects and influences that we choose to investigate. A system can be open or closed.

## T

**technology** — designed world, including all inventions, devices, processes, and products developed by humans. Technology is the output of engineering.

**temperature** — (1) measurement quantifying the degree of hot or cold of a substance;
(2) average kinetic energy per molecule in a substance. Usually measured using either a relative scale (such as degrees Celsius or Fahrenheit) or an absolute scale (such as kelvins).

**tensile strength** — stress value at which a solid material typically fails in tension.

**tension** — force that pulls or stretches a material. Since tension is a force it has units of newtons (N).

**terminal velocity** — speed at which air friction and weight become equal and downward acceleration ceases. Terminal velocity can also apply to motion through water or any other condition where friction increases with speed.

**theory** — well-verified set of explanations that typically are supported by a large body of scientific evidence. Theories generally start as hypotheses. A hypothesis is elevated to the status of a theory when it becomes supported by a large enough body of evidence.

**thermal conductivity** — measure of a material's ability to conduct heat. Thermal conductivity has units of watts per meter per degree Celsius ($W\ m^{-1}\ °C^{-1}$). A thermal conductivity of 1 W/(m °C) means one watt of heat flows through a one meter cube when the temperature difference is 1°C.

**thermal energy** — kinetic energy that comes from the random motion of atoms and/or molecules due to the effects of temperature.

**thermal equilibrium** — condition of equal temperature at which no heat flows.

**thermodynamics** — study of heat and its transformation to and from mechanical energy.

**thermometer** — instrument that measures or senses temperature.

**third law of thermodynamics** — absolute zero is the minimum possible temperature at which the entropy of a system is zero. The implication of the third law is that there exists no temperature below absolute zero.

**threshold frequency** — minimum frequency of light for a given metal that causes electrons to be ejected through the photoelectric effect. Related to the work function of the metal.

**torque** — product of force and the perpendicular distance between its point of application and the center of rotation. Units of torque are newton meters (N m).

**total internal reflection** — complete reflection of an incident light ray back into the incident media. This occurs at a boundary with a material that has a lower index of refraction. Total internal reflection occurs at angles of incidence greater than a critical angle that depends on the difference in refractive indices.

**trajectory** — path a projectile follows. In two dimensions, the trajectory is a curved path due to a constant horizontal velocity and an accelerating vertical velocity.

**transformer** — device that increases or decreases voltage in AC circuits in order to minimize energy lost during transmission.

**translation** — linear motion that results in a change in location. Other types of motion include rotation and oscillation.

**translucent** — optical device that allows only some light to pass through it. The light may change direction many times as it passes through a translucent material, creating a distorted image.

**transparent** — optical device that allows all the light to pass through it without distorting the image.

**transverse** — direction perpendicular to the long direction or direction of motion. Transverse wave: type of wave in which the oscillations are perpendicular to the wave's direction of motion.

**trough** — minimum value or lowest amplitude of a wave.

## U

**ultraviolet light** — form of electromagnetic radiation at somewhat higher frequencies than visible light.

**uncertainty principle** — momentum and position of a quantum particle cannot be known simultaneously better than $\Delta h \Delta p = h/4\pi$, nor is possible to know energy and time simultaneously better than $\Delta E \Delta t = h/4\pi$.

## V

**vacuum** — void or absence of matter.

**variable** — quantity that may take on different values. All possible values of a particular variable have the same dimensions. For example, speed is a variable that can have many different values, but each value has units of length over time and expresses the rate of change in position.

**vector** — quantity that includes both magnitude (or value) and direction. Force, displacement, and velocity are vectors.

**vector diagram** — graphical representation of a vector in which the coordinate axes are scaled to represent the components of the vector in perpendicular directions.

**velocity** — vector quantity that describes the rate at which position changes. Like speed, velocity has units of length over time, such as m/s.

**virtual image** — image formed by light rays that do not converge at the location of the image. Virtual images cannot be projected or displayed on a screen.

**viscosity** — resistance of a fluid to flow or shear deformation. For example, honey has a high viscosity and water has a low viscosity.

**voice coil** — cylindrical coil of wire attached to the loudspeaker cone and moves in and out in response to a varying electrical signal passing through the wire. In a microphone, the equivalent device called an induction coil moves in response to sound waves, inducing a varying electrical signal in the wire.

**volt (V)** — unit of electric potential and electric potential difference.

**voltage** — amount of electrical potential energy that each unit of electric charge has. Also called potential difference. Measured in volts (V).

**voltmeter** — instrument to measure the voltage, or potential difference, across one or more components of a closed circuit. A multimeter usually includes a voltmeter as one of its instrument settings.

**volume** — number of cubic units that completely fill the interior space of an object.

## W

**watt (W)** — SI unit of power. One watt corresponds to one joule per second.

**wave** — oscillation that travels. Waves are a traveling form of energy. They have properties of frequency, wavelength, speed, and amplitude.

**wave function** — quantum mechanical description of a particle's probability amplitude as a function of time and space, denoted by $\Psi$.

**wavefront** — two- or three-dimensional shape of the crest of a wave. A spatial curve or surface that is the locus of all points on a given cycle of a wave that have the same phase.

**wavelength** — spatial extent (length) of one cycle of a wave.

**weak nuclear force** — one of the four fundamental forces of nature that governs the process of radioactive decay.

**wedge** — piece of material, thick at one end and tapered at the other end, typically inserted into a crevice in order to split an object. One of six types of simple machines.

**weight** — downward-acting force created by gravity acting on the mass of an object. Equal to $mg$, where $g$ is the local strength of gravity, such as 9.8 N/kg at the Earth's surface. Weight has the units of force, such as newtons (N) or pounds.

**wheel and axle** — lifting machine comprised of a rope attached to a wheel such that applying a torque to the wheel winds the rope onto the axle. One of six types of simple machines.

**work** — form of energy equal to one newton of force exerted for one meter in the direction of the force. Measured in joules (J).

**work function** — energy needed to liberate an electron from a metal atom in the photoelectric effect.

**work–energy theorem** — change in the kinetic energy of an object equals the net work done on it.

## X

**x-rays** — high frequency form of electromagnetic radiation.

## Z

**zeroth law of thermodynamics** — if two systems are in thermal equilibrium with a third system, then they are also in thermal equilibrium with each other. The implication of the zeroth law is that no heat will flow between two objects at the same temperature.

# Index

## C

# Index

# Index

## Q



## R

# Index

# Index

## X



## Y



## Z

## Fundamental constants

| Quantity | Symbol | Value | Units |
|---|---|---|---|
| Speed of light in a vacuum | $c$ | $3.00\times10^{8}$ | m/s |
| Gravitational constant | $G$ | $6.67\times10^{-11}$ | N $m^2/kg^2$ |
| Boltzmann's constant | $k_B$ | $1.38\times10^{-23}$ | J/K |
| Stefan–Boltzmann constant | $\sigma$ | $5.67\times10^{-8}$ | W/($m^2$ $K^4$) |
| Planck's constant | $h$ | $6.63\times10^{-34}$ | J s |
| Avogadro's number | $N_A$ | $6.022\times10^{23}$ | $mol^{-1}$ (or particles/mol) |
| Gas constant | $R$ | 8.315 | J/(mol K) |
| Permittivity of free space | $\varepsilon_0$ | $8.85\times10^{-12}$ | $C^2$/(N $m^2$) |
| Permeability of free space | $\mu_0$ | $4\pi\times10^{-7}$ | T m/A |
| Mass of the electron | $m_e$ | $9.11\times10^{-31}$ | kg |
| Mass of the proton | $m_p$ | $1.6726\times10^{-27}$ | kg |
| Mass of the neutron | $m_n$ | $1.6749\times10^{-27}$ | kg |
| Charge of the electron | $e$ | $-1.60\times10^{-19}$ | C |
| Coulomb's law constant | $k_e$ | $8.99\times10^{9}$ | N $m^2/C^2$ |
| Absolute zero | 0 K | −273.15 | °C |

## Useful values constants

| Quantity | Symbol | Value | Units |
|---|---|---|---|
| Strength of gravity at Earth's surface | $g$ | 9.81 | $m/s^2$ or N/kg |
| Speed of sound at 20°C and 1 atm | $c_s$ | 343 | m/s |
| Standard atmospheric pressure | atm | $1.01\times10^{5}$ | Pa or $N/m^2$ |

## Scientific notation and powers of ten

| Description | Scientific notation | Equivalent value | SI prefix | Abbreviation |
|---|---|---|---|---|
| One quadrillion | $1\times10^{15}$ | 1,000,000,000,000,000 | peta | P |
| One trillion | $1\times10^{12}$ | 1,000,000,000,000 | tera | T |
| One billion | $1\times10^{9}$ | 1,000,000,000 | giga | G |
| One hundred million | $1\times10^{8}$ | 100,000,000 | | |
| Ten million | $1\times10^{7}$ | 10,000,000 | | |
| One million | $1\times10^{6}$ | 1,000,000 | mega | M |
| One hundred thousand | $1\times10^{5}$ | 100,000 | | |
| Ten thousand | $1\times10^{4}$ | 10,000 | | |
| One thousand | $1\times10^{3}$ | 1,000 | kilo | k |
| One hundred | $1\times10^{2}$ | 100 | hecto | h |
| Ten | $1\times10^{1}$ | 10 | | da |
| One | $1\times10^{0}$ | 1 | | |
| One tenth | $1\times10^{-}$ | 0.1 | deci | d |
| One hundredth | $1\times10^{-2}$ | 0.01 | centi | c |
| One thousandth | $1\times10^{-3}$ | 0.001 | milli | m |
| One ten thousandth | $1\times10^{-4}$ | 0.0001 | | |
| One hundred thousandth | $1\times10^{-5}$ | 0.00001 | | |
| One millionth | $1\times10^{-6}$ | 0.000001 | micro | µ |
| One ten millionth | $1\times10^{-7}$ | 0.0000001 | | |
| One hundred millionth | $1\times10^{-8}$ | 0.00000001 | | |
| One billionth | $1\times10^{-9}$ | 0.000000001 | nano | n |
| One trillionth | $1\times10^{-12}$ | 0.000000000001 | pico | p |
| One quadrillionth | $1\times10^{-15}$ | 0.000000000000001 | femto | f |

# Periodic table colored by element groups

**metals**
- alkali metals
- alkaline earth metals
- transition metals
- rare earth metals
- other metals

**nonmetals**
- halogens
- noble gases
- other nonmetals

**metalloids**
- metalloids

Key: 1 — Atomic number; H — Symbol; 1.0079 — Atomic mass; hydrogen — Name

| group 1 1A | group 2 2A | group 3 3B | group 4 4B | group 5 5B | group 6 6B | group 7 7B | group 8 | group 9 8B | group 10 | group 11 1B | group 12 2B | group 13 3A | group 14 4A | group 15 5A | group 16 6A | group 17 7A | group 18 8A |
|---|---|---|---|---|---|---|---|---|---|---|---|---|---|---|---|---|---|
| 1 H 1.0079 hydrogen | | | | | | | | | | | | | | | | | 2 He 4.0028 helium |
| 3 Li 6.941 lithium | 4 Be 9.0122 beryllium | | | | | | | | | | | 5 B 10.811 boron | 6 C 12.011 carbon | 7 N 14.007 nitrogen | 8 O 15.999 oxygen | 9 F 18.998 fluorine | 10 Ne 20.180 neon |
| 11 Na 22.990 sodium | 12 Mg 24.305 magnesium | | | | | | | | | | | 13 Al 26.982 aluminum | 14 Si 28.086 silicon | 15 P 30.974 phosphorous | 16 S 32.065 sulfur | 17 Cl 35.453 chlorine | 18 Ar 39.948 argon |
| 19 K 39.098 potassium | 20 Ca 40.078 calcium | 21 Sc 44.956 scandium | 22 Ti 47.867 titanium | 23 V 50.942 vanadium | 24 Cr 51.996 chromium | 25 Mn 54.938 manganese | 26 Fe 55.845 iron | 27 Co 58.933 cobalt | 28 Ni 58.693 nickel | 29 Cu 63.546 copper | 30 Zn 65.41 zinc | 31 Ga 69.723 galium | 32 Ge 72.61 germanium | 33 As 74.922 arsenic | 34 Se 78.96 selenium | 35 Br 79.904 bromine | 36 Kr 83.80 krypton |
| 37 Rb 85.468 rubidium | 38 Sr 87.62 strontium | 39 Y 88.906 yttrium | 40 Zr 91.224 zirconium | 41 Nb 92.906 niobium | 42 Mo 95.94 molybdenum | 43 Tc (98) technetium | 44 Ru 101.07 ruthenium | 45 Rh 102.91 rhodium | 46 Pd 106.42 palladium | 47 Ag 107.87 silver | 48 Cd 112.41 cadmium | 49 In 114.82 indium | 50 Sn 118.71 tin | 51 Sb 121.76 antimony | 52 Te 127.60 tellurium | 53 I 126.90 iodine | 54 Xe 131.29 xenon |
| 55 Cs 132.91 cesium | 56 Ba 137.33 barium | 71 Lu 174.97 lutetium | 72 Hf 178.49 hafnium | 73 Ta 180.95 tantalum | 74 W 183.84 tungsten | 75 Re 186.21 rhenium | 76 Os 190.23 osmium | 77 Ir 192.22 iridium | 78 Pt 195.08 platinum | 79 Au 196.97 gold | 80 Hg 200.59 mercury | 81 Tl 204.38 thallium | 82 Pb 207.2 lead | 83 Bi 208.98 bismuth | 84 Po (209) polonium | 85 At (210) astatine | 86 Rn (222) radon |
| 87 Fr (223) francium | 88 Ra (226) radium | 103 Lr (262) lawrencium | 104 Rf (261) rutherfordium | 105 Db (262) dubnium | 106 Sg (266) seaborgium | 107 Bh (264) bohrium | 108 Hs (277) hassium | 109 Mt (278) meitnerium | 110 Ds (281) darmstadtium | 111 Rg (272) roentgenium | 112 Uub (285) ununbium | 113 Uut (284) ununtrium | 114 Uuq (289) ununquadium | 115 Uup (288) ununpentium | 116 Uuh (292) ununhexium | | 118 Uuo (294) ununoctium |

| | | | | | | | | | | | | | |
|---|---|---|---|---|---|---|---|---|---|---|---|---|---|
| 57 La 138.91 lanthanum | 58 Ce 140.12 cerium | 59 Pr 140.91 praseodymium | 60 Nd 144.24 neodymium | 61 Pm (145) promethium | 62 Sm 150.36 samarium | 63 Eu 151.96 europium | 64 Gd 157.25 gadolinium | 65 Tb 158.93 terbium | 66 Dy 162.50 dysprosium | 67 Ho 164.93 holmium | 68 Er 167.26 erbium | 69 Tm 168.93 thulium | 70 Yb 173.06 ytterbium |
| 89 Ac (227) actinium | 90 Th 232.04 thorium | 91 Pa 231.04 protactinium | 92 U 238.03 uranium | 93 Np (237) neptunium | 94 Pu (244) plutonium | 95 Am (243) americium | 96 Cm (247) curium | 97 Bk (247) berkelium | 98 Cf (251) californium | 99 Es (252) einsteinium | 100 Fm (257) fermium | 101 Md (258) mendelevium | 102 No (259) nobelium |